Design of Machine Elements

Design of Machine Elements

H.G. Patil
Retired Professor
Formerly HOD
Department of Mechanical Engineering
Gogte Institute of Technology, Belagavi

Dr. S.C. Pilli
Professor
Department of Mechanical Engineering
Acharya Institute of Technology, Bengaluru
Formerly Principal
KLECET (MSSCET), Belagavi

Dr. Ravindra R. Malagi
Professor
Department of Product Design and Manufacturing
Center for P.G. Studies, VTU, Belagavi

Dr. M.S. Patil
Professor
Department of Mechanical Engineering
Gogte Institute of Technology, Belagavi

TECHSAR™
NEW DELHI

Published and Printed by

Techsar Pvt. Ltd.
4435-36/7, nsari Road, Daryaganj
New Delhi–110 002 (India)
E-mail: tech_info@techsarworld.com
Website: www.techsarworld.com

ISBN: 978-93-84588-50-2

Reprint 2025

Preface

The subject of 'machine design' is important to mechanical engineers. Thorough understanding of the subject is essential to the practising engineers for preparation of workable designs in engineering applications.

The main purpose of the book is to develop skills in preparation of designs of simple machine elements. Each chapter contains a brief theory and a large number of solved examples. The chapters one to five deal with the basic principles of machine design. Topics such as design philosophy, engineering materials, mechanics of materials and failure prediction methods are discussed. Each of chapters six to twenty four covers the design procedures for individual machine elements. The details of the forces acting, stresses induced, materials used and design procedures prescribed by the ISO, ISI AGMA standards are discussed. A large number of problems are solved in SI units. In all solved problems extensive reference is made to the data from our Machine Design Data Handbook, indicated by the equation number, figure number, table number followed by MDHB. The probabilistic approach to design is introduced in the last chapter.

We used the data from various books and standards. We sincerely express our gratitude to the authors and publishers of the books. We place on record our gratitude and indebtedness to Indian Standards Institution for the extracts from ISI publications reproduced in this book

Authors

Contents

CHAPTER 1

INTRODUCTION TO MACHINE DESIGN

"Engineering is a profession concerned with primarily the application of a certain body of knowledge, a set of skills, and point of view, in the creation of devices, structures and processes to transform resources to forms which satisfy the needs of society." —E V Krick

Human beings began to modify the natural resources to fulfil the needs. Man became a designer as soon as he began to modify the form of natural objects. The engineering designers tend to regard design as achieving solutions to technical requirements. The objective of all designers is to optimise fitness for purpose, based on market requirements and criterion of quality, performance and cost effectiveness. Quality is used to include attributes based on design, the appearance, visual coherence, compatibility, reliability, safe, easy to maintain and environment-friendly. Performance is in relation to function leading to qualitative and quantities assessment. Cost effectiveness is an aspect for commercial success. The designer generally works in one of the following environment, namely: i) Certainty: each alternative's outcome is known in advance, ii) Risk: probabilities can be assigned to each outcome, iii) Uncertainty: lack of information or knowledge makes the outcome of each alternative unpredictable such that no probabilities can be determined.

Engineering design which encompasses all the disciplines, in literature has been defined as:

"Engineering design is a purposeful activity directed towards the goal of fulfilling human needs, particularly those which can be met by the technological factors of our culture." —M Asimov, 1962

"A creative activity - it involves bringing into being something new and useful that has not existed before." —J B Reswick, 1965

"A goal directed problem solving activity." —B Archer, 1965

"Engineering design is problem solving science based problem solving with social-human awareness." —J R Dixon, 1966

"The design process inherently is an iterative decision-making activity, to produce plans by which resources are converted, preferably optimally with due consideration for environment, into systems and devices (products) to meet human needs." —T T Woodson, 1966.

1.1 WHAT IS DESIGN?

The word 'Design' can either be a noun or a verb. As a noun, 'Design' can be referred to an object's appearance functionally and aesthetically. Design is the act of formalizing an idea or concept into system.

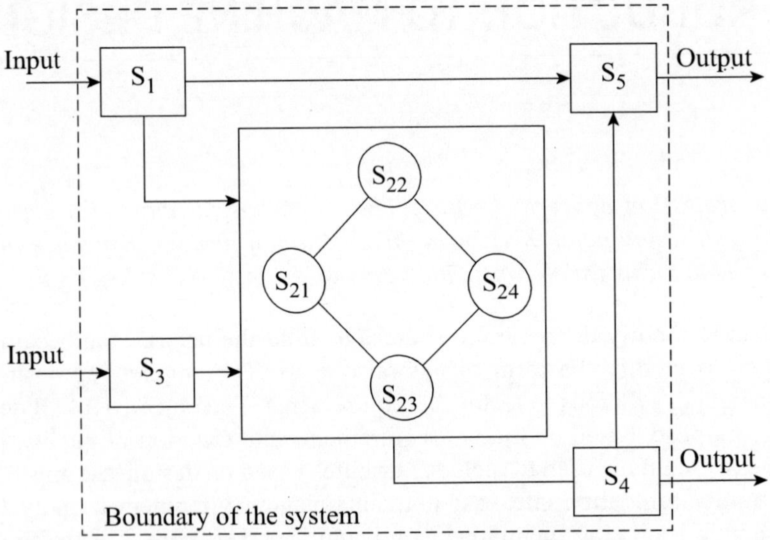

Figure 1.1. Schematic representation of system

A system (product) consists of subsystems (sub-assembly) which together are interconnected with cause and effect relations, converting an input to an output. Important features being (i) Breaking down complex system into a sequence of simpler design activities and (ii) Element of iteration is not a shortcoming. In fact it is essential for getting creative solutions as shown in Fig. 1.1.

In the current scenario a product is looked at in all its entities in its product life cycle. The product life cycle encompasses the use of resources for creation of a product to reclamation of resources on retirement. The key components of product life cycle are shown in Fig. 1.2. The product life cycle includes the processes of conception, invention, visualization, specifications, product design (analysis, synthesis and optimization), technology, manufacturing, marketing, service and disposal that determine the form of a product.

Design is as an iterative decision-making process to formulate a plan by which resources are optimally converted into systems, processes or devices in order to solve a specific problem or to fulfil a specific need. It begins with a need, a problem, an idea or a concept and ends in the form of drawings, and/or a document, which helps in manufacturing and realisation of a product. The process of design, needs creativity resulting in new or modified product, decision making on the variables or parameters to resolve complexity, making choice between many possible solutions and compromise in balancing multiple and some time conflicting requirements. It is a creative activity to produce a plan or system, or product that is functional, manufacturable, safely usable, reliable,

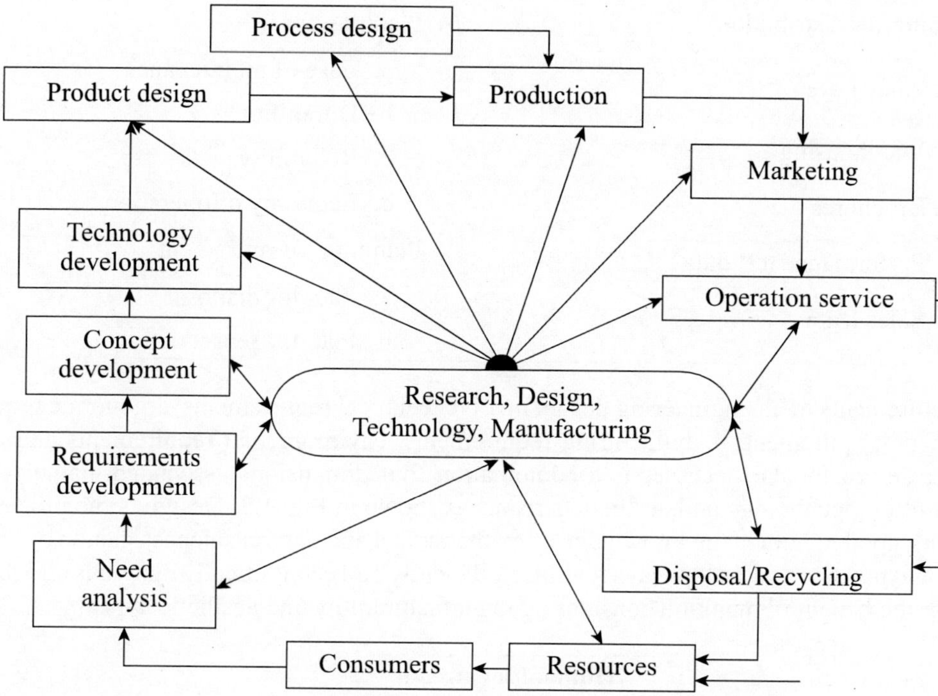

Figure 1.2. Product life cycle

competitive, marketable. Asimov was among the first to give a design morphology consisting of seven phases:

1) Conceptual design
   a. Define problem
   b. Gather information
   c. Concept generation
   d. Evaluation of concepts
2) Embodiment of design
   a. Product architecture
   b. Configuration design
   c. Parametric design
3) Detailed design
   a. Part drawings suitable for manufacture
   b. Assembly drawings
   c. Product specifications
   d. Verification testing of prototypes
   e. Decision on manufacturing or subcontracting
   f. Cost estimation
4) Planning for manufacture
   a. Designing tooling and fixtures
   b. Production planning
   c. Planning quality assurance
   d. Establishing time and cost
   e. Establishing the process flow and control

5) Planning for distribution
    a. Shelf-life
    b. Warehousing
    c. Brochures
    d. Performance test data
    e. Sales team

6) Planning for use
    a. Ease of maintenance
    b. Durability
    c. Reliability
    d. Economy of operation

7) Planning for retirement
    a. Plan for disposal
    b. Plan for reuse

The requirements of an engineering design involve technical requirements, ergonomic requirement, aesthetic requirement, manufacturing requirements, environmental requirements and cost. Engineering design for its successes is a combination four dimensions associated, namely: science, humanities , technology and aesthetic streams as shown in Fig. 1.3. Design is the process of applying with good judgment the knowledge of mathematical and natural sciences gained by study, experience and practice to develop ways to utilize efficiently and economically materials and forces of nature for the benefit of mankind considering manufacturability and aesthetic appeal.

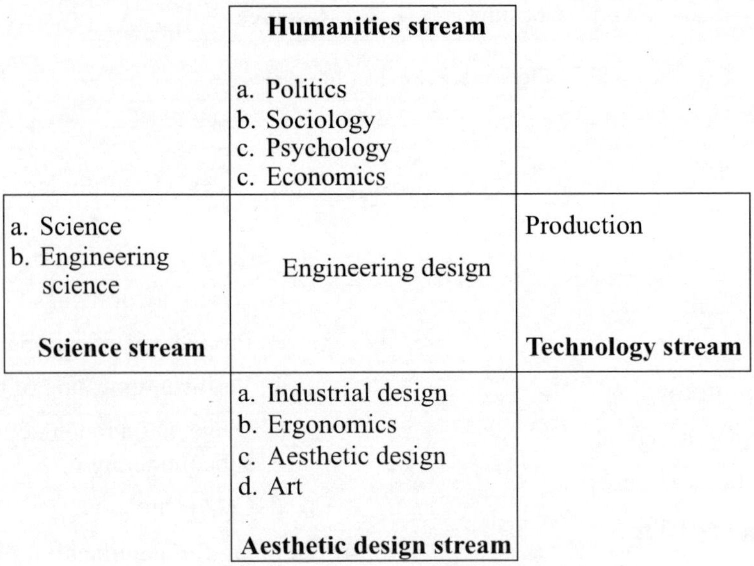

Figure 1.3. Dimensions of engineering design

Mainly there are three types of designs i) Adaptive design ii) Development design iii) New design. In adaptive design,a designers makes modifications or improvements in an existing design. In developmental design the designer adopts the existing design concepts for other uses by modifying the system, materials and manufacturing techniques. New designs are based on research and innovations and is a creation of a new device altogether. In all these types of designs the design may be based on rational design depending of the formulae of mechanics, empirical design based on practice and industrial design considering the production aspects. optimum design maximizes the benefit and minimizes the undesirable effects. If computer systems are used for creation, modification, analysis and optimization, it is called computer aided design.

1.2 MECHANICAL DESIGN

> *The process is the use of scientific principles and technical information along with innovations, ingenuity or imagination in the definition of a machine, mechanical device or system (product) to perform, prespecified functions with maximum economy and efficiency.*
>
> *—Engineering Design Council, UK*

Design is the process of applying the knowledge of science and technology to devise new or improved products. Design process begins with identification of a need and a decision to do something about it. After a number of iterations, the process ends with the presentation of the plans for satisfying the need. Depending on the nature of the design task, several phases may be repeated before the design is finalized. Complete design process can be explained with the help of Fig. 1.4, which shows the phases of design process and iterations.

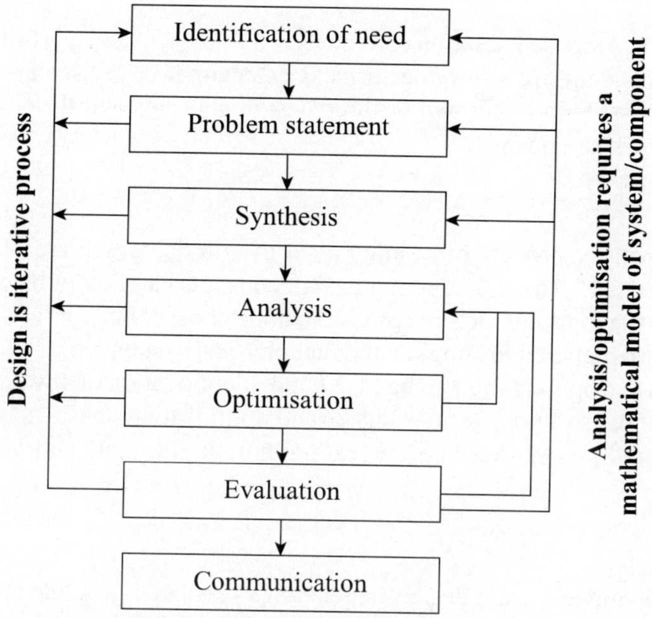

Figure 1.4. Phases of design process

1.3 IDENTIFICATION OF NEED

Design process generally starts with the identification of need, a problem or with the recognition of a potential market for a product, device or process. It is understanding of customer needs. Recognition of the need requires a lot of imagination and creativity, survey because the need is generally not clear and evident and can be vague.

Problem Statement

The statement describes what has to be accomplished to satisfy the customer needs. It involves assessment of competing products. The statement of problem is more specific and includes all the specifications for the object to be designed, which include the input, i.e., resources to be utilized, expected output, cost limitations, quantity required, expected life and reliability, limitations, maximum range and expected variations in the variables, dimensional and weight limitations, manufacturing limitations, etc. Anything that limits the designer's freedom of choice is a constraint and all the constraints should be clearly defined in the problem definition.

Synthesis

Synthesis is the process of concept generation, the purpose of which is to develop as many ideas as possible to offer potential solutions to the problem stated in previous phase. Primary need is identified for initial focus and then all feasible design alternatives are developed. Different proposed alternatives should not be compared with each other, in this phase, and each alternative should be separately evaluated on the basis of established criteria.

Analysis

After synthesis, different proposed solutions are analyzed to assess their performance and can be ranked. Alternatives having unsatisfactory performance can either be revised and improved or discarded. It is always advisable to select two or three design concepts and develop them instead of selecting presumably the best one only.

Optimization

Optimization is the repetitive process of refining a set of criteria, which are often conflicting, to achieve the best compromise. These competing design concepts can finally be compared again so that the path leading to the most competitive product can be chosen. For analysis and optimization, mathematical models are prepared to simulate the real physical system.

Synthesis, analysis and optimization go hand in hand. Identification of any kind of deficiency or inadequacy in the proposed solution, in the analysis and optimization phase, may require synthesis of some new solution and process keeps on repeating until an adequate and optimum solution is obtained.

Evaluation

In this phase, prototype model of the design is generally simulated or made and tested, which is the final evaluation of the design to make sure that it satisfies the original requirements. In addition to this, design is evaluated for its reliability, competitiveness, economic viability, maintenance requirements, profitability and so on.

Communication

Final phase of the design process is to communicate the design to others. This presentation can be in the form of drawings, computer models or in any other form that would help in manufacturing and utilization of the product.

1.4 MACHINE DESIGN

Machine design is creation of plans for components of a system to perform the desired functions. Machine design can be defined as creation of combination of correctly proportioned moving and stationary components so constructed and joined as to enable the liberation, transformation and utilization of energy. Machine design requires the knowledge of basic and engineering sciences such as Physics, Mathematics, Engineering Mechanics, Strength of Materials, Theory of Machines, Thermodynamics and Heat Transfer, Vibrations, Fluid Mechanics, Metallurgy, Manufacturing Processes and Engineering Drawing.

In machine design, the designer's task is to choose the material, determine the motion, forces and energy transfer involved so as to determine the sizes, shapes for each element of the machine. Generally, the machine elements are designed on the basis of strength and rigidity so that they are able to withstand the applied load with permissible deformation or stress. In addition to strength and stiffness, other factors that are considered in design of machine element are weight, cost, wear, safety, reliability, etc. Design work may involve concentrating on one component at a time, but it is very important to simultaneously consider its relation with the other components and the product as a whole. The foundation course in machine design focuses on designing the components or assembly for specified operating conditions. This involves mechanics, kinematics, dynamics, heat transfer, vibration analysis, The course activity summarized in Fig. 1.5 which is an iterative process.

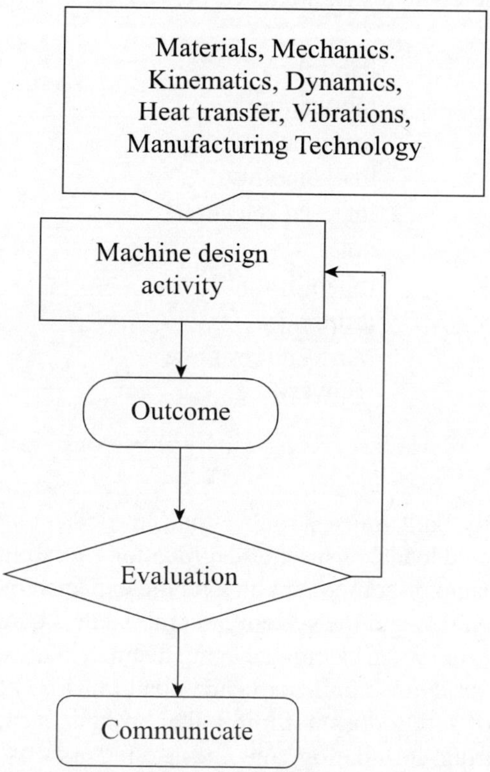

Figure 1.5. Schematic representation of foundation course in machine design

The following steps are generally followed in machine design:

i) Problem to be solved is completely and clearly stated.
ii) Possible mechanisms that will provide the desired motion or set of motions are selected.
iii) Forces acting on and energy transmitted by each element of the machine are determined.
iv) Suitable material is selected for each of the machine element.
v) Allowable values of stress and deflection are determined for each machine element, depending upon its material and functional requirements.
vi) Size and shape of each machine element is determined so that it can withstand the applied loads without failure.
vii) Dimensions of the machine elements are modified considering manufacturing aspects.
viii) Assembly and detailed drawings of the machine are made with complete specification of materials and manufacturing methods.

1.4.1 Design Considerations

The strength and stiffness are primary factors that are considered in machine design but there are a number of other considerations that the designer has to consider, depending on the problem. Some of the important factors considered in design are listed below:

availability	liability	safety
control	lubrication	standards and codes
cost	manufacturability	strength and rigidity
ease of assembly	marketability	shape, and size
ease of maintainability	material selection	styling
ease of manufacture	noise	thermal stability
choice of theory of	reliability	utility
failure,	retirement	wear and corrosion
friction and lubrication	reuse and resource	characteristics
functionality	recovery	weight

1.4.2 Factors of Safety

Many systems are purposefully built much stronger than needed for normal usage to allow for emergency situations, unexpected loads, misuse, or degradation (reliability). Factor of safety is the load carrying capacity of a system or component beyond the expected or actual loads. Essentially, the factor of safety is how much stronger the system or component is than it usually needs to be for an intended load. Safety factors are often calculated using detailed analysis because comprehensive testing is impractical, but the structure's ability to carry load must be determined to a reasonable accuracy.The use of a factor of safety does not imply that an item, structure, or design is ”safe”. Quality assurance, manufacturing, installation, and end-use factors may influence whether or not something is safe in any particular situation.

Factors which affect the factor of safety:

1. Material strength basis:
    a. Brittle Materials - use ultimate strength,
    b. Ductile Materials - use yield strength
2. Manner of loading:
    a. Static - applied slowly; remains applied or is infrequently removed.
    b. Repeated - fatigue failure may occur at stresses lower than static load failure.
    c. Impact - high initial stresses are developed.
3. Possible misuse - designer must consider any reasonable foreseeable misuse of product.
4. Complexity of stress analysis - the actual stress in a part isn't always known.
5. Environment - temperature, weather, radiation, chemical, etc.

There are two types of the factor of safety: i) ratio of absolute strength to actual applied load. This is a measure of realized factor of safety indicating the reliability of a particular design. ii) a design factor of safety or required factor of safety, which is a constant value imposed by law, standard, specification, contract .

There are two ways to estimate the value of an acceptable factor of safety (i) rule of thumb and (ii) probabilistic method or statistical method, based on knowledge of material, loading, geometric property, the failure mode and reliability. For better known properties the factors must be closer to unity and for less known factors a larger value may be assumed.

For rule of thumb, factor of safety = f (material, stress, geometry, type of failure, reliability). For the probabilistic, or statistical method FS $= \frac{\bar{\sigma}_{allowed}}{\bar{\sigma}_{applied}}$. The average allowed and applied stress are indicated in Fig. 1.6.

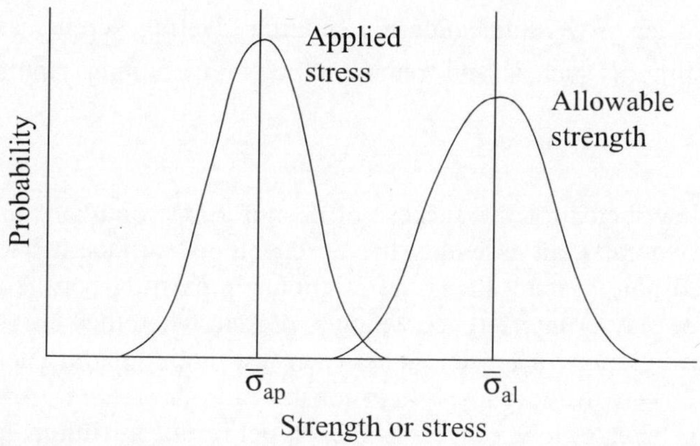

Figure 1.6. Average values of stress and strain statistically

The factor of safety from stress point of view are calculated as:

i) Based on yield strength for ductile materials $N = \dfrac{S_y}{S_d}$

ii) Based of ultimate strength for brittle material $N = \dfrac{S_u}{S_d}$

iii) Based on yield strength in shear $N = \dfrac{S_{sy}}{S_{sd}}$

iv) Design safety factor $= \dfrac{\text{Failure load}}{\text{Design load}}$

v) Margin of safety factor $= \dfrac{\text{Failure load}}{\text{Design load}} - 1$

Fundamentally, all these measure the extra load beyond what an intended structure will actually take (or will be required to withstand).

1.4.3 Manufacturing Considerations in Design

The knowledge of manufacturing process is important for designer engineer. A conventional manufacturing can be divided into i) Process engineering ii) Tool engineering iii) work standards iv) plant engineering and v) production planning and control. When computers are interfaced in manufacturing, it is called computer integrated manufacturing (CIM). There are four types of manufacturing systems: shop, batch, assembly/line and continuous flow. The factors that influence the selection of a process to make a part are: quantity of parts required, complexity-shape, size and features, materials, quality, cost to manufacture, availability, lead time and delivery schedule.

The manufacturing processes can be classified as:

i) Primary shaping process involving casting, forging, extrusion, rolling, drawing, shearing, powder metallurgy, composites.
ii) Machining process involving turning, milling, drilling, shaping, grinding.
iii) Joining processes such as welding, soldering, brazing, riveting, screw fastening.
iv) Surface finishing process such as heat treatment, polishing, honing, lapping, electroplating.

1.4.4 Fits

Interchangeability is a key factor for the success of design. Size control of the finished product is called limit system. Generally, an assembly has an enveloped surface called shaft system and enveloping system called hole system. Fit is the relationship existing between two mating parts with respect to amount of play or interference which is present when they are assembled together. It indicates the degree of tightness or looseness between two mating parts. There are two types of fits (i) Hole basis system where the hole size is kept constant and shaft size is varied to give different types of fits. Basic size is taken as lower limit of hole. Upper limit (maximum hole size-basic size) and lower limit (minimum hole size-basic size) deviation of shaft give desired fit which may be clearance fit, transition fit, and interference fit. (ii) Shaft basis system shaft size is kept fixed and hole size is varied to obtain different types of fit. Basic size is taken as the maximum limit size of

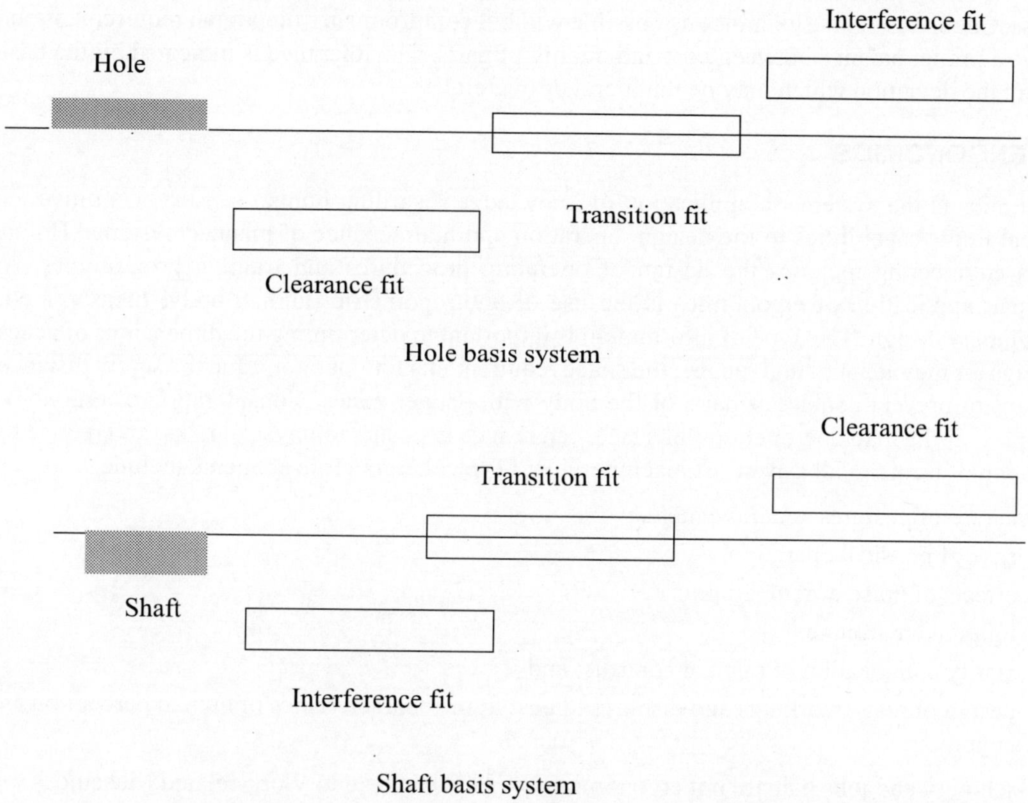

Figure 1.7. Types of fits based on hole basis and shaft basis

the shaft. Lower limit (minimum shaft size-basic size) of shaft upper limit (maximum shaft size-basic size) of hole give desired fit which may be clearance fit, transition fit, and interference fit. Permissible variation in size or dimension is called tolerance.

ISO system of limits and fits covers holes and shafts up to 3150 mm. For any basic size there are 28 different holes progressively oversize and undersize. They are A, B, C, CD, D, E, EF, F, FG, G, H, J, JS, K, M, N, P, R, S, T, U, V, X, Y, Z, ZA, ZB, and ZC. Each hole has a choice of 18 grades of tolerance, i.e., IT01, ITO, IT1 to IT16. Similarly for shafts. Tolerance grades decide the accuracy of manufacture (values of tolerances.)

Then tolerance is represented by a symbol consisting of basic size letter symbol, e.g., 45 E8/e7 (45 is basic size E8/e7 freezes one limit of hole/shaft)

1.4.5 Tolerance

Permissible variation in size or dimension is called tolerance. It is an amount by which a job is allowed to go away from accuracy without any functional trouble. The tolerance is decided upon i) Functional requirements of mating parts, ii) Cost of production and iii) Available manufacturing process. Choose as coarse tolerance as possible without compromising functional requirements and achieve a proper balance between cost and quality of parts. The tolerance is indicated by the basic size and the deviation which may be unilateral or bilateral.

1.5 ERGONOMICS

Ergonomics is the systematic application of knowledge regarding human sensory, cognitive and physical limits/capabilities to the design, operation and maintenance of manned systems. Human factors engineering includes the design of operating procedures and training programmes. An important application of ergonomics is the use of anthropometric (human body) measurements in machinery design. This type of information is important in determining the dimensions of access openings for maintenance and repairs, the space requirements for operators, and the safety distances and gaps to prevent contact of parts of the body with danger zones. Consideration of ergonomic principles, or the machine-operator interface, can reduce exposure to physical risks, and reduce the likelihood of errors in all aspects of machinery use. Typical ergonomic elements include:

i) nature of postures and movements;
ii) ease of physical operation;
iii) effects of noise or temperature;
iv) lighting environment;
v) clarity and location of manual controls; and
vi) design of dials, markings and displays to best fit the characteristics of human perception and cognition.

Machinery and jobs that are not ergonomically sound can lead to Work-related Musculo Skeletal Disorders (WMSDs). WMSDs may be caused, aggravated or precipitated by intense, repeated or sustained work activities with insufficient recovery. These are low back pain, tension neck syndrome, carpal tunnel syndrome, rotator cuff syndrome, DeQuervain's syndrome, Trigger finger; and tarsal tunnel syndrome. Less serious, but still significant WMSD injuries include muscle strains and tears, ligament sprains, joint and tendon inflammation, pinched nerves and spinal disc degeneration. Another important consideration is human factor engineering, which uses sciences of biomechanics, ergonomics and psychology to assure that the design can be safely operated by humans.

1.6 COMPUTER-AIDED DESIGN AND COMPUTER-AIDED ENGINEERING

Computer-aided design (CAD) is the use of computers in the design process. A CAD system consists of hardware softwares for different applications and peripherals. The software's have the capability of carrying out the analysis, optimization and drafting through graphic capability. The data bases stores the information of product model. The CAD helps the designer to reduces the cycle time by:

i) Accurately generated and easily modifiable graphical representation of the product.
ii) The user can nearly view the product on screen, make any modifications to it, and present his/her ideas without any prototype.
iii) Perform complex design analysis in short time.

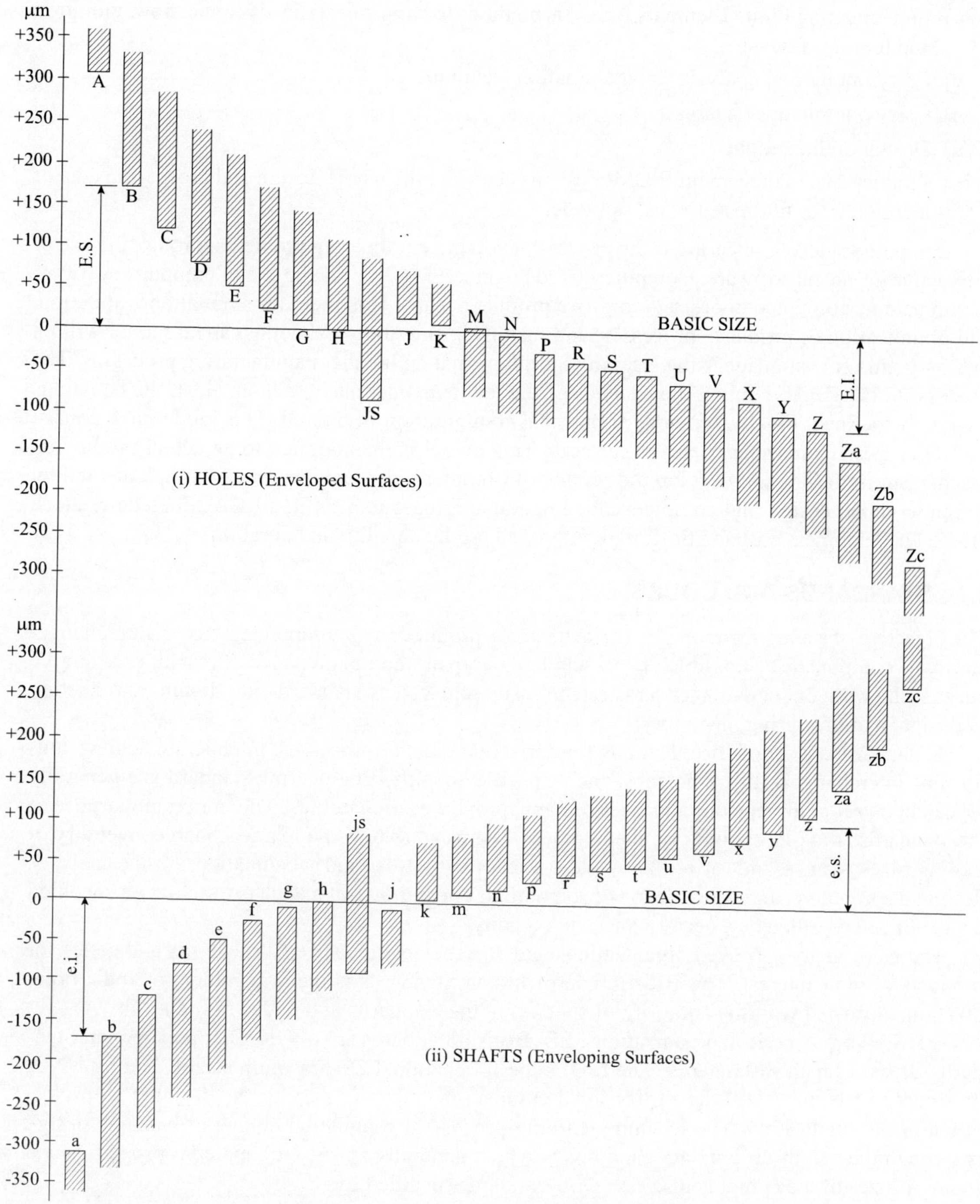

Figure 1.8. Types of fits based on ISO system

iv) Implementing Finite Elements Analysis methods to carry out static, dynamic, flow, vibration and thermal analysis.
v) Carry out motion analysis using animation techniques.
vi) Carry out tolerance analysis.
vii) Design optimization.
viii) Product data management (PDM) systems can store the whole design and processing data of a product, for future reuse and upgrade.

Computer-aided engineering is the process of solving engineering problems through the use of interactive graphical software. Computer-aided engineering (CAE) is the use of computer software to simulate performance in order to improve product design. This includes simulation, validation, and optimization of products, processes, and manufacturing tools. A mathematical model written in a programming language using a set of algorithms that define the manufacturing processes. The process starts by first defining the analysis of the mathematical phenomenon. Next, the equations have to be defined. Finally, a model of physical configuration is created. This model may consist of 2-D or 3-D figures/shapes/curves/surfaces. This model is than applied to an actual production mechanism to design and develop the product. Computer aided engineering is complemented by computer-aided design and computer-aided manufacturing.The benefits of CAE include reduced product development cost and time, with improved product quality and durability.

1.7 STANDARDS AND CODES

The objective of standardization is to make the mass production of components easier. Interchangeability of components is possible due to standards. Repair and maintenance of machines is simple because the worn out or damaged parts can be easily replaced by standard ones. It aims at reduction of design, manufacturing, inventory.

A standard is a set of specifications for parts, materials or processes, intended to achieve uniformity, interchangeability, efficiency, and a specified quality. Engineering standard is a generally agreed upon set of dimensions and sizes of parts, properties of materials. The important purpose of the standards is to place a limit on the number of items so as to provide a reasonable inventory of tooling, sizes, shapes, and varieties. A code is a set of laws, rules and specifications for the analysis, design, manufacture, and construction of something. The purpose of a code is to achieve a specified degree of safety, efficiency, performance, and quality.

There are many approved organizations and societies, which specify standards and codes for products of their interest. The BIS-Bureau of Indian Standards is the government of India organization, entrusted with the authority of specifying the standards and codes in India. The ASME, ASTM, SAE are American organizations. BS- British Standards are of UK. DIN are German Standards. JIS are Japan's Standards. The ISO is the International Organization established by UNO. The codes of ethics of all the professional engineering societies require the designer to protect public health and safety. The teaching learning process of foundation course in machine design is through the use of design data hand books which generally specify the material properties and standard sizes of the components. There are standards specified for–

i) engineering drawing of components.
ii) materials, their chemical compositions, mechanical properties and heat treatment.

iii) shapes and dimensions of commonly used machine elements such as bolts, screws and nuts, rivets, belts and chains, bearings, wire ropes, keys, gears, etc.
iv) fits, tolerances and surface finish of components.
v) testing of products such as pressure vessels, boilers, overhead travelling cranes.
vi) engineering drawing of components.

1.7.1 Advantages of Standardization

i) The standardization reduces the types and number of identical components.
ii) The standardized components can be produced on mass scale at a few selected specialized production centres.
iii) The quality and reliability of products is better.
iv) The availability of standard components, reduces the time and effort of the designer. Now he has to only select from the manufacturer's catalogue.
v) The infrastructure needed to produce the components is not necessary and the cost of the machine is less.
vi) The standardization leads to interchangeability. The servicing and maintenance are easy and economical.

SHORT ANSWER QUESTIONS

1.1. Define machine design.
1.2. Give the block diagram of machine design process.
1.3. What is synthesis?
1.4. What is analysis?
1.5. What is design report?
1.6. What are the types of design?
1.7. What is standardization?
1.8. What are the standards followed in design?
1.9. What is tolerance?
1.10. How many IT grades are there?
1.11. What are the different type of fits?
1.12. What are the manufacturing considerations in design?
1.13. Define ergonomics.
1.14. What are the types of controls?
1.15. What are the types of displays?

CHAPTER 2

MATERIAL SELECTION

2.1 MATERIAL PROPERTIES

The mechanical components perform various functions in a mechanical system. To serve these functions the components have to possess certain characteristics. These are referred to as "design considerations". The following is the list of the factors to be considered during the design-process.

1. Strength
2. Stiffness
3. Toughness
4. Creep
5. Surface finish
6. Wear
7. Lubrication
8. Corrosion
9. Friction
10. Weight
11. Shape
12. Size
13. Volume
14. Cost
15. Reliability
16. Utility
17. Safety
18. Flexibility
19. Processing
20. Noise
21. Thermal properties
22. Maintenance

The factors to be considered in a particular case, depend upon the application. In some cases, a few of these are critical and when they are satisfied, others no longer need to be considered. Some important factors are as follows:

(A) Mechanical Properties

In majority of the design situations, strength, stiffness, toughness, wear, creep, corrosion and friction are important factors influencing the design. They depend on the following mechanical properties.

1. Ultimate strength or ultimate stress – S_u
2. Yield strength or yield stress – S_y
3. Endurance strength or endurance limit stress – S_{en}
4. Modulii of elasticity and rigidity, i.e., E & G
5. Hardness—BHN, VHN, etc.

We look for the materials having a desirable combination of these properties to impart the desired functional characteristics to the components.

(B) Manufacturing Properties

The components are shaped or formed by manufacturing processes like easting, forging, extrusion, welding and machining. The properties like—castability, ductility, weldability, machnability are important. The materials which can be processed economically by these methods will be the natural choice of the designer.

(C) Availability

Material should be available in required quantities, at reasonable price.

(D) Cost

The cost analysts consists of two components (a) the cost of the material, and (b) the cost of processing. It is important that the overall cost must be minimum. The materials selected should have the optimum combination of these factors. Some of the commonly used materials are discussed in the following sections.

2.2 CAST IRONS

The cast irons are alloys of iron and carbon. The carbon content is 2% to 4.5%. Some trace elements are also present. Cast irons are the most widely used materials. They have the following characteristics.

(i) They can be cast easily into a variety of shapes—they have very good castability.
(ii) They can be machined with available tools, i.e., they have good machinability.
(iii) They have high compressive strength and are very much suitable for components subjected to compressive loads, e.g., machine beds
(iv) They have good wear resistance combined with vibration damping capacity. They are suitable for machine bed, guides and frames.
(v) Cast irons are not affected by small variations in temperatures.
(vi) They have low notch sensivity.
(vii) The major drawback of cast irons is their brittleness. They fail without any plastic deformation.
(viii) They are weak in tension and are not suitable for components subjected to tensile loads.

There are four types of cast-irons based on the distribution of carbon particles.

1. Grey cast iron
2. White cast iron
3. Malleable cast iron
4. Nodular cast iron.

1. *Grey cast iron.* It is the most commonly used form of cast iron. It has 2% carbon in the form of pearlite ferrite and graphite. Graphite flakes are uniformly distributed throughout the structure and give grey appearance. It is easy to pour and easy to machine. It has good damping property. This is popularly chosen for machine frames, engine blocks, brake drums

and rotors. It has good wear resistance and lubricity. Specifications for grey-cast-iron are covered in IS210–1993. They are designated by symbol FG followed by tensile strength in MPa. For example, the grey cast iron with tensile strength of 150 MPa is designated by FG150.

2. *White cast iron*. It has cementite and pearlite and no graphite. This is produced by chilling process. The structure is very hard and wear resistant. By selective chilling of a component hard structure can be produced at the surface, where the wear resistance is required as in ploughs and other agricultural equipment.
3. *Malleable cast iron*. It has graphite in the nodular form. The structure is produced by prolonged annealing. It has better tensile strength and ductility. It is used in parts subjected to bending stresses. The specifications for malleable cast iron are contained IS14329–1995. There are three basic types of malleable cast irons-black-heart pearlite and white heart and are designated by symbols BM, PM, and WM respectively and followed by the tensile strength in MPa.
   For example–
   BM350 refers to black heart malleable-cast iron with minimum tensile strength of 350 MPa.
   PM700 refers to Pearlite malleable-cast iron with minimum tensile strength of 700 MPa.
   WM400 refers to white-heart-malleable cast iron with minimum tensile strength of 400 MPa.
4. *Nodular cast iron*. In this structure the graphite particles are spheroidal in shape. The nodular structure is same as in the cast condition. This is achieved by the addition of magnesium and cesium during the casting process. This structure has the highest tensile strength and high modulus of elasticity. It is tougher, stronger, more ductile and less porous than grey cast iron. It is chosen for fatigue-loaded parts such as crankshafts, pistons and cams. They

Table 2.1. Physical constants of metals

Material	Specific mass, kg/lt	Modulus of elasticity, E Gpa	Modulus of rigidity, G Gpa	Thermal conductivity, watts/mK	Coeff. of linear expn. (a), μm/m°K	Poisson's ratio μ
Aluminium	2.7	67.5	26.0	201	23.8	0.34
Brass	8.45	97.0	35.0	110	16.7	0.34
Bronze	8.73	113.0	46.5	180	17.3	0.35
Cast iron	7.20	100.0	40.0	075	11.0	0.218
Copper	8.96	123.0	39.0	385	17	0.34
Lead	11.34	160	7.6	035	29	0.45
Monel metal	8.83	180.0	67.0	210	14.0	0.32
Steel carbon	7.84	210.0	84.5	065	15	0.30
Steel alloy	7.75	210.0	84.5	65	11.1	0.30
Tungsten	19.30	410.0	177.0	200	4.5	0.17

Table 2.2. Mechanical properties of grey cast iron

Properties	Grade						
	FG 150	**FG 200**	**FG 220**	**FG 260**	**FG 300**	**FG 350**	**FG 400**
Tensile strength (N/mm^2)	150	200	220	260	300	350	400
Compressive strength (N/mm^2)	600	720	768	864	960	1080	1200
Shear strength (N/mm^2)	173	230	253	299	345	403	460
Modulus of elasticity (GPa)	100	114	120	128	135	140	145
Modulus of rigidity (GPa)	40	46	48	51	54	56	58
Hardness (HB)	130–180	160–220	180–220	180–230	180–230	207–241	207–27
Relative density (specific gravity)	7.05	7.10	7.15	7.20	7.25	7.30	7.30

are also called ductile-cast iron or spheroidal cast iron. Their specification are covered in IS1865-1991. They are designated by the symbol SG followed by the minimum tensile strength and minimum elongation. For example, SG900/2 refers to spheroidal cast iron with tensile strength of 900 MPa and elongation of 2%.

Alloy cast irons. They are produced by the addition of alloying elements like nickel, chromium and molybdenum. They have higher tensile strength and impact resistance along with good machinability.

The major applications of cast irons are:

(i) Machine-tool beds, frames, tables, guides gears, levers, sprockets.

(ii) I.C. engine cylinders cylinder heads, piston and piston rings, crankcase, exhaust manifolds flywheels, brake-drum, brake shoes, etc.

(iii) Hydraulic cylinders, pipe fittings, valve bodies, dome covers.

Table 2.3. Mechanical properties of some cast iron alloys

Cast-iron alloy	Condition	Tensile yield strength (2% offset), S_y MPa	Ultimate tensile strength, S_{ut} MPa	Compressive strength, S_{uc} MPa	Brinell hardness, HB
Grey Cast Iron—Class 20	as cast	–	152	572	156
Grey Cast Iron—Class 30	as cast	–	221	752	210
Grey Cast Iron—Class 40	as cast	–	290	965	235
Grey Cast Iron—Class 50	as cast	–	359	1131	262
Grey Cast Iron—Class 60	as cast	–	427	1289	302
Ductile Iron 60-40-18	annealed	324	448	359	160
Ductile Iron 65-45-12	annealed	331	462	365	174
Ductile Iron 80-55-06	annealed	365	565	386	228
Ductile Iron 120-90-02	*Q & T*	827	965	924	325

2.3 STEELS

The steels are alloys of iron and carbon with carbon content less than 0.8%, i.e., less than that in cast iron. Depending upon the process, they may be classified into two groups.

1. Cast steels
2. Wrought steels

Cast steels are produced by sand casting or investment casting. The mechanical properties of cast steels are better than the cast irons. The tensile strengths range from 450 MPa to 1350 MPa. They can be cast into intricate shapes. They can be heat treated. They are classified into two groups.

1. Carbon steel casting
2. High tensile steel casting

Carbon steel castings (IS1762-1974) are used for heavy duty machinery, gears and wheels. They are classified into five groups; each group being represented by Grade $x - y$ where x represents 1/10 of the yield-strength and y represents 1/10 of the ultimate strength in MPa. They are Grade 20–40, Grade 23–45, Grade 26–32, Grade 27–54, Grade 30–57.

High tensile steel castings have higher strength, good toughness, high wear resistance. They are used for transportation and agricultural machinery. They are classified into five groups, namely, CS640; CS700; CS840; CS1030; CS1230. The number represents the ultimate tensile strength in MPa.

Wrought steels. (IS1570-1978) They are produced by processes that manipulate the shape of the material without melting it. Hot-rolling and cold-rolling are the two commonly used methods. Wrought steels have better mechanical properties compared to cast steels.

Table 2.4. Properties of carbon steels (IS:1570 and IS:1871)

IS Code	Ultimate tensile strength (MPa)	Yield strength (MPa)	BHN	Applications
C07	320–400	200	–	(1) Cold-drawing, deep drawing, forging and heat-treating application: Auto parts-Cams-camshaft-light-duty gears; worms, gudgeon pins-selector fork pawl and rachet; chain wheels; spindles, rivets, etc.
C10	340–420	210	–	
C14	370–450	220	137	
C15	370–490	240	137	
C15 Mn 75	420–500	250	163	
C20	440–520	260	156	
C25	440–540	280	170	
C25 Mn 75	470–570	280	207	
C30	500–600	300	179	(2) Levers, tie-rods-cables, sprockets, hubs, bushes fasteners, machine structure–motorcycle frames—fish plates, etc.
C35	520–620	310	187	
C35 Mn 75	550–650	320	223	
C40	580–680	330	217	(3) Shaft spikes, bolts, crankshaft auto axles-push rods—connecting rods gears, worms, bolts, cylinder liners, nails spike-bolts, gearshaft, rocket—gears.
C45	630–710	360	229	
C50	660–780	380	241	
C50 Mnl	720 min	400	255	(4) Rails spick-bolts, gears shaft rocker—gears, cylinder liners.
C55 Mn 75	720 min	400	265	
C60	750 min	420	255	(5) Machine—tool spindles hardened screws, nuts—coupling—crankshafts, pinions, locomotive—carriage and wagon tyres and springs—valve springs washers, stamped parts.
C65	750 min	430	255	

Hot-rolled steels are produced by forcing the hot billets through a set of rollers or dies which progressively change their shape into I-beams, channel sections, angle sections, flats, square rods, round rods, tubes, sheets, plates, etc. The surface of the hot-rolled shapes is rough and the mechanical properties are also low, but can be improved by treatment process. They are used for building and machine frame construction.

The cold: rolled steels are produced from billets or hot-rolled shapes. The final shape is achieved by rolling between hardened steel rollers or drawing through dies at room temperatures. The rolls or dies burnish the surface, increasing its strength and reducing its ductility. The material is hardened

Table 2.5. Properties of typical carbon steels

Material SAE code (ISI)	Ultimate strength, MPa: Tension S_{ut}	Ultimate strength, MPa: Shear S_{us}	Yield strength, MPa: Tension S_y	Yield strength, MPa: Shear S_{sy}	Endurance limit stress S'_{en} MPa	BHN	Suggested use
Cast Steel Soft	420	300	200	100	185	110	General purpose casting
Medium	490	350	220	135	210	120	
Hard	560	395	255	150	240	130	
SAE 1025 (C25)							Machinery and general-purpose steel
Annealed	470	290	240	140	205	120	
Water quenched	550	390	290	170	300	159	
	630	440	410	240	350	183	
SAE 1045 (C45)							
Annealed	600	420	315	185	300	140	Large forgings and axle, shaft
Water quenched	670	470	420	245	370	197	
	845	590	630	365	470	248	
Oil quenched	675	475	435	245	372	192	
	810	560	560	315	455	235	
SAE 1095 (C95)							
Annealed	775	530	390	230	370	200	Springs; cutting instruments
Oil quenched	915	595	450	475	475	300	
	1320	845	915	530	705	280	

Upper figures—steel quenched and drawn at 5.50°C; lower figures—steel quenched and drawn at 430°C.

and there are built in strains in the material. Hence, the process should be followed by a suitable heat-treatment process.

Based on the alloying elements the steels are classified as follows:

1. Plain carbon steels
2. Alloy steels
3. Strainless steels
4. Free-cutting steels.

Plain carbon steels: (IS1570–1978) They contain carbon from 0.2% to 0.8% and a small percentage of manganese, silicon and phosphorus. They are designated by the letter C, followed by a number representing the average percentage of carbon multiplied by 100. A steel with 0.35% to 0.45% carbon is represented by C40. In the new BIS-system, the designation has three parts. (i) A figure (number) indicating 100 times the average percentage of carbon; (ii) letter C (iii) a figure (number) indicating 10 times the average percentage of manganese.

For example, 55C8 represents a plain carbon steel with 0.55% carbon and 0.8% manganese. They are classified into three groups.

(i) Low carbon steel with carbon percentage less than 0.2%
(ii) Medium carbon steels with carbon percentage—range from 0.2% to 0.55%
(iii) High-carbon steels with carbon-percentage greater 0.55% but less than 0.8%

The applications are

Low-carbon steels (C10, C20, C30). They are used for automobile bodies hoods, cam and cam-shaft-gudgeon pin, chain-wheel, spindle, rocker arm, lever, etc.

Medium carbon steels (C40, C45, C50, C55) are used for transmission shafts crank-shafts, spindles, connecting rods, gears, worms, bolts, studs, cylinders, cams sprockets.

High carbon steels (C60, C65) are used for machine-tool-spindles high-tensile bolts, springs, etc.

2.4 ALLOY-STEELS (IS1570-1978)

They are steels with one or two alloying elements in sufficiently large quantities to modify its properties satisfactorily. The symbol for the alloy steels consists of three parts: (i) The number indicating the 100 times the carbon percentage; (ii) the symbol of the significant alloying element; and (iii) number indicating the average percentage of the alloying element multiplied by a factor for the element given in the table below.

Element	Multiplying factor
Cr, Ca, Ni, Mn, Si, W	4
Al, Be, V, Pb, Cu, Mb Ti, Ta, Zs, Mo	10
P, S, N	100

Examples:

(i) 40Cr4—is an alloy containing 0.35%–0.45% carbon 1% chromium and is called chromium alloy steel.

(ii) 50Cr4V23—is chrome-vanadium steel with 0.45%–0.55% carbon, 1% chromium 23% vanadium.

(iii) 27Mn2—is manganese-steel with 0.27% carbon and 2% manganese.

The effects of various alloying elements are explained below.

Chromium (Cr). The presence of chromium increases toughness, hardness, resistance to wear and corrosion.

Manganese (Mn). It is a deoxidizing and desulfurizing agent. It increases hardness and strength. It counteracts the brittleness due to sulphur. It prolongs the transformation time so that oil quenching becomes possible.

Nickel (Ni). Its presence increase the strength, toughness without losing ductility. It prevents distortion during heat treatment.

Silicon. It is a deoxidizing agent and stabilizes the carbides of other alloying elements. It reduces the hysteresis loss and improves magnetic permeability.

Molybdenum. It improves hardness, toughness and hardenability. It also improves the creep properties.

Vanadium. It is a strong deoxidizing agent. It increases the hardening range and retains hardness at elevated temperatures. It increases the endurance strength.

Tungstem. It produces a fine, dense structure, increases both hardness and toughness. It is widely used in tool steels.

The applications of alloy steels are as follows.

Chromium and nickel steels > [40Cr4; 40Ni3; 35Ni1Cr60; 30Ni4Cr4] are used for gears connecting rods, crankshafts, camshaft, gear shafts and high strength bolts.

Manganese steels > [27Mn2; 37Mn2] are used for transmission shaft, axles and welded parts.

Chrome-vanadium steels > [50Cr4V23] are used for leaf and coil springs.

The high alloy steels contain more than 10% of the alloying element. They are designated as follows:

(i) A letter 'X' followed by (ii) a figure indicating 100% of average carbon percentage

(iii) The chemical symbol of the alloying element followed by the average percentage of the alloying element

(iv) Chemical symbol of the specially added element to acquire the desired properly.

For example, X20Cr18Ni2 is a high alloy steel containing 0.20% Carbon; 18% Chromium; 2% Nickel.

Stainless steels. They are alloy steels with more than 12% chromium, nickel, molybdenum and additional alloying elements to improve the corrosion and heat-resistant properties.

Chromium steels with 12% to 16% Cr are magnetic and hardenable. They are used for hydraulic-pumps, valves and steam valves.

Chromium steels with more than 16% chromium are magnetic but non-hardenable. They have good ductility but have low strength at high temperatures. They are used for cold-press work.

Chrome-nickel and chrome-vanadium steels have austenitie structure. They are non-magnetic, resistant to corrosion, hardenable and weldable. They are used for springs and high temperature applications.

Method of designation free-cutting steels. They are medium to high carbon-manganese steels with small percentage of sulphur. They have good machinability. The free-cutting steels are designated by a number indicating 100 times the average percentage of carbon followed by letter 'C' then a number indicating 10 times the manganese content followed by the symbol of the alloying element along with a number 100 times the average percentage. For example, 20C12Pb15 is a free cutting steel with 20% Carbon, 12% Manganese; 0.15% Lead.

14C12S14, 10C8S10—steels are used for bushes, tappets gudgeon pins, pawls, pinions, etc., 25C12S14, 40C10S18, 40C15S12 are used for bolts and studs.

Table 2.6. Properties of alloy steels (IS:1570 and IS:1871)

IS Code	Ultimate tensile strength, S_u MPa	Yield strength, S_y MPa	BHN	Applications
20 Mn2 27 Mn2 37 Mn2 35 Mn2 Mo28 35 Mn2 Mo45	600–850 600–850 600–1050 700–1150 800–1150	440–500 440–500 440–700 540–800 600–800	170–248 170–248 170–311 200–341 229–341	Welded structure, shafts, crankshafts-axles, spindles, levers, connecting rod, gears, bolts and stud.
40 Crl 40 Crl Mo28 15 Cr3 Mo55 25 Cr3 Mo55	700–1050 700–1150 700–1250	540–700 540–800 540–880	201–311 201–341 201–363	Parts requiring high surface handness and wear resistance bearings, gears.
40 Cr3 Mol V20 40 Cr2 All Mol 18	1350–1520 700–1050	1120–1275 540–700	363–444 201–311	Heavy duty forgings-turbine blades—highly stressed screws—bolts and nuts—components of refrigerators and compressors.
40 Ni3 31 Ni3 Cr65 Mo55 35 Nil Cr60 40 Ni Crl Mo15 40 Ni2 Crl Mo28 40 Ni3 Cr65 Mo55	800–1050 900–1350 700–1050 800–1250 800–1350 1000–1350	600–700 700–1000 540–700 600–880 600–1000 800–1000	299–311 255–401 201–311 229–363 229–401 285–401	Aircraft, heavy vehicle components, highly stressed gears, collets, spindles, bolts, etc.

Table 2.7. Mechanical properties of high tensile steel castings

Grade	S_{ut}, MPa (Min)	S_y, MPa (Min)	Elongation %	Brinell Hardness (HB)
CS 640	640	390	15	190
CS 700	700	560	14	207
CS 840	840	700	12	248
CS 1030	1030	850	8	305
CS 1230	1230	1000	5	355

Table 2.8. Properties of typical alloy steels (SAE Code)

Material SAE code	Ultimate strength, MPa		Yield strength, MPa		Endurance limit in reversed bending, S'_{en} MPa	BHN	Suggested use
	Tension, S_{ut}	Shear, S_{us}	Tension, S_y	Shear, S_{sy}			
Nickel Steel **SAE 2330**							
Water quenched	540 985	390 690	350 775	210 455	345 480	143 277	Case hardening stock for heavy parts
Oil quenched	530 915	380 635	340 705	205 420	325 350	140 262	Gears
SAE 2340							
Water quenched	670 1250	420 775	460 1050	270 610	375 530	183 340	Forgings, Axles
Oil quenched	655 1160	415 745	435 1050	253 600	390 530	183 330	

Material SAE code	Ultimate strength, MPa		Yield strength, MPa		Endurance limit in reversed bending, S'_{en} MPa	BHN	Suggested use
	Tension, S_{ut}	Shear, S_{us}	Tension, S_y	Shear, S_{sy}			
Nickel-Chromium-Steels **SAE 3120** Water quenched	605 985	420 690	400 810	240 480	390 410	174 269	Heavy section required in medium depth case hardening
Water quenched	540 845	390 575	340 670	205 400	320 340	163 241	
SAE 3220 Water quenched	610 1160	420 810	420 1000	250 585	410 510	187 331	Massive section required in medium depth case hardening
Oil quenched	560 990	395 750	375 915	220 670	360 460	174 311	
SAE 3240 Oil quenched	775 1400	515 915	615 1260	365 740	590 635	229 388	Rollers, sprockets gears
Chrome-Molybdenum-Steels **SAE 4140** Oil quenched	845 1335	635 1050	705 1160	390 640	420 670	240 380	-do-

(Contd.)

Table 2.8. Contd.

Material SAE code	Ultimate strength, MPa		Yield strength, MPa		Endurance limit in reversed bending, S'_{en} MPa	BHN	Suggested use
	Tension, S_{ut}	Shear, S_{us}	Tension, S_y	Shear, S_{sy}			
Chrome-Vanadium-Steels **SAE 6145**							
Oil quenched	735	525	690	380	660	220	Roller sprockets gears
	1600	1260	1470	815	735	425	
Silicon-Manganese-Steels **SAE 9260** Oil quenched	1100	845	705	420	435	240	Springs, gears
Stainless steel 0.12C, 18.0Cr 8.0 Ni	635 1400	420 1050	245 1230	140 705	280 635	135 380	-do-

upper figures–*Q* & *T* at 730°C
lower figures–*Q* & *T* at 425°C

2.5 ALUMINIUM AND ITS ALLOYS (IS5047)

Aluminium is the most widely used nonferrous metal. Its usage is second only to steels. Aluminium is produced in both pure and alloyed form. The most common alloying elements are copper, silicon magnesium, manganese and zinc. Favourable characteristics of aluminium and its alloys are:

(i) High strength to weight ratio
(ii) Resistance to corrosion
(iii) High thermal conductivity and electrical conductivity
(iv) Excellent workability—castability and weldability
(v) Low cost.

Pure aluminium is rarely used in engineering applications. The aluminium alloys are used extensively in aircraft and automobile industries. The alloys are available in two forms: (i) castings and (ii) wrought.

The alloys are designated by a digital-number system. The alloying elements are represented by the following numbers.

Aluminium – 1	Magnesium – 5
Copper – 2	Magnesium Silicide – 6
Manganese – 3	Zinc – 7
Silicon – 4	Other elements – 8

The cast aluminium alloys are specified by a four-digit number system and the wrought aluminium alloys are specified by a five-digit number system. The meaning of the digits is as follows.

First digit represents major alloying element.

The second digit represents the average percentage of the alloying element halved and rounded off.

Third, fourth, fifth digits indentify the minor alloying elements in the order of decreasing percentage.

Example, **4683** has silicon as major alloying element with average of 12%. Fe and Mn are minor alloying elements.

24345 has copper as the major alloying element with average of 8% manganese, silicon and magnesium are minor alloying elements in the decreasing order of percentage.

Aluminium alloys are generally used for:

(1) Rapidly reciprocating or oscillating parts such as pistons and connecting rods of high-speed engines.

(2) High rotating parts—like pulleys, bearing cages, engine components.

(3) Intricate castings of engines, and thin walled structures.

(4) Covers, hoods and guards for hydraulic machines.

(5) Welded cylinders, pressure vessels, and structure of aircraft.

(6) Forged and deep-drawn parts.

Important applications of specific aluminium alloys are given below.

Cast Alloys:

4450 – Cylinder block valve body, fan blade.

4600 – Water cooled manifolds of engines and pumps.

4685 – I.C. engine pistons.

2280 – Connecting rod and fly wheel.

2285 – Piston and air cooled cylinders.

2550 – Hydraulic cylinder and bearing cap.

Wrought Alloys:

24345 – Heavy duty forgings and structures.
24534 – Critical parts of aircraft.
54300 – Welded structures like car tank.
64430 – Roof trusses and deep-drawn containers.
74530 – Welded pressure vessels.

Table 2.9. Mechanical properties of some wrought aluminium alloys

Wrought aluminium alloy	Condition	Tensile yield strength (2% offset), MPa	Ultimate tensile strength, MPa	Fatigue strength at 5E8 Cycles, MPa	Elongation over 2 in %- MPa	Brinell hardness, HB
1100	Sheet					
	annealed	34	90	–	35	23
	cold rolled	152	165	–	5	44
2024	Sheet					
	annealed	76	179	–	20	–
	heat treated	290	110	138	19	–
3003	Sheet					
	annealed	41	110	–	30	28
	cold treated	186	200	–	4	55
5052	Sheet					
	annealed	90	193	–	25	47
	cold treated	255	290	–	7	77
6061	Sheet					
	annealed	55	124	–	25	30
	heat treated	276	310	97	12	95
7075	Bar					
	annealed	103	228	–	16	60
	heat treated	503	572	97	11	150

Table 2.10. Typical uses and characteristic properties of aluminium alloys for general engineering purposes

Designation	Tensile strength, MPa		Use and characteristic properties
	Chill	Send	
A-1	126	157	Medium stressed castings such as timing gears, general-purpose machine parts, moulding flask and pattern casting possess, moderate strength. Non-heat treatable alloy.
A-2	126	150	Used for positions and similar parts, combines high wear resistance with a good strength at elevated temperatures. A low coefficient of thermal expansion.
A-6	165	189	Suitable for sand casting, gravity and pressure die cast and for casting intricate patterns. It has good corrosion resistance and medium strength.
A-9	244	299	Used for cylinder blocks, cylinder heads, valve bodies, large fan blades, pneumatic tools, etc. Has good castability, resistance to chemical attack and good machining and welding characteristics, can be fully heat treated.
A-14 (Y-Alloy)	220	283	Used mainly for piston and cylinder heads. Has high strength at elevated temperatures and good machinability.
A-11	118	142	Used for electrical and household appliances. Has good castability and resistance to corrosion.
A-24	–	184	Used mainly for pressure dye casting.

2.6 COPPER ALLOYS (IS3288; IS2378-1974; 191-2007)

They have the following favourable properties:

(1) Excellent bearing properties
(2) High thermal conductivity
(3) Good corrosion resistance
(4) Excellent machinability
(5) Ease of fabrication.

The copper alloys are grouped under two broad categories:

(a) Brasses
(b) Bronzes

Brasses. They are copper and zinc alloys with varying proportions. The complex brasses contain additional elements like lead-silicon, manganese, aluminium, iron, nickel and tin. The brasses are available as (i) cast brasses and (ii) wrought brasses. The cast brasses have excellent foundry properties and can be casted into complicated shapes easily.

The wrought-brasses can be cold-worked. They can be rolled into thin sheets and can be drawn into thin tubes. The brasses have good corrosion resistance, good-electrical conductivity and good machinability.

The brasses are used for (i) tubings (ii) wires (iii) electric fittings and machines (iv) instruments.

The various brasses are:

Gilding brasses—used for gold plated articles—in jewellery.

Red-brass—used for tubings and piping's in radiators and condensors.

Low brass—used for deep-drawn articles.

Cartidge brass—It has better ductility combined with strength. It is used for machine screw parts, instruments, lock and watch parts.

Admiralty brass—It has good corrosion resistance. It is used for power plant and chemical equipments.

Aluminium brass—It is used for piping in chemical plants.

Bronzes are mainly copper-tin alloys but other elements like lead, aluminium, beryllium, silicon may be added as alloying elements. The bronzes in general have high antifriction properties, good corrosion resistance and good processing properties.

They are widely used for (i) friction joints like sleeve bearings, guides, worm wheels, helical gears and power screws (ii) for water, steam and oil fittings.

The various types of bronzes are as follows:

Silicon bronze. It has good strength and corrosion resistance. It can be cold or hot worked, machined and welded.

Phospher bronze. It is resistant to fatigue and corrosion. It has good tensile strength and high capacity to absorb energy. It is used for nut, worm wheel, springs, bearings pumps, etc.

Aluminium bronze. It has high strength and toughness, high corrosion resistance. It is used for hydraulic valve, bearings, cam followers and gears.

Beryllium bronze. It has corrosion resistance, high strength, hardness, and resistance to wear. It is used for springs subjected to fatigue.

Lead bronze. It is very good bearing material, but has low hardness and requires hard backing material.

Babbit metals. They are alloys of soft metals like tin, lead and copper. They have good antifriction properties and are used for bearings operating at high speeds and under high pressures.

Table 2.11. Mechanical properties for some casting-aluminium alloys

Casting aluminium alloys	Condition	Tensile yield strength (2% offset) MPa	Ultimate tensile strength, MPa	Fatigue strength, MPa
43	Permanent mould casting – as cast	62	159	45
195	Sand casting – as cast	165	248	–
220	Sand casting – solution heat treated	179	331	75
380	Die casting – as cast	165	331	–
A1132	Permanent mould casting heat treated @ 170°C	296	324	125
A142	Sand casting – heat treated @ 345°C	207	221	85

Table 2.12. Mechanical properties for some wrought and cast-copper alloys

Copper alloys	Condition	Tensile yield strength (2% offset) MPa	Ultimate tensile strength, MPa	Elongation over 2 in %	Brinell or Rockwell hardness, HB
CA-110- Pure Copper	Strip annealed Spring temper	69 345	221 379	45 4	40 HRF 60 HRB
CA 170- Beryllium Copper	Strip annealed plus age Hard plus age	1000 1172	1138 1310	7 3	35 HRF 40 HRC
CA 220- Commercial Bronze	Strip annealed Spring temper	69 427	255 496	45 3	53 HRF 78 HRB
CA-230 Red Brass	Strip annealed Spring temper	103 414	276 517	50 7	50 HB 135 HB
CA-260- Cartridge Brass	Strip annealed Spring temper	76 448	303 648	66 3	54 HB 91 HB

(Contd.)

Table 2.12. Contd.

Copper alloys	Condition	Tensile yield strength (2% offset) MPa	Ultimate tensile strength, MPa	Elongation over 2 in %	Brinell or Rockwell hardness, HB
CA-270-Yellow Brass	Strip annealed Spring temper	97 427	317 627	65 30	58 HRF 90 HRF
CA 510-Phosphor Bronze	Annealed Spring temper	131 552	324 689	64 4	73 HRF 95 HRF
CA 614-Aluminium Bronze	Soft Hard	310 414	565 614	40 32	84 HRB 87 HRB
CA 655-High Silicon Bronze	Annealed Spring temper	145 427	386 758	63 4	76 HRF 97 HRB
CA 675-Manganese Bronze	Soft Half-hard	207 414	448 579	33 19	65 HRB 90 HRB
Leaded-Tin Bronze	As cast	131	234	18	60 HB
Nickle-Tin Bronze	As cast Cast and heat treated	138 379	345 586	40 10	85 HB 180 HB

Table 2.13. Properties of non-ferrous materials

Materials	Ultimate strength, MPa	Yield point tension, MPa	E GPa	G GPa	Uses
1. Copper					Electrical industries, pipe, tubes, wire rods wires, pipes, tubes, strips, plates, sheets
(a) Annealed	224	–	108.5	44.30	
(b) Cold drawn	392	182	108.5	44.30	
(c) Cold rolled	364	–			
2. Brass					Valves, fittings, pumplines, general parts requiring corrosion resistance, general building hardware, tank
(a) Cast	280	126	–	–	
(b) Annealed	378	343	–	–	
(c) Cold drawn	670	–	105.0	42.0	
3. Broze					Bearings of all kinds, general building hardware
(a) Cast	150	–	113	–	
(b) Cold drawn	595	–	–	–	
4. Aluminium					Airplane and automobile parts exposed to wear. Automobile building hardware
(a) Cast	413	–	67	26	
(b) Extruded	546	–	–	–	
5. Monel metal					General purpose under corrosion condition: abrasive resistant
(a) Cast	490	525	175	68	
(b) Hard rolled	665	–	–	–	
6. Babbit	840	–	–	–	Lining for bearing, can be dye cast

2.7 PLASTICS

They are materials of high-molecular organic compounds, generally processed synthetically by heat and pressure. The plastics can be processed easily. They can be moulded, extruded, machined and bounded. They have a wide range of mechanical and physical properties.

The favourable characteristics are:

(i) Low density

(ii) High thermal and electrical resistance (insulator)

(iii) High damping capacity

(iv) Low coefficient of friction and self-lubrication

(v) Wear resistance

(vi) Ease of fabrication into complex shapes.

The unfavourable characteristics are:

(i) Low strength

(ii) Poor heat resistance

(iii) High thermal expansion

(iv) High creep

There are various types of plastics with different properties. They are grouped under two categories:

(a) Thermoplastics, which soften under heat, harden on cooling and can be resoflened again by heating, to be formed again.

(b) Thermosetting plastics which undergo a chemical change due to high temperature during manufacturing and cure to a infusible shape, not permitting any reforming.

Some of the commonly used plastics are as follows:

(i) *Nylon (Polymide)*. It has good mechanical strength, resistance to heat, and abrasion. It has self-lubricating property. It is used for bearings, cams, valve seats, gears.

(ii) *Polythene*. It is used for gaskets and washers.

(iii) *Delrin (Acetal)*. It is used for bearings cams and gears.

(iv) *Texin (Polyurethane)*. It is used for bearings gears, gaskets and seals.

(v) *Teflon (Poly tetra fluoro ethylene)*. It is used for piston rings, seals for hydraulic cylinders, bearings, gaskets and belows.

Table 2.14. Properties of some engineering plastics

Material	Approximate modules of elasticity, E GPa	Ultimate tensile strength, MPa	Ultimate compressive strength, MPa	Elongation over 2 in %	Maximum temperature, °C	Specific gravity, ρ
ABS 20–40% glass filled	2.1 4.1	41.4 68.9	68.9 82.7	5 to 25 3	160 – 200 200 – 230	1.05 1.30
Acetal 20–30% glass filled	3.4 6.9	60.7 68.9	124.1 124.1	60 7	220 185 – 220	1.41 1.56
Acrylic	2.8	68.9	103.4	5	140 – 190	1.18
Fluoroplastic (PTFE)	1.4	34.5	41.4	100	350 – 330	2.10
Nylon 6/6	1.4	68.9	68.9	60	180 – 300	1.14
Nylon 11 20–30% glass filled	1.3 2.5	55.2 88.3	55.2 88.3	300 4	180 – 300 250 – 340	1.04 1.26

(Contd.)

Table 2.14. Contd.

Material	Approximate modules of elasticity, *E* GPa	Ultimate tensile strength, MPa	Ultimate compressive strength, MPa	Elongation over 2 in %	Maximum temperature, °C	Specific gravity, ρ
Polycarbonate	2.4	62.1	82.7	100	250	1.20
10–40% glass filled	6.9	117.2	117.2	2	275	1.35
HMW polyethylene	0.7	17.2	–	525	–	0.94
Polyphenylene oxide	2.4	62.2	113.1	20	212	1.06
20–30% glass filled	7.8	106.9	120.7	5	260	1.231
Polyphenylene	1.4	34.5	48.3	500	250 – 320	0.90
20–30% glass filled	4.8	51.7	42.7	2	300 – 320	1.10
Impact polystryrene	2.1	27.6	41.4	2 to 80	140 – 175	1.07
20–30% glass filled	0.7	82.7	110.3	1	180 – 200	1.25
Polysufone	2.5	70.3	95.8	5	300 – 345	1.24

2.8 RUBBER

The rubber is a soft and elastic material. It may be natural or synthetic. It has:

(a) hardness and elasticity over a wide range;
(b) high energy absorption capacity and good damping capacity;
(c) good resistance to attrition and the action of many aggressive media; and
(d) very high dielectric properties.

Rubber can be hardened by the addition of sulphur. The rubber can be reinforced by textile and steel elements to impart high-tensile strength.

It is used in the manufacture of tire shock absorbers, clutches, belts, hoses, seals and packings, electric-insulation applications.

2.9 ECONOMIC CHOICE OF MATERIALS

The principal methods of economising the material are:

1. Choice of optimal types of parts and their design features. The fish-belied beam has less mass than constant cross section. The welded designs require less material than riveted designs.
2. Choice of optimum parameters. Higher design speeds reduce the mass of the parts.
3. Choice of optimal material and their heat-treatments and use of bimetallic or non-metallic parts. By hardening, the strength can be increased and mass of the parts can be reduced. The use of bimetallic liners with a thin antifriction layer reduces the cost of piston.
4. Choice of smaller factor of safety and more accurate methods of calculations.
5. Choice of simplest and cheapest shapes for blanks. The blanks in form of tubings, rolled shapes, finish drawn material means smaller machining allowance and reduced wastage in the form of chips.

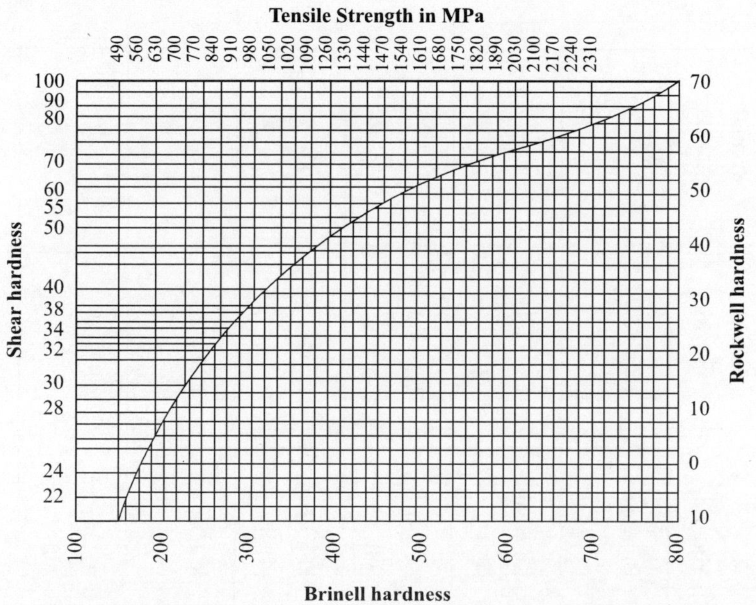

Figure 2.1. Hardness to tensile strength conversion chart

Table 2.15. Physical properties of rubber-like material

Material	Specific gravity	Compressive strength, S_c MPa	Tension strength, S_y MPa	Flexural strength, S_b MPa	Hardness shore durometer	Maximum temperature for use°C	Effect of heat
Duperence	1.27 – 3.00	–	15 – 28	–	15 – 95	150	Stiffens slightly
Koreseal hard	1.3 – 1.4	–	14 – 63.5	–	80 – 100	100	Softens
Koreseal soft	1.2 – 1.3	–	3.5 – 11.5	–	30 – 80	88	Softens
Philoform plastic	1.06	60 – 78	28 – 35	49 – 63	–	70 – 120	Softens
Rubber hard	1.12 – 2	14 – 105	7 – 70	63 – 105	50 – 80	54 – 70	Softens
Rubber soft	0.97 – 1.25	–	37 – 42	–	–	65 – 93	Softens
Rubber	0.98 – 1.35	–	–	–		–88	Softens

EXERCISES FOR PRACTICE

2.1. Explain the desired mechanical properties, the material for a machine component, should possess.
2.2. Enumerate various factors influencing selection of materials.
2.3. What is cast iron?
2.4. What are the elements in cast iron?
2.5. How are the cast irons classified?
2.6. Which are the important properties of cast irons from design consideration?
2.7. Enumerate the applications of cast irons.
2.8. How are the cast irons designated?
2.9. What are the steels? How are they classified?
2.10. What are plain carbon steels? How are they designated?
2.11. What are the advantages and disadvantages of plain carbon steels?
2.12. What are the applications of plain carbon steels?
2.13. What are the alloy steels and how are they designated?
2.14. Name the alloying elements in alloy steels and explain their effect on the properties.
2.15. What are the advantages and disadvantages of alloy steel?
2.16. What are applications of alloy steels?
2.17. What are the advantages and disadvantages of alluminium and its alloys?
2.18. How are the alluminium alloys designated?
2.19. Enumerate the applications of aluminium alloys.
2.20. What are the copper alloys? What are advantages of copper alloys?
2.21. What are brasses? What are their applications?
2.22. What are bronzes? What are their applications?
2.23. What are plastics and how are they classified?
2.24. Enumerate the advantages and disadvantages of plastics?
2.25. What are the applications of plastics?
2.26. What are the favourable properties of rubber?
2.27. What are the applications of rubber?
2.28. What are the methods for economizing the material usage?

SHORT ANSWER QUESTIONS

2.1. What are the important material properties considered in design?
2.2. What are the materials in design?
2.3. What is cast iron?
2.4. What are the types of cast irons?
2.5. How are the cast irons designated?
2.6. What are the advantages of cast irons?
2.7. What are the disadvantages of cast irons?
2.8. What are the applications cast irons?
2.9. What are the steels?

2.10. What are types of steels?
2.11. How are the plain carbon steels designated?
2.12. How are the alloy steels designated?
2.13. What are the advantages of steels?
2.14. What are the disadvantages of steels?
2.15. What are the applications of steels?
2.16. What are alloying elements used in alloy steels?
2.17. What are the brasses?
2.18. What are the types of brasses?
2.19. What are the bronzes?
2.20. What are types of bronzes?
2.21. What are the advantages of aluminium alloys?
2.22. What are the applications of aluminium alloys?
2.23. What are plastics?
2.24. What are types of plastics?
2.25. What are the applications of plastics?

CHAPTER 3

STRESS AND STRAIN ANALYSIS

3.1 DEFINITION OF STRESS AND NOTATIONS

When a machine part or an elastic body is subjected to external forces called surface tractions it is deformed and intermolecular forces are developed to resist this deformation. These forces are referred to as internal tractions. These tractions are distributed throughout the body.

They are different on different planes. The intensity of these tractions on a particular plane through a point is called stress.

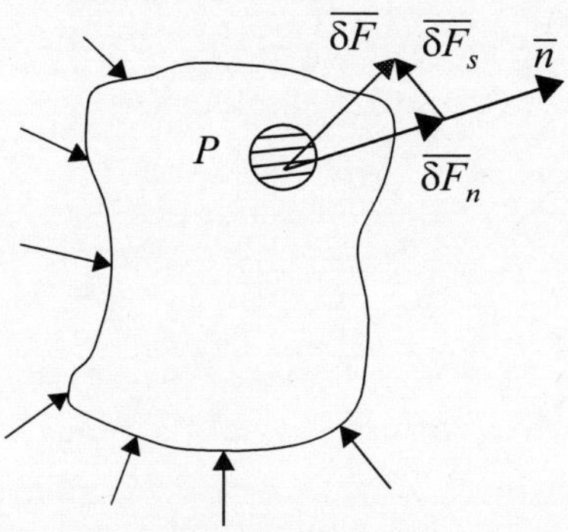

Figure 3.1

Let (Fig. 3.1)

δF = the total of fractions on elemental area δA of a plane though point 'P'

n = normal to the plane

$\overline{\delta F_n}$ = component of this force along the normal

$\overline{\delta F_s}$ = component along the plane

The normal stress is defined as

$$\sigma_n = \lim_{\delta A \to 0} \left(\frac{\overline{\delta F_n}}{\delta A} \right) = \left(\frac{dF_n}{dA} \right) \tag{3.1}$$

The shear stress is defined as

$$\tau_{ns} = \lim_{\delta A \to 0} \left(\frac{\overline{\delta F_s}}{\delta A} \right) = \left(\frac{dF_s}{dA} \right) \tag{3.2}$$

For a simple bar subjected to a force F, distributed uniformly on its cross section,

Normal stress is

$$\sigma_n = (F_n/A)$$

Shear stress is

$$\tau_{ns} = (F_s/A)$$

Figure 3.2

where F_n, F_s, are the normal and tangential components of F (Fig. 3.2).

For mathematical analysis of the distribution of the stress they are referred to a coordinate system more commonly a rectangular system. One of the axis is taken perpendicular to the plane and other two lying in the plane, the shear stress being resolved parallel to these two directions.

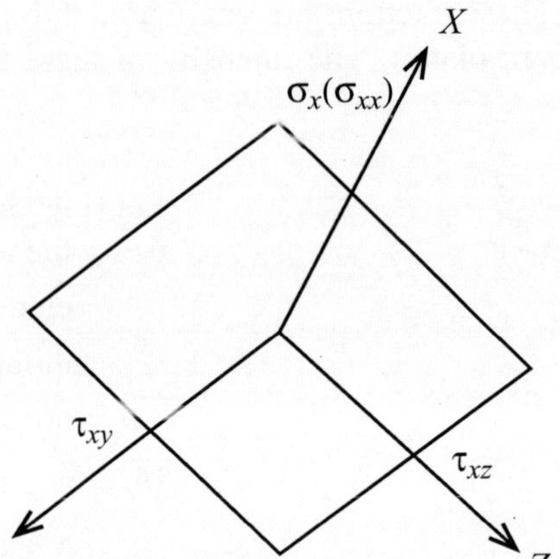

(a)

Figure 3.3

The normal component is represented by σ_{xx} and the shear componenets by τ_{xy} and τ_{xz}. The first suffix represents the plane and the second represents the direction of the stress.

They are written in the matrix form as follows.

$$\begin{bmatrix} \sigma_{xx} & \tau_{xy} & \tau_{xz} \\ \tau_{yx} & \sigma_{yy} & \tau_{yz} \\ \tau_{zx} & \tau_{zy} & \sigma_{zz} \end{bmatrix}$$

This is called stress-tensor with

$$\tau_{xy} = \tau_{yx}, \quad \tau_{yz} = \tau_{zy} \quad \text{and} \quad \tau_{zx} = \tau_{xz}$$

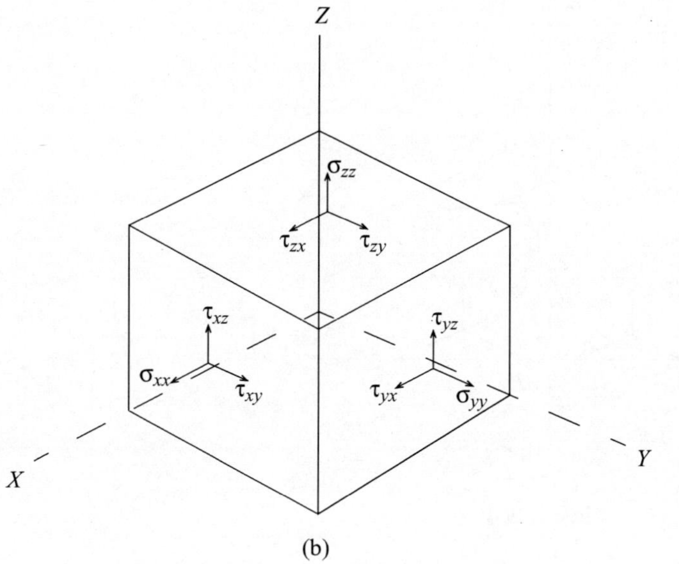

(b)

Figure 3.3

It is customary to describe the distribution of the stresses by the stresses acting on the faces of an elemental cube around the point, as shown in Fig. 3.3. The stresses acting on any other plane can be expressed in terms of these stresses and hence they represent the state of stress at the point.

For a two-dimensional stress analysis,

$$\tau_{zz} = \tau_{zx} = \tau_{zy} = 0$$

Then the state of stress is given by

$$\begin{bmatrix} \sigma_{xx} & \tau_{xy} & 0 \\ \tau_{yx} & \sigma_{yy} & 0 \\ 0 & 0 & 0 \end{bmatrix}$$

3.2 EQUILIBRIUM EQUATIONS AND STRESS TRANSFORMATION

When body subjected to a system of body and surface forces the stress of variable magnitude and direction are produced throughout the body. The distribution of the stresses must satisfy the overall equilibrium of the body and also of each element in the body. Let us consider the equilibrium at an elemental level. An element is shown in Fig. 3.4. The stresses on the negative and positive faces are different. The incremental changes are shown by differentials. The stresses in the x-direction are indicated for simplicity. We consider the equilibrium of the element in x-direction.

$$\left(\sigma_{xx} + \frac{\partial \sigma_{xx}}{\partial x} \,.\, dx - \sigma_{xx}\right) dy \,.\, dz + \left(\tau_{yx} + \frac{\partial \tau_{yx}}{\partial y} \,.\, dy - \tau_{yx}\right) dxdz$$
$$+ \left(\tau_{zx} + \frac{\partial \tau_{zx}}{\partial z} \,.\, dz - \tau_{zx}\right) dy \,.\, dx - F_x \,.\, dx \,.\, dy \,.\, dz = 0$$

where F_x is the body force per unit volume.

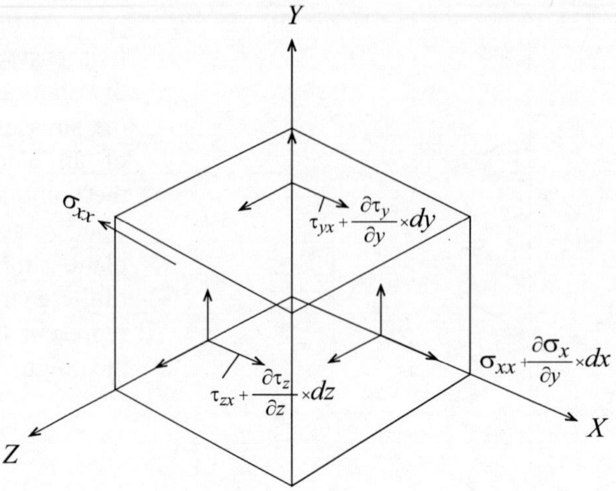

Figure 3.4

Dividing throughout by $dx \, . \, dy \, . \, dz$, we get

$$\frac{\partial \sigma_{xx}}{\partial x} + \frac{\partial \tau_{yx}}{\partial y} + \frac{\partial \tau_{zx}}{\partial z} + F_x = 0. \tag{3.3a}$$

Similarly, considering the equilibrium of forces in y and z directions, we get

$$\frac{\partial \tau_{xy}}{\partial x} + \frac{\partial \sigma_{yy}}{\partial y} + \frac{\partial \tau_{zy}}{\partial z} + F_y = 0 \tag{3.3b}$$

$$\frac{\partial \tau_{zx}}{\partial x} + \frac{\partial \tau_{yz}}{\partial y} + \frac{\partial \sigma_{zz}}{\partial z} + F_z = 0 \tag{3.3c}$$

These equations can be generalized and the generalized form is given below.

$$\frac{\partial \sigma_{ij}}{\partial_{xi}} + F_i = 0 \tag{3.3d}$$

$$i = x, y, z \quad (\text{or } 1, 2, 3)$$

$$j = x, y, z \quad (\text{or } 1, 2, 3)$$

The repeated index i means it is summation for $j = x, y, z$ or $1, 2, 3$

3.2.1 Two-Dimensional Stress System

The stresses on any other plane (AC) with normal n, at an angle ϕ to the x-direction can be found as follows:

The element ABC has unit thickness in the z-direction. The stresses acting on the faces are as shown in the Fig. 3.5.

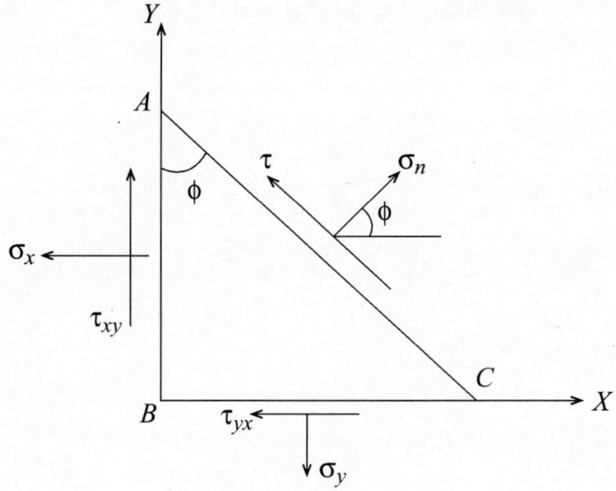

Figure 3.5

Considering the equilibrium of forces along the normal direction.

1. $$AC \quad \sigma_n = 1\,.\,AB.\,\sigma_x\,\cos\phi + 1.\,AB\,.\,\tau_{xy}\,\sin\phi + 1.\,BC.\,\sigma_y\,\sin\phi + 1.\,BC.\,\tau_{yx}\,\cos\phi.$$

$$\therefore \quad \sigma_n = \frac{AB}{AC}\,.\,\sigma_x\,\cos\phi + \frac{AB}{AC}\,\tau_{xy}\,\sin\phi + \frac{BC}{AC}\,.\,\sigma_y\,\sin\phi + \frac{BC}{AC}\,.\,\tau_{xy}\,\cos\phi$$

$$\sigma_n = \sigma_x\,\cos^2\phi + \tau_{xy}\,.\sin\phi\,.\,\cos\phi + \sigma_y\sin^2\phi + \tau_{xy}\sin\phi\cos\phi$$

$$= \sigma_x\cos^2\phi + \sigma_y\sin^2\phi + 2\tau_{xy}\sin\phi\cos\phi$$

$$= \sigma_x\frac{\cos 2\phi + 1}{2} + \sigma_y\,.\,\frac{1-\cos 2\phi}{2} + \tau_{xy}\sin 2\phi$$

$$= \frac{\sigma_x+\sigma_y}{2} + \frac{\sigma_x-\sigma_y}{2}\cos 2\phi + \tau_{xy}\sin 2\phi \tag{3.4}$$

Considering the equilibrium of the forces along the parallel direction.

$$AC\;\tau = 1\,.\,AB\,.\,\sigma_x\sin\phi + 1.\,AB\,.\,\tau_{xy}\cos\phi + 1\,.\,BC\,.\,\sigma_y\cos\phi - 1\,.\,BC\,.\,\tau_{yx}\sin\phi$$

$$\tau = -\frac{AB}{AC}\,.\,\sigma_x\sin\phi + \frac{AB}{AC}\,.\,\tau_{xy}\,.\,\cos\phi + \frac{BC}{AC}\,.\,\sigma_y\,.\,\cos\phi - \frac{BC}{AC}\,.\,\tau_{xy}\sin\phi$$

$$= -\sigma_x\sin\phi\cos\phi + \sigma_y\sin\phi\cos\phi + \tau_{xy}\cos^2\phi - \tau_{yx}\sin^2\phi$$

$$= \left(\frac{\sigma_y - \sigma_x}{2}\right) 2\sin\phi\cos\phi + \tau_{yx}(\cos^2\phi - \sin^2\phi)$$

$$= \left(\frac{\sigma_y - \sigma_x}{2}\right) \sin 2\phi + \tau_{yx}\cos 2\phi \tag{3.5}$$

For maximum values of direct stress σ_n

$$\frac{\delta\sigma_n}{\delta\phi} = 0 = 0 - 2\left(\frac{\sigma_x - \sigma_y}{2}\right)\sin 2\phi_n + 2\tau_{xy}\cos 2\phi_n$$

$$\tan 2\phi_n = \left(\frac{2\tau_{xy}}{\sigma_x - \sigma_y}\right) \tag{3.6}$$

Values of stresses are (by using the relation 3.6)

$$\sigma_1, \sigma_2 = \left(\frac{\sigma_x + \sigma_y}{2}\right) \pm \left(\frac{\sigma_x - \sigma_y}{2}\right)\left[\frac{\sigma_x - \sigma_y}{\sqrt{(\sigma_x - \sigma_y)^2 + 4\tau_{xy}^2}}\right]$$

$$\pm\,\sigma_{xy}\frac{2\sigma_{xy}}{\sqrt{(\sigma_x - \sigma_y)^2 + 4\tau_{xy}^2}}$$

$$= \left(\frac{\sigma_x + \sigma_y}{2}\right) \pm \sqrt{\left(\frac{\sigma_x - \sigma_y}{2}\right)^2 + \tau_{xy}^2} \tag{3.7}$$

These are called principal stresses and the planes are called principal planes. Incidently, the shear stress on these planes $\tau = 0$ from the relation (3.5).
For maximum shear stress

$$\frac{\delta\tau}{\delta\phi} = 0 = -\left(\frac{\sigma_x - \sigma_y}{2}\right) 2\cos 2\phi - 2\tau_{xy}\sin 2\phi$$

Then the location of maximum shear plane is given by

$$\cot 2\phi_s = \left(\frac{2\tau_{xy}}{\sigma_y - \sigma_x}\right) \text{ OR } \tan 2\phi_s = \left(\frac{\sigma_y - \sigma_x}{2\tau_{xy}}\right) \tag{3.8}$$

Substituting the value of ϕ in Eq. (2) the maximum shear stress is

$$\tau_{\max} = \frac{(\sigma_y - \sigma_x)(\sigma_y - \sigma_x)/4}{\sqrt{\left(\frac{\sigma_y - \sigma_x}{2}\right)^2 + \tau_{xy}^2}} \pm \tau_{xy}.\frac{\tau_{xy}}{\sqrt{\left(\frac{\sigma_y - \sigma_x}{2}\right)^2 + \tau_{xy}^2}}$$

$$\tau_{\max} = \sqrt{\left(\frac{\sigma_x - \sigma_y}{2}\right)^2 + \tau_{xy}^2} \tag{3.9}$$

The direct stresses on the planes

$$\sigma = \frac{\sigma_x + \sigma_y}{2} + 0 \tag{3.10}$$

The angle between the principal planes and maximum shear planes is 45° as indicated by the expressions (3.6) and (3.8).

3.2.2 Mohr's Circle Method

These solutions are obtained graphically by the Mohr's circle diagram constructed as follows (Fig. 3.6).

Let,

$$OP = \sigma_x;\ OM = \sigma_y;\ PR = \sigma_{xy};\ MS = \sigma_{yx};\ OO_1 = \frac{\sigma_x + \sigma_y}{2}$$

$$\text{Radius,} \qquad R = O_1R = O_1C = O_1D = O_1A = \sqrt{\left(\frac{\sigma_x - \sigma_y}{2}\right)^2 + \tau_{xy}^2}$$

with O_1 as the centre and O_1R as radius draw the circle. The intercepts are $OA = \sigma_1$ and $OB = \sigma_2; O_1C = O_1D = \tau_{\max}$

$\angle RO_1A = 2\phi_n$ gives the orientation of the principal plane w.r.t. x-plane.

Let R_1 be the reflection of R

RR_1 is the x-plane and the element is shown in Fig. 3.6(a).

Then R_1A is the principal plane of σ_1

R_1B is the plane of σ_2

The principal element is shown at Fig. 3.6(b)

R_1C and R_1D are the planes of maximum shear stress. The max shear element in shown in Fig. 3.6(c).

Angle between AR_1 and CR_1 is 45°.

That is the principal element and the max shear element are oriented at 45° to each other.

$$OP = \sigma_x, \quad PR = \tau_{xy}, \quad OM = \sigma_y, \quad OO_1 = (\sigma_x + \sigma_y)/2, \quad OA = \sigma_1, \quad O_1C = \tau_{\max}$$

$$OB = \sigma_z, \angle RR_1A = \phi_n, \angle DO_1R = 2\phi_s, \quad \angle QO_1R = 2\phi$$

The stresses on the element with $u - v$ axis at an angle ϕ w.r.t. x-axis can be found by locating Q, with $\angle RO_1Q = 2\phi$ or $\angle RR_1Q = \phi$. The coordinates of Q give the stresses. The stress element is shown in Fig. 3.6(d).

$$OT = \sigma_u, OT_1 = \sigma_v. \qquad TQ = \tau_{uv}.$$

$$OT = OO_1 + O_1T$$

$$= \frac{\sigma_x + \sigma_y}{2} + R\cos(2\phi_n + 2\phi) = \frac{\sigma_x + \sigma_y}{2} + R[\cos 2\theta_n . \cos 2\phi + \sin 2\phi_n . \sin 2\phi]$$

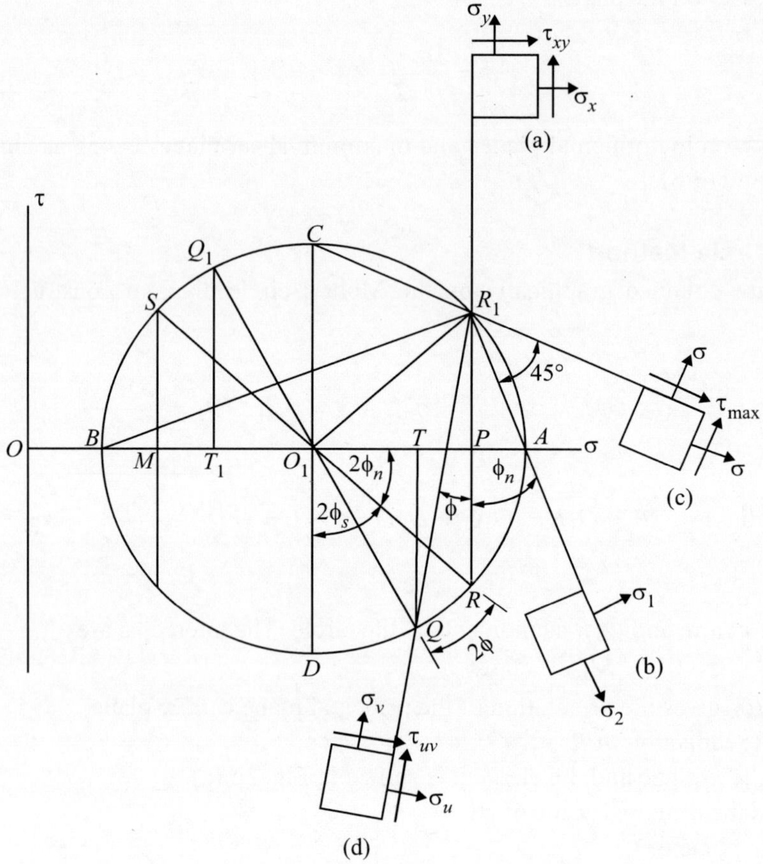

Figure 3.6

$$\therefore \quad \sigma_u = \frac{\sigma_x + \sigma_y}{2} + \frac{\sigma_x - \sigma_y}{2} . \cos 2\phi + \tau_{xy} \sin 2\phi$$

$$TQ = R \,.\, \sin\,(2\phi_n + 2\phi)$$

$$= R \sin\,(2\phi_n) . \cos 2\phi + R \cos 2\phi_n \sin 2\phi$$

$$\therefore \quad \tau_{uv} = \tau_{xy} \cos 2\phi + \frac{\sigma_x - \sigma_y}{2} . \sin 2\phi .$$

3.2.3 For a Three-dimensional System

The stress on any plane can be found in the following way. Refer Fig. 3.7.

The plane *ABC* is inclined with normal *n*, a_{nx}, a_{ny}, a_{nz} are direction cosines of the normal '*n*' to the plane *ABC*.

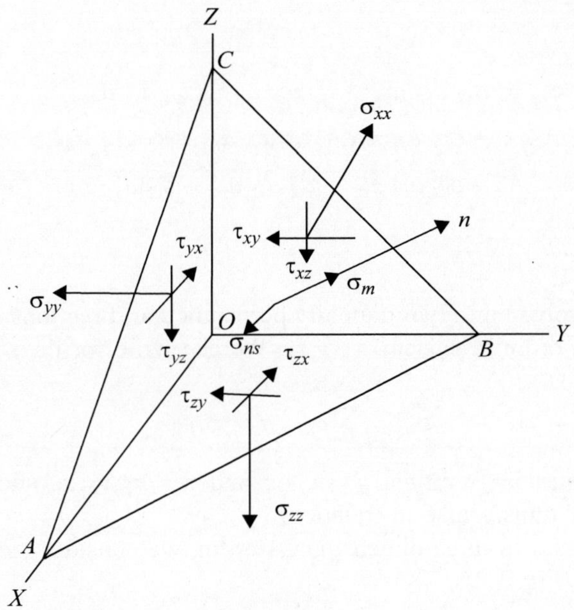

Figure 3.7

The stresses on the faces are shown in Fig. 3.7. Considering equilibrium the components along the normal.

$$
\begin{aligned}
ABC\,.\,\sigma_n = {} & OBC.\,\sigma_{xx}\,.\,a_{nx} + AOC.\,\tau_{yx}\,.\,a_{nx} + AOB.\,\tau_{zx}\,.\,a_{nx} \\
& + AOC\,.\,\sigma_{yy}\,.\,a_{ny} + COB.\,\tau_{xy}\,.\,a_{ny} + AOB.\,\tau_{zy}\,.\,a_{ny} \\
& + AOB.\,\sigma_{zz}\,.\,a_{nz} + BOC.\,\tau_{xz}\,.\,a_{nz} + AOC.\,\tau_{yz}\,.\,a_{nz}
\end{aligned}
$$

$$
\begin{aligned}
\therefore \sigma_n = {} & \sigma_{xx}\,.\,a_{nx}\frac{OBC}{ABC} + \tau_{yx}\,.\,a_{nx}\frac{AOC}{ABC} + \tau_{zx}\,.\,a_{nx}\frac{AOB}{ABC} \\
& + \sigma_{yy}\,.\,a_{ny}\frac{AOC}{ABC} + \tau_{xy}\,.\,a_{ny}\frac{COB}{ABC} + \tau_{zy}\,.\,a_{ny}\frac{AOB}{ABC} \\
& + \sigma_{zz}\,.\,a_{nz}\frac{AOB}{ABC} + \tau_{xz}\,.\,a_{nz}\frac{BOC}{ABC} + \tau_{yz}\,.\,a_{nz}\frac{AOC}{ABC}
\end{aligned}
$$

$$
\begin{aligned}
\therefore \sigma_n = {} & \sigma_{xx}\,a_{nx}^2 + \tau_{yx}\,.\,a_{nx}\,a_{nx} + \tau_{zx}\,a_{nx}\,a_{nz} \\
& + \sigma_{yy}\,a_{ny}^2 + \tau_{xy}\,.\,a_{ny}\,a_{nx} + \tau_{zy}\,a_{ny}\,a_{nz} \\
& + \sigma_{zz}\,a_{xz}^2 + \tau_{yz}\,.\,a_{nz}\,a_{nx} + \tau_{yz}\,a_{nz}\,a_{nx}
\end{aligned}
\tag{3.11}
$$

Shear stress is

$$\begin{aligned}\sigma_{ns} = {} & \sigma_{xx}\, a_{sx}\, a_{nx} + \tau_{yx}\, a_{sx}\, a_{ny} + \tau_{zx}\, a_{sx}\, a_{nz} \\ & + \sigma_{yy}\, a_{sy}\, a_{ny} + \tau_{xy}\, a_{sy}\, a_{nx} + \tau_{zy}\, a_{sy}\, a_{nz} \\ & + \sigma_{zz}\, a_{sz}\, a_{nz} + \tau_{xz}\, a_{sz}\, a_{nx} + \tau_{yz}\, a_{sz}\, a_{ny}. \end{aligned} \tag{3.12}$$

Generalizing $\sigma_{ns} = \Sigma(a_{ni} \,.\, a_{sj} \,.\, \sigma_{ij})$
where $i = x, y.z; j = x, y, z.$

The shear stress is resolved into two mutually perpendicular directions in the plane, the three axes represent another co-ordinate system, x_1, y, z_1. We can write for the stresses in the new co-ordinate system as

$$\sigma_{ns} = \sum a_{ni} \,.\, a_{si} \,.\, \sigma_{ij} \tag{3.13}$$

where the indices take values $i = x, y, z$ and $j = x, y, z$ and $n = x_1, y_1, z_1$ and $s = x_1, y_1, z_1$

This equation is called transformation equation.

For the principal stresses in three-dimensional system, we consider the principal plane *ABC* with only normal stresses (Fig. 3.8).

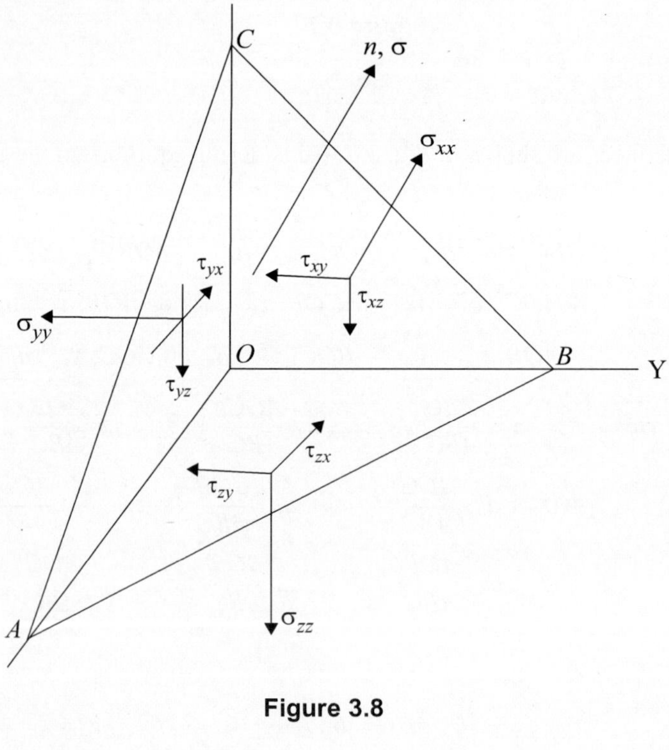

Figure 3.8

For equilibrium in x-direction,

$$ABC.\sigma.a_{nx} = \sigma_{xx}.COB + \tau_{yx}.AOC + \tau_{zx}AOB$$

$$\sigma . a_{nx} = \sigma_{xx}\frac{COB}{ABC} + \tau_{yx}\frac{AOC}{ABC} + \tau_{zx}\frac{AOB}{ABC}.$$
$$= \sigma_{xx}.a_{nx} + \tau_{yx}\, a_{ny} + \tau_{zx}.a_{nz}$$
$$= (\sigma_{xx} - \sigma)\, a_{nx} + \tau_{yz}\, a_{ny} + \tau_{zx}\, a_{nz} = 0$$

Similarly, we can write for y and Z directions

$$\tau_{xy}\, a_{nx} + (\sigma_{yy} - \sigma)\, a_{ny} + \tau_{zy.}\, a_{nz} = 0$$

$$\tau_{zx}\, a_{nx} + \tau_{yz}\, a_{ny} + (\sigma_{zz} - \sigma)\, a_{nz} = 0$$

These equations are written in matrix form as

$$\begin{bmatrix} (\sigma_{xx} - \sigma) & \tau_{yx} & \tau_{zx} \\ \tau_{xy} & (\sigma_{yy} - \sigma) & \sigma_{zy} \\ \tau_{xz} & \tau_{yz} & (\sigma_{zz} - \sigma) \end{bmatrix} \begin{bmatrix} a_{nx} \\ a_{ny} \\ a_{nz} \end{bmatrix} = 0 \tag{3.14}$$

Since $a_{nx}^2 + a_{ny}^2 + a_{nz}^2 = 1$

The necessary condition is

$$\begin{vmatrix} \sigma_{xx} - \sigma & \tau_{yx} & \tau_{zx} \\ \tau_{xy} & \sigma_{yy} - \sigma & \sigma_{zy} \\ \tau_{xz} & \tau_{yz} & \sigma_{zz} - \sigma \end{vmatrix} = 0 \tag{3.15a}$$

which reduces to

$$\sigma^3 - C_1\sigma^2 + C_2\sigma - C_3 = 0 \tag{3.15b}$$

where,

$$C_1 = \sigma_{xx} + \sigma_{yy} + \sigma_{zz}$$

$$C_2 = \sigma_{xx}\,\sigma_{yy} + \sigma_{yy}\,\sigma_{zz} + \sigma_{zz}.\sigma_{xx} - \tau_{xy}^2 - \tau_{yz}^2 - \tau_{zx}^2$$

$$\begin{vmatrix} \sigma_{xx} & \tau_{yx} \\ \tau_{xy} & \sigma_{yy} \end{vmatrix} + \begin{vmatrix} \sigma_{yy} & \tau_{zy} \\ \tau_{yz} & \sigma_{zz} \end{vmatrix} + \begin{vmatrix} \sigma_{xx} & \tau_{zx} \\ \tau_{xz} & \sigma_{zz} \end{vmatrix}$$

$$C_3 = \sigma_{xx}\,\sigma_{yy}\,\sigma_{zz} - \sigma_{xx}.\tau_{yz}^2 - \sigma_{yy}\,\tau_{zx}^2 - \sigma_{zz}.\tau_{xy}^2 + 2\tau_{xy}\tau_{yz}\,\tau_{zx}$$

$$= \begin{vmatrix} \sigma_{xx} & \tau_{yx} & \tau_{zx} \\ \tau_{xy} & \sigma_{yy} & \sigma_{zy} \\ \tau_{xz} & \tau_{yz} & \sigma_{zz} \end{vmatrix}$$

For two-dimensional stress system (Fig. 3.9),

$$\sigma_{zz} = \tau_{xz} = \tau_{yz} = 0$$

Then equation reduces to

$$\sigma_{nn} = \sigma_{xx}\, a_{nx}^2 + \tau_{yx}\, a_{nx}\, a_{ny} + \tau_{xy}\, a_{nx}\, a_{ny} + \sigma_{yy}\, a_{ny}^2$$

$$= \sigma_{xx}\, a_{nx}^2 + \sigma_{yy}\, a_{ny}^2 + 2\tau_{xy}\, a_{nx}\, a_{ny}$$

where,
$a_{nx} = \cos\theta$
$a_{ny} = \sin\theta.$

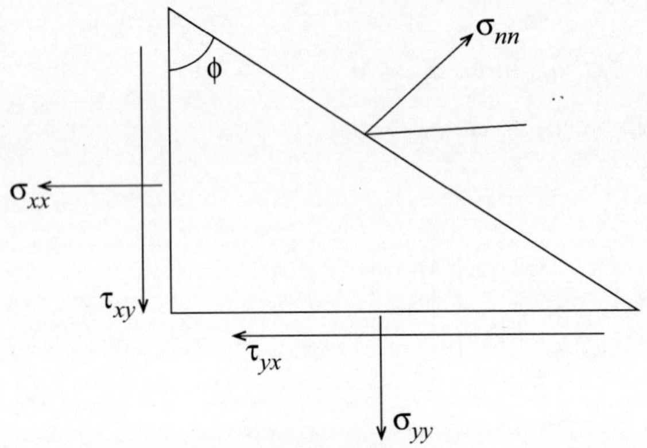

Figure 3.9

$$\therefore \quad \sigma_{nn} = \sigma_{xx}\cos^2\phi + \sigma_{yy}\sin^2\phi + 2\cos\phi\sin\phi.\tau_{xy}$$

$$= \frac{\cos 2\phi + 1}{2}.\sigma_{xx} + \frac{\cos 2\phi - 1}{2}\sigma_{yy} + \tau_{xy}\sin 2\phi$$

$$= \frac{\sigma_{xx} + \tau_{xy}}{2} + \frac{\sigma_{xx} - \sigma_{yy}}{2}\cos 2\phi + \tau_{xy}\sin 2\phi$$

This is the same as proved earlier.
Similarly, for the principal stresses

$$C_1 = \sigma_{xx} + \sigma_{yy};\ C_2 = \sigma_{xx}.\sigma_{yy} - \tau_{xy}^2 \text{ and } C_3 = 0$$

Then Eq. (3.15a) reduces to

$$\sigma^3 - (\sigma_{xx} + \sigma_{yy})\sigma^2 + (\sigma_{xx}\,\sigma_{yy} - \tau_{xy}^2)\sigma - 0 = 0$$

$$\sigma^2 - (\sigma_{xx} + \sigma_{yy})\sigma + (\sigma_{xx}\,\sigma_{yy} - \tau_{xy}^2) = 0$$

The roots are

$$\sigma_1, \sigma_2 - \left(\frac{\sigma_{xx} + \sigma_{yy}}{2}\right) \pm \sqrt{\left(\frac{\sigma_{xx} + \sigma_{yy}}{2}\right)^2 - (\sigma_{xx}\,\sigma_{yy} - \tau_{xy}^2)}$$

$$= \frac{\sigma_{xx} + \sigma_{yy}}{2} \pm \sqrt{\left(\frac{\sigma_{xx} - \sigma_{yy}}{2}\right)^2 + \tau_{xy}^2}$$

As proved earlier.

SOLVED EXAMPLES

Example 3.1: The stress element has the stresses as shown in Fig. 3.10, $\sigma_x = 10$ MPa, $\sigma_y = 5$ MPa, $\tau_{xy} = 2$ MPa *CW*. Evaluate the principal stresses and their orientations.

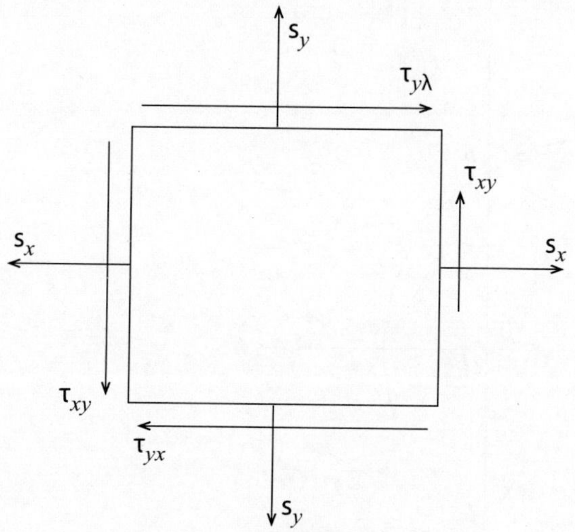

Figure 3.10

SOLUTION

Step-1: Principal stress

$$\sigma_1, \sigma_2 = \frac{\sigma_x + \sigma_y}{2} \pm \sqrt{\left(\frac{\sigma_x - \sigma_y}{2}\right)^2 + \tau_{xy}^2}$$

$$= \frac{10+5}{2} \pm \sqrt{\left(\frac{10-5}{2}\right)^2 + 2^2}$$

$$= 7.5 \pm \sqrt{6.25 + 4}$$

$$= 7.5 \pm \sqrt{10.25}$$

$$= 7.5 \pm 3.2 \text{ MPa.}$$

$$\sigma_1 = 10.7 \text{ MPa}; \sigma_2 = 4.3 \text{ MPa}$$

Step-2: Principal directions

$$\tan 2\theta_n = \frac{-2\tau_{xy}}{\sigma_x - \sigma_y} = \frac{2 \times 2}{10 - 5} = \frac{4}{5} = 0.8$$

$$2\theta_n = \tan^{-1}(0.8) = 38.66^\circ$$

$$\theta_n = 19.33^\circ; \theta_{n_2} = -70.67^{(\circ)}$$

Step-3: Maximum shear stress

$$\tau_{\max} = \sqrt{\left(\frac{\sigma_x - \sigma_y}{2}\right)^2 + \tau_{xy}^2}$$

$$= \sqrt{\left(\frac{10-5}{2}\right)^2 + 2^2} = \sqrt{10.25}\text{ MPa}$$

$$= 3.2\text{ MPa}$$

Step-4: Mohr's circle methods

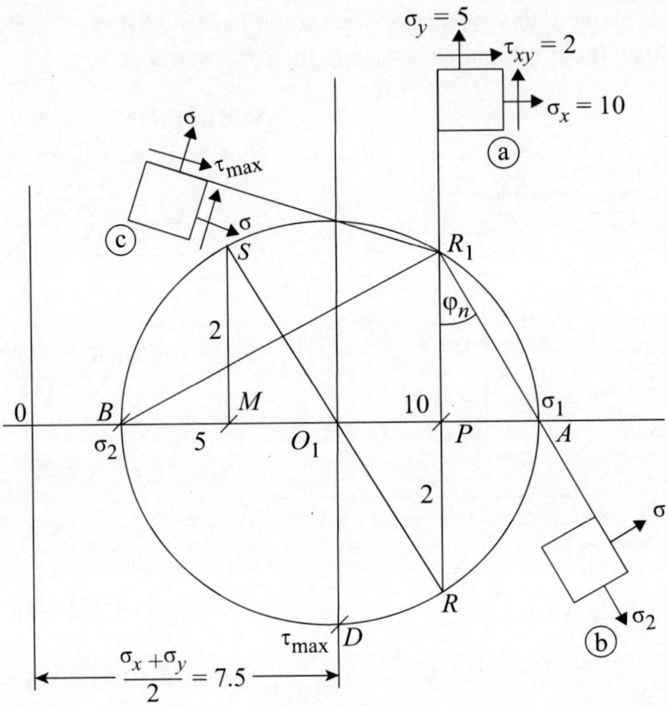

Figure 3.11

$$OP = 10, OM = 5, PR = MS = 2$$

$$OO_1 = \frac{10+5}{2} = 7.5, O_1P = \frac{10-5}{2} = 2.5$$

$$O_1R = \sqrt{O_1P^2 + PR^2} = \sqrt{2.5^2 + 2^2} = \sqrt{10.25} = 3.2$$

$$\sigma_1 = OA = OO_1 + O_1A = OO_1 + O_1R$$

$$= 7.5 + \sqrt{10.25} = 7.5 + 3.2 = 10.7\text{ MPa}$$

$$\sigma_2 = OO_1 - O_1B = 7.5 - 3.2 = 4.3\text{ MPa}$$

$$\tau_{max} = O_1D = O_1R = \sqrt{10.25} = 3.2 \text{ MPa.}$$

$$\tan\phi_n = \left(\frac{PA}{PR_1}\right) = \left(\frac{O_1A - O_1P}{PR_1}\right) = \frac{3.2 - 2.5}{2} = 0.35$$

$$\phi_n = \tan^{-1}(0.35) = 19.3^\circ \text{ and } \phi_{n_2} = 19.3^\circ - 90^\circ = -70.7^\circ$$

Example 3.2: The stresses on an element are given as follows: $\sigma_x = 15$ MPa, $\sigma_y = 10$ MPa, $\tau_{xy} = 5$ MPa. Find the principal stresses and their orientation.

SOLUTION

Step-1: Principal stress

$$\sigma_1, \sigma_2 = \frac{10+15}{2} \pm \sqrt{\left(\frac{15-10}{2}\right)^2 + 5^2} = 12.5 \pm \sqrt{31.25}$$

$$= 18.09 \text{ MPa and } 6.91 \text{ MPa}$$

Step-2: The orientations of the principal stresses are

$$\tan 2\phi_n = \frac{2 \times (-5)}{15 - 10} = -2$$

$$\tan(\pi - 2\phi_n) = 2$$

$$\pi - 2\phi_2 = \tan^{-1} 2$$

$$\phi_n = \frac{\pi - \tan^{-1}(2)}{2} = 58.3^\circ \quad \text{and} \quad 31.7^\circ.$$

Step-3: Mohr's circle method

$$OO_1 = \frac{10+15}{2} = 12.5,\ O_1P = \frac{15-10}{2} = 2.5, PR = 5$$

$$O_1A = O_1R$$

$$= \sqrt{OP^2 + PR^2} = \sqrt{2.5^2 + 5^2} = 5.59$$

$$\sigma_1 = OA = OO_1 + O_1A = 12.5 + 5.59 = 18.09 \text{ MPa}$$

$$\sigma_2 = OB = OO_1 - O_1B = 12.5 - 5.59 = 6.91 \text{ MPa}$$

$$\phi_n = \tan^{-1}\left(\frac{PA}{PR_1}\right) = \tan^{-1}\left(\frac{O_1A - O_1P}{PR_1}\right) = \tan^{-1}\left(\frac{5.59 - 2.5}{5}\right) = 31.7^\circ.$$

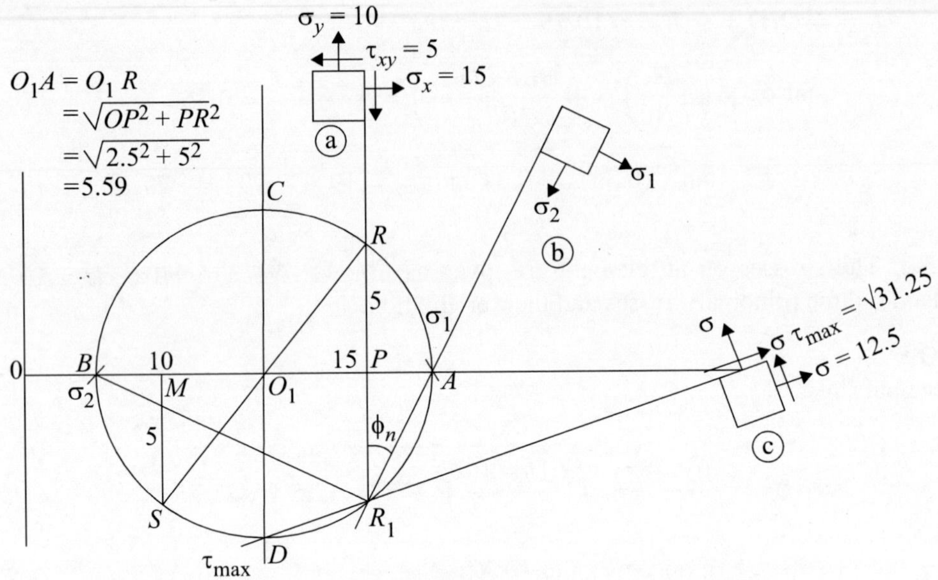

Figure 3.12

Example 3.3: The stresses are shown in Fig. 3.13a. Find the principal stresses and their orientation.

SOLUTION

Step-1: Principal stresses

$$\sigma_1, \sigma_2 = \frac{15-10}{2} \pm \sqrt{\left(\frac{10+15}{2}\right)^2 + 3^2}$$

$$= 2.5 \pm \sqrt{(12.5)^2 + 3^2} = 2.5 + 12.85$$

$$= 15.35 \text{ MPa and } - 10.35 \text{ MPa}$$

Step-2: Principal directions

$$\tan 2\phi_n = \left(\frac{2 \times 3}{15+10}\right) = \frac{2 \times 3}{25} = 0.24$$

$$2\phi_n = \tan^{-1}(0.24) = 13.5°.$$

$$\phi_n = 6.75°, \phi_{n_2} = 6.75 - 90 = -83.25°$$

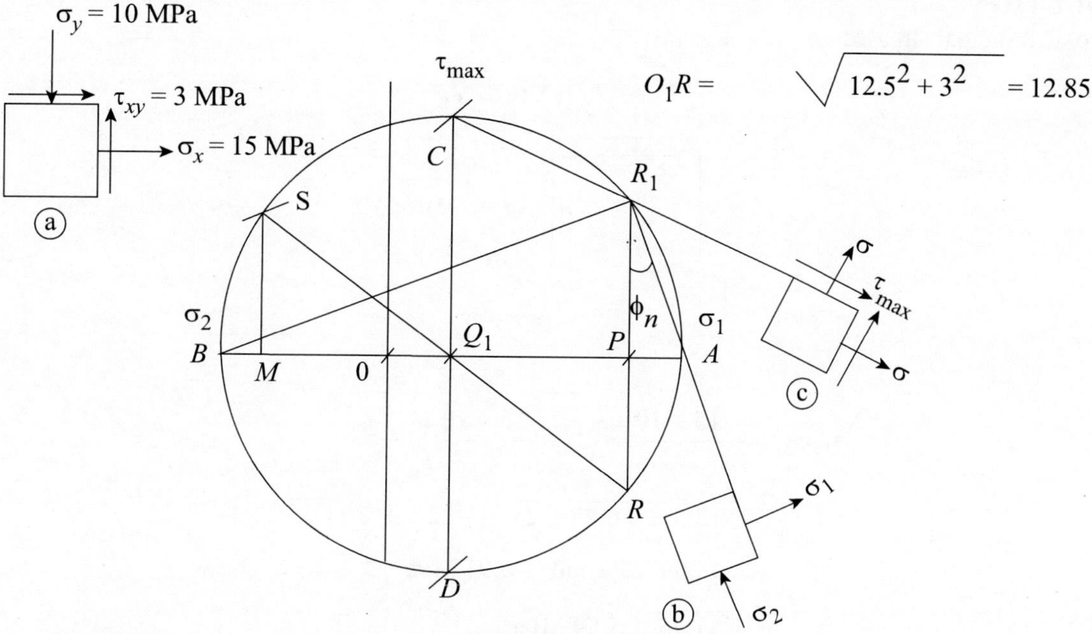

Figure 3.13

Step-3: Mohr's circle method

$$OO_1 = \frac{15-10}{2} = 2.5; O_1P = \frac{15+10}{2} = 12.5; O_1R = \sqrt{12.5^2+3^2} = 12.85.$$

$$\sigma_1 = OA = OO_1 + O_1A = 2.5 + 12.85 = 15.35 \text{ MPa}$$

$$\sigma_2 = OB = OO_1 - O_1B = 2.5 - 12.85 = -10.35 \text{ MPa}$$

$$\phi_n = \tan^{-1}\left(\frac{PA}{PR_1}\right) = \tan^{-1}\left(\frac{O_1A - O_1P}{PR_1}\right) = \tan^{-1}\left(\frac{12.85-12.5}{5}\right) = 6.75°$$

$$\tau_{max} = O_1R = \sqrt{12.5^2+3^2} = 12.85$$

Example 3.4: The stresses at a point in a load body are shown in Fig. 3.14. Find the principal stresses, maximum shear stress and their orientation.

SOLUTION

Step-1: Principal stresses

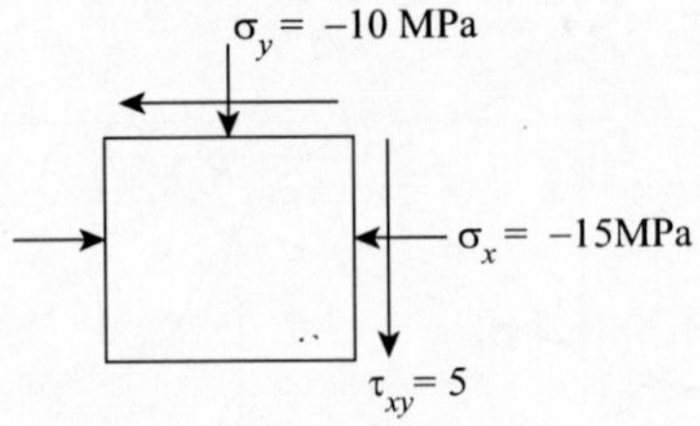

$$\sigma_1, \sigma_2 = \frac{-15-10}{2} \pm \sqrt{\left(\frac{-15+10}{2}\right)^2 + 5^2}$$

$$= -12.5 \pm \sqrt{6.25+25} = -12.5 \pm \sqrt{31.25}$$

$$= -18.09 \text{ MPa and } -6.91 \text{ MPa.}$$

$$\tau_{max} = \sqrt{31.25} = 5.59 \text{ MPa.}$$

Step-2: Their orientation

$$\tan 2\phi_n = \frac{2\times(-5)}{10-15} = 2$$

$$2\phi_n = \tan^{-1}(2) = 63.43$$

$$\therefore\ \phi_n = 31.7^\circ; \phi_{n_2} = 31.7 - 90^\circ = -58.3^\circ$$

Step-3: Mohr's circle method

$$OO_1 = \frac{-15-10}{2} = -12.5$$

$$OP = -15, OM = -10, PR = 5$$

$$O_1P = -15 - (-12.5) = -2.5$$

$$O_1R = \sqrt{5^2 + 2.5^2} = 5.59$$

$$\sigma_2 = OB = OO_1 - O_1B = 12.5 - 5.59 = 6.91 \text{ MPa (Compressive)}$$

$$\sigma_1 = OA = OO_1 - O_1A = 12.5 + 5.59 = 18.09 \text{ MPa (Compressive)}$$

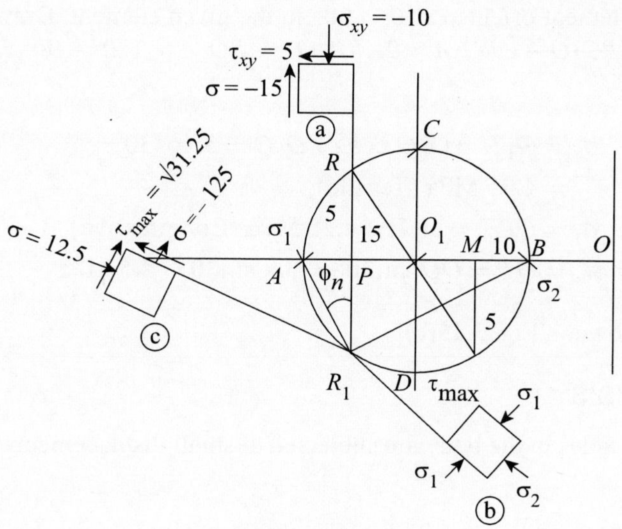

Figure 3.14

Example 3.5: Pure shear exists at a point with $\tau = 5$ MPa. Find the principal stresses. Find also the stresses on an element inclined at 60° to the given element.

SOLUTION

Mohr's Circle Method (Refer Fig. 3.15)

O_1 is the centre, $O_1C = 5$ as the radius, draw the circle, A,B,C & D are intersecting points.

Step-1: Principal stresses and orientations are (Fig. 3.15b)

$$\sigma_1 = OA = OR = 5 \text{ MPa (Tensile)}$$
$$\sigma_2 = OB = OR = 5.0 \text{ MPa (Compressive)}$$
$$\phi_n = \angle RR_1A = \angle O_1CA = 45°$$

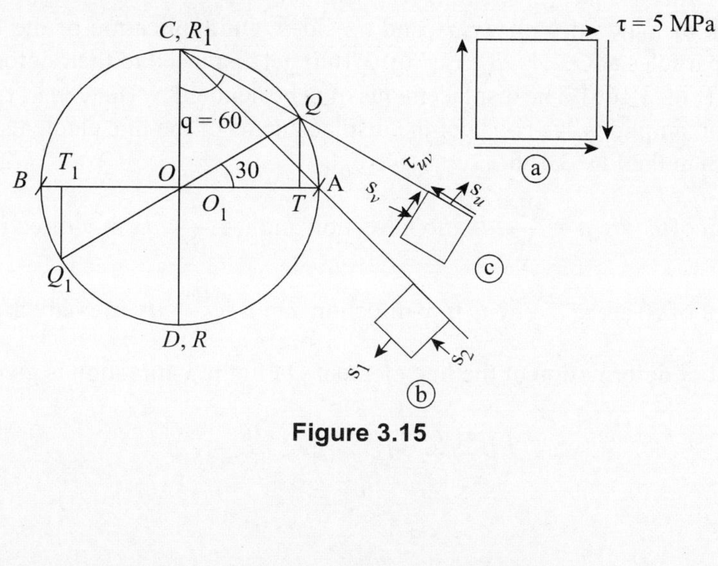

Figure 3.15

Step-2: Stresses on an element oriented at $\theta = 60°$ to the given element. Draw line R_1Q at 60° to RR_1. Then $\angle TO_1Q = \angle RO_1Q - \angle RO_1A = 2\angle RR_1Q - \angle RO_1A = 120° - 90° = 30°$
The stresses are

$$\sigma_u = O_1T = O_1Q \cos\angle TO_1Q = 5.\cos 30°.$$
$$= 4.35 \text{ MPa (Tensile)}.$$
$$\sigma_v = O_1T_1 = O_1T = 4.35 \text{ MPa (Compressive)}$$
$$\sigma_{uv} = QT = OQ.\sin 30° = 5 \times \sin 30 = 2.5 \text{ MPa}$$

The stress element is shown in Fig. 3.15(c).

3.3 STRAIN ANALYSIS

In a loaded body the particles in the body are subjected to small displacements but there is no body movement.

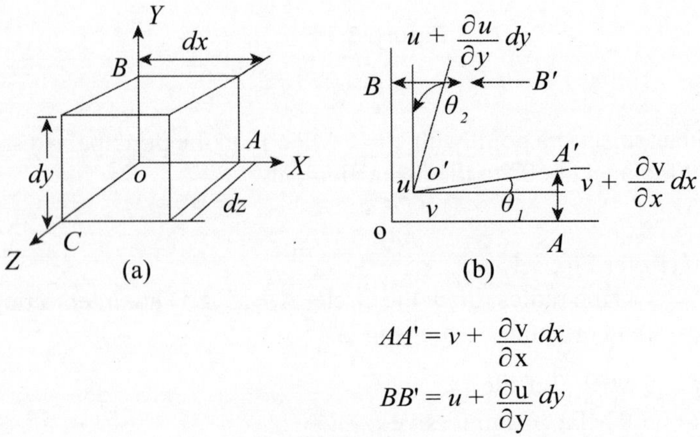

Figure 3.16

Consider a block of material with dx, dy and dz sides. On application of the load the block gets deformed. The particles at $O,\ A,\ B,\ C$ (Fig. 3.16a) get displaced to their deformed positions $O',\ A',\ B'$ and C' (Fig. 3.16b). The displacements of O be denoted by (u, v, w) in $x, y,$ and z directions respectively. For simplicity we consider the displacements of the line elements OA and OB in $X - Y$ plane as shown in the Fig. 3.16b.

The displacement of A' are $u + \dfrac{\partial u}{\partial y}.dx$ in x-direction and $v + \dfrac{\partial v}{\partial x}.dx$ in y-direction.

The displacement of B' are $u + \dfrac{\partial u}{\partial y}.d_y$ in x-direction and $v + \dfrac{\partial v}{\partial x}.dy$ in y-direction.

Then the amount of deformation of the line element OA in the x-direction is given by

$$\Delta l_x = \left(u + \frac{\partial u}{\partial x}.dx - u\right) = \frac{\partial u}{\partial x}.dx$$

The strain in x-direction is given by

$$\varepsilon_x = \frac{\Delta l_x}{dx} = \frac{\partial u}{\partial x} \tag{3.16a}$$

Similarly, strain in y-direction is

$$\varepsilon_y = \frac{\partial v}{\partial y} \tag{3.16b}$$

and strain in z-direction is

$$\varepsilon_z = \left(\frac{\partial w}{\partial z}\right) \tag{3.16c}$$

The included angle between the line elements OA and OB, $\angle AOB$ gets changed to $\angle A'O'B'$. The change is shear deformation.
The notation of the line element OA is

$$\theta_1 = \frac{v + \dfrac{\partial v}{\partial x}.dx - v}{d_x} = \frac{\partial v}{\partial x}.$$

The rotation of the line element OB is

$$\theta_2 = \frac{u + \dfrac{\partial u}{\partial y}.dy - u}{dy} = \frac{\partial u}{\partial y}.$$

The total change in the included angle,

$$\theta_{xy} = \theta_1 + \theta_2 = \frac{\partial v}{\partial x} + \frac{\partial u}{\partial y}.$$

This is called the shear strain in $x - y$ plane denoted by

$$\gamma_{xy} = \frac{\partial u}{\partial y} + \frac{\partial v}{\partial x}. \tag{3.16d}$$

Similarly, we get the shear strain component in YZ plane as

$$\gamma_{yz} = \frac{\partial w}{\partial y} + \frac{\partial u}{\partial z} \tag{3.16e}$$

The shear strain component in XZ plane is

$$\gamma_{zx} = \frac{\partial u}{\partial z} + \frac{\partial w}{\partial x}. \tag{3.16f}$$

γ_{xy}, γ_{yz} and γ_{zx} are called engineering strains. The shear strain components in the shear tensor are given by

$$\varepsilon_{xy} = \frac{1}{2}\gamma_{xy} = \frac{1}{2}\left(\frac{\partial u}{\partial y} + \frac{\partial v}{\partial x}\right) = \varepsilon_{yx}$$

$$\varepsilon_{yz} = \frac{1}{2}\gamma_{yz} = \frac{1}{2}\left(\frac{\partial u}{\partial z} + \frac{\partial w}{\partial y}\right) = \varepsilon_{zy} \quad (3.17a)$$

$$\varepsilon_{zx} = \frac{1}{2}\gamma_{zx} = \frac{1}{2}\left(\frac{\partial w}{\partial x} + \frac{\partial u}{\partial z}\right) = \varepsilon_{xz}$$

The strain tensor is

$$\left[\varepsilon_{ij}\right], \; i = x,y,z \text{ and } j = x,y,z \quad (3.17b)$$

This is similar to the stress tensor

$$\left[\sigma_{ij}\right], \; i = x,y,z \text{ and } j = x,y,z.$$

The equations derived for the stress transformation and principal stress hold good for the strain also. The strain transformation is given by

$$\varepsilon'_{ns} = a_{ni}\, a_{sj}\, \varepsilon_{ij} \quad (3.18)$$

where $i = x,y,z;\; j = x,y,z$ and $n = x'y'z', s = x', y', z'\, a_{ni}$ & a_{si} are directions cosines of the new system.
The principal strains are given by the equations,

$$\left|\varepsilon_{ij} - \varepsilon\ \delta_{ij}\right| = 0. \quad (3.19)$$

which reduces to $\varepsilon^2 - J_1\varepsilon^2 + J_2\varepsilon - J_3 = 0$
The strain invariants are

$$J_1 = \varepsilon_{xx} + \varepsilon_{yy} + \varepsilon_{zz}$$

$$J_2 = \varepsilon_{xx}.\varepsilon_{yy} + \varepsilon_{yy}\varepsilon_{zz} + \varepsilon_{zz}\varepsilon_{xx} - \varepsilon_{xy}^2 - \varepsilon_{yz}^2 - \varepsilon_{zx}^2$$

$$J_3 = \varepsilon_{xx}\varepsilon_{yy}\varepsilon_{zz} - \varepsilon_{xx}\varepsilon_{yz}^2 - \varepsilon_{yy}\varepsilon_{xz}^2 - \varepsilon_{zz}\varepsilon_{xy}^2 + 2\ \varepsilon_{xy}\varepsilon_{yz}\varepsilon_{zx}$$

Compatibility - equations

From the displacement field, i.e., the three equations for u, v, w expressed as functions of x, y & z an unique strain field can be obtained by the definitions of strain components, but the reverse process of obtaining displacement field from strain field may yield an impossible displacement field with voids in the body. To avoid this situation the strain field must satisfy the following compatibility conditions.

$$\frac{\partial^2 \gamma_{xy}}{\partial x.\partial y} = \frac{\partial^2 \varepsilon_{xx}}{\partial y^2} + \frac{\partial^2 \varepsilon_{yy}}{\partial x^2} \quad (3.20a)$$

$$\frac{\partial^2 \gamma_{yz}}{\partial y . \partial z} = \frac{\partial^2 \varepsilon_{yy}}{\partial_{zz}^2} + \frac{\partial^2 \varepsilon_{zz}}{\partial y^2} \tag{3.20b}$$

$$\frac{\partial^2 \gamma_{xz}}{\partial z . \partial x} = \frac{\partial^2 \varepsilon_{xx}}{\partial x^2} + \frac{\partial^2 \varepsilon_{xx}}{\partial_{zz}^2} \tag{3.20c}$$

$$2 . \frac{\partial^2 \varepsilon_{xx}}{\partial y \partial z} = \frac{\partial}{\partial x}\left(-\frac{\partial \gamma_{yz}}{\partial x} + \frac{\partial \gamma_{zx}}{\partial y} + \frac{\partial \gamma_{xy}}{\partial z}\right) \tag{3.20d}$$

$$2 . \frac{\partial^2 \varepsilon_{yy}}{\partial z . \partial x} = \frac{\partial}{\partial y}\left(\frac{\partial \gamma_{yz}}{\partial x} - \frac{\partial \gamma_{zx}}{\partial y} + \frac{\partial \gamma_{xy}}{\partial z}\right) \tag{3.20e}$$

$$2 . \frac{\partial^2 \varepsilon_{zz}}{\partial x \partial y} = \frac{\partial}{\partial z}\left(\frac{\partial \gamma_{yz}}{\partial x} + \frac{\partial \gamma_{zx}}{\partial y} - \frac{\partial \gamma_{xy}}{\partial z}\right) \tag{3.20f}$$

3.4 STRESS-STRAIN RELATIONSHIPS

From Hooke's law we know that stress is proportional to strain.

$$\sigma = E . \varepsilon$$

where E is the constant of proportionality and is called Young's modulus. But as per Poisson's law there is strain due to lateral stress given by

$$(\varepsilon_x)_y = \mu . \varepsilon_y = \frac{\mu}{E} . \sigma_y .$$

Considering the Poisson effect, we can write an expression for the total strain in x-direction.

$$\varepsilon_{xx} = \frac{1}{E}\left[\sigma_{xx} - \mu(\sigma_{yy} + \sigma_{zz})\right] \tag{3.21a}$$

Similarly, we can write

$$\varepsilon_{yy} = \frac{1}{E}\left[\sigma_{yy} - \mu(\sigma_{zz} + \sigma_{xx})\right] \tag{3.21b}$$

$$\varepsilon_{zz} = \frac{1}{E}\left[\sigma_{zz} - \mu(\sigma_{yy} + \sigma_{zz})\right] \tag{3.21c}$$

These are the relations connecting the direct strains to direct stresses. By applying Hooke's law for shear load, we can write

$$\tau = G . \gamma$$

where G is called modulus of rigidity in shear or modulus of rigidity. The relations connecting shear stresses to shear strain can be written as

$$\tau_{xy} = G.\gamma_{xy} \tag{3.21d}$$

$$\tau_{yz} = G.\gamma_{yz} \tag{3.21e}$$

$$\text{and} \quad \tau_{zx} = G.\gamma_{zx} \tag{3.21f}$$

Considering the case of pure shear we can establish the following relation for the elastic constant as

$$G = \frac{E}{2(1+\mu)} \tag{3.22}$$

By adding the shear strain relations, we can write

$$e = \left(\frac{1-2\mu}{E}\right)\theta \tag{3.23}$$

where, $e = \varepsilon_{xx} + \varepsilon_{yy} + \varepsilon_{zz}$ and

$\theta = \sigma_{xx} + \sigma_{yy} + \sigma_{zz}$

Applying this to the case of hydrostatic pressure

$$\sigma_{xx} = \sigma_{yy} = \sigma_{zz} = -P.$$

$$\text{Volume strain} \quad e = \frac{-3(1-2\mu)}{E}P = -KP.$$

where $K = \dfrac{+3(1-2\mu)}{E}$ is called the bulk modulus.

By rearranging the terms, we can express the stress components in terms of strain components as follows:

$$\sigma_{xx} = \lambda.e + 2G\,\varepsilon_{xx} \tag{3.24a}$$

$$\sigma_{yy} = \lambda.e + 2G\,.\,\varepsilon_{yy} \tag{3.24b}$$

$$\sigma_{zz} = \lambda.e + 2G\,.\,\varepsilon_{zz} \tag{3.24c}$$

$$\tau_{xy} = 2G\,.\,\varepsilon_{xy} \tag{3.24d}$$

$$\tau_{yz} = 2G\,.\,\varepsilon_{yz} \tag{3.24e}$$

$$\tau_{zx} = 2G\,.\,\varepsilon_{zx} \tag{3.24f}$$

where $\lambda = \dfrac{\mu E}{(1+\mu)(1-\mu)}$ is called Lame's constant. Summing up we can write a general equation as

$$\sigma_{ij} = \lambda \cdot \sigma_{kk}\,\delta_{ij} + 2G\,\varepsilon_{ij} \tag{3.25}$$

where δ_{ij} is knocker $\delta = 1$, when $i = j$ $\delta = 0$, when $i \neq j$.

3.5 APPLIED LOADS AND STRESSES

The applied loads create or induce stresses. They can be referred to the co-ordinate system chosen to suit the particular geometry of the object. The components of the stress tensor so defined are called the applied stresses. The principal stresses are the three normal stresses and the three shear stresses on the planes inclined to the chosen co-ordinate system at the particular angles determined by the stress tensor.

Many of the stress components in the stress-tensor may be zero. For example, in a tensile stress specimen the only non-zero component is σ_x which is unidirectional and normal but the principal stresses are both normal and shear.

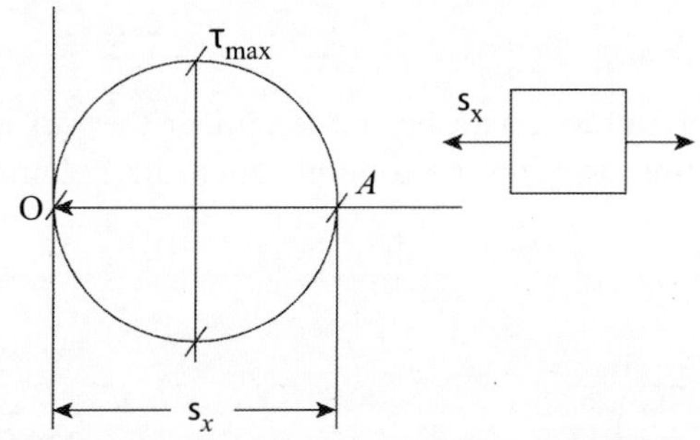

Figure 3.17. Mohr's circle for UTM-test

Normal principal stresses are

$$\sigma_1 = \sigma_x, \sigma_2 = \sigma_3 = 0$$

Principal shear stresses are,

$$\tau_{12} = \frac{\sigma_x}{2}; \tau_{23} = 0; \tau_{13} = \frac{\sigma_x}{2}.$$

Principal shear stresses are present at an planes inclined to the principal normal stress planes at an angle of 45°. They are present even if the applied shear stress is zero. Similarly, when the applied stresses only shear, the principal normal stresses do exist. Figure 3.18 represents the principal normal stress and shear stress for torsion test.

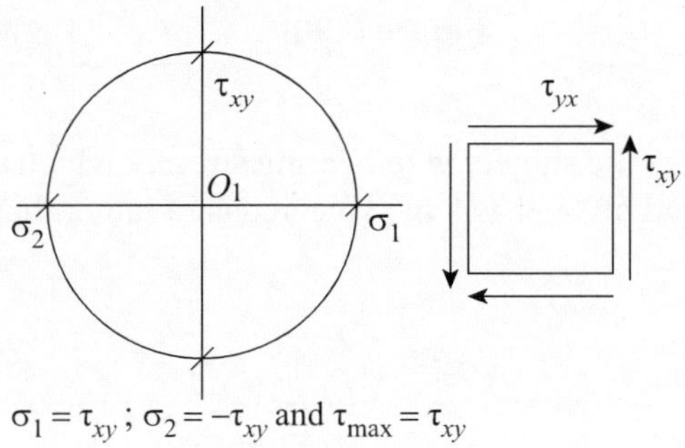

Figure 3.18. Mohr's circle for torison test

We will consider the stresses induced by the particular types of loadings.

3.5.1 Axial Load

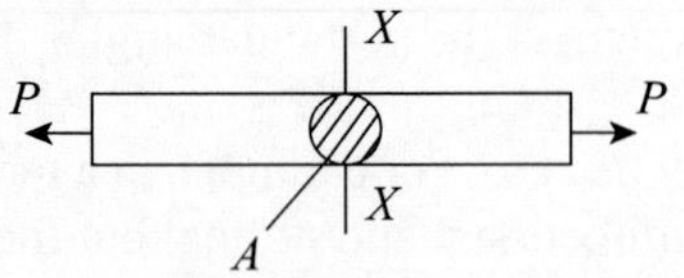

Figure 3.19

The load applied is coaxial with the road cross section. The distribution of stress on a section $X - X$ away from the ends is uniform and is given by

$$\sigma = \frac{P}{A} \tag{3.26}$$

This expression holds good both in tension and compression when the length of the rod is short, for longer lengths under compression this equation does not hold good. Then the column action equations are to be used.

The elongation caused by the load is given by

$$\delta = \left(\frac{\sigma}{E}\right)L = \left(\frac{PL}{AE}\right) \tag{3.27}$$

Normally, the load application is through a pin joint as shown in Fig. 3.20.

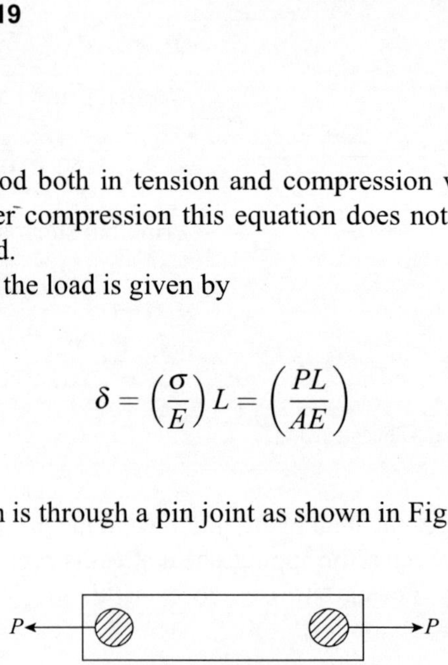

Figure 3.20

The pin and the hole surfaces are subjected to bearing stresses which are compressive in nature. The stress induced are calculated on the basis of projected area rather than the circumferential area

$$\sigma_c = \left(\frac{P}{ld}\right) \tag{3.28}$$

3.5.2 Direct Shear Stress

Direct shear stress is induced when the force is applied parallel to the surface with no gap between the forces and the surface. A pair of shears, i.e., scissors designed to cut material causes direct shear in the metal. The edges of the scissors are aligned and produce a force P on the material along the common plane of the blades. This force creates uniform stresses on the plane and the material is sheared into two pieces when the shear stress induced is more than ultimate strength in shear of the material. The shear stress induced is given by the expression

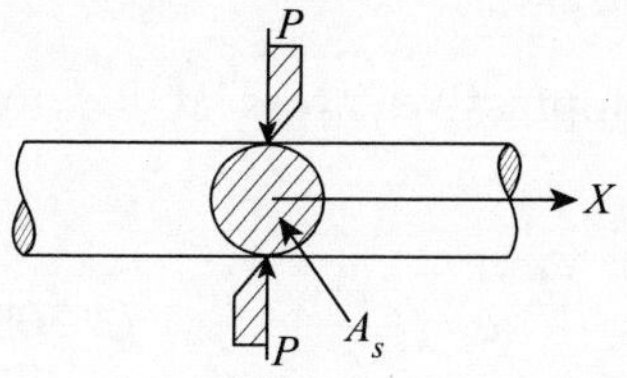

(a)

$$\tau_{xy} = \frac{P}{A_s} \tag{3.29a}$$

where, A_s is area under shear.

The pure shear stress is induced in a pin-joint (closely fitted) at the common surfaces between the rod and the fork, i.e., planes XX and YY.

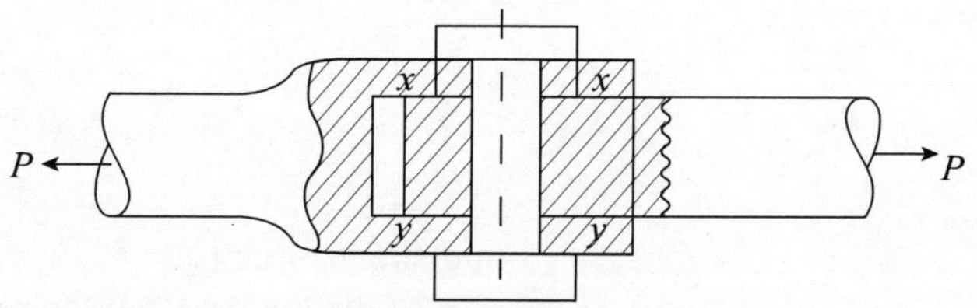

(b)

Figure 3.21

This is the case of double shear which is given by

$$\tau_{xy} = \frac{P}{2A_s} \tag{3.29b}$$

where A_s is the cross-sectional area of the pin.

3.5.3 Straight Beams and Bending Stresses

Pure Bending:

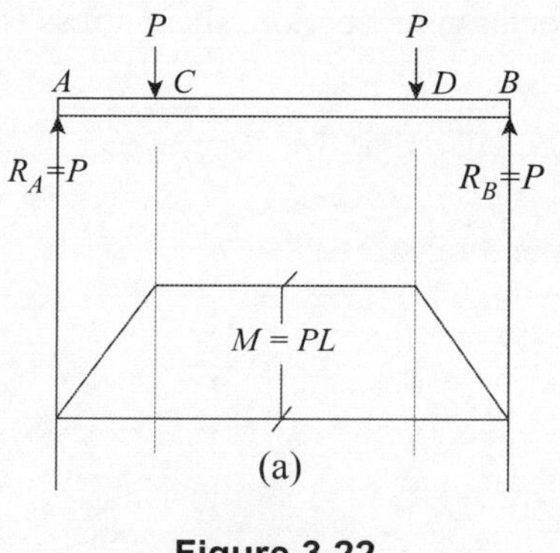

(a)

Figure 3.22
(Contd.)

In a beam loaded as shown in Fig. 3.22a the section between CD is subjected to pure bending $M = PL$.

This induces bending stresses in the cross section of the beam which has a maximum tensile stress at the bottom fibre.

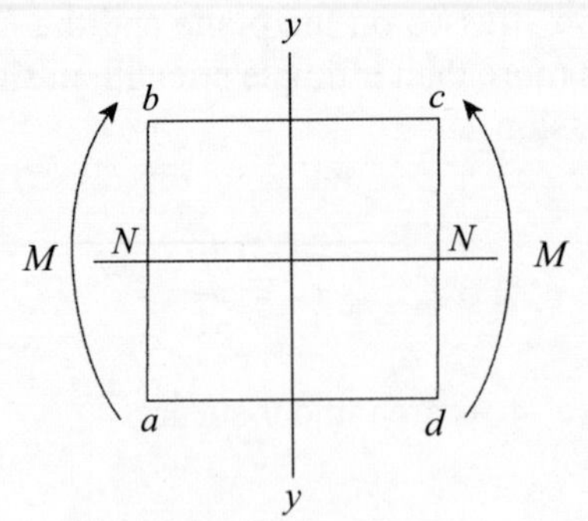

(b) BM at any section between *C&D*

$$\sigma_x = \frac{M}{I} \cdot c = \frac{M}{Z} \tag{3.30a}$$

The maximum compressive stress at the top fibre.

$$\sigma_x = \frac{M}{I} \cdot c = \frac{-M}{Z} \tag{3.30b}$$

It varies linearly from bottom fibre to the top fibre with zero value at centre fibre called neutral axis.

At any section between A and C there is bending moment M_x.

$$M_x = P \cdot x$$

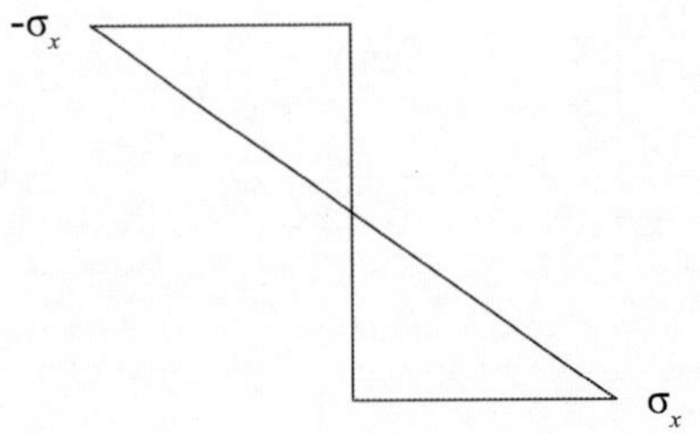

(c) Bending stress-distribution across section

and shear force $F_s = P$.

At distance y from the neutral axis the shear stress induced is given by

$$\tau_{xy} = \frac{FQ}{Ib} \tag{3.31}$$

where F is shear force and

$$Q = \int_y^c y \cdot dA = A\bar{y} \tag{3.32}$$

For rectangular section, shear stress is

$$\tau_{xy} = \frac{FA\bar{y}}{Ib} \tag{3.33}$$

The bending stress is,

$$\sigma_x = \frac{Mx}{I} \cdot y \tag{3.34}$$

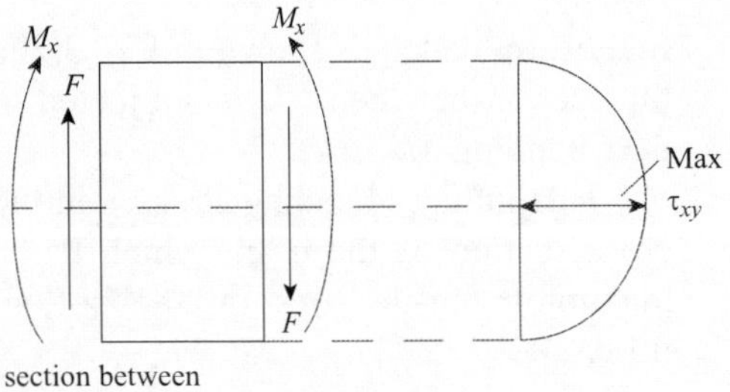

section between $A \times C$

(d) Transverse shear stress

Figure 3.22

The maximum transverse shear stress occurs at the centre and depends upon the cross section. For rectangular section,

$$\tau_{\max} = \left(\frac{3}{2} \cdot \frac{F}{A}\right) \tag{3.35a}$$

For circular section,

$$\tau_{\max} = \left(\frac{4}{3} \cdot \frac{F}{A}\right) \tag{3.35b}$$

For hollow circular cross section,

$$\tau_{\max} = 2\left(\frac{F}{A}\right) \tag{3.35c}$$

For I-Section,

$$\tau_{\max} = \frac{F}{A_w} \tag{3.35d}$$

where A_w is area of the web.

3.5.4 Curved Beam Theory

The curved beam is a beam with large curvature causing variation in the fibre length from inner to outer fibre. The basic assumptions (due to Bernoulli-Euler) are:

1. The beam is initially curved.
2. The plane cross sections initially perpendicular to the neutral axis remains plane and perpendicular to the neutral axis, even after loading.
3. The beam bends without twisting, i.e., it is loaded and deflected in the longitudinal plane of symmetry.
4. The loading is within the elastic limit and deflections are small.
5. Material of the beam is homogeneous and obeys Hooke's law.

The stresses induced under the action of pure bending moment are derived as follows. Consider an element *abcd* of the beam. By application of moment M, the section *cd* rotates through an angle $d\theta$ to fg, about the netural axis with respect to *ab*. The reduction in the length of the fiber at a distance y from the neutral axis is.

$\delta l = -y \cdot d\theta$

The strain in the fibre,

$$\varepsilon = \frac{\delta 1}{1} = \frac{-y d\theta}{(r_n + y)\theta} \tag{3.36}$$

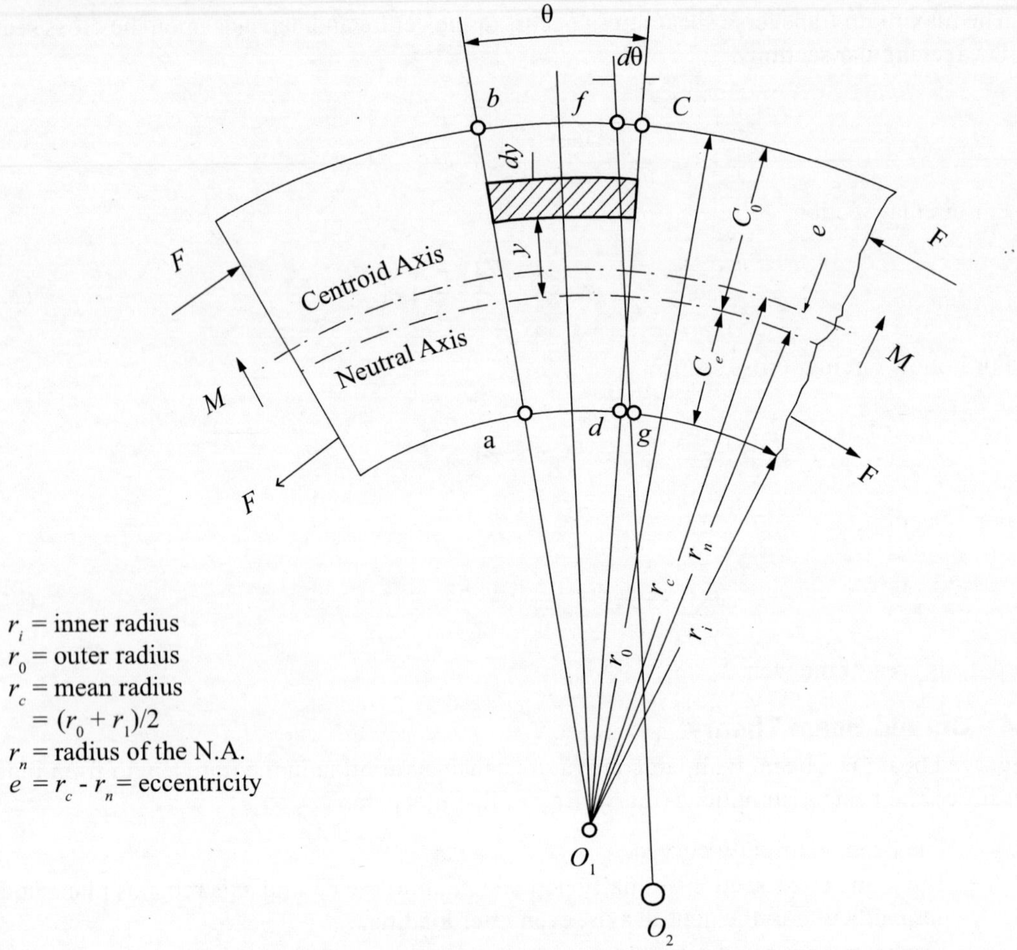

Fig 3.23. Curved Beam Nomenclature

The stress induced,

$$\sigma = \varepsilon . E = \frac{-yd\theta E}{(r_n + y)\theta} \tag{3.37}$$

Since there is no net axial force on the cross section
$\int \sigma dA = 0$

i.e.,

$$\int \frac{yd\theta E}{(r_n + y)\theta} dA = \left(\frac{Ed\theta}{\theta}\right) \int \frac{ydA}{(r_n + y)} = 0 \tag{3.38}$$

$$\therefore \int \frac{ydA}{(r_n + y)} = 0 \tag{3.39}$$

The moment due to the induced stresses (resisting moment) must be equal to the applied moment.

$M = -\int y\sigma dA$

$$= \left(\frac{Ed\theta}{\theta}\right) \int \frac{y^2 dA}{(r_n + y)}$$

$$= \left(\frac{Ed\theta}{\theta}\right) \int \frac{y(y + r_n - r_n)dA}{(r_n + y)}$$

$$= \left(\frac{Ed\theta}{\theta}\right) [\int ydA - r_n \int \frac{ydA}{(r_n + y)}] = \left(\frac{Ed\theta}{\theta}\right) \int ydA \text{ (using(Eqn.3.39))}$$

$$= +(\frac{Ed\theta}{\theta})(\text{eA}) \tag{3.40}$$

where $e = r_c - r_n$ = eccentricity of neutral axis.

From Eqn. 3.37, $\frac{Ed\theta}{\theta} = -(\frac{r_n + y}{y})\sigma$

Substituting in Eqn. 3.40,

$$M = \left[-(\frac{r_n + y}{y})\sigma eA\right]$$

Then stress is given by

$$\sigma = -\frac{M.y}{Ae(r_n + y)} \tag{3.41}$$

The stress at the outer fibre,

$$\sigma_0 = -(\frac{MC_o}{A.e.r_o}) \tag{3.42}$$

And at the inner fibre

$$\sigma_i = +(\frac{MC_i}{A.e.r_i}) \tag{3.43}$$

where $C_o = r_o - r_n$ and $C_i = r_n - r_i$

The Eqn. (3.39) used to determine the radius of the neutral axis as shown below.

$$\int \frac{y.dA}{r_n + y} = 0$$

$$\text{i.e.,} \int \frac{(y + r_n - r_n)(dA)}{(r_n + y)} = 0$$

Table 3.1. Expressions of radius of neutral axis and centroidal axis for curved beams (Plate 1.5 MDHB)

Section	Radius of mean neutral surface r_n & Radius of mean surface r_c
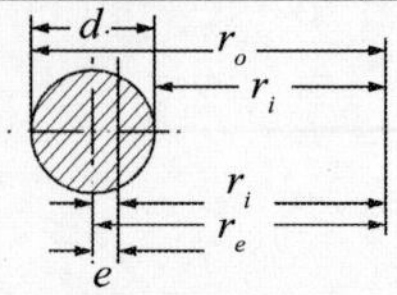	1) $r_n = \left[\frac{\sqrt{r_o}+\sqrt{r_i}}{2}\right]^2 \quad r_c = \left(\frac{r_i+r_o}{2}\right)$ $e = r_e - r_n = \frac{d^2}{4[4r_o - \sqrt{4r_c^2 - d^2}]}$
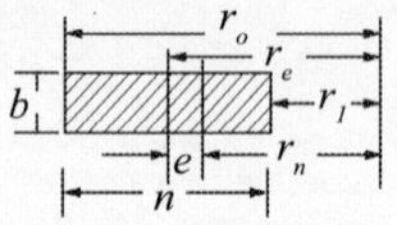	2) $r_n = \left[\frac{h}{\ln(r_o/r_i)}\right]; \quad r_c = \left(\frac{r_i+r_o}{2}\right) \quad e = r_e - r_n$
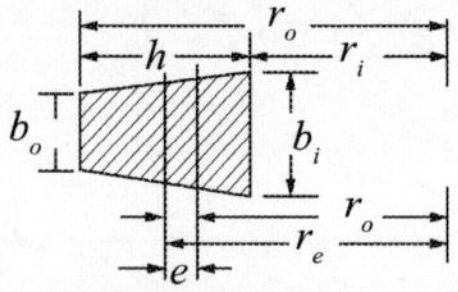	3) $r_n = \frac{(b_i+b_o)h/2}{\left(\frac{b_i r_o - b_o r_i}{h}\right)\ln(r_o/r_i) - (b_i - b_o)}. \quad r_c = r_i + \frac{h}{3}\left[\frac{b_i+2b_o}{b_i+b_o}\right]$ if $b_o = 0$, this section reduces to a triangle.
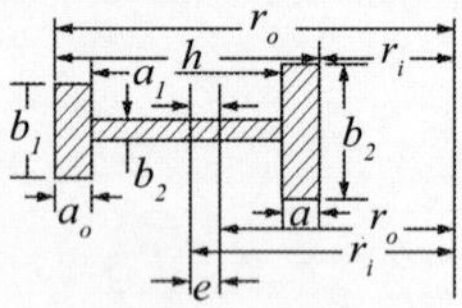 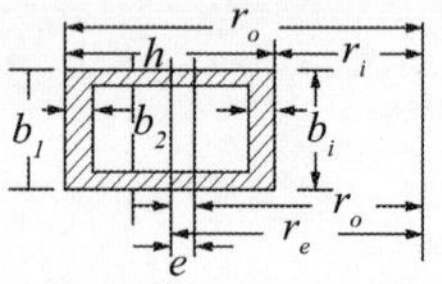	4) $r_n = \frac{A}{b_1\ln\left[\frac{r_i+a_i}{r_i}\right] + b_2\ln\left[\frac{r_o+a_o}{r_i+a_i}\right] + b_o\ln\left[\frac{r_o}{r_o+a_o}\right]}$ $r_c = r_i + \frac{0.5b_i a_i^2 + b_2 a_2(a_i+0.5a_2) + b_o a_o(h-0.5a_o)}{A}$ where $A = b_i a_i + b_2 a_2 + b_o a_o \quad e = r_c - r_n$ The equation for r_n is the same for a box section with each side panel of thickness $= 0.5b_2$. If $a_o = 0$, the section reduces to a T section.

$$\therefore \int dA - r_n \int \frac{dA}{r_n+y} = 0$$

$$\therefore r_n = \frac{A}{\int \frac{dA}{(r_n+y)}} = \frac{A}{\int \frac{dA}{r}} \tag{3.44}$$

The values of r_n for common sections are given in Table 3.1.

3.5.4a Chain-Ring (Fig 3.24)

Due to symmetry, only quarter portion is considered. The moment on the Section AA,

$$M = M_2 + \frac{FR}{2}(1 - \sin\theta)$$

The strain energy between section 1-1 and 2-2 is

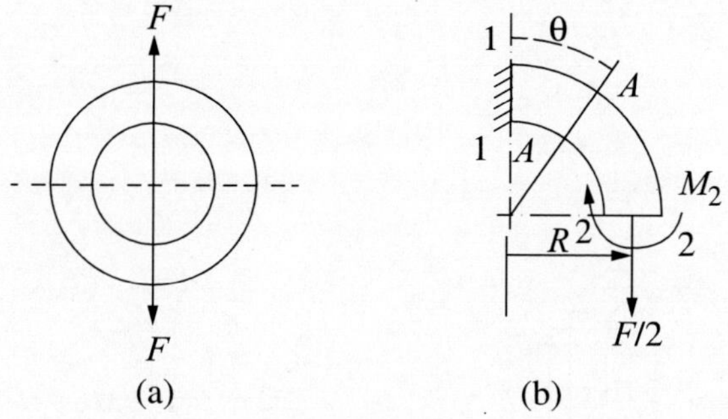

Figure 3.24

$$U = \int_o^{\pi/2} \frac{M^2 R d\theta}{2EI} = \int_o^{\pi/2} \frac{\left[M_2 + \frac{FR}{2}(1 - \sin\theta)\right]^2 R d\theta}{2EI}$$

By Castiglione theorem, because orientation of section 2-2 does not change,

$$\frac{\delta U}{\delta M_2} = \text{rotation under } M_2 = 0$$

$$\therefore \int_o^{\pi/2} \frac{[M_2 + \frac{FR}{2}(1 - \sin\theta)]R d\theta}{EI} = 0$$

$$\therefore \int_o^{\pi/2} M_2 d\theta + \frac{FR}{2} \int_o^{\pi/2} (1 - \sin\theta)] d\theta = 0$$

$$\therefore \frac{\pi}{2} M_2 \cdot + \frac{FR}{2}(\frac{\pi}{2} - 1) = 0$$

$$M_2 = -\frac{FR}{\pi}(\frac{\pi}{2} - 1)$$

$$M_2 = \frac{FR}{\pi}\left(\frac{\pi}{2}\right) - \frac{FR}{\pi} = -0.182 \; FR \tag{3.45}$$

The negative sign indicates that the moment is anticlockwise.
Then the moment at section 1-1 under load, i.e., $\theta = 0$.

$$M_1 = -0.182\,FR + \frac{FR}{2} = 0.318\,FR \tag{3.46}$$

3.5.4b Chain-link (Fig 3.25)

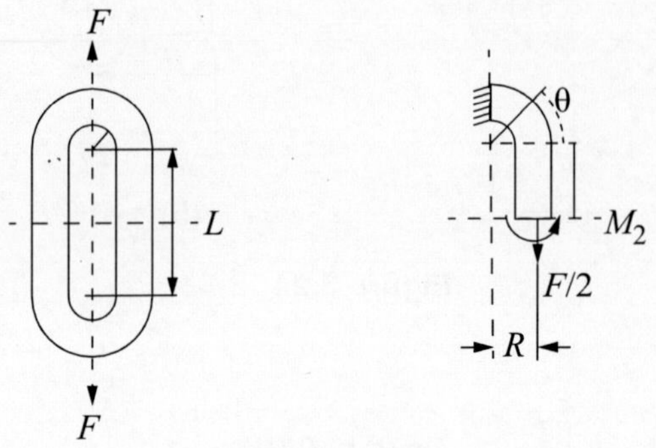

Figure 3.25

Due to symmetry, only quarter portion is considered.
The moment along the straight portion,

$$M = +M_2$$

Moment along the curved portion,

$$M = M_2 - \frac{FR}{2}(1 - \cos\theta)$$

The strain energy is given by,

$$U = \int \frac{M^2 dx}{2EI}$$

$$U = \int_o^{L/2} \frac{M^2 d1}{2EI} + \int_o^{\pi/2} \frac{M^2 R d\theta}{2EI}$$

$$\text{Since } \frac{\delta U}{\delta M_2} = 0$$

$$\int_o^{L/2} \mathrm{M_2 d1} + \int_o^{\pi/2} [M_2 - \frac{FR}{2}(1 - \cos\theta)] R d\theta = 0$$

$$\therefore \quad M_2\frac{L}{2} + M_2\frac{\pi R}{2} - \frac{FR^2}{2}(\frac{\pi}{2} - 1) = 0$$

$$\therefore \quad M_2 = \frac{FR^2(\frac{\pi}{2} - 1)}{(L + \pi R)} = \frac{FR}{2}\left[\frac{\pi R - 2R}{L + \pi R}\right] \tag{3.47}$$

The moment is clockwise.
The moment under the load,

$$M_1 = M_2 - \frac{FR}{2} = \frac{FR}{2}(\frac{\pi R - 2R}{L + \pi R}) - \frac{FR}{2}$$

$$= (\frac{FR}{2})[\frac{\pi R - 2R - L - \pi R}{L + \pi R}] = -\frac{FR}{2}\left[\frac{2R + L}{L + \pi R}\right] \tag{3.48}$$

The negative sign indicates that the moment is anticlockwise.

3.5.5 Deflections in Beams

The deflection of the beams is given by the controlling equation,

$$\frac{M}{EI} = \frac{d^2y}{dx^2}$$

This equation is applicable for the deflections caused by the bending moment. The bending moment varies along the length of the beam and is the function of loading. The successive differentiation gives the shear load function and load intensity function. The successive integrations give the slope function and deflection function. All these equations are given below.

The load intensity function : $\frac{q}{EI} = \frac{d^4y}{dx^4}$ (3.49a)

The shear load function : $\frac{v}{EI} = \frac{d^3y}{dx^3}$ (3.49b)

The bending moment function : $\frac{M}{EI} = \frac{d^2y}{dx^2}$ (3.49c)

The slope of beam central line : $\theta = \frac{dy}{dx}$ (3.49d)

The deflection of beam centre line $y = f(x)$. The results of these equations for various loading patterns are given in the Plate 1.8 MDHB.

The maximum deflections for concentrated load:

(i) Cantilever beam, $y_{max} = \dfrac{WL^3}{3EI}$ (3.50)

(ii) Centrally loaded beam, $y = \dfrac{WL^3}{48EI}$ (3.51)

If the load is uniformally distributed:

(iii) Cantilever beam, $y_{max} = \dfrac{WL^4}{8EI}$ (3.52)

(iv) Simply supported beam, $y_{max} = \dfrac{5WL^4}{384EI}$ (3.53)

3.5.6 Torsional Load

The shaft of machines transmit power. They are subjected to torsional shear stress. For circular shafts the governing equation is

$$\frac{T}{J} = \frac{\tau}{r} = \frac{G\theta}{L} \tag{3.54}$$

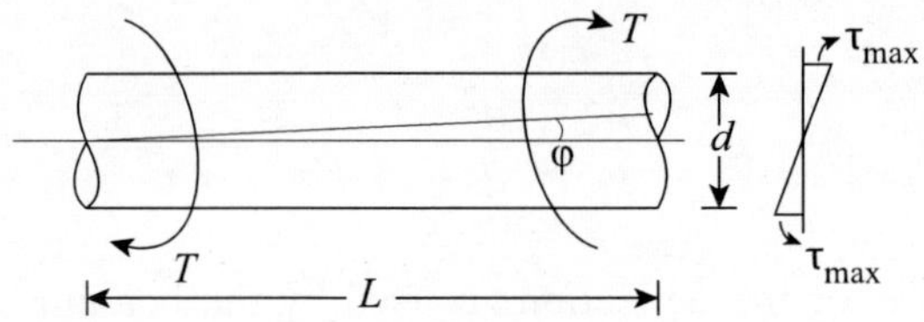

Figure 3.26

The maximum shear stress is given by

$$\tau_{max} = \frac{T}{J} \cdot \frac{d}{2} = \frac{16T}{\pi d^3} \tag{3.55}$$

The angular twist is given by

$$\theta = \frac{TL}{GJ} = \frac{32TL}{\pi d^3 G} \tag{3.56}$$

The torsional rigidity of the shaft is defined as

$$K = \frac{T}{\theta} = \frac{GJ}{L} \tag{3.57}$$

where T = polar moment $= \dfrac{\pi}{32} d^4$

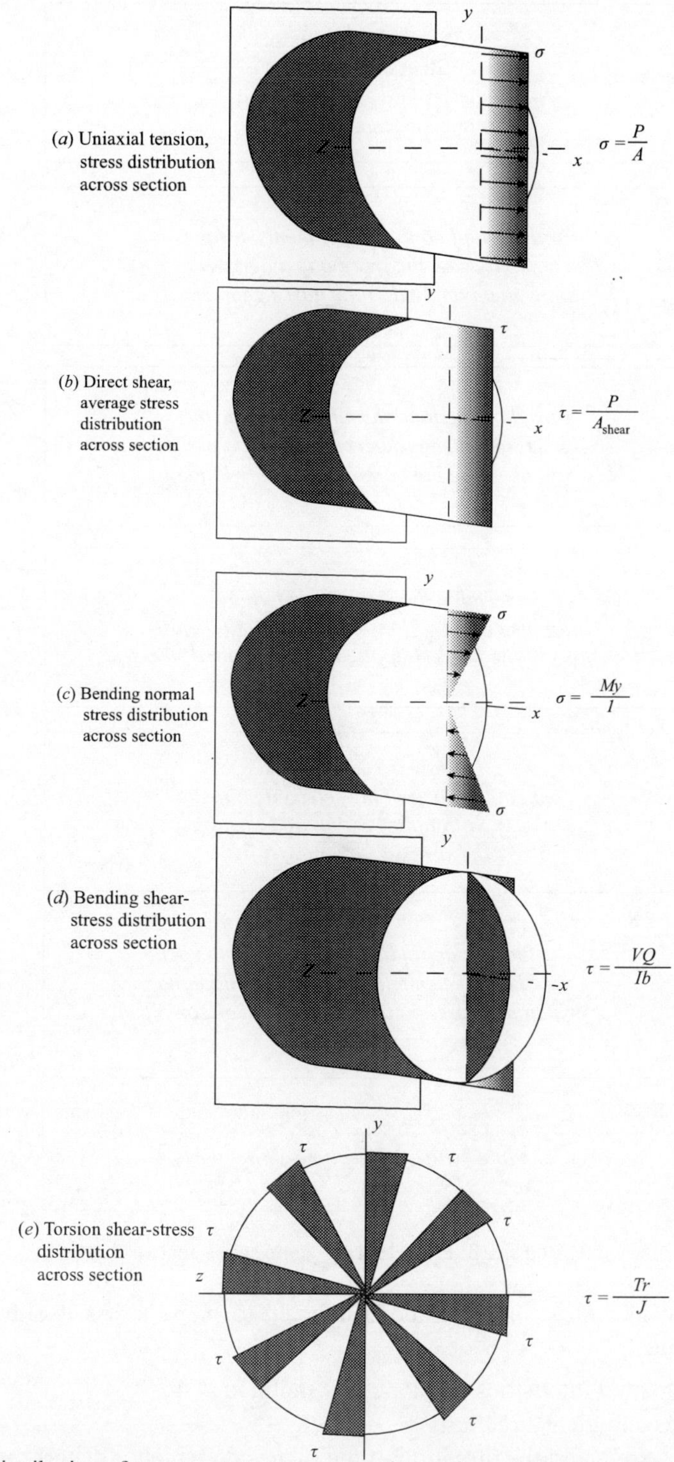

Figure 3.27. Distribution of stresses across a cross section under various types of loading conditions

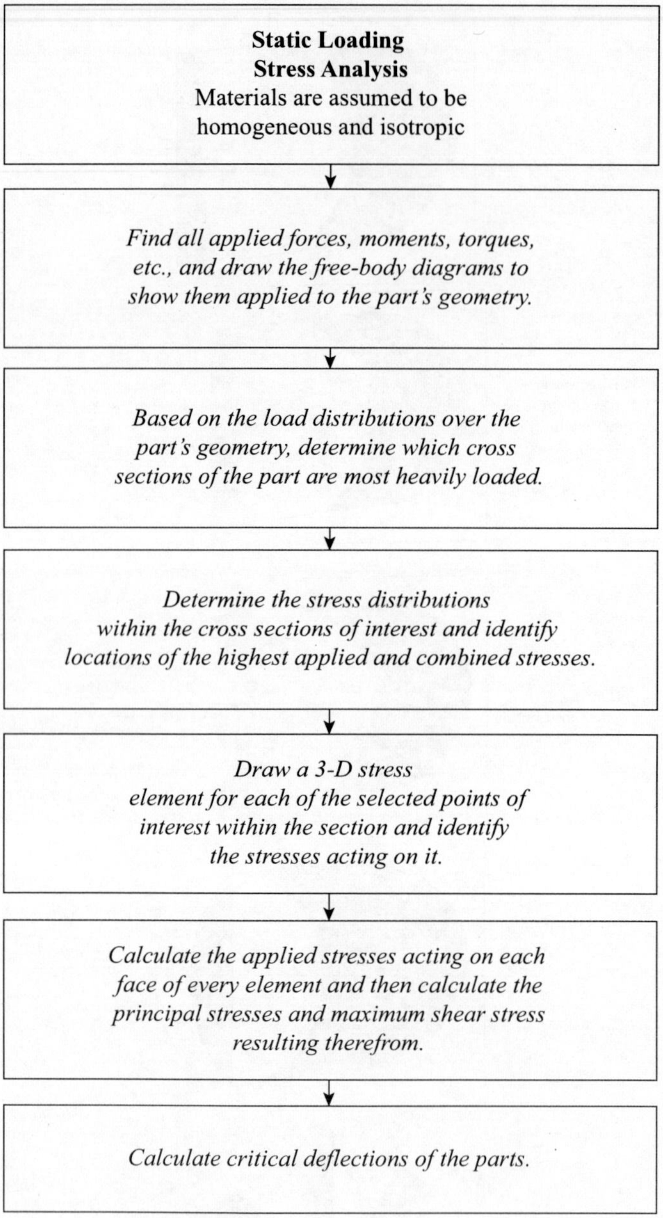

Figure 3.28. Flowchart for static stress analysis

Example 3.6: A point in a structural member subjected to plane stress is shown in Fig. 3.27. Determine the following:

(i) Normal and tangential intensities on plane *MN* inclined at 45°
(ii) Principal stresses and their directions
(iii) Maximum shear stress and the direction of the planes on which it occurs.

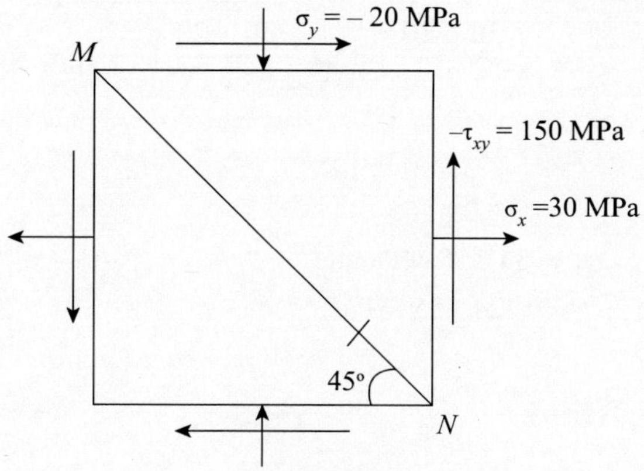

Figure 3.29

SOLUTION

Given Data

$$\sigma_x = 30 \text{ MPa}, \sigma_y = -20 \text{ MPa}, \tau_{xy} = 15 \text{ MPa CCW}$$

Step-1: Stresses on plane MN

Normal stress on plane *MN* (Eqn. 1.18 MDHB)

$$\sigma_\phi = \frac{\sigma_x + \sigma_y}{2} + \left(\frac{\sigma_x - \sigma_y}{2}\right)\cos 2\phi + \tau_{xy}\sin 2\phi$$

$$= \frac{30-20}{2} + \left(\frac{30+20}{2}\right)\cos\,(2\times 45^\circ) + 15\sin\,(2\times 45^\circ)$$

$$= 20 \text{ MPa}$$

Shear stress on the plane *MN* (Eqn. 1.19 MDHB)

$$\tau_\phi = -\left(\frac{\sigma_x - \sigma_y}{2}\right)\sin 2\phi + \tau_{xy}\,\cos 2\phi$$

$$= \frac{30+20}{2}\sin\,(2\times 45^\circ) + 15\cos\,(2\times 45^\circ)$$

$$= 25 \text{ MPa}$$

Step-2: Principal stresses

Principal stresses (Eqn. 1.21 MDHB)

$$\sigma_1, \sigma_2 = \left(\frac{\sigma_x + \sigma_y}{2}\right) \pm \sqrt{\left(\frac{\sigma_x - \sigma_y}{2}\right)^2 + \tau_{xy}^2}$$

$$= \left(\frac{30-20}{2}\right) \pm \sqrt{\left(\frac{30+20}{2}\right)^2 + 15^2}$$

$$= 5 \pm \sqrt{25^2 + 15^2}$$

$$= 5 \pm 29.155$$

$$\sigma_1 = 34.155 \text{ MPa and}$$

$$\sigma_2 = -24.155 \text{ MPa}$$

Step-3: Principal directions
The principal directions are given by (Eqn. 1.20 MDHB)

$$\tan 2\phi_n = \frac{-2\tau_{xy}}{\sigma_x - \sigma_y}$$

$$\phi_n = \frac{-1}{2}\tan^{-1}\left(\frac{2\times 15}{30+20}\right)$$

$$\phi_{n1} = \frac{-1}{2}\tan^{-1}(0.6) = -15.48^\circ$$

$$\phi_{n2} = -15.48^\circ + 90 = 74.51^\circ$$

Principal stress element

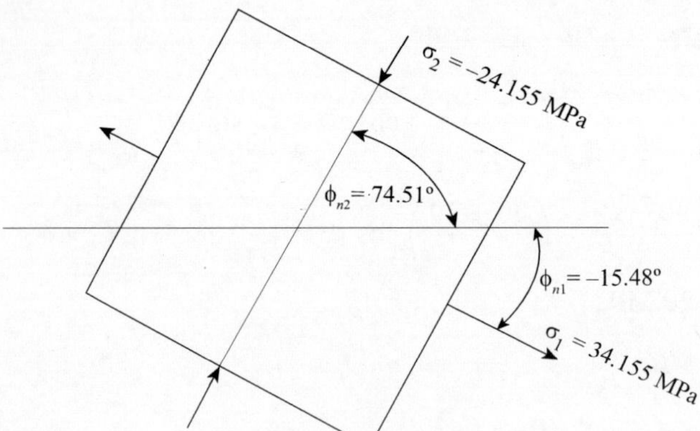

Figure 3.30

Step-4: Maximum shear stress and direction
Maximum shear stress (Eqn. 1.23 MDHB)

$$\tau_{\max} = \sqrt{\left(\frac{\sigma_x - \sigma_y}{2}\right)^2 + \tau_{xy}^2}$$

$$= \sqrt{\left(\frac{30+20}{2}\right)^2 + 15^2} = 29.155 \text{ MPa}$$

The directions are (Eqn. 1.22 MDHB)

$$\tan 2\phi_s = \frac{\sigma_x - \sigma_y}{2\tau_{xy}} = \left(\frac{30+20}{2\times 15}\right)$$

$$\phi_{s_1} = \frac{1}{2}\tan^{-1}\left(\frac{50}{30}\right) = \frac{1}{2}\times 59.036 = 29.5181$$

$$\phi_{s_2} = 90 + 29.5181 = 119.52$$

$$\text{OR, } \phi_s = \phi_n + 45^\circ$$

$$= -15.48 + 45^\circ$$

$$= 29.52^\circ$$

Maximum shear element

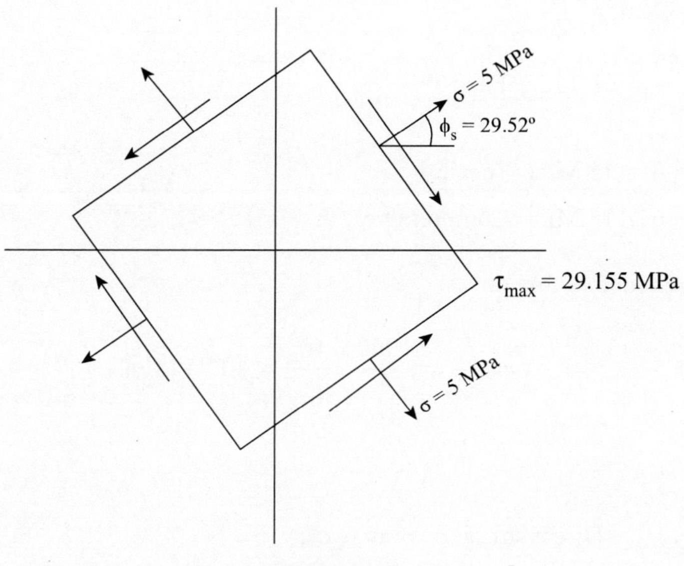

Figure 3.31

Example 3.7: A cantilever beam of circular cross section is loaded as shown in the Fig. 3.31. Determine the maximum and minimum normal stresses and maximum shear stresses at points *A* and *B*.

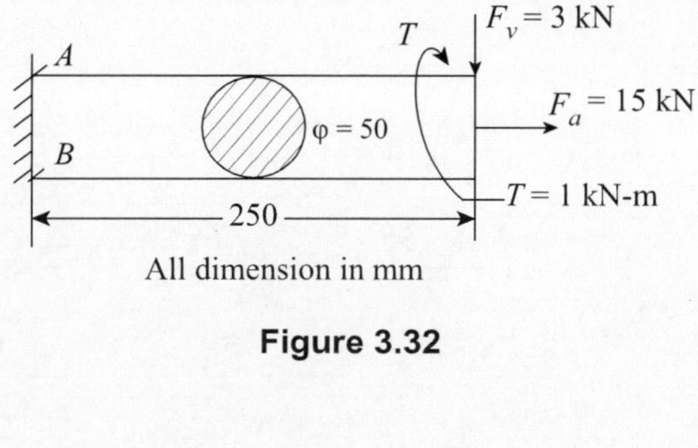

All dimension in mm

Figure 3.32

SOLUTION

Given Data :

Axial force, $F_a = 15$ kN
Lateral force, $F_v = 3$ kN
Torsional moment, $T = 1$ kN-m
The stresses are induced by these loads are as follows:

Step-1: Direct tensile stress due to axial load

$$\sigma_t = \frac{F_a}{A} = \frac{15 \times 10^3}{\frac{\pi}{4} \times (50)^2} = 7.64 \text{ MPa}$$

Step-2: Bending stress due to lateral force

$$\sigma_b = \frac{M}{I}.c = \frac{F_v L}{I}.c = \frac{F_v L}{Z}$$

$$= \frac{3000 \times 250}{\frac{\pi}{32} \times 50^3} = 61.115 \text{ MPa}$$

At $A, \sigma_{bA} = +61.115$ MPa (Tensile)

At $B, \sigma_{bB} = -61.115$ MPa (Compressive).

Step-3: The shear stress due to torsion

$$\tau = \frac{T}{Z_P} = \frac{1000 \times 10^3}{\frac{\pi}{16} \times 50^3} = 40.74 \text{ MPa}$$

Step-4: The stresses at A

$$\text{Direct stress, } \sigma_x = \sigma_t + \sigma_b$$
$$= 7.64 + 61.115 = 68.755 \text{ MPa}$$
$$\sigma_y = 0$$
$$\text{Shear stress, } \tau_{xy} = 40.74 \text{ MPa}$$

Step-5: Principal stresses

$$\sigma_1, \sigma_2 = \left(\frac{\sigma_x}{2}\right)^2 \pm \sqrt{\left(\frac{\sigma_x}{2}\right)^2 + \tau_{xy}^2}$$

$$= \left(\frac{68.755}{2}\right)^2 \pm \sqrt{\left(\frac{68.755}{2}\right)^2 + 40.74^2}$$

$$- 34.377 \pm 53.31$$

$$\text{Maximum } \sigma_1 = 87.687 \text{ MPa}$$
$$\text{Minimum } \sigma_2 = -18.933 \text{ MPa}$$
$$\text{Maximum shear} = \frac{\sigma_1 - \sigma_2}{2} = \frac{87.687 + 18.933}{2} = 53.31 \text{ MPa}$$

Step-6: Stresses at B

$$\text{Direct stress, } \sigma_x = 7.64 - 61.115 = -53.475 \text{ MPa}$$
$$\text{Shear stresses, } \tau_{xy} = 40.74 \text{ MPa}$$

Step-7: Principal stresses

$$\sigma_1, \sigma_2 = \frac{53.475}{2} \pm \sqrt{\left(\frac{-53.475}{2}\right)^2 + 40.74^2}$$
$$= -26.7375 \pm 48.738$$
$$\text{Maximum } \sigma_1 = 21.996 \text{ MPa (Tensile)}$$
$$\text{Minimum } \sigma_2 = -75.471 \text{ MPa (Compressive)}$$
$$\text{Maximum shear stress, } \tau_{max} = \frac{21.996 + 75.471}{2} = 48.734 \text{ MPa}$$

Example 3.8: A point in a structural member is subjected to plane stress as shown in Fig. 3.32. Determine the following:

(i) Normal and tangential stresses on a plane at an angle of 45° to x-plane
(ii) Principal stresses and their directions
(iii) Maximum shear stress and its plane.

SOLUTION

Given Data : $\sigma_x = 40$ MPa, $\sigma_y = -30$ MPa and $\tau_{xy} = 25$ MPa.
Inclination of the plane MN $\phi = 45°$
Step-1: The normal stress on the plane (Eqn. 1.18 MDHB)

$$\sigma_\phi = \frac{\sigma_x + \sigma_y}{2} + \left(\frac{\sigma_x - \sigma_y}{2}\right) \cos 2\phi + \tau_{xy} \sin 2\phi$$
$$= \frac{40 - 30}{2} + \left(\frac{40 + 30}{2}\right) \cos\,(2 \times 45°) + 25 \sin\,(2 \times 45°)$$
$$= 30 \text{ MPa}$$

Step-2: Shear stress on the plane (Eqn. 1.19 MDHB)

$$\tau_\phi = \left(\frac{\sigma_x - \sigma_y}{2}\right) \sin 2\phi + \tau_{xy} \cos 2\phi$$

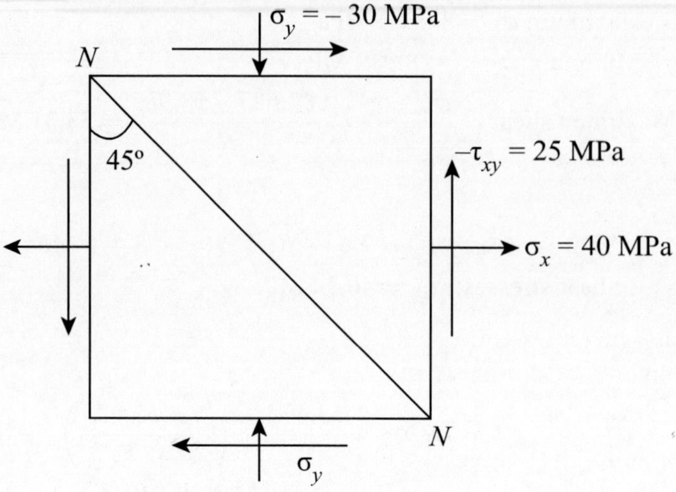

Figure 3.33

$$= \left(\frac{40+30}{2}\right) \sin\,(2\times 45^\circ) + 25 \cos\,(2\times 45^\circ)$$
$$= 35 \text{ MPa}$$

Step-3: Principal stresses

$$\sigma_1, \sigma_2 = \left(\frac{\sigma_x+\sigma_y}{2}\right) \pm \sqrt{\left(\frac{\sigma_x-\sigma_y}{2}\right)^2 + \tau_{xy}^2}$$
$$= \left(\frac{40-30}{2}\right) \pm \sqrt{\left(\frac{40+30}{2}\right)^2 + 25^2}$$
$$= 5 \pm 43.011$$
$$\sigma_1 = 48.011 \text{ MPa (Tensile)}$$
$$\sigma_2 = -38.011 \text{ MPa (Compressive)}$$

Step-4: Directions are given by

$$\tan 2\phi_n = \frac{-2\tau_{xy}}{\sigma_x - \sigma_y}$$
$$\phi_n = \frac{-1}{2}\tan^{-1}\left(\frac{2\times 25}{40+30}\right)$$
$$\phi_{n_1} = -17.769^\circ$$
$$\text{and} \quad \phi_{n_2} = 90 - 17.769 = 72.231^\circ.$$

Element is

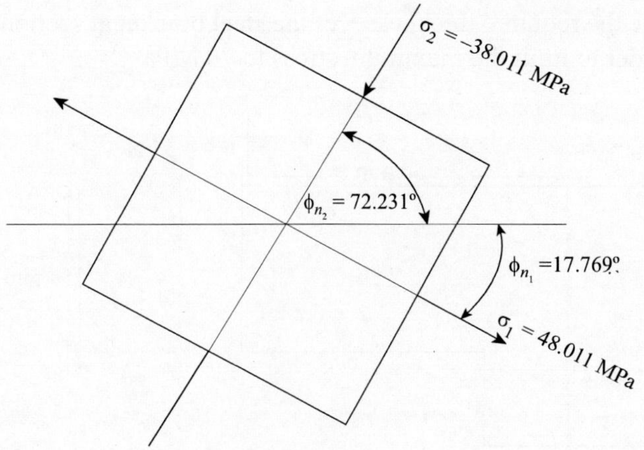

Figure 3.34

Step-5: Maximum shear stress is given by

$$\tau_{\max} = \pm\sqrt{\left(\frac{\sigma_x - \sigma_y}{2}\right) + \tau_{xy}^2}$$

$$= \pm\sqrt{\left(\frac{40+30}{2}\right)^2 + 25^2}$$

$$= 43.011 \text{ MPa}$$

Step-6: Directions are

$$\phi_{s_1} = -17.769° + 45° = 27.231°$$

$$\text{and } \phi_{s_2} = 72.231 + 45° = 117.231°$$

Element is

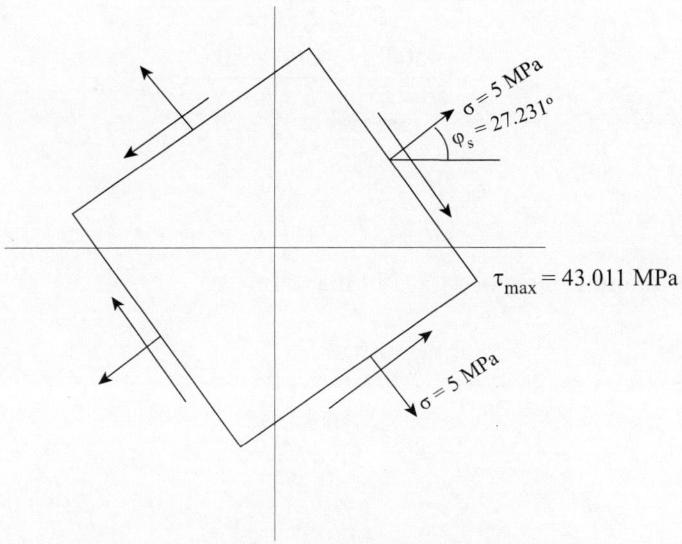

Figure 3.35

Example 3.9: Determine the required thickness h of the steel bracket at section AB when loaded as shown in Fig. 3.35 in order to limit the maximum stress to 70 MPa.

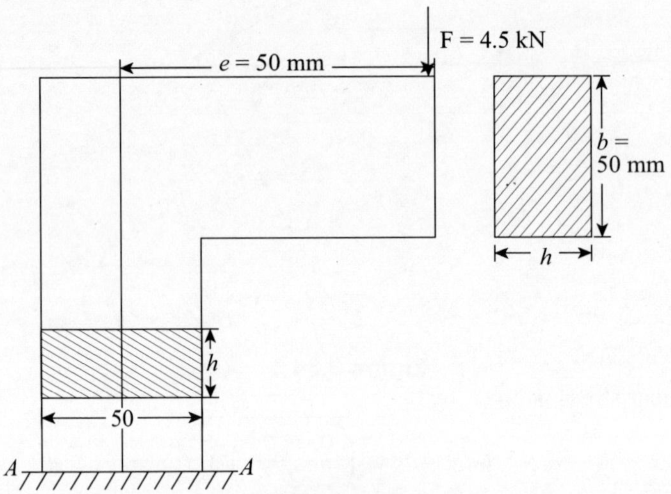

Figure 3.36

SOLUTION

Step-1: This is a problem on eccentric loading. The stress at section AA is given by

$$\sigma = \frac{F}{A} + \frac{M}{Z} = \frac{F}{A} + \frac{F \cdot e}{Z}$$

$$= \frac{4500}{50 \times h} + \frac{4500 \times 50}{\left(\frac{h \times 50^2}{6}\right)}$$

$$= \frac{90}{h} + \frac{540}{h} = \frac{630}{h}$$

Step-2: Thickness for limiting value of $\sigma = 70$ MPa

$$\therefore \quad 70 = \frac{630}{h}$$

$$h = \frac{630}{70} = 9 \text{ mm}$$

Example 3.10: Determine the extreme fibre stresses at the critical section of a machine member loaded as shown in the Fig. 3.36. Also show the distribution of stresses at this section.

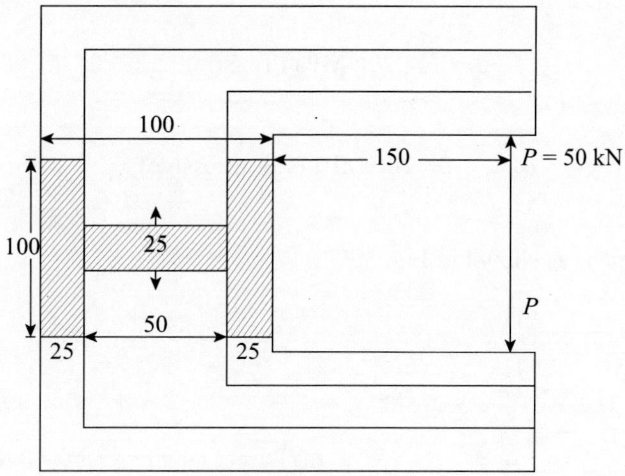

Figure 3.37

SOLUTION

Step-1: The clamp is eccentrically loaded. The stresses in the extreme fibres are given by

$$\sigma = \frac{P}{A} \pm \frac{M}{Z} = \frac{P}{A} \pm \frac{P.e}{Z}$$

where,

$$P = 50 \times 10^3 \text{N}$$
$$A = 2 \times 100 \times 25 + 50 \times 25 = 6250 \text{ mm}^2$$
$$e = 150 + 50 = 200 \text{ mm}$$

Section modulus of area,

$$Z = \frac{BH^3 - bh^3}{6H} = \frac{100 \times 100^3 - 75 \times 50^3}{6 \times 100}$$
$$= 151041.667 \text{ mm}^3$$

Stresses in the fibres,

$$\sigma = \frac{50 \times 10^3}{6250} \pm \frac{50 \times 10^3 \times 200}{151041.667}$$
$$= 8 \pm 66.207$$
$$\sigma_1 = 74.206 \text{ MPa and } \sigma_2 = -58.206 \text{ MPa}$$

Stress in the inner fibre,

$$\sigma_i = 74.206 \text{ MPa (Tensile)}$$

Stress in the outer fibre,

$$\sigma_o = -58.206 \text{ MPa (Compressive)}$$

Step-2: Total stresses,
The stress variation is linear as shown in Fig. 3.37.

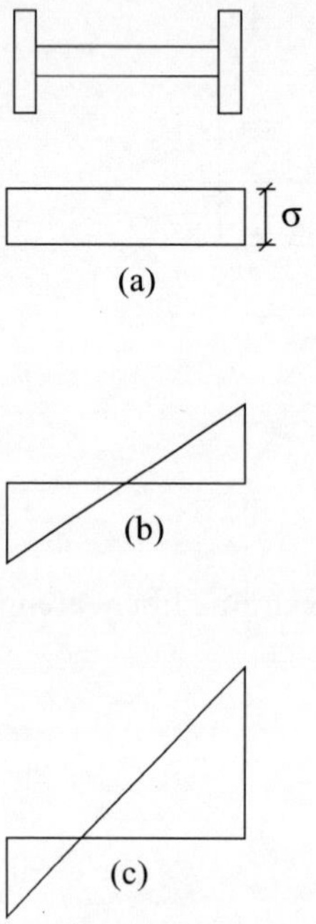

(a) Direct tensile stress = 8 MPa (σ)

(b) Bending stress distribution with extreme values, $\sigma_b = \pm 66.206$ MPa

(c) Total distribution with inner stress $\sigma_i = 74.206$ MPa and outer stress $\sigma_o = -58.206$ MPa.

Figure 3.38

Example 3.11: A hollow shaft of 40 mm diameter and 25 mm inner diameter is subjected to a twisting moment of 118 N-m and axial thrust of 9806 N and a bending moment of 79 N-m. Calculate the maximum compressive and shear stresses.

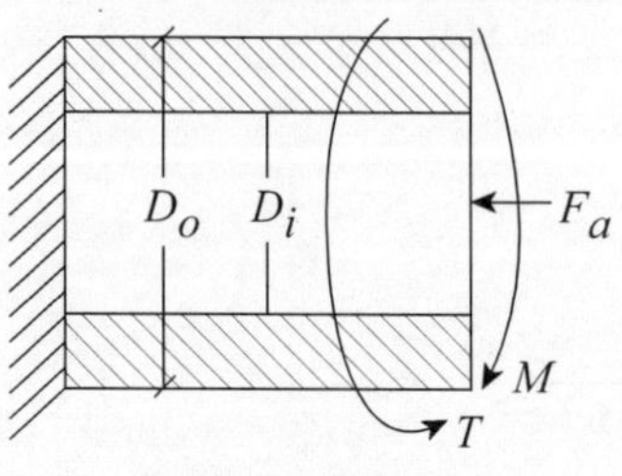

Figure 3.39

Given Data:

$D_o = 40$ mm $D_i = 25$ mm
F_a = Axial thrust = 9806 N
T = Torsional moment = 118 N-m
M = Bending moment = 79 N-m

SOLUTION

Step-1: The direct compressive stress due to axial thrust

$$\sigma_c = \frac{F_a}{A} = \frac{9806}{\frac{\pi}{32}(40^2 - 25^2)} = \frac{9806}{765.76} = 12.8 \text{ MPa}$$

Step-2: The compressive stress due to bending

$$\sigma_c = \frac{M}{Z} = \frac{79 \times 10^3}{\frac{\pi}{32} \cdot 40^3 \left[1 - \left(\frac{25}{40}\right)^4\right]} = 14.84 \text{ MPa}$$

Step-3: The torsional shear stress

$$\tau = \frac{T}{Z_P} = \frac{16 \times 118 \times 10^3}{\pi \times 40^3 \left[1 - \left(\frac{25}{40}\right)^4\right]} = 11.08 \text{ MPa}$$

Step-4: Total normal compressive stress

$$\sigma = \sigma_c + \sigma_b = 12.8 + 14.84 = 27.64 \text{ MPa}$$

Step-5: The principal stresses

$$\sigma_1, \sigma_2 = \frac{-27.64}{2} \pm \sqrt{\left(\frac{27.64}{2}\right)^2 + 11.08^2}$$

$$= -13.82 \pm 17.7$$

$$\sigma_1 = +3.89 \text{ MPa (Tensile)}$$

$$\sigma_2 = -31.52 \text{ MPa (Compressive)}$$

Step-6: Maximum shear stress

$$\tau_{\text{max}} = \sqrt{\left(\frac{27.64}{2}\right)^2 + 11.08^2}$$

$$= 17.7 \text{ MPa}$$

Example 3.12: The maximum force P acting on the frame is 20 kN. The plates are made of steel W5C8 with a tensile yield strength of 380 MPa. The factor of safety is 2.5. Determine the plate thickness.

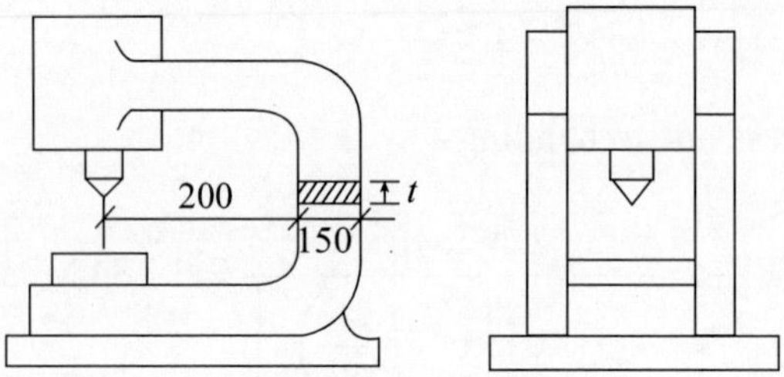

Figure 3.40

SOLUTION

Given data

Applied load, $P = 20$ kN
Material width, $S_y = 380$ MPa
Factor of safety, $n = 2.5$

$$\text{Design Stress, } S_d = \frac{380}{2.5} = 152 \text{ MPa}$$

This is the case of eccentric loading with eccentricity,

$$e = 200 + 75 = 275 \text{ mm}$$

The stress induced is the sum of the direct stress and the bending stress.

$$\sigma = \frac{P}{A} + \frac{p.e}{\frac{th^2}{6}}$$

$$152 = \frac{20 \times 10^3}{150 \times t} + \frac{20 \times 10^3 \times 275}{\frac{t \times (150)^2}{6}}$$

$$\therefore \quad t = \frac{20 \times 10^3}{152 \times 150}\left[1 + \frac{275 \times 6}{150}\right] = \frac{20 \times 10^3 \times 12}{152 \times 150}$$

$$t = 10.5 \text{ mm.}$$

Example 3.13: A bracket shown in Fig. 3.40 is subjected to a pull of 15 kN at 60° to the vertical. Determine the maximum tensile stress in the bracket.

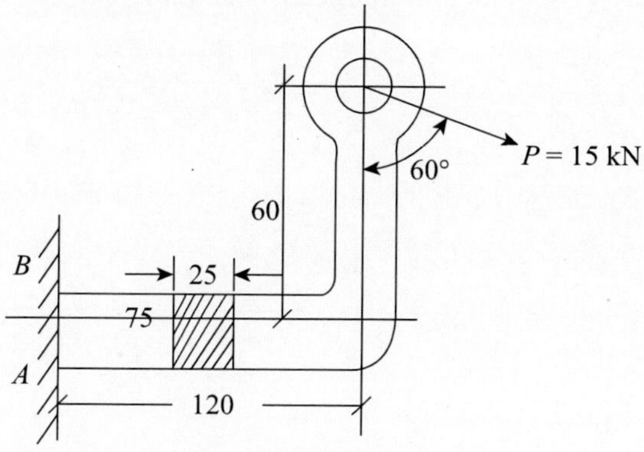

Figure 3.41

SOLUTION

Step-1: The effect of the pull on the Section AB is

(i) Direct tensile force, $F_a = P \cdot \cos 30° = 15 \cdot \cos 30° = 13$ kN

(ii) Transverse force, $F_t = P \cdot \cos 60° = 15 \cdot \cos 60° = 7.5$ kN

(iii) Bending moment, $M_b = 13 \times 60 + 7.5 \times 120 = 1680$ kN-mm

(iv) Direct tensile stress, $\sigma_1 = \dfrac{F_a}{A} = \dfrac{13 \times 10^3}{25 \times 75} = 6.933$ MPa

(v) Bending stress, $\sigma_b = \dfrac{M}{Z} = \dfrac{1680 \times 10^3}{\left(\dfrac{25 \times 75^2}{6}\right)} = 71.68$ MPa

(vi) The total tensile stress at extreme fibre at B,

$$\sigma = 6.933 + 71.68 = 78.61 \text{ MPa}$$

Considering the effect of transverse shear.

(vii) Transverse shear (average),

$$\tau = \frac{F_t}{A} = \frac{7.5 \times 10^3}{25 \times 75} = 4 \text{ MPa}$$

Step-2: Maximum principal stress

$$\sigma_1 = \frac{\sigma}{2} + \sqrt{\left(\frac{\sigma}{2}\right)^2 + \tau^2}$$

$$= \frac{78.61}{2} + \sqrt{\left(\frac{78.61}{2}\right)^2 + 4^2} = 39.305 + 39.51$$

$$\sigma_1 = 78.813 \text{ MPa}$$

Example 3.14: A bolt in an assembly is subjected to a pull of 1000 N along its axis and a shear force of 500 N. What will be the maximum stress induced in the bolt. If the bolt is made of SAE 1045 annealed; is the bolt safe given that bolt diameter is 12 mm?

Given data

Axial pull, $P = 1000$ N

Transverse shear, $F = 500$ N

Bolt diameter, $d = 12$ mm

Bolt material SAE 1045

SOLUTION

The yield strength (Table 2.2 MDHB)

$S_y = 315$ MPa in tension

$S_{sy} = 185$ MPa in shear

Step-1: Stress induced

The direct tensile stress, $\sigma = \frac{P}{A} = \frac{1000}{\frac{\pi}{4} \times 12^2} = 8.84$ MPa

Transverse shear, $\tau = \frac{F}{A} = \frac{500}{\frac{\pi}{4} \times 12^2} = 4.42$ MPa

Step-2: Principal stresses

$$\sigma_1, \sigma_2 = \left(\frac{\sigma}{2}\right)^2 \pm \sqrt{\left(\frac{\sigma}{2}\right)^2 + \tau^2}$$

$$= \frac{8.84}{2} \pm \sqrt{\left(\frac{8.84}{2}\right)^2 + 4.42^2}$$

$$= 4.42 \pm 6.25$$

$$\sigma_1 = \sigma_{max} = 10.67 \text{ MPa (Tensile)} < S_y = 315 \text{ MPa}$$

$$\sigma_2 = \sigma_{min} = -1.83 \text{ (Compressive)}$$

Maximum stress is less than S_y and hence the bolt is safe.

Example 3.15: A 50 mm diameter steel rod supports 9000 N axial load and in addition is subjected to torsional moment of 100 N-m. Determine the maximum normal and shear stresses.
Given data

Axial load, $F = 9000$ N

Twist, $T = 100$ N-m

Rod diameter, $d = 50$ mm

SOLUTION
Step-1: Stresses induced
The direct stress

$$\sigma = \frac{F}{A} = \frac{9000}{\frac{\pi}{4} \times 50^2} = 4.58 \text{ MPa}$$

The torsional shear stress

$$\tau = \frac{16\,T}{\pi d^3} = \frac{16 \times 100 \times 10^3}{\pi \times 50^3} = 4.074 \text{ MPa}$$

Step-2: Maximum principal stresses are given by

$$\sigma_{\max} = \left(\frac{\sigma}{2}\right) + \sqrt{\left(\frac{\sigma}{2}\right)^2 + \tau^2}$$

$$= \left(\frac{4.58}{2}\right) + \sqrt{\left(\frac{4.58}{2}\right)^2 + (4.074)^2}$$

$$= 2.29 + 4.673 = 6.963 \text{ MPa}$$

Step-3: Maximum shear stress

$$\tau_{\max} = \sqrt{\left(\frac{\sigma}{2}\right)^2 + \tau^2}$$

$$= \sqrt{\left(\frac{4.58}{2}\right)^2 + (4.074)^2}$$

$$= 4.673 \text{ MPa}$$

Example 3.16: A portable hydraulic rivetor has a maximum riveting force of 70 kN. The U-frame is made of cast steel with an ultimate tensile stress of 480 MPa and yield point in tension of 240 MPa. Find the stresses at sections AA and BB.

Given Data : Load $F = 70$ kN, $R = 100$ mm, $b = 75$ mm, $h = 125$ mm
$S_u = 480$ MPa, $S_y = 240$ MPa

SOLUTION

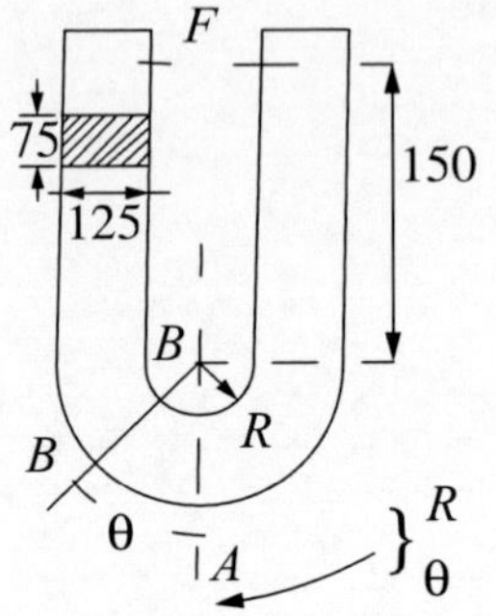

Figure 3.42

Step-1: Radius of neutral axis

For the curved portion

$r_1 = 225$ mm, $\quad r_i = 100$ mm

$$r_c = 100 + \frac{125}{2} = 162.5 \text{ mm}$$

$$r_n = \frac{h}{\ln\,(r_o/r_i)}$$

$$= \frac{125}{\ln\left[\frac{225}{100}\right]} = 154.144 \text{ mm}$$

Then eccentricity, $e = r_c - r_n = 162.5 - 154.144 = 8.356$ mm

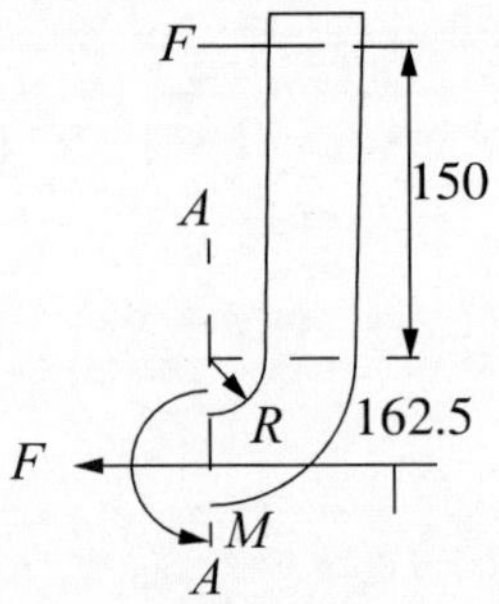

Figure 3.43

Step-2: The distances of extreme fibres $C_i = r_n - r_i = 154.144 - 100 = 54.144$ mm

$C_o = r_o - r_n = 225 - 154.144 = 70.856$ mm

Step-3: Stresses at section AA

For section AA, the loads are shown

Tensile force, $F = 70$ kN

Bending moment

$M = 70 \times (150 + 162.5) = 21875$ kN mm

The stress at the inner fibre,

$$\sigma_i = \frac{MC_i}{Aer_i} + \frac{F}{A} \qquad \text{(Eqn. 1.31 MDHB)}$$

$A = 125 \times 75 = 9375$ mm^2

$e = 8.36$ mm

$C_i = 54.144$

$$\sigma_i = \frac{21875 \times 10^3 \times 54.144}{75 \times 125 \times 8.356 \times 100} + \frac{70 \times 10^3}{125 \times 75} = 151.19 + 7.47 = 158.66 \text{ MPa}$$

At the outer fibre,

$$\sigma_o = \frac{-MC_o}{Aer_o} + \frac{F}{A} \qquad \text{(Eqn. 1.31 MDHB)}$$

$$A = 9375 \text{ mm}$$

$$e = 8.36 \text{ mm}$$

$$C_o = 70.836 \text{ mm}$$

$$\sigma_o = \frac{-21875 \times 10^3 \times 70.856}{9375 \times 8.36 \times 225} + \frac{70 \times 10^3}{9375}$$

$$= -87.89 + 7.47 = -80.42 \text{ MPa (Compressive)}$$

Step-4: Stresses at section BB
For the section *BB*, the loads are
Bending moment,
$M = 70 \times 10^3 \ (150 + 162.5 \cos 45°)$
$= 70 \times 10^3 \times 264.9 = 18543.3 \times 10^3$ N.mm

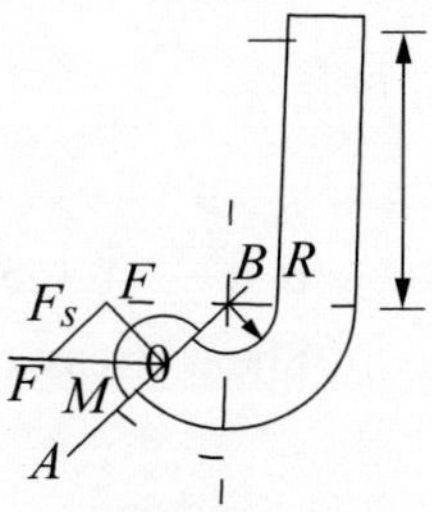

Figure 3.44

Direct tensile load,

$$F_t = F \cos \theta = 70 \times \cos 45° = 49.497 \text{ kN}$$

Shear load,

$$F_s = F \sin \theta = 70 \times \sin 45° = 49.497 \text{ kN}$$

$$\text{Direct tensile stress } \sigma_i = \frac{49.947}{125 \times 75} = 5.28 \text{ MPa}$$

$$\text{Shear stress } \tau = F_s / A = 5.28 \text{ MPa}$$

Bending stresses are,
Stress at inner fibre,

$$\sigma_i = \frac{MC_i}{Aer_i} = \frac{18543.3 \times 10^3 \times 54.144}{9375 \times 8.36 \times 100} = 128.1 \text{ MPa}$$

Stress at outer fibre,

$$\sigma_i = -\frac{MC_o}{Aer_o} = \frac{18543.3 \times 10^3 \times 70.856}{9375 \times 8.36 \times 225} = -74 \text{ MPa}$$

Maximum stress at the inner fibre
Total direct stress,

$$\sigma = 128.1 + 5.28 = 133.38 \text{ MPa}$$

Shear stress, $\tau = 5.28$ MPa

Step-5: Maximum stresses
Maximum direct stress,

$$\sigma_{max} = \frac{\sigma}{2} + \sqrt{\left(\frac{\sigma}{2}\right)^2 + \tau^2} = \frac{133.38}{2} + \sqrt{\left(\frac{133.38}{2}\right)^2 + 5.28^2} = 133.5 \text{ MPa}$$

$$\text{Maximum shear stress, } \tau_{max} = \sqrt{\left(\frac{133.38}{2}\right)^2 + 5.28^2} = 66.898 \text{ MPa}$$

Maximum stress at the outer fibre
Direct stress,

$$\sigma = -74. + 5.28 = -68.72 \text{ MPa}$$

Shear stress $\tau = 5.28$ MPa

Maximum direct stress,

$$\sigma_{max} = \frac{\sigma}{2} + \sqrt{\left(\frac{\sigma}{2}\right)^2 + \tau^2} = \frac{-68.72}{2} + \sqrt{\left(\frac{-68.75}{2}\right)^2 + 5.28^2}$$

$$= -69.123 \text{ MPa}$$

$$\text{Maximum shear stress, } \tau_{max} = \sqrt{\left(\frac{-68.75}{2}\right)^2 + 5.28^2} = 34.76 \text{ MPa}$$

Example 3.17: A ring is made from a bar of 75 mm diameter. The inside diameter of the ring is 100 mm. A load of 20 kN acts as shown in Fig. 4.44. Calculate the maximum stresses in the bar and the location.

Given Data : Load $F = 20$ kN, $D_1 = 100$ mm, $d = 75$ mm

SOLUTION Any section of the left portion of the ring is subjected a direct load and moment. The stress due to bending moment is predominating stress. Since the section AA is subjected maximum moment, it is the critical section. The compressive stress due to the direct load.

$$\sigma_1 = \frac{20 \times 10^3}{\frac{\pi}{4}(75)^2} = 4.53 \text{ MPa}$$

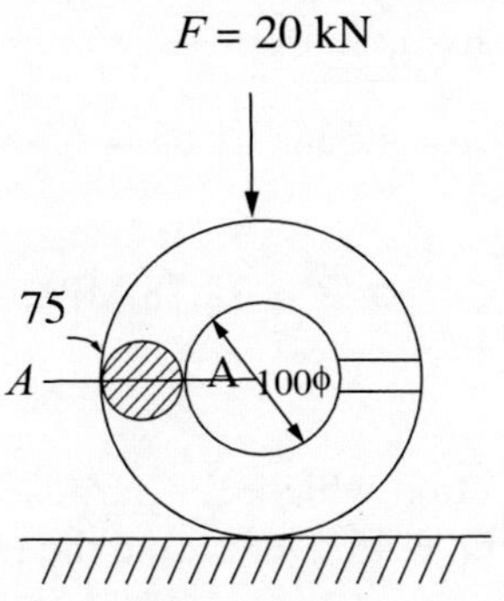

Figure 3.45

Step-1: Cross section details

The bending moment,

$$M = 20 \times 10^3 \times \left(50 + \frac{75}{2}\right) = 175 \times 10^4 \text{ N-mm}$$

$$r_i = 50 \text{ mm}, \; r_o = 50 + 75 = 125 \text{ mm}$$

$$\text{The mean radius } r_c = \frac{50 + 125}{2} = 87.5 \text{ mm}$$

$$\text{The radius of neutral axis, } r_n = \left(\frac{\sqrt{r_1} + \sqrt{r_o}}{2}\right)^2$$

$$r_n = \left(\frac{\sqrt{50} + \sqrt{125}}{2}\right)^2 = 83.28 \text{ mm}$$

$$\text{Eccentricity } e = r_c - r_n = 87.5 - 83.28 = 4.22 \text{ mm}$$

$$A = \frac{\pi}{4} d^2 = \frac{\pi}{4} \times (75)^2 = 4417.86 \text{ mm}^2$$

$$C_i = r_n - r_i = 83.28 - 50 = 33.28 \text{ mm}$$

$$C_o = r_o - r_n = 125 - 83.28 = 81.72 \text{ mm}$$

Step-2: The bending stresses are

$$\sigma_{bi} = \frac{MC_i}{Aer_i} = \frac{175 \times 10^4 \times 33.28}{4417.86 \times 4.32 \times 50} = 61.03 \text{ MPa (Compressive)}$$

At the outer fibre,

$$\sigma_{bo} = \frac{MC_o}{Aer_o} = \frac{175 \times 10^4 \times 81.72}{4417.86 \times 4.32 \times 125} = 30.6 \text{ MPa (Tensile)}$$

Total stress at the inner fibre (Compressive),

$$\sigma = \sigma_{bi} + \sigma_i = 4.53 + 61.03 = 65.56 \text{ MPa}$$

Maximum shear stress,

$$\tau_{\max} = \sigma/2 = 32.78 \text{ MPa}$$

Example 3.18: A crane hook has a section which, for purpose of analysis, is considered a trapezoidal. The dimensions are shown in Fig. 3.45. Determine the maximum stresses, and the location for the work load of 90 kN.

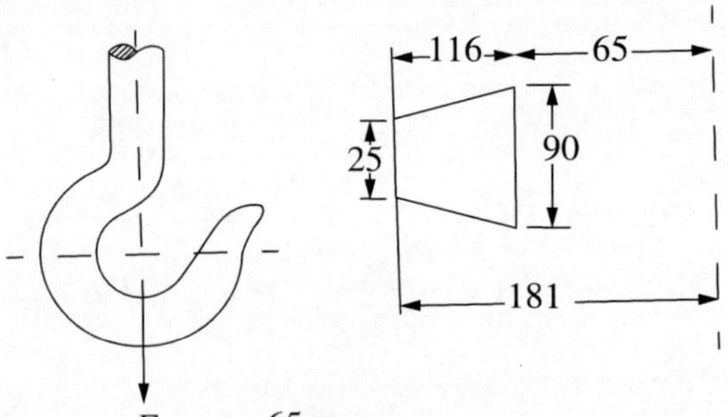

F $r_i = 65$ mm
$r_o = 181$ mm
$h = r_o - r_i = 181 - 65 = 116$ mm

Figure 3.46

Data: $r_i = 65$ mm, $r_o = 181$ mm, $h = r_o - r_i = 181 - 65 = 116$ mm

SOLUTION

Step-1: Cross section properties

The radius of the central axis,

$$r_c = r_i + \frac{h}{3}\left[\frac{b_i + 2b_o}{b_i + b_o}\right] \qquad \text{(Plate 1.5 MDHB)}$$

$$= 65 + \frac{116}{3}\left[\frac{90 + 50}{90 + 25}\right] = 112.07 \text{ mm}$$

The radius of N. A.,

$$r_n = \frac{(b_i + b_o)h/2}{\dfrac{b_i r_o - b_o r_i}{h}\ \ln\dfrac{r_o}{r_i} - (b_i - b_o)} \qquad \text{Table 3.1 (Plate 1.5 MDHB)}$$

$$= \frac{(90+25) \times \dfrac{116}{2}}{\dfrac{90 \times 181 - 25 \times 65}{116} \ln\left(\dfrac{181}{65}\right) - (90-25)}$$

$$= \frac{6670}{126 - 42 \times 1.024 - 65} = 103.46 \text{ mm}$$

Eccentricity, $e = r_c - r_n = 112.07 - 103.46 = 8.611$ mm

Distances from natural axis to extreme fibres,

$$C_i = r_n - r_i = 103.46 - 65 = 38.46 \text{ mm}$$

$$C_o = r_o - r_n = 181 - 103.46 = 77.54 \text{ mm}$$

Step-2: Stresses induced

Bending moment,

$$M = F.r_c = 90 \times 10^3 \times 112.07 = 100863 \times 10^3 \text{N-mm}$$

$$\text{Area } A = \frac{90+25}{2} \times 116 = 6670 \text{ mm}^2$$

The stress at the inner fibre due to M,

$$\sigma_i = \frac{MC_i}{Aer_i} = \frac{10086 \times 10^3 \times 38.46}{6670 \times 8.61 \times 165} = 40.94 \text{ MPa (Compressive)}$$

At the outer fibre,

$$\sigma_o = -\frac{MC_o}{Aer_o} = -\frac{10086 \times 10^3 \times 77.54}{6670 \times 8.6 \times 181} = -75.32 \text{ MPa (Tensile)}$$

The direct tensile stress,

$$\sigma_a = \frac{90 \times 10^3}{6670} = 13.49 \text{ MPa}$$

Total stress at the inner fibre,

$$\sigma = \sigma_a + \sigma_i = 40.94 + 13.49 = 54.433 \text{ MPa}$$

Total stress at the outer fibre,

$$\sigma = \sigma_a + \sigma_o = -75.32 + 13.49 = -61.83 \text{ MPa}$$

Example 3.19: A chain ring 200 mm mean diameter is made from 50 mm steel bar. It is subject to be load 25 kN. Find the maximum stresses and the location.

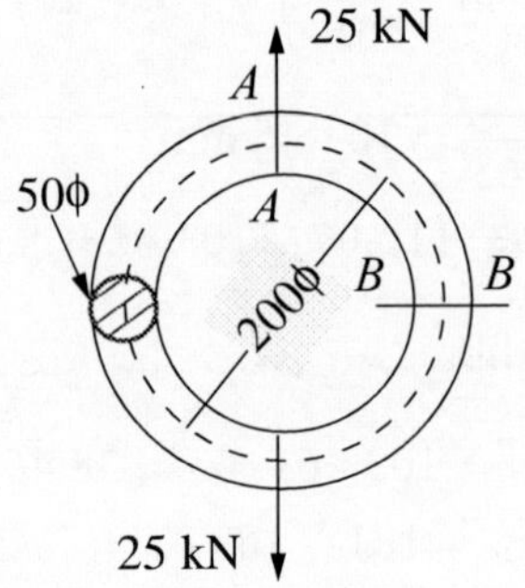

Figure 3.47

SOLUTION

Step-1: Geometric relations

$$R_i = \frac{200-50}{2} = \frac{150}{2}\text{ mm} = 75\text{ mm}$$
$$R_o = \frac{200+50}{2} = \frac{250}{2}\text{ mm} = 125\text{ mm}$$
$$R_c = 100\text{ mm}$$
$$R_n = \left(\frac{\sqrt{R_o}+\sqrt{R_i}}{2}\right)^2 = \left(\frac{\sqrt{125}+\sqrt{75}}{2}\right)^2 = 98.41\text{ mm}$$
$$\text{Eccentricity,}\quad e = R_c - R_n = 100 - 98.41 = 1.59\text{ mm}$$

Step-2: Bending stress at AA

The bending moment under the load at section AA by Eqn. 3.46 (Eqn. 23.9 MDHB)

$$M_1 = 0.318\,FR = 0.318 \times 25 \times 10^3 \times 100 = 7.95 \times 10^5\text{ N-mm}$$
$$\text{Area } A = \frac{\pi}{4}d^2 = \frac{\pi}{4}(50)^2 = 1963.495\ mm^2$$
$$C_i = R_n - R_i = 98.41 - 75 = 23.41$$
$$C_o = 125 - 98.41 = 26.59\text{ mm}$$

Maximum bending stress (Compressive)

$$\sigma_i = \frac{MC_i}{AeR_i} = \frac{7.95 \times 10^5 \times 23.41}{1963.495 \times 1.59 \times 75} = 79.48\text{ MPa}$$

Step-3: Stresses at section BB

At section *BB* - 90° away from the load direction,

$$BM = M_2 = 0.182\ FR \qquad \text{Eqn. 3.45 (Eqn. 23.9 MDHB)}$$
$$= 0.182 \times 25 \times 10^3 \times 100 = 4.55 \times 10^5 \text{ N-mm}$$

Maximum bending stress (Tensile)

$$\sigma_i = \frac{4.55 \times 10^5 \times 23.41}{19.63.495 \times 1.59 \times 75} = 45.49 \text{ MPa}$$

$$\text{Direct tensile stress, } \sigma_i = \frac{25 \times 10^3}{1963.495} = 12.732 \text{ MPa}$$

$$\text{Total tensile stress, } \sigma = 12.732 + 45.49 = 58.22 \text{ MPa}$$

$\therefore$ Maximum stress occurs at the inner fibre under the load.

Example 3.20: A chain link made of 40 mm diameter rod is semicircular at each end, the mean diameter of which is 80 mm. The straight sides of the link length are also equal to 80 mm. If the link carries a load of 90 kN, estimate the tensile and compressive stresses in the link along the section of load line. Also find the stresses at the section 90° away from the load line.
Given Data : Refer Fig. 3.48

$$l = 80 \text{ mm}; \ d_c = 80 \text{ mm} \ \therefore r_c = 40 \text{ mm};$$

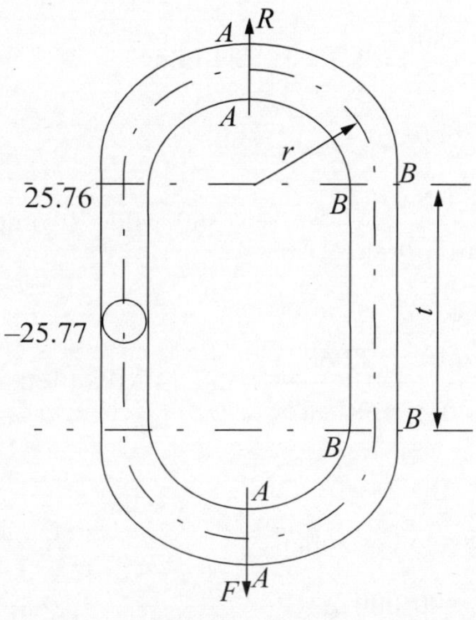

Figure 3.48

SOLUTION

Step-1: Geometric relations

For the critical section

Inner radius, $r_i = 40 - \dfrac{40}{2} = 20$ mm

Outer radius, $r_o = 40 + \dfrac{40}{2} = 60$ mm

Radius of centroidal axis, $r_c = 40$ mm

Radius of neutral axis, $r_n = [\dfrac{\sqrt{r_o} + \sqrt{r_i}}{2}]^2 = \dfrac{(\sqrt{60} + \sqrt{20})^2}{4} = 37.32$ mm

Eccentricity, $e = r_c - r_n = 40 - 37.32 = 2.68$ mm

Distance to inner fibre, $c_i = r_n - r_i = 37.32 - 20 = 17.32$ mm

Distance to outer fibre, $c_o = r_o - r_n = 60 - 37.32.68$ mm

Direct stress at any cross section at an angle θ with horizontal, $\sigma_1 = \dfrac{F\cos\theta}{2A}$

Applied force $F = 90$ kN $= 90,000$ N.

Step-2: Stress at the section A - A

At section $A - A$, $\theta = 90°$ with respect to horizontal

$\therefore$ Direct stress, $\sigma_1 = \dfrac{F\cos\theta}{2A} = 0$

Bending moment, $M_{b_A} = \dfrac{Fr(2r+1)}{2(\pi r + l)}$

$$\therefore \quad M_{b_A} = \frac{90,000 \times 40[2 \times 40 + 80]}{2[\pi \times 40 + 80]} = 1.4 \times 10^6 \text{ N-mm}$$

$\therefore$ Maximum stress at the inner fibre,

$$\sigma_{r_{i_A}} = \sigma_d - \frac{M_{b_A} c_i}{Aer_i} = 0 - \frac{1.4 \times 10^6 \times 17.32}{(\frac{\pi}{4} \times 40^2)(2.68)(20)} = -360 \text{ MPa (Compressive)}$$

Maximum stress at the outer fibre,

$$\sigma_{r_{o_A}} = \sigma_d - \frac{M_{b_A} c_o}{Aer_o} = 0 + \frac{1.4 \times 10^6 \times 22.68}{(\frac{\pi}{4} \times 40^2)(2.68)(60)} = 157.14 \text{ MPa (Tensile)}$$

Step-3: Stresses at the section B - B

At section B-B, $\theta = 0$ with respect to horizontal

$$\therefore \quad \text{Direct stress, } \sigma_1 = \frac{F\cos\theta}{2A} = \frac{90,000\cos\theta}{2 \times \frac{\pi}{4} \times 40^2} = 35.81 \text{ MPa (Tensile)}$$

Bending moment, $M_{b_B} = \dfrac{Fr(2r - \pi r)}{2(\pi r + l)}$

$$\therefore \quad M_{b_B} = \frac{90000 \times 40[2 \times 40 - \pi \times 40]}{2[\pi \times 40 + 80]} = -399655.7 \text{ N-mm}$$

$\therefore$ Maximum stress at the inner fibre,

$\sigma_{r_{i_B}}$ = Direct stress + Bending stress

$$= \sigma_d + \frac{M_{b_B} c_i}{Ae r_i} = 35.81 + \frac{399655.7 \times 17.32}{(\frac{\pi}{4} \times 40^2)(2.68)(20)} = 138.578 \text{ MPa (Tensile)}$$

Maximum stress at the outer fibre,

$$\sigma_{r_{i_B}} = \sigma_d + \frac{M_{b_B} c_o}{Ae r_o} = 35.81 - \frac{399655.7 \times 22.68}{(\frac{\pi}{4} \times 40^2)(2.68)(60)} = -9.047 \text{ MPa (Compressive)}$$

3.6 IMPACT LOADING

Generally, the load is applied gradually but there are situations when the load is applied suddenly with impact or shock.

Examples:

(i) Hammer falling on anvil
(ii) The cylinder of internal combustion engine
(iii) Momentary high loads induced during accidents.

The impact loads induce transient stresses (and deformations) which are much higher than the stresses when the load is applied gradually, we consider some cases of impact load.

Linear Impact

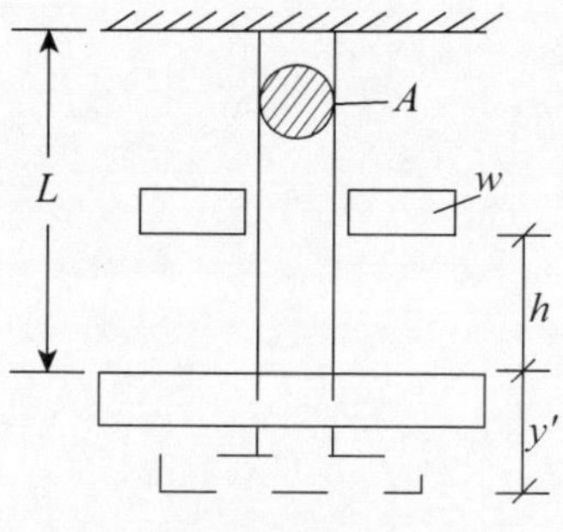

Figure 3.49

Consider a vertical bar of length L and cross-sectional area A with its top end fixed and a collar at the bottom end. A weight W falls through a height h on to the collar.

Let y' be the instantaneous deformation

Step-1: Energy balance
The potential energy of the falling is

$$E_P = W(y' + h) \tag{3.58a}$$

The strain energy stored in the rod is

$$E_s = \frac{\sigma'^2}{2E} \times \text{Vol} \tag{3.58b}$$

where σ^1= instantaneous stress induced
For energy balance

Step-2: Stress induced

$$\left(\frac{\sigma'^2}{2E}\right) \times \text{Vol} = W(y' + h)$$

$$\therefore \quad \frac{\sigma'^2}{2E} \times AL = W\left[h + \frac{\sigma'}{E}.L\right]$$

$$\therefore \quad \sigma'^2 - 2\left(\frac{W}{A}\right)\sigma' - \frac{2hEW}{LA} = 0$$

Solving

$$\sigma' = \left(\frac{W}{A}\right) + \sqrt{\left(\frac{W}{A}\right)^2 + \frac{2hEW}{LA}}$$

$$= \frac{W}{A}\left[1 + \sqrt{1 + \frac{2AEh}{WL}}\right] \tag{3.59a}$$

$$= \sigma\left[1 + \sqrt{1 + \frac{2Eh}{\sigma L}}\right] \tag{3.59b}$$

Substituting static deflection, $y = \dfrac{\sigma}{E} \cdot L$

$$\therefore \quad \sigma' = \sigma\left[1 + \sqrt{1 + \frac{2h}{y}}\right] \tag{3.59c}$$

Step-3: Deflection

The dynamic deflection, $y' = \dfrac{\sigma'}{E} \cdot L$

$$\therefore \quad \sigma' = \left(y' \cdot \frac{E}{L}\right)$$

Similarly, $\sigma = \left(\dfrac{yE}{L}\right)$

Then substituting for σ' & σ in (3.59c)

$$y_i' \frac{E}{L} = \frac{yE}{L}\left[1 + \sqrt{1 + \frac{2h}{y}}\right]$$

$$\therefore \text{Dynamic deflection, } y_i' = y\left[1 + \sqrt{1 + \frac{2h}{y}}\right] \tag{3.60}$$

3.6.1 Bending Impact

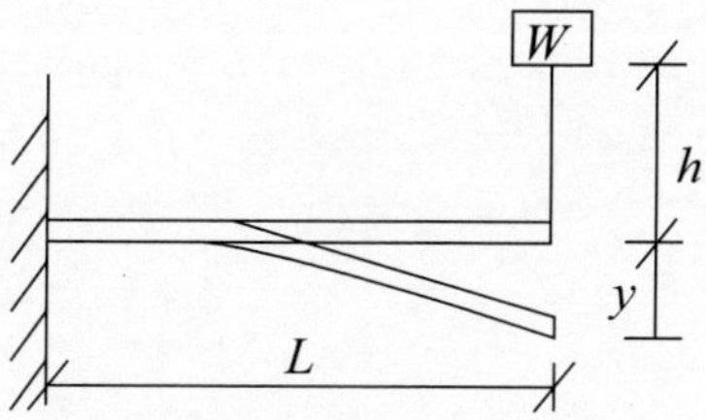

Figure 3.50

Energy of the fall,

$$E = W(h + y')$$

W' is equivalent dynamic load

The work done, $U = \dfrac{1}{2} W' y'$

$$\therefore \quad \frac{1}{2} W' y' = W(h + y') \tag{i}$$

$$\text{Dynamic deflection,} \quad y' = \frac{W' L^3}{3EI} = K_2 W' \tag{ii}$$

where, $\quad K_2 = \dfrac{L^3}{3EI}$

The stress in the beam is

$$\sigma = \frac{Mc}{I} = \frac{WLc}{I} = K_1 W \tag{iii}$$

Using the relations (ii) and (iii) where $K_1 = \dfrac{WL \cdot C}{I}$

$$\therefore \; \frac{1}{2}W' \cdot W' \cdot K_2 = \frac{\sigma}{k_1}[h + W'K_2]$$

$$\frac{1}{2}\sigma_2' \frac{K_2}{K_1^2} = \frac{\sigma}{K_1}\left[h + \sigma' \frac{K_2}{K_1}\right]$$

$$\therefore \; \sigma'^2 = 2\sigma \frac{K_1}{K_2}\left[h + \sigma' \frac{K_2}{K_1}\right] = 2\frac{K_1}{K_2}\sigma h + 2\sigma\sigma'$$

$$\therefore \; \sigma'^2 - 2\sigma\sigma' - \left(2\frac{K_1}{K_2}\right)\sigma h = 0 \qquad \text{(iv)}$$

Solving

$$\sigma_1' = \sigma + \sqrt{\sigma^2 + \left(2 \cdot \frac{K_1}{K_2}\right)\sigma h}$$

$$= \sigma\left[1 + \sqrt{1 + \left(2 \cdot \frac{K_1}{K_2}\right)\left(\frac{h}{\sigma}\right)}\right]$$

$$= \sigma\left[1 + \sqrt{1 + \frac{2h}{y}}\right] \qquad \text{(v)}$$

Multiply by $\left(\frac{K_2}{K_1}\right)$

$$\left(\frac{K_2}{K_1}\sigma'\right) = \frac{K_2}{K_1} \cdot \sigma\left[1 + \sqrt{1 + \frac{2h}{y}}\right]$$

$$\therefore \qquad y' = y\left[1 + \sqrt{1 + \frac{2h}{y}}\right] \qquad \text{(vi)}$$

These equations are same as (3.59) and (3.60)
For simply supported beams,

$$K_1 = \frac{Lc}{4I} \quad \text{and} \quad K_2 = \frac{L^3}{48EI} \quad \text{and} \quad \frac{K_2}{K_1} \cdot \sigma = y \quad \text{holds good.}$$

The energy equation,

$$\frac{1}{2}W'Y'Z = W(h + y') \qquad \text{holds good.}$$

$\therefore$ We get the same solutions as in (v) and (vi)

3.6.2 Torsional Impact

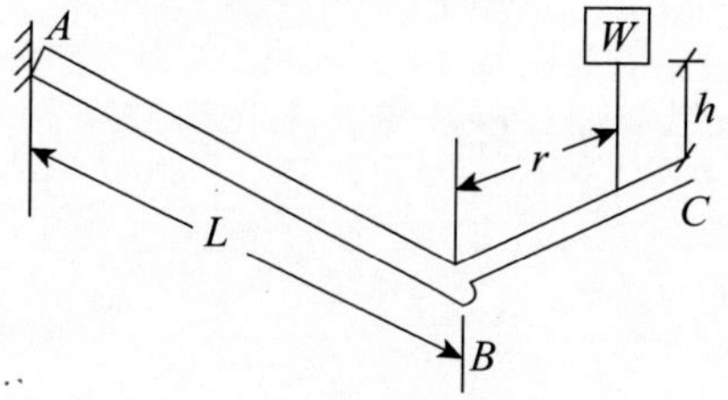

Figure 3.51

The rod is fixed at end A and an arm is attached at the end B. A weight W falls through a height h onto the end of the arm. The dynamic deflection = Y'

Then the potential energy delivered is

$$PE = W(h+Y') \tag{i}$$

If the effect can be represented by the application of equivalent load W' to the end of the rod to produce a deflection Y'. Then the strain energy is

$$U = \frac{1}{2}W'Y' \tag{ii}$$

For energy balance,

$$U = \text{P.E}$$

$$\therefore \quad \frac{1}{2}W'Y' = W(h+Y') \tag{iii}$$

Then the stress produced in the bar is

$$\tau' = \frac{T}{J}\cdot C = W'\left(\frac{rC}{J}\right) = K_1 W' \tag{iv}$$

where, $K_1 = \left(\frac{rC}{J}\right)$

The dynamic deflection (twist),

$$Y' = \theta' r = \left(\frac{T'J}{GL}\right) r \qquad \left[\because \frac{T}{J} = \frac{\tau}{r} = \frac{G\theta}{l}\right]$$

$$= W'\left(\frac{r^2 L}{GJ}\right) = K_2 W' \tag{v}$$

where $K_2 = \left(\frac{r^2 L}{GJ}\right)$

Substituting the relations (iv) and (v) in relation (iii)

$$\therefore \quad \frac{1}{2}W'W'K_2 = W[h+W'K_2]$$

$$\therefore \frac{1}{2}\frac{\tau'^2}{K_1^2}\cdot K_2^1 = \frac{\tau}{K_1}\left[h+\sigma'\frac{K_2}{K_1}\right]$$

$$\therefore \tau'^2 = 2\frac{K_1}{K_2}\tau\cdot h + 2\tau\tau'$$

$$\tau_2'^2 - 2\tau\tau' - 2\frac{K_1}{K_2}\tau\cdot h = 0 \qquad \text{(vi)}$$

The solution is

$$\tau' = \tau + \sqrt{\tau^2 + 2\frac{K_1}{K_2}\tau\cdot h} = \tau\left[1+\sqrt{1+2\frac{K_1}{K_2}\cdot\frac{h}{\tau}}\right]$$

$$= \tau\left[1+\sqrt{1+\frac{2h}{y}}\right] \qquad \text{(vii)}$$

$$= \tau\left[1+\sqrt{1+\frac{2h}{r\theta}}\right] \qquad \text{(viii)}$$

Multiply (viii) by $\frac{K_2}{K_1}$ and simply,

$$\theta' = \theta\left[1+\sqrt{1+2h/r\theta}\right] \qquad \text{(ix)}$$

These equations are similar to equations (3.59) and (3.60).

Example 3.21: A circular rod of 0.30 m length is subjected to an impact due to a weight of 500 N through a height of 30 mm onto end-plate. Determine the diameter of the rod taking maximum stress as 100 MPa. The modulus of elasticity as 207 GPa.

SOLUTION

Step-1: Static stress, $\sigma = \frac{500}{A}$ MPa

Static deflection,

$$\delta = \varepsilon L = \frac{\sigma}{E}.L = \frac{500}{A}\times\frac{300}{207\times 10^3} = \frac{0.7246}{A}\ \text{mm}$$

Step-2: Dynamic stress and the cross section

$$\sigma^1 = \sigma\left[1+\sqrt{1+\frac{2h}{\delta}}\right] = \frac{500}{A}\left[1+\sqrt{1+\frac{2\times 30}{\left(\frac{0.7246}{A}\right)}}\right]$$

$$\therefore \quad 100 = \frac{500}{A}\left[1+\sqrt{1+\frac{60\times A}{0.7246}}\right]$$

$$\therefore \quad \left(\frac{A}{5}-1\right)^2 = 1+\frac{60\times A}{0.7246} = 1+82.8\,A$$

$$A^2 - 10\,A + 25 = 25 + 2070\,A$$

$$A^2 - 2080\,A = 0$$

$$A = 2120 \text{ mm}^2$$

$$\text{Diameter, } d = \sqrt{\frac{4\times 2080}{\pi}} = 51.4 \text{ mm.}$$

Example 3.22: A weight falls through a height of 20 mm onto a collar attached to the lower end of a vertical bar 3 m long and 500 mm^2 in cross section. It causes an instantaneous extension of 2.5 mm. The modulus of elasticity is 210 GPa. Determine the magnitude of the weight and the stress induced.

SOLUTION

Step-1: Static deflection

The dynamic deflection,

$$y^1 = y\left[1+\sqrt{1+\frac{2h}{y}}\right]$$

$$2.5 = y\left[1+\sqrt{1+\frac{2\times 20}{y}}\right]$$

$$\therefore \quad \left(\frac{2.5}{y}-1\right)^2 = 1+\frac{40}{y}$$

$$\therefore \quad (2.5-y)^2 = y^2 + 40\,y$$

$$\therefore \quad 6.25 - 5y + y^2 = y^2 + 40\,y$$

$$\therefore \quad 45y = 6.25$$

$$\therefore \quad y = 0.1389 \text{ mm}$$

but,

$$y = e \cdot L = \frac{\sigma}{E} \cdot L$$

Step-2: Static stress

$$\therefore \sigma = y \cdot \frac{E}{L} = \frac{0.1389 \times 210 \times 10^3}{3000} = 9.7 \text{ MPa}$$

The weight, $W = \sigma \cdot A = 9.7 \times 500 = 4861 \text{ N} = 4.861 \text{ kN}$

Example 3.23: A steel rod 1.5 m long has to resist longitudinally an impact of 2.5 kN falling under gravity at a velocity 1.0 m/s. The stress induced is limited to 150 MPa. The modulus of elasticity is 210 GPa. Find the diameter of the rod.

SOLUTION

Step-1: Height of fall

The height of fall causing a velocity of 1.0 m/s is found by the relation,

$$v^2 - u^2 = 2\,ah,$$

where $v = 1 \text{ m/s}$, $u = 0$ and $a = g =$ acceleration dues to gravity

$$\text{Equivalent height} - h = \frac{v^2}{2g} = \frac{1.0}{2 \times 9.81} = 0.05097 \text{ m}$$

$$h = 50.97 \text{ mm}$$

Step-2: Cross section for dynamic strength

Dynamic stress,

$$\sigma^1 = \sigma\left[1 + \sqrt{1 + \frac{2h}{y}}\right]$$

$$= \frac{W}{A}\left[1 + \sqrt{1 + \frac{2\,hAE}{W \cdot L}}\right]$$

$$150 = \left(\frac{2500}{A}\right)\left[1 + \sqrt{1 + \frac{2 \times 50.97 \times A \times 210 \times 10^3}{2500 \times 1500}}\right]$$

$$(0.06\,A - 1)^2 = 1 + 5.71\,A.$$

$$0.0036\,A^2 - 0.12\,A + 1 = 1 + 5.71\,A$$

$$0.0036\,A^2 - 5.83\,A = 0$$

$$A = \frac{5.83}{0.0036} = 1619.44\ \text{mm}^2$$

$$\text{Then diameter, } d = \sqrt{\frac{4 \times 1619.44}{\pi}} = 45.41\ \text{mm}$$

Example 3.24: A cantilever beam of section 50 mm width and 150 mm depth. A weight of 1.5 kN falls through a height of 10 mm onto the free end of the beam. The modulus of elasticity is $E = 210$ GPa.

Find:

(a) Dynamic stress
(b) Dynamic deflection
(c) Dynamic load
(d) Impact factor

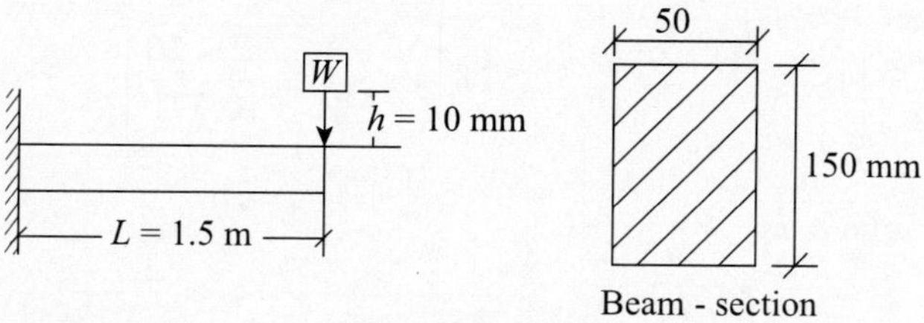

Figure 3.52

Given Data

Load $W = 1.5\ \text{kN} = 1500\ \text{N}$

$$E = 210\ \text{GPa} = 210 \times 10^3\ \text{MPa}$$

SOLUTION

$$\text{Moment of inertia, } I = \frac{bd^3}{12} = \frac{50 \times 150^3}{12} = 14062500\ \text{mm}^4$$

$$\text{Section modulus, } Z = \frac{I}{(d/2)} = \frac{14062500}{75} = 187500\ \text{mm}^3$$

Step-1: Static stress and deflection
Static bending stress,

$$\sigma = \frac{M}{Z} = \frac{W \cdot L}{Z} = \frac{1500 \times 1500}{187500} = 12 \text{ MPa}$$

Static deflection,

$$y = \frac{WL^3}{3 \cdot EI} = \frac{1500 \times 1500^3}{3 \times 210 \times 10^3 \times 14062500} = 0.571 \text{ mm}$$

Step-2: Dynamic stress and deflection

(a) Dynamic stress,

$$\sigma^1 = \sigma\left[1 + \sqrt{1 + \frac{2h}{y}}\right] = 12\left[1 + \sqrt{1 + \frac{2 \times 10}{0.571}}\right] = 84.03 \text{ MPa}$$

(b) Dynamic deflection,

$$y^1 = y\left[1 + \sqrt{1 + \frac{2h}{y}}\right] = 0.571\left[1 + \sqrt{1 + \frac{2 \times 10}{0.571}}\right] = 3.998 \text{ mm}$$

Step-3: Equivalent dynamic load

$$W^1 = \frac{\sigma^1 \times Z}{L} = \frac{84.03 \times 187500}{1500} = 10503.75 \text{ N}$$

$$W^1 = 10.504 \text{ kN}$$

Step-4: Impact ratio

$$R = \frac{\sigma^1}{\sigma} = \frac{84.03}{12} = 7.0025$$

Example 3.25: A weight of 500 N is dropped from a height of 25 mm at the free end of a cantilever beam of length 200 mm. Determine the size of the cross section (square section) taking the allowable stress as 80 MPa and modulus of elasticity as 210 GPa.

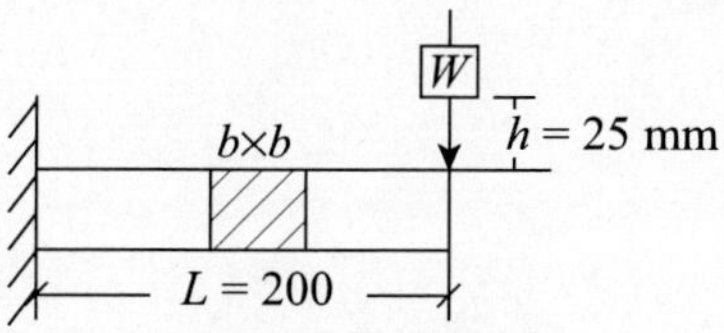

Figure 3.53

SOLUTION

Step-1: State stress and deflection

$S_d = 80$ MPa and $E = 210 \times 10^3$ MPa

Moment of Inertia, $I = \dfrac{bd^3}{12} = \dfrac{b^4}{12}$

Section modulus, $z = \dfrac{bd^2}{6} = \dfrac{b^3}{6}$

Static stress,

$$\sigma = \frac{M}{Z} = \frac{WL}{Z} = \frac{500 \times 200}{b^3/6} = \frac{6 \times 10^5}{b^3} \text{ MPa}$$

Static deflection,

$$y = \frac{WL^3}{3EI} = \frac{500 \times 200^3}{3 \times 210 \times 10^3 \times \left(\dfrac{b^4}{12}\right)} = \frac{76190.5}{b^4} \text{ mm}$$

Step-2: Dynamic stress

$$\sigma^1 = \sigma\left[1 + \sqrt{1 + \frac{2h}{y}}\right]$$

$$80 = \left(\frac{6 \times 10^5}{b^3}\right)\left[1 + \sqrt{1 + \frac{2 \times 25 \times 6^4}{76190.5}}\right]$$

$$\left(\frac{80 \times b^3}{6 \times 10^5} - 1\right)^2 = 1 + 0.656 \times 10^{-3} \cdot b^4.$$

Step-3: Dimension of the beam

Let $\frac{b}{10} = x$

$$\left(\frac{80 \times x^3}{6 \times 10^2} - 1\right)^2 = 1 + 6.56\,x^4$$

$$\therefore\ (0.133\,x^3 - 1)^2 = 1 + 6.56\,x^4$$

$$\therefore\ 0.01778\,x^6 - 0.266\,x^3 + 1 = 1 + 6.56\,x^4$$

$$\therefore\ 0.01778\,x^6 - 6.56\,x^4 - 0.266\,x^3 = 0$$

$$\therefore\ x^3 - 368\,x - 14.96 = 0$$

Solving,

$x = 19.2$
$\therefore b = 192$ mm

Example 3.26: A machine member is considered a simply supported beam of one metre length. The cross section of the beam is 60 mm × 60 mm square. Determine the instantaneous maximum deflection and bending stress. If a mass of 15 kg falls from a height of 250 mm at the midpoint of the beam made of steel.
Given Data

Simply supported beam length, $L = 1000$ mm

Beam cross section, $b \times h = 60$ mm × 60 mm

Falling mass, 15 kg = 147.15 N

Height of fall, $h = 250$ mm

Material-steel, $E = 210 \times 10^3$ MPa

SOLUTION
Step-1: Static deflection

$$y = \frac{WL^3}{48\,EI}$$

where,

$W = 147.15$ N

$L = 1000$ mm

$E = 210 \times 10^3$ MPa for steel

Moment of inertia, $Z = \dfrac{bh^3}{12} = \dfrac{60 \times 60^3}{12} = 108 \times 10^4 \text{mm}^4$

Substituting the values

$$\therefore \quad y = \frac{147.15 \times 1000^3}{48 \times 210 \times 10^3 \times 108 \times 10^4} = 0.0135 \text{ m}$$

Step-2: Static stress

$$\sigma = \frac{M}{Z}$$

where, $$M = \frac{WL}{4} = \frac{147.15 \times 1000}{4} = 36.7875 \times 10^3 \text{ N-mm}$$

$$Z = \frac{bh^2}{6} = \frac{60 \times 60^2}{6} = 36 \times 10^3 \text{ mm}^3.$$

Substituting the values

$$\sigma = \frac{36.7875 \times 10^3}{36 \times 10^3} = 1.02 \text{ MPa}$$

Step-3: The dynamic (instantaneous) strees due to the falling weight

$$\sigma_b' = \sigma_b \left[1 + \sqrt{1 + \frac{2h}{y}}\right] \qquad \text{(Eqn. 1.48 MDHB)}$$

$$= 1.02 \left[1 + \sqrt{1 + \frac{2 \times 250}{0.0135}}\right]$$

$$= 197.32 \text{ MPa}$$

Step-4: The dynamic (instantaneous) deflection

$$y' = y \left[1 + \sqrt{1 + \frac{2h}{y}}\right]$$

$$= 0.0135 \left[1 + \sqrt{1 + \frac{2 \times 250}{0.0135}}\right]$$

$$= 2.61 \text{ mm}$$

Example 3.27: The brasses of an automobile engine connecting rod have worn so as to allow play which gives shock loading equivalent to a weight of 5886 N falling through a height of 0.2 mm. The connecting rod is 250 mm long and has a cross section of $3 \times 10^{-4} \text{m}^2$. Determine the stress induced in the connecting rod and compare it with the static stress.

Given Data

Connecting rod length, $L = 250$ mm

Cross-sectional area, $A = 3 \times 10^{-4}\ \text{m}^2 = 300\ \text{mm}^2$

Falling weight, W = 5886 N

Height of fall, $h = 0.2$ mm

For brass, $E = 97 \times 10^3$ MPa (Table 2.1 MDHB)

SOLUTION

The static stress,

$$\sigma = \frac{W}{A} = \frac{5886}{300} = 19.62 \text{ MPa}$$

Dynamic stress,

$$\sigma' = \sigma\left[1 + \sqrt{1 + \frac{2hEA}{WL}}\right] \quad \text{(Eqn. 1 1.47 MDHB)}$$

$$= 19.62\left[1 + \sqrt{1 + \frac{2 \times 0.2 \times 97 \times 10^3 \times 300}{5886 \times 250}}\right]$$

$$= 19.62 \times 3.986$$

$$= 78.186 \text{ MPa}$$

Impact-ratio is,

$$\frac{\sigma'}{\sigma} = \frac{78.186}{19.62} = 3.986.$$

Example 3.28: A weight of 6000 N falls from a height of 50 mm on to a box-section beam of 5 m length at its midpoint. The beam is simply supported at its end and is 400 mm deep. The moment of inertia for the box section is 10^8 mm. Determine the maximum stress induced and compare it with the static stress.

Given Data

Simply supported beam length, $L = 5$ m

Moment of Inertia, $I = 10^8$ mm^4

Depth of the beam, $H = 400$ mm

Falling weight, $W = 6000$ N

Height of fall, $h = 50$ mm

Young's modulus, $E = 210 \times 10^3$ MPa

SOLUTION

Step-1: Static stress and deflection

Bending moment, $M = \dfrac{WL}{4} = \dfrac{6000 \times 5 \times 10^3}{4} = 75 \times 10^5$ N-mm

Section modulus, $Z = \dfrac{I}{C} = \dfrac{10^8}{200} = 5 \times 10^5$ mm^3

Static bending stress, $\sigma_b = \dfrac{M}{Z} = \dfrac{75 \times 10^5}{5 \times 10^5} = 15$ MPa

Static deflection, $Y = \dfrac{WL^3}{48EI} = \dfrac{6000 \times 5000^3}{48 \times 210 \times 10^3 \times 10^8} = 0.744$ mm

Step-2: Dynamic stress

$$\text{Instantaneous stress,} \quad \sigma_b' = \sigma_b \left[1 + \sqrt{1 + \frac{2h}{y}}\right]$$

$$= 15 \left[1 + \sqrt{1 + \frac{2 \times 50}{0.744}}\right]$$

$$= 15 \times 12.6365$$

$$= 189.55 \text{ MPa}$$

$$\text{Instantaneous deflection,} \quad y' = y \left[1 + \sqrt{1 + \frac{2h}{y}}\right]$$

$$= 0.744 \left[1 + \sqrt{1 + \frac{2 \times 50}{0.744}}\right]$$

$$= 9.4 \text{ mm}$$

Step-3: Ratio of the stresses (Impact ratio)

$$\frac{\sigma_b'}{\sigma_b} = \frac{189.55}{15} = 12.6365$$

Example 3.29: An unknown weight falls through a height of 10 mm onto a collar rigidly attached to the lower end of rod 3000 mm long and 600 mm^2 in cross section. If the max instantaneous extension of the bar is 2 mm what is the corresponding stress and the value of the unknown weight. Take Young's modulus $E = 200 \times 10^3$ MPa.

Given Data

Height of fall, $h = 10$ mm
Length of the rod, $L = 3000$ mm
Cross-section area, $A = 600$ mm^2
Instantenous deflection, $y' = 2$ mm
Young's modulus, $E = 200$ kN/mm$^2 = 200 \times 10^3$ MPa

SOLUTION

Step-1: Instantaneous stress

$$\sigma' = \frac{Y'}{L}.E = \frac{2 \times 200 \times 10^3}{3000} = 133.33 \text{ MPa}$$

Step-2: For energy balance, the falling weight W

$$\text{Strain energy} = \text{Potential energy}$$

$$\frac{1}{2}\frac{\sigma_2'}{E}.AL = W(h+y')$$

$$\frac{1}{2}\frac{(133.33)^2 \times 600 \times 3000}{200 \times 10^3} = W(10+2)$$

$$\therefore\ W = \frac{1}{2} \times \frac{133.33^2 \times 600 \times 3000}{200 \times 10^3 \times 12} = 6666.33 \text{ N.}$$

Example 3.30: A steel rod 1.5 m long resists an impact load of 2 kN dropped through a distance of 50 mm along its axis. The limiting max stress in the rod is 150 MPa. Determine the diameter of the rod and the impact factor. Use $E = 200$ GPa.

Given Data

Falling weight, $W = 2$ kN $= 2000$ N
Height of fall, $h = 50$ mm
Length of the rod, $L = 1.5$ m $= 1500$ mm

Young's modulus, $E = 200$ GPa
Limiting stress, $\sigma' = 150$ MPa

SOLUTION

Step-1: Instantaneous deflection

$$y' = \frac{\sigma'}{E} \cdot L = \frac{150 \times 1500}{200 \times 10^3} = 1.125$$

Step-2: The cross section for energy balance

$$\frac{1}{2}\frac{\sigma'^2}{E} \cdot A \cdot L = W(h + y')$$

$$\frac{1}{2} \times \frac{150^2 \times A \times 150}{200 \times 10^3} = 2000(50 + 1.125)$$

$$\therefore A = \frac{2000 \times 51.125 \times 200 \times 10^3 \times 2}{150^2 \times 1500}$$

$$\therefore A = 1211.85 \text{ mm}^2$$

Diameter of the rod,

$$d = \sqrt{\frac{4A}{\pi}} = \sqrt{\frac{4 \times 1211.85}{\pi}}$$

$$d = 39.28 \text{ mm} \cong 40 \text{ mm.}$$

EXERCISES FOR PRACTICE

3.1. What are the methods for economizing the material usage?
3.2. Define stress and state of stress (stress tensor).
3.3. Derive the equations for the direct and shear stresses on any plane in terms of the stress components on the rectangular reference planes for two-dimensional stress system.
3.4. Derive the equations for principal stresses for plane stress system.
3.5. Derive the equations for the stresses on any plane in terms of the stress components of the stress tensor for three-dimensional stress system.
3.6. Derive the expressions for the principal stresses in three-dimensional stress system.
3.7. Explain Mohr's circle method for finding the principal stresses for plane stress system.
3.8. For each of the following plane stress states, determine the principal normal and shear stresses, both analytically and graphically,

(i) $\sigma_x = 18$ MPa, $\sigma_y = 9$ MPa, $\tau_{xy} = 6$ MPa.c

(ii) $\sigma_x = 15$ MPa, $\sigma_y = 36$ MPa, $\tau_{xy} = 9$ MPa.ccw

(iii) $\sigma_x = 12$ MPa, $\sigma_y = -10$ MPa, $\tau_{xy} = 16$ MPa, ccw

(iv) $\sigma_x = 18$ MPa, $\sigma_y = -12$ MPa, $\tau_{xy} = 6$ MPacw

(v) $\sigma_x = -24$ MPa, $\sigma_y = 44$ MPa, $\tau_{xy} = 24$ MPacw

(vi) $\sigma_x = 45$ MPa, $\sigma_y = -15$ MPa, $\tau_{xy} = 15$ MPa, ccw.

3.9. For each of the following states of stress, determine the principal stresses, and draw the Mohr's three-circle diagram, labelling all important points.

(i) $\sigma_x = -4$ MPa, $\sigma_y = 12$ MPa, $\sigma_z = -8$ MPa, $\tau_{xy} = 6$ MPa, $\tau_{yz} = 4$ MPa, $\tau_{yz} = -10$ MPa.

(ii) $\sigma_x = 15$ MPa, $\sigma_y = 0$ MPa, $\sigma_z = 15$ MPa, $\tau_{xy} = 30$ MPa, $\tau_{yz} = -22$ MPa, $\tau_{zx} = 0$ MPa.

(iii) $\sigma_x = 5$ MPa, $\sigma_y = 20$ MPa, $\sigma_z = 20$ MPa, $\tau_{xy} = 10$ MPa, $\tau_{yz} = -20$ MPa, $\tau_{zx} = -10$ MPa.

(iv) $\sigma_x = 40$ MPa, $\sigma_y = 0$ MPa, $\sigma_z = -30$ MPa, $\tau_{xy} = 20$ MPa, $\tau_{yz} = \tau_{zx} = 0$ MPa.

3.10. Determine the stresses at the section $A - B$ of the clamp, loaded as shown in the figure below.

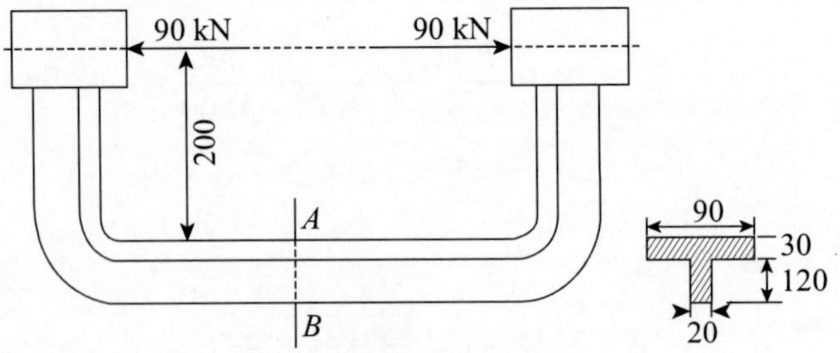

Figure 3.54

3.11. A 45 mm dia steel rod carries a 12 kN load and a twisting moment of 180 Nm as shown in the figure below.

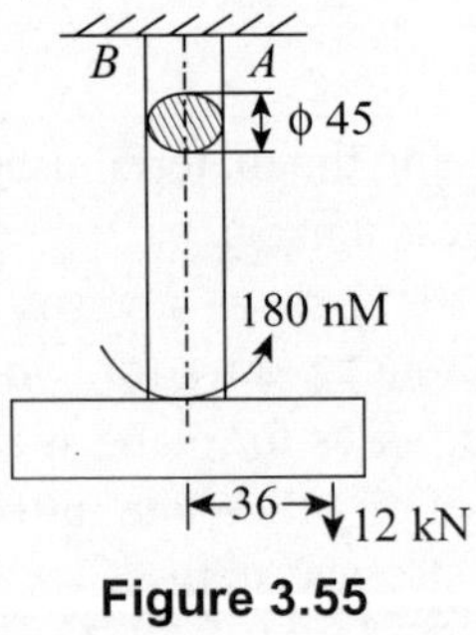

Figure 3.55

3.12. a) State the assumptions made in the theory of curved beams.

b) The section of crane hook is trapezium, whose inner and outer sides are 50 mm and 25 mm respectively and the depth is 60 mm. The centre of curvature is 65 mm from inside of the section and the load line is 50 mm from the same point. Find the greatest load the hook will carry if the maximum stress is not to exceed 50 MPa.

3.13. a) Derive the equation for the stress distribution in a curved beam subjected to pure bending.

b) A crane hook used to lift a load 90 kN has trapezoidal section as shown in Fig. 3.56. Calculate the stresses induced in the inner fibre and outer fibre of the section.

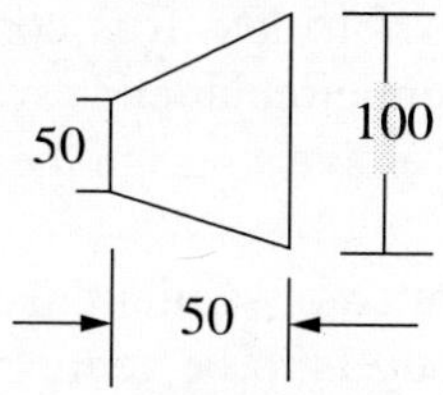

Figure 3.56

3.14. a) Derive an equation for bending stresses in a curved beam.

b) A ring is made of 75 mm diameter round bar. The inside diameter is 100 mm. Calculate the maximum stresses induced in the ring if the load acting is 5 kN.

3.15. a) Why are unsymmetrical sections recommended for crane hooks?

b) The frame of a 10 tonnes steel punch press is shown in Fig. 3.57. Determine the maximum stress in the frame.

c) If holes must be put on curved beam, they should be located on the neutral axis rather than on centroidal axis. Discuss with sketch.

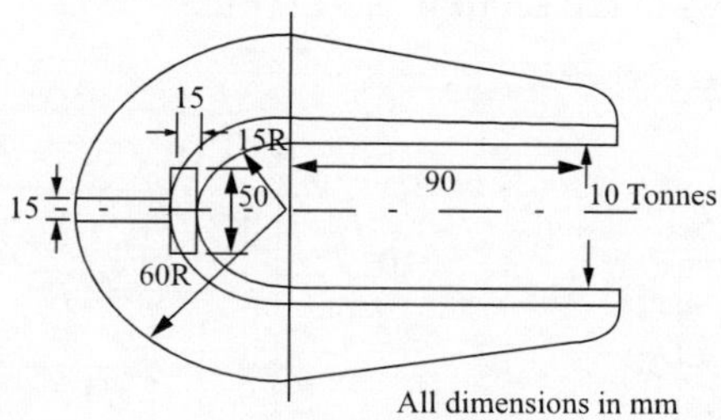

Figure 3.57

3.16. A curved bar of rectangular section initially unstressed is subjected to a bending movement of 1500 Nm which tends to straighten the bar. The section is 40 m wide by 50 mm deep in the plane of bending and the mean radius of curvature is 100 mm. Find the position of the neutral axis and magnitude of the greatest bending stress and draw the diagram to show approximately how the stress varies across the section.

3.17. The section of a crane hook is a rectangle 60 mm × 40 mm. The centre of curvature of the section is at a distance of 80 mm from the centroid of the section. A load of 15 kN is acting through the centre of curvature. Determine minimum and maximum stresses induced in the hook.

3.18. A chain link is subjected to a pull of 15 kN. It is composed of steel bar 20 mm diameter and has a mean radius 25 mm. Its semicircular ends are connected by straight pieces 25 mm long. Estimate the maximum compressive stress in the link and the tensile stress at the same section.

3.19. Design and make a dimensioned sketch of semicircular C-clamp to text a clamping force of 1 tonne. The gap is 20 cm. The body is to be semicircular with rectangular cross section. Select suitable materials for body, spindle, nut handle, etc.

3.20. A crane hook carries a load of kN. The line of the load being at a horizontal distance of 32.5 mm from the inside edge of a horizontal section though the centre of curvature. The centre of curvature being 40 mm from the same edge. The horizontal section is a trapezium whose parallel sides are 15 mm and 30 mm and perpendicular distance apart by 32.5 mm. Find the greatest tensile and compressive stresses in the hook.

3.21. a) A cranke hook carries a load of 5,000 N. The line of action of load is at horizontal distance of 52 mm from the inside edge of a horizontal section through the centre of curvature and centre of curvature is 40 mm from the same edge. The horizontal section is I-section whose outer and inner flanges are 25 mm and 45 mm wide and flange and web thicknesses are 10 mm and depth is 50 mm. Determine the greatest tensile and compressive stresses in the hook. Make a neat sketch of the hook giving all dimensions.

b) A closed circular ring is made of 20 mm diameter steel bar and is subjected to a pull of 10 kN whose line of action through the centre of curvature of the ring. Find the ratio of maximum tensile stress and maximum compressive stress in the ring. The dia of the ring is 120 mm.

3.22. Determine the force F that will produce a maximum tensile stress of 50 MPa in section $A-B$ and the corresponding stress at the section $C-D$ (Fig. 3.58).

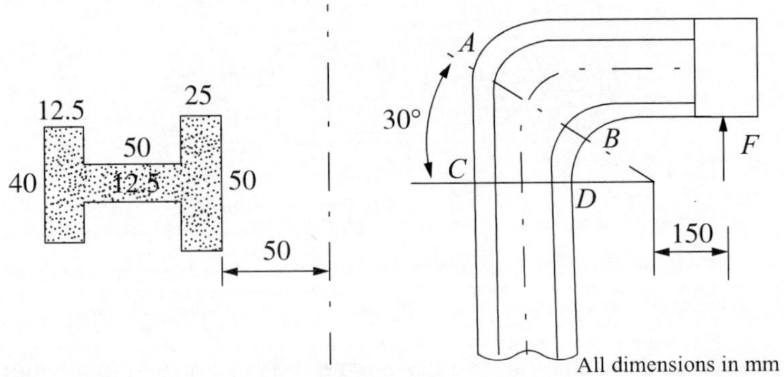

Figure 3.58

3.23. Determine a safe value for load P for a machine element loaded as shown in Fig. 3.59, limiting the maximum normal stress induced on the cross section XX to 120 MPa.

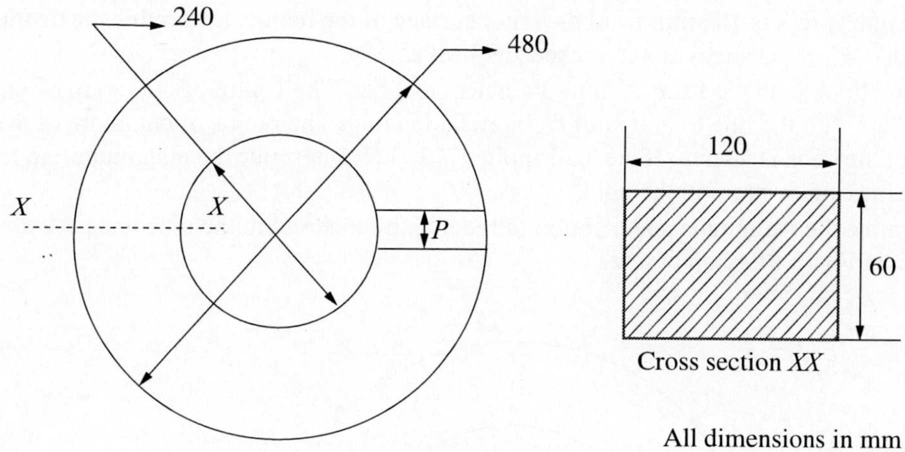

Figure 3.59

3.24. Calculate the stresses at points A and B for a circular beam as as shown in Fig. 3.60, the circular beam is subjected to a compressive load of 6 kN.

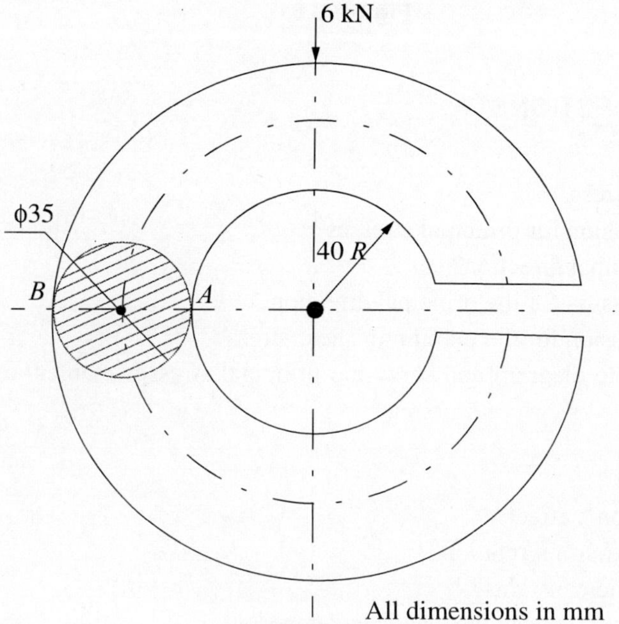

Figure 3.60

3.25. A small punch has a capacity to punch 16 mm hole in a 6 mm thick steel plate, whose ultimate strength in tension is 600 MPa. The punch frame is shaped like C-clamp with T-section. The dimensions are $t_1 = 25\,\text{mm}, t_2 = 30\,\text{mm}$, and $b = 150\,\text{mm}$. The central line of the puch press is 165 mm from the inner surface of the flange. Determine the depth of the section - 'h', if the stress is not to exceed 95 MPa.

3.26. A crane hook is made from 75 mm diameter steel bar. The centre of curvature of the hook coincides with the line of action of the vertical loading. The radius of curvature of the inner fibre of the hook is 50 mm. If the load applied is 20 kN, determine the magnitude and location of the maximum stress in the hook.

3.27. Determine the magnitude and location of the maximum stress in the machine part, loaded as shown in Fig. 3.61 given below.

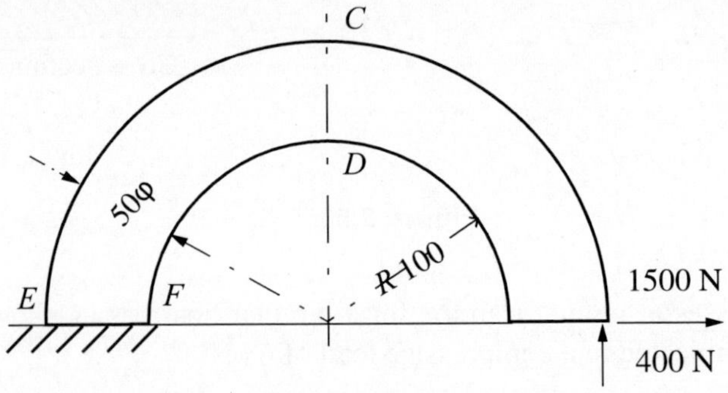

Figure 3.61

SHORT ANSWER QUESTIONS

3.1. Define stress.
3.2. Define principal stress.
3.3. What is the expression for principal stresses?
3.4. What are the principal direction?
3.5. What is the expression for the principal directions?
3.6. What is the expression for the maximum shear stress?
3.7. Draw Mohr's circle diagram and show the principal stresses element and maximum shear stress-element?
3.8. Define strain.
3.9. State Hooke's law.
3.10. What is the Poisson's effect?
3.11. What are the stress-strain relations?
3.12. What are the elastic constants?
3.13. What is the relation between the elastic constants?
3.14. What are the principal strains?
3.15. What is the relation between principal stress planes and principal strain planes?
3.16. What is equation for bending?
3.17. What is the equation of torsion?

3.18. What is the expression for bending stress in a curved beam?
3.19. What is the expression for the radius of the neutral axis?
3.20. What is the eccentricity?
3.21. What is a column?
3.22. How are the columns classified?
3.23. What are the buckling load equations?
3.24. What the expression for the stress under suddenly applied load?
3.25. What is the expression for the stress under the impact load?

CHAPTER 4

DESIGN FOR STATIC STRENGTH

4.1 DESIGN-STRESS

Machine parts are subjected to different types of loads. The loads may be constant or variable with time. The magnitude of variation may be small and can be neglected. The steady loads and loads with insignificant variation are treated as static loads. The maximum value of such load a component can withstand is called static strength. Evaluation of this strength is not easy and exact, because of many uncertainties involved. The following are factors to be considered in evaluating the static strength.

A. The material properties are influenced by:

   (i) Statical nature of properties. The tabulated values are average values only. The actual values may be different.

   (ii) The size of the component

   (iii) Surface finish of the component

   (iv) Shape of the component

   (v) The actual environmental conditions like temperature and corrosive atmosphere.

B. The load acting on a component is to be estimated taking into consideration—the actual working conditions like the load may be treated as concentrated or distributed. The magnitude may not be determined exactly as in case of wind load.

C. The evaluation of stresses induced is based on the theoretical model of analysis. The assumption made for the model may not hold good in the actual situation. The abrupt change in shape—and cross section induces stress-concentration. The actual stress induced is much larger than given by normal equations.

D. Severity of the failure also influences the design procedure. A more severe situation, where human life is involved, requires a larger margin of safety.

A margin of safety has to be provided for the possible increase in the actual stress induced in the component. The strength value used for design is much less than the tabulated value.

Then the margin safety = Maximum experimental stress – Stress used for design calculation

= Maximum stress – Design stress

Then a factor of safety is defined as the ratio of the maximum value of the strength property to the actual value of the stress induced.

$$\text{Factor of safety N} = \frac{\text{Ultimate tensile strength}}{\text{Actual stress induced}} = \frac{S_u}{\sigma_{max}} \quad (4.1)$$

$$\text{OR} \quad N = \frac{\text{Yield strength}}{\text{Actual stress induced}} = \frac{S_y}{\sigma_{max}}. \quad (4.1a)$$

For brittle materials, the factor of safety is based on ultimate strength and for ductile materials, it is based on the yield strength.

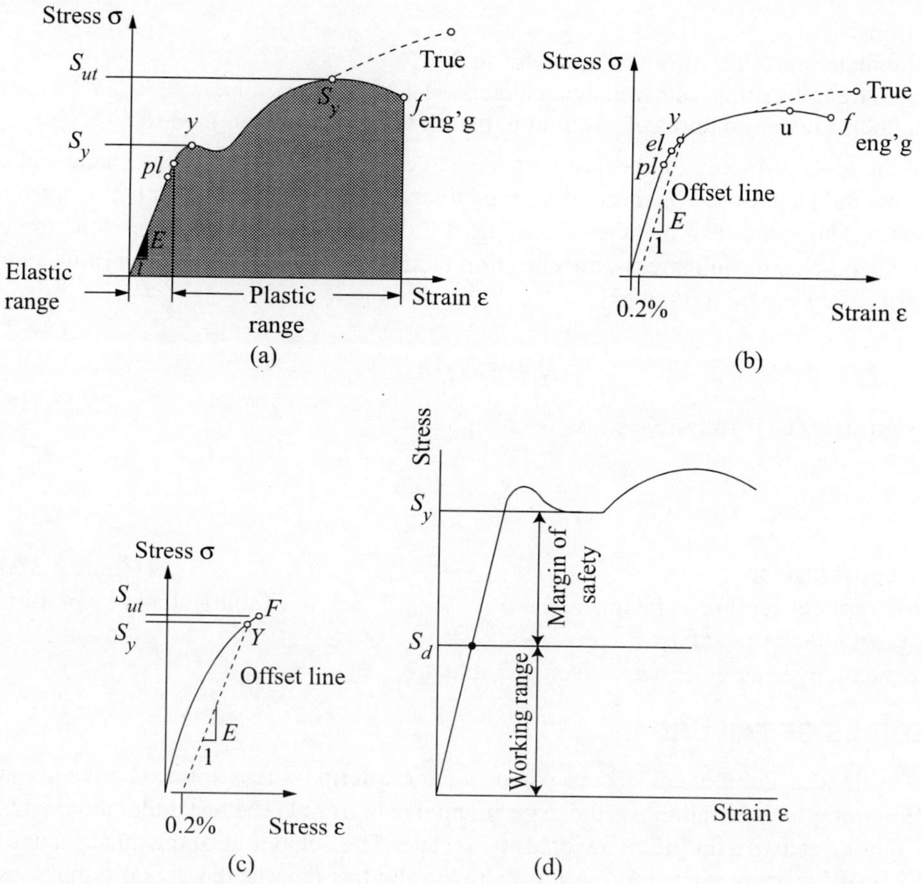

Figure 4.1. Engineering and True Stress-Strain Curves for Ductile Materials: (*a*) Low-Carbon Steel (*b*) Annealed High-Carbon Steel (*c*) Stress-Strain Curve of a Brittle Material (*d*) Typical elastic material with design stress marked in it

Theoretically, the factor of safety may be factorized as follows:

$$N = n_m \cdot n_l \cdot n_s \cdot n_d \quad (4.2)$$

where,

n_m is the factor accounting for the uncertainties of material strength
n_l is the factor accounting for the uncertainties in load evaluation
n_s is the factor accounting for the uncertainties in stress evaluation
n_d is the factor for the severity of the damage

This method helps in reducing the factor of safety without reducing the safety considerations.

Many engineering societies and government agencies have developed safety codes for specific area of engineering design. For example, steam boilers, pressure vessels, elevators escalator, cranes and hoists.

The material based values of factor of safety are as follows:

Cast iron—3 to 5
Ductile materials (steel, aluminium)—1.5 to 2
Surface strength (cams followers, gears, etc.)—1.8 to 2.5
Components under compression (column, piston rod, power screw)—3 to 6

The test and the tools of theory of elasticity are used to evaluate the effect of the shape and size on the actual stress induced. Based on the studies, the effect of size is accounted for by the introduction of size factor B. The values can be taken from Fig. 5.7. Extensive work has been done to evaluate the effect of shape-change and stress-concentration factors. K_c are reported for various situations. Then factor of safety can be written as

$$N = n \cdot K_t / B \tag{4.3}$$

The design stress or permissible stress is given by

$$S_d = \frac{S_y}{N} = \frac{S_y \cdot B}{n \cdot K_t} \tag{4.4}$$

where, B is the size factor (Figure 3.3 MDHB)
K_t is stress-concentration factor (Table 4.4 and Fig 4.14 to 4.28)
n is the net factor of safety.

The representative values of n are given in Table 4.1.

4.2 THEORIES OF FAILURE

The failure of loaded component occurs whenever the internal stress induced exceeds material strength. The strength data available is the experimental data from UTM test under uniaxial loading. The actual loading causes a multidimensional stress state. The behaviour of the material under such a situation is difficult to predict. In an effort to relate the test data to the actual failure stress the following theories of failure are formulated.

(1) Maximum Normal Stress Theory (Rankine Theory)
(2) Maximum Principal Strain Theory (Saint-Venant Theory)
(3) Maximum Shear Stress Theory (Guest Theory)
(4) Maximum Strain Energy Theory (Haigh's Theory)
(5) Maximum Distortion Energy Theory (Von-Mises-Hencky Theory)

We discuss the theories in details in the following paragraphs.

Table 4.1. Net factor of safety

S. No.	Particulars	Factor of safety
(1)	When close dimensional control is exercised in production and materials have been purchased to close tolerance specifications such as in the design of aircraft and space vehicles.	1.0
(2)	For reliable high grade materials when loads and stresses can be determined very accurately and a low weight is desired.	1.1 to 1.25
(3)	For the same above condition 2 when a low weight is not important of critical.	1.5 to 2.0
(4)	For ordinary materials where overload is possible and reliability is important.	2.0 to 3.0
(5)	For cast iron and similar materials where loading is uncertain in design applications and where the stresses cannot be determined accurately.	3 and over
(6)	In mechanical design of steel and aluminium elements.	1.25 to 1.5
(7)	If the material (steel or aluminium) is not of best quality or if there is no definite information about its quality.	2.00

4.2.1 Maximum Normal Stress Theory (Rankine Theory)

The theory states that failure in a loaded member occurs whenever the maximum principal stress in the member exceeds the yield strength value as predicted by the UTM-test.

If $\sigma_1 > \sigma_2 > \sigma_3$ are the principal stresses, the failure according to this theory occurs only if

$$\text{Max.} \quad (|\sigma_1|, |\sigma_2|, |\sigma_3|) \geq S_{yt} \tag{4.5}$$

$$\text{OR} \quad \sigma_1 \geq S_{yt} \text{ or } \sigma_3 \leq S_{yc} \tag{4.5a}$$

For design, equations are

$$|\sigma_1| \leq \frac{S_y}{N} \quad \text{or} \quad |\sigma_2| < \frac{S_y}{N} \quad \text{or} \quad |\sigma_3| \leq \left(\frac{S_y}{N}\right)$$

$$\text{OR} \quad \text{Max.} \quad (|\sigma_1|; |\sigma_2|; |\sigma_3|) \leq \left(S_d = \frac{S_y}{N}\right) \tag{4.6}$$

For two-dimensional case, the failure equation becomes

$$|\sigma_1| = \text{Max.}\left[\frac{\sigma_x - \sigma_y}{2} \pm \sqrt{\left(\frac{\sigma_x + \sigma_y}{2}\right)^2 + \tau x_y^2}\right] \geq S_y. \tag{4.7}$$

The design equation is,

$$|\sigma_1| = \text{Max.}\left[\frac{\sigma_x - \sigma_y}{2} \pm \sqrt{\left(\frac{\sigma_x + \sigma_y}{2}\right)^2 + \tau x_y^2}\right] < \left(S_d = \frac{S_y}{N}\right) \tag{4.8}$$

The failure diagram can be drawn by plotting equations

$$\sigma_1 \geq S_y \quad \text{or} \quad \sigma_2 \geq S_y$$

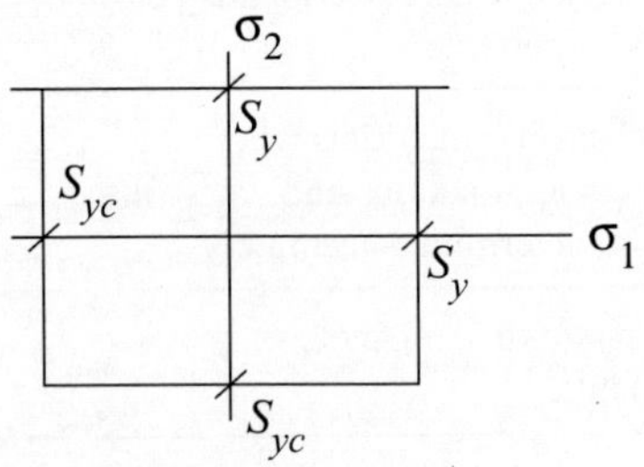

Figure 4.2

For ductile materials,

$$|S_{yc}| = |S_y|$$

In case of pure torsion test, the Mohr's circle and the principal stress elements are shown in Fig. 4.3(a) and Fig. 4.3(b).

As per torsion test, the failure occurs when $\tau_{max} = S_{sy}$, yield stress in shear. The principal stresses in the test specimen at failure are $\sigma_1 = \tau_{max} = S_{sy}$ and $|\sigma_2| = \tau_{max} = S_{sy}$. As per the normal stress theory, at failure $\sigma_1 = S_y$

This means

$$S_y = S_{sy}$$

This conclusion contradicts the observed fact that

$$\boxed{S_{sy} = 0.57 S_y}$$

Hence, the theory fails to predict the failure correctly, whenever the principal stresses are unlike, i.e., one is tensile and the other compressive. Therefore, if the loading results in the stress combination in the second or fourth quadrant of the failure diagram, theory is unreliable.

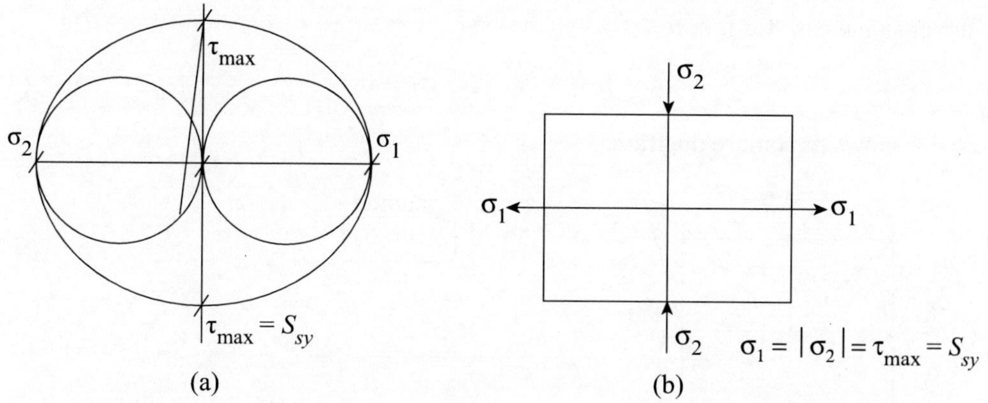

Figure 4.3. (a) Mohr's circle diagram for torsion test (b) Principal stress element

4.2.2 Maximum Principal Strain Theory (Saint-Venant Theory)

According to this theory, the failure at any point in a loaded body, occurs only if the maximum strain at that point exceeds the limiting value of the strain at yield in an uniaxial test (UTM). Assuming that $\varepsilon_1 > \varepsilon_2 > \varepsilon_3$, where ε_1, ε_2, ε_2 are the principal strains.

$$\varepsilon_{\max} = \varepsilon_1 = \frac{1}{E}[\sigma_1 - \mu(\sigma_2 + \sigma_3)]$$

The limiting value of strain at yield in UTM,

$$\begin{aligned} \varepsilon_y &= \frac{\sigma_1}{E} \qquad \because \quad \sigma_2 = \sigma_3 = 0 \\ &= \frac{S_y}{E} \qquad \because \quad \sigma_1 = S_y \text{ at yield in UTM} \end{aligned}$$

For failure to occur,

$$\varepsilon_{\max} \geq \varepsilon_y$$

$$\therefore \quad \sigma_1 - \mu(\sigma_2 + \sigma_3) \geq S_y \tag{4.9a}$$

Generalizing we can write,

$$|\sigma_2 - \mu(\sigma_1 + \sigma_3)| \geq S_y \tag{4.9b}$$

$$\text{OR} \quad |\sigma_3 - \mu(\sigma_1 + \sigma_2)| \geq S_y \tag{4.9c}$$

The design equations are

$$|\sigma_1 - \mu(\sigma_2 + \sigma_3)| \leq \left(S_d = \frac{S_y}{N}\right) \tag{4.10a}$$

$$\text{OR} \quad |\sigma_2 - \mu(\sigma_1 + \sigma_3)| \leq S_d \tag{4.10b}$$

$$\text{OR} \quad |\sigma_3 - \mu(\sigma_1 + \sigma_2)| \leq S_d \tag{4.10c}$$

For two-dimensional case for failure,

$$\sigma_1 - \mu\sigma_2 \geq S_y \quad \text{for } \sigma_3 = 0 \tag{4.11}$$

Figure 4.4 shows the failure diagram.

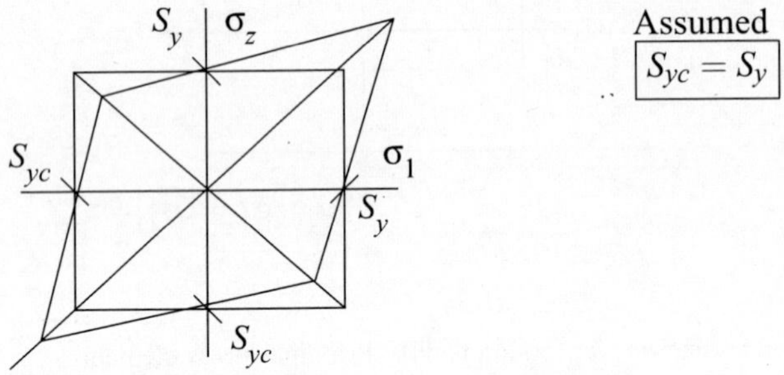

Figure 4.4

For design equation is

$$|\sigma_1 - \mu\sigma_2| \leq \left(S_d = \frac{S_y}{N}\right) \tag{4.12}$$

4.2.3 Maximum Shear Theory (Guest-Theory)

The theory states that the failure by yielding at any point in a loaded body occurs when the maximum shear stress induced in the member, becomes equal to or exceeds the maximum shear stress in the specimen of the same material, under tension test (UTM) when the yielding starts.

Maximum shear stress in the specimen under UTM test is given by Mohr's circle diagram in Fig. 4.5.

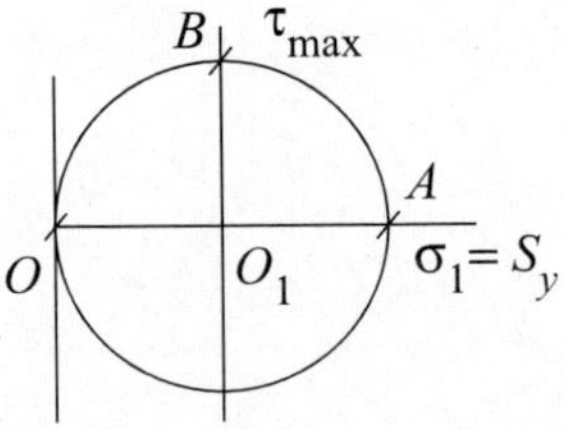

Figure 4.5. Mohr's circle for UTM

Maximum shear stress

$$\tau_{max} = O_1B = O_1A = \frac{1}{2}OA = \frac{1}{2}\sigma_1$$

$$\text{but} \quad \sigma_{max} = \sigma_1 = S_y$$

$$\therefore \quad \tau_{max} = \frac{S_y}{2}$$

Then as per the shear theory failure occurs when

$$\tau_{\max}\ (\text{induced}) \geq \left(\frac{S_y}{2}\right)$$

From the Mohr's circle for torsion test,

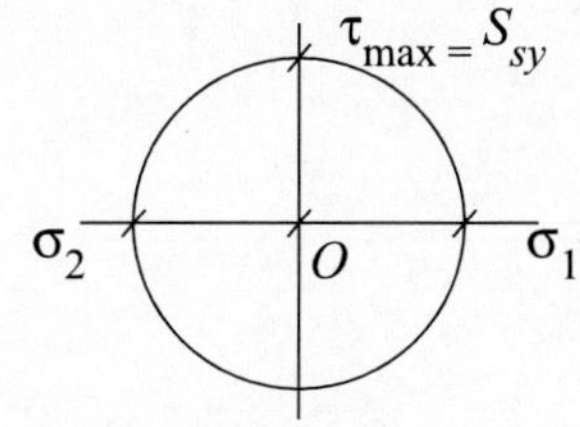

Figure 4.6

$$\sigma_1 = \tau_{\max} = S_{sy}$$
$$\sigma_2 = -\tau_{\max} = -S_{sy}$$

But we know

$$\tau_{\max} = \left(\frac{\sigma_1 - \sigma_2}{2}\right) = \frac{S_{sy} + S_{sy}}{2} = S_{sy}$$

As per shear theory at failure,

$$\therefore \qquad \tau_{\max} = \frac{S_y}{2}$$

Then we conclude

$$\boxed{S_{sy} = \frac{S_y}{2}}$$

This result is very much nearer to the observed result,

$$\boxed{S_{sy} = 0.57S_y}$$

Thus the shear theory can be used with confidence in all situations. The general equations for failure can be written in as follows:

$$|\sigma_1 - \sigma_2| \geq S_y \quad \text{or} \quad |\sigma_2 - \sigma_3| \geq S_y \quad \text{or} \quad |\sigma_1 - \sigma_3| \geq S_y$$
$$\text{OR} \quad \text{Max}\left\{|\sigma_1 - \sigma_2|; |\sigma_2 - \sigma_3|; |\sigma_1 - \sigma_3|\right\} \geq S_y. \tag{4.13}$$

For design equation, we can write

$$\text{Max.}\left\{|\sigma_1 - \sigma_2|; |\sigma_2 - \sigma_3|; |\sigma_1 - \sigma_3|\right\} \leq \left(S_d = \frac{S_y}{N}\right). \tag{4.14}$$

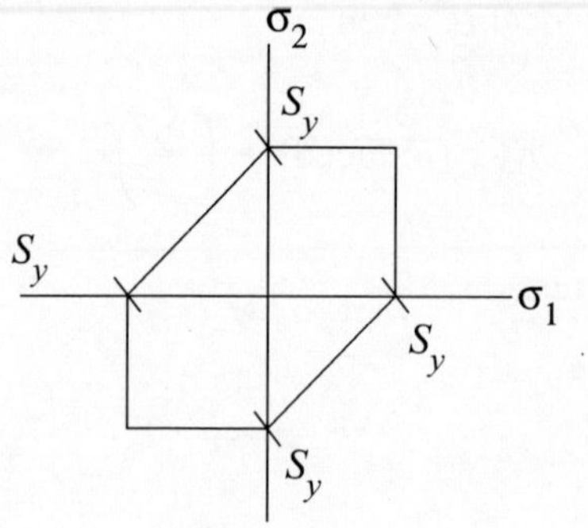

Figure 4.7

For two-dimensional stress system, we can write for failure equations as

$$|\sigma_1 - \sigma_2| \geq S_y \quad \text{or} \quad |\sigma_1| \geq S_y \quad \text{or} \quad |\sigma_2| \geq S_y$$
$$\text{OR} \quad \text{Max.} \quad \left\{|\sigma_1 - \sigma_2|; |\sigma_1|; |\sigma_2|\right\} \geq S_y. \tag{4.15}$$

The failure diagram can be plotted as shown in Fig. 4.7.

In the first and third quadrants, it is similar to maximum normal stress theory. But in the second and fourth quadrants it is much different from the normal stress theory. It predicts the failure more correctly in all the quadrants. The test failure points are much beyond the failure lines. Thus, the theory is more conservative and is used more widely for design.

The design equation can be written as Max. $\left\{|\sigma_1 - \sigma_2|;\ |\sigma_1|;\ |\sigma_2|\right\} \leq S_d$.

In terms of the general stress terms the equation is

$$\sqrt{(\sigma_x - \sigma_y)^2 + 4\tau_{xy}^2} \leq \left(S_d = \frac{S_y}{N}\right) \tag{4.16}$$

4.2.4 Maximum Strain Energy Theory (Haigh's Theory)

According to this theory, the failure at any point in loaded member occurs only if the maximum strain energy per unit volume equals or exceeds that at yield in the UTM test.
The strain energy in loaded body is

$$U = \sum_{n=1}^{3} \frac{\sigma_n \varepsilon_n}{2}$$
$$= \frac{1}{2E}\left[\sigma_1^2 + \sigma_2^2 + \sigma_3^2 - 2\mu(\sigma_1\sigma_2 + \sigma_2\sigma_3 + \sigma_3\sigma_1)\right]$$

where μ is Poisson's ratio.

The strain energy at yield in UTM test,

$$U_y = \frac{1}{2E}\left[S_y^2\right] \quad \because \quad \sigma_1 = S_y,\ \sigma_2 = \sigma_3 = 0.$$

For failure, $U \geq U_y$

$$\therefore \qquad \sigma_1^2 + \sigma_2^2 + \sigma_3^2 - 2\mu(\sigma_1\sigma_2 + \sigma_2\sigma_3 + \sigma_1\sigma_3) > S_y^2 \tag{4.17}$$

For design,

$$\sigma_1^2 + \sigma_2^2 + \sigma_3^2 - 2\mu(\sigma_1\sigma_2 + \sigma_2\sigma_3 + \sigma_1\sigma_3) \leq \left(S_d = \frac{S_y}{N}\right)^2 \leq \left(\frac{S_y}{N}\right)^2 \tag{4.18}$$

For two-dimensional stress system the failure equation is

$$\sigma_1^2 + \sigma_2^2 - 2\mu(\sigma_1\sigma_2) \geq S_y^2 \tag{4.19}$$

4.2.5 Maximum Distortion-Energy Theory (Von-Mises-Hencky Theory)

The results of hydraulic test on a cube of material have shown that the material can withstand loads larger than those predicted by UTM test. Based on the results, it is postulated that the failure is not a stress related phenomenon but distortion energy related phenomenon. Mr Von-Mises and Mr Hencky concluded independently as above based on their experiments and the theory is named after them. It is also called shear energy theory.

The theory states that the failure in a loaded member occurs when the distortion energy per unit volume becomes equal to or exceeds the distortion energy per unit volume in the simple tension test specimen (UTM) at yield.

The strain energy per unit volume for a loaded member is given by

$$U = \sum_{n=1}^{3}\left(\frac{\sigma_n \varepsilon_n}{2}\right) = \frac{1}{2E}\left[\sigma_1^2 + \sigma_2^2 + \sigma_3^2 - 2\mu(\sigma_1\sigma_2 + \sigma_2\sigma_3 + \sigma_3\sigma_1)\right]$$

where μ is Poisson's n ratio.

The deformation of a stressed member may be split into two parts (i) volume-change plus and (ii) angular distortion. The volume-change is due to hydraulic loading and the distortion is due to the balance load as shown below.

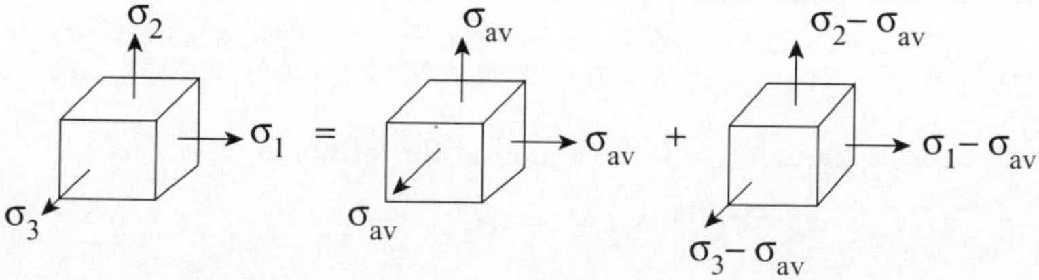

where $\sigma_{av} = \dfrac{\sigma_1 + \sigma_2 + \sigma_3}{3}$

By using the theorem of superposition
Total strain energy,

$$U = U_v + U_d$$
$$= \text{Energy for volume-change } (U_v) + \text{Energy for distortion } (U_d)$$

The distortion energy,

$$\begin{aligned} U_d &= U - U_v \\ &= \frac{1}{2E}\left[\sigma_1^2 + \sigma_2^2 + \sigma_3^2 - 2\mu(\sigma_1\sigma_2 + \sigma_2\sigma_3 + \sigma_3\sigma_1)\right] - \frac{1}{2E}\left[3\sigma_{av}^2 - 6\mu\sigma_{av}^2\right] \\ &= \frac{1}{2E}\left[\sigma_1^2 + \sigma_2^2 + \sigma_3^2 - 2\mu(\sigma_1\sigma_2 + \sigma_2\sigma_3 + \sigma_3\sigma_1)\right] - \frac{1}{2E}\left[(1-2\mu)(3\sigma_{av}^2)\right] \\ &= \frac{1}{2E}\left[\sigma_1^2 + \sigma_2^2 + \sigma_3^2 - 2\mu(\sigma_1\sigma_2 + \sigma_2\sigma_3 + \sigma_3\sigma_1)\right] \\ &\quad - \left(\frac{1-2\mu}{6E}\right)\left[\sigma_1^2 + \sigma_2^2 + \sigma_3^2 + 2\sigma_1\sigma_2 + 2\sigma_2\sigma_3 + 2\sigma_3\sigma_1\right] \\ &= \left(\frac{1+\mu}{3E}\right)\left[\frac{(\sigma_1 - \sigma_2)^2 + (\sigma_2 - \sigma_3)^2 + (\sigma_3 - \sigma_1)^2}{2}\right] \end{aligned}$$

For a simple-tension test (UTM) at yield

$$\sigma_1 = S_y, \quad \sigma_2 = \sigma_3 = 0$$

Distortion energy at yield,

$$\therefore \quad U_{dy} = \left(\frac{1+\mu}{3E}\right)(S_y^2)$$

By the statement of the theory, at failure

$$U_d \geq U_{dy}$$

$$\therefore \quad \left(\frac{1+\mu}{3E}\right)\left[\frac{(\sigma_1 - \sigma_2)^2 + (\sigma_2 - \sigma_3)^2 + (\sigma_3 - \sigma_1)^2}{2}\right] \geq \left(\frac{1+\mu}{3E}\right) S_y^2$$

$$\therefore \quad (\sigma_1 - \sigma_2)^2 + (\sigma_2 - \sigma_3)^2 + (\sigma_3 - \sigma_1)^2 \geq 2S_y^2 \tag{4.20}$$

In terms of general stress components, the equation for failure is

$$\left[(\sigma_x - \sigma_y)^2 + (\sigma_y - \sigma_z)^2 + (\sigma_z - \sigma_x)^2 + 6(\tau_{xy}^2 + \tau_{yz}^2 + \tau_{zx}^2)\right] \geq 2S_y^2 \tag{4.21}$$

For two-dimensional stress system, $\sigma_3 = 0$ the equation for failure is

$$\sigma_1^2 + \sigma_2^2 - \sigma_1 \cdot \sigma_2 \geq S_y^2 \tag{4.22}$$

This represents an ellipse shown below.

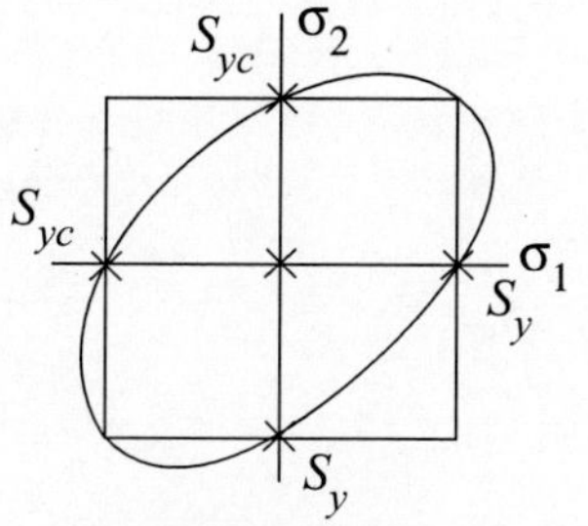

Figure 4.8

Assume

$$\boxed{S_{yc} = S_y}.$$

In terms of the general stress, the failure equation is

$$\left(\sigma_x^2 + \sigma_y^2 - \sigma_x \cdot \sigma_y + 3\sigma_{xy}^2\right) \geq S_y^2. \tag{4.23}$$

This curve passes through the majority of failure points, observed experimentally.
For pure torsion test,

$$\sigma_1 = -\sigma_2 = S_{sy}\sigma_3 = 0.$$

Substituting these values in failure equation, we get

$$S_y^2 = 3S_{sy}^2$$

$$\therefore \quad S_{sy} = \frac{1}{V3} \cdot S_{sy} = 0.577S_y$$

This value agrees with experimentally observed value. Hence, theory predicts failure more correctly. But the calculations are more cumbersome for use. Hence, theory is used for checking the design done by the use of Shear-Stress theory.
The design check equation is

$$(\sigma_1 - \sigma_2)^2 + (\sigma_2 - \sigma_3)^2 + (\sigma_3 - \sigma_1)^2 \leq 2\left(\frac{S_y}{N}\right)^2 \tag{4.24}$$

for 2D state of stress,

$$\sigma_1^2 - \sigma_1\sigma_2 + \sigma_2^2 \leq 2\left(\frac{S_y}{N}\right)^2 \tag{4.25}$$

4.3 FAILURE OF DUCTILE MATERIALS

The failure envelopes of the various theories are shown in Fig. 4.9. The experimental failure points for ductile materials are shown by dotted points. The distortion energy theory-envelope passes through these points. Hence, we can conclude that the theory predicts the failure more correctly.

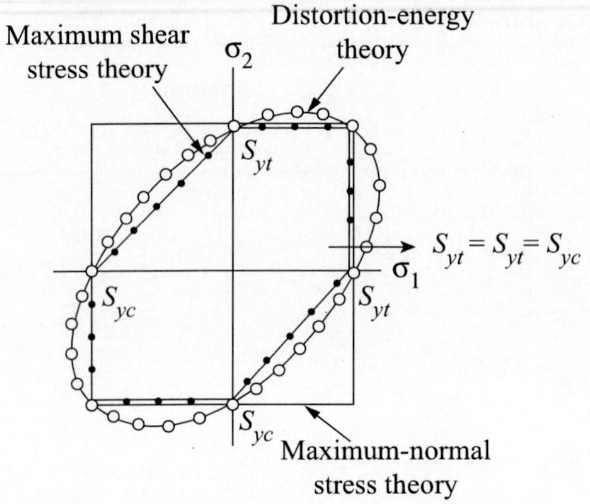

Figure 4.9. Comparison of the static failure theories for ductile materials

The shear-stress theory envelope is within the failure points. It is more conservative, and safer to use. The maximum normal stress theory is the same as shear theory in the first and third quadrants. In the second and fourth quadrants, the failure points are within the failure-envelope and the failure-predictions are totally unreliable. Hence, it is better to avoid using this theory for ductile materials. The maximum strain theory is an improvement on the normal stress theory but the predicted stress-values are much higher than the experimental values. Maximum strain energy theory envelope also includes all the failure-points within it and hence the predictions may not be correct. From the above comparision of the experimental data with the various theories of failure, we conclude that shear-theory is the best choice for design. The predictions are conservative and the design-equations can be easily applied. The distortion energy theory is more accurate but the computations are more complicated. Hence, the theory is used only for checking and modifying the preliminary design done by shear theory. The method is adopted only in case of costly materials.

Table 4.2. Comparison of theories of failure

	Theories of failure			
Type	**Rankine Theory**	**Saint–Vanant's Theory**	**Guest's Theory**	**Hencky Mises Theory**
1. For pure tension, i.e., $\sigma_x = \sigma$, $\sigma_y = 0$, $\sigma_{xy} = 0$	$\sigma = S_y$	$\sigma = S_y$	$\sigma = S_y$	$\sigma = S_y$
2. For pure shear, i.e., $\sigma_x = 0$, $\sigma_y = 0$, $\sigma_{xy} = \tau_{max}$	$\tau_{max} = S_y$	$\tau_{max} = 0.77S_y$	$\tau_{max} = 0.5S_y$	$\tau_{max} = 0.58S_y$

4.4 FAILURE OF BRITTLE MATERIALS

The brittle materials under static loads, fail by fracture rather than yielding. The brittle fracture in tension is due to normal tensile stress. The brittle fracture under compressive loads, is due to excessive compressive stress and shear stress. The compressive strength is much more than the tensile strength for cast irons. The compressive strength is three to four times the tensile strength. For ceramics, the ratio is much larger. Another characteristic feature of the brittle materials is that their shear strength is greater than the tensile strength but less than the compressive strength.

To account for all these features, the normal stress theory is modified. The modification is the Coulomb-Mohr Theory. For two-dimensional stress system the failure plot is shown in Fig. 4.10.

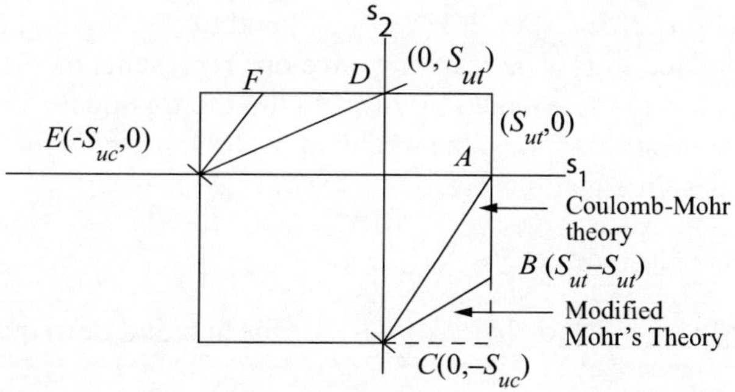

Figure 4.10

For tensile stresses, it is similar to normal stress, theory with the tensile strength S_{ut} as the limiting value. For the compressive-stresses, (third quadrant) it is similar to the normal stress-theory with the compressive strength S_{uc} as the limiting-value. In the second and fourth quadrants, i.e., the stresses are a combination of tensile and compressive stresses, the lines joining S_{ut} to S_{uc}, form the limiting lines (lines AC & ED). These lines can be represented by the equations.

$$\frac{\sigma_2}{S_{ut}} - \frac{\sigma_1}{S_{uc}} = 1 \quad \text{for the second quadrant} \tag{4.26a}$$

$$\frac{\sigma_1}{S_{ut}} - \frac{\sigma_2}{S_{uc}} = 1 \quad \text{for the fourth quadrant.} \tag{4.26b}$$

The design equations for the first and third quadrants are,

$$\sigma_1 \leq \frac{S_{ut}}{N} \quad \text{or} \quad \sigma_2 \leq \frac{S_{ut}}{N} \quad \text{for } 1^{\text{st}} \text{ quadrant} \tag{4.27a}$$

$$|\sigma_1| \leq \left(\frac{S_{uc}}{N}\right) \quad \text{or} \quad |\sigma_2| \leq \left(\frac{S_{uc}}{N}\right) \quad \text{for } 3^{\text{rd}} \text{ quadrant.} \tag{4.27b}$$

For the second and fourth quadrants the design equations are

$$\frac{\sigma_2}{S_{ut}} - \frac{\sigma_1}{S_{uc}} = \frac{1}{N} \quad \text{for the 2}^{\text{nd}} \text{ quadrant} \tag{4.28a}$$

$$\frac{\sigma_1}{S_{ut}} - \frac{\sigma_2}{S_{uc}} = \frac{1}{N} \quad \text{for the 4}^{\text{th}} \text{ quadrant.} \tag{4.28b}$$

This theory is similar to shear stress theory with unequal values for tensile strength and compressive strength.

Modified Mohr's Theory

It is observed that in the fourth and second quadrants, the failure points are well outside the Coulomb-Mohr lines *AC* and *ED*. To predict the failure more correctly the normal stress theory lines are extended from $A(S_{ut}, o)$ to $B/1(S_{uc}, -S_{ut})$ and from $D(O, S_{ut})$ to $F(-S_{ut}, S_{ut})$. Then *B* is joined to *C* and *F* is joined to *D*. The resulting envelope represents modified Mohr theory. For pure torsion test, $(\sigma_1/\sigma_2 = -1)$ represented by *B*(or *F*) the the torsional strength $S_{us} = S_{ut}$ is the characteristic of the brittle materials. Thus, the modified Mohr's theory is the more realistic theory for brittle materials. The design equations are,

When $|\sigma_1| < S_{ut}$ and $|\sigma_3| \leq S_{ut}$.

$$|\sigma_1| \leq \left(\frac{S_{ut}}{N}\right) \text{ or } |\sigma_2| \leq \left(\frac{S_{ut}}{N}\right) \quad \text{for first and third quadrants} \tag{4.29}$$

In the second and fourth quadrants for $\sigma_1 < -S_{ut}$ and $\sigma_2 > 0$

$$\left|\sigma_2 - \frac{S_{ut}}{S_{uc}}(\sigma_1 + \sigma_2)\right| \leq \frac{S_{ut}}{N} \quad \text{second quadrant} \tag{4.30a}$$

For $\sigma_1 > 0$ and $\sigma_2 < -S_{ut}$

$$\left|\sigma_1 - \frac{S_{ut}}{S_{uc}}(\sigma_1 + \sigma_2)\right| \leq \frac{S_{ut}}{N} \quad \text{fourth quadrant} \tag{4.30b}$$

For three-dimensional stress system. The Dowling-effective stresses are defined as follows:

$$C_1 = \frac{1}{2}\left[|\sigma_1 - \sigma_2| + \frac{S_{uc} + 2S_{ut}}{S_{uc}}(\sigma_1 + \sigma_2)\right] \tag{4.31a}$$

$$C_2 = \frac{1}{2}\left[|\sigma_2 - \sigma_3| + \frac{S_{uc} + 2S_{ut}}{S_{uc}}(\sigma_2 + \sigma_3)\right] \tag{4.31b}$$

$$C_3 = \frac{1}{2}\left[|\sigma_3 - \sigma_1| + \frac{S_{uc} + 2S_{ut}}{S_{uc}}(\sigma_3 + \sigma_1)\right] \tag{4.31c}$$

The effective maximum stress is given by $\bar{\sigma} = \text{Max.}[C_1, C_2, C3, \sigma_1, \sigma_2, \sigma_3]$

$$= 0 \quad \text{if Max.} \leq 0 \tag{4.32}$$

Then the factor of safety is $\quad N = \left(\dfrac{S_{ut}}{\bar{\sigma}}\right)$ (4.33)

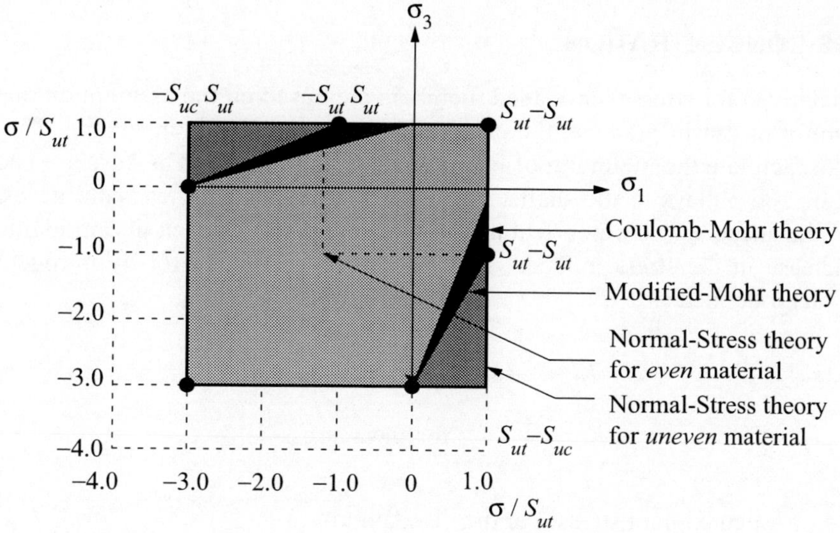

Figure 4.11. Coulomb-Mohr, Modified Mohr, and Maximum Normal stress theories for uneven, brittle materials

Design Equation,

$$\frac{\sigma_1}{S_{ut}} - \frac{\sigma_3}{S_{uc}} = \frac{1}{N} \quad \text{for second and fourth quadrants}$$

Torsional-Shear Strength,

$$S_{sy} = \frac{S_{yt} S_{yc}}{S_{yt} + S_{yc}}$$

Modified Mohr's Theory

$$C_1 = \frac{1}{2}\left[|\sigma_1 - \sigma_2| + \frac{S_{uc} + 2S_{ut}}{S_{uc}}(\sigma_1 + \sigma_2)\right]$$

$$C_2 = \frac{1}{2}\left[|\sigma_2 - \sigma_3| + \frac{S_{uc} + 2S_{ut}}{S_{uc}}(\sigma_2 + \sigma_3)\right]$$

$$C_3 = \frac{1}{2}\left[|\sigma_3 - \sigma_1| + \frac{S_{uc} + 2S_{ut}}{S_{uc}}(\sigma_3 + \sigma_1)\right]$$

The largest of the set six values (C_1, C_2, C_3, plus the three principal stresses) is the desired effective stress as suggested by Dowling:

Dowling Maximum effective stress.

$$\bar{\sigma} = \text{Max.}[C_1, C_2, C3, \sigma_1, \sigma_2, \sigma_3]$$
$$= 0 \quad \text{if Max.} \leq 0$$

$$F.S. = \frac{S_{ut}}{\bar{\sigma}}$$

4.5 STRESS CONCENTRATION

The equations derived for stresses in loaded members are based on the assumption that the cross sections are uniform. But in practice, the change in shape and size is inevitable. The shafts are to be stepped to facilitate the mounting of bearing, gears and pulleys. The keys are necessary for connecting gears and pulleys to the shafts. For bolts, the threads and heads are necessary parts. The change in the shape are common features in machine parts. At such discontinuities, there is a localized increase in the stress induced. The increase is accounted for by introducing stress-concentration factor defined as

$$K_t = \frac{\sigma_{\max}}{\sigma} \quad \text{for direct stress}$$

$$K_{ts} = \frac{\tau_{\max}}{\tau} \quad \text{for shear stress}$$

where, $\sigma_{\max}$, $\tau_{\max}$ are maximum stresses at the discontinuity;

σ, τ are the nominal stresses based on the uniform cross section;

K_t is the geometric stress-concentration factor for normal stress; and

K_{ts} is the geometric stress-concentration factor for shear stress.

The increase of stress is due to the change in the geometry of the part. The material of the part has no effect. Hence, it is called geometric stress-concentration factor. Extensive work, both theoretical and experimental, has been done for the evaluation of the maximum stress for various types of discontinuities under different kinds of loading situations. The stress-concentration factors are evaluated and presented in the form of charts. The exhaustive collection is "stress-concentration factors" by R. E. Peterson. A few of these are given in Fig. 4.14 to Fig. 4.28.

The reduction in the stress-concentration factor can be achieved by properly modifying the geometry of the parts. A few of the remedies are shown in Fig. 4.12 and Fig. 4.13.

(i) In case of stepped shafts, a proper fillet radius is provided (Fig. 4.13a)

(ii) Providing one or two grooves by the side of the step. (Fig. 4.12b) and (Fig. 4.13b)

(iii) Providing an under-cutting groove at the step. (Fig. 4.13c and 4.13d)

(iv) Providing smaller holes on both sides of the existing hole (Fig. 4.12a)

(v) Reducing the shank diameter to the root diameter of threaded bolt.

The method to be adopted can be easily visualized by flow analogy method.

Stress Concentration and Methods to Reduce Stress Concentration

Figures 4.12 and 4.13 show the stress concentration and some methods to reduce stress concentration.

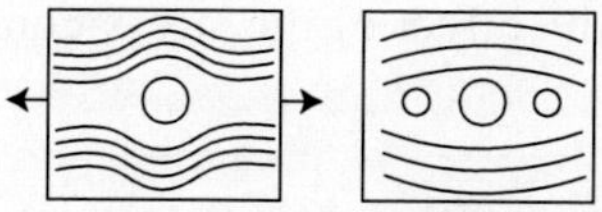

Better

By providing two small holes
on either side of big hole

(a) Plate with hole

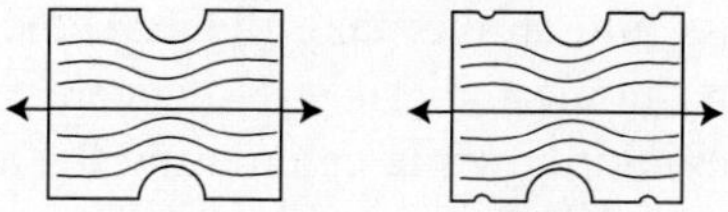

Better

Small notches are provided
on both sides of big notch

(b) Notched plate

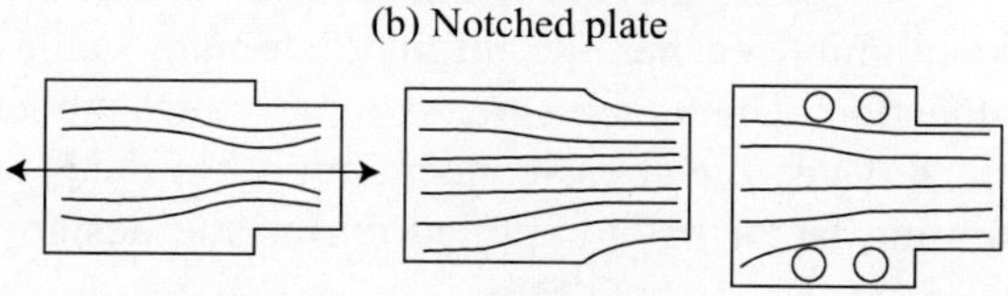

Better
by providing a fillet

Better
by drilling hole

(c) Stepped plate

Figure 4.12. Plates

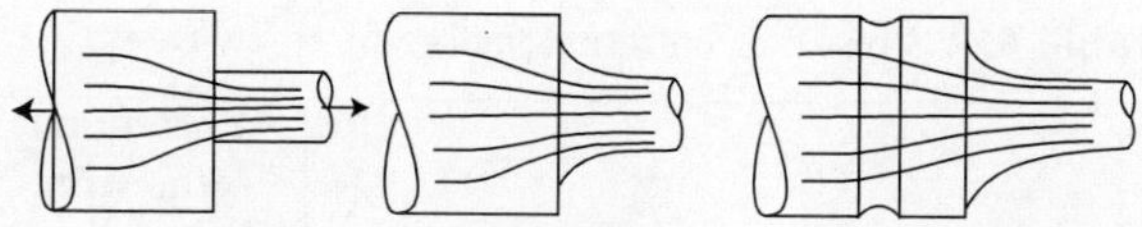

(a) By providing a fillet
stepped shaft

(b) By providing a
grooved shaft

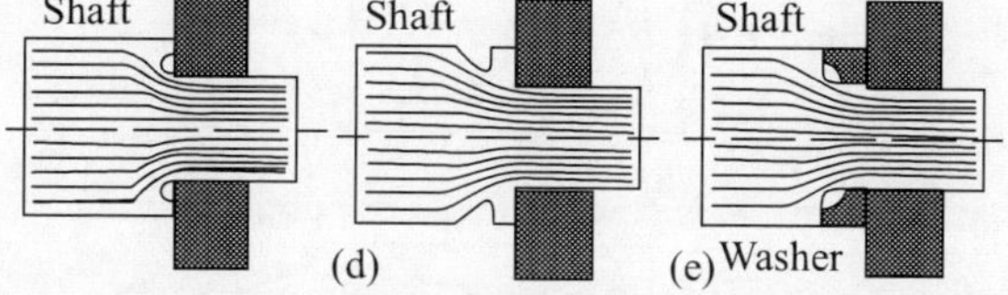

(c)
Stress concentration
reduced with radius

(d)
Stress concentration
reduced with groove

(e)
Stress concentration
reduced with washer

Figure 4.13. Shafts

Stress-Concentration for Static-loading

The influence of material properties on the effect of the stress-concentration factor has to be considered during the design process.

Ductile Materials

The stress at the stress-raiser reaches the yield limit much before the stress at the remaining portion. The localized yielding occurs at the stress-raiser and the stress is relieved. The further increase is halted. This continues until yielding has occurred in the complete section. Only after reaching this condition, further increases in stress and strains take place. Hence, it is very common to ignore the effect of stress-concentration in ductile materials. However, it is necessary to account for the reduced cross-sectional area due to yielding while calculating the actual stress induced.

Brittle Materials

In case of brittle materials, the stress-strain curve is almost straight line upto fracture (Fig. 4.1c). When the stress at the discontinuity reaches the ultimate strength values, the localized deformation causes crack at the location. This crack spreads rapidly throughout the body and causes sudden fracture of the member. Thus, the stress-concentration has dangerous effect and the full values of the stress-concentration factor has to applied during the design process for the brittle materials.

Table 4.3. Chart for selecting stress concentration factor for various types of loading and materials

Loading	**Brittle materials**		**Ductile materials**	
Static	K_t	K_{ts}	$K_t = 1$	$K_{ts} = 1$
Cyclic	K_f	K_{fs}	K_f	K_{fs}

Table 4.4. Stress concentration factors for a few cases

Type of form Irregularity or stress riser	**Manner of loading**		**Equation for the stress concentration factor**	**Experimental Value of the Factor K_t or K_{ft}**
1. A round hole in a plate of finite width	Tension	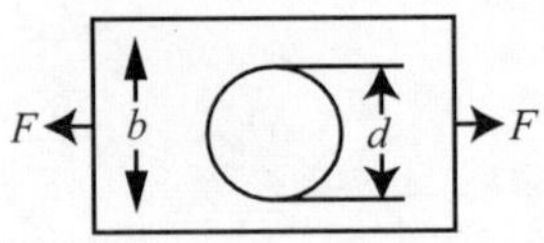	$K_t = 0.5$ $[(2+d/2x)^2$ $+3(d/2x)^4]$	$K_t = 3$
2. Round filled hole	Tension	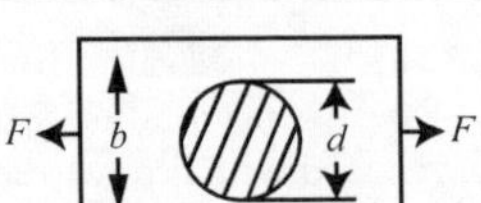		$K_t = 2.5$

Contd.

Table 4.4. Contd.

Type of form Irregularity or stress riser	Manner of loading	Equation for the stress concentration factor	Experimental Value of the Factor K_t or K_{ft}
3. The pin in a hole transmits the load	Tension 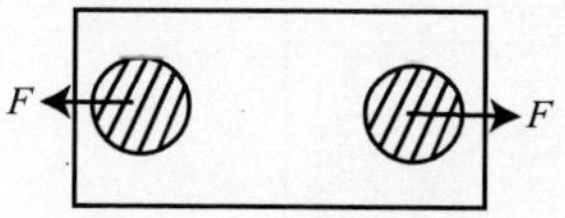		$K_t = 2.5$
4. An elliptical hole in a plate with major axis normal to the load	Tension Shear 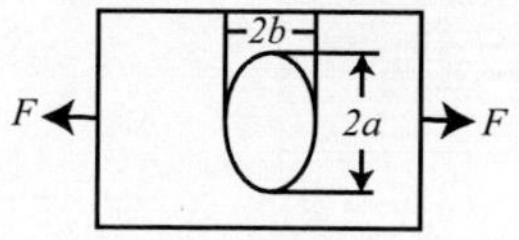	$K_t = 1 + 2a/b$ $K_{ts} = 1 + a/b$	
5. An elliptical hole in a plate with major axis along the load	Tension 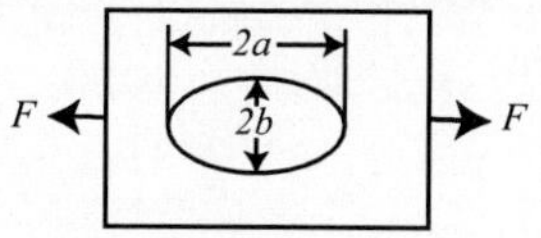	$K_t = 1 + 2b/a$ a = semi-major-axis b = semi-minor-axis	
6. For a small V-notch at the edge of the plate	Tension 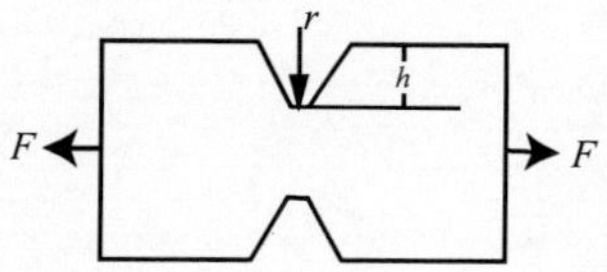	$K_t = 1 + 2\sqrt{h/r}$ $K_{ts} = 1 + \sqrt{h/r}$ h = depth of the notch r = radius at the notch	
7. Key ways with $b = d/4$ and $h = b/2$ (i) Profile key-way(ii) Sled runner	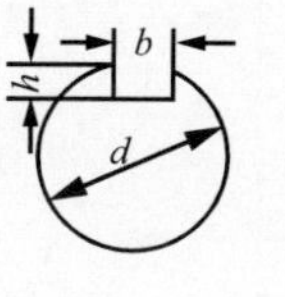 Shear Shear		(i.e.) $K_{ts} = 1.68$ (i.e.) $K_{ts} = 1.44$
8. Screw thread (i) UNF thread (ii) Whitworth	Tension Tension		$K_f = 2.5 - 3.75$ $K_f = 1.5 - 3.0$
9. Press and shrink fit (i) Plain member (ii) Grooved member			$K_t = 1.95$ $K_t = 1.34$
10. Rotating disc with hole (i) For $R_i/R_o \cong 0$ (ii) Thin ring			$K_t = 2$ $K_t = 1$

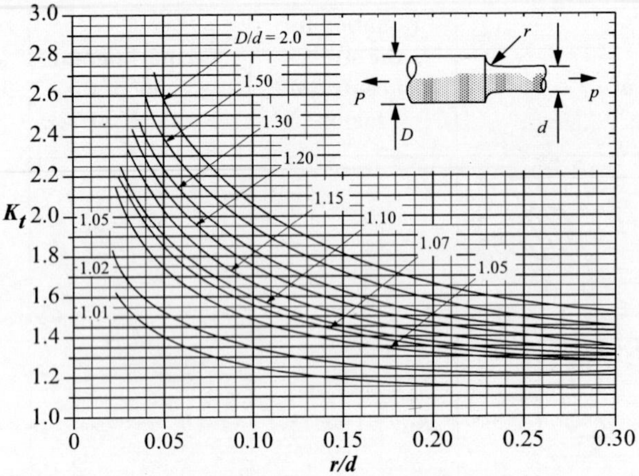

Figure 4.14. Geometric stress-concentration factor K_t for a shaft with a shoulder fillet in axial torsion

$$K_t \cong A\left(\frac{r}{d}\right)^b$$

where:

D/d	A	b
2.00	1.01470	−0.30035
1.50	0.99957	−0.28221
1.30	0.99682	−0.25751
1.20	0.96272	−0.25527
1.15	0.98084	−0.22485
1.10	0.98450	−0.20818
1.07	0.94898	−0.19548
1.05	1.00480	−.0.17076
1.02	1.01220	−0.12474
1.01	0.98413	−0.10474

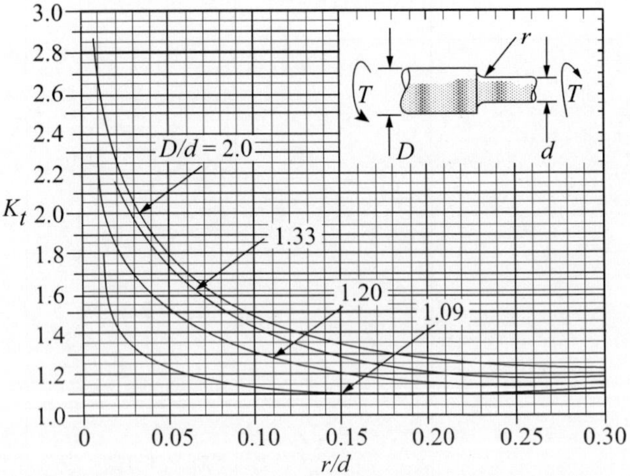

Figure 4.15. Geometric stress-concentration factor K_t for a shaft with a shoulder fillet in bending

$$K_t \cong A\left(\frac{r}{d}\right)^b$$

where:

D/d	A	b
6.00	0.87868	−0.33243
3.00	0.89334	−0.30860
2.00	0.90879	−0.28598
1.50	0.93836	−0.257.59
1.20	0.97098	−0.21796
1.10	0.95120	−0.23757
1.07	0.97527	−0.20958
1.05	0.98137	−0.19653
1.03	0.98061	−0.18381
1.02	0.96048	−0.17711
1.01	0.91938	−0.10732

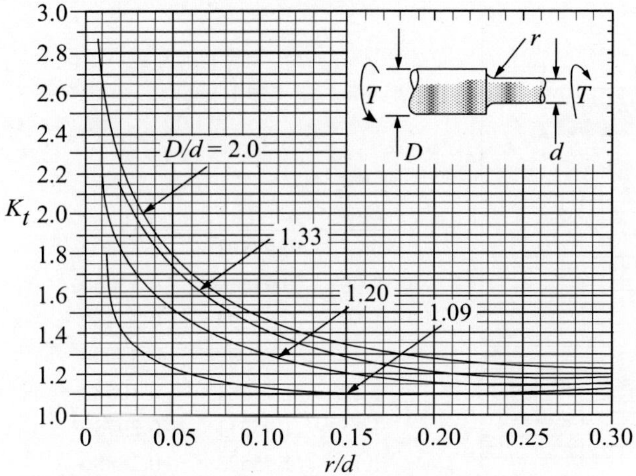

$$K_t \cong A\left(\frac{r}{d}\right)^b$$

where:

D/d	A	b
2.00	0.86331	−0.23865
1.33	0.84897	−0.23161
1.20	0.83425	−0.21649
1.09	0.90337	−0.12692

Figure 4.16. Geometric stress-concentration factor K_t for a shaft with a shoulder fillet in torsion

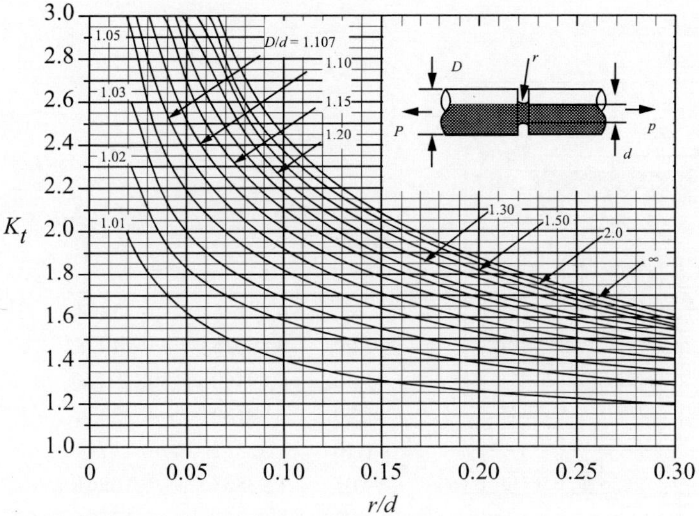

$$K_t \cong A\left(\frac{r}{d}\right)^b$$

where:

D/d	A	b
∞	0.99372	−0.39352
2.00	0.99383	−0.38231
1.50	0.99808	−0.36955
1.30	1.00490	−0.35545
1.20	1.01070	−0.33765
1.15	1.02630	−0.31673
1.10	1.02720	−0.29484
1.07	1.02380	−0.27618
1.05	1.02720	−0.25256
1.03	1.03670	−0.21603
1.02	1.03790	−0.18755
1.01	1.00030	−0.15609

Figure 4.17. Geometric stress-concentration factor K_t for a grooved shaft in axial tension

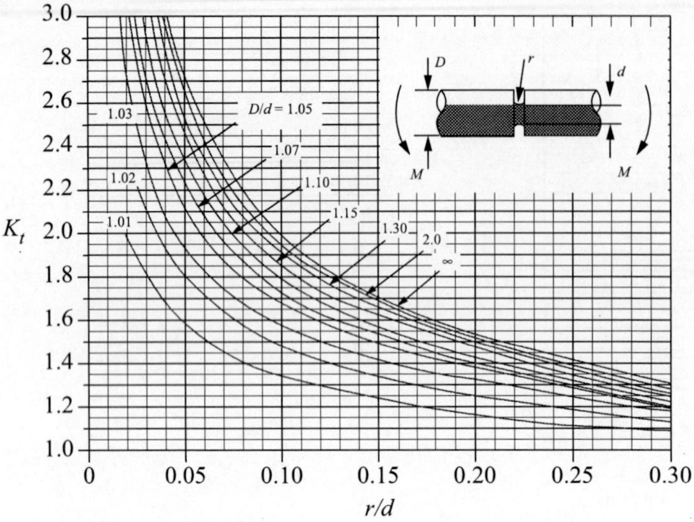

Figure 4.18. Geometric stress-concentration factor K_t for a grooved shaft in bending

$$K_t \cong A\left(\frac{r}{d}\right)^b$$

where:

D/d	A	b
∞	0.94801	−0.33302
2.00	0.93619	−0.33066
1.50	0.93894	−0.32380
1.30	1.94299	−0.31504
1.20	1.94681	−0.30582
1.15	1.95311	−0.29739
1.12	1.95573	−0.28886
1.10	1.95454	−0.28268
1.07	1.96774	−0.26452
1.05	1.98755	−0.24134
1.03	1.99033	−0.21517
1.02	0.97753	−0.19793
1.01	0.99393	−0.15238

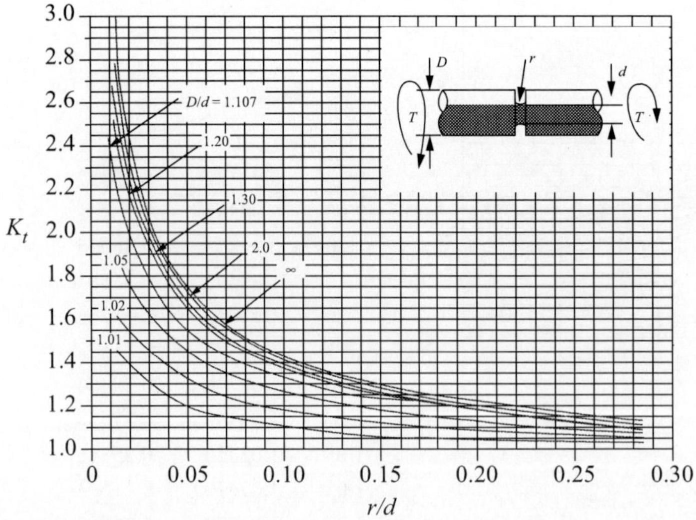

Figure 4.19. Geometric stress-concentration factor K_t for a grooved shaft in torsion

$$K_t \cong A\left(\frac{r}{d}\right)^b$$

where:

D/d	A	b
∞	0.88126	−0.25204
2.00	0.89035	−0.24075
1.30	0.89460	−0.023267
1.20	1.90182	−0.22334
1.10	1.92311	−0.19740
1.05	1.93853	−0.16941
1.02	1.96877	−0.12605
1.01	1.97245	−0.10162

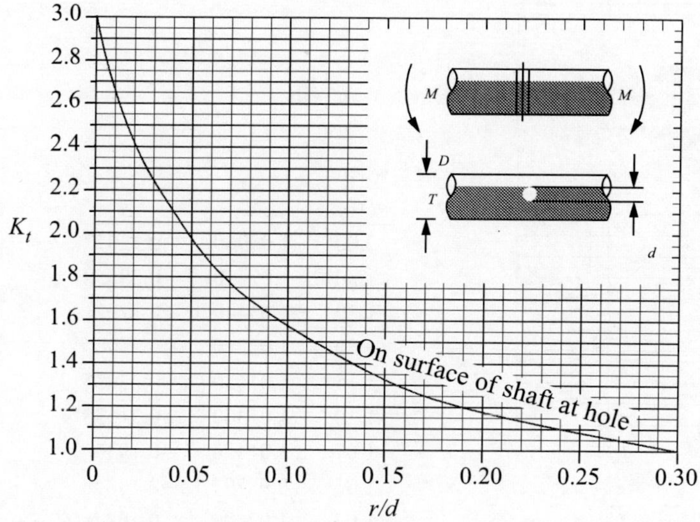

Figure 4.20. Geometric stress-concentration factor K_t for a shaft with a transverse hole in bending

$$K_t \cong 1.58990 - 0.63550 \log\left(\frac{d}{D}\right)$$

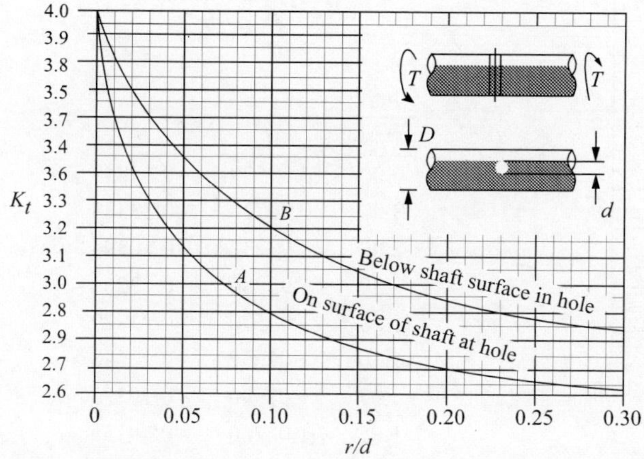

Figure 4.21. Geometric stress-concentration factor K_t for a width a transverse hole in torsion

$$K_{t_B} \cong 3.9702 - 9.292\left(\frac{d}{D}\right) + 27.159\left(\frac{d}{D}\right)^2 + 30.231\left(\frac{d}{D}\right)^3 - 393.19\left(\frac{d}{D}\right)^4 + 650.39\left(\frac{d}{D}\right)^5 + 15.451\left(\frac{d}{D}\right)^6$$

$$K_{t_A} \cong 3.92150 - 24.435\left(\frac{d}{D}\right) + 234.06\left(\frac{d}{D}\right)^2 - 1.200.5\left(\frac{d}{D}\right)^3 + 3059.5\left(\frac{d}{D}\right)^4 - 3042.4\left(\frac{d}{D}\right)^5$$

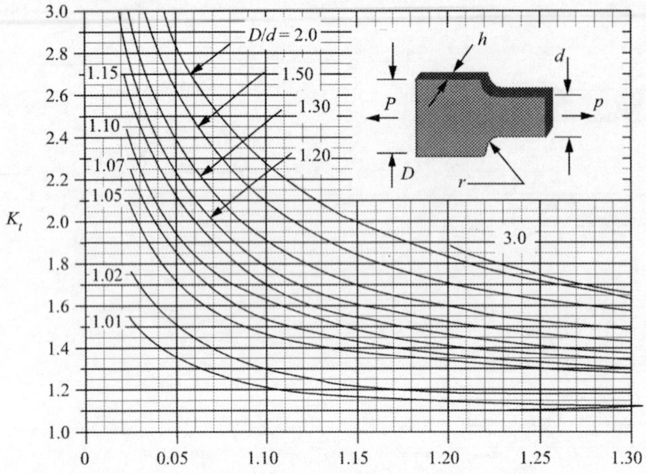

Figure 4.22. Geometric stress-concentration factor K_t for a filleted flat bar in axial tension

$$K_t \cong A\left(\frac{r}{d}\right)^b$$

where:

D/d	A	b
2.00	1.09960	−0.32077
1.50	1.07690	−0.29558
1.30	1.05440	−0.27021
1.20	1.03510	−0.25084
1.15	1.01420	−0.23935
1.10	1.01300	−0.21535
1.07	1.01450	−0.19366
1.05	0.98797	−0.13848
1.02	1.02590	−0.16978
1.01	1.97662	−0.10656

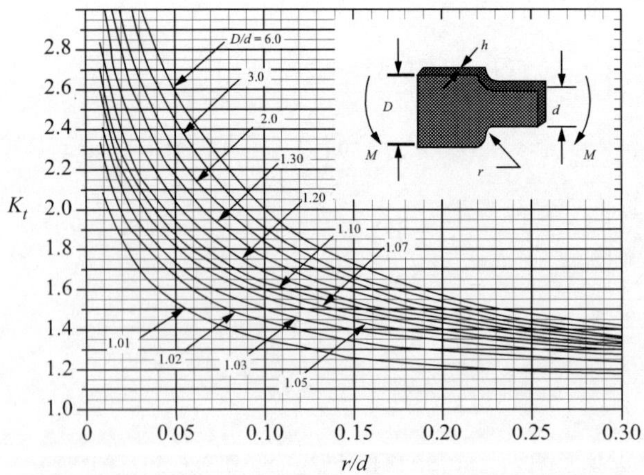

Figure 4.23. Geometric stress-concentration factor K_t for a filleted flat bar in bending

$$K_t \cong A\left(\frac{r}{d}\right)^b$$

where:

D/d	A	b
6.00	0.89579	−0.35847
3.00	0.90720	−0.33333
2.00	1.93232	−0.30304
1.30	1.95880	−0.27269
1.20	1.99590	−0.23829
1.10	1.01650	−0.21548
1.07	1.01990	−0.20333
1.05	1.02260	−0.19156
1.03	1.01660	−0.17802
1.02	1.99528	−0.17013
1.01	1.96689	−0.15417

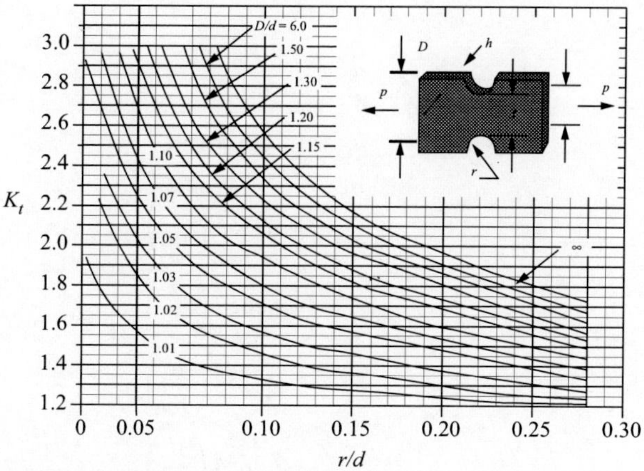

Figure 4.24. Geometric stress-concentration factor K_t for a notched flat bar in axial tension

$$K_t \cong A\left(\frac{r}{d}\right)^b$$

where:

D/d	A	b
∞	1.19050	−0.41712
3.00	1.11390	−0.40923
2.00	0.13390	−0.38586
1.50	1.13260	−0.36592
1.30	1.15860	−0.33260
1.20	1.14750	−0.31507
1.15	1.09520	−0.32517
1.10	1.08510	−0.29997
1.07	1.09120	−0.26857
1.05	1.09060	−0.24163
1.03	1.05180	−0.22216
1.02	1.05400	−0.18879
1.01	1.04260	−0.14145

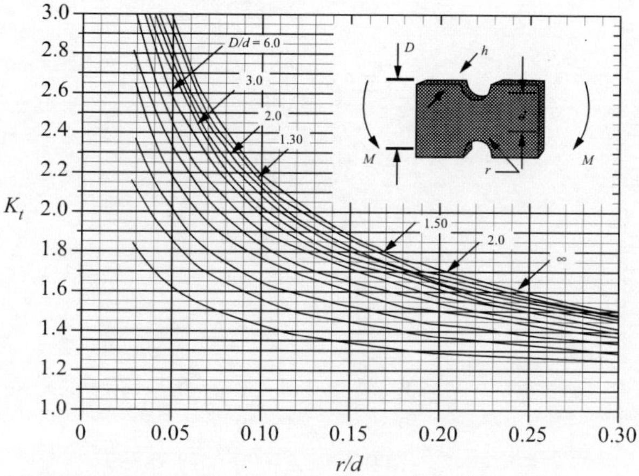

Figure 4.25. Geometric stress-concentration factor K_t for a notched flat bar in bending

$$K_t \cong A\left(\frac{r}{d}\right)^b$$

where:

D/d	A	b
∞	0.97079	−0.35672
3.00	0.97079	−0.35047
2.00	0.96801	−0.34915
1.50	0.98315	−0.33395
1.30	0.98288	−0.32606
1.20	0.99055	−0.31319
1.15	0.99304	−0.30263
1.10	1.00710	−0.28379
1.07	1.01470	−0.26145
1.05	1.02500	−0.24008
1.03	1.02940	−0.21161
1.02	1.03740	−0.18428
1.01	1.06050	−0.13369

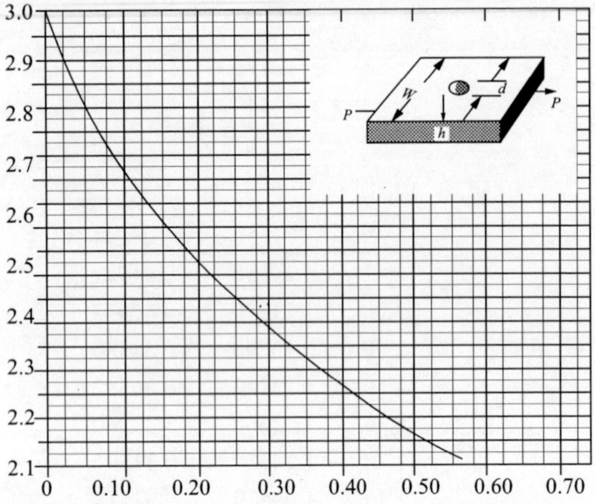

Figure 4.26. Geometric stress-concentration factor K_t for a flat bar with transverse hole in axial tension

for $\frac{d}{W} \leq 0.65$:

$$K_1 \cong 3.0039 - 3.753\frac{d}{W} + 7.9735\left(\frac{d}{W}\right)^2 - 9.2659\left(\frac{d}{W}\right)^3 + 1.8145\left(\frac{d}{W}\right)^4 + 2.9684\left(\frac{d}{W}\right)^5$$

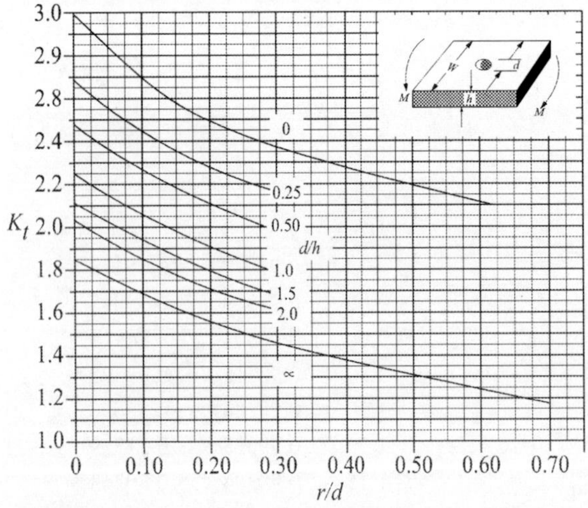

Figure 4.27. Geometric stress-concentration factor K_t for a flat bar with transverse hole in bending

For $\frac{d}{h} \Rightarrow 0$ and $\frac{d}{W} \leq 0.65$:

$$K_t \cong 2.9947 - 3.4833\frac{d}{W} + 5.8268\left(\frac{d}{W}\right)^2 - 4.1986\left(\frac{d}{W}\right)^3$$

for $\frac{d}{h} \geq 0.25$: $K_t \cong Ae[b(d/w)]$

where:

d/h	A	b
0.25	2.68750	-0.75128
0.50	2.46620	-0.77215
1.00	2.24000	-0.78739
1.50	2.02430	-0.80821
2.00	2.10560	-0.79878
∞	1.80820	-0.66702

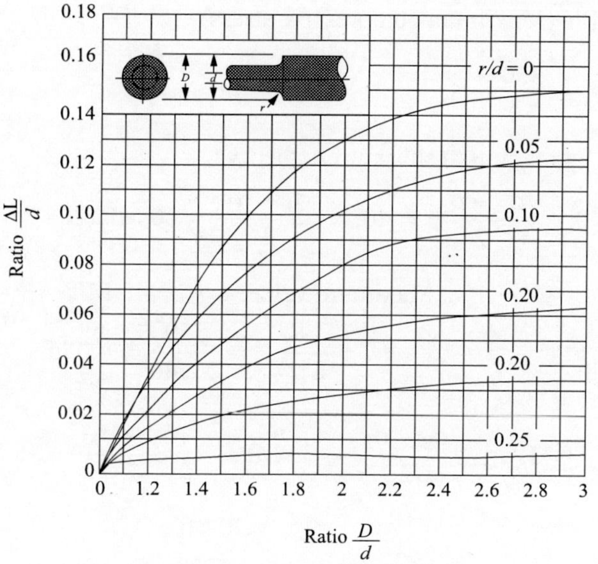

Figure 4.28. Junction effect for cylindrical shafts

SOLVED EXAMPLES

Example 4.1: A machine component made of mild steel having yield point stress of 350 MPa is subjected to the following stresses. Calculate the factor of safety using (a) Maximum normal stress theory, (b) Maximum shear stress theory, (c) Maximum distortion energy theory and (d) Maximum strain theory.

(i) $\sigma_1 = 80$ MPa; $\sigma_2 = 80$ MPa; $\sigma_3 = 0$
(ii) $\sigma_1 = 70$ MPa; $\sigma_2 = -70$ MPa; $\sigma_3 = 0$
(iii) $\sigma_1 = 45$ MPa; $\sigma_2 = 40$ MPa; $\sigma_3 = -45$ MPa
(iv) $\sigma_1 = 0$ MPa; $\sigma_2 = -30$ MPa; $\sigma_3 = -70$ MPa

SOLUTION

Step-1: Factor of safety by maximum normal stress theory

(i) $\sigma_1 = \sigma_2 = 80$ MPa

$$\sigma_1 = \frac{S_y}{F_s} = F_s = \frac{350}{80} = 4.375$$

(ii) $\sigma_1 = 70$ MPa (Max.)

$$F_s = \frac{350}{70} = 5$$

(iii) $\sigma_1 = 45$ MPa (Max.)

$$F_s = \frac{350}{45} = 7.78$$

(iv) $\sigma_3 = -70$ MPa (Maximum compressive stress)

$$\sigma_3 = \frac{S_y}{F_s} = F_s = \frac{350}{70} = 5$$

Step-2: Factor of safety by maximum shear stress theory

(i) $\left|\frac{\sigma_1 - \sigma_2}{2}\right| = 0; \quad \left|\frac{\sigma_2 - \sigma_3}{2}\right| = 40; \quad \left|\frac{\sigma_1 - \sigma_3}{2}\right| = 40$ MPa

$$\text{Maximum value, } \tau_{\max} = 40 \text{ MPa}$$

$$\therefore \quad F_s = \frac{S_y}{2\tau_{max}} = \frac{350}{2 \times 40} = 4.375$$

(ii) $\left|\frac{\sigma_1 - \sigma_2}{2}\right| = 70$ MPa; $\left|\frac{\sigma_2 - \sigma_3}{2}\right| = 35$ MPa; $\left|\frac{\sigma_1 - \sigma_3}{2}\right| = 35$ MPa

$$\tau_{\max} = 70 \text{ MPa}$$

$$F_s = \frac{S_y}{2\tau_{max}} = \frac{350}{2 \times 70} = 2.5$$

(iii) $\left|\frac{\sigma_1 - \sigma_2}{2}\right| = 2.5; \quad \left|\frac{\sigma_2 - \sigma_3}{2}\right| = 42.5; \quad \left|\frac{\sigma_1 - \sigma_3}{2}\right| = 45$ MPa

$$\tau_{\max} = 45 \text{ MPa}$$

$$F_s = \frac{S_y}{2\tau_{max}} = \frac{350}{2 \times 45} = 3.89$$

(iv) $\left|\frac{\sigma_1 - \sigma_2}{2}\right| = 15; \quad \left|\frac{\sigma_2 - \sigma_3}{2}\right| = 20; \quad \left|\frac{\sigma_1 - \sigma_3}{2}\right| = 35$ MPa

$$\tau_{\max} = 35 \text{ MPa}$$

$$F_s = \frac{S_y}{2\tau_{max}} = \frac{350}{2 \times 35} = 5$$

Step-3: Factor of safety by maximum distortion energy theory

(i) $\sigma_1{}^2 + \sigma_2{}^2 - \sigma_1\sigma_2 = \left(\frac{S_y}{F_s}\right)^2 \quad (\sigma_3 = 0)$

$$80^2 + 80^2 - 80 \times 80 = \left(\frac{350}{F_s}\right)^2$$

$F_s = 4.375$

(ii) $(70)^2 + (-70)^2 - (70)(-70) = \left(\frac{350}{FS}\right)^2 \quad (\sigma_3 = 0)$

$F_s = 2.89$

(iii) $\sigma_1{}^2+\sigma_2{}^2+\sigma_3{}^2-\sigma_1\sigma_2-\sigma_2\sigma_3-\sigma_1\sigma_3=\left(\frac{S_y}{F_s}\right)^2$

$$(45)^2+(40)^2+(-45)^2-45\times 40-40\times(-45)-45\times(-45)=\left(\frac{350}{FS}\right)^2$$

$F_s = 4.11$

(iv) $\sigma_2{}^2+\sigma_3{}^2-\sigma_2\sigma_3=\left(\frac{S_y}{F_s}\right)^2 \qquad (\sigma_=0)$

$$(-30)^2+(-70)^2-(-30)(-70)=\left(\frac{350}{F_s}\right)^2$$

$F_s = 5.75$

Step-4: Factor of safety by maximum strain theory
Assume Poisson's rate $= 0.3$ and $E = 2\times 10^5$ MPa

(i) $\in_1=\frac{1}{E}(\sigma_1-v\sigma_2) \qquad (\sigma_3=0)$

$$=\frac{1}{2\times 10^5}[80-0.3\times 80]$$

$$=280\times 10^{-6}$$

$$\in_2=\frac{1}{E}(\sigma_2-v\sigma_1) \qquad (\sigma_1=\sigma_2)$$

$$=280\times 10^{-6}$$

$$\text{Maximum strain, } \in_{\max} = 280\times 10^{-6}$$

$$=\frac{S_y}{E\cdot F_S}=\frac{350}{2\times 10^5\times F_s}$$

$$\therefore \quad F_s = 6.25$$

(ii) $\in_1=\frac{1}{2\times 10^5}[70-0.3(-70)] \quad = 455\times 10^{-6}$

$$\in_2=\frac{1}{2\times 10^5}(-70-0.3\times 70) \quad = -455\times 10^{-6}$$

$$\in_{\max} = 455\times 10^{-6}=\frac{S_y}{E(F_s)}$$

$$F_s = \frac{350}{2\times 10^5\times 455\times 10^{-6}}=3.85$$

(iii) $\in_1=\frac{1}{E}[\sigma_1-v(\sigma_2+\sigma_3)]$

$$=\frac{1}{2\times 10^5}\left\{45-0.3[40+(-45)]\right\}=232.5\times 10^{-6}$$

$$\in_2=\frac{1}{E}[\sigma_2-v(\sigma_1+\sigma_3)]$$

$$= \frac{1}{2\times 10^5}\left\{40 - 0.3[45 + (-45)]\right\} = 200\times 10^{-6}$$

$$\in_3 = \frac{1}{E}[\sigma_3 - \nu(\sigma_1 + \sigma_2)]$$

$$= \frac{1}{2\times 10^5}\left\{-45 - 0.3[45 + (40)]\right\} = -325.5\times 10^{-6}$$

$$\therefore \quad \text{Maximum strain, } \in_{max} = 352.5\times 10^{-6} = \frac{S_y}{E\times F_s}$$

$$\therefore \quad F_s = 4.96$$

(iv) $$\in_1 = \frac{1}{2\times 10^5}\{0 - 0.3[-30 + (-70)]\} = -150\times 10^{-6}$$

$$\in_2 = \frac{1}{2\times 10^5}\{-30 - 0.3[0 + (-70)]\} = -45\times 10^{-6}$$

$$\in_3 = \frac{1}{2\times 10^5}\{-70 - 0.3[0 + (-30)]\} = -305\times 10^{-6}$$

$$\in_{max} = 305\times 10^{-6} = \frac{S_y}{F_s\times E}$$

$$\therefore \quad F_s = \frac{350}{2\times 10^5\times 305\times 10^{-6}} = 5.74$$

Example 4.2: A shaft having circular cross section of diameter 'd' and made of still having yield point strength 300 MPa is subjected to a twisting moment of 1.5 kNm and simultaneously a bending moment of 1 kNm. Determine the diameter using (a) Maximum shear strain theory and (b) Maximum distortion energy theory. Assume the shear strength as 0.4 times the yield point stress and factor of safety as 2.5.

SOLUTION

Given Data

$T = 1.5$ kNm $= 1.5\times 10^6$ Nmm;'d' in mm

$M = 1$ kNm $= 1\times 10^6$ Nmm

Step-1: Principal stresses

$$\text{Torsional shear stress, } \tau = \frac{16T}{\pi d^3} = \frac{7.64\times 10^6}{d^3}\ \text{MPa}$$

$$\text{Bending stress, } \sigma_b = \frac{32M}{\pi d^3} = \frac{10.19\times 10^6}{d^3}\ \text{MPa}$$

$$\text{Maximum Principal stress, } \sigma_1 = \frac{10.19\times 10^6}{2d^3} + \sqrt{\left(\frac{10.19\times 10^6}{2d^3}\right)^2 + \left(\frac{7.64\times 10^6}{d^3}\right)^2}$$

$$\sigma_1 = \frac{14.27 \times 10^6}{d^3} \text{ MPa}$$

$$\text{Maximum Principal stress, } \sigma_2 = \frac{10.19 \times 10^6}{2d^3} - \sqrt{\left(\frac{10.19 \times 10^6}{2d^3}\right)^2 + \left(\frac{7.64 \times 10^6}{d^3}\right)^2}$$

$$\sigma_2 = \frac{-4.09 \times 10^6}{d^3} \text{ MPa}$$

$$\text{For bi-axial stress system, } \sigma_3 = 0$$

Step-2: The diameter of the shaft by maximum shear stress theory

$$\tau_{max} = \text{Maximum of } \left(\left|\frac{\sigma_1 - \sigma_2}{2}\right|, \left|\frac{\sigma_2 - \sigma_3}{2}\right| \text{ and } \left|\frac{\sigma_1 - \sigma_3}{2}\right| \right)$$

$$= \text{Maximum} \left(\frac{9.18 \times 10^6}{d^3}, \frac{2.045 \times 10^6}{d^3}, \frac{7.135 \times 10^6}{d^3} \right)$$

$$\tau_{max} = \frac{9.18 \times 10^6}{d^3}$$

$$= \frac{Ssy}{Fs} \qquad (Ss_y = 0.4 \times S_y = 0.4 \times 300 = 120 \text{ MPa})$$

$$\frac{9.8 \times 10^6}{d^3} = \frac{120}{2.5}$$

$$\therefore \quad d = 57.61 \text{ mm} = 58 \text{ mm}$$

Step-3: The diameter of the shaft by maximum distortion energy theory

$$\sigma_1{}^2 + \sigma_2{}^2 - \sigma_1\sigma_2 = \left(\frac{S_y}{F_s}\right)^2$$

$$\left(\frac{14.27 \times 10^6}{d^3}\right)^2 + \left(\frac{-4.09 \times 10^6}{d^3}\right)^2 - \left(\frac{14.27 \times 10^6}{d^3}\right)\left(\frac{-4.09 \times 10^6}{d^3}\right) = \left(\frac{300}{2.5}\right)^2$$

$$d = 51.81 \text{ mm} = 52 \text{ mm}$$

Example 4.3: A machine element is loaded so that $\sigma_1 = 120$ MPa, $\sigma_2 = 0$ MPa and $\sigma_3 = -90$ MPa, the material has a minimum yield strength in tension and compression of 360 MPa; Find the factor of safety for each of the following failure theories:

(a) Maximum normal stress theory
(b) Maximum shear stress theory
(c) Maximum distortion stress theory

SOLUTION

Given Data

$\sigma_1 = 120$ MPa, $\sigma_2 = 0$ MPa, $\sigma_3 = -90$ MPa
$S_{yt} = S_{yc} = 360$ MPa

Step-1: Factor of safety by maximum normal stress theory

$\sigma_{max} = \text{Max.}\ (\sigma_1, \sigma_2, \sigma_3) = \text{Max.}\ (120, 0, -90) = 120$ MPa
The factor of safety is

$$N = \frac{S_y}{\sigma_{max}} = \frac{360}{120} = 3.0$$

Step-2: Factor of safety by maximum shear stress theory

$$\tau_{max} = \text{Max}\left(\left|\frac{\sigma_1 - \sigma_2}{2}\right|, \left|\frac{\sigma_2 - \sigma_3}{2}\right|, \left|\frac{\sigma_1 - \sigma_3}{2}\right|\right)$$
$$= \left|\frac{120 - (-90)}{2}\right| = \left|\frac{210}{2}\right| = 105 \text{ MPa}$$

Yield stress in shear $S_{sy} = S_y/2 = 360/2 = 180$ MPa
Factor of safety

$$N = \frac{S_{sy}}{\tau_{max}} = \frac{180}{105} = 1.71$$

Step-3: FS by maximum distortion stress theory

Von-Mises stress,

$$\sigma^1 = \sqrt{\frac{(\sigma_1 - \sigma_2)^2 + (\sigma_2 - \sigma_3)^2 + (\sigma_3 - \sigma_1)^2}{2}}$$
$$= \sqrt{\frac{(120 - 0)^2 + (0 + 90)^2 + (-90 - 120)^2}{2}}$$
$$= \sqrt{\frac{14,400 + 8,100 + 44,100}{2}}$$
$$= 182,483 \text{ MPa}$$

$$\text{Factor of safety, } N = \frac{S_y}{\sigma} = \frac{350}{182.483} = 1.97$$

Example 4.4: A machine part is statistically loaded and has a yield strength of 350 MPa. For each state of stress, indicated below, find the factor of safety using each of the three static failure theories;

a) $\sigma_A = 70$ MPa, $\sigma_B = 70$ MPa
b) $\sigma_A = 70$ MPa, $\sigma_B = 35$ MPa
c) $\sigma_A = 70$ MPa, $\sigma_B = -70$ MPa
d) $\sigma_A = -70$ MPa, $\sigma_B = 0$ MPa

SOLUTION

(a) (i) Maximum stress theory $\sigma_1 = \sigma_A, \sigma_2 = \sigma_B \;\&\; \sigma_3 = 0$

$$\sigma_{\max} = \text{Max.}\,(70, 70) = 70 \text{ MPa}$$

The factor of safety, $N = \dfrac{S_y}{\sigma_{max}} = \dfrac{350}{70} = 5.0$

(ii) Maximum shear stress theory $\sigma_1 = \sigma_A, \sigma_2 = \sigma_B \;\&\; \sigma_3 = 0$

$$\tau_{\max} = \text{Max.}\left[\frac{70-70}{2}; \frac{70}{2}\right] = 35 \text{ MPa}$$

Yield in shear,

$$S_{sy} = \frac{S_y}{2} = \frac{350}{2} = 175 \text{ MPa}$$

$\therefore$ Factor of safety, $N = \dfrac{175}{35} = 5.0$

(iii) Maximum distortion energy theory

Von-Mises stress,

$$\sigma^1 = \sqrt{\frac{(\sigma_1-\sigma_2)^2+(\sigma_2-\sigma_3)^2+(\sigma_3-\sigma_1)^2}{2}}$$

$$\text{For } \sigma_1 = \sigma_A = 70 \text{ MPa}; \sigma_2 = \sigma_B = 70 \text{ MPa}; \sigma_3 = 0$$

$$\sigma^1 = \sqrt{\sigma_A{}^2 - \sigma_A \times \sigma_B + \sigma_A{}^2}$$

$$= \sqrt{(70)^2 - 70 \times 70 + (70)^2} = 70 \text{ MPa}$$

$\therefore$ Factor of safety, $N = \dfrac{S_y}{\sigma^1} = \dfrac{350}{70} = 5.0$

(b) (i) Maximum normal stress theory

$\sigma_{\max} = \text{Max. } (70,\ 35,\ 0) = 70 \text{ MPa}$

Factor of safety, $N = \dfrac{350}{70} = 5.0$

(ii) Maximum shear theory

$\tau_{\max} = \text{Max.}\left(\dfrac{70-35}{2}, \dfrac{70}{2}, \dfrac{35}{2}\right) = 35 \text{ MPa}$

Yield stress in shear, $S_{sy} = \dfrac{S_y}{2} = \dfrac{350}{2} = 175 \text{ MPa}$

Factor of safety, $N = \dfrac{S_{sy}}{\tau_{\max}} = \dfrac{175}{35} = 5$

(iii) Distortion energy theory

Von-Mises stress

$\sigma^1 = \sqrt{\sigma_A{}^2 - \sigma_A\sigma_B + \sigma_B{}^2} = \sqrt{70^2 - 70 \times 35 + 35^2} = 60.62$

Factor of safety, $N = \dfrac{350}{60.62} = 5.77$

(c) (i) Maximum normal stress theory

$$\sigma_{max} = \text{Max}\ (70, -70) = 70 \text{ MPa}$$

Factor of safety, $N = \dfrac{350}{70} = 5.0$

(ii) Maximum shear theory

$$\tau_{max} = \frac{(70-(-70))}{2} = 70 \text{ MPa}$$

Factor of safety, $N = \dfrac{S_{sy}}{\tau_{max}} = \dfrac{175}{70} = 2.5$

(iii) Maximum distortion energy theory

Von-Mises stress

$$\begin{aligned} \sigma^1 &= \sqrt{\sigma_A{}^2 - \sigma_A\sigma_B + \sigma_B{}^2} \\ &= \sqrt{70^2 + 70 \times 70 + (70)^2} \\ &= 121.24 \text{ MPa} \end{aligned}$$

Factor of safety, $N = \dfrac{350}{121.24} = 2.88$

(d) (i) Maximum normal stress theory $|\sigma_{max}| = 70$ MPa

F.S. - N$=\dfrac{350}{70} = 5.0$

(ii) Maximum shear theory, $\tau_{max} = \dfrac{0-(-70)}{2} = 35$ MPa

$N = \dfrac{175}{35} = 5.0$

(iii) Maximum distortion energy $\sigma^1 = \sqrt{(70)^2 - 70 \times 0 + 0^2} = 70;$

$N = \dfrac{350}{70} = 5.0$

Graphical Representation

a) Factor of safety for all theories

$N = \dfrac{OA}{Oa} = \dfrac{OP}{Oa_1} = 5$

b) Factor of safety by (i & ii) theories

$$N = \frac{OB_1}{ob} = \frac{OP}{Ob_1} = \frac{350}{70} = 5.0$$

by (iii) theory

$$N = \frac{OB_3}{Ob} = 5.77$$

c) Factor of safety by (i) theory

$$N = \frac{OC_1}{Oc} = 5.0$$

by (ii) theory

$$N = \frac{OC_2}{Oc} = 2.5$$

by (iii) theory

$$N = \frac{OC_3}{Oc} = 2.85$$

d) Factor of safety by all theory

$$N = \frac{OR}{Od} = \frac{350}{70} = 5$$

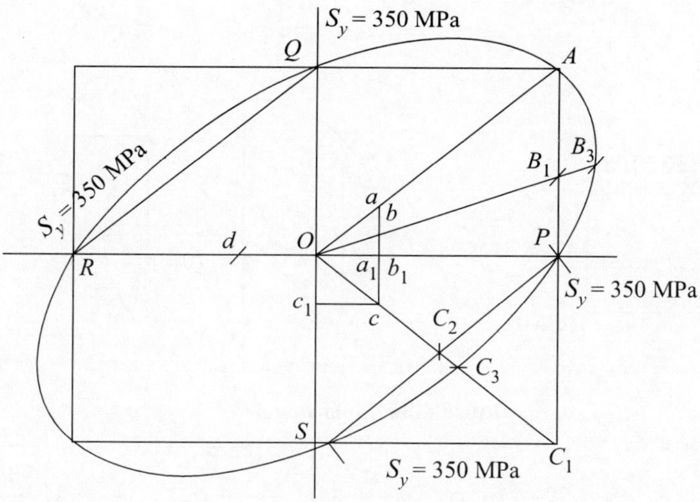

Figure 4.29

Example 4.5: Based on the use of CA614 Aluminium bronze soft rod as the material, find factors of safety for each of the three static-failure theories for the following stress states:

(a) $\sigma_x = 70$ MPa, $\sigma_y = 30$ MPa

(b) $\sigma_x = 70$ MPa, $\tau_{xy} = 30$ MPa cw

(c) $\sigma_x = -10$ MPa, $\sigma_y = -60$ MPa, $\tau_{xy} = 30$ MPa ccw
(d) $\sigma_x = 50$ MPa, $\sigma_y = 20$ MPa, $\tau_{xy} = 40$ MPa cw

SOLUTION

For CA614 Aluminium bronze soft (Table 2.12 MDHB)

$S_y = 310$ MPa $S_u = 565$ MPa $\therefore S_{sy} = 155$ MPa

(a) $\sigma_x = 70$ MPa $\quad \sigma_y = 30$ MPa

$$\sigma_{max} = 70 \text{ MPa}, \ \tau_{max} = \frac{70-0}{2} = 35 \text{ MPa}$$

Factor of safety by

(i) Maximum normal stress theory, $N = \dfrac{S_y}{\sigma_{max}}$

$$N = \frac{310}{70} = 4.43$$

(ii) Maximum shear stress theory, $N = \dfrac{S_{sy}}{\tau_{max}} = \dfrac{155}{35} = 4.43$

(iii) Maximum distortion - energy theory

Von-Mises stress

$$\sigma^1 = \sqrt{(70)^2 - 70 \times 30 + (30)^2} = \sqrt{3700} = 60.8 \text{ MPa}$$

$$N = \frac{S_y}{\sigma^1} = \frac{310.0}{60.8} = 5.1$$

(b) $\sigma_x = 70$ MPa, $\tau_{xy} = 30$ MPa cw

Mohr's circle,

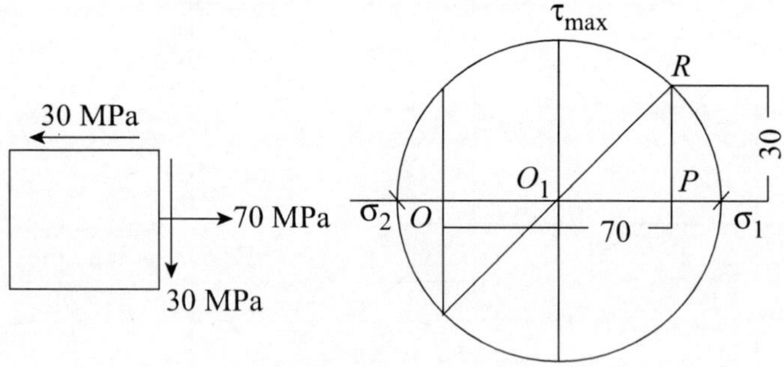

Figure 4.30. Mohr's circle

Principal stresses,

$$\sigma_1, \sigma_2 = \frac{\sigma_x}{2} + \sqrt{\frac{\sigma_x{}^2}{2} + \tau^2{}_{xy}} = \frac{70}{2} \pm \sqrt{\frac{70^2}{2} + 30^2}$$

$$= 35 \pm 46.1$$

$$\sigma_1 = 81.1 \text{ MPa}, \ \sigma_2 = -11.1 \text{ MPa}, \ \sigma_3 = 0$$

$$\sigma_{max} = 81.1 \text{ MPa}$$

Maximum shear stress, $\tau_{max} = \dfrac{81.1 + 11.1}{2} = 46.1$ MPa

Factor of safety by

(i) Maximum normal stress theory, $N = \dfrac{S_y}{\sigma_{max}} = \dfrac{310}{81.1} = 3.82$

(ii) Maximum shear stress theory, $N = \dfrac{S_{sy}}{\tau_{max}} = \dfrac{155}{46.1} = 3.36$

(iii) Maximum distortion energy theory

Von-Mises stress,

$$\begin{aligned} \sigma^1 &= \sqrt{(81.1)^2 - (81.1)(-11.1) + (11.1)^2} \\ &= 87.18 \text{ MPa} \\ N &= 310/87.18 = 3.55 \end{aligned}$$

(c) $\sigma_x = -10$ MPa, $\sigma_y = -60$ MPa, $\tau_{xy} = 30$ MPa ccw

Mohr's circle,

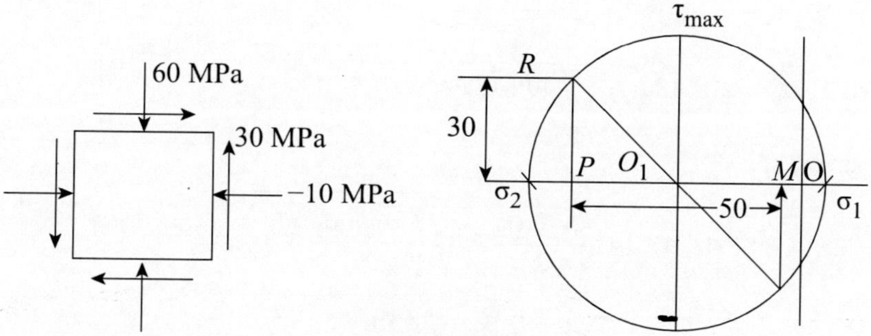

Figure 4.31

Principal stresses,

$$\begin{aligned} \sigma_1, \sigma_2 &= \frac{\sigma_x + \sigma_y}{2} \pm \sqrt{\left(\frac{\sigma_x - \sigma_x}{2}\right)^2 + \tau^2{}_{xy}} \\ &= \frac{-10 + (-60)}{2} \pm \sqrt{\left(\frac{-10 + 60}{2}\right)^2 + 30^2} \\ &= -35 \pm 39.05 \text{ MPa} \\ \sigma_1 &= 4.05 \text{ MPa}, \ \sigma_2 = -74.05 \text{ MPa} \\ \tau_{max} &= 39.05 \text{ MPa} \end{aligned}$$

Numerically, maximum normal stress,

$$-\sigma_{max} = 74.05 \text{ MPa} \quad \text{(compressive)}$$

Maximum shear stress, $\tau_{max} = \dfrac{\sigma_1 - \sigma_2}{2} = \dfrac{4.05 - (-74.05)}{2} = 39.05\text{MPa}$

Factor of safety

(i) By maximum normal stress theory,

$$N = \frac{S_y}{\sigma_{max}} = \frac{310}{74.05} = 4.186$$

(ii) By maximum shear stress theory,

$$N = \frac{S_{sy}}{\tau_{max}} = \frac{155}{39.05} = 3.97$$

(iii) By maximum distortion energy theory,

$$\begin{aligned} N &= \frac{S_y}{\sqrt{\sigma_1{}^2 - \sigma_1\sigma_2 + \sigma_2{}^2}} = \frac{310}{\sqrt{4.05^2 + 4.05 \times 74.05 + 74.05^2}} \\ &= 4.07 \end{aligned}$$

(d) $\sigma_x = 50$ MPa, $\sigma_y = 20$ MPa, $\tau_{xy} = 40$ MPa ccw

Principal stresses,

$$\begin{aligned} \sigma_1, \sigma_2 &= \frac{\sigma_x + \sigma_y}{2} \pm \sqrt{\left(\frac{\sigma_x - \sigma_y}{2}\right)^2 + \tau_{xy}{}^2} \\ &= \frac{50 + 20}{2} \pm \sqrt{\left(\frac{50 - 20}{2}\right)^2 + 40^2} \\ &= 35 \pm 42.72 \\ \sigma_1 &= 77.72 \text{ MPa}, \ \sigma_2 = -7.72 \text{ MPa}, \ \sigma_3 = 0 \\ \sigma_{max} &= 77.72 \text{ MPa}, \\ \tau_{max} &= \frac{77.72 + 7.72}{2} = 42.7 \text{ MPa} \end{aligned}$$

Factor of safety

(i) By maximum normal stress theory,

$$N = \frac{S_y}{\sigma_{max}} = \frac{310}{77.72} = 3.9887$$

(ii) By maximum shear stress theory,

$$N = \frac{S_{sy}}{\tau_{max}} = \frac{155}{42.72} = 3.63$$

(iii) By maximum distortion energy theory,

$$N = \frac{S_y}{\sqrt{\sigma_1{}^2 - \sigma_1\sigma_2 + \sigma_2{}^2}} = \frac{310}{\sqrt{77.72^2 + 77.72 \times 7.72 + 7.72^2}}$$
$$= 3.787$$

Graphical Representation

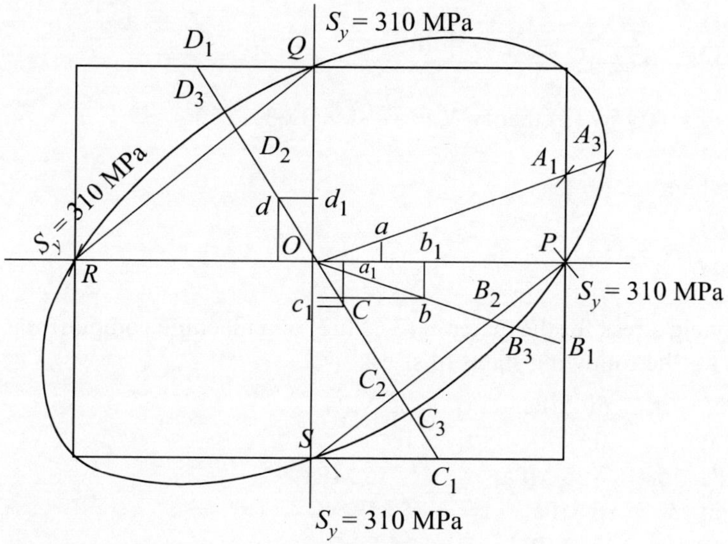

Figure 4.32

a) Factor of safety by (i) and (ii)

$$N = \frac{OA_1}{Oa} = \frac{OP}{Oa_1} = 4.43$$

Factor of safety by (iii) $N = \frac{OA_3}{Oa} = 5.1$

b) Factor of safety by (i) theory

$$N = \frac{OB_1}{Ob} = \frac{OP}{Ob_1} = \frac{310}{81.1} = 3.82$$

Factor of safety by (ii) theory

$$N = \frac{OB_2}{Ob} = 3.36$$

by (iii) theory

$$N = \frac{OB_3}{Ob} = 3.55$$

c) Factor of safety by (i) $N = \dfrac{OC_1}{Oc} = \dfrac{OS}{Oc_1} = 4.186$

Factor of safety by (ii) $N = \dfrac{OC_2}{Oc} = 3.97$

Factor of safety by (iii) $N = \dfrac{OC_3}{Oc} = 4.07$

d) Factor of safety by (i) theory

$$N = \frac{OD_1}{Od} = \frac{OQ}{Od_1} = \frac{310}{77.72} = 4.00$$

Factor of safety by (ii) theory $N = \dfrac{OD_2}{Od} = 3.63$

by (iii) theory

$$N = \frac{OD_3}{Od} = 3.87$$

Example 4.6: The yield stress for the material (37 Mn2) of a machine component is 500 MPa. Find the factor of safety for the following states of stress:

(a) $\sigma_x = 75$ MPa; $\sigma_y = 50$ MPa; $\sigma_z = 25$ MPa
$\tau_{xy} = 50$ MPa; $\tau_{yz} = 0$; $\tau_{xz} = 0$

(b) $\sigma_x = 100$ MPa; $\sigma_y = -25$ MPa; $\sigma_{zz} = -75$ MPa
$\tau_{xy} = 25$ MPa; $\tau_{yz} = 50$ MPa; $\tau_{xz} = 75$ MPa

(c) $\sigma_x = -50$ MPa; $\sigma_y = 75$ MPa; $\sigma_z = 25.0$ MPa
$\tau_{xy} = 10.0$ MPa; $\tau_{yz} = 25$ MPa; $\tau_{zx} = 100$ MPa

SOLUTION

a) $C_1 = \sigma_x + \sigma_y + \sigma_z = 75 + 50 + 25 = \mathbf{15.0}$

$$C_2 = \begin{vmatrix} 75 & 50 \\ 50 & 50 \end{vmatrix} + \begin{vmatrix} 50 & 0 \\ 0 & 25 \end{vmatrix} + \begin{vmatrix} 75 & 0 \\ 0 & 25 \end{vmatrix} = 1250 + 1250 + 1875 = \mathbf{4375}$$

$$C_3 = \begin{vmatrix} 75 & 50 & 0 \\ 50 & 50 & 0 \\ 0 & 0 & 25 \end{vmatrix}$$

$$= 75\begin{vmatrix} 50 & 0 \\ 0 & 25 \end{vmatrix} - 50\begin{vmatrix} 50 & 0 \\ 0 & 25 \end{vmatrix} + 0\begin{vmatrix} 50 & 0 \\ 0 & 25 \end{vmatrix}$$

$$= 93750 - 62500 + 0 = \mathbf{31250}$$

The equation for principal stresses is

$\sigma^3 - C_1\sigma^2 + C_2\sigma^2 - C_3 = 0$

$\sigma^3 - 150\sigma^2 + 4375\sigma - 31250 = 0$

Normalizing $x = \sigma/10$

$x^3 - 15.0x^2 + 43.75x - 31.25 = 0$

$$\begin{aligned}\text{Try}\quad x &= 12,\ f(x) = 91.75\\ x &= 11,\ f(x) = -34.0\\ x &= 11.5,\ f(x) = 9.0\\ x &= 11.4,\ f(x) = 0.356\\ x &= 11.4 \text{ is one of the roots}\end{aligned}$$

$\therefore \quad (x - 11.4)(x^2 - 3.6x + 2.71) = 0$

$(x - 11.4)(x - 2.5)(x - 1.1) = 0$

$\therefore x_1 = 11.4;\ x_2 = 2.5;\ x_3 = 1.1$

The solutions are

$\therefore \sigma_1 = 114.0$ MPa; $\sigma_2 = 25$ MPa; $\sigma_1 = 11$ MPa

Maximum normal stress,

$\therefore \quad \sigma_{\max} = 114.0$ MPa

Factory safety by normal stress theory,

$N = 500/114 = 4.386$

Maximum shear stress-theory,

$$\tau_{\max} = \frac{114 - 11}{2} = 51.5 \text{ MPa}$$

Yield stress in shear $S_{sy} = S_y/2 = 250.0$ MPa

Factor of safety by maximum shear theory, $N = 250/(51.5) = 4.85$

Von-Mises stress,

$$\begin{aligned}\sigma^1 &= \sqrt{\frac{(\sigma_1 - \sigma_2)^2 + (\sigma_2 - \sigma_3)^2 + (\sigma_3 - \sigma_1)^2}{2}}\\ &= \sqrt{\frac{(114 - 25)^2 + (25 - 11)^2 + (11 - 114)^2}{2}}\\ &= \sqrt{\frac{7921 + 1961 + 10609}{2}} = 96.76 \text{ MPa}\end{aligned}$$

Factor of safety by distortion energy theory,

$N = (500/96.76) = 5.16$

b) $\sigma_x = 100$ MPa; $\sigma_y = -25$ MPa, $\sigma_z = -75$ MPa

$\tau_{xy} = 25$ MPa; $\tau_{yz} = 50$ MPa; $\tau_{xz} = 75$ MPa

$$C_1 = \sigma_x + \sigma_y + \sigma_z = 100 - 25 - 75 = 0$$

$$C_2 = \begin{vmatrix} 100 & 25 \\ 25 & -25 \end{vmatrix} + \begin{vmatrix} -25 & 50 \\ 50 & -75 \end{vmatrix} + \begin{vmatrix} 100 & 75 \\ 75 & -75 \end{vmatrix}$$

$$= -3125 - 625 - 13125 = \mathbf{-16875}$$

$$C_3 = \begin{vmatrix} 100 & 25 & 75 \\ 25 & -25 & 50 \\ 75 & 50 & -75 \end{vmatrix}$$

$$= 100\begin{vmatrix} -25 & 50 \\ 50 & -75 \end{vmatrix} - 25\begin{vmatrix} 25 & 50 \\ 75 & -75 \end{vmatrix} + 75\begin{vmatrix} 25 & -25 \\ 75 & 50 \end{vmatrix} = -62,500 + 1,40,625 + 2,34,375$$

$$= \mathbf{3,12,500}$$

Then equation for principal stresses is

$$\therefore \quad \sigma^3 - 0x\sigma^2 - 16875\sigma - 312500 = 0$$

put $\sigma = 10x$

$$\therefore \quad x^3 - 168.750x - 312.5 = 0$$

Try $x = 10$, $f(x) = -1000$

$x = 15$, $f(x) = 531.25$

$x = 12$, $f(x) = -609.5$

$x = 13$, $f(x) = -309.25$

$x = 14$, $f(x) = 69$

$x = 13.8$, $f(x) = -13.18$

$x = 13.9$, $f(x) = 27.5$

$\therefore x = 13.833$ is the root

$$\therefore \quad (x - 13.83)(x^2 + 13.83x + 22.5) = 0$$

$$(x - 13.83)(x + 11.945)(x + 1.883) = 0$$

$x_1 = 13.83$; $x_2 = -11.945$; $x_3 = -1.883$

$\therefore \quad \sigma_1 = 138.3$ MPa; $\sigma_2 = -119.45$ MPa; $\sigma_3 = -18.83$ MPa

Maximum normal stress,

$\sigma_{max} - 138.3$ MPa

Factor of safety by normal stress theory,

$$N = \frac{500}{138.3} = 3.6$$

Maximum shear stress,

$$\tau_{\max} = \frac{138.3 - (-119.45)}{2} = 128.875 \text{ MPa}$$

Factor of safety by maximum shear stress theory,

$$N = \frac{500/2}{128.875} = 1.94$$

Von-Mises stress,

$$\begin{aligned}\sigma^1 &= \sqrt{\frac{(138.3+18.83)^2+(119.45-18.83)^2+(138.3+119.45)^2}{2}}\\ &= \sqrt{\frac{24689.8+10124.4+66435}{2}}\\ &= 225 \text{ MPa}\end{aligned}$$

Factor of safety by maximum distortion energy theory,

$$N = \frac{500}{225} = 2.22$$

(c) $\sigma_x = -50$ MPa; $\sigma_y = 75$ MPa, $\sigma_z = 25.0$ MPa

$\tau_{xy} = 10.0$ MPa; $\tau_{yz} = 25$ MPa; $\tau_{xz} = 100$ MPa

Stress invariants,

$$\begin{aligned}C_1 &= \sigma_x + \sigma_y + \sigma_z = -50+75+25 = 50\\ C_2 &= \begin{vmatrix}-50 & 10\\ 10 & 75\end{vmatrix} + \begin{vmatrix}75 & 25\\ 25 & 25\end{vmatrix} + \begin{vmatrix}-50 & 100\\ 100 & 25\end{vmatrix}\\ &= -3850+1250-11250 = \mathbf{-13850}\\ C_3 &= \begin{vmatrix}-50 & 10 & 100\\ 10 & 75 & 25\\ 100 & 25 & 25\end{vmatrix}\\ &= -50\begin{vmatrix}75 & 25\\ 25 & 25\end{vmatrix} - 10\begin{vmatrix}10 & 25\\ 100 & 25\end{vmatrix} + 100\begin{vmatrix}10 & 75\\ 100 & 25\end{vmatrix}\\ &= -62500+22500-72500 = \mathbf{-765000}\end{aligned}$$

Then the equation for principal stresses is

$\sigma^3 - 50\sigma^2 - 13850\sigma - 765000 = 0$

put $\sigma = 10x$,

$\therefore \quad x^3 - 5x^2 - 138.5x + 765 = 0$

$$
\begin{aligned}
\text{Try } x &= 10,\ f(x) = -120 \\
\text{Try } x &= 15,\ f(x) = +937.5 \\
\text{Try } x &= 11,\ f(x) = -32.5 \\
\text{Try } x &= 11.5,\ f(x) = 31.875 \\
\text{Try } x &= 11.25,\ f(x) = -2.11
\end{aligned}
$$

$\therefore$ One of the roots is $x = 11.26$

$$
\begin{aligned}
\therefore \quad (x - 11.26)(x^2 + 6.26x - 68.00) &= 0 \\
\therefore \quad (x - 11.26)(x - 5.69)(x + 11.95) &= 0
\end{aligned}
$$

$\therefore \quad x_1 = 11.26;\ x_2 = 5.69,\ x_3 = -11.95$

$\therefore \quad \sigma_1 = 112.6\ \text{MPa};\ \sigma_2 = 56.9\ \text{MPa},\ \sigma_3 = -119.5\ \text{MPa}$

Maximum normal stress.

$\sigma_{max} = 119.5\ \text{MPa (Compressive)}$

$\therefore$ Factor of safety by normal stress theory,

$N = 500/119.5 = 4.184$

Maximum shear stress,

$$\tau_{max} = \frac{112.6 - (-119.5)}{2} = 116.05\ \text{MPa}$$

$\therefore$ Factor of safety by stress theory, $N = 250/116.05 = 2.15$

Von-Mises stress,

$$
\begin{aligned}
\sigma_1{}^1 &= \sqrt{\frac{(112.6 - 56.9s)^2 + (56.9 + 119.5)^2 + (112.6 + 119.5)^2}{2}} \\
&= \sqrt{\frac{3102.5 + 31116.96 + 53870.41}{2}} = 209.86
\end{aligned}
$$

$\therefore$ Factor of safety by distortion energy theory, $N = (500/209.86) = 2.382$

Example 4.7: Tests on a particular melt of cast iron gave $S_{ut} = 150$ MPa and $S_{uc} = 600$ MPa. Find the factor of safety for each of the three failure theories for brittle materials for the following stress states:

a) $\sigma_x = 50$ MPa; $\tau_{xy} = 30$ MPa, cw
b) $\sigma_x = -80$ MPa; $\sigma_y = -40$ MPa; $\tau_{xy} = 20$ MPa, ccw
c) $\sigma_x = 40$ MPa; $\sigma_y = 30$ MPa; $\tau_{xy} = 10$ MPa
d) $\sigma_x = 30$ MPa; $\sigma_y = -60$ MPa; $\tau_{xy} = 30$ MPa cw

SOLUTION
Graphical Representation:

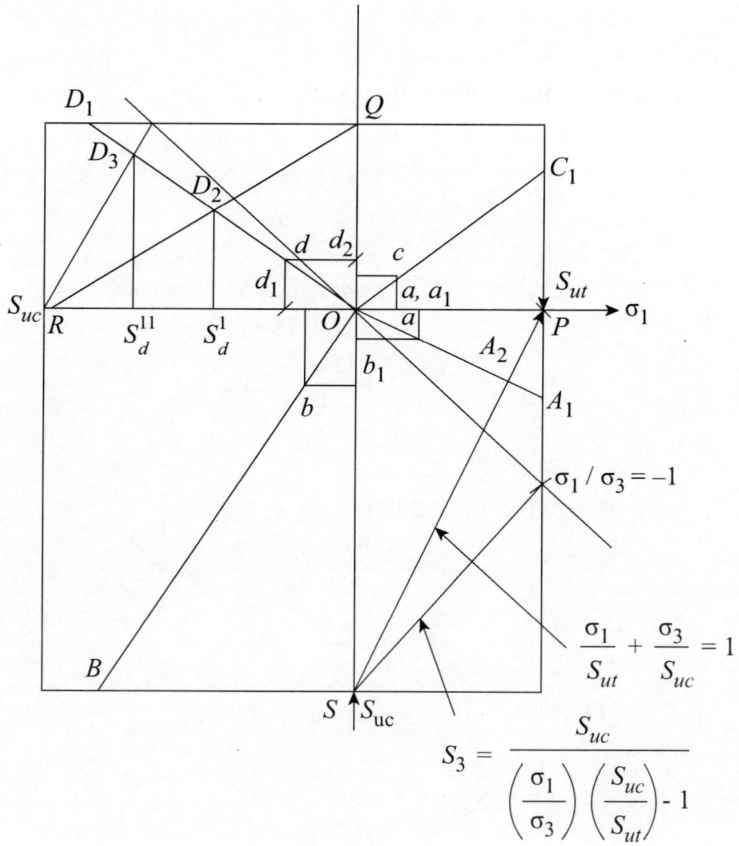

Figure 4.33

a) $\sigma_x = 50$ MPa; $\tau_{xy} = 30$ MPa, cw

Principal stresses are

$$\sigma_1, \sigma_2 = \frac{50}{2} \pm \sqrt{\left(\frac{50}{2}\right)^2 + 30^2} = 25 \pm 39.0$$

$\sigma_1 = 64.0$ MPa; $\sigma_2 = -14.0$ MPa
On the graph
$a(64, -14)$ is the state of stress.

(i) Factor of safety by normal stress theory and Mohr's theory

$$N = \frac{OA_1}{Oa} = \frac{OP}{Oa_1} = \frac{150}{64} = 2.4$$

(ii) Factor of safety by Columb - Mohr theory

$$N = \frac{OA_2}{Oa} = \frac{S_3}{\sigma_2}$$

$$\text{Where } S_3 = \frac{600}{\frac{64}{-14}\left(\frac{600}{150}\right) - 1} = -31.1$$

$$\therefore N = \frac{-31.1}{-14} = 2.22$$

b) $\sigma_x = -80$ MPa; $\sigma_y = -40$ MPa; $\tau_{xy} = 20$ MPa, ccw

Principal stresses,

$$\sigma_1, \sigma_2 = \frac{-80 - 40}{2} \pm \sqrt{\left(\frac{40 - 80}{2}\right)^2 + 20^2} = -60 \pm 28.28$$

$\sigma_1 = -31.72$ MPa; $\sigma_2 = -88.28$ MPa,
on graph **b is the point.**

The point lies in the third quadrant Factor of safety by all theories,

$$N = \frac{OB}{Ob} = \frac{OS}{Ob_1} = \frac{-600}{-88.28} = 6.8$$

c) $\sigma_x = 40$ MPa; $\sigma_y = 30$ MPa; $\tau_{xy} = 10$ MPa, ccw

Principal stresses,

$$\sigma_1, \sigma_2 = \left(\frac{\sigma_x + \sigma_y}{2}\right) \pm \sqrt{\left(\frac{\sigma_x - \sigma_y}{2}\right)^2 + \tau_{xy}^2}$$

$$\frac{(40+30)}{2} \pm \sqrt{\left(\frac{40-30}{2}\right)^2 + 10^2} = 35 \pm 14.$$

$\therefore \quad \sigma_1 = 49.14$ MPa; $\sigma_2 = 20.86$ MPa,
on graph **c is the point.**

The point lies in the first quadrant. Factor of safety by all theories,

$$N = \frac{OC_1}{OC} = \frac{OP}{OC_1} = \frac{150}{49.14} = 3.05$$

d) $\sigma_x = 30$ MPa; $\sigma_y = -60$ MPa; $\tau_{xy} = 30$ MPa cw

Principal stresses,

$$\sigma_1, \sigma_2 = \frac{30-60}{2} \pm \sqrt{\left(\frac{30+60}{2}\right)^2 + 30^2}$$

$$= -15 \pm 54.08$$

$\therefore \quad \sigma_2 = 39.08$ MPa; $\sigma_1 = -69.08$ MPa,
on graph **d is the point.**

Factor of safety by normal stress theory,

$$N = \frac{OD_1}{Od} = \frac{OQ}{Od_2} = \frac{S_{ut}}{\sigma_2} = \frac{150}{39.08} = 3.84$$

By Coulomb - Mohr Theory,

$$s_d^1 = \frac{S_{uc}}{\frac{S_{uc}}{S_{ut}} \cdot \frac{\sigma_2}{\sigma_1} - 1}$$

$$= \frac{600}{\left(\frac{600}{150}\right)\left(\frac{39.08}{-69.08}\right) - 1} = 183.88 \text{ MPa}$$

$$\text{Factor of safety} = \frac{s_d^1}{\sigma_1} = \frac{-183.88}{-69.08} = 2.66$$

By Modified Mohr's Theory,

$$\text{Factor of safety} = \frac{S_d^{11}}{\sigma_1}$$

$$\text{where } S_d^{11} = \frac{S_{uc}}{\frac{S_{uc} - S_{ut}}{S_{ut}} \cdot \frac{\sigma_2}{\sigma_1} - 1}$$

$$= \frac{600}{\left(\frac{600 - 150}{150}\right)\left(\frac{39.08}{-69.08}\right) - 1} = -222.46 \text{ MPa}$$

$$\text{Factor of safety} = \frac{S_d^{11}}{\sigma_1} = \frac{-222.46}{-69.08} = 3.22$$

Example 4.8: A round stepped shaft is made of brittle material cast-iron FG260 and subjected to a bending moment of 15 Nm as shown in Fig. 4.34. The stress concentration factor at the fillet is 1.5. Determine the following:

(i) Step diameter
(ii) Magnitude of stress at fillet
(iii) Factor of Safety

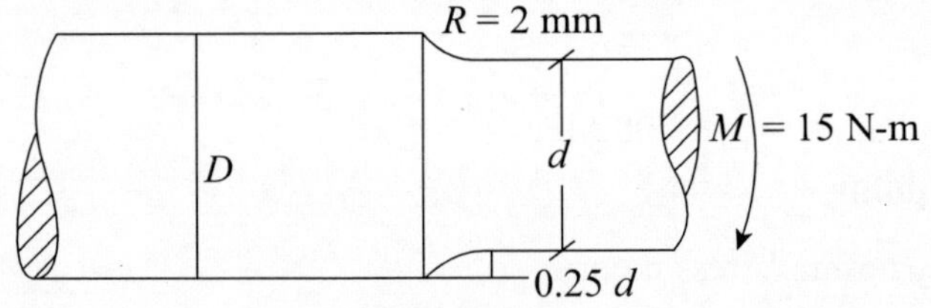

Figure 4.34

(i) Step diameter,

$$D = d + 0.25d + 0.25d$$
$$= 1.5d$$

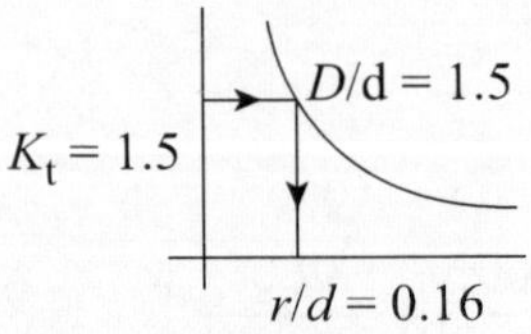

For $\frac{D}{d} = 1.5$ & $K_t = 1.5$ from MDHB Fig 3.6,

$r/d = 0.16 \quad d = \frac{r}{0.16} = \frac{2}{0.16} = 12.5$ mm

(ii) Magnitude of stress at the fillet bending stress, $\sigma = \dfrac{M}{Z} = \dfrac{15 \times 10^3}{\dfrac{\pi}{32}(12.5)^3} = 78.28$ MPa

The stress at the fillet,

$\sigma_{max} = K_t \sigma = 1.5 \times 78.28 = 117.34$ MPa

(iii) Material of the shaft is FG260

$S_u = 260$ MPa

$\therefore$ Factor of safety $= \dfrac{S_u}{\sigma_{max}} = \dfrac{260}{117.34} = 2.216$

Example 4.9: A round rod of diameter 60 mm is subjected to an axial tensile load of 10 kN and a twisting moment of 3 kN-m. The rod is made of steel *C30*. Factor of safety is 3. Determine whether the design is safe according to

(i) Maximum principal stress theory of failure

(ii) Maximum shear stress theory

Given Data

Rod diameter, $d = 60$ mm
Axial force, $F_a = 10$ kN (Tensile)
Twisting Moment, $T = 3$ kN-m
Material, $C30$
Factor of safety, $N = 3$.

SOLUTION

For material C30 (Table 2.6 MDHB)

$S_y = 300$ MPa

Allowable stress in tension,

$$S_d = S_y/N = \frac{300}{3} = 100 \text{ MPa}$$

Allowable stress in tension,

$$S_{sd} = \frac{1}{2} S_d = 50 \text{ MPa}$$

The stresses induced,

Axial tensile stress, $\sigma_t = \dfrac{F}{A} = \dfrac{10 \times 10^3}{\dfrac{\pi}{4}(60)^2} = 3.537$ MPa

Shear stress due to torsion,

$$\tau = \frac{16 \cdot T}{\pi d^3} = \frac{16 \times 3 \times 10^6}{\pi \times (60)^3} = 70.7355 \text{ MPa}$$

The principal stresses are

$$\begin{aligned}\sigma_1, \sigma_2 &= \left(\frac{3.537}{2}\right) \pm \sqrt{\left(\frac{3.537}{2}\right)^2 + (70.7355)^2}\\ &= 1.7685 \pm \sqrt{1.7685^2 + 70.7355^2}\\ &= 1.7685 \pm 70.76\\ \sigma_1 &= 72.526 \text{ MPa}\\ \sigma_2 &= -68.989 \text{ MPa}\end{aligned}$$

$$\text{Maximum shear stress, } \tau_{max} = \left(\frac{\sigma_1 - \sigma_2}{2}\right) = \frac{72.526 + 68.989}{2} = 70.76 \text{ MPa}$$

(i) As per maximum principal stress theory,

$\sigma_{max} = 72.526 < S_d \quad \therefore$ Safe

(ii) As per maximum shear theory,

$\tau_{max} = 70.76 \text{ MPa} > (S_{sd} = 50 \text{ MPa}) \quad \therefore$ Unsafe

Example 4.10: Determine the maximum stress induced in the following cases taking stress concentration into account.

(i) A rectangular plate 50 mm wide; and 8 mm thick with a central hole of 10 mm is loaded with axial tension of 14.7 kN.

(ii) A stepped shaft stepped down from 45 mm to 30 mm with fillet radius of 6 mm is subjected to twisting moment of 98 Nm.

SOLUTION

(i) Rectangular plate

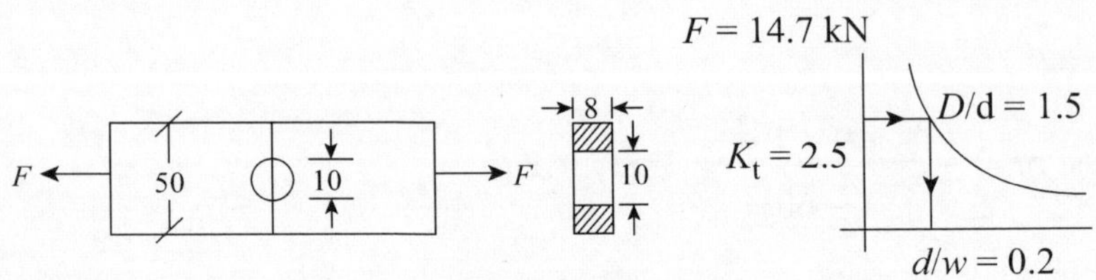

Figure 4.35

The stress concentration factor from Fig 3.17 MDHB for $d/w = 10/50 = 0.2$, is

$K_t = 2.5$

The maximum stress,

$$\sigma_{\max} = K_t \cdot \frac{F}{A} = 2.5 \times \frac{14.7 \times 10^3}{8 \times (50 - 10)} = 114.84 \text{ MPa}$$

(ii) Stepped shaft under twisting moment, $T = 98$ N-m.

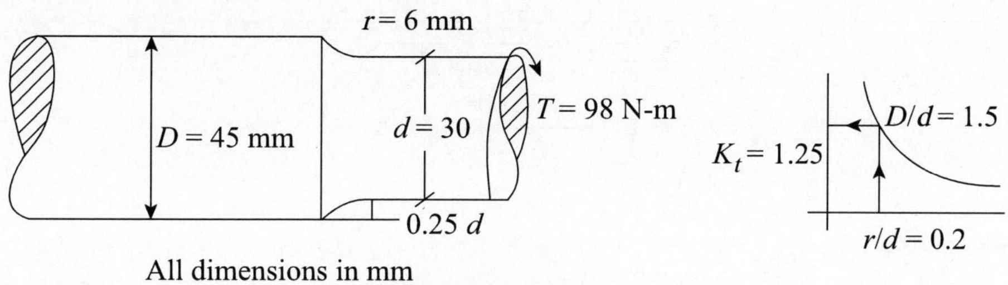

Figure 4.36

The stress concentration factor due to step for $D/d = 1.5$ & $\frac{r}{d} = 0.2$

$K_t = 1.25$ (Fig. 3.7 MDHB)

The maximum shear stress,

$$\tau_{\max} = K_t \frac{T}{Z_p} = \frac{1.25 \times 98 \times 10^3}{\frac{\pi}{16}(30)^3} = 23.107 \text{ MPa}$$

Example 4.11: A steel shaft subjected to a bending moment of 9 kNm and twisting moment of 12 kNm yield strength $S_y = 360$ MPa and the Poisson's ratio is 0.3 with a factor of safety of 2 determine the shaft diameter according to

(i) Maximum shear theory

(ii) Maximum normal stress theory

(iii) Maximum distortion energy theory

Given Data

Bending moment, $M_b = 9$ kN-m $= 9 \times 10^6$ N-mm
Twisting moment, $T = 12$ kN-m $= 12 \times 10^6$ N-mm
Yield strength, $S_y = 360$ MPa

Poisson's ratio = 0.3
Factor of safety, $N = 2$

SOLUTION

The design stress in tension,

$$S_d = S_y/N = \frac{360}{2} = 180 \text{ MPa}$$

(i) The bending stress,

$$\sigma = \frac{32M_b}{\pi d^3} = \frac{32 \times 9 \times 10^6}{\pi \times d^3}$$
$$= \left(\frac{91.67 \times 10^6}{d^3}\right) \text{MPa}$$

(ii) The shear stress,

$$\tau = \frac{16T}{\pi d^3} = \frac{16 \times 12 \times 10^6}{\pi \times d^3}$$
$$= \left(\frac{61.115 \times 10^6}{d^3}\right) \text{MPa}$$

The maximum principal stress,

$$\sigma_{max} = \left(\frac{\sigma}{2}\right) + \sqrt{\left(\frac{\sigma}{2}\right)^2 + \tau^2}$$

$$\sigma_{max} = \frac{91.67 \times 10^6}{2d^3} + \sqrt{\left(\frac{91.67 \times 10^6}{d^3}\right)^2 + \left(\frac{61.115 \times 10^6}{d^3}\right)^2}$$

$$= \frac{45.835 \times 10^6 \times 2.667}{d^3}$$

$$= \left(\frac{122.24 \times 10^6}{d^3}\right)$$

But, $\sigma_{max} = S_d = 180$ MPa

$$\therefore \quad d^3 = \frac{122.24 \times 10^6}{180} = 679.12 \times 10^3$$

$\therefore \quad d = 87.89$ mm

Choose $d = 90$ mm

(iii) Maximum shear stress theory

$$S_d = \sqrt{\sigma^2 + 4\tau^2}$$

$$180 = \sqrt{\left(\frac{91.67 \times 10^6}{d^3}\right)^2 + 4\left(\frac{61.115 \times 10^6}{d^3}\right)^2}$$

$$d = \left(\frac{91.67 \times 10^6 \times 1.667}{180}\right)^{\frac{1}{3}}$$

$$= 94.7 \text{ mm}$$

$$\text{Choose } d = 95 \text{ mm}$$

(iv) By distortion energy theory

$$S_d = \sqrt{\sigma^2 + 3\tau^2}$$

$$180 = \sqrt{\left(\frac{91.67 \times 10^6}{d^3}\right)^2 + 3\left(\frac{61.115 \times 10^6}{d^3}\right)^2}$$

$$= \left(\frac{91.67 \times 10^6}{d^3}\right)\sqrt{1 + 3\left(\frac{61.115}{91.67}\right)^2}$$

$$d = 92.07 \text{ mm} = 92.0 \text{ mm}$$

EXERCISES FOR PRACTICE

4.1. A hollow annealed steel propeller shaft has an external diameter of 340 mm and an internal diameter of 170 mm; it transmits 7500 kW at 200 rpm. (a) Compute the torsional stress in the shaft (stress from bending and propeller thrust are not considered). (b) Compute the factor of safety. Does it look risky?

4.2. Determine the maximum normal stress and maximum shear stress at section A-A for the crank shown in Fig. 4.37. A load of 10 kN is applied at the centre of the crank pin. Neglect the effect of transverse shear.

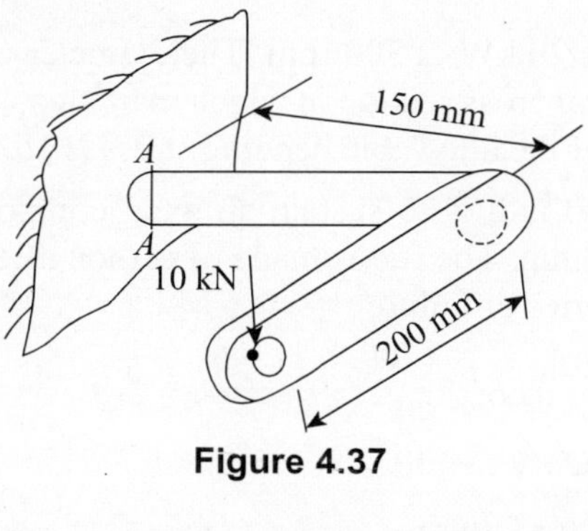

Figure 4.37

4.3. A C–Clamp carries a load P = 20000 N. The cross section of the clamp shown in figure below at x-x is rectangular having width equal to twice thickness. Assuming that the clamp is made of steel casting with an allowable stress of 120 MPa, find its dimensions. Also determine the stress at section zz.

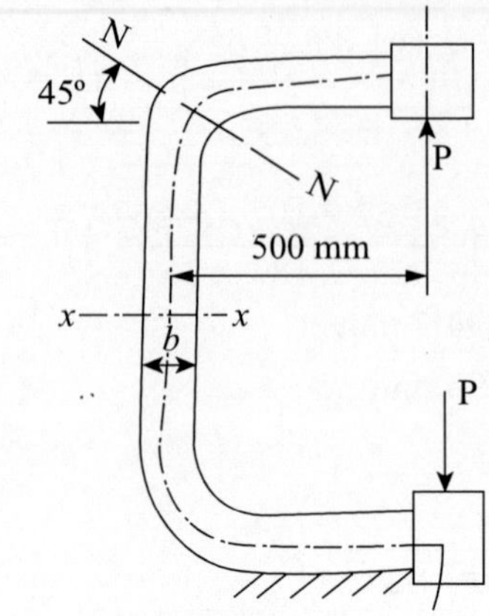

Figure 4.38

4.4. A wall bracket as shown in Fig. 4.39 is subjected to a pull of 7.5 KN at 60° to the vertical. The cross section of the bracket is rectangular having $b = 4t$. Determine the dimensions 'b' and't', if the tensile stress in the material of the bracket is limited 50 MPa.

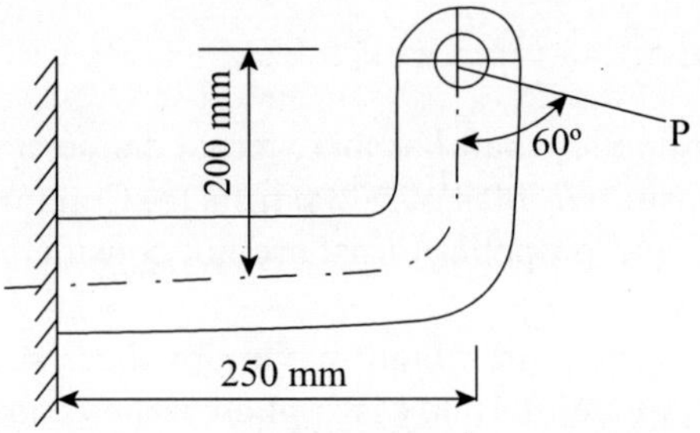

Figure 4.39

4.5. A cast iron pulley transmits 20 kW at 300 rpm. The diameter of pulley is 550 mm and has four straight arms of elliptical cross section in which major axis is twice the minor axis. Find the dimensions of the arm, if the allowable bending stress is 15 MPa.

4.6. A round rod of diameter 30 mm is to sustain an axial compressive load of 20 kN and a twisting moment of 150 kNmm. The rod is made of carbon steel C40. Determine factors of safety as per following theories of failure

(i) Maximum principal stress theory

(ii) Maximum shear stress theory

(iii) Distortion energy theory for failure

4.7. A shaft is designed based on maximum distortion energy theory with a factor of safety of 2. The material used is 30C8 steel with a yield stress of 310 MPa. It is subjected to an axial load of 40 kN. Determine the maximum torque capacity. Diameter of the shaft is 20 mm.

4.8. A mild steel shaft of 60 mm diameter is subjected to a bending moment of 3000 Nm and a torque T. If the yield point of steel in tension is 200 MPa, find the maximum value of this torque without yielding of the shaft according to: i) maximum principal stress; ii) maximum shear stress; and iii) maximum distortion energy theories of failure.

4.9. A rotating shaft carries a 18 kN pulley at the center of a 0.75 m simply supported span. The average torque is 230 Nm. Assume the torque range to be 10 % of the average torque. The material has yield point of 770 MPa and the endurance limit of 450 MPa. Determine the required diameter of the shaft based on: (i) maximum stress theory; and (ii) distortion energy theory. Factor of safety may be taken as 2.

4.10. A machine part is statically loaded and has yield strength of 350 MPa. For the following stresses calculate the factor of safety using the following theories of failure:

(i) Maximum normal stress theory.

(ii) Maximum shear stress theory

(iii) Von Mises-Hanky theory

(a) $\sigma_1 = -70$ MPa, $\sigma_2 = 0$ MPa
(b) $\sigma_1 = 70$ MPa, $\sigma_2 = -70$ MPa
(c) $\sigma_1 = 70$ MPa, $\sigma_2 = 70$ MPa

4.11. A stepped shaft has maximum diameter 45 mm and minimum diameter 30 mm. The fillet radius is 6 mm. If the shaft is subjected to (i) an axial load of 10 kN, (ii) twisting moment of 0.25 kNm find the maximum stress induced, taking stress concentration into account.

4.12. A flat bar shown in Fig. 4.40 is subjected to an axial load of F equal to 50 kN. Assuming that the stress in the bar is limited to 200 N/mm^2. Considering the effect of stress concentration, determine the thickness of flat bar.

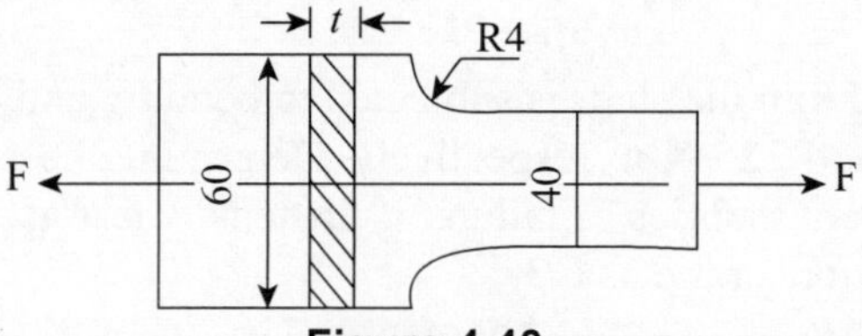

Figure 4.40

4.13. A weight of 1400 N is dropped onto a collar at the lower end of a vertical steel shaft of 3 m long and 25 mm diameter, calculate the height of drop if the maximum instantaneous stress produced is not to exceed 120 MPa. Take $E = 2.1 \times 10^5$ MPa.

4.14. A weight of 1 kN is dropped from a height of 50 mm at the free end of a cantilever beam of effective length 300 mm. Determine the cross section of the cantilever beam of square cross section if the allowable stress in the material of the beam is limited to 80 MPa.

4.15. A rectangular plate 15 mm thick made of brittle material is shown in Fig. 4.41. Calculate the stresses at each of three holes.

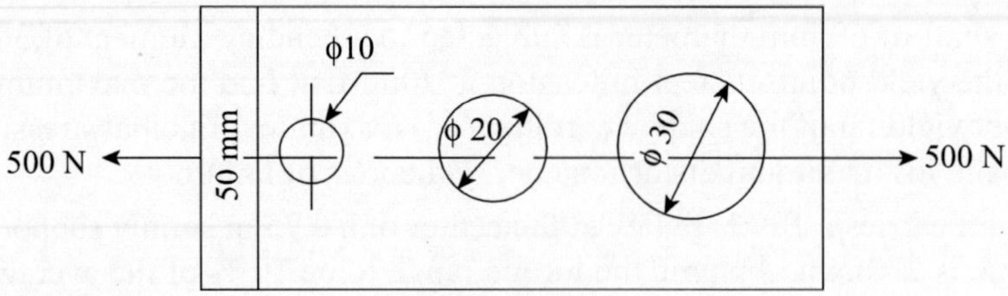

Figure 4.41

4.16. A cantilever beam of span 800 mm has a rectangular cross section of depth 200 mm. The free end of the beam is subjected to a transverse load of 1 KN, dropped onto it from a height of 40 mm. Selecting a suitable material and assuming an appropriate value for the factor of safety, determine the width of the rectangular cross section.

4.17. A plate of C45 steel is subjected to the following stresses:
$\sigma_x = 25$ MPa, $\sigma_y = 20.0$ MPa, and $\tau_{xy} = 0$
Find the factor of safety by,

i) Maximum shear theory

ii) Maximum principal stress theory

iii) Mises-Henky theory.

4.18. a) What is factor of safety? Why is it used in design ?
b) The load on a C-clamp is 25 kN. Design its frame. The frame is made of steel casting having an allowable stress of 90 MPa. Assume width equals to half the length of the cross section and eccentricity of 150 mm.

4.19. At a critical point in a machine member made of ductile material, the principal stresses are 50 MPa, 40 MPa. Find the safe tensile stress for the member according to each of the three theories of failure.

4.20. A critical section in a 40 mm dia shaft is subjected to bending and twisting moments simultaneously of 350 N-m and 225 N-m respectively. Determine the factor safety according to each of the three important theories of failure, if the tensile test gives a proportional limit of 285 MPa. Assume Poission's ratio as 0.3.

4.21. A shaft is loaded by twisting moment of 200 N-m and a bending moment of 100 N-m. The shaft is made from rolled steel with an allowable shear stress of 100 MPa and a tensile stress of 180 MPa. Determine the diameter of the shaft using the maximum shear stress theory of failure.

4.22. A 50 mm diameter shaft is subjected to a bending moment of 400 N-m and a twisting moment of 350 N-m. The yield stress for the material of the shaft is 285 MPa and Poisson's ratio is 0.3. Determine the factor of safety according to the three major theories of failure.

4.23. A 45 mm dia steel shaft with 10 mm dia transverse hole is subjected a combined loading consisting of bending stress of +400 MPa and an alternating torsional stress of 210 MPa. They are in phase. Find the factor of safety for infinite life, if the ultimate tensile stress for the shaft material is 1.2 GPa.

4.24. A bolt is subjected to a force with an axial tensile component of 15 kN and a transverse shear force of 10 kN. The bolt material is steel with yield strength of 350 MPa. The factor of safety is to be 2.0. Determine the diameter of the bolt using the three theories of failure.

4.25. A 5 mm thick flat bar has its width reduced fromm 35 mm to 30 mm with a fillet radius of 6 mm determine the maximum stress induced when

i) it is subjected to tensile load of 15 kN,

ii) it is subjected to a bending moment of 5 kN-m

SHORT ANSWER QUESTIONS

4.1. What is yield strength?

4.2. What is ultimate strength?

4.3. Draw stress-strain diagrams for ductile and brittle materials and show the important points.

4.4. What is factor of safety?

4.5. What are the factors that influence the choice of factor of safety?

4.6. What are the modes of failure of machine components?

4.7. Which are the theories of elastic-failure?

4.8. What is the allowable stress?

4.9. State the maximum principal stress theory.

4.10. State the maximum shear stress theory.

4.11. When do you use the maximum shear stress theory?

4.12. State the maximum distortion energy theory.

4.13. Compare the distortion energy theory with other theories.

4.14. When do you use the distortion energy theory?

4.15. Name the failure theories for uneven brittle materials.

4.16. State the Coulomb-Mohr theory.

4.17. Explain Modified Mohr theory.

4.18. What is Dowling effective stress?

4.19. What is the factor of safety by Modified Mohr theory?

4.20. Define stress concentration factor.

4.21. State the reasons for stress concentration.

4.22. How do you determine the stress concentration factor?

4.23. How do you reduce the stress concentration?

4.24. When is the stress concentration more dangerous?

4.25. Why the stress concentration factor is not considered for elastic materials?

CHAPTER 5

DESIGN FOR FATIGUE STRENGTH

For evaluating material strength properties by tension test or torsion test, the load is applied gradually providing sufficient time for the strains to develop in phase. The specimen elongates gradually, yields and breaks eventually. This is called the static loading condition. The strength values obtained by these tests, will not hold good for variable loads. Majority of the machine members are rotating and are subjected to variable stresses, even though the load apparently remains steady. Such members are found to fail at stresses much less than those predicted by tensile test. The physical examination of the failed parts, exhibits two regions of failure.

(i) smooth surface due to the progressive crack propagation, and

(ii) a 'rough surface' due to the sudden brittle fracture as shown in Fig. 5.1(a) & 5.1(b). The failure is called fatigue failure.

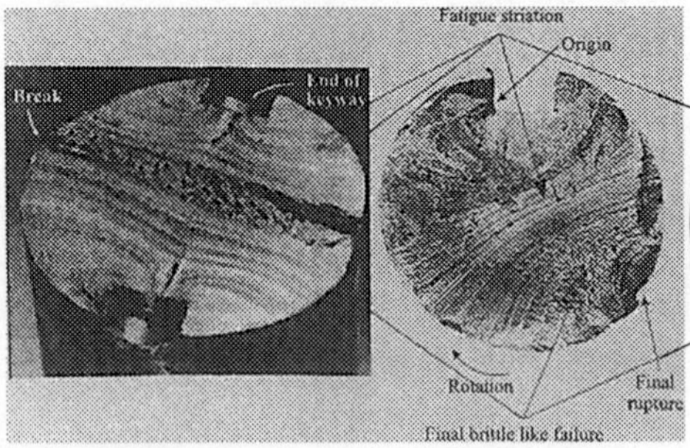

Figure 5.1(a). Typical fatigue failure of a ductile material

5.1 THE MECHANISM OF FATIGUE FAILURE

The examination of failed parts, shown in Fig. 5.1(a) & Fig. 5.1(b), shows that the fatigue failure starts at a crack. The crack may be present due to the manufacturing process. Or may be developed due to the cyclic loading. The micro-cracks due to cyclic-loading, coalesce to form a bigger crack.

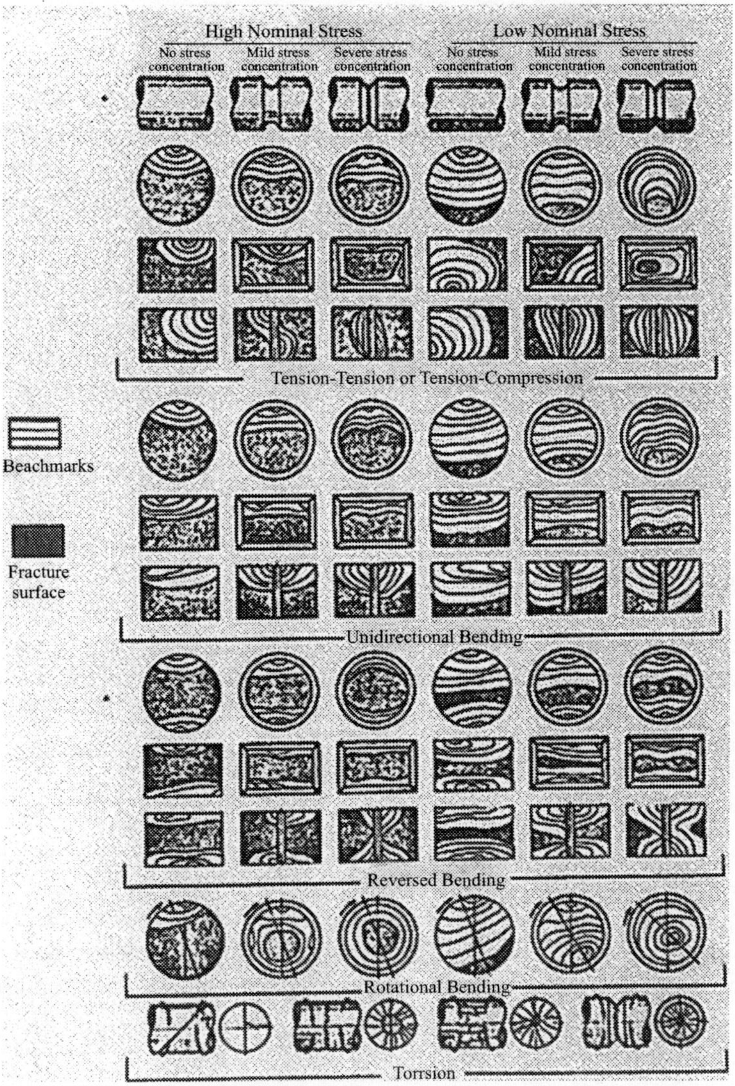

Figure 5.1(b). Schematic representations of fatigue-fracture surfaces of various smooth and notched cross sections under various loading conditions and stress levels (From Metals Handbook, 1975, Am. Soc. for Metals, Metals Park, Ohio, Vol.10, 8th ed., p.102, with permission)

This crack spreads gradually by a small amount in each cycle. The successive growth at every cycle of variation leads to critical size and finally the part fails.

Thus, there are three stages of fatigue failure:

(i) Crack initiation
(ii) Progressive crack propagation
(iii) Fracture.

Crack Initiation

The discontinuities, notches, voids and inclusions serve as – stress risers and cause stress concentration. The local stress at such points reach yield stress even though the average stress is much less, causing local yielding. Under fatigue loading, regions of intense deformation are created due to the shear motion along the grain boundaries and creation of slip bands. The bands coalesce and give rise to a microscopic crack. In most of the cases the crack begins at the free surface and grows at a very small speed.

Crack propagation

The crack-tip so formed increases the stress concentration. The crack opens when the stresses become tensile and becomes blunt. When the stresses decrease and become compressive, the crack tip closes and becomes sharp. The stress concentration increases and crack opens when the stress becomes tensile again. Thus, at every cycle of stress variation, the crack spreads by a small amount. A striation is formed for every cycle of stress variation. The progressive propagation of the crack, leaves a series of striations on the cracked surface. The surface looks smooth to the bare eyes.

Fracture

As the crack continues to grow under fatigue loading, it attains critical size when the crack tip stress is too large and the stress in the remaining portion also becomes large, leading to sudden fracture of the remaining section. The fractured surface is rough.

This kind of failure is called fatigue failure. This is quite dangerous because it occurs without warning. Hence, great care is to be taken during the design of the components subjected to the fluctuating stresses.

5.2 FATIGUE LOADS

Any load that varies with time can potentially cause fatigue failure. The character of variation may vary from one application to another. In rotating machinery, the load variation is a regular in amplitude and frequency with time. In service equipment, the load variation can be quite random with variable amplitude and frequency. Any load variation can be represented by a combination of sinusoidal functions for analysis purposes. Depending upon the sinusoidal variations, they can be classified as

(i) Reversed
(ii) Repeated
(iii) Fluctuating.

The reversed loads vary with equal extreme values in tension and compression (Fig. 5.2a).
The repeated loads vary with from zero to an extreme value in tension (Fig. 5.2b).
The extreme values in fluctuating loads may be both tensile or tensile and compressive (Fig. 5.2c).

Any of these waveforms can be characterized by the parameters like maximum and minimum values, mean and alternating (amplitude) components or the ratio of the extreme values. They are defined as follows:

$$\sigma_{max} = \text{Maximum value of stress}$$
$$\sigma_{min} = \text{Minimum value of stress}$$

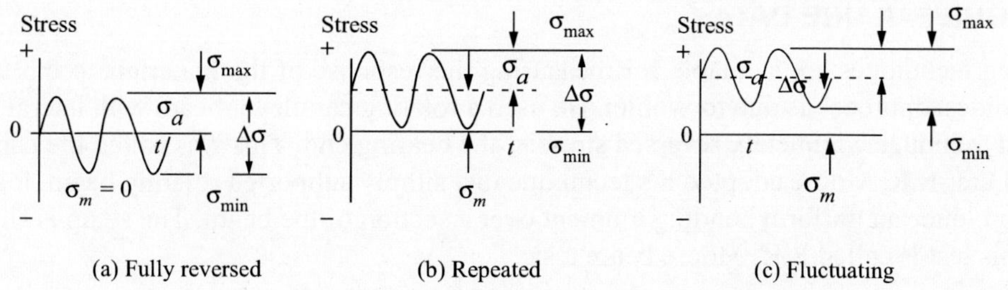

Figure 5.2. Alternating, mean, and range values for fully reversed, repeated, and fluctuating cyclic stresses

Stress variation range,

$$\Delta\sigma = (\sigma_{\max} - \sigma_{\min})$$

Stress amplitude,

$$\sigma_a = \frac{\Delta\sigma}{2} = \left(\frac{\sigma_{\max} - \sigma_{\min}}{2}\right) \tag{5.1}$$

Mean stress,

$$\sigma_m = \left(\frac{\sigma_{\max} + \sigma_{\min}}{2}\right) \tag{5.2}$$

Stress ratio,

$$R = \frac{\sigma_{\min}}{\sigma_{\max}}. \tag{5.3}$$

Amplitude ratio,

$$A = \left(\frac{\sigma_a}{\sigma_m}\right) \tag{5.4}$$

For reversed loads,

$$\begin{aligned} \text{Mean stress, } \sigma_m &= 0 \\ \text{Stress ratio, } R &= -1 \\ \text{Amplitude ratio, } A &= \infty \end{aligned}$$

For repeated loads,

$$\begin{aligned} \text{Minimum stress, } \sigma_{\min} &= 0 \\ \text{Stress ratio, } R &= 0 \\ \text{Amplitude ratio, } A &= 1 \end{aligned}$$

For fluctuating loads

The extreme values,

$$\sigma_{\max} > 0$$
$$\sigma_{\min} \gtrless 0$$

The stress-ratio R may be positive or negative.

5.3 FATIGUE FAILURE DATA

Many testing techniques are available for measuring the response of the materials to the fatigue-loads. The oldest approach is due to Wohler. He used a rotating cantilever beam with lateral load at the free end to induce completely reversed stress at the bearing end. This was called the cantilever beam test. Later R.R. Moore adopted his technique to a simply supported rotating beam, loaded in a manner to induce an uniform bending moment over a section of the beam. The setup is shown in Fig. 5.3. This test is called R.R. Moore beam test.

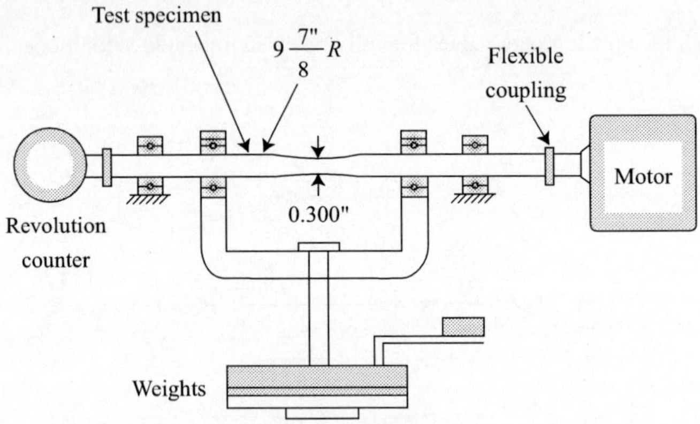

Figure 5.3. Fatigue testing and standard test specimen

In this test a highly polished specimen of standard dimensions is mounted in a fixture which allows a constant magnitude of pure bending moment to be applied. The specimen is rotated at 1725 rpm. The load is adjusted to induce a bending stress less than the ultimate tensile stress for the material of the specimen. The specimen is rotated until it is fractured. The number of cycles (rotations) are noted down. The load is reduced progressively and the corresponding number of rotation to failure are noted down, until the specimen does not fail. The data is plotted on a graph with stress amplitude as ordinate and number of cycles to failure as abscissa. The resulting graph is a drooping curve reaching asymptotically a constant value. This curve is shown in Fig. 5.4(a) and is called *S–N* curve.

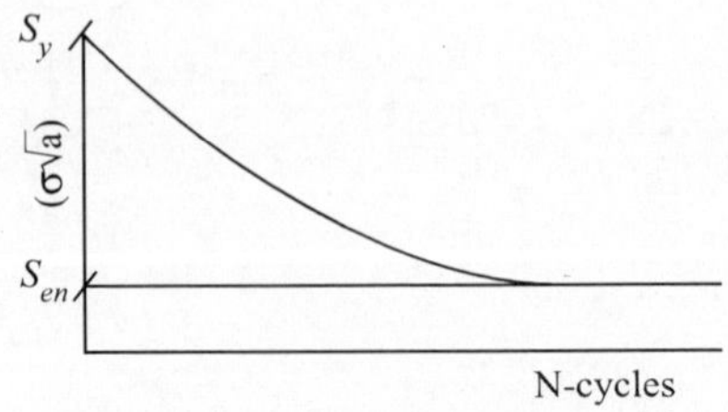

Figure 5.4(a). *S–N* diagram

Stress life Approach

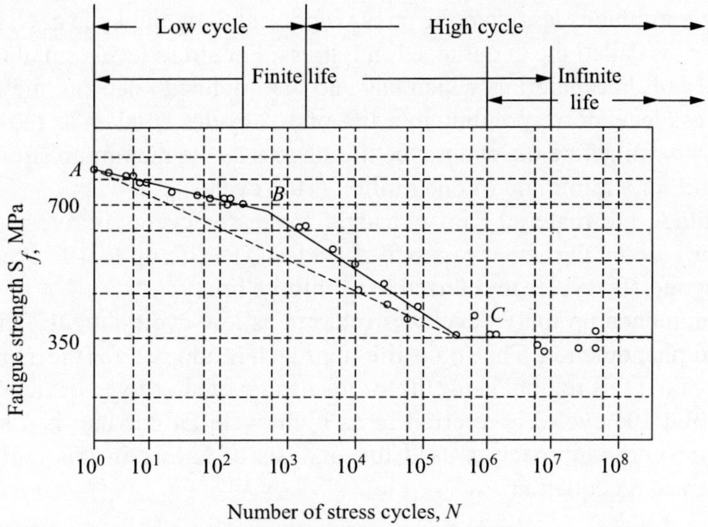

Figure 5.4(b). Stress-life-approach

Low-cycle fatigue in the range of $1 \leq N \leq 10^3$ cycles) it is given by [*AB*]

$$S_f \geq S_{ut} N^{(\log f)/3}. \quad 1 \leq N \leq 10^3$$

Basquin equation for high cycle fatigue N> 10^3 cycles [*BC*]

$$S_f = aN^b$$

where,

$$a = \frac{(fS_{ut})^2}{S_e}$$

$$b = -\frac{1}{3}\log\left(\frac{fS_{ut}}{S_{en}}\right) \quad \text{where, } f = 0.8 - 0.9$$

Approximate equation has, [*AC*]

$$S_f = AN^B$$

where,

$$A = S_{ut}$$

$$B = -\frac{1}{Z}\log\left(\frac{S_{ut}}{S_{en}}\right)$$

$$Z = 6 \text{ for } N_f = E6$$

$$= 8.699 \text{ for } N_f = 5E8$$

The *S–N* data can also be plotted on log-log graph Fig. 5.4(b). Then the plot consists of a combination of straight lines. *AB* from 10^0 to 10^3 cycles; *BC*-from 10^3 to 10^6 cycles and *CD*-beyond 10^6 cycles. At stress-amplitudes less than the stress represented by the knee at *C*, the material has infinite life. This stress is called the endurance limit-stress. For stress levels smaller than this value the material does not fail. For materials which have no well-defined knee and the failure line continues to fall, the stress level corresponding to a life of 10^6 cycles is taken as the endurance limit stress. Generalizing we can define the endurance limit stress as the maximum stress-amplitude the material can withstand for a minimum of one million (10^6) cycles.

The stress-amplitude the material can withstand for lesser cycles is referred to as fatigue-strength corresponding to the particular number of cycles. The life up to 10^6 cycles is called the finite life and life beyond 10^6 cycles in referred to as infinite life.

The fatigue phenomenon up to 10^3 cycles is referred as low-cycle-fatigue. The failure in this region is strain-based phenomenon. The strain-life-approach is adopted for the analysis of the failure. The knowledge of the this region, is useful for the design of shortlived devices like missiles.

The fatigue beyond 10^3 cycles is referred to as high-cycle fatigue and is a stress controlled phenomenon. The stress-life-approach is used for analysis of failure in this region. The fatigue strength is given by Basquin equation

$$S_f = aN^b \tag{5.5}$$

where, S_f is the fatigue strength for *N* no. of fatigue cycles.

a and *b* are constants determined by the relations

$$a = \left(\frac{fS_{ut}}{S_{en}}\right)$$

$$b = \frac{1}{3}\log\left(\frac{fS_{ut}}{S_{en}}\right)$$

where, $f = 0.8$ to 0.9

S_{ut} – Ultimate tensile strength

S_{en} – Corrected value of endurance limit stress

For approximate analysis, line *AC* can be taken as failure line and the modified.

Basquin equation is

$$S_f = AN_f^B \tag{5.6}$$

where,

$$A = S_{ut}$$

and

$$B = -\frac{1}{Z}\log\left(\frac{S_{ut}}{S_{en}}\right)$$

$Z = 6$ for $N_f = E6$ for steels

$= 8.699$ for $N_f = 5E8$ for aluminium and other ductile materials.

The conduction of fatigue test is not a simple affair. The set up is more complicated and the test duration is very long. A quick method of estimating the endurance limit stress is based on the correlation between the results of UTM-test and fatigue test. The endurance-limit stresses S'_{en} v/s ultimate strength value plots for different material are shown in Fig. 5.5.

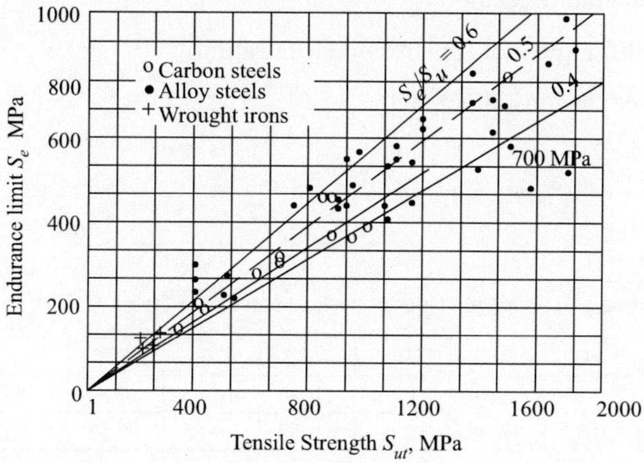

Figure 5.5

It can be observed that there are linear relations between S'_{en} and S_{ut} for various materials. They are given below.

For steels, $S'_{en} = 0.5\ S_{ut}$ for $S_{ut} \leq 1400$ MPa

$= 700$ MPa for $S_{ut} > 1400$ MPa

For cast irons and cast steels, $S'_{en} = 0.45\ S_{ut}$ for $S_{ut} \leq 600$ MPa

$= 275$ MPa for $S_{ut} > 600$ MPa

For aluminium, $S'_{en} = 0.4\ S_{ut}$ for $S_{ut} \leq 330$ MPa

$= 130$ MPa for $S_{ut} > 330$ MPa

For copper alloys, $S'_{en} = 0.4 S_{ut}$ for $S_{ut} \leq 280$ MPa

$= 100$ MPa for $S_{ut} > 280$ MPa

5.4 MODIFYING FACTORS

The actual values of endurance limit stress used in design, are different from the test value s'_{en} for various conditions of material, manufacturing and design. To account for these, modifying factors are introduced as follows:

$$S_{en} = K_a \cdot K_b \cdot K_c \cdot K_d \cdot K_e \cdot K_i\ S'_{en} \tag{5.7}$$

where, K_a = Surface factor (Fig. 5.6)

K_b = Size factor (Fig. 5.7)

K_c = Reliability factor (Table 5.2 MDHB be reproduced as Table 5.2)

K_d = Temperature factor

K_e = Modifying factor for stress-concentration

K_i = Miscellaneous effect factor

5.4.1 Surface Finish Factor, K_a

Table 5.1. Average values of surface correction factor, $C(Ka)$

Ultimate Stress S_u (MPa)	418	489	561	632	704	775	836	1050	1397
C for machined surface	0.91	0.90	0.88	0.86	0.85	0.84	0.82	0.78	0.72
C for hot rolled surface	0.72	0.68	0.62	0.58	0.55	0.52	0.48	0.38	0.30

The surface of the test specimen is highly polished but the actual surface finish of the machine components is not that good. There may be surface scratches. They are the possible locations for fatigue-crack initiation. A rough surface means reduced strength. The surface factor K_a accounts for this effect. The values are chosen from Fig. 5.6.

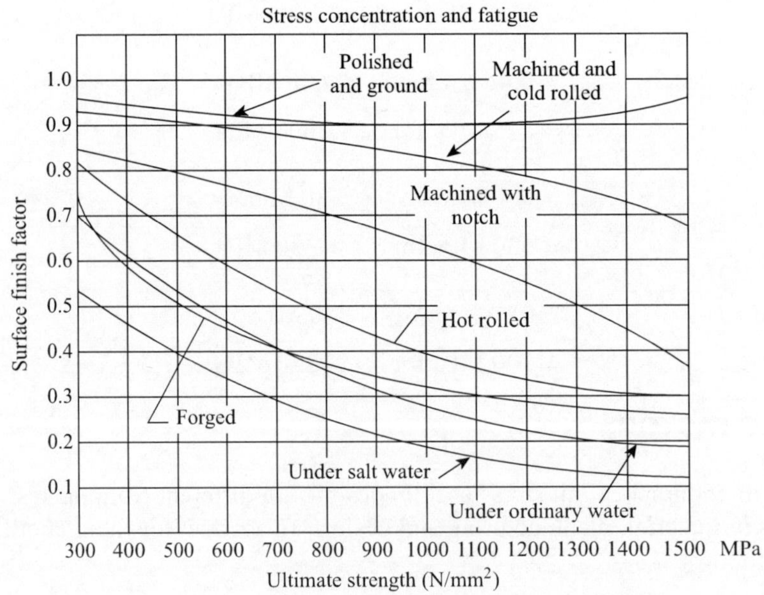

Figure 5.6. Surface factors for various finishes on steel

Shigley and Mischke expression for surface factor

$$K_a = A(S_{ut})^b \tag{5.8}$$

Table 5.2. Coefficients for the surface factor equation

	MPa	
Surface finish	A	b
Ground	1.58	−0.085
Machined or cold-drawn	4.51	−0.265
Hot-rolled	57.70	−0.781
As-forged	272.00	−0.995

5.4.2 Size Factor, K_b

The number of deflects in the material increase with volume. The larger the number of defects, lesser is the strength. Members with larger cross sections are weaker than the members with smaller cross sections. The size factor is introduced to take into account the effect of size on the strength of a component. The value may be selected from (Fig. 5.7) or from the following relation,

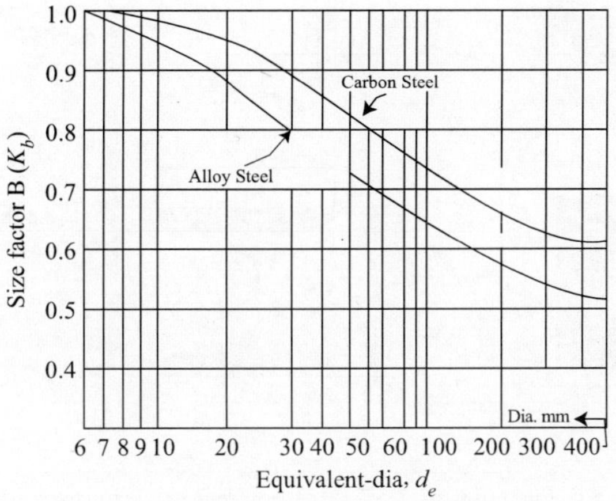

Figure 5.7. Size effect on endurance limit stress, K_b

Shigley and Mischke expression for size effect

$$K_b = 1 \text{ for } d_e < 8 \text{ mm}$$

$$K_b = 1.189 d_e^{-0.097} \text{ for } 8 \text{ mm } d_e < 250 \text{ mm}$$

For rectangular section, the equivalent diameter is given by

$$d_e = 0.808\sqrt{h.b} \tag{5.8a}$$

The equivalent diameter rotating-beam specimen for any cross section is

$$d_e = \sqrt{\frac{A_{95}}{0.0766}} \tag{5.8b}$$

where A_{95} is the portion of the cross-sectional area of the non-round part that is stressed between 95% and 100% of its maximum stress. A_{95} varies for various sections. (Fig 5.8)

5.4.3 The Reliability Factor, K_c

It is introduced to account for the statistical distribution of test data and the values are chosen from table (Table 5.3) depending upon the desired reliability.

Table 5.3. Reliability factors K_c corresponding to an 8 per cent standard deviation of the endurance limit

Reliability	0.50	0.90	0.95	0.99	0.999	0.999 9	0.999 99	0.999 999	0.999 999 9	0.999 999 99
Z_r	0	1.288	1.645	2.326	3.090	3.719	4.265	4.753	5.199	5.997
K_c	1.000	0.897	0.868	0.814	0.753	0.702	0.659	0.620	0.584	0.520

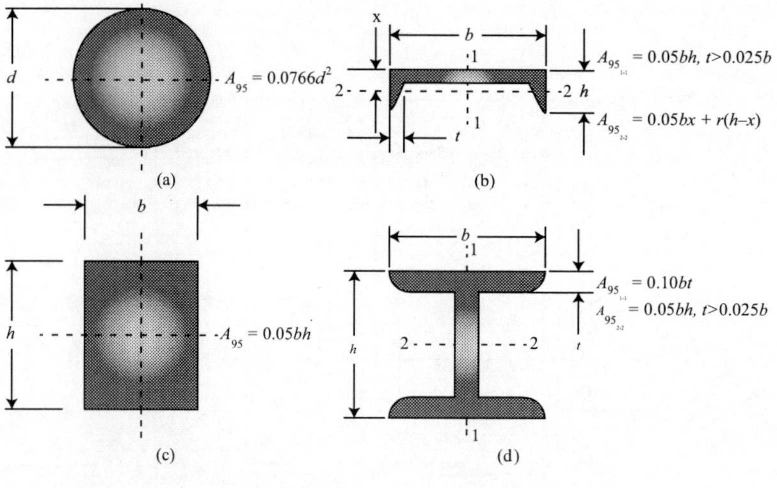

Figure 5.8. Formulas for 95% stressed areas of various sections

5.4.4 Temperature Factor, K_d

Like any other property, fatigue strength also depends upon the temperature. Normally, higher the temperature, lower is strength. The approximate value of this factor is chosen from the following relation

$$K_d = 1 \text{ for } T < 350°\text{C}$$
$$= 0.5 \text{ for } 350°\text{C} < T \leq 500°\text{C}$$

This factor is to be applied to both ends of S–N diagram.

Modifying factor for fatigue stress concentration factor, K_e

The stress concentrations at discontinuities are most likely cause crack initiation due to local cyclic plastic deformation. The effect depends upon the geometry of discontinuity and the material property called notch sensitivity. The factor is determined as follows:

$$K_e = \frac{1}{K_f}$$

The fatigue stress concentration factor K_f is given by,

$$K_f = 1 + q(K_t - 1) \tag{5.9}$$

where,

q – is the notch sensitivity of the material (Fig. 5.9a, 5.9b & 5.9c)

K_t – Geometrical stress concentration factor Fig (Fig. 4.14 to 4.28)

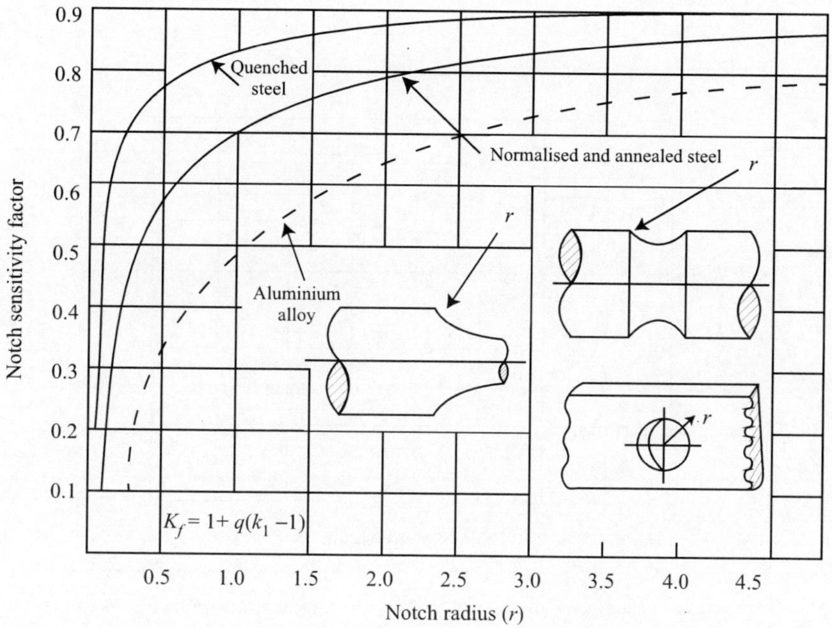

Figure 5.9(a). Notch sensitivity factor (q) for bending and axial load

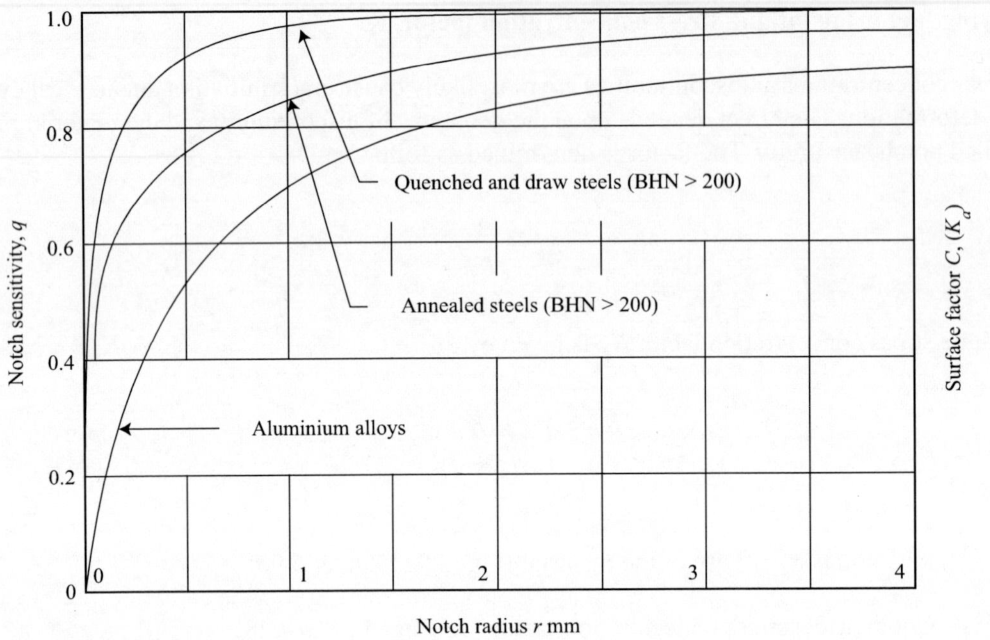

Figure 5.9(b). Notch sensitivity curves for material in reversed torsion. For large notch radii use the values of q corresponding to $r = 4$ mm

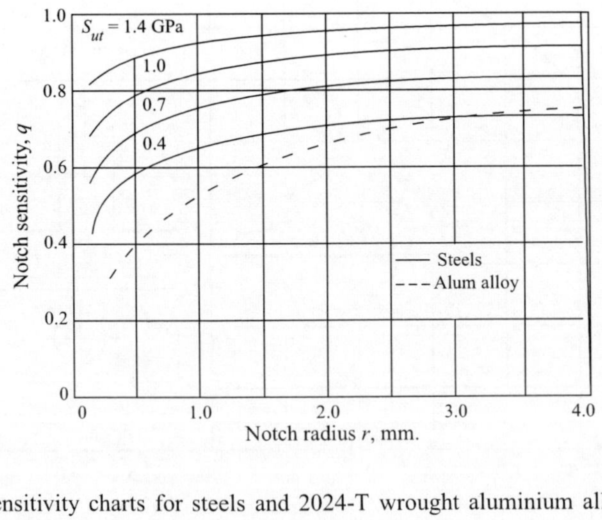

Figure 5.9(c). Notch sensitivity charts for steels and 2024-T wrought aluminium alloys subjected to reversed bending or reversed axial loads

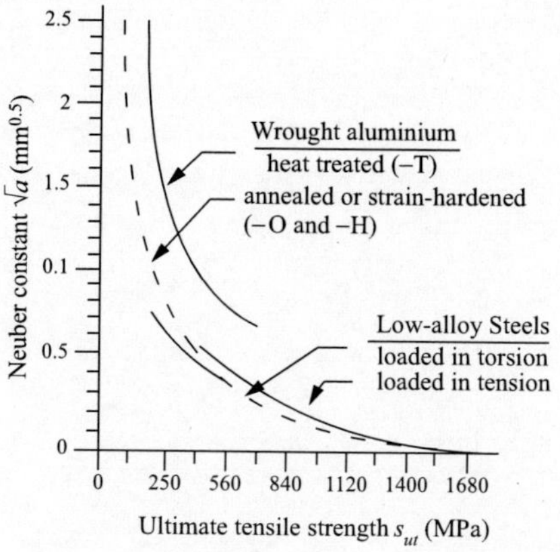

Figure 5.9(d). Neuber constant

Notch sensitivity (Neuber's formula)

$$q = \frac{1}{1 + \sqrt{\frac{a}{r}}}$$

$a =$ Neuber constant [Fig. 5.9(d)]

Peterson formula

$$q = \frac{1}{1 + \frac{a}{r}}$$

where $a = 0.0254 \left(\frac{2070}{S_u}\right)^{1.8}$ for steels.

Miscellaneous effect factor, K_i

This is introduced to account for the effect of the following factors:

(i) Residual stresses
(ii) Surface treatment
(iii) Corrosion

The value of K_i is chosen approximatly from the knowledge of the above factors.

5.5 FATIGUE FAILURE THEORIES

The $S-N$ diagram gives fatigue-strength under completely reversed loading conditions. In practice, loading is fluctuating between two extreme values. The behaviour of the materials under such loads is different. To predict the failure under the fluctuating loads, two theories are formulated.

1. Goodman theory
2. Soderberg theory

5.5.1 Goodman Theory

The theory is described by the Goodman diagram shown in Fig. 5.10. This is a plot of varying stress as ordinate and the mean stress as abscissa. OA is the line drawn at 45° to the x-axis. It represented the mean-stress-line. The amplitude values are plotted above and below this line $OC = OD = S_{en}$ – represents the endurance limit stress. This corresponds to the completely reversed loading condition in rotating beam test, $A(S_{ut}, S_{ut})$ represents the situation of zero amplitude, i.e., UTM test condition.

CA – represents the maximum stresses.

DA – represents the minimum stresses.

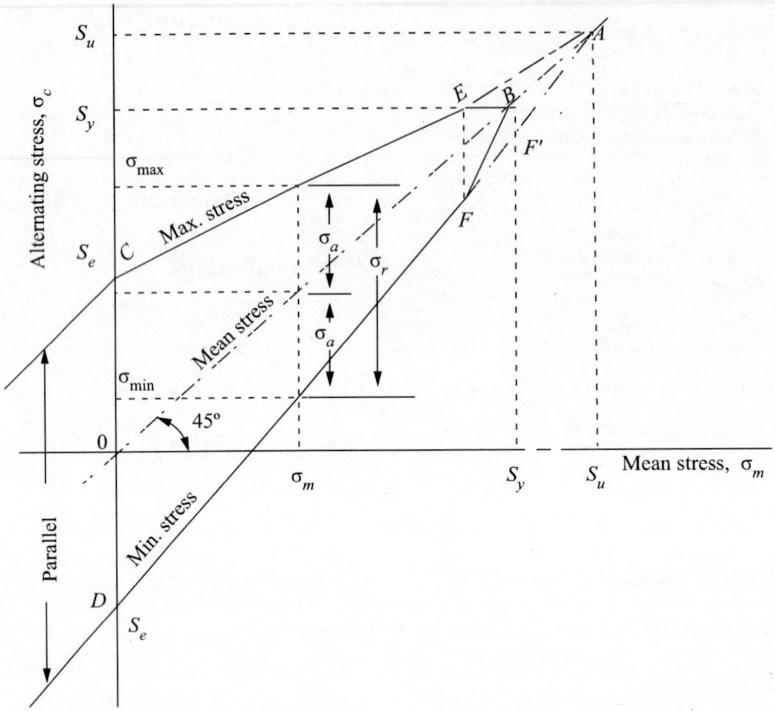

Figure 5.10. Modified Goodman diagram showing all the strengths and the limiting values of all stress components for a particular mean stress

From *C* and *D* two parallel lines, parallel to mean line are drawn backward. The diagram *CAD* is called Goodman diagram.

Any combination of mean and amplitude of stress, resulting in a point within the lines *CA* and *DA*, means the member is safe. If the point lies outside this range, the fatigue failure occurs.

For ductile materials yield point is the failure point. It is plotted as $B(S_y, S_y)$ on the mean-stress line *OA*. Then the modified Goodman diagram is *CEBFD*.

A more simpler diagram can be drawn to represent theory. This is the plot of the data with stress amplitude σ_a as ordinate and mean-stress (σ_m) as abscissa. (Fig 5.11)

The UTM test data $A(S_{ut}, o)$ is plotted on the *x*-axis.
The rotating beam test data $B(O, S_{en})$ is plotted on *y*-axis.
AB represents fatigue failure line.
If the point $P(\sigma_{m1}, \sigma_{a1})$ within the line is safe.
If the point $Q(\sigma_{m2}, \sigma_{a2})$ above the line is unsafe.
The point $R(\sigma_m, \sigma_a)$ on the line *AB* failure has started.
An equation for the failure-line *AB* can be written as follows:

$$\frac{\sigma_a}{S_{en}} + \frac{\sigma_m}{S_{ut}} = 1 \tag{5.10}$$

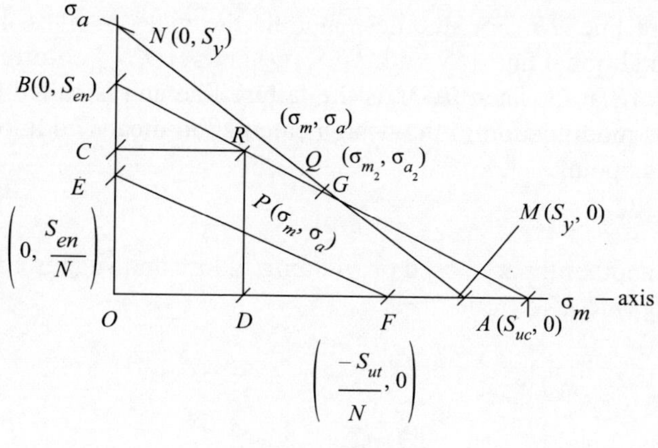

Figure 5.11

The design equation is (line EF)

$$\frac{\sigma_a}{S_{en}} + \frac{\sigma_m}{S_{ut}} = \frac{1}{N} \tag{5.11}$$

where,

S_{en} is the corrected value of the endurance limit stress
N is factor of safety.

Equation can be derived by considering the point $R(\sigma_m, \sigma_a)$ on the failure line. From similar triangles BOA and RDA,

$$\frac{RD}{BO} = \frac{DA}{OA}$$

$$\therefore \quad \frac{RD}{BO} = \frac{OA - OD}{OA} = 1 - \frac{OD}{OA}$$

$$\therefore \quad \frac{\sigma_a}{S_{en}} = 1 - \frac{\sigma_m}{S_{ut}}$$

$$\text{i.e.,} \quad \boxed{\frac{\sigma_a}{S_{en}} + \frac{\sigma_m}{S_{ut}} = 1}$$

Similarly, we consider the design line EF with co-ordinates

$$E\left(O, \frac{S_{en}}{N}\right) \text{ and } F\left(\frac{S_{ut}}{N}, O\right).$$

For the design line EF, we can derive the design equation as

$$\boxed{\frac{\sigma_a}{S_{en}} + \frac{\sigma_m}{S_{ut}} = \frac{1}{N}}$$

This is parallel to failure line AB. For ductile materials the yield point is the failure point. To incorporate this fact an additional line MN with $M(S_y, O)$ and $N(O, S_y)$ co-ordinate is drawn.

It intersects the line AB at G. Then BGM is the failure line and is called the modified Goodman line. To include this modification in the design calculation means additional calculations for locating G the intersecting point.

5.5.2 Soderberg Theory

Soderberg formulated another theory based on yielding failure and suggested that the failure line is the line joining $B(O, S_{en})$ to $M(S_y, O)$.

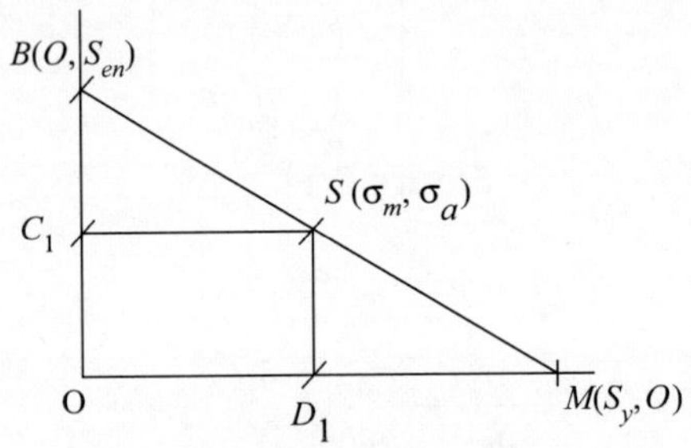

Figure 5.12

Considering a point $S(\sigma_m, \sigma_a)$ on this line. From the similar Δs, BOM and SDM, we can write

$$\frac{BO}{SD_1} = \frac{OM}{D_1M}$$

i.e.,

$$\frac{SD_1}{BO} = \frac{D_1M}{OM} = \frac{OM - OD_1}{OM} = 1 - \frac{OD_1}{OM}$$

$\therefore$

$$\frac{\sigma_a}{S_{en}} = 1 - \frac{\sigma_m}{S_y}$$

$\therefore$

$$\frac{\sigma_a}{S_{en}} + \frac{\sigma_m}{S_y} = 1 \tag{5.12}$$

This is the Soderberg equation for fatigue failure.

The design equation is

$$\frac{\sigma_m}{S_y} + \frac{\sigma_a}{S_{en}} = \frac{1}{N} \tag{5.13}$$

The Soderberg theory is more conservative and can be used for all materials. The theory which confirms the experimental data is the Gerber theory represented by the Gerber parabola.

$$\frac{\sigma_a}{S_{en}} + \left(\frac{\sigma_m}{S_{ut}}\right)^2 = 1 \tag{5.14}$$

Comparison of Failure Theories for Ductile Materials

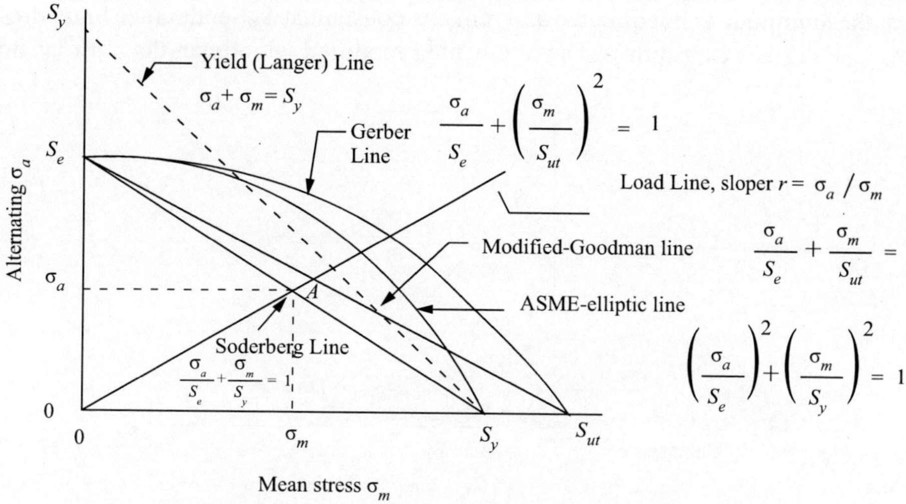

Figure 5.13. Various fatigue failure criteria

Fatigue failure equations

Gerber parabola:

$$\sigma_a = S_{en}\left(1 - \frac{\sigma_m^2}{S_{ut}^2}\right)$$

Modified-Goodman line:

$$\sigma_a = S_{en}\left(1 - \frac{\sigma_m}{S_{ut}}\right)$$

Soderberg line:

$$\sigma_a = S_{en}\left(1 - \frac{\sigma_m}{S_y}\right)$$

Langer line:

$$S_a + S_m = S_y$$

Design equations are

Soderberg equation

$$\frac{\sigma_a}{S_{en}} = \frac{\sigma_m}{S_y} = \frac{1}{N}$$

Modified-Goodman equation

$$\frac{\sigma_a}{S_{en}} + \frac{\sigma_m}{S_{ut}} = \frac{1}{N}$$

The Gerber equation

$$\frac{N\sigma_a}{S_{en}} + \left(\frac{N\sigma_m}{S_{ut}}\right)^2 = 1$$

ASME-elliptic equation

$$\left(\frac{N\sigma_a}{S_{en}}\right)^2 + \left(\frac{N\sigma_m}{S_y}\right)^2 = 1$$

Langer static yield equation

$$\sigma_a + \sigma_m = \frac{S_y}{N}$$

N - factor of safety

The experimental failure points lie around this curve. It is the mean curve of the failure data. Because of the quadratic terms involved, it is seldom used for design calculations.

In the precious discussions it is assumed that the mean stress σ_m is tensile. For compressive mean stresses, the amplitude is not affected and remains constant at S_{en}-endurance limit stress. But the failure by yielding is a possibility. The augmented modified. Goodman diagram is shown in Fig. 5.14.

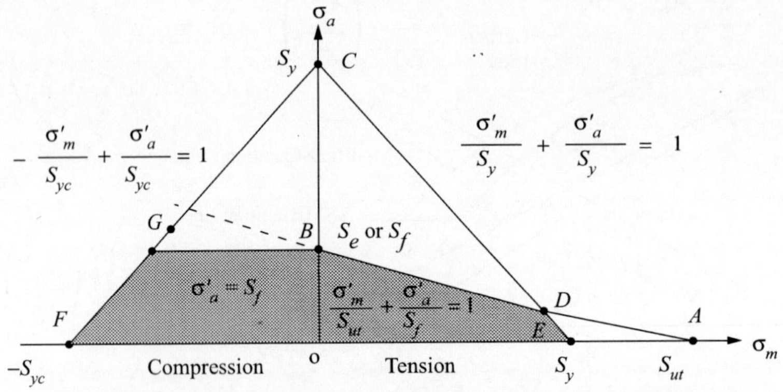

Figure 5.14. An "Augmented" Modified-Goodman diagram

As shown in Fig. 5.14, the failure line on the compressive mean stress side is a combination of a line *CF* having intercepts S_{y_c} on the *x*-axis and S_y on the *y*-axis. and the horizontal line *BG*, passing through *B* having intercept 'S_{en}' on the *y*-axis.

For Soderberg criterion, the improvement in strength due to the higher yield strength in compression, is neglected. The modification is done by replacing 'S_{y_c}' by 'S_y'. Then the failure equation for the compressive mean stress is given by

$$\frac{\sigma_a}{S_{en}} + \frac{\sigma_m}{S_y} = 1$$

and for design,

$$\frac{\sigma_a}{S_{en}} + \frac{\sigma_m}{S_y} = \frac{1}{N}$$

5.5.3 Fluctuating-Torsional Stresses

As in case of direct stresses, we define the following quantities.

$$\tau_{max} = \text{Maximum shear stress,}$$

$$\tau_{min} = \text{Minimum shear stress,}$$

$$\text{Mean shear stress, } \tau_m = \frac{\tau_{max} + \tau_{min}}{2} \tag{5.15}$$

$$\text{Amplitude of shear stress, } \tau_a = \frac{\tau_{max} - \tau_{min}}{2} \tag{5.16}$$

We make a plot of shear amplitude τ_a (along y-axis) versus mean shear stress τ_m (along x-axis). Mark $A(O, S_{sen})$ on y-axis and $C\ (S_y, O)$ an x-axis. Draw a horizontal line through A. Draw a line @45° to x-axis through 'C' to intersect horizontal line at B. Then $OABC$ is the fatigue failure diagram for shear stress. Plot the experimentally observed failure points on this diagram. These points lie above the horizontal line AB. It is interesting to note that the torsional mean stress has no effect on the torsional fatigue strength S_{sen}. For larger value of mean stress, the failure by yield is predominant. The inclined line BC with intercepts S_{sy} on both axes represent failure criteria.

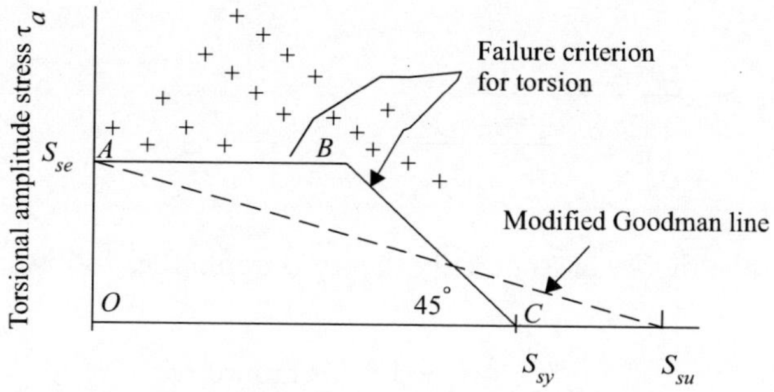

Figure 5.15. Torsional mean stress, τ_m

The failure criteria can be expressed in equations form are as follows:

$$\text{AB - The fatigue failure if } \tau_a \geq S_{se} \tag{5.17}$$

and

$$\text{BC - Static failure if } \tau_{\max} = \tau_a + \tau_m \geq S_{sy} \tag{5.18}$$

The values of S_{se} and S_y can be estimated by the following relations:
Yield stress in shear, $S_{sy} = 0.577\ S_y$ and
Endurance limit stress in shear, $S_{se} = 0.577\ S_e$.

5.5.4 Combined Fatigue Loads

We have discussed the fatigue failure under uniaxial fluctuating loads either direct loads or shear loads. But in practice the loads are multiaxials and are varying simultaneously. To analyse such combined loading, we make an assumption that the loads vary with same frequency and are in-phase, i.e., the maximum and minimum values occur at the same instant. Then we draw two stress elements — one for mean stresses and the second for the alternating components, i.e., stress amplitudes. By applying the Mohr cycles, the principal mean stresses, $(\sigma_{1m}, \sigma_{2m})$ and the principal alternating stresses $(\sigma_{1a}, \sigma_{2a})$ are determined. Then applying Von-Mises distortion theory, the equivalent Von-Mises stresses are calculated as follows:

$$\sigma_m^1 = \sqrt{\sigma_{1m}^2 - \sigma_{1m} \cdot \sigma_{2m} + \sigma_{2m}^2} = \sqrt{\sigma_m^2 + 3\tau_m^2} \tag{5.19}$$

$$\sigma_a^1 = \sqrt{\sigma_{1a}^2 - \sigma_{1a} \cdot \sigma_{2a} + \sigma_{2a}^2} = \sqrt{\sigma_a^2 + 3\tau_a^2} \tag{5.20}$$

By Sine Method

$$\sigma_m^1 = (\sigma_{xm} + \sigma_{ym} + \sigma_{zm}) \tag{5.21}$$

$$\sigma_a^1 = \sqrt{\frac{(\sigma_{xa}-\sigma_{ya})^2+(\sigma_{ya}-\sigma_{za})^2+(\sigma_{za}-\sigma_{xa})^2+6(\sigma_{xya}^2+\sigma_{yza}^2+\sigma_{zxa}^2)}{2}} \tag{5.22}$$

For biaxial stress system,

$$\sigma_m^1 = \sigma_{xm} + \sigma_{ym} \tag{5.23}$$

$$\sigma_a^1 = \sqrt{\sigma_{xa}^2 + \sigma_{ya}^2 - \sigma_{xa} \cdot \sigma_{ya} + 3\sigma_{xya}^2} \tag{5.24}$$

Then the modified Goodman theory or Soderberg theory is applied as follows:

$$\frac{\sigma_m^1}{S_{ut}} + \frac{\sigma_a^1}{S_{en}} = 1 \quad \text{for failure}$$

or

$$\frac{\sigma_m^1}{S_y} + \frac{\sigma_a^1}{S_{en}} = 1 \quad \text{for failure}$$

For design using Soderberg theory, we can write the equation as

$$\frac{\sigma_m^1}{S_y} + \frac{\sigma_a^1}{S_{en}} = \frac{1}{N}$$

or

$$\sigma_m^1 + \left(\frac{S_y}{S_{en}}\right)\sigma_a^1 \leq \left(\frac{S_y}{N} = S_d\right) \tag{5.25}$$

Another approach is to convert the variable components into their static equivalents and apply a static theory of failure. The equivalent static values based on Soderberg theory is given as follows:

Equivalent direct stress,

$$\sigma_e = \sigma_m + \left(\frac{S_y}{S_{en}}\right) \cdot \sigma_a \tag{5.26}$$

Equivalent shear stress,

$$\tau_e = \tau_m + \left(\frac{S_y}{S_{en}}\right) \cdot \tau_a \tag{5.27}$$

Then the design equation using shear theory is

$$\sqrt{\sigma_e^2 + 4\tau_e^2} \leq \left(S_d = \frac{S_y}{N}\right) \tag{5.28}$$

5.6 CUMULATIVE FATIGUE DAMAGE

In practice, the loading may not remain the same throughout the life cycles. It may have different alternating values for different periods. The component is supposed to be damaged partially at each loading period. For failure, the sum of all these damages should add up to unity. This is given by the equation,

$$\frac{n_1}{N_1} + \frac{n_2}{N_2} + \ldots + \frac{n_n}{N_n} = \sum_{i=1}^{n} \frac{n_i}{N_i} = 1 \tag{5.29}$$

where, n_1 – number of load cycles at stress, σ_1

N_1 – life cycles or stress life, σ_1

n_2 – number of load cycles at stress, σ_2

N_2 – life cycles for stress, σ_2

This is called the Miner's rule.

The effect of the stress σ_1 for n_1, cycles is shown on S–N as in Fig. 5.16 below.

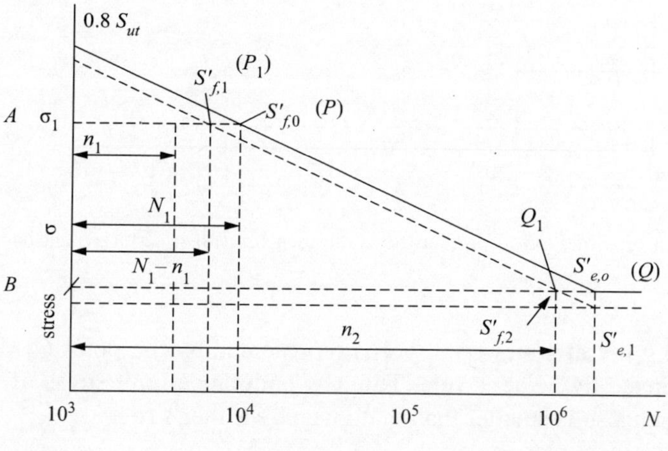

Figure 5.16. Use of Miner's method to predict the endurance limit of a material that has been stressed for a number of cycles

At stress σ_1, the member has a life of N_1 cycles. Since it is loaded for n_1 cycles the remaining life is $(N_1 - n_1)$ cycles. It is represented by P. The new failure line is a parallel line through P_1 intersecting the endurance stress line at Q_1. The remaining life @ stress $\sigma_2 = S_{en}$ is given by the similarity.

$$\frac{AP_1}{P} = \frac{BQ_1}{BQ}$$

$$\frac{n_2}{10^6} = \frac{N_1 - n_1}{N_1}$$

$$\therefore \quad \frac{n_1}{N_1} + \frac{n_2}{10^6} = 1 \tag{5.30}$$

or $\quad n_2 = \left(1 - \frac{n_1}{N_1}\right) \times 10^6$ is the remaining life at stress, $\sigma_2 = S_{en}$.

The shifting of the failure line from PQ to P_1Q_1 means that the damaged specimen has lower endurance limit stress, S_{en_1} and reduced static strength. The reduction in the static strength is not confirmed by experimental results. To account for this fact, Manson suggested to retain the static strength point C at $\sigma = 0.8\, S_u$ and to join this point to P_1 to get the new failure line. This is shown in Fig. 15.17.

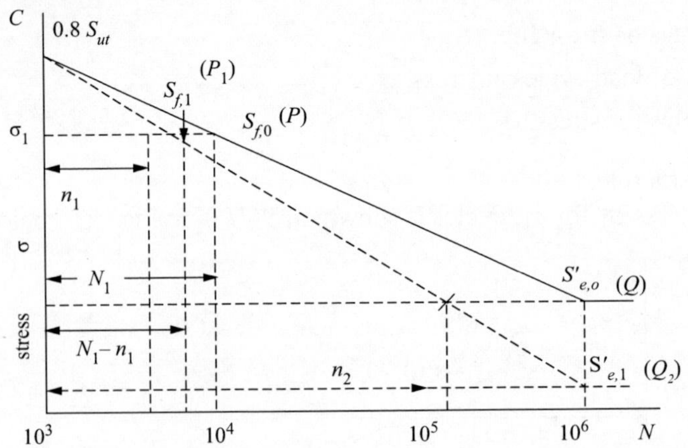

Figure 5.17. Use of Manson's method to predict the endurance limit of a material that has been stressed for a number of cycles

Then the remaining life at stress $\sigma_2 = S_{en}$ is represented by the point Q_1. This life is much smaller than that predicted by Miner's rule. The new endurance limit stress of 10^6 life cycles is represented by Q_2 which also is smaller than that given by Miner's rule.

5.7 FACTORS OF SAFETY

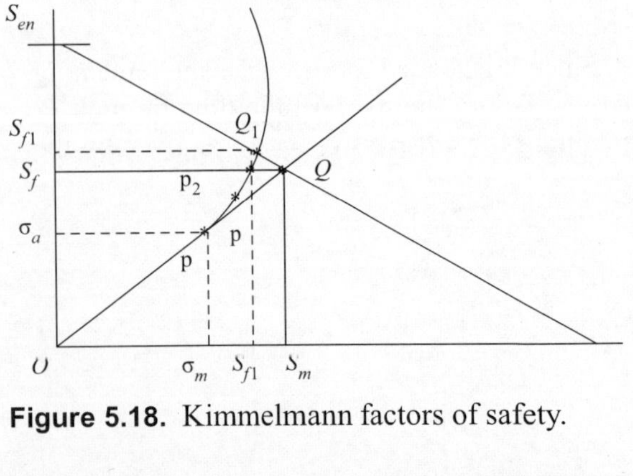

Co-ordinates of Q are (S_m, S_f)
Co-ordinates of Q_1 are (S_{m1}, S_{f1})

Figure 5.18. Kimmelmann factors of safety.

The factor of safety for a given combination σ_m and σ_a, is found by plotting $P(\sigma_m, \sigma_a)$ on the diagram and extending OP to intersect the failure line at Q. The co-ordinates of $Q(S_m, S_f)$ are the strength value for the loading situation. The factor of safety is defined by the relation,

$$N = \frac{OQ}{OP} = \left(\frac{S_f}{\sigma_a}\right) \tag{5.31}$$

It is assumed that loading conditions follow the rule $\sigma_m/\sigma_a = \text{Const.}$

If the loading condition does not follow the linear relation but follows a curved path Pp_1, p_2, Q_1, then the strength values as represented by $Q_1(S_{m1}, S_{f1})$ are to be considered. The factors of safely are:

$$N_a = \frac{S_{f_1}}{\sigma_a} \quad \text{for amplitude stress.} \tag{5.32a}$$

$$N_m = \frac{S_{m_1}}{\sigma_m} \quad \text{for mean stress.} \tag{5.32b}$$

They are called Kimmelmann factors of safety.

SOLVED EXAMPLES (Fatigue Failure)

Example 5.1: What is the fatigue strength of a rotating beam specimen made of C30 hot rolled steel corresponding to life of 250E3 cycles of stress reversals? What would be the life of the specimen, if the alternating stress were 220 MPa

Given Data

Material: C30 hot rolled steel

(i) Life, $N_{f_1} = 250E3 = 250 \times 10^3$

Fatigue strength, $S_{f_1} = ?$

(ii) Fatigue strength, $S_{f_2} = 220$ MPa

Life of $N_{f_2} = ?$

SOLUTION

Material properties (Table 2.6 MDHB)

$$S_u = 500 \text{ MPa}$$
$$S_y = 300 \text{ MPa}$$
$$S_{en} = 0.5\, S_u = 0.5 \times 500 = 250 \text{ MPa}$$
$$S_{en} = S'_{en} = 250 \text{ MPa} \quad \text{(No correction factors)}$$

Basquin equation for (S–N curve) fatigue strength

$$S_f = aN_f^b \quad \text{– Equation (3.44a)}$$
$$\log S_f = \log a + b \cdot \log N_f$$
$$= b \cdot \log N_f + C$$

where,

$$b = -\frac{1}{3}\log\left[\frac{0.8\,S_{ut}}{S_{en}}\right]$$

$$= -\frac{1}{3}\log\left[\frac{0.8\times 500}{250}\right]$$

$$= -0.06804$$

and $$C = \log\left[\frac{(0.8\,S_{ut})^2}{S_{en}}\right]$$

$$= \log\left[\frac{(0.8\times 500)^2}{250}\right]$$

$$= 2.80618$$

Then

$$\log S_f = -0.06804\log N_f + 2.80618$$

For $N_f = 250\text{E}3 = 250\times 10^3$,

$$\log S_f = -0.06804\log(250\times 10^3) + 2.80618$$

$$= -0.06804(4 + \log 25) + 2.80618$$

$$= 2.4389$$

$\therefore$ $$S_f = 274.0 \text{ MPa}$$

$S_f = 220 \text{ MPa} \le (S_{en} = 250 \text{ MPa})$

The applied stress amplitude is less than the endurance stress-limit stress and hence the life N_f is infinite, $N_f = \infty$.

Example 5.2: A portion of a machine member is shown in Fig. 5.19. It is loaded by completely reversed axial load F which are uniformly distributed across the width. The material is cold-drawn steel with yield stress, $S_y = 385$ MPa and ultimate stress, $S_u = 530$ MPa. For 90% reliability and infinite life, determine the maximum force F that can be applied.

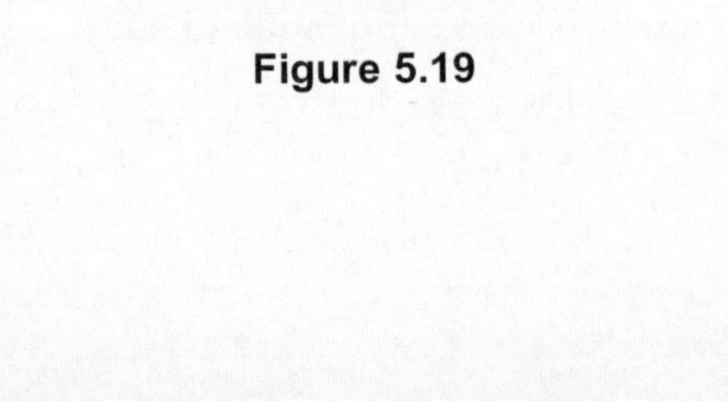

Figure 5.19

Given Data

Machine-member plate with a hole

Plate width, $w = 60$ mm

Thickness, $t = 10$ mm

Hole diameter, $d = 12$ mm

Material properties = cold-drawn steel

Ultimate stress, $S_{ut} = 530$ MPa

Yield stress, $S_y = 385$ MPa

To find $F = ?$

SOLUTION

$$\text{The endurance limit stress, } S'_{en} = 0.5\, S_{ut} = 0.5 \times 530 = 265 \text{ MPa}$$

The corrected value of endurance limit stress,

$$S_{en} = K_a \cdot K_b \cdot K_c \cdot K_d \cdot K_e \cdot K_f\, S'_{en} \qquad \text{(Eqn. 3.7 MDHB)}$$

where,

$K_a = 0.78$ (Fig. 3.2 MDHB)

$K_b = 0.94$ (Fig. 3.3 MDHB)

$K_c = 0.897$ (Table 3.2 MDHB)

$K_d = 1.0$ assumed

$$K_e = \frac{1}{K_f} = \frac{1}{1 + q(K_t - 1)}$$

For $K_t = 2.5$ (Fig. 3.17 MDHB)

$q = 0.9$ (Fig. 3.1a MDHB)

$$K_e = \frac{1}{1 + 0.9(2.5 - 1)} = 0.425$$

$K_i = 1.0$ assumed

Then the corrected value of endurance limit stress,

$$S_{en} = 0.78 \times 0.94 \times 0.897 \times 1 \times 0.425 \times 1 \times 265 = 105.2 \text{ MPa}$$

Stress amplitude,

$$\sigma_a = F/b_i \cdot h = \frac{F}{48 \times 10} = \frac{F}{450}$$

For infinite life with factor of safety = 1,

$$\sigma_a = S_{en}$$

$$\therefore \quad \frac{F}{450} = 105.2$$

$$F = 105.2 \times 480 = 50,496 \text{ N}$$

$$= \mathbf{50.496 \ kN}$$

Example 5.3: Figure is an idealized representation of a machine member subjected to the action of an alternating force *F* which places the member in completely reversed bending. The material is 50C-steel hardened and tempered to 190 BHN. All surfaces are ground, based on 50% reliability, infinite-life and no margin of safety, determine the max. value of alternating force *F* which can be applied.

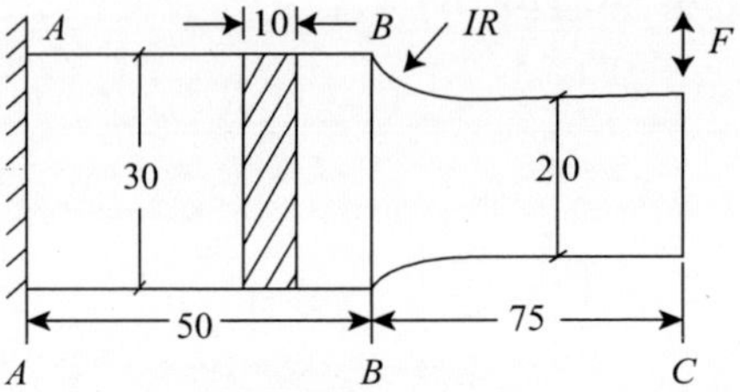

Figure 5.20

Given Data
Material-50C-steel-hardened and tempered
BHN = 190 BHN
Ground surface
Reliability, $R = 50\%$
Factor of safety = 1.0

SOLUTION
The ultimate tensile stress,

$$S_{ut} = 3.5 \times \text{BHN} = 3.5 \times 190 = 665 \text{ MPa}$$

Endurance limit stress,

$$S'_{en} = 1.76 \times \text{BHN} = 1.76 \times 190 = 334.5 \text{ MPa}$$

Step-1: For section -*A*

$$\text{Bending moment, BM} = 125F$$

$$\text{Bending stress, } \sigma_b = \frac{\text{BM}}{Z} = \frac{125F}{\left(\frac{10 \times 30^2}{6}\right)} = 0.083F$$

The corrected value of endurance limit stress,

$$S_{en} = K_a \cdot K_b \cdot K_c \cdot K_d \cdot K_e \cdot K_f \, S'_{en}$$

where,

$$K_a = 0.9 \qquad \text{(Fig. 3.2 MDHB)}$$

$$K_b = 0.95 \qquad \text{(Fig. 3.3 MDHB)}$$

$$K_c = 1.00$$

$$K_e = \frac{1}{1+q(K_t-1)} = \frac{1}{1+q(1-1)} = 1.0$$

$$K_i = 1 \quad \text{assumed}$$

$$K_d = 1 \quad \text{assumed}$$

$$S_{en} = 0.9 \times 0.95 \times 1 \times 1 \times 1 \times 1 \times 334.5 = 286 \text{ MPa}$$

For infinite life and factor of safety = 1,

$$\sigma_b = S_{en}$$

$$0.083F = 286$$

$$F = 286/.083 = 3445.9 \text{ N}$$

Step-2: For section -*B*

From (Fig. 3.14 MDHB) for $D/d = 30/20 = 1.5$ & $r/d = 1/20 = 0.05$

$$K_t = 2.2$$

and

$$q = 0.85 \qquad \text{(Fig. 3.1a MDHB)}$$

$$K_f = 1 + q(K_t - 1) = 1 + 0.85(2.2 - 1)$$

$$= 2.020$$

$$K_e = \frac{1}{K_f} = \frac{1}{2.02} = 0.495$$

$$K_b = 0.98 \qquad \text{(Fig. 3.3 MDHB)}$$

Then corrected endurance limit stress,

$$\begin{aligned} S_{en} &= K_a \cdot K_b \cdot K_c \cdot K_d \cdot K_e \cdot K_f \times S'_{en} \\ &= 0.9 \times 0.98 \times 1.0 \times 1.0 \times 0.495 \times 1.0 \times 334.5 \\ &= 145 \text{ MPa} \end{aligned}$$

The bending moment,

$$\text{BM} = 75F$$

The bending stress,

$$\sigma_b = \frac{\text{BM}}{Z} = \frac{75F}{\left(\frac{10 \times 20^2}{6}\right)} = 0.1125F$$

For infinite life and factor of safety = 1,

$$\sigma_b = S_{en}$$

$$0.1125F = 145$$

$$\therefore \; F = 1284.0\ N$$

This is smaller and hence it is the answer.

Example 5.4: The connecting rod shown in the Fig. 5.21 is machined from a 10 mm thick bar of C50 cold-drawn steel. The bar is axially loaded by completely reversed forces acting on the pins through two-drilled holes. Based on infinite life, 99% reliability and factor of safety of 2 estimate the maximum safe axial load that can be employed.

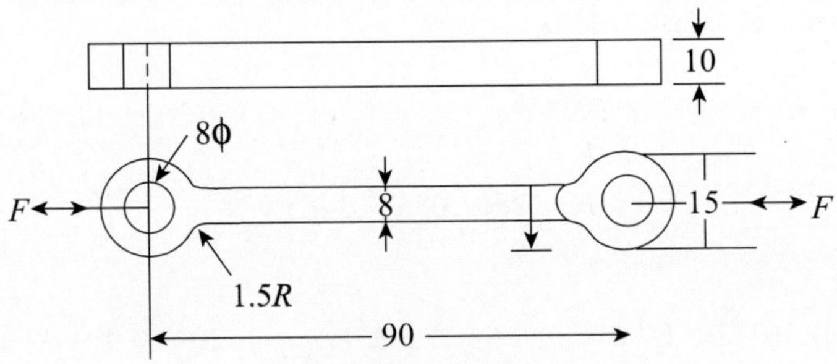

All dimensions in mm.

Figure 5.21

Given Data

Completely reversed axial load F

Material C50 steel cold-drawn

$$\begin{aligned}
\text{Rod section } &= 8 \times 10 \text{ mm} \\
\text{Length, } L &= 90 \text{ mm} \\
\text{Pin hole diameter, } d &= 8 \text{ mm} \\
\text{Boss diameter, } D &= 15 \text{ mm} \\
\text{Reliability, } R &= 99\% \\
\text{Factor of safety, } FS &= 2.
\end{aligned}$$

SOLUTION

Material properties (Table 2.6 MDHB)

$$\begin{aligned}
\text{Ultimate stress, } S_u &= 660 \text{ MPa} \\
\text{Yield stress, } S_y &= 380 \text{ MPa} \\
\text{The endurance limit stress, } S'_{en} &= 0.5\, S_u = 0.5 \times 660 \\
&= 330 \text{ MPa}
\end{aligned}$$

Corrected value of endurance stress,

$$S_{en} = K_a \cdot K_b \cdot K_c \cdot K_d \cdot K_e \cdot K_f\, S'_{en}$$

where,

$K_a = 0.85$ (Fig. 3.2 MDHB)

$K_b = 0.96$ (Fig. 3.3 for $de = \sqrt{8 \times 10} = 9.0$)

$K_c = 0.814$ (Table 3.2 MDHB)

$K_d = 1.0$ assumed

$$K_e = \frac{1}{[1 + q(K_b - 1)]}$$

For hole $K_t = 2.15$ (Fig. 3.17 MDHB), $d/w = \dfrac{8}{15}$ and notch sensitivity $q = 0.9$ ((Fig. 3.1a MDHB), for $r = 8$)

$$K_e = 1/[1 + 0.9(2.15 - 1)] = 0.491$$

For fillet,

$$K_t = 1.85 \quad \left(\text{(Fig. 3.13 MDHB); } r/d = 0.1875 \text{ and } D/d = \frac{15}{8}\right)$$

$$q = 0.85 \quad \text{(Fig. 3.1a for } r = 1.5\text{)}$$

$$K_e = 1/[1 + q(K_t - 1)] = 1/[1 + 0.85(1.85 - 1)] = 0.58$$

Smaller of the two $K_e = 0.491$.

Then corrected value of endurance limit stress,

$$S_{en} = 0.85 \times 0.96 \times 0.814 \times 1.0 \times 0.491 \times 1 \times 330$$
$$= 107.6 \text{ MPa}$$

The design stress,

$$S_{dr} = S_{en}/FS = \frac{107.6}{2} = 53.81 \text{ MPa}$$

The load that can be employed

$$F = S_{dr}\,A = 53.81 \times 8 \times 10$$
$$= \mathbf{4304.96\ N}.$$

Example 5.5: Figure 5.22 shows a formed round wire cantilever spring subjected to a varying force F. A hardness test made on 25 springs gave a minimum hardness of 380 BHN. It is apparent from the mounting details that there is no stress concentration. A visual inspection of the surface finish indicates a hot-rolled finish. Based on a 50% reliability, find the number of cycles of load applications likely to cause failure.

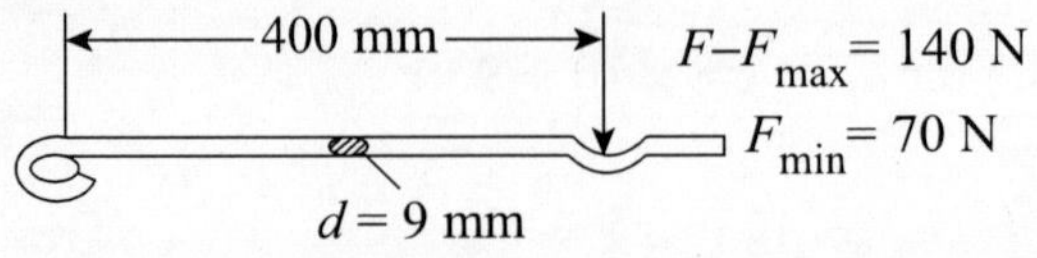

Figure 5.22

Given Data

Spring wire hardness $= 380$ BHN

Surface finish – hot rolled

Reliability, $R = 50\%$

Load applied, $F_{max} = 140$ N

$F_{min} = 70$ N

SOLUTION

Step-1: Material properties

Ultimate tensile strength for BHN = 380

$S_{ut} = 1400$ MPa (Fig. 2.1 MDHB)

Endurance stress,

$$S'_{en} = 0.5\, S_{ut} = 0.5 \times 1400 = 700 \text{ MPa}$$

Correction factors are

$$\begin{aligned} K_a &= 0.32 && \text{[Fig. 3.2 MDHB]} \\ K_b &= 0.97 && \text{[Fig. 3.3 MDHB]} \\ K_c &= 1.0 && \text{[Table 3.2 MDHB]} \\ K_d &= 1.0 \quad \text{assumed} \\ K_e &= 1.0 \quad \text{for} \quad K_t = 1 \\ K_i &= 1.0 \quad \text{assumed} \end{aligned}$$

Therefore, the corrected value of the endurance limit stress

$$\begin{aligned} S_{en} &= K_a \cdot K_b \cdot K_c \cdot K_d \cdot K_e \cdot K_i\, S'_{en} \\ &= 0.32 \times 0.97 \times 1 \times 1 \times 1 \times 1 \times 700 \\ &= 217.3 \text{ MPa} \end{aligned}$$

Step-2: Stresses induced

The bending moment,

$$M = FL$$

Maximum bending moment, $M_{\max} = 140 \times 400 = 56000$ N-mm

Minimum bending moment, $M_{\min} = 70 \times 400 = 28,000$ N-mm

Section module, $Z = \dfrac{\pi}{32} \cdot d^3 = \dfrac{\pi}{32} \times (9)^3 = 71.57 \text{ mm}^3$

Maximum bending stress, $\sigma_{\max} = \dfrac{M_{\max}}{Z} = \dfrac{56000}{71.57} = 782.45$ MPa

Minimum bending stress, $\sigma_{\min} = \dfrac{M_{\min}}{Z} = \dfrac{28000}{71.57} = 391.225$ MPa

Mean stress,

$$\sigma_m = \frac{\sigma_{\max} + \sigma_{\min}}{2} = \frac{782.45 + 391.225}{2} = 586.85 \text{ MPa}$$

Stress amplitude,

$$\sigma_a = \frac{\sigma_{\max} - \sigma_{\min}}{2} = \frac{782.45 - 391.225}{2} = 195.61 \text{ MPa}$$

Step-3: Applying Goodman theory, the permissibe amplitude for infinite life

$$\sigma_a' = \left[1 - \frac{\sigma_m}{S_{ut}}\right] \cdot S_{en}$$

$$\sigma_a' = \left[1 - \frac{586.85}{1400}\right] \times 217.3$$

$$= 126.9 \text{ MPa}$$

This is less than the actual amplitude $\sigma_a = 195.61$ MPa. Then the spring will have a life lesser than 10^6 cycles.

Step-4: The actual life can be found by considering the Basquin equation for fatigue strength.

$$S_f = aN^b \qquad \text{(Eqn. 3.14 MDHB)}$$

For approximate form,

$$a = 0.9\ S_u = 0.9 \times 1400 = 1260$$

and

$$b = -\frac{1}{6}\log\left(\frac{1260}{217.3}\right)$$

$$= -0.1272$$

The fatigue strength S_f for this situation is calculated from Goodman equation as follows:

$$\frac{\sigma_a}{S_f} + \frac{\sigma_m}{S_u} = 1, \quad \text{for} \quad FS = 1$$

$$\frac{195.61}{S_f} + \frac{586.85}{1400} = 1$$

$$S_f = \frac{195.61}{\left[1 - \frac{586.85}{1400}\right]} = 336.78 \text{ MPa}$$

Substituting the values in Basquin equation, we get

$$336.78 = 1260\,N^{-0.1272}$$

$$\therefore \qquad \text{Solving } N = \left(\frac{1260}{336.78}\right)^{\frac{1}{0.1272}} = (3.74)^{7.86} = 31{,}825 \text{ cycles}$$

Example 5.6: Figure 5.23 is the drawing of a 18 × 3 mm latching spring. The spring is assembled by deflecting at 2 mm initially and then deflecting it an additional 4 mm during each latching operation. The material is machined high-carboon-steel hardened and tampered to 490 BHN. The stress concentration factor at the bend is 1.7 corresponding to a fillet radius of 3 mm and uncorrected for notch sensitivity.

(a) Find the maximum and minimum latching forces.

(b) Will the spring fail in fatigue? Why?

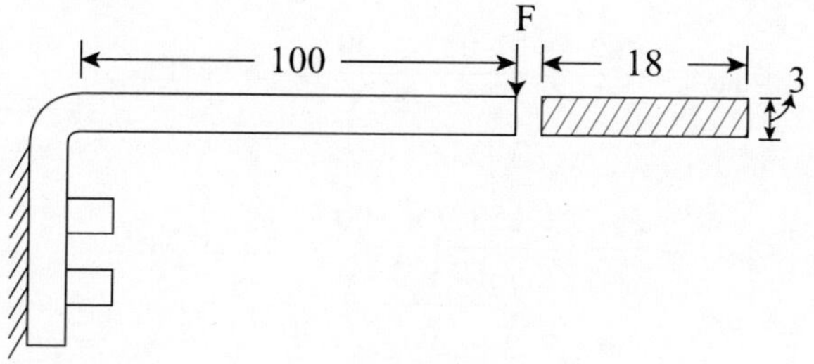

Figure 5.23

Given Data

Latch material: high-carbon steel BHN = 490

Latch cross section, $A = 18 \times 3\ \text{mm}^2$

Length = 100 mm

Stress-concentration factor, $K_t = 1.7$

Initial deflection, $y_1 = 2$ mm

Additional deflection, $y_2 = 4$ mm

SOLUTION

Material properties

$E = 210\ \text{GPa} = 0.21\times = 0.21 \times 10^6\ \text{MPa}.$ (Table 2.1 MDHB)

$S_u = 1500$ MPa for BHN = 490 (Fig. 2.1 MDHB)

$S'_{en} = 700\ \text{MPa} \quad \because \quad S_u > 1400\ \text{MPa}$

The deflection of spring,

$$y = \frac{FL^3}{3EI}$$

$$I = \frac{bt^3}{12} = \frac{18 \times 3^3}{12} = 40.5 \text{ mm}^4.$$

Initial force for initial deflection, $y_1 = 2.0$ mm

$$F_1 = \frac{3EI}{L^3} \quad y_1 = \frac{3 \times 0.21 \times 10^6 \times 40.5 \times 2}{100^3} = 51.03 \text{ N}$$

Latching force, for $y_{max} = 6$ mm

$$F_2 = \frac{3EI}{L^3} \quad y_2 = \frac{3 \times 210 \times 10^3 \times 40.5 \times 6}{100^3} = 153.9 \text{ N}$$

The stresses induced in the spring,

$$\sigma_{\min} = \frac{F_1 L}{Z} = \frac{51.03 \times 100}{\left(\frac{18 \times 3^2}{6}\right)} = 189 \text{ MPa}$$

$$\sigma_{\max} = \frac{F_2 L}{Z} = \frac{153.9 \times 100}{\left(\frac{18 \times 3^2}{6}\right)} = 567.0 \text{ MPa}$$

Mean stress,

$$\sigma_m = \frac{\sigma_{\max} + \sigma_{\min}}{2} = \frac{567.0 + 189}{2} = 378 \text{ MPa}$$

Stress amplitude,

$$\sigma_a = \left(\frac{\sigma_{\max} - \sigma_{\min}}{2}\right) = \frac{567 - 189}{2} = 189 \text{ MPa}$$

The corrected value of the endurance limit stress,

$$S_{en} = K_a.K_b.K_c.K_d.K_e.K_i S'_{en}$$

$$K_a = 0.62 \quad \text{(cold rolled – Fig. 3.2 MDHB)}$$

$$K_b = 0.99 \quad \text{(Fig. 3.3. for } de\sqrt{00766} = 7.3\text{)}$$

$$K_c = 1.0 \quad \text{(assumed)}$$

$$K_d = 1.0 \quad \text{assumed}$$

$$q = 0.95 \qquad \text{(Fig. 3.1a MDHB)}$$

$$K_e = \frac{1}{1+q(k_t-1)} = \frac{1}{1+0.95(1.7-1)} = 0.6$$

$$K_i = 1 \quad \text{assumed}$$

$$\therefore \quad S_{en} = 0.62 \times 0.99 \times 1 \times 1 \times 0.6 \times 1 \times 700 = 258.05 \text{ MPa}$$

By Goodman theory, the permissible amplitude

$$\sigma_a' = \left[1 - \frac{\sigma_m}{S_{ut}}\right].S_{en} = \left[1 - \frac{378}{1800}\right] \times 258.0 = 203.86 \text{ MPa} > (\sigma_a = 189 \text{ MPa})$$

The actual amplitude $\sigma_a = 189$ MPa is less than permissible value. Hence, no fatigue failure occurs.

Example 5.7: Figure 5.24 shows a short rectangular link-rod which is loaded by F acting on the pins at each end. The forces F vary so as to produce axial tension ranging from 70 kN to 30 kN. A nickel-chrome alloy steel (40 Nil Cr4) is used for the link. It is hardened and tempered to have $S_u = 1000$ MPa and is finished by grinding all over. Determine the dimension t to the nearest mm basing design on infinite life at 99.99% reliability and factor of safety of 1.35.

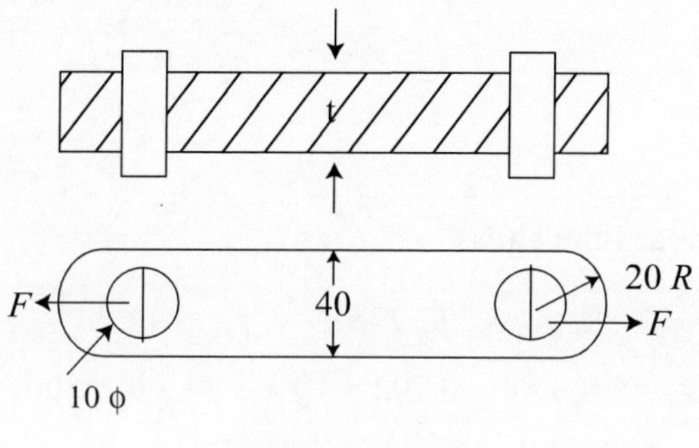

Figure 5.24

Given Data

Maximum force, $F_{max} = 70$ kN

Minimum force, $F_{min} = 30$ kN

Reliability, $R = 99.99\%$

Factor of safety, $FS = 1.35$

Material nickel steel alloy with $S_u = 1000$ MPa

SOLUTION

Step-1: Permissible stress

Test value of endurance limit stress,

$$S'_{en} = 0.5\,S_{ut} = 0.5 \times 1000 = 500 \text{ MPa.}$$

The stress concentration factor,

$$K_t \text{ for } \quad d/w = \frac{10}{40} = 0.25 \qquad \text{(Fig. 3.17 MDHB)}$$

$$K_t = 2.42.$$

The notch sensitivity factor,

$$q = 0.9 \qquad \text{(Fig 3.1a MDHB)}$$

The fatigue stress concentration factor,

$$K_f = l + q(K_f - 1) = 1 + 0.9(2.42 - 1) = 2.28$$

$$K_e = \frac{1}{K_f} = \frac{1}{2.28} = 0.44$$

Surface factor, $K_a = 0.9$ (Fig. 3.2 MDHB)

Size factor, $K_b = 0.88$ (Fig. 3.3 MDHB)

Reliability factor, $K_c = 0.702$ (Table. 3.2 MDHB)

Assume $K_d = 10$ & $K_j = 1.0$

The corrected value of endurance limit stress,

$$\begin{aligned} S_{en} &= K_a.K_b.K_c.K_d.K_e.K_iS'_{en} \\ &= 0.9 \times 0.88 \times 0.702 \times 10 \times 0.44 \times 1 \times 500 \\ &= 122.3 \text{ MPa} \end{aligned}$$

Step-2: The stresses induced

$$\sigma_{\max} = \frac{70 \times 10^3}{(40 - 10)t} = \frac{2.33 \times 10^3}{t} \text{ MPa}$$

$$\sigma_{\min} = \frac{30 \times 10^3}{(40 - 10)t} = \frac{10^3}{t} \text{ MPa}$$

Mean stress,

$$\sigma_m = \frac{\sigma_{\max} + \sigma_{\min}}{2} = \frac{2.33 \times 10^3 + 10^3}{2t} = \frac{3.33 \times 10^3}{2t} \text{ MPa}$$

Stress amplitude,

$$\sigma_a = \frac{\sigma_{\max} - \sigma_{\min}}{2}$$

$$= \frac{2.33 \times 10^3 - 10^3}{2t} = \frac{1.33 \times 10^3}{2t} \text{ MPa}$$

Step-3: Thickness for infinite-life

Using Goodman theory,

$$\frac{\sigma_m}{S_u} + \frac{\sigma_a}{S_{en}} = \frac{1}{N}$$

$$\therefore \quad \frac{\left(\frac{3.33 \times 10^3}{2t}\right)}{1000} + \frac{\left(\frac{1.33 \times 10^3}{2t}\right)}{122.3} = \frac{1}{1.35}$$

$$\therefore \quad t = \frac{1.35}{2}[3.33+10.87] = 9.588 \text{ mm}$$

Adopt $t = 10$ mm

Example 5.8: A rotating shaft is made of steel with material properties $S_{ut} = 620$ MPa, $S_y = 480$ MPa, $S'_{en} = 207$ MPa, $S_{su} = 462.0$ MPa, $S_{sy} = 276$ MPa, $S_{sen} = 117$ MPa The shaft is subjected to steady torsional shear stress of 62 MPa and to completely reversed torsional stress of 41 MPa due to vibration. Find the factor of safety against fatigue failure as well as static failure.

Give Data

Material properties

In tension,

Ultimate stress, $S_{ut} = 620$ MPa

Yield stress, $S_y = 480$ MPa

Endurance limit stress, $S_{en} = 207$ MPa

In shear,

Ultimate stress, $S_{su} = 462.0$ MPa

Yield stress, $S_{sy} = 276$ MPa

Endurance limit stress, $S'_{sen} = 117$ MPa

The load

Steady torsional stress, $\tau_m = 62$ MPa

Torsional stress amplitude, $\tau_a = 41.0$ MPa

SOLUTION

Maximum shear stress,

$$\tau_{max} = \tau_m + \tau_a = 62.0 + 41.0 = 103 \text{ MPa}$$

Factor of safety for static failure,

$$N_s = \frac{S_{sy}}{\tau_{max}} = \frac{276}{103} = 2.68$$

Factor of safety for fatigue failure,

$$N_f = \frac{S_{sen}}{\tau_a} = \frac{117}{41.0} = 2.85$$

Example 5.9: The shaft shown in Fig. 5.25 rotates at 1720 rpm and to have a life of 3 min at 50% reliability. The steel used has the following properties $S_{ut} = 610$ MPa $E = 207$GPa; $HB = 178$. The shaft is finished by grinding. It is simply supported in bearing at A and B and is loaded with static forces $F_1 = 6.0$ kN and $F_2 = 9$ kN. Find the factor of safety.

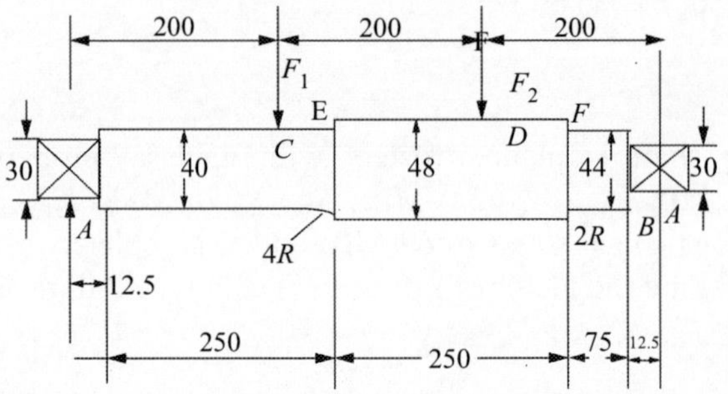

Figure 5.25

Given Data

Shaft material steel with

Ultimate stress, $S_{ut} = 610$ MPa

Young's modulus, $E = 207$ GPa

Hardness, $HB = 178$ BHN

The shaft speed, $N = 1720$ rpm

Life of shaft, $L = 3$ min $= 5160$ cycles

Lateral loads, $F_1 = 6.0$ kN

$F_2 = 9.0$ kN

SOLUTION

Reaction at A

Taking moment about B,

$$0R_n = \frac{6 \times 400 + 9 \times 200}{600} = 7.0 \text{ kN}$$

For reaction at B taking moment about A,

$$R_B = \frac{6 \times 200 + 9 \times 400}{600} = 8.0 \text{ kN}$$

Bending moment (BM) at E,

$$\begin{aligned} M_E &= R_A \times 262.5 - F_1 \times 62.5 \\ &= 7 \times 262.5 - 6 \times 62.5 \\ &= 1462.5 \text{ kN-mm.} \end{aligned}$$

BM at C, $\quad M_C = R_A \times 200 = 7 \times 200 = 1400$ kN-mm

BM at D, $\quad M_D = R_B \times 200 = 8.0 \times 200 = 1600$ kN-mm

BM at F, $\quad M_F = 8 \times 87.5 = 700.0$ kN-mm

Even though M_D is maximum there is no stress concentration. At E, stress concentration occurs and hence the section is more critical.

Stress concentration at E for $r/d = 4/40 = 0.1$ and $D/d = \frac{48}{40} = 1.2$ from Fig. 3.6 MDHB, $K_t = 1.62$.

This increases the stress and failure is more likely to occur at E.

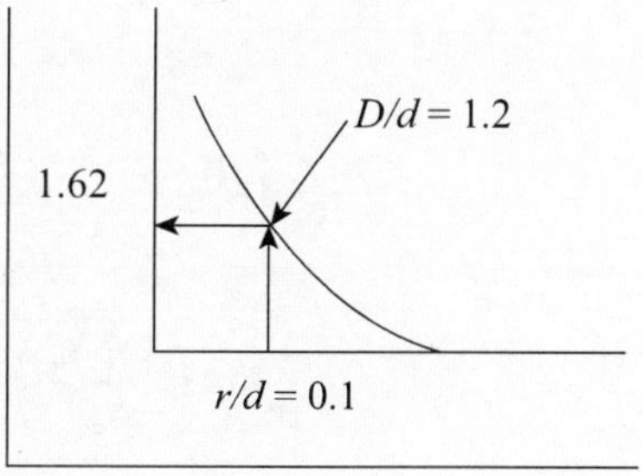

Bending stress at E,

$$\sigma_b = \frac{M_E}{2} = \frac{1462.5}{\frac{\pi}{32} \times 40^3} = (232.67) \text{ MPa}$$

Test value endurance limit stress, $S'_{en} = 0.5\,S_{ut} = 0.5 \times 610 = 305$ MPa

Corrections are

Surface factor, $K_a = 0.75$ (Fig. 3.2 MDHB)

Size factor, $K_b = 0.75$ (Fig. 3.3 MDHB)

Reliability, $K_c = 1.0$

Temperature factor, $K_d = 1.0$

Notch sensitivity factor, $q = 0.85$ (Fig. 3.1a MDHB)

Fatigue stress concentration factor, $K_f = 1 + q(K_t - 1)$

$$= 1 + 0.85(1.62 - 1) = 1.527$$

$\therefore$ Stress factor, $K_e = \frac{1}{K_f} = \frac{1}{1.527} = 0.655$

Miscellaneous factor, $K_i = 1.0$ assumed

The corrected endurance stress,

$$S_{en} = K_a \cdot K_b \cdot K_c \cdot K_d \cdot K_e \cdot K_i \cdot S'_{en}$$
$$= 0.75 \times 0.75 \times 1 \times 1 \times 0.655 \times 1 \times 305 = (112.37)\text{ MPa}$$

The fatigue strength equation is

$$S_f = AN^b = 10^c \cdot N^b$$

where,

$$c = \log\left[\frac{(0.8S_u)^2}{S_{en}}\right] = \log\left[\frac{(0.8 \times 610)^2}{112.37}\right] = 3.326$$

$$b = -\frac{1}{3}\log\left(\frac{0.8S_u}{S_{en}}\right) = -\frac{1}{3}\log\frac{(0.8 \times 610)}{112.37} = -0.2126$$

Then permissible-stress by S-N equation,

$$S_f = 10^{3.326}N^{-0.2126} = 2118.4(5160)^{-0.2126} = 344.11\text{ MPa}$$

Then the factor of safety, $FS = \frac{S_f}{\sigma_b} = \frac{344.11}{232.67}$

$$= \mathbf{1.49}$$

Example 5.10: A steel shaft is subjected to the following stresses:

(a) A steady torsional shear stress of 100 MPa and alternating bending stress of 120 MPa.
(b) A steady tensile stress of 100 MPa and alternating torsional stress of 75 MPa.
(c) A torsional stress varying from 70 MPa to 150 MPa and an alternating bending stress of 75 MPa.

The shaft material has the following properties:

Ultimate tensile stress, $S_{ut} = 550$ MPa
Yield stress, $S_y = 420$ MPa
Endurance stress, $S_{en} = 275$ MPa

Find the factor of safety for

(1) static failure
(2) fatigue failure in each case

Given Data

Shaft material properties

Ultimate tensile stress, $S_{ut} = 550$ MPa
Yield stress, $S_y = 420$ MPa
Endurance limit stress, $S'_{en} = 275$ MPa

SOLUTION

Assume there is no stress concentration

$$K_t = 1 \text{ and}$$
$$K_a = K_b = K_c = K_d = K_e = K_i = 1$$
$$\therefore\ S_{en} = S'_{en} = 275 \text{ MPa}$$

(a) Torsional mean stress, $\tau_m = 100$ MPa.

Torsional amplitude, $\tau_a = 0$.

Mean direct stress, $\sigma_m = 0$

Stress amplitude, $\sigma_a = 120$ MPa

Equivalent-direct stresses are as follows:

Von-Mises stress amplitude (Eqn. 3.21 MDHB)

$$\sigma_a^1 = \sqrt{\sigma_a^2 + 3\tau_a^2} = \sqrt{120^2 + 3 \times 0} = 120 \text{ MPa}$$

Von-Mises mean stress,

$$\sigma'_m = \sqrt{\sigma_m^2 + 3\tau_m^2}$$

$$= \sqrt{0 + 3 \times 100^2}$$

$$= 173.2 \text{ MPa}$$

Equivalent static stress,

$$\sigma_e = \sigma'_m + \frac{S_y}{S_{en}} \sigma'_a \quad \text{(Eqn. 3.18 MDHB)}$$

$$= 173.2 + \left(\frac{420}{275}\right) \times 120 = 356.4 \text{ MPa}$$

Factor of safety by normal stress theory,

$$\sigma_e = S_d = \left(\frac{S_y}{N}\right)$$

$$\therefore \quad 356.4 = \frac{420}{N}$$

$$\therefore \quad FS = N = \frac{420}{356.4} = 1.178$$

By shear theory,

$$\sqrt{\sigma^2 + 4\tau^2} = S_d = \frac{S_y}{N.}$$

$$N = \frac{S_y}{\sqrt{\sigma^2 + 4\tau^2}} = \frac{420}{\sqrt{120^2 + 4 \times 100^2}}$$

$$= \frac{420}{233.2} = 1.8$$

(b) Stresses are

$$\sigma_m = 100 \text{ MPa and } \sigma_a = 0.0$$

$$\tau_m = 0 \text{ and } \tau_a = 75.0 \text{ MPa}$$

(i) By static failure

Maximum direct stress,

$$\sigma_{\max} = \sigma_m + \sigma_a = 100 \text{ MPa}$$

Maximum shear stress,

$$\tau_{\max} = \tau_m + \tau_a = 0 + 75.0 = 75 \text{ MPa}$$

By shear theory,

$$\sqrt{\sigma_{\max}^2 + 4.\tau_{\max}^2} = \frac{S_y}{N}$$

$$\sqrt{100^2 + 4. \times 75^2} = \frac{(420)}{N}$$

$$N = \frac{420}{\sqrt{100^2 + 4 \times 75^2}} = \frac{(420)}{180.2}$$

$$= 2.33$$

(ii) By fatigue failure

Von-Mises equivalent stress, (Eqn. 3.22 MDHB)

$$\sigma_m' = \sqrt{\sigma_m^2 + 3\tau_m^2} = \sqrt{100^2 + 0} = 100 \text{ MPa}$$

Mean stress, (Eqn. 3.21 MDHB)

$$\sigma_a' = \sqrt{\sigma_a^2 + 3\tau_a^2} = \sqrt{0 + 3 \times 75^2} = 129.9 \text{ MPa}$$

Equivalent static stress, (Eqn. 3.18 MDHB)

$$\sigma_e = \sigma_m' + \left(\frac{S_y}{S_{en}}\right).\sigma_a' = 100 + \left(\frac{420}{275}\right) \times 129.9 = 298.4 \text{ MPa}$$

Equating to design stress,

$$\sigma_e = \frac{S_y}{N}$$

$$298.4 = \frac{420}{N}$$

$$\text{Factor of safety, } N = \frac{420}{298.4} = 1.4$$

(c) Stresses are

$$\sigma_m = 0 \text{ and } \sigma_a = 75 \text{ MPa}$$

$$\tau_m = \frac{150 + 70}{2} = 110 \text{ MPa}$$

$$\tau_a = \frac{150 - 70}{2} = 40 \text{ MPa}$$

(i) Factor of safety for static failure

$$\sigma_{\max} = \sigma_m + \sigma_a = 0 + 75$$
$$= 75 \text{ MPa}$$
$$\tau_{\max} = \tau_m + \tau_a = 110 + 40$$
$$= 150 \text{ MPa}$$

By shear theory, (Eqn. 3.20 MDHB)

$$\sqrt{\sigma_{\max}^2 + 4\tau_{\max}^2} = S_d$$
$$\sqrt{75^2 + 4 \times 150^2} = \frac{420}{N}$$
$$N = \frac{420}{\sqrt{75^2 + 4 \times 150^2}} = \frac{420}{309.2}$$
$$= 1.36$$

By principal stress theory,

$$\frac{\sigma}{2} + \sqrt{\left(\frac{\sigma}{2}\right)^2 + \tau^2} = S_d = \frac{S_y}{N}$$
$$\frac{75}{2} + \sqrt{\left(\frac{75}{2}\right)^2 + 150^2} = \frac{420}{N}$$
$$N = \frac{420}{\frac{75}{2} + \sqrt{\left(\frac{75}{2}\right)^2 + 150^2}} = \frac{420}{192.1} = 2.86$$

(ii) Factor of safety for fatigue failure

Equivalent Von-Mises stresses,

$$\sigma_m^1 = \sqrt{\sigma_m^2 + 3\tau_m^2} = \sqrt{0 + 3 \times 110^2} \quad \text{(Eqn. 3.22 MDHB)}$$
$$= 190.5 \text{ MPa}$$
$$\sigma_a^1 = \sqrt{\sigma_a^2 + 3\tau_a^2} \quad \text{(Eqn. 3.21 MDHB)}$$
$$= \sqrt{75^2 + 3 \times 40^2}$$
$$= 102.1 \text{ MPa}$$

Equivalent static stress,

$$\sigma_e = \sigma_m^1 + \frac{S_y}{S_{en}}\sigma_a^1 = S_d = \frac{S_y}{N}$$

$$= 190.5 + \left(\frac{420}{275}\right) 102.1 = \frac{420}{N}$$

$$\therefore N = \frac{420}{346} = 1.2$$

Example 5.11: A spherical vessel 600 mm in dia is made from steel and is to withstand a fluctuating pressure of 0 to $p_{\max}$. Find the value of the pressure for:

(a) static failure
(b) fatigue failure

Material properties are $S_{ut} = 440$ MPa, $S_y = 370$ MPa, shell thickness is 3.0 mm.

Given Data

Spherical vessel diameter, $D = 600$ mm.
Material steel with properties
Ultimate stress, $S_{ut} = 440$ MPa
Yield stress, $S_y = 370$ MPa
Shell thickness, $t = 3$ mm

SOLUTION

Assume that the joints do not cause any stress concentration. The maximum stress in the shell,

$$\sigma_{\max} = \frac{p_{\max} D}{4t} = \frac{600}{4 \times 3} \cdot p_{\max}$$

$$= 50\, p_{\max}$$

(a) For static failure,

$$\sigma_{\max} = S_y$$

$$\therefore \quad 50 p_{\max} = 370$$

$$p_{\max} = 370/50 = 7.5 \text{ MPa}$$

(b) For fatigue failure,

$$\sigma_m = \frac{\sigma_{\max} + \sigma_{\min}}{2} = \frac{50 p_{\max} + 0}{2} = 25 p_{\max}$$

Stress amplitude,

$$\sigma_a = \frac{\sigma_{\max} - \sigma_{\min}}{2} = \frac{50p_{\max} - 0}{2} = 25p_{\max}.$$

Endurance limit,

$$S_{en} = S'_{en} = 0.5S_{ut} = 0.5 \times 440 = 220 \text{ MPa}.$$

Then by Soderberg theory,

$$\frac{\sigma_m}{S_y} + \frac{\sigma_a}{S_{en}} = 1$$

$$\therefore \quad \frac{25p_{\max}}{370} + \frac{25p_{\max}}{220} = 1$$

$$\therefore \quad p_{\max} = \frac{370 \times 220}{25(220+370)} = 5.52 \text{ MPa}.$$

Example 5.12: A shaft supported on bearings 450 mm apart carries a vertical load of 10 kN at its centre. It transmits a torque varying between 800 N-m to 1600 N-m. The material properties are $S_{ut} = 1200$ MPa and $S_y = 900$ MPa. Assume a stress-concentration factor $K_t = 1.2$. the surface finish factor $K_a = 0.9$ and size factor $K_b = 0.9$, notch sensitivity factor $q = 0.9$ and the factor of safety $N = 1.5$. Design a suitable diameter for the shaft.

Given Data

Length of shaft, $L = 450$ mm

(simply supported)

Torque transmitted, $T_{\max} = 1600$ N-m

$T_{\min} = 800$ N-m.

The lateral central load, $W = 10$ kN.

Material properties

Ultimate stress, $S_{ut} = 1200$ MPa

Yield stress, $S_y = 900$ MPa

Correction factors

$K_a = 0.9, K_b = 0.9,\ q = 0.9,\ K_t = 1.2$

Factor of safety, $N = 1.2$.

SOLUTION

Step-1: Permissible (design) stress

Endurance limit stress (test),

$$S'_{en} = 0.5S_{ut} = 0.5 \times 1200 = 600 \text{ MPa}$$

Fatigue stress concentration factor,

$$K_f = 1 + q(K_{t-1}) = 1 + 0.9(1.2 - 1) = 1.18$$

Correction factors, $K_e = 1/K_f = \dfrac{1}{1.18} = 0.85$

Corrected value of endurance limit stress,

$$\begin{aligned} S_{en} &= K_a \cdot K_b \cdot K_c \cdot K_d \cdot K_e \cdot K_f \cdot S'_{en} \\ &= 0.9 \times 0.9 \times 1 \times 1 \times 0.85 \times 1 \times 600 \\ &= 413 \text{ MPa} \end{aligned}$$

Step-2: Bending moment

$$\begin{aligned} M = \pm\frac{WL}{4} &= \frac{10 \times 450}{4} \\ &= 1125 \text{ kN-mm} = 1125 \text{ N-m} \\ &= \pm 1125 \times 10^3 \text{ N-mm} \end{aligned}$$

Maximum bending moment, $M_{\max} = 1125 \times 10^3$ N-mm

Minimum bending moment, $M_{\min} = -1125 \times 10^3$ N-mm

Mean bending moment, $M_m = \dfrac{1125 \times 10^3 + (-1125 \times 10^3)}{2} = 0.0$

Amplitude of bending moment

$$M_a = \frac{1125 \times 10^3 - (-1125 \times 10^3)}{2} = \mathbf{1125 \times 10^3 \text{ N-mm}}$$

Step-3: Torsional moment

Maximum torque, $T_{\max} = 1600$ N-m

Minimum torque, $T_{\min} = 800$ N-m.

Then mean torque, $T_m = \dfrac{1600 + 800}{2} = \dfrac{2400}{2} = 1200$ N-m.

Torque amplitude,

$$T_a = \frac{1600 - 800}{2} = 400 \text{ N-m}$$

Step-4: Equivalent static values

The equivalent static value of the bending moment,

$$M_e = M_m + \left(\frac{S_d}{S_{dr}}\right) \cdot M_a \qquad \text{(Eqn. 3.16 MDHB)}$$

$$= M_m + \left(\frac{S_y}{S_{en}}\right) M_a$$

$$= 0 + \left(\frac{900}{413}\right) \times 1125 \times 10^3$$

$$= 2450 \times 10^3 \text{ N-mm.}$$

Equivalent static value of torque,

$$T_e = T_m + \left(\frac{S_d}{S_{dr}}\right) T_a \qquad \text{(Eqn. 3.17 MDHB)}$$

$$= T_m + \left(\frac{S_y}{S_{en}}\right) \cdot T_a$$

$$= 1200 + \left(\frac{900}{413}\right) \times 400$$

$$= 2017.67 \text{ N-m} = 2017.67 \times 10^3 \text{ N-mm.}$$

Step-5: Diameter of shaft

Using maximum shear theory, the shaft diameter is

$$d = \left[\frac{16}{\pi S_{sd}} \left\{\sqrt{M_e^2 + T_e^2}\right\}\right]^{1/3} \qquad \text{Eqn. 4.8a MDHB for } K = 0$$

where,

$$S_{sd} = \frac{1}{2} S_d = \frac{1}{2} \cdot \frac{S_y}{N} = \frac{1}{2} \times \frac{900}{1.5} = 300 \text{ MPa}$$

$$d = \left[\frac{16}{\pi \times 300}\left\{\sqrt{(2450 \times 10^3)^2 + (2071.67 \times 10^3)^2}\right\}\right]^{1/3}$$

$$= 37.9 \text{ mm}$$

Choose $d = 40$ mm.

Example 5.13: Estimate the life in hours according to Miner's rule of a piston rod of a reciprocating pump that operates at 30 rpm. The expected life schedule is as follows:

(i) 20 min at $\sigma_{a1} = 550$ MPa
(ii) 30 min at $\sigma_{a2} = 600$ MPa
(iii) 10 min at $\sigma_{a3} = 530$ MPa

The material of the piston rod is carbon steel C45 with hardness = 310 BHN.
Given Data

Rotating speed, $n = 30$ rpm.

Loads, $\sigma_{a1} = 550$ MPa for 20 min
$\sigma_{a2} = 600$ MPa for 30 min
$\sigma_{a3} = 530$ MPa for 10 min

Material of the rod - C45

Hardness = 310 BHN

SOLUTION
Step-1: Permissible stress

For C45, the ultimate stress

$$S_u = 3.2 \text{ BHN} = 3.2 \times 310 = 992.0 \text{ MPa}$$

Test value of endurance limit stress,

$$S'_{en} = 0.5\, S_u = 0.5 \times 992 = 496.0 \text{ MPa}$$

Assuming no correction,

$$S_{en} = S'_{en} = 496.0 \text{ MPa}$$

Step-2: The fatigue strength by Basquin equation

$$S_f = aN_f^b$$

i.e.,

$$\log S_f = \log a + b \, \log N_f$$
$$= c + b. \log N_f.$$

With S_f

$$S_f = S_{en} \text{ for } N_f = 10^6$$
$$S_f = 0.8\ S_u \text{ for } N_f = 10^3$$

We get,

$$b = -\frac{1}{3}\log\left(\frac{0.8\ S_u}{S_{en}}\right)$$
$$c = \log\left[\frac{(0.8\ S_u)^2}{S_{en}}\right]$$

For this case,

$$b = -\frac{1}{3}\log\left(\frac{0.8\times 992}{496}\right) = -0.068$$
$$c = \log\left[\frac{(0.8\ S_u)^2}{S_{en}}\right] = \log\left[\frac{(0.8\times 992)^2}{496}\right] = 3.10$$
$$a = 10^c = 10^{3.10} = 1258.9$$

Then Basquin equation is

$$\boxed{S_f = 1258.9\ N_f^{-0.068}}$$

Step-3: Life cycles for individual loads

$$N_f = 10^{4558}\,(S_f)^{-\frac{1}{0.068}}$$
$$= 3.8\times 10^{4.5}\ S_f^{-14.71}$$
$$= \left(\frac{1258.9}{S_f}\right)^{14.71}$$

The life cycles for various loads are

For $\sigma_{a1} = 550$ MPa

$$N_1 = \left(\frac{1258.9}{550}\right)^{14.71} = 195045.5 \text{ cycles.}$$

For $\sigma_{a2} = 600$ MPa

$$N_2 = \left(\frac{1258.9}{600}\right)^{14.71} = 54{,}223.2 \text{ cycles.}$$

For $$\sigma_{a3} = 530 \text{ MPa}$$

$$N_3 = \left(\frac{1258.9}{530}\right)^{14.71} = 3{,}36{,}335.7 \text{ cycles.}$$

Step-4: Miner's rule for total life cycles

$$\frac{n_1}{N_1} + \frac{n_2}{N_2} + \frac{n_3}{N_3} = 1$$

$$\therefore \quad \frac{t_1.n}{N_1} + \frac{t_2.n}{N_2} + \frac{t_3.n}{N_3} = 1$$

If T is the total time,

$$\frac{(t_1/T)}{N_1} + \frac{(t_2/T)}{N_2} + \frac{(t_3/T)}{N_3} = \frac{1}{n.T} = \frac{1}{N_T}$$

$$\frac{\alpha_1}{N_1} + \frac{\alpha_2}{N_2} + \frac{\alpha_3}{N_3} = \frac{1}{N_T}.$$

But $$\alpha_1 = 20/60 = 0.333$$

$$\alpha_2 = 30/60 = 0.500$$

$$\alpha_3 = 10/60 = 0.167$$

$$\therefore \quad \frac{1}{N_T} = \frac{0.333}{195045.5} + \frac{0.5}{54233.3} + \frac{0.167}{336335.7}$$

$$= 11.428 \times 10^{-6}$$

Total life cycles, $$N_T = 10^6/(11.428)$$

$\therefore$ Life in time duration, $$T = \frac{N_T}{n} = 10^6/(11.428 \times 30) = 2916.8 \text{ min} = 48.6 \text{ hours.}$$

Example 5.14: The working cycle of a component subjected to completely reversed loading is consisting of load schedule as follows:

(i) at $\sigma_{a1} = 350$ MPa for 85% of the time

(ii) at $\sigma_{a2} = 400$ MPa for 12% of the time

(iii) at $\sigma_{a3} = 500$ MPa for 3% of the time.

The material for the component is carbon steel 50 C4 with $S_u = 660$ MPa and $S_{en} = 280$ MPa. Determine the life of the component.

SOLUTION

For material 50 C4,

$$S_{ut} = 660 \text{ MPa} \quad \text{and} \quad S_{en} = 280 \text{ MPa}$$

Step-1: *S-N* line (Basquin equation) is

$$\log S_f = c + b\log N_f.$$

where,

$$b = -\frac{1}{3}\log\left(\frac{0.8S_u}{S_{en}}\right) = -\frac{1}{3}\log\left(\frac{0.8\times 660}{280}\right) = -0.092$$

$$c = \log\left[\frac{(0.8S_{ut})^2}{S_{en}}\right] = \log\left[\frac{(0.8\times 660)^2}{280}\right] = 3.0$$

Basquin equation is

$$S_f = 10^c \cdot N_f^b = 10^3 N^{-0.092}$$

or by rearranging, the life cycles in terms of stress amplitude is given as

$$N_f = 10^{c/b}.S_f^{1/b} = 10^{3/.092}.(S_f)^{-\frac{1}{0.092}}$$

$$= \left(\frac{1000}{S_f}\right)^{10.87}$$

Step-2: Life cycles for individual loads
For load $\sigma_{a1} = 350$ MPa

$$N_1 = \left(\frac{1000}{350}\right)^{10.87} = 90360.86 \text{ cycles.}$$

For load $\sigma_{a2} = 400$ MPa

$$N_2 = \left(\frac{1000}{400}\right)^{10.87} = 21164.5 \text{ cycles.}$$

For load $\sigma_{a3} = 500$ MPa

$$N_3 = \left(\frac{1000}{500}\right)^{10.87} = 1871.5 \text{ cycles.}$$

Step-3: Total life cycles by Miner's rule

$$\frac{n_1}{N_1} + \frac{n_2}{N_2} + \frac{n_3}{N_3} = 1$$

i.e.,

$$\frac{\alpha_1}{N_1} + \frac{\alpha_2}{N_2} + \frac{\alpha_3}{N_3} = \frac{1}{N_T}$$

where $\alpha_1 = 0.85; \alpha_2 = 0.12; \alpha_3 = 0.03$

$$\therefore \qquad \frac{0.85}{90360} + \frac{0.12}{21164.5} + \frac{0.03}{1871.5} = \frac{1}{N_T}$$

$$\therefore \qquad \frac{1}{N_T} = 3.11 \times 10^{-5}$$

$$N_T = 32154.34 \text{ cycles}$$

Example 5.15: A circular shaft of 500 mm length is supported freely at its two ends. It is acted upon by a central concentrated cyclic load having minimum value of 20 kN and a maximum value of 50 kN. Determine the diameter of bar by taking a factor of safety of 2, size effect of 0.85 surface finish factor of 0.9. The material properties of the bar are given as ultimate strength of 650 MPa, yield strength of 500 MPa and the endurance limit stress of 350 MPa.

Given Data

Shaft length, $L = 500$ mm; simply supported

Material properties $S_u = 650$ MPa; $S_y = 500$ MPa; $S'_{en} = 350$ MPa

Modifying factors

Surface finish factor, $K_a = 0.9$

Size-effect factor, $K_b = 0.85$

Factor of safety, $n = 2$

Cyclic-load, $F_{\min} = 20$ kN, $F_{\max} = 50$ kN

SOLUTION

Step-1: Bending loads

The bending moment at the centre,

$$M = \frac{Fl}{4}$$

Minimum bending moment,

$$M_{\min} = \frac{F_{\min} \cdot L}{4} = \frac{20 \times 10^3 \times 500}{4} = 25 \times 10^5 \text{ N-mm}$$

Maximum bending moment,

$$M_{max} = \frac{F_{max} \cdot L}{4} = \frac{50 \times 10^3 \times 500}{4} = 62.5 \times 10^5 \text{ N-mm}$$

Mean bending moment,

$$M_m = \frac{M_{max} + M_{min}}{2} = \frac{62.5 \times 10^5 + 25 \times 10^5}{2} = 4375 \times 10^3 \text{ N-mm}$$

Amplitude of bending moment,

$$M_o = \frac{M_{max} - M_{min}}{2} = \frac{62.5 \times 10^5 - 25 \times 10^5}{2} = 1875 \times 10^3 \text{ N-mm}$$

Step-2: Bending stresses,

Mean bending stress,

$$\sigma_m = \frac{M_m}{Z} = \frac{32 \times 4375 \times 10^3}{\pi \times d^3} = \frac{44563.4 \times 10^3}{d^3} \text{ MPa}$$

Bending stress amplitude,

$$\sigma_a = \frac{M_a}{Z} = \frac{32 \times 1875 \times 10^3}{\pi \times d^3} = \frac{19098.6 \times 10^3}{d^3} \text{ MPa}$$

Step-3: Design stress

Corrected endurance limit stress,

$$S_{en} = K_a \cdot K_b \cdot K_c \cdot K_d \cdot K_e \cdot K_i \cdot S'_{en}$$

$$= 0.9 \times 0.85 \times 1 \times 1 \times 1 \times 1 \times 350$$

$$= 267.75 \text{ MPa}$$

Step-4: Soderberg equation for design (Eqn. 3.12 MDHB)

$$\frac{\sigma_a}{S_f} + \frac{\sigma_m}{S_y} = \frac{1}{n} \quad \text{where } S_f = S_{en} \text{ for infinite life}$$

$$\frac{(19098.6 \times 10^3/d^3)}{267.75} + \frac{(44563.4 \times 10^3/d^3)}{500} = \frac{1}{2}$$

$$d^3 = 320.91 \times 10^3$$

$$d = 68.46 \text{ mm} = 70 \text{ mm}$$

Example 5.16: A cold drawn steel cantilever member shown in Fig. 5.26 is subjected to a transverse load at its end that varies from 50 N upward to 150 N downward and axial load that varies from 100 N compressive to 500 N tensile. Determine the required diameter at the step for infinite life using a factor of safety of 2. The material has an ultimate strength of 550 MPa and yield strength of 470 MPa. Take the notch sensitivity factor for the fillet as 0.9.

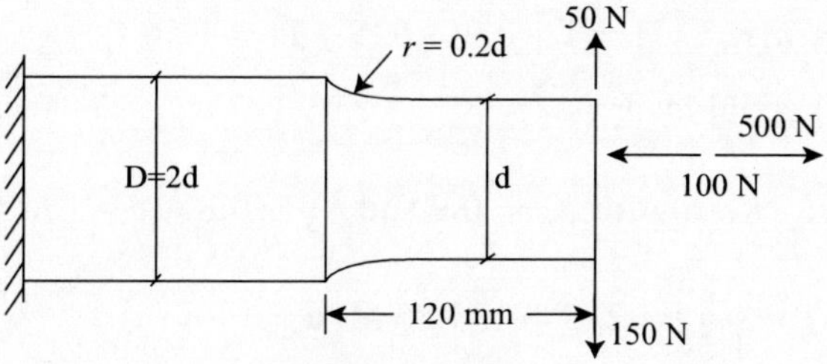

Figure 5.26

Given Data

Axial load $= -100$ N to $+500$ N

Lateral load $= 50$ N $\uparrow$ $+150$ N $\downarrow$

Step $= 2d$ to d

Fillets radius, $r = 0.2d$

Length $= 120$ mm

Material properties

$S_u = 550$ MPa

$S_y = 470$ MPa

$S'_{en} = \frac{1}{2}550 = 275$ MPa

Notch factor, $q = 0.9$

SOLUTION

Step-1: Axial stresses due to axial load

Axial load,

$$\sigma_1 = \frac{F_{\min}}{A} = -\frac{100}{\frac{\pi}{4}d^2} = -\left(\frac{127.32}{d^2}\right) \text{ MPa}$$

$$\sigma_2 = \frac{F_{\max}}{A} = \frac{500}{\frac{\pi}{4}d^2} = \left(\frac{636.62}{d^2}\right) \text{ MPa}$$

$$\text{Mean stress, } \sigma_{m_1} = \left(\frac{636.62}{d^2} - \frac{127.32}{d^2}\right)/2 = \left(\frac{254.65}{d^2}\right) \text{ MPa}$$

$$\text{Stress amplitude, } \sigma_{a_1} = \left(\frac{636.62}{d^2} + \frac{127.32}{d^2}\right)/2 = \left(\frac{381.97}{d^2}\right) \text{ MPa}$$

The stress concentration factor for $D/d = 2r$ and $\frac{r}{d} = 0.2$ (Fig. 3.5 MDHB)

$K_t = 1.625$

Fatigue factor, $K_f = 1 + q(K_t - 1) = 1 + 0.9(1.625 - 1) = 1.56$

Corrected endurance for axial load,

$$S_{en} = K_a \cdot K_b \cdot K_c \cdot K_d \cdot K_e \cdot K_i \cdot S'_{en} \text{ for } K_a = 0.9 \text{ and } K_b = 0.85$$

$$= 0.9 \times 0.85 \times 1 \times 1 \times \frac{1}{1.56} \times 275 = 134.85 \text{ MPa}$$

Equivalent static stress, (Eqn. 3.18 MDHB)

$$\sigma_{e_1} = \sigma_{m_1} + \left(\frac{S_y}{S_{en}}\right)\sigma_{a_1} = \frac{254.65}{d^2} + \left(\frac{470}{13464}\right)\left(\frac{381.97}{d^2}\right)$$

$$= \left(\frac{1588.02}{d^2}\right) \text{MPa}$$

Step-2: Bending stresses due to lateral loads

Due to upward bending moment,

$$M_u = 50 \times 120 \text{ N-mm} = 6000 \text{ N-m}$$

Bending stresses at the upper fibre,

$$\sigma_{u_1} = \frac{M_u}{Z} = -\frac{32 \times 6000}{\pi \times d^3} = -\frac{61115.44}{d^3} \text{ MPa (Compressive)}$$

at the bottom fibre,

$$\sigma_{b_1} = \frac{32 \times 600}{\pi \times d^3} = +\frac{61115.44}{d^3} \text{ MPa (Tensile)}$$

Due to downward moment, $M_d = 150 \times 120$ N-mm
At the upper fibre,

$$\sigma_{u_2} = \frac{32 \times 150 \times 120}{\pi \times d^3} = \frac{183346.49}{d^3} \text{ MPa (Tensile)}$$

At the bottom fibre,

$$\sigma_{b_2} = -\frac{183346.49}{d^3} \text{ MPa (Compressive)}$$

The stress amplitude, $\sigma_{a_2} = (183346.49 + 61115.44)/2d^3 = \dfrac{122230.96}{d^3}$

The mean stress, $\sigma_{m_2} = (183346.49 - 61115.44)/2d^3 = \dfrac{61115.525}{d^3}$

Considering the stress concentration,

$$\text{For } D/d = 2 \text{ and } r/d = 0.2 \qquad \text{(Fig. 3.6 MDHB)}$$

$$K_t = 1.5$$

$$\text{Fatigue factor, } K_f = 1 + q(K_t - 1) = 1 + 0.9(1.5 - 1) = 1.45$$

Corrected endurance stress for bending,

$$\begin{aligned} S_{en} &= K_a \cdot K_b \cdot K_d \cdot K_d \cdot K_e \cdot K_i \cdot S'_{en} \\ &= 0.9 \times 0.55 \times 1 \times 1 \times \frac{1}{1.45} \times 275 \\ &= 145.08 \text{ MPa} \end{aligned}$$

Equivalent static stress,

$$\begin{aligned} \sigma_{e_2} &= \sigma_{m_2} + \left(\frac{S_y}{S_{en}}\right)\sigma_{a_2} = \frac{61115.525}{d^3} + \left(\frac{470}{145.08}\right)\frac{12223096}{d^3} \\ &= \frac{457094.06}{d^3} \end{aligned}$$

Step-3: Total equivalent stress

$$\sigma_e = \sigma_{e_1} + \sigma_{e_2} = \frac{1588.027}{d^2} + \frac{457094.06}{d^3}$$

Step-4:

But $\sigma_e = s_d = \dfrac{S_y}{FS} = 470/2 = 235$ MPa

$$\therefore \qquad \text{Substituting the } K \text{ values, } 235 = \frac{1588.027}{d^2} + \frac{457094.06}{d^3}$$

$$\therefore \qquad d^3 - \frac{1588.027}{235}d - \frac{457094.06}{235} = 0$$

$$\therefore \qquad d^3 - 6.7575d - 1945.08 = 0$$

By trial and error method solution is,

$$d = 12.7 \text{ mm}$$

$$\text{Adopt} \quad d = 13.00 \text{ mm}$$

Example 5.17: A cantilever beam made of 35C8 steel ($S_u = 540$ MPa) is subjected to a completely reversed load of 1000 N as shown in Fig. 5.27 the notch sensitivity factor q at the fillet is 0.85 and the expected reliability is 90%. Determine the diameter of the beam for life cycle of 10,000 cycles.

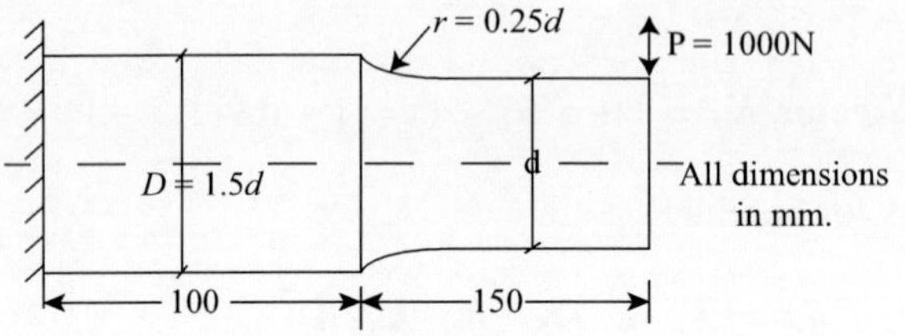

Figure 5.27

Given Data

The applied load, $P = \pm 1000$ N
$D/d = 1.5$ fillet radius, $r = 0.25d$
$l = 150$ mm

SOLUTION
Step-1: Material properties
Beam material 35C8-steel

$$S_u = 540 \text{ MPa}$$

$$S'_{en} = \frac{1}{2} \times 540 = 270 \text{ MPa}$$

$$\text{Notch sensitivity, } q = 0.85$$

$$\text{Reliability, } R = 90\%$$

From MDHB, Fig. 3.6 for $D/d = 1.5$ $r/d = 0.25$

$$K_t = 1.35$$

∴ Fatigue factor,

$$K_f = 1 + q(K_t - 1) = 1 + 0.85(1.35 - 1)$$
$$= 1.2975$$

$$\therefore \quad K_e = \frac{1}{K_f} = \frac{1}{1.2975} = 0.7707$$

Surface factor K_a (Fig. 3.2 MDHB)

$$K_a = 0.8 \quad \text{for machined with notch}$$

The size factor assumed,

$$K_b = 0.85$$

and reliability factor (Table 3.2 MDHB)

$$K_c = 0.897$$

Assume $K_d = 1$ and $K_i = 1$.

Then, the corrected endurance limit stress

$$\begin{aligned} S_{en} &= K_a \cdot K_b \cdot K_c \cdot K_d \cdot K_e \cdot K_i \cdot S'_{en} \\ &= 0.8 \times 0.85 \times 0.897 \times 1 \times .7707 \times 1 \times 270 \\ &= 126.926 \text{ MPa} \end{aligned}$$

Step-2: Bending stresses

Mean load, $P_m = \dfrac{p_{\max} + p_{\min}}{2} = \frac{1000-1000}{2} = 0$

$$\text{Load amplitude,} \quad P_a = \frac{p_{\max} - p_{\min}}{2} = \frac{1000 + 1000}{2} = 1000 \text{ N}$$

Stress amp[illegible]

$$\sigma_a = \frac{M_a}{Z} = \left(\frac{P_a \cdot L}{\frac{\pi}{32} d^3} \right)$$

$$\sigma_a = \frac{1000 \times 150}{\frac{\pi}{32} d^3} = \frac{1527887.5}{d^3} \text{ MPa}$$

Step-3: Diameter for fatigue strength

Using Soderberg equation (Eqn. 3.12 MDHB)

$$\frac{\sigma_a}{S_f} + \frac{\sigma_m}{S_y} = \frac{1}{n}$$

Assume factor of safety, $n = 1.5$

Fatigue stress (Eqn. 3.14 MDHB)

$$\begin{aligned} S_f &= aN^b \text{ for } N = 1000 \\ &= 0.9 \times 540 \times N^{-.0855} \\ &= 221.124 \text{ MPa} \end{aligned}$$

Substituting in Soderberg equation,

$$\frac{(1527887.5)/d^3}{221.129} + \frac{0}{S_y} = \frac{1}{1.5}$$

$$d^3 = \frac{1527887.5 \times 1.5}{221.129} = 10364.227$$

$$\therefore \quad d = 21.8 \text{ mm}$$

$$\therefore \quad \text{Adopt}\, d \cong \mathbf{22.0\ mm}$$

Example 5.18: A cantilever beam of cold drawn C 30 steel ($S_u = 550$ MPa, and $S_y = 470$ MPa) Subjected to load which varies from $-F$ to $3F$ as shown in Fig. 5.28. Determine the maximum load that this member can withstand for an indefinite life using a factor of safety of 2. The stress-concentration effect has to be considered with notch sensitivity of 0.9. Analyse at the fillet section only.

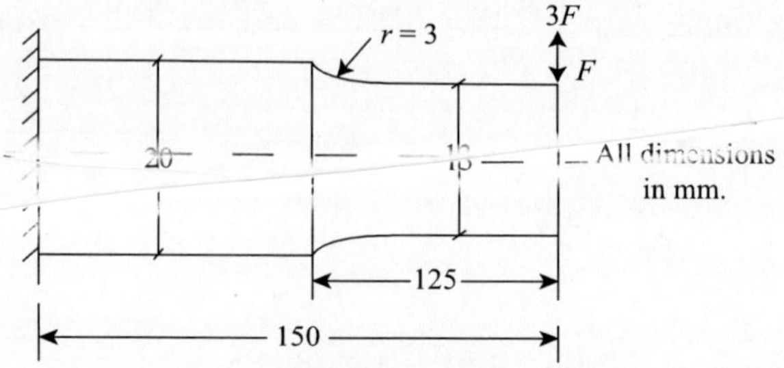

Figure 5.28

Given Data
$D = 20$ mm, $d = 13$ mm $r = 3$ mm.
Length, $L = 125$ m
$S_u = 550$ MPa and $S_y = 470$ MPa, $S'_{en} = \frac{1}{2}$, $S_u = 275$ MPa
$q = 0.9$ and $n = 2$

SOLUTION
Step-1: Bending stresses
Bending moment,

$$M_{\min} = -F \times L = -125F \text{ N-mm}$$

$$M_{\max} = 3F \times L = 3 \times 125F$$

$$= 375F \text{ N-mm}$$

Bending stresses,

$$\sigma_{\min} = \frac{M_{\min}}{\frac{\pi}{32}d^3} = \frac{-125F}{\frac{\pi}{32}\times(d)^3} = (-0.5795F)\text{ MPa}$$

$$\sigma_{\max} = \frac{125\times 3F}{\frac{\pi}{32}\times(13)^3} = (1.7386F)\text{ MPa}$$

Mean stress,

$$\sigma_m = \frac{\sigma_{\max}+\sigma_{\min}}{2} = \left(\frac{1.7386F-0.5795F}{2}\right) = 0.5795F$$

Stress amplitude,

$$\sigma_a = \frac{1.7386F+0.5795F}{2} = (1.15925F)\text{ MPa}$$

Step-2: Design-stress
Correction factors for endurance limit
Surface factor, $K_a = 0.87$ (Fig. 3.2 MDHB)
The size factor, $K_b = 0.85$ (Approx.) (Fig. 3.3 MDHB)
Reliability factor, $K_c = 1$ assumed
Temperature factor, $K_d = 1$ assumed
The stress concentration-effect factor, $K_e = \frac{1}{K_f}$
where, $K_f = 1+q(K_t-1)$
and stress concentration factor,

$$K_t \text{ for } D/d = 20/13 = 1.54 \text{ and } r/d = 3/13 = 0.23$$

From Fig. 3.6 MDHB, $K_t = 1.35$

$$\therefore \quad K_f = 1+0.9(1.35-1) = 1.315$$

$$\therefore \quad K_e = 1/1.315 = 0.76045$$

and miscellaneous factor, $K_j = 1$ assumed

Then corrected endurance limit stress is

$$S_{en} = K_a\cdot K_b\cdot K_c\cdot K_d\cdot K_i\cdot S'_{en}$$
$$= 0.88\times 0.85\times 1\times 1\times 0.76045\times 1\times 275$$
$$= 156.426\text{ MPa}$$

Step-3: Permissible load for infinite life
Using Soderberg equation (Eqn. 3.12 MDHB)

$$\frac{\sigma_u}{S_f}+\frac{\sigma_m}{S_y}=\frac{1}{n} \quad \text{where } S_f=S_{en} \text{ for infinite life}$$

$$\frac{1.15905F}{156.42}+\frac{0.5795F}{470}=\frac{1}{2}$$

$$F=\mathbf{57.85\ N}$$

and

$$3F=\mathbf{173.55\ N}$$

Example 5.19: A pulley is keyed to a shaft midway between two bearings. The shaft is made of steel (with yield stress $S_y = 380$ MPa and ultimate stress $S_u = 660$ MPa). The bending moment at the pulley varies from -300 N-m to $+500$ N-m and the torque varies from -100 N-m to 200 N-m The fatigue stress concentration for the keyway in bending and torsion are 1.6 and 1.3 respectively. The factor of safety is 1.5. Determine the diameter of the shaft.

Given Data

Torque, $T = 100$ N-m to $+200$ N-m
Bending moment, $M = 300$ N-m to $+500$ N-m

Material properties,

$$S_u = 660 \text{ MPa, and } S_y = 380 \text{ MPa}$$

$$S'_{en} = \frac{1}{2}S_u = 330 \text{ MPa}$$

Fatigue factors

in bending, $K_{f_b} = 1.6$
in torsion, $K_{ft} = 1.3$
Factor of safety, $n = 1.5$

SOLUTION

Step-1: Equivalent static bending stress
Mean value of bending moment,

$$M_m = \frac{M_{\max}+M_{\min}}{2} = \frac{500-300}{2} = 100 \text{ N-m}$$

Amplitude of bending moment,

$$M_a = \frac{M_{\max}-M_{\min}}{2} = \frac{500+300}{2} = 400 \text{ N-m}$$

Static design stress,

$$S_d = \frac{S_y \cdot B}{nK_t} \quad \text{(Eqn. 3.1 MDHB)}$$

where $S_y = 380$ MPa and $K_t = 1$ for ductile material,

$$B = 1.0 \text{ (assumed) and } n = 1.5$$

$$S_d = \frac{380 \times 1}{1.5 \times 1} = 253.33 \text{ MPa}$$

Design stress for variable load,

$$S_{dr} = \frac{S'_{en}ABC}{K_f \cdot n}$$

$$= \frac{330 \times 1 \times 0.85}{1.6 \times 1.5}$$

$$= 116.875 \text{ MPa}$$

where, $S'_{en} = 330$ MPa
$A = 1$
$B = 1.0$ (assumed)
$C = 0.85$ (assumed)

Equivalent static moment, (Eqn. 3.16 MDHB)

$$M_e = M_m + \left(\frac{S_d}{S_{dr}}\right) M_a$$

$$= 100 + \left(\frac{253.33}{116.875}\right) \times 400$$

$$= 967.0 \text{ N-m}$$

Then equivalent direct stress,

$$\sigma_e = \frac{M_c}{Z} = \frac{967 \times 10^3}{\frac{\pi}{32}d^3}$$

$$= \frac{9849.9 \times 10^3}{d^3} \text{ MPa}$$

Step-2: Equivalent static torsional stress
Torsional variation,

$$\text{Mean torque, } T_m = \frac{T_{\max} + T_{\min}}{2} = \frac{200 - 100}{2} = 50 \text{ N-m}$$

$$\text{Torque amplitude, } T_a = \frac{T_{\max} - T_{\min}}{2} = \frac{200 + 100}{2} = 150 \text{ N-m}$$

The static design stress in shear,

$$S_{sd} = \frac{1}{2} S_d = \frac{1}{2} \times 253.33 = 126.665 \text{ MPa}$$

The design stress for variable torque loads,

$$S_{sdr} = \frac{S'_{sen}ABC}{K_f \cdot n} \quad \text{(Eqn. 3.8 modified for shear stress)}$$

where $A =$ load factor $= 0.7$

$B = 1$ assumed

$C = 0.85$ assumed

$K_{fs} = 1.3$ given

FS $n = 1.5$

$S'_{sen} = \frac{1}{2}S'_{en} = \frac{330}{2} = 165.0$ MPa

$\therefore$ Design stress,

$$S_{sdr} = \frac{165.0 \times 0.7 \times 1 \times 0.85}{1.3 \times 1.5} = 50.35 \text{ MPa}$$

Equivalent static torque, (Eqn. 3.17 MDHB)

$$T_e = T_m + \left(\frac{S_{sd}}{S_{sdr}}\right) \cdot T_a$$

$$= 50 + \left(\frac{126.665}{50.35}\right) \times 150$$

$$= 427.38 \text{ N-m}$$

Equivalent static shear stress,

$$\tau_e = \frac{16T_e}{\pi d^3} = \frac{16 \times 427.38 \times 10^3}{\pi \cdot d^3} = \left(\frac{2176.63}{d^3}\right) \text{ MPa.}$$

Step-3: The shaft-diameter by static theory
Applying shear theory, (Eqn. 3.20 MDHB)

$$S_d = \sqrt{\sigma_e^2 + 4\tau_e^2}$$

$$253.33 = \sqrt{\left(\frac{9849.9 \times 10^3}{d^3}\right)^2 + 4\left(\frac{2176.63 \times 10^3}{d^3}\right)^2}$$

$$d = \sqrt[3]{\frac{9849.9 \times 10^3 \times 1.093}{253.33}} = 34.89$$

$\therefore$ Choose $d = 35$ mm.

Example 5.20: A hot rolled steel shaft is subjected to a torsional moment that varies from 300 N-m clockwise to 100 N-m counterclockwise as the bending moment at the critical section varies from 400 N-m to −200 N-m. Neglecting the stress concentration effect, find the required shaft diameter. The material has an ultimate strength of 550 MPa and yield strength of 410 MPa, take the endurance limit as half the ultimate strength and a factor of safety as 2. Assume surface, size and load factor for bending as 1.111, 1.1765 and 1 and that for torsion as 1.05263, 1.1765 and 1.7 respectively.

Given Data

Torsional moment, $T_{max} = 300$ N-m CW to

$T_{min} = -100$ N-m CCW

Bending moment, $M_{max} = 400$ N-m

$M_{min} = -200$ N-m

Material properties, $S_u = 550$ MPa and $S_y = 410$ MPa

$S'_{en} = \frac{1}{2} \times 550 = 275$ MPa

Modifying factor for bending load

Surface finish factor, $C = 1/1.111$

Size factor, $B = \dfrac{1}{1.765}$

Load factor, $A = 1$

For torsional load

Surface factor, $C = \dfrac{1}{1.05263}$

Size factor, $B = \dfrac{1}{1.1765}$

Load factor, $A = \dfrac{1}{1.7}$

SOLUTION

Step-1: Equivalent static bending stress

Mean moment,

$$M_m = \frac{M_{max} + M_{min}}{2} = \frac{400 - 200}{2} = 100 \text{ N-m}$$

Moment amplitude,

$$M_a = \frac{M_{max} + M_{min}}{2} = \frac{400 + 200}{2} = 300 \text{ N-m.}$$

Design stress for static load,

$$S_d = \frac{S_y \cdot B}{n \cdot K_t} = \frac{410}{1.1765 \times 2 \times 1} \qquad \text{(Eqn. 3.1 MDHB)}$$

$$= 174.25 \text{ MPa}$$

Design stress for variable load,

$$S_{dr} = \frac{S'_{en} \cdot ABC}{n\,K_t} \quad \text{[Eqn. 3.8 MDHB]}$$

$$= \frac{275}{2 \times 1} \times \frac{1}{1} \times \frac{1}{1.1765} \times \frac{1}{1.111}$$

$$= 105.2 \text{ MPa}$$

∴ Equivalent static moment, (Eqn. 3.16 MDHB)

$$M_e = M_m + \left(\frac{S_d}{S_{Sdr}}\right) M_a$$

$$= 100 + \left(\frac{174.25}{105.2}\right) \times 300$$

$$= 596.91 \text{ N-m.}$$

Then equivalent static stress,

$$\sigma_e = \frac{M_e}{Z} = \frac{596.91 \times 10^3}{\left(\frac{\pi}{32} \times d^3\right)} = \left(\frac{6080.075 \times 10^3}{d^3}\right) \text{ MPa}$$

Step-2: Equivalent static torsional shear stress
Static design stress,

$$S_{sd} = \frac{S_{sy} \cdot B}{n \cdot K_t} = \frac{1}{2} \cdot \frac{S_y \cdot B}{n.K_t} = \frac{1}{2} S_d = \frac{174.25}{2}$$

$$= 87.125 \text{ MPa}$$

Design stress for variable load,

$$S_{sdr} = \frac{S'_{sen} \cdot ABC}{n \cdot K_t} = \frac{1}{2} \frac{S'_{sen} \cdot ABC}{n \cdot K_t}$$

$$= \frac{1}{2} \times \frac{275}{2 \times 1} \times \frac{1}{1.7} \times \frac{1}{1.1765} \times \frac{1}{1.05263}$$

$$= 32.655 \text{ MPa}$$

Equivalent static torque, [Eq. 3.17 MDHB]

$$T_e = T_m + \left(\frac{S_{sd}}{S_{sdr}}\right) T_a$$

$$= \left(\frac{300 - 100}{2}\right) + \left(\frac{87.125}{32.655}\right) \left(\frac{300 + 100}{2}\right)$$

$$= 633.61 \text{ N-m}$$

Equivalent static shear stress,

$$\tau_e = \frac{16 \cdot T_e}{\pi \cdot d^3} = \frac{16 \times 633.61 \times 10^3}{\pi \times d^3}$$
$$= \left(\frac{3226.94}{d^3}\right) \times 10^3 \text{ MPa}$$

Step-3: Shaft diameter
By maximum shear theory (Eqn. 3.20 MDHB)

$$S_d = \sqrt{\sigma_e^2 + 4 \cdot \tau_e^2}$$
$$= \sqrt{\left(\frac{6080.075 \times 10^3}{d^3}\right)^2 + 4\left(\frac{3226.94 \times 10^3}{d^3}\right)^2}$$

$$\therefore \qquad 174.25 = \left[\frac{6080.075 \times 10^3}{d^3}\right]\left[\sqrt{1 + 4\left(\frac{3226.94}{6080.075}\right)^2}\right]$$

$$d^3 = \frac{6080.075 \times 10^3 \times 1.4583}{174.25}$$
$$= 50.8854 \times 10^3$$
$$\therefore \qquad d = 37.05 \text{ mm}$$

Adopt $d = 37.0$ mm

Example 5.21: A component machined from a plate, made of 45C8 ($S_{ut} = 630$ MPa) is shown in Fig. 5.29. It is subjected to a completely reversed axial force of 50 kN. The expected reliability is 90% and factor of safety is 2. Determine the plate thickness t for infinite life, if the notch sensitivity factor is 0.8.

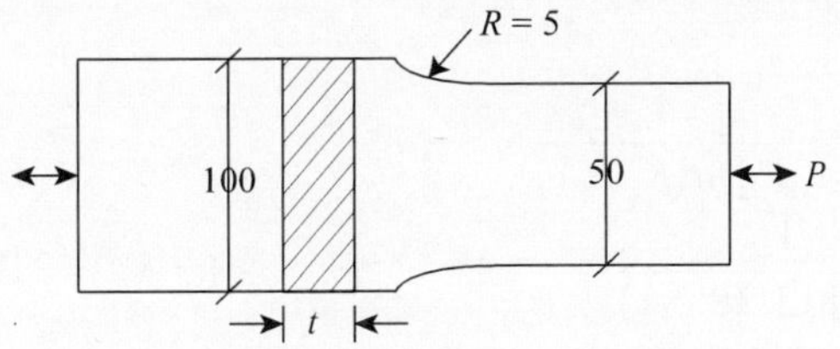

All dimensions in mm.

Figure 5.29

Given Data

Axial pull, $P = \pm 50$ kN (completely reversed)
Plate material 45C8
Ultimate strength, $S_u = 630$ MPa
Expected reliability, $R = 90\%$
Notch sensitivity, $q = 0.8$

SOLUTION

Step-1: Loads

$$\text{Mean load, } P_{\text{mean}} = 0$$

$$\text{Amplitude, } P_a = 50 \text{ kN}$$

Stress amplitude,

$$\sigma_a = P_a/A = \left(\frac{50 \times 10^3}{A}\right) \text{ MPa}$$

$$= \frac{50 \times 10^3}{50\,t} = \left(\frac{1000}{t}\right) \text{ MPa}$$

Step-2: Design stress

Endurance limit stress,

$$S'_{en} = \frac{1}{2} S_u = \frac{630}{2} = 315 \text{ MPa}$$

$$\text{Surface factor, } K_a = 0.8 \quad \text{Assumed}$$

$$\text{Size factor, } K_b = 0.8 \quad \text{Assumed}$$

$$\text{Reliability factor, } K_c = 0.897 \qquad \text{(Table 3.2 MDHB)}$$

$$\text{Temperature factor, } K_d = 1.0 \quad \text{Assumed}$$

$$\text{Miscellaneous factor, } K_i = 1.0 \quad \text{Assumed}$$

Fatigue stress concentration factor,

$$K_e = 1/K_f = \frac{1}{1 + q(K_t - 1)}$$

$$= \frac{1}{1 + 0.8(2.28 - 1)} \qquad (\because K_t = 2.28 \text{ from Fig. 3.13 MDHB})$$

$$= 0.494$$

Then the corrected endurance limit stress,

$$S_{en} - K_a \cdot K_b \cdot K_c \cdot K_d \cdot K_e \cdot K_t \cdot S'_{en}$$

$$= 0.8 \times 0.8 \times 0.897 \times 1 \times 0.494 \times 1 \times 315$$
$$= 89.33 \text{ MPa}$$

Step-3: Thickness of the plate

Then using Soderberg equation (Eqn. 3.12 MDHB)

$$\frac{\sigma_a}{S_f} + \frac{\sigma_m}{S_y} = \frac{1}{n}$$

where $\sigma_m = 0, \sigma_a = \left(\frac{1000}{t}\right)$; $S_f = S_{en} = 89.33$; $n = 2$

$$\therefore \quad \frac{(1000/t)}{89.33} + \frac{0}{S_y} = \frac{1}{2}$$

$$\therefore \quad t = \frac{1000 \times 2}{89.33} = 22.39 \text{ mm} = \mathbf{23\ mm}$$

EXERCISES FOR PRACTICE

5.1. Define 'geometric stress concentration factor' and notch sensitivity.

5.2. What are the various factors influencing endurance strength?

5.3. Find the diameter of a shaft to transmit twisting moments varying from 800 Nm to 1600 Nm. The ultimate tensile strength for the material is 600 MPa, and yield stress is 400 MPa. Assume the stress concentration factor = 1.2; surface finish factor = 0.8 and size factor = 0.85.

5.4. Determine the size of a piston rod subjected to a total fluctuation between 15 kN compression and 25 kN in tension. The endurance limit is 360 MPa and yield strength 400 MPa. The FS = 1.5; surface factor = 0.88; stress concetration factor = 2.25.

5.5. A cantilever beam of rectangular cross section has a span of 800 mm. The rectangular cross section of the beam has a depth of 200 mm. The free end of the beam is subjected to a transverse load that fluctuates between 8 kN down and 5 kN up. Selecting carbon steel C30 as material for the beam and a value of 2.5 for the factor of safety, determine the width of rectangular cross section. Surface factor and size factor are respectively 0.95 and 0.90. Stress concentration factor is 1.65.

5.6. A cantilever beam of circular cross section is subjected to an alternating bending stress at a point on the outer fibre in the plane of the support that varies from 20 MPa (compression) to 32 MPa (tension). At the same time there is an alternative stress due to axial loading that varies from 12 MPa (tension) to 28 MPa (compression). The material has an ultimate strength of 400 MPa, yield strength of 320 MPa. Assume that fatigue stress concentration factor = 1.2, size correction factor = 0.85, and surface correction factor = 0.9. Determine:

i) the equivalent normal stress due to axial loading;

ii) the equivalent normal stress due to bending; and

iii) the total equivalent normal stress due to axial loading and bending.

5.7 A round rod of diameter '1.2d' is reduced to a diameter 'd' with a fillet radius of '0.1d'. This stepped rod is to sustain a twisting moment that fluctuates between + 2.5 kNm and + 1.5 kNm together with a bending moment that fluctuates between + 1.5 kNm and – 1.5 kNm. The rod is made of carbon steel C40. Determine a suitable value for 'd'.

5.8. A shaft of circular cross section is subjected to a twisting moment that fluctuates between 1000 kNm and 800 kNm and also a bending moment that fluctuates between + 600 kNm and – 300 kNm. The material selected for the shaft has a shear stress value of 90 MPa as endurance limit and a shear stress value of 110 MPa as the yield limit. Determine the diameter of the solid circular shaft taking a value of 2.50 for the factor of safety. Surface factor, size factor and load factor can be taken as 0.90, 0.85 and 1.0 respectively. Shear stress concentration factor is 1.80 and the notch sensitivity is 0.95.

5.9. A hot rolled square bar of 500 mm length is supported freely at its two ends. It is acted upon by a central concentrated cyclic load having a minimum value of 20 kN and a maximum value of 50 kN. Determine the size of the bar by taking a factor of safety of 1.5, size factor of 0.85, surface finish factor of 0.9. The material properties of the bar are given as: ultimate strength of 650 MPa, yield strength of 500 MPa and endurance strength of 350 MPa.

5.10. A steel rod is subjected to a reversed axial load of 180 kN. Find the diameter of the rod for a factor of safety of 2. Neglect column action. The material has an ultimate tensile strength of 1070 MPa and yield strength of 910 MPa. The endurance limit is reversed, bending may be assumed to be one half of the ultimate tensile strength. The correction factors are as follows: Load factor = 0.7; surface finish factor = 0.8; size factor = 0.85; stress concentration factor = 1.

5.11. A beam is subjected to load causing a bending stress of 200 MPa. The material of the beam is 2320 steel, oil quenched. The stress concertation factor is 1.2. Find the factor of safety, when,

(i) the stress is completely reversed;

(ii) the stress is varying between – 150 MPa and 200.0 MPa;

(iii) the load is steady.

5.12. (a) Define the following terms:

i) Geometric stress concentration factor

ii) Notch sensitivity factor

iii) Fatigue stress concertation factor.

iv) Factor of safety

5.13. (a) Explain the Soderberg theory for fatigue failure.

(b) A pulley is keyed to a shaft midway between two antifriction bearings. The bending moment at the pulley varies from – 170 N-m to + 510 N-m as the torsional varies from 55 N-m to 165 N-m; the frequency of variation is the same as the shaft speed. The shaft is made of cold drawn steel having an ultimate strength of 540 MPa and yield strength of 400 MPa. Determine the required diameter for infinite life. The stress concentration factors for the keyway in bending and torsion may be taken as 1.6 and 1.3 respectively. Use design factor N = 1.5.

5.14. (a) Define (i) stress concentration factor (ii) notch sensitivity (iii) fatigue failure (iv) endurance limit stress.

(b) Determine the diameter of a solid shaft that is subjected to alternating torque of 160 to 200 N-m. The shaft is made of C40 hot rolled steel. Assume a factor of safety of 2, and yield stress of 240 MPa, endurance limit as 140 MPa.

5.15. A bar of circular cross section is subjected to alternating tensile forces varying from a minimum of 200 kN to a maximum of 500 kN. It is to be manufactured from a material with an ultimate tensile strength of 900 MPa, and an endurance limit of 500 MPa. Determine the diameter of the bar using a factor of safety of 3.5, related to ultimate tensile strength and 4 related to endurance limit and stress concentration factor of 1.65 for fatigue load. Use Goodman straight line as basis for design.

5.16. (a) State and explain the criteria of fatigue design for infinite life of machine parts.

(b) Find the maximum fatigue load (completely reversed) using Soderberg criterion that can be applied to a plate with a 5 mm dia hole in the middle. The width and thickness of the plate are 30 mm and 10 mm respectively. The plate supports a steady load of 25 kN. The yield point stress for the plate material is 350 MPa and its endurance limit is 140 MPa. Use a factor of safety of 2.

5.17. A fuel pump pusher rod is to be designed for a life of 10^5 cycles, while it is being subjected to a released load of 7.8 kN. The test data indicate that the material of the rod has a yield point stress of 380 MPa and endurance limit stress of 260 MPa. For reversed bending, but has an endurance strength of 345 MPa for 10^5 cycles. For a factor of safety of 2, determine the diameter of the push rod for both finite life of 10^5 cyles and for infinite life.

5.18. (a) Derive the Soderberg design equation.

(b) A section of a shaft of diameter 'd' is joined to a section of diameter '2d' with a fillet radius of '0.2d'. The shaft is supported at the larger end and is free at the smaller end. The length from free end to the change of cross section is 125 mm, to the support is 180 mm. The shaft is subjected to a transverse load at its free end that varies from 45 N up to 135 N down, as an axial load that varies from 450 N compressive to 250 N tensile. Find the diameter 'd' for a factor of safety of 2. The properties of the shaft material are, ultimate tensile stress $S_u = 550$ MPa, yield stress $S_y = 470$ MPa, and endurance limit stress $S_{en} = 275$ MPa.

5.19. A cold drawn steel rod of circular cross section, made of C30 steel, is subjected to variable bending moment of 500 kN-m to 1000 kN-m as the axial load varies from 5 kN to 10 kN, both the maximum values occur at the same instant every time. The rod has a discontinuity, that causes stress concentration of 1.6 in bending and 1.5 in axial loading. Assume the notch sensitivity to be 0.8 and a factor of safety to be 2. Determine the diameter of the 200 for infinite life.

5.20. (a) Define endurance limit strength of a machine component and list the factors on which it depends.

(b) A connecting rod for a steam engine is made of SAE1045 anealed steel. The rod is subjected to completely reversed axial load of 7.5 kN. Determine the diameter of the rod, for a factor of safety of 2. Consider the endurance correction factors.

5.21. (a) Explain the Goodman theory of fatigue failure.

(b) A solid circular shaft, 20 mm in diameter, is subjected to torsional shear stress varying from 0 to 35 MPa and at the same time to an axial stress varying from – 15 MPa to + 30 MPa. The frequency of variation of these stresses is same as the shaft speed. The shaft material has an ultimate strength of 540 MPa, yield strength of 400 MPa and endurance strength of 270 MPa. Determine the factor of safety by the Soderberg theory of fatigue failure.

5.22. (a) Explain the types of fatigue stresses with sketches.

(b) A piston rod is subjected to a completely reversed axial load of 115 kN. The material for the rod has an ultimate strength of 900 MPa. Assume that the surface is machined and there is no discontinuity. Determine the rod diameter for a factor of safety equal to 1.5.

5.23. (a) Explain the various theories of fatigue failure.

(b) A steel shaft transimits 100 kW of power at 100 rpm, the torque varies over a range of 40%. This is also subjected to a completely reversed bending moment of 100 kN-m. The material of the shaft has a yield strength of 400 MPa and endurance strength of 300 MPa. Assume the size factor as 1.2, surface factor as 0.9, stress concentration factor as 1.14 and notch sensivity factor as 0.9. Determine the shaft diameter for a factor of safety of 2.5.

5.24. (a) Explain the cumulative fatigue damage theories.

(b) A grooved shaft of diameter D has a semicircular groove of radius = 0.1D. The material of the shaft has a torsional yield strength of 300 MPa and torsional endurance strength of 210 MPa. The shaft is subjected a twisting moment varying between – 300 N-m and + 900 N-m. Find the diameter of the shaft for a factor of safety of 3. Assume the following size factor = 0.85, surface factor = 0.87, notch sensitivity factor = 0.95.

5.25. A thin walled cylindrical pressure vessel of diameter 600 mm is subjected to an internal pressure varying between 3 MPa and 6 MPa. Find the thickness of the pressure vessel for a factor of safety = 2.5. The material properties are, yield stress = 360 MPa, endurance stress = 240 MPa.

SHORT ANSWER QUESTIONS

5.1. How are the variable stresses classified?
5.2. What are the fluctuating stresses?
5.3. What are the repeated stresses?
5.4. What are reversed stresses?
5.5. What is fatigue failure?
5.6. What is endurance limit stress?
5.7. How do you determine the endurance limit stress?
5.8. What is S-N curve?
5.9. What is low cycle fatigue?
5.10. What is high cycle fatigue?
5.11. What is fatigue life?
5.12. What is Basquin equation?
5.13. Which are factors affecting the endurance limit stress?
5.14. What is surface finish factor?
5.15. What is size factor?
5.16. What is reliability factor?
5.17. What is fatigue stress concentration factor?
5.18. What is notch sensitivity factor?
5.19. What is temperature factor?
5.20. What is Goodman equation?
5.21. What is Soderberg equation?
5.22. What is static equivalent stress?
5.23. What is cumulative fatigue damage?
5.24. What is Miner's rule cumulative fatigue damage?
5.25. What is Manson's method for cumulative fatigue damage?

CHAPTER 6

SHAFTS

A shaft is a rotating or stationary member. Usually, it is circular in cross section and supports gears, pulleys, flywheels, sprockets, cranks. It is supported in bearings. It transmits power from driving machine (motor or engine) to driven machines. It is subjected to torsion and bending. An axle is a stationary shaft, subjected to bending only. In some cases, there may be axial force also. The design of the shaft should consider the effect of these loads. There are three basic considerations, namely, (a) strength, (b) torsional rigidity and (c) lateral rigidity.

The shaft diameter is normally determined by strength considerations and then checked for rigidity.

6.1 DESIGN FOR STRENGTH

Design based on static theories of failure

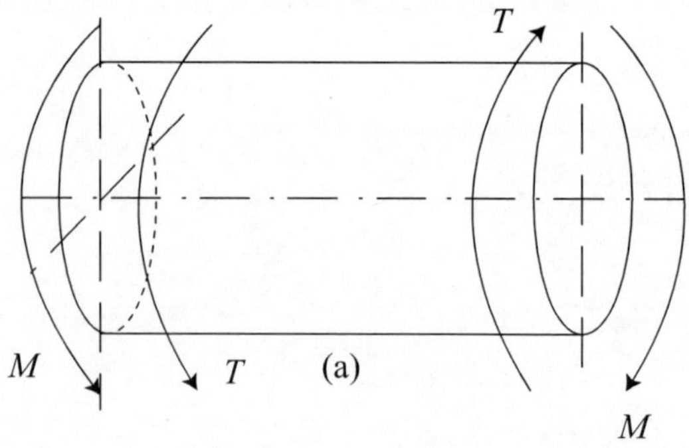

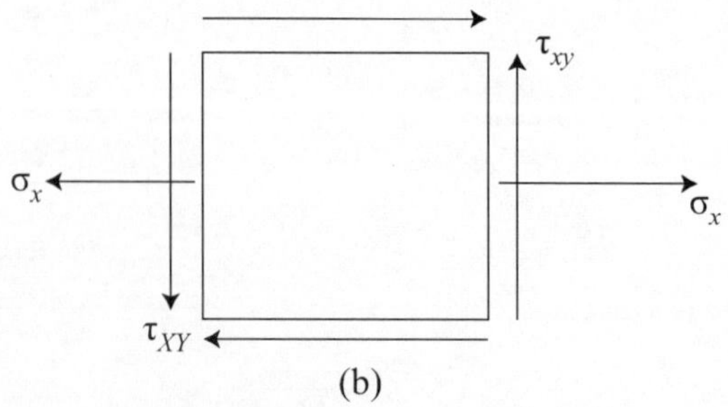

Figure 6.1

We consider the bending and torsional loads acting on the shaft.

The stresses acting on an element on surface of the shaft are:

$$\text{Bending stress, } \sigma_x = \frac{32M}{\pi d^3}$$

$$\text{Torsional shear, } \tau_{xy} = \frac{16T}{\pi d^3}$$

where,

M = bending moment at the critical section

T = Torsional moment at the critical section

The maximum principal stress,

$$\sigma_{\max} = \frac{\sigma_x}{2} + \sqrt{\left(\frac{\sigma_x}{2}\right)^2 + \tau_{xy}^2}$$
$$= \frac{16\,M}{\pi d^3} + \sqrt{\left(\frac{16M}{\pi d^3}\right)^2 + \left(\frac{16\,T}{\pi d^3}\right)^2}$$

By principal stress theory,

$$\sigma_{\max} = \frac{S_y}{n} = S_d$$

where, n is the factor of safety.

$$\therefore \quad S_d = \frac{16}{\pi d^3} \times \left[M + \sqrt{M^2 + T^2}\right]$$
$$= \frac{32\,M_{eq}}{\pi d^3}$$

where, $M_{eq} = \left[\left(M + \sqrt{M^2 + T^2}\right)\right]/2$ is called equivalent bending moment.

Then the diameter is

$$\therefore d = \left[\frac{32\,M_{eq}}{\pi S_d}\right]^{1/3}$$
$$= \left[\frac{16}{\pi S_d}\left(M + \sqrt{M^2 + T^2}\right)\right]^{1/3} \tag{6.1}$$

The maximum shear stress,

$$\tau_{\max} = \sqrt{\left(\frac{\sigma_x}{2}\right)^2 + \tau_{xy}^2}$$
$$= \frac{16}{\pi d^3}\sqrt{M^2 + T^2}$$
$$= \left(16\,T_{eq}/\pi d^3\right)$$

where $T_{eq} = \sqrt{M^2 + T^2}$ is called the equivalent torque.

By maximum shear stress theory,

$$\tau_{\max} = S_{sd} = \frac{S_{sy}}{n} \qquad \text{where, } S_{sy} = \frac{S_y}{2}$$
$$S_{sd} = \left(\frac{16\,T_{eq}}{\pi d^3}\right)$$
$$\therefore d = \left[\frac{16\,T_{eq}}{\pi S_{sd}}\right]^{1/3} = \left[\left(\frac{16}{\pi\,S_{sd}}\right)\left(\sqrt{M^2 + T^2}\right)\right]^{1/3} \tag{6.2}$$

In the above expressions, the average values of M and T are to be modified to account for the shock and fatigue effect and are replaced by the effective values M_e, T_e as given below.

$$M_e = C_m M \text{ for bending} \tag{6.2(i)}$$

$$\text{or} = C_m M + \frac{aFd_o(1+K^2)}{8} \text{ for bending and axial load} \tag{6.2(ii)}$$

$$T_e = C_t\, T \text{ for torsion} \tag{6.2(iii)}$$

Where C_m and C_t are shock and fatigue factors.(refer Table 6.2)
The values of C_m and C_t are as per the recommendations of ASME.
Then general equations for hollow shafts are as follows:
According to maximum normal stress theory,

$$d_o = \left[\frac{16}{\pi\, S_{sd}}\left(M_e + \sqrt{M_e^2 + T_e^2}\right)\left(\frac{1}{1-K^4}\right)\right]^{1/3} \tag{6.3}$$

According to shear theory,

$$d_o = \left[\frac{16}{\pi\, S_d}\left(\sqrt{M_e^2 + T_e^2}\right) \times \left(\frac{1}{1-K^4}\right)\right]^{1/3}. \tag{6.4}$$

where $K = d_i/d_o$ is the ratio of inside diameter to outside diameter. The variable nature of bending moment and torque can be considered by defining the effective values M_e and T_e as follows:

$$M_e = M_m + \frac{K_f \cdot S_y}{S_{en}} M_a$$

$$T_e = T_m + \frac{K_f \cdot S_y}{S_{en}} T_a$$

where M_m and T_m are the mean values and M_a and T_a are amplitude values.
The other approach is to apply the Von-Mises-Henky Theory as follows:
Then the diameter is given by

$$d_o = \left[\frac{16}{\pi S_d}\left\{M_{me} + \left(\frac{K_f S_y}{S_{en}}\right) M_{ae}\right\}\left(\frac{1}{1-K^4}\right)\right]^{1/3} \tag{6.5}$$

where,

$$M_{me} = \sqrt{4M_m^2 + 3T_m^2}$$

$$M_{ae} = \sqrt{4M_a^2 + 3T_a^2}$$

The ASME code equation for variable loads assumes that bending moment is fully reversed and the torque is steady, and is based on the Gough ellipse.

$$\left(\frac{\sigma_a}{S_{en}}\right)^2 + \left(\frac{\tau_m}{S_{sy}}\right)^2 = 1$$

The code equation for shaft diameter is

$$d_o = \left\{\frac{32n}{\pi}\left[\sqrt{\left(\frac{K_f M_a}{S_{en}}\right)^2 + \frac{3}{4}\left(\frac{T_m}{S_y}\right)^2} \times \left(\frac{1}{1-K^4}\right)\right]\right\}^{1/3}$$

$$d_o = \left\{\frac{32}{\pi\, S_d}\left[\sqrt{\left(\frac{K_f S_y}{S_{en}} M_a\right)^2 + \frac{3}{4} T_m^2} \times \left(\frac{1}{1-K^4}\right)\right]\right\}^{1/3} \qquad (6.6)$$

6.2 MATERIAL FOR SHAFTS

From low to medium carbon steels in either cold rolled or hot rolled form are used for the shaft. The alloy steels are used for higher strength. The stainless steel or bronze are sometimes used for marine and other corrosive environments.

The permissible or design stress is given by

$$S_{sd} = 0.18\ S_{ut}$$

$$\text{OR} \qquad S_{sd} = 0.3\ S_{yt}$$

The lower of the two is used.

Refer Tables 6.3 and 6.6 for design stresses.

6.3 RIGIDITY

Torsional rigidity

The torsional equation,

$$\frac{T}{J} = \frac{\tau}{r} = \frac{G\theta}{L}$$

provides for the calculation of angular twist the shaft experiences due to the torsion (T).

$$\theta = \frac{T.L}{G.J} \qquad (6.7)$$

The angle of twist for a line shaft is limited to 1° in a length, $L = 20\,d$.

In specific application the max value of the angle may be specified.

$$\theta = 0.3\,\text{degree/m} \quad \text{for machine tool shaft}$$

$$\theta = 3\,\text{degree/m} \quad \text{for line shaft}$$

Lateral Rigidity

The shaft is supported in bearings. For lateral deflections the shaft may be treated as a simply supported beam or cantilever beam with applied lateral loads. Using the beam equations the deflection can be calculated. The amount of max deflection is specified for particular application.

6.4 COMPARISON OF SOLID AND HOLLOW SHAFTS

Solid Shaft

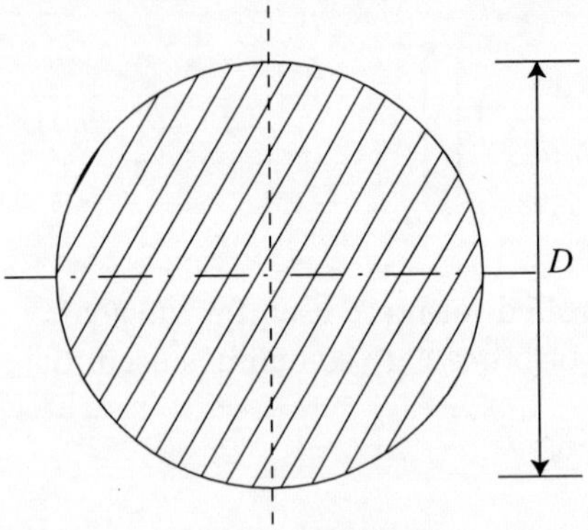

Figure 6.2

Cross-sectional area, $A_s = \frac{\pi}{4}D^2$

Polar moment of inertia, $J_s = \frac{\pi}{32}D^4$.

Hollow Shaft

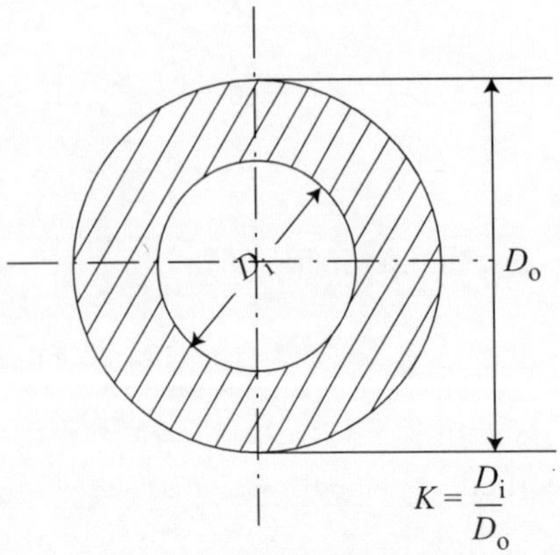

Figure 6.3

Cross-sectional area,

$$A_H = \frac{\pi}{4}(D_o^2 - D_i^2) = \frac{\pi}{4}D^2(1 - K^2)$$

Polar moment of inertia,

$$J_H = \frac{\pi}{32}(D_o^4 - D_i^4) = \frac{\pi}{32}D_o^4(1 - K^4)$$

where, $K = \frac{D_i}{D_o}$.

(a) For same diameter and length

Weight, $W = A.L.\rho.$

$$\text{For solid shaft,} \quad W_s = \frac{\pi}{4} D^2.L.\rho.$$

$$\text{For hollow shaft,} \quad W_H = \frac{\pi}{4} D_o^2(1-K^2).\text{L}.\rho.$$

$$\frac{W_H}{W_S} = \frac{\frac{\pi}{4} D_o^2(1-K^2).\text{L}.\rho}{\frac{\pi}{4} D^2\text{L}.\rho} = (1-K^2) < 1. \qquad (\because D = D_0) \tag{6.8}$$

Hollow shaft is lighter.

Torsional Strength

$$\text{For solid shaft,} \quad T_s = \frac{J_s}{r} \cdot S_{sd}$$

$$\text{For hollow shaft,} \quad T_H = \frac{J_H}{r} \cdot S_{sd}$$

$$\text{Then} \quad \frac{T_H}{T_S} = \frac{J_H}{J_S} = \frac{\frac{\pi}{32} D_o^4(1-K^4)}{\frac{\pi}{32} D^4} = (1-K^4) < 1 \qquad (\because D = D_0) \tag{6.9}$$

The torsional strength of hollow shaft is less than that of solid shaft.

Lateral Stiffness (M_θ),

$$M_\theta = \frac{G \cdot J}{L}$$

$$\text{For solid shaft,} \ M_{\theta S} = \frac{G}{L} \cdot J_S$$

$$\text{For hollow shaft,} \ M_{\theta H} = \frac{G}{L} \cdot J_H$$

$$\text{Then} \ \frac{M_{\theta\, H}}{M_{\theta\, S}} = \frac{J_H}{J_S} = (1-K^4) < 1 \tag{6.10}$$

The solid shaft is stronger.

(b) For same weight and length

For equal weight,

$$W_S = W_H$$

$$\frac{\pi}{4} D^2.\text{L}.\rho. = \frac{\pi}{4} D_o^2(1-K^2)\, L.\rho.$$

$$D^2 = D_o^2(1-K^2)$$

$$D = D_o\sqrt{1-K^2} \tag{6.11}$$

$$\text{OR } D_o = \frac{D}{\sqrt{1-K^2}} \tag{6.11a}$$

The hollow shaft has larger diameter.

Torsional Strength

For solid shaft,

$$T_S = \frac{J_S}{r} \cdot S_{sd} = \frac{\pi}{16} D^3 \cdot S_{sd}$$

For hollow shaft,

$$T_H = \frac{J_H}{r} \cdot S_{sd} = \frac{\pi}{16} D_o^3 (1-K^4) S_{sd}$$

$$\frac{T_H}{T_S} = \frac{D_o^3(1-K^4)}{D^3} = \frac{(1-K^4)}{(1-K^2)^{3/2}}$$

$$= \frac{(1+K^2)}{(1-K^2)^{1/2}} > 1 \tag{6.12}$$

The hollow shaft is stronger.

6.5 THE FORCES ACTING ON THE SHAFT IN BELT DRIVE

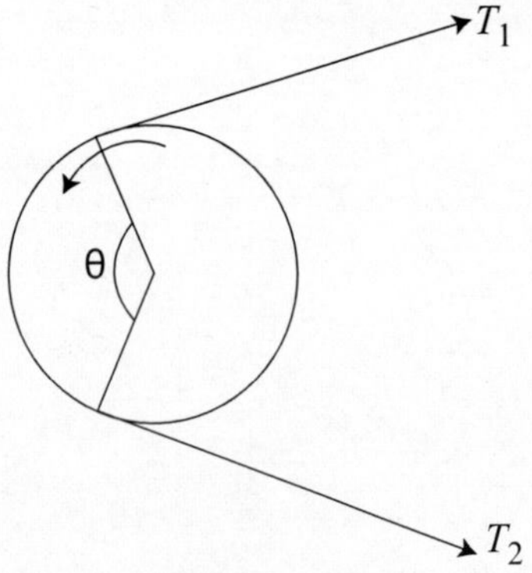

Figure 6.4

T_1 – Tension on tight side, N
T_2 – Tension on slack side, N
μ = Coefficient of friction
θ = Angle of lap (rad)

The ratio of tensions,

$$\frac{T_1}{T_2} = e^{\mu\theta} \tag{1}$$

Power transmitted,

$$P = \frac{(T_1 - T_2)V}{1000} \text{ kW}$$

$$T_1 - T_2 = \frac{1000P}{V} = \frac{60 \times 10^6 \cdot P}{\pi DN} \tag{2}$$

where, P in kW
D in mm
N in rpm
Using Eqns. (1) and (2), T_2 and T_1 can be evaluated.

$$T_2 = \frac{60 \times 10^6 P}{\pi DN(1 - e^{\mu\theta})}$$

$$T_1 = \left(\frac{60 \times 10^6 P}{\pi\, DN}\right)\left(\frac{e^{\mu\theta}}{1 - e^{\mu\theta}}\right)$$

The pull on the shaft $= T_1 + T_2$

$$F = \left(60 \times \frac{10^6\, P}{\pi\, DN}\right)\left(\frac{1 + e^{\mu\theta}}{1 - e^{\mu\theta}}\right) \quad (6.13)$$

6.6 FORCES ACTING ON THE SHAFT DUE TO GEAR DRIVE

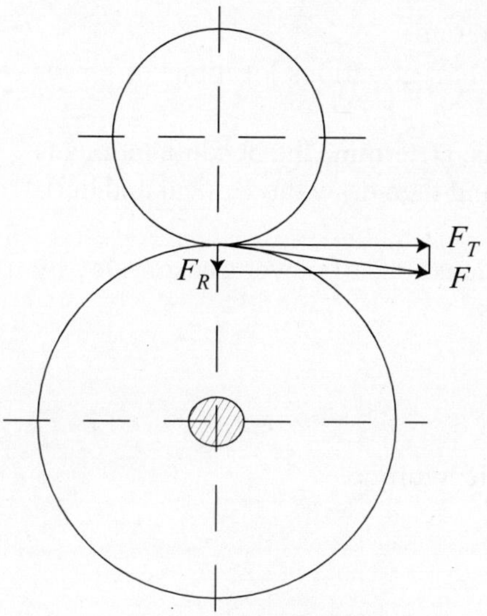

Figure 6.5

F_T = Tangential force along the common tangent
F_R = Radial force at pitch point
= $F_T \tan\psi$. 1 ψ = Pressure angle for the tooth system
= 20° or 14.5°.

Tangential force is evaluated as follows:

$$F_T = \frac{\text{Torque on gear}}{\text{Radius of the gear}}$$

$$\left.\begin{aligned} &= \frac{\left(\dfrac{955 \times 10^4 \times (\text{kW})}{N}\right)}{R_g} \\ &= \frac{955 \times 10^4 \times (\text{kW})}{N \cdot R_g} \end{aligned}\right\} \quad (6.14)$$

Resultant force on shaft,

$$F = \sqrt{F_T^2 + F_R^2} = F_T \cdot \sec\psi. \quad (6.15)$$

6.7 STRESS CONCENTRATION

In order to mount and locate several types of machines – elements on the shaft properly, a final design contains several diameters keyseats, ring grooves and other geometric discontinuities that create stress concentration. The effects of these stress concentrations have to be accounted for during the design process. But the values of the stress concentration factors are not known in the beginning because of unknown diameters and geometries. The preliminary values of K_t may be chosen approximately as follows.

Key seats, $K_t = 2.0$ for profile keyway

$= 1.6$ for weld runner keyway

Shoulder fillets, $K_t = 2.5$ for sharp fillet

$= 1.5$ for weld rounded fillet

Retaining ring, $K_t = 30$ for bending stresses

6.8 SHAFT DESIGN PROCEDURE

1. Determine the torque transmitted and draw the torque diagram.
2. Determine the forces and their directions acting on the shaft at various locations of power transmission by belt drive gear drive, etc.
3. Resolve the forces in the vertical and horizontal directions. Determine the bending moments due to these forces in the vertical and horizontal planes and then draw the vertical and horizontal bending moment diagrams.
4. Determine the resultant bending moments at the various locations of power transmission by adding the vertical and horizontal components vectorially.

$$M_R = \sqrt{M_V^2 + M_H^2}$$

5. Locate the critical section subjected to maximum equivalent torque.

$$T_{eq} = \sqrt{M_R^2 + T^2}$$

6. Select the material and design stress for the shaft using that data (Tables 4.3 and 4.6 MDHB Page 67).
7. Determine the diameter using the Eqns. 6.3 or 6.4 using the appropriate values of effective moments M_e and effective forgue T_e depending upon the approach selected, i.e., steady load approach using ASME code or variable load approach or by Von-Mises-Henky theory using Eqn. 6.5.

Table 6.1. Stress concentration factor

Particulars	K_t **factor**
Key-way	1.68 - 1.44
Straight sided splines	2
Involute splines	1.2 - 1.3

Table 6.2. Combined shock and fatigue factor

Type of loading	**Rotating shaft**		**Stationary shaft**	
	C_m	C_t	C_m	C_t
Gradually applied and steady load	1.5	1.0	1.0	1.0
Suddenly applied load with minor shock	1.5 - 2.0	1.0 - 1.5	1.5 - 2.0	1.5 - 2.0
Suddenly applied load with heavy shock	2.0 - 3.0	1.5 - 3.0	1.5 - 2.0	1.5 - 2.0

Table 6.3. Permissible stress for shafts

Grade of shafting	**Simple bending** S_d MPa	**Simple torsion** S_{sd} MPa	**Combined stress** S_{sd} MPa
1) Commercial steel shafting without allowance for key-way	110	55	55
2) Commercial shafting with allowance for key-way	85	42	42
3) Specified steel S_y-Test value of elastic limit stress S_u-Ultimate stress	0.6 S_y or 0.3 S_u	0.3 S_y or 0.18 S_u	0.3 S_y or 0.18 S_u

Table 6.4. Standard size shaft sizes in mm

6	8	10	12	14	16	18	20	22	25	28	32	36	40
45	50	56	63	71	80	100	110	125	140	160	180	200	

Table 6.5. Values of (a) in (Eqn. 4.8 (iii) MDHB)

Slenderness ratio (L/k)	0	25	50	75	100	115
Factor (a)	1.00	1.12	1.28	1.49	1.78	2.02

Table 6.6. Properties of shafting materials

Material	Percentage carben	Ultimate strength S_u MPa			Elastic limit S_y MPa			Percentage elongation
		Tension S_{ut}	Compression	Shear S_{us}	Tension S_{yt}	Compression	Shear S_{ys}	
Commercial cold rolled	0.10-0.25	492	492	246	246	246	125	35
Commercial turned	0.10-0.25	420	420	210	210	210	105	35
Hot rolled/forged	0.15-0.25	460	460	230	250	250	115	26
	0.25-0.35	492	492	246	280	280	123	24
	0.35-0.45	530	530	265	320	320	132.5	22
3.5% Nickel	0.45-0.55	564	564	282	352	352	141	20
Chrome steel	0.15-0.25	600	600	300	390	390	150	26
vanadium	0.25-0.35	632	632	316	422	422	158	25

SOLVED EXAMPLES

Example 6.1: A machinery shaft is supported on bearings 2.5 m apart and transmits 150 kW at 250 rpm. It is subjected to a lateral load of 4.0 kN located at a distance of 600 mm from the left bearing. Determine the shaft diameter for (i) steady loads (ii) steady bending load and suddenly applied torsional load.
The allowable stress in shear and bending are 40 MPa and 80 MPa respectively.

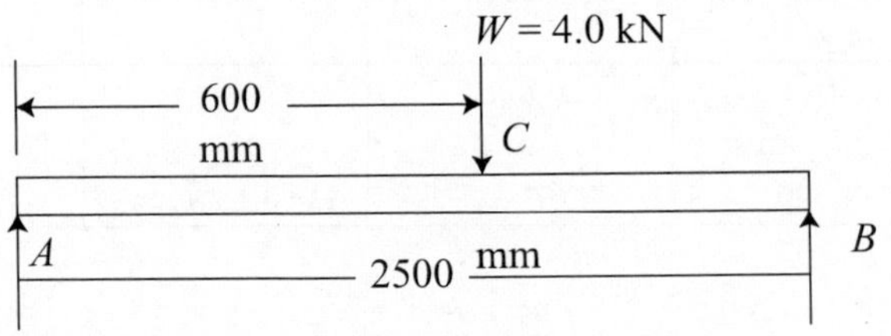

Figure 6.6

SOLUTION

Step-1: The bending moment at C

$$M = \frac{Wab}{L} = \frac{4000 \times 600 \times 1900}{2500} \text{N.mm} = 1824 \times 10^3 \text{ N-mm}$$

Step-2: The average torque

$$T = \frac{955 \times 10^4 \times 150}{250} = 5730 \times \ 10^3 \text{ N-mm}$$

Step-3: (i) For steady loads (corrected values from Table 4.2 MDHB)

$$\begin{aligned} C_m &= 1.5, \quad C_t = 1.0 \\ \therefore \qquad M_e &= C_m \quad M = 1.5 \times 1824 \times 10^3 = 2736 \times 10^3 \text{ N-mm} \\ T_e &= C_t \quad T = 5730 \times 10^3 \text{ N-mm} \end{aligned}$$

Step-4: The shaft diameter by Rankine theory

$$d = \left[\frac{16}{\pi S_d}\left(M_e + \sqrt{M_e^2 + T_e^2}\right)\right]^{1/3}$$

where,

$$S_d = 80 \text{ MPa}, \quad M_e = 2736 \times 10^3 \text{ N-mm}; \quad T_e = 5730 \times 10^3 \text{ N-mm}$$

$$\begin{aligned} d &= \left[\frac{16}{\pi \times 80}\left\{2736 \times 10^3 + \sqrt{(2736 \times 10^3)^2 + (5730 \times 10^3)^2}\right\}\right]^{1/3} \\ &= 73.79 \text{ mm} \end{aligned}$$

Step-5: The shaft diameter by shear stress theory

$$\begin{aligned} d &= \left[\frac{16}{\pi S_{sd}}\sqrt{(M_e^2 + T_e^2)}\right]^{1/3} \\ &= \left[\frac{16}{\pi \times 40}\sqrt{(2736 \times 10^3)^2 + (3730 \times 10^3)^2}\right]^{1/3} \\ &= 92.95 \text{ mm} \end{aligned}$$

Step-6: Standardise the diameter (Table 4.4 MDHB)

∴ Adopt the larger of the two

i.e., $d = 92.95 \text{mm}$

Standardizing $d = 100 \text{mm}$

Step-7: The stock and fatigue factors for the case (ii) are

$$C_m = 1.5 \quad C_t = 1.5$$

$$M_e = 1.5 \times 1824 \times 10^3 \text{ N-mm} = 2736 \times 10^3 \text{ N-mm}$$
$$T_e = 1.5 \times 5730 \times 10^3 = 8595 \times 10^3 \text{ N-mm}$$

Step-8: Since the load is predominantly torsion, we adopt shear theory, then

$$d_o = \left[\frac{16}{\pi\, S_{sd}}\sqrt{(M_e^2 + T_e^2)}\right]^{1/3}$$

$$d_o = \left[\frac{16}{\pi \times 40}\sqrt{(2736 \times 10^3)^2 + (8595 \times 10^3)^2}\right]^{1/3}$$

$$= 104.7 \text{ mm}$$

Standardizing, $d_o = 110.0$ m. (Table 4.4 MDHB)

Example 6.2: A shaft 90 cm long between bearings supports a 60 cm pulley, located at 30 cm to the right of left hand bearing and the belt drives a pulley directly below it. Another pulley of 45 cm in diameter is located at 20 cm to the left of the right hand bearing and the belt is driven from a pulley horizontally to the right. The angle of contact for both the pulleys is 180° and the tension ratio is 2.25. The maximum tension in the belt on the 60 cm pulley is 2250 N. Determine the suitable diameter for the solid shaft, with permissible stresses of 75 MPa in tension and 45 MPa in shear.

SOLUTION

Step-1: Load diagram

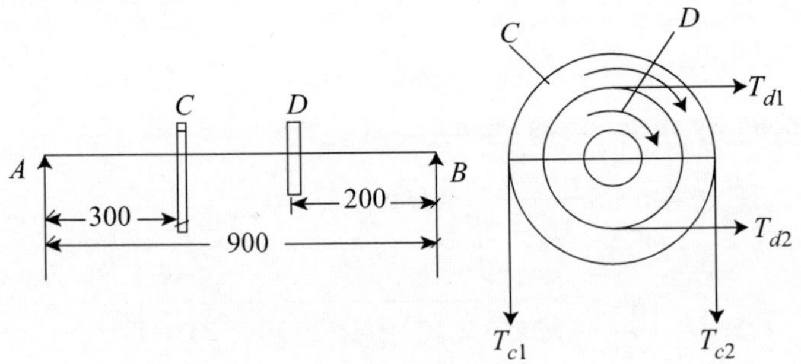

Figure 6.7

$$\frac{T_{c1}}{T_{c2}} = \frac{T_{d1}}{T_{d2}} = 2.25$$
$$T_{c1} = 2250 \text{ N}$$
$$T_{c2} = T_{c1}/2.25 = 1000 \text{ N}$$

The driving torque, $T = (T_{c1} - T_{c2})R_c$

$$= (2250 - 1000) \times 300 \text{ N-mm} = 375 \times 10^3 \text{ N-mm}$$

Then for pulley at D,

$$(T_{d1} - T_{d2})\, R_d = 375 \times 10^3$$
$$T_{d1} - T_{d2} = 375 \times 10^3/225 = 1666.7\,\text{N}$$
$$2.25\, T_{d2} - T_{d2} = 1666.7$$
$$T_{d2} = 1666.7/1.25 = 1333.3\ \text{N}$$
$$T_{d1} = 3000\ \text{N}$$

The vertical pull at C,

$$F_{cv} = 2250 + 1000 = 3250\ \text{N}$$

The horizontal pull at D,

$$F_{DH} = 3000 + 1333.3 = 4333.3\ \text{N}$$

Step-2: The vertical bending moment diagram

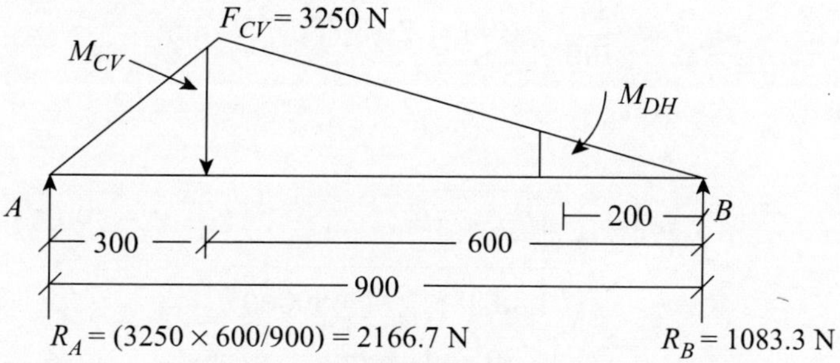

Figure 6.8

Vertical reaction at A, $R_{AV} = \dfrac{3250 \times 600}{900} = 2166.7\ \text{N}$

Vertical reaction at at B, $R_{BV} = 3250 - 2166.7 = 1083.3\ \text{N}$

The vertical moment at C,

$$M_{CV} = 2166.7 \times 300 = 6,50,000.0\ \text{N-mm}$$

The vertical moment at D,

$$M_{DV} = 1083.3 \times 200 = 2,16,660.0\ \text{N-mm}$$

Step-3: Horizontal bending moment diagram

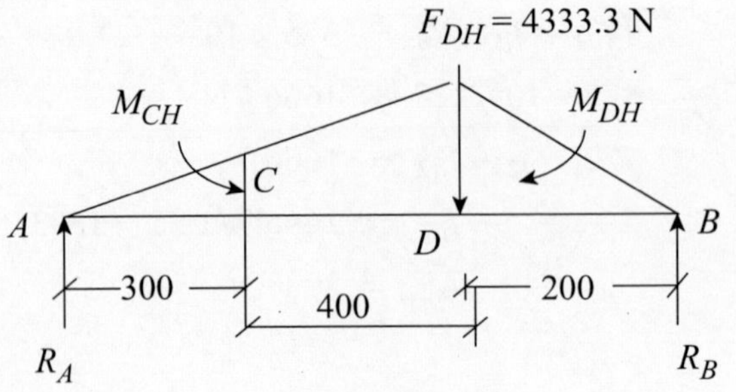

Figure 6.9

Horizontal moment at D, $M = \dfrac{Wab}{L}$

$$M_{DH} = \frac{4,333.3 \times 200 \times 700}{900} = 6,74,068.9 \text{ N-mm}$$

Horizontal moment at C,

$$M_{CH} = \frac{M_{DH}}{700} \times 300 = 2,88,886.7 \text{ N-mm}$$

Step-4: Total moments
Total moment at C,

$$\begin{aligned} M_c &= \sqrt{M_{CH}^2 + M_{CV}^2} \\ &= \sqrt{2,88,886.7^2 + 650000^2} \\ &= 7,11,305.5 \text{ N-mm} \end{aligned}$$

Total moment at D,

$$\begin{aligned} M_D &= \sqrt{M_{DH}^2 + M_{DV}^2} \\ &= \sqrt{6,74,068.9^2 + 2,16,666.7^2} \\ &= 7,08,034.8 \text{ N-mm} \end{aligned}$$

The bending moment is maximum at C. The torque being the same for both sections the section at C is critical. Step-5: Then diameter by principal stress theory.

$$d = \left[\frac{16}{\pi S_d}(M_e + \sqrt{M_e^2 + T_e^2})\right]^{1/3}$$

where, for $C_m = 1.5$ and $C_t = 1.0$ (Table 4.2 MDHB)

$$M_e = C_m \cdot M = 1.5 \times 711305.5 \text{ N-mm} = 10,66,958.25 \text{ N-mm}$$

$$T_e = C_t T = 1.0 \times 375 \times 10^3 \text{ N-mm}$$

and $S_d = 75$ MPa. Then the diameter is given by,

$$d = \left[\frac{16}{\pi\, 75}\left(1066958.25 + \sqrt{1066958.25^2 + (375 \times 10^3)^2}\right)\right]^{1/3} = 53.04 \text{ mm}$$

Step-6: The diameter by shear theory

$$d = \left[\frac{16}{\pi\, S_{sd}}\sqrt{M_e^2 + T_e^2}\right]^{1/3}$$

$$d = \left[\frac{16}{\pi \times 45}\sqrt{1066958.25^2 + (375 \times 10^3)^2}\right]^{1/3}$$

$$= 50.31 \text{ mm}$$

Step-7: Standardize the diameter
Shaft diameter, $d = 53.04$ mm is the larger of the two and hence adopt standard diameter, $d = 56.0$ mm (Table 4.4 MDHB)

Example 6.3: A machine shaft, supported on bearings having their centres 750 mm apart, transmits 185 kW at 600 rpm. The power is received through a gear of 200 mm diameter, located at 200 mm to the right of left hand bearing. The teeth are 20° involute full depth. The gear is driven by a pinion with downward tangential force. The power is delivered through a horizontal belt drive from a pulley of 450 mm diameter and located at 200 mm to the right of the right hand bearing. The pulley weighs 100 N and the ratio of belt tension is 3:1. Find the diameter of the shaft for an allowable stress of 65 MPa in shear.

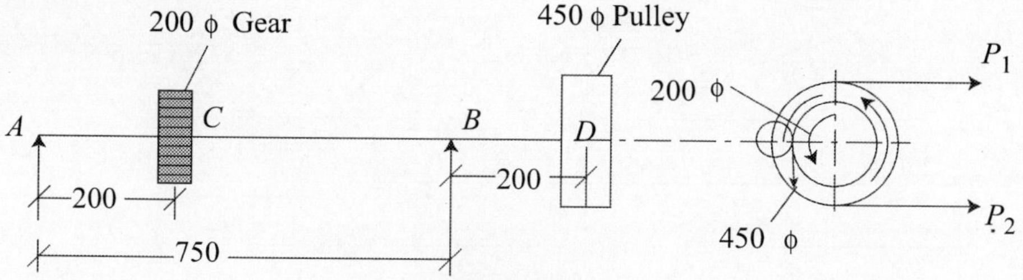

Figure 6.10

SOLUTION

Step-1: Loads on the shaft

$$\text{Torque transmitted} = \frac{955 \times 10^4 \times \text{kW}}{\text{N}}$$

$$= \frac{955 \times 10^4 \times 185}{600} = 294.46 \times 10^4 \text{ N-mm}$$

Tangential force F_T at the gear,

$$F_T = \frac{T}{r} = \frac{294.46 \times 10^4}{200/2} = 29446 \text{ N}$$

Radial force, F_r

$$F_r = F_T \tan \phi = 29446 \tan 20^\circ = 10717.5 \text{ N}$$

Pull in the belt on pulley at D.

$$\text{Torque } (P_1 - P_2)\, R = T$$

$$\therefore \quad P_1 - P_2 = T/R = 294.46 \times 10^4/(450/2) = 13087.1 \text{ N} \qquad \text{(i)}$$

But $\quad$ Tension ratio $(P_1/P_2) = 3.0 \qquad$ (ii)

$$2P_2 = 13087.1$$

$$P_2 = 13087.1/2 = 6543.0 \text{ N}$$

$$\therefore \quad P_1 = 3P_2 = 19630.0 \text{ N}$$

$\therefore$ Horizontal pull on the shaft,

$$P_1 + P_2 = 19630 + 6543.0 = 26173.6 \text{ N}$$

Step-2: Horizontal bending moment diagram

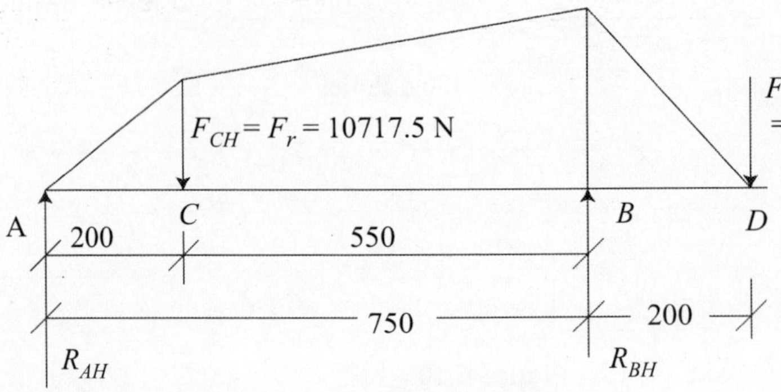

Figure 6.11

Taking moment about A,

$$26173.6 \times 950 + 10717.5 \times 200 - R_{BH} \times 750 = 0$$

$$R_{BH} = \frac{26173.6 \times 950 + 10717.5 \times 200}{750} = 36011.23 \text{N}$$

Taking moment about *B*,

$$R_{\text{AH}} = \frac{10717.5 \times 550 - 26173.6 \times 200}{750} = 879.87 \text{ N}$$

Moment at *C*,

$$M_{CH} = 879.87 \times 200 = 175974.0 \text{ N-mm}$$

Moment at *B*,

$$M_{BH} = 26173.6 \times 200 = 5234720.0 \text{ N-mm}$$

Step-3: Vertical bending moment diagram

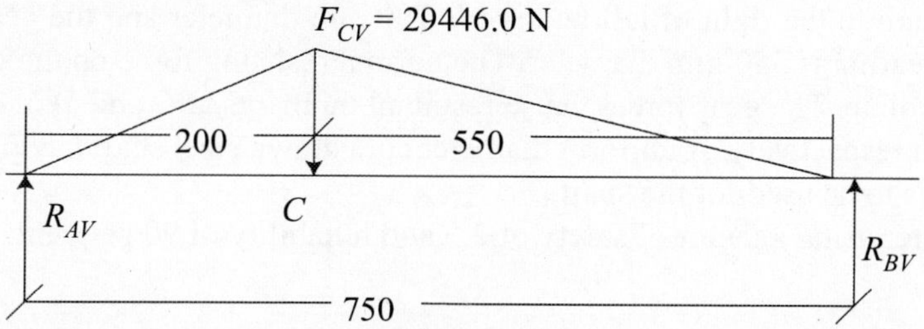

Figure 6.12

$$\begin{aligned}\text{Moment at } C, M_{CV} &= (F_T \cdot a \cdot b)/L) \\ M_{CV} &= \frac{29446.0 \times 200 \times 550}{750} = 4318746.7 \text{ N-mm} \\ \text{Moment at } B, M_{BV} &= 0\end{aligned}$$

Step-4: Total moments
∴ Resultant moment at *C*,

$$\begin{aligned}M_C &= \sqrt{M_{CV}^2 + M_{CH}^2} \\ &= \sqrt{43{,}18{,}746.7^2 + 175974.0^2} = 43{,}22{,}330.346 \text{ N-mm}\end{aligned}$$

Resultant moment at *B*,

$$M_B = \sqrt{M_{BH}^2 + M_{BV}^2} = M_{BH} = 5234720.0 \text{ N-mm}$$

Step-5: Diameter at critical section
∴ The torque being the same at both the sections, section *B* is critically loaded. Assuming steady load conditions,

$$C_m = 1.5, C_t = 1.0 \text{ (Table 4.2 MDHB)}$$

Then the diameter of the shaft,

$$d = \left[\frac{16}{\pi S_{sd}}\sqrt{(CmM)^2 + (CtT)^2}\right]^{1/3}$$

$$= \left[\frac{16}{\pi \times 65}\sqrt{(1.5 \times 5234720)^2 + (2944600)^2}\right]^{1/3}$$

$$= 86.95 \text{ mm} = 87.00 \text{ mm}$$

The next standard diameter = 100 mm (Table 4.4 MDHB)

Example 6.4: A countershaft 1.2 mm long between the supports has two spur gears mounted on it. The gear C at 450 mm to the right of left bearing is 700 mm diameter and the gear D at 450 mm to the left of right bearing is 350 mm diameter. The normal resisting force on the gear C is 12 kN. The teeth are 20 involute. The gear forces cause resultant moments at C and D of $M_c = 2720$ N.m and $M_d = 5150$ N.m respectively. A uniform diameter cold drawn steel shaft having $S_y = 325$ MPa and $S_u = 386$ MPa is to be used for the shaft.
Find the safe diameter using a factor of safety of 2.5 and reliability of 99 per cent.

SOLUTION

Step-1: Loads on the shaft
The normal force on the gear C,

$$F_n = 12 \text{ kN}.$$

∴ Tangential force on the gear C,

$$F_t = F_n \cos\phi = 12 \times \cos 20° = 11.276 \text{ kN}$$

∴ Torque on the shaft,

$$T = F_t \cdot r = 11.276 \times 350 = 3946.7 \text{ kN-mm}$$

$$= 3946.7 \text{ Nm}$$

This steady torque induces steady torsional shear stress and components of torque are

$$\therefore \quad T_m = 3946.7 \text{ Nm} \quad \text{and} \quad T_a = 0$$

The bending at D is more than at C and hence the max. bending stress is induced at D. This stress is alternating, and the components of moment are

$$M_m = 0; \quad M_a = 5150 \text{ N.m}$$

Step-2: Design stress
Endurance limit stress (test value),

$$S'_{en} = 0.5\,S_u = 0.5 \times 386.0 = 193.0 \text{ MPa}$$

Corrected value of the endurance stress,

$$S_{en} = K_a K_b K_c K_d K_e K_j S_{en}^1$$
$$K_a = 0.89 \text{ (Fig. 3.2)}$$
$$K_b = 1.00 \text{ (Assumed)}$$
$$K_c = 0.814 \qquad \text{(Table 3.2 MDHB)}$$
$$K_a = 1.0 \text{ (Assumed)}$$
$$K_e = 1.0 \text{ (Uniform diameter, } K_t = 0)$$
$$K_j = 1.0 \text{ (Assumed)}$$

$$\therefore \quad S_{en} = 0.89 \times 1 \times 0.814 \times 1.0 \times 1.0 \times 1.0 \times 193.0$$
$$= 139.8 \text{ MPa}$$

Step-3: Diameter by code equation
Then using Westing House code formula,

$$d = \left[\left(\frac{32.n}{\pi}\right)\left\{\left(\frac{T_a}{S_{en}} + \frac{T_m}{S_y}\right)^2 + \left(\frac{M_a}{S_{en}} + \frac{M_m}{S_y}\right)^2\right\}^{1/2}\right]^{1/3}$$

$$d = \left[\left(\frac{32 \times 2.5 \times 10^3}{\pi}\right)\left\{\left(\frac{3946.7}{325}\right)^2 + \left(\frac{5150}{139.8}\right)^2\right\}^{1/2}\right]^{1/3}$$

$$d = \left[\left(\frac{32 \times 2.5 \times 10^3}{\pi}\right)(147.4 + 1357.0)^{1/2}\right]^{1/3}$$
$$= 99.36 \text{ mm} \cong 100 \text{ mm}$$

The standard diameter, $d = 100$ (Table 4.4 MDHB)

Example 6.5: Figure 6.13 shows a countershaft subjected to alternating bending together with steady torsion and axial loads because of the action of the two gears. The part (b) shows all shafts dimensions. Sled runner keyways are used at A and C but these do not extend completely to the shaft shoulders. Note that the shaft design is such that only the bearing at O can take the axial loads. The bevel gear force is $F_D = 0.242 F_D i - 0.242 F_D j + 0940 F_D k$. The force at the helical gear $F_A = -1.05i + 4j - 1.47k$.

The shaft is made of medium carbon steel heat treated to a hardness of 310 BHN and all important surfaces are finished by grinding. The shaft is to be analysed to determine how safe is, using a reliability of 99 per cent and the Von-Mises-Henkey-Goodmann approach.

(a) Determine the factor of safety at shoulder G.
(b) Determine the factor of safety at A, where the keyway is located
(c) Find the factor of safety at B.

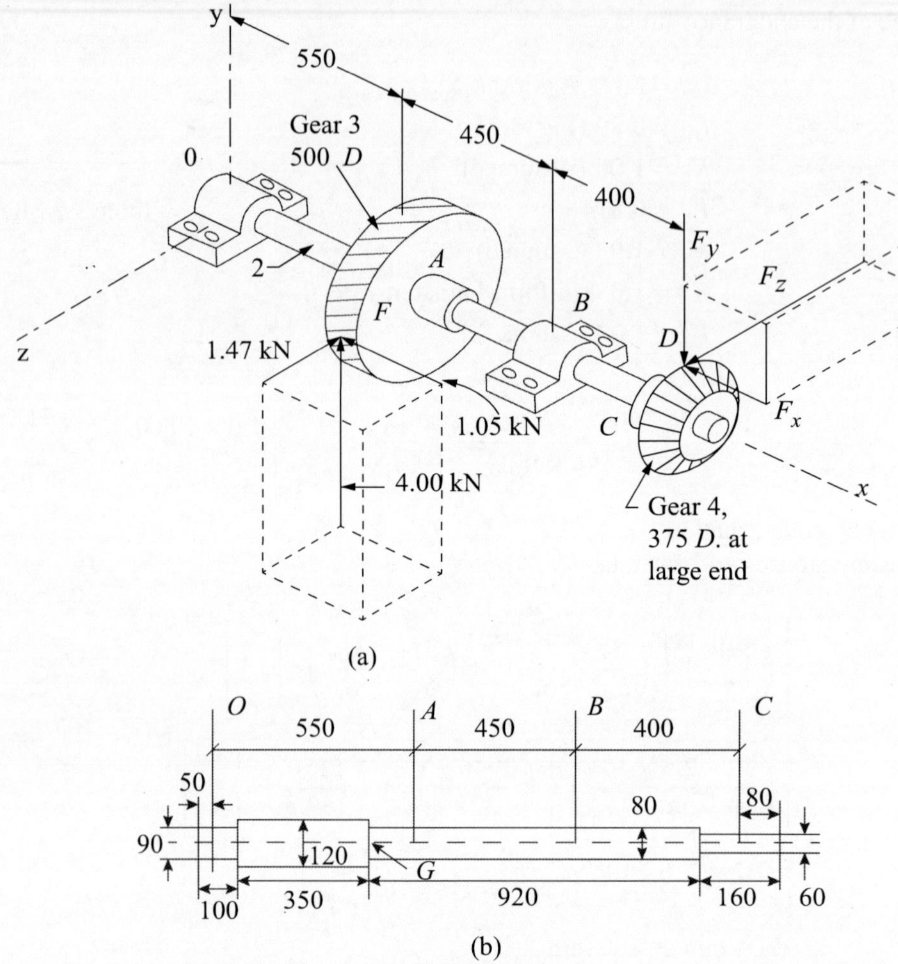

Figure 6.13

SOLUTION

Step-1: Loads on the shaft

Tangential force on the gear F, $F_t = 4.00$ kN $= 4000$ N

$\therefore$ Torque on the shaft,

$$T = F_t \times R = 4000 \times 250 = 100 \times 10^4 \text{ N-mm}$$

$\therefore$ Tangential force on the bevel gear,

$$F_z = \frac{T}{r} = \frac{100 \times 10^4}{(375/2)} = 53334 \text{ N}$$

$$\therefore \quad 0.94\, F_D = 5333 \text{ N}$$

$$\therefore \quad F_D - 5333/0.94 - 5674 \text{ N}$$

$$\therefore \qquad F_x = 0.242\ F_D = 1373.0\text{ N}$$
$$F_y = 0.242\ F_D = 1373.0\text{ N}$$

Step-2: Vertical bending moment diagram

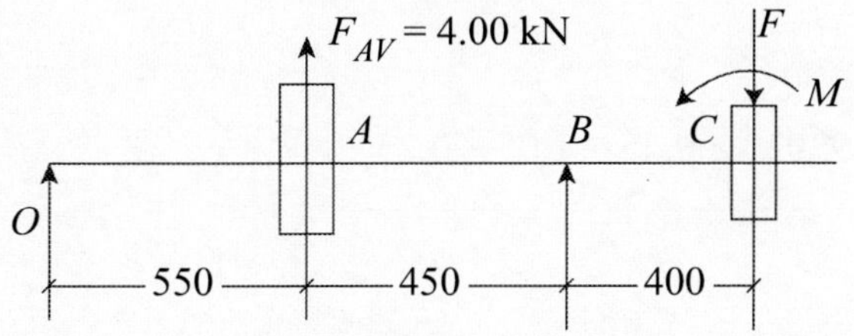

$$\therefore \quad \begin{aligned} F &= 1373.0\text{ N} = 1.373\text{ kN} \\ M &= 1.373 \times \frac{375}{2} \\ &= 0.2574\,\text{kNm} \end{aligned}$$

Figure 6.14

Reaction at B, taking moment about O

$$R_{BV} = \frac{1.373 \times 1.4 - 4 \times 0.55 - 0.2574}{1.0}$$
$$= -0.5352375\text{ kN (Downward)}$$

Moment at B,
$$M_{BV} = 1.373 \times 0.4 - 0.2574 =$$
$$= 0.2918\,\text{kN m}$$

Moment at A,
$$M_{AV} = (1.373 \times 0.85 + 0.5352375 \times 450) - 0.2574$$
$$= 1.15049\text{ kNm}$$

Step-3: Horizontal bending moment diagram (Top)

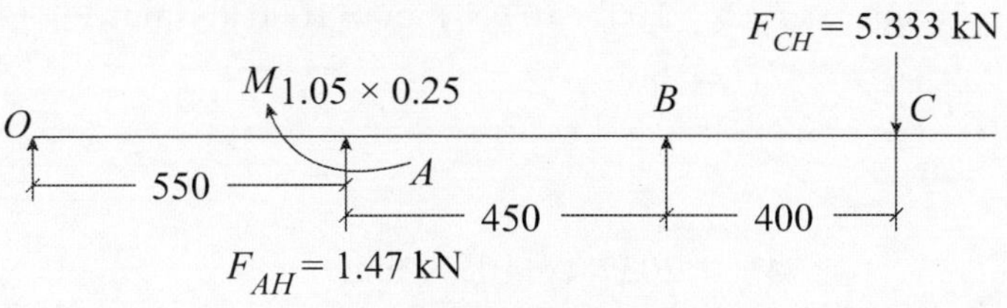

Figure 6.15

Taking moment about O,

$$R_{BH} = \frac{5.333 \times 1.4 + 1.050 \times 0.25 - 1.47 \times 0.55}{1.0}$$
$$= 6.92020\text{ kN}$$

$$R_{OH} = -\frac{5.333 \times 0.4 + 1.05 \times 0.25 + 1.47 \times 0.450}{1.0}$$
$$= -3.0572 \text{ kN (Downward)}$$

Moment at B,

$$M_{BH} = 5.333 \times 0.4 = 2.13332 \text{ kNm}$$

Moment at A,

$$M_{AH} = 5.333 \times 0.85 - 692020 \times 0.45 + 1.05 \times 0.25$$
$$= 1.68 \text{ kNm}$$

Step-4: Resultant moments

∴ Resultant moment at A,

$$M_A = \sqrt{M_{AH},^2 + M_{AV}^2}$$
$$= \sqrt{1.68^2 + 1.15049^2}$$
$$= 2037.4 \text{ kN-mm}$$

Resultant moment at B,

$$M_B = \sqrt{M_{BH}^2 + M_{BV}^2}$$
$$= \sqrt{2.1333^2 + 0.2918^2}$$
$$= 2.148 \text{ kNm}$$

Step-5: Stress in the shaft at the section A

At section A

$$\text{Torque, } T_m = 100 \times 10^4 \text{ N-mm} = 1000 \text{ kN-mm}$$
$$T_a = 0.0$$

Bending moment, $M_m = 0$

$$M_a = 2037.4 \times 10^3 \text{ N-mm}$$

Axial load, $P_A = 1.05 + 1.373 = 2.423$ kN = **2423 N**

∴ Torsional shear stress components are

$$\tau_m = \frac{16\,T_m}{\pi d^3} = \frac{16 \times 100 \times 10^4}{\pi \times 80^3} = 9.78 \text{ MPa}$$
$$\tau_a = 0$$

Components of the direct stresses are

$$\sigma_m = 2423/\left(\pi(80)^2/4\right) = 0.48195\,\text{MPa}$$

$$\sigma_a = \frac{2037.4 \times 10^3}{(\pi/32) \times (80)^3} = 40.53 \text{ MPa}$$

Von-Mises stress,

$$\sigma_m^1 = \sqrt{\sigma_m^2 + 3\tau_m^2} = \sqrt{(0.48195)^2 + 3 \times 9.78^2} = 16.95 \text{ MPa}$$

$$\sigma_a^1 = \sqrt{\sigma_a^2 + 3\tau_a^2} = \sqrt{40.53^2 + 3 \times 0^2} = 40.53 \text{ MPa}$$

Step-6: Design stress

$$S_u = 1050 \text{ MPa} \qquad \text{(Fig. 2.1)}$$

$$S_{en}^1 = 0.5\,S_u = 525 \text{ MPa}$$

Corrected value of S_{en}

$$S_{en} = K_a \cdot K_b \cdot K_c \cdot K_d \cdot K_e \cdot K_j \cdot S_{en}^1$$

$$K_a = 0.9 \qquad \text{(Fig. 3.2 MDHB)}$$

$$K_b = 0.75 \qquad \text{(Fig. 3.2 MDHB)}$$

$$K_c = 0.814 \qquad \text{(Table 3.2 MDHB)}$$

$$K_d = 1.0 \text{ (Assumed)}$$

$$K_c = \frac{1}{1 + q(K_t 1)} = \frac{1}{1 + 0.9(1.44 - 1)}$$

$$K_j = 1.0 \text{ (Assumed)}$$

$\therefore$

$$S_{en} = 0.9 \times 0.75 \times 0.814 \times 1.0 \times 0.716 \times 1.0 \times 525$$

$$= 206.5 \text{ MPa}$$

Step-7: Factor of safety by applying Goodman equation,

$$\frac{1}{n} = \frac{\sigma_m^1}{S_u} + \frac{\sigma_a^1}{S_{en}} = \frac{16.95}{1050} + \frac{40.53}{206.5}$$

$$= 0.2124, \text{ i.e., } \mathbf{n = 4.71}$$

Step-8: Stresses in Section at G

$$\text{B.M.} - M_G = \frac{MA}{550} \times 400 = 2037.4 \times \frac{400}{550} = 1481.7 \text{ N.m}$$

∴ Bending stresses are

$$\sigma_a = \frac{M_a}{Z} = \frac{1481.7 \times 10^3}{(\pi/32)(80)^3} = \mathbf{29.5\ MPa}$$

$$\text{and} \quad \sigma_m = \mathbf{0.48195\ MPa}\ \text{(axial force is the same)}$$

Shear stress

The driving torque acts between sections *A* and *C*, and there is no torsion in the portion *O*–*A* and hence no shear stress.

$$\tau_a = 0, \tau_m = 0$$

∴ Stress concentration factor,

$$K_t = 1.8\ (\text{for } D/d = 1.5 \text{ and } r/d = 0.08 \text{ in (Fig. 3.13 MDHB)})$$

Notch sensitivity,

$$q = 0.9\,(\text{Fig. 3.1a MDHB})$$

$$\therefore \quad K_f = (1+0.9)(1.8-1) = 1.72$$

$$K_e = 1/K_f = 1/1.72 = 0.58$$

All factors remain the same and the corrected endurance stress,

$$S_{en} = 206.5 \times \frac{0.58}{0.716} = 167.7\ \text{MPa}$$

Step-9: Factor of safety
By Goodman equation,

$$\frac{1}{n} = \frac{\sigma_m}{S_u} + \frac{\sigma_a}{S_{en}}$$

$$= \frac{0.48915}{1050} + \frac{29.5}{167.7} = 0.17624$$

$$\therefore \mathbf{n} = \mathbf{5.67}$$

Step-10: Stresses in Section at B
Moment at *B*,

$$M_B = 2148\ \text{N-m}$$

Bending stress,

$$\sigma_a = \frac{2148 \times 10^3}{\pi/32 \times (80)^3} = 42.73\ \text{MPa}$$

Axial load, $P_a = 1373$ N

$$\therefore \quad \text{Stress } \sigma_m = \frac{1373}{(\pi/4)\times(80)^2} = 0.273\text{ MPa}$$

Torsion, $T_m = 100 \times 10^4$ N-m

$$\tau_a = 0.0$$
$$\tau_m = 9.78\text{ MPa}$$

$\therefore$ Von-Mises stress,

$$\sigma_m^1 = \sqrt{\sigma_m^2 + 3\tau_m^2}$$
$$= \sqrt{0.273^2 + 3\times 9.78^2} = 16.94\text{ MPa}$$
$$\sigma_a^1 = \sqrt{\sigma_a^2 + 3\tau_a^2} = 42.73\text{ MPa}$$

Since, $K_t = 1$, $K_e = 1.0$
Corrected value of S_{en} is given by,

$$S_{en} = 206.5 \times \frac{1.0}{0.716} = 288.4\text{ MPa}$$

Step-11: Factor of safety
By Goodman equation

$$\frac{1}{n} = \frac{16.94}{1050} + \frac{42.873}{288.4} = 0.16295$$
$$\mathbf{n = 6.086}$$

The lowest value of the factor of safety is $\mathbf{n = 4.71}$

Example 6.6: An axial shaft of SAE 1045 annealed steel, 1.5 m long between bearings has a 1 kN pulley at its mid-point. The pulley is keyed to the shaft and receives 20 kW power at 200 rpm. The belt drive is horizontal and ratio of belt tension is 3:1. The diameter of the pulley is 600 mm. Calculate the necessary shaft diameter. Take $K_b = 1.75$ and $K_t = 1.25$ and factor of safety = 2.

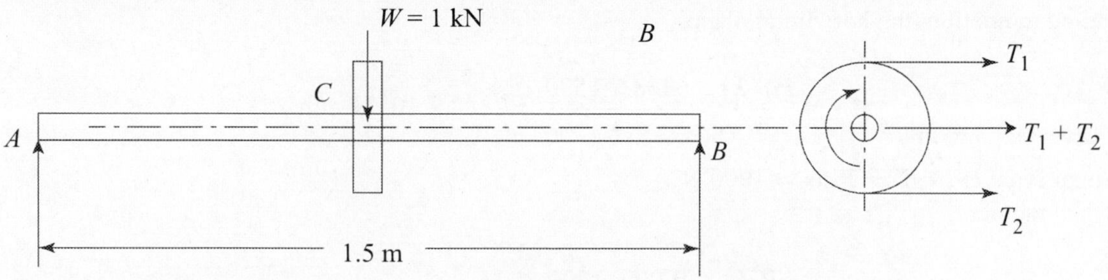

Figure 6.16

Given Data

Pulley diameter = 600 mm
Power input, P = 20 kW
Speed, N = 200 rpm
$K_b = 1.75, K_t = 1.25$ and Factor of safety = 2

SOLUTION

Step-1: Loads on the shaft
Torque at the pulley,

$$T = \frac{955 \times 10^4 \cdot P}{\text{N}} = \frac{955 \times 10^4 \times 20}{200} = 955 \times 10^3 \text{ N-mm}$$

Belt tension and torque relation

$$(T_1 - T_2)\, r = T \qquad \text{r = Radius of pully}$$

$$T_1 - T_2 = \frac{T}{r} = \frac{955 \times 10^3}{300} = 3183.3\,\text{N}$$

$$\text{Ratio of belt tension} = \frac{T_1}{T_2} = 3$$

$$\therefore \quad 3T_2 - T_2 = 3183.3 \text{ N}$$

$$\therefore \quad T_2 = 3183.3/2 = 1.592 \times 10^3\,\text{N} = 1592\,\text{N}$$

$$\therefore \quad T_1 = 1592 \times 3 = 4776\,\text{N}.$$

Pull on the shaft,

$$P_H = T_1 + T_2 = 4776 + 1592 = 6368 \text{ N}.$$

Step-2: Bending moment
Bending moment in the horizontal plane,

$$M_H = \frac{P_H \cdot L}{4} = \frac{6368 \times 1500}{4} = 2388 \times 10^3 \text{ N-mm}$$

Vertical force $P_V = W = 1 \text{ kN} = 1000 \text{ N}$
Vertical moment,

$$M_V = \frac{P_V L}{4} = \frac{1000 \times 1500}{4} = 375 \times 10^3 \text{ N-mm}$$

Resultant bending moment,

$$M = \sqrt{M_H^2 + M_V^2}$$
$$= \sqrt{(2388 \times 10^3)^2 + (375 \times 10^3)^2}$$
$$= 2417.26 \times 10^3 \text{ N-mm}$$

Step-3: Design stress
For SAE 1045 (Table 2.2 MDHB)
Yield stress $S_y = 315$

Design stress in tension, $S_d = S_y/N = \frac{315}{2} = 157.5$ MPa.

Design stress in shear, $S_{sd} = S_d/2 = \frac{157.5}{2} = 78.75$ MPa.

Step-4: The diameter by shear theory

According to maximum shear theory the shaft diameter is given by Eqn. 4.8 MDHB with $K = 0$

$$d_o = \left\{\frac{16}{\pi \cdot S_{sd}}\left[\sqrt{M_e^2 + T_e^2}\right]\right\}^{1/3}$$

where, $M_e = C_m M = 1.75 \times 2417.26 \times 10^3$

$$= 4230 \times 10^3 \text{ N-mm}$$

$$T_e = C_t T = 1.25 \times 955 \times 10^3 = 1193.75 \times 10^3 \text{ N-mm}$$

$$d_o = \left\{\frac{16}{\pi \times 78.75}\left[\sqrt{(4230 \times 10^3)^2 + (1193.75 \times 10^3)^2}\right]\right\}^{1/3}$$
$$= \left\{64.67\left[\sqrt{4230^2 + 1193.75^2}\right]\right\}^{1/3}$$
$$= [64.67 \times 4230 \times 1.039]^{1/3}$$
$$= 65.75 \text{ mm}$$

Standardize $d_o = 71$ mm

Example 6.7: In an axial flow rotary compressor the hollow shaft is subjected to 1500 N-m torque and a maximum bending moment of 3000 N-m. The inner diameter is 0.4 times the outer diameter. The shear stress in the shaft diameter is limited to 50 MPa. Determine the diameter of the hollow shaft. What will be the shear stress if an axial load of 10 kN is taken into account.

SOLUTION

Given Data

Maximum torque, $T_{max} = 1500$ Nm

Maximum bending moment, $M_{max} = 3000$ N-m
and ratio of diameters,

$$d_i/d_o = 0.4$$

Design stress in shear, $S_{sd} = 50$ MPa
Step-1: Corrected values

Assume, $C_m = 1.5$ and $C_t = 1.0$ (Table 4.2 MDHB)
Then effective values of moments are,

$$\begin{aligned} M_e &= C_m \cdot M = 1.5 \times 3000 = 4500 \text{ N-mm} \\ T_e &= C_t \cdot T = 1 \times 1500 = 1500 \text{ N-mm} \end{aligned}$$

Step-2: Diameter of the shaft by shear theory

$$\begin{aligned} d_o &= \left[\frac{16}{\pi \cdot S_{sd}}\left\{\sqrt{M_e^2 + T_e^2}\right\}\left(\frac{1}{1-K^4}\right)\right]^{1/3} \\ &= \left[\frac{16}{\pi \times 50}\left\{\sqrt{(4500 \times 10^3)^2 + (1500 \times 10^3)^2}\right\}\left(\frac{1}{1-0.4^4}\right)\right]^{1/3} \\ &= \left[\left(\frac{16 \times 4500 \times 10^3}{\pi \times 50}\right)\left\{\sqrt{1 + \left(\frac{1500}{4500}\right)^2}\right\}(1.02627)\right]^{1/3} \\ &= 79.15 \text{ mm} \end{aligned}$$

Choose $d_0 \cong 80$ mm

$$\begin{aligned} \therefore \quad d_i &= 0.4\, d_o \\ &= 0.4 \times 80 = 32 \text{ mm} \end{aligned}$$

Step-3: Stresses in the shaft with axial load
Axial load, $F_a = 10$ kN
Axial stress,

$$\begin{aligned} \sigma = \frac{F_a}{A} &= \frac{4(10 \times 10^3)}{\pi\,(80^2 - 32^2)} \\ &= 2.368 \text{ MPa} \end{aligned}$$

Bending stress,

$$\sigma_b = \frac{32.M}{\pi(80^3)(1-K^4)} = \frac{32 \times 4500 \times 10^3}{\pi \times 80^3 \times 0.9744}$$

$$\sigma_b = 91.88\,\text{MPa}.$$

Total direct stress,

$$\sigma_1 = 91.88 + 2.368$$
$$= 94.248\,\text{MPa}$$

Shear stress,

$$\tau = \frac{16.T}{\pi d_o^3(1-K^4)} = \frac{16 \times 1500 \times 10^3}{\pi \times 80^3 \times (1-0.4^4)}$$

$$\tau = 15.31\,\text{MPa}.$$

Maximum shear stress,

$$\tau_{max} = \sqrt{\left(\frac{94.248}{2}\right)^2 + (15.31)^2}$$

$$\tau_{max} = 49.55\,\text{MPa}.$$

Example 6.8: A steel shaft (C45) transmitting 15 kW at 210 rpm is supported between two bearings 1000 mm apart. On this two spur gears are mounted. The gear having 80 teeth of module 6 mm is located 200 mm to the left of right hand bearing and receives power from a driving gear such that the tangential force acts vertical. The pinion having 24 teeth and module of 6 mm is located 100 mm to the right of left bearing and delivers power to a gear mounted behind it. Taking combined shock and fatigue factors as 1.75 in bending and 1.25 in torsion, determine the shaft diameter.

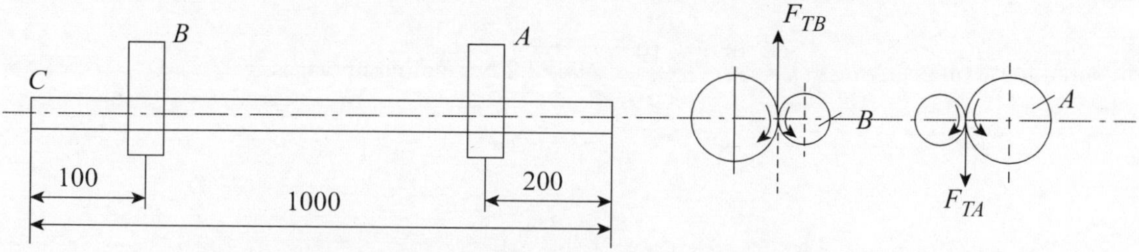

Figure 6.17

Given Data

Power transmitted = 15 kW

Speed of operation, $N = 210$ rpm

The size of the gear at A, $Z_A = 80$ and module, $m = 6$ mm

The size of the gear at B, $Z_B = 24$ and module, $m = 6$ mm

Shock and fatigue factors, $C_m = k_m = 1.75$ and $c_t = k_t = 1.25$,

Material C-45 with $s_y = 360$ MPa (Table 2.8 MDHB)

Factor of safety = 2, $S_d = 360/2 = 180$ MPa and $S_{sd} = 90$ MPa.

SOLUTION

Step-1: Loads on the shaft

$$\text{Torque transmitted, } T = \frac{955 \times 10^4 \cdot \text{kW}}{} \text{ N-mm}$$

$$= \frac{955 \times 10^4 \times 15}{210} = 68.21 \times 10^4 \text{ N-mm}$$

Diameter of gear at A, $d_A = 80 \times 6 = 480$ mm

Hence, radius $r_A = 240$ mm

$$\text{Tangential force on the gear, } F_{TA} = \frac{68.21 \times 10^4}{240} = 2842 \text{ N} \quad \text{acting downwards.}$$

Diameter of gear, at B, $d_B = 24 \times 6 = 144$ mm, and radius, $r_B = 72$ mm

$$\text{Tangential force on gear, } B.\ F_{TB} = \frac{68.2 \times 10^4}{72} = 9472.2 \text{ N} \quad \text{acting upwards.}$$

Step-2: Bending moment diagram

Vertical load diagram

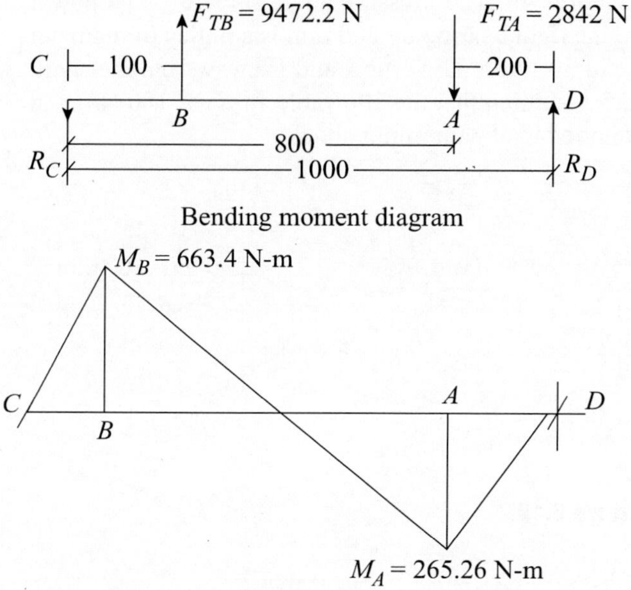

Figure 6.18

The bending reactions,

$$R_D = \frac{2842 \times 800 - 9472.2 \times 100}{1000}$$

$R_D = 1326.3$ N acting upwards.

$$R_C = 9472.2 + 1326.3 - 28.42$$
$$= 6634.2 \text{ acting downwards.}$$

Moments:

$$M_B = 6634.2 \times 100$$
$$= 663.42 \times 10^3 \text{ N-mm}$$
$$M_A = 1326.3 \times 200$$
$$= 265.26 \times 10^3 \text{ N-mm}$$

Maximum moments,

$$M_{\max} = M_B = 663.4 \text{ Nm} = 663.4 \times 10^3 \text{ N-mm}$$

Effective moment,

$$M_e = C_m M_{\max} = 1.75 \times 663.4 \times 10^3 = 1160.95 \times 10^3 \text{ N-m}$$

Step-3: Torque on the shaft maximum torque, $T_{\max} = 682.1 \times 10^3$ N-mm

Effective torque considering shock and fatigue factor,

$$T_e = 1.25 \times 682.1 \times 10^3$$
$$= 852.625 \times 10^3 \text{ N-mm}$$

Step-4: Shaft diameter is given by shear theory

$$d = \left[\frac{16}{\pi . S_{sd}} \left\{\sqrt{M_e^2 + T_e^2}\right\} \left(\frac{1}{1-0}\right)\right]^{1/3}$$

$$= \left[\frac{16}{\pi .90} \left\{\sqrt{(1160 \times 10^3)^2 (852.625 \times 10^3)^2}\right\}\right]^{1/3} = 43.35 \text{ mm}$$

Adopt standard value, $d = 45$ mm

Example 6.9: A shaft supported between bearings 400 mm apart gets its drive through a gear drive. A gear is mounted 200 mm to the right of left bearing and is driven by a pinion just above it. The gear has a module of 10 mm the number of teeth is 40. The pressure angle $\phi = 20°$. The power received is 20 kW at 500 rpm. Overhanging to the right bearing by 200 mm is a pulley of diameter 200 mm. The belt drive is inclined at an angle of 30° with the vertical and is away from the shaft. The ratio of belt tension is 3:1. Design the shaft assuming that the allowable stress as 100 N/m^2 in tension and 40 N/mm^2 in shear for suddenly applied load with minor shock.

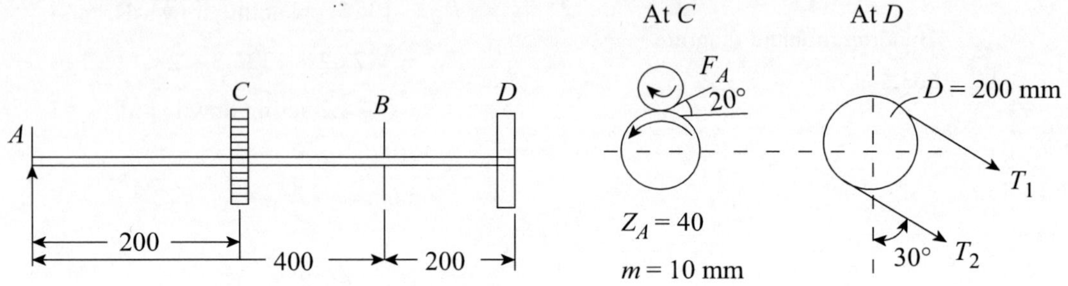

Figure 6.19

Given data

Power input at $C = 20$ kW

Speed, N = 500 rpm

Load suddenly applied with minor shock

Allowable stress, $S_d = 100$ MPa tension

$S_{sd} = 40$ MPa in shear.

SOLUTION

Step-1: Loads on the shaft

Torque transmitted, (Eqn. 4.1a MDHB)

$$T = \frac{955 \times 10^4.\text{kW}}{\text{N}} = \frac{955 \times 10 \times 20}{500} = 382 \times 10^3 \text{ N-mm}$$

Tangential force on the gear,

$$F_T = \frac{T}{r} = \frac{382 \times 10^3}{\frac{Z_C \times m}{2}} = \frac{382 \times 10^3}{\frac{40 \times 10}{2}} = 1910\,\text{N}$$

Radial force,

$$F_R = F_T \cdot \tan\phi = 1910 \cdot \tan 20 = 695.18 \text{ N}$$

Belt tensions are

$$T_1/T_2 = 3 \quad \text{(i)}$$
$$\text{and } (T_1 - T_2)R = T \quad \text{(ii)}$$
$$T_1 - T_2 = \frac{382 \times 10^3}{100} = 3820 \text{ N.}$$
$$\therefore \quad 3T_2 - T_2 = 3820$$
$$\therefore \quad T_2 = 1910 \text{ N}$$
$$\therefore \quad T_1 = 3 \times 1910 = 5730 \text{ N}$$

Pull on the shaft,

$$P_D = T_1 + T_2 = 5730 + 1910 = 7640 \text{ N}$$

Components:

Vertical component,

$$P_{dV} = 7640 \times \cos 30 = 6616.43 \text{ N}$$

Horizontal component,

$$P_{dH} = 7640 \times \sin 30 = 3820 \text{ N}$$

Step-2: Vertical bending moment diagram
Load diagram in vertical place

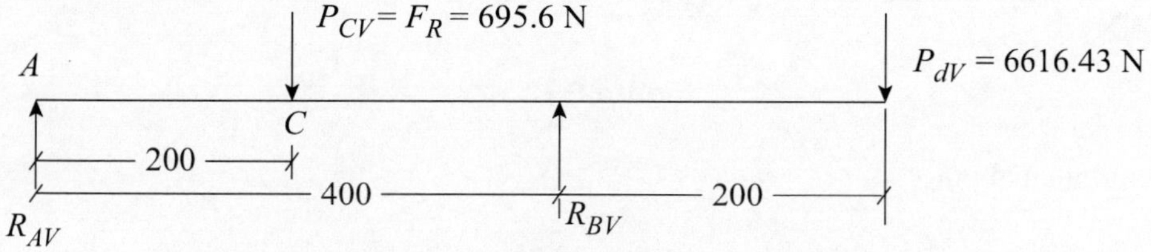

Figure 6.20

Taking moment about A,

$$695.6 \times 200 - R_{B_V} \times 400 + 6616.43 \times 600 = 0$$
$$R_{B_V} = \frac{695.6 \times 200 + 6616.43 \times 600}{400}$$
$$= 10272.45 \text{ N}$$

Taking moment about B,

$$R_{A_V} \times 400 - 695.6 + 6616.43 \times 200 = 0$$

$$R_{A_V} = \frac{695.6 \times 200 - 6616.43 \times 200}{400}$$
$$= -2960.41 \text{ N.}$$

Bending moment diagram

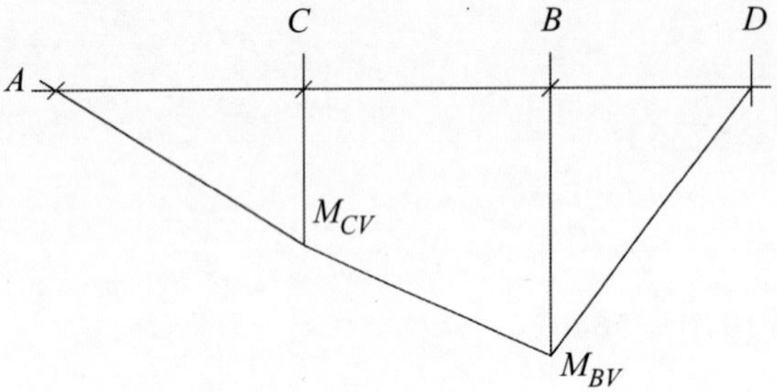

Figure 6.21

Moment at C,

$$M_{CV} = 2960.41 \times 200$$
$$= -592083 \text{ N-m}$$

Moment at B,

$$M_{BV} = 6616.43 \times 200$$
$$= 1323286 \text{ N-m}$$

Step-3: Horizontal bending moment diagram
Load diagram in horizontal plane

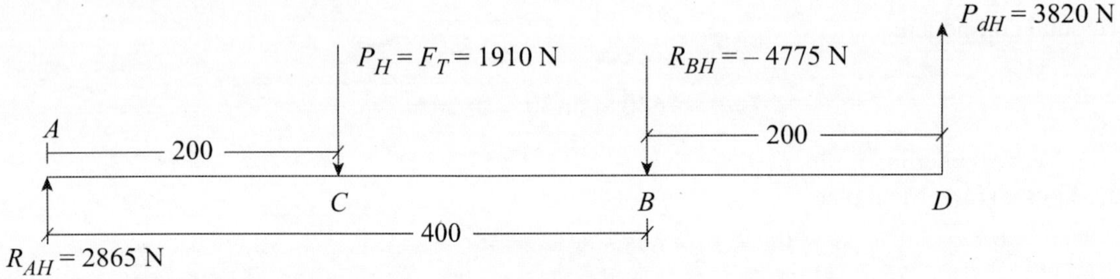

Figure 6.22

Taking moment about A,

$$1910 \times 200 - R_{BH} \times 400 - 3820 \times 600 = 0$$

$\therefore$ Reaction at B, $R_{BH} = \left(\dfrac{1910 \times 200 - 3820 \times 600}{400}\right)$

$$= -4775 \text{ N}$$

Reaction at A, $R_{AH} = 1910 + 4775 - 3820$

$$= 2865 \text{ N}$$

Moment at C,

$$M_{CH} = 2865 \times 200 = 573 \times 10^3 \text{ N-mm}$$

Moment at B,

$$M_{BH} = 3820 \times 200 = 764 \times 10^3 \text{ N-m}$$

Bending Moment diagram

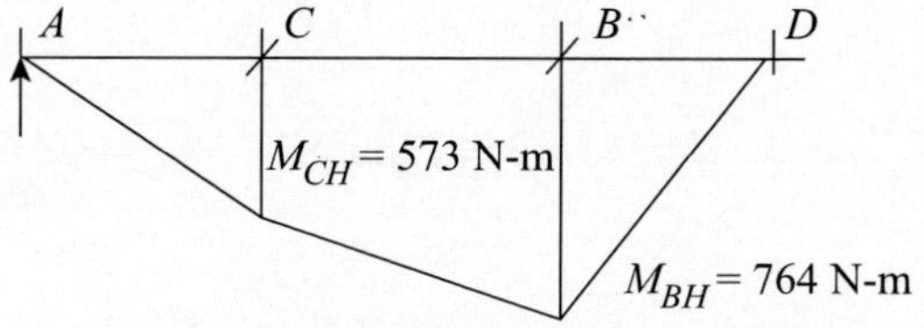

Figure 6.23

Step-4: Resultant moment
Resultant moment at B,

$$\begin{aligned} M_B &= \sqrt{M_{BH}^2 + M_{BV}^2} \\ &= \sqrt{(764 \times 10^3)^2 + (1323.28 \times 10^3)^2} \\ &= 1528 \times 10^3 \text{ N-mm} \end{aligned}$$

This is maximum and the section at B is subjected to maximum torque and maximum moment and hence is critical.
Step-5: The shaft diameter by normal stress (Eqn. 4.7 MDHB with $K = 0$)

$$d_o = \left[\frac{16}{\pi.\, S_d}\left\{M_e + \sqrt{M_e^2 + T_e^2}\right\}\right]^{1/3}$$

where $M_e = C_m M$ and $T_e = C_t T$

$C_m = 2$ and $C_t = 1.5$ (Table 4.2 MDHB)

$M_e = 2 \times 1528 \times 10^3 = 3056 \times 10^3$ N-mm

$T_e = 1.5 \times 382 \times 10^3 = 573 \times 10^3$ N-mm

Then, by principal stress theory,

$$\begin{aligned} d_o &= \left[\frac{16 \times 3056 \times 10^3}{\pi \times 100}\left\{1 + \sqrt{1 + \left(\frac{2573}{3056}\right)^2}\right\}\right]^{1/3} \\ &= 67.96 \text{ mm} \end{aligned}$$

By shear theory,

$$d_o = \left[\frac{16}{\pi \cdot S_{sd}}\left\{\sqrt{M_e^2 + T_c^2}\right\}\right]^{1/3}$$

$$= \left[\frac{16}{\pi \times 40}\left\{\sqrt{(3056\times 10^3)^2 + (573\times 10^3)^2}\right\}\right]^{1/3}$$

$$= 73.42 \text{ mm}$$

Adopt the larger of the two

$$d = 73.42 \cong 75 \text{ mm}.$$

Adopt $d = 75$ mm

Example 6.10: A shaft is supported between two bearings located 0.6 m apart. Gear *A* of pitch circle diameter 0.1 m is keyed to the shaft 0.1 m to the right of the left bearing. Gear *B* of 0.15 m diameter is keyed to the shaft 0.3 m to the right of the left bearing. Another gear *C* of pitch circle diameter 0.08 m is keyed to the shaft 0.1 m to the left of the right bearing. Gear *B* receives 10 kW power at 500 rpm from a mating gear mounted directly below it. Gear *A* delivers 6 kW power to another gear mounted directly in front of it, such that the tangential force acts vertically upwards. Gear *C* delivers the remaining power to its mating gear mounted directly behind it, such that the tangential force acts directly vertically downwards. All gears are of 20° full depth involute form. The shaft is made of steel which has an ultimate strength of 510 MPa and a yield strength of 330 MPa. Determine the required diameter of the shaft under study load condition using ASME code.

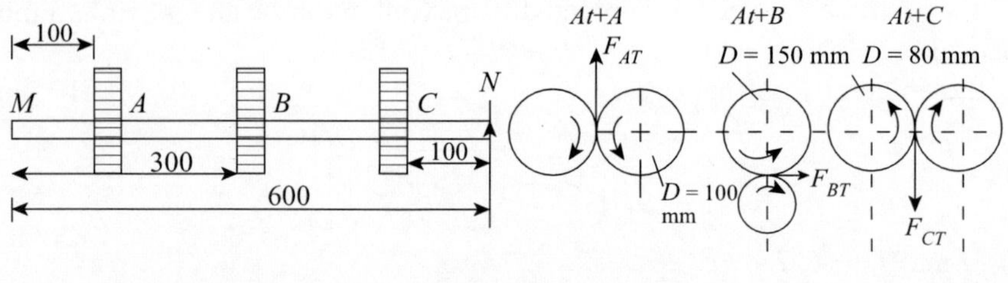

Figure 6.24

Given Data

Power input of 10 kW at $N = 500$ rpm

$D_A = 100$ mm, $D_B = 150$ mm, $D_C = 80$ mm

Ultimate strength, $S_u = 510$ MPa

Yield strength, $S_y = 330$ MPa

Factor of safety = ?

SOLUTION

Assume Factor of safety, $n = 2$

$$S_d = \frac{330}{2} = 165 \text{ MPa}$$

Step-1: Loads on the shaft
The power input at $B = 10$ kW

Torque input, $T = \dfrac{955 \times 10^4 \times 10}{500} = 191 \times 10^3$ N-mm

Tangential force F_{BT} on the gear (horizontal),

$$\therefore \quad F_{BT} = \frac{T}{r} = \frac{191 \times 10^3}{150/2} = 2546.7 \text{ N}$$

Radial force, F_{Br} on the gear (vertical)

$$F_{Br} = F_{BT} \cdot \tan 20^\circ = 2546.7 \times \tan 20^\circ = 926.9 \text{ N}$$

At gear A power delivered is 6 kW.

Torque output,

$$T_A = \frac{955 \times 10^4 \times 6}{500} = 114.6 \times 10^3 \text{ N-mm}$$

Tangential force F_{AT} on the gear (vertical),

$$F_{AT} = \frac{T}{r_a} = \frac{114.6 \times 10^3}{100/2} = 2292 \text{ N}$$

Radial force F_{Ar} on the gear (horizontal),

$$F_{Ar} = F_{AT} \tan 20^\circ = 2292 \times \tan 20^\circ = 834.22 \text{ N}$$

Power delivered at C is 4 kW.
Torque output,

$$T_c = \frac{955 \times 10^4 \times 4}{500} = 76400 \text{ N-mm}$$
$$= 76.4 \times 10^3 \text{ N-mm}$$

Tangential force on the gear F_{CT} (vertical down),

$$F_{CT} = \frac{T}{r_c} = \frac{76.4 \times 10^3}{80/2} = 1910 \text{ N}$$

Radial force on the gear (horizontal),

$$F_{Cr} = F_{CT} \tan 20^\circ = 1910 \tan 20^\circ$$
$$= 695.18 \text{ N.}$$

Step-2: Vertical moment diagram

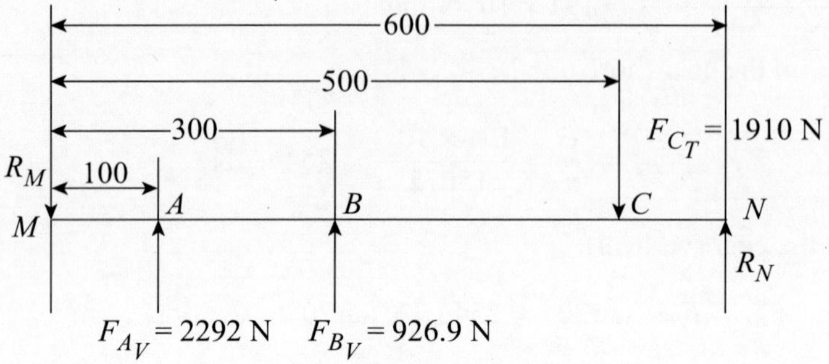

Figure 6.25

Taking moment about M,

$$2292 \times 100 + 926.9 \times 300 - 1910 \times 500 - R_N \times 600 = 0$$

$$R_N = \frac{229200 + 92690 \times 3 - 191000 \times 5}{600}$$

$$= -746.22 \text{ N(upward)}$$

Taking moment about N,

$$1910 \times 100 - 926.9 \times 300 - 2292 \times 500 + R_M \times 600 = 0$$

$$R_M = \frac{2292 \times 500 + 926.9 \times 300 - 1910 \times 100}{600}$$

$$= 2055.12 \text{ N}$$

Moments

$$\text{At } A,\ M_{AV} = 2055.12 \times 100 = 205.512 \times 10^3 \text{ N-mm}$$

$$\text{At } B,\ M_{BV} = 2055.12 \times 300 - 2292 \times 200 = 158136$$

$$= 158.136 \times 10^3 \text{ N-mm}$$

$$\text{At } C,\ M_{CV} = 746.22 \times 100 = 74.622 \times 10^3 \text{ N-mm.}$$

Bending moment diagram

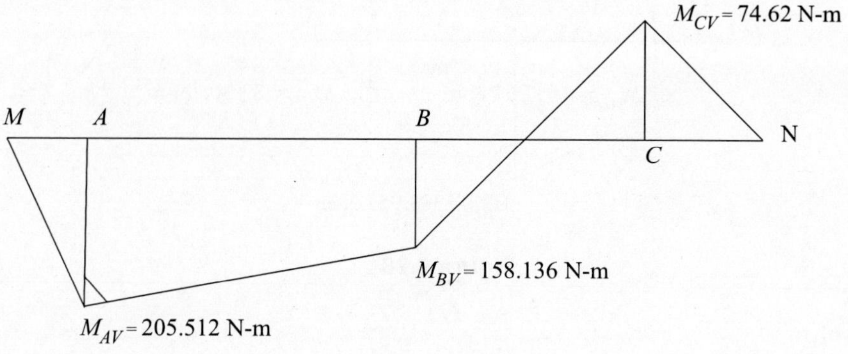

Figure. 6.26

Step 3: Horizontal moment diagram (from top)

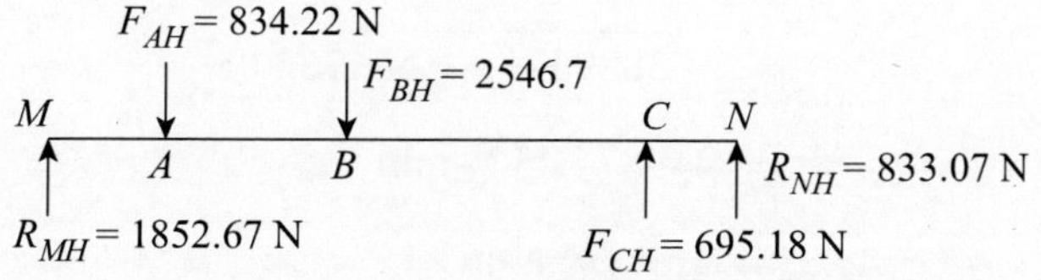

Figure 6.27

Taking moment about M,

$$834.22 \times 100 + 2546.7 \times 300 - 695.18 \times 500 - R_{N_H} \times 600 = 0$$

$$R_{N_H} = \frac{834.22 \times 100 + 2546.7 \times 300 - 695.18 \times 500}{600}$$

$$= 833.07 \text{ N}$$

Reaction at M,

$$R_{M_H} = 834.22 + 2546.7 - 695.18 - 833.07$$

$$= 1852.67 \text{ N}$$

Bending moments,

$$\text{At } A, \quad M_{AH} = 1852.67 \times 100 = 185267 \text{ N-mm}$$

$$= 185.267 \text{ N-m}$$

$$\text{At } B, \quad M_{BH} = 1852.67 \times 300 - 834.22 \times 200$$

$$= 388957 \text{ N-mm}$$

$$= 388.957 \text{ N-m}$$

$$\text{At } C, \quad M_{CH} = 833.07 \times 100 = 83307 \text{ N-mm}$$

$$= 83.307 \text{ N-m}$$

Bending moment diagram

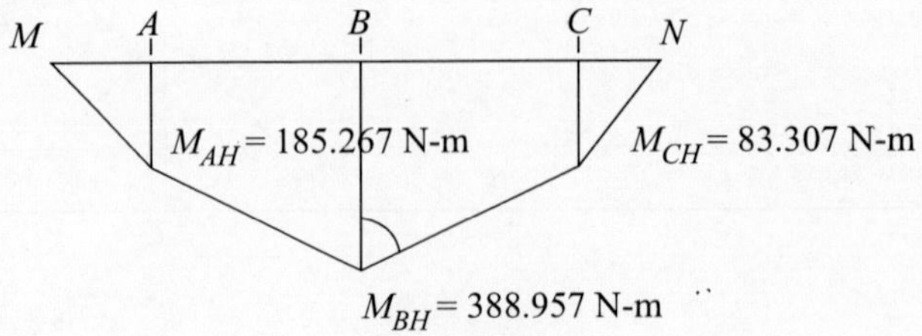

Figure 6.28

Step-4: Resultant moment

Maximum moment occurs at B,

$$\begin{aligned} M_B &= \sqrt{M_{BH}^2 + M_{BV}^2} \\ &= \sqrt{388957^2 + 1581362} \\ &= 419874.43 \text{ N-mm} \\ &= 419.874 \text{ N-m} \end{aligned}$$

The section at B is subjected to max torque and bending moment and hence is critical.

Step-5: Diameter of shaft

The shock and fatigue factors from Table 4.2 MDHB are:

For steady load (assumed),

$$C_m = 1.5 \quad \text{and} \quad C_t = 1.0$$

Effective moment,

$$\begin{aligned} M_e &= C_m.M = 1.5 \times 419874.43 \\ &= 629811.645 \text{ N-mm} \end{aligned}$$

Effective torque,

$$T_e = C_t.T = 1 \times 191 \times 10^3$$

Using normal stress theory [Eqn. 4.8 MDHB with $K = 0$]

$$d_o = \left[\frac{16}{\pi \cdot S_d}\left\{M_e + \sqrt{M_e^2 + T^2}\right\}\left(\frac{1}{1-0}\right)\right]^{1/3}$$

$$= \left[\frac{16}{\pi \times 165}\left\{629811.645 + \sqrt{(629811.645)^2 + \left(191 \times 10^3\right)^2}\right\}\right]^{1/3}$$

$$= 34.13 \text{ mm}$$

Adopt $d - 35$ mm

Example 6.11: A horizontal steel shaft supported on bearings A at the left end and B at the right end, carries two gears C and D, located at distances 250 mm and 400 mm respectively, from the centre lines of left and right end bearings. The pitch diameter of gear C is 600 mm and that of gear D is 200 mm. The pressure angle is 20°. The distance between the centre lines of the bearings is 2400 mm. The shaft transmits 20 kW power at 120 rpm. The power is delivered to shaft at gear C and is taken out at gear D in such a manner that the tooth pressures F_T and F_D of gears C and D act vertically downwards. Find the diameter of the shaft, if the working stresses are 100 MPa in tension and 56 MPa in shear. The gears C and D weigh 950 N and 350 N respectively. Take $C_m = 1.5$ and $C_T = 1.2$.

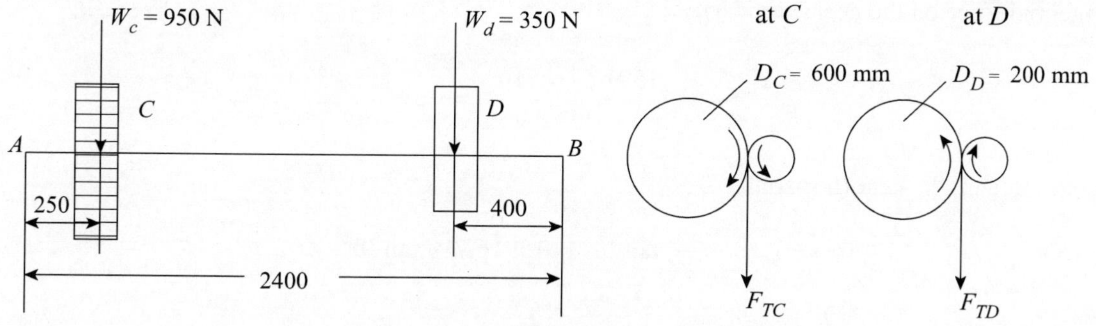

Figure 6.29

Given Data
Power input at C
$P = 20$ kW at $N = 120$ rpm
Working stress, $S_d = 100$ MPa and $S_{sd} = 56$ MPa.

SOLUTION
Step-1: Loads on the shaft
Torque at C (Eqn. 4.1 a MDHB)

$$T_c = \frac{955 \times 10^4 \text{kW}}{\text{N}} = \frac{955 \times 10^4 \times 20}{120} = 1591.67 \times 10^3 \text{ N-mm}$$

Tangential force on the gear (vertical),

$$F_{rc} = \frac{T_c}{r_c} = \frac{1591.67 \times 10^3}{600/2} = 5305.55 \text{ N}$$

Radial force on the gear (horizontal),

$$F_{rc} = F_T \times \tan 20^\circ = 5305.55 \times \tan 20^\circ$$
$$= 1931.06 \text{ N}$$

Total vertical load at C,

$$\because F_{cv} = W_c + F_T = 950 + 5305.55$$
$$= 6255.55 \text{ N}$$

At D,
Tangential force on the gear (vertical),

$$F_{TD} = \frac{T_C}{r_d} = \frac{1591.67 \times 10^3}{200/2} = 15916.7 \text{ N}$$

Radial force on the gear (horizontal),

$$F_{rD} = F_{TD} \tan 20^\circ = 15916.7 \times \tan 20^\circ$$
$$= 5793.21 \text{ N}$$

Total vertical load at D,

$$F_{DV} = 15916.7 + 350$$
$$= 16266.7 \text{ N}$$

Step-2: Vertical load moment diagram

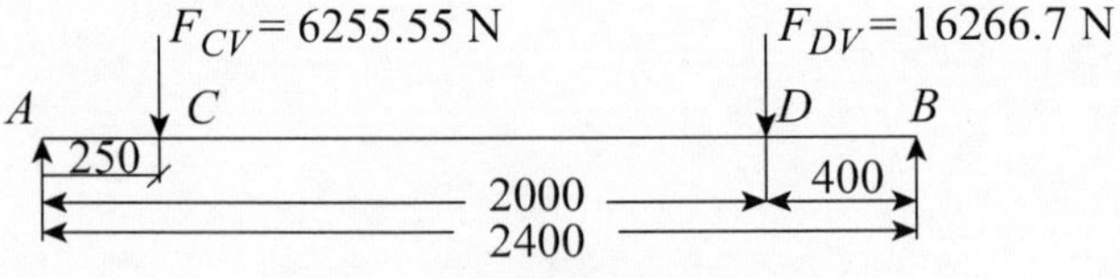

All dimensions in mm

Figure 6.30

Taking moment about A the reaction at B is given by

$$R_{BV} = \frac{6255.55 \times 250 + 16266.7 \times 2000}{2400} = 14207 \text{ N}$$

Similarly, taking moments about B,

$$R_{AV} = \frac{16266.7 \times 400 + 6255 \times 2150}{2400}$$
$$= 8315.05 \text{ N}$$

Bending moment at C,

$$M_{CV} = 8315.05 \times 250 = 20{,}78{,}762.5 \text{ N-mm} = 2078.762 \text{ Nm}$$

Moment at D,

$$M_{DV} = 14207 \times 400 = 56{,}82{,}800 \text{ N-mm} = 5682.8 \text{ Nm}$$

Bending Moment diagram

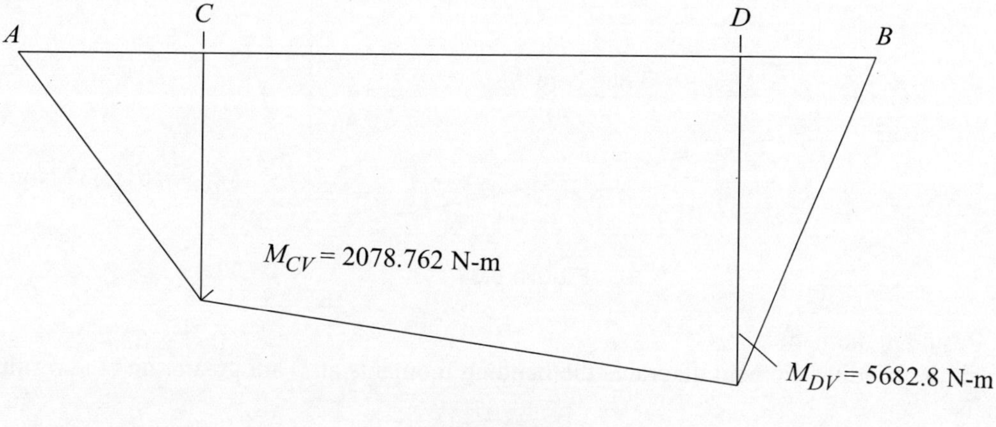

Figure 6.31

Step-3: Horizontal load moment diagrams

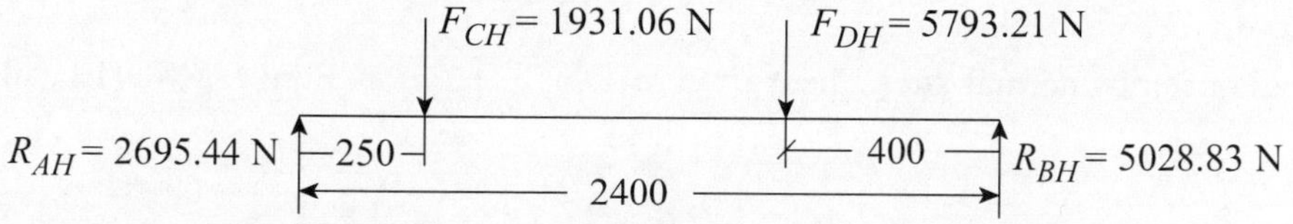

Figure 6.32

Taking moments about A, the reaction at B,

$$R_{BH} = \frac{1931.06 \times 250 + 5793.21 \times 2000}{2400} = 5028.83 \text{ N}$$

Similarly, taking moment about B,

$$R_{AH} = \frac{5793.21 \times 400 + 1931.06 \times 2150}{2400} = 2695.44 \text{ N}$$

Bending moment at C,

$$M_{CH} = 2695.44 \times 250 = 6,73,860.73 \text{ N-mm} = 673.86 \text{ Nm}$$

Bending moment at D,

$$M_{DH} = 5028.83 \times 400 = 2011532 \text{ N-mm} = 2011.532 \text{ Nm}$$

Bending moment diagram

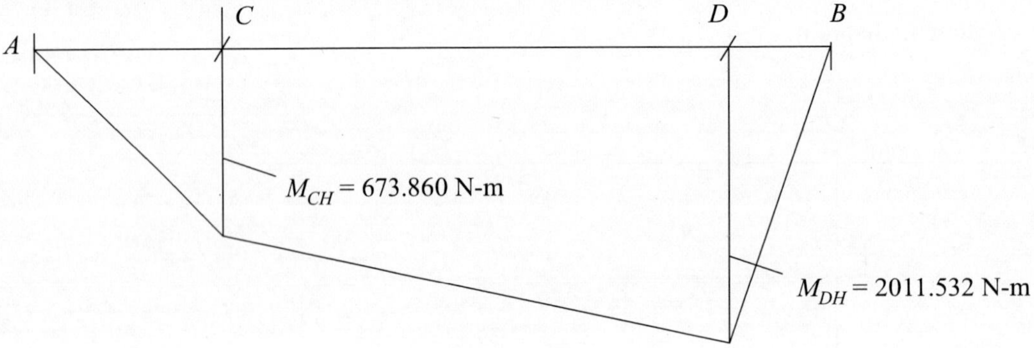

Figure 6.33

Step-4: Resultant moment
From both the bending moment diagrams the bending moments at D are greater and the resultant moment is

$$\begin{aligned} M_D &= \sqrt{M_{DH}^2 + M_{DV}^2} \\ &= \sqrt{2011532^2 + 5682800^2} \\ &= 6028306.3 \text{ N-mm} \end{aligned}$$

Step-5: The diameter by normal-stress theory [Eqn. 4.7 MDHB with $K = 0$]

$$d_o = \left[\frac{16}{\pi S_d}\left\{M_e + \sqrt{M_e^2 + T_e^2}\right\}\left(\frac{1}{1-0}\right)\right]^{1/3}$$

where,

$$M_e = C_m M = 1.5 \times 6028306.3$$

$$= 9042459.45 \text{ N-mm}$$

$$T_e = C_T\, T = 1.2 \times 1591.67 \times 10^3$$

$$= 1910 \times 10^3 \text{ N-mm}$$

$$S_d = 100 \text{ MPa}$$

$$d_o = \left[\frac{16}{\pi \times 100}\left\{9042459 + \sqrt{9042459^2 + (1910 \times 10^3)^2}\right\}\right]^{1/3}$$

$$= 97.65 \text{ mm}$$

Adopt $d_o = 100$ mm.

Example 6.12: A hoisting drum 0.5 m in diameter is keyed to a shaft which is supported in two bearings and driven through a 12:1 reduction gear by an electric motor. Determine the power of the driving motor if the max load of 8 kN is hoisted at a speed of 50 m/min and the efficiency of the drive is 80%. Also determine the torque on the drum shaft and the speed of the motor in rpm. Determine the diameter of the shaft made of machinery steel, working stresses of which are 115 MPa in tension and 50 MPa in shear. The drive gear whose diameter is 450 mm is mounted at the end of the shaft such that it overhangs the nearest bearing by 150 mm. The combined shock and fatigue factors for bending and torsion maybe taken as 2 and 1.5 respectively.

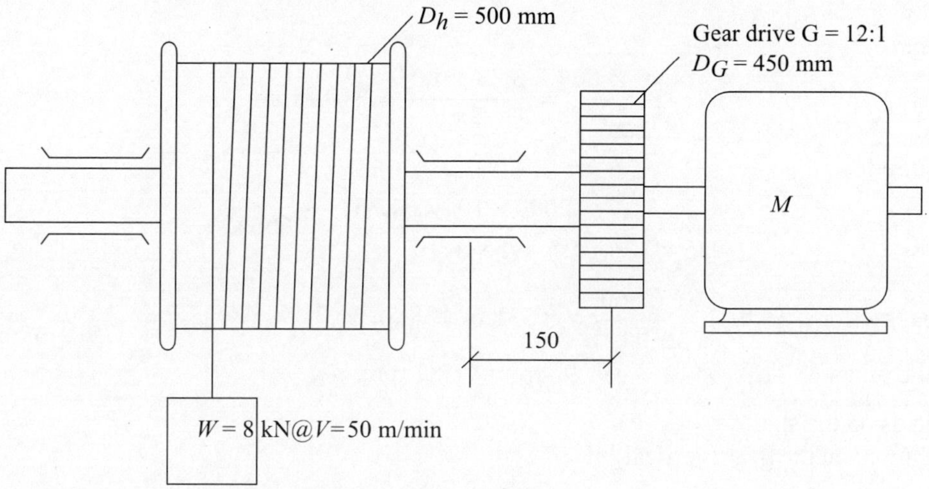

Figure 6.34

Given Data

Hoist diameter, $D_h = 0.5 \text{ m} = 500 \text{ mm}$

Load lifted = 8 kN

Speed of hoisting, $V = 50 \text{ m/min} = 5/6 \text{ m/sec}$

Efficiency of the drive, $\eta = 80\%$

Drive-gear diameter, $D_G = 450$ mm

Gear overhang, $L = 150$ mm

Material properties: Design stress in tension, $S_d = 115$ MPa

Design stress in shear, $S_{sd} = 50$ MPa

Fatigue factor: $C_m = 2$ and $C_t = 1.5$.

SOLUTION

Step-1: Power of motor

Load lifted, $W = 8$ kN.

Radius of the drum, $R = \dfrac{D}{2} = \dfrac{500}{2} = 250$ mm.

Torque applied,

$$T = W \cdot R = 8 \times 10^3 \times 250 = 2000 \times 10^3 \text{ N-mm.}$$

Speed of the hoisting,

$$\begin{aligned} V &= \omega \cdot R \\ \omega &= \frac{V}{R} = \frac{5}{6} \times \frac{1}{0.25} \\ \omega &= 3.33 \text{ rad/ sec.} \end{aligned}$$

Speed in rpm,

$$N = \frac{\omega \times 60}{2\pi} = \frac{3.33 \times 60}{2\pi} = 31.83 \text{ rpm}$$

Power required,

$$P = \frac{T \cdot N}{955 \times 10^4} = \frac{2000 \times 10^3 \times 31.83}{955 \times 10^4} = 6.666 \text{ kW.}$$

Motor capacity (for $\eta = 0.8$) $= \dfrac{6.666}{0.8} = 8.33\,\text{kW} \cong 8.5\,\text{kW}$.

Speed of the motor $= 31.83 \times 12 = 381.96$ rpm $\cong 382$ rpm

Step-2: Loads on the shaft

Tangential force on the gear (vertical),

$$F_T = \frac{T}{r_g} = \frac{2 \times 10^6}{450/2} = \frac{2 \times 10^6 \times 2}{450} = 8888.9 \text{ N.}$$

Radial force on the gear (horizontal),

$$\begin{aligned} F_r &= F_T \tan\phi = 8888.9 \times \tan 20^\circ \\ &= 3235.3 \text{ N} \end{aligned}$$

Total force,

$$F = \sqrt{F_t^2 + F_r^2}$$
$$= \sqrt{(8888.9)^2 + (3235.32)^2}$$
$$= 9459.37 \text{N}$$

Step-3: Bending moment on the shaft

$$M_b = F \times l = 9459.37 \times 150$$
$$= 1418905 \text{ N-mm}$$

Step-4: The shaft diameter by maximum shear stress theory [Eqn. 4.8 MDHB with $K = 0$]

$$d_o = \left[\frac{16}{\pi.\, S_{sd}}\left\{\sqrt{M_e^2 + T_e^2}\right\}\right]^{1/3}$$

where,

$$M_e = 2 \times 1418905 = 2837810 = 2837.8 \times 10^3$$

$$T_e = 1.5 \times 2 \times 10^6 = 3 \times 10^6 = 3000 \times 10^3 \text{ N.}$$

$$d_o = \left[\frac{16}{\pi \times 50}\left\{\sqrt{(2837.8 \times 10^3)^2 + (3000 \times 10^3)^2}\right\}\right]^{1/3}$$

Solving, $d_o = 74.926$

Adopt, $d_o = 75$ mm

Example 6.13: A horizontal piece of shafting is supported by two bearings 1.5 m apart. A keyed gear, 20° involute and 175 mm diameter is located 400 mm to the left of the right bearing and is driven by a gear directly behind it. A 600 m diameter pulley is keyed to the shaft 600 mm to the right of the left bearing and drives another pulley by means of a belt drive inclined at 45° to the horizontal below the shaft and in front of it. The tension ratio of the belt is 3:1. The gear transmits 45 kW at 330 rpm CW when viewed from right side. The allowable shear stress for shaft material is 40 MPa. The combined shock and fatigue factor for torsion and bending are 1.5 and 2.0 respectively. Draw the moment diagrams and calculate the necessary shaft diameter.

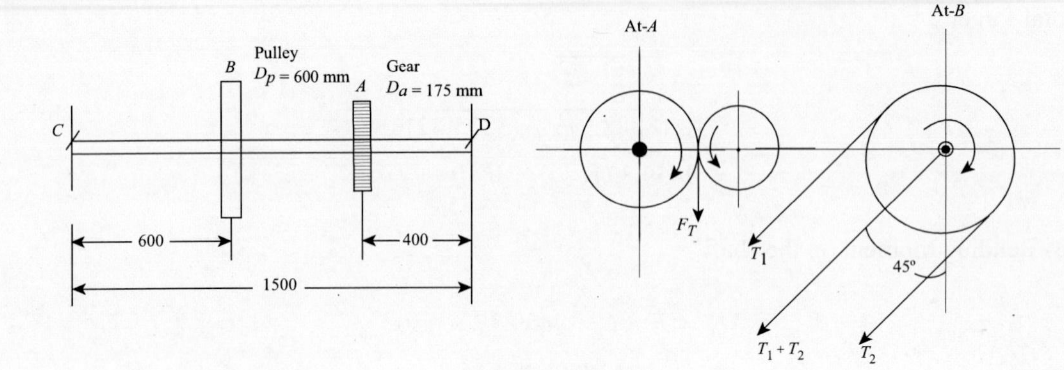

Figure 6.35

Power input at A, $P = 45$ kW at $N = 330$ rpm

Allowable stress in $S_{sd} = 40$ MPa

Shock and fatigue factors, $C_m = 2.0$ and $C_t = 1.5$

SOLUTION

Step-1: Loads on the shaft

The input torque, $T_A = \dfrac{955 \times 10^4 \times 45}{330} = 1302.27 \times 10^3$ N-mm

Tangential force on the gear (vertical),

$$F_{AT} = \frac{T_A}{r_A} = \frac{1302.27 \times 10^3}{175/2} = 14883.1 \text{ N}$$

Radial force on the gear (horizontal),

$$F_{Ar} = F_{AT} \times \tan 20° = 5417.00 \text{ N}$$

At pulley B, the ratio of belt tensions

$$\frac{T_1}{T_2} = 3 \qquad \text{(i)}$$

Effective driving force,

$$T_1 - T_2 = \frac{T_B}{r_B} = \frac{1302.27 \times 10^3}{600/2} = 4340.9 \text{ N} \qquad \text{(ii)}$$

From Eqns. (i) and (ii),

$$2T_2 = 4340.9$$
$$T_2 = \frac{4340.9}{2} = 2170.45\text{ N}$$
$$T_1 = 3 \times 2170.45 = 6511.35$$

Total belt pull on the shaft,

$$F_B = T_1 + T_2 = 6511.35 + 2170.45$$
$$= 8681.8\text{N}$$

Horizontal component,

$$F_{BH} = F_B \cdot \cos 45^\circ = 8681.8 \times \cos 45^\circ$$
$$= 6138.96\text{ N}$$

Vertical component,

$$F_{BV} = F_B \cdot \sin 45^\circ = 8681.8 \times \sin 45^\circ$$
$$= 6138.96\text{ N}$$

Step-2: Vertical load diagram and moment diagram

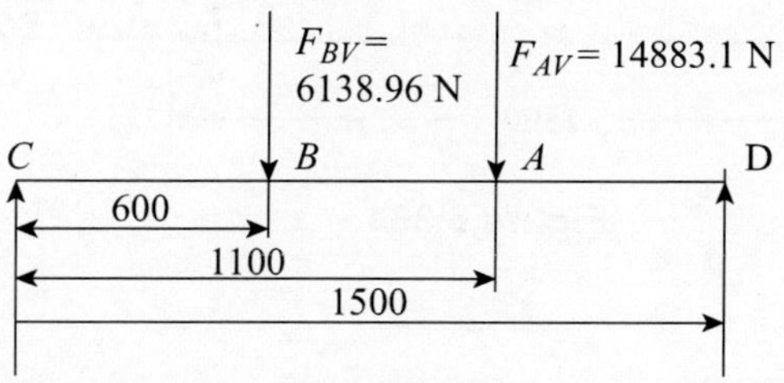

Figure 6.36

Reaction at D is evaluated by taking moments about C:

$$6138.96 \times 600 + 14883.1 \times 1100 - R_{DV} \times 1500 = 0$$
$$\therefore \quad R_{DV} = \frac{6138.96 \times 600 + 14883.1 \times 1100}{1500} = 13369.85\text{ N}$$

Similarly, reaction at C $\quad R_{CV} = \dfrac{6138.96 \times 900 + 14883.1 \times 400}{1500} = 7652.2\text{ N}$

Bending moments diagram

$$\text{At } B,\ M_{BV} = 7652.2 \times 600 = 4591320\text{ N-mm}$$
$$\text{At } A,\ M_{AV} = 13369.85 \times 400 = 5347940\text{ N-mm}$$

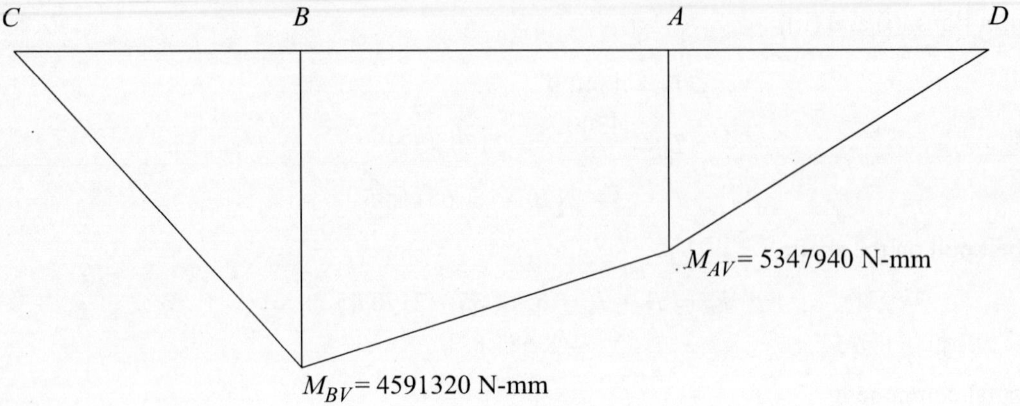

Figure 6.37

Step-3: Horizontal load diagram and moment diagram

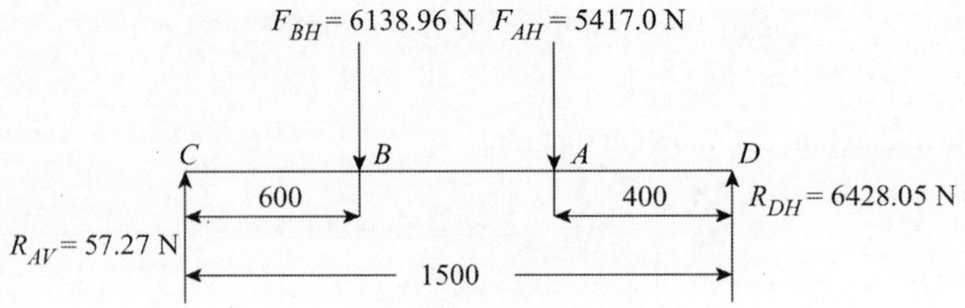

Figure 6.38

Reaction at D,

$$R_{DH} = \frac{6138.96 \times 600 + 5417 \times 1100}{1500} = 6428.05 \text{ N}$$

Reaction at C,

$$R_{CH} = \frac{5417 \times 400 + 6138.96 \times 900}{1500} = 5127.0 \text{ N}$$

Bending moment diagram

Moment at B,

$$M_{BH} = 5127 \times 600 = 3076200 \text{ N-mm}$$

Moment at A,

$$M_{AH} = 6428.05 \times 400 = 2571220 \text{ N-mm}$$

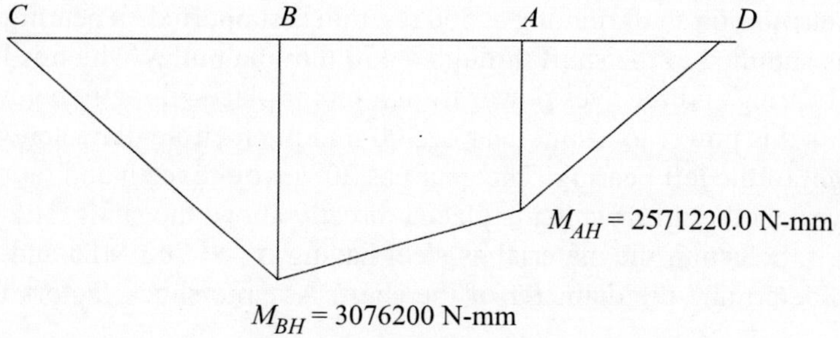

Figure 6.39

Step-4: Resultant moments
Resultant moment at A,

$$M_A = \sqrt{M_{AH}^2 + M_{AV}^2}$$
$$= \sqrt{(2571220)^2 + (5347940)^2}$$
$$= 5933939.21 \text{ N}$$

Resultant moment at B,

$$M_B = \sqrt{(3076200)^2 + (4591320)^2}$$
$$= 5526592.0 \text{ N-mm}$$

The section at A carries maximum torque and maximum moment and hence is critical.
Step-5: The design of shaft by shear theory is [Eqn. 4.8 MDHB with $K = 0$]

$$d_o = \left[\frac{16}{\pi \cdot S_{sd}} \left\{\sqrt{M_e^2 + T_e^2}\right\} \left(\frac{1}{1-0}\right)\right]^{1/3}$$

where,

$$M_e = C_m M_{\max} = 2 \times 5933939.21 = 11867878.42 \text{ N-mm}$$
$$= 11867.9 \times 10^3 \text{ N-mm}$$
$$T_e = C_t T = 1.5 \times 1302.27 \times 10^3 = 1953.405 \times 10^3 \text{ N-mm}$$

Design stress, $S_{sd} = 40$ MPa.
Substituting the values,

$$d_o = \left[\left(\frac{16 \times 1953.405 \times 10^3}{\pi \times 40}\right) \left\{\left(\frac{11867878.4}{1953405}\right)^2 + 1\right\}^{1/2}\right]^{1/3}$$
$$= 115.26 \text{ mm}$$

Adopt, $d_o = 125$ mm next available diameter (Table 6.4)

Example 6.14: A transmission shaft running at 500 rev/min is supported on bearings 800 mm apart. Twenty kW power is supplied to the shaft through a 450 mm dia pulley which is located 400 mm to the right of right bearing and receives power from a motor placed directly below the shaft. The shaft further transmits this power to a spur gear of 300 mm pitch circle diameter, which is located at 400 mm to the right of the left bearing. The gear has 20° involute teeth and ratio of belt tension is 3:1. The gear drives another gear which is placed directly above the shaft. The gear and pulley are keyed to the shaft. Selecting the material as steel having $\sigma_{ut} = 700$ MPa and $\sigma_{yt} = 460$ MPa as per ASME code, determine the diameter of the shaft. Assume shock factors for bending and torsion as 1.5.

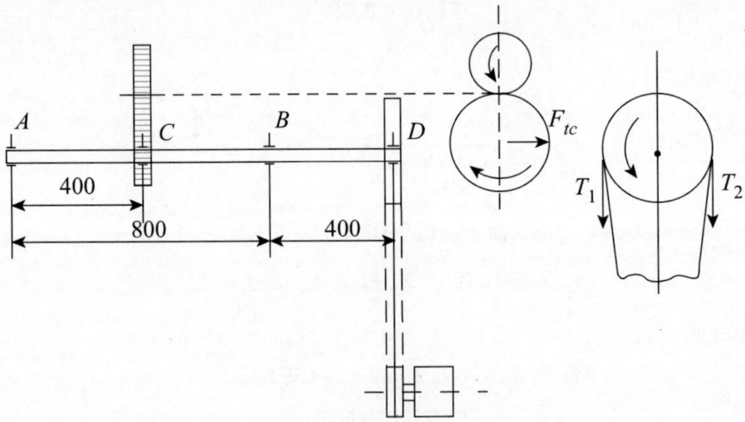

Figure 6.40

Given Data

Power = 20 kW

Speed, $N = 500$ rpm

Gear diameter, $D_G = 300$ rpm

Pulley diameter, $D_P = 450$ mm and $S_y = 460$ MPa.

$$C_m = C_t = 1.5 \quad \text{Factor of safety} = 2 \text{ (assumed)}$$

$$S_d = \frac{S_y}{2} = \frac{460}{2} = 230 \text{ MPa}$$

$$S_{sd} = 115 \text{ MPa.}$$

SOLUTION

Step-1: Loads on the shaft

Torque transmitted (Eqn. 4.1 MDHB)

$$T = \frac{955 \times 10^4 \times \text{kW}}{\text{N}} = \frac{955 \times 10^4 \times 20}{500} = 382 \times 10^3 \text{ N-mm}$$

Tangential force on the gear at C (horizontal),

$$F_{Tc} = \frac{T}{r_c} = \frac{382 \times 10^3}{150} = 2546.67\,\text{N}.$$

Radial force on the gear (vertical),

$$F_{rc} = F_{TC} \times \tan 20^\circ = 2546.67 \times \tan 20^\circ = 926.911\,\text{N}.$$

At D,
Ratio of belt tensions,

$$T_1/T_2 = 3 \tag{i}$$

Effective belt tension,

$$T_1 - T_2 = \frac{T}{r_d} = \frac{382 \times 10^3}{450/2}$$

$$T_1 - T_2 = 1697.78\,\text{N} \tag{ii}$$

From Eqns. (i) and (ii),

$$2T_2 = 1697.78$$

$$\therefore \quad T_2 = 848.9$$

$$\therefore \quad T_1 = 3T_2 = 3 \times 848.9 = 2546.67\ \text{N}.$$

Total pull on the shaft (vertical),

$$F_{DV} = T_1 + T_2 = 2546.67 + 848.9$$
$$= 3395.57\ \text{N}.$$

Step-2: Vertical load diagram and moment diagram

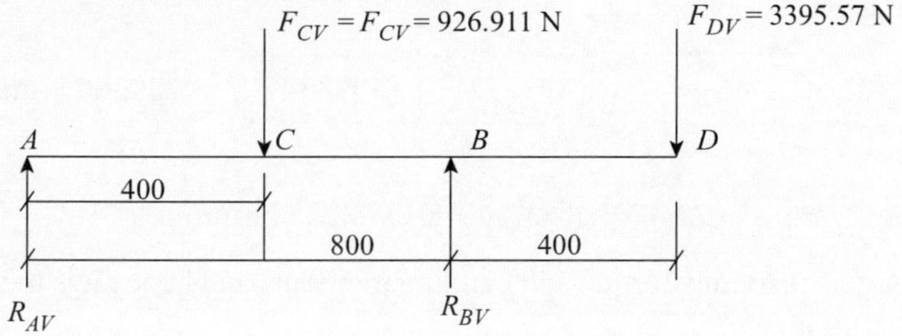

Figure 6.41

Reaction at B, taking moments about A

$$R_{BV} = \frac{926.911 \times 400 + 3395 \times 1200}{800} = 5556.81\,\text{N}$$

The reaction at A, $R_{AV} = \dfrac{3395.57 \times 400 - 926.911 \times 400}{800} = 1234.33\,\text{N}$

Then the moment at B,

$$M_{BV} = 3395.57 \times 400$$
$$= 1358228 \text{ N-mm}$$

Moment at C,

$$M_{CV} = 1234.33 \times 400$$
$$= 493732 \text{ N-mm}$$

Step-3: Horizontal load diagram

Reactions, $R_{AH} = R_{BH} = \dfrac{F_{CH}}{2} = \dfrac{2546.67}{2} = 1273.3354$

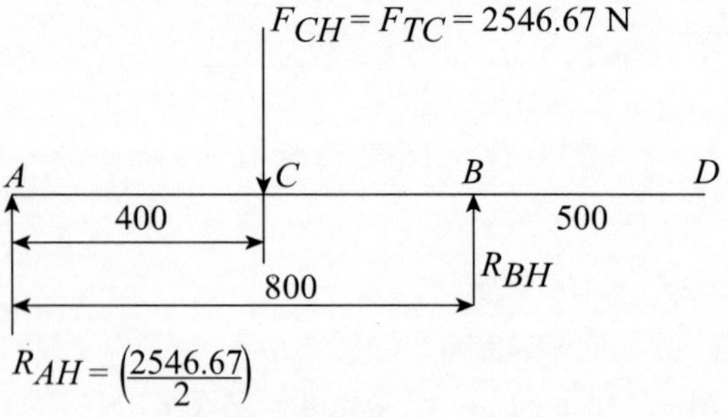

Figure 6.42

Bending moment at C,

$$M_{CH} = \frac{FL}{4} = \frac{2546.67 \times 800}{4} = 509334 \text{ N-mm.}$$

Step-4: Resultant moments
Resultant moment at C,

$$M_c = \sqrt{{M_{CH}}^2 + {M_{CV}}^2} = \sqrt{(493732)^2 + (509334)^2 ll} = 709360 \text{ N-mm.}$$

Resultant moment at B,

$$\therefore \quad M_{BH} = 0 \quad M_{BV} = 1{,}358{,}228 \text{ N-mm}$$

The section at B is subjected to max torque and bending moments and hence it is the critical section.
Step-5: Design of shaft diameter
Effective moment and torque,

$$M_e = C_m \cdot M = 1.5 \times 1358228 = 2037342 \text{ N-mm} = 2037.342 \times 10^3 \text{ N-mm}$$
$$T_e = C_t T = 1.5 \times 382 \times 10^3 = 578 \times 10^3 \text{ N-mm.}$$

Shaft diameter by shear theory (Eqn. 4.8 MDHB with $K = 0$),

$$d_o = \left[\frac{16}{\pi \cdot S_{sd}} \left\{\sqrt{M_e^2 + T_e^2}\right\} \left(\frac{1}{1-0}\right)\right]^{1/3}$$

$$= \left[\frac{16}{\pi \times 115} \left\{\sqrt{(2037342)^2 + (578 \times 10^3)^2}\right\}\right]^{1/3}$$

$$= 45.43 \text{ mm}$$

Diameter by normal stress theory, (Eqn. 4.7 with $K = 0$)

$$d_o = \left[\frac{16}{\pi S_d} \left\{M_e + \sqrt{M_e^2 + T_e^2}\right\}\right]^{1/3}$$

$$d_o = \left[\frac{16 \times 2037342}{\pi \times 230} \left\{1 + \sqrt{1 + \left(\frac{578000}{2037342}\right)^2}\right\}\right]^{1/3}$$

$$d_o = 45.14 \text{ mm}$$

Adopt the larger value

$$d = 45.43 \text{ mm}$$

Choose next standard value (Table 4.4 MDHB)

$$d = 50.0 \text{ mm}$$

Example 6.15: (a) Briefly explain advantages of hollow shafts over solid shafts.
(b) A power transmission shaft 1300 mm long is supported in bearings at its extreme ends *A* and *B*. A power of 30 kW is received at 500 rpm through a gear drive located at 400 mm to the right of the left extreme end of the shaft. The gear mounted on the shaft has a pitch diameter of 300 mm and weighs 800 N. This gear receives power from a gear located exactly behind. The power is delivered through a belt drive located 500 mm to the left of right bearing. The pulley mounted on the shaft has a diameter of 400 mm and weighs 1 kN. The belt is directed towards the observer below the horizontal and inclined at 45°. Ratio of belt tensions is 3, material of the shaft is C40 steel. Assuming a FOS of 2.5 and loading to be with minor shock, determine the diameter of solid shaft.

SOLUTION
(a) Hollow shafts have greater strength and stiffness than solid shafts for the same weight and length.
For same weight and length

$$A = \frac{\pi}{4} d_o^2 (1 - K^2) = \frac{\pi}{4} d^2$$

$$d_o = \frac{d}{\sqrt{1 - K^2}} \qquad \text{(i)}$$

Then, the ratio of polar moment of inertias is

$$\frac{J_H}{J_s} = \frac{\frac{\pi}{32} d_o^4 (1-K^4)}{\frac{\pi}{32} d^4} = \frac{1-K^4}{(1-K^2)^2} = \left(\frac{1+K^2}{1-K^2}\right) > 1$$

$J_H > J_s$ => hollow shaft is stronger.

The stiffness, $K_S = \frac{T}{\theta} = \frac{GJ}{L}$

$$\frac{(K_s)_h}{(K_s)_s} = \left(\frac{J_H}{J_s}\right) > 1 \quad \text{(ii)}$$

Torsional rigidity for hollow shafts is more than that for solid shaft.
Torque capacity, $T = \tau_{max} \cdot J_p$

$$\frac{T_H}{T_s} = \frac{(J_p)_H}{(J_p)_S} = \frac{\frac{\pi}{16} d_o^3 (1-K^4)}{\frac{\pi}{16} d^3} = \frac{1-K^4}{(1-K^2)^{3/2}}$$

$$= \frac{1+K^2}{\sqrt{(1-K^2)}} > 1$$

$$\text{i.e.,} \quad T_h > T_s \quad \text{(iii)}$$

Torque capacity for hollow shaft is more than solid shaft.

(b)

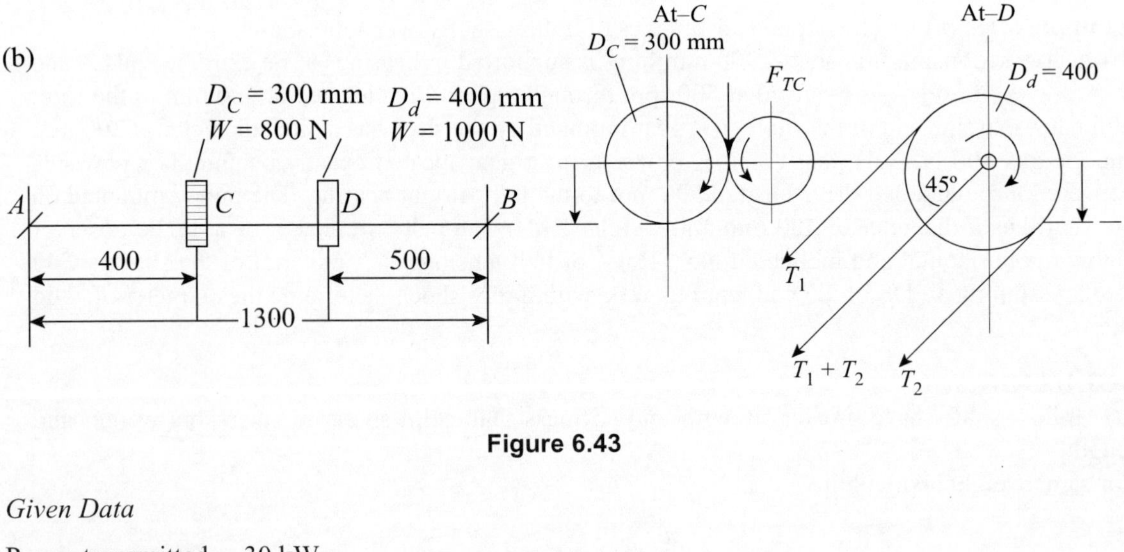

Figure 6.43

Given Data

Power transmitted = 30 kW

Speed, N = 500 rpm

Material C40, $S_y = 360$ MPa.

$$\text{Factor of safety} = 2.5, \therefore\ S_d = \frac{360}{2.5} = 144 \text{ MPa.}$$

SOLUTION

Step-1: Loads on the shaft

The torque transmitted (Eqn. 4.1 MDHB)

$$T = \frac{955 \times 10^4 \times 30}{500} = 573 \times 10^3 \text{ N-mm}$$

Forces at C

The tangential force on the gear (vertical),

$$F_{TC} = \frac{T}{R_c} = \frac{573 \times 10^3 \times 2}{300} = 3820 \text{ N}$$

Total vertical force $= F_{TC} + W_c = 3820 + 800 = 4620\ N$ The radial force on the gear (horizontal)

$$F_{rc} = F_{TC}.\tan 20^o = 3820 \tan 20^o = 1390.37 \text{ N}$$

Belt tensions at D

Ratio of belt tensions, $T_1/T_2 = 3$ (i)

Effective drive tension,

$$T_1 - T_2 = \frac{T}{R_D} = \frac{573 \times 10^3}{400/2} = 2865 \text{ N} \quad \text{(ii)}$$

From (i) and (ii),

$$2T_2 = 2865 \text{ N}$$
$$\therefore\quad T_2 = 2865/2 = 1437.5 \text{ N}$$
$$\therefore\quad T_1 = 3T_2 = 4312.5 \text{ N}$$

Total pull on the shaft,

$$P_D = T_1 + T_2 = 4312.5 + 1437.5$$
$$= 5750 \text{ N.}$$

Horizontal component,

$$P_{DH} = P_D \cos 45^o = 5750.\ \cos 45^o = 4065.86 \text{ N}$$

Vertical component,

$$P_{DV} = P_D \sin 45^o = 5750.\ \sin 45^o = 4065.86 \text{ N}$$

Total vertical force,

$$F_{DV} = P_{DV} + W = 4065.86 + 1000 = 5065.86 \text{ N}$$

Step-2: Vertical load diagram and moment diagram

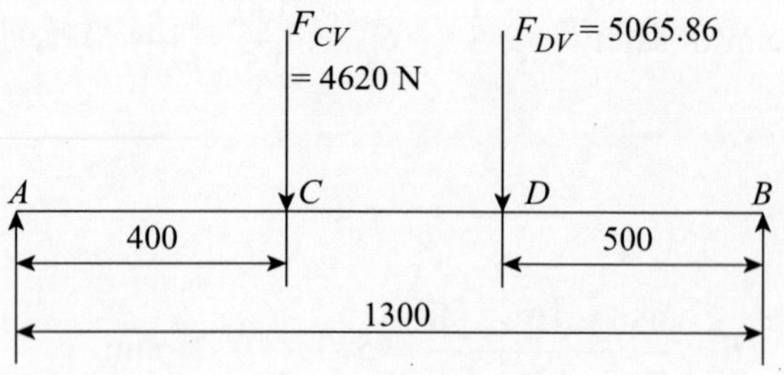

Figure 6.44

Bending moment stress for individual loads

Using Fig. (7) in Plate 1.8 MDHB

Moment due to the load at D,

$$M = \frac{Pab}{L}$$

$$M_{DV_1} = \frac{5065.86 \times 800 \times 500}{1300}$$

$$= 1558726.15 \text{ N-mm}$$

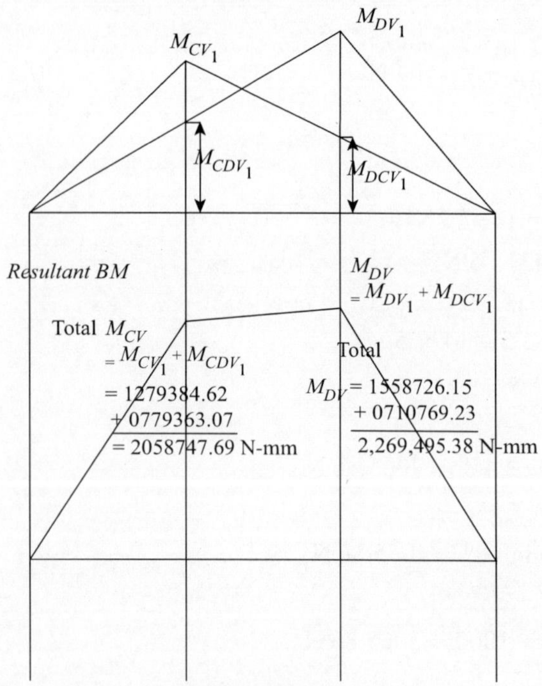

Figure 6.45

Moment at C due to this load,

$$M_{CDV_1} = \frac{M_{DV}}{AD} \cdot AC$$

$$= \frac{1558726.15 \times 400}{800}$$

$$= 779363.07 \text{ N-mm}$$

Moment due to load at C,

$$M_{CV_1} = \frac{4620 \times 400 \times 900}{1300}$$

$$= 1279384.22 \text{ N-mm}$$

Moment at D due to this,

$$M_{DCV_1} = \frac{1279384.62 \times 500}{900}$$

$$= 710769.23 \text{ N-mm}$$

Step-3: Horizontal loads and moments

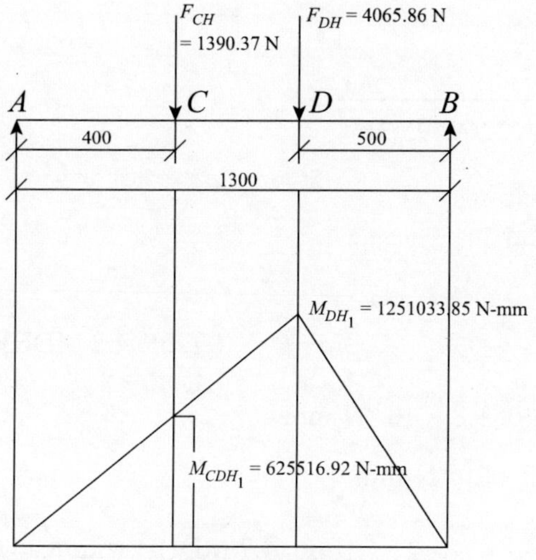

BM. due to load at *D*

Figure 6.46

Using the Eqn., $M = \dfrac{Pab}{L}$ in Fig. (7) in Plate 1.8 MDHB

Moment at *D*,

$$M_{DH_1} = \frac{4065.86 \times 500 \times 800}{1300}$$
$$= 1251033.85 \text{ N-mm}$$

Moment at *C* due to this load,

$$M_{C_{DH_1}} = \frac{M_{DH} \times 400}{800} = \frac{1251033.85 \times 400}{800}$$
$$= 625516.92$$

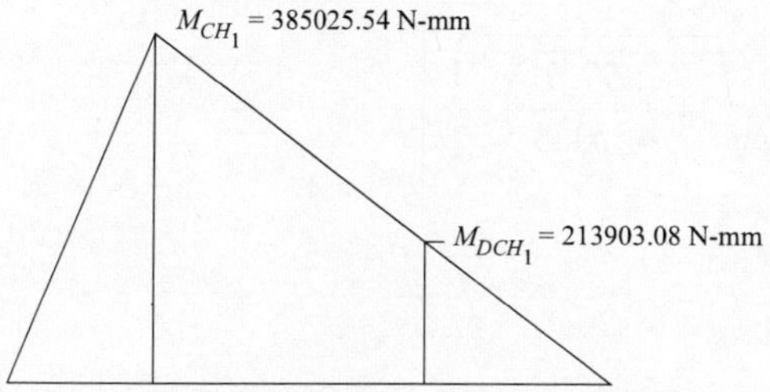

BM due to load at C

Moment at *C* due to load at *C*,

$$M_{CH_1} = \frac{1390.37 \times 400 \times 900}{1300}$$
$$= 385025.54 \text{ N-mm}$$

Moment at *D* due to this load,

$$M_{D_{CH_1}} = \frac{M_{CH}}{900} \times 500 = 213903.08 \text{ N-mm}$$

Total moment diagram,

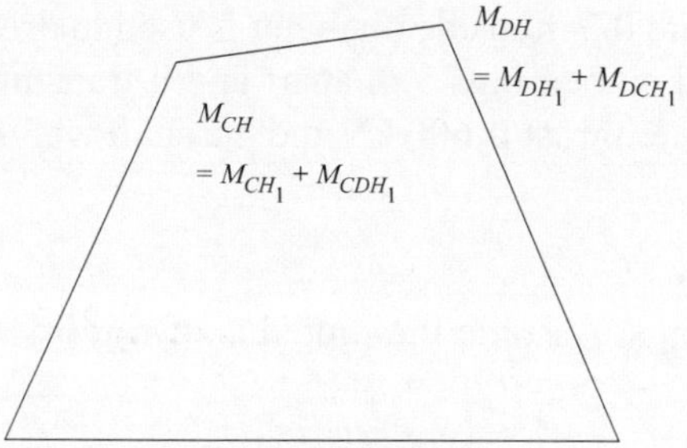

Figure 6.47

$$M_{DH} = M_{DH_1} + M_{DH_1}$$
$$= 1251033.55$$
$$+ 213903.08$$
$$1,46,4,936.93 \text{ N-mm}$$

$$M_{CH} = M_{CH_1} + M_{CDH_1}$$
$$= 385025.54$$
$$+ 625516.92$$
$$1,010,542.46 \text{ N-mm}$$

Step-4: Resultant bending moment diagram
Maximum moment occurs at D,

$$M_D = \sqrt{M_{DH}^2 + M_{DV}^2}$$
$$= \sqrt{(1464936.93)^2 + (2269495.38)^2}$$
$$= 2701227.20 \text{ N-mm}$$
$$= 2701.23 \times 10^3 \text{ N-mm}$$

Step-5: Design of shaft
Then using shock factors $C_m = 1.5$ and $C_t = 1.5$, (Table 4.2 MDHB)

$$M_e = 1.5 \times 2701.23 \times 10^3 = 4051.8 \times 10^3 \text{ N-mm}$$
$$T_e = 1.5 \times 573 \times 10^3 = 859.5 \times 10^3 \text{ N-mm.}$$

Then using normal stress theory, [Eqn. 4.7 MDHB with $K = 0$]

$$d_o = \left[\frac{16}{\pi S_d}\left\{M_e + \sqrt{M_e^2 + T_e^2}\right\}\left(\frac{1}{1-0}\right)\right]^{1/3}$$
$$= \left[\frac{16 \times 4051.8 \times 10^3}{\pi \times 144}\left\{1 + \sqrt{1\left(\frac{859.5}{4051.8}\right)^2}\right\}\right]^{1/3}$$
$$= 66.17 \text{ mm}$$

Adopt $d = 70$ mm

Example 6.16:

(a) Prove that a hollow shaft is stronger and stiffer than a solid shaft of same length, weight and material.

(b) A hollow propeller shaft of 0.6 m outside diameter and 0.3 m inside diameter is used to drive a propeller of a marine vessel. The shaft is mounted on bearings 5 m apart and it transmits 6 MW power at 100 rpm. The maximum axial propeller thrust is 600 kN and the shaft weighs 60 kN. Determine:

(i) Maximum shear stress developed in the shaft.

(ii) Angle of twist of the shaft between the bearings. Assume the modulus of rigidity as 84 GPa.

SOLUTION

(a) Weight $= (A \cdot L \cdot \rho)_h = (A \cdot L \cdot \rho)_s$

$$A_h = A_s$$

$$\frac{\pi}{4}d_o^2\left(1-K^2\right)=\frac{\pi}{4}d^2$$

$$\therefore\ d_o=\frac{d}{\sqrt{1-K^2}} \tag{i}$$

The torsion equation is

$$\because\ \frac{T}{J}=\frac{\tau}{R}=\frac{G\theta}{L} \qquad \text{(R : radius)}$$

For strength, torque capacity

$$T=\tau\left(\frac{J}{R}\right)$$

For hollow shaft,

$$T_h=\tau\left[\frac{\pi}{16}d_o^3\left(1-K^4\right)\right] \tag{ii}$$

For solid shaft,

$$T_s=\tau\left[\frac{\pi}{16}d^3\right] \tag{iii}$$

Ratio of strengths,

$$\frac{T_h}{T_s}=\frac{d_o^3\left(1-K^4\right)}{d^3}=\frac{d^3}{\left(1-K^2\right)^{3/2}}\times\frac{\left(1-K^4\right)}{d^3}$$

$$=\frac{1-K^4}{\left(1-K^2\right)^{3/2}}=\frac{1-K^2}{\sqrt{1+K^2}}>1$$

i.e., $\quad T_h>T_s$

Hollow shaft is stronger than the solid shaft.
Stiffness from the torsion equation,

$$K_s=\frac{T}{\theta}=\frac{GJ}{L}$$

Ratio of stiffness,

$$\frac{(K_s)_h}{(K_s)_s}=\frac{J_h}{J_s}=\frac{\frac{\pi}{32}d_o^4\left(1-K^4\right)}{\frac{\pi}{32}d^4}=\frac{1-K^4}{\left(1-K^2\right)^2}=\frac{1+K^2}{1-K^2}>1$$

$$(K_s)_h>(K_s)_s$$

Hollow shaft is stiffer than solid shaft.

(b) *Marine shaft*

Given Data

$d_o = 0.6$ m = 600 mm

$d_i = 0.3$ m = 300 mm

Length of the shaft = 5 m = 5000 mm

Power transmitted, $P = 6$ MW = 6000 kW

Speed, $N = 100$ rpm

Axial thrust, $F_a = 600$ kN

Weight of the shaft, $W = 60$ kN

Modulus of rigidity, $G = 84$ GPa = 84×10^3 MPa

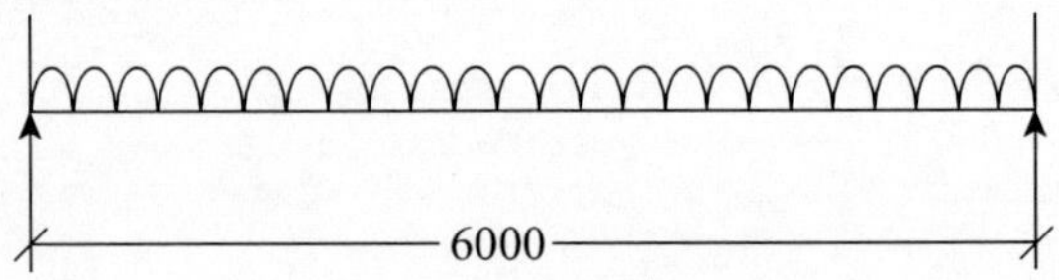

Figure 6.48

Step-1: Loads on the shaft

The shaft weight is uniformly distributed

$$w = \frac{60 \times 10^3}{6000} = 10 \text{ N/mm}$$

Maximum bending moment (Fig. 10 in Plate 1.8 MDHB)

$$M_b = \frac{wl^2}{8} = \frac{10 \times 6000^2}{8} = 45 \times 10^6 \text{ N-mm}$$

Torque transmitted,

$$T = \frac{955 \times 10^4 \times \text{kW}}{N} = \frac{955 \times 10^4 \times 6000}{100} = 5730 \times 10^5 \text{ N-mm}$$

$$T = 573 \times 10^6 \text{ N-mm}$$

Step-2: Stresses induced in the shaft

Moment of inertia of the shaft in bending,

$$I = \frac{\pi}{64}\left(d_o^4 - d_i^4\right) = \frac{\pi}{64} d_o^4 \left(1 - K^4\right)$$

$$I = \frac{\pi}{64} \times 600^4 \times \left(1 - 0.5^4\right)$$

$$I = 59.64 \times 10^8 \, \text{mm}^4$$

$$\begin{aligned} \text{Polar moment of intertia, } J &= 2 \times I \\ &= 2 \times 59.64 \times 10^8 \\ &= 119.28 \times 10^8 \, \text{mm}^4 \end{aligned}$$

Bending stress,

$$\sigma_b = \frac{M}{\text{I}} \cdot r_0 = \frac{45 \times 10^6 \times (600/2)}{59.64 \times 10^8} = 2.26 \text{ MPa}$$

Axial stress,

$$\sigma_1 = \frac{F_a}{A} = \frac{600 \times 10^3}{\frac{\pi}{4} \times (600)^2 (1 - 0.5^2)} = 2.83 \text{ MPa}$$

Total direct stress,

$$\sigma = \sigma_b + \sigma_1 = 2.26 + 2.83 = 5.09 \, \text{MPa}$$

Torsional shear stress,

$$\tau = \frac{T}{J} \cdot r_0 = \frac{573 \times 10^6 \times 300}{119.28 \times 10^8} = 14.411 \, \text{MPa}$$

Step-3: Maximum shear stress in the shaft

$$\begin{aligned} \tau_{\max} &= \sqrt{\left(\frac{\sigma}{2}\right)^2 + \tau^2} \\ &= \sqrt{\left(\frac{5.09}{2}\right)^2 + (14.411)^2} \\ &= 14.634 \, \text{MPa} \end{aligned}$$

Step-4: Angle of twist of the shaft

$$\theta = \frac{T}{J} \cdot \frac{L}{G} = \frac{573 \times 10^6 \times 6000}{119.28 \times 10^8 \times 84 \times 10^3}$$

$$\theta = 0.003431 \, \text{rad}$$

$$\theta = 0.1966^\circ$$

EXERCISES FOR PRACTICE

6.1. Find the diameter of a solid steal shaft to transmit 20 kW at 200 rpm. The ultimate shear stress for the steel may be taken as 360 MPa and a factor of safety of 5. If a hollow shaft is used in place of solid shaft, find the inside and outside diameters if their ratio is 0.6.

6.2. A line shaft is driven by means of a motor placed vertically below it. The pulley on the line shaft is 1.5 metre in diameter and has belt tensions 5.8 kN and 1.2 kN on the tight side and slack side of the belt respectively. Both these tensions may be assumed vertical. The pulley is overhung on a shaft with a distance of 500 mm. Find the diameter of shaft. The shear stress is limited to 42 MPa.

6.3. A mild steel shaft transmits 20 kW at 200 rpm. It carries a central load of 1.5 kN and supported between bearings 3 m apart. Determine the size of the shaft, if the allowable stress in shear is 42 MPa and in tension is 56 MPa. What is the size of the shaft if it is subjected to gradually applied load. Take the maximum torque to be 20% more than the full-load torque.

6.4. A shaft, 2 m long between bearings, carries a 900 N pulley at its midpoint. Through a belt drive, the shaft receives 25 kW at 180 r.p.m. The belt drive is horizontal, and the sum of the belt tensions is 7 kN. Determine the shaft diameter and angle of twist the shaft undergoes. Take G = 80 GPa

6.5. A shaft is supported on bearings *A* and *B*, 1.5 m between centres. A 20° straight tooth spur gear having 0.75 m pitch diameter is located 0.35 m to the right of the left hand bearing *A* and a 0.85 m diameter pulley is mounted 0.22 m towards left of bearing *B*. The gear is driven by a pinion with a downward tangential force while the pulley drives a horizontal belt having 180o angle of wrap. The pulley also serves as a flywheel and weighs 2.2 kN. The maximum belt tension is 3.5 kN and the tension ratio is 3:1. Determine the maximum bending moment and the necessary shaft diameter if allowable shear stress of the material is 45 MPa. Assume that the power transmission takes place with minor shocks.

6.6. A power transmission shaft made of C40 steel is supported on two bearings 2 m apart. The shaft receives a power of 40 kW through a belt drive situated, at a distance of 600 mm to the right of the left bearing. The weight and diameter of the pulley are respectively 800 N and 400.0 mm. The belt moves towards the observer below the horizontal, inclined at 60° to it. The ratio of the belt tensions is 3. The power is transmitted out of the shaft through a gear drive located on the shaft at a distance of 500 mm to the left of the right bearing. The weight and pitch diameter of the gear mounted on the shaft are respectively 600 N and 300 mm. The gear which receives from this gear is located exactly behind. The teeth are of involute profile with a pressure angle of 20°. Determine the diameter of the solid circular shaft assuming 2.5 for the factor of safety and 25% over load.

6.7. A hollow shaft 0.5 m outside diameter and 0.3 m inside diameter is supported by two bearings 6 m apart, the shaft is driven by a flexible coupling at one shaft end and drives a ship's propeller at 10.5 rad/sec. The maximum thrust on the propeller is 540 kN when the shaft is transmitting 5885 kW. The shaft weight is 67.5 kN. Determine the maximum shear stress in the shaft considering the weight of the shaft and the column effect. Assume $K_b = 1.5$ and $K_s = 1.0$.

6.8. A hollow shaft of 0.6 m outside diameter and 0.4 m inside diameter is used to drive a propeller of a marine vessel. The shaft is mounted on bearings 6 m apart and transmits 5000 kW at 150 rpm. The thrust on propeller is 500 kN and shaft weighs 70 kN. Determine:

(a) Maximum shear stress in shaft.

(b) Angular twist between bearings.

6.9. A machine shaft is supported in ball bearings placed at a distance of 750 mm. The shaft carries a 450 mm diameter pulley at a distance of 200 mm on the right of right hand bearing and a straight tooth spur gear 200 mm pitch circle diameter on the right of left hand bearing at a distance of 250 mm from it. 15 kW of power is supplied at spur gear at 600 rpm and taken off at the pulley on which belt is mounted making an angle of 60° with the horizontal. The ratio of the belt tensions is 3:1 and pulley weighs 800 N. The gear meshes with another gear directly located above the shaft. The permissible bending stress is 100 MPa and permissible shearing stress is 55 MPa. The pressure angle of gear is 20°. Determine the diameter of the shaft as per ASME code.

6.10. A horizontal piece of commercial shafting is supported by two bearings 1.5 m apart. A keyed gear 20° involute and 200 mm in diameter is located 400 mm to the left of right bearing and is driven by a gear directly behind it. A 600 mm diameter pulley is keyed to the shaft 600 mm to the right of left bearing and drives a pulley with a horizontal belt directly behind it. The weight of the gear and pulley are 200 N and 150 N respectively. The tension ratio of the belt is 2.5:1, with the slack side on top. The driver transmits 50 kW at 350 rpm, $K_b = K_t = 1.5$.

a) Draw the moment diagram

b) Calculate the diameter of the solid shaft required

c) Calculate the torsional deflection in degrees.

6.11. Design a shaft to transmit power from motor to a lathe head stock through a pulley by means of a belt drive. The pulley weighs 250 N and is located at 300 mm from the centre of bearing. The diameter of pulley is 200 mm and maximum power transmitted is 1.5 kW at 120 rpm. The angle of lap of belt is180° and coefficient of friction between belt and pulley is 0.25. The shock and fatigue factors for bending and twisting are 1.5 and 2.0 respectively. The allowable shear stress in the shaft is 42 MPa.

6.12. A shaft 1.2 mm long is supported at the ends by simple bearings. A vertical load of 10 kN is applied 0.25 m from the left end : a load of 15 kN acting down and forward at an angle of 60° with the vertical is applied at 0.6 m from the left hand bearing and a load of 12.5 kN acting down and backward at an angle of 30° with the vertical is applied at 0.95 m from the left end. A torque of 500 N-m is taken off at the second load and a torque of 300 N-m is taken of at the third load. Find the maximum resulting bending moment and the required shaft diameter if the allowable stress in the material is 45 MPa.

6.13. A horizontal piece of commercial shafting is supported by two bearings 1.5 m apart. A keyed gear 20° involute and 175 mm diameter is located 400 mm to the left of right-hand bearing and is driven by a gear directly behind it. A 600 mm pulley is keyed to the shaft at 600 mm to the right of left-hand bearing and drives a pulley with a horizontal belt, directly behind it. The tension ratio of belts is 3 to 1 with slack side on the top. The drive transmits 50 kW at 330 rpm. Calculate (i) the shaft diameter and (ii) angular thrust.

6.14. Three identical pulleys of 50 mm diameter and weighing 500 kN each are mounted on a line shaft supported on two bearings 4000 mm apart. The pulley *A* is mounted at 300 mm to the right of left bearing and receives 40 kW at 200 rpm from a pulley vertically below it. The pulley B is mounted 1,000 mm. to the right of left bearing and delivers 8 kW to a pulley through a belt-drive inclined backward at 45° to the vertical. The remaining power is taken out through another pulley C, which is mounted at 3000 mm to the right of left bearing and drives a planning machine, the drive being 30° to the front of the vertical. The angle of lap for all pulleys may be taken as 180° and the coefficient of friction is 0.3. The working stress in shear for the shaft material is 80 MPa. Determine the diameter of the shaft.

6:15. A shaft is supported by two bearings placed 1.2 m. Apart. A 600 mm pulley is mounted at a distance of 300 mm to the right of left hand bearing. This drives a pulley directly below it with the help of a belt having a maximum tension of 2500 N. Another pulley 400 mim diameter is placed 200 mm to the left of right hand bearing and is driven with the belt of electric motor, which is placed horizontally to the right. The angle of contact for both the pulleys is 180°. Coefficient of friction 0.24. Determine a suitable diameter of shaft.

6.16. The allowable stresses are :
in tension =102 N/mm^2
in shear = 62 N/mm^2 and Take $K_b = 1.5$; $K_t = 1$.

6.17. A horizontal piece of commercial shafting is supported by two bearings 2.0 m apart. A keyed gear, 20° involutes and 200 mm in diameter, is located 500 mm to the left of the right bearing and is driven by a gear directly behind it. A 600 mm diameter pulley is keyed to the shaft 500 mm to the right of the left bearing and drives a pulley with horizontal belt directly behind it. The tension ratio of the belt is 3 to 1 with the slack side on top. The drive transmits 50 kW at 500 rpm. Assume $C_m = C_t = 1.5$.

(a) Draw the moment diagrams showing the values at the clange points.

(b) Calculate the necessary shaft diameter.

(c) Calculate the angular deflection in degrees.

6.18. A mild steel shaft is to be designed to transmit 15 kW at 200 rpm. It is supported on two bearings 600 mm apart. A 20 teeth 20° involute, 8 mm module gear is mounted at 100 mm to the right of right hand bearing and delivers power to a gear directly below it. Another gear with 80 teeth 6 mm module is located 1250 mm to the right of the left-hand bearing and receives power from a gear directly behind it. The permissible stress are 85 MPa in tension and 60 MPa in shear. Determine the diameter of the shaft and the twist between the end sections.

SHORT ANSWER QUESTIONS

6.1. What is the shaft?

6.2. What are the types of shafts?

6.3. What is a transmission shaft?

6.4. What is an axle?

6.5. What is a spindle?

6.6. What is line shaft?

6.7. What is crankshaft?

6.8. What are the materials used for shaft?
6.9. What are the kinds of stresses induced in the shafts?
6.10. What are the correction factors applied to the bending and torsional moments?
6.11. What is the equivalent bending moment by maximum principal stress (Rankine) theory?
6.12. What is the equivalent torsional moment by maximum shear (Guest) theory?
6.13. What is the design for torsional rigidity?
6.14. What is the design for lateral rigidity?
6.15. What the the ASME code equation for shaft design?
6.16. What are the Westing house code equations for equivalent bending and torsion moments in case of fluctuating loads?
6.17. What is the equation for shaft diameter by Von-Mises-Hencky theory?
6.18. What are the advantages of hollow shaft over the solid shaft?
6.19. What is the relation between the diameters of hollow and solid shafts for equal torional rigidity ?
6.20. How do their strengths compare for equal weight?
6.21. Why are the shafts provided with steps?
6.22. What are the forces that cause bending moments in shafts?
6.23. Are the bending stresses steady or variable?
6.24. Are the torsional shear stresses steady or variable?
6.25. How do you account for the effect of the variable stresses in design of the shaft?

CHAPTER 7

KEYS, COTTER AND KNUCKLE JOINTS

7.1 INTRODUCTION

The key is a simple device used to connect a hub to a shaft. Keys can be grouped under two classes, (1) sunk key (2) saddle key.

1. *Sunk keys*. The sunk keys are placed in the keyways formed in the shaft and hub. These are generally rectangular or square in cross section and may be parallel or tapered.

2. *Saddle keys*. There is no keyway in the shaft. The key is inserted in the keyway in the hub and rests against the flattened or round surface of the shaft. The friction between the surfaces is responsible for the torque transmitted. Normally, these keys are tapered in order to create pressure between the surfaces as the key is driven in.

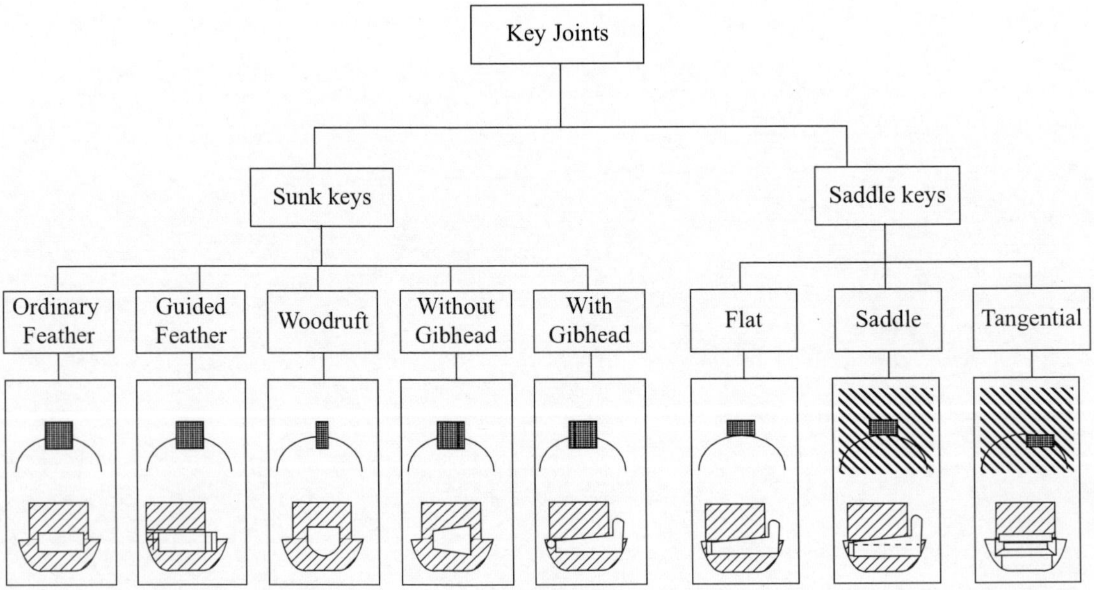

Figure 7.1

The standard sizes of square and rectangular sunk keys are

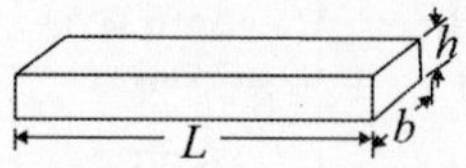

$$b = h = \frac{d}{4} \quad \text{for square key}$$

$$b = \frac{d}{4} \quad \text{and} \quad h = \frac{d}{6} \quad \text{for rectangular key}$$

The standard values are given in Table 7.1.

7.2 STRENGTH OF KEYS

The key may fail due to crushing failure or shear failure. Torque capacity for strength against crushing failure,

$$T = \frac{1}{4} \; Lhd\, S_{cd} \tag{7.1}$$

Torque for strength against failure due to shearing,

$$T = \frac{1}{2} \; L \cdot b \cdot d \cdot S_{sd} \tag{7.2}$$

7.3 KEY MATERIAL

Medium carbon steel with ultimate tensile stress $S_u \geq 600$ MPa is commonly used for keys. A factor of safety 1.5 to 2 may be used. It is good practice to make the key a weaker member than the shaft and the hub. The allowable pressure between the key and hub may be taken in the range of 80 MPa to 150 MPa. Lower values are used for cast iron hubs and higher value for steel. If the hub is to slide along the shaft, the pressure is limited to 10 to 20 MPa.

Table 7.1. Dimension of keys (All dimensions in mm)

Shaft dia., *d* in mm range	Parallel keys			Taper Keys		
	Width (*b*)	Height (*h*)	Length (L) min-max	Width (*b*)	Height (*h*)	Length (L) min-max
6-8	2	2	6-20	2	2	6-20
8-10	3	3	6-36	3	3	8-36
10-12	4	4	8-45	4	4	10-45
12-17	5	5	10-56	5	5	12-56
17-22	6	6	14-70	6	6	16-70
22-30	8	7	18-90	8	7	20-90
30-38	10	8	22-110	10	8	25-110
38-44	12	8	28-140	12	8	32-140
44-50	14	9	36-180	14	9	40-160
50-58	16	10	45-180	16	10	45-180
58-65	18	11	50-200	18	11	50-200
65-75	20	12	56-220	20	12	56-220
75-85	22	14	63-250	22	14	63-250
85-95	25	14	70-280	25	14	70-280
95-110	28	16	80-320	28	16	80-315
110-130	32	18	90-360	32	18	90-355
130-150	36	20	100-400	36	20	100-400
150-170	40	22	110-400	40	22	110-400
170-200	45	25	125-400	45	25	125-400
200-230	50	28	140-400	50	28	140-400
230-260	56	32	160-400	56	32	–
260-290	63	32	180-400	63	32	180-400
260-330	70	36	200-400	70	36	200-400
330-380	80	40	220-400	80	40	220-400
380-440	90	45	250-400	90	45	250-400
440-500	100	50	280-400	100	50	–

7.4 MULTIPLE SPLINES AND SERRATION JOINTS

The splines and serrations are the projectors on the shaft which fit into the reccess of the corresponding shape in the hub.

These joints can be treated as multiple key joints in which the keys are integral with the shaft.

The splined joints have the following advantages over keyed joints:

(a) They have greater load carrying capacity due to larger working surface and more uniform pressure distribution.

(b) The mounted parts are more accurately centred and guided.

(c) They have higher fatigue strength for the shaft due to reduced stress concentration factor.

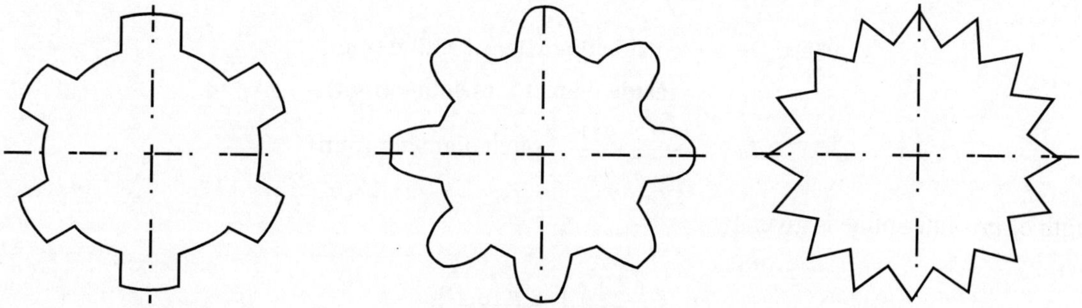

(a) Parallel splines (b) Involute splines (c) Serrations

Figure 7.2. Splines and serrations

There are two types of splines.

(a) *Parallel sided splines*. The profile is formed by major and minor circles and straight parallel sides. The locating feature may be (i) the sides or (ii) major or minor diameter.

For high torque capacity, the locating on sides is adopted. When accurate centring, major or minor diameter based locating feature is used.

The splines are designated as $Z \times d \times D$, where Z is the number of splines, d is the minor diameter and D is the major diameter.

Involute Splines

The profile consists of major and minor circles and involute sides. They are like gear tooth profile. The pressure angle is in the range of 20° to 30° and depth of the spline is 0.9 to 1.0 times module. The involute splines have higher strength and can be easily manufactured. They are increasingly becoming and more popular.

7.5 DESIGN PROCEDURE

(1) Choose the standard size from the Table 7.2 for parallel splines and Tables 7.3 and 7.4 for involute splines.

The length of the splines for parallel spline is given by the relation,

$$L = \frac{4T}{d_p(D_1 - d_1)ZP_b} \tag{7.3}$$

Check for shear strength,

$$\tau = \frac{2T}{d.w.L.Z} \leq S_{sd} \tag{7.4}$$

where D_1 = outside diameter of spline (mm)

d_1 = inside diameter of spline (mm)

$d_p = \frac{(D_1 + d_1)}{2}$ = pitch diameter (mm)

Length of involute spline is given by

$$L = \left(\frac{10}{3}\right)\left(\frac{2T}{d_p m Z P_b}\right) \tag{7.5}$$

where m is the module.

Check for shear strength,

$$\tau = \frac{T}{0.7854Ld^2} \leq S_{sd} \tag{7.6}$$

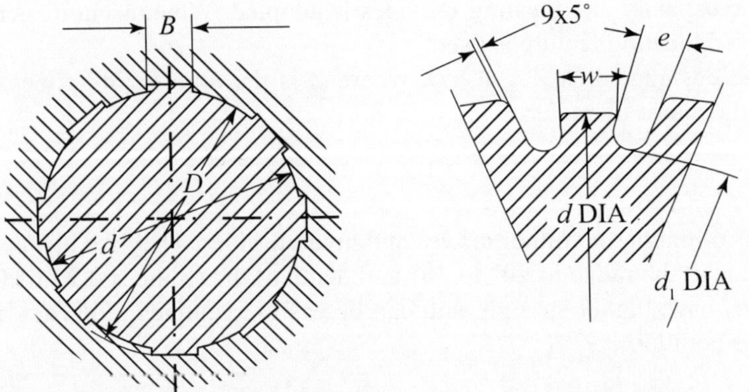

(a) Splined shaft and hub profile (b) Splined shaft

Figure 7.3

Table 7.2. Dimensions for straight sided splines for general engineering use (IS 2317-1963) (All dimension in mm) (Fig 7.3)

Normal size, $Z \times d \times D$	**Number of splines,** Z	**Minor diameter,** d	**Major diameter, D**	**Width,** W	$\mathbf{d_1^*}$ **min.**
Light duty series					
$6 \times 23 \times 26$	6	23	26	6	22.1
$6 \times 26 \times 30$	6	26	30	6	24.6
$6 \times 28 \times 32$	6	28	32	7	26.7
$8 \times 32 \times 36$	8	32	36	6	30.4
$8 \times 36 \times 40$	8	36	40	7	34.5
$8 \times 42 \times 46$	8	42	46	8	40.4
$8 \times 46 \times 50$	8	46	50	9	44.6
$8 \times 52 \times 58$	8	52	58	10	49.7
$8 \times 56 \times 62$	8	56	62	10	53.6
$8 \times 62 \times 68$	8	62	68	12	59.8
$10 \times 72 \times 78$	10	72	78	12	69.6
$10 \times 82 \times 88$	10	82	88	12	79.3
$10 \times 92 \times 98$	10	92	98	14	89.4
$10 \times 102 \times 108$	10	102	108	16	99.9
$10 \times 112 \times 120$	10	112	120	18	108.8
Medium duty series		[t]			
$6 \times 11 \times 14$	6	11	14	3	9.9
$6 \times 13 \times 16$	6	13	16	3.5	12.0
$6 \times 16 \times 20$	6	16	20	4	14.5
$6 \times 18 \times 22$	6	18	22	5	16.7
$6 \times 21 \times 25$	6	21	25	5	19.5
$6 \times 23 \times 28$	6	23	28	6	21.3
$6 \times 26 \times 32$	6	26	32	6	23.4
$6 \times 28 \times 34$	6	28	34	7	25.9

Table 7.2. (Contd.)

Normal size, $Z \times d \times D$	**Number of splines,** Z	**Minor diameter,** d	**Major diameter,** **D**	**Width,** W	$\mathbf{d_1^*}$ **min.**
$8 \times 32 \times 38$	8	32	38	6	29.4
$8 \times 36 \times 42$	8	36	42	7	33.5
$8 \times 42 \times 48$	8	42	48	8	39.5
$8 \times 46 \times 54$	8	46	54	9	42.7
$8 \times 52 \times 60$	8	52	60	10	48.7
$8 \times 56 \times 65$	8	56	65	10	52.2
$8 \times 62 \times 72$	8	62	72	12	57.8
$10 \times 72 \times 82$	10	72	82	12	67.4
$10 \times 82 \times 92$	10	82	92	12	77.1
$10 \times 92 \times 102$	10	92	102	14	87.3
$10 \times 102 \times 112$	10	102	112	16	97.7
$10 \times 112 \times 125$	10	112	125	18	106.3

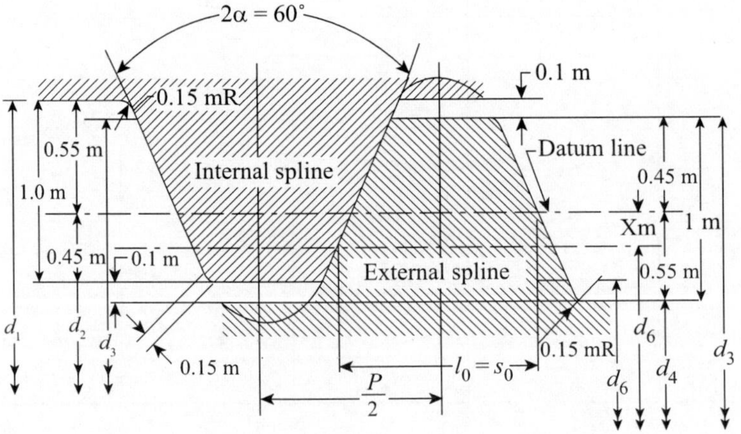

Figure 7.4. Reference profile

Table 7.3. Dimension for involute sided splines of module 2. All dimensions in mm (Fig 7.4)

Normal size							
$d_1 \times d_2$	z	d_0	d_b	d_3	d_4	d_5 **Min**	d_6 **Max**
15 × 11	6	12	10.392	14.6	10.6	14.68	10.92
17 × 13	7	14	12.124	16.6	12.6	16.68	12.92
18 × 14	7	14	12.124	17.6	13.6	17.68	13.92
20 × 16	8	16	13.856	19.6	15.6	19.68	15.62
22 × 18	9	18	15.588	21.6	17.6	21.68	17.92
25 × 21	11	22	19.053	24.6	20.6	24.68	20.92
28 × 24	12	24	20.785	27.6	23.6	27.68	23.92
30 × 26	14	26	24.249	29.6	25.6	29.69	25.91
32 × 28	14	28	24.249	31.6	27.6	31.69	27.91
35 × 31	16	32	27.713	34.6	30.6	34.91	30.91
37 × 33	17	34	29.445	36.6	32.6	39.69	32.91
38 × 34	18	36	31.177	37.6	33.6	37.69	33.91
40 × 36	18	36	31.177	39.6	35.6	39.69	35.91
42 × 38	20	40	34.641	41.6	37.6	41.69	37.91
45 × 41	21	42	36.373	44.6	40.6	44.69	40.91
47 × 43	22	44	38.105	46.6	42.6	46.69	42.91
48 × 44	22	44	38.105	47.6	43.6	47.69	43.91
50 × 46	24	48	41.569	49.6	45.6	49.69	45.91
(52 × 48)	24	48	41.569	51.6	47.6	51.69	47.91
55 × 51	26	52	45.033	54.6	50.6	54.70	50.90
(58 × 54)	28	56	48.497	57.6	53.6	57.70	53.90
60 × 56	28	56	48.497	59.6	55.6	59.70	55.90
(62 × 58)	30	60	51.962	61.6	57.6	61.70	57.90
65 × 61	31	62	53.694	64.6	60.6	64.70	60.90
(68 × 64)	32	64	55.426	67.6	63.6	67.70	63.90
70 × 66	34	68	58.890	69.6	65.6	69.70	65.90
(72 × 68)	34	68	58.890	71.6	67.6	71.70	67.90
75 × 71	36	72	62.354	74.6	70.6	74.70	70.90
(78 × 74)	38	76	65.818	77.6	73.6	77.70	73.90
80 × 76	38	76	65.818	79.6	75.6	79.70	75.90
(82 × 78)	40	80	69.282	81.6	77.6	81.70	77.90

Note: Values in bracket are not preferred.

Table 7.4. Dimension for involute sided splines of module 2.5. All dimensions in mm notations as in Fig. 7.4

Normal size							
$d_1 \times d_2$	z	d_0	d_b	d_3	d_4	d_5 **Min**	d_6 **Max**
20 × 15	6	15.0	12.990	19.5	14.5	19.58	14.92
22 × 17	7	17.5	15.155	21.5	16.5	21.58	16.92
25 × 20	8	20.0	17.321	24.5	19.5	24.58	19.92
28 × 23	10	25.0	21.651	27.5	22.5	27.58	22.92
30 × 25	10	25.0	21.651	29.5	24.5	529.58	24.92
32 × 27	11	27.5	23.816	31.5	26.5	31.59	26.91
35 × 30	12	30.0	25.981	34.5	29.5	34.59	29.91
37 × 32	13	32.5	28.146	36.5	31.5	36.59	31.91
38 × 33	14	35.0	30.311	37.5	32.5	37.59	32.91
40 × 35	14	35.0	30.311	39.5	34.5	39.59	34.91
42 × 37	15	37.5	32.476	41.5	36.5	41.59	36.91
45 × 40	16	40.0	36.641	44.5	39.5	44.59	39.91
47 × 42	17	42.5	36.806	46.5	41.5	46.59	41.91
48 × 43	18	45.0	38.971	47.5	42.5	47.59	42.91
50 × 45	18	45.0	38.971	49.5	44.5	49.59	44.91
(52 × 47)	19	47.5	41.136	51.5	46.5	51.59	46.91
55 × 50	20	50.0	43.301	54.5	49.5	54.59	49.91
(58 × 53)	22	55.0	47.631	57.5	52.5	57.60	52.90
60 × 55	22	55.0	47.631	59.5	54.5	59.60	54.90
(62 × 57)	23	57.5	49.796	61.5	56.5	61.60	56.90
65 × 60	24	60.0	51.962	64.5	59.5	64.60	59.90
(68 × 63)	26	65.0	56.292	67.5	62.5	67.60	62.90
70 × 65	26	65.0	56.292	69.5	64.5	69.60	64.90
(72 × 67)	27	67.5	58.457	71.5	66.5	71.60	66.90
75 × 70	28	70.0	60.622	74.5	69.5	74.60	69.90
(78 × 73)	30	75.0	64.952	77.50	72.50	77.90	72.90
80 × 75	30	75.0	64.952	79.50	74.5	79.60	74.90
(82 × 77)	31	77.5	67.117	81.5	76.5	81.60	76.90
85 × 80	32	80.0	69.282	84.5	79.5	84.60	79.90
(88 × 83)	34	85.0	73.612	87.5	82.5	87.60	82.90
90 × 85	34	85.0	73.612	89.5	84.5	89.60	84.90

(Contd.)

Table 7.4. (Contd.)

Normal size							
$d_1 \times d_2$	z	d_0	d_b	d_3	d_4	d_5 **Min**	d_6 **Max**
(92 × 87)	35	87.50	75.777	91.5	86.5	91.60	86.90
95 × 90	36	90.0	77.942	94.5	89.5	94.60	89.90
(98 × 93)	38	95.0	82.272	97.5	92.5	97.60	92.90
100 × 95	38	95.0	82.272	99.5	94.5	99.60	94.90
105 × 100	40	100.0	86.603	104.5	99.5	104.60	99.90

Note: Values in bracket are not preferred.

Table 7.5. Dimensions (mm) of woodruff keys (IS229)

Shaft diameter d		**Key size**			**Depth in shaft/hub**	
Min	**Max**	l/b	b_1	D	t_1	t_2
3	4	1.0	1.4	4.0	1.0	0.6
4	5	1.5	2.6	7.0	2.0	0.8
5	6	2.0	2.6	7.0	1.8	1.0
6	7	2.0	3.7	10.0	2.9	1.0
7	8	2.5	3.7	10.0	2.7	1.2
8	10	3.0	5.0	13.0	3.8	1.4
10	12	3.0	6.5	16.0	5.3	1.4
12	14	4.0	6.5	16.0	5.0	1.8
14	16	4.0	7.5	19.0	6.0	1.8
16	18	5.0	6.5	16.0	4.5	2.3
18	20	5.0	7.5	19.0	5.5	2.3
20	22	5.0	9.0	22.0	7.0	2.3
22	25	6.0	9.0	22.0	6.5	2.8
25	28	6.0	10.0	25.0	7.5	2.8
28	32	8.0	11.0	28.0	8.0	3.3
32	38	10.0	13.0	32.0	10.0	3.3

7.6 COTTER JOINT

Cotter is a tapered metal plate or strip having rectangular cross section which is used to connect two rod ends. The two ends are specially formed to provide slots to insert the cotter. The spigot and socket type is shown in the Fig. 7.5. It is used to transmit axial tensile and compressive loads. The cotter joint dimensions are normally fixed by proportions. The design is checked for the strength against various failures.

Design of Cotter Joint

1. For tensile failure of the rod ends, the rod diameter is

$$d = \sqrt{\frac{4P}{\pi S_d}} \tag{7.7}$$

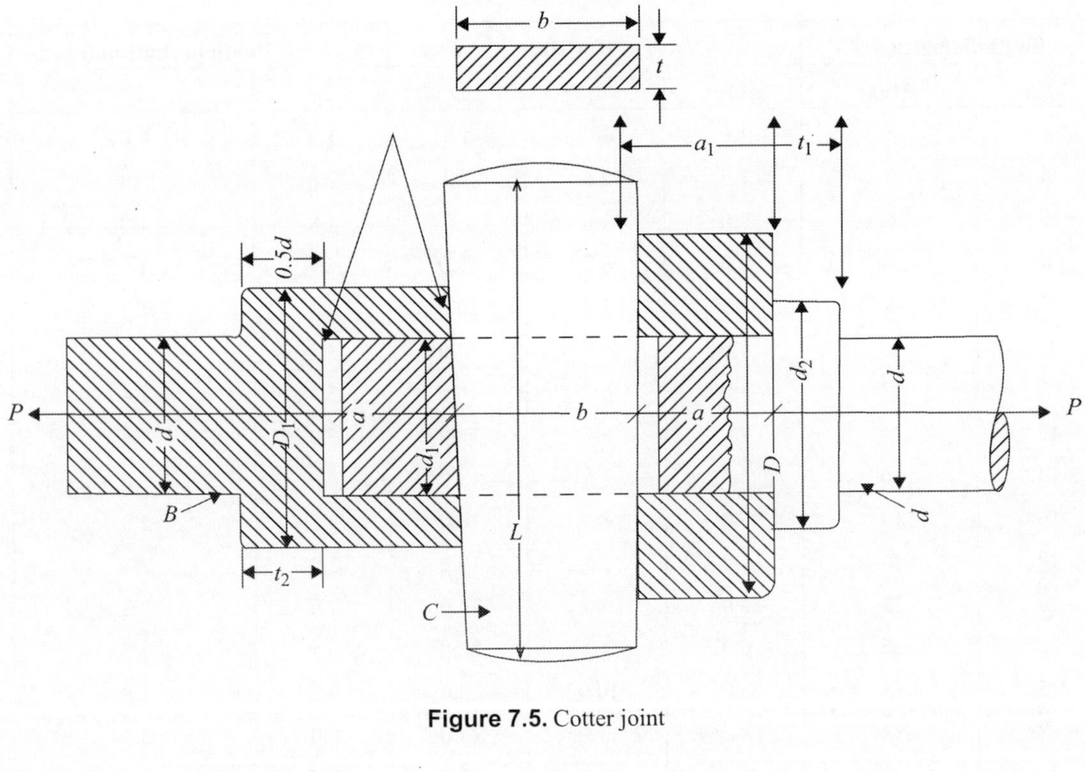

Figure 7.5. Cotter joint

Spigot

2. Tensile failure of the spigot across the cotter hole,

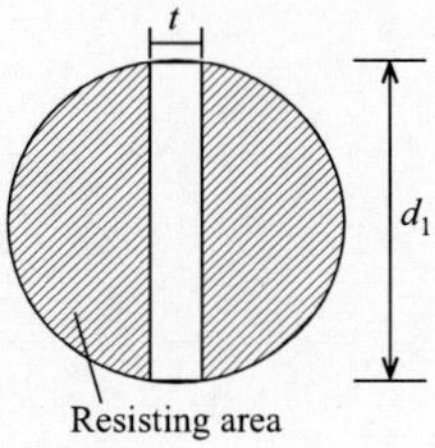

$$P = \left(\frac{\pi}{4}d_1^2 - d_1 t\right) \cdot S_d \tag{7.8}$$

3. Crushing failure of the spigot under the cotter,

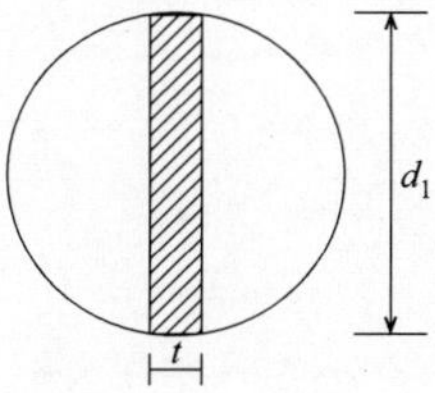

$$P = d_1 t \, S_{cd} \tag{7.9}$$

4. Shearing of the spigot across the cotter hole,

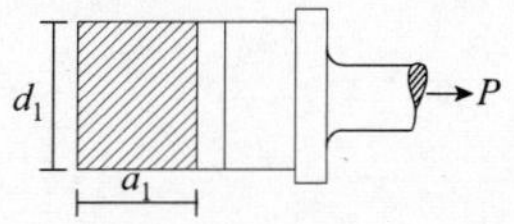

$$P = 2 d_1 a \, S_{sd} \tag{7.10}$$

5. Crushing failure of the spigot collar,

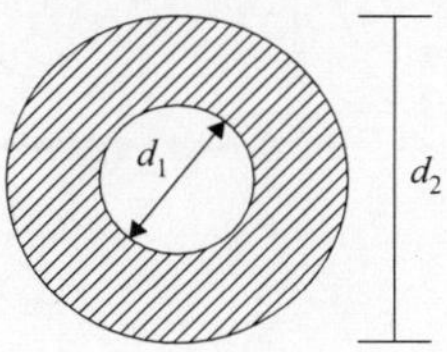

$$P = \frac{\pi}{4}(d_2^2 - d_1^2) \cdot S_{cd} \tag{7.11}$$

6. Shearing failure of the spigot collar,

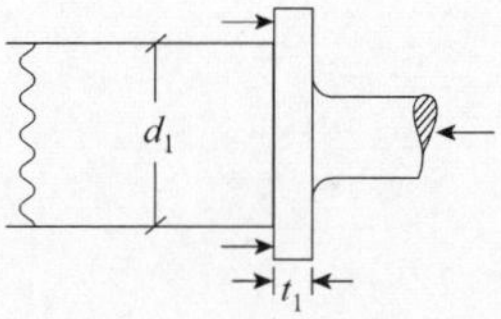

$$P = (\pi d_1 t_1) \, S_{sd} \tag{7.12}$$

Cotter

1. Double shear failure of the cotter,

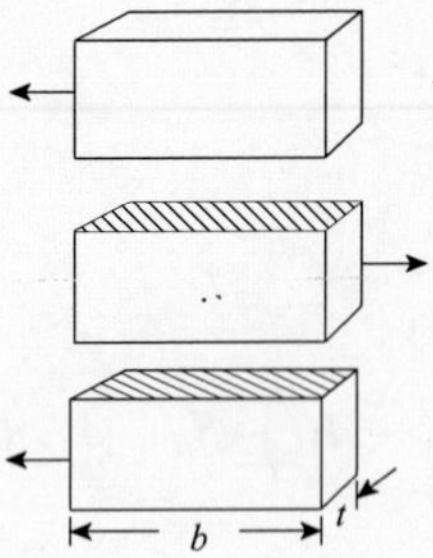

$$P = 2btS_{sd} \tag{7.13}$$

2. Crushing failure of the cotter,

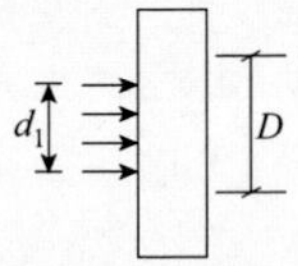

$$P = d_1 t\, S_{cd} \tag{7.14}$$

3. Bending failure of the cotter,

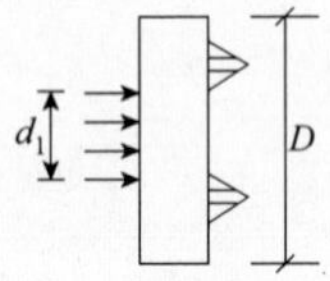

$$\text{Bending stress} \quad \sigma_b = \frac{5\,PD}{8b^2t} < S_d \tag{7.15}$$

Socket

1. Tensile failure of socket across,

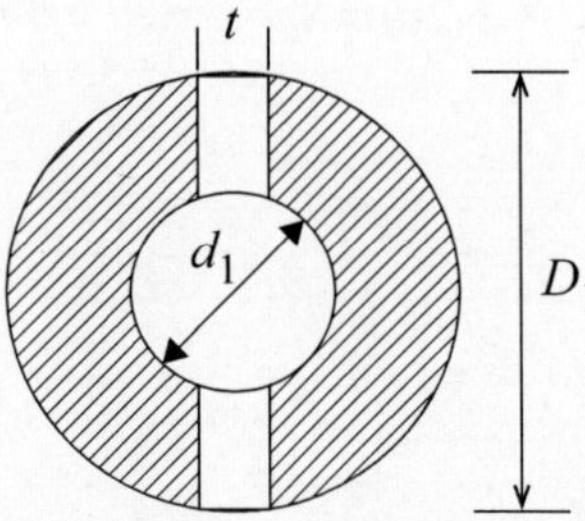

$$P = \left[\frac{\pi}{4}\left(D_1^2 - d_1^2\right) - (D_1 - d_1)t\right] S_d \tag{7.16}$$

2. Crushing failure of socket under the cotter (near collar),

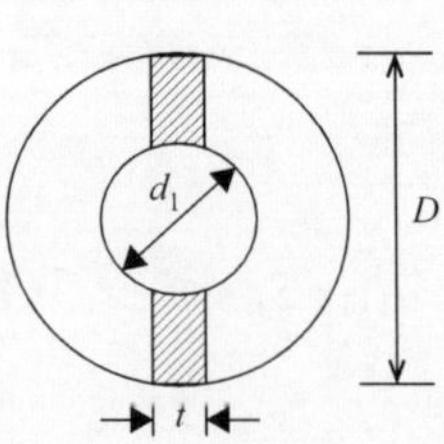

$$P = (D - d_1)t\, S_{cd} \tag{7.17}$$

3. Shearing of the socket across the cotter

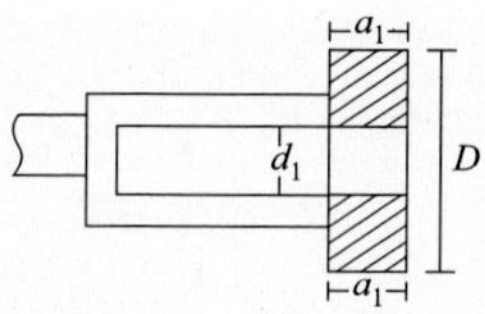

$$P = 4 \times \text{shear resistance across} \times S_{sd}$$
$$= 4 \times \left(\frac{D - d_1}{2}\right) \cdot a_1 S_{sd}$$
$$= 2(D - d_1) a_1 S_{sd} \tag{7.18}$$

4. Shearing of the socket near rod end

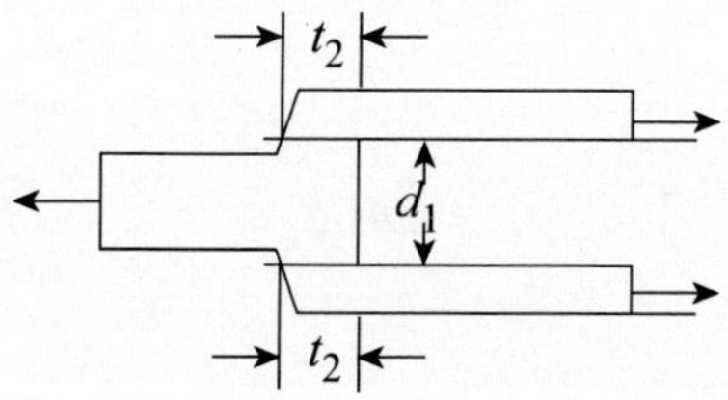

$$P = \pi d_1 t_2 S_{sd} \tag{7.19}$$

7.7 KNUCKLE JOINT

It is used to connect two rods subjected to axial tensile or compressive load. It also allows the rotation of the rods about the axis of the pin. The knuckle joint is as shown in Fig. 7.6.

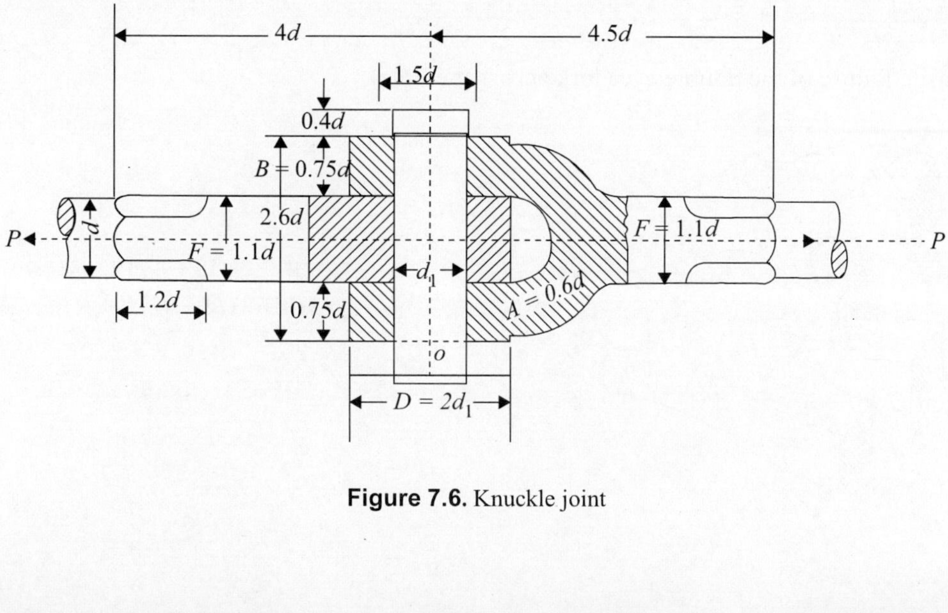

Figure 7.6. Knuckle joint

Design of knuckle joint

(1) Tensile failure of the rod,

$$P = \frac{\pi}{4} d^2 S_d$$

$$\therefore \quad d = \sqrt{\frac{4P}{\pi S_d}} \tag{7.20}$$

(2) Tensile failure of square section,

$$P = F^2 S_d \tag{7.21}$$

(3) Tensile failure of the two fork section,

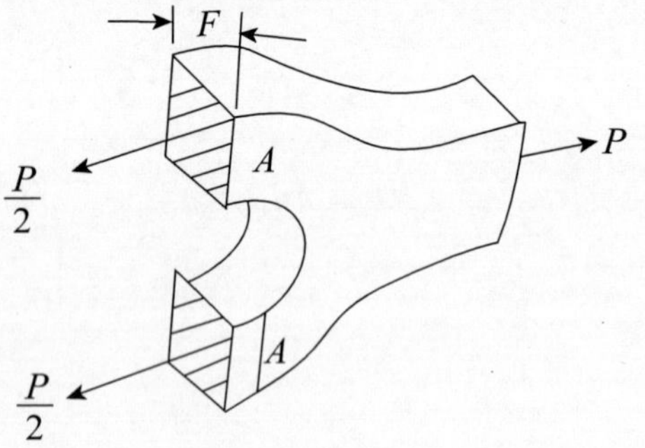

$$P = 2AF S_d \tag{7.22}$$

(4) Shear failure of the knuckle pin,

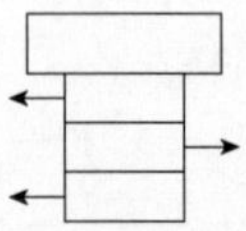

$$P = 2 \times \frac{\pi}{4} d_1^2 S_{sd}$$

$$P = \frac{\pi}{2} d_1^2 S_{sd} \tag{7.23}$$

(5) Tensile failure of the double eye (fork across the hole),

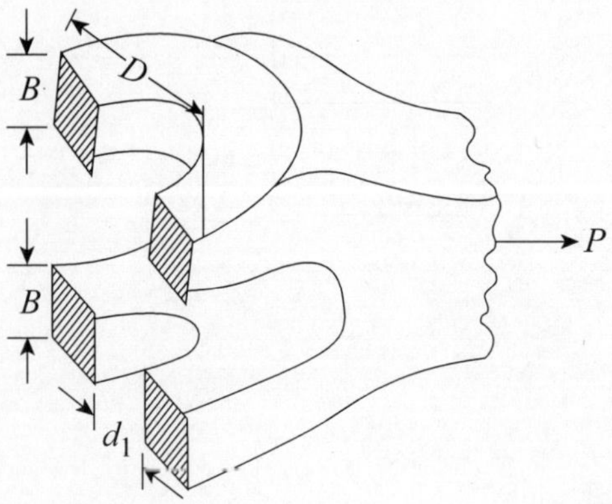

$$P = 2\,(D - d_1)\,B\,S_d \tag{7.24}$$

(6) Shearing of the fork across the hole,

$$P = 4 \times \text{resisting area} \times S_{sd}$$
$$= 4 \times \left(\frac{D - d_1}{2}\right) \cdot B \cdot S_{sd}$$
$$= 2(D - d_1)BS_{sd} \tag{7.25}$$

(7) Crushing of the pin in the rod,

$$P = Fd_1S_{cd} \quad \text{(area is based on projected area)} \tag{7.26}$$

(8) Tensile failure of the single eye (rod across the hole),

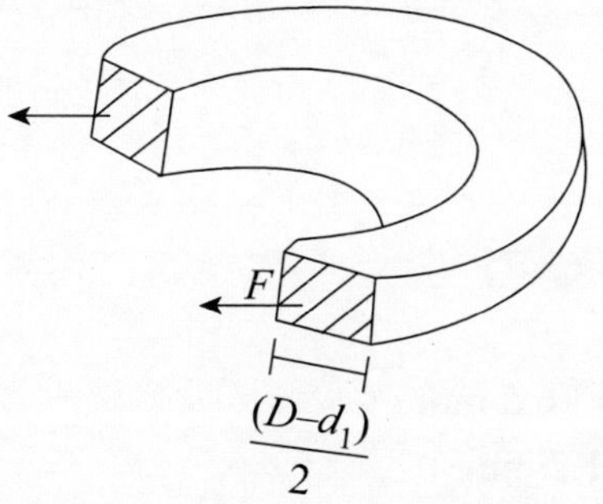

$$P = (D - d_1)F\,S_d \tag{7.27}$$

(9) Shearing of the single eye (rod across the hole),

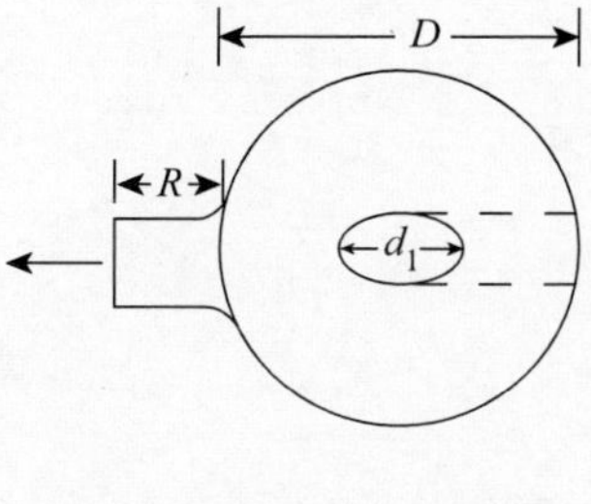

$$P = (D - d_1)F\,S_{sd} \tag{7.28}$$

Note: Joints can be designed by two methods–

1. Obtain diameter of the rod d, use proportion and check for the load or the stresses induced.
2. Obtain the diameter of the rod, d. Also calculate all the other dimensions using the load and the design stress.

SOLVED EXAMPLES

Example 7.1: Two rod ends of a pump are joined by means of a spigot and socket type of cotter joint. Design the joint for an axial load of 100 kN, which alternatively changes from tensile to compressive. The allowable stress for the materials used are 50 MPa in tension, 40 MPa in shear and 100 MPa in crushing.

Given Data

Axial load, $P = 100$ kN
Design stresses in tension, $S_d = 50$ MPa
In compression, $S_{cd} = 100$ MPa
In shear, $S_{sd} = 40$ MPa

SOLUTION

Refer Fig. 7.5
Step-1: The rod diameter for tearing (Eqn. 5.18 MDHB)

$$d = \sqrt{\frac{4P}{\pi S_d}} = \sqrt{\frac{4 \times 100 \times 10^3}{\pi \times 50}} = 50.4 \text{ mm}$$

$$\text{Choose} \quad d = 51 \text{ mm}$$

Step-2: Then the dimensions by proportion
For spigot,

$$d_1 = 1.25d = 1.25 \times 51 = 63.75$$
$$\text{Choose} \quad d_1 = 65 \text{ mm}$$
$$d_2 = 1.5 \times d = 1.5 \times 51 = 76.5 \text{ mm}$$
$$a = 1.2d = 1.2 \times 51 = 61.2 \text{ mm}$$
$$\text{Choose} \quad a = 60 \text{ mm}$$
$$t_1 = 0.5 \times d_1 = 0.5 \times 51 \cong 25.0 \text{ mm}$$

For cotter size,

$$\text{Thickness,} \quad t = 0.4d = 0.4 \times 51 = 20.4$$
$$\text{Choose} \quad t = 20 \text{ mm}$$
$$\text{Width,} \quad b = 4t = 4 \times 20 = 80 \text{ mm}$$

For socket,

$$\text{Socket end diameter,} \quad D = 2d_1 = 2 \times 65 = 130 \text{ mm}$$
$$\text{Socket sleeve diameter,} \quad D_1 = 2d = 2 \times 51 = 102 \text{ mm}$$
$$\text{Choose} \quad D_1 = 100 \text{ mm}$$

Step-3: Check for strengths of spigot
(i) Spigot tearing strength across the cotter hole, (Eqn. 5.19 MDHB)

$$P_1 = \left(\frac{\pi}{4}d_1^2 - d_1 t\right) \cdot S_d$$
$$= \left[\frac{\pi}{4} \times 65^2 - 65 \times 20\right] \times 50$$
$$= 100915 \cdot 0 \text{ N} > 100 \text{ kN} \quad \therefore \quad \text{Safe}$$

Crushing of cotter in the rod, (Eqn. 5.20 MDHB)

$$P_1 = d_1 t S_{cd} = 65 \times 20 \times 100$$
$$= 130{,}000 \text{ N} > 100 \text{ kN} \quad \therefore \quad \text{Safe}$$

Shearing of the rod across the cotter hole, (Eqn. 5.21 MDHB)

$$P_1 = 2d_i a \cdot S_{sd}$$
$$= 2 \times 65 \times 60 \times 40$$
$$= 312{,}000 \text{ N} > 100 \text{ kN} \quad \therefore \quad \text{Safe}$$

Step-4: Check for cotter strength
Shearing strength of the cotter (Eqn. 5.24 MDHB)

$$P_1 = 2btS_{sd}$$
$$= 2 \times 80 \times 20 \times 40$$
$$= 128{,}000 \text{ N} > 100 \text{ kN} \quad \therefore \quad \text{Safe}$$

Bending strength (Eqn. 5.25 MDHB)

$$\sigma_b = \left(\frac{5}{8} P \times D\right) / b^2 \cdot t$$
$$= \frac{5}{8} \times \frac{100 \times 10^3 \times 130}{80^2 \times 20}$$
$$= 63.5 \text{ MPa} > S_d = 50 \text{ MPa} \quad \text{Unsafe}$$

Choose $b = 100$ mm

$$\sigma_b = \frac{5}{8} \times \frac{100 \times 10^3 \times 130}{100^2 \times 20}$$
$$= 40.625 \text{ MPa} < S_d = 50 \text{ MPa} \quad \therefore \quad \text{Safe}$$

Step-5: Check for socket

Tearing across-cotter hole (Eqn. 5.26 MDHB)

$$\begin{aligned} P_1 &= \left[\frac{\pi}{4}(D_1^2 - d_1^2) - (D_1 - d_1)t\right] S_d \\ &= \left[\frac{\pi}{4}(100^2 - 65^2) - (100 - 65) \times 20\right] \times 50 \\ &= 191,783.72 \text{ N} > 100 \text{ kN} \quad \therefore \quad \text{Safe} \end{aligned}$$

Crushing of cotter in socket (Eqn. 5.27 MDHB)

$$\begin{aligned} P_1 &= (D - d_1)tS_{cd} \\ &= (130 - 65) \times 20 \times 100 \\ &= 130,000 = 00 \text{ N} > 100 \text{ kN} \quad \therefore \quad \text{Safe} \end{aligned}$$

Shearing of the socket across the cotter hole (Eqn. 5.28 MDHB)

$$\begin{aligned} P_1 &= 2(D - d_1) \cdot a_1 \cdot S_{sd} \\ &= 2(130 - 65) \times 60 \times 40 \\ &= 312,000 \text{ N} > 100 \text{ kN} \quad \therefore \quad \text{Safe} \end{aligned}$$

Shearing of the socket across the rod end (Eqn. 5.28a MDHB)

$$\begin{aligned} P_1 &= \pi d_1 t_2 S_{sd} \\ &= \pi \times 65 \times 26 \times 40 \quad (t_2 = 0.5\, d = 26 \text{ mm}) \\ &= 212371.66 \text{ N} > 100 \text{ kN} \quad \therefore \quad \text{Safe} \end{aligned}$$

Example 7.2: Design a knuckle joint to connect two mild steel rods under tensile load of 25 kN. The permissible stresses are 80 MPa in tension; 50 MPa in shear and 100 MPa in crushing.
Given Data

Load transmitted, $P = 25$ kN

Permissible stresses

In tension, $S_d = 80$ MPa

In shear, $S_{sd} = 50$ MPa

In crushing, $S_{cd} = 100$ MPa

SOLUTION

Refer Fig. 7.6

Step-1: Diameter of the rod (Eqn. 5.29 MDHB)

$$\begin{aligned} d &= \sqrt{\frac{4P}{\pi S_d}} \\ &= \sqrt{\frac{4 \times 25000}{\pi \times 80}} \\ &= 19.94 \text{ mm} \\ &\cong 20.00 \text{ mm} \end{aligned}$$

Step-2: Diameter of the pin for shearing (Eqn. 5.32 MDHB)

$$\begin{aligned} d_1 &= \sqrt{\frac{2P}{\pi S_{sd}}} \\ &= \sqrt{\frac{2 \times 25000}{\pi \times 50}} \\ &= 17.84 \text{ mm} \\ \text{Choose} \quad d_1 &= 20 \text{ mm} \end{aligned}$$

Step-3: Dimensions as per standard proportions are

(i) For eye,

Inner diameter, $d_1 = 20$ mm

Outer diameter, $D = 2d_1 = 40$ mm

Thickness of the square-portion, $F = 1.1\ d = 1.1 \times 20 = 22.0$ mm

(ii) For fork,

Inner diameter, $d_1 = 20.0$ mm

Outer diameter, $D = 2d_1 = 2 \times 20 = 40$ mm

(iii) Thickness of each fork,

$\text{B} = 0.75d = 0.75 \times 20 = 15$ mm

Fork width at the junction,

$$F = 1.1d = 1.1 \times 20 = 22.0 \text{ mm}$$

Fork thickness, $A = 0.6\,d = 0.6 \times 20 = 12$ mm check for strengths

Step-4: Check for strengths

Tearing of the fork at the junction (Eqn. 5.30 MDHB)

$$\begin{aligned} F_1 &= 2AFS_d \\ &= 2 \times 12 \times 22 \times 80 \\ &= 42{,}240 \text{ N} > P = 25000 \text{ N} \quad \therefore \quad \text{Safe} \end{aligned}$$

Tearing of the square section (Eqn. 5.31 MDHB)

$$\begin{aligned} F_1 &= F^2 \cdot S_d = 22^2 \times 80 \\ &= 38720 \text{ N} > P = 25 \text{ kN} \quad \therefore \quad \text{Safe} \end{aligned}$$

Tearing of the eye across the hole (Eqn. 5.33 MDHB)

$$\begin{aligned} F_1 &= F(D - d_1)S_d \\ &= 22 \times (40 - 20) \times 80 \\ &= 35200 \text{ N} > P = 25 \text{ kN} \quad \therefore \quad \text{Safe} \end{aligned}$$

Tearing of the forks across the hole (Eqn. 5.34 MDHB)

$$\begin{aligned} F_1 &= 2(D - d_1)B\,S_d \\ &= 2(40 - 20) \times 15 \times 80 \\ &= 480{,}000 \text{ N} > P = 25 \text{ kN} \quad \therefore \quad \text{Safe} \end{aligned}$$

Crushing of the pin in the eye (Eqn. 5.35 MDHB)

$$\begin{aligned} F_1 &= F \cdot d_1 S_{cd} \\ &= 22 \times 20 \times 100 \\ &= 44000 \text{ N} > P = 25 \text{ kN} \quad \therefore \quad \text{Safe} \end{aligned}$$

Crushing of the pin in forks (Eqn. 5.36 MDHB)

$$\begin{aligned} F_1 &= 2B\,d_1 S_{cd} \\ &= 2 \times 15 \times 20 \times 100 \\ &= 60{,}000 \text{ N} > P = 25 \text{ kN} \quad \therefore \quad \text{Safe} \end{aligned}$$

Shearing of the rod across the hole (Eqn. 5.37 MDHB)

$$
\begin{aligned}
F_1 &= F(D - d_1)S_{sd} \\
&= 22 \times (40 - 20) \times 50 \\
&= 22000 \text{ N} < P\ 25 \text{ kN}
\end{aligned}
$$

Increase thickness $F = 25$ mm

$$
\begin{aligned}
F_1 &= 25 \times (40 - 20) \times 50 \\
&= 25000 \text{ N} = P = 25 \text{ kN} \quad \therefore \quad \text{Safe}
\end{aligned}
$$

Shearing of the forks across the hole (Eqn. 5.38 MDHB)

$$
\begin{aligned}
F_1 &= 2B(D - d_1)S_{sd} \\
&= 2 \times 15 \times (40 - 20) \times 50 \\
&= 30,000 \text{ N} > P = 25 \text{ kN} \quad \therefore \quad \text{Safe}
\end{aligned}
$$

Hence, the design is acceptable.

Example 7.3: Design a spigot and socket joint to connect two rods of 30 C8 steel to carry axial tensile and compressive load of 10 kN. Allowable stresses are $\sigma_t = 90$ MPa, $\sigma_c = 100$ MPa and $\tau = 50$ MPa.

Refer Fig. 7.5

Given Data

$$P = 10 \text{ kN and } S_{cd} = \sigma_c = 100 \text{ MPa}; \ S_{td} = 90 \text{ MPa and } S_{sd} = 50$$

SOLUTION

Design of spigot and socket joint

Step-1: Design of rod for tearing (Eqn. 5.18 MDHB)

$$d = \sqrt{\frac{4P}{\pi S_d}} = \sqrt{\frac{4 \times 10 \times 10^3}{\pi \times 90}} = 14.14 \text{ mm}$$

$$\text{Choose} \quad d = 15 \text{ mm}$$

Step-2: Dimensions as per standard proportions

$$
\begin{aligned}
D &= 2d = 2 \times 15 = 30 \text{ mm} \\
D_1 &= 1.75d = 1.75 \times 15 = 26.25 \text{ mm} \\
d_1 &= 1.25d = 15 \times 1.25 = 18.75 \text{ mm} \\
d_2 &= 1.5d = 1.5 \times 15 = 22.5 \text{ mm} \\
t_1 &= 0.5 \times 15 = 7.5 \text{ mm} \\
t_2 &= 0.5 \times 15 = 7.5 \text{ mm} \\
a_1 &= 1.25d = 1.25 \times 15 = 18.75 \text{ mm}
\end{aligned}
$$

Cotter thickness,

$$t = 0.4d = 0.4 \times 15 = 6.0 \text{ mm}$$

Width of cotter,

$$b = 4t = 4 \times 6 = 24 \text{ mm}$$

Step-3: Check for strength of cotter
Shearing strength of cotter (Eqn. 5.24 MDHB)

$$\begin{aligned} F &= 2bt\,S_{sd} = 2 \times 24 \times 6 \times 50 \\ &= 14400 > (P = 10{,}000 \text{ N}) \quad \therefore \quad \text{Safe} \end{aligned}$$

Bending strength (Eqn. 5.25 MDHB)

$$\begin{aligned} \sigma_D &= \left(\frac{5}{8} P \times D\right) / b^2 t \\ &= \frac{5}{8} \times \frac{10{,}000 \times 30}{(24)^2 \times 6} = 54.25 \text{ MPa} < S_d = 90 \text{ M} \end{aligned}$$

The cotter is safe against bending.
Crushing strength of the cotter (Eqn. 5.22 MDHB)

$$\begin{aligned} P_1 &= \frac{\pi}{4}(d_2^2 - d_1^2) S_{cd} \\ &= \frac{\pi}{4}(22.5^2 - 18.75^2) \times 100 \\ &= 12{,}149.13 \text{ N} > (P = 10 \text{ kN}) \quad \therefore \quad \text{Safe} \end{aligned}$$

Step-4: Check for strength of spigot
Shearing strength of the cotter (Eqn. 5.22a MDHB)

$$\begin{aligned} P &= \pi d_1 t_1 S_{sd} = \pi \times 18.75 \times 7.5 \times 50 \\ &= 22089.32 \text{ N} > (P = 10 \text{ kN}) \quad \therefore \quad \text{Safe} \end{aligned}$$

Tearing resistance across the cotter hole (Eqn. 5.19 MDHB)

$$\begin{aligned} P_1 &= \left\{\frac{\pi}{4} d_1^2 - d_1 t\right\} S_d \\ &= \left\{\frac{\pi}{4}(18.75)^2 - 18.75 \times 6\right\} \times 90 \\ &= 14725 \text{ N} > (P = 10000 \text{ N}) \quad \therefore \quad \text{Safe} \end{aligned}$$

Crushing strength between cotter and rod (Eqn. 5.20 MDHB)

$$\begin{aligned} P_1 &= d_1 t S_{cd} = 18.75 \times 6 \times 100 \\ &= 11250 \text{ N} > (P = 10 \text{ kN}) \quad \therefore \quad \text{Safe} \end{aligned}$$

Shearing strength of the rod across cotter (Eqn. 5.21 MDHB)

$$P_1 = 2d_1a \cdot S_{sd} = 2 \times 18.75 \times 18.75 \times 50$$
$$= 35156.25 \text{ N} > (P = 10 \text{ kN}) \quad \therefore \quad \text{Safe}$$

Step-5: Check for strength of socket

Tearing strength of the socket across the hole (Eqn. 5.26 MDHB)

$$P_1 = \left\{\frac{\pi}{4}(D_1^2 - d_1^2) - (D_1 - d_1)t\right\} S_d$$
$$= \left\{\frac{\pi}{4}(26.25^2 - 18.75^2) - (26.25 - 18.75) \times 6\right\} \times 90$$
$$= \{265.07 - 45\} \times 90$$
$$= 19806.3 \text{ N} > (P = 10 \text{ kN}) \quad \therefore \quad \text{Safe}$$

Crashing strength of socket under the cotter (Eqn. 5.27 MDHB)

$$P_1 = (D - d_1)t\, S_{cd}$$
$$= (30 - 18.75) \times 6 \times 100$$
$$= 6750 < (P = 10 \text{ kN}) \quad \therefore \quad \text{Unsafe}$$

Then putting $P_1 = 10,000$ N

$$10,000 = (D - 18.75) \times 6 \times 100$$
$$D = 35.41 \text{ mm}$$
$$\cong 35.5 \text{ mm} \quad \text{(Adopt)}$$

Shearing strength of the socket across the cotter hole (Eqn. 5.28 MDHB)

$$P_1 = 2(D - d_1)a_1S_{sd}$$
$$= 2(35.5 - 18.75) \times 18.75 \times 50$$
$$= 31406.25 \text{ N} > (P = 10 \text{ kN}) \quad \therefore \quad \text{Safe}$$

Shearing strength of the socket across the rod end (Eqn. 5.28a MDHB)

$$P_1 = \pi d_1 \cdot t_2 \cdot S_{sd}$$
$$= \pi \times 18.75 \times 7.5 \times 50$$
$$= 22089.32 \text{ N} > P = (10 \text{ kN}) \quad \therefore \quad \text{Safe}$$

Example 7.4: Design a knuckle joint to connect two mild steel rods to sustain an axial pull of 150 kN. The pin and rods are made of same material. Assume the working stresses in the material as 80 MPa in tension; 40 MPa in shear and 120 MPa in crushing.

Given Data (Refer Fig. 7.6)
Pull in the rod, $P = 150$ kN
Design stresses

$$S_d = 80 \text{ MPa};\ S_{sd} = 40 \text{ MPa and } S_{cd} = 120 \text{ MPa.}$$

SOLUTION

The rod diameter for tearing strength (Eqn. 5.29 MDHB)

$$P_1 = \frac{\pi}{4} d^2 \cdot S_d$$

$$\therefore \quad d = \sqrt{\frac{4P}{\pi S_d}} = \sqrt{\frac{4 \times 150 \times 10^3}{\pi \times 80}}$$

$$= 48.8 \text{ mm}$$

$$\text{Adopt,} \quad d = 50 \text{ mm}$$

Tearing strength of the two fork sections

$$P_1 = 2AFS_d$$

where $A = 0.6d = 0.6 \times 50 = 30$ mm
$F = 1.1d = 1.1 \times 50 = 55$ mm
∴ Tearing strength of the fork sections

$$P_1 = 2 \times 30 \times 55 \times 80$$

$$= 264000 \text{ N} > (P = 150 \text{ kN}). \quad \therefore \quad \text{Safe}$$

Tearing strength of square section (Eqn. 5.31 MDHB)

$$P = F^2 \cdot S_d$$

$$= 55^2 \times 80$$

$$= 242000 \text{ N} > (P = 150 \text{ kN}). \quad \therefore \quad \text{Safe}$$

Shearing resistance of the pin (Eqn. 5.32 MDHB)

$$P = \frac{\pi}{2} d_1^2 \cdot S_{sd}$$

$$150 \times 10^3 = \frac{\pi}{2} (d_1^2) \cdot \times 40$$

$$d_1 = \sqrt{\frac{150 \times 10^3 \times 2}{p \times 40}} = 48.8 \text{ mm}$$

$$\text{Adopt,} \quad d_1 = 50 \text{ mm}$$

Tearing strength of rod, across the hole (Eqn. 5.33 MDHB)

$$P_1 = F(D - d_1) \cdot S_d$$

where, $F = 55$ mm

and $D = 2d_1 = 100$ mm

$$P_1 = 55(100 - 50) \times 80$$
$$= 220,000 \text{ N} > (P = 150 \text{ kN}). \quad \therefore \quad \text{Safe}$$

Tearing strength of the fork across the hole (Eqn. 5.34 MDHB)

$$P_1 = 2(D - d_1)B \cdot S_d$$

where, $B = 0.75 \times 50 = 37.5$ mm

$\therefore$

$$P_1 = 2(100 - 50) \times 37.5 \times 80$$
$$= 300,000 \text{ N} > (P = 150 \text{ kN}). \quad \therefore \quad \text{Safe}$$

Crushing strength of the pin in the rod (Eqn. 5.35 MDHB)

$$P = Fd_1S_{cd}$$

$$P_1 = 55 \times 50 \times 120$$
$$= 330,000 \text{ N} > (P = 150 \text{ kN}). \quad \therefore \quad \text{Safe}$$

Crushing strength of pin in the fork (Eqn. 5.36 MDHB)

$$P_1 = 2Bd_1S_{cd}$$
$$= 2 \times 37.5 \times 50 \times 120$$
$$= 450,000 \text{ N} > (P = 150 \text{ kN}). \quad \therefore \quad \text{Safe}$$

Shearing strength of rod across the hole (Eqn. 5.37 MDHB)

$$P_1 = F(D - d_1)S_{sd}$$
$$= 55(100 - 50) \times 40$$
$$= 110,000 \text{ N} < (P = 150 \text{ kN}). \quad \therefore \quad \text{Unsafe}$$

Then modify F as follows:

$$F = \frac{150 \times 10^3}{50 \times 40} = 75 \text{ mm}$$

or modify D

$$D = \frac{150 \times 10^3}{55 \times 40} + 50 = 118.18 \text{ mm}$$

To keep the portions, variations small,

Adopt

$D = 120$ mm.
and $F = 55$ mm
Check for shearing strength of fork across the hole (Use $D = 120$ mm)

$$P = 2B(D - d_1)S_{sd} \quad \text{(Eqn. 5.38 MDHB)}$$
$$= 2 \times 37.5(120 - 50) \times 40$$
$$= 210{,}000 \text{ N} > (P = 150 \text{ kN}). \quad \therefore \quad \text{Safe}$$

Example 7.5: Design a knuckle joint for a tie rod of circular cross section to sustain a maximum tensile load of 70 kN. The ultimate strength of the rod against tension is 420 N/mm^2. The ultimate tensile and shearing stresses for the pin material are 500 N/mm^2 and 360 N/mm^2 respectively. Take a factor of safety of 6.

Given Data

The pull, $P = 70$ kN
Material properties
Ultimate strength in tension for the rod material, $S_u = 420$ N/mm^2
For the pin material
Ultimate tensile strength, $S_u = 500$ N/mm^2
Shear strength, $S_{su} = 360$ N/mm^2
Factor of safety, $n = 6$.

SOLUTION

Design stress for the rod,

$$S_d = 420/6 = 70 \text{ MPa}$$
$$S_{sd} = 0.57 \times 70 = 39.9 \rightarrow 40 \text{ MPa}$$

Design stress for the pin,

$$S_d = 500/6 = 83.33 \text{ MPa}$$
$$S_{sd} = 360/6 = 60 \text{ MPa}$$

Tearing resistance of rod (Eqn. 5.29 MDHB)

$$P = \frac{\pi}{4}d^2 \cdot S_d$$

Diameter,

$$d = \sqrt{\frac{4 \cdot P}{\pi S_d}} = \sqrt{\frac{4 \times 70 \times 10^3}{\pi \times 70}}$$
$$= 35.7 \text{ mm} \cong 40 \text{ mm}$$

The pin diameter for shearing strength (Eqn. 5.32 MDHB)

$$P_1 = \frac{\pi}{2} d_1^2 \cdot S_{sd}$$

$$\therefore \quad d_1 = \sqrt{\frac{2 \times P}{\pi \cdot S_{sd}}} = \sqrt{\frac{2 \times 70 \times 103}{\pi \times 60}}$$

$$= 27.26 \text{ mm}$$

$$\text{Choose} \quad d_1 = 30 \text{ mm}$$

Tearing of the square section (Eqn. 5.33 MDHB)

$$P_1 = F^2 \cdot S_d$$

$$\text{But} \quad F = 1.1d = 1.1 \times 40 = 44$$

$$\text{Choose} \quad F = 45 \text{ mm}$$

Then

$$P_1 = 45 \times 45 \times 70$$

$$= 141750 \text{ N} > (P = 70 \text{ kN}) \quad \therefore \quad \text{Safe}$$

Tearing strength of the two forks (Eqn. 5.32 MDHB)

$$P = 2.A.F.S_d$$

$$\text{Where} \quad A = 0.6d = 0.6 \times 40 = 24 \text{ mm}$$

$$\text{Choose} \quad A = 25 \text{ mm}$$

∴ Tearing strength,

$$P_1 = 2 \times 25 \times 45 \times 70$$

$$= 163800 \text{ N} > (P = 70 \text{ kN}) \quad \therefore \quad \text{Safe}$$

Tearing strength of the rod across the hole (Eqn. 5.33 MDHB)

$$P_1 = F(D - d_1)S_d$$

where, $D = 2d_1 = 2 \times 30 = 60$ mm

$$\therefore \quad P = 45 \times (60 - 30) \times 70$$

$$= 94500 \text{ N} > (P = 70 \text{ kN}) \quad \therefore \quad \text{Safe}$$

Tearing strength of forks across the hole (Eqn. 5.34 MDHB)

$$P = 2(D - d_1)\, BS_d$$

where $B = 0.75 \times d = 0.75 \times 40 = 30$ mm

$$P = 2 \times (60 - 30) \times 30 \times 70$$
$$= 126000 \text{ N} > (P = 70 \text{ kN}) \quad \therefore \quad \text{Safe}$$

Check for shearing strength of the rod across the hole (Eqn. 5.37 MDHB)

$$P = F(D - d_1)S_{sd}$$
$$= 45 \times (60 - 30) \times 40$$
$$= 54000 \text{ N} < (P = 70 \text{ kN}) \quad \therefore \quad \text{Unsafe}$$

Check for shearing strength of the fork across the hole (Eqn. 5.38 MDHB)

$$P = 2 \times 30 \times (60 - 30) \times 40$$
$$= 72000 \text{ N} > (P = 70 \text{ kN}) \quad \therefore \quad \text{Safe}$$

To satisfy the Eqn. 5.37 (MDHB), modify D as

$$D = \frac{70 \times 10^3}{45 \times 40} + 30 = 68.89$$

Choose $D = 70$ mm

Example 7.6: Design and sketch the assembly of a cotter joint to connect two rods, subjected to an axial pull of 600 kN. The material selected for the joint has the following permissible stresses 300 MPa in tension, 220 MPa in shear and 450 MPa in crushing.

Given Data

Axial pull, $P = 600$ kN
Allowable stresses
Tension, $S_d = 300$ MPa, Shear, $S_{sd} = 220$ MPa
In crushing, $S_{cd} = 450$ MPa

SOLUTION

Refer Fig. 7.5

Step-1: Rod diameter for tensile strength (Eqn. 5.18 MDHB)

$$d = \sqrt{\frac{4P}{\pi S_d}}$$
$$= \sqrt{\frac{4 \times 600 \times 10^3}{\pi \times 300}} = 50.46 \text{ mm}$$

Choose $d - 51$ mm

Step-2: Design of cotter
Thickness of cotter (Eqn. 5.23 MDHB)

$$t = 0.4d = 0.4 \times 51 = 20.4 \text{ mm}$$

Choose $t = 20$ mm
Width of cotter,

$$b = 4\,t = 4 \times 20 = 80 \text{ mm}$$

Shearing strength of cotter, (Eqn. 5.24 MDHB)

$$\begin{aligned} P_1 &= 2 \cdot 6tS_{sd} \\ &= 2 \times 80 \times 20 \times 220 \\ &= 704000 \text{ N} > (P = 600 \text{ kN}) \quad \therefore \quad \text{Safe} \end{aligned}$$

Bending strength of the cotter,

Take $D = 2d = 2 \times 51 = 102$ mm

$$\sigma_b = \left(\frac{5}{8} \times P \times D\right) / (b^2 t) = \frac{5}{8} \times \frac{600 \times 10 \times (2 \times 51)}{(80^2 \times 20)} \quad (\because \quad D = 2d)$$

$$= 298.83 \text{ MPa} < (S_d = 300 \text{ MPa}) \quad \therefore \quad \text{Safe}$$

Step-3: Design of the spigot
Crushing between cotter and rod,

$$P = d_1 t S_{cd}$$

$$\therefore \qquad d_1 = \frac{P}{tS_{cd}} = \frac{600 \times 10^3}{20 \times 450} = 66.67 \text{ mm}$$

Choose $d_1 = 67$ mm
Tearing strength across the cotter hole,

$$\begin{aligned} P_1 &= \left[\frac{\pi}{4} d_1^2 - d_1 t\right] \cdot S_d \\ &= \left[\frac{\pi}{4} (67)^2 - 67 \times 20\right] \times 300 \\ &= 655695.7 \text{ N} > (P = 600 \text{ kN}) \quad \therefore \quad \text{Safe} \end{aligned}$$

Shearing strength of the rod across the cotter hole (Eqn. 5.21 MDHB)

$$P_1 = 2d_1 \cdot a \cdot S_{sd}$$

$$\therefore \qquad a = \frac{P}{2d_1 \cdot S_{sd}} = \frac{600 \times 10^3}{2 \times 67 \times 220} = 20.35 \text{ mm}$$

Choose $a = 21$ mm and $a_1 = 21$ mm

The crushing strength of the collar (Eqn. 5.22 MDHB)

$$P_1 = \pi d_1 \cdot t_1 S_{sd}$$

$$\therefore \quad t_1 = \frac{P}{\pi d_1 \cdot S_{sd}} = \frac{600 \times 10^3}{\pi \times 67 \times 220} = 12.9 \text{ mm}$$

Choose $t_1 = 15$ mm

Step-4: Design of the socket

Choose $D_1 = 1.75d = 1.75 \times 51 = 89.25 \cong 90.0$ mm

Tearing strength of the socket across the collar hole (Eqn. 5.26 MDHB)

$$P_1 = \left[\frac{\pi}{4}(D_1^2 - d_1^2) - (D_1 - d_1)t\right] S_d$$

$$= \left[\frac{\pi}{4}(90^2 - 67^2) - (90 - 67) \times 20\right] \times 300$$

$$= 712821 \text{ N} > (P = 600 \text{ kN}) \quad \therefore \quad \text{Safe}$$

Crushing strength of socket under cotter (Eqn. 5.27 MDHB)

$$P_1 = (D - d_1)tS_{cd} = (102 - 67) \times 20 \times 450$$

$$= 315000 \text{ N} < (P = 600 \text{ kN}) \quad \therefore \quad \text{Unsafe}$$

Step-5: Redesign of the spigot and cotter

Then modifying the design as

$$D = 2d \text{ and } d_1 = 1.1d, \text{ and } t = 0.4d$$

Then by equation 5.27 MDHB,

$$600 \times 10^3 = (2d - 1.1d) \times 4d \times 450$$

$$d = \sqrt{\frac{600 \times 10^3}{0.9 \times .4 \times 450}} = 60.86 \text{ mm}$$

Adopt

$$d = 61 \text{ mm and } d_1 = 1.1d = 67.0 \text{ mm}$$

$$D = 122 \text{ mm}, \ D_1 = 1.75 \times 61 = 106.75 = 107 \text{ mm}$$

Cotter size,

$$t = 0.4 \times 61 = 24.4 \cong 25 \text{ mm}$$

$$b = 4t = 4 \times 25 = 100 \text{ mm}$$

All other dimensions may be retained as they are.

Example 7.7: Design a knuckle joint to connect two mild steel rods. The joint has to transmit a tensiles load of 80 kN. The materials for the rod and pin has the working stress as $\sigma_c = 80$ MPa, $\sigma_c = 120$ MPa $\times$ $\tau = 40$ MPa.

Given Data

Pull $P = 80$ kN
Working stress tension $S_d = 80$ MPa; Compression, $S_{cd} = 120$ MPa and Shear, $S_{sd} = 40$ MPa

SOLUTION

Refer Fig. 7.6

Design of rod

Diameter of the rod for tensiles strength (Eqn. 7.20)

$$d = \sqrt{\frac{4P}{\pi S_d}} = \sqrt{\frac{4 \times 80 \times 10^3}{\pi \times 80}} = 35.68$$

Choose $d = 40$ mm

The design of pin

Diameter for shear strength (Eqn. 7.23)

$$d_1 = \sqrt{\frac{2 \cdot P}{\pi S_{sd}}} = \sqrt{\frac{2 \times 80 \times 10^3}{\pi \times 40}}$$

$$= 35.68 \text{ mm}$$

Choose $d_1 = d = 40$ mm

Tearing strength of the square section of the rod, (Eqn. 7.21)

$$P = F^2 \cdot S_d$$

i.e.,

$$\therefore \quad F = \sqrt{\frac{P}{S_d}} = \sqrt{\frac{80 \times 10^3}{80}} = 31.63$$

But by proportions,

$$F = 1.1d = 40 \times 1.1 = 44.$$

Choose $F = 45$ mm

For tearing strength of the fork, (Eqn. 7.22)

$$P = 2 \cdot AF\, S_d$$

$$A = \left(\frac{P}{2F\, S_d}\right) = \left[\frac{80 \times 10^3}{2 \times 45 \times 80}\right] = 11.11$$

By proportions,

$$\text{Choose} \qquad A = 0.6d = 0.6 \times 40 = 24 \text{ mm}$$

For tearing strength of the rod, across the pin hole, (Eqn. 7.27)

$$P = F(D - d_1)\, S_d$$

$$D = \frac{P}{F\, S_d} + d_1$$

$$= \frac{80 \times 10^3}{45 \times 80} + 40 = 62.222$$

But by proportions,

$$\text{Choose} \qquad D = 2d_1 = 2 \times 40 = 80 \text{ mm}$$

For tearing strength of the fork across the pin hole, (Eqn. 7.24)

$$P = B(D - d_1)\, S_d$$

$$B = \frac{P}{2(D - d_1)\, S_d} = \frac{80 \times 10^3}{2 \times (80 - 40) \times 80}$$

$$= 12.5 \text{ mm}$$

By proportions, choose $B = 0.75 \times 40 = 30$ mm

Crushing strength of the pin in the rod, (Eqn. 7.26)

$$P_1 = F d_1 \cdot S_{cd}$$

$$= 45 \times 40 \times 120$$

$$= 216000 \text{ N} > (P = 80 \text{ kN}) \quad \therefore \quad \text{Safe}$$

Crushing strength in fork,

$$P_1 = 2B \cdot d_1 \cdot S_{cd}$$

$$= 2 \times 30 \times 40 \times 120$$

$$= 288000 \text{ N} > (P = 80 \text{ kN}) \quad \therefore \quad \text{Safe}$$

Shearing strength of the rod across the hole, (Eqn. 7.28)

$$\begin{aligned} P_1 &= F(D-d_1)\cdot S_{sd} \\ &= 30\times(80-40)\times 40 \\ &= 48,000\text{ N} < P = 80\text{ kN} \quad \therefore \quad \text{Unsafe} \end{aligned}$$

$\therefore$ Modify D,

$$80\times 10^3 = F(D-40)\times 40$$

$$D = \frac{80\times 10^3}{45\times 40} + 40$$

$$= 44.44 + 40 = 84.4$$

Choose $D = 85$ mm

Shearing strength of fork across hole, (Eqn. 7.25)

$$\begin{aligned} P &= 2\cdot B(D-d_1)S_{sd} \\ &= 2\times 30\times(85-40)\times 40 \\ &= 108000\text{ N} > (P = 80\text{ kN}) \quad \therefore \quad \text{Safe} \end{aligned}$$

Example 7.8: A shaft 80 mm diameter transmits power at a maximum shear stress of 65 MPa. Design a rectangular key for a design stress of 42 MPa.

SOLUTION

Standard proportions of a rectangular key are $b = d/4$ and $h = d/6$

$$\therefore \quad b = 80/4 = 20\text{ mm}$$

$$h = 80/6 = 13.33\text{ mm} \quad \cong \quad \text{Choose} \quad h = 14\text{ mm.}$$

Torque transmitted by the shaft,

$$T_d = \frac{\pi}{16}d^3 S_{sd} = \frac{\pi}{16}\times(80)^3\times 65.0 = 6534512.7\text{ N-mm}$$

Torque capacity of the key by shear strength, (Eqn. 7.2)

$$\begin{aligned} T_d &= b.L.S_d.d/2 \\ &= 20\times L\times 42\times 80/2 \end{aligned}$$

Equating the two,

$$20\times L\times 42\times 40 = \frac{\pi}{16}(80)^3\cdot 65.0$$

$$\therefore \quad L = \frac{\pi\times 80^3\times 65.0}{16\times 20\times 42\times 40} = 194.5\text{ mm}$$

Choose $L = 195$ mm

Check for crashing stress, (Eqn. 7.1)

$$\sigma_c = \frac{T_d}{\frac{h}{2} \cdot L \cdot \frac{d}{2}} = \frac{4 \times T_d}{h.L.d}$$

$$= \frac{4 \times 6534512.7}{14 \times 95 \times 80} = 119.0 \text{ MPa} < S_{cd}$$

This value is acceptable.

Example 7.9: A shaft 30 mm diameter is transmitting power at a maximum shear stress of 80.0 MPa. If a pulley is connected to the shaft by a key, find the dimensions of the key so that the shear stress in the key does not exceed 50 MPa and the length of the key is 8 times the width.

SOLUTION

Torque transmitted by the shaft,

$$T_d = \frac{\pi}{16} d^3 \cdot S_{sd} = \frac{\pi}{16} \times 30^3 \times 80 \text{ N-mm}$$
$$= 424115.01 \text{ N-mm}$$

Shear strength of the key,

$$T = b.L.\, S_{sd} \cdot d/2$$
$$= b(8b) \cdot S_{sd} \cdot d/2 \quad (\because L = 8b)$$
$$= 8b^2 \times 50 \times 30/2$$
$$= 6000b^2 \text{ N-mm}$$

$\therefore$ Equating the two,

$$6000b^2 = 424115.01$$

$$b = \sqrt{\frac{424115.01}{6000}} = 8.4 \text{ mm}$$

Choose the standard size, $b \times h = 8 \text{ mm} \times 7 \text{ mm}$

Length, $L = 8 \times b = 8.4 \times 8 = 672 \text{ mm}$

Rounded value of $L = 70$ mm may be adopted.

Example 7.10: A gear 650 mm in diameter is used to transmit 100 kW at 500 rpm on a 75 mm diameter shaft. The design stress for the key in shear is 45.0 MPa and in bearing (crushing) is 100 MPa. Design the square key for the joint.

SOLUTION

Standard dimensions of the square key are $b = h = d/4 = 75/4 = 18.75$

Choose $b = h = 18.0$ mm.

Torque to be transmitted,

$$T_d = \frac{955 \times 10^4 \times \text{kW}}{\text{N}} = \frac{955 \times 10^4 \times 100}{500}$$

$$= 191 \times 10^4 \text{ N-mm}$$

For shearing strength of the key, (Eqn. 7.2)

$$T_d = b.L.(d/2)S_{sd}.$$

$\therefore$ Length, $L = 2T_d/b.d.S_{sd}$

$$= \frac{2 \times 191 \times 10^4}{18 \times 75 \times 45.0} = 62.88 \text{ mm}$$

Choose $L = 65$ mm

Check for crushing stress, (Eqn. 7.1)

$$\sigma_c = \left(\frac{4T}{hdL}\right) \frac{191 \times 10^4 \times 4}{18 \times 75 \times 65} = 87.06 \text{ MPa}$$

It is less than the maximum value of 100 MPa and the design is safe.

Example 7.11: A wood-ruf key is to be used to fit on a 40 mm diameter shaft to transmit 15 kW, through a gear at 500 rpm. Calculate the number of keys needed for maximum shear stress of 65.0 MPa and crushing stress of 120 MPa.

SOLUTION

Standard dimensions of the key,

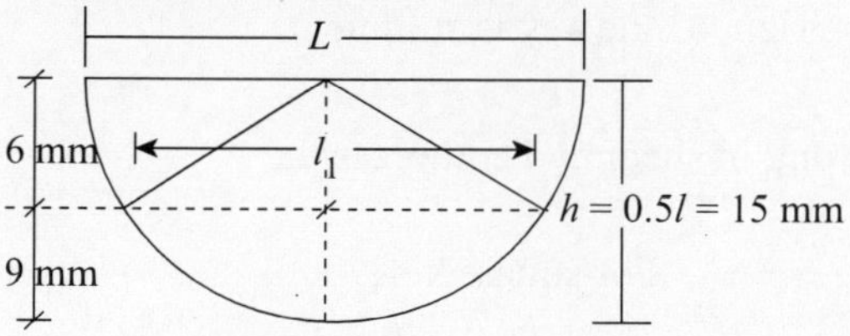

Figure 7.7

$$L = 0.75d$$

$$\therefore \quad L = 0.75 \times 40 = 30 \text{ mm}$$

$$h = 0.5L = 5 \times 30$$

$$= 15 \text{ mm}$$

$$\text{Thickness, } t = \frac{d}{4} = \frac{40}{4} = 10 \text{ mm}$$

Key depth in hub = 6.0 mm

in shaft = 9.0 mm

Length of the key at the junction,

$$l_1 = 2\left(\sqrt{15^2 - 6^2}\right) = 2\left(\sqrt{9 \times 21}\right) = 27.5 \text{ mm}$$

$$\text{Area of the full key, } A = \frac{\pi r^2}{2} = \frac{\pi 15^2}{2} = 353.3 \text{ mm}^2$$

$$\text{Area in the hub, } A_1 = \left(\frac{30 + 27.5}{2}\right) \times 6 = 172.5 \text{ mm}^2$$

$\therefore$ Area in the shaft, $A_2 = (353.3 - 172.5) = 180.8 \text{ mm}^2$

Area in hub is less and crushing stress is more in hub. The torque transmitted,

$$T = \frac{(955 \times 10^4 \times 15)}{500} = 286500 \text{ N-mm}$$

Tangential force on the key,

$$F_t = \frac{286500}{20} = 14325 \text{ N}$$

Shear stress in the key,

$$\tau = \frac{F_t}{a} = \frac{14325}{27.5 \times 10} = 52.1 \text{ MPa} < (S_{sd} = 65 \text{ MPa}) \quad \therefore \quad \text{Safe}$$

Crushing stress on key in hub,

$$\sigma_c = \frac{F_1}{A_1} = \frac{14375}{172.5} = 83.3 \text{ MPa.} < (S_{sd} = 120 \text{ MPa}) \quad \therefore \quad \text{Safe}$$

Both the stresses are less than the permissible values. Hence, one key is sufficient.

Example 7.12: Prove that a square key is equally strong in shear and compression.

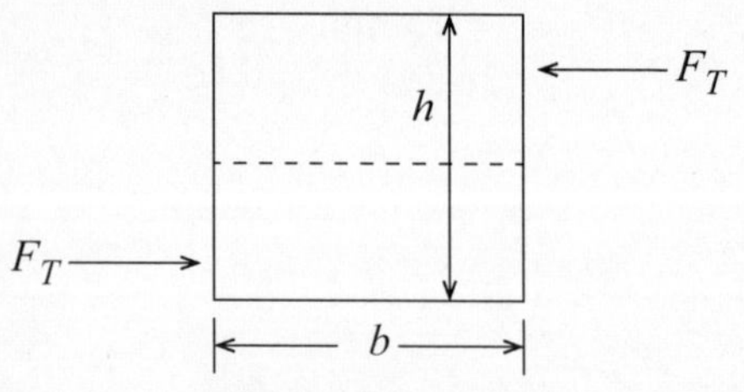

Figure 7.8

For square key,

$$b = h$$

For steel,

$$S_d = 2S_{sd}$$

SOLUTION

Key in subjected to shear at the junction of shaft and hub.
Torque capacity by shear,

$$T_s = \frac{b.L.S_{sd}.d}{2}$$
$$= \frac{bLd\ S_{sd}}{2} \qquad \text{(i)}$$

Key faces are subjected to compressive stress in the hub and the shaft.
Torque capacity by compression,

$$T_c = \frac{hL}{2} \cdot S_d \cdot \frac{d}{2} = \frac{hLd}{4} \cdot S_d \qquad \text{(ii)}$$
$$= \frac{b.L.d}{2} \cdot S_{sd} = T_s. \quad (\text{Submitting } S_d = 2S_{sd})$$

$$\therefore \qquad T_c = T_s$$

The key is equally strong in shear and compression.

Example 7.13: A 45 mm diameter shaft is made of steel with a yield strength of 400 MPa. A parallel key of size 14 mm wide and 9 mm thick is made of steel with a yield-stress of 350 MPa. Find the required length of the key, if the shaft is loaded to transmit permissible torque. Design based on maximum shear theory and take factor of safety as 2.

Given Data

Shaft diameter = 45 mm
Yield strength = 400 MPa
Factor of safety = 2
Size of the parallel key = 14 × 9 mm
Yield strength of key material, S_y = 350 MPa
Factor of safety = 2

SOLUTION

Maximum permissible stress in shear for the shaft,

$$S_{sd} = \frac{S_d}{2} = \frac{S_y}{2n} = \frac{400}{2 \times 2}$$
$$= 100 \text{ MPa}$$

Maximum torque transmitted by the shaft,

$$T = \frac{\pi}{16} \cdot d^3 \cdot S_{sd} = \frac{\pi}{16}(45)^3 \times 100$$
$$= (1789.235 \times 10^3) \text{ N-mm}$$

Torque capacity of the key for shear,

$$T_s = \frac{bL.d}{2} S_{sd}$$

where, $T_s = 1789.235 \times 10^3$ N-mm

$b = 14$ mm

$$S_{sd} = \frac{1}{2} \cdot S_d = \frac{1}{2} \cdot \frac{350}{2} = 87.5$$

Then the length of the key,

$$L = \frac{1789235 \times 2}{14 \times 45 \times 87.5} = 64.9$$

Choose $L = 65$ mm

Example 7.14: Design a square key for fixing a gear on a shaft of 25 mm diameter 15 kW power at 720 rpm is transmitted from the shaft to the gear. The allowable compressive stress in the material is 150 MPa and allowable shear stress is 88 MPa.

Given Data

Power transmitted = 15 kW
Speed, $N = 720$ rpm
Allowable stresses for key material
In compression, $S_d = 150$ MPa
In shear, $S_s d = 88$ MPa
Shaft diameter, $d = 25$ mm

SOLUTION

Torque transmitted,

$$T = \frac{955 \times 10^4 \text{ kW}}{\text{N}} = \frac{955 \times 10^4 \times 15}{720}$$
$$= 199.0 \times 10^3 \text{ N-mm}$$

Standard size of the square key,

$$b = h = \frac{d}{4} = \frac{25}{4} = 6.25 \text{ mm}$$

From Table 7.1 choose the standard size, $b = h = 6.0$ mm.

Crushing strength of the key, $T_c = \frac{1}{4} S_{cd} . h \cdot d \cdot L$ (Eqn. 7.1)

$$\therefore \quad T_c = \frac{1}{4} \times 150 \times 6 \times 25 \times L_c = 199 \times 10^3$$

$$\therefore \quad L_c = \frac{199 \times 10^3 \times 4}{150 \times 6 \times 25} = 35.37 \text{ mm}$$

Shear strength, (Eqn. 7.2)

$$T_s = \frac{1}{2} \cdot S_{sd} \times b.d.L$$

$$= \frac{1}{2} \times 88 \times 6 \times 25 \times L_s = 199 \times 10^3$$

$$\therefore \quad L_s = \frac{199 \times 10^3 \times 2}{88 \times 6 \times 25} = 30.15 \text{ mm}$$

Choose the larger of the two,

$$L = 35.37 \cong 36 \text{ mm}$$

Example 7.15: A splined connection in an automobile transmission consists of 10 splines cut in a 58 mm diameter shaft. The height of each spline is 5.5 mm and the keyways in the hub are 45 mm long. Determine the power that may be transmitted at 2500 rpm, of the allowable normal pressure in limited to 4.8 MPa.

Given Data

Number of splines = 10
Diameter of shaft, $D_1 = 58$ mm
Height of splines, $h = 5.5$ mm
Length of splines, $L = 45$ mm
Allowable pressure, $P_b = 4.8$ MPa

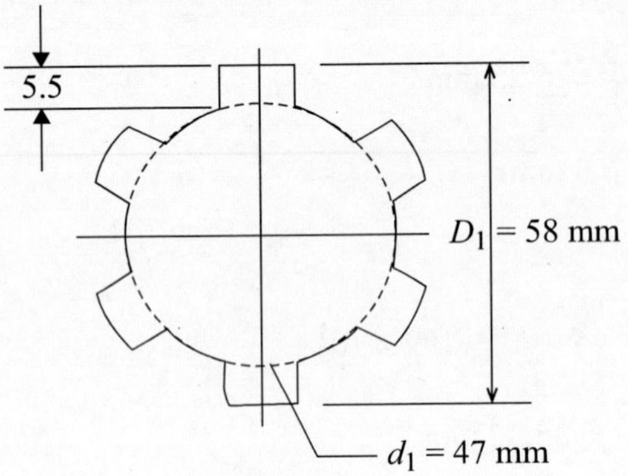

Figure 7.9

$$D_1 = 58 \text{ mm}$$
$$d_1 = 58 - 2 \times 5.5$$
$$= 47 \text{ mm}$$

SOLUTION

The bearing pressure, (Eqn. 7.3)

$$P_b = \frac{4T}{d_P(D_1 - d_l)Z.L}$$

Rearranging, the torque capacity

$$T = \frac{d_P(D_i - d_i)Z.L \times P_b}{4}$$

where $d_P = \dfrac{58+47}{2} = 52.5$

$$\therefore \quad T = \frac{52.5 \times 11 \times 10 \times 4.5 \times 4.8}{4}$$
$$= 311850 \text{ N-mm}$$

$$\text{Power, } \quad \text{kW} = \frac{TN}{955 \times 10^4} = \left(\frac{311850 \times 2500}{955 \times 10^4}\right) = 81.6 \text{ kW}$$

Example 7.16: The standard cross section of a flat key which is fitted on a 50 mm diameter shaft is 16×10 mm. The key is trasmitting 475 N-m torque from the shaft to the hub. The key is made of commercial steel for which yield strength in both tension and compression may be taken as 230 N-mm^2. Determine the length of the key required, if the factor of safety is 3.

Given Data

The flat key size = 16 × 10 mm
Shaft diameter, $d = 50$ mm
Torque transmitted, $T = 475$ N.m = 475×10^3 N-mm
Yield strength, $S_{yt} = S_{yc} = 230$ MPa
Factor of safety, $n = 3$.

SOLUTION

The design stress in tension or compression,

$$S_d = \frac{S_y}{n} = \frac{230}{3} = 76.67 \text{ MPa}$$

Length of the key by crushing consideration, (Eqn. 7.1)

$$L_c = \frac{4T}{S_{cd} \times h \times d}$$

$$L_c = \frac{4 \times 475 \times 10^3}{76.67 \times 10 \times 50} = 49.56 \text{ mm}$$

Allowable shear stress,

$$S_{sd} = S_{cd}/2 = \frac{76.67}{2}$$
$$= 38.335 \text{ MPa}$$

The length of the key by shear strength, (Eqn. 7.2)

$$L_s = \frac{2T}{S_{sd} \cdot b \cdot d} = \frac{2 \times 475 \times 10^3}{38.335 \times 16 \times 50}$$
$$= 30.97 \text{ mm}$$

Larger of the two, is to be used

$$\text{Adopt} \quad L = 49.56 \rightarrow 50.0 \text{ mm}$$

Example 7.17: If a shaft and key are made of same material, determine the length of the key required in terms of shaft diameter taking key with $b = \frac{d}{4}$ and key thickness $h = \frac{3d}{16}$. Assume key way factor as 0.75.

Given Data

Material for shaft and key in the same S_y is the same.

Key size, $b = \frac{d}{4}$ and $h = \frac{3d}{16}$

Key way factor, $K_e = 0.75$

SOLUTION

Torque that can be transmitted by the shaft,

$$T = \frac{\pi d^3}{16} \cdot S_{sd} \times K_e \quad \text{(i)}$$

Length of the key by crushing strength, (Eqn. 7.1)

$$L_c = \frac{4 \cdot T}{S_{cd} \cdot h \cdot d} = \frac{4 \cdot \frac{\pi}{16} d^3 \cdot S_{sd} \cdot K_e}{S_{cd} \cdot \left(\frac{3d}{16}\right) \cdot d}$$

$$= \frac{2\pi}{3} \cdot d \cdot K_e \quad \text{(Assuming } S_{cd} = 2S_{sd}) \quad \text{(ii)}$$

Length of the key by shearing strength of the key, (Eqn. 7.2)

$$L_s = \frac{2T}{S_{sd} \cdot bd} = \frac{2 \times \frac{\pi}{16} d^3 \cdot S_{sd} \cdot K_e}{S_{sd} \cdot \frac{d}{4} \cdot d}$$

$$= \frac{\pi}{2} d \cdot ke \quad \text{(iii)}$$

Choosing larger of the two values,

$$L_c = \frac{2\pi}{3} \cdot K_e \cdot d = \left(\frac{2\pi}{3}\right)(0.75)d$$

$$L = 1.57d$$

EXERCISES FOR PRACTICE

7.1. Design a spigot and socket type cotter joint to carry a load of 25 kN. The stress induced is limited to 50 MPa in tension; 35 MPa in shear and 90 MPa in crushing.

7.2. Design a socket and spigot cotter joint to connect two rods subjected to load ± 15 kN. The ultimate strengths of the material for the joint are: tensile at 200 MPa; shear at 140 MPa and crushing at 300 MPa. Consider a factor of safety of 4.

7.3. Design a knuckle joint to connect two rods to carry a load of 100 kN. The design stresses may be taken as 75 MPa in tension; 60 MPa in shear; and 100 MPa in crushing.

7.4. A belt pulley is keyed to a 70 mm diameter shaft running at 240 rpm. The key is 20 mm wide and 100 mm long. Permissible shear stress for the shaft and key materials is 45 MPa and crushing stress for key material is 90 MPa. Determine the power that can be transmitted at 240 rpm. Also determine the thickness of the key.

7.5. A 50 mm diameter shaft, of cold-drawn C40 steel, has a pulley keyed to it. Compute the length of the square key and the length of flat key, if the materials for the key is same as shaft.

7.6. Design a key for a 100 mm diameter shaft transmitting 750 kW at 1000 rpm. The allowable shear stress in the key is 75 MPa and allowable compressive stress is 200 MPa.

7.7. (a) Explain the various types of failures of keys.

(b) A splined joint in an automobile has a specification of $8 \times 42 \times 48$. Allowable bearing pressure at the surface of contact is 10 MPa. Determine the torque that can be transmitted over a contract length of 60.0 mm. Also determine the power that can be transmitted at 750 rpm.

7.8. A 100 mm shaft rotating at 100 rpm transmits 300 kW; power is taken off through a gear whose hub is 200 mm long. The key is made of steel having an ultimate shearing stress of 350 MPa. Using a factor of safety of 5, determine the width of key required.

7.9. A hand lever keyed to the spindle is 0.8 m long from the centre of the spindle to the point of application of load of 300 N. The effective overhang of the spindle from the nearest bearing is 100 mm. If the permissible stress in tension, shear and crushing is not to exceed 66 MPa, design the spindle, key and lever. Assume the arm of the lever to be rectangular having width twice of its thickness

7.10. (a) Prove that a square key is equally strong in shear and compression.

(b) A square key has a width equal to 0.3 times the diameter of the shaft. The shaft and key are made of materials, which are equally strong with yield point value in shear equal to one half of the yield point value in tension. Find the required length of the key in terms of shaft dimater necessary to transmit the shaft torque.

7.11. (a) Discuss the various types of keys and their applications.

(b) A $8 \times 36 \times 50$ mm spline fitting in an automobile transmission has a length of 50 mm. The normal pressure in spline fitting on the active surface is limited to 7 MPa. Determine the power transmitted at 800 rpm.

7.12. a) A shaft and key are made of the same material and the key width is one fourth the diameter of shaft. Considering shear only determine the minimum key length in terms of shaft diameter.

b) Select a rectangular parallel key for transmitting a power of 50 kW at a rated speed of 500 rpm, to mount a hub of length 60.0 mm on a solid circular shaft of diameter 50.0 mm.

7.13. (a) (a) Explain the various types of failures in knuckle joint.

(b) Design fully a socket and spigot type of cotter joint for an axial load of 80 kN, which alternately changes from tensile to compressive. Allowable stresses for the material of all the parts are 50 N/mm^2, 35 N/mm^2 and 90 N/mm^2 in tension, shear and compression respectively. Draw a neat sketch.

7.14. (a) Design a cotter joint to withstand an axial load varying from 60 kN in tension to 60 kN in compression. Allowable stresses for the steel used in the joint are 60 MPa in tension, 75 MPa in crushing and 48 MPa in shear. Give neat proportioned sketch of the joint.

(b) Design a kunckle joint to transmit tensile load of 50 kN. The design stresses for the material for the components of the joint are as follows: $S_d = 80$ MPa in tension; $S_{dc} =$ 100 MPa in crushing; $S_{ds} = 50$ MPa in shear pressure dimensioned sketch of the joint.

(c) Design a cotter joint to carry an alternating force of 25 kN. The allowable stresses are

Rod: Tension $\sigma_t = 60$ MPa; Compression, $\sigma_c = 20$ MPa; Shear, $\tau = 40$ MPa

Cotter: Shear, $\tau = 50$ MPa; Bending, $\sigma_b = 80$ MPa

SHORT ANSWER QUESTIONS

7.1. What is a key?
7.2. What are the functions of keys?
7.3. What are the types of keys?
7.4. What are the materials used for keys?
7.5. What is a saddle key?
7.6. When is the saddle preferred?
7.7. What is a sunk key?
7.8. What are the types of sunk keys?
7.9. What are the standard proportions of square keys?
7.10. What are the standard proportions of rectangular key?
7.11. What is a tapered sunk key?
7.12. What is the taper provided?
7.13. What are the advantages of the taper key?
7.14. What is the feather key? What are its applications?
7.15. What is woodruff key? What are its applications?
7.16. What is Kennedy key? Where is it used?
7.17. What are splines? Where are they used?
7.18. What are the advantages of splines over keys?
7.19. What are the types of splines?
7.20. What are the advantages of involute splines over the straight sided splines?
7.21. What is cotter joint?
7.22. What are the applications of cotter joints?
7.23. What is knuckle joint?
7.24. Where are the knuckle joints used?

CHAPTER 8

COUPLINGS

8.1 DEFINITION OF COUPLING

The coupling are the machine components that join two rotating shafts permanently.

The shaft couplings are used:

1. To connect the shafts of units that are manufactured separately such as a motor and a generator and to provide for disconnection for repair or alteration.
2. To provide for misalignment of the shafts or to introduce flexibility.
3. To reduce the transmission of shock loads from one shaft to another.
4. To introduce protection against overloads.

Requirements of good coupling

1. It should be easy to connect and disconnect.
2. It should transmit the full torque (power) from one shaft to the other without loses.
3. It should hold the shafts in perfect alignment.
4. It should reduce the transmission of shock loads from one shaft to another.
5. It should have no projecting parts.

8.2 TYPES OF COUPLINGS

1. Rigid couplings.
2. Flexible couplings.

1. *Rigid couplings* are used to connect two shafts which are perfectly aligned. The various types are:

   (a) Sleeve or muff coupling

   (b) Clamp or split-muff or compression coupling

   (c) Protected type flange couplings

   (d) Claw coupling (star coupling)

2. *Flexible couplings* are used to connect two shafts having both lateral and angular misalignment. The various types are:
   (a) Bushed pin type coupling
   (b) Universal coupling
   (c) Oldham coupling
   (d) Spring coupling
   (e) Tyre coupling

8.2.1 Sleeve or Muff Coupling

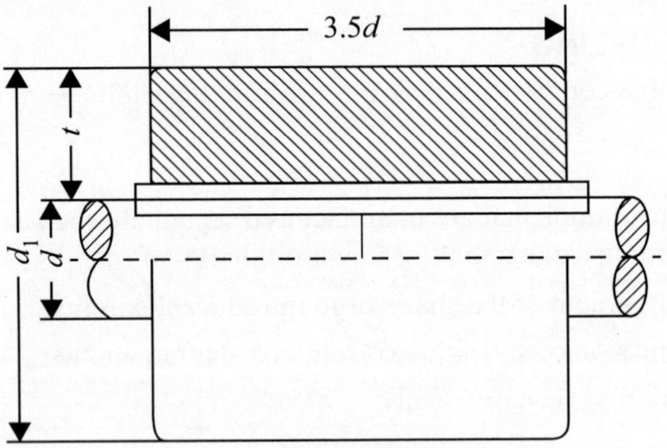

Proportions

Thickness of the sleeve.
$t = d/3 + 10$ mm
Length of the sleeve
$L = 3.5d$ to $4\,d$

Figure 8.1. Sleeve or muff coupling

It is the simplest type of rigid coupling made of cast iron. It consists of a hollow cylinder with inner diameter equal to the shaft diameter. It is fitted over the ends of the two shafts by means of a gib-headed key. The power is transmitted from the driving shaft to the sleeve through the key and then from the sleeve to the driven shaft through the key.

Dimensions of the sleeve are determined by usual proportions.

$$\text{Outside diameter, } d_1 = 1.7d + 20 \text{ mm to } 2d + 15 \text{ mm}$$

$$\text{Length, } L = 3.5d \text{ to } 4d$$

Torque capacity of the sleeve: It is designed as a hollow shaft.

$$T = \frac{\pi d_1^3}{16}\left(1 - K^4\right) S_{sd} \geq T_d \tag{8.1}$$

where, $K = \dfrac{d}{d_1}$ and design stress, $S_{sd} = 14$ MPa for cast iron.

Design of the key

The key is a rectangular key with $h = d/4$ and $h = d/6$
The length of the key equals length of the coupling $= 3.5d$
The length of the key in each shaft, $l = \dfrac{L}{2} = 1.75d$.
The shearing strength and the crushing strength of the key are checked by the following relations.

$$\text{Shearing strength, } T_s = \frac{bld\,S_{sd}}{2} \geq T_d \tag{8.2}$$

$$\text{Crushing strength, } T_c = \frac{hld\,S_{cd}}{4} \geq T_d \tag{8.3}$$

8.2.2 Clamp or Compression Coupling

In this coupling the muff is split into two halves one half is placed below and the other half is placed from above. The shaft ends are made to butt each other. A single key is placed in keyways of both the shafts. The halves of the sleeves are held together by means of mild steel stud bolts and nuts. The number of bolts may be two, four or six. The advantage of the coupling is that the position of the shaft need not be changed for assembling or disassembling of the coupling.

Design of the Muff

The dimensions are determined by usual proportions.
Outside diameter of the sleeve, $d_1 = 2d + 13$ mm to $2.5d$
Length of the sleeve, $L = 1.5\ d_1, = 3.75d$
The torque capacity of the sleeve is evaluated by treating it as hollow shaft,

$$T = \frac{\pi d_1^3}{16}\left(1 - K^4\right) S_{sd} \tag{8.4}$$

The permissible shear stress, $S_{sd} = 14$ MPa for CI.
The key is designed in the same way as in the solid-muff coupling.

Bolt design

The fixing bolt diameter is determined by the following relations:

$$d_b = 0.2d + 10 \text{ mm for } d \leq 55 \text{ mm}$$

$$= 0.15d + 15 \text{ mm for } d > 55 \text{ mm}$$

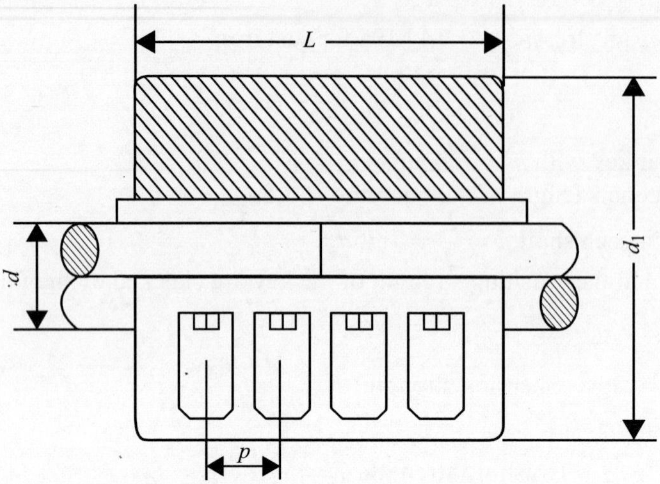

Proportions

Bolt diameter
$d_b = 0.2d + 10$ mm for $d < 55$ mm $= 0.15d + 15$ for $d > 55$ mm Pitch of the bolts, $p = 2d_b + 20$ mm Outside diameter of the sleeve $d_1 = 2.5\ d$ Length of the sleeve, $L = 1.5\ d_1 = 3.75\ d$

Figure 8.2. Split muff or clamp coupling

The torque capacity based on the bolt tension,

$$T = \frac{n.f.\pi^2\ d_b^2\ S_d\ .d}{16} \geq T_d \tag{8.5}$$

This relation may be used to find the number of bolts required.

SOLVED EXAMPLES

Example 8.1: A sleeve coupling, used to transmit a torque of 120 N-m, has the following details–

Diameter of the shaft, $d = 30$ mm
Outside diameter of sleeve, $d_1 = 75$ mm
Length of the sleeve, $L = 100$ mm
Key used $b \times h = 8 \times 7$ mm

Find the stresses induced.

SOLUTION

Shear stress in the shaft,

$$\tau = \frac{16T}{\pi d^3} = \frac{16 \times 120000}{\pi \times 30^3}$$

$$= 22.6 \text{ MPa}$$

Shear stress in sleeve,

$$\tau = \frac{16T}{\pi d_1^3(1 - K^4)}, \text{ where } K = \frac{30}{75} = 0.4$$

$$= \frac{16 \times 120 \times 1000}{\pi \times 75^3(1 - 0.4^4)}$$

$$= 1.49 \text{ MPa}$$

Shear stress in the key,

$$\tau = \frac{2T}{bdL} = \frac{2 \times 120000}{8 \times 30 \times 50} = 20.00 \text{ MPa}$$

Compressive stress in the key,

$$\sigma_c = \frac{4.T}{hdL} = \frac{4 \times 120000}{7 \times 30 \times 50} = 45.7 \text{ MPa}$$

Example 8.2: Design a sleeve coupling to transmit 15 kW at 1000 rpm. The design stresses for C.I. sleeve are $S_{sd} = 10$ MPa, for steel shaft $S_{sd} = 40$ MPa and for key $S_{sd} = 55$ MPa and $S_{cd} = 110$ MPa.

SOLUTION

$$\text{Torque transmitted, } T = \frac{955 \times 10^4 \text{ kW}}{\text{N}} \text{ N-mm}$$

$$= \frac{955 \times 10^4 \times 15}{1000}$$

$$= 143250 \text{ N-mm}$$

$$\text{Shaft diameter, } d = \left[\frac{16T}{\pi S_{sd}}\right]^{1/3}$$

$$= \left[\frac{16 \times 143250}{\pi \times 40}\right]^{1/3}$$

$$= 26.322 \text{ mm}$$

Choose $d = 27$ mm.

Sleeve-design

Outside diameter of sleeve

$$d_1 = d + 2t = d + \frac{2}{3}d + 20 = \frac{5}{3}d + 20$$

$$= \frac{5}{3} \times 27 + 20$$

$$= 65 \text{ mm}$$

$$\text{Length of the sleeve, } L = 3.5d$$

$$= 3.5 \times 27$$

$$= 94.5 \text{ mm}$$

$$\text{Choose } L = 95 \text{ mm}$$

Check for the shear stress in the sleeve,

$$\tau = \frac{16T}{\pi d_1^3 (1 - K^4)} \quad \text{where, } K = \frac{d}{d_1} = \frac{27}{65} = 0.415$$

$$= \frac{16 \times 143250}{\pi \times 65^3 (1 - 0.415^4)}$$

$$= 2.74 \text{ MPa} < (S_{sd} = 10 \text{ MPa})$$

Hence, the design is safe.

Design of key

$$\text{Length of the key, } l = \frac{L}{2} = \frac{95}{2}$$

$$= 47.5 \text{ mm}$$

$$\text{Choose } l = 47.0 \text{ mm}$$

Choose the standard key size from Table 5.1 MDHB

$$b = 8 \text{ mm and } h = 7 \text{ mm}$$

Check for shear stress in key (Eqn. 5.7 MDHB)

$$S_s = \frac{2T}{b.d.l} = \frac{2 \times 143250}{8 \times 27 \times 47}$$

$$= 28.22 \text{ MPa} < S_{sd} = 55 \text{ MPa} \quad \therefore \quad \text{Safe}$$

It is safe.

Check for crashing stress (Eqn 5.6 MDHB)

$$\sigma_c = \frac{4T}{h.dl} = \frac{4 \times 143250}{7 \times 27 \times 47}$$

$$= 64.5 \text{ MPa} < (S_{cd} = 110 \text{ MPa})$$

Hence, it is safe.

Example 8.3: Design a split muff coupling to transmit 40 kW at 150 rpm. The material for the shaft key and bolts is C40 steel. The factor of safety is to be 5. The number of clamping bolts are 8. The coefficient of friction between the sleevs and shaft is 0.3.

SOLUTION

Material properties from Table 2.6 MDHB;

Ultimate strength, $S_u = 580$ MPa

Yield strength, $S_y = 330$ MPa

Yield strength in shear, $S_{sy} = 0.5 \times S_y = 165$ MPa

Design stress in

$$\text{Tension } S_d = 330/5 = 66 \text{ MPa}$$
$$\text{Shear } S_{sd} = 165/5 = 33 \text{ MPa}$$

Torque transmitted,

$$T = \frac{955 \times 10^4 \text{kW}}{\text{N}} = \frac{955 \times 10^4 \times 40}{150}$$
$$= 2546666.7 \text{ N-mm}$$

Diameter of the shaft,

$$d = \left[\frac{16.T}{\pi S_{sd}}\right]^{1/3} = \left[\frac{16 \times 2546666.7}{\pi \times 33}\right]^{1/3}$$
$$= 73.25 \text{ mm}$$
$$\text{Choose } d = 75 \text{ mm}$$

Outside diameter of the sleeve,

$$d_1 = 2.5d = 2.5 \times 75$$
$$= 187.5 \text{ mm}$$
$$\text{Choose } d_1 = 190 \text{ mm}$$

Length of the sleeve,

$$L = 1.5d_1 = 1.5 \times 190$$
$$= 285 \text{ mm}$$

Clamping bolt diameter,

$$d_b = 0.15d + 15$$
$$= 0.15 \times 75 + 15$$
$$= 26.25 \text{ mm}$$

Choose the standard size, (Table 8.7 MDHB)

$$d_b = 27 \text{ mm and } p = 2 \text{ mm}$$

i.e., M27 × 2 with $A_s = 496 \text{ mm}^2$
Torque capacity due to friction (Eqn. 6.5 MDHB)

$$T = \left(n f \pi^2 d_b^2 . S_d . d\right) / 16$$
$$= (8 \times 0.3 \times \pi^2 \times 27^2 \times 66 \times 75)/16$$
$$= 5342244.14 \text{ N-mm} > (T_d = 2546666.7)$$

The design is safe.

Example 8.4: Design a split muff coupling for transmitting 25 kW of power at 150 rpm. The material for the shaft-key and bolt is C30 steel. The factor of safety is 3. The coefficient of friction is 0.3.

SOLUTION

Material properties from Table 2.6 MDHB

$$\text{Yield strength in tension, } S_{yt} = 300 \text{ MPa}$$

$$\text{Yield strength in shear, } S_{sy} = 0.5.\ S_{yt} = 0.5 \times 300 = 150 \text{ MPa}$$

Design stress in

$$\text{Tension, } S_d = S_y/\text{N} = 300/3 = 100 \text{ MPa}$$

$$\text{Shear, } S_{sd} = \frac{S_{sy}}{\text{N}} = 150/3 = 50 \text{ MPa}$$

$$\text{Compression, } S_{cd} = 1.5 S_d \text{ (assumed)} = 1.5 \times 100 = 150 \text{ MPa}$$

Torque transmitted,

$$T = \frac{955 \times 10^4 \text{kW}}{\text{N}} = \frac{955 \times 10^4 \times 25}{150} = 1591.67 \times 10^3 \text{ N-mm}$$

Shaft diameter,

$$d = \left[\frac{16.T}{\pi S_{sd}}\right]^{1/3}$$

$$d = \left[\frac{16 \times 1591.67 \times 10^3}{\pi \times 50}\right]^{1/3} = 54.53 \text{ mm}$$

$$\text{Choose } \quad d = 55 \text{ mm}$$

Outside diameter of the sleeve,

$$d_1 = 2.5d = 2.5 \times 55 = 137.5 \text{ mm}$$

$$\text{Choose } \quad d_1 = 140 \text{ mm}$$

Length of the sleeve,

$$L = 1.5\ d_1 = 1.5 \times 140 = 210 \text{ mm}$$

Length of the key,

$$l = \frac{L}{2} = \frac{210}{2} = 105 \text{ mm}$$

Size of the key from Table 5.1 MDHB

$$b = 16 \text{ mn and } h = 10 \text{ mm}$$

Torque capacity based on shear (Eqn. 5.7 MDHB)

$$T = \frac{1}{2} S_{sd}\, b \cdot d \cdot L$$

$$= \frac{1}{2} \times 50 \times 16 \times 55 \times 105$$

$$= 2310 \times 10^3 \text{ N-m} > (T_d = 159.167 \times 10^3) \quad \therefore \quad \text{Safe}$$

Torque capacity based on crushing (Eqn. 5.6 MDHB)

$$T = \frac{1}{4} S_{cd} \cdot b \cdot d \cdot L$$

$$= \frac{1}{4} \times 150 \times 10 \times 55 \times 105$$

$$= 2165.625 \times 10^3 \text{ N-mm} > (T_d = 1591.6710^3 \text{ N-mm}) \;\therefore \quad \text{Safe}$$

The key is safe in both shear and crushing.
Torque capacity based on the friction between shaft and sleeve (Eqn. 6.5 MDHB)

$$T = \left(n \cdot f \cdot \pi^2 \cdot d_b^2 \cdot S_d \cdot d\right) / 16$$

where, $T = 1591.67 \times 10^3$ N-mm
Diameter of bolt,

$$d_b = 0.15d + 15$$

$$= 0.15 \times 55 + 15$$

$$= 23.25 \text{ mm}$$

Choose the standard size (Table 9.7 MDHB)

$$\boxed{\text{M24} \times 2 \text{ mm}} \; A_s = 384 \text{ mm}^2$$

Then the number of bolts required by rearranging the above equation

$$n = 16T/\left(f \cdot \pi^2 \cdot d_b^2 \cdot d S_d\right)$$

$$= \frac{16 \times 1591.67 \times 10^3}{0.3 \times \pi^2 \times 24^2 \times 55 \times 100} = 2.7$$

Choose $n = 4$

8.2.3 Rigid Flange Coupling

The coupling is so widely used that the dimensions are standardized in terms of the shaft diameter. The check is made for the various kinds of failures.

The shaft diameter is found by torsional strength consideration, i.e., by the equation

$$d = \sqrt[3]{\frac{16T}{\pi S_{sd}}} \tag{8.6}$$

The major dimensions are calculated by standard proportions, namely,

$$\text{Hub diameter, } d_1 = 1.75d \text{ to } 2d$$

$$\text{Hub length, } L = 1.5d \text{ to } 1.75\,d$$

$$\text{Bolt circle diameter, } D_b = 2(d+25)$$

Flange thickness, $t = 0.5\,d$
Outside diameter of flage, $D = 2.5d + 75$
Number of fixing bolts, $n = d/50 + 3.0$
Diameter of the bolt, $d_b = (0.5d/\sqrt{n})$

Check for shear strength of the bolts

$$T = (\pi d_b^2 \cdot n \cdot D_b \cdot S_{sd})/8 > T_d \tag{8.7}$$

where T_d is design torque.

Check for the bearing pressure on the bolts

$$T = n \cdot d_b \cdot t \cdot D_b \cdot S_{cd}/2 > T_d \tag{8.8}$$

Check for the shearing strength of the flange

$$T = \pi d_1^2 \cdot t \cdot S_{cd}/2 \geq T_d \tag{8.9}$$

Torsional strength of the hub is quite high, and no check is required.

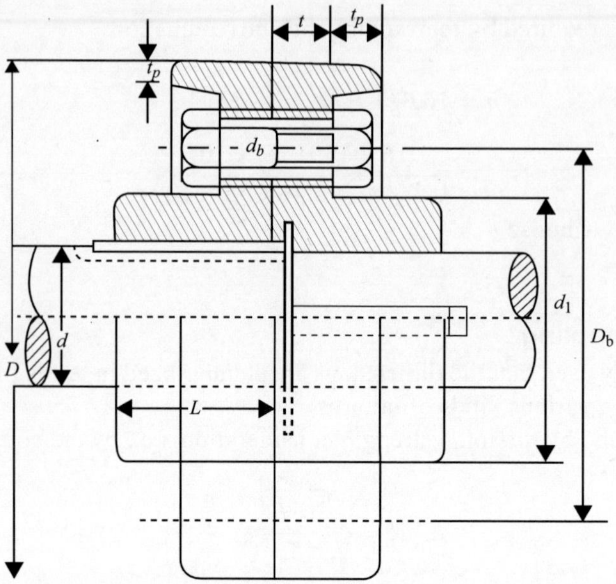

Proportions

The hub diameter, $d_1 = 1.75d$ to $2d$
Hub length, $L = 1.5d$ to $1.75d$
Bolt circle didmeter, $D_b = 2\,(d+25)$
Flange thickness, $t = (0.25d + 10)$ to $0.5d$

Outside diameter of the flange,
$D = 2.5d + 75$ to $4d$
Flange width, $tp = 0.25d$

Figure 8.3. Protected type flange coupling

Key design

Rectangular keys are used with proportions,

$$b = d/4 \text{ and } h = d/6$$

Length of the key is calculated by shearing strength consideration,

$$L_1 = (2T/S_{sd} b.d) \leq \text{hub length } L \tag{8.10}$$

Check for the crushing stress

$$S_c = (4T/h \cdot d \cdot \mathrm{L}_1) \leq S_{cd} \tag{8.11}$$

If this condition is not satisfied, the key length is increased to reduce the crushing S_e below S_{cd}.

8.2.4 Flexible Couplings

Pin and Bush Type Flexible Flange Coupling

This is similar to flange coupling, with modified connecting bolts. The bolts have enlarged diameter in the driven flange with a rubber bush on it. The rubber bush provides for slight misalignments of the shafts and absorbs shock due the load variations.

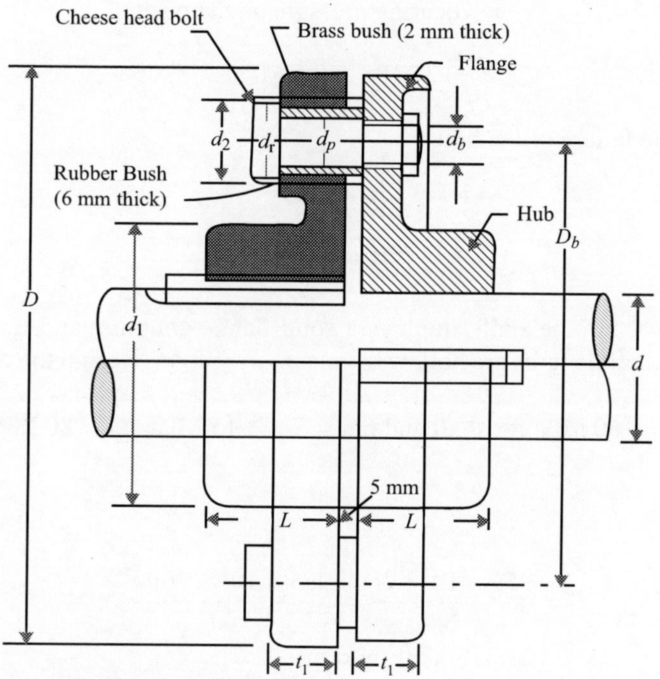

Proportions – **(additional)**

Pitch circle diameter, $D_b = 3.5d$
Outside diameter of the flange, $D = 5d$
Enlarged dia of the pin (bolt), $d_p = 1.5d_b$
Dia of the rubber bush, $d_r = d_p + 15$
Length of the bush, $L_r = 1.5d_r$ to $1.8d_r$

Figure 8.4. Pin and bush type flange coupling

The design procedure is the same as for rigid flange coupling.
Equations 8.7 to 8.11 are applicable to this coupling also. The change is that each bolt has an enlarged diameter d_p, with a rubber bush on it, in the driven flange. There will be two additional failure considerations as given below.
Considering the crushing failure of the rubber bush in the driven flange,

$$T = \frac{nd_r L_r D_b p}{2} \geq T_d \tag{8.12}$$

where, p = bearing pressure on the bush

$= 0.3$ MPa to 0.5 MPa

Considering the bending failure of the bolt,

$$\sigma = \frac{32 T_d L_r}{\pi n D_b d^3} \leq S_d \tag{8.13}$$

Example 8.5: A vertical turbine shaft employs a solid flange coupling and transmits 3 MW of power at 120 rpm. The shaft lengths are hollow having $d_i/d_o = 0.45$. Design the coupling using the following stresses:

$S_{sd} = 80$ MPa, $S_d = 160$ MPa for shaft and bolts, $S_{sd} = 15$ MPa, $S_d = 80$ MPa for the flange.

SOLUTION
Torque transmitted,

$$T = \frac{955 \times 10^4 \text{ kW}}{\text{N}} = \frac{955 \times 10^4 \times 3000}{120}$$
$$= 23.875 \times 10^7 \text{ N-mm}$$

Diameter of the shaft (Eqn. 4.5 MDHB)

$$d_o = \sqrt[3]{\frac{16.T}{\pi S_{sd}(1 - K^4)}}$$
$$= \sqrt[3]{\frac{16 \times 23.875 \times 10^7}{\pi \times 80(1 - 0.45^4)}}$$
$$= 248.88 \cong 250 \text{ mm}$$

Hub diameter, $d = 1.5d + 25$
$= 1.5 \times 250 + 25$
$= 400$ mm

Hub length, $L = 1.75\, d = 1.75 \times 250 = 437.5 = 440$ mm

Rectangular key dimensions are

$$b = d/4 = 250/4 = 62.5 \text{ mm}$$
$$h = d/6 = 250/6 = 41.7 \text{ mm}$$

Choose size is 63 × 42 mm.
Length of the key for shear strength (Eqn. 5.7 MDHB)

$$\begin{aligned} L_1 &= 2T/(S_{sd} \cdot b \cdot d) \\ &= 2 \times 23.875 \times 10^7/(80 \times 63 \times 250) \\ &= 378 \cong 400 \text{ mm} \end{aligned}$$

Check for the crushing stress (Eqn. 5.6 MDHB)

$$\begin{aligned} S_c &= (4T/hdL_1) \\ &= 4 \times 23.875 \times 10^7/42 \times 250 \times 400 \\ &= 245.0 \text{ MPa} \end{aligned}$$

It is more than the permissible and unsafe.

and use a square key $b = h = d/4 = \dfrac{250}{4} \cong 63$ mm
Then shear stress in the key,

$$\begin{aligned} \tau &= 2T/bdL_1 = 2 \times 23.875 \times 10^7/(63 \times 250 \times 400) \\ &= 75.7 \text{ MPa} < 80 \text{ MPa}. \quad \therefore \quad \text{Safe} \end{aligned}$$

Crushing stress,

$$\begin{aligned} \sigma_c &= 4T/h.d.L_1 \\ &= 4 \times 23.875 \times 10^7/(63 \times 250 \times 400) \\ &= 151.6 \text{ MPa} < S_d = 160 \text{ MPa}. \quad \therefore \quad \text{Safe} \end{aligned}$$

Flange thickness, $t = 0.5\, d = 0.5 \times 250 = 125$ mm
Shearing stress in the flange (Eqn. 6.13 MDHB)

$$\begin{aligned} \tau &= (2T/\pi d_1^2.t) \\ &= 2 \times 23.875 \times 10^7/(\pi \times 400^2 \times 125) \\ &= 7.6 \text{ MPa} < 15 \text{ MPa}. \quad \therefore \quad \text{Safe} \end{aligned}$$

$$\begin{aligned} \text{Bolt circle diameter, } D_b &= 2.0(d + 25) \\ &= 2.0(250 + 25) \\ &= 550 \text{ mm} \end{aligned}$$

Number of bolts, (Eqn. 6.9 MDHB)

$$n = d/50 + 3 = 250/50 + 3 = 8$$

Bolt diameter, $d_b = 0.5\,d/\sqrt{n}$

$$= 0.5 \times 250/\sqrt{8} = 44.19$$

The standard size is **M45 × 3.0**
Shearing stress in the bolt (Eqn. 6.11 MDHB)

$$\begin{aligned}\tau &= 8T_d/(\pi d_b^2 \times n\,D_b)\\ &= 8 \times 23.875 \times 10^7/(\pi \times 45^2 \times 8 \times 550)\\ &= 0.8 \text{ MPa} < (S_{sd} = 80 \text{ MPa}). \quad \therefore \quad \text{Safe}\end{aligned}$$

Bearing (crushing) stress on bolt is the flange.

$$\begin{aligned}\sigma_c &= 2T_d/(n\,d_b t\,D_b)\\ &= (2 \times 23.875 \times 10^7)/(8 \times 45 \times 125 \times 550)\\ &= 19.3 \text{ MPa} \le (S_{cd} = 160 \text{ MPa}). \quad \therefore \quad \text{Safe}\end{aligned}$$

Outer diameter of the flange,

$$D = 2.5\,d + 75 = 2.5 \times 250 + 75 = 700 \text{ mm}$$

Flange width, $t_p = 0.25 \times d = 0.25 \times 250 = 625$ mm
Rim flange thickness $= t_p/4 = 15$ mm

Example 8.6: Design a bush type flexible flange coupling to transmit 25 kW at 250 rpm. The design stress are:
For shaft in shear, $S_{sd} = 60$ MPa
For key in shear, $S_{sd} = 50$ MPa in crushing, $S_{cd} = 120$ MPa
For bolt in shear, $S_{sd} = 60$ MPa, bending, $S_d = 120$ MPa
Bearing pressure for bush, $P_b = 3.0$ MPa

SOLUTION
For diagram and standard proportions refer Fig. 6.4a in MDHB.
Torque to be transmitted,

$$\begin{aligned}T &= \frac{955 \times 10^4 \text{ kW}}{\text{N}} = \frac{955 \times 10^4 \times 25}{250}\\ &= 955 \times 10^3 \text{ N-mm}\end{aligned}$$

Assume a load factor, $K_L = 1.2$

$$\therefore \text{ Design torque, } T_d = 1.2 \times 955 \times 10^3$$
$$= 1146 \times 10^3 \text{ N-mm}$$

Shaft diameter for torsional strength (Eqn. 4.5 MDHB)

$$d = \left[\frac{16T}{\pi S_{sd}}\right]^{1/3} \quad \text{with } K = 0 \text{ for solid shaft}$$
$$= \left[\frac{16 \times 1146 \times 10^3}{\pi \times 60}\right]^{1/3} = 45.9 \text{ mm}$$

Choose $d = 50.0$ mm

Hub: Hub diameter, $d_1 = 2\,d = 100$ mm
Length of the hub, $L_1 = 1.75\,d = 87.5$ mm
Choose $L_1 = 90.0$ mm
Check for torsional stress in hub (Eqn. 4.5 MDHB)

$$\tau = \left(\frac{16T}{\pi d_1^3}\right)\left(\frac{1}{1-K^4}\right)$$

where, $T = 1146 \times 10^3$ N-mm

$d_o = 100.0$ mm

$K = 50/100 = 0.5$

$$\tau = \left(\frac{16 \times 1146 \times 10^3}{\pi \times 100^3}\right) \times \left(\frac{1}{1-0.50^4}\right)$$

$= 6.225$ MPa. This is less than the permissible and safe.

Flange thickness, $t = 0.5\,d = 25$ mm
Shearing stress in the flange (Eqn. 6.13 MDHB)

$$\tau = 2T_d/\pi d_1^2 t = 2 \times 1146 \times 10^3/(\pi \times 100^2 \times 25)$$
$$= 2.9 \text{ MPa. Safe}$$

Key: The rectangular key be used. The standard dimensions from Table 5.1 MDHB are $b = 14$ mm, $h = 9$ mm
The length of the key by shear strength (Eqn. 5.7 MDHB)

$$L = 2T_d/(S_{sd}.b.d)$$
$$= 2 \times 1146 \times 10^3/(50 \times 14 \times 50) = 65.5 \text{ mm}$$

Choose $L = 70.0$ mm.

Then the crushing stress (Eqn. 5.6 MDHB)

$$\sigma = 4T/h.d.L. = (4 \times 1146 \times 10^3)/(9 \times 50 \times 70)$$
$$= 145.5 \text{ MPa} > S_{cd} = 120 \text{ MPa}. \quad \therefore \quad \text{Unsafe}$$

Choose length of the key $L = 90.0$ mm

Then crushing stress,

$$\sigma = 4 \times 1146 \times 10^3/(9 \times 50 \times 90)$$
$$= 1132 \text{ MPa} < (S_{cd} = 120.0 \text{ MPa}). \quad \therefore \quad \text{Safe}$$

Bolts:

The bolt circle diameter,

$D_b = 3.5\ d = 3.5 \times 50 = 175.0$ mm

Number of bolts, $n = d/50 + 3 = 50/50 + 3 = 4.$

Diameter of the bolt, $d_b = 0.5\ d/\sqrt{n}$

$= 0.5 \times 50/\sqrt{4}$

$= 12.5$ mm

Choose bolt size **M14** $\times$ **1.5** (Table 9.7 MDHB)

Diameter of the enlarged pin portion, $d_p = 1.5\ d_b = 1.5 \times 14 = 21.00$ mm

Diameter of rubber bush,

$$d_r = d_p + 15 \text{ mm} = 21 + 15 = 36.0 \text{ mm}$$

Length of the bush, $l_r = 1.5\ d_r = 54.0$ mm

Torque capacity of bolts for shear (Eqn. 6.11 MDHB)

$$T = \pi d_b^2 n \cdot D_b \cdot S_{sd}/8$$
$$= \pi \times 14^2 \times 4 \times 175 \times 60/8$$
$$= 3232698 = 3232.698 \times 10^3 \text{ Nm} > T_d. \quad \therefore \quad \text{Safe}$$

Torque capacity for bearing in flange (Eqn. 6.12 MDHB)

$$T = n \cdot d_b \cdot t\ D_b \cdot S_{cd}/2$$
$$= 4 \times 14 \times 25 \times 175 \times 120/2$$
$$= 147 \times 10^5 \text{ N-mm} > T_d. \quad \therefore \quad \text{Safe}$$

Crushing strength of rubber bush (Eqn. 6.14 MDHB)

$$T = nd_r \cdot L_r \cdot D_b \cdot p/2$$
$$= 4 \times 36 \times 54 \times 175 \times 3/2$$
$$= 2041200 \text{ N-mm} > T_d. \quad \therefore \quad \text{Safe}$$

Bending stress in the bolt (Eqn. 6.15 MDHB)

$$\sigma = \frac{32\, T_d \cdot L_r}{\pi \cdot n \cdot D_b \cdot d_b^3}$$

$$= \frac{32 \times 1146 \times 10^3 \times 54}{\pi \times 4 \times 175 \times 21^3} = 97.23 \text{ MPa} < S_d = 120. \quad \therefore \quad \text{Safe}$$

It is within the design stress,

$$S_d = 120 \text{ MPa}$$

Outer diameter of the flange,

$$D = 5d = 5 \times 50 = 250 \text{ mm}$$

Minimum required, $D_{min} = D_b + d_r + \text{clearance}$

$$= 175 + 36 + 25$$
$$= 236.0 \text{ mm}$$

$\therefore$ $D = 250$ mm is acceptable.

Example 8.7: Design a bushed pin type flexible coupling for connecting a motor shaft to a pump shaft for the following service conditions:

Power to be transmitted	= 40 kW
Speed of the motor	= 1000 rpm
Diameter of motor shaft	= 50 mm
Diameter of pump shaft	= 45.0 mm
Bearing pressure in the bush	= 0.45 MPa
Allowable shear stress in the pin	= 25 MPa

SOLUTION

Torque $= 955 \times 10^4 \times 40/1000 = 382000$ N-mm

Choose design stress for shaft and bolt

$S_{sd} = 40$ MPa, $S_d = 60$ MPa, $S_{cd} = 100$ MPa

Choose design stress for key as

$S_{sd} = 30$ MPa, $S_{cd} = 100$ MPa

Hub Design

The proportional dimensions based on the motor shaft diameter.

Hub diameter $d_1 = 1.75\, d = 1.75 \times 50 = 87.5$ mm

Choose $d_1 = 90.0$ mm

Length of the hub, $L_1 = 1.75\, d = 1.85 \times 50 = 87.5$ mm

Choose $L_1 = 90.0$ mm

Shear stress in the hub for $K = \frac{50}{90} = 0.55$,

$$\tau = \left(\frac{16T}{\pi . d_1^3}\right)\left(\frac{1}{1-K^4}\right) = \left(\frac{16 \times 382 \times 10^3}{\pi \times 90^3}\right) \times \left(\frac{1}{1-0.55^4}\right) = 2.07 \text{ MPa} \quad \therefore \quad \text{Safe}$$

Design of Key

Dimensions of the key (Table 5.1 MDHB)

$b = 14$ mm; $h = 9$ mm; $L = 36$ - 180 mm

By shear strength consideration, the length of the key for

$S_{sd} = 30.0$ MPa (Eqn. 5.7 MDHB)

$$\begin{aligned} L &= 2T/S_{sd} . b.d \\ &= 2 \times 382 \times 10^3/(30 \times 14 \times 45) \text{ (consider smaller diameter)} \\ &= 40.0 \text{ mm} \end{aligned}$$

By the crushing strength consideration, the length of the key for $S_{cd} = 100.0$ MPa (Eqn. 5.6 MDHB)

$$\begin{aligned} L &= 4T/(S_{cd}\; h.d) \\ &= 4 \times 382 \times 10^3/(100 \times 9 \times 45) \\ &= 37.7 \text{ mm} \end{aligned}$$

Choose larger of the two, i.e., L = 40 mm

The length required is less than the hub length, $L_1 = 90$ mm. Hence, the hub length may be reduced to

$$L_1 = 50.0 \text{ mm}$$

Design of flange

Flange thickness,

$$t = 0.5\, d = 0.5 \times 50 = 25 \text{ mm}$$

Then the shear stress in the flange (Eqn. 6.13 MDHB)

$$[\tau = 2\,Td/\pi d_1^2 \cdot t]$$

$$= 2 \times 382 \times 10^3/(\pi \times 90^2 \times 25)$$

$= 1.2$ MPa. This is very small, hence safe.

Design of bolts

Number of bolts, $n = d/50 + 3 = 50/50 + 3.0 = 4.0$

Bolt diameter, $d_b = 0.5d/\sqrt{n} = 0.5 \times 50/\sqrt{4}$

$$= 12.5 \text{ mm}$$

Choose the standard size **M 14 × 2.0** (Table 9.7 MDHB)

Bolt circle diameter, $D_b = 3.5 \times 50 = 175.0$ mm

Torque by shearing strength of the bolt for $S_{sd} = 40.0$ MPa (Eqn. 6.11 MDHB)

$$T = \pi d_b^2 \cdot n \cdot D_b \cdot S_{sd}/8$$
$$= \pi \times 14^2 \times 4 \times 175 \times 40/8$$
$$= 2155132 \text{ N-mm} > (T_d = 382 \times 10^3)$$

Hence, safe.

Torque capacity based on crushing stress, $S_{cd} = 100$ MPa (Eqn. 6.12 MDHB)

$$T = n \cdot d_b \cdot tD_b \cdot S_{cd}/2$$
$$= 4 \times 14 \times 25 \times 175 \times 100/2$$
$$= 1{,}22{,}50{,}000 \text{ N-mm}, > T_d, \quad \therefore \text{Safe}$$

Enlarged diameter, $d_p = 1.5\,d = 1.5 \times 14 = 21.0$ mm

Diameter of the rubber bush,

$$d_r = d_p + 15 = 21 + 15 = 36 \text{ mm}$$

Choose $d_r = 40.0$ mm

Length of the rubber bush,

$$L_r = 1.5\,d_r = 1.5 \times 40 = 60.0 \text{ mm}$$

Choose $L_r = 60.0$ mm

Bearing pressure on the bush (Eqn. 6.14 MDHB)

$$p = 2T/nd_r \cdot L_r \cdot D_b$$
$$= 2 \times 382 \times 10^3/(4 \times 40 \times 60 \times 175)$$
$$= 0.455 \text{ MPa}$$

This is equal to the permissible value of $p = 0.45$ MPa. Hence, safe.
Bending stress in the bolt (Eqn. 6.15 MDHB)

$$\sigma = \frac{32T_d \cdot L_r}{\pi n D_b \cdot d_p^3}$$
$$= \frac{32 \times 382 \times 10^3 \times 60}{\pi \times 4 \times 175 \times 21^3} = 36.0 \text{ MPa}$$

This is less than the permissible stress $S_d = 80.0$ MPa. Hence, safe.
The outside diameter of the flange,

$$D = 5d$$
$$= 5 \times 50 = 250 \text{ mm}$$

This should be greater than $D_{min} = D_b + d_r + \text{clearance}$

$$= 175 + 40 + 20 = 235 \text{ mm}$$

Hence, $D = 250.0$ mm is OK.

Example 8.8: Design a flexible flange coupling for a diesel engine, developing 75 kW at 500 rpm. The permissible stresses are as follows:

For the shaft material in shear, $S_{sd} = 50$ MPa

For the cast iron flange in shear, $S_{sd} = 15$ MPa

For the bolt and key material(steel) in shear = 35 MPa

For the bolt and key in crushing = 65 MPa

Bearing pressure for the rubber bush = 0.5 MPa

Assume a load factor of 1.25.

SOLUTION

Shaft design

Average torque, T = 9550000 kW/N

$= 9550000 \times 75/500 = 143.25 \times 10^4$ N-mm

Design torque, $T_d = 1.25 \times 143.25 \times 10^4$ N-mm = 1790625 N-mm

Shaft diameter, $d = \sqrt[3]{16T_d/\pi S_{sd}} = \sqrt[3]{16 \times 1790625/(\pi \times 50)} = 56.6$ mm

Choose $d = 60$ mm

Key design

Choose a square key with size $b \times h = 15 \times 15$ mm

Length of the key for shear strength,

$$L = 2T_d/(b.d.S_{sd})\text{(Eqn. 5.7 MDHB)}$$

$$= 2 \times 1790625/(15 \times 60 \times 35) = 113.69 \text{ mm}$$

Choose $L = 120$ mm

Crushing stress $= 4T_d/(hdL)$

$$= 4 \times 1790625/(15 \times 60 \times 120) = 66.32 \text{ MPa}$$

This is slightly more than the permissible value of 65 MPa and OK.

Hub design

Hub diameter, $d_1 = 1.75\ d = 1.75 \times 60 = 105$ mm

Flange thickness, $t = 0.5\ d = 0.5 \times 60 = 30$ mm

Shearing stress in the flange,

$$\tau = (2T_d)/\left(\pi d_1^2 t\right)$$

$$= 2 \times 1790625/\left(\pi \times 105^2 \times 30\right) = 3.44 \text{ MPa} < \text{permissible. Hence, OK.}$$

Bolt design

Number of bolts, $n = d/50 + 3 = 60/50 + 3 = 4.2$

Choose $n = 6$

Bolt diameter, $d_b = 0.5/\sqrt{n} = 0.5 \times 60/\sqrt{6} = 12.24$ mm

Choose M14 × 1.5 mm bolts
Pitch circle diameter, $D_b = 3.5d = 3.5 \times 60 = 210$ mm
Torque capacity based on the shearing strength of bolts (Eqn. 6.11 MDHB)

$$T = \left(\pi d_b^2 \cdot n \cdot D_b S_{sd}\right)/8 = \left(\pi \times 14^2 \times 6 \times 210 \times 35\right)/8 = 3394,333.7 \text{ N-mm}$$

This is more than the design torque and hence, OK.
Torque capacity based on bearing pressure (Eqn. 6.12 MDHB)

$$T = nd_b \cdot t \cdot D_b \cdot S_{cd}/2 = 6 \times 14 \times 30 \times 210 \times 65/2 = 17199 \times 10^3 \text{ N-mm}$$

This is more than the design torque and hence, OK.
Enlarged diameter of the bolt, $d_p = 1.5d = 1.5 \times 14 = 21$ mm
Rubber bush, diameter $d_r = d_p + 15 = 21 + 15 = 36$ mm
Take $d_r = 40$ mm
Length of the bush, $L_r = 1.5\, d_r = 1.5 \times 40 = 60$ mm
Torque capacity based on the bearing pressure, $p = 0.5$ MPa

$$T = (n \cdot d_r \cdot L_r \cdot D_b \cdot p)/2 = (6 \times 40 \times 60 \times 210 \times 0.5)/2 = 756,000 \text{ N-mm}$$

This value is very much smaller than the design torque, hence modify the dimensions as follows.
Increase the pin diameter, $d_p = 2d_b = 2 \times 14 = 28$ mm
Rubber bush diameter, $d_r = 28 + 20 = 48$ mm, choose $d_r = 50$ mm
Length the bush, $L_r = 2d_r = 2 \times 50 = 100$ mm
P C D $D_b = 250$ mm

Torque capacity based on bearing pressure,

$$T = (n \cdot d_r \cdot L_r \cdot D_b \cdot p)/2 = (6 \times 50 \times 100 \times 250 \times 0.5)/2 = 1875000 \text{ N-mm}$$

The value is more than the design torque and hence the modified dimensions be accepted.
The outside diameter of the flange,

$$D_o = D_b + d_r + \text{thickness of the flange cover}$$
$$= 250 + 50 + 20 = 320 \text{ mm}$$

Example 8.9: Design a rigid flange coupling to transmit 18 kW at 1440 rpm. The allowable shear stress in the flange is 4 MPa. The shaft and the key are made of AISI 1040 annealed steel with ultimate strength and yield strength values as 518.8 MPa and 353.4 MPa respectively. Use ASME code to design the shaft and key.

SOLUTION

Power transmitted, $P = 18$ kW

Speed, $N = 1440$ rpm

For shaft and key material AISI 1040

$S_u = 518.8$ MPa and $S_y = 353.4$ MPa

Assume the factor of safety = 2.5

Design of shaft

$$\text{Design stresses in tension, } S_d = \frac{353.4}{2.5} \cong 140 \text{ MPa}$$

$$\text{Design stresses in shear, } S_{sd} = 0.5\,S_d \cong 70 \text{ MPa}$$

The torque transmitted,

$$T = \frac{955 \times 10^4 \times \text{kW}}{\text{N}} = \frac{955 \times 10^4 \times 18}{1440} = 119375 \text{ N-mm}$$

The ASME code eqn. (Eqn. 4.8 MDHB)

$$d = \left[\frac{16}{\pi S_{sd}}\left\{\sqrt{M_c^2 + T_c^2}\right\}\left(\frac{1}{1-K^4}\right)\right]^{1/3}$$

for $M_e = 0,\ K = 0$ and $C_t = 1$ assumed

$T_c = C_t T = 1.0 \times 119375$

$S_{sd} = 70$ MPa.

$$d = \left[\frac{16}{\pi \times 70}\{1.0 \times 119375\}\left(\frac{1}{1-0}\right)\right]^{1/3}$$

$$= 20.55 \text{ mm}$$

Allowing for key-way weakening effect choose, $d = 28$ mm.

Key design

Adopting parallel key the suitable standard size from Table 5.1 MDHB

$b = 8$ mm and $h = 7$ mm

Length of the key for shearing strength of the key (Eqn. 5.7 MDHB)

$$T = \frac{1}{2S_{sd}}\, b \cdot d \cdot L$$

$$\therefore\ L = \frac{2T}{b \cdot d \cdot S_{sd}} = \frac{2 \times 119375}{8 \times 28 \times 70} = 15.23 \text{ mm}$$

For crushing strength,

$$L = \frac{4T}{h \cdot d \cdot S_{cd}} = \frac{4 \times 119375}{7 \times 28 \times 140} = 17.40 \text{ mm}$$

Choose $L = 20$ mm.

Hub design

Hub diameter, $d_1 = 2d = 2 \times 28 = 56$ mm
Length of the hub, $L = 1.5 \times d = 1.5 \times 28 = 42.0$ mm
Check for shear stress in the hub

$$\tau = \frac{16.T}{\pi d_1^3(1-K^4)} = \frac{16 \times 119375}{\pi \times 56^3(1-0.5^4)}$$

$$= 3.69 \text{ MPa} < S_{sd} = 4 \text{ MPa for hub.}$$

Design is safe.
Flange design
Thickness of the flange, $t = 0.5 \times d = 14$ mm
Torque capacity for shearing stress in the flange,

$$T = \pi d_1 t\, S_{sd} \frac{d_1}{2}$$
$$= \pi \times 56 \times 15 \times 4 \times \frac{56}{2}$$
$$= 295561 \text{ N-mm} > T_d = 119375 \text{ N-mm}$$

Hence, flange is safe.
Outside diameter of the flange,

$$D = 2.5d + 75 = 2.5 \times 28 + 75 = 145 \text{ mm}$$

Choose $D = 150$ mm

Bolt design

Number of bolts, $n = \dfrac{d}{50} + 3 = \dfrac{28}{50} + 3 = 4$
Diameter of the bolt,

$$d_b = \frac{0.5d}{\sqrt{n}} = \frac{0.5 \times 28}{\sqrt{4}} = 7.0 \text{ mm}$$

Bolt circle diameter,

$$D_d = 2(d+25) = 2(28+25) = 106 \text{ mm}$$

Choose $D_d = 110$ mm
Torque capacity based on the shearing strength of the bolts,

$$T = \frac{\pi d_b^2\, n\, D_b\, S_{sd}}{8} \quad \text{(Eqn. 6.11 MDHB)}$$

$$= \frac{\pi \times 7^2 \times 4 \times 110 \times 70}{8} = 677327.0 \text{ N-mm} > T_d = 119375 \text{ N-mm}$$

Torque capacity based on the bearing strength of the bolts in the flange.

$$T = \frac{\pi d_b\, t.\pi D_b.S_{sd}}{2} \qquad \text{(Eqn. 6.12 MDHB)}$$

$$= \frac{\pi \times 7 \times 15 \times 110 \times 140}{2} = 2539977.7 \text{ N-mm} > T_d = 119375 \text{ N-mm}$$

Hence, the bolts are safe.

Example 8.10: A mild steel shaft has to transmit 40 kW power at 600 rpm. The maximum torque to be transmitted is 30% greater than the average torque. Design the rigid flange coupling for this application.

SOLUTION

Design torque

Power, $P = 40$ kW
Speed, $N = 600$ rpm
Load factor, $K_a = 1.3$

Average torque, $T_a = \dfrac{955 \times 10^4 \times 40}{600} = 636666.7$ N-mm

Design torque,

$$T_d = \mathrm{T}_a \cdot \mathrm{K}_a = 1.3 \times 636666.7$$
$$= 827666.7 \text{ N-mm}$$

Design stresses

The material properties are not given hence assume the following.
Material for shaft key and bolt –C40 steel with $S_u = 580$ MPa $S_y = 330$ MPa (Table 2.6 MDHB), factor of safety, $n = 2.5$
Design stresses are

$$\text{Tension, } S_d = S_y/n = 330/2.5 = 132 \text{ MPa}$$
$$\text{Shear, } S_{sd} = S_d/2 = 132/2 = 66 \text{ MPa}$$
$$\text{Crushing, } S_{cd} = 1.2 \times S_d = 1.2 \times 132 = 158.4 \text{ MPa}$$

Shaft design

Shaft diameter (Eqn. 4.5 MDHB)

$$d = \left[\frac{16.T_d}{\pi S_{sd}}\right]^{1/3} = \left[\frac{16 \times 827666.7}{\pi \times 66}\right]^{1/3} = 39.97 \text{ mm}$$

Accounting for keyway weakening effort adopt $d = 45$ mm

Key design

Use parallel key, the key size from (Table 5.1 MDHB) is $b = 14$ mm and $h = 9$ mm
Length of the key for shearing strength (Eqn. 5.7 MDHB)

$$L = \frac{2T}{bd.S_{sd}} = \frac{2 \times 827666.7}{14 \times 45 \times 66} = 39.81 \text{ mm}$$

Length of the key for crushing strength (Eqn. 5.6 MDHB)

$$L = \frac{4T}{hd.S_{sd}} = \frac{4 \times 827666.7}{9 \times 45 \times 15834} = 51.606 \text{ mm}$$

The larger one is to be used
$\therefore$ Choose $L = 52$ mm

Flange design

Assume allowable shear stress, $S_{sd} = 10$ MPa
Hub diameter, $d_1 = 1.75\,d = 1.75 \times 45 = 78.75$
Adopt $d_1 = 80$ mm
Length of the hub, $L = 1.5\,d = 1.5 \times 45 = 67.5$
Adopt $L = 65$ mm
Bolt circle diameter, $D_b = 2(d + 25) = 2(45 + 25) = 140$ mm
Adopt $D_b = 140$ mm
Outside diameter, $D = 2.5\,d + 75 = 2.5 \times 45 + 75 = 187.5$ mm
Adopt $D = 190$ mm
Flange thickness, $t = 0.5\,d = 0.5 \times 45 = 225$
Adopt $t = 20$ mm

Torque capacity for shearing strength of the flange,

$$T_d = \frac{\pi d_1^2\, t.S_{sd}}{2} \text{ (Assume } S_{sd} = 10 \text{ MPa)}$$

$$= \frac{\pi \times 80^2 \times 20 \times 10}{2} = 2010619.3 \text{ N-mm}$$

$$= 2010619.3 \text{ N-mm} > (T_d = 827666.7) \quad \therefore \text{Safe}$$

Check for shear stress in the hub,

$$\tau = \frac{16T}{\pi d_1^3(1 - K^4)} = \frac{16 \times 827666.7}{\pi \times 80^3(1 - 0356244^4)} = 9.15 \text{ MPa} < 10 \text{ MPa (Safe).}$$

Bolt design

Number of bolts,

$$n = \frac{d}{50} + 3 = \frac{45}{50} + 3 = 4 \qquad \text{(Eqn. 6.9 MDHB)}$$

Diameter of the bolts,

$$d_b = \frac{0.5d}{\sqrt{n}} = \frac{0.5 \times 45}{\sqrt{4}} = 11.25 \text{ mm.}$$

Then adopt standard diameter, $d_b = 12.00$ mm
From table bolt size M12 × 1 (Table 9.7 MDHB)
Torque capacity based on shearing strength (Eqn. 6.11 MDHB)

$$T_d = \frac{\pi d_b^2\, n\, D_b\, S_{sd}}{8} = \frac{\pi \times 12^2 \times 4 \times 140 \times 66}{8}$$
$$= 2,090,880 \text{ N-mm} >> T_d = 827,666.7 \quad \therefore \text{Safe}$$

Torque capacity based on bearing pressure (crushing of bolt) in flange (Eqn. 6.12 MDHB)

$$T_d = \frac{n d_s t\, D_b . S_{sd}}{2} = \frac{4 \times 12 \times 20 \times 140 \times 158.4}{2}$$
$$= 21288960 \text{ N-mm} >> T_d = 827666.7 \text{ N-mm}$$

∴ Design is safe.

Example 8.11: It is required to design a rigid type flange coupling to connect two shafts. The input shaft transmits 37.5 kW at 180 rpm to the shaft through the coupling. The starting torque is 50% higher than the rated torque. Select material for flanges as cast iron FG-200 ($S_{ut} = 400$ MPa) with a factor of safety of 6. The material for shaft may be taken as carbon steel with $S_{yt} = 380$ MPa with factor of safety of 2.5. The material for key and bolt as carbon steel with $S_{yt} = 400$ MPa (in tension) and $S_{yc} = 600$ MPa (in compression) respectively and factor of safety of 2.5. Design the coupling and follow major dimensions.

SOLUTION

Given data

Power transmitted = 37.5 kW
Speed of the shaft = 180 rpm
Maximum torque, $T_{max} = 1.5\ T_a$ (rated torque)
Material for flange CI-FG200, $S_{ut} = 200$ MPa
Factor of safety, $n_1 = 6$
Material for the shaft *cs* with $S_{yt} = 380$ MPa

Factor of safety, $n_2 = 2.5$
Material for the key and bolt steel with $S_{yt} = 400$ MPa, $S_{yc} = 600$ MPa
Factor of safety, $n_3 = 2.5$

Design torque
The average torque transmitted,

$$T = \frac{955 \times 10^4 \times 37.5}{180} = 1989583.3 \text{ N-m}$$

Design torque,

$$T_d = K_a\, T = 1.5 \times 1989583.3 = 2984375 \text{ N-mm}$$

Shaft design
Design stress for the shaft,

$$S_{sd} = \frac{S_d}{2} = \frac{S_y}{2n_2} = \frac{380}{2 \times 2.5} = 76 \text{ MPa}$$

Shaft diameter, $d = \left[\frac{16\, T_d}{\pi S_{sd}}\right]^{1/3}$ (Eqn. 4.5)

$$d = \left[\frac{16 \times 2984375}{\pi \times 76}\right]^{1/3} = 58.479$$

Choose $d = 63$ mm. (Table 4.4 MDHB)

Key design
The parallel key is suitable for this shaft (Table 5.1 MDHB) with
$b = 18$ mm and $h = 11$ mm
Key material properties
Design stress in shear,

$$S_{sd} = \frac{S_d}{2} = \frac{S_y}{2n_3} = \frac{400}{2 \times 2.5} = 80 \text{ MPa}$$

Design stress in compression,

$$S_{cd} = \frac{S_{yc}}{n_3} = \frac{6000}{2.5} = 240 \text{ MPa}$$

Length of the key for shear strength (Eqn. 5.7 MDHB)

$$L = \frac{2T}{b \cdot d \cdot S_{sd}} = \frac{2 \times 2984375}{18 \times 63 \times 80} = 65.79 \text{ mm}$$

Key length for crushing strength (Eqn. 5.6 MDHB)

$$L = \frac{4T}{h \cdot d \cdot S_{sd}} = \frac{4 \times 2984375}{11 \times 63 \times 240} = 71.77 \text{ mm}$$

Choose the length, $L = 72$ mm

Flange design (Fig. 6.4 MDHB)

Hub diameter, $d_1 = 1.75d = 1.75 \times 63 = 110.25$

Choose $d_1 = 110$ mm

Hub length, $L_1 = 1.5d = 1.5 \times 63 \cong 95$ mm

Flange thickness, $t = 0.25\, d + 10 = 0.25 \times 63 + 10 = 25.75$ mm

Choose, $t = 25$ mm.

Torque capacity based on shearing strength of the flange (Eqn. 6.13 MDHB)

$$T_d = \frac{\pi d_1^2 t S_{sd}}{2}$$

For flange material design stress, $S_d = \frac{200}{6} = 33.3$ MPa

Design stress for shearing, $S_{sd} = \frac{S_d}{2}$ (assumed)

$= 33/2 = 16.7$ MPa

$$T_d = \frac{\pi \times 63^2 \times 25 \times 16.7}{2} = 5205799 \text{ N-mm}$$

$$5205799 \text{ N-mm} > (T_d = 2984375 \text{ N-mm}) \quad \therefore \text{Safe}$$

Check the shearing stress in the hub

$$\tau = \frac{16T}{\pi d_1^3 (1 - K^4)} = \frac{16 \times 2984375}{\pi \times 110^3 (1 - 0.57^4)} = 12.78 \text{ MPa}$$

$$12.78 \text{ MPa} < S_{sd} = 16.7 \text{ MPa} \quad \therefore \text{Safe}$$

Bolt design

The number of bolts, $n = \dfrac{d}{50} + 3 \quad = \dfrac{63}{50} + 3 \quad = 4.26$

Choose $n = 6$ ($\because$ n is 3, 4, 6, 8 in this series)

Bolt diameter, $d_b = \dfrac{0.5d'}{\sqrt{6}} = \dfrac{0.5 \times 63}{\sqrt{6}} = 12.86$

Choose bolt size as M14 × 1.5

Bolt circle diameter, $D_b = 2(d + 25) = 2 \times (63 + 25) \cong 176$ mm

$$\text{Choose } D_b = 180\,\text{mm}$$

Torque capacity based on shearing strength,

$$T_d = \frac{\pi\, d_b^2\, n\, D_b\, S_{sd}}{8} \qquad \text{(Eqn. 6.11 MDHB)}$$

where $S_{sd} = 80$ MPa

$$T_d = \frac{\pi \times 14^2 \times 6 \times 180 \times 80}{8} = 6650123.3 \text{ N-mm}$$

$$> (T_d = 2984375 \text{ N-mm}) \quad \therefore \text{Safe}$$

Torque capacity based crushing in flange,

$$T_d = \frac{n\, d_b t\, D_b\, S_{cd}}{2}$$

$$= \frac{6 \times 14 \times 25 \times 180 \times 240}{2}$$

$$= 45360000 \text{ N-mm} > T_d = 2984375 \text{ N-mm}$$

The bolt is strong in shear and crushing and the design is safe.

Example 8.12: Design a bushed pin type flexible coupling to connect a pump shaft to a motor shaft transmitting 32 kW at 960 rpm. The overall torque is 20% more than mean torque. The material properties are as follows:

(i) Allowable share and crushing stresses for shaft and key material is 40 MPa and 180 MPa respectively.

(ii) Allowable shear stress for cast iron is 15 MPa.

(iii) Allowable bearing pressure for rubber bush is 0.8 MPa.

(iv) Material of the pin is same as that of shaft and key.

SOLUTION

Given Data

Power transmitted = 32 kW

Speed of shaft, $N = 960$ rpm

Load factor, $K_L = 1.2$

Material for shaft, key and pin are same with allowable stress in shear $S_{sd} = 40$ MPa and allowable in crushing $S_{cd} = 180$ MPa.

Material for flange CI with $S_{sd} = 15$ MPa

Bearing pressure for bush, $P_b = 0.8$ MPa

[Refer Fig. 6.4a MDHB]

Design of shaft

Average torque (Eqn. 4.1)

$$T = \frac{955 \times 10^4 \times \text{kW}}{\text{N}} = \frac{955 \times 10^4 \times 32}{960} = 318333.3 \text{ N-mm}$$

Design torque,
$$T_d = K_L T = 1.2 \times 318333.3$$
$$= 382000 \text{ N-mm}$$
$$= 382 \times 10^3 \text{ N-mm}$$

Shaft diameter,

$$d = \left[\frac{16 T_d}{\pi S_{sd}}\right]^{1/3} = \left[\frac{16 \times 382 \times 10^3}{\pi \times 40}\right]^{1/3} = 36.5 \text{ mm}$$

Allowing for key-way weakening effect

Choose $d = 40$ mm (Table. 4.4 MDHB)

Key design

Standard key size (Table 5.1 MDHB)

$b = 12$ mm and $h = 8$ mm

Length of the key for shearing strength, (Eqn. 6.7 MDHB)

$$L = \frac{2\ T_d}{b.d.\ S_{sd}} = \frac{2 \times 382 \times 10^3}{12 \times 40 \times 40} = 39.8 \text{ mm}$$

Length of the key for crushing strength,

$$L = \frac{4 T_d}{hd\ S_{cd}} = \frac{4 \times 382 \times 10^3}{8 \times 40 \times 80} = 59.68 \text{ mm}$$

Choose the larger one

Let $L = 60$ mm

Flange design

Hub diameter, $d_1 = 1.75\,d$ (same as right flange coupling)

$= 1.75 \times 40$

$= 70$ mm

Length of the hub, $L = 1.5\,d$

$= 1.5 \times 40$

$= 60$ mm

Flange thickness, $t = 0.5\,d$

$= 0.5 \times 40$

$= 20$ mm

The torque capacity of the flange in shear,

$$T_d = \frac{\pi d_1^2 t s_{sd}}{2} = \frac{\pi \times 70^2 \times 20 \times 15}{2} = 2309070$$

$$= 2309.07 \times 10^3 \text{ N-m} > T_d = 382 \times 10^3 \quad \therefore \text{Safe}$$

Check for shearing stress in the hub,

$$\tau = \frac{16.T}{\pi d_1^3\,(1 - K^4)} = \frac{16 \times 38 \times 10^3}{\pi \times 70^3\,(1 - 0.57^4)} = 6.34 \text{ MPa} \quad \therefore \text{Safe}$$

Bolt circle diameter,

$$D_6 = 3.5\,d = 3.5 \times 40$$
$$= 140 \text{ mm}$$

Outside diameter of flange, $D = 5\,d = 5 \times 40 = 200$ mm

Design of the bolt (pin)

$$\text{Number of bolts, } n = \frac{d}{50} + 3 = \frac{40}{50} + 3 = 4$$

$$\text{Diameter of bolt, } d_b = \frac{0.5d}{\sqrt{n}} = \frac{0.5 \times 40}{\sqrt{4}} = 10 \text{ mm}$$

Choose bolt size M10 × 1.5

Torque capacity for shearing strength of bolt, (Eqn. 6.11 MDHB)

$$T_s = \frac{\pi\, d_b^2\, n D_b\, S_{sd}}{8} = \frac{\pi \times 10^2 \times 4 \times 140 \times 40}{8}$$

$$= 879{,}645.9 \text{ N-m} > \left(T_d = 382 \times 10^3\right) \quad \therefore \text{Safe.}$$

Torque capacity for crushing strength strength of bolt, (Eqn. 6.12 MDHB)

$$T_c = \frac{nd_b tD_b S_{cd}}{2} = \frac{4 \times 10 \times 20 \times 140 \times 80}{2}$$

$$= 4480 \times 10^3 > \left(T_d = 3882 \times 10^3\right) \therefore \text{Safe.}$$

Diameter of the pin,

$$d_p = 1.5\ d_b = 1.5 \times 10 = 15 \text{ mm}$$

Diameter of the rubber bush,

$$d_r = d_p + 15 = 15 + 15 = 30 \text{ mm}$$

Length of the rubber bush,

$$L_r = 1.5\ d_r = 1.5 \times 30 = 45 \text{ mm}$$

Torque capacity based bearing pressure on the bush, (Eqn. 6.14 MDHB)

$$T_b = \frac{nd_r L_r Dp.P}{2} = \frac{4 \times 30 \times 45 \times 140 \times 0.8}{2} \tag{8.14}$$

$$= 302.4 \times 10^3 \text{ N-mm} < T_d = 382 \times 10^3 \text{ Unsafe} \tag{8.15}$$

Increase bush diameter and let $d_r = 40$ mm
Then the torque capacity for bearing,

$$T_b = \frac{302.4 \times 10^3 \times 40}{30} = 403.2 \times 10^3 = \text{N-mm} > (T_d = 382 \times 10^3 \text{ N-mm}) \tag{8.16}$$

Check for bending stress in the bolt

$$\sigma = \frac{32 \times 382 \times 10^3 \times 45}{\pi \times 4 \times 140 \times 15^3} = 92.6 \text{ MPa} > (s_d = 80 \text{ MPa}). \therefore \text{Unsafe} \qquad \text{(Eqn. 6.15 MDHB)}$$

Increase the number of bolts

$$n = 6.$$

Then the bending stress,

$$\sigma = \frac{92.6 \times 4}{6} = 61.76 \text{ MPa} < s_d = 80 \text{ MPa} \quad \therefore \text{Safe}$$

EXERCISES FOR PRACTICE

8.1. Design a protective type rigid flange coupling for steel shaft transmitting 20 kW power at 250 rpm. Maximum torque is 30% greater than full load torque. Material properties are as follows:

(a) Allowable shear stress for shaft and key = 40 MPa

(b) Allowable shear stress for bolts = 30 MPa

(c) Allowable crushing stress for shaft and key = 80 MPa

(d) Allowable shear stress for flange = 14 MPa

(e) Allowable compressive stress for bolts = 60 MPa

(f) Take four bolts on P.C.D.

8.2. In a rigid flange shaft coupling having 37.5 mm bore, it is desired that torsional stress in the shaft will not exceed 25 N/mm^2. The outside diameter of the coupling limited by space is 200 mm. There are three 15 mm bolts on a bolt circle diameter of 140 mm. The radial flange thickness is 18 mm. Determine the following:

(a) The power that may be transmitted at 600 rpm.

(b) The shearing stress in the bolts.

(c) The bearing pressure on the bolts.

8.3. Design a cast iron protective type flange coupling to transmit 15 kW at 900 rpm. The maximum load is 35% more than mean value on shaft. The limiting stresses are 40 MPa in shear for shaft, bolt and key material; 80 MPa in crushing for bolt and key; and 8 MPa in shear for cast iron.

8.4. Design a bushed-pin type flexible coupling for connecting a motor shaft to a pump shaft, with the following service conditions:

(a) Power to be transmitted = 40 kW

(b) Speed of the motor shaft = 1000 rpm

(c) Diameter of motor and pump shafts = 45 mm

(d) Bearing pressure on the rubber bush = 0.7 MPa

(e) Allowable stress in the pins = 60 MPa

8.5. Design a bushed-pin type flexible coupling to transmit 25 kW at 900 rpm. Use the following stress values:

(a) Shear stress = 50 MPa (shaft and key), 30 MPa (pin)

(b) Crushing stress = 90 MPa (key)

(c) Bearing pressure = 0.45 MPa (rubber bush).

8.6. a) What is the function of a coupling? Name and describe the various types of couplings used in industry.
b) Two 50 mm shafts running at 350 rpm and transmitting a torque of 1200 N-m are connected by a flange coupling. The space limitations restrict the outside diameter of the coupling to 250 mm. Design the coupling and make the dimensioned sketch of the coupling.

8.7. Why are shaft couplings used in machineries? Explain briefly.

8.8. Design a bushed-pin type flexible coupling for connecting a motor shaft to a pump shaft for the following service conditions:

(a) Power to be transmitted = 30 kW

(b) rpm of motor shaft = 960

(c) Diameter of motor shaft = 50 mm

(d) Diameter of pump shaft = 45 mm

(e) The bearing pressure in the rubber bush = 0.45 MPa and allowable stress in the pins = 25 MPa

8.9. Design a C.I. flange coupling to connect shaft with 100 mm diameter running at 300 rpm and transmitting 5000 N-m of torque. The design stress in shear for shaft. bolt and key is 50 MPa and crushing stress is 120 MPa and shear stress for C.I. is 8 MPa.

8.10. a) Explain the effect of keyway cut in to the shaft.
b) Design a cast iron flange coupling to connect two shafts of 80 mm diameter. The shaft runs at 250 rpm and transmits a torque of 4300 N-m.

The following permissible stresses may be used

(i) Shear stress for shaft, and key material = 75 MPa

(ii) Crushing stress for bolt and key material = 150 MPa

(iii) Shear stress for cast iron = 80 MPa

SHORT ANSWER QUESTIONS

8.1. What is coupling?
8.2. What are the types of couplings?
8.3. Which are rigid couplings?
8.4. What are the disadvantages of rigid couplings?
8.5. What is the flexible coupling?
8.6. Which are the flexible couplings?

8.7. What are the advantages of flexible couplings over the rigid couplings?
8.8. When are the flexible couplings used?
8.9. What are the requirements of a good coupling?
8.10. What is sleeve (muff) coupling?
8.11. When is muff used?
8.12. What is clamp coupling?
8.13. When is clamp coupling used?
8.14. What are the advantages of clamp coupling over muff coupling?
8.15. What is rigid flange coupling?
8.16. What are the components of rigid flange coupling?
8.17. What are the possible failures in the rigid flange coupling?
8.18. What are the advantages of protected flange coupling?
8.19. What is the additional feature in bush type flexible flange coupling?
8.20. What are the advantages of the flexible flange coupling?
8.21. Why is there a clearance between the flanger in the flexible flange coupling?
8.22. Why is a portion of the bolt diameter enlarged?
8.23. What are the types of failures in the bolts?
8.24. How do you determine the number of bolts required?
8.25. How do you fix pitch circle diameter of the bolts?

CHAPTER 9

POWER SCREWS

9.1 THEORY OF POWER SCREWS

The power screw consists of a threaded spindle and a nut. The thread forms are square or Acme. They are used for lifting loads, for exerting pressure as in a press or clamp, for transmitting power and converting rotary motion into linear motion as in the lead screw of a lathe and linear activators.

Terminology used in power screw: Refer Fig. 9.1 (a).

(i) Major diameter (d_o)

(ii) Minor diameter (d_i)1

(iii) Mean diameter $(d_m) = \left(\dfrac{d_o + d_i}{2}\right)$

(iv) Pitch (p)

(v) Lead $(L) =$ pitch $\times$ number of starts (N)

(vi) Helix (lead) angle $(\alpha) = \tan^{-1}\left(\dfrac{L}{\pi d_m}\right)$

(vii) Coefficient of friction between nut and screw (f)

(viii) Friction angle $\phi = \tan^{-1}(f)$

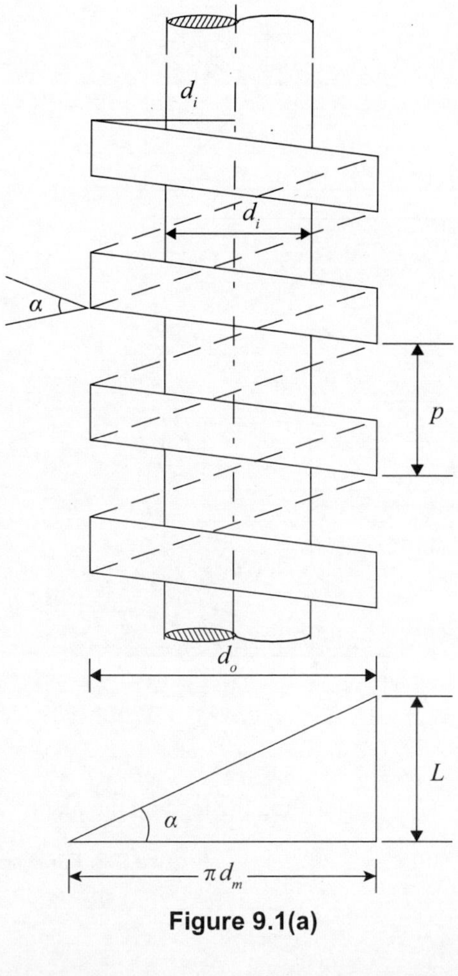

Figure 9.1(a)

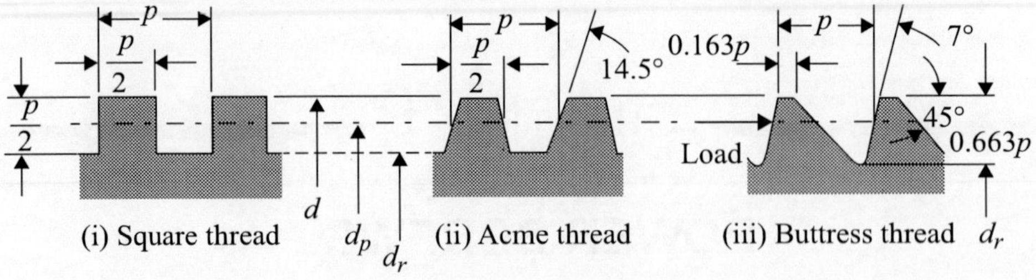

(b) Square, Acme, and Buttress threads

Figure 9.1(b)

Expression for the torque required to
(a) Raise the load: We consider a square threaded screw.

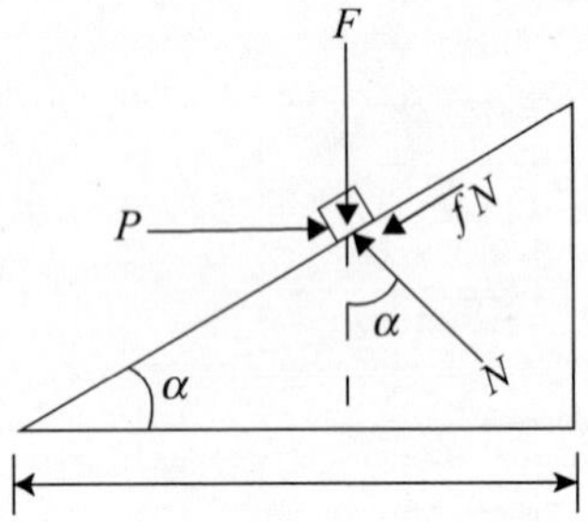

(a)

For the mechanics of operation, the thread is treated as an inclined plane with a horizontal force '*P*' applied at the mean diameter. For the equilibrium of forces; resolving the forces in the horizontal direction and taking $\sum F_H = 0$, we get

$$P = N(f\cos\alpha + \sin\alpha) \tag{9.1}$$

$N =$ normal force

$f =$ coefficient of friction

$\alpha =$ helix angle

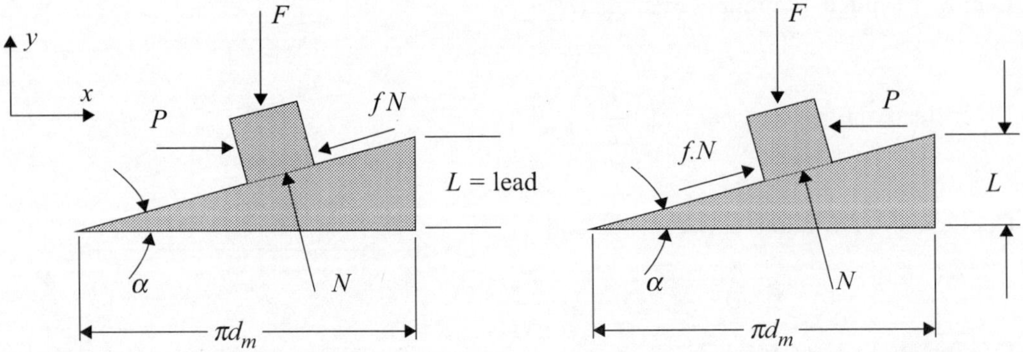

(b) Lifting the load up the plane (c) Lowering the load down the plane

Figure 9.2. Force analysis at the screw-nut interface

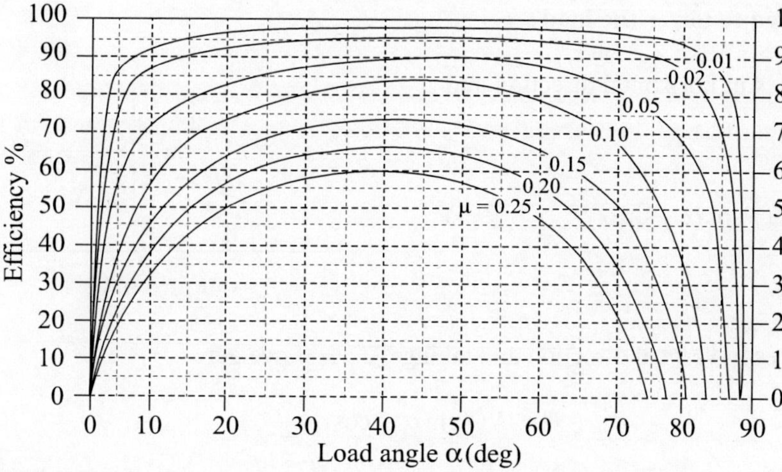

Figure 9.3. Efficiency for an acme-thread power screw (neglecting thrust-collar friction)

Resolving the forces in the vertical direction and taking $\sum F_v = 0$, we get

$$F = N(\cos\alpha - f\sin\alpha)$$

$$\therefore \quad N = \frac{F}{(\cos\alpha - f\sin\alpha)} \tag{9.2}$$

Substituting Eqn. (9.2) in Eqn. (9.1), we have

$$P = \frac{F}{(\cos\alpha - f\sin\alpha)}(\sin\alpha + f\cos\alpha)$$

$$= F\frac{(\tan\alpha + f)}{(1 - f\tan\alpha)}$$

$$= F\frac{(\tan\alpha + \tan\phi)}{(1 - \tan\phi\tan\alpha)} \quad [\because \quad f = \tan\phi]$$

Effort required, $P = F\tan(\phi + \alpha)$ (9.3)

Torque required $T = P\dfrac{d_m}{2}$

$$T = F\frac{d_m}{2}\tan(\phi + \alpha) \tag{9.4}$$

(b) Torque required to lower the load

Resolving the forces in horizontal direction and taking $\sum F_H = 0$,

$$P = N(f\cos\alpha - \sin\alpha) \tag{9.5}$$

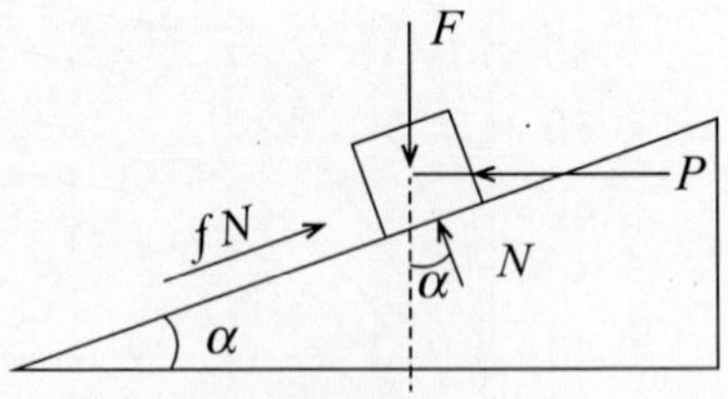

Figure 9.4

Resolving the forces in vertical direction and taking $\sum F_v = 0$, we get

$$F = N(f\sin\alpha + \cos\alpha)$$

$$N = \frac{F}{(f\sin\alpha + \cos\alpha)} \tag{9.6}$$

Substituting Eqn. 9.6 in Eqn. 9.5, we get

$$P = \frac{F}{(f\sin\alpha + \cos\alpha)}(f\cos\alpha - \sin\alpha)$$

$$= F\frac{(f - \tan\alpha)}{(f\tan\alpha + 1)} = F\frac{(\tan\phi - \tan\alpha)}{(\tan\phi\tan\alpha + 1)}$$

$$P = F\tan(\phi - \alpha)$$

$\therefore$ Effort required, $P = F\tan(\phi - \alpha)$ (9.7)

$$\text{Torque required, } T = P\frac{d_m}{2}$$

$$= F\frac{d_m}{2}\tan(\phi - \alpha) \tag{9.8}$$

From Eqn. 9.8, it can be seen that, if $\alpha > \phi$, then torque is negative. This means that the load will lower without the application of any effort. Such a screw is called overhauling screw. If $\alpha < \phi$, then torque is positive. This means that effort is required to lower the load. Such a screw is called self-locking screw.

Table 9.1. Safe bearing pressures in power screw

Service	Material		Safe bearing pressure (mpa)	Remarks
	Screw	Nut		
Hand press	Steel	Bronze	17.5 to 24.5	Low speed, well lubricated
.. Jack screw	Steel	C.I.	12.6 to 17.5	Low speed, not over 2.4 m/mm.
Jack screw	Steel	Bronze	11.2 to 17.5	Low speed, not over 3 m/min.
Hoisting screw	Steel	C.I.	4.2 to 7.1	Medium speed, (6 to 12 m/min)
Hoisting screw	Steel	Bronze	5.6 to 9.8	Medium speed (6 to 12 m/min)
Lead screw	Steel	Bronze	10.5 to 16.8	(High speed 15 m/min and over)
Lead screw	Steel	C.I.	3.0	Precision feed
			8	Normal accuracy
			12	General purpose

Table 9.2. Coefficient of friction for screws and nuts

Material of Screw	Lubrication	Material of nut		
		Steel	Brass	Cast iron
Steel-Milld	Dry	0.15-0.25	0.15-0.23	0.15-0.25
or case hardened	Machinery oil	0.11-0.17	0.10-0.16	0.11-0.17
or Bronze	Oil with graphite	0.08-0.12	0.04-0.06	0.06-0.09

Table 9.3. Coefficient of friction on thrust collar

Material	Coefficient of running friction	Coefficient of starting friction
Soft steel on cast iron	0.121	0.170
Hardened steel on C.I.	0.092	0.147
Soft steel on bronze	0.084	0.101
Hardened steel on bronze	0.063	0.081

For acme threads or trapezoidal threads,

$$F' = \frac{F}{\cos\theta}$$

where θ is the thread angle.

$$\text{Friction force} = f.F' = f\frac{F}{\cos\theta}$$

$$= \left(\frac{f}{\cos\theta}\right).F$$

$$= f'F$$

where, f' = Modified coefficient of friction

$f' = f\sec\theta$

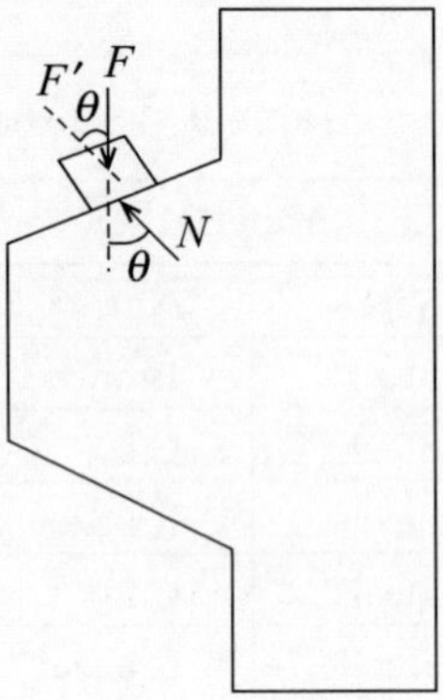

Figure 9.5

$$\therefore \quad \text{Friction angle, } \phi = \tan^{-1}(f')$$

$$= \tan^{-1}(f\sec\theta) \tag{9.9}$$

Hence, for power screw having acme threads or trapezoidal threads, Eqn. 9.4 and 9.8 can be used with the friction angle calculated using Eqn. 9.9.

9.2 EFFECT OF COLLAR FRICTION

Torque required to raise the load = Torque required to overcome friction in both screw and nut + Torque required to overcome friction at the collar

$$T = T_1 + T_c$$

$$= F\frac{d_m}{2}\tan(\theta+\alpha) + f_c F\frac{d_c}{2}.$$

where, f_c = Coefficient of collar friction

d_c = Mean collar diameter

$$d_c = \frac{d_{co}+d_{ci}}{2} \text{ for uniform wear theory}$$

$$d_c = \frac{2}{3}\frac{(d_{co}^3 - d_{ci}^3)}{(d_{co}^2 - d_{ci}^2)} \text{ for uniform pressure theory}$$

$$\text{Total torque, } T = F\frac{d_m}{2}\left[\tan(\phi+\alpha) + f_c.\frac{d_c}{d_m}\right] \tag{9.10}$$

Torque required to lower the load,

$$T = F\frac{d_m}{2}\tan(\phi-\alpha) + f_c \cdot F\frac{d_c}{2}$$

$$= F\frac{d_m}{2}\left[\tan(\phi-\alpha) + f_c\frac{d_c}{d_m}\right] \tag{9.11}$$

9.3 EFFICIENCY OF POWER SCREW

$$\text{Efficiency} = \frac{\text{Torque required without the friction}}{\text{Torque required with friction}}$$

$$\eta = \frac{F(d_m/2).\tan\alpha}{F(d_m/2)\tan(\phi+\alpha)}$$

$$\therefore \eta = \frac{\tan\alpha}{\tan(\phi+\alpha)} \tag{9.12}$$

$$\text{Collar friction, } \eta = \frac{\tan\alpha}{[\tan(\phi+\alpha)+f_c d_c/d_m]} \tag{9.12a}$$

9.4 STRESSES IN THE SCREW AND NUT

Stresses induced in the screw spindle:

(a) Compressive stress, $\sigma_c = \dfrac{4F}{\pi d_i^2}$ (9.13)

(b) Torsional shear stress, $\tau = \dfrac{16T}{\pi d_i^3}$. (9.14)

$$\text{Maximum principal stress, } \sigma_1 = \frac{\sigma_c}{2} + \sqrt{\left(\frac{\sigma_c}{2}\right)^2 + \tau^2}$$

$$\text{Minimum principal stress, } \sigma_2 = \frac{\sigma_c}{2} - \sqrt{\left(\frac{\sigma_c}{2}\right)^2 + \tau^2}$$

$$\text{Maximum shear stress, } \tau_{\max} = \sqrt{\left(\frac{\sigma_c}{2}\right)^2 + \tau^2}$$

Stresses in the nut

Let N be the number of threads, then

Bearing pressure, $p_b = \dfrac{4F}{\pi(d_0^2 - d_i^2)N}$ (9.15)

Shear stress in the threads of the nut,

$$\tau = \frac{F}{\pi d_o t N}; \quad \text{where } t = \text{thickness of the threads} \tag{9.16}$$

Shear stress in the thread of the bolt, $\tau = \dfrac{F}{\pi d_i t N}$. (9.17)

Height of nut, $H = N \cdot p$. (9.18)

9.5 DESIGN OF SCREW JACK

Stress area, $A_s = F/S_{cd}$
(Choose the screw dimensions from Tables 9.4 or 9.5 or 9.6 to give a stress area almost twice the value A_s subject to the condition)

$$\frac{L}{\pi d_m} \leq f$$

Check for stresses
The buckling stress,

$$\sigma_c = \frac{F}{A_s}\left[1 + \frac{a}{n}\left(\frac{L_1}{K}\right)^2\right] \quad (9.19)$$

where, L_1 = lift of screw jack
a (Table 1.1 MDHB)
n (Table 1.2 MDHB)
$K = d_c/4$

Shear stress, $\tau = \dfrac{16T}{\pi d_i^3}$
(T-from Eqns. 9.7 or 9.10)
The resultant principal stress,

$$\sigma_1 = \frac{\sigma_c}{2} + \sqrt{\left(\frac{\sigma_c}{2}\right)^2 + \tau^2} \leq S_d$$

Maximum shear stress,

$$\tau_{max} = \sqrt{\left(\frac{\sigma_c}{2}\right)^2 + \tau^2} \leq S_{sd}$$

Number of threads
for bearing,

$$N_1 = \frac{4F}{p_b \cdot \pi(d_0^2 - d_i^2)} \quad (9.20)$$

for shearing,

$$N_2 = \frac{2F}{\pi(d_i \cdot S_{sd} \cdot p)} \quad (9.21)$$

Height of nut, $H = N.p$
(N-from Eqns. 9.20 or 9.21 larger of the two)
Nut collar thickness,

$$t = \frac{F}{\pi D_1 S_s} \quad (9.22)$$

Check for crushing of collar

$$\sigma_c = \frac{4F}{\pi(D_2^2 - D_1^2)} \leq S_{cd} \quad (9.23)$$

Handle design
Length of the handle
$L_2 = T/P$ where T-from Eqn. 9.19
P-effort applied by an operator $\leq$ 400 N
Diameter of the handle,

$$d_1 = \sqrt[3]{\frac{32T}{\pi S_d}} \quad (9.24)$$

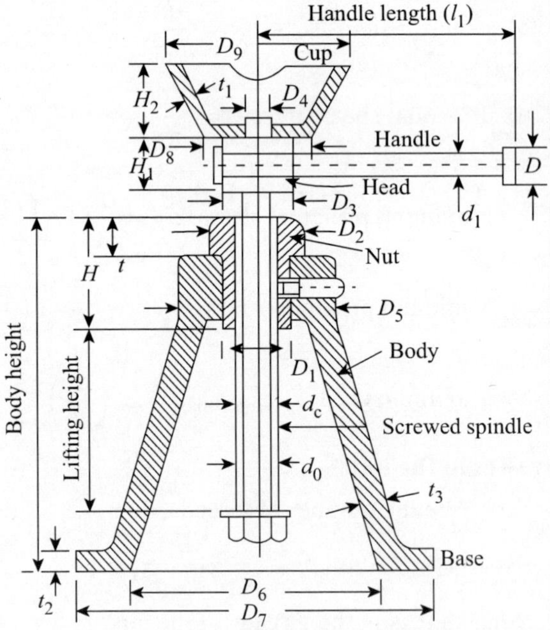

Screw jack

Figure 9.6

$D_1 = 1.7d_0$, $D_4 = D_3/4$, $D = 1.25d_1$, $H = pxn$,
$H_1 = 2D$, $D_5 = 1.5D_2$, $t_3 = 0.25d_0$, $D_6 = 2.25D_2$,
$D_7 = 1.75D_6$, $I_1 = T/P$, $t_1 = p/2$, $D_8 = 2d$,
$D_9 = 2.5d_0$, $H_2 = 1.2H$, $D_1 = 2d_c$ $t_2 = t_3$
$D_2 = 1.25D_1$

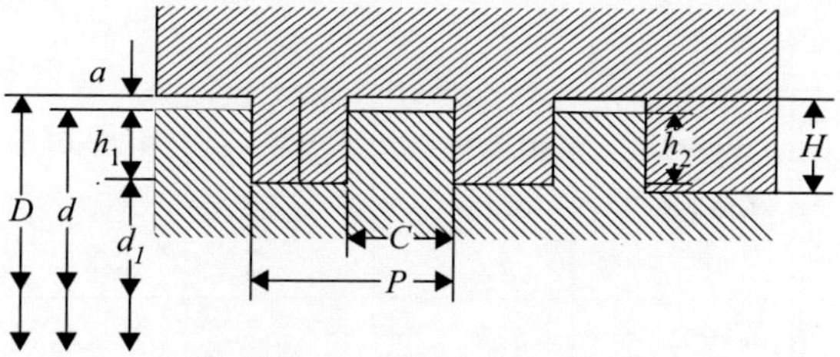

Figure 9.7. Basic square thread

Designation

SQ 30 × 6-IS-4694

SQ diameter in mm × pitch in mm

$h_1 = 0.5P \qquad C = 0.5P$

$H = 0.5P + a \quad D = d + 2a$

$h_2 = 0.5P - b \quad d_1 = d - 2h_1$

$$\text{Area of core} = \frac{\pi d_1^2}{4}$$

(1) Fine series = 10 to 640 mm $\phi - 2 - 24$ mm pitch
(2) Normal series = 22 to 300 mm $\phi - 5.26$ mm pitch
(3) Coarse series = 22 to 400 mm $\phi, 8 - 43$ mm pitch

Table 9.4. Basic dimensions for square threads: fine series (all dimensions in mm)

Normal dia., d	Minor dia., d_1	Pitch, p	Area of core, A_s mm^2	Normal dia., d	Minor dia., d_1	Pitch, p	Area of core, A_s mm^2
(1)	(2)	(3)	(4)	(1)	(2)	(3)	(4)
10	08		50.3	(115)	109		9331
12	10		78.5	120	114		10207
14	12	2	113	(125)	119	6	11122
16	14		154	130	124		12076
18	16		201	(135)	129		13070
20	18		254	140	134		14103
22	19		284	(145)	139		15175
24	21		346	150	144		16286
26	23	3	415	(155)	149	6	17437
28	25		491	160	154		18627
30	27		573	(165)	159		19856
32	29		661	170	164		21124
(34)	31		755	(175)	169		22432
36	33	3	855	180	172		23235
(38)	35		962	(185)	177		24606
40	37		1075	190	182	8	26016

(Contd.)

Table 9.4. (Contd.)

Normal dia., d	Minor dia., d_1	Pitch p	Area of core, A_s mm^2	Normal dia., d	Minor dia., d_1	Pitch, p	Area of core, A_s mm^2
				(195)	187		27465
42	39		1195	200	192		28953
44	41		1320	210	202		32047
(46)	43	3	1422	220	212	8	35299
48	45		1590	230	222		38708
50	47		1735	240	232		42272
52	49		1886	250	238		44488
55	52		2124	260	248		48305
(58)	55	3	2376	270	258	12	52279
60	57		2552	280	268		56410
(62)	59		2734	290	278		60699
65	61		2922	300	288		65144
(68)	64		3217	320	308		74506
70	66	4	3421	340	328		84496
(72)	68		3632	360	348	12	95115
75	71		3959	380	368		106362
(78)	74		4301	400	388		118237
80	76		4538	420	402		126923
(82)	78	4	4778	440	422		139867
(85)	81		5153	460	442	18	153439
(88)	84		5542	480	462		167639
				500	482		182467
90	86		5809	520	496		193221
(92)	88		6082	540	516		209117
95	91	4	6504	560	536		225642
(98)	94		6940	580	556	24	242795
100	96		7288	600	576	24	260576
(105)	101		8018	620	596		278986
110	106		8825	640	616		298024

Table 9.5. Basic dimensions for square threads: normal series (all dimensions in mm)

Normal dia d	Minor dia d_1	Pitch p	Area of core A_s mm^2	Normal dia d	Minor dia d_1	Pitch p	Area of core A_s mm^2
(1)	(2)	(3)	(4)	(1)	(2)	(3)	(4)
				95	83	12	5411
22	17		227	(98)	86		5809
24	19	5	284	(98)	86		5809
26	21		346	(105)	93		6793
28	23		415	110	98		7543
30	24		452	(115)	101		8012
32	26	6	531	120	106		8825
(34)	28		616	125	111		9677
36	30		707	130	116	14	10568
(38)	31		755	(135)	121		11499
40	33	7	855	140	126		12499
(42)	35		962	145	131		13478
44	37		1075	150	134		14103
(46)	38		1134	(155)	139		15175
48	40	8	1267	160	144	16	16286
50	42		1385	(165)	149		17437
52	44		1521	170	154		18627
55	46	9	1662	(175)	159		19856
(58)	49		1886	180	162		20612
60	51		2043	(185)	167		21904
(62)	53		2206	190	172	18	23235
				(195)	177		24606
65	55		2376	200	182		26016
(68)	58		2642	210	190		28353
70	60		2829	220	200	20	31416
(72)	62	10	3019	230	210		34636

(Contd.)

Table 9.5. (Contd.)

Normal dia d	Minor dia d_1	Pitch p	Area of core A_s mm^2	Normal dia d	Minor dia d_1	Pitch p	Area of core A_s mm^2
(1)	(2)	(3)	(4)	(1)	(2)	(3)	(4)
75	65		3318	240	218		37325
(78)	68		3632	250	228	22	40828
80	70		3848	260	238		44488
(82)	72		4072				
85	73		4185	270	246		47529
(88)	76	12	5436	280	256	24	51472
90	78		4778	290	266		55572
(92)	80		5027	300	274	26	58965

Table 9.6. Basic dimensions for square threads: coarse series (all dimensions in mm)

(1)	(2)	(3)	(4)	(1)	(2)	(3)	(4)
22	14		164				
24	16		201	(115)	93		6793
26	18	8	254	120	98	22	7543
28	20		314	(125)	103		9161
30	20		314	130	108		9161
32	22		360	135	111		9577
(34)	24		452	140	116		10568
36	26	10	531	(145)	121	24	11499
(38)	28		616	150	126		12469
40	28		616	(155)	131		13478
(42)	30		707	160	132		13685
44	32		804	(165)	137		14741
(46)	34		908	170	142	28	15837
48	36	12	1018	(175)	147		16972
50	38		1134	180	152		18146

(Contd.)

Table 9.6. (Contd.)

(1)	(2)	(3)	(4)	(1)	(2)	(3)	(4)
52	40		1157	(185)	153		18385
55	41		1321	190	158		19607
(53)	44	14	1521	(195)	163	32	20867
60	46		1662	200	168		22157
(62)	48		1810	210	174		23779
65	49		1886	220	184	36	26590
(68)	52		2124	230	194		29659
70	54	16	2290	240	204		32685
(72)	56		2463				
75	59		2734	250	210		34636
(78)	62		3019	260	220		38013
80	64		3217	270	230	40	41548
(82)	66		3421	280	240		45239
85	67		3526	290	246		47529
(88)	70		3848	300	256	44	51772
90	72	18	4072	320	276		59828
(92)	74		4301	340	296		68813
95	77		4657				
(98)	80		5027				
100	80		5027	360	312		76454
(105)	85	20	5675	380	332	43	86570
110	90		6362	400	352		97314

SOLVED EXAMPLES

Example 9.1: Find the power required to drive a 40 mm power screw having double start square thread with a pitch of 6.0 mm. The frictional coefficients are 0.10 for the threads and collar. The frictional diameter of the collar is 80 mm, the nut is to rotate at a velocity of 48 mm / sec and move a load $F = 10$ kN.

SOLUTION

Step-1: Geometric relations

Lead, $L = 2p = 2 \times 6.0 = 12$ mm

Minor diameter, $d_i = d_o - p = 40 - 6.0 = 34.0$ mm

Mean diameter, $d_m = (34 + 40)/2 = 37.0$ mm or $d_m = d_o - p/2 = 40 - 3 = 37$ mm

$$\text{Helix angle, } \alpha = \tan^{-1}\left(\frac{L}{\pi d_m}\right) = \tan^{-1}\left(\frac{12}{\pi \times 37}\right) = 5.89^\circ$$

$$\text{Friction angle, } \phi = \tan^{-1}(f) = \tan^{-1}(0.1) = 5.7^\circ$$

Step-2: Torque to lift the load (Eqn. 9.10)

$$T = \left(\frac{F d_m}{2}\right)\left[\tan(\alpha + \phi) + f_c \frac{d_c}{d_m}\right]$$

$$T = \left(\frac{10 \times 10^3 \times 37}{2}\right)\left[\tan(5.87 + 5.7) + 0.1\left(\frac{80}{37}\right)\right]$$

$$= 77874.11 \text{ N-mm} = 77.874 \text{ N.m}$$

Step-3: Power requirement

$$\text{Nut speed} = 48 \text{mm/s}$$

$$\therefore \quad \text{Spindle rotation, } N_s = \frac{48}{12} = 4.0 \text{rps}$$

$$\text{Angular speed, } \omega = 2\pi \text{ Ns} = 2\pi \times 4$$

$$= 25.13 \text{ rad/sec.}$$

$\therefore$ Power required $= T.\omega.$ watts $= 77.874 \times 25.13$

$$= 1957.189 \text{ watts}$$

$$= 1.957 \text{ kW}$$

Step-4: Efficiency

T_o = Torque with zero friction

$$= \frac{F d_m}{2} \tan \alpha = \left(\frac{10 \times 10^3 \times 37}{2}\right) \tan(5.87)$$

$$= 1.9 \times 10^4 \text{ N-mm}$$

$$= 19.0 \text{ Nm}$$

$\therefore$ Efficiency of the screw

$$\eta = \frac{T_o}{T} = \frac{19.0}{77.874} = 0.244 = 24.4\%$$

Example 9.2: The lead screw of a lathe having square threads requires to exert an axial force of 15 kN to the carriage. The thrust is carried by the collar. The length of the screw is 1500 mm. The coefficients of friction at the threads and collar are 0.15 and 0.10 respectively. Assume a factor of safety = 6.0 and allowable bearing pressure for the threads = 4.0 MPa. Design the screw and nut. If the screw rotates at 30 rpm, find the power required to drive it.

SOLUTION

Step-1: Material selection and design stresses

Material for screw–hardened steel-C80 with $S_u = 750.0$ MPa;

$S_y = 420.0$ MPa (Table 2.6 MDHB)

For nut – bronze with $S_u = 595.0$ MPa (Table 2.8 MDHB)

Design stress for the screw

$S_d = 750/6 = 125.0$ MPa, $S_{sd} = 0.5\ S_d = 62.5$ MPa

for nut $S_d = 595/6 = 100$ MPa and $S_{sd} = 0.5\ S_d = 50.0$ MPa

Step-2: Design of the screw.

The core area of screw for compression or tension

$$A_c = F/S_d = 15 \times 10^3/125 = 120.0\ \text{mm}^2.$$

Choosing square thread of normal series, and making allowance for the additional torsional stress, choose **SQ 26 × 5.0** screw with $A_s = 346.0\ \text{mm}^2$ and $d_o = 26.0$ mm $d_i = 21mm$ and $p = 5.0$ mm. The mean diameter, $d_m = d_o - p/2 = 26 - 2.5 = 23.5$ mm. For single start thread, $L = p = 5.0$ mm

Step-3: Torque required

Helix angle, $\alpha = \tan^{-1}\left(\dfrac{L}{\pi d_m}\right) = \tan^{-1}\left(\dfrac{5.0}{\pi \times 23.5}\right) = 3.87^\circ$

Friction angle, $\phi = \tan^{-1}(0.15) = 8.53^\circ$

$\phi > \alpha \Longrightarrow$ This is self locking screw.

Assuming collar diameter, $d_o = 2d = 2 \times 26 = 52$ mm

The mean collar diameter, $d_c = \left(\dfrac{2 \times 26 + 26}{2}\right) = 39.0$ mm

Driving torque required,

$$T = \frac{Fd_m}{2}\left[\tan(\alpha + \phi) + f_c\left(\frac{d_e}{d_m}\right)\right]$$

$$= \left(\frac{15 \times 10^3 \times 23.5}{2}\right)\left[\tan(3.87 + 8.53) + 0.1 \times \left(\frac{39}{23.5}\right)\right]$$

$$= 68001.0\ \text{N-mm}$$

Step-4: Check for the stress induced
Shear stress in the screw,

$$\tau = \frac{16T}{\pi d_c^3} = \frac{16 \times 68001}{\pi \times 21^3} = 37.39 \text{ MPa}$$

Direct stress, $\sigma = F/A_s = 15 \times 10^3/346 = 43.35$ MPa
Then maximum principal stress

$$\sigma_{\max} = \frac{\sigma}{2} + \sqrt{\left(\frac{\sigma}{2}\right)^2 + \tau^2}$$
$$= \left(\frac{43.35}{2}\right) + \sqrt{\left(\frac{43.35}{2}\right)^2 + 37.39^2}$$
$$= 21.67 + \sqrt{469.86^2 + 1398.012}$$
$$= 21.67 + 43.21 = 64.89 \text{ MPa} < S_d (= 125 \text{ MPa})$$

Hence, it is safe. The chosen dimensions for the spindle are O.K.

Step-5: The check for column action

The critical load (Eqn. 1.39 MDHB)

$$F_{cr} = \frac{\sigma_c \cdot A_s}{\left[1 + \frac{a}{n}\left(\frac{L}{K}\right)\right]^2} \quad \text{where} \quad K - \text{Radius of gyration} = \frac{d_i}{4}$$

$$= \frac{750 \times 346}{\left[1 + \frac{1}{7500}\left(\frac{1500}{21/4}\right)^2\right]}$$

$$= 21835.4 \text{ N}$$

$$= 21.8354 \text{ kN} > (F = 15 \text{ kN}) \quad \text{Hence, safe.}$$

The design is acceptable.

Step-6: Design of nut
Number of threads on the nut for bearing pressure (Eqn. 9.24 MDHB)

$$n = 4F/\left[p_b \pi (d_o^2 - d_i^2)\right]$$
$$= 4 \times 15 \times 10^3/\left[4\pi(26^2 - 21^2)\right]$$
$$= 20.3$$

Number of threads in the nut for shearing strength (Eqn. 9.25 MDHB)

$$n = 2F/(\pi d_i S_{sd} p)$$
$$= 2 \times 15 \times 10^3/(\pi \times 21 \times 50 \times 5)$$
$$= 1.82$$

∴ The required no. of threads is the larger of two, i.e, **n** = **20**

∴ Height of the nut,

$$H = n.p = 20 \times 5.0 = \mathbf{100\ mm}$$

Step-7: Power requirement

Speed of the screw, $\omega = \dfrac{2\pi \times 30}{60} = \pi \text{ rad/sec.}$

Power to drive the screw, $P = T.\omega = (6800 \times \pi/1000)$ watts

$$= 213.6 \text{ watts} = 0.2136 \text{ kW.}$$

Example 9.3: Design a C-clamp for exerting force of $F = 7.5$ kN. The gap between ends in 200 mm and distance between the screw centre line and inner face is 150 mm. The materials used are 30C8 for screw and handle, 200 FG C.I. for the frame and pad; P-bronze for the nut, the coefficient of friction at the nut and collar are 0.1 and 0.12 respectively.

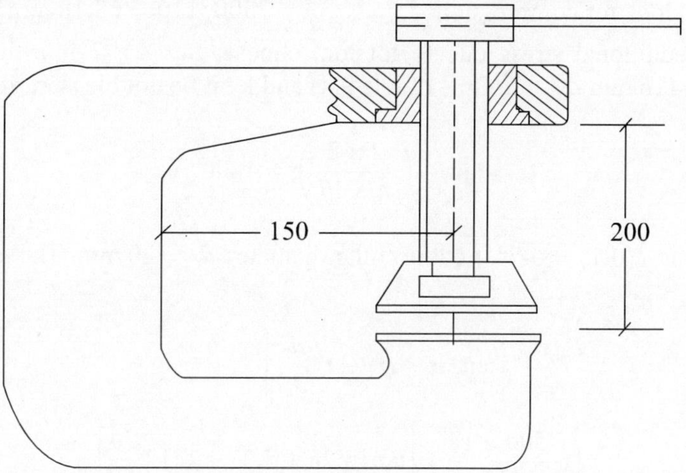

Figure 9.8

SOLUTION

Step-1: Material properties

for 30C8 (Table 2.6 MDHB for C30) $S_u = 550$ MPa, $S_y = 300$ MPa

FG 200 C.I. (Table 2.9 MDHB) $S_u = 200$ MPa, $S_{en} = 100$ MPa, $S_{uc} = 720$ MPa, $S_{su} = 230$ MPa

P bronze – (Table 2.12 MDHB), $S_y = 131$ MPa, $S_u = 324$ MPa

Step-2: Design stresses
Factor of safety based ultimate stress = 5.0 (Assumed)
∴ Design stress for spindle bar [30C8]

$$S_d = 550/5 = 110 \text{ MPa}$$
$$S_{sd} = 0.56 S_d = 60.0 \text{ MPa}$$

Design stress for the frame [FG 200 C1]

$$S_d = 200/5 = 40.0 \text{ MPa}$$
$$S_{cd} = 720/5 = 144.0 \text{ MPa}$$

Design stress for the nut [P-bronze]

$$S_d = 324/5 = 65.0 \text{ MPa}$$
$$S_{sd} = 32.5 \text{ MPa}$$

Step-3: Screw design

The core area of the screw for compression $A_c = F/S_d = 7500/110 = 68.18$ mm^2.
To account for the additional stress due to torsion, choose, SQ18 × 2 fine thread series with $A_s = 201$ mm^2 , ($d_i = 16$ mm $\alpha_p = 2$ mm) (Table 9.4) and let it be double start, then helix angle

$$\alpha = \tan^{-1}\left(\frac{2 \times 2}{\pi \times 17}\right) = 4.28°$$

Friction angle, $\phi = \tan^{-1}(0.1) = 5.7°$. Let the collar diameter, $d_c = 20$ mm. The torque required to exert a force of 7500 N is

$$T = \frac{F \cdot d_m}{2}\left[\tan(\alpha + \varphi) + f_c \frac{d_c}{d_m}\right] \qquad \text{(Eqn. 9.10)}$$

$$= \left(\frac{7500 \times 17}{2}\right)\left[\tan(4.28 + 5.7) + 0.12\frac{20}{17}\right]$$

$$= 20218.0 \text{ N-mm.}$$

Step-4: Check for the stresses
Shear stress in the screw,

$$\tau = \frac{16\,T}{\pi\, d_i^3} = \frac{16 \times 20218}{\pi \times 16^3} = 25.14 \text{ MPa}$$

Direct stress compressive stress,

$$\sigma = F/A_s = 7500/201 = 37.3 \text{ MPa}$$

Maximum shear stress,

$$\tau_{\max} = \sqrt{\left(\frac{\sigma}{2}\right)^2 + \tau^2} = \sqrt{\left(\frac{37.3}{2}\right)^2 + 25.13^2}$$
$$= 31.3 \text{ MPa} < (S_{sd} = 60.0 \text{ MPa})$$

Maximum principal stress,

$$\sigma_1 = \sigma/2 + \sqrt{(\sigma/2)^2 + \tau^2}$$
$$= \left(\frac{37.3}{2}\right) + \sqrt{\left(\frac{37.3}{2}\right)^2 + 25.13^2}$$
$$= 49.95 \text{ MPa} < (S_d = 110 \text{ MPa})$$

Both stresses are within the permissible values and hence safe.

Step-5: Nut Design

The number of threads N for bearing pressure, (Eqn. 9.15)

$$N = 4F/[p_b\pi(d_o^2 - d_i^2)]$$

with $p_b = 20.0$ MPa (Table 9.1)

$$N = 4 \times 7500/[20 \times \pi \times (18^2 - 16^2)]$$
$$= 7.02$$

Number of threads for shearing strength (Eqn. 9.16)

$$N = 2F/[\pi(d_i S_s p]$$
$$= 2 \times 7500/[\pi \times 16 \times 32.5 \times 2]$$
$$= 4.59$$

∴ Choose the large of the two **N = 7.0**.
Then height of the nut (Eqn. 9.18)

$$H = Nb = 7 \times 2 \cong \mathbf{15.0\ mm}$$

Step-6: The handle design bending moment on the handle

= Torque to be applied to the spindle = 20218 N-mm

Design stress for the handle, $S_d = 110$ MPa

$\therefore$ Section modules, $z = 20218/110 = 183.8$ mm^3

Diameter of the handle,

$$d = \sqrt[3]{\frac{32 \times 183.8}{\pi}} = 12.29 \text{ mm, choose } d = \mathbf{12.5\ mm}$$

Length of the handle, $l_2 = \dfrac{M}{P} = 20218/100 = 202.18$ mm $\quad (\therefore p = \text{effort} = 100N)$

Allowing for the grip, choose $l_2 =$ **300 mm**

Step-7: Frame design

Section of the frame. Section may be (i) rectangular (ii) trapezoidal (iii) T-section (iv) I-section. We select I-section with following proportions.

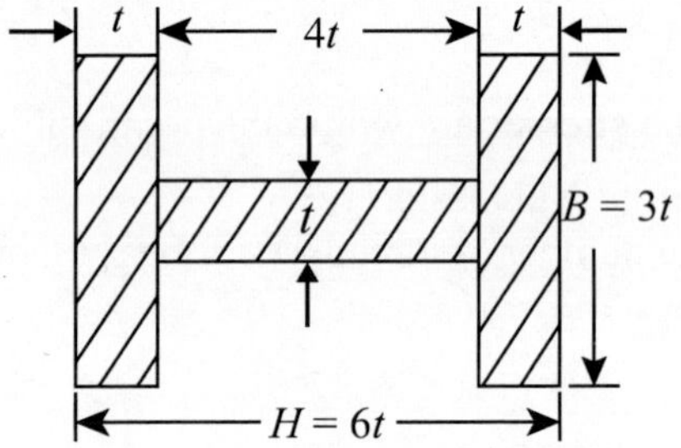

Figure 9.9

Area, $A = 2 \times t \times 3t + t \times 4t = 10t^2$

$$\text{Moment of Interia} = \frac{3t(6t)^3}{12} - \frac{2t(4t)^3}{12} \quad \frac{520t^4}{12}$$

$$= 43.3t^4$$

Section modulus, $Z = I/C = 43.33t^4/3t$

$$= (14.44t^3) \text{ mm}^3$$

Bending stress, $\quad \sigma_b = \dfrac{M}{Z} = \dfrac{7500(150+3t)}{14.44t^3}$

Direct stress, $\quad \sigma_1 = \dfrac{7500}{10t^2} = \dfrac{750}{t^2}$

$\therefore$ Total stress, $\quad \sigma = \sigma_b + \sigma_t \leq \sigma_d$

$\therefore$
$$\frac{7500\,(150+3t)}{14.44t^3} + \frac{750}{t^2} = 40.0$$

$\therefore$
$$40t^3 - 2308t - 77908.6 = 0$$

i.e., $t^3 - 57.7t - 1947.7 = 0$

By trial the solution is $t = 14$ mm

$\therefore$ Choose $t = 15$ mm

$\therefore$ Section is $B = 3t = 45$ mm, $H = 6t = 90$ mm

$$\text{Area, } A = 10 \times 15^2 = 2250\ \text{mm}^2$$

$$= 2250\ \text{mm}^2$$

$$\text{Section modulus, } Z = \left[\frac{45 \times 90^3}{12} - \frac{30 \times 60^3}{12}\right] / 45$$

$$= 48750\ \text{mm}^3$$

$$\therefore \quad \text{Direct stress, } \sigma_c = 7500/2250 = 3.33\ \text{MPa}$$

$$\text{Bending stress, } \sigma_b = \frac{7500 \times 195}{48750} = 30\ \text{MPa}$$

$$\therefore \quad \text{Total stress, } \sigma = (30 + 3.33)$$

$$= 33.33\ \text{MPa} < (S_d = 40\ \text{MPa}) \text{ and hence, safe.}$$

The I section is $(90 \times 45 \times 15$ mm)

The other dimensions are chosen proportionately.

Example 9.4: Design a screw jack for lifting a load of 100 kN through a height of 500 mm. The material for the screw is 30C8 steel, *P*-bronze for nut FG 200 for the body.

SOLUTION

Step-1: Material properties and design stress for factor of safety = 5

Screw 30C8, (Table 2.6 MDHB)

$$S_u = 550\ \text{MPa}, \ S_y = 300\ \text{MPa}$$
$$S_d = 110\ \text{MPa}$$

Nut P – bronze, (Table 2.12 MDHB)

$$S_u = 324\ \text{MPa}, \ S_d = 65.0\ \text{MPa}$$
$$S_{sd} = 32.5\ \text{MPa}$$

Body CI – FG 200, (Table 2.9 MDHB)

$$S_{ut} = 200\ \text{MPa}, S_{uc} = 720.0\ \text{MPa}, \ S_{su} = 230\ \text{MPa}$$

$$S_{cd} = 720/5 = 144.0\ \text{MPa}$$

Step-2: Design of screw
Screw Stress area required

$$A_s = F/S_d = 10 \times 10^3/110.0 = 90.9 \text{ mm}^2$$

Choose normal square thread, (Table 9.5)

$$\text{SQ } 22 \times 5 \text{ with } A_s = 227.0 \text{ mm}^2 \; d_o = 22 \text{ mm, and } d_i = 17 \text{ mm and } P = 5 \text{ mm}$$

Check for buckling stress

$$\sigma_c = \frac{F}{A_s}\left[1 + \frac{a}{n}\left(\frac{L}{k}\right)^2\right]$$

$$\sigma_c = \left(\frac{10 \times 10^3}{227}\right)\left[1 + \frac{1}{7500}\left(\frac{500}{17/4}\right)^2\right]$$

$$= 125.0 > S_d \,(= 110.0 \text{ MPa}) \text{ hence, unsafe.}$$

∴ Choose next size **SQ.24 × 5** with A_s = 284; mm^2; d_o = 24 mm; d_i = 19 mm; α_p = 5 mm; then buckling stress

$$\sigma_c = \left(\frac{10 \times 10^3}{284}\right)\left[1 + \frac{1}{7500}\left(\frac{500}{19/4}\right)^2\right]$$

$$= 87.23 \text{ MPa} < S_d = 110.0 \text{ MPa}$$

Hence, it is strong against buckling.
Step-3: Torque on the screw (Eqn. 9.4)

$$T = \frac{F \cdot d_m}{2}(\tan(\alpha + \phi))$$

Helix angle for $d_m = (24 + 19)/2 = 21.5$ mm

$$\alpha = \tan^{-1}\left(\frac{L_1}{\pi d_m}\right) = \tan^{-1}\left(\frac{5}{\pi \times 21.5}\right) = 4.2°$$

For coefficient of friction, $f = 0.1$, friction angle

$$\phi = \tan^{-1}(0.1) = 5.7°$$

$$\therefore \quad \text{Torque, } T = \left(\frac{10 \times 10^3 \times 21.5}{2}\right)[\tan(4.2 + 5.7)]$$

$$= 1.8782 \times 10^4 \text{ N-mm}$$

Step-4: Check for the stresses
Shear stress induced,

$$\tau = \frac{16 \times 1.8782 \times 10^4}{\pi \times (19)^3} = 13.9 \text{ MPa}$$

Principal stress,

$$\sigma_1 = \frac{\sigma_c}{2} + \sqrt{\left(\frac{\sigma_c}{2}\right)^2 + \tau^2}$$

$$\sigma_1 = \frac{87.23}{2} + \sqrt{\left(\frac{87.23}{2}\right)^2 + (13.9)^2}$$

$$= 43.615 + \sqrt{2095.5}$$

$$= 43.615 + 45.776 = 89.39 \text{ MPa}$$

This is less than the design stress ($S_d = 110$ MPa. Hence, it is safe.)
Step-5: Nut design
Number of threads in the nut
For bearing (Eqn. 9.15)

$$N = 4F \,/\, [p_b \pi \, (d_o^2 - d_i^2)]$$

where, $p_b = 15.0$ MPa (Table 12.1)
$d_o = 24$ mm and $d_i = 19$ mm
$N = (4 \times 10 \times 10^3) \,/\, [(15 \times \pi (24^2 - 19^2)]$

$$\mathbf{N = 3.95 \cong 4.0}$$

For shearing strength, (Eqn. 9.17)

$$N = F/(\pi d_i t s_{sd})$$
$$N = 2 \times (10 \times 10^3)/[\pi \times 19 \times 32.5 \times 5]$$
$$= 2.06$$
$$\therefore \quad \text{Choose} \quad n = 4.0$$

$\therefore$ Height of the nut, $H = N \cdot p = 4 \times 5 = 20$ mm
other dimensions of the nut,

$$D_1 = 2d_i = 2 \times 19 = 38.0 \text{ mm} \cong 40.0 \text{ mm}$$

Collar diameter, $D_2 = 1.25,\ D_1 = 50.0$ mm

Step-6: Dimensions of the screw head,

$$D_3 = 1.7d_o = 1.7 \times 24 = 40.8 \text{ mm}$$

Choose, $D_3 = 40.0$ mm
Pin diameter, $d_4 = D_2/4 = 40/4 = 10$ mm

Step-7: Torque for collar friction
For collar friction, equivalent collar diameter

$$d_i = \left(\frac{2}{3}\right)\left[\frac{40^3 + 10^3}{40^2 - 10^2}\right] = \left(\frac{2}{3}\right)\left[\frac{63000}{1500}\right]$$
$$= 28.0 \text{ mm}$$

Torque due to collar friction for $f_c = 0.15$

$$T_c = \frac{f_c F\, d_c}{2} = \frac{0.15 \times 10 \times 10^3 \times 28}{2}$$
$$= 2.1 \times 10^4 \text{ N-mm}$$

Step-8: Design of handle
Total torque to be exerted by the handle,

$$T = 1.8782 \times 10^4 + 2.1 \times 10^4 = 3.9782 \times 10^4 \text{ N-mm}$$

Then diameter of the handle,

$$d_i = \sqrt[3]{\frac{32T}{\pi S_d}} = \sqrt[3]{\frac{32 \times 3.9782 \times 10^4}{\pi \times 110}} = 15.4 \text{ mm}$$

$$\text{Choose} \quad d_i = 16 \text{ mm}$$

Height of the screw head, $H_1 = 2d_1 = 32.0$ mm

Choose $H_1 = 35.0$ mm
For handle effort, $P = 250$ N,
Length of the handle, $L_1 = T/P$

$$L_1 = 39782/250 = 159.0 \text{ mm}$$

Allowing for the grip, let $L_1 = 200.0$ mm
Step-9: Cup design
Dimensions of the cup,

$$D_8 = 2d_0 = 2 \times 24 = 48.0 \text{ mm}$$
$$= 50.0 \text{ mm}$$
$$D_9 = 2.5 \times d_0 = 2.5 \times 24 = 60.0 \text{ mm}$$

Thickness, $t_i = p/2 = 5.0/2 = 2.5$ mm

Compressive stress in the cup,

$$\sigma_c = 10 \times 10^3/(\pi \times 50 \times 2.5) = 25.5 \text{ MPa}$$

$\therefore$ Safe

$$\therefore \quad \text{Height of the cup} = H_2 = 1.2 \times H = 1.2 \times 35.0 = 42.0 \text{ mm}$$

Step-10: Body design
Dimensions of the body by proportions

$$D_5 = 1.5\,D_2 = 1.5 \times 50 = 75.0 \text{ mm}$$

$$D_6 = 2.25\,D_2 = 2.25 \times 50 = 112.5$$

$$D_7 = 1.75\,D_6 = 1.75 \times 112.5 = 196.8$$

$$\text{Thickness,} \quad t_3 = 0.25\,d_o = 0.25 \times 24 = 6.0 \text{ mm}$$

$$\text{Choose} \quad t_3 = 10 \text{ mm (min)}$$

Compressive stress in the body,

$$\sigma_c = (10 \times 10^3/\pi \times 40 \times 10) = 7.9 \text{ MPa}$$

It is very small and hence, safe.

Example 9.5: The lead screw of a lathe has trapezoidal threads. To drive the carriage the screw has to exert an axial force of 20 kN. The thrust is carried by the collar. The length of the lead-screw is 1.5 m. Coefficient of friction at the collar and nut are 0.1 and 0.15 respectively. Determine dimensions of the screw and the nut, if the permissible bearing pressure is 4 MPa.
Given Data

Axial load on the screw, $P = 20$ kN

Coefficient of friction for collar, $f_c = 0.1$; for nut $f = 0.15$

Safe bearing pressure, $p_b = 4$ MPa

Length of the screw, $L = 1.5$ m

SOLUTION
Step-1: Data to be selected or assumed

Material for the screw –M.S.-C40 (Table 9.2 MDHB)
Yield strength in tension, $S_y = 316$ MPa

Nut material – CI

From Table 2.4 MDHB

$$\text{CI class} - 20 - S_{ut} = 221 \text{ MPa}$$
$$S_{uc} = 752 \text{ MPa}$$

Assume factor of safety, $N = 2$

Design stress for screw, $S_d = 316/2 = 158.0$ MPa

For nut,

$$S_d = 220/2 = 110 \text{ MPa}$$
$$S_{cd} = 752/2 = 376.0 \text{ MPa}$$

Step-2: Design of screw

Considering the screw in simple compression, area required

$$P = A_s \cdot S_d$$

$$A_s = \frac{P}{S_d} = \frac{20 \times 10^3}{158} = 126.58 \text{ mm}^2$$

Required core diameter

$$d_c = \sqrt{\frac{4A_c}{\pi}} = 122.69 \text{ mm}$$

To allow for torison and bukling choose (Table 9.9 MDHB)

Choose $T_r 24 \times 5$ threads

Nominal diameter, $d = 24$ mm; pitch $p = 5$ mm

Core diameter, $d_i = 19$ mm and $A_s = 284.0 \text{ mm}^2$

Included angle $2\theta = 29°$

$\therefore \quad \theta = 14.5°$

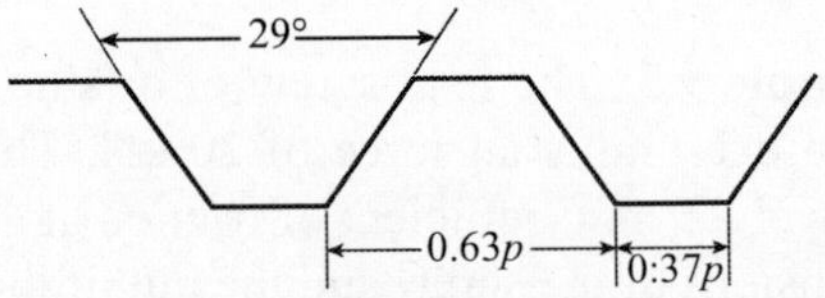

Figure 9.10

Step-3: Torque on the screw

$$T = \frac{Fd_m}{2}\left[\tan(\alpha+\phi)+fc\times\frac{d_c}{d_m}\right] - \qquad \text{(Eqn. 9.10)}$$

where, $F = 20$ kN

$$d_m = \frac{24+19}{2} = 21.5 \text{ mm}$$

$$\text{Helix angle, } \alpha = \tan^{-1}\left(\frac{p}{\pi d_m}\right)$$

$$\tan^{-1} = \left(\frac{5}{\pi \times 21.5}\right)$$

$$= 4.23^\circ$$

$$\text{Friction angle, } \quad \phi = \tan^{-1}(f \sec\theta) \qquad \text{(Eqn. 9.21)}$$

$$= \tan^{-1}\left(\frac{0.1}{\cos 14.5}\right)$$

$$= \mathbf{5.9^\circ}$$

$$\text{Collar diameter, } \quad d_c = 1.5\, d_m = 1.5 \times 21.5$$

$$= 32.25 \text{ mm}$$

Then substituting the values; the torque required

$$T = \frac{20 \times 10^3 \times 21.50}{2}\left[\tan(4.23 + 5.9) + 0.15\,\frac{32.25}{21.5}\right]$$

$$= 86788.5 \text{ N-mm}$$

Step-4: Check for the stresses
Torsional shear stress,

$$\tau = \frac{16T}{\pi d_c^3} = \frac{16 \times 86788.5}{\pi \times 19^3}$$

$$= 64.44 \text{ MPa}$$

Compressive stress considering buckling,

$$\sigma_c = \frac{F}{A_s}\left[1 + \frac{a}{n}\left(\frac{l}{k}\right)^2\right]$$

where, $F = 20 \times 10^3$

$A_s = 284.0 \text{ mm}^2$ (Table 12.5)

$a = \dfrac{1}{7500}$ (Table 1.1 MDHB)

$n = 4$ (Table 1.2) assuming fixed ends (Table 1.2 MDHB)

$$L = 1.5 \text{ m} = 1500 \text{ mm}$$

$$K = 19/4 = 4.75$$

$$\therefore \quad \sigma_c = \frac{20 \times 10^3}{284}\left[1 + \frac{1}{7500}\left(\frac{1500}{4.75}\right)^2 \times \frac{1}{4}\right]$$

$$= 304.5 \text{ MPa}$$

i.e., too large and unsafe.

Step-5: Redesign the screw
Choose $T_r 32 \times 6$ threads

$$d = 32 \text{ mm and } d_i = 26 \text{ mm and } p = 6 \text{ mm}$$

and
$$A_s = 531 \text{ mm}^2$$

Then

$$d_m = \frac{32+26}{2} = 29 \text{ mm}$$

$$\text{Helix angle,} \quad \alpha = \tan^{-1}\left(\frac{p}{\pi d_m}\right) = \tan^{-1}\left(\frac{6}{\pi \times 29}\right) = 3.77^\circ$$

$$\text{Friction angle, } \phi = \tan^{-1}(0.1 \times \sec 14.5) = 5.9^\circ$$

Collar diameter, $d_c = 1.5\, d_m = 1.5 \times 29 = 43.5$ mm
Then torque,

$$T = \frac{F d_m}{2}\left[\tan(\alpha+\phi) + f_c \frac{d_c}{d_m}\right] = \frac{20 \times 10^3 \times 29}{2}\left[\tan\,(3.77+5.9) + 0.15 \times \frac{43.5}{29}\right] = 114664.34 \text{ N-mm}$$

Then torsional shear stress,

$$\tau = \frac{16T}{\pi d_i^3} = \frac{16 \times 114664.34}{\pi \times 29^3} = 23.94 \text{ MPa}$$

Compressive stress,

$$\sigma_c = \frac{F}{A_s}\left[1 + \frac{a}{n}\left(\frac{L}{k}\right)^2\right] = \frac{20 \times 10^3}{531}\left[1 + \frac{1}{7500} \times \frac{1}{4} \times \left(\frac{1500}{6.5}\right)^2\right] - 104.52 \text{ MPa}$$

Then maximum shear stress,

$$\tau_{max} = \sqrt{\left(\frac{\sigma}{2}\right)^2 + \tau^2} = \sqrt{\left(\frac{104.52}{2}\right)^2 + 23.94^2}$$

$$= 57.5 \text{ MPa. Safe}$$

Maximum normal stress

$$\sigma_{max} = \frac{\sigma}{2} + \sqrt{\left(\frac{\sigma}{2}\right)^2 + \tau^2}$$

$$= \frac{104.52}{2} \pm \sqrt{\left(\frac{104.52}{2}\right)^2 + 23.94^2}$$

$$= 109.75 \text{ MPa} < S_d = 158.0 \text{ MPa.} \quad \therefore \text{Safe.}$$

Step-6: Nut design

Nut dimensions

$$D_i = d_i = 26 \text{ mm}$$

$$D_0 = d_0 = 32 \text{ mm}$$

$$D = 2.d_i = 2 \times 26 = 52 \text{ mm}$$

The number of threads in the nut for bearing pressure consideration,

$$N = \frac{4F}{p_b \cdot \pi (d_o^2 - d_i^2)}$$

$$= \frac{4 \times 20 \times 10^3}{4 \times \pi \times (32^2 - 26^2)}$$

$$= 18.29 \simeq 19$$

Height,

$$H = n.p = 19 \times 6 = \mathbf{114.0 \text{ mm}}$$

Example 9.6: A triple threaded power screw is used in a screw jack, has a nominal diameter of 50 mm and a pitch of 8 mm. The threads are square in shape and the length of the nut is 48 mm. The screw jack is used to life a load of 7.5 kN. The coefficient of friction at the threads is 0.12 and the collar friction is negligible. Calculate

(i) Principal shear stresses in the rod
(ii) Transverse shear stress in the rod and nut
(iii) Bearing pressure for threads and
(iv) State whether the screw is self-locking.

Given Data

Screw diameter, $d_o = 50$ mm

Pitch, $p = 8$ mm

Square threads: triple threaded

Nut length, $H = 48$ mm

Load lifted, $F = 7.5$ kN

SOLUTION

Step-1: Torque on screw

$$\text{Inner diameter, } d_i = d_o - p = 50 - 8 = 42 \text{ mm}$$

$$\text{Mean diameter, } d_m = \frac{d_o + d_i}{2} = \frac{50 + 42}{2} = 46 \text{ mm}$$

Helix angle of thread,

$$\alpha = \tan^{-1}\left(\frac{3 \times p}{\pi d_m}\right) = \tan^{-1}\left(\frac{3 \times 8}{\pi \times 46}\right) = 9.43°$$

Friction angle,

$$\phi = \tan^{-1}(f) = \tan^{-1}(0.12) = 6.84°$$

The screw is overhauling ($\phi < \alpha$).
Torque applied to raise the load,

$$T = \frac{F d_m}{2} \tan(\alpha + \phi) \qquad \text{(Eqn. 9.4)}$$

$$= \frac{7500 \times 46}{2} \tan(9.43 + 6.84)$$

$$= \mathbf{50344.5 \text{ N-mm}}$$

Step-2: Stresses in screw
Torsional shear in the screw,

$$\tau = \frac{16 \cdot T}{\pi d_i^3} = \frac{16 \times 50344.5}{\pi \times 42^3} = 3.46 \text{ MPa}$$

Compressive stress,

$$\sigma = \frac{F}{A_s} \cdot = \frac{4F}{\pi d_i^2} = \frac{4 \times 7500}{\pi \times 42^2}$$

$$= 5.41 \text{ MPa}$$

Principal shear stress,

$$\tau_{\max} = \sqrt{\left(\frac{\sigma}{2}\right)^2 + \tau^2}$$

$$= \sqrt{\left(\frac{5.41}{2}\right)^2 + 3.46^2}$$

$$= 4.393 \text{ MPa}$$

Step-3: Stresses in threads
(i) Shear stress in the threads in screw,

$$\tau = \frac{F}{n \cdot \pi d_i \cdot p/2} = \frac{2F}{n \cdot \pi d_i p}$$

$$= \frac{2 \times 7500}{48 \times \pi \times 42} \qquad (\because n \cdot p = H = 48)$$

$$= 2.37 \text{ MPa}$$

(ii) Shear stress at the root of thread in nut,

$$\tau = \frac{2 \cdot F}{n \cdot p \cdot \pi d} = \frac{2 \times 7500}{48 \times \pi \times 50}$$

$$= 1.99 \text{ MPa}$$

Number of threads, $n = H/P = \frac{48}{8} = 6$
(iii) Bearing pressure between the thread,

$$p_b = \frac{F}{n \cdot \frac{\pi}{4}(d_o^2 - d_i^2)} \qquad \text{(Eqn. 9.15)}$$

$$= \frac{4 \times 7500}{6 \times \pi \times (50^2 - 42^2)}$$

$$= \mathbf{2.16 \text{ MPa}}$$

Step-4: Since friction angle $\phi = 6.84°$ and helix angle $\alpha = 9.43°$. There is no self-locking. The load is lowered automatically.

Example 9.7: A single start square threaded power screw is used to raise a load of 120 kN. The screw has a mean diameter of 24 mm and four threads per 24 mm length. The mean collar diameter is 40 mm. The coefficient of friction is estimated as 0.1 for both thread and collar.

(i) Determine the major diameter of the screw.
(ii) Estimate the screw torque required to raise the load.
(iii) Estimate the overall efficiency.
(iv) If the collar friction is eliminated what minimum value of thread coefficient is needed to prevent the screw from overhauling.

Given Data

Load raised, $F = 120$ kN

Mean diameter, $d_m = 24$ mm

Pitch of thread, $p = 24/4 = 6$ mm

Collar diameter, $d_c = 40$ mm

Coefficient of friction, $f = 0.1$

SOLUTION

Step-1: Geometric relations

$$\text{Major diameter, } d_o = d_m + p/2 = 24 + 3 = 27 \text{ mm}$$

$$\text{Minor diameter, } d_i = d_m - p/2 = 24 - 3 = 21 \text{ mm}$$

$$\text{Helix angle, } \alpha = \tan^{-1}\left(\frac{p}{\pi d_m}\right) = \tan^{-1}\left(\frac{6}{\pi \times 24}\right) = \mathbf{4.55^\circ}$$

$$\text{Friction angle, } \phi = \tan^{-1}(f) = \tan^{-1}(0.1) = \mathbf{5.71^\circ}$$

Step-2: Torque required to raise the load

$$T = \frac{F \cdot d_m}{2}\left[\tan(\alpha + \phi) + f_c \frac{d_c}{d_m}\right]$$

$$= \left(\frac{120 \times 10^3 \times 24}{2}\right)\left[\tan(4.55 + 5.71) + 0.1 \times \frac{40}{24}\right]$$

$$= 500654 \text{ N-mm}$$

Step-3: Overall efficiency,

$$\eta = \frac{\tan\alpha}{\left[\tan(\alpha+\phi) + f_c \dfrac{d_c}{d_m}\right]}$$

$$= \frac{\tan(4.55)}{\tan(4.55 + 5.71) + 0.1 \times \dfrac{40}{24}}$$

$$= 0.229 = \mathbf{22.9\%}$$

Step-4: Minimum coefficient of friction
Condition for overhauling,

$$\tan\alpha > \frac{f d_m + f_c d_c}{d_m - f f_c d_c}$$

For preventing the overhauling,

$$\tan\alpha < \frac{f d_m + f_c d_c}{d_m - f f_c d_c}$$

For $f_c = 0$

$$f_{\min} \geq (\tan\alpha = \tan 4.55)$$

$\therefore$ Minimum friction coefficient $\therefore \quad f_{\min} \geq 0.0795$

Example 9.8:

(a) Explain the term overhauling and self-locking with respect to power screws.

(b) The screw of a shaft-straighter exerts a load of 30 kN as shown in Fig. 9.11. The screw is square threaded of outside diameter 75 mm and 6 mm pitch. Determine:

  (i) Force required at the rim of 300 mm diameter hand-wheel assuming the coefficient of friction for threads as 0.12.

  (ii) Maximum compressive stress in screw, bearing pressure on the threads and the maximum shear stress in the thread.

  (iii) Efficiency of the straightner.

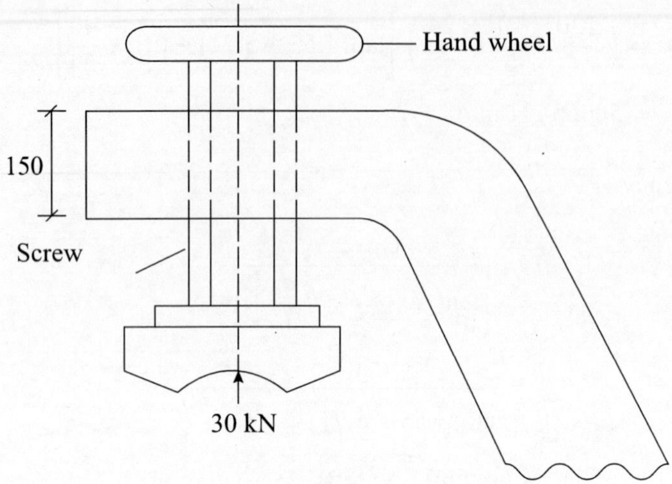

Figure 9.11

(a) *Overhauling*. It is a condition where there is no need to apply a force to lower the load. The helix angle of the threads is so large that screw winds down on its own. The condition is

$$\tan\alpha > \left(\frac{f\,d_m + f_c\,d_c}{d_m - f\,f_c\,d_c}\right)$$

self-locking is the condition where there is a need for the application of force to lower the load. The condition is

$$\tan\alpha < \left(\frac{f\,d_m + f_c\,d_c}{d_m - f\,f_c\,d_c}\right)$$

(b) *Given Data*

Diameter of the screw, $d = 75$ mm

Pitch of the threads, $p = 6.0$ mm

Hand wheel diameter $d_1 = 300$ mm

Coefficient of friction, $f = 0.12$.

SOLUTION

Step-1: Geometric relations

$$\text{Inner diameter of the screw, } d_i = d - p$$
$$= 75 - 6 = 69 \text{ mm}$$

$$\text{Mean diameter, } d_m = \frac{d_o + d_c}{2}$$
$$= \frac{75 + 69}{2} = 72 \text{ mm}$$

Helix angle of the thread,

$$\alpha = \tan^{-1}\left(\frac{p}{\pi\, d_m}\right)$$
$$= \tan^{-1}\left(\frac{6}{\pi \times 72}\right)$$
$$= \mathbf{1.52^\circ}$$

Friction angle,

$$\phi = \tan^{-1}(f) = \tan^{-1}(0.12)$$
$$= 6.84^\circ$$

Step-2: Torque requirement
Torque required neglecting collar friction,

$$T = \frac{F\, d_m}{2} \tan(\alpha + \phi)$$
$$= \frac{30 \times 10^3 \times 72}{2} \tan(1.52 + 6.84)$$
$$= 158710.18 \text{ N-mm}$$

Force at the hand wheel,

$$P = \frac{T}{r} = \frac{158710.18}{150}$$
$$= 1058.06 \text{ N}$$

Step-3: Stresses in screw
Compressive stress in the screw,

$$\sigma = \frac{F}{A_c} = \frac{30 \times 10^3}{\frac{\pi}{4} \times 69^2}$$
$$= 8.023 \text{ MPa}$$

Torsional shear,

$$\tau = \frac{16 \cdot T}{\pi\, d_i^3} = \frac{16 \times 158710.8}{\pi \times 69^3}$$
$$= 2.46 \text{ MPa}$$

Maximum shear stress,

$$\tau_{\max} = \sqrt{\left(\frac{\sigma}{\alpha}\right)^2 + \tau^2}$$

$$= \sqrt{\left(\frac{8.023}{2}\right)^2 + 2.46^2}$$

$$= 4.706 \text{ MPa}$$

Maximum compressive stress,

$$\sigma_c = \frac{\sigma}{2} + \sqrt{\left(\frac{\sigma}{2}\right)^2 + \tau^2}$$

$$= \frac{8.023}{2} + \sqrt{\left(\frac{8.203}{2}\right)^2 + 2.46^2}$$

$$= 8.72 \text{ MPa}$$

Step-4: Efficiency of the straightner

$$\eta = \frac{\tan\alpha}{\tan(\alpha + \phi)}$$

$$= \frac{\tan(1.52)}{\tan(1.52 + 6.84)}$$

$$= 0.1806 = \mathbf{18.06\%}$$

Example 9.9: The cutter of a broaching machine is pulled by square threaded screw of 55 mm external diameter and 10 mm pitch. The operating nuts take the axial load of 400 N on a flat surface of 60 mm and 90 mm internal and external diameters respectively. If the coefficient of friction is 0.15 for all contact surfaces determine the power required to rotate the nut when the cutting speed is 6 m/min. Also find efficiency of the screw.

Given Data

External diameter of the screw, $d_o = 55$ mm

Pitch of the thread, $p = 10$ mm

Axial force, $F = 400$ N

Collar diameter, $d_{c_o} = 90$ mm

$d_{ci} = 60$ mm

Coefficient of friction, $f = 0.15$
Cutting speed, $v = 6$ m/min

SOLUTION

Step-1: Geometric relations

Inside diameter of screw, $d_i = 55 - 10 = 45$ mm.

Mean diameter, $d_m = \dfrac{d_o + d_i}{2} = \dfrac{55+45}{2} = 50$ mm

Helix angle,

$$\alpha = \tan^{-1}\left(\frac{p}{\pi d_m}\right)$$

$$= \tan^{-1}\left(\frac{10}{\pi \times 50}\right) = \mathbf{3.64^\circ}$$

Friction angle, $\phi = \tan^{-1}(f) = \tan^{-1}(0.15) = \mathbf{8.53^\circ}$

Mean diameter of the collar,

$$d_c = \frac{90+60}{2} = \frac{150}{2} = 75 \text{ mm}$$

Step-2: Torque required

$$\text{T} = \frac{Fd_m}{2}\left[\tan(\alpha+\phi) + t_c\frac{d_c}{d_m}\right] \qquad \text{(Eqn. 9.10)}$$

$$= \frac{400 \times 50}{2}\left[\tan(3.64 + 8.53) + 0.15\frac{75}{50}\right]$$

$$= 4406.135 \text{ N-mm} = 4.4066 \text{ N.m}$$

Step-3: Power requirement

Cutting speed, $v = N.p$ mm/min

where $N =$ rpm of the screw

$6000 = \text{N} \times 10$

$N = 600$ rpm

Angular speed,

$$\omega = \frac{2\pi N}{60} \text{ rad/rec} = \frac{2\pi \times 600}{60}$$

$$= 62.83 \text{ rad/rec}$$

$$\text{Power} = T.\omega \text{ watts}$$
$$= 4.406 \times 62.83 = 276{,}856.31 \text{ watts}$$
$$= 0.277 \text{ kW}$$

Step-4: Efficiency of the screw

$$\eta = \frac{\tan(\alpha)}{\tan(\alpha+\phi)}$$
$$= \frac{\tan(3.64)}{\tan(3.64+8.53)} = 0.295 = 29.5\%$$

Example 9.10: A sluice gate weighing 600 kN is raised by means of two square threaded screws. The coefficient of collar friction is 0.03 and coefficient of thread friction is 0.14. The outer diameter of the collar is 100 mm and inner diameter is 50 mm. The gate is raised at a rate of 0.6 m/min. The permissible stress of the screw material in tension and compression is 80 MPa and in shear is 60 MPa. Design the screw and nut, Check for the stresses induced also determine the speed of the screw and the power required at the motor to raise the gate, assuming an efficiency of 75% for the reduction drive. Permissible bearing pressure is 15 MPa.
Given Data

Weight of the gate, $W = 600$ kN

Number of screws used $= 2$

Collar friction, $f_c = 0.03$

Thread friction, $f = 0.14$

Collar dimensions, $d_{co} = 100$ mm

$d_{ci} = 50$ mm

Gate speed, $V = 0.6$ m/min

Permissible stresses for screw, $S_d = 80$ MPa

$S_{sd} = 60$ MPa

Bearing pressure, $p_y = 15$ MPa

Efficiency of the drive, $\eta_d = 75\%$

SOLUTION

Step-1: Design of screw

$$\text{Mean collar diameter, } d_c = \frac{100+50}{2} = 75 \text{ mm}$$

$$\text{Load per screw} = \frac{600}{2} \text{ kN}$$
$$F = 300 \text{ kN}$$

Core area of the screw considering direct tensile stress,

$$A_s = \frac{F}{S_d} = \frac{300 \times 10^3}{80} = 3750 \text{ mm}^2$$

Allowing for torsional shear,

$$\text{Let} \quad A_s = 2 \times 3750 \text{ mm}^2 = 7500 \text{ mm}^2$$

From Table 9.5
Suitable screw size is $d = 110$ mm, $d_i = 98$ mm and pitch 12 mm

$$A_s = 7543 \text{ mm}^2$$

Step-2: Torque on screw
Mean diameter of the screw,

$$d_m = \frac{d + d_i}{2} = \frac{110 + 98}{2} = 104 \text{ mm}$$

Assuming single start threads the helix angle,

$$\alpha = \tan^{-1}\left(\frac{p}{\pi d_m}\right) = \tan^{-1}\left(\frac{12}{\pi \times 104}\right) = \mathbf{2.103^\circ}$$

Then friction angle,

$$\begin{aligned} \varphi &= \tan^{-1}(f) = \tan^{-1}(0.14) \\ &= 7.97^\circ \end{aligned}$$

Then the torques required,

$$\begin{aligned} \text{T} &= \frac{F d_m}{2}\left[\tan(\alpha + \varphi) + f_c \frac{d_c}{d_m}\right] \\ &= \frac{300 \times 10^3 \times 104}{2}\left[\tan(2.103 + 7.97) + 0.03 . \frac{75}{104}\right] \\ &= \left(3108 \times 10^3\right) \text{ N-mm} \end{aligned}$$

Step-3: Check for stresses in the screw
The torsional shear stress,

$$\tau = \frac{16T}{\pi d_c^3} = \frac{16 \times 3108 \times 10^3}{\pi \times 98^3} = 16.82 \text{ MPa}$$

Direct tensile stress,

$$\sigma = F/A_s = \frac{300 \times 10^3}{7543} = 39.77 \text{ MPa}$$

Maximum shear stress,

$$\tau_{\max} = \sqrt{\left(\frac{\sigma}{2}\right)^2 + \tau^2}$$
$$= \sqrt{\left(\frac{39.77}{2}\right)^2 + 16.82^2}$$
$$= 26.04 \text{ MPa} < S_{sd} = 60 \text{ MPa.} \quad \therefore \text{Safe}$$

Maximum normal stress,

$$\sigma_{\max} = \frac{\sigma}{2} + \sqrt{\left(\frac{\sigma}{2}\right)^2 + \tau^2}$$
$$= \frac{39.77}{2} + \sqrt{\left(\frac{39.77}{2}\right)^2 + 16.82^2}$$
$$= 45.95 \text{ MPa} < S_d = 80 \text{ MPa.} \quad \therefore \text{Safe}$$

Step-4: Nut design
Considering the bearing pressure the height of the nut,

$$H = n.p$$
$$= \frac{4F\,p}{\pi(d^2 - d_e^2)p_b} \times$$
$$= \frac{4 \times 300 \times 10^3 \times 12}{\pi\,(110^2 - 98^2) \times 15}$$
$$= 122.43\,\text{mm} = 125\text{ mm}$$

Number of threads, $n = 125/12 = 10.42 \cong 11$
Shear stress in the root,

$$\tau = \frac{2F}{n.\pi d_c p}$$

$$= \frac{2 \times 300 \times 10^3}{10.42 \times \pi \times 98 \times 12}$$

$$= 15.58 < S_{sd} = 60 \text{ MPa.} \quad \therefore \text{Safe}$$

Step-5: Power requirement
Rotation of the screw (N rpm),

$$V = N.p$$

$$0.6 = N \times \frac{12}{1000}$$

$$N = \frac{0.6 \times 1000}{12} = 50 \text{ rpm}$$

Then power required at the screw,

$$NP_1 = \frac{TN}{955 \times 10^4} \qquad \text{(Eqn. 4.1a MDHB)}$$

$$= \frac{3108 \times 10^3 \times 50}{955 \times 10^4}$$

$$= 16.27 \text{ kW}$$

Power at the motor,

$$\text{NP} = \frac{NP_1}{\eta_d} = \frac{16.27}{0.75} = 21.69$$

$$\cong 21.7 \text{ kW}$$

Example 9.11: The lead screw of a lathe has single start 150 metric trapezoidal threads of 52 mm nominal diameter and 8 mm pitch. The screw is required to exert and axial force of 2 kN in order to drive the tool carriage during turning operation. The thrust is carried on a collar of 100 mm outer diameter and 60 mm inner diameter. The values of coefficient of friction at the screw threads and collar are 0.15 and 0.12 respectively. The lead screw rotates at 30 rev/min. Calculate:

1. Power required to drive the screw
2. Efficiency of the screw

Given Data

Diameter of the screw, $d = 52$ mm

Pitch of the screw, $p = 8$ mm

Axial force, $F = 2$ kN

Collar diameter, $d_{co} = 100$ mm
$d_{ci} = 60$ mm

Coefficient of friction

for screw $= 0.15$

for collar, $f_c = 0.12$

screw rotation, $N = 30$ rpm

SOLUTION

Step-1: Geometric relations

Inner diameter screw, $d_i = 52 - 8 = 44$ mm

Mean diameter of the screw,

$$d_m = \frac{d_o + d_i}{2} = \frac{52+44}{2} = 48 \text{ mm}$$

Helix angle of the thread,

$$\alpha = \tan^{-1}\left(\frac{p}{\pi d_m}\right) = \tan^{-1}\left(\frac{8}{\pi \times 48}\right) = 3.037^\circ$$

Thread angle, $\theta = \dfrac{29^\circ}{2} = 14.5^\circ$

Friction angle, $\phi = \tan^{-1}(f \cdot \sec\theta) = \tan^{-1}(0.15 \cdot \sec 14.5^\circ) = 8.81^\circ$

Mean diameter of the collar,

$$d_c = \frac{d_{co} + d_{ci}}{2} = \frac{100+60}{2} = 80 \text{ mm}$$

Step-2: Torque required

$$\text{T} = \frac{F d_m}{2}\left[\tan(\alpha+\phi) + f_c \frac{d_c}{d_m}\right]$$

$$= \frac{2000 \times 48}{2}\left[\tan(3.037 + 8.81) + 0.12 \cdot \frac{80}{48}\right]$$

$$= 19668.8 \text{ N-mm}$$

Step-3: Power required

$$\text{NP} = \frac{TN}{955 \times 10^4} \text{kW} \qquad \text{(Eqn. 4.1a MDHB)}$$

$$= \frac{19668.8 \times 30}{955 \times 10^4}$$

$$= 0.06179 \text{ kW}$$
$$\text{NP} = 61.79 \text{ watts}$$

Step-4: Efficiency of screw

$$\eta = \frac{\tan\alpha}{\tan(\alpha+\phi)} = \frac{\tan(3.037)}{\tan(3.037+8.81)}$$
$$= 0.253 = 25.3\%$$

Example 9.12: Design a screw jack to lift a load of 30 kN with the following data. The allowable compressive stress is 160 MPa. Coefficient of friction in threads $f = 0.14$. Coefficient of collar friction $f_c = 0.2$. Height of lift $L = 150$ mm.

Given Data

Load lifted, $F = 30$ kN

Weight of lift, $L = 150$ mm

Compressive stress, $S_{cd} = 160$ MPa

Thread friction, $f = 0.14$

Collar friction, $f_c = 0.20$

SOLUTION

Step-1: Design of screw

Core area based on the axial load only,

$$A_s' = \frac{F}{S_{cd}} = \frac{30 \times 10^3}{160} = 187.5 \text{ mm}^2$$

Considering the torsional shear and buckling, let

$$A_s = 3 \cdot A_s = 3 \times 187.5 \text{ mm}^3 = 562.5 \text{ mm}^2$$

From Table 9.5, the suitable screw size is SQ 34 × 6 with $A_s = 616$ mm

Minor diameter, $d_i = 34 - 6 = 28$ mm

Mean diameter,

$$d_m = \frac{34+28}{2} = 31 \text{ mm}$$

Helix angle of the thread,

$$\alpha = \tan^{-1}\left(\frac{p}{\pi d_m}\right) = \tan^{-1}\left(\frac{6}{\pi \times 31}\right) = 3.525^\circ$$

Friction angle,

$$\phi = \tan^{-1}(f) = \tan^{-1}(0.14) = 7.97^\circ$$

Choose collar diameter, $d_c = 1.5\, d_m = 1.5 \times 31 = 46.5$ mm

Step-2: Torque required

$$T = \frac{F d_m}{2}\left[\tan(\alpha + \phi) + f_c \frac{d_c}{d_m}\right]$$

$$= \frac{30 \times 10^3 \times 31}{2}[\tan(3.525 + 7.97) + 0.2 \times 1.5]$$

$$= 234063.06 \text{ N-mm}$$

Step-3: Check for stresses induced

Shear stress,

$$\tau = \frac{16T}{\pi d_c^3} = \frac{16 \times 234063.06}{\pi \times 28^3} = 54.30 \text{ MPa}$$

Compressive stress considering the column action,

$$\sigma_c = \frac{F}{A_s}\left[1 + \frac{\alpha}{n}\left(\frac{L}{k}\right)^2\right]$$

$$= \left(\frac{30 \times 10^3}{616}\right)\left[1 + \frac{1}{7500} \times \frac{1}{0.25}\left(\frac{150}{28/4}\right)^2\right] \qquad (\because n = 0.25)$$

$$= 60.63 \text{ MPa}$$

Maximum shear stress,

$$\tau_{max} = \sqrt{\left(\frac{\sigma_c}{2}\right)^2 + \tau^2}$$

$$= \sqrt{\left(\frac{60.63}{2}\right)^2 + 54.30^2}$$

$$= 62.19 \text{ MPa}$$

Maximum normal stress,

$$\sigma_{max} = \left(\frac{60.63}{2}\right) + \sqrt{\left(\frac{60.63}{2}\right)^2 + 54.3^2}$$

$$= (92.5) \text{ MPa} < S_{cd} = 160 \text{ MPa}. \quad \therefore \text{Safe.}$$

Step-4: Nut design

Choosing CI as nut material, safe bearing pressure (Table 9.1)

$$p_b = 15 \text{ MPa}.$$

Then the number of threads in the nut,

$$n = \frac{4F}{p_b\,\pi\left(d_o^2 - d_i^2\right)}$$

$$= \frac{4 \times 30 \times 10^3}{15 \times \pi \times (34^2 - 28^2)}$$

$$= 6.85$$

Choose $n = 7$.

Nut height, $H = n.p = 7 \times 6 = 42$ mm.
Assuming safe shear stress for threads,

$$S_{sd} = 60 \text{ MPa}$$

Number of threads,

$$n = \frac{2F}{\pi d_i S_s p} = \frac{2 \times 30 \times 10^3}{\pi \times 28 \times 60 \times 6}$$

$$= 1.89 < 7$$

$$\therefore \quad H = 42 \text{ mm is OK.}$$

Example 9.13: Design completely the screw handle and the nut of a screw jack of capacity 40 kN. The maximum lift is limited to 0.2 m. The screw and handle are made of C40 [40C8] steel and the nut and the cup are made of cast iron. Also find the efficiency of the screw. Check the screw for bucking load.

Given Data

Loaded lifted $F = 40 \text{ kN} = 40 \times 10^3$ N

Maximum left $L = 0.2 \text{ m} = 200$ mm

Materials for Screw and Handle C 40 steel

for Nut and Cup – C.I.

SOLUTION
(Refer plate 9.1 MDHB for equations)
Step-1: Selection of materials and design stresses
Allowable stresses for factor of safety = 2 (assumed)

Screw and Handle-C40.

Yield stress, $S_y = 330$ MPa
Design stress, $S_d = S_{y/FS} = 330/2$
S_d in tension compression $= 165$ MPa

Nut and cup-C.l-class 40

Ultimate tensile stress, $S_{ut} = 290$ MPa

Ultimate compressive stress, $S_{uc} = 965$ MPa

Design stress in tension, $S_d = S_{ut}/Fs = \dfrac{290}{2} = 145$ MPa

Design stress in compression, $S_{cd} = 965/2 = 482.5$ MPa

Step-2: Design screw
Screw under compression and the core area,

$$A_s = \frac{F}{S_d} = \frac{40 \times 10^3}{165} = 242.42 \text{ mm}^2$$

To allow for torsional shear and buckling let $A_s = 2 \times 242.42 = 484.84 \text{ mm}^2$
Suitable screw size is,

i.e.,

SQ 32 × 6 with $A_s = 531 \text{ mm}^2$

Major diameter $d_o = 32$ mm

Inner (minor) diameter $d_i = 32 - 6 = 26$ mm

Mean diameter of the screw,

$$d_m = \frac{d_o + d_i}{2} = \frac{32 + 26}{2} = 29 \text{ mm}$$

Assuming single start square thread the helix angle,

$$\alpha = \tan^{-1}\left(\frac{P}{\pi d_m}\right) = \tan^{-1}\left(\frac{6}{\pi \times 29}\right)$$

$$= 3.768°$$

Coefficient of friction

for thread, $f = 0.15$ (Table 12.2 MDHB)

For collar, $f_c = 0.121$ (Table 12.3 MDHB)

Friction angle,

$\phi = \tan^{-1}(f) = \tan^{-1}(0.15) = 8.53°$

Screw head diameter, $D_3 = 1.7 \cdot d_o$

$$= 1.7 \times 3.2 = 54.4 \text{ mm}$$

Diameter of the pin, $D_4 = D_{3/4} = 54.4/4 = 13.6$ mm

Mean diameter of the contact area,

$$d_c = \frac{54.4 + 13.6}{2} = 34.0 \text{ mm}$$

Step-3: Torque at the handle

$$T = \frac{F d_m}{2}\left[\tan(\alpha + \phi) + f_c \frac{d_c}{d_m}\right]$$

$$= \left(\frac{40 \times 10^3 \times 29}{2}\right)\left[\tan(3.768 + 8.53) + 0.121 \times \frac{34}{29}\right]$$

$$= 208719.24 \text{ N-mm}$$

Step-4: Check for stresses in the screw
Torsional shear stress in the screw,

$$\tau = \frac{16 \cdot T}{\pi d_c^3} = \frac{16 \times 208719.24}{\pi \times 26^3}$$

$$= 60.48 \text{ MPa}$$

Compressive stress considering buckling (Assuming lift $l = 200$ mm),

$$\sigma_c = \frac{F}{A_c}\left[1 + \frac{a}{n}\left(\frac{l}{k}\right)^2\right]$$

$$= \left(\frac{40 \times 10^3}{531}\right)\left[1 + \frac{1}{7500} \times \frac{1}{0.25}\left(\frac{200}{26/4}\right)^2\right]$$

$$= 113.37 \text{ MPa}$$

Maximum shear stress,

$$\tau_{\max} = \sqrt{\left(\frac{\sigma_c}{2}\right)^2 + \tau^2} = \sqrt{\left(\frac{113.37}{2}\right)^2 + 60.48^2} = 82.9 \text{ MPa}$$

Maximum normal stress,

$$\sigma_{\max} = \left(\frac{\sigma_c}{2}\right) + \sqrt{\left(\frac{\sigma_c}{2}\right)^2 + \tau^2}$$

$$= \frac{113.37}{2} + \sqrt{\left(\frac{113.37}{2}\right)^2 + 60.48^2}$$

$$= 139.59 \text{ MPa} < (S_d = 165 \text{ MPa})$$

Step-5: Nut design

Assume the bearing pressure $p_b = 15$ MPa
Number of threads for bearing pressure,

$$n = \frac{4F}{p_b \cdot \pi(d_o^2 - d_i^2)}$$

$$= \frac{4 \times 40 \times 10^3}{15 \times \pi \times (32^2 - 26^2)}$$

$$= 9.757$$

Number of threads

$$\text{for shearing,} \quad n = \frac{2F}{\pi d_i \cdot S_s \cdot p} = \frac{2 \times 40 \times 10^3}{\pi \times 26 \times \left(\frac{165}{2}\right) \times 6}$$

$$= 1.98$$

Choose the larger of the two

$$n = 9.575 \cong 10$$

Height of the nut $H = np = 10 \times 6 = 60$ mm.
Nut diameter, $D_1 = 2\, d_c = 2 \times 26 = 52$ mm
Collar diameter, $D_2 = 1.25 D_1 = 65$ mm
Nut collar thickness,

$$t = \frac{F}{\pi D_1 S_s} = \frac{40 \times 10^3}{\pi \times 52 \times 116} = 2.11$$

Choose $t = 3$ mm (minimum thickness)

$$\text{Check for crushing} \quad \sigma_c = \frac{4F}{\pi(D_2^2 - D_1^2)} = \frac{4 \times 40 \times 10^3}{\pi \times 52^2\,(1 - 25^2 - 1)} = 33.5 \text{ MPa} \quad \therefore \quad \text{Safe}$$

Step-6: Handle design
Assuming the operator can exert a force of 350 N.
Length of the handle,

$$l = \frac{T}{F} = \frac{208719.24}{350}$$
$$= 596.34 \text{ mm}$$
$$\therefore \simeq 600 \text{ mm}$$

Diameter of the handle for bending,

$$d = \sqrt[3]{\frac{32 \cdot M}{\pi S_d}} = \sqrt[3]{\frac{32 \times 208719.24}{\pi \times 165}}$$
$$= 23.44 \text{ mm}$$

Let, $d = 24$ mm

EXERCISES FOR PRACTICE

9.1. A shaft straightner is designed to exert a load of 20 kN. It utilizes a square threaded screw having outside diameter of 75 mm and pitch of 6 mm. Determine the force required to operate the hand wheel of 300 mm diameter if the coefficient of friction for threads is 0.12. Also determine efficiency of straightner.

9.2. A split nut used with a lead screw is propelled at a speed of 5 m/min, against a load of 20 kN, along the spindle of a square thread (single start), having nominal diameter of 30 mm and a pitch of 6 mm. The axial thrust is absorbed by a collar of 100 mm outside diameter and 70 mm inside diameter. Assuming coefficient of friction for threads is 0.14 and at the collar as 0.10, determine:

(i) Power required to drive
(ii) Height of the bronze nut required if allowable bearing pressure is 17 MPa.

9.3. A two-start trapezoidal is used in a screw jack to raise a load of 300 kN. The screw has nominal diameter as 90 mm, pitch as 12 mm and helix angle of 15°. Coefficient of friction is screw thread is 0.15. Neglecting collar friction, calculate:

a) Torque required to raise the load
b) Torque required to lower the load
c) Screw efficiency
d) Diameter of the 400 mm long lever assuming an effort of 300 N of force is applied at the end of the lever

9.4. A single-start square threaded screw having outside diameter 32 mm and pitch 3.5 mm is used to raise 15 kN load at a speed of 25 mm/s. The coefficient of friction at the threads is 0.08 and at the collar is 0.12. The collar mean diameter is 55 mm. Determine the power required to raise the load. Check the screw for self-locking.

9.5. A 25 kN load is to be raised by two single-start square threaded power screws at a speed of 35 mm/s. The power supplied is 1750 W/screw. Due to space limitations the screw diameter should not be larger than 45 mm. The collar diameter is 65 mm. The coefficient of friction for threads and collar is 0.09. Assuming the load is distributed uniformly; suggest the size of the screw to be used and its efficiency.

9.6. A screw press is required to exert a force of 50 kN when the applied torque is 560 Nm. The unsupported length of the screw is 450 mm and a thrust bearing of hardened steel on cast iron is provided at the power end.
Permissible stresses in the steel screw are:
Tension and compression – 85 MPa, Shear – 55 MPa, the permissible bearing pressure is 13.5 MPa for steel screw and C.I nut, the permissible shearing stress in the CI is 20 MPa, yield strength of steel of screw, $\sigma_Y = 260$ MPa.
The coefficient of friction in screw and nut is 0.15. Determine the dimensions of screw and nut, and efficiency.

9.7. Double start square threaded screw rotating at 60 rpm pulls a broaching cutter through a job. The tensile force in the screw is supported by a collar whose internal and external diameters are respectively 60 mm and 90 mm. Coefficient of friction for all contact surfaces is 0.15. The requirement of the tool displacement demands that the pitch of the thread should be 10 mm and corresponding major screw diameter is 55 mm. If the power consumed by the machine is 0.39 kW, calculate the axial load exerted upon the tool.

9.8. Design a screw jack to lift a load of 100 kN through a height of 300 mm. Assume $S_u = 400$ MPa, $S_{su} = 200$ MPa, $S_Y = 300$ MPa, $p_b = 10$ MPa. Coefficient of friction between collar on screw and CI is 0.2. Coefficient of friction between steel screw and bronze nut is 0.15. Take a factor of safety of 5 for screw and nut but take a factor of safety of 4 for operating lever.

9.9. Design a turn buckle to take an axial load of 100 kN. The material of which the turn buckle is to be made, has a design normal stress of 165 MPa and design shear stress of 100 MPa.

9.10. Design the following parts of a 10 tonne (100 kN) screw press with a travel of 300 mm. Select suitable materials for each of the components to be designed: a) screw rod b) nut and c) hand lever.

9.11. a) Discuss the equations for buckling load for column action of screw spindle.
b) Design a screw and nut for a screw jack suitable for a maximum load of 5 tonnes and having a lift of 225 mm with a ground clearance of 400 mm. The steel for the screw has the following design stresses. $S_t = 94$ MPa, $S_s = 63$ MPa, $S_c = 120$ MPa. The stresses for prosperous bronze nut are $S_t = 63$ MPa $S_s = 47$ MPa $S_c = 55$ MPa. The bearing pressure between the nut and screw is 16 MPa.

9.12. Design the following parts of a screw press of 5 tonnes capacity, selecting suitable materials: (i) screw rod, (ii) nut, and (iii) the hand lever.

9.13. A turn buckle with right and left hand threads is used to draw two railway wagons nearer to each other. The axial force required to pull the wagon is 50 kN. Design the turn buckle with square threads. Safe tensile and shear stresses are 50 MPa and 40 MPa respectively.

9.14. Design a screw jack to lift a load of 50 kN through a height of 300 mm. Select the suitable materials for the components. Prepare a dimensioned sketch.

9.15. The power screw of a screw press is to be designed to exert a force of 55 kN and has 8 mm pitch single start square thread. The screw is to be made of medium carbon steel and to be driven by a grey C.I. nut. Efficiency of the screw is estimated to be 15%. Determine the necessary dimensions of screw and sketch the same.

9.16. a) Obtain an expression for the torque required to raise a load on square threaded screw.

b) Estimated the maximum 'Wrench torque' which can be applied in tightening a 20 mm bolt if the shear stress in the body of the bolt is not to exceed 140 MPa. Following data are given. Outside diameter 20 mm; Root diameter = 16.72 mm Thread section has 60° included angle; pitch 2.5 mm Effective friction radius under nut r_c= 12 mm. Thread and collar friction coefficient is estimated as 0.10.

9.17. a) What do you understand by overhauling and working screw?

b) Design and sketch toggle jack to lift a load of 6000 N. The jack is to be so designed that the distance between the centre lines of nuts varies from 80 mm to 200 mm. The eight links are symmetrical and 120 mm long. The link pins in the base one set 50 mm apart. The links, screw and pins are made from mild steel for which the stresses are 90 N/mm^2 in tension and 50 N/mm^2 in shear. The bearing pressure at the pins is 20 N/mm^2 and at the threads is 8 N/mm^2. Assume $\mu = 0.15$.

9.18. a) What are the power screws? What do you know about self-locking property of the screw?

b) Lead screw of a lathe has square threads of 50 mm outside diameter and 10 mm pitch. It drives a tool carriage and exerts an axial pressure of 25 kN. The collar bearing with outside diameter = 10 cm and inside diameter = 5 cm, is provided to take up the thrust. If the lead screw rotates at 30 rpm. Find the efficiency and power required to drive the screw. Take coefficient of friction as 0.3 for screw threads and 0.12 for collar.

9.19. a) What is self-locking screw?

b) A sluice gate weighing 300 kN is raised and lowered by means of two 40 mm standard square threaded screws. The screws are operated by 570 rpm motor. A ball thrust bearing is used at the collar, reducing the coefficient of collar friction to 0.03 at a diameter of 100 mm. The coefficient of thread friction may be assumed to be 0.14: If the gate is to be raised at a rate of 0.6 mpm, determine the following i) The rpm of the screws ii) The KW of motor required to raise the gate, asuming a mechanical efficiency of 0.85 for the speed reducing equipment. Sketch the arrangement.

SHORT ANSWER QUESTIONS

9.1. What is a power screw?

9.2. What are the applications of power screws?

9.3. What are types of threads used in power screws?

9.4. Give the proportions of square-threads

9.5. Give the proportions of trapezoidal-threads
9.6. What are the advantages of square-threads over the trapezoidal-threads?
9.7. What are the advantages of trapezoidal-threads over square-threads?
9.8. What is acme thread?
9.9. What are the differences between acme and trapezoidal-threads?
9.10. What is a buttress thread?
9.11. Why are the buttress threads not popular?
9.12. What is the pitch of the screw?
9.13. What are the multi-start screws?
9.14. What is lead of the screw?
9.15. What are the advantages of multi-start screws?
9.16. What is the relation between the pitch and lead of a screw?
9.17. What is the nominal diameter of the screw?
9.18. What is the helix angle of screw-thread?
9.19. What is friction angle?
9.20. What is the effort P to lift a load W?
9.21. What is torque to be applied to lift the load W?
9.22. What is the lifting efficiency of the screw?
9.23. What is the effort P to lower the load W?
9.24. What is overhauling of power screw? What is the condition for overhauling?
9.25. What is self-locking of power screw? What is the condition for self-locking?

CHAPTER 10

THREADED JOINTS

10.1 INTRODUCTION

The threaded fasteners are cylindrical members with threads formed on part of its length. They are used to connect or assemble two members.

Advantages of bolted connections

1. The processes of joining is noiseless.
2. No skilled labour is required.
3. Needs less labour.
4. The connection can be made easily and quickly.
5. Accommodates minor discrepancies in dimensions.
6. Alteration can be done easily.

Disadvantages of bolted connections

1. Reduction in the tensile strength due to the reduced cross-sectional area at the root of the threads and the stress concentration.
2. Reduction in the rigidity of the joint due to loose-fit leading to excessive deflections.
3. Loosening of the nut due to vibration leading to reduced safety.

10.2 TYPES OF SCREW FASTENERS

Fasteners can be classified by the intended use, by thread-types, heads-type and their strength. The most widely known classification is by their intended use.

There are three categories, namely:

1. Bolts, 2. Machine screw and 3. Studs.

Bolt (Fig. 10.1 and Fig. 10.2a) It is a cylindrical member with a head and a shank being fully or partially threaded. It is inserted through the drilled holes in the parts to be joined and a nut is tightened on the protruded part of the threaded shank. This is the most commonly used fastener.

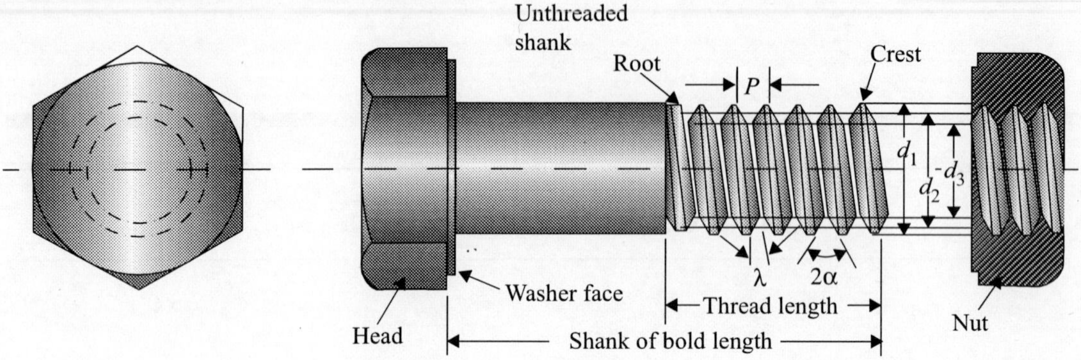

Figure 10.1. A threaded fastener

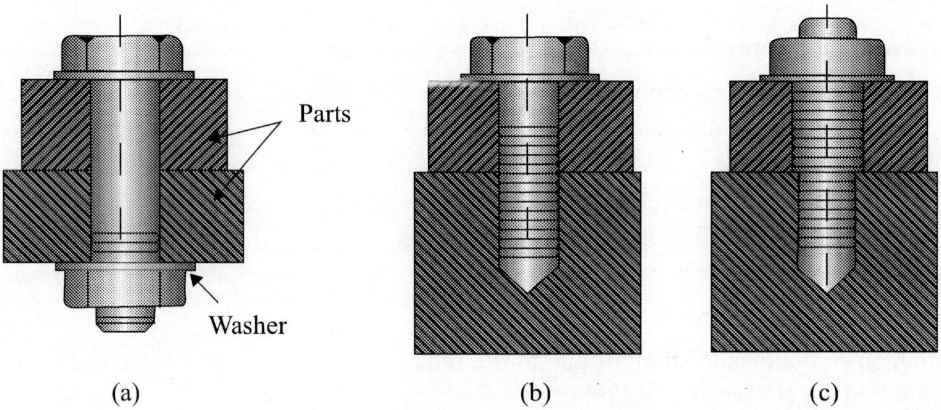

Figure 10.2. Bolt and nut, screw and stud

Machine-screw (Fig. 10.2b) It is similar in shape to bolt. In this case one of the parts to be joined has a tapped hole and other has drilled hole. The screw is passed through the drilled hole and tightened into the tapped hole. Sometimes it is in the form of a hook with threaded shank. It is tightened into the tapped hole in the part to lifted or moved.

Stud (Fig. 10.2c) It is a headless bolt with both ends of the shank threaded. It is screwed tightly into the tapped hole of one part and the other part with drilled hole is placed over it. The two are clamped together by a nut tightened on the projected portion of the stud it is used in cases where the other end of the bolt cannot be made accessible or when the assembly is dismantled less frequently.

Thread form (Fig. 10.3)

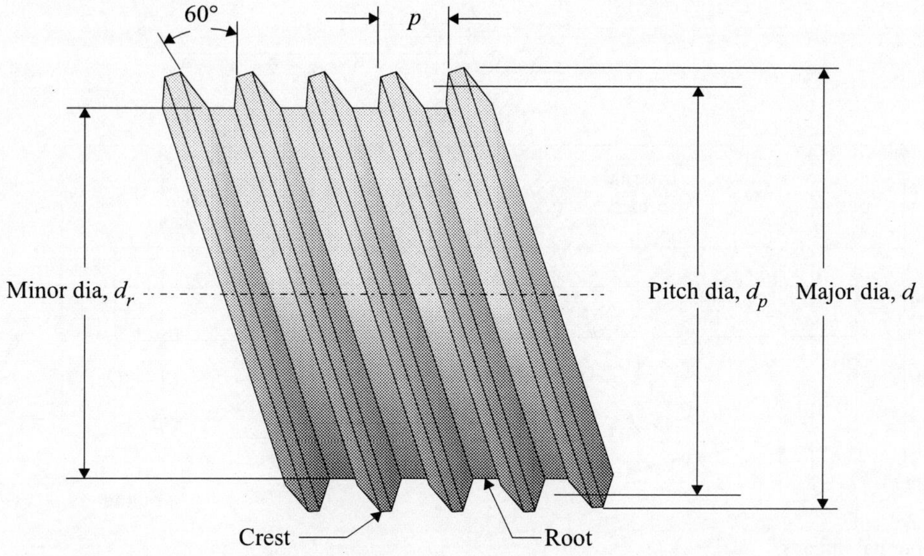

Figure 10.3(a). ISO standard thread form

$$D_1 = d_2 - 2\left[\frac{H}{2} - \frac{H}{4}\right] = d - 2H_1 = d - 1.08253P$$

$$d_3 = d_2 - 2\left[\frac{H}{2} - \frac{H}{6}\right] = d - 1.22687P$$

$$P = \text{Pitch}$$

$$H = 0.86603P$$

$$D = d = \text{Major diameter}$$

$$D_2 = d_2 = d - \frac{3}{4}H$$

$$h_3 = \frac{d - d_3}{2} = \frac{17}{24}H = 0.61343P$$

$$R = \frac{H}{6} = 0.14434P$$

$$H_1 = \frac{D - D_1}{2} = \frac{5}{8}H = 0.54127P$$

Designation: M10 × 1.25

M10 mm diameter × 1.25 mm pitch

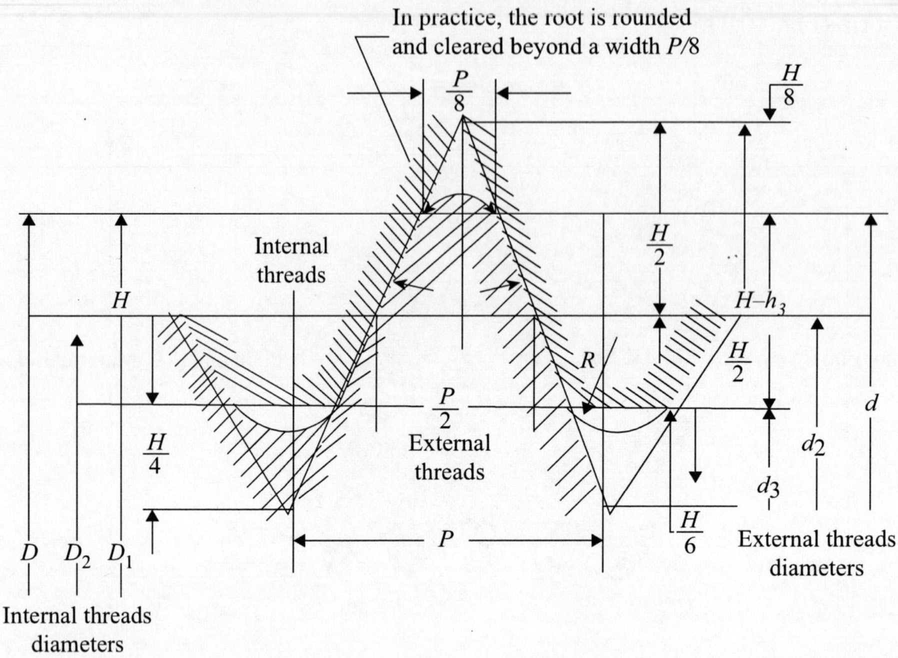

Figure 10.3(b). ISO metric screw thread-design profiles of external and internal threads

Usually, the ISO metric threads are used.

The standard proportions are given in Table 10.4. (Table 9.7 MDHB)

10.3 PRELOADED BOLTS

This kind of joints are used to fix the cover plate (head) onto the cylinder of engines, pressure vessels, to connect pipes conveying fluid under pressure. A soft gasket is used between the flanges of the cylinder and cover plate to make the joint leakproof.

The stiffness of a bolt is evaluated as follows:

$$\frac{1}{K_b} = \frac{l_t}{A_r \cdot E_b} + \frac{l_s}{A_s E_b} \cong \frac{l}{A_r \cdot E_b} \tag{10.1}$$

where, l_t – Length of threaded portion of bold

l_s – Length of unthreaded shank

A_r – The root area of the bolt

A_s – The area of unthreaded shank

l – Total length of the bolt

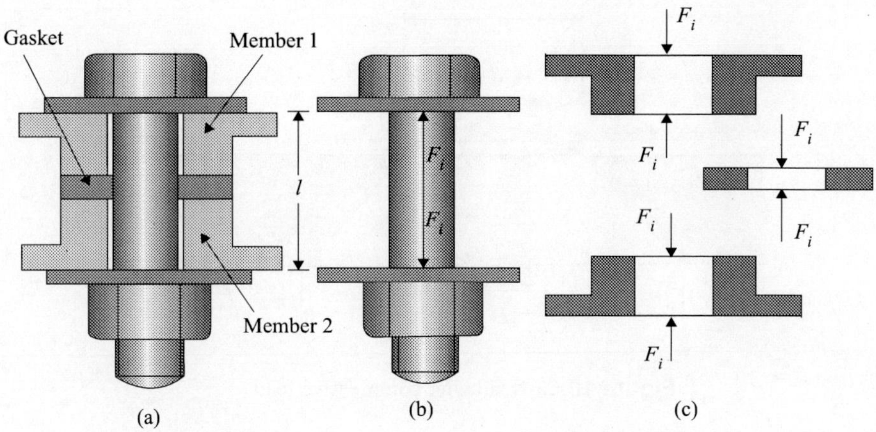

Figure 10.4. An assembly under preload only

The stiffness of the connected parts is given by

$$\frac{1}{K_c} = \frac{t_1}{Ac_1 E_1} + \frac{t_2}{Ac_2 E_2} + \frac{t_g}{A_g \cdot E_g} \tag{10.2}$$

t_1, t_2 – are thicknesses of the flanges of cylinder and cover plate

E_1, E_2 – Elastic modulii of the flange material

Ac_1, Ac_2 – Area of the flanges under compression (calculated as follow)

t_g – Thickness of the gasket

E_g – Elastic modulus for gasket material

A_g – Area of gasket under compression

Referring Fig. 10.5, the area under compression under the bolt-head,

$$Ac = \frac{\pi}{4}\left[\left(\frac{dw + D}{2}\right)^2 - d^2\right] \tag{10.3}$$

where, $D = dw + l \tan 45$

In majority of the cases, the flange materials and thicknesses are same and hence

$$t_1 = t_2 \text{ and } E_1 = E_2 = E_c$$

$$A_{c_1} = A_{c_2} = A_c$$

$$= \frac{\pi}{4}\left(d_g - d\right)^2 \quad \text{if} \quad d_g < D \tag{10.4a}$$

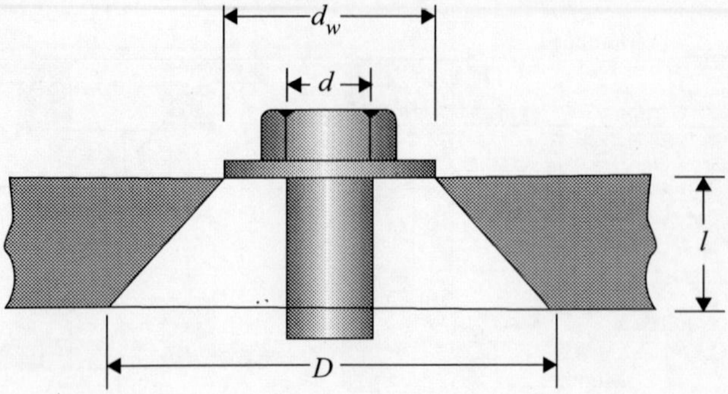

Figure 10.5. Area under compressive load

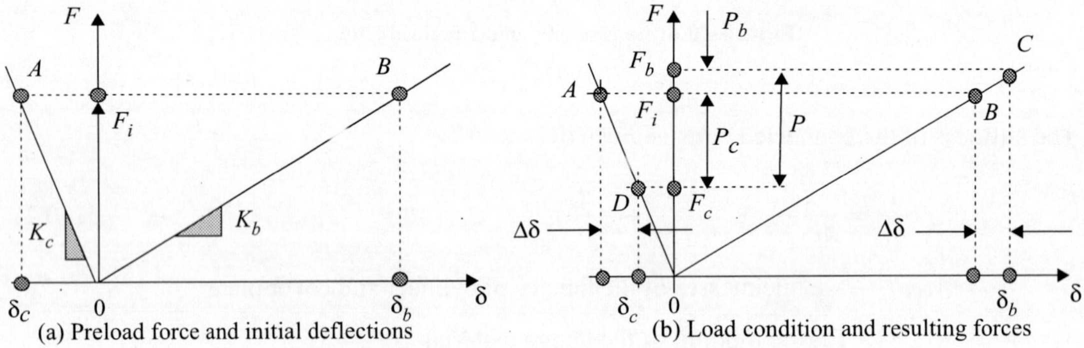

Figure 10.6. Effects on bolt and material from preload

$$= \frac{\pi}{4}(D-d)^2 \quad \text{if} \quad d_g > D \tag{10.4b}$$

Therefore,

$$k_c = \frac{E_c A_c}{l_c} \quad \text{and} \quad k_g = \frac{E_g A_g}{l_g}$$

When the assembly is made the nut is tightened by a torque-wrench to create desired tension in the bolt and compression in the gasket. Under the action of this initial load, the bolt elongates by δ_b and the connected parts (gasket) get compressed δ_c. This is shown in Fig. 10.6.

The initial tension, $F_i = 2805\, d$ for metal to metal joint

$= 1400\, d$ for gasket joints

Deflection of bolt,

$$\delta_b = F_i/k_b$$

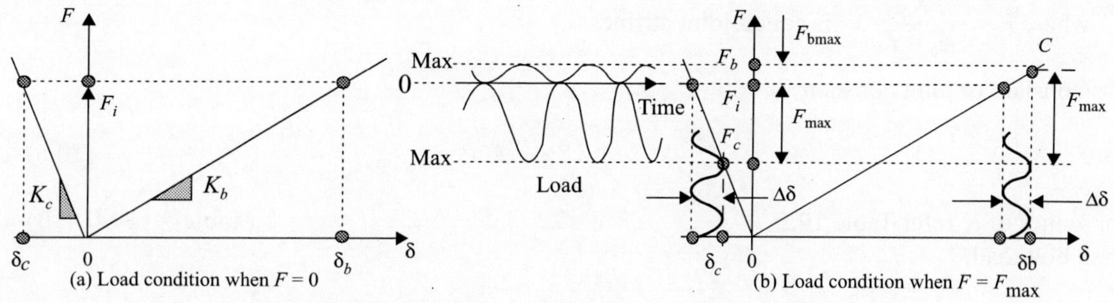

Figure 10.7. Effects on bolt and material of a load fluctuating from zero to P_{max}

Compression of connected parts (gasket),

$$\delta_c = \frac{F_i}{k_c}$$

When an external load P is applied on the coverplate, (due to fluid pressure) it is shared between the bolt and the connected parts. There is additional tension in the bolt and the bolt elongates by $\Delta\delta_b$. The compression in the connected parts is reducced and the amount of the compression removed is $\Delta\delta$. The corresponding reduction is P_c.

Then $P = P_b + P_c$

The load on the bolt,

$$F = F_i + P_b \tag{10.5}$$

The load on the connected parts,

$$F_c = F_i - P_c$$

The extension in bolt length,

$$\Delta\delta_b = \frac{P_b}{K_b}$$

The compression removed,

$$\Delta\delta_c = \frac{P_c}{K_c}$$

Equating the two,

$$\therefore \quad \Delta\delta = \frac{P_b}{K_b} = \frac{P_c}{K_c}$$

$$\therefore \quad P_b = \frac{K_b}{K_c} P_c = \frac{K_b}{K_c}(P - P_b)$$

$$\therefore \quad P = \frac{K_c}{K_b} P_b + P_b = \left(\frac{K_b + K_c}{K_b}\right) P_b$$

$$= P_b / K$$

where $K = \dfrac{K_b}{K_b + K_c}$ is called joint stiffness.

Constant or joint constant

$$\therefore \qquad P_b = K\,P \tag{10.6}$$

For values of K refer Table 10.5 (Table 9.11 MDHB)

Then bolt load,

$$\begin{aligned} F &= F_i + P_b \\ &= F_i + K\,P. \end{aligned} \tag{10.7}$$

The load on the connected parts,

$$\begin{aligned} F_c &= F_i - P_c \\ &= F_i - (P - P_b) \\ &= F_i - (1 - K)P \end{aligned} \tag{10.8}$$

These relations are true as long as the contact between the parts is maintained. When the compression in the connected parts is reduced to zero, the separation occurs and the entire applied load P is carried by the bolt.

At separation,

$$F_c = 0 = F_i - (l - K)P$$

$$\therefore \qquad F_i = (l - K)\,P. \tag{10.9}$$

For no separation, pre-load is given by

$$\boxed{F_i > (l - K)P}. \tag{10.9a}$$

Thus, the amount of pre-load is an important factor in bolt design. The torque necessary to induce the predetermined initial load F_i is given by

$$T_i = K_i F_i \,.\, d \qquad \text{(Eqn. 9.7 MDHB)} \tag{10.10}$$

where,

$$K_i = \frac{d_m}{2d}\tan(\alpha + \phi') + 0.625 f_c$$

The values are given in Table 10.7.

The average values is $K_i = 0.21$.

Then initial torque is $T_i = 0.21\,F_i d$. (10.10a)

When the applied load is fluctuating, the distribution of load is shown in Fig. 10.7.
The equivalent static load can be evaluated by the relation,

$$F_{eq} = F_{am} + K_f \left(\frac{S_y}{S_{en}}\right) . F_{aa} \tag{10.11}$$

where mean load,

$$F_{am} = \left(\frac{F_{\max} + F_{\min}}{2}\right)$$

Load amplitude,

$$F_{aa} = \left(\frac{F_{\max} - F_{\min}}{2}\right)$$

The bolt load,

$$F = F_i + K F_{eq} \tag{10.12}$$

$$= F_i + K \left[F_{am} + K_f \left(\frac{S_y}{S_{en}}\right) . F_{aa}\right] \tag{10.12a}$$

10.4 STRESSES IN THE BOLT

Axial stress-due to the axial force.

The conservative approach is to consider to the minimum cross-section area corresponding to the root diameter (d_r).

The stress area is

$$A_s = \frac{\pi}{4} d_r^2$$

However, for more correct approach, the strengthening effect of the threads is to be considered

$$A_s = \frac{\pi}{4}\left(\frac{d_p + d_r}{2}\right)^2$$

where, for ISO metric threads

$$d_p = d - 0.649519\, p$$

$$d_r = d - 1.226869\, p$$

$$d = \text{outside diameter in mm}$$

$$p = \text{pitch in mm}$$

The stress induced by axial load F,

$$\sigma = (F/A_s) \le S_d \text{ for safe design} \tag{10.13}$$

Torsional shear in bolt

When a nut is tightened on the bolt, the torsional shear stresses are developed in the shank of the bolt. The maximum shear stress is given by the expression,

$$\tau = \frac{16T}{\pi d_r^3} \tag{10.14}$$

The combined maximum resultant stress,

$$\sigma_{max} = \frac{\sigma}{2} + \sqrt{\left(\frac{\sigma}{2}\right)^2 + \tau^2} \leq S_d \text{ for safe design} \tag{10.15}$$

The maximum shear stress,

$$\tau_{max} = \sqrt{\left(\frac{\sigma}{2}\right)^2 + \tau^2} \leq S_{sd} \text{ for safe design} \tag{10.16}$$

Nut-Height

There is shear in the threads due to axial load. The effective area under shear for one screw thread is

$$A_s = \pi d_r w_i p$$

The total area is

$$A_{st} = \pi d_r w_i pN \tag{10.17}$$

where,

w_i – a factor depending upon the thread form
= 0.80 for ISO metric threads
N is number of threads in the nut

Then the shear stress in threads,

$$\tau = \left(\frac{F}{A_{Sc}}\right) = \frac{F}{\pi d_r w_i p \cdot N} \leq S_{sd} \tag{10.18}$$

i.e., Number of threads

$$N = \left(\frac{F}{\pi d_r w_i p S_{sd}}\right) \tag{10.19}$$

Then the nut height H is

$$H = N.p. \tag{10.20}$$

The nut is made long enough so that the load required to ship the threads exceeds the load needed to fail the bolt in tension for ISO threads, the nut height for safe design, is given as

$$H \geq 0.5d \text{ for } d < 25 \text{ mm}$$

$$= 0.6d \text{ for } d < 25 \text{ mm}$$

This is true only if the material for screw and nut are the same minimum tap-hole engagement

When the screw is threaded into a tapped hole (as in case of studs) longer threads engagement is required. For the same material combination,

$$H = d$$

For steel screw in cast iron, brass or bronze

$$H = 1.5d$$

For steel screw in aluminium,

$$H = 2d$$

10.5 UNIFORM – STRENGTH BOLTS

The bolts are subjected to shock and impact loads in many applications. The bolts used to connect the cover plate of an IC engine cylinder and the bolts used in connecting rod are the examples. In such situations the resilience of the bolts becomes an important design consideration. The resilience is proportional to the square of the stress induced in material and the volume of the material under the stress.

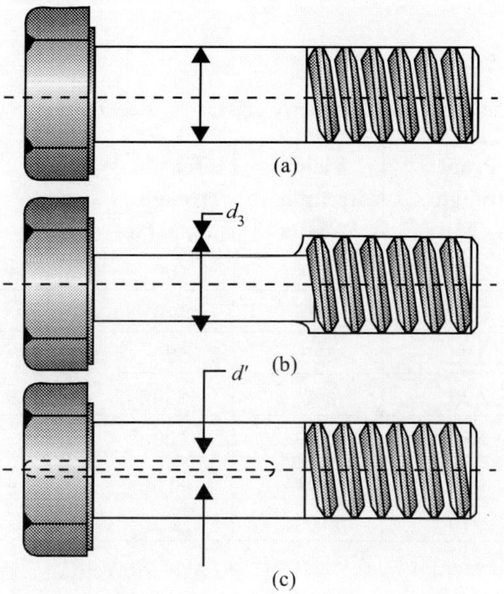

Figure 10.8. Bold of uniform strength

In the normal bolt Fig. 10.8(a), the energy is stored in the bolts is in two parts:

(i) Energy stored in threaded portion

(ii) Energy stored in the shank portion.

In the threaded portion, the stress-induced is dependent on the core area, i.e., core diameter of the bolt. The core diameter is smaller than shank diameter. The stress induced is more. There is likely stress concentration at the root of the threads. Hence, the average stress is much less than the design stress.

The shank diameter is larger than the core diameter and there is no stress concentration. Thus, average stress in the shank is much less than that in the thread portion and the energy stored is also much smaller.

The overall energy stored is less than the possible capacity. One optimum storage capacity is when the stress induced in the entire body of the bolt is the same as the design stress. This condition is achieved by two methods as shown is Fig. 10.8(b) and Fig. 10.8(c).

In the first method the shank diameter is reduced to the core diameter of the threaded portion. This method removes the possibility of stress concentration and ensures uniform stress equal to the design stress throughout the body of the bolt and is called uniform strength bolt. This bolt can store maximum possible strain energy.

In the second method the hole is drilled in the shank portion of the bolt. The diameter of the drilled hole is such that the net cross-sectional area of the shank is equal to the core area. The stress induced in the entire length is the same in the magnitude but it can be smaller than the design stress due to the possible stress concentration at the junction of shank and threaded portion.

Thus, the resilience capacity of the reduced shank bolt is optimum. This solution is better.

Table 10.1. Metric-mechanical property classes for steel bolts screws and studs

Property class	Size range	Proof strength, S_p MPa	Yield strength, S_y MPa	Tensile strength, S_u MPa	Material
4.6	M5 to M36	225	250	400	Low or medium carbon steel
4.8	M1.6 to M16	310	340	420	-do-
5.8	M5 to M24	380	420	520	-do-
8.8	M1.6 to M36	600	660	830	Medium carbon steel Q & T
9.8	M1.6 to M36	650	720	900	-do-
10.9	M5 to M36	830	940	1040	Low carbon-martensite steel Q & T
12.9	M1.6 to M36	970	1100	1220	Alloy steel (Q & T)

Table 10.2. Properties of some bolt materials

Materials	Ultimate strength, MPa	Yield point, MPa	Elastic limit, MPa	Endurance limit, MPa
Low carbon steel (Hot rolled) C–0.10 to 0.20	412	268	212	172
Low carbon steel (Hot rolled) C–0.20 to 0.30	460	295	240	256
Medium carbon steel (Annealed) C–0.30 to 0.40	480	316	288	240
Ni–Steel (Heat treated drawn at 400°C) Ni–3.25 to 3.75 C–0.35 to 0.45	865	715	652	343
Ni–Cr. Steel (Heat treated drawn at 540°C) Ni–1.00 to 1.5 C–0.35 to 0.45 Cr–0.45 to 0.75	1030	961	652	550
Cr–Va steel (Heat treated drawn at 430°C) Cr–0.80 to 1.10, C–0.45 to 0.55, Va–0.18	1566	1442	1167	550

Table 10.3. Working stress and load for Indian metric coarse thread

Major diameter (mm)	Stress area, A_S (sq.mm)	Design stress, (MPa)	Permissible load, (N)
16	157	19	2972
20	245	23	5590
24	353	27	9575
30	561	32	18050
36	817	38	30900
42	1120	43	48363
48	1472	48	71122
56	2030	55	111835

Table 10.4. Basic dimension for design profiles of ISO metric screw threads

Basic diameter, *d* mm	Pitch, *p* mm	Stress area, A_S mm^2	Basic diameter, *d* mm	Pitch, *p* mm	Stress area, A_S mm^2
1	2	3	1	2	3
1	0.25	0.46	3	0.50	5.03
	0.20	0.53		0.35	5.61
1.1	0.25	0.59	3.5	0.60	6.78
	0.20	0.67		0.35	7.90
1.2	0.25	0.73	4	0.70	8.78
	0.20	0.82		0.50	9.79
1.4	0.30	0.98	4.5	0.75	11.3
	0.20	1.17		0.50	12.8
1.6	0.35	1.27	5	0.88	14.2
	0.20	1.59		0.50	16.1
1.8	0.35	1.70	6	1.00	20.1
	0.20	2.66		0.75	22.0
2	0.40	2.07	7	1.00	28.9
	0.25	2.45		0.75	31.3
2.2	0.45	2.48	8	1.25	36.6
	0.25	3.03		1.00	39.2
				0.75	41.8
2.5	0.45	3.39			
	0.35	3.70	Note: Basic diameter = Major diameter		
			33	3.5	694
10	1.5	58.0		3	716
	1.25	61.2		2	761
	1.00	64.5		1.5	785
	0.75	67.9	35	1.5	886
12	1.75	84.3			
	1.5	88.1	36	4	817
	1.25	92.1		3	865
	1.0	96.1		2	915

(Contd.)

Table 10.4. (Contd.)

Basic diameter d mm	Pitch, p mm	Stress area, A_S mm^2	Basic diameter, d mm	Pitch, p mm	Stress area, A_S mm^2
14	2	115	36	1.5	940
	1.5	125	39	4	976
	1.25	129		3	1030
	1	134		2	1080
16	2	157		1.5	1110
	1.5	167			
	1	178	42	4.5	1120
18	2.5	192		4	1150
	2	204		3	1210
	1.5	216		2	1260
	1	229		1.5	1290
20	2.5	245	45	4.5	1300
	2.0	258		4	1340
	1.5	272		3	1400
	1	285		2	1460
22	2.5	303		1.5	1490
	2	318			
	1.5	333	48	5	1470
	1	348		4	1550
24	3	353		3	1600
	2	384		2	1670
	1.5	401		1.5	1710
	1	418	52	5	1760
25	3	385		4	1830
27	3	459		3	1900
	2	496	52	2	1970
	1.5	514		1.5	2010
	1	533			

(Contd.)

Table 10.4. (Contd.)

Basic diameter, *d* mm	Pitch, *p* mm	Stress area, A_S mm^2	Basic diameter, *d* mm	Pitch, *p* mm	Stress area, A_S mm^2
30	3.5	561	56	5.5	2030
	3	581		4	2140
	2	621		3	2220
	1.5	642		2	2300
	1	663		1.5	2340
60	5.5	2360	80	6	4340
	4	2490		4	4570
	3	2570			
	2	2650		3	4680
	1.5	2700		2	4790
64	6	2680		1.5	4850
	4	2850	85	6	4950
	3	2940		4	5190
	2	3030		3	2310
	1.5	3080			
68	6	3060		2	5430
	4	3240	90	6	5590
	3	3340		4	5840
	2	3430		3	5940
	1.5	3480			
72	6	3460		2	6100
	4	3660	95	6	6270
	3	3760		4	6540
	2	3860		3	6670
	1.5	3910		2	6810
76	6	3890	100	6	7000
	4	4100		4	7280
	3	4210		3	7420
	2	4320		2	7560
	1.5	4370			

Table 10.5. Value of K for Eqns. 10.7 & 10.8

Types of joints	K
Soft packing with studs	1.00
Soft packing with through bolts	0.75
Asbestos	0.60
Soft copper gasket with long through bolts	0.50
Hard copper gasket with long through bolts	0.25
Soft copper corrugated gasket	0.04
Hard copper gasket with long through bolts	0.01
Metal-to-metal joints through bolts	0.00

Table 10.6. Modulus of elasticity for gasket materials

Materials	E_c-MPa
Cork	86
Compressed asbestos	480
Copper asbestos	93E3
Copper (pure)	121E3
Plain rubber	69
Spiral wound	280
Teflon	240
Vegetable fibre	120

Table 10.7. Torque factors K_i for Eqn. 10.10

Bolt conditions	K_1
Non-plated black finish	0.3
Zinc plated	0.2
Lubricated	0.18
Cadmium plated	0.16
With Bowman Anti-seize	0.12
With Bowman Grip nuts	0.09

An average value of $K_1 = 0.2$ is generally used.

Table 10.8. Fatigue strength reduction factors K_f for threaded elements

Metric grade	Rolled threads	Cut threads	Fillet
3.6 to 5.8	2.2	2.8	2.1
3.6 to 10.9	3.0	3.8	2.3

SOLVED EXAMPLES

Example 10.1: Figure 10.9 illustrates the connection of a cylinder head to a pressure vessel using 10 bolts and a confined gasket. The dimensions in mm are $A = 100, B = 200, C = 300, D = 20$ and $E = 25.0$ mm. The cylinder is used to store gas at a static pressure of 6.0 MPa. Metric grade 5.8 bolts are to be used with a factor of safety of at least 3.0. What size of the bolts are to be used?

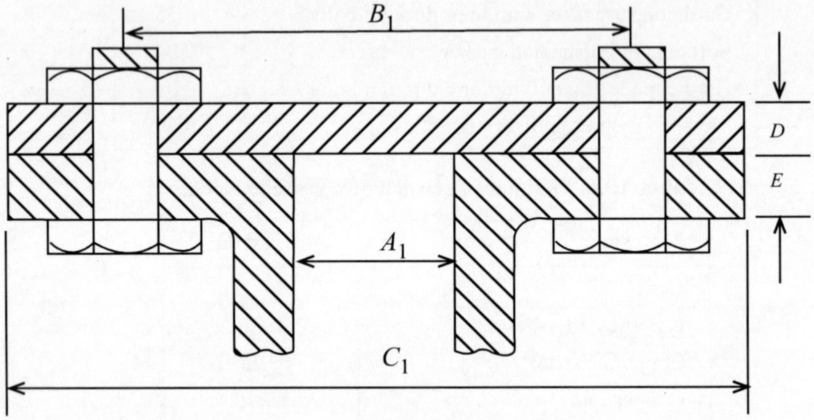

Figure 10.9

SOLUTION

Step-1: Load per bolt
Gas load on the cover,

$$P_g = \frac{\pi}{4} A^2 \cdot p = \frac{\pi}{4}(100)^2 \times 6.0 = 4.7 \times 10^4 \text{ N}$$

Applied load per bolt,

$$F_a = P_g / n = \frac{4.7 \times 10^4}{10} = 4.7 \times 10^3 \text{ N}$$

Stiffness constant for hard copper gasket, (Table 10.5)

$$K = 0.25$$

Minimum initial load, $F_i = (1 - K) F_a$

$$\text{Let } F_i = 1.5(1 - K)F_a = 1.5 \times (1 - 0.25) \times 4.7 \times 10^3 = 5287.5 \text{N}$$

Then the total bolt load,
$F = F_i + KF_a$

$$F = 0.25 \times 4.7 \times 10^3 + 5.2875 \times 10^3 = 6.4625 \times 10^3 \text{N}$$

Step-2: Bolt design
For property class 5.8 tensile strength, S_u = 520.0 MPa (Table 10.1)
Design stress, $S_d = S_u/FS = 520/3 = 173.3$ MPa
Stress area, $A_s = F/S_d$ = 6462.5 / 173.3 = 37.29 mm^2
Bolt size from Table 10.4 is

$$\text{M8} \times 1.0 \text{ with } A_s = 39.2 \text{ mm}^2$$

Circumferential pitch, $P_c = \pi \times 200/10 = 62.8 \text{ mm} > (6 \times d = 6 \times 8 = 48 \text{ mm})$ Not acceptable
Hence, choose M12 × 1.5 bolts with $A_s = 88.1$ mm^2
Then $P_c = \pi \times 200/10 = 62.8 \text{ mm} < 6d\ (= 6 \times 12 = 72 \text{ mm})$ Acceptable

Example 10.2: The cylinder head of a 200 mm × 350 mm compressor is secured by means of 12 studs of rolled mild steel. The gas pressure is 1.5 N/mm^2 gauge. The initial tension in the bolts, assumed to be equally loaded such that a cylinder pressure of 3.0 N/mm^2 gauge is required for the joint to open, suggest the suitable size of studs in according with Soderberg equation, assuming equivalent diameter of the compressed parts to be twice the size of the bolt and factor of safely of 2.0. The stress concentration factor may be taken as 2.8. The endurance limit stress is $S_{en} = 0.5 S_u$.

SOLUTION
Step-1: Stiffness constant
Stiffness of bolt, $$K_b = \frac{AE}{l}$$

Then stiffness of compressed parts, $$K_c = \frac{A_c \cdot E}{l} = \frac{4AE}{l} = 4K_b$$

Assuming E same for both bolt and compressed flanges.

Stiffness constant for the joint,

$$K = \frac{K_b}{K_b + K_c} = K_b/(K_b + 4K_b) = 0.20$$

Step-2: Then the initial load for opening the joint at $p = 3.0$ MPa

$$F_i = (1 - K) F_a$$

where,

$$F_a = \frac{\pi}{4} D^2 p/n = \frac{\pi}{4}(200)^2 \times 3/12 = 7854 \text{ N}$$

$$\therefore F_i = (1-0.2)\,7854 = 6283.18 \text{ N}$$

Step-3: The applied load, F_a for normal gas pressure

$$p = 1.5 \text{ N/mm}^2$$

$$F_a = \frac{\pi}{4} \times (200)^2 \times 1.5/12 = 3927 \text{ N}$$

This is the maximum load and minimum load is nil.
Then mean load, $F_{am} = 3927/2 = 1963.5$ N
Load amplitude, $F_{am} = 3927/2 = 1963.5$ N
Step-4: Equation for resultant load

$$F = F_i + K\left[F_{am} + K_f \frac{S_y}{S_{en}} F_{aa}\right]$$

where, $K_f = 1 + q\,(K_t - 1) = 1+0.9(2.8\text{-}1) = 2.62$

$$S_y = 295 \text{ MPa}, S_u = 460 \text{ MPa}$$

$$S_{en} = 0.5\,S_u = 0.5 \times 460 = 230 \text{ MPa}$$

$$\therefore \quad F = 6283.18 + 0.2\left[1963.5 + 2.62\,\frac{295}{230} \times 1963.5\right] = 7995.5 \text{ N}$$

Step-5: Bolt design
The design stress, $S_d = S_y/FS = 295/2 = 147.5$ MPa

$$\therefore \quad \text{Stress area, } A_s = 7995.5\ /\ 147.5 = 54.2 \cong 55 \text{ mm}^2$$

The suitable bolt size (Table 10.4) is, M10 × 1.5 mm with A_s = 58 mm^2

10.6 ECCENTRICALLY LOADED BOLT JOINTS

Case 1 : The load is parallel but away from the plane of bolts (Fig. 10.10)
The bracket has tendency to turn about the bottom edge.
The bolts are subjected to a combination of tensile and shear loads. The tensile load is proportional to the distance of the bolt from the edge and shear load is uniformly distributed.
If F_1, F_2, F_3, F_4 are tensile forces in the bolts.

$$\frac{F_1}{l_1} = \frac{F_{i2}}{l_1} = \frac{F_3}{l_2} = \frac{F_4}{l_2} = K \text{ (a constant)}$$

$$F_1 = F_2 \text{ and } F_3 = F_4$$

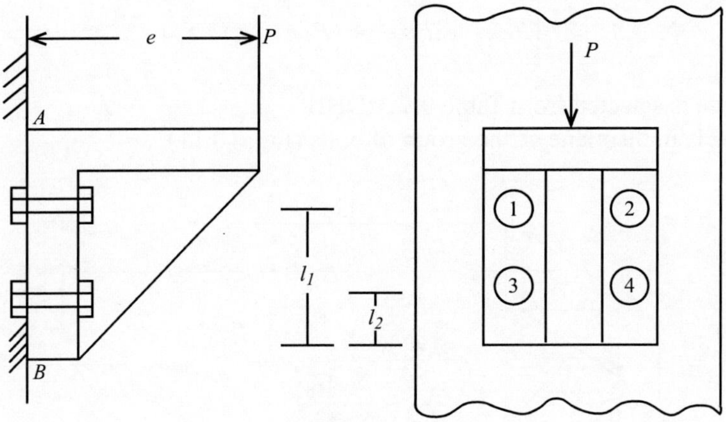

Figure 10.10

The moment of these forces about the bottom edge of the bracket is

$$\begin{aligned} M_r &= l_1 F_1 + l_1 F_2 + l_2 F_3 + l_2 F_4 \\ &= (l_1^2 + l_1^2 + l_2^2 + l_2^2) K \end{aligned}$$

The overturning moment due to the applied load P,

$$M_a = P.e$$

Equating the two, i.e., $M_r = M_a$

$$K = \frac{P.e.}{l_1^2 + l_1^2 + l_2^2 + l_2^2} = \frac{P.e}{\Sigma l_i^2}$$

$$\therefore \quad F_1 = F_2 = \frac{P.e}{\varepsilon l^2} l_1 \tag{10.21}$$

These forces are maximum

The tensile stress, $\sigma = F_1 / A_s$

$$A_s = \text{Cross-sectional area of the bolt}$$

$$\text{The shear stress, } \tau = \frac{P/n}{A_s}$$

By the maximum principal stress theory,

$$\begin{aligned} \sigma_i &= \frac{\sigma}{2} + \sqrt{\left(\frac{\sigma}{2}\right)^2 + \tau^2} \\ &= \left(\frac{1}{2A_s}\right)\left[F_1 + \sqrt{F_1^2 + 4(P/n)^2}\right] = S_d \end{aligned}$$

$$\therefore \quad A_s = \left[F_1 + \sqrt{F_1^2 + 4\,(P/n)^2}\right] / (2\,S_d) \tag{10.22}$$

The required bolt size is selected from Table 9.7 MDHB.

Case 2: The loading is in the plane of the group of bolts (Fig. 10.11)

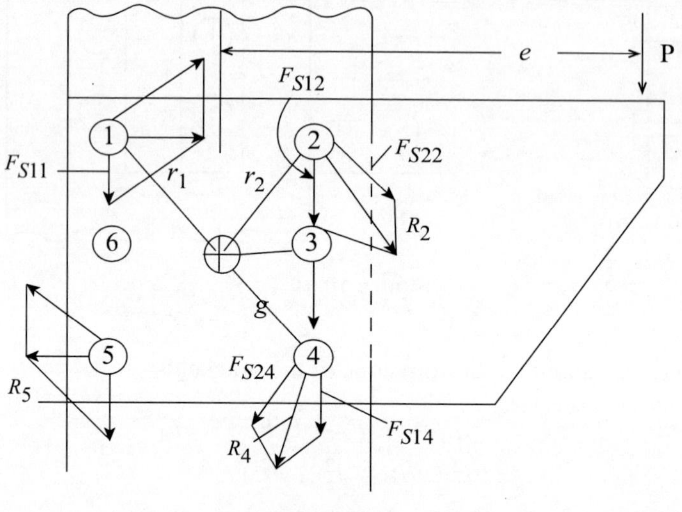

Figure 10.11

The load line has an eccentricity, e w.r.t. the C.G. of the bolts. The moment due to the eccentricity causes a secondary shear force F_{s2} perpendicular to the line joining the bolt centre to the C.G. This force is proportional to the distance r from C.G., i.e, F_{s2} and r are connected by the relation $F_{s2i} = k.r_i$

i.e.,

$F_{S21} = kr_1$	$F_{S23} = kr_3$	$F_{S25} = kr_s$
$F_{S22} = kr_2$	$F_{S24} = kr_4$	$F_{S26} = kr_6$

The value of k constant of proportionality is found by equating the resisting moment due to these forces to the external moment = $P.e$

$$\therefore \qquad k = \frac{P.e}{\Sigma r_i^2}$$

$$F_{S21} = \frac{P.e}{\Sigma r_i^2} r_i \,;\; F_{S22} = \frac{P.e}{\Sigma r_i^2} r_2 \text{ ate} \tag{10.23}$$

This force is added vectorially to the primary shear force.

$F_{si} = P/n$ to get the resultant force R_1

The maximum resultant force R_{max} is considered for design.

Example 10.3: A wall brake carries an inclined load $P = 16$ kN at angle of $\theta = 30°$ with vertical as shown in Fig. 10.12(a). The braket is fixed to the vertical column by 6 bolts spaced as shown in Fig. 10.12(b). Calculate the diameter of the bolts if permissible tensile stress is 60 MPa.

SOLUTION

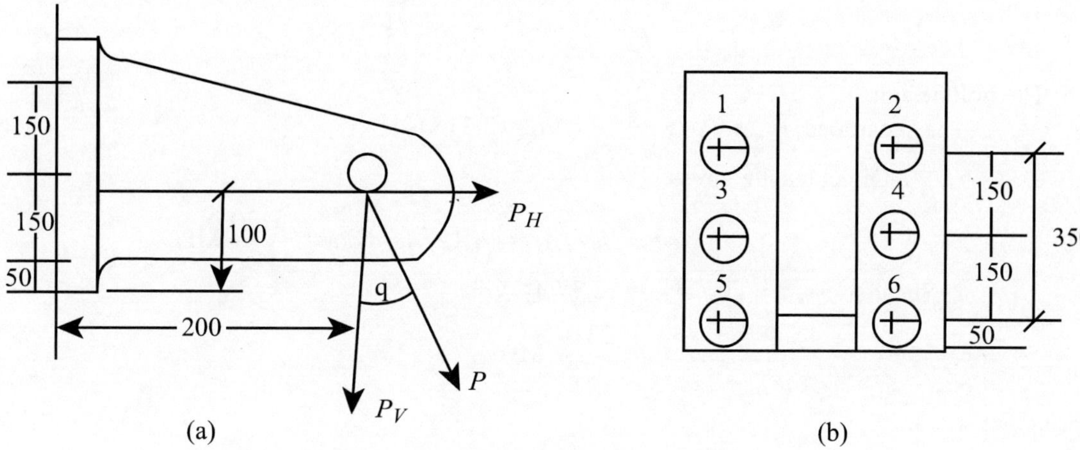

Figure 10.12

Step-1: Loads on the bolts

$$P_H = P\sin\theta = 16\sin 30° = 8 \text{ kN}$$

$$P_V = P\cos 30° = 16\cos 30° = 13.87 \text{ kN.}$$

$$\Sigma l^2 = \Sigma(2 \times 350^2 + 2 \times 200^2 + 2 \times 50^2)$$

$$= (2 \times 165000)$$

Step-2: Moment of the applied force about the bottom edge

$$M = 100 \times P_H + 200 \times P_V$$

$$= 800 + 2774$$

$$= 3574 \text{ kN. mm}$$

Step-3: The maximum bending force in bolt 1 or 2

$$F_1 = F_2 = \frac{M \cdot l_1}{\Sigma l_i^2}$$

$$F_1 = F_2 = \frac{3574 \times 350}{2 \times 165000}$$

$$= 3.79 \text{ kN}$$

Step-4: Maximum combined loads on bolts

The direct tensile load due to P_H,

$$F_{11} = P_H/6 = 8/6 = 1.33 \text{ kN}$$

$\therefore$ Total tensile load in bolts 1, and 2,

$$F_{1T} = F_{2T} = 1.33 + 3.79 = 5.12 \text{ kN}$$

Step-5: The bolt design

The shear load, $F_{s1} = P_V/6 = 13.87/6 = 2.31$ kN

$\therefore$ Direct tensile stress,

$$\sigma = F_{iT}/A_s = (5.12/A_s) \text{ GPa} = \frac{5120}{A_s} \text{ MPa}$$

$$\text{Shear stress, } \tau = F_{si}/A_s = (2.3/A_s) \text{ GPa}$$

$$= \frac{2310}{A_s} \text{ MPa}$$

The principal stress,

$$\sigma_1 = \frac{\sigma}{2} + \sqrt{\left(\frac{\sigma}{2}\right) + \tau^2} < S_d$$

$$\therefore \quad \frac{5.12}{2A_s} + \sqrt{\left(\frac{5.120}{2A_s}\right)^2 + \left(\frac{2.31}{A_s}\right)^2} = 60 \times 10^{-3}$$

$$\therefore \quad A_s = \left[2.56 + \sqrt{2.56^2 + 2.31^2}\right] / 60.0 \times 10^{-3}$$

$$= 100 \text{ mm}^2$$

Refer Table 10.4 and choose M14 × 2 bolt with $A_s = 115 \text{ mm}^2$.

Example 10.4: A bracket is bolted to a column by 6 bolts of equal size as shown in Fig. 10.13 It carries a load of 60 kN at a distance of 100 mm from the centre of the column. If the max. stress in the bolt is limited to 150 MPa, determine the bolt dia?

SOLUTION

Step-1: Loads on the bolts

The primary shear on each bolt is $F_{si} = P/n = 60/6 = 10$ kN

The bolt 1, 2, 4, 5 are at the farthest distance of $r_1 = r_2 = r_4 = r_5$

$$= \sqrt{75^2 + 50^2} = 90.14 \text{ mm}$$

The maximum secondary force,

$$F_{s26} = F_{s22} = \frac{P.e}{\Sigma r_i^2} r_i \qquad \text{(Eqn. 10.23)}$$

$$= \frac{60 \times 200 \times 90.14}{\Sigma (4 \times 90.14^2 + 2 \times 50^2)} = 28.85 \text{ kN}$$

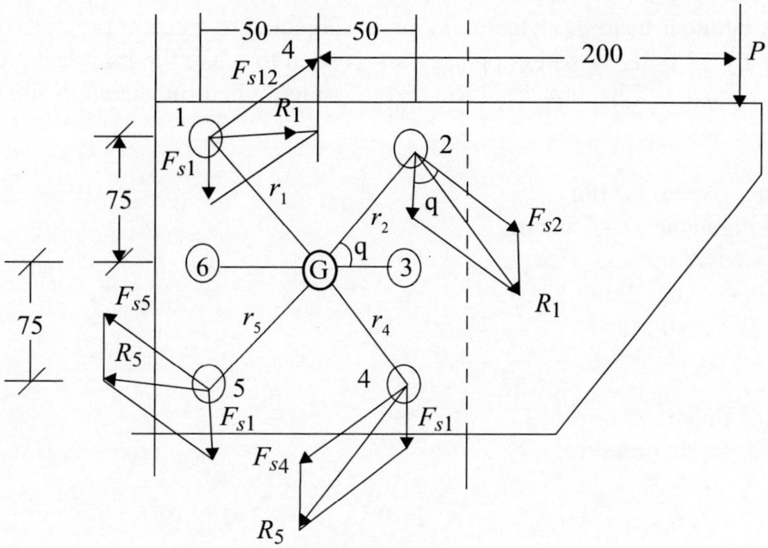

Figure 10.13

The angle between these two forces,

$$\theta = \tan^{-1}\left(\frac{75}{50}\right) = 56.31^\circ$$

Step-2: The resultant force

$$F_{SR} = \sqrt{F_{s1}^2 + F_{s2}^2 + 2F_{si} \times F_{s2}\cos\theta}$$

$$= \sqrt{\sqrt{10^2 + 28.85^2 + 2 \times 10 \times 28.85 \cos 56.31^\circ} = 35.39\text{ kN}}$$

Step-3: Bolt design
The permissible tensile stress = 150 MPa
The permissible shear stress = 75 MPa
∴ The bolt area required, $A_s = (35.39 \times 10^3/75) = 471.85\,\text{mm}^2$
The diameter of the bolt,

$$d = \sqrt{\frac{4 \times 471.85}{\pi}} = 24.5\text{ mm}$$

Choose $d = 25.0$ mm

Example 10.5: A cylinder head of a steam engine is subjected to a steam pressure of 0.8 MPa. It is held in position by 12 bolts. A soft copper gasket is used to make the joint leakproof. The bore diameter of the cylinder is 250 mm. Find the size of bolts, when the stress in the bolt is not to exceed 110 MPa.

Given Data

The steam pressure, $p_a = 0.8$ MPa
The cylinder bare diameter, $D = 250$ mm
Number of bolts used, $N = 12$
Soft copper gasket, $K = 05$ (Table 10.5)
Allowable stress, $S_d = 110$ MPa.

SOLUTION

Step-1: Load on the bolts
Applied load due to steam pressure,

$$F_a = \frac{\pi}{4} D^2 p_a = \frac{\pi}{4} \times (250)^2 \times 0.8 = 39269.91 \text{ N}$$

Practical value of initial tightening load (Eqn. 10.9)

$$\begin{aligned} F_i &\geq F_a \left[\frac{K_c}{K_b + K_c} \right] = F_a \left[1 - \frac{K_b}{K_b + K_c} \right] \\ &\geq F_a [1 - K] = 39269.91(1 - 0.5) \\ &\geq 19634.95 \text{ N} \end{aligned}$$

Let $F_i = 20,000$ N.
Then the total load on the bolts (Eqn. 10.7)

$$\begin{aligned} F &= F_i + KF_a \\ &= 20,000 + 0.5 \times 39269.91 \\ &= 39634.955 \text{ N} \end{aligned}$$

Force on each bolt,

$$\begin{aligned} F' &= F/N \\ &= 39634.955/12 \\ &= \mathbf{3302.91\ N} \end{aligned}$$

Step-2: Bolt design
Then for $S_d = 110$ MPa, the bolt diameter

$$d = \sqrt{\frac{4 \times 3302.91}{\pi \times 110}} = 6.18.$$

Choose the next suitable diameter (from Table 10.4) designated as $\boxed{\text{M7} \times 0.75}$ or $\boxed{\text{M8} \times 1.25}$

OR stress-area required,

$$A_s = \frac{F'}{S_d} = \frac{3302}{110} = 30.018 \text{ mm}^2$$

The bolt size from Table 10.4 M8 × 1.2 with $A_s = 36.6$ is suitable or M7 × 0.75 may be chosen.

Example 10.6: A bracket is fixed to the wall by means of four bolts and loaded as shown in Fig. 10.14. Calculate the size of the bolt if the load is 25 kN and yield stress is 380 N/mm^2. The factor of safety is taken as 2.5. Use maximum shear theory.

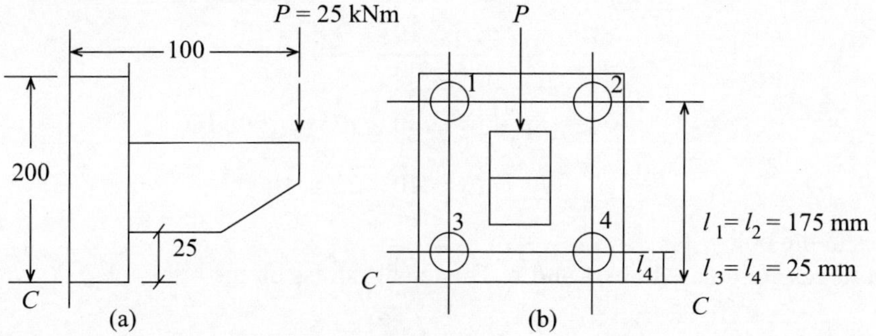

Figure 10.14

Given Data

Applied load, $P = 25$ kN = 25×10^3 N

Load eccentricity, $e = 100$ mm

Yield stress for bolt, $S_y = 380$ MPa

Factor of safety, $n = 2.5$

SOLUTION

Step-1: Forces on the bolts

Shear force on each bolt, $F = \dfrac{25 \times 10^3}{4} = 6250$ N

The bending moment on the bolts is due to eccentricity;

$$M = P.e = 25 \times 10^3 \times 100 = 25 \times 10^5 \text{ N-mm}$$

The bracket overturns about bottom edge C due to the moment and the bolts are subjected to tensile loads, F_1, F_2, F_3 & F_4. Their magnitudes are proportional to the distances from edge $C - C$.

$$F_1 = Kl_1 \text{ and } F_2 = Kl_2$$
$$F_3 = Kl_3 \text{ and } F_4 = Kl_4$$

The resisting moment due to these forces,

$$\begin{aligned} M_r &= F_1.l_1 + F_2.l_2 + F_3.l_3 + F_4.l_4 \\ &= K\left[l_1^2 + l_2^2 + l_3^2 + l_4^4\right] \\ &= K\left[175^2 + 175^2 + 25^2 + 25^2\right] \\ &= [62500] \text{ kN-mm} \end{aligned}$$

Equating to the applied moment,

$$25 \times 10^5 = 62500 \text{ K}$$

$$\therefore \qquad K = \frac{25 \times 10^5}{62500} = 40$$

Then

$$F_1 = F_2 = 40 \times 175 = 7000 \text{ N}$$

$$F_3 = F_4 = 40 \times 25 = 1000 \text{ N}$$

Step-2: Stresses in the bolt

The maximum forces are on the bolts 1 and 2. The tensile stress on the bolts,

$$\sigma = \left(7000 \Big/ \frac{\pi}{4}\, d^2\right) = \left(\frac{8912.68}{d^2}\right) \text{ MPa}$$

The transverse shear stress,

$$\tau = \left(6250 \Big/ \frac{\pi}{4}\, d^2\right) = \left(\frac{7957.75}{d^2}\right) \text{ MPa}$$

The maximum shear stress,

$$\begin{aligned} \tau_{\max} &= \sqrt{\left(\frac{\sigma}{2}\right)^2 + \tau^2} \\ &= \sqrt{\left(\frac{8912.68}{2d^2}\right)^2 + \left(\frac{7957.75}{d^2}\right)^2} \\ &= \frac{4456.34}{d^2}\sqrt{1 + \left(\frac{7957.75}{4456.34}\right)^2} \\ &= \frac{9120.57}{d^2} \end{aligned}$$

Step-3: Bolt design

Design stress in shear,

$$S_{sd} = \frac{S_{sy}}{n} = \frac{S_y}{2n} = \frac{1}{2} \times \frac{380}{2.5} = 76 \text{ MPa}$$

∴ The bolt diameter,

$$d^2 = \frac{9120.57}{76} = 120.00$$

$$\therefore \quad d = 10.955 \text{ mm}.$$

Choose M12 × 1.0 bolts

Example 10.7: The cylinder head of a reciprocating air compressor is held in place by ten bolts. The total joint stiffness is four times the total bolt-stiffness. Each bolt is tightened to an initial tension of 5 kN. The total extreme force acting to separate the joint is 20 kN. Find the size of the bolts so that the stress in bolts does not exceed 100 MPa.

Given Data

Initial bolt-load, $F_i = 5 \text{ kN} = 5 \times 10^3$ N

Applied load, $F_a = 20 \text{ kN} = 20 \times 10^3$ N

Stiffness, $K_T = 4K_b = K_b + K_c$

Allowable stress, S_d = 100 MPa

Number of bolts, N = 10

SOLUTION

Step-1: Joint stiffness factor

Total stiffness,

$$K_T = K_b + K_c = 4K_b$$
$$\therefore K_c = 3K_b$$

∴ Stiffness factor,

$$K = \frac{K_b}{K_b + K_c} = \frac{K_b}{K_b + 3K_b} = \frac{1}{4} = 0.25$$

Step-2: Total to bolt load on each bolt (Eqn. 10.7)

$$F_b = F_i + KF_a$$
$$= 5000 + 0.25 \times \frac{20000}{10}$$
$$= 5500 \text{ N}$$

Step-3: Bolt design

The bolt area,

$$A_s = {}^{P}/_{S_d} = \frac{5500}{100} = 55 \text{ mm}^2$$

The bolt size (from Table 10.4) is

M 10 × 1.5

Example 10.8: A radial drilling machine with circular base is minuted to a base plate by means of three steel bolts equally spaced on a bolt circle diameter of 0.3 m. The diameter of the circular plate is 0.4 m. The spindle is subjected to a force of 4.5 kN. Determine the size of the bolts, if the allowable stress in bolt material is 100 MPa and the spindle is positioned at radial distance of 0.335 m.

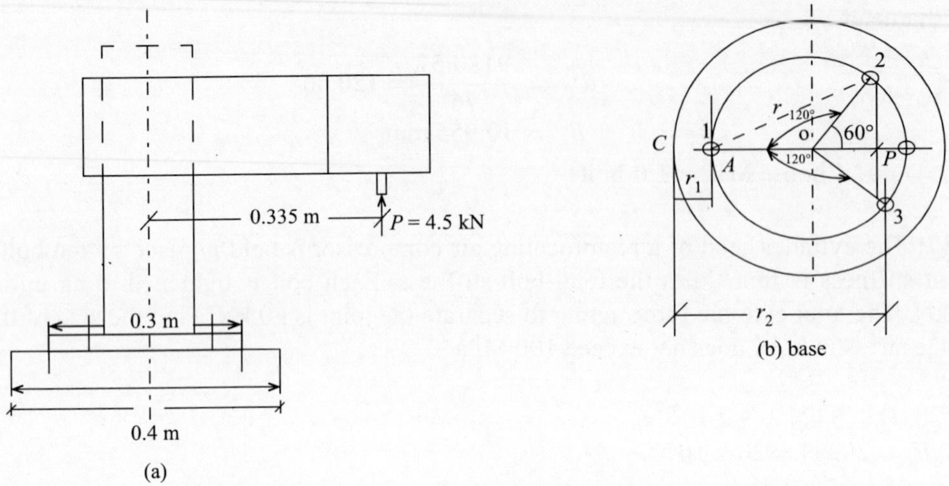

Figure 10.15

SOLUTION

Step-1: Distances of bolts from C

Under the load the plate over turn about C

The distance of bolt from C,

$$\begin{aligned} r_1 &= \text{CA} = \text{CO} - \text{AO} \\ &= \frac{0.4}{2} - \frac{0.3}{2} = 0.05 \text{ m} \\ r_2 &= r_3 = \text{CO} + \text{OP} \\ &= 0.2 + 0.3\cos 60^\circ \\ &= 0.275 \text{ m} \end{aligned}$$

Step-2: Forces on bolts

The forces on bolts are

$$F_1 = Kr_1 = 0.05\,\text{K}$$
$$F_2 = F_3 = Kr_2 = Kr_3 = 0.275\,\text{K}$$

K is a constant of proportionality.

Then applied moment = Resisting moment

$$4.5 \times 0.335 = K[0.05^2 + 2 \times 0.275^2] = 0.15375\text{ K}$$

$$K = \frac{4.5 \times 0.335}{0.15375} = 9.8$$

$$F_2 = Kr_2 = 9.8 \times 0.275 = 2.7 \text{ kN}$$

Primary tension,

$$P = 4.5/3 = 1.5 \text{ kN}$$

Maximum tension,

$$F_{max} = 1.5 + 2.7 = 4.2 \text{ kN}$$

Step-3: Bolt design
The bolt area required,

$$A_s = \frac{4.2 \times 10^3}{100} = 42.0 \text{ mm}^2$$

The suitable bolt size from Table 10.4 is

$$\text{M10} \times 1.5 \text{ with } A_s = 58.0 \text{ mm}^2$$

Example 10.9: Suggest the suitable size of bolts. If the bracket shown in Fig. 10.16 is subjected to a load of 20 kN at 60° from the vertical. The allowable tensile stress in the bolts is 120 MPa. The bracket is fixed using four bolts.

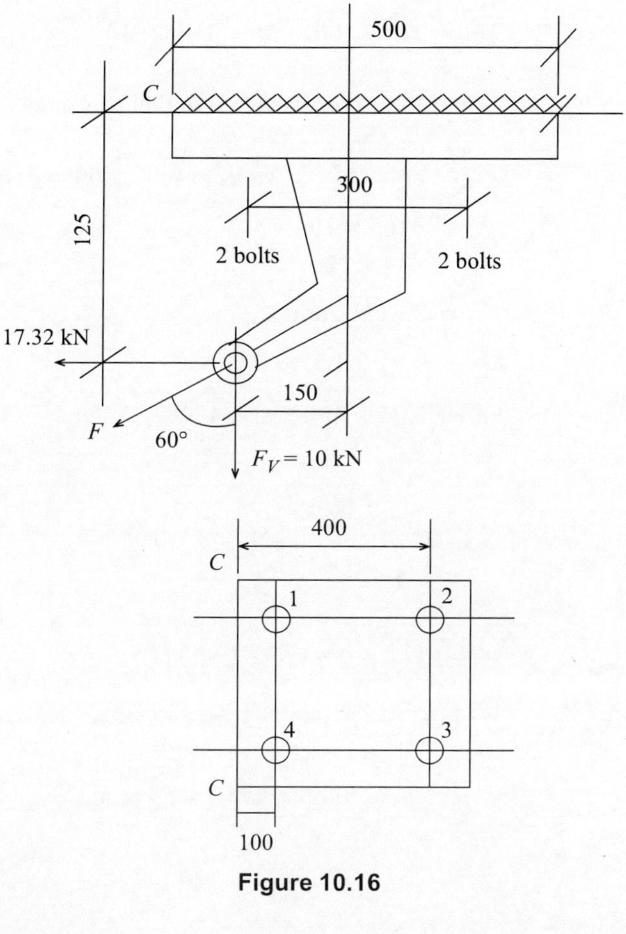

Figure 10.16

SOLUTION

Step-1: Force components

$$\begin{aligned} F &= 20 \text{ kN} \\ F_V &= 20 \cos 60^\circ = 10 \text{ kN} \\ F_H &= 20 \sin 60^\circ = 17.32 \text{ kN} \end{aligned}$$

Step-2: Forces on the bolts

Bending moment on the bracket about edge $C-C$,

$$\begin{aligned} M &= 17.32 \times 125 + 10 \times (250 - 150) \\ &= 3165 \text{ kN.mm} \\ &= 3165 \times 10^3 \text{ N.mm} \end{aligned}$$

The secondary tensile forces due to bending,

$$2F_1 \times 100 + 2F_2 \times 400 = M = 3165 \times 10^3$$

because $F_1 = F_4$ and $F_2 = F_3$ and F_1 and F_2 are proportional to their distance.

$$2K \times 100 \times 100 + 2 \times K \times 400 \times 400 = 3165 \times 10^3$$

$$\begin{aligned} K &= \frac{3165 \times 10^3}{34 \times 10^4} = 9.31 \\ &= 9.31 \text{ N/mm} \end{aligned}$$

$$\therefore \quad \begin{aligned} F_1 &= 9.31 \times 100 = 931 \text{ N} \\ F_2 &= 9.31 \times 400 = 3724 \text{ N} \end{aligned}$$

Primary shear force/bolt,

$$P_s = 17.32/4 = 4.33 \text{ kN} = 4330 \text{ N}$$

Primary tensile force,

$$P_s = 10/4 = 2.5 \text{ kN} = 2500 \text{ N}$$

Total maximum tensile force,

$$F_{max} = P_1 + F_2 = 2500 + 3724 = 6224 \text{ N}$$

Step-3: Stresses in bolts
Tensile stress,

$$\sigma = F_{max}/A_s = \left(\frac{6224}{A_s}\right) \text{ MPa} \quad \text{where } A_s = \text{area of bolt}$$

Transverse shear force,

$$F_s = F_h = 4330 \text{ N}$$

Transverse shear stress,

$$\tau = {}^{F_s}/_{A_s} = \left(\frac{4330}{A_s}\right) \text{ MPa}$$

Maximum normal stress,

$$\begin{aligned}\sigma_{max} &= \frac{\sigma}{2} + \sqrt{\left(\frac{\sigma}{2}\right)^2 + \tau^2} \\ &= \frac{6224}{2A_s} + \sqrt{\left(\frac{6224}{2A_s}\right)^2 + \left(\frac{4330}{A_s}\right)^2} \\ &= \frac{3112}{A_s} + \frac{3112}{A_s}\sqrt{\left[1 + \left(\frac{4330}{3112}\right)^2\right]} \\ &= \frac{3112}{A_s} + \frac{5332}{A_s} = \frac{8444.3}{A_s} \text{ MPa}\end{aligned}$$

Step-4: Design of bolt
By normal stress theory [neglecting the initial lightening, stress],

$$\begin{aligned}S_d &= \sigma_{max} \\ 120 &= \frac{8444.3}{A_s} \\ \therefore \quad A_s &= \frac{8444.3}{120} = 70.37 \text{ mm}^2\end{aligned}$$

Maximum shear stress,

$$\tau_{max} = \sqrt{\left(\frac{\sigma}{2}\right)^2 + \tau^2} = \left(\frac{5332}{A_s}\right)$$

By shear theory,

$$\begin{aligned}S_{sd} &= \tau_{max} \\ 120/2 &= \frac{5332}{A_s} \\ A_s = \frac{5332}{60} &= 88.87 \text{ mm}^2\end{aligned}$$

This is greater & hence A_s required $= 88.87\,\text{mm}^2$
The suitable bolt size is M 12×1.25 with $A_s = 92.1\,\text{mm}^2$ (Table 10.4)

Example 10.10: Estimate the maximum wrench torque which can be applied in tightening a 20 mm bolt if the shear stress in the body of the bolt is not to exceed 140 MPa. Outside bolt diameter = 20 mm; root diameter = 16.72 mm; the thread section has 60° included angle ($\theta = 30°$) pitch 2.5 mm; effective friction radius $r = 12$ mm; thread and collar friction coefficient $\mu = 0.10$.

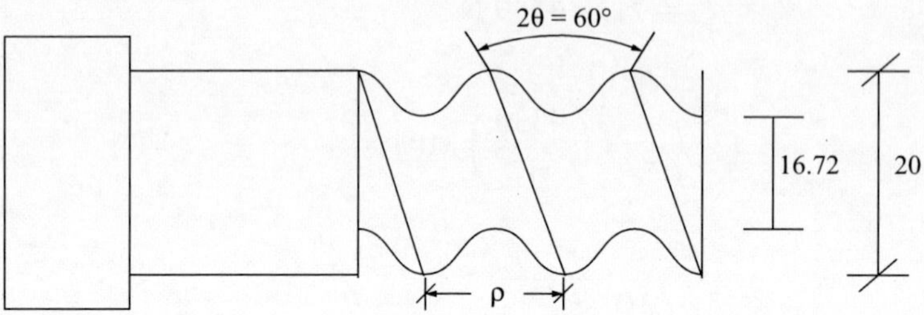

Figure 10.17

Given Data

Effective friction radius, $f_c = 12$ mm
Coefficient of friction, $\mu = 0.10$

SOLUTION

Step-1: The tightening torque applied to induce initial tension in the bolt. It can be evaluated from Eqn. 9.19 MDHB.

$$T_i = \frac{F_i d_m}{2}\left[\tan c\alpha + \varphi + fc\frac{d_l}{d_m}\right]$$

where,

$$\alpha = \tan^{-1}\left(\frac{L}{\pi d_m}\right)$$

$$L = \rho = 2.5 \text{ mm}$$

$$d_m = \frac{20 + 16.72}{2} = 18.36 \text{ mm}$$

$$\therefore \quad \alpha = \tan^{-1}\left[\frac{2.5}{\pi \times 18.16}\right] = 2.51°$$

The friction angle, $\varphi = \tan^{-1}(f \sec\theta)$ (Eqn. 9.21 MDHB)

$$= \tan^{-1}(f/\cos\theta) = \tan^{-1}\left(\frac{0.1}{\cos 30}\right) = 6.587$$

$$\therefore \quad \alpha + \varphi = 2.48 + 6.587 = 9.066°$$

Collar mean, $d_c = 2r_c - 2 \times 12 = 24$ mm

Then the torque is

$$\begin{aligned} T_i &= \frac{F_i \times 18.36}{2}\left[\tan(9.066) + \frac{0.1 \times 24}{18.36}\right] \\ &= 2.665\,F_i \end{aligned}$$

Step-2: Stresses induced
Shear stress induced,

$$\therefore \quad \tau = \frac{16.T}{\pi d_c^3} = \frac{16 \times 2.665\,F_i}{\pi \times (16.72)^3} = (0.0029\,F_i)\ \text{MPa}$$

Direct tensile stress,

$$\sigma = F_i/A_c = \frac{4F_i}{\pi d_c^2} = \frac{4 \times F_i}{\pi(16.72)^2} = 0.00455\,F_i$$

Maximum shear stress,

$$\begin{aligned} \tau_{\max} &= \sqrt{\left(\frac{\sigma}{2}\right)^2 + \tau^2} \\ &= \sqrt{(0.0029\,F_i)^2 + \left(\frac{0.00455\,F_i^2}{2}\right)} \\ &= 0.0037\,F_i \end{aligned}$$

Step-3: Initial force
Maximum allowable is limited 140 MPa.

$$\begin{aligned} \therefore \quad 140 &= 0.0037\,F_i \\ F_i &= 140/(0.0037) \\ &= 37983.0\ \text{N} \end{aligned}$$

Step-4: Applied torque

$$\begin{aligned} T_i &= 2.665\,F_i = 2.665 \times 37983.0\,\text{N} \\ &= 101224.45\ \text{N-mm} \\ &= 101.22445\ \text{Nm} \end{aligned}$$

Example 10.11: A bolt carries a tensile load of 8 kN and tightening load is 3 kN. It is made of steel having an allowable tensile stress of 120 MPa. Find its size. A soft-copper gasket is used.
Given Data
Tightening load, $F_i = 3$ kN
Applied load, $F_a = 8$ kN
Soft gasket is used
Allowable tensile stress, $S_d = 120$ MPa

SOLUTION
Step-1: Bolt load
For soft gasket the joint stiffness from Table 10.5.

$K = 0.5$

The load on the bolt, (Eqn. 10.7)

$$\begin{aligned} F &= K \cdot F_a + F_i \\ &= 0.5 \times 8 + 3 \\ &= 7\ \text{kN} = 7 \times 10^3\ \text{N} \end{aligned}$$

Step-2: Bolt design

For allowable stress $S_d = 120$, the bolt stress area required is

$$A_s = \frac{7 \times 10^3}{120} = 58.33\ \text{mm}^2$$

The suitable bolt size is M 10×1.25 $A_s = 61.2\ \text{mm}^2$. (Table 10.4)

Example 10.12: A bracket is bolted as shown in Fig 10.18. All bolts are identical and have a yield strength of 400 MPa. Determine the size of the bolts assuming factor of safety.

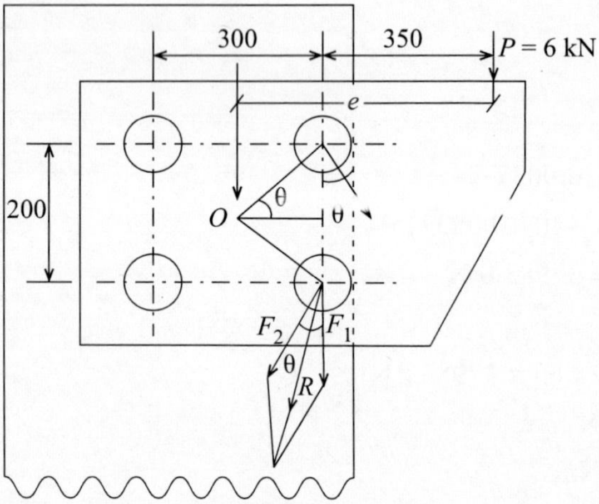

Load eccentricity,

$$e = 150 + 350 = 500\ \text{mm}$$

Figure 10.18

SOLUTION

Step-1: Loads on the bolts [Refer Fig. 10.18]

The effect of the applied load P is

(i) Primary shear load through C.G. '0' equally distributed an all the four bolts,

$$F_1 = P/n = \frac{6 \times 10^3}{4} = 1500\ \text{N}$$

(ii) Moment about *O*,

$$M = P.e = 6 \times 10^3 \times 500 \text{ N-mm}$$
$$= 30 \times 10^5 \text{ N-mm}$$

This moment creates a secondary shear force an each bolt. This force is proportional the distances from the C.G. The distances are equal to

$$r = \sqrt{100^2 + 150^2} = 180.277 \text{ mm}$$

The secondary force (Eq. 10.23)

$$F_2 = \frac{P.e.r_{\max}}{\varepsilon r_2} = \frac{30 \times 10^5 \times 180.277}{4 \times 180.277^2} = 4160.3 \text{ N}$$

The angle between the forces θ

$$\tan\theta = 100/150 = 0.667$$
$$\therefore \theta = 33.69^\circ$$

Step-2: The resultant shear force

$$F = \sqrt{F_1^2 + F_2^2 + 2F_1F_2\cos\theta}$$
$$= \sqrt{1500^2 + 4160.3^2 + 2 \times 1500 \times 4160.3\cos 33.69} = 5472.0 \text{ N}$$

Step-3: Bolt design
The allowable stress in shear,

$$S_{sd} = \frac{1}{2}S_d = \frac{S_y}{2n} = \frac{1}{2} \times \frac{400}{3} = 66.67 \text{ MPa}$$

The bolt stress area required,

$$A_s = F/S_{sd} = \frac{5472.0}{66.67} = 82.076 \text{ mm}^2$$

The suitable bolt size is (Table 10.4)
M12 × 1.75 with $A_s = 84.3 \text{ mm}^2$

Example 10.13: A wall bracket is attached to a wall by means of 4 identical bolts to *A* and two at *B* as shown in Fig. 10.19. Assuming that the bracket is held against the wall and prevented from tipping about C by all four bolts and using allowable stress of 35 MPa, determine the size of the bolts on the basis of maximum principal stress theory, selecting ISO metric threads of not more than 15 mm pitch.

Given Data

Applied load, $P = 25$ kN
Eccentricity of load, $e = 500$ mm
Allowable stress, $S_d = 35$ MPa

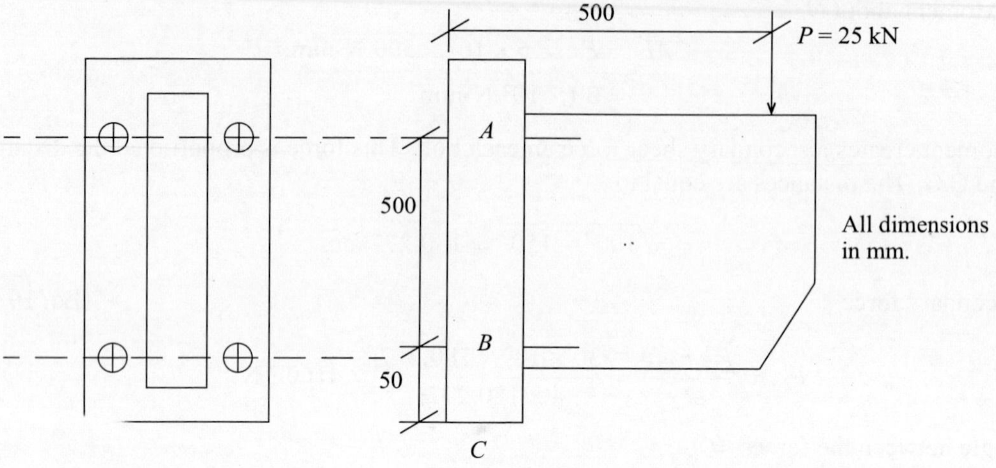

Figure 10.19

SOLUTION

Step-1: Loads on the bolts

The primary shear load per bolt,

$$F_S = \frac{25 \times 10^3}{4} = 6250\ \text{N}$$

There is secondary load on each bolt due to eccentric loading and evaluated by equalling the external moment to resisting moment

F_A = force in the bolts at A

F_B = force in the bolts at B.

Then

$$2 \times 550 \times F_A + 2 \times 50 \times F_B = 25 \times 10^3 \times 500$$

But $F_A = 550$ K and $F_B = 50$ K, where, K is a constant.

$$\therefore \quad 2 \times 550^2 \cdot K + 2 \times 50^2 \cdot K = 25 \times 10^3 \times 500$$

$$K = \frac{25 \times 10^3 \times 500}{2 \times 305 \times 10^3} = 20.49\ \text{N/mm}$$

$$\therefore \quad F_A = 20.49 \times 550 = 11270.49\ \text{N}$$

Step-2: Stresses in the bolt

The normal stress to this tensile load,

$$\sigma = F_A/A_s = [11270.49/A_s]$$

The shear stress due to shear load,

$$\tau = \frac{F_S}{A_s} = (6250/A_s)$$

Maximum principal stress,

$$\sigma_1 = \left(\frac{\sigma}{2}\right) + \sqrt{\left(\frac{\sigma}{2}\right)^2 + \tau^2}$$

$$= \left(\frac{11270.49}{2 \times A_s}\right) + \sqrt{\left(\frac{11270.49}{2A_s}\right)^2 + \left(\frac{6250}{A_s}\right)^2}$$

$$= \frac{5635.25}{A_s}\left[1 + \sqrt{1 + \left(\frac{6250}{5635.25}\right)^2}\right]$$

$$35 = \frac{14050.625}{A_s} \qquad \because [\sigma_1 = S_d]$$

$$\therefore \quad A_s = \frac{14050.625}{35} = 401.446 \text{ mm}^2$$

Step-3: Bolt design

The suitable size (Table 10.4)

M24 × 1 with $A_s = 418$ mm^2

Example 10.14: A cylinder head is fastened to the cylinder of a compressor using 6 bolts of M20 size. Bolt material is C20 steel. The maximum fluid pressure is 35 MPa; cylinder diameter is 75 mm. A soft gasket is used. Assuming the initial tension required in each bolt is 40 kN, determine the factor of safety.

Given Data

Number of bolts = 6

Bolt size = M20

Bolt material = C20

Fluid-pressure, $p = 3.5$ MPa

Cylinder diameter, $D = 75$ mm

Packing material = soft gasket

Initial tension = $F_i = 40$ kN.

SOLUTION

Step-1: Bolt load

Total fluid force on the cover,

$$F_T = \frac{\pi D^2}{4} \cdot p = \frac{\pi \times 75^2}{4} \times 3.5 = 15462.53 \text{ N}$$

$\therefore$ Applied load on each bolt,

$$F_a = F_{T/N} = 15462.53/6 = 2577.09 \text{ N}$$

The joint factor (Table 10.5)

$$K = 0.75$$

The total load on each bolt,

$$\begin{aligned} F &= K \cdot F_a + F_i \\ &= 0.75 \times 2577.07 + 40 \times 10^3 \\ &= 41932.816 \text{ N} \end{aligned}$$

Step-2: Factor of safety
The stress area of M20 bolt (Table 10.5)

$$A_s = 245 \text{ mm}^2 \quad \text{for pitch} \quad 2.5 \text{ mm}$$

The stress induced

$$\sigma = F/A_s = \frac{41932.816}{245} = 171.15 \text{ MPa}$$

For C20, S_y = 260 MP. (Table 2.6 MDHB)

$$\therefore \quad FS = \frac{S_y}{\sigma} = 260/171.15 = 1.52$$

Example 10.15: An M20 × 2 steel bolt of 100 mm long is subjected to impact load. The energy absorbed by the bolt is 2 N-m:

(i) Determine the stress in the shank of the bolt, if there is no threaded portion between the nut and bolt head.
(ii) Determine the stress in the shank, if the entrie length of the bolt is the threaded. Assume modulus of elasticity for steel as 206 GPa.

Given Data

Bolt size = M20 × 2
Bolt length, L = 100 mm
Impact energy, L = 2 N-m = 2000 N-m
Modulus of elasticity, E = 206 GPa

SOLUTION
Impact energy, U = 2000 N-m
Strain energy in the bolt,

$$E_S = \frac{1}{2}\frac{\sigma^2}{E} \times \text{vol.} = \frac{1}{2} \cdot \frac{\sigma^2}{E} \times A.L.$$

(i) $A = \frac{\pi}{4} d_0^2 = \frac{\pi \times 20^2}{4} = 314.16 \text{ mm}^2$

Since $E_s = U$

$$\frac{1}{2} \times \frac{\sigma^2}{206 \times 10^3} \times 314.16 \times 100 = 2000$$

$$\sigma = \sqrt{\frac{2000 \times 2 \times 206 \times 10^3}{314.16 \times 100}} = 161.95 \text{ MPa}$$

(ii)

$$A = \frac{\pi}{4} d_i^2 = \left(\frac{\pi}{4} \times 17.55^2\right) \quad \left[\begin{array}{ll} \therefore \quad d_i & = d - 1.22687\, p \\ & = 17.55 \text{ mm} \end{array}\right]$$

$$= 241.90 \text{ mm}^2$$

For energy balance,

$$\frac{1}{2} \frac{\sigma^2 \times 241.9 \times 100}{206 \times 10^3} = 2000$$

$$\sigma = \sqrt{\frac{2000 \times 2 \times 206 \times 10^3}{241.9 \times 100}} = 184.56$$

$$= 184.56 \text{ MPa}$$

Example 10.16: Determine the size of the bolts for the loaded bracket shown in Fig. 10.20, if the allowable tensile stress is limited to 80 MPa.

SOLUTION

Step-1: Loads on the bolts

The applied force is resolved in vertical and horizontal components

The vertical component,

$$P_V = P . \sin 30 = 20 \times \sin 30 = 10 \text{ kN}$$

The horizontal component,

$$P_H = P \cos 30^\circ = 20 \times \cos 30^\circ = 17.320 \text{ kN}$$

The overturning moment about the hollow edge CC is

$$\begin{aligned} M &= P_V \times 200 + P_H \times 80 \\ &= 10 \times 10^3 \times 200 + 17.320 \times 10^3 \times 80 \\ &= 3385.6 \times 10^3 \text{ N-mm} \end{aligned}$$

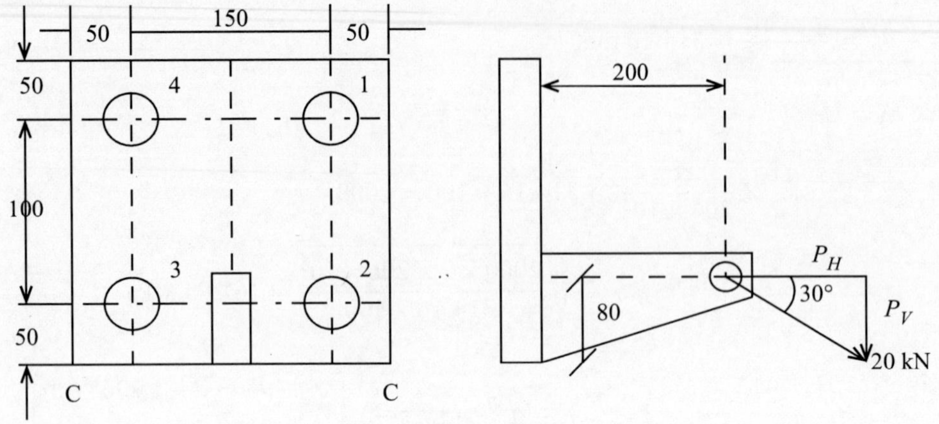

Figure 10.20

The maximum tensile force occurs on the bolts 1 and 4 and is evalualated as follows (Eqn. 10.23-modified),

$$F_{max} = \frac{M \cdot r_{max}}{\sum r^2} = \frac{3385.6 \times 10^3 \times 150}{(2 \times 50^2 + 2 \times 150^2)} = 10156 \text{ N}$$

Primary tensile force,

$$F_1 = \frac{P_H}{n} = \frac{17.320 \times 10^3}{4} = 4.33 \times 10^3 \text{N}$$

Total direct (normal) force,

$$F_T = 10156 + 4330 = 14486 \text{ N}$$

Step-2: Stresses in the bolts
Normal stress,

$$\sigma = F_T / A_s = \left(\frac{14486}{A_s}\right) \text{MPa}$$

Transverse shear stress,

$$\tau = \frac{P_v / n}{A_s} = \frac{10/4}{A_s} = \frac{2.5 \times 10^3}{A_s} \text{MPa}$$
$$= \left(\frac{2500}{A_s}\right) \text{MPa}$$

Maximum normal stress,

$$\sigma_{\max} = \frac{\sigma}{2} + \sqrt{\left(\frac{\sigma}{2}\right)^2 + C^2}$$

$$= \left(\frac{14486}{2A_s}\right) + \sqrt{\left(\frac{14486}{2A_s}\right)^2 + \left(\frac{2500}{A_s}\right)^2}$$

$$= \left(\frac{7243}{A_s}\right)\left[1 + \sqrt{1 + \left(\frac{2500}{7243}\right)^2}\right]$$

$$= (14905/A_s)$$

Step-3: Bolt design

Equality $\sigma_{\max} = S_d = 80$ MPa

$$\therefore\ A_s = \left(\frac{14905}{80}\right) = 186.312\,\text{mm}^2.$$

Suitable bolt size (Table 10.4)

$$\text{M}18 \times 2.5 \quad \text{with} \quad A_s = 192\ \text{mm}^2$$

If the pitch is limited to 1.5 m

$$\text{M}18 \times 1.5 \quad \text{with} \quad A_s = 216\ \text{mm}^2$$

EXERCISES FOR PRACTICE

10.1. An eye bolt is to be used for lifting a load of 80 kN. Find the nominal diameter of the bolt, if the tensile stress is not to exceed 90 MPa.

10.2. Two shafts are connected by means of a flange coupling to transmit torque of 25 Nm. The flanges of the coupling are fastened by four bolts of the same material at a radius of 30 mm. Find the size of the bolts, if the allowable shear stress for the bolt material is 30 MPa.

10.3. A bolt in a steel structure is subjected to a tensile load of 9 kN. The initial tightening load on the bolt is 5 kN. Determine the size of the bolt taking allowable stress for the bolt material to be 80 MPa.

10.4. A mild steel cover plate is to be designed for an inspection hole in the shell of a pressure vessel. The hole is 120 mm in diameter and the pressure inside the vessel is 6 MPa. Design the bolts. Assume the allowable tensile stress for bolt material as 50 MPa.

10.5. A steam engine of effective diameter of 300 mm is subjected to a steam pressure of 1.5 MPa. The cylinder head is connected by 8 bolts having yield point 330 MPa and endurance limit at 240 MPa. The bolts are tightened with an initial preload of 1.5 times the steam load. A soft copper gasket is used to make the joint leakproof. Assuming factor of safety 2, find the size of the bolt required. The stiffness factor for copper gasket may be taken as 0.5.

10.6. The cylinder head of a steam engine is subjected to a steam pressure of 0.7 MPa. It is held in position by means of twelve bolts. A soft copper gasket is used to make the joint leakproof. The effective diameter of cylinder is 300 mm. Find the size of the bolt so that the stress in the bolts is not to exceed 110 MPa.

10.7. The cylinder head of a 200 mm bore compressor is secured by means of 12 studs of rolled mild steel. The gas pressure is 1.5 N/mm^2 gauge. The initial tension in the bolts, assumed to be equally loaded such that a cylinder pressure of 1.7 N/mm^2 gauge is required for the joint to be on the point of opening. Suggest the suitable size of the studs assuming suitable material and factor of safety of 2. The joint should be leakproof.

10.8. A steam engine cylinder of 350 mm diameter is subjected to steam pressure of 1.5 MPa. The cylinder cover is held by means of eight bolts of size M 20. The joint is made leakproof by means of suitable gaskets. Determine the stress induced in the bolts.

10.9. A wall bracket subjected to vertical load is shown in Fig. 10.21. It is fixed to a wall by means of four bolts. Find the size of the bolts if working stress in tension for the bolt may be assumed as 75 MPa.

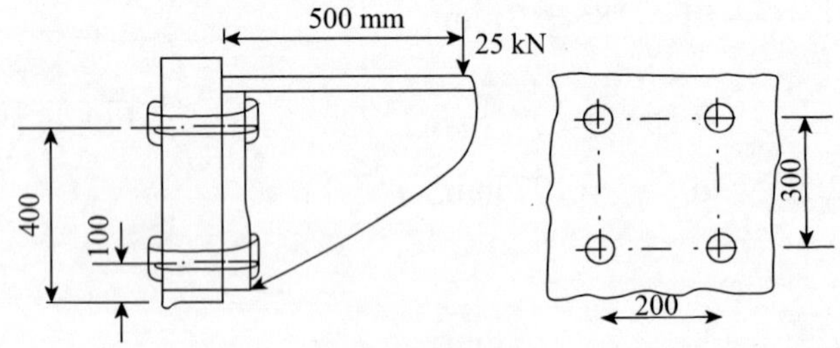

Figure 10.21

10.10. A plate subjected to load is bolted to a column by four bolts of equal size as shown below. Determine the minimum strength required for the bolt material if the factor of safety is considered to be 1.5.

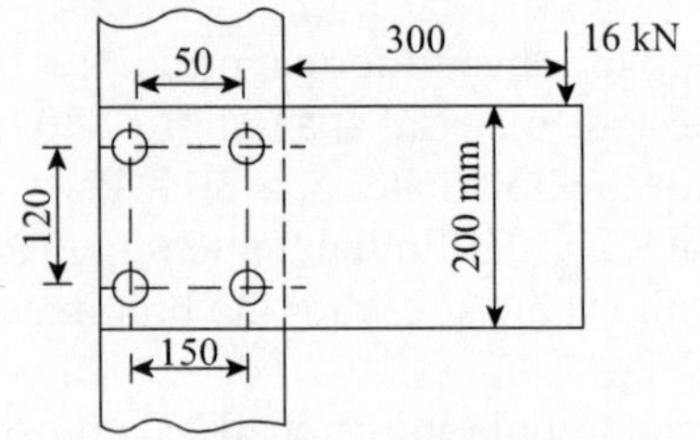

Figure 10.22

10.11. (a) Explain the various causes for stress concentration in bolts.

The cylinder head of a steam engine is subjected to a steam pressure of 0.75 MPa. It is held in position by twelve bolts. A soft copper jacket is used to make the joint leakproof. The effective diameter of the cylinder is 300 mm. Find the size of bolts if the stress is limited to 110 N/mm^2.

10.12. Discuss the effect of initial tightening of fasteners on the design load. How is it influenced by the gasket put between the parts connected together?

10.13. (a) What are the advantages and disadvantages of screwed joints?

(b) A steam engine cylinder of effective diameter of 30 cm is subjected to a steam pressure of 1.5 MPa. The cylinder head is connected by means of 8 bolts having yield point stress of 330 MPa and endurance limit of 240 MPa. The bolts are tightened with an initial preload of 1.5 times that of steam load. A soft copper gasket is used to make the joint leak proof. Assuming a stress concentration factor of 2.8 and factor of safety as 2, find the size of the bolts required.

10.14. (a) Discuss the effect of preloading in bolted joints.

(b) A steam engine cylinder of size 300 mm diameter × 400 mm stroke, operates at 1.5 MPa. The cylinder head is connected by eight bolts having an yield point stress of 350 MPa and an endurance limit stress of 240 MPa. The joint is made leakproof by using a cooper gashed with K = 0.5. With stress concentration factor of 3 and factor of safety of 2.0, determine the bolt diameter.

10.15. A steam engine cylinder of 300 mm effective diameter is subjected to a steam pressure of 1.8 MPa. The cylinder head is connected by means of 8 bolts having yield point stress of 330 MPa and endurance limit of 240 MPa. The bolts are tightened with an initial preload of 1.5 times that of steam load. A soft copper gasket is used to make the joint leakproof. Assuming a stress concentration factor of 2.5 and a factor of safety as 2.5, find the size of bolts required.

10.16. Determine the size of the bolts required for the head of cylinder of 350 mm diameter containing steam at a pressure of 1.5 MPa. 12 number of bolts are used over a pitch circle diameter of 400 mm for fastening. Assume that a hard copper gasket is to be used in the assembly.

10.17. A wall bracket of rectangular base is fixed to a vertical column and supports a travelling crane. The maximum load on the bracket is 18 kN and acts at a distance of 450 mm from the face of the column. The bracket is secured to the column by 6 bolts in three rows at distances of 60 mm, 110 mm and 380 mm from the lower edge of the bracket. Determine the size of the bolts if the permissible value of the tensile stress for the bolt material is 100 MPa.

10.18. The cover of a pressure vessel is held in place of ten M20 bolts. The pressure is 1.4 MPa and the effective area of the cover exposed to the pressure is O.2 m_2. The ratio of the stiffness of the bolts to the connected parts is 1/2. Each bolt is tightened initially, before the pressure is applied, to 27 kN. (i) Will the joint separate as result of the load due to internal pressure? (ii) If the joint does not separate, determine the tensile stress in each bolt due to axial load (stiffness here is defined as meters of deflection per Newton of load).

10.19. A bracket is fixed to a vertical support by means of 4 bolts in two rows at distance of 50 mm and 500 mm from the lower edge of the bracket. The load on the bracket is 15 kN, which acts at a distance of 500 mm from the face of the column. Determine the size of the bolts if perrnissible value of the tensile stress for the bolt material is 80 MPa. Derive the formula used.

SHORT ANSWER QUESTIONS

10.1. What is a threaded joint?

10.2. What are the advantages of threaded joints?

10.3. What are the disadvantages of threaded joints?
10.4. What are the components of threaded joint?
10.5. What are the types of bolts?
10.6. What is a through bolt?
10.7. What is a stud?
10.8. What is a tap bolt?
10.9. What is a setscrew?
10.10. What is cap screw?
10.11. What is the difference between setscrew and cap screw?
10.12. Why is a hexagonal shape preferred for nut and bolt head?
10.13. What is locking device?
10.14. What are the various locking devices?
10.15. What is lock nut?
10.16. What is a castle nut?
10.17. What are the types of threads used for bolts and screws?
10.18. What are the advantages of V-threads?
10.19. What is profile of ISO metric threads?
10.20. How do you designate ISO metric fine threads?
10.21. How do you designate ISO metric coarse threads?
10.22. What is tensile stress are of screw thread?
10.23. What are the materials used for bolts and screws?
10.24. What is initial tightening load?
10.25. What is bolt of uniform strength? What are the methods to obtain the uniform strength?

CHAPTER 11

RIVETED JOINTS

11.1 INTRODUCTION

The riveted joints are a type of permanent joints which connect two plates. As in tanks and pressure vessels ship building or two structural members as in bridges, roof structures, etc. The rivet is a cylindrical member with a head. It is passed through the drilled and reamed holes in the members to be joined. The protruding tail end is hammered to form another head. This is done by cold forging or hot forging. In cold forging process the rivet expands and fills the holes in the plates. In case of hot forging process the diameter and length both contract and create additional compressive pressure between the plates. This induces friction between the plates and prevents the relative sliding. The various forms of heads are shown in Fig. 11.1. The snap head is the most commonly used form.

11.2 TYPES OF JOINTS

Depending upon the manner the plate ends are placed, the joint may be:

(a) Lap joint
(b) Butt joint

Lap joint

In lap joint the plate ends are placed one over the other such that they overlap. The drilled holes are aligned and the rivets are placed into the holes then the tail ends are forged to form head. The rivets may be arranged in single, double and triple rows. Accordingly, we have single, double or triple or quadruple riveted lap joint. The arrangement of rivets in each row may be the same with same central lines for all rows resulting in chain riveting or the rivets in the adjacent rows may be displaced by half-a-pitch, then the arrangement is called zig-zag riveting. These arrangements are shown in Fig. 11.2(a) and 11.2(b). The joints are used for circumferential joints in pressure vessels.

Butt joint

In case of butt joints the plate ends are butted with each other and cover plates are placed on one side or both sides. The holes are drilled through the plates. Thus, there are two sets of rivets one set for each main plate. The arrangement may be chain or zig-zag. These are shown in Fig. 11.3(a), 11.3(b), 11.3(c) and 11.3(d). The butt joints are used for longitudinal joints in pressure vessels.

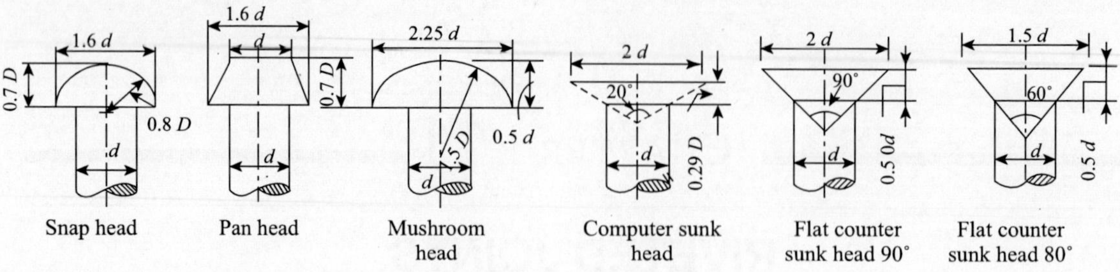

General purpose rivets (below 12 mm diameter)

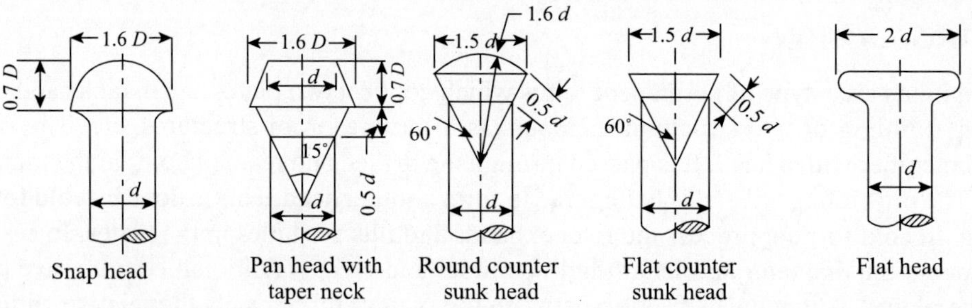

General purpose rivets (above 12 mm diameter)

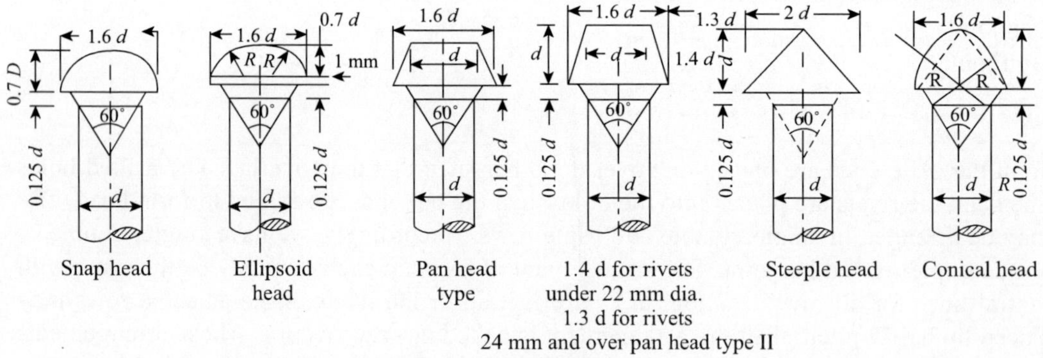

Rivets for boilers

Figure 11.1. Types of rivet heads

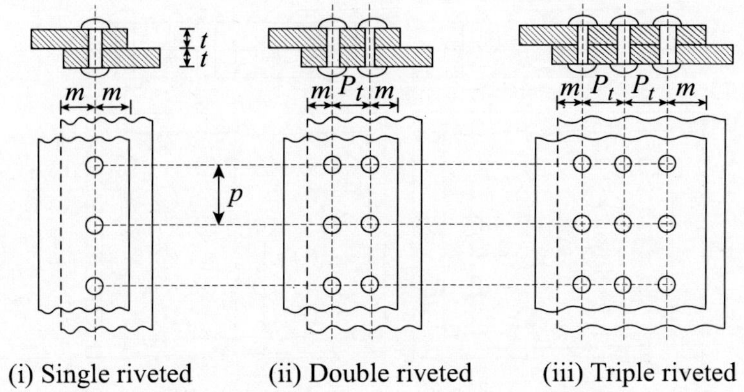

Figure 11.2(a). Lap joint with chain rivetting

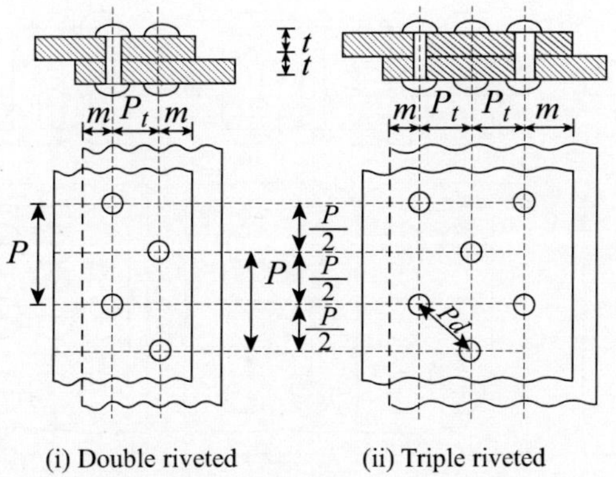

Figure 11.2(b). Lap joint with zig-zag rivetting

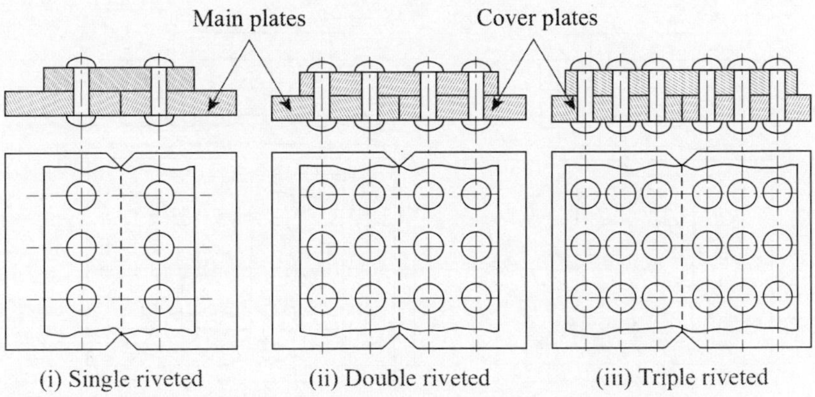

Figure 11.3(a). Single cover butt joints with chain riveting

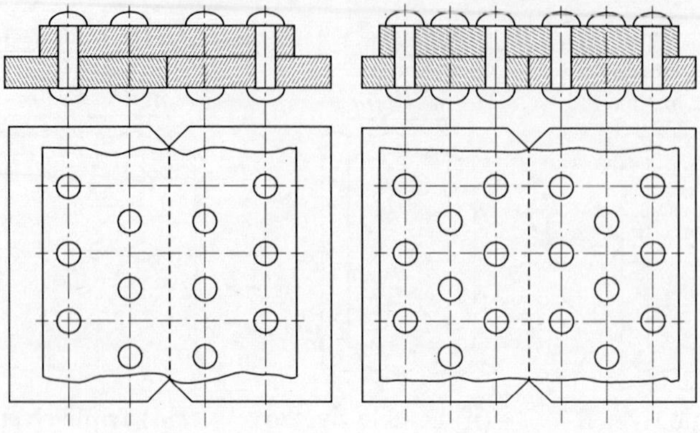

(i) Double riveted (ii) Triple riveted

Figure 11.3(b). Single cover butt joints with zig-zag riveting

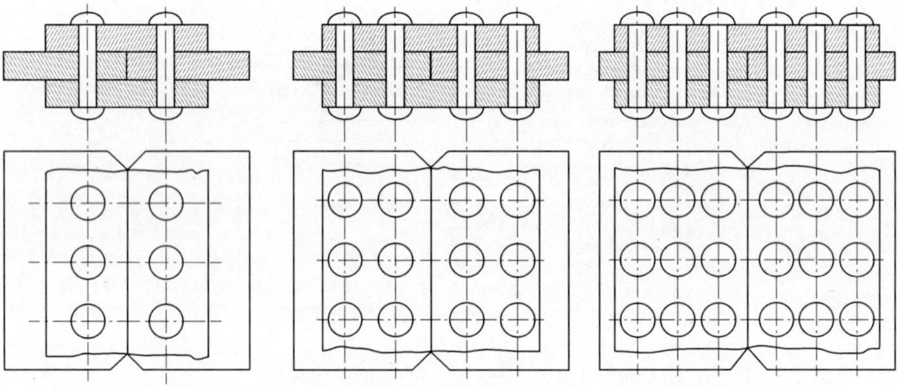

(i) Single riveted (ii) Double riveted (iii) Triple riveted

Figure 11.3(c). Double cover butt joints (equal covers)

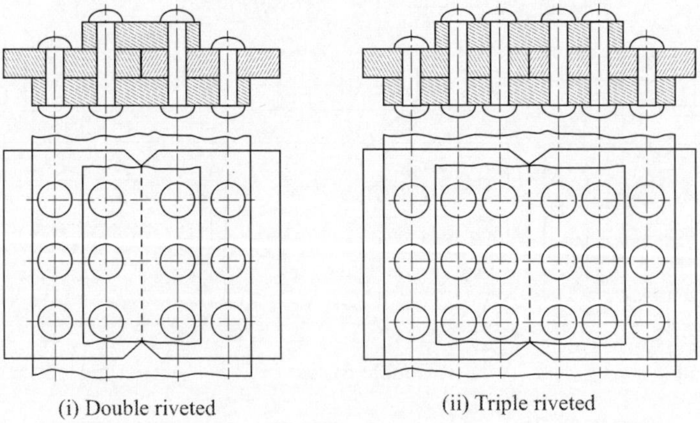

(i) Double riveted (ii) Triple riveted

Figure 11.3(d). Double cover butt joints (unequal covers)

Terminology (Fig 11.2b)

Rivet hole diameter: $d = 6\sqrt{t}$ (11.1)

Pitch: The distance between the centres of the consecutive rivets in a row is called pitch.

$$P_{\min} = 2.25d \text{ and } P_{\max} = 6d \quad (11.2a \text{ and } 11.2b)$$

Transverse pitch, P_t: The distance between the consecutive rows of the rivets is called transverse pitch.

$$P_t \geq 2d \quad \text{for chain rivetting} \quad (11.3a)$$

$$P_t \geq 0.33P + 0.67d \quad \text{for zig-zag rivetting} \quad (11.3b)$$

P_d. The distance between rivets in the adjacent rows is called diagonal pitch.
Margin, m. It is the distance between the end of the plate and the first row of the rivets. Normally, $m = 1.5d$.

Materials for rivets

The rivets are made of the same material as the plates are being joined. The steel rivets are made for steel plates. Aluminium rivets are used for aluminium plates. The qualities of the material are specified by ISI standards IS: 1148–1982(re–1992), IS: 1149–1982(re–1992), IS: 1990–1973(re–1992).

11.3 FAILURE OF RIVETED JOINTS

A riveted joints can fail in one or more ways as mentioned below:

(i) Tearing of plate
(ii) Shearing of rivets
(iii) Crushing of rivets

(i) **Tearing of plate**. Joint may fail due to tearing of plate at margin or tearing of plate along line of rivet holes as shown in Fig. 11.4(a) and (b) respectively.

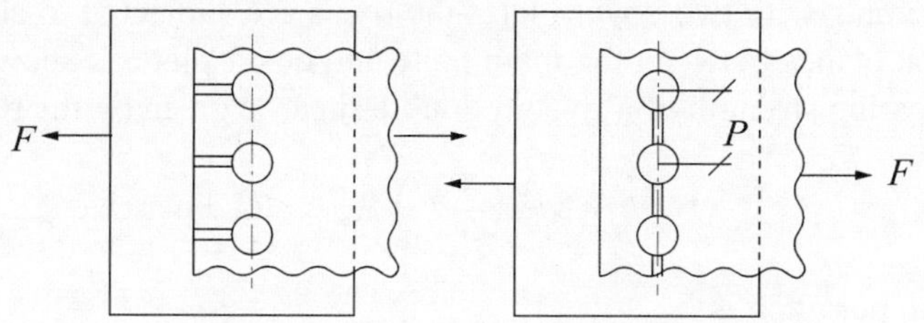

(a) At margin (b) At the rivet holes

Figure 11.4. Failure of plate in tearing

Failure-Modes of riveted joints

The tearing of plates

(a) Due to insufficient margin the plate between the first row of rivets and plates end is likely to tearing failure. This is safeguarded by having the minimum margin.

$$m = 1.5d \tag{11.4}$$

(b) The tearing of the plates between the first row of rivets is likely to occur due to the reduced cross-sectional area. This is prevented by providing a proper pitch to have sufficient strength per pitch as given by the relation

$$F = (P - d)t \cdot S_t \tag{11.5}$$

(ii) **The shearing of the rivets**
In case of lap joints and butt joints with a single cover plate the rivets are subjected to shearing at the common plane between the overlapping plate. The strength per rivet in single shear is given by

$$F_{s1} = \frac{\pi}{4}d^2 \cdot S_{sd} > \text{load per rivet} \tag{11.6}$$

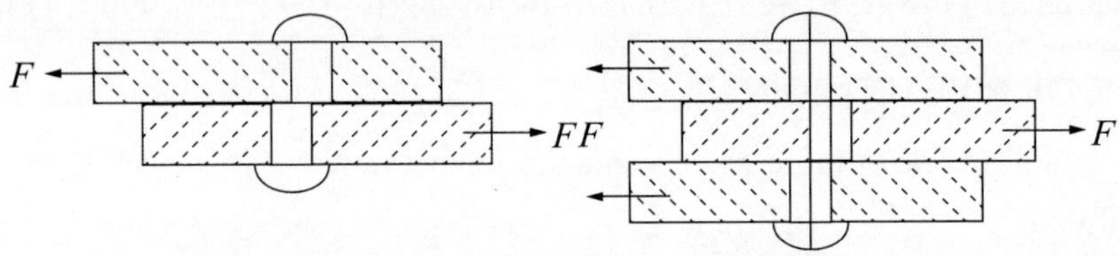

(a) Single shear (b) Double shear

Figure 11.5. Shearing of rivets

In butt joints with two cover plates the rivets are subjected to shearing at the two common planes between the main plate and cover plate as shown in Fig. 11.5(b). The shearing strength of a rivet in double shear is given by the relation,

$$F_{s_2} = 1.875\left(\frac{\pi}{4}d^2\right)S_{sd} > \text{ load per rivets.} \tag{11.7}$$

Strength per pitch is

$$F_s = \left(n_1 \cdot \frac{\pi}{4}d^2 + 1.875\,n_2\frac{\pi}{4}d\right)S_{sd}$$
$$= (n_1 + 1.875\,n_2)\frac{\pi}{4}d^2 \cdot S_{sd} \tag{11.8}$$

n_1 = number of rivets in single shear per pitch
n_2 = number of rivets in double shear per pitch

(iii) **Crushing of the rivets in the plates**
There are contact stresses induced between the plate and rivet surface due to the pull in the plate. The rivet surface may fail due to the excessive contact stress.

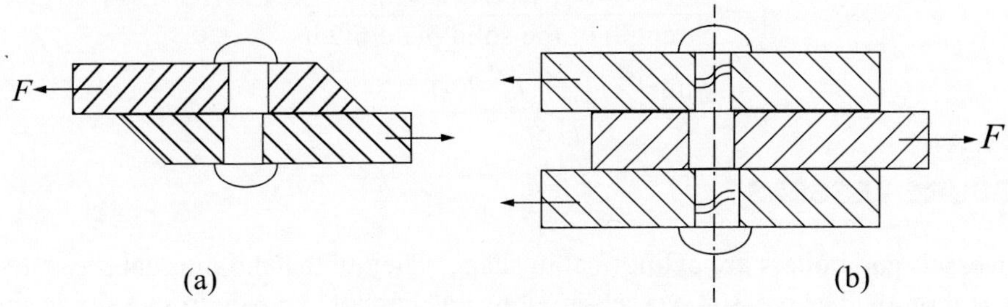

Figure 11.6. Crushing of rivets

In lap joint, the crushing strength per rivet is

$$F_{c_1} = dt \cdot S_c > \text{load per rivet} \tag{11.9a}$$

In butt joint with double cover plates. The crushing area is less in the main plate and hence the crushing strength per rivet is given by

$$F_{c_1} = dt \cdot S_c > \text{load per rivet} \tag{11.9b}$$

In case of butt joint with single cover plate, the contact area is less in the cover plate and crushing strength per rivet is given by

$$F_{c_2} = h \cdot d \cdot S_c > \text{load per rivet} \tag{11.9c}$$

Crushing strength per pitch,

$$F_{c_1} = (n_1 h + n_2 t) d S_c > \text{load per pitch} \tag{11.10}$$

In addition the tearing of the plate at the inner row combined with shearing or crushing of the rivets in the outer row is another possibility of failure in case the alternate rivet is omitted in the outer row. Then the strength values for these possibilities are given by
For tearing combined with shearing,

$$F_{st} = (P - 2d) S_t + \frac{\pi}{4} d^2 \cdot S_s > \text{load/pitch} \tag{11.11}$$

For tearing combined with crushing,

$$F_{ct} = (P - 2d) S_t + hd \cdot S_c > \text{load/pitch} \tag{11.12}$$

Strength of solid-plate,

$$F_t = P \cdot t \cdot S_t \tag{11.13}$$

Usually, these strength values are compared with the strength of the solid plate and efficiency is defined as follows:

$$\eta = \frac{\text{Min}(F_t, F_s, F_c, F_{st}, F_{ct})}{\text{Strength of the solid plate/pitch}}$$

$$\therefore \quad \eta = \frac{\text{Min}(F_t, F_s, F_c, F_{st}, F_{ct})}{P \cdot t \cdot S_t} \tag{11.14}$$

11.4 PRESSURE VESSELS

The pressure vessels and boilers are cylindrical in shape. They withstand internal pressure and are required to be leakproof. The pressure vessels are long and cannot be made from a single sheet. The number of cylindrical sections are made of by bending the sheets and joining the edges by riveted butt joints along the length. The cylinders are joined by riveted lap joint along the circumference. Circumferential stress,

$$\sigma_t = \frac{P \cdot r}{t} \quad \text{or} \quad \sigma_c = \frac{PD}{2t} \tag{11.15}$$

Longitudinal stress,

$$\sigma_l = \frac{P \cdot r}{2t} \text{ or } \sigma_l = \frac{PD}{4t} \tag{11.16}$$

where, P = fluid pressure

r = radius of the drum

D = diameter of the drum

t = thickness of the plate

We observe that

$$\sigma_t = 2\sigma_l$$

This means that the longitudinal joint which resists the circumferential stress is to be stronger than the circumferential joint which resists the longitudinal stress. The butt joint is used for the longitudinal joint and lap joint is used for circumferential joint.

Design procedure for longitudinal joint

1. Thickness of the pressure vessel is calculated by the relation,

$$t = \frac{D \cdot p_f}{2\eta S_t} + 1 \text{ mm} \tag{11.17}$$

(1 mm for wear due to corrosion, etc). Efficiency is choosen from the Table 11.2. The choice of the type of the joint is guided by the Table 11.4.

2. The rivet diameter is calculated by using the Unwin's formula $d = 6\sqrt{t}$ (Eqn. 11.1). The diameter is standardized by using the Table 11.8. Generally used thickness of the plate is more than 8 mm.
3. Determine the pitch of the rivets with restrictions.

$$P_{\min} > 2.25d$$

and I B R eqn for maximum pitch

$$P_{\max} \leq C \cdot t + 41.28 \text{ mm} \tag{11.18}$$

The value of C is chosen from the Table 11.2.

The pitch can also be calculated based on the optimum strength consideration by using the equation,

$$P = \frac{(n_1 + 1.875n_2)\pi d^2 \cdot S_s}{4tS_t} + d \tag{11.19}$$

where, n_1 = number of rivets in single shear/pitch

n_2 = number of rivets in double shear/pitch.

The value is accepted if it is smaller than the value given by IBR equation.

4. Determine the transverse pitch p_t by using appropriate equations.

Equal number of rivets in all rows

$$\text{Chain riveting,} \quad P_t \geq 2d \tag{11.20a}$$

$$\text{Zig-zag riveting,} \quad P_t = 0.33P + 0.37d \tag{11.20b}$$

Alternate rivet in the outer row is omitted (unequal cover plates). Chain riveting between the outer row and inner row,

$$P_t \leq 3d \quad \text{or} \quad P_t = 0.33P + 0.67d$$

between inner rows,

$$P_t \geq 2d$$

Zig-zag riveting between outer row and inner row,

$$P_t = 0.2P + 1.15d \tag{11.20c}$$

between inner rows,

$$P_t = 0.165P + 0.67d \tag{11.20d}$$

5. Determine the thickness of the cover plates.

(a) Single cover plate, $h = 1.125t$ (11.21a)

(b) Equal double cover plates,

$$h_1 = h_2 = h = 0.625t \tag{11.21b}$$

(c) Unequal double cover plates,

$$\text{Wider cover plate,} \quad h_1 = 0.75t \tag{11.21c}$$

$$\text{Thin cover plate,} \quad h_2 = 0.625t \tag{11.21d}$$

6. Checking by Swet formula for caulking pitch.

$$P_c = d + 14\sqrt[4]{\frac{h_1^3}{p_f}} > \text{inner pitch} \tag{11.22}$$

The inner pitch chosen should be smaller than this value.

7. Calculation of strength values.

(i) Tearing strength of the plate at the outer row,

$$F_t = (P - d)t\ S_t \qquad \text{(Eqn. 11.5)}$$

(ii) Shearing strength of the rivets/pitch,

$$F_s = (n_1 + 1.875\,n_2)\frac{\pi}{4}d^2 \cdot S_s \qquad \text{(Eqn. 11.8)}$$

(iii) Crushing strength of the rivets per pitch,

$$F_c = (n_2 t + h_1 n_1) d S_c \qquad \text{(Eqn. 11.10)}$$

(iv) Tearing strength of the plate at the second row combined with shearing of the rivets in the outer row (unequal cover plates),

$$F_{ts} = (P - 2d)t \cdot S_t + \frac{\pi}{4}d^2 S_s \qquad \text{(Eqn. 11.11)}$$

(v) Tearing strength of the plate at the second row combined with crushing of rivets in the outer row,

$$F_{tc} = (P - 2d)t \cdot S_t + d h_1 S_c \qquad \text{(Eqn. 11.12)}$$

(vi) Strength of the solid plate/pitch,

$$F = P_t \cdot S_t \qquad \text{(Eqn. 11.13)}$$

(8) Calculate the efficiency of the joint.

$$\eta = \frac{\text{Min}(F_t, F_s, F_c, F_{ts}, F_{tc})}{\text{Strength of the solid plate}} \qquad \text{(Eqn. 11.14)}$$

The value should be more than the value choosen at step (1).

Design procedure for circumferential lap joint

The shell thickness and rivet diameter already calculated for butt joint are used.

1. Calculate the number of rivets.

   The rivets are in single shear under fluid pressure P_f, acting on the circular end plate of the cylinder which tends to separate the two length sections of the cylinder.

   Equating the shearing strength of the rivets to the bursting force,

$$N \cdot \frac{\pi}{4} d^2 \cdot S_s = \frac{\pi}{4} D^2 \cdot P_f$$

$$N = \left(\frac{D}{d}\right)^2 \left(\frac{p_f}{S_s}\right) \tag{11.23}$$

2. Calculate the permissible pitch.

   The efficiency of the lap joint η_l can be half of the efficiency of the longitudinal butt joint

   i.e.,
$$\eta_l = \frac{\eta}{2} \tag{11.24}$$

   Then the permissible pitch,

$$P_e = \left(\frac{d}{1-\eta_l}\right) \tag{11.25}$$

3. Calculate the number of row of rivets. The number of rivets in one row,

$$n_1 = \frac{\pi(D+t)}{P_e} \tag{11.26}$$

   Then the number of rows of rivets,

$$N_1 = \frac{N}{n_1} = \frac{N \cdot P_e}{\pi(D+t)} \tag{11.27}$$

   The value of N_1 determines whether the joint is to be single riveted or multiple riveted, since the number of rows are to be a whole number the pitch has to be readjusted. In any case the pitch selected will have to satisfy the condition of caulking pitch as specified in step (6) for butt joint.

11.5 RIVETED JOINTS FOR STRUCTURE

(a) Lozenge joint or Diamond joint

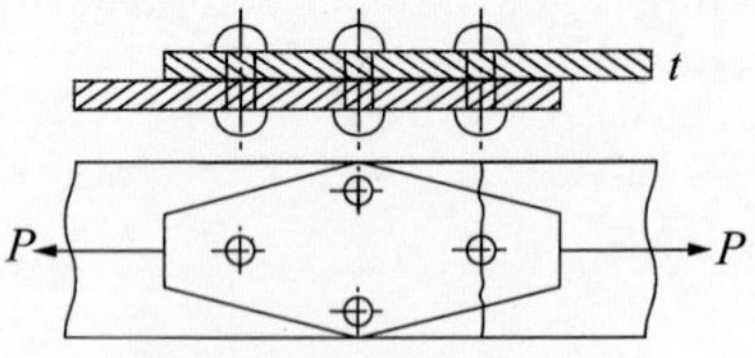

Figure 11.7. Lap joint

Number of rivets C (greater for the following)

$$\text{For shear strength, } n = \frac{4F}{n \cdot d^2 \cdot S_{sd}} \quad (11.28a)$$

$$\text{For crushing strength, } n = \frac{F}{dt \cdot S_{cd}} \quad (11.28b)$$

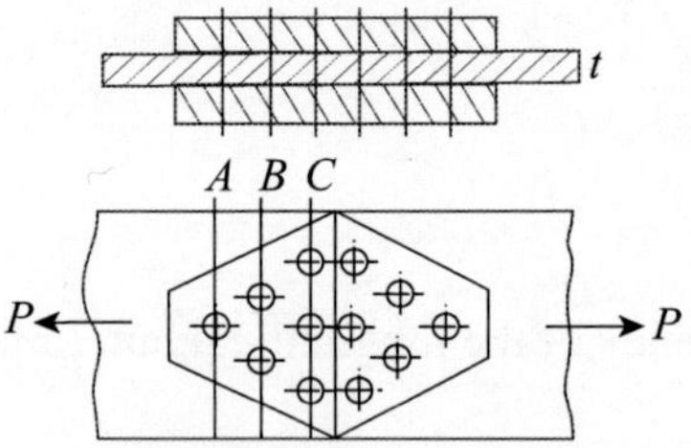

Figure 11.8. Butt joint

Number of rivets (greater for the following)

$$\text{For shear strength, } n = \frac{4F}{1.75\pi d^2 \cdot S_{sd}} \quad (11.29a)$$

$$\text{For crushing strength, } n = \frac{F}{dt \cdot S_{cd}} \quad (11.29b)$$

For structures like trusses bridges or girders, etc., the width of the members to be joined is known. The Lozenges or diamond-shaped joints are preferred. The joints are usually double cover butt type with rivets so arranged that there is only one rivet in the outermost row and their number increases towards the inner rows. The load carrying capacity in tension is given by the relation,

$$F = (b - d)t \cdot S_t$$

where, b = width of the plate in mm
t = thickness of the plate in mm
d = diameter of the rivet in mm
S_t = design stress in tension in MPa

These joints have uniform strengths in all modes of failure. Marginal variation in the calculated dimensions may cause slight variation in strength in a particular mode.

Procedure

1. Calculate the diameter using the Unwin's equation $d = 6\sqrt{t}$ and standardize it using Table 11.7.
2. Shearing strength of each rivet, $F_s = \frac{\pi}{4} d^2 \cdot S_s$
3. Crushing strength of each rivet, $F_c = dt \cdot S_c$

4. Strength of each rivet, $F_r = \text{Min}(F_s, F_c)$
5. Total number of rivets required, $N = (F/F_r)$
6. Arrange the rivets in diamond form or shape as shown in Table 11.8.

(b) Eccentrically loaded group of rivets

In many structural applications like columns, eccentric loads are applied through a bracket fixed to the column, by a group of rivets. The eccentric load is equivalent to parallel load P through the centre of gravity of the group of rivets and moment $M = P.c$ about an axis through CG. The parallel is shared equally by all rivets. The direct shearing load in each rivet is

$$F_P = \frac{P}{n}$$

where n is the number of rivets in the group.

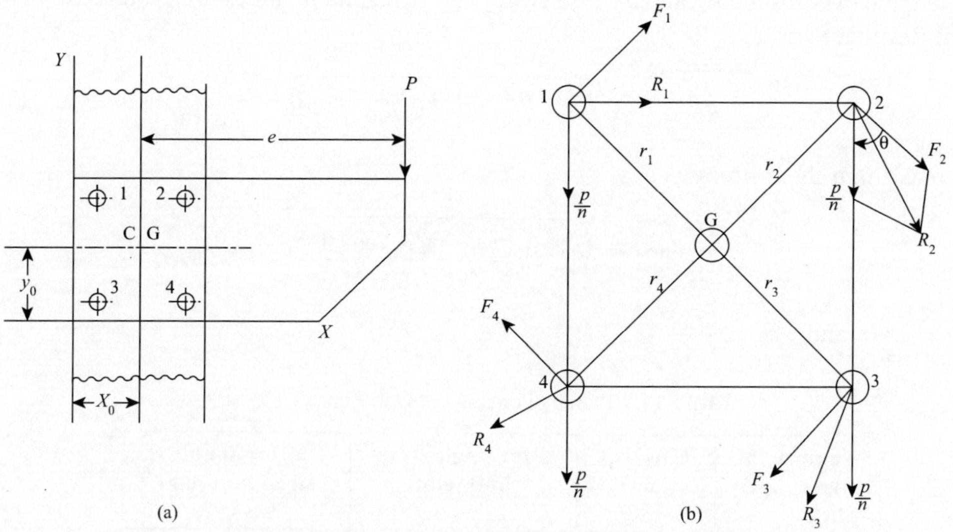

Figure 11.9. Group of eccentrically loaded rivets

The moment, M induces in each rivet a secondary force F_i which is proportional to its distance from the CG of the group of rivets.

$$F_i = Kr_i$$

The moment due to this force about the axis through CG is

$$M_i = Kr_i r_i = Kr_i^2$$

The sum of all such moments must be equal to externally applied moment.

$$M = P \cdot e = \sum Kr_i^2$$

$$K = \frac{P \cdot e}{\Sigma r_i^2} \tag{11.30}$$

Then the secondary shear force is given by

$$F_i = Kr_i = \left(\frac{P \cdot e \cdot r_i}{\Sigma r_i}\right) \tag{11.31}$$

This force is perpendicular to the radius r_i, line joining the rivet centre to the CG. Thus, there is an angle θ_i between the primary force and the secondary force. The resultant force is

$$F_{Ri} = \sqrt{F_P^2 + F_i^2 + 2F_P F_C \cos\theta_i} \tag{11.32}$$

The max value depends upon the value F_i and θ_i. Normally, the farthest rivets (i.e., $r_i = r_{\max}$) carry max F_i and the resultant is maximum for minimum θ_i. By inspection the rivet with max load can be identified, the max resultant values chosen after all the resultant values are evaluated.
Maximum Resultant Force

$$F_{R_{\max}} = \sqrt{F_P^2 + F_{i_{\max}}^2 + 2F_P F_{i_{\max}} \cdot \cos\theta_i} \tag{11.33}$$

Then the maximum shear stress is

$$\tau_{\max} = \sqrt{\tau_1^2 + \tau_2^2 + 2\tau_1\tau_2 \cdot \cos\theta_i} \tag{11.34}$$

where, $\tau_1 = \dfrac{F_P}{A}$ and $\tau_2 = \dfrac{F_{i_{\max}}}{A}$

Table 11.1. Value of coefficient C in Eqn. 11.17

Number of rivets per pitch	C for lap joint	C for single strap butt joint	C for double strap butt joint
1	1.31	1.53	1.75
2	2.61	3.06	3.50
3	3.47	4.05	4.63
4	4.17	–	5.52
5	–	–	6.00

Table 11.2. Standard proportions

Types of joints	$c = \frac{D_{P_f}}{2}$	Rivet diameter, d mm	Pitch, P mm	Efficiency, $\eta = \frac{P-d}{P}$
(1) Single riveted lap joint	500	$t+8$ mm	$2d+8$ mm	0.56 – 0.60
(2) Double riveted lap joint	350 – 950	$t+8$ mm	$2.6d+15$ mm	0.70
(3) Triple riveted lap joint	430 – 1350	$t+6$ to 8 mm	$3d+22$ mm	0.75
(4) Double riveted two strap bolt joint	450 to 1350	$t+5$ to 6 mm	$3.5d+15$ mm	0.75
(5) Triple riveted two strap bolt joint	450 to 2300	$t+5$ mm	$6d+20$ mm	0.85

Table 11.3. Efficiency of riveted joints

Types of joints	% efficiency		Number of rivets per pitch
	Normal	Maximum	
Lap joints			
Single riveted	50 – 60	63	1
Double riveted	60 – 72	77	2
Triple riveted	72 – 80	86.6	3
Butt joint two strap			
Single riveted	55 – 60	63.3	1
Double riveted	76 – 84	87	2
Triple riveted	80 – 88	95	3
Quadruple riveted	86 – 94	98	4

Weakening factor due to varying load for low carbon steels

$$v = \frac{1}{1 - 0.3\dfrac{F_{\min}}{F_{\max}}}$$

Table 11.4. Suggested types of joints

Diameter of shell, mm	Thickness of the shell, mm	Type of joint
600 – 1800	6 – 12	Double riveted
900 – 2100	8 – 25	Triple riveted
1500 – 2700	10 – 40	Quadraple riveted

Table 11.5. Working for rivets in structural work for buildings

Types of rivets	Shear, N/mm^2		Bearing, N/mm^2
	Double shear	Single shear	
Power driven rivets in reamed holes	280	225	105
Hand driven rivets	175	140	70

Table 11.6. Working stress in rivets used in railway bridges

Shear Stress, S_s	MPa
Power driven rivets	95
Hand driven rivet	77
Bearing Stress, S_c	
Power driven rivets	190
Hand driven rivets	140

Table 11.7. Commercially available rivet sizes

Basic rivet size, mm	12	14	16	18	20	22	24	27
Hole dia. (pressure vessel)	13	15	17	19	21	23	25	28.5
Hole dia. (general purpose)	13.5	15.5	17.5	19.5	21.5	23.5	25.5	29.0
Basic rivet size, mm	30	33	36	39	42	48		
Hole dia. (pressure vessel)	31.5	34.5	37.5	37.5	44	50.0		
Hole dia. (general purpose)	32.0	35.0	38.0	38.0	44	50.0		

Table 11.8. Arrangement of rivets for structural joints

Number of rivets	Lap joint	Butt joint
3	—	
4	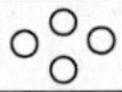	—
5	—	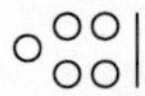
6	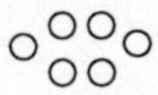	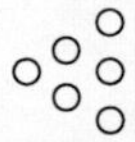
8	—	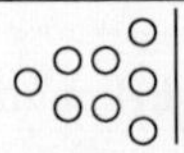
9	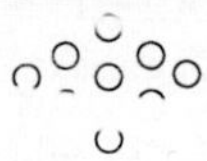	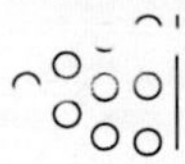
10	—	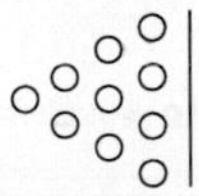
11	—	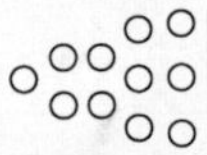
12	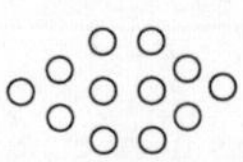	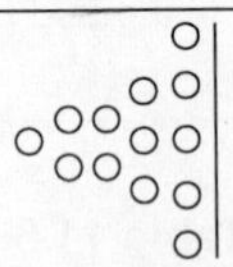
15	—	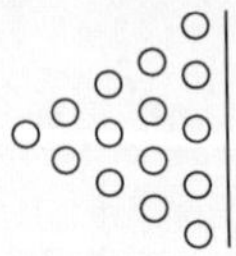
16	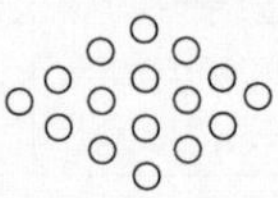	—

SOLVED EXAMPLES

Example 11.1: Design a double riveted double strap but joint for the longitudinal seam of a boiler shell 750 mm in diameter to carry a maximum stream pressure of 1.5 N/mm^2 gauge. The allowable stresses are $S_{td} = 50$ MPa; $S_{sd} = 30.0$ MPa and $S_{cd} = 75.0$ MPa. Assume the joint efficiency as 75%.

SOLUTION

Step-1: Design of joint details

Thickness of the shell (Eqn. 11.17)

$$t = \frac{D \cdot p_f}{2\eta\ S_{td}} = \frac{750 \times 1.5}{2 \times 75 \times 50} = 15.0 \text{ mm}$$

Rivet diameter,

$$d = 6\sqrt{t} = 6\sqrt{15} = 23.24 \text{ mm}$$

Choose the available diameter (rivet hole diameter) (Table 11.7)

$$d = 24.0 \text{ mm}$$

Cover plate thickness,

$$h_1 = h_2 = 0.625t = 9.375 \text{ mm} \cong \mathbf{10.0\ mm}$$

Pitch of the rivets

As per BR specifications (Table 11.2 and Eqn. 11.18)

$$\begin{aligned} P &= Ct + 40 = 3.5 \times 15 + 40 \\ &= 92.5 \text{ mm} \end{aligned}$$

Pitch for optimum strength, (Eqn. 11.19)

$$\begin{aligned} P &= \frac{(n_1 + 1.875n_2)\pi d^2\, S_s}{4t\, S_t} + d \\ &= \frac{(0 + 1.875 \times 2)\pi \times 24^2 \times 30}{4 \times 15 \times 50} + 24 = 91.8 \text{ mm} \end{aligned}$$

Then choose P = 92.0 mm (choose the lowest among the two)
It satisfies the condition set by Eqns. 11.2a and 11.2b

$$P > P_{\text{min}} = 2.5d = 2.5 \times 24 = 60 \text{ mm}$$

and

$$P < P_{\text{max}} = 6d = 6 \times 24 = 144 \text{ mm}$$

Transverse pitch for chain riveting,

$$P_t \geq 2d = 2 \times 24 = 48 \text{ mm}$$

Choose $P_t = 50.0$ mm

Step-2: Check for strengths
Strengths of solid plate (Eqn. 11.13)

$$F = P \cdot t \cdot S_t = 92 \times 15 \times 50.0 = 69000 \text{ N}$$

Shearing strength of rivets (Eqn. 11.8)

$$F_s = (0 + 1.875 \times 2)\pi \times 24^2 \times 30/4$$
$$= 50893.0 \text{ N}$$

Tearing strength of the plate (Eqn. 11.5)

$$F_t = (P - d)tS_t = 92 - 24 \times 15 \times 50$$
$$= 51000 \text{ N}$$

Crushing strength of the rivets (Eqn. 11.10)

$$F_c = (2 \times 15 + 0) \times 24 \times 75.0$$
$$= 54000 \text{ N}$$

Other strengths will be greater.
Minimum strength $= F_s = 50893.0$ N

Step-3: Efficiency of the joint

$$\eta = \frac{\text{Minimum strengths}\,(F_s, F_t, F_c)}{\text{Strength of solid plate}\,(F)}$$

$$\eta = 50893/69000 = 0.74$$

This is nearly equal to the given value of $\eta = 75\%$
Hence, the design is acceptable.

Example 11.2: Design longitudinal and circumferential joints for a steam drum of 1500 mm inside diameter. The drum is to store steam at pressure of 1.6 MPa gauge. The allowable stresses for the plate and rivets are $S_t = 80$ MPa, $S_s = 62.5$ MPa and $S_c = 125$ MPa.

SOLUTION

Step-1: Design of joint details
Choose triple riveted butt joint for the longitudinal joint with alternate rivet in the outer row omitted. The approximate efficiency of the joint $\eta = 80\%$ (Table 11.3).

Then thickness of the shell (Eqn. 11.17)

$$t = \frac{p_f D}{2\eta S_t} = \frac{1.6 \times 1500}{2 \times .8 \times 80} = 18.75 \text{ mm}$$

Choose **t = 20.0 mm**
The rivet diameter (Eqn. 11.1)

$$d = 6\sqrt{t} = 6\sqrt{20} = 26.8 \cong 27.0 \text{ mm}$$

Choose **d = 27 mm** (Rivet hole diameter)
The cover plate thickness (Eqn. 11.21)

$$\text{Wider cover,} \quad h_1 = 0.75\,t = 0.75 \times 20 \times 15 \text{ mm}$$
$$\text{Narrow cover,} \quad h_2 = 0.625\,t = 0.625 \times 20 = 12.5 \text{ mm}$$

Optimum strength pitch, (Eqn. 11.19)

$$P = \frac{(n_1 + 1.875\,n_2)\pi d^2 S_s}{4t\,S_t} + d$$

For $n_1 = 1$ and $n_2 = 4$

$$P = \frac{(1 + 1.875 \times 4)\pi (27)^2 \times 62.5}{4 \times 20 \times} + 27$$
$$= 217.1 \text{ mm}$$

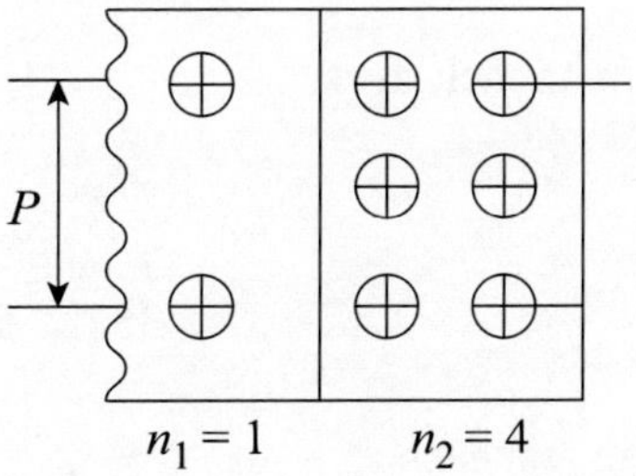

Pitch recommended by IBR, (Eqn. 11.18 and Table 11.2)

$$P \le Ct + 40 \le 6 \times 20 + 40 = 160.0 \text{ m}$$

Then choose pitch, $P = 160.0 \text{ mm} < P_{max} = 6d = 6 \times 27 = 162$ mm
Transverse pitch, $P_t \le (3d = 81.0 \text{ mm})$
Choose $P_t = 80.0$ mm

Step-2: Check for strength and efficiency
Strength of solid plate, (Eqn. 11.13)

$$F = 160 \times 20 \times 80 = 256000 \text{ N}$$

Tensile strength at the outer row, (Eqn. 11.5)

$$F_t = (160 - 27) \times 20 \times 80 = 212800 \text{ N}$$

Shearing strength of rivets per pitch, (Eqn. 11.8)

$$F_s = (1 + 1.875 \times 4)\pi/4(27)^2 \times 62.5 = 304169.98 \text{ N}$$

Crushing strength of rivets, (Eqn. 11.10)

$$F_c = (4 \times 20 + 1 \times 15) \times 27 \times 125 = 320625.0\text{ N}$$

The other strength will be larger than these.
The efficiency, (Eqn. 11.14)

$$\begin{aligned}\eta &= \frac{(\min F_s, F_c, F_t)}{F}\\ &= 212800/256000 = 0.8312 = 83.125\%\end{aligned}$$

This is more than the assumed value of 80% and the design is OK.

Step-3: Circumferential joint
It is a lap joint.

Shearing strength of each rivet, (Eqn. 11.6)

$$\begin{aligned}F_s &= \frac{\pi}{4}d^2 \cdot S_s = \frac{\pi}{4} \times 27^2 \times 62.5\\ &= 35785.0\text{ N}\end{aligned}$$

Crushing strength and each rivet, (Eqn. 11.9)

$$\begin{aligned}F_c = d \cdot t \cdot S_c &= 27 \times 20 \times 125\\ &= 67500\text{ N}\end{aligned}$$

Shear strength is smaller and the design is based on the shear strength.
The total force on the end cover.

$$\begin{aligned}F_s &= \frac{\pi}{4}D^2 \cdot p_f = \frac{\pi}{4} \times (1500)^2 \times 1.6\\ &= 2827433.4\text{ N}\end{aligned}$$

Number of rivets in the lap joint, (Eqn. 11.22)

$$n = \frac{2827433.4}{35785} = 79.0$$

$$\text{Choose } n = 80$$

Pitch for double row,

$$P = \frac{\pi D}{n/2} = \frac{\pi \times 1500}{40} = 117.8\text{ mm}$$

As per IBR recommendation, (Eqn.11.18)

$$P = 2.61 \times 20 + 40 \times 92.2 \text{ mm}$$

$\therefore$ Choose $P = 92.0$ mm

$$\therefore \quad \text{No. of rivets,/row} = \frac{\pi D}{P} = \frac{\pi \times 1500}{92} = 51.22$$

Choose total number of rivets, $n = 2 \times 51 = 102$

Transverse pitch, $P_t \rightarrow 2d = (2 \times 27 = 54)$ (Eqn. 11.20)

Choose $P_t = 60$ mm

Step-4: Check for the strengths

Strength of solid plate, (Eqn. 11.13)

$$F = t \cdot P \cdot St = 20 \times 92 \times 80 = 1{,}47{,}200 \text{ N}$$

Shearing strength, (Eqn. 11.8)

$$F_s = n\frac{\pi}{4}d^2 \cdot S_s = 2 \times 35{,}785 = 71{,}570 \text{ N}$$

Crushing strength, (Eqn. 11.9)

$$F_c = n_1 \cdot dt\,S_c = 2 \times 67500 = 1{,}35{,}000 \text{ N}$$

Tearing strength of the plate, (Eqn. 11.5)

$$F_c = (P - d)\,t \cdot S_t = (92 - 27) \times 20 \times 80$$
$$= 1{,}04{,}000 \text{ N}$$

$\therefore$

Step-5: Efficiency

$$\eta = \frac{\min F_t, F_s, F_c}{F} \quad \text{(Eqn. 11.14)}$$
$$= 71{,}570/1{,}47{,}200 = 0.486 = 48.6\%$$

This is more than half of the longitudinal efficiency of 83%, hence, OK.

Example 11.3: Design a double-rivetted zig-zag butt joint with equal cover plates to connect two plates 20 mm thick. Calculate the efficiency of the joint. Take $S_d = 100$ MPa; $S_{sd} = 60$ MPa and $S_{cd} = 120$ MPa.

Given data

Plate thickness, $t = 20$ mm
Material strength values
Design stress in tension, $S_d = 100$ MPa
Design stress in compression, $S_{cd} = 120$ MPa
Design stress in shear, $S_{sd} = 60$ MPa

SOLUTION

Step-1: Design of joint details

$$\text{Rivet diameter, } d = 6\sqrt{t} = 6\sqrt{20} = 26.83 \text{ mm.} \qquad \text{(Eqn. 7.2 MDHB)}$$

Standard rivet hole diameter,

$$\text{Choose d} = 27 \text{ mm} \qquad \text{(Table 7.7 MDHB)}$$

Thickness of the cover plates,

$$h = h_1 = h_2 = 0.625t = 0.625 \times 20 = 12.5 \text{ mm} \qquad \text{(Eqn. 7.6 MDHB)}$$

Pitch of riveting,

(i) Based on strength, $p = \dfrac{(n_1 + 1.875n_2)\pi d^2 S_{sd}}{4t \cdot S_d} + d$ (Eqn. 7.11 MDHB)

where, $n_1 = 0$ and $n_2 = 2$

$$P = \frac{1.875 \times 2\pi \times 27^2 \times 60}{4 \times 20 \times 100} + 27 = 91.41 \text{ mm}$$

(ii) By IBR code, (Eqn. 7.10a MDHB)

$$P \le Ct + 41.28 \quad \text{where} \quad C = 3.5 \qquad \text{(Table 7.1 MDHB)}$$
$$\le 3.5 \times 20 + 41.28$$
$$P = 110 \text{ mm}$$

(iii) Limitations

$$P_{\min} \geq 2.5d = 2.5 \times 27 = 67.5 \text{ mm}$$
$$P_{\max} \leq 6d = 6 \times 27 = 162 \text{ mm}$$
$$\text{Choose} \quad P = 95 \text{ mm}$$

Transverse pitch for zig-zag riveting,

$$\begin{aligned} P_t &\geq 0.165P + 0.67d \\ &\geq 0.165 \times 95 + 0.67 \times 27 \\ &= 33.765 \text{ mm} \end{aligned} \qquad \text{(Eqn. 7.18 MDHB)}$$

$$\text{Choose } P_t = 35 \text{ mm}$$

Step-2: Check for strength of the butt joint
Tensile strength of the solid plate,

$$\begin{aligned} F_t &= P_t.S_d \times 100 \\ &= 95 \times 20 \times 100 \\ &= 190 \times 10^3 \text{ N} \end{aligned} \qquad \text{(Eqn. 7.19 MDHB)}$$

Tensile strength of the outer row,

$$\begin{aligned} F_t &= (p-d)t\,S_t = (p-t)t \cdot S_d \\ &= (95-27) \times 20 \times 100 \\ &= 136000 \text{ N} \\ &= 136 \times 10^3 \text{ N} \end{aligned}$$

Shear strength of the rivets,

$$F = (n_1 + 1.875n_2)\frac{\pi}{4}d^2 \cdot S_{sd} \qquad \text{(Eqn. 7.21 MDHB)}$$

where, $n_1 = 0$ and $n_2 = 2$

$$\begin{aligned} F_s &= (0 + 1.875 \times 2)\frac{\pi}{4} \times 27^2 \times 60 \\ &= 128824.93 \text{ N} \\ &= 128.824 \times 10^3 \text{ N} \end{aligned}$$

Crushing strength of the rivets,

$$\begin{aligned} F_c &= (n_2 t + h_1 n_1)d\,S_{cd} \\ &= (2 \times 20 + 0) \times 27 \times 120 \\ &= 129600 \text{ N} \\ &= 129.6 \times 10^3 \text{ N} \end{aligned} \qquad \text{(Eqn. 7.22 MDHB)}$$

Step-3: Efficiency of the joint

$$\eta = \frac{\text{Min}(F_t \cdot F_s \cdot F_c)}{\text{Solid plate strength}(F)} \qquad \text{(Eqn. 7.25 MDHB)}$$

$$= \frac{128.825}{190} = 0.68$$

$$\eta = 68\%$$

Example 11.4: Design a double rivetted lap joint with zig-zag rivetting for 13 mm thick plates. The working stresses to be used are $S_d = 80$ MPa, $S_{sd} = 60$ MPa and $S_{cd} = 120$ MPa state how the joint will fail and find the efficiency of the joint.

Given data

Plate thickness, $t = 13$ mm
Working stresses in

Tension, $S_d = 80$ MPa
Compression, $S_{cd} = 120$ MPa
Shear, $S_{sd} = 60$ MPa

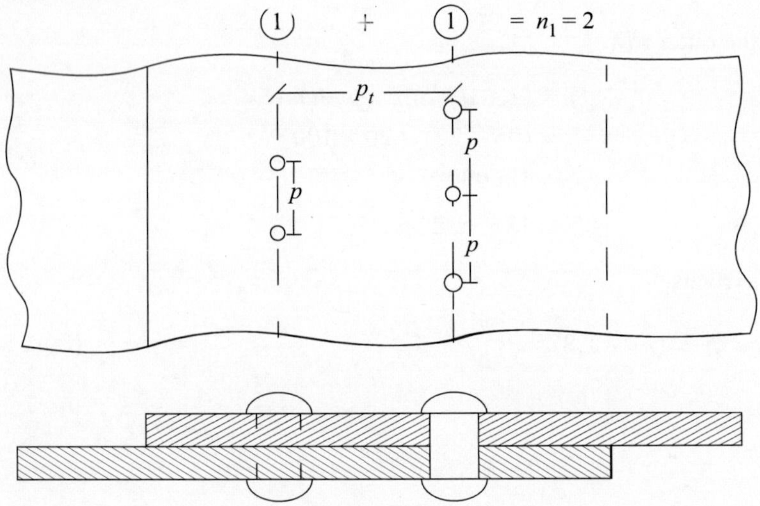

Figure 11.10

SOLUTION

Step-1: Design of the joint details

Rivet diameter, $d = 6\sqrt{t}$ (Eqn. 7.2 MDHB)

$$= 6\sqrt{13} = 21.63$$

Choose $d = 22$ mm (Rivet hole diameter) (Table 7.7 MDHB)

Pitch of riveting,

(i) For strength consideration,

$$P = \frac{(n_1 + 1.875n_2)\pi d^2 \cdot S_s}{4t \cdot s_t} + d \qquad \text{(Eqn. 7.11 MDHB)}$$

$$= \frac{(2 + 1.875 \times 0)\pi \times 22^2 \times 60}{4 \times 13 \times 30} + 22 = 65.86 \text{ mm}$$

(ii) By IBR code,

$$P \leq Ct + 41.28 \quad \text{where} \quad C = 2.61 \qquad \text{(Table 7.1)}$$

$$\leq 2.61 \times 13 + 41.28 = 75.21 \text{ mm}$$

(iii) Limitations

$$P_{\text{min}} = 3d = 3 \times 22 = 66 \text{ mm}$$

$$P_{\text{max}} = 6d = 6 \times 12 = 132 \text{ mm}$$

$$\text{Choose} \quad P = 70 \text{ mm}$$

Step-2: Check for strength of the joint
Strength of the solid plate,

$$F = Pt \cdot S_d \qquad \text{(Eqn. 7.31 MDHB)}$$

$$= 70 \times 13 \times 80$$

$$= 72800 \text{ N}$$

Tensile strength at the outer row,

$$F_t = (P - d)t \cdot S_d \qquad \text{(Eqn. 7.32 MDHB)}$$

$$= (70 - 22) \times 13 \times 80$$

$$= 49920 \text{ N}$$

Shear strength of rivets,

$$F_s = n_1 \frac{\pi}{4} \cdot d^2 S_s \qquad \text{(Eqn. 7.33 MDHB)}$$

$$= 2 \times \frac{\pi}{4} \times 22^2 \times 60$$

$$= 45615.93 \text{ N}$$

Crushing strength of the rivets,

$$F_c = n_1 t d S_c \qquad \text{(Eqn. 7.34 MDHB)}$$

$$= 2 \times 13 \times 22 \times 120$$

$$= 68640 \text{ N}$$

Step-3: Efficiency of the joint

$$\eta = \frac{\text{Min}(F_t, F_s, F_c)}{F}$$
$$= \frac{45615.93}{72800} = 0.627$$
$$= 62.7\%$$

The failure may occur by shearing of the rivets.

Example 11.5: Design a double rivetted butt joint with two cover plates for the longitudinal seam of a boiler shell 1.5 m in diameter subjected to a steam pressure of 0.9 MPa. Assume joint efficiency as 75%, allowable stress in tension for the plate as 83 MPa, in compression 138 MPa and shear stress in rivets may be assumed as 55 MPa. Assume chain rivetted joint.

Given Data
Joint type: Double rivetted double cover butt joint

Drum diameter = 1.5 m = 1500 mm

Fluid pressure, $p_f = 0.9$ MPa

Joint efficiency, $\eta = 75\%$

Allowable stresses

Tension, $S_D = 83$ MPa

Compression, $S_{cd} = 138$ MPa

Shear, $S_{sd} = 55$ MPa

SOLUTION
Step-1: Design of joint details
Plate thickness,

$$t = \frac{DP_f}{2\eta\, S_t} + 1 \quad \text{(Eqn. 7.1 MDHB)}$$
$$= \frac{1500 \times 0.9}{0.75 \times 2 \times 83} + 1$$
$$= 10.84 + 1 = 11.84 \text{ mm}$$

Choose $t = 12$ mm

Rivet diameter,

$$d = 6\sqrt{t} \quad \text{(Eqn. 7.2 MDHB)}$$
$$= 6 \times \sqrt{12} = 20.78$$

Choose, $d = 22$ mm (Standard rivet diameter) (Table 7.7 MDHB)
Thickness of the cover plate (Eqn. 7.6 MDHB)

$$h = h_1 = h_2 = 0.625t$$
$$= 0.625 \times 12$$
$$= 7.5 \text{ mm}$$

Pitch of the rivets IBR code,

$$P \leq Ct + 41.28 \qquad \text{(Eqn. 7.10a MDHB)}$$

where $C = 3.5$ (Table 7.1 MDHB)

$$\therefore \quad P = 3.5 \times 12 + 41.28$$
$$= 83.28$$
$$\text{Choose} \quad P = 80 \text{ mm} > 3d \text{ and } < 6d$$

Optimum pitch,

$$P = \frac{(n_1 + 1.875n_2)\pi d^2 S_s}{4t \cdot S_t} + d$$
$$P = \frac{(0 + 1.875 \times 2)\pi \times 22^2 \times 55}{4 \times 12 \times 83} + 22$$
$$= 78.72 + 22$$
$$= 100.7 \text{ mm } [< 6d = 132]$$

Then choose $P = 80$ mm
Transverse pitch,

$$P_t = 2d > 2 \times 22$$
$$= 2 \times 22 = 44 \text{ mm}$$
$$\text{Choose} \quad P_t = 50 \text{ mm}$$

Step-2: Check for strength
Strength of the plate per pitch,

$$F = P_t S_d \qquad \text{(Eqn. 7.19 MDHB)}$$
$$= 80 \times 12 \times 83$$
$$= 79680 \text{ N}$$

Tearing strength of the plate at the first row,

$$F_t = (P - d)tS_t \qquad \text{(Eqn. 7.20 MDHB)}$$
$$= (80 - 22) \times 12 \times 83$$
$$= 57768 \text{ N}$$

Shearing strength of rivets,

$$F_s = (n_1 + 1.875n_2)\frac{\pi}{4}d^2 \cdot S_s$$
$$= (0 + 1.875 \times 2)\frac{\pi}{4} \times 22^2 \times 55$$
$$= 78402.4 \text{ N}$$

Crushing strength of rivets,

$$F_c = (n_2 t + n_1 h_1)dS_c$$
$$= (2 \times 12 + 0) \times 22 \times 138$$
$$= 72864 \text{ N}$$

Others will be more and hence need not be calculated.
Step-3: Joint efficiency,

$$\eta = \frac{\text{Min}(F_t, F_s, F_c)}{F} = \frac{57768}{79680} = 0.725$$
$$\eta = 72.5\%$$

Less than given value, $\eta = 75\%$
Increase $P = 83$ mm
Then $F_t = (83 - 22) \times 12 \times 83 = 60756$
Now, efficiency is

$$\eta = \frac{60756}{79680} = 0.7625$$
$$\eta = 76.25\% > \eta = 75\% \text{ specified value}$$

Hence, design is acceptable.

Example 11.6: Design a double riveted (longitudinal) double strap butt joint with unequal cover plates for the pressure vessel. The internal diameter for the pressure is 1 m and is subjected to an internal pressure of 2.2 MPa. The pitch of the rivets in the outer row is to be double the pitch in the inner row. The allowable tensile stress in the plate is 124 N/mm^2. The allowable shear and crushing stress in the rivets are 93 N/mm^2 and 165 N/mm^2 respectively. The resistance of the rivets in double shear is to be taken as 1.875 times that of single shear.

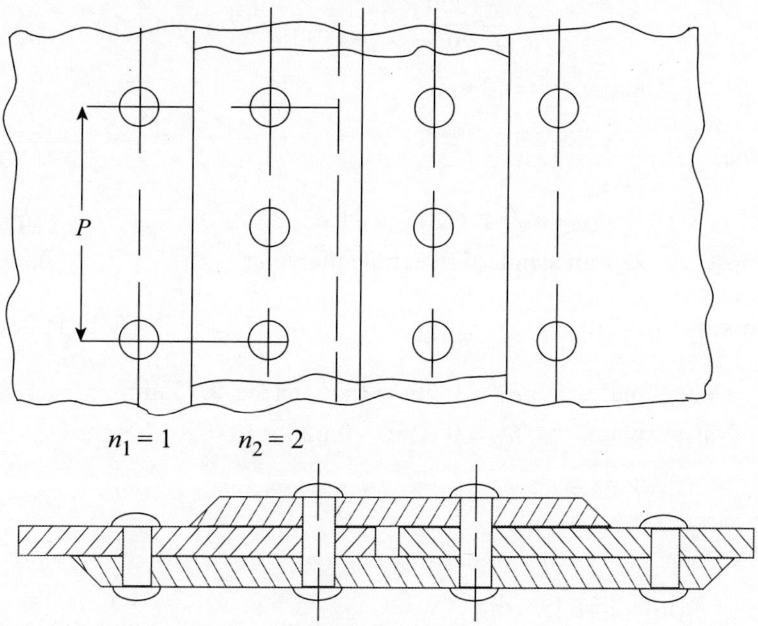

Figure 11.11

Given Data

Joint type: Double riveted double strap with unequal straps

Diameter of the drum, $D = 1$ m $= 1000$ mm

Fluid pressure, $P_f = 2.2$ MPa

Plate strength, $S_t = 124.0$ MPa

Rivet strength in shear, $S_{sd} = 93$ MPa

Rivet strength in crushing, $S_{cd} = 165$ MPa

Assume the efficiency of the joint, $\eta = 75\%$ (Table 7.3 MDHB)

SOLUTION

Step-1: Design of joint details

Plate thickness,

$$t = \frac{DP_f}{2\eta S_t} + 1 \qquad \text{(Eqn. 7.1 MDHB)}$$

$$= \frac{1000 \times 2.2}{2 \times 0.75 \times 124} + 1 = 12.8 \text{ mm}$$

$$\text{Choose} \quad t = 13 \text{ mm}$$

Diameter of the rivet,

$$d = 6\sqrt{t} = 6\sqrt{13} = 21.63 \qquad \text{(Eqn. 7.2 MDHB)}$$

Choose $d = 22$ mm standard rivet hole diameter (Table 7.7 MDHB)

Cover plate thicknesses (Eqn. 11.21c and 11.21d)

$$\text{Wider plate,} \quad h_1 = 0.75t = 0.75 \times 13 = 9.75 \text{ mm}$$

$$\text{Narrow plate,} \quad h_2 = 0.625t = 0.625 \times 13 = 8.125 \text{ mm}$$

The pitch of riveting,

(i) Based on strength (Eqn. 11.19)

$$P = \frac{(n_1 + 1.1875n_2)\pi d^2 \cdot S_s}{4t \cdot S_t} + d \quad \text{where} \quad n_1 = 1 \quad \text{and} \quad n_2 = 2$$

$$= \frac{(1 + 1.875 \times 2)\pi \times 22^2 \times 93}{4 \times 13 \times 124} + 22 = 126.17 \text{ mm}$$

(ii) By IBR code equation (Eqn. 7.11a MDHB)

$P \leq Ct + 41.28$ where $C = 4.63$ (Table 11.1)

$$\leq 4.63 \times 13 + 41.28 = 101.47 \text{ mm}$$

(iii) Limitations,

$$P_{\text{min}} = 3d = 3 \times 22 = 66 \text{ mm}$$

$$P_{\text{max}} = 6d = 6 \times 22 = 132 \text{ mm}$$

$$\text{Choose } P = 100 \text{ mm}$$

Inner pitch,

$$P_1 = \frac{1}{2}P = \frac{1}{2} \times 100 = 50 \text{ mm}$$

Transverse pitch,

$$P_t \geq 0.33P + 0.67d \qquad \text{(Eqn. 7.15 MDHB)}$$
$$\geq 0.33 \times 100 + 0.67 \times 22$$
$$= 47.7 \text{ mm}$$
$$\text{Choose } P_t = 50 \text{ mm}$$

Step-2: Check for strength of the joint
Strength of the solid plate,

$$F = P \cdot t \cdot S_t \qquad \text{(Eqn. 11.13)}$$
$$= 100 \times 13 \times 124$$
$$= 161200 \text{ N}$$

Tearing strength at the outer row,

$$F_t = (P - d)tS_t \qquad \text{(Eqn. 11.5)}$$
$$= (100 - 22) \times 13 \times 124$$
$$= 125736 \text{ N}$$

Shearing strength of rivets,

$$F_s = (n_1 + 1.875n_2)\frac{\pi}{4}d^2 \cdot S_{sd} \qquad \text{(Eqn. 11.8)}$$
$$= (1 + 1.875 \times 2)\frac{\pi}{4} \times 22 \times 93$$
$$= 167923 \text{ N}$$

Crushing strength of rivets,

$$F_c = (n_2 t + n_1 h_1)d \cdot S_{cd} \qquad \text{(Eqn. 11.10)}$$
$$= (2 \times 13 + 1 \times 9.25) \times 22 \times 165$$
$$= 129772.0 \text{ N}$$

Tearing strength at inner two with shearing of rivets in outer row,

$$F_{st} = (p - 2d)tS_d + \frac{\pi}{4}d^2 \cdot S_{sd} \qquad \text{(Eqn. 11.11)}$$
$$= (100 - 2 \times 22) \times 13 \times 124 + \frac{\pi}{4} \times 22^2 \times 93$$
$$= 125624 \text{ N}$$

Tearing strength at inner row with crushing of rivets in outer row,

$$F_{ct} = (p - 2d)tS_t + dh_1 S_{cd} \qquad \text{(Eqn. 11.12)}$$
$$= (100 - 2 \times 22) \times 13 \times 124 + 22 \times 9.75 \times 165$$
$$= 125.664.5 \text{ N}$$

Step-3: Efficiency of the joint

$$\eta = \frac{\text{Min}(F_t\ F_s\ F_c\ F_{st}\ F_{ct})}{F} = \frac{125624}{161200} = 0.7793$$

$$\eta = 77.93\%$$

This is more than the assumed value of $\eta = 75\%$ hence, the design is acceptable.

Example 11.7: Two length of mild-steel that the bars 200 mm × 10 mm are to be connected by a double cover joint using 24 mm diameter rivets. Design the joint, of the allowable working stresses are 110 MPa in tension, 85 MPa in shear and 200 MPa in crushing.

SOLUTION

Strength of the solid plate,

$$F = 200 \times 10 \times 10 \times 110 = 2,20,000 \text{ N} = 220 \text{ kN}$$

Shear strength of each rivet,

$$F_s = 1.75 \left(\frac{\pi}{4}\right) d^2 \cdot S_s = \pi \times 24^2 \times 85)/2 = 67292.9 \text{ N} = 67.29 \text{ kN}$$

Tearing strength of the plate/pitch,

$$F_t = (P - d)tS_t = (200 - 24) \times 10 \times 110.0$$

$$= 1,93,600 \text{ N} = 193.6 \text{ kN}$$

The crushing strength of the rivets, $F_e = dtS_c = 24 \times 10 \times 200 = 48000 \text{ N} = 48 \text{ kN}$

Equating the crushing strength and tearing strength the number of rivets required,

$$n = \frac{193600}{48000} = 4$$

The rivets are arranged in the diamond shape with one rivet in the last row and three in the inner row.

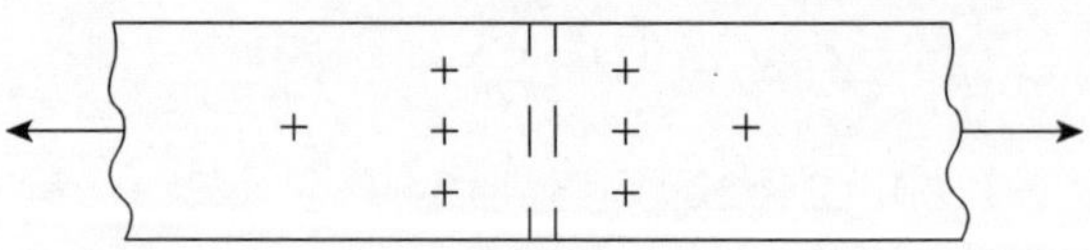

Figure 11.12

Efficiency of the joint for strength,

$$\eta = \frac{p - d}{p} = \frac{200 - 25}{200} = 0.88$$

Example 11.8: Two mild steel tie rods having a width of 225 mm and a thickness of 15 mm are held together by a double strap Lozenge joint. Design the joint if the permissible stresses as given below:

Permissible tensile strength, $S_t = 110$ MPa
Permissible shear strength, $S_s = 75$ MPa
Permissible crushing strength, $S_c = 185$ MPa

Determine the efficiency of the joint.

Given Data

Width of the plate = 225 mm
Thickness of the plate = 15 mm
Joint used: Lozenge joint with double cover

Permissible stresses:

Tension, $S_{td} = 110$ MPa
Shear, $S_{sd} = 75$ MPa
Compression, $S_{cd} = 185$ MPa

SOLUTION

Step-1: Design load
Refer Fig. 7.1b MDHB.
The load is not given hence design is to be based on the load the plate can support.
The plate strength at the last row,

$$F_t = (w - d)t \cdot S_t$$

Rivet diameter, $d = 6\sqrt{t} = 6\sqrt{15} = 23.24$

$$\text{Choose} \quad d = 24 \text{ mm} \qquad \text{(Table 7.7 MDHB)}$$

$$\therefore \quad F_t = (225 - 24) \times 15 \times 110$$
$$= 331650 \text{ N}$$

This is the design load.

Step-2: Minimum strength of a rivet

Shear strength of each rivet in double shear,

$$F_s = 1.75 \times \frac{\pi}{4} d^2 S_{sd}$$
$$= 1.75 \times \frac{\pi}{4} \times 24^2 \times 75$$
$$= 59375 \text{ N}$$

Crushing strength of each rivet,

$$F_c = tdS_{cd} = 15 \times 24 \times 185 = 66600 \text{ N}.$$

$$\text{Minimum strength, } F_{\text{min}} = 59376 \text{ N}$$

Step-3: Then the number of rivets required,

$$n = \frac{F_t}{F_{\text{min}}} = \frac{331650}{59376} = 5.58$$

Adopt $n = 6$.

Step-4: Arrangement of the rivets is shown in the figure below.

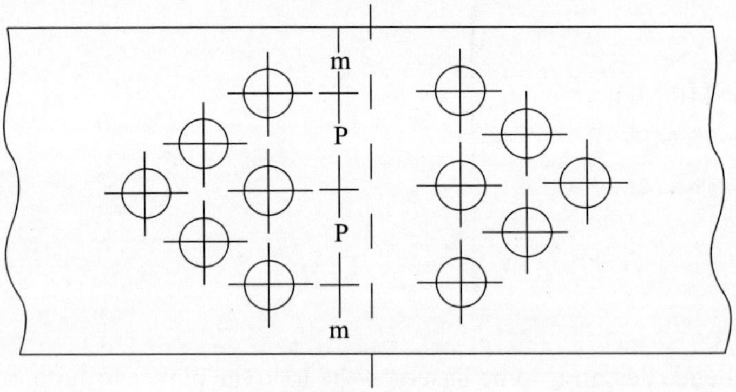

Figure 11.13

Pitch of the rivets for the innermost row is evaluated by the relation,

$$\begin{aligned} w &= 2m + 2P \\ &= 2 \times 1.5d + 2P \\ 225 &= 2 \times 1.5 \times 24 + 2P \\ P &= \frac{225 - 2 \times 1.5 \times 24}{2} \\ P &= 76.5 \text{ mm} \end{aligned}$$

The transverse pitch,

$$\begin{aligned} P_t &= 0.165P + 0.67d \qquad \text{(Eqn. 11.20d)} \\ &= 0.165 \times 76.5 + 0.67 \times 24 \\ &= 28.7 \text{ mm} \end{aligned}$$

$$\text{Choose} \quad P_t = 30 \text{ mm}$$

Step-5: Check for strengths
Shearing strength of all the rivets,

$$F_s = n \times 1.75\frac{\pi}{4}d^2 \cdot S_{sd}$$
$$= 6 \times 1.75 \times \frac{\pi}{4} \times 24^2 \times 75$$
$$= 356256.7 \text{ N}$$

Crushing strength of all the rivets,

$$F_c = n \times dt \cdot S_{sd}$$
$$= 6 \times 24 \times 15 \times 185$$
$$= 399600 \text{ N}$$

Tearing strength at the last row,

$$F_t = (w - d)t \cdot S_d$$
$$= (225 - 24) \times 15 \times 110$$
$$= 331,650 \text{ N}$$

Tearing strength at the second row,

$$F_{t2} = (w - 2d)tS_d + 1.75 \times \frac{\pi}{4}d^2 S_{sd}$$
$$= (225 - 2 \times 24) \times 15 \times 110 + 1.75 \times \frac{\pi}{4} \times 24^2 \times 75$$
$$= 351426.1 \text{ N}$$

The tearing strength at the innermost row,

$$F_{t3} = (w - 3d)tS_d + 3 \times 1.75 \times \frac{\pi}{4}d^2 S_{sd}$$
$$= (225 - 3 \times 24) \times 15 \times 110 + 3 \times 1.75 \times \frac{\pi}{4} \times 24^2 \times 75$$
$$= 430578.3 \text{ N}$$

Strength of the solid plate,

$$F = wtS_d$$
$$= 225 \times 15 \times 110$$
$$= 371250 \text{ N}$$

Step-6: Efficiency of the joint

$$\eta = \frac{\text{Min}(F_t\ F_s\ F_c\ F_{t2}\ F_{t3})}{F}$$

$$= \frac{331426}{371250} = 0.89$$

$$\eta = 89\%$$

Example 11.9: A bracket is fixed to a column by a group of rivets as shown in Fig. 11.14. The bracket is to support load of 40 kN. The shear stress in the rivet is not to exceed 60 MPa. Determine the size of the rivets.

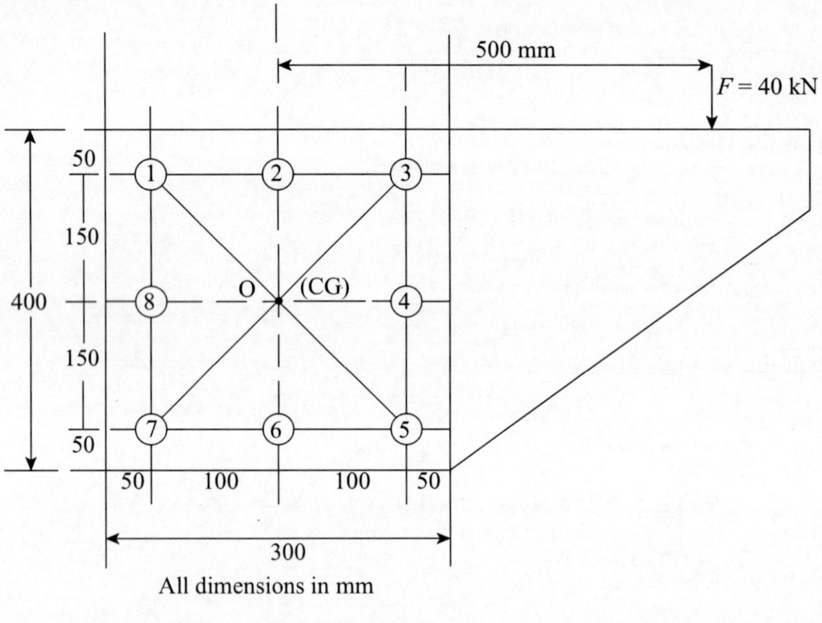

Figure 11.14

SOLUTION

Determine the CG

$$\bar{x} = \frac{0+100+200+0+200+0+100+200}{8} = \frac{800}{8} = 100\text{ m}$$

$$\bar{y} = \frac{0+150+300+300+300+150+0+0}{8} = \frac{1200}{8} = 150\text{ mm}$$

The primary shear force in each rivet,

$$F_{st} = 40000/8 = 5000\text{ N}$$

Distance of the rivets from C.G.,

$$r_1 = r_3 = r_5 = r_7 = \sqrt{150^2 + 100^2} = 180.28 \text{ mm}$$

$$r_2 = r_6 = 150 \text{ mm, and } r_4 = r_8 = 100 \text{ mm}$$

Secondary shear in the rivets 1, 3, 5, and 7 is the same, each perp. to the respective radial lines (Eqn. 7.37),

$$F_1 = F_3 = F_5 = F_7 = \frac{40 \times 10^3 \times 500 \times 180.28}{\Sigma(4 \times 180.28^2 + 2 \times 150^2 + 2 \times 100^2)}$$

$$= 102.56 \times 180.28 = 18490.0 \text{ N}$$

Force in rivets (2) and (6) $F2 = F6 = 102.56 \times 150 = 153840$ N
Forces in rivets (4) and (8),

$$F_4 = F_8 = 102.56 \times 100 = 10256 \text{ N}$$

The forces in rivets (4), the primary and secondary are oriented in the same direction and add together. The resultant force is

$$R_4 = 5000 + 10256 = 15256.0 \text{ N}$$

For the rivets (2) and (6) the forces is right angles and the resultant is smaller. Hence, no need to calculate. The angle between the primary and secondary forces for rivets (1) and (7) are abtuse and resultant is smaller. Hence, no need to calculate. For rivets (3) and (5) the angle between the primary and secondary shears is acute and resultant is larger. Hence, it is to be calculated.

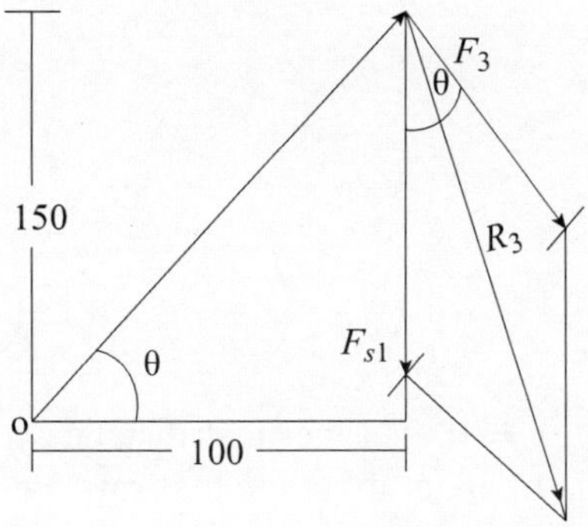

The angle $\theta = \tan^{-1}\left(\frac{150}{100}\right) = 56.31°$

Resultant on rivet, (3) (Eqn. 10.31)

$$R_3 = \sqrt{F_{st}^2 + F_3^2 + 2F_{st}\, F_3 \cos 56.31°}$$

$$= \sqrt{(5000)^2 + 18490^2 + 2 \times 5000 \times 18490 \times \cos 56.31}$$

$$= 21666.7 \text{ N}$$

For rivet (5) also it is the same.

∴ Maximum force, $F_{max} = 21666.7$ N

$\therefore$ Rivet area, $A_s = F_{\max}/S_s = 21666.7/60 = 361.1 \text{ mm}^2$

$\therefore$ The diameter of rivet, $d = \sqrt{\dfrac{4\,A_s}{\pi}}$

$$= \sqrt{\frac{4 \times 361.1}{\pi}} = 21.44 \text{ mm}$$

$\therefore$ Choose $d = 21$ mm (Table 11.7)

Example 11.10: A bracket plate 25 cm thick is riveted to a vertical column. The spacing of the rivets is shown in Fig. 11.15 The load $P = 50$ kN is applied with an eccentricity of 400 mm. Find the rivet size for permissible shear stress of 65 MPa and crushing stress of 120 MPa.

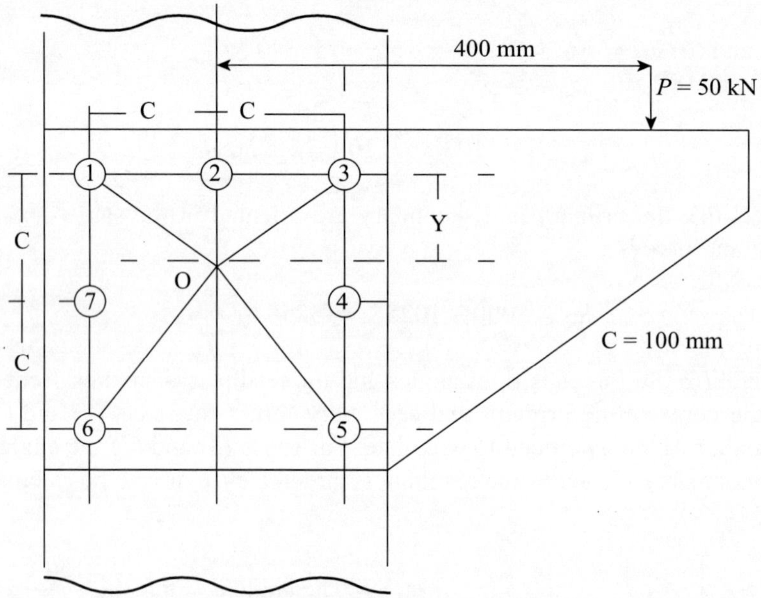

Figure 11.15

SOLUTION

Step-1: Geometric relations

The CG of the group of rivets is on the vertical central line, i.e., $\bar{x} = 0$. The vertical distance y is to be determined. Let A be the area of each rivet. Taking moment of areas about the first row,

$$7 \times A \times Y = 3A \times 0 + 2A \times C + 2A \times 2C$$

$$\therefore Y = \frac{2 \times 100 + 4 \times 100}{7} = 85.7 \text{ mm}$$

Distance of the rivets from CG are

$$r_1 = r_3 = \sqrt{100^2 + 85.7^2} = 131.71$$

$$r_2 = 85.7 \text{ mm}$$

$$r_4 = r_7 = \sqrt{100^2 + 14.3^2} = 101.0 \text{ mm}$$

$$r_5 = r_6 = \sqrt{100^2 + 114.3^2} = 151.87 \text{ mm}$$

Step-2: Forces on the bolts
Rivet (5) is at the farthest distance and the angle between the primary shear and secondary shear is acute. Hence, the resultant force on this rivet is maximum.

Primary shear, $F_{st} = \dfrac{50}{7} = 7.143$ kN
Secondary shear, (Eqn. 11.3)

$$F_{S_2} = \frac{50 \times 400 \times 151.87}{[2 \times 131.7^2 + 85.7^2 + 2 \times 101^2 + 51.87^2]}$$
$$= 27.98 \text{ kN}$$

Angle θ between the forces,

$$\theta = \tan^{-1}\left(\frac{114.3}{100}\right) = 48.8^\circ$$

Resultant force,

$$R_5 = \sqrt{7.143^2 + 27.98^2 + 2 \times 7.14^3 \times 27.98 \cos 48.8}$$
$$= 32.12 \text{ kN}$$

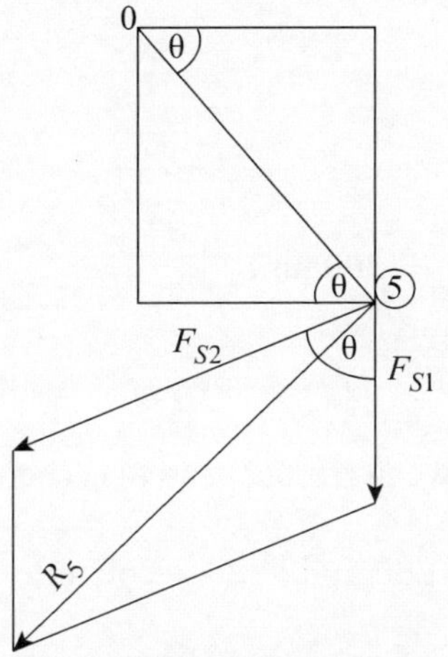

Figure 11.16

Step-3: Rivet design

Shearing stress for the rivet = 65 MPa

$$\therefore \quad \text{The rivet diameter for shearing strength} = \sqrt{\frac{4 \times 33.12 \times 10^3}{\pi \times 65}}$$

$$= 25.47 \text{ mm}$$

Crushing stress = 120 MPa

$$\therefore \quad \text{The rivet diameter for crushing strength, } d = \frac{33.12 \times 10^3}{25 \times 120} = 11.04 \text{ mm}$$

∴ Choose the larger of the two as the diameter of rivet $d = 27$ mm (Table 11.7)

Example 11.11: A bracket is supported by means of four rivets of same size as shown in Fig. 11.17. Determine the diameter of the rivet if the maximum shear stress in the rivet is 90 N/mm^2

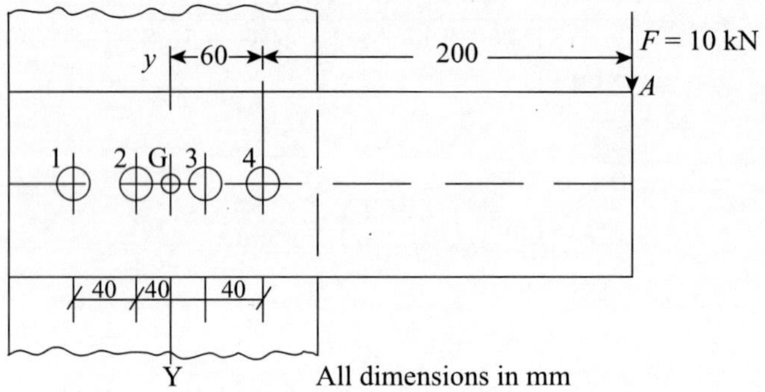

Figure 11.17

Given Data

Four rivets placed at 40 mm each

Load applied to the bracket = 10 kN

Distance of the load from the last rivet = 200 mm

Permissible shear stress, $S_{sd} = 90$ N/mm^2

SOLUTION

The C.G. of the rivets-G is placed between rivet (2) and (4). Then the eccentricity of the load,

$$e = 200 + 40 + 20 = 260 \text{ mm}$$

Load applied at '*A*' end of the bracket is equivalent a load '*F*' through *G* and twisting moment,

$$T = F.e = 10 \times 10^3 \times 260 \text{ N-mm}$$
$$= 2600 \times 10^3 \text{ N-mm}$$

Primary shear force per rivet,

$$F_{S_1} = \frac{10 \times 10^3}{4} = 2.5 \times 10^3 \text{ N}$$

Maximum secondary shear force in rivet (4),

$$F_{S_2} = \frac{F \cdot e}{\Sigma r_1^2} \cdot r_{\text{max}}$$
$$= \frac{2600 \times 10^3 \times 60}{\Sigma[2(20^2 + 60^2)]} = \frac{156 \times 10^6}{8000} = 19.5 \times 10^3 \text{ N-m}$$

Total shear force,

$$F_t = 2.5 \times 10^3 + 19.5 \times 10^3$$
$$= 22 \times 10^3 \text{ N}$$

Area of the rivet required,

$$A = \frac{22 \times 10^3}{90} = 244.4 \text{ mm}^2$$

Diameter of the rivet,

$$d = \sqrt{\frac{4.A}{\pi}} = \sqrt{\frac{4 \times 244.4}{\pi}}$$
$$= 17.64 \text{ mm}$$

Choose $d = 18$ mm

Example 11.12: Determine the size of the rivets for a 20 mm thick flat plate riveted to a column as shown in the Fig. 11.18. The allowable stresses are 56 MPa in shear and 100 MPa in crushing.

Given Data

Plate thickness, $t = 20$ mm
Applied load, $F = 20$ kN

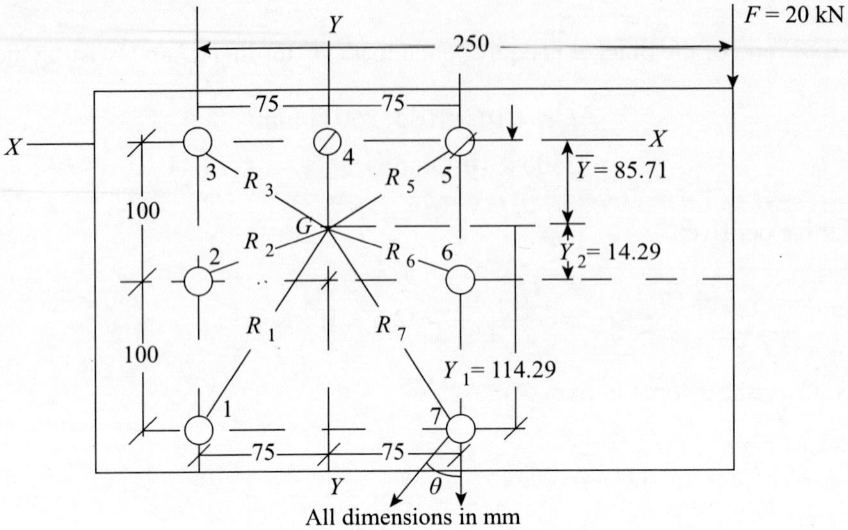

Figure 11.18

The distance of the load from the first column of rivets = 200 mm
Permissible stresses,

$$\text{Shear, } S_{sd} = 56 \text{ MPa}$$
$$\text{Crushing, } S_{cd} = 100 \text{ MPa}$$

SOLUTION

Step-1: Geometric relations

The CG of the rivets is placed on the $Y - Y$ axis by symmetry. The vertical distance $\bar{y}$ is given by

$$\bar{y} = \frac{2[100A + 200A]}{7A} = 85.71 \text{ mm}$$

The distance of the rivets from CG are

$$R_1 = R_7 = \sqrt{75^2 + 114.29^2} = 136.7 \text{ mm}$$

$$R_2 = R_6 = \sqrt{75^2 + 14.29^2} = 76.35 \text{ mm}$$

$$R_3 = R_5 = \sqrt{75^2 + 85.71^2} = 113.89 \text{ mm}$$

$$R_4 = 85.71 \text{ mm}$$

Eccentricity of the load,

$$e = 200 - 75 = 125 \text{ mm}.$$

Step-2: Forces on the rivet

Maximum secondary force on the rivets (1) and (7) because they are farthest from C.G.

$$F_{S_2} = \frac{F \cdot e}{\Sigma r_1^2} \cdot R_{\max}$$

$$= \frac{20 \times 10^3 \times 125 \times 136.7}{2 \times (136.7^2 + 113.89^2 + 76.35^2) + 85.71^2}$$

$$= 4151.46 \text{ N}$$

Primary shear, F_{S_1}

$$F_{S_1} = \frac{20 \times 10^3}{7} = 2857.14 \text{ N}$$

Angle between the forces,

$$\theta = \tan^{-1}\left(\frac{114.29}{75}\right) = 56.73^\circ$$

Then the resultant force,

$$F_R = \sqrt{F_{S_1}^2 + F_{S_2}^2 + 2F_{S_1}F_{S_2}\cos\theta} \qquad \text{(Eqn. 7.38 MDHB)}$$

$$= \sqrt{2857.14^2 + 4151.5^2 + 2 \times 2857.14 \times 4151.5 \times \cos 56.73}$$

$$= 6197.76 \text{ N}$$

Step-3: Design of rivet

Rivet diameter for crushing strength $[F = dt\, S_c]$

$$\therefore d = \frac{6197.76}{20 \times 100} = 3.1 \text{ mm}$$

Rivet diameter for shearing strength $\left[F = \frac{\pi}{4} d^2 \cdot S_{sd}\right]$

$$\therefore d = \sqrt{\frac{4 \times 6197.76}{\pi \times 56}} = 11.87 \text{ mm}$$

Choose diameter of rivet = 12 mm

EXERCISES FOR PRACTICE

11.1. Two plates 16 mm thick are joined by a double riveted lap joint. The pitch of each row of rivets is 90 mm. The rivets are 25 mm in diameter. The permissible stresses are 140 MPa in tension, 80 MPa in shear and 160 MPa in crushing. Find the efficiency of the joint.

11.2. A double riveted lap joint with zig-zag riveting is to be designed for 13 mm plates. Assume permissible stresses in tension, shear and compression are 80 MPa, 60 MPa and 120 MPa respectively. Find the efficiency of the joint and state how the joint will fail.

11.3. Two plates of 8 mm thickness each are to be joined by a single riveted double strap butt joint. Design the joint with working stresses in tension and shearing and compression are 80 MPa, 60 MPa and 100 MPa respectively.

11.4. A double riveted lap joint is made between 15 mm thick plates. The rivet diameter and pitch are 25 mm and 75 mm respectively. If the ultimate stresses are 400 MPa in tension and 320 MPa in shear and 640 MPa in crushing, find the minimum force per inch, which will rupture the joint. If the above joint is subjected to a load such that the factor of safety is 4, find out the actual stresses developed in the plates and the rivets.

11.5. Design the longitudinal joint for a 1.25 m diameter steam boiler to carry a steam pressure of 2.5 MPa. The ultimate strength of the boiler plate may be assumed as 420 MPa in tension, crushing strength as 650 MPa and shear strength of 300 MPa. The joint efficiency as 80%. FOS = 2.

11.6. Design a chain type double riveted butt joint with two cover plates, for the longitudinal seam of a boiler shell 2 m diameter subjected to a steam pressure of 1.5 MPa. Assume the joint efficiency as 75%, allowable tensile stress in plate 85 MPa, compressive stress 130 MPa; and shear stress in the rivets is 52 MPa.

11.7. A double riveted butt joint in which the pitch of the rivets in the outer rows is twice that in the inner rows, connects two 16 mm thick plates with two cover plates each 12 mm thick. The diameter of rivets is 22 mm. Determine the pitches of the rivets in the two rows if the working stresses are not to exceed the following limits. Tensile stress in plates = 100 MPa; shear stress in rivets = 75 MPa; bearing stress in rivets and plates = 150 MPa.

11.8. Design a triple riveted longitudinal butt joint with zig-zag riveting having unequal equal widths of cover plates for a pressure vessel of diameter 1200 mm subjected to an internal pressure of 0.90 MPa. A joint efficiency of 75% can be assumed. The pitch of rivets in the extreme rows, which are in single shear, is twice the pitch of rivets in the inner rows which are in double shear. The design stresses of the materials of the main plate and the rivets are as follows:

(i) plate material in tension, $\sigma_t = 120$ MPa

(ii) rivet material in compression, $\sigma_c = 160$ MPa

(iii) rivet material in shear, $\tau = 80$ MPa

11.9. Two plates of mild steel tie rod are connected with width 200 mm and thickness 12.5 mm are to be connected by means of a butt joint with double cover plates. Design the joint if safe stresses are 80 MPa in tension; 65 MPa in shear; and 160 MPa in crushing.

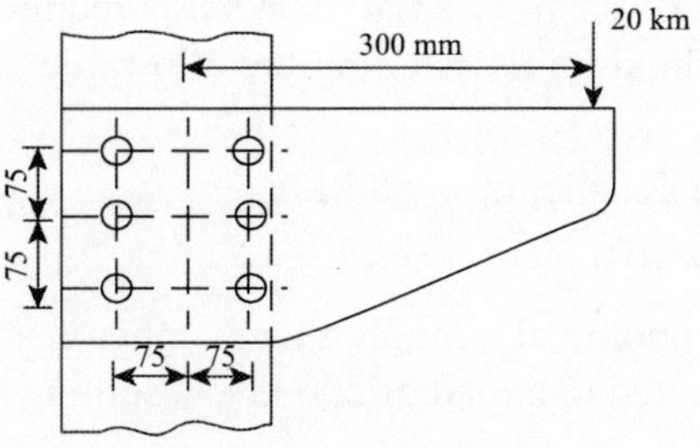

Figure 11.19

11.10. A bracket is riveted to a column by 6 rivets of equal size as shown in Fig. 11.19. It carries a load of 20 kN at a distance of 300 mm from the column centre. If the maximum shear stress in the rivet is limited to 63 MPa, find the diameter of the rivet.

11.11. A bracket is riveted to a column by 5 rivets of equal size as shown in Fig. 11.20. If the maximum shear stress in the rivet is limited to 50 MPa, determine the diameter of the rivet used.

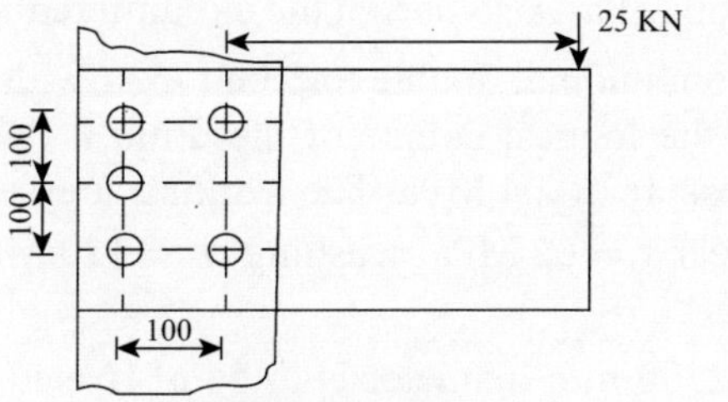

Figure 11.20

11.12. a) Explain the various types of rivets and riveted joints. Compare the efficiencies of different type of joints.

b) Design a double riveted butt joint to join plates of 20 mm thick.

11.13. Design the riveted joints for 2000 mm diameter boiler subjected to an internal pressure of 1.2 MPa. The ultimate strength for the plate material is 400 MPa and the factor of safety is 5. Allowable stresses for the rivet material are:
Shear, $\tau_d = 60$ MPa; Crushing, $\sigma_{cd} = 100$ MPa

11.14. A triple-riveted butt joint with unequal cover straps are used to connect two 2 mm thick plates. Design the joint completely if allowable stresses are 80 N/mm^2 in tension, 50 N/mm^2 in shear, and 100 N/mm^2 in crushing. Find the joint efficiency. Draw a neat sketch.

11.15. Design both longitudinal and circumferential joints for boiler drum (efficiency of 80%) for steam pressure of 2.5 MPa. The inside diameter of the drum is 1,350 mm The safe working stresses are: $\sigma_t = 77$ N/mm^2; $\tau = 61.6$ N/mm^2; $\sigma_c = 133$ N/mm^2. Show the schematic arrangement of rivets with freehand sketches. Prepare dimensioned sketches of the junction of the two joints.

11.16. a) A Lancashire boiler of 1.2 m internal diameter is required to produce steam at 2 MPa. Design the longitudinal joint assuming the allowable stresses in the material as follows:

Shear, $\tau = 60$ MPa; Tensile, $\sigma_t = 80$ MPa;

Crushing, $\sigma_c = 140$ MPa

b) Design and draw economical joint for a tie member in steel structure having thickness of 12 mm and subjected to a load of 200 kN, assuming suitable values for stresses.

11.17. Design triple riveted butt joint for two plates of 12 mm thickness. The butt joint should have two cover plates with inner cover plate being wider than the outer plate. The allowable stresses are $\sigma_t = 80$ MPa, $\sigma_c = 160$ MPa and $\tau = 60$ MPa. Find the efficiency of the joint.

11.18. A bracket is fixed to a vertical support by means of 4 bolts in two rows at distance of 50 mm and 500 mm from the lower edge of the bracket. The load on the bracket is 200 kN, which acts at a distance of 500 mm from the face of the column. Determine the size of the bolts if permissible value of the tensile stress for the bolt material is 90 MPa. Derive the formula used.

11.19. Design a riveted joint suitable for longitudinal Seam of a boiler 2 metre diameter to withstand a maximum pressure of 1.5 N/mm^2. The allowable stresses are $\sigma_t = 80$ MPa, $\sigma_c = 120$ MPa and $\tau = 60$ MPa. Also design the circumferential joint for the same boiler.

11.20. Design a triple riveted longitudinal double trap butt joint with unequal strap's for a boiler. The inside diameter of the longest course of the drum is 1.3 metres. The joint is to be designed for a steam pressure of 2.4 MPa. The working stresses to be used are:
tensile $\sigma_t = 77$ MPa, shear $\tau = 62$ MPa, crushing $\sigma_c = 120$ MPa. Assume the efficiency of the joint as 81%.

11.21. A cylindrical boiler of 2500 mm diameter is to be of riveted steel plate construction and is to operate at a maximum steam pressure of 1.5 MPa. Design the main longitudinal and circumferential riveted joints using triple riveted butt joint with two equal covers straps for longitudinal joints and double riveted lap joint for circumferential joints. The working stress are:

$$S_T = 240 \text{ MPa}; S_S = 110 \text{ MPa and } S_C = 220 \text{ MPa}.$$

11.22. Design the longitudinal and circumferential joint for a boiler whose diameter is 2.4 m and is subjected to a pressure of 1.0 MPa. The longitudinal joint is a triple riveted butt joint with an efficiency of 84% and the circumferential joint is a double riveted lap joint with an efficiency of 70%. The widths of the cover plates are not equal and the outer pitch is double the inner pitch. Draw the complete joint.

(a) Explain different types of failure of the riveted joint.

(b) Design and make a neat dimensioned sketch of longitudinal seam of a boiler with the following data. (i) Boiler shell diameter = 200 cm (ii) Steam pressure = 1.5 MPa; (iii) Allowable stress in tension, compression, and shear are 100 MPa; 150 MPa; and 70 MPa respectively. (iv) Joint efficiency = 85%.

SHORT ANSWER QUESTIONS

11.1. What is a rivet?
11.2. What is a riveted joint?
11.3. What are the advantages of riveted joint over the welded joint?
11.4. What are the disadvantages of riveted joint compared to welded joint?
11.5. How is the riveted joint made?
11.6. How do you determine the rivet diameter?
11.7. What is the pitch of the rivets?
11.8. What is longitudinal pitch?
11.9. What is trasvese pitch?
11.10. What is diagonal pitch?
11.11. How do you determine the longitudinal pitch?
11.12. What is optimum pitch?
11.13. What are the types of riveted joints?
11.14. Where do you use the lap joint?
11.15. Where do you use the butt joint?
11.16. What are the types of failures in a riveted lap joint?
11.17. What are the types of failures in a butt joint?
11.18. What is the efficiency of a riveted joint?
11.19. Where do you use the butt joint?
11.20. Where do use the lap joint?
11.21. What is Lozenge joint? Where is it used?
11.22. What is the efficiency of Lozenge joint?
11.23. What is an eccentrically group of rivets?
11.24. What are the stresses induced in the eccentrically loaded rivets?

CHAPTER 12

WELDED JOINTS

12.1 INTRODUCTION

A welded joint is obtained by the fusion of the edges of two metal parts to be joined together with or without the application of pressure and heat, and filler material. In most commonly used method of welding, the two ends are heated to fusion temperature and the molten filler material is added in the gap formed between the two ends. Welding has become more popularly used technique for joining components, both structural and machinery which are subjected to heavy loads.
Advantages of this technique:

1. Welded joints have higher strength and efficiency.
2. Results in a light weight joint.
3. Less expensive.
4. Ease of joining at different locations.
5. Tight and leakproof joint.
6. Curved plates can be joined.
7. Easy to inspect.
8. Can be fully automated.
9. The operation is less noisy.

Disadvantages of this technique:

1. It cannot be used for impact and vibration loads.
2. Higher temperature may affect the mechanical properties. Loss of ductility is a major problem.
3. Residual stress are induced resulting is reduced fatigue strength.
4. Post-weld processing is necessary.
5. Skilled operator required for good weld joint.
6. Fumes and the gases generated during arc welding cause health hazard.

The welded joints are used in construction, piping, pressure vessels, storage tanks, boilers, shipbuilding, aerospace, automobile, etc.

12.2 TYPES OF WELD JOINTS

There are five basic types of welded joints. They are shown in Fig. 12.1.

1. **Butt joint**. A joint between two members aligned approximately in the same plane.
2. **Lap joint**. A joint between two parallel numbers having overlap.
3. **Corner joint**. A joint between two members at right angles to each other in the form of T.
4. **T-joint**. A joint between two members located at right angles to each other in the form of T.
5. **Edge-joint**. A joint between two members whose edges are placed parallel by bending the end of one of the members.

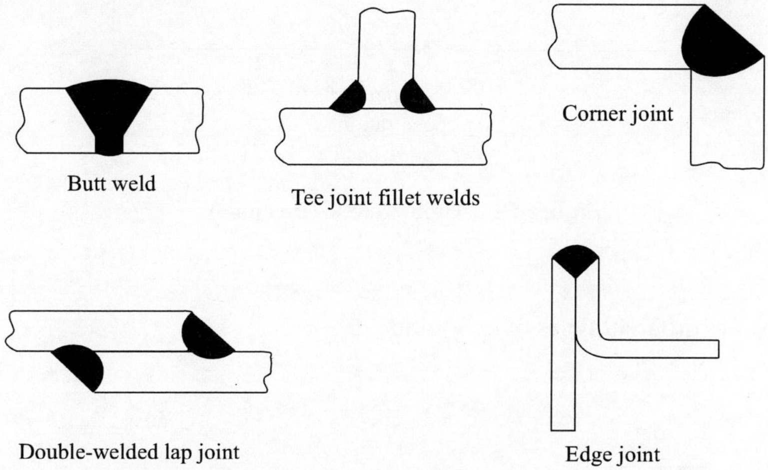

Figure 12.1. Welded joints

12.3 TYPES OF WELDS

There are two types of welds:

(i) Butt weld or groove weld
(ii) Fillet weld

(i) ***Butt Weld or Groove Weld***

Figures 12.2 and 12.3 show the details of the butt weld.

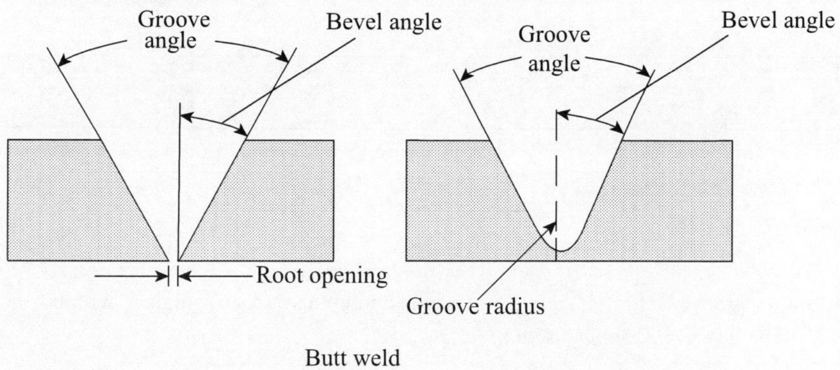

Figure 12.2. Weld nomenclature

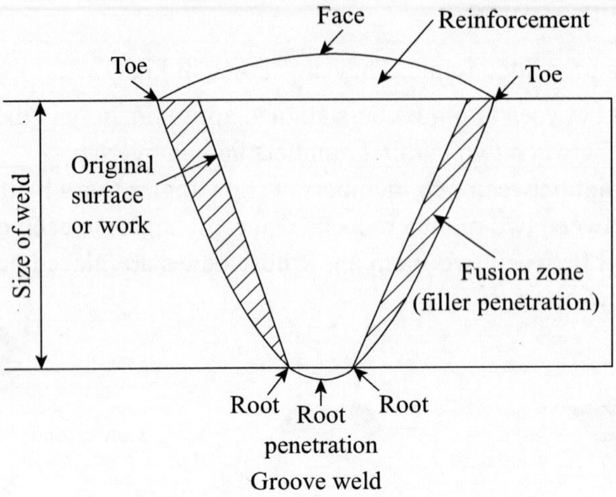

Figure 12.3. Grooved weld (Butt joint)

Figure 12.4 shows different types of butt joints.

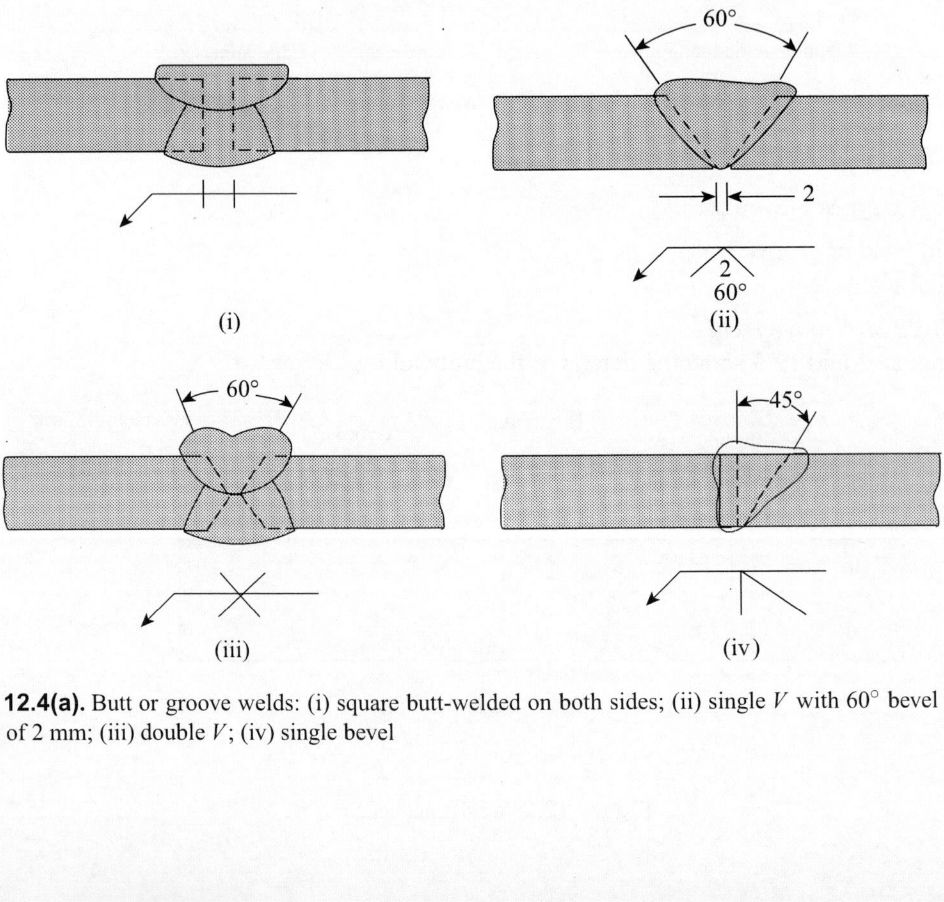

Figure 12.4(a). Butt or groove welds: (i) square butt-welded on both sides; (ii) single V with 60° bevel and root opening of 2 mm; (iii) double V; (iv) single bevel

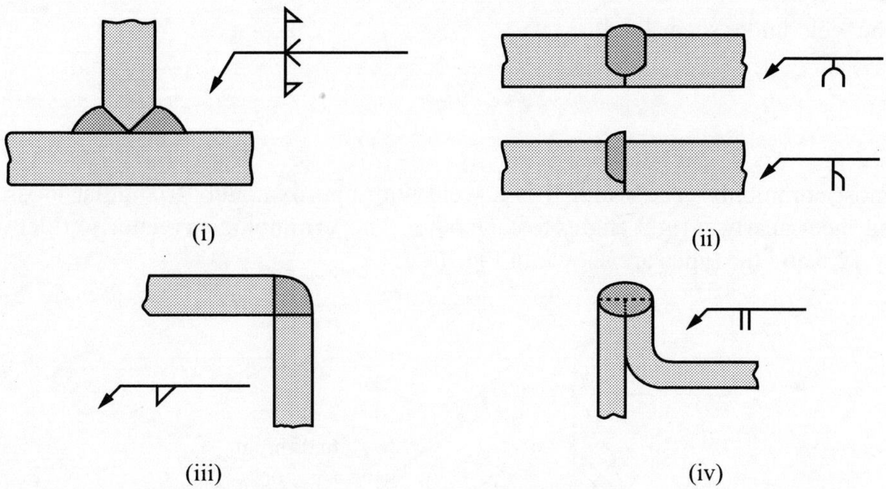

Figure 12.4(b). Special groove welds: (i) T joint for thick plates; (ii) U and J welds for thick plates; (iii) corner weld (may also have a bead weld on inside for greater strength but should not be used for heavy loads); (iv) edge weld for sheel metal and light loads

Strength of Butt Weld

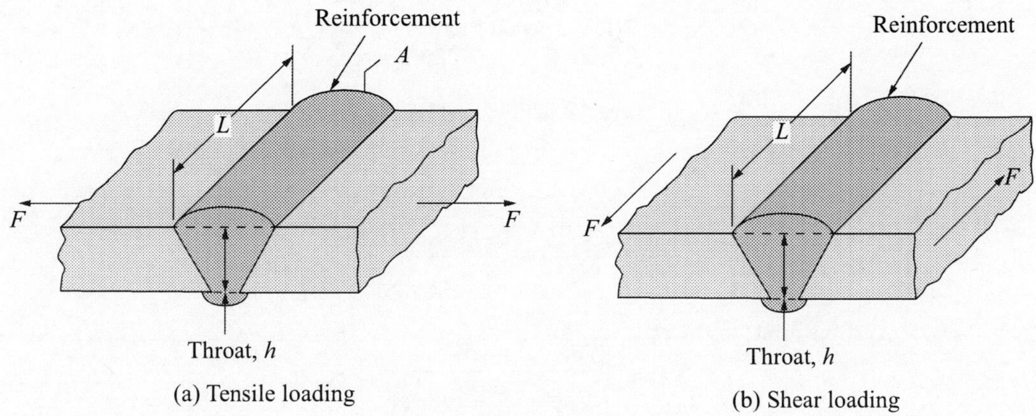

Figure 12.5. A typical butt joint

For determining the strength of the weld, the thickness of the plate is considered to be the size of the weld.

Strength of the weld under tensile load,

$$F_t = hLS_t \tag{12.1}$$

Strength of the weld under shear load,

$$F_s = hLS_s \tag{12.2}$$

(ii) ***Fillet Weld***

This is the most commonly used weld. It is a weld with approximately triangular cross section joining two surfaces nearly at right angles to each other. The terminologies related to fillet weld are shown in Fig. 12.6 and the types are shown in Fig. 12.7.

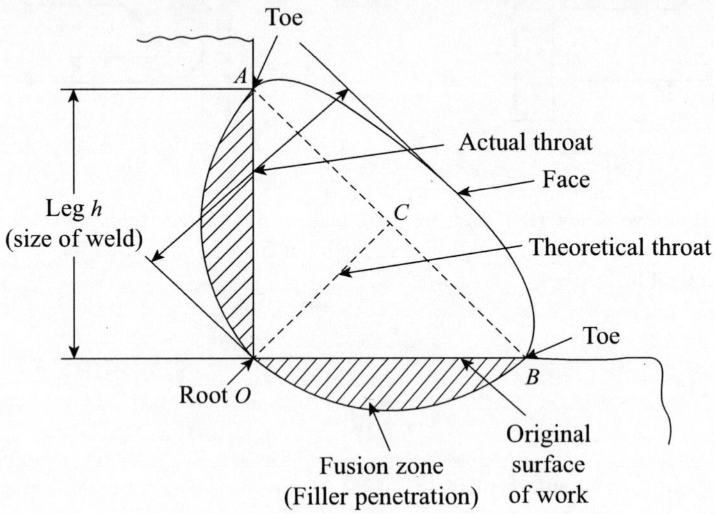

Figure 12.6. Fillet weld

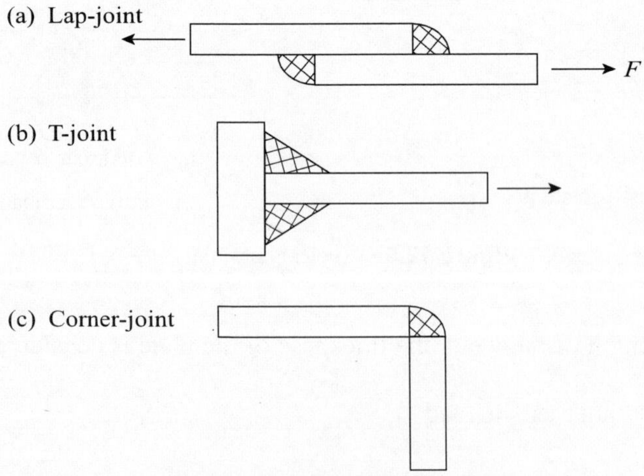

Figure 12.7. Types of fillet welds

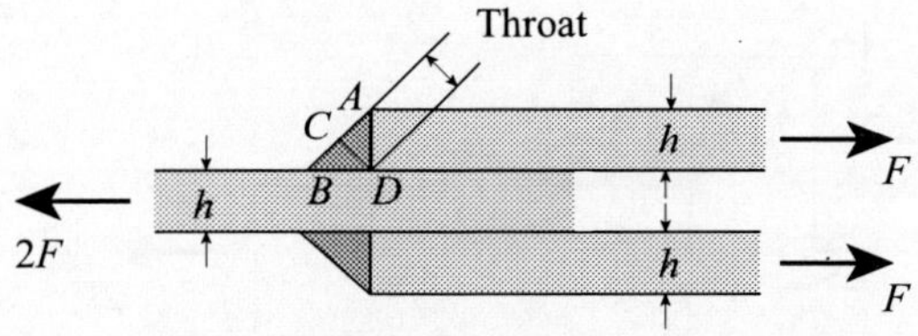

Figure 12.8(a). A transverse fillet weld

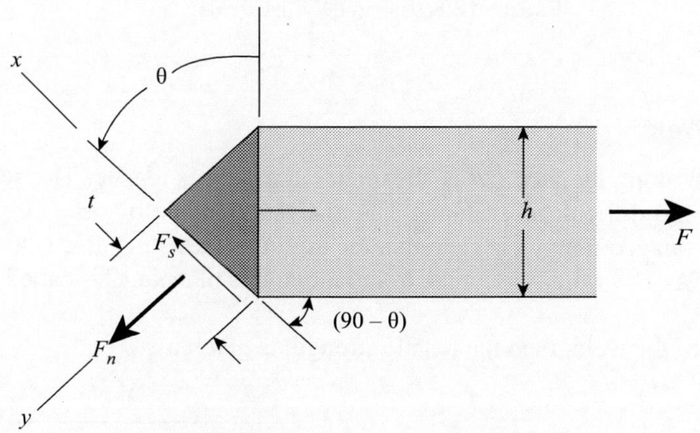

Figure 12.8(b). Free body from Fig. 12.8(a)

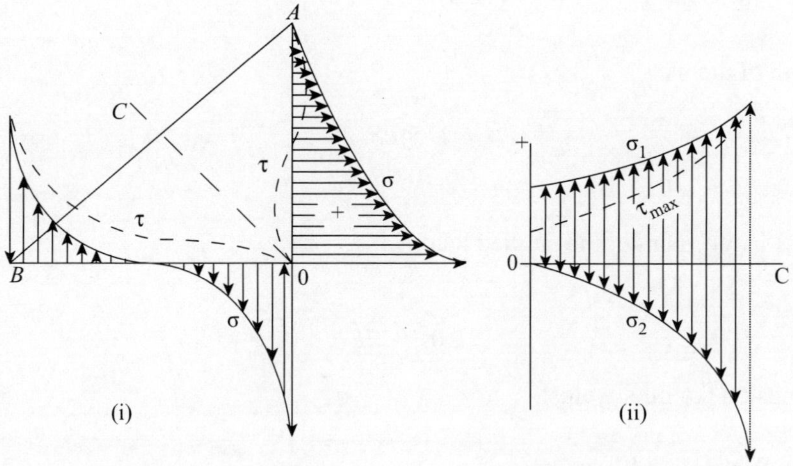

Figure 12.8(c). Stress distribution in fillet welds: (i) stress distribution on the legs as reported by Norris; (ii) distribution of principal stresses and maximum shear stress as reported by Salakian

Contd.

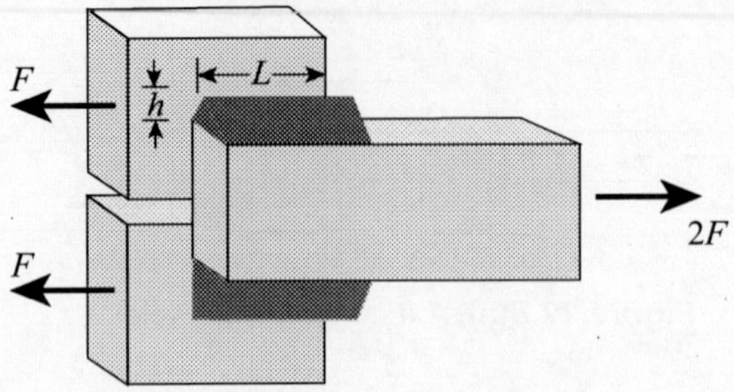

Figure 12.8(d). Parallel fillet welds

Strength of Fillet Weld

The stress distribution along the face *OB* is shear and along *OA* is tensile. The actual distribution is more complicated and difficult to evaluate. The analysis is made by assuming the weld to be triangular in shape. From geometry, it is clear that the minimum resisting area is located at an angle of 45° from *OB* and *OA*. The weld is assumed to fail along this plane. *OC* is called the throat of the weld (Fig. 12.6).

If L is the length of the weld, then the tensile strength is given by

$$F_t = (OC)LS_t$$
$$= (h\cos 45)LS_d$$
$$Ft = 0.707hLS_d \qquad (12.3)$$

Shear strength of the weld,

$$F_s = (OC)LS_{sd}$$
$$= 0.707hLS_{sd} \qquad (12.4)$$

or the shear stress induced under the applied load F is

$$\tau = \frac{F}{0.707hL} \qquad (12.4a)$$

The basic loadings on the fillet welds:

1. Transverse weld (Double)

(i) $\tau = \dfrac{1.414F}{hL}$

with $K_f = 1.5$ (12.5)

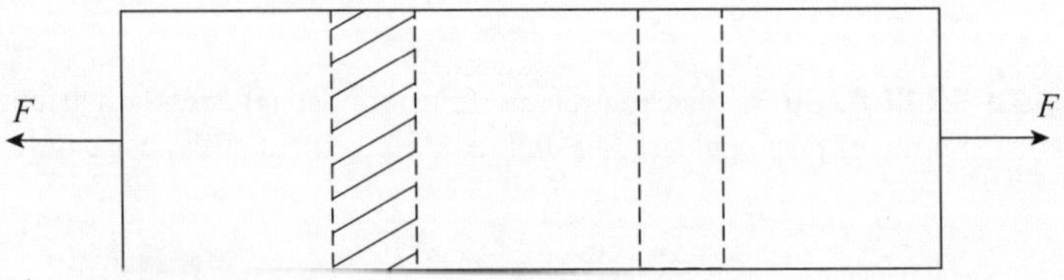

Figure 12.9

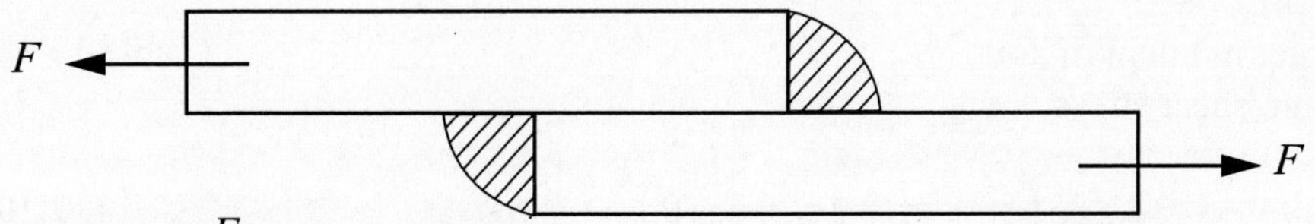

(ii) $\tau = \dfrac{F}{1.414hL}$

with $K_f = 2$ (12.6)

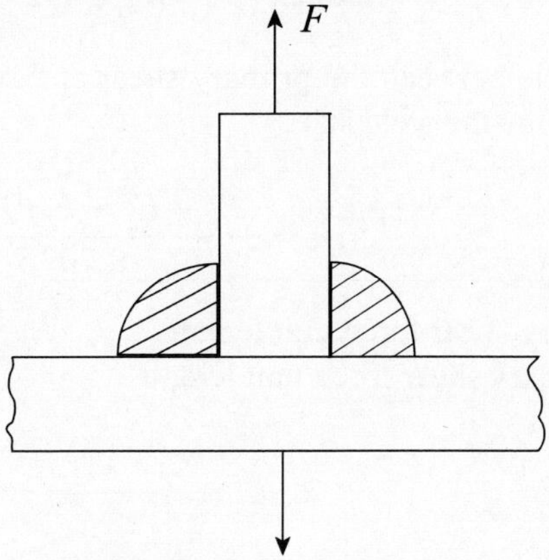

Figure 12.10

2. Parallel weld

$$\tau = \frac{F}{1.414hL}$$

$$K_f = 2.7 \qquad (12.7)$$

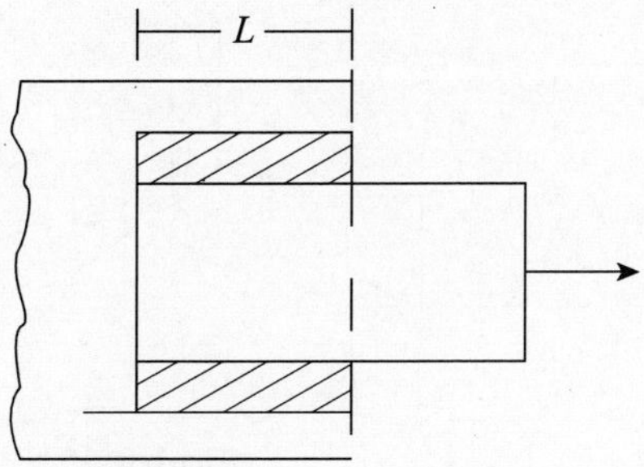

Figure 12.11

3. Torsion

Primary shear stress,

$$\tau_1 = \frac{F}{0.707hL_{\text{total}}}$$

$$K_f = 2.7 \qquad (12.8)$$

Secondary shear stress,

$$\tau_2 = \frac{M\gamma_{\max}}{J} = \frac{F.e.\gamma_{\max}}{J} \qquad (12.9)$$

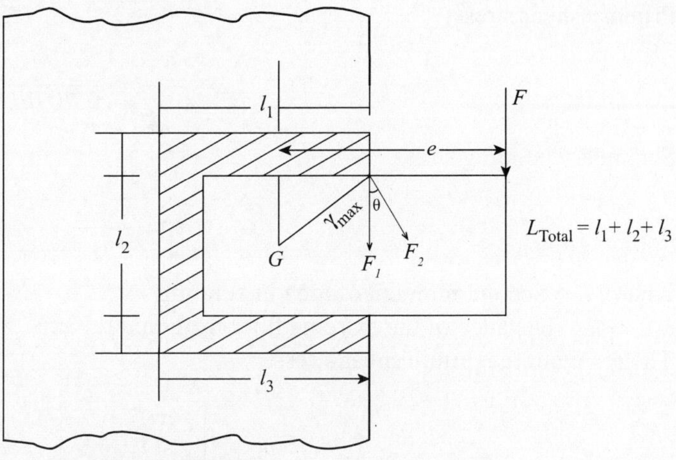

Figure 12.12

where $J =$ Second polar moment of area (Table 12.13)
The resultant maximum shear stress,

$$\tau_{\max} = \left(\tau_1^2 + \tau_2^2 + 2\tau_1\tau_2\cos\theta\right)^{1/2} \leq S_{sd} \tag{12.10}$$

where θ is the angle between the primary shear and secondary shear stress. The equation can be modified to determine the weld size,

$$h = \frac{\left(F_1^2 + F_2^2 + F_1 F_2 \cos\theta\right)^{1/2}}{0.707\, S_{sd}} \tag{12.11}$$

where, $F_1 =$ Primary shear force/unit length
$F_2 =$ Secondary shear force/unit length

4. Bending

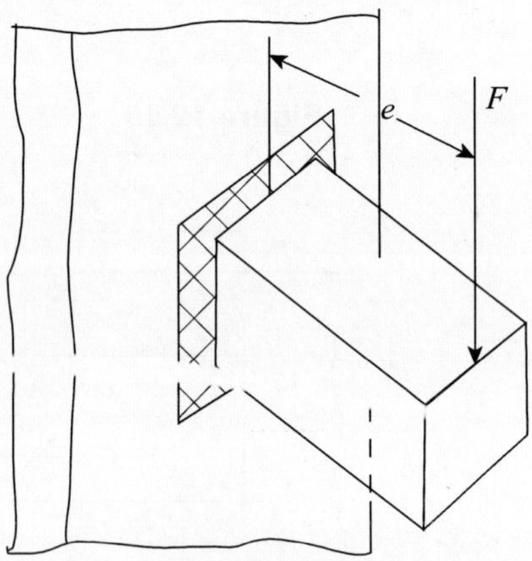

Figure 12.13

Primary shear stress,

$$\tau = \frac{F}{A_{\text{total}}} = \frac{F}{0.707hL} \tag{12.12}$$

Secondary bending stress,

$$\tau = \frac{F.e.C}{I}; \tag{12.13}$$

where, $I =$ Second moment of area in bending (Table 12.12)
$C =$ Distance of the extreme fiber from neutral axis
The resultant maximum shear stress,

$$\tau_{max} = \left[\left(\frac{\sigma}{2}\right)^2 + \tau^2\right]^{1/2} \leq S_{sd} \tag{12.14}$$

SOLVED EXAMPLES

Example 12.1: A mild steel plate 100 mm wide and 10 mm thick is to be connected to another steel plate with normal welds at the ends and has to carry a tensile load of 68 kN. Determine the size of the weld required, if the permissible shear stress is 70 MPa.

SOLUTION

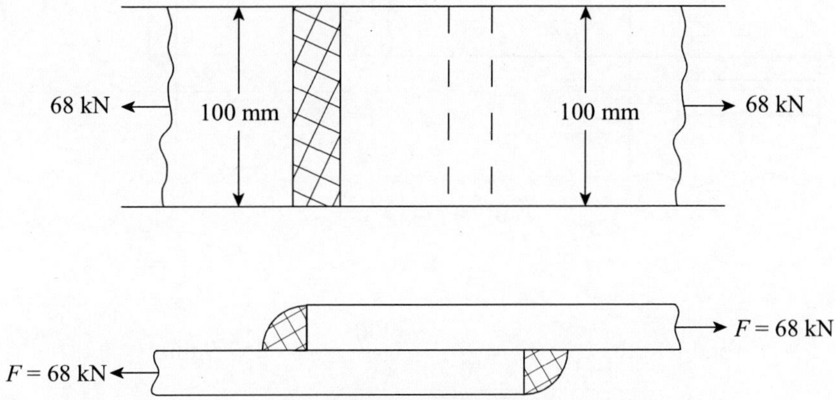

Figure 12.14

Length of each fillet welds, L = 100 mm

The strength of the weld,

$$F = 2 \times 0.707\, hLS_s$$

$$68 \times 10^3 = 2 \times 0.707 \times h \times 100 \times 70$$

$$h = 6.87 \text{ mm}$$

Recommended size of the weld = 10 mm (Table 12.9)

Example 12.2: A 100 mm × 100 mm × 6 mm angle is to be welded to a plate to carry a tensile load of 93 kN. Estimate the length of the weld and its distribution. The permissible shear stress is 85 MPa.

SOLUTION

The CG of the section, $\bar{y} = \dfrac{(100 \times 6 \times 3 + 100 \times 6 \times 50)}{(100 \times 6 + 100 \times 6)} = 26.5$ mm

Assume the weld size = thickness of the angle = 6 mm

Weld strength,

$\therefore \quad F = 0.707h\,(L_a + L_b)S_s$

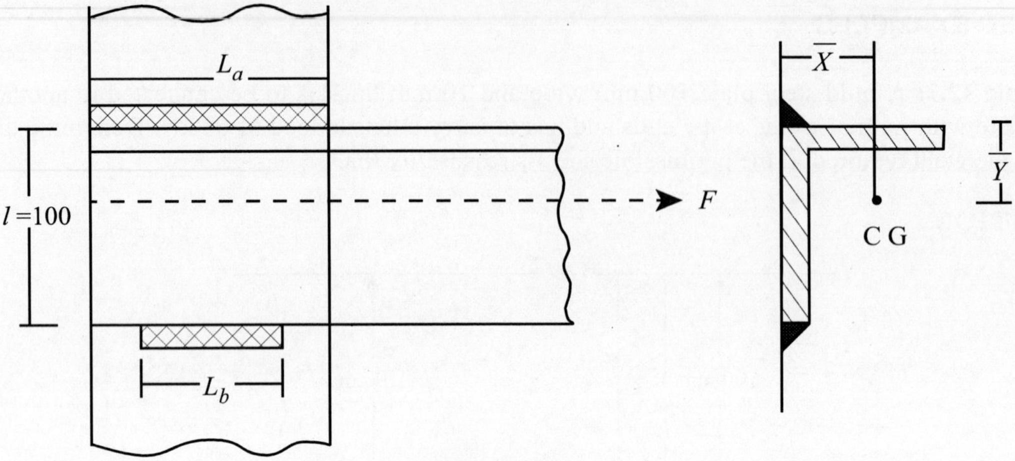

Figure 12.15

$$\text{Total length, } L_a + L_b = \frac{93000}{0.707 \times 6 \times 85} = 257.9 \text{ mm} \quad \text{(i)}$$

Distribution of weld

If the line of action of load passes through the CG of the weld which coincides with the CG of the angle, then

$$(0.707 \times h \times L_a \times S_s) \times 26.5 = (0.707 \times h \times L_b \times S_s) \times (100 - 26.5)$$

$$\therefore \frac{L_a}{L_b} = 2.773$$

$$\therefore L_a = 2.773\, L_b \quad \text{(ii)}$$

$\therefore$ We know

$$L_a + L_b = 257.9$$

$$2.773\, L_a + L_b = 257.9$$

$$L_b = 68.35 \text{ mm}$$

$$L_a = 189.55 \text{ mm}$$

Example 12.3: A 70 mm diameter solid shaft is to be welded to a flat plate around the circumference of the shaft. Determine the size of the weld if the torque on the shaft is 3 kN-m. The permissible shear stress in the weld is 70 MPa.

SOLUTION

Length of the weld,

$$L = \pi D = \pi \times 70 = 219.9 \text{ mm}$$

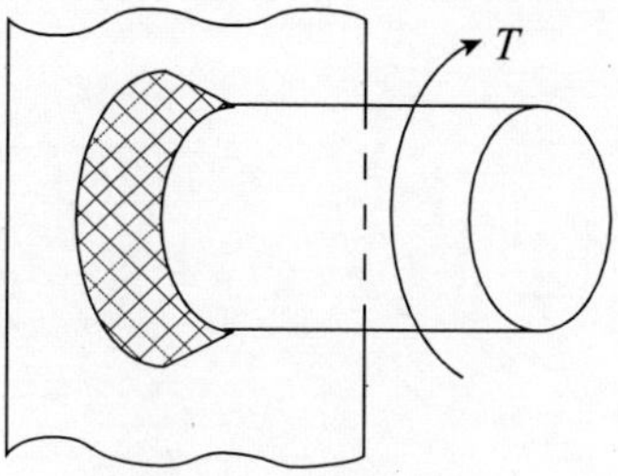

Figure 12.16

Let h be size of the weld.

Resisting torque of the weld,

$$T_r = 0.707\ h.L.S_s\ D/2$$

$$= 0.707 \times h \times 219.9 \times 70 \times 70/2$$

$$= 380919.68\ h \text{ N-mm}$$

Resisting moment = Applied moment

$$\therefore \quad 380919.68h = 3 \times 10^3 \times 10^3$$

$$h = 3 \times 10^6/380919.68 = 7.87 \text{ mm}$$

The recommended size, h = 10 mm

Example 12.4: A bracket carrying a load of 25 kN is to be welded on to a stanchion by fillet weld as shown in Fig. 12.17. Calculate the size of the weld if the permissible shear in the welded is limited to 85 MPa.

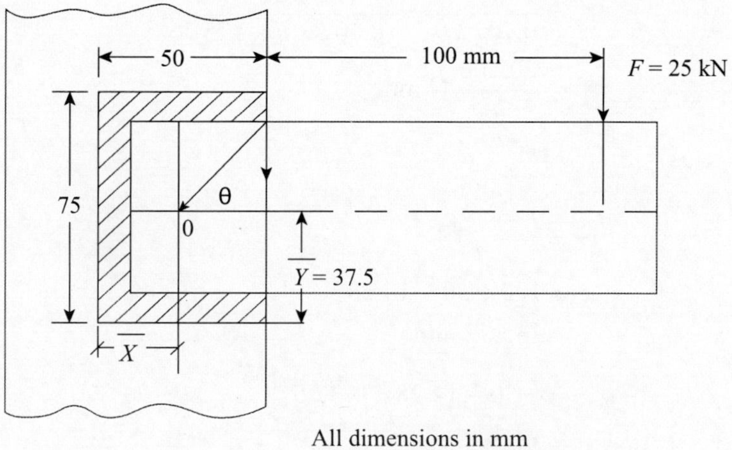

Figure 12.17

SOLUTION

Step-1: Properties of the section

CG of the weld section (weld subjected to torsion), (Table 12.13)

$$\bar{x} = \frac{b^2}{2b+d} = \frac{50^2}{2 \times 50 + 75} = 14.28 \text{ mm}$$

$$\bar{y} = \frac{d}{2} = \frac{75}{2} = 37.5 \text{ mm}$$

Unit MI,

$$J_u = \left(\frac{8b^3 + 6bd^2 + d^3}{12}\right) - \left(\frac{b^4}{2b+d}\right)$$

$$= \left(\frac{8 \times 50^3 + 6 \times 50 \times 75^2 + 75^3}{12}\right) - \left(\frac{50^4}{(2 \times 50 + 75)}\right)$$

$$= 223400.29 \text{ mm}^3$$

$$\therefore \quad J = 0.707h \times J_u$$

$$= 157944.01\, h \text{ mm}^4$$

Step-2: Stresses

Primary shear stress,

$$\tau_1 = \frac{F}{0.707\, hL} = \frac{25 \times 10^3}{(0.707 \times h \times 175)}$$

$$= \frac{202.06}{h} \text{ N/mm}^2$$

Secondary shear stress,

$$\tau_2 = \frac{F \cdot e \cdot r_{\max}}{J}$$

$$e = 100 + 50 - 14.28$$

$$= 135.72 \text{ mm}$$

$$r_{\max} = \left[(50 - 14.28)^2 + 37.5^2\right]^{1/2}$$

$$= 51.78 \text{ mm}$$

$$\tau_2 = \frac{25 \times 10^3 \times 135.72 \times 51.78}{157944.01h}$$

$$= \frac{1112.35}{h} \text{ N/mm}^2$$

$$\theta = \tan^{-1}\left(\frac{37.5}{(50 - 14.28)}\right) = 46.392^\circ$$

Resultant shear stress,

$$\tau_R = \sqrt{\tau_1^2 + \tau_2^2 + 2\tau_1\tau_2\cos\theta}$$

$$= \tau_h\sqrt{\left(\frac{202.06}{h}\right)^2 + \left(\frac{1112.35}{h}\right)^2 + 2\left(\frac{202.06}{h}\right)\left(\frac{1112.35}{h}\right)\cos(46.392)}$$

$$\tau_R = \frac{1260.24}{h}$$

Step-3: Weld design

$$\text{but} \quad \tau_R = S_s = 85 \text{ MPa}$$

$$\therefore \quad \frac{1260.24}{h} = 85$$

$$h = 14.826 \text{ mm}$$

$$\text{Choose} \quad h = 15 \text{ mm}$$

Example 12.5: For the eccentrically loaded welded joint shown in Fig. 12.18, determine the size of the weld required. Take the permissible stress for the weld as 75 MPa.

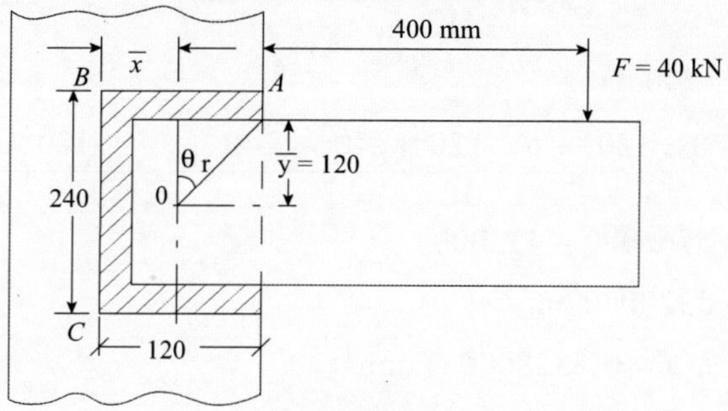

All dimensions in mm

t = Throat of weld

$= 0.707\ h$

h = weld size

Figure 12.18

SOLUTION

Step-1: Geometric properties of the section

Locate the CG of the weld

Taking moments about vertical the edge BC,

$$\bar{x} = \frac{2 \times 120 \times t \times \left(\frac{120}{2}\right) + 240 \times t \times 0}{2 \times 120t + 240 \times t}$$

$$= 30 \text{ mm}$$

By symmetry

$$\bar{y} = 120 \text{ mm}$$

The load eccentrically,

$$e = 400 + (120 - 30) = 490 \text{ mm}$$

Step-2: Stresses in the weld
Primary shear stress,

$$\tau_1 = \frac{F}{t_L} = \frac{40 \times 10^3}{480t} = \left(\frac{83.33}{t}\right) \text{ MPa}$$

Secondary shear stress,

$$\tau_2 = \frac{F \cdot e \cdot r_{\max}}{J}$$

where $r_{\max} = \sqrt{120^2 + 90^2} = 150$ mm
and $J = t \cdot J_u$

$$J_u = \frac{8b^2 + 6bd^2 + d^3}{12} - \frac{b^4}{2b + d} \quad \text{(Table 12.13)}$$

where, $b = 120$ mm and $d = 240$ mm

$$\therefore \quad J_u = \frac{8 \times 120^3 + 6 \times 120 \times 240^2 + 240^3}{12} - \frac{120^4}{2 \times 120 + 240}$$

$$= 5760000 - 432000$$

$$= 5328000 \text{ mm}^3$$

$$J = t \cdot J_u = (5328000\, t) \text{ mm}^4$$

Secondary shear stress at A,

$$\tau_2 = \frac{40 \times 10^3 \times 490 \times 150}{5328000\, t}$$

$$= \left(\frac{551.8}{t}\right) \text{ MPa}$$

Angle between τ_1 and τ_2,

$$\tan\theta = \frac{120}{90} = 0.75$$

$$\theta = \tan^{-1}(0.75) = 53.13^\circ$$

Resultant shear stress at A is

$$\tau_{\max} = \sqrt{\tau_1^2 + \tau_2^2 + 2\tau_1\tau_2\cos\theta}$$

$$= \sqrt{\left(\frac{83.33}{t}\right)^2 + \left(\frac{551.8}{t}\right)^2 + 2\left(\frac{83.33}{t}\right)\left(\frac{551.8}{t}\right)\cos 53.13}$$

$$= \left(\frac{605.48}{t}\right) \text{ MPa}$$

Step-3: Design of weld size
Since design stress, $S_{sd} = 75$ MPa

$$\tau_{\max} = 75 = \frac{605.48}{t} = \frac{605.48}{0.707\,h}$$

$$\therefore \qquad h = \frac{605.48}{0.707 \times 75} = 11.41 \text{ mm} \cong 12 \text{ mm}$$

Choose the weld size as 12 mm × 12 mm

Example 12.6: Determine the size of the fillet weld required for the flat plate loaded as shown in Fig. 12.19.

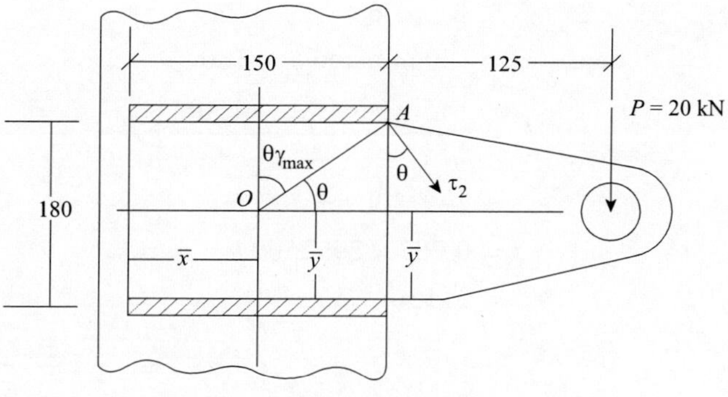

Figure 12.19

SOLUTION
Step-1: Geometric relations
Throat thickness, $t = 0.707\,h$
The CG of the weld is located at O by symmetry, with

$$\bar{x} = 75 \text{ mm and } \bar{Y} = 90 \text{ mm}$$

Maximum distance,

$$r_{max} = \sqrt{75^2 + 90^2} = 117.15 \text{ mm}$$

Angle between the primary and secondary stress,

$$\theta = \tan^{-1}\left(\frac{90}{75}\right) = 50.2°$$

Step-2: Primary shear stress, (Eqn. 8.3 MDHB)

$$\tau_1 = \frac{F}{A2} = \frac{20 \times 10^3}{300\,h \times 707} = \left(\frac{94.3}{h}\right) \text{ MPa}$$

where, A_2 is the area of the weld $= 0.707$ h.L $=$ total length of weld.

Step-3: Secondary shear stress (Eqn. 8.4 MDHB)

$$\tau_2 = \frac{F \cdot e \cdot r_{\max}}{J}$$

where, $F = 20 \times 10^3$ N

$e = 125 + 75 = 200$ mm

$r_{\max} = 117.15$ mm

and $J = 0.707\, h.J_u$

and $J_u = \left[\frac{d(3b^2 + d^2)}{6}\right].$ (Table 12.13)

where, $d = 150$ mm and $b = 180$

$$J_u = \left[\frac{150(3 \times 180^2 + 150^2)}{6}\right]$$

$$= 2992500 \text{ mm}^3$$

$$J = 0.707 \times 2992500\, h$$

$$= 2115697.5h$$

Substituting the values,

$$\tau_2 = \frac{20 \times 10^3 \times 200 \times 117.15}{2115697.5\, h}$$

$$= \left(\frac{221.48}{h}\right) \text{ MPa}$$

Step-4: Resultant shear stress

$$\tau_{\max} = \sqrt{\tau_1^2 + \tau_2^2 + 2\tau_1\tau_2\cos\theta}$$

$$= \sqrt{\left(\frac{94.30}{h}\right)^2 + \left(\frac{221.48}{h}\right)^2 + 2\left(\frac{94.30}{h}\right)\left(\frac{22148}{h}\right) \times \cos 51.2^\circ}$$

$$= (\frac{290}{h}) \text{ MPa}$$

Step-5: Design of weld

Assume the design stress $S_{sd} = 100$ MPa, then

$$\tau_{\max} = S_{sd} = 100$$

$$\frac{290}{h} = 100$$

$$h = \frac{290}{100} = 2.9 \text{ mm}$$

Choose the weld size as 3×3 mm

Example 12.7: A pulley has been fabricated by welding the rim of pulley to the annular web plate by a weld of size 3 mm × 3 mm whereas hub is welded to the web plate by 5 mm × 5 mm weld. Determine safe power that can be transmitted by this welded pulley considering any welded joint.

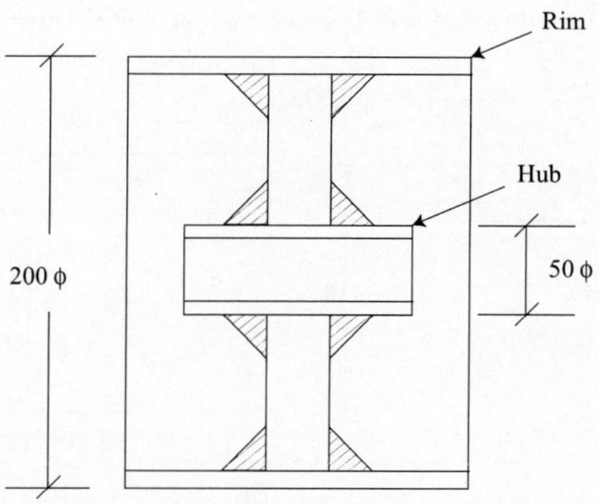

Figure 12.20

Assume, the permissible shear stress $S_{sd} = 75$ MPa.

SOLUTION

Step-1: The torque capacity considering the shearing of the weld between hub and web plate,

$$T_1 = A_1 \,.\, S_{sd} \,.\, \frac{d}{2} = 0.707\, h \times \pi d \times S_{sd} \times \frac{d}{2}$$

$$= 0.707 \times 5 \times \pi \times 50 \times 75 \times \frac{50}{2}$$

$$= 1{,}041{,}143.44 \text{ N-mm}$$

Step-2: The torque capacity considering the shearing of the weld between the rim and plate,

$$T_2 = A_2 \, . \, S_{sd} \, . \, \frac{D}{2} = 0.707\, h \times \pi \times D \times S_{sd} \times \frac{D}{2}$$

$$= 0.707 \times 3 \times \pi \times 200 \times 75 \times \frac{200}{2}$$

$$= 9994977.0 \text{ N-mm}$$

Step-3: Permissible torque capacity

The smaller of the two is the safe torque capacity

$$= 1,041,143.44 \text{ N-mm}$$

$$= 1.041 \times 10^6 \text{ N-mm}$$

Example 12.8: An eccentrically loaded connection is shown in Fig. 12.21. Determine the size of weld if the maximum shear induced in the weld is not exceed 75 N/mm^2.

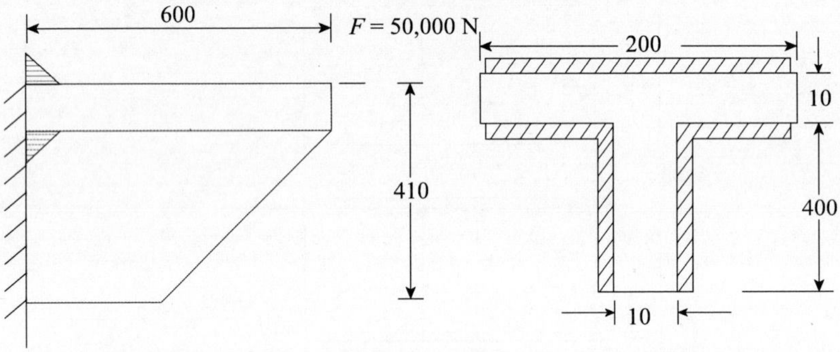

Figure 12.21

SOLUTION

The applied load F is equivalent to

(i) Transverse shear force F at the weld $= 50 \times 10^3$ N

(ii) Bending moment $M = F \cdot e = 50 \times 10^3 \times 600 = 30 \times 10^6$ mm

Step-1: Primary shear stress

$$\tau = \frac{F}{A} = \frac{50 \times 10^3}{(2 \times 400 + 190 + 200) \times (.707h)}$$

$$= (59.43/h) \text{ MPa}$$

Step-2: Secondary bending stress

$$\sigma = \frac{M}{I} \times \mathrm{r}_{\max}$$

where $M = 50 \times 10^3 \times 600 = 30 \times 10^3$ N-mm

$$r_{max} = \bar{y} = \frac{d^2}{2b + 2d + b_1} \qquad \text{(Table 12.12)}$$

$$= \frac{410^2}{200 + 2 \times 400 + 190} = 141.26$$

$$= 141.26 \text{ mm}$$

$$I = 0.707 \cdot hI_u$$

$$I_u = \frac{2 \cdot d^3}{3} - 2d^2 \cdot \bar{y} + (b + 2d)\bar{y}^2 + b_1\bar{y}^2 \qquad \text{(Table 12.12)}$$

$$= \frac{2 \times 400^3}{3} - 2 \times 400^2 \times 141.26 + (200 + 2 \times 400) \times 141.26^2 + 190 \times 141.26^2$$

$$= 21{,}209187.91$$

$$I = 0.707\, hI_u$$

$$= 14994895.85\, h$$

$$\therefore \quad \sigma = \frac{30 \times 10^6 \times 141.26}{14994895.85}$$

$$= (282.6/h) \text{ MPa}$$

Step-3: The maximum shear stress

$$\tau_{max} = \sqrt{\left(\frac{\sigma}{2}\right)^2 + \tau^2} = \sqrt{\left(\frac{282.6}{h}\right)^2 + \left(\frac{59.43}{h}\right)^2} = (153.29/h) \text{ MPa}$$

Step-4: Design of weld

$$\text{But} \quad \tau_{max} = 75 \text{ MPa}$$

$$\therefore \quad 75 = \frac{153.29}{h}$$

$$h = 2.04$$

Choose the weld size 3 mm $\times$ 3 mm

Example 12.9: One end of a rectangular bar of 120 mm × 70 mm cross section is welded to a vertical support by four fillet welds along its circumference. A steady transverse load of 10 kN is applied at the free end of the bar of length 160 mm and is parallel to 120 mm side. Determine the size of weld, if the allowable stress in the material is limited to 15 MPa.

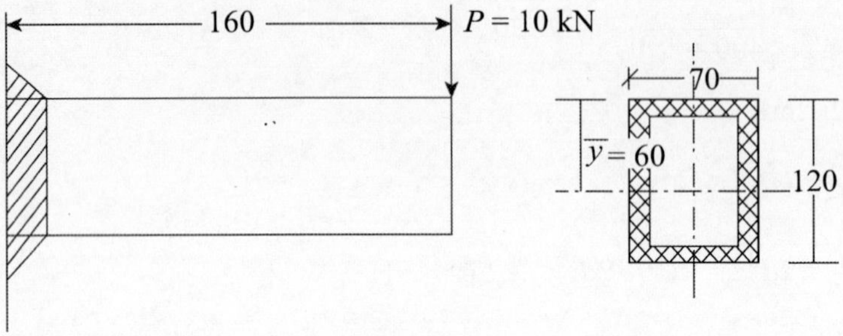

Figure 12.22

SOLUTION

Step-1: Primary shear stress

$$\tau = \frac{F}{A} = \frac{10 \times 10^3}{0.707 \times h \times 190 \times 2}$$

$$= \left(\frac{37.22}{h}\right) \text{MPa}$$

Step-2: Secondary sending stress

$$\sigma = \frac{M \cdot \text{r}_{\max}}{I}$$

where,

$$\text{r}_{\max} = \bar{y} = 60.0 \text{ mm} \quad \text{(Symmetry)}$$

$$I = 0.707h \cdot I_u$$

$$\text{and} \quad I_u = \frac{d^2}{6}[3b + d] \qquad \text{(Table 12.12)}$$

$$= \frac{120^2}{6}[3 \times 70 + 120]$$

$$= 792000$$

$$I = 0.707 \times 79200 \times h$$

$$= (559944\,h) \text{ mm}^4$$

$$M = 10 \times 10^3 \times 160 \text{ N.mm}$$

$$\therefore \quad \sigma = \left(\frac{1600 \times 10^3 \times 60}{559944 \times h}\right)$$

$$= \left(\frac{171.45}{h}\right) \text{ MPa}$$

Step-3: Maximum shear stress

$$\tau_{max} = \sqrt{\left(\frac{\sigma}{2}\right)^2 + \tau^2}$$

$$= \sqrt{\left(\frac{171.45}{2h}\right)^2 + \left(\frac{37.22}{h}\right)^2}$$

$$= \left(\frac{93.47}{h}\right) \text{ MPa}$$

Step-4: Weld size

$$\text{Since} \quad \tau_{max} = 155 \text{ MPa}$$

$$h = \frac{93.47}{115} = 0.812$$

Choose the minimum size as 3 mm × 3 mm

Example 12.10: Determine the size of the weld required for a flat-plate welded to a steel column and loaded as shown in Fig. 12.23. The allowable shear stress in the weld is limited to 80 MPa at the throat section.

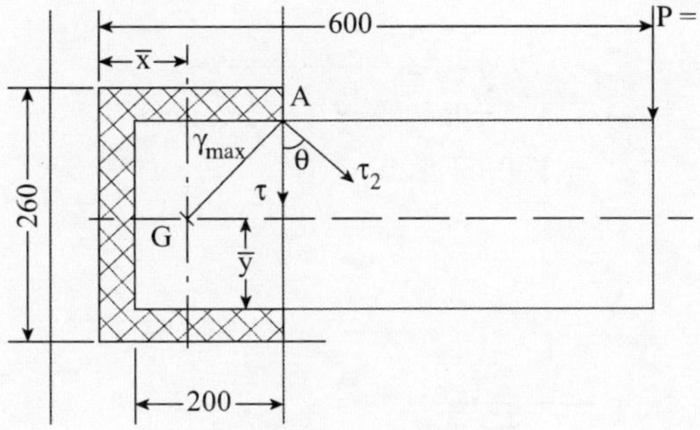

Figure 12.23

Allowable shear stress, $S_{sd} = 80$ MPa

SOLUTION

Step-1: Weld section properties

From Table 12.13, the properties of the fillet weld

$$A = 0.707\,h(2b+d) = 0.707(400+260)\cdot h$$
$$= (466.62\,h)\text{ mm}$$

The CG is given by

$$\bar{x} = \frac{b^2}{2b+d} = \frac{200^2}{(2\times 200+260)}$$
$$= 60.60\text{ mm}$$
$$\bar{y} = d/2 = 260/2 = 130\text{ mm}$$

Unit moment of inertia,

$$J_u = \left[\frac{8b^3+6\times b\times d^2+d^3}{12}\right] - \left[\frac{b^4}{2b+d}\right]$$
$$= \left[\frac{8\times 200^3+6\times 200\times 260^2+260^3}{12}\right] - \left[\frac{200^4}{2\times 200+260}\right]$$
$$= 13558000 - 2424242.42 = 1133757.58$$
$$J = 0.707\,h\cdot J_u = 0.707\times h\times 11133757.58$$
$$= 7871566.609\,h\text{ mm}^4$$

Eccentrically,

$$e = 600 - 60.6 = 539.4\text{ mm}$$

Maximum distance,

$$r_{\max} = \sqrt{(200-60.6)^2+130^2}$$
$$= 190.61\text{ mm}$$

Step-2: Stresses in the weld

Secondary shear stress of A,

$$\tau_2 = \frac{F\cdot e\cdot r_{\max}}{J}$$
$$= \frac{40\times 10^3\times 539.4\times 190.61}{7871566.609\,h}$$
$$= \left(\frac{522.46}{h}\right)\text{ MPa}$$

Primary shear at A,

$$\tau_1 = F/A = \frac{40 \times 10^3}{466.62\, h}$$
$$= \left(\frac{85}{h}\right) \text{ MPa}$$

Angle between τ_1 and τ_2 is

$$\theta = \tan^{-1}(30.0/139.4)$$
$$= \mathbf{43^\circ}$$

Step-3: Weld size for maximum shear
Maximum shear stress,

$$\tau_{\max} = \sqrt{\tau_1^2 + \tau_2^2 + 2\tau_1 \cdot \tau_2 \cos\theta} \qquad = \sqrt{\left(\frac{85}{h}\right)^2 + \left(\frac{522.46}{h}\right)^2 + 2 \times \left(\frac{85}{h}\right)\left(\frac{522.46}{h}\right)\cos 43^\circ}$$

$$= \left(\frac{583.75}{h}\right) = S_d = 80 \text{ MPa}$$

$\therefore h = 7.297$

Choose the weld size as 8 mm × 8 mm

Example 12.11: A bracket carrying a load of 15 kN is to be welded as shown in Fig. 12.24. Find the size of the weld required if the allowable shear stress is not to exceed 80 MPa.

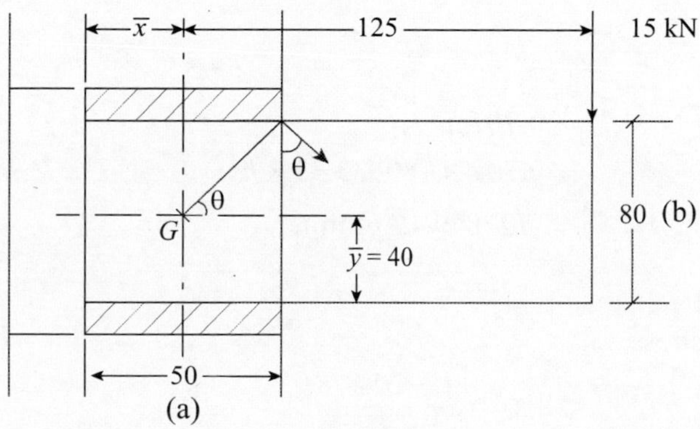

Figure 12.24

SOLUTION

Step-1: Torsional properties of the weld section

The C.G. of the weld (Table 12.13)

$$\bar{x} = 50/2 = 25 \text{ mm}$$

$$\bar{y} = 80/2 = 40 \text{ mm}$$

The area,

$$\begin{aligned} A &= 1.414\, h \cdot d \\ &= 1.414 \times 50 \times h \\ &= (70.7\, h) \text{ mm}^2 \end{aligned}$$

The unit moment of intertia (Table 12.13)

$$J_u = \left[\frac{d(3b^2 + d^2)}{6}\right]$$

where,

$$d = 50 \text{ mm} \quad \text{and} \quad b = 80 \text{ mm}$$

$$\therefore \quad J_u = \left[\frac{50 \times (3 \times 80^2 + 50^2)}{6}\right]$$

$$= 180833.33 \text{ mm}^3$$

Moment of inertia,

$$\begin{aligned} J &= 0.707\, h \cdot J_u \\ &= 0.707 \times 180833.33 \times h \\ &= (127849.17\, h) \text{ mm}^4 \end{aligned}$$

Step-2: Stresses in the weld

Maximum secondary shear stress,

$$\tau_2 = \frac{F \cdot e \cdot r_{\max}}{J}$$

where, $e = 125$ mm

$r_{\max} = \sqrt{40^2 + 25^2} = 47.17$ mm

$F = 15 \times 10^3$ N

$J = 127849.17$ h

$\therefore$ Substituting the value, $$\tau_2 = \frac{15 \times 10^3 \times 125 \times 47.17}{127849.17 \times h}$$

$$= \left(\frac{691.78}{h}\right) \text{ MPa}$$

Primary shear stress,

$$\tau_1 = \frac{F}{A} = \frac{15 \times 10^3}{70.7\, h}$$
$$= \left(\frac{212.16}{h}\right) \text{ MPa}$$

Angle between the primary and secondary shear stresses,

$$\theta = \tan^{-1}\left(\frac{40}{25}\right) = 58^\circ$$

Then the combined maximum shear stress,

$$\tau_{\max} = \sqrt{\tau_2^1 + \tau_2^2 + 2 \cdot \tau_1 \cdot \tau_2 \cdot \cos\theta} \qquad \text{(Eqn. 8.5 MDHB)}$$
$$= \sqrt{\left(\frac{212.16}{h}\right)^2 + \left(\frac{691.78}{h}\right)^2 + 2 \times \frac{212.16}{h} \times \frac{691.78}{h} \times \cos 58^\circ}$$
$$= \left(\frac{824.1}{h}\right) \text{ MPa} \leq (Sd = 80 \text{ MPa})$$

Step-3: Weld size $\therefore$ $$h = \frac{824.1}{80} = 10.3 \text{ mm}$$

Choose the weld size as 12 mm $\times$ 12 mm

Example 12.12: A circular shaft 50 mm diameter is welded to a support by means of a fillet weld and loaded as shown in Fig. 12.25. Determine the size of weld if permissible shear stress in the weld is limited to 100 MPa.

Given Data

Applied load, F = 10 kN
Eccentricity, e = 200 mm
Diameter of the rod, d = 50 mm
Allowable stress, S_{sd} = 100 MPa

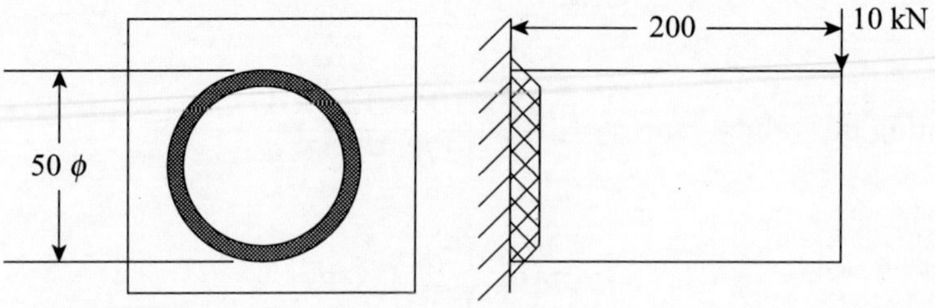

Figure 12.25

SOLUTION

Step-1: Bending properties of weld section

From Table 12.12,

$$\text{Area of the weld, } A = 1.414\pi \cdot h \cdot r$$
$$= 1.414 \times \pi \times \frac{50}{2} h$$
$$= (111.055h) \text{ mm}^2$$

Unit moment of inertia,

$$I_u = \pi r^3 = \pi \times 25^3$$
$$= 49087.4 \text{ mm}^3$$

Moment of inertia,

$$I = 0.707\, h \cdot I_u$$
$$= 0.707 \times 49087.4 \times h$$
$$= 34704.78 \text{ mm}^4$$

Step-2: Stresses in the weld

Primary shear,

$$\tau = \frac{F}{A} = \frac{10 \times 10^3}{111.055h} = \left(\frac{90.045}{h}\right) \text{ MPa}$$

Secondary bending stress,

$$\sigma = \frac{M}{I} \cdot r_{\max} = \frac{F \cdot e \cdot r_{\max}}{I}$$
$$= \frac{10 \times 10^3 \times 200 \times 25}{34704.78\, h}$$
$$= \left(\frac{1440.72}{h}\right) \text{ MPa}$$

Then the maximum shear stress,

$$\tau_{\max} = \sqrt{\left(\frac{\sigma}{2}\right)^2 + \tau^2}$$
$$= \sqrt{\left(\frac{1440.72}{2 \times h}\right)^2 + \left(\frac{90.45}{h}\right)^2}$$
$$= \left(\frac{726.016}{h}\right) \text{ MPa}$$

Step-3: Weld size

$$100 = \frac{726.016}{h}$$
$$h = 7.26016 \text{ mm}$$

Choose weld size as 8 mm × 8 mm

Example 12.13: Determine the maximum normal stress and maximum shear stress in the weld shown in Fig. 12.26.

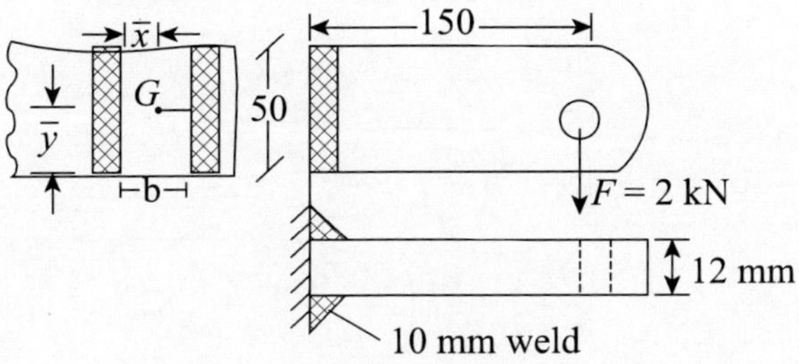

Figure 12.26

SOLUTION

Step-1: Bending properties of the weld section
From Table 12.12
Area of the weld,

$$A = 1.414\ hd = 1.414 \times 50 \times h$$
$$= 70.7h \text{ mm}^2$$

The CG is

$$\bar{x} = b/2 = 12/2 = 6.0 \text{ mm}$$
$$\bar{y} = d/2 = 50/2 = 25 \text{ mm}$$

Unit moment inertia,

$$I_u = \frac{d^3}{6} = \frac{50^3}{6} = 20833.3 \text{ mm}^3$$

Moment of inertia,

$$\begin{aligned} I &= 0.707hI_u \\ &= 0.707 \times 20833.3 \times 10 \\ &= 14729.17\,h \text{ mm}^4 \end{aligned}$$

Step-2: Stresses in the weld
Primary shear,

$$\tau = F/A = \frac{2000}{70.7h} = \left(\frac{28.29}{10}\right) \text{MPa} = 2.83 \text{ MPa}$$

Bending stress,

$$\sigma = \frac{M}{I}y = \frac{2000 \times 150 \times 25}{14729.17 \times 10} = 50.92 \text{ MPa}$$

Maximum normal stress,

$$\begin{aligned} \sigma_1 &= \left(\frac{\sigma}{2}\right) + \sqrt{\left(\frac{\sigma}{2}\right)^2 + \tau^2} \\ &= \left(\frac{50.92}{2}\right) + \sqrt{\left(\frac{50.92}{2}\right)^2 + 2.83^2} \\ &= 25.46 + 25.62 \\ &= 51.08 \text{ MPa} \end{aligned}$$

Maximum shear stress,

$$\begin{aligned} \tau_{\max} &= \sqrt{\left(\frac{\sigma}{2}\right)^2 + \tau^2} \\ &= \sqrt{\left(\frac{50.92}{2}\right)^2 + 283^2} \\ &= 2562 \text{ MPa} \end{aligned}$$

Table 12.1. Weld symbols

Form of weld	Sectional representation	Appropriate symbol	Form of weld	Sectional representation	Appropriate symbol
Fillet	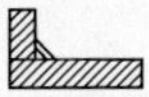	◺	Plug or slot		
Square butt			Backing strip		=
Single-V butt			Spot		
Double-V butt			Seam		
Single-U butt			Machined Seam	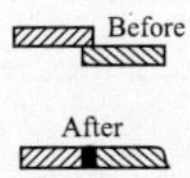	
Double-U butt					
Single bevel butt			Stitch		
Double bevel butt	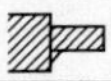		Mashed Stitch		
Single-J butt	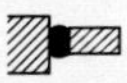				
Double-J butt	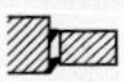		Projection	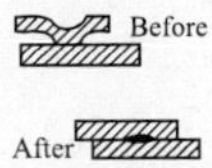	△
Stud					
Bead (edge or seal)			Flash		
Sealing run		○	Butt resistance or pressure (upset)		|

Table 12.2. Drawing representation and symbols

S. No.	Particulars	Drawing representation and symbols
1.	Weld all round	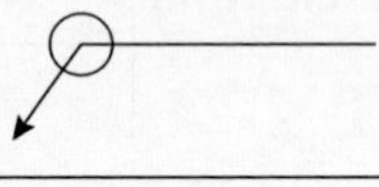
2.	Field or site weld	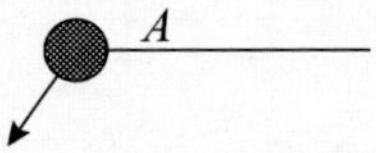
3.	Flash contour	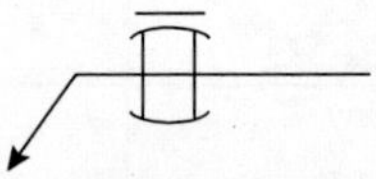
4.	Convex contour	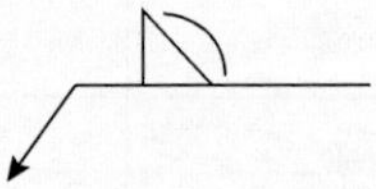
5.	Concave contour	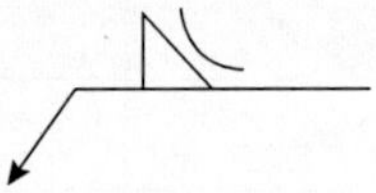
6.	Grinding finish	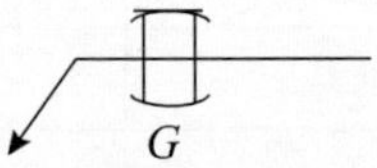
7.	Machining finish	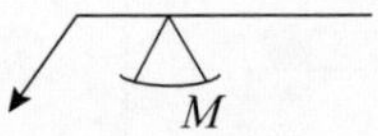
8.	Chipping finish	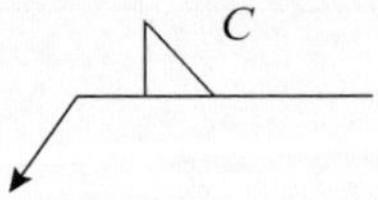

Table 12.3. Properties of deposited weld metal

Property	Bare electrode		Coated electrode	
	Minimum	**Maximum**	**Minimum**	**Maximum**
Yield point, MPa	240	280	295	387
Ultimate strength, MPa	316	387	422	492
Endurance strength, MPa	112.5	140	183	211
% Reduction of area	15	20	45	65
Impact strength, Izod, N-mm	7000	21000	55000	69000
Density, gm/cc (specific gravity)	7.5	7.6	7.81	7.85

Table 12.4. Recommended design stresses for welds made with mild steel electrodes in MPa

Type of load	For bare electrodes		For coated electrodes	
	Steady loads	**Reversed loads**	**Steady loads**	**Reversed loads**
A.W.S. Recommendations				
Tension	91	36	-	-
Compression	126.5	49	-	-
Shear	79.5	24	-	-
C.W. Jennings Recommendations				
Butt welds				
Tension	91	35	112	56
Compression	105	35	126	56
Shear	56	21	70	35
Fillet welds (all)	80	21	98	35

Table 12.5. Permissible weld stresses

Type of loading	Type of weld	Permissible stress	Factor of safety
Tension	Butt	0.60 S_y	1.67
Bearing	Butt	0.90 S_y	1.11
Bending	Butt	0.60 – 0.66 S_y	1.52 – 1.67
Simple compression	Butt	0.60 S_y	1.67
Shear	Butt or fillet	0.40 S_y	1.44

Table 12.6. Allowable loads on mild steel fillet welds

Size of weld, mm	Allowable static load per unit length of weld N/mm			
	Bare welding rod		Covered electrode	
	Normal weld	Parallel weld	Normal weld	Parallel weld
3 by 3	168	135	210	168
5 by 5	280	225	350	280
10 by 10	560	450	700	560
12 by 12	672	540	840	672
15 by 15	840	675	1050	840
20 by 20	1120	900	1400	1120

Table 12.7. Allowable static load on mild steel butt weld per linear mm of weld, N

Plate thickness, mm	Bare electrode		Shield arc	
	Normal weld	Parallel weld	Normal weld	Parallel weld
3	300	190	380	240
5	450	285	570	360
10	900	570	1140	720
12	1240	775	1550	970
16	1545	960	1930	1200
20	1800	1140	2280	1440
25	2472	1544	3090	1930

Table 12.8. Strength of shielded arc welds subjected to repeated loads

Types of Joints	Kind of stress	Recommended stress (MPa)	
		Load varies from 0 to F	Load varies from + F to –F
Butt welded from both sides	Tension or compression	95	63
Butt welded from both sides	Shear	63	42
Fillet welded	Tension, compression or shear	50	34

Table 12.9. Recommended minimum size of welds

Plate thickness (mm)	3-5	6-8	10-16	18-24	25-35	Over 38
Minimum weld size (mm)	3	5	6	10	14	20

Table 12.10. Stress concentration factors

Types of welds	Stress concentration factors
Reinforced butt weld	1.2
Toe of transverse fillet weld	1.5
T-butt joint with short corners	2.0
End of longitudinal fillet weld for lap weld	2.7

Table 12.11. Weld stress formulae

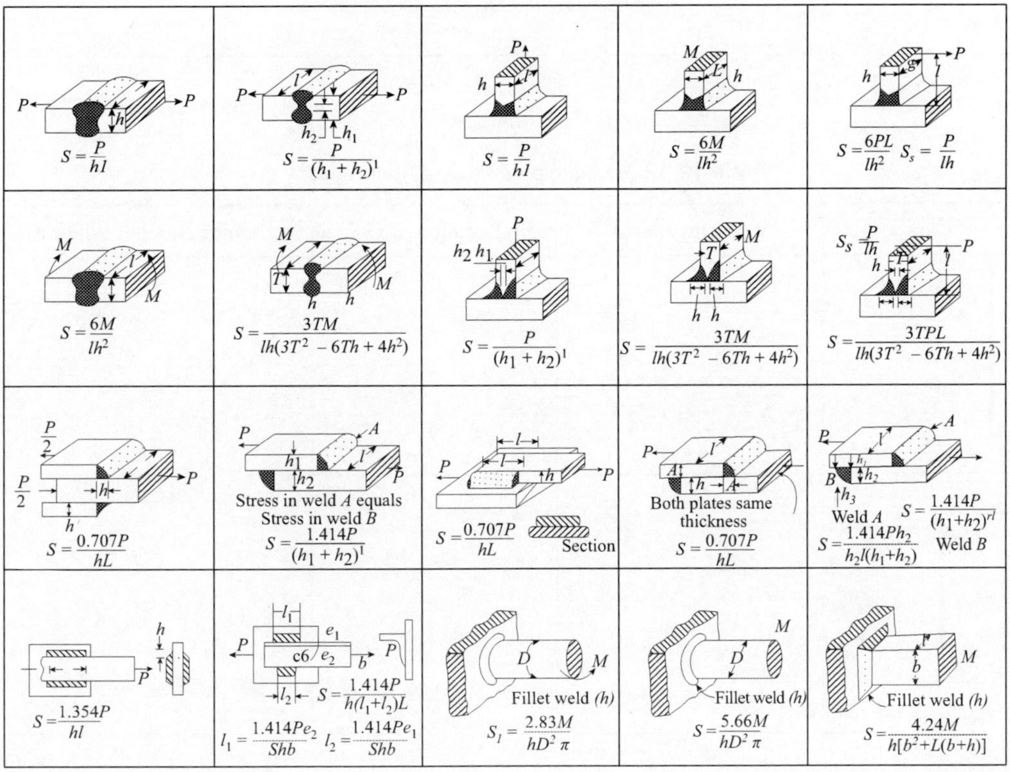

(*Contd*)

Table 12.11. Contd.

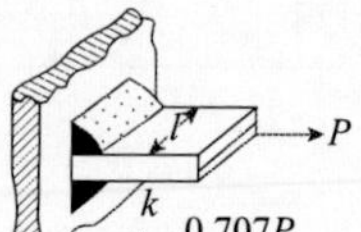 $S=\dfrac{0.707P}{hl}$	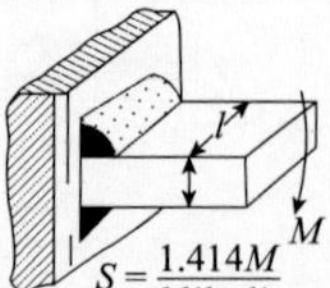 $S=\dfrac{1.414M}{hl(b+h)}$	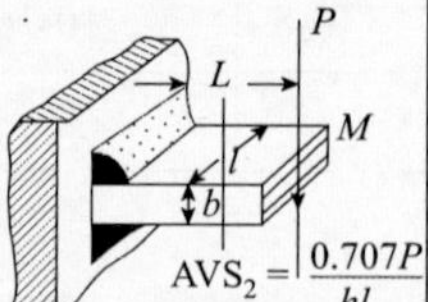 $AVS_2=\dfrac{0.707P}{hl}$ $S_{Max}=\dfrac{1.414PL}{hL(b+h)}$	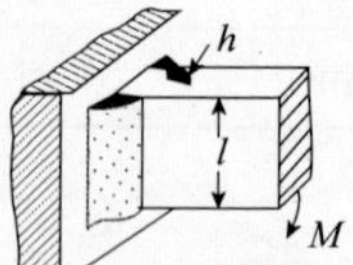 $S=\dfrac{4.24M}{hl^2}$	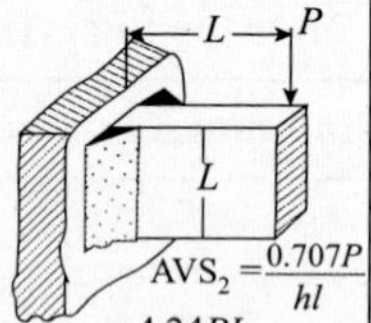 $AVS_2=\dfrac{0.707P}{hl}$ $S_{Max}=\dfrac{4.24PL}{hl^2}$
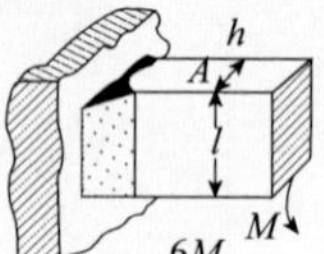 $S=\dfrac{6M}{hl^2}$	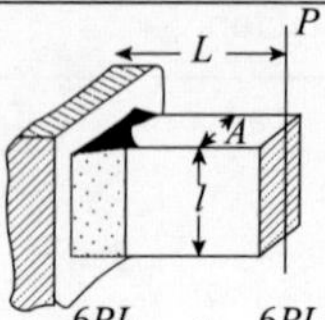 $S=\dfrac{6PL}{hl^2}$ $S_5=\dfrac{6PL}{hl^2}$	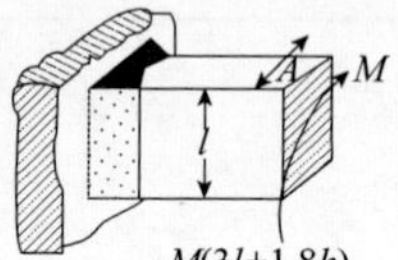 $S_s=\dfrac{M(3l+1.8h)}{h^2l^2}$	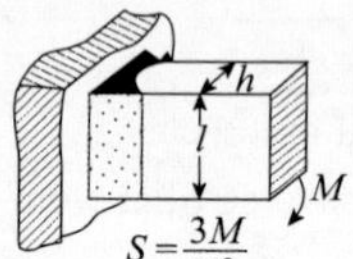 $S=\dfrac{3M}{hl^2}$	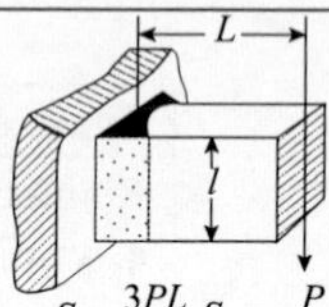 $S=\dfrac{3PL}{hl^2}$ $S_5=\dfrac{P}{2hl}$
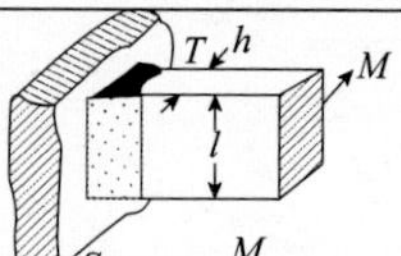 $S=\dfrac{M}{2(T-h)(l-h)h}$	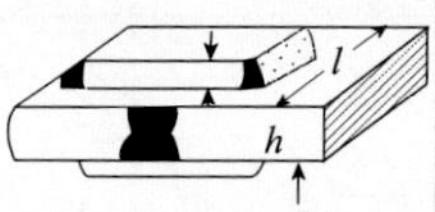 Butt weld $S=\dfrac{P}{2hl+h_1+l_1}$ Fillet weld $S=\dfrac{1.414P}{2hl+h_1+l_1}$	S = Normal stress, MPa S_s = Normal stress, MPa M = Bending moment N, mm	L = Linear distance, mm h = Size of weld, mm l = Length of weld, mm P = External load, N	

Table 12.12. Bending properties of fillet welds*

	Weld throat area	Location of G	Unit moment of inertia
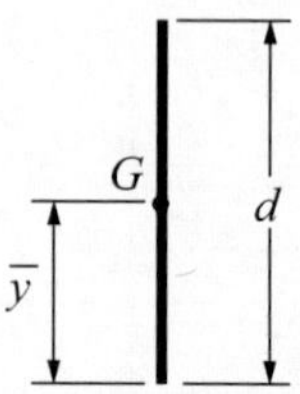	$A = 0.707\ hd$	$\bar{x}=0$ $\bar{y}=d/2$	$I_u=\dfrac{d^3}{12}$
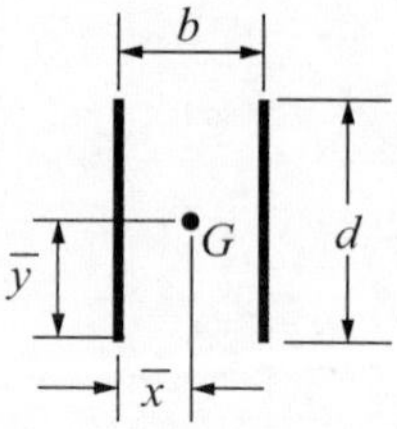	$A = 1.414\ hd$	$\bar{x}=b/2$ $\bar{y}=d/2$	$I_u=\dfrac{d^3}{6}$

Contd.

Table 12.12. Contd.

	Weld throat area	Location of G	Unit moment of inertia
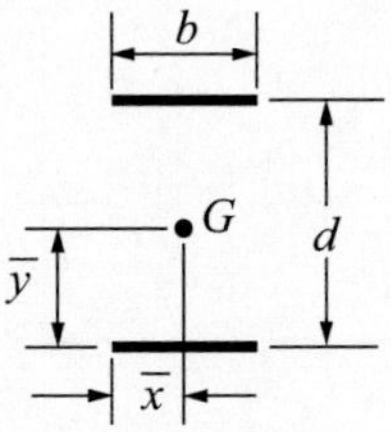	$A = 1.414\ hd$	$\bar{x} = b/2$ $\bar{y} = d/2$	$I_u = \dfrac{bd^2}{2}$
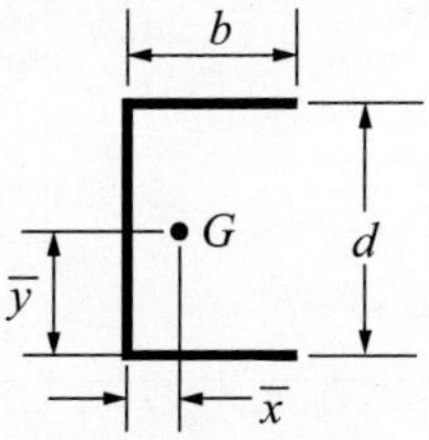	$A = 0.707\ h(2b+d)$	$\bar{x} = \dfrac{b^2}{2b+d}$ $\bar{y} = d/2$	$I_u = \dfrac{d^2}{12}(6b+d)$
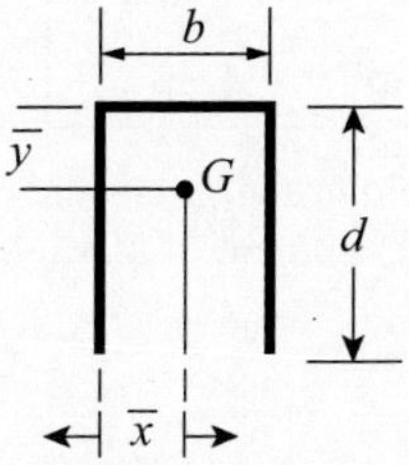	$A = 0.707\ h(b+2d)$	$\bar{x} = b/2$ $\bar{y} = \dfrac{d^2}{b+2d}$	$I_u = \dfrac{2d^3}{3} - 2\bar{y}d^2 + (b+2d)\bar{y}^2$
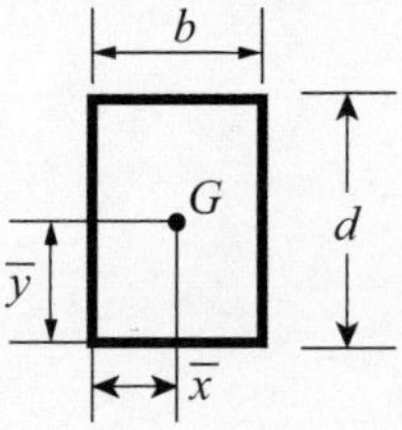	$A = 1.414\ h(b+d)$	$\bar{x} = b/2$ $\bar{y} = d/2$	$I_u = \dfrac{d^2}{6}(3b+d)$

Contd.

Table 12.12. Contd.

	Weld throat area	Location of G	Unit moment of inertia
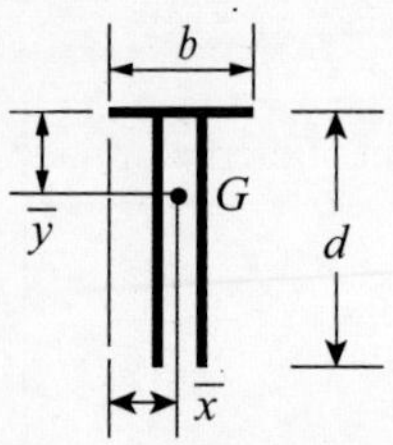	$A = 0.707\,h(b+2d)$	$\bar{x} = b/2$ $\bar{y} = \dfrac{d^2}{b+2d}$	$I_u = \dfrac{2d^3}{3} - 2d^2\bar{y} + (b+2d)\bar{y}^2$
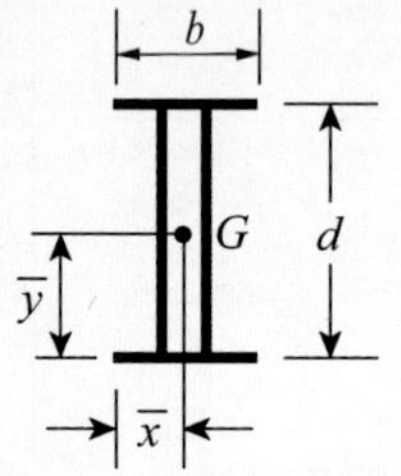	$A = 1.414h(b+d)$	$\bar{x} = b/2$ $\bar{y} = d/2$	$I_u = \dfrac{d^2}{6}(3d+b)$
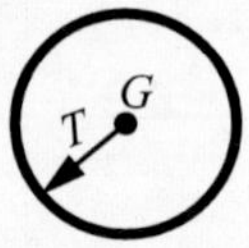	$A = 1.414\,\pi hr$		$I_u = \pi r^3$

*I_u Unit moment of inertia, is taken about a horizontal axis through G, the centroid of the weld group; h is weld size; the plane of the bending couple is normal to the paper; all welds are of the same size. The moment of inertia of the weld $I = 0.707hI$.

Table 12.13. Torsional Properties of fillet welds*

	Weld throat area	Location of G	Unit moment of inertia
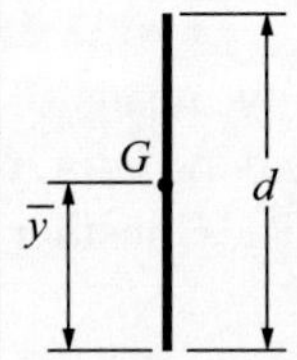	$A = 0.707\ hd$	$\bar{x} = 0$ $\bar{y} = d/2$	$J_u = d^3/12$
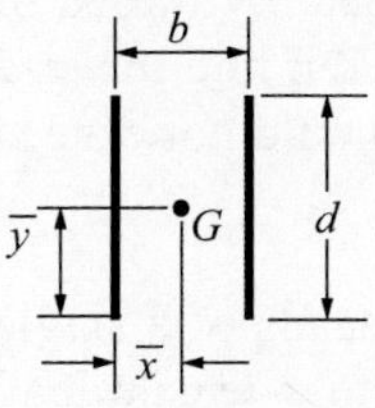	$A = 1.414\ hd$	$\bar{x} = b/2$ $\bar{y} = d/2$	$J_u = \left[\dfrac{d(3b^2 + d^2)}{6}\right]$
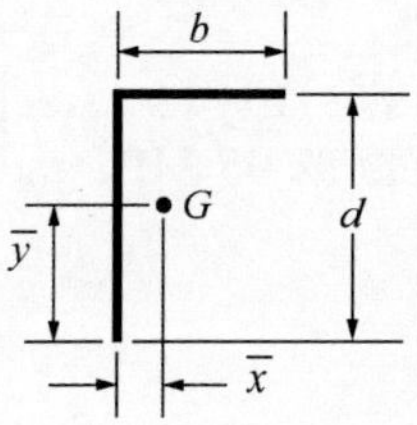	$A = 0.707\ h(b+d)$	$\bar{x} = \dfrac{b^2}{2(b+d)}$ $\bar{y} = \dfrac{2bd + d^2}{2(b+d)}$	$J_u = \left[\dfrac{(b+d)^4 - 6b^2d^2}{12(b+d)}\right]$
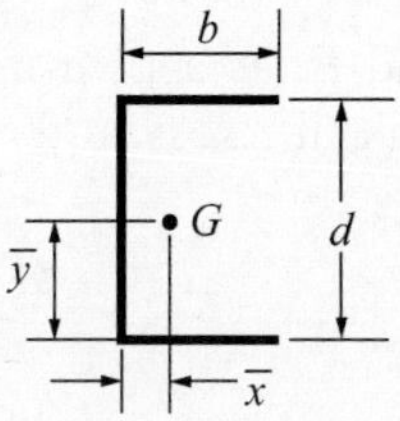	$A = 0.707\ h(2b+d)$	$\bar{x} = \dfrac{b^2}{2b+d}$ $\bar{y} = d/2$	$J_u = \left[\dfrac{8b^3 + 6bd^2 + d^3}{12}\right] - \left[\dfrac{b^4}{2b+d}\right]$
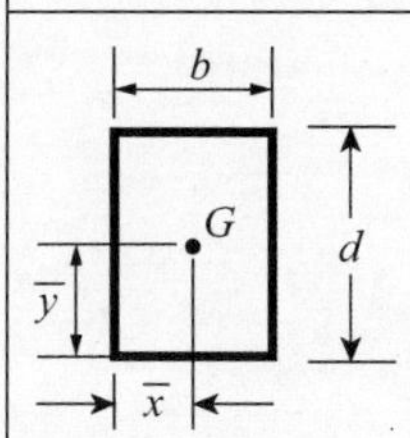	$A = 01.414h(b+d)$	$\bar{x} = b/2$ $\bar{y} = d/2$	$J_u = \left[\dfrac{(b+d)^3}{6}\right]$
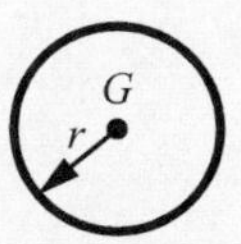	$A = 1.414\ \pi hr$		$J_u = 2\pi r^3$

*G is the C.G. of the group; h is weld size; the plane of the couple is the plane of the paper; all welds are of the same size. The relation between the polar moment of inertia of the weld and unit polar moment of inertia $J = 0.707\ h.J_u$.

EXERCISES FOR PRACTICE

12.1. a) Explain different types of failure of the riveted joint.

b) A plate 80 mm wide and 10 mm thick is to be welded to another plate by means of double parallel fillets. The plates are subjected to a static load of 70 kN. Find the length of weld if the permissible shear stress in the weld does not exceed 45 MPa. Consider the joint first under static loading and then under fatigue loading.

12.2. A plate 100 mm wide and 12 mm thick is to be welded with another plate by means of transverse welds at the ends. If the plates are subjected to a load of 100 kN, find the size of weld for static as well as fatigue load. The permissible tensile stress should not exceed 70 MPa.

12.3. A plate 150 mm wide and 12 mm thick is welded to another plate by means of a single transverse weld and double parallel fillet welds. The joint is subjected to a tensile load 120 kN. Calculate the length of the weld required.

12.4. A circular steel bar 50 mm diameter and 150 mm long is welded perpendicularly to a steel plate to form a cantilever. Assume the allowable tensile stress in the weld as 95 MPa and shear stress as 50 MPa. Determine the size of the weld if:

(a) it is subjected to transverse load of 5 kN at the free end.

(b) twisting moment of 2 kNm.

12.5. A bracket is welded to the side of a column and carries a vertical load P, as shown in Fig. 12.27. Determine the load P it can support so that the maximum shear stress in the 10 mm fillet welds is 80 MPa.

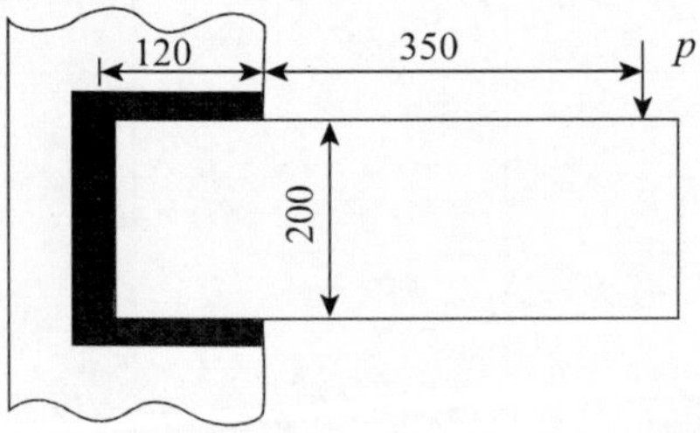

Figure 12.27

12.6. A bracket is welded to the side of a column and carries a vertical load of 15 kN is shown in Fig. 12.28. Determine the size of the weld so that the maximum shear stress does not excced 80 MPa.

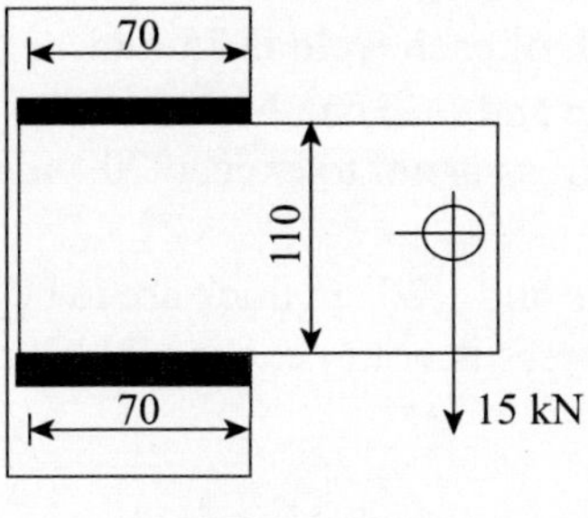

Figure 12.28

12.7. A solid rectangular bar of cross section 80 mm × 50 mm is welded by a fillet weld on all sides to a flat plate with axis perpendicular to the plate surface is as shown in Fig. 12.29. Calculate the size of the weld required to support the load of 15 kN.

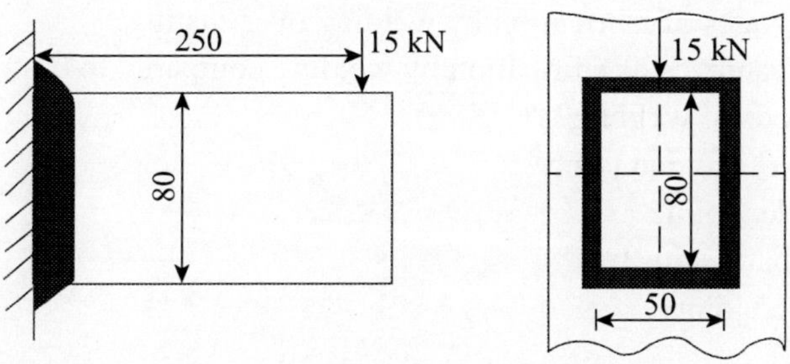

Figure 12.29

12.8. a) Mention the different types of welding, and explain with the help of a neat sketch the application of these.

b) A circular bar 50 mm diameter and 200 mm long is welded to a steel plate and supports a vertical load of 10 kN applied at its free end. Determine the size of the weld required.

12.9. a) Explain the procedure of designing an axially loaded unsymmetrical welded joint.

b) A 125 × 95 × 10 mm unequal angle is to be welded to a steel plate by two parallel fillet welds along the edge and back of the 125 mm leg. The angle is subjected to a tensile load of 120 kN, acting through the centre of gravity of the angle. Load is parallel to both the sides of the angle. The position of the centre of gravity is 36 mm above the 95 mm leg. Assume the weld size as 9 mm and allowable static load /mm length of weld is 500 N/mm.

12.10. A 150 × 100 × 12 mm angle is to be welded to a steel plate by fillet welds along the edges of 1 50 mm leg. The angle is to support a tension load of 100 kN. Determine the lengths of shielded arc welds.

12.11. A 150 × 100 × 12.5 mm angle is welded to its supports by two 12.5 mm fillet welds along the back and edge of 150 mm leg. A load of 20 kN is applied normal to the gravity axis of the angle at a distance of 400 mm from the CG of welds. Determine the maximum shearing stress in the welds if the length of each weld is 75 mm.

12.12. Two steel plates of 10 cm, wide and 1.25 cm thick are to be joined by double transverse fillet weld. The maximum tensile stress is not to exceed 70 MPa. Find the length of the weld for static and dynamic loading.

12.13. Two steel plates of 10 cm, wide and 1.25 cm thick are to be joined by double transverse fillet weld. The maximum tensile stress is not to exceed 70 MPa. Find the length of the weld for static and dynamic loading.

SHORT ANSWER QUESTIONS

12.1. What is a welded joint?
12.2. What is it made?
12.3. What are the advantages welded joints over riveted joints?
12.4. What are the disadvantages of welded joints compared to riveted joints?
12.5. What are the advantages of fabrication by welding over casting?
12.6. What are the disadvantages of fabrication by welding compared to casting?
12.7. What are the methods of welding?
12.8. What are the types of welded joints?
12.9. What is a butt welded joint?
12.10. What are types of butt welds?
12.11. What is a lap welded joint?
12.12. What is fillet weld?
12.13. What is transverse fillet weld?
12.14. What is parallel fillet weld?
12.15. What is leg of fillet weld?
12.16. What is throat of fillet weld?
12.17. What is the relationship between leg and throat of weld?
12.18. How is weld designated?
12.19. What is the cross section for considering the strength of the weld?
12.20. What is the strength of a parallel weld?
12.21. What is the strength of a transverse weld?
12.22. What are primary and secondary shear stresses in eccentrically loaded welded joints?
12.23. Why are the residual stresses induced in welded joints? How are they relieved?
12.24. What are the four basic elements of weld symbol?
12.25. What are the methods of testing of welded joints?

CHAPTER 13

CYLINDERS AND CYLINDER HEADS

Cylinders are classified as thin cylinders and thick cylinders, depending on the ratio of the thickness to the diameter. If the ratio, $t/d \leq 0.05$, the cylinder is treated as thin cylinder. If the ratio, $t/d \geq 0.05$, the cylinder is treated as thick cylinder.

13.1 THIN CYLINDERS

They are used as pressure vessels. The stresses in the cylinder are assumed to be uniform across the thickness of the cylinder wall. There are only two types of stresses:

(i) Circumferential (tangential) stress, σ_t

(ii) Longitudinal stress, σ_l

Consider the half portion of a cylinder

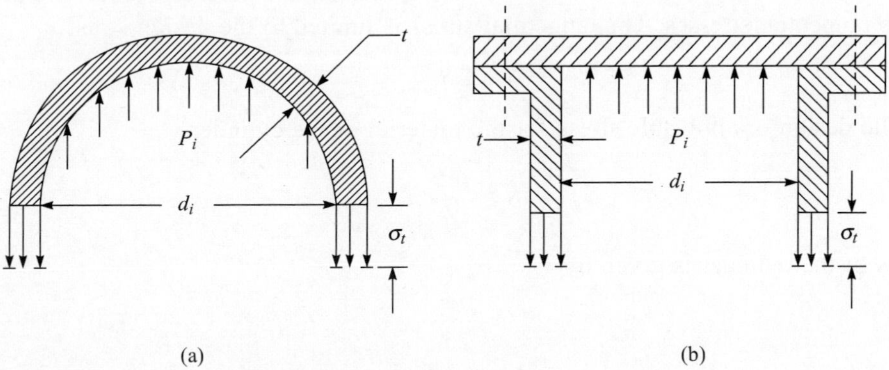

Figure 13.1 Stresses in thin cylinder

Bursting force due to the internal pressure (p) on the half of the cylinder is given by

$$F_b = \text{Pressure} \times \text{projected area} = p(d_i l) \qquad \text{(i)}$$

where, d_i = inside dia. of the cylinder,

l is the length of the cylinder.

This is resisted by the tangential stresses developed in the cylinder wall given by

$$F_r = 2\sigma_t (tl) \qquad \text{(ii)}$$

For equilibrium, $F_r = F_b$

∴ $2\sigma_t tl = pd_i l$

∴ $$\sigma_t = \left(\frac{pd_i}{2t}\right) \qquad \text{(iii)} \qquad (13.1)$$

Consider the cross section of the cylinder with end cover plate (Fig. 2.1 b). There is bursting force on the cover, given by

$$F_{b_1} = \frac{\pi}{4} d_i^2 p \qquad \text{(iv)}$$

This is resisted by the axial or longitudinal stresses, developed in the cylinder wall. The resisting force is given by

$$F_{l_1} = (\pi d_i t)\sigma_l \qquad \text{(v)}$$

For equilibrium, $F_{l_1} = F_{b_1}$

∴ $$\pi d_i t \sigma_l = \frac{\pi}{4} d_i^2 p$$

∴ $$\sigma_l = \frac{pd_i}{4t} \qquad \text{(vi)} \qquad (13.2)$$

This shows that the longitudinal stresses are half the tangential stresses. Hence, the design is based on the tangential stresses. The tangential stress is limited to the design stress,

$$\sigma_t \le S$$

where S is the design or allowable stress for the material of the cylinder.

i.e., $$S \ge \frac{pd_i}{2t}$$

Thickness of the cylinder is given by

$$t = \frac{pd_i}{2S} \qquad \text{(vii)} \qquad (13.3)$$

The pressure vessels are usually made by bending the plate and joining the ends by riveting or welding. The joint strength may be less than the plate strength. The above expression for the thickness is corrected for the joint efficiency as follows:

$$t = \frac{pd_i}{2\eta \cdot S} \qquad \text{(viii)} \qquad (13.3a)$$

where, η is the joint efficiency in Table 13.1 (Table 7.3 MDHB)

The design code for the pressure vessels specifies the following equation,

$$t = \frac{pd_i}{2SJ - p} \qquad \text{(ix)} \qquad (13.3b)$$

where J is joint factor given in Table 13.2. (Table 11.1 MDHB)

Table 13.1. Efficiency of riveted joints

Type of joint	% efficiency		No. of rivets per pitch
	Normal	Max.	
Lap joints			
Single riveted	50-60	63	1
Double riveted	60-72	77	2
Triple riveted	72-80	86.6	3
Butt joint two strap			
Single riveted	55-60	63.3	1
Double riveted	76-84	87	2
Triple riveted	80-88	95	3
Quadruple riveted	86-94	98	4

Table 13.2. Classification of pressure vessels and weld joint efficiency factor (J)

Sl. No.	Requirement	Class I	Class II	Class III		
1	2	3	4	5	6	7
1.	Weld joint efficiency (J)	1.00	0.85	0.70	0.60	0.50
2.	Shell or end plate thickness	No limitation on thickness	Maximum thickness 38 mm, after adding corrosion allowance	Maximum thickness 16 mm, before corrosion allowance is added	Maximum thickness 16 mm, before corrosion allowance is added	Maximum thickness 16 mm, before corrosion allowance is added
3.	Type of joint	Single welded butt joints with backing strip $J = 0.9$	Single welded butt joints with backing strip $J = 0.80$	Single welded butt joints with backing strip $J = 0.65$	Single welded butt joints with backing strip $J = 0.55$	-

13.1.1 Thick Cylinders: Lame's Equations

When the thickness is large, there is variation in the radial and tangential stresses induced across the thickness but the longitudinal or axial stress is assumed to be constant given by

$$\sigma_l = p\left(\frac{d_i^2}{d_0^2 - d_i^2}\right) \qquad \text{(i)}$$

where,

p is the internal fluid pressure

d_i is the internal diameter

d_0 is the outer diameter

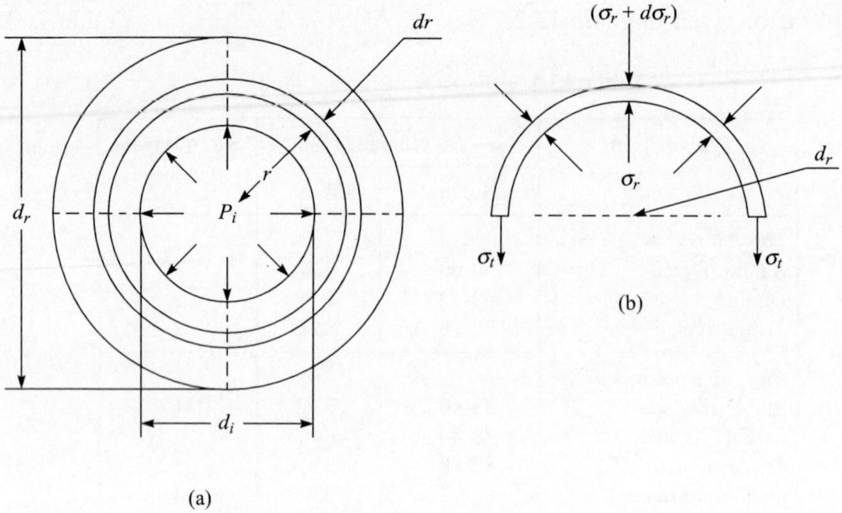

Figure 13.2 Stresses in thick cylinder

For finding the radial and tangential stress, we consider a thin ring of material of thickness *dr*, at a radius *r*. (Fig. 13.2-b)

The thin cylinder is subjected to a radial stress 'σ' at the inner surface and radically inward stress '$\sigma_r + d\sigma_r$' at the outer surface. There is tangential stress σ_t on the thinness *dr*. Considering the equilibrium of the upper half of cylinder, we can write the balances of the forces as follows:

$$2\sigma_t drl + (\sigma_r + d\sigma_r)(2r + 2dr)l - \sigma_r(2rl) = 0$$

Simplifying, we get

$$\therefore \quad \sigma_t + \sigma_r + r\left(\frac{d\sigma_r}{dr}\right) = 0 \qquad \text{(ii)}$$

The axial stress and strain are assumed to be constant

i.e., $$\varepsilon_a = \frac{\sigma_l}{E} + \mu\frac{\sigma_r}{E} - \mu\frac{\sigma_t}{E} = \text{Const.}$$

where μ is the Poisson's ratio

i.e., $$\sigma_r - \sigma_t = \frac{E}{\mu}\left(\varepsilon_a - \frac{\sigma_l}{E}\right) = \text{Const.}$$

i.e., $$\sigma_r - \sigma_t = \text{Const.} = -2a \qquad \text{(iii)}$$

where '*a*' is a constant to be determined

Adding the equations (ii) and (iii)

$$2\sigma_r + r\frac{d\sigma_r}{dr} = -2a$$

Multiplying throughout by '*r*'

$$2\sigma_r r + r^2\frac{d\sigma_r}{dr} = -2ar$$

$$\therefore \qquad \frac{d}{dr}(r^2\sigma_r) = -2ar$$

Integrating, we get

$$r^2\sigma_r = -ar^2 + b$$

$$\sigma_r = -a + b/r^2 \qquad \text{(iv)}$$

where '*b*' is a constant of integration to be determined from the boundary conditions

Substituting (iv) in (iii)

$$-a + \frac{b}{r_2} - \sigma_t = -2a$$

$$\therefore \qquad \sigma_t = a + \frac{b}{r^2} \qquad \text{(v)} \qquad (13.4)$$

If σ_r compressive is taken with negative sign then

$$\sigma_r = \left(a - \frac{b}{r_2}\right) \qquad \text{(vi)} \qquad (13.5)$$

These equations are called Lame's equations. The constants, *a* and *b*, are to be determined by the boundary conditions. If the internal pressure = p_i at $r = r_i$ and the outer pressure = p_o at $r = r_o$

Then,

$$-p_i = a - \frac{b}{r_i^2} \qquad \text{(vii)}$$

$$-p_o = a - \frac{b}{r_0^2} \qquad \text{(viii)}$$

By subtracting (vii) from (viii)

$$(p_i - p_o) = \frac{b}{r_i^2} - \frac{b}{r_0^2}$$

$$\therefore \qquad b = \left(\frac{r_o^2 \cdot r_i^2}{r_o^2 - r_i^2}\right)(p_i - p_o) = \frac{d_i^2 \cdot d_o^2}{4(d_o^2 - d_i^2)}(p_i - p_o) \qquad \text{(ix)}$$

Substituting (ix) in (vii)

$$-p_i = a - \frac{1}{r_i^2} \times \left(\frac{r_o^2 \cdot r_i^2}{r_o^2 - r_i^2}\right)(p_i - p_o)$$

Simplifying we get,

$$a = \left(\frac{p_i r_i^2 - p_o r_o^2}{r_o^2 - r_i^2}\right) = \left(\frac{p_i d_i^2 - p_o d_o^2}{d_o^2 - d_i^2}\right) \qquad \text{(x)}$$

13.1.2 Cylinder with Internal Pressure Only

For cylinder subjected to internal pressure only, the Lame's equations (with $p_o = 0$) are

The radial stress is

$$\sigma_r = \left(\frac{p_i d_i^2}{d_o^2 - d_i^2}\right)\left[1 - \frac{d_o^2}{4r^2}\right] \qquad \text{(xi)} \qquad (13.6)$$

The tangential stress is

$$\sigma_t = \left(\frac{p_i d_i^2}{d_o^2 d_i^2}\right)\left[1 + \frac{d_o^2}{4r^2}\right] \qquad \text{(xii)} \qquad (13.7)$$

The maximum radial stress is at $r = r_{\min} = r_i$

$$\sigma_{r\max} = -p_i \text{ (compression)} \qquad \text{(xiii)}$$

The minimum radial stress is at $r = r_{\max} = r_o$

$$\sigma_r \min = 0 \qquad \text{(xiv)}$$

The maximum tangential stress occurs at $r = r_{\min} = r_i$

$$\sigma_{t\max} = \left(\frac{p_i d_i^2}{d_o^2 d_i^2}\right)\left[1 + \frac{d_o^2}{4r_i^2}\right] = \left(\frac{d_0^2 + d_i^2}{d_0^2 - d_i^2}\right)p_i \qquad \text{(xv)}$$

Minimum tangential stress occurs at $r = r_{\max} = r_0$

$$\sigma_{t\min} = \left(\frac{2p_i d_i^2}{d_0^2 - d_i^2}\right) \qquad \text{(xvi)}$$

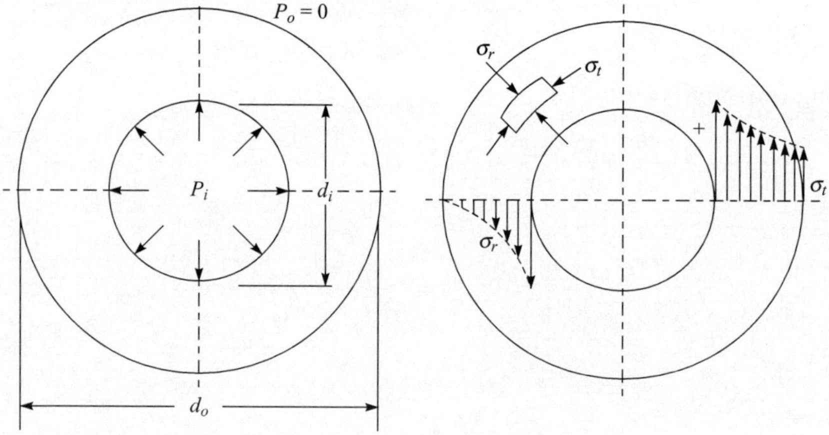

(a) (b)

Figure 13.3 Variation of principal stresses

(Cylinder with internal pressure)

13.1.3 Cylinder with External Pressure Only

When the cylinder is subjected to external pressure only, i.e., $p_i = 0$. The Lame's equations are:

Radial stress is

$$\sigma_r = -\frac{p_0 d_0^2}{d_0^2 - d_1^2}\left[1 - \frac{d_i^2}{4r^2}\right] \qquad \text{(xvii)} \qquad (13.8)$$

Tangential stress is

$$\sigma_t = +\left(\frac{p_0 d_0^2}{d_0^2 - d_i^2}\right)\left(1 + \frac{d_i^2}{4r^2}\right) \qquad \text{(xviii)} \qquad (13.9)$$

For maximum radial stress at $r = r_{max} = r_o$

$$(\sigma_r)_{max} = -p_o \qquad \text{(xix)}$$

The maximum tangential stress at $r = r_{min} = r_i$

$$\sigma_{t\,max} = \left(\frac{2p_0 d_0^2}{d_0^2 - d_i^2}\right) \qquad \text{(xx)}$$

The minimum tangential stress at $r = r_{max} = r_0$

$$\sigma_{t\,min} = \left(\frac{d_0^2 + d_i^2}{d_0^2 - d_i^2}\right)p_o \qquad \text{(xxi)}$$

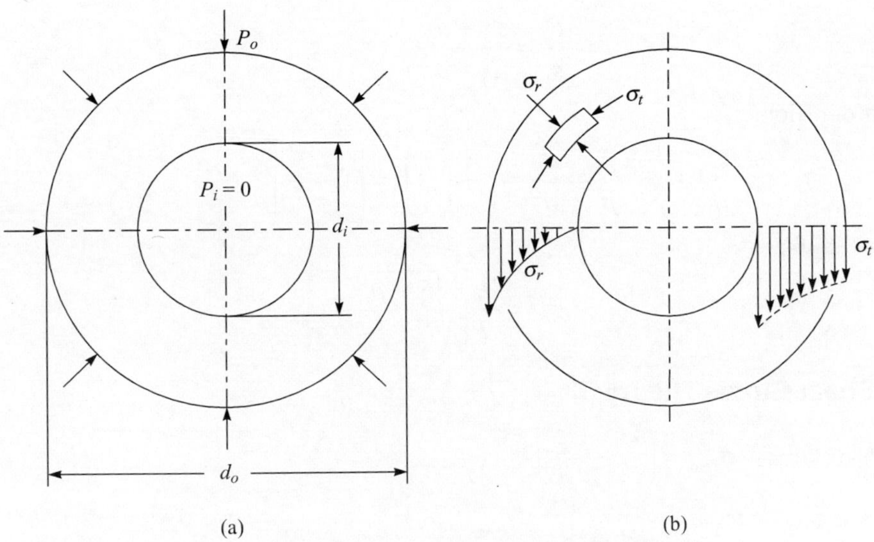

Figure 13.4 Variation of principal stresses (Cylinder with external pressure)

13.1.4 Thickness of the Cylinder Subjected to Internal Pressure Only

The stresses at the inner layer of the cylinder are, (at $r = r_i$)

Radial stress, $\sigma_r = -p_i$

Tangential stress, $\sigma_t = \left[\dfrac{d_0^2 + d_i^2}{d_0^2 - d_i^2}\right] p_i$

The stresses at the outer layer are (at $r = r_o$)

Radial stress, $\sigma_r = 0$

Tangential stress, $\sigma_t = \dfrac{2p_i d_i^2}{(d_0^2 - d_i^2)}$

The axial stress is the same at both layers and is given by

$$\sigma_a = \left(\frac{p_i d_i^2}{d_0^2 - d_i^2}\right)$$

Maximum Principal Stress Theory

The maximum principal stress is the tangential stress at the inner layer. Then by the max principal theory

$$(\sigma_t)_{\max} = S \quad \text{(design stress)}$$

i.e.,

$$\frac{d_0^2 + d_i^2}{d_0^2 - d_i^2} p_i = S$$

$\therefore$

$$\frac{S}{p_i} = \frac{d_0^2 + d_i^2}{d_0^2 - d_i^2}$$

By ratio proportions

$$\frac{S + p_i}{S - p_i} = \frac{d_0^2}{d_i^2} = \left(\frac{d_i + t}{d_i}\right)^2 = \left[1 + \frac{2t}{d_i}\right]^2$$

Solving for thickness,

$$t = \frac{d_i}{2}\left[\sqrt{\frac{S + p_i}{S - p_i}} - 1\right] \tag{13.10}$$

Maximum Shear Stress Theory

Maximum shear stress, $\sigma_{\max} = \left(\dfrac{\sigma_1 - \sigma_2}{2}\right)$

$$S_{s\max} = \frac{\left[\dfrac{d_0^2 + d_i^2}{d_0^2 - d_i^2} p_i + p_i\right]}{2}$$

$$= \frac{p_i}{2}\left[\frac{d_0^2 + d_i^2 + d_0^2 - d_i^2}{d_0^2 - d_i^2}\right] = \frac{p_i}{2}\left[\frac{2d_0^2}{d_0^2 - d_i^2}\right]$$

But $$S_{s\max} = \frac{S}{2}$$

$$\therefore \quad \frac{S}{2p_i} = \frac{d_0^2}{d_0^2 - d_i^2}$$

i.e., $$\frac{S}{S - 2p_i} = \frac{d_0^2}{d_i^2} = \left(1 + \frac{2t}{d_i}\right)^2$$

Solving for thickness,

$$t = \frac{d_i}{2}\left[\sqrt{\frac{S}{S - 2p_i}} - 1\right] \tag{13.11}$$

13.2 COMPOUND CYLINDERS: SHRINK-FIT

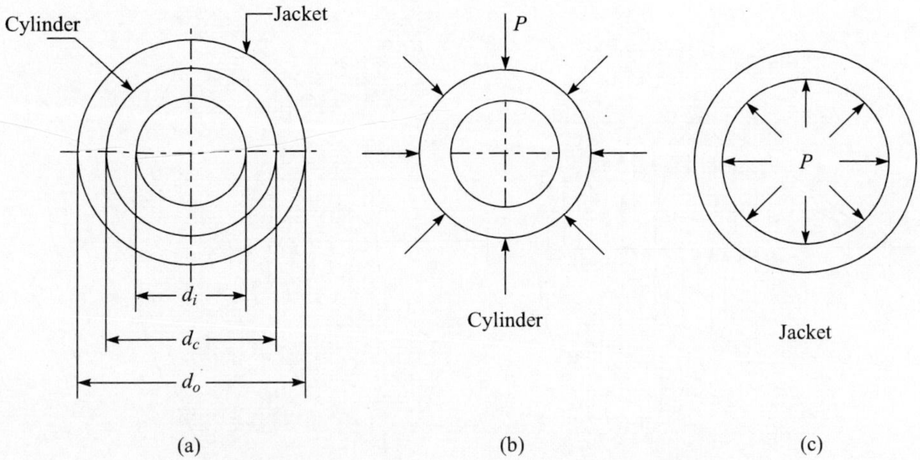

Figure 13.5 Stresses in compound cylinder

The outside cylinder is heated and shrunk on the inner cylinder. There is a shrinkage pressure p_c at the common interface with radius $r = r_c$, the outer cylinder is subjected to internal pressure p_c and the stresses induced are:

At the inner layer,

Radial stress, $\sigma_{roi} = -p_c$

$$\text{The tangential stress, } \sigma_{toi} = \frac{d_0^2 + d_c^2}{d_0^2 - d_c^2}\, p_c \tag{13.12}$$

At the outer layer,

Radial stress, $\sigma_{roo} = 0$

$$\text{Tangential stress, } \sigma_{too} = \left(\frac{2d_c^2}{d_0^2 - d_c^2}\right) p_c \tag{13.13}$$

The inner cylinder is subjected to external pressure 'p_c' and hence the stresses induced are:

At the inner layer,

Radial stress, $\sigma_{r_{ii}} = 0$

$$\text{Tangential stress, } \sigma_{t_{ii}} = -\frac{2p_c d_c^2}{d_c^2 - d_i^2} \tag{13.14}$$

At the outer layer,

Radial stress, $\sigma_{r_{i_o}} = -p_c$

$$\text{Tangential stress, } \sigma_{t_{i_o}} = -p_c\left[\frac{d_c^2 + d_i^2}{d_c^2 - d_i^2}\right] \tag{13.15}$$

The distribution of stresses is shown in Fig. 13.6.

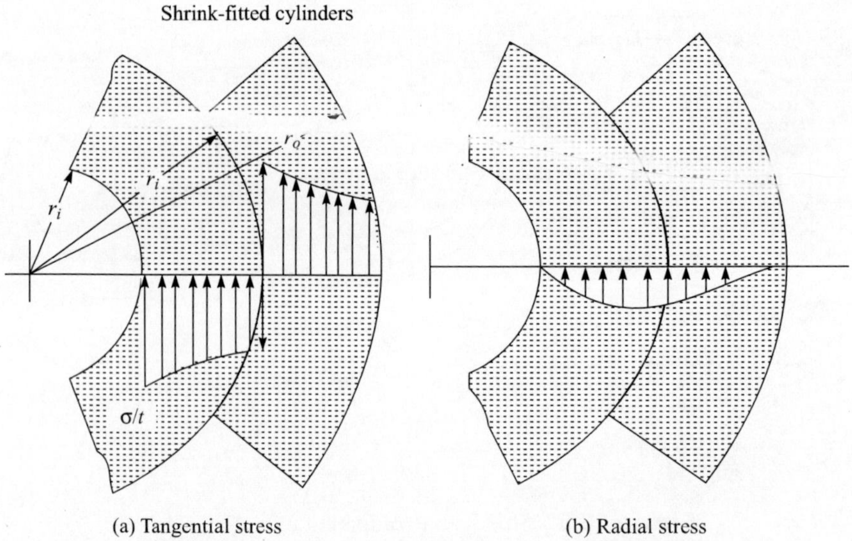

Figure 13.6 Stresses due to shrink-fit

where the fluid pressure, 'p_a' is applied inside the cylinder, the stresses induced are:

At the inner layer $r = r_i$

Radial stress $\sigma_{r_{a_i}} = -p_a$

$$\text{Tangential stress } \sigma_{t_{a_i}} = p_a\left[\frac{d_0^2 + d_i^2}{d_0^2 - d_i^2}\right] \tag{13.16}$$

At the common surface $r = r_c$

Radial stress, $\sigma_{r_{a_c}} = -p_a\left(\frac{d_i}{d_c}\right)^2\left[\frac{d_o^2 - d_c^2}{d_o^2 - d_i^2}\right]$ (13.17)

Tangential stress,

$$\sigma_{t_{a_c}} = p_a\left(\frac{d_i}{d_c}\right)^2\left[\frac{d_o^2 + d_c^2}{d_o^2 - d_i^2}\right] \quad (13.18)$$

At the outer layer, $r = r_o\ (d = d_o)$

The radial stress, $\sigma_{r_{a_o}} = 0$

Tangential stress, $\sigma_{t_{a_o}} = \left[\frac{2p_a d_i^2}{d_o^2 - d_i^2}\right]$ (13.19)

The resultant stress distribution is the sum of the two distributions as shown below.

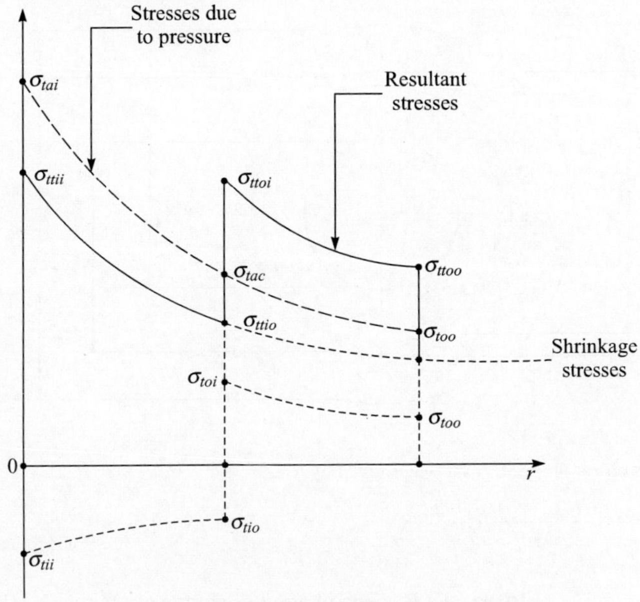

Figure 13.7 Distribution of tangential stresses

The increase in the internal diameter of external cylinder,

$$\Delta d_{co} = \left(\frac{p_a d_c}{E_o}\right)\left[\frac{d_o^2 + d_c^2}{d_o^2 - d_c^2} + \mu_o\right] \quad (13.20)$$

The decrease in the external diameter of the internal cylinder,

$$\Delta d_{ci} = -\left(\frac{p_c d_c}{E_i}\right)\left[\frac{d_c^2 + d_i^2}{d_c^2 - d_i^2} - \mu_i\right] \quad (13.21)$$

The total shrinkage allowance, $\Delta B = |\Delta d_{co}| + |\Delta d_{ci}|$

$$= \left(\frac{p_c d_c}{E_i}\right)\left[\frac{d_c^2 + d_i^2}{d_c^2 - d_i^2} - \mu_i\right] + \left(\frac{p_c d_c}{E_o}\right)\left[\frac{d_o^2 + d_c^2}{d_o^2 - d_c^2} + \mu_o\right] \quad (13.22)$$

If the material for the two cylinders is same, then

$$\Delta B = \left(\frac{p_c d_c}{E}\right)\left[\frac{d_c^2 + d_i^2}{d_c^2 - d_i^2} + \frac{d_o^2 + d_c^2}{d_o^2 - d_c^2}\right] \quad \because E_i = E_o = E \text{ and } \mu_i = \mu_o \quad (13.23)$$

Then the shrinkage pressure 'p_c', due to the shrinkage allowance 'ΔB',

$$p_c = \frac{\Delta B \cdot E \cdot (d_o^2 - d_c^2)(d_c^2 - d_i^2)}{2d_c^3 (d_o^2 - d_i^2)} \quad (13.24)$$

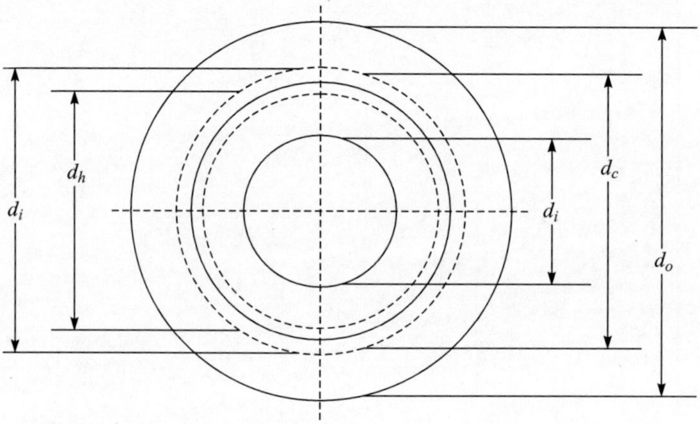

$\Delta B = (d_i - d_h)$

Pressure at the contact surface is given by

$$p_c = \frac{\Delta BE\,(d_c^2 - d_o^2)\,(d_c^2 - d_i^2)}{2d_c^3\,(d_o^2 - d_i^2)}$$

Figure 13.8 Shrinkage allowance

SOLVED EXAMPLES

Example 13.1: A gas cylinder with an internal diameter of 200 mm is subjected to an operating pressure of 12 MPa. It is made of plain carbon steel C40 and the factor of safety is to be 2.5. Calculate the cylinder thickness (neglect the joint efficiency).

Given Data: Dia. of the cylinder, $d_i = 200$ mm; Fluid pressure, $p = 12$ MPa

Material C40; FOS = 2.5

SOLUTION For C40 from Table 2.6 MDHB

Yield strength, $S_y = 330$ MPa

The design stress, $S = S_y/\text{FOS} = 330/2.5 = 132$ MPa

The thickness of the cylinder,

$$t = \frac{pd_i}{2\eta s} \text{ Eqn. 2.3a} \qquad \text{(Eqn. 11.4 MDHB)}$$

where, $p = 12$ MPa, $d_i = 200$ mm; $\eta = 100\%$ and $S = 132$ MPa

$$t = \frac{12 \times 200}{2 \times 1 \times 132} = 9.09 \text{ mm}$$

Example 13.2: An air receiver consisting of a 500 mm cylinder closed by hemispherical ends is made of C30 steel and the factor of safety is 2.5; operating pressure is limited to 5MPa. Calculate the thickness of the cylinder and hemispherical end. Neglect the effect of weld joint.

Given Data: Cylinder dia. $d_i = 500$ mm; Air pressure, $p = 5$ MPa

Material C30; Factor of safety = 2.5

SOLUTION Material property from Table 2.6 MDHB, $S_y = 300$ MPa

The design stress, $S = Sy/\text{FOS} = 300/2.5 = 120$ MPa

The thickness of the cylinder,

$$t = \frac{pd_i}{2\eta s} \quad \text{where } p = 5 \text{ MPa}, d_i = 50 \text{ mm}; \eta = 1 \text{ and } S = 120 \text{ MPa}$$

$$= \frac{5 \times 500}{2 \times 1 \times 120} = 10.42 \text{ mm}$$

Thickness of end-cover,

$$t = \frac{p_i d_i}{4\sigma_t \eta - 04 p_i} + CA \qquad \text{(Eqn. 11.39 MDHB)}$$

$$= \frac{5 \times 500}{4 \times 120 \times 1 - 0.4 \times 5} + 0 = 5.23 \text{ mm} \quad (CA = 0 \text{ assumed})$$

Example 13.3: Derive the expression for the thickness of a cylinder with closed ends using distortion energy theory.

SOLUTION The design equation according to the distortion energy theory

$$\sqrt{\frac{(\sigma_1 - \sigma_2)^2 + (\sigma_2 - \sigma_3)^2 + (\sigma_3 - \sigma_1)^2}{2}} \le S_d \qquad \text{(i)} \qquad \text{(Eqn. 3.4 MDHB)}$$

where, $\sigma_1 =$ tangential stress $= \left(\dfrac{d_o^2 + d_i^2}{d_o^2 - d_i^2}\right) p$

$\sigma_2 =$ radial stress $= -p$

$$\sigma_3 = \text{ axial stress } = \left(\frac{d_i^2}{d_o^2 - d_i^2}\right)p$$

$$S_d = \text{design stress} = S$$

Substituting these values in (i)

$$\left(\frac{d_o^2 + d_i^2}{d_o^2 - d_i^2}p + p\right)^2 + \left(p - \frac{d_i^2}{d_o^2 - d_i^2}p\right)^2$$

$$+\left[\frac{d_i^2}{d_o^2 - d_i^2}p - \left(\frac{d_o^2 + d_i^2}{d_o^2 - d_i^2}\right)p\right]^2 = 2 \cdot S^2$$

$$\therefore \quad 2\frac{S^2}{p^2} = \left(\frac{2 \cdot d_o^2}{d_o^2 - d_i^2}\right)^2 + \left(\frac{d_o^2}{d_o^2 - d_i^2}\right)^2 + \left(\frac{d_o^2}{d_o^2 - d_i^2}\right)^2 = 6\left(\frac{d_o^2}{d_o^2 - d_i^2}\right)^2$$

$$\therefore \quad \frac{S}{p} = \sqrt{3}\left(\frac{d_o^2}{d_o^2 - d_i^2}\right)$$

$$\therefore \quad \frac{S}{S - \sqrt{3}p} = \left(\frac{d_o^2}{d_i^2}\right) = \left(1 + \frac{2t}{d_i}\right)^2$$

$$\therefore \quad t = \frac{d_i}{2}\left[\sqrt{\frac{S}{S - \sqrt{3}p}} - 1\right]$$

Example 13.4: A hydraulic cylinder with closed ends is subjected to an internal pressure of 15 MPa. The inner and outer diameters are 200 mm and 240 mm respectively. The cylinder is made of cast iron FG 300. Determine the factor of safety used in design.

Given Data: Inner diameter, $d_i = 200$ mm; Outer diameter, $d_o = 240$ mm
Internal pressure, $p = 15$ MPa; Material – Cast Iron FG – 300

SOLUTION Material property (Table 2.9 MDHB)

Tensile strength, $S_u = 300$ MPa

The maximum stress in the cylinder, $\sigma_{t\max} = \left(\frac{d_0^2 + d_i^2}{d_0^2 - d_i^2}\right)p_i$

which can be written as

$$\frac{\sigma_{t\max}}{p_i} = \frac{d_o^2 + d_i^2}{d_o^2 - d_i^2} = \frac{240^2 + 200^2}{240^2 - 200^2} = 5.5454$$

$\therefore \quad \sigma_{t\max} = 5.5454 \times 15 = 83.18 \text{ MPa}$

$\therefore \quad \text{FOS} = \dfrac{S_y}{\sigma_{t\max}} = \dfrac{300}{83.18} = 3.6$

Example 13.5: A high pressure cylinder consists of an inner cylinder of inner and outer diameters as 200 mm and 300 mm respectively. It is jacketed by an outer cylinder of outside diameter 400 mm. The difference between the outer diameter of the inner cylinder and inner diameter of the outer cylinder before the assembly is 0.25 mm. If the modulus of elasticity for the material of cylinders, E = 207 GPa, determine the contact pressure and maximum stress in each cylinder.

Given Data: The inner diameter, $d_i = 200$ mm; Common diameter, $d_c = 300$ mm
Outer diameter, $d_o = 400$ mm; Shrinkage allowance, $\Delta B = 0.25$ mm
Modulus of elasticity, $E = 207 \text{ GPa} = 207 \times 10^3 \text{ MPa}$

SOLUTION The contact pressure,

$$P_c = \frac{\Delta BE\,(d_o^2 - d_c^2)\,(d_c^2 - d_i^2)}{2d_c^3\,(d_o^2 - d_i^2)} \qquad \text{(Eqn. 12.29 MDHB)}$$

$$= \frac{0.25 \times 207 \times 10^3 \times (400^2 - 300^2) \times (300^2 - 200^2)}{2 \times 300^3 \times (400^2 - 200^2)} = 27.95 \text{ MPa}$$

The maximum tangential stress in the outer cylinder,

$$\sigma_{t_o} = \frac{d_o^2 + d_c^2}{d_o^2 - d_c^2}\,p_c \qquad \text{(Eqn. 12.24 MDHB)}$$

$$= \frac{(400^2 + 300^2)}{(400^2 - 300^2)} \times 27.95 = 99.826 \text{ MPa}$$

The maximum tangential stress in the inner cylinder,

$$\sigma_{t_i} = -\frac{2p_c d_c^2}{d_c^2 - d_i^2} = -\frac{2 \times 27.95 \times 300^2}{(300^2 - 200^2)} = -99.54 \text{ MPa} \qquad \text{(Eqn. 12.22 MDHB)}$$

Example 13.6: A shrink-fit assembly formed by shrinking one cylinder over another, is subjected to an internal fluid pressure 60 MPa. Before the fluid is admitted, the internal and external diameters of the assembly are 120 mm and 200 mm respectively and the diameter at the junction is 160 mm. If after shrinking, the contact pressure at the junction is 8 MPa, determine using Lame's equations, the tangential stresses at inner, mating and outer surfaces of the assembly after the fluid has been admitted.

Given Data: $p_i = 60$ MPa; $d_i = 120$ mm; $d_o = 200$ mm; $d_c = 160$ mm; $p_c = 8$ MPa.

SOLUTION

Step-1: Stresses due to internal fluid pressure

Lame's equation for tangential stress is

$$\sigma_t = \frac{p_i d_i^2}{(d_o^2 - d_i^2)}\left[1 + \frac{d_o^2}{4r^2}\right] \qquad \text{Eqn. 13.7 (Eq. 12.6 MDHB)}$$

Tangential stress at the inner surface of inner member at $r = \frac{d_i}{2}$

$$\sigma_{t_i} = \frac{p_i d_i^2}{(d_o^2 - d_i^2)}\left[1 + \frac{d_o^2}{4r^2}\right] = \frac{60 \times 120^2}{(200^2 - 120^2)}\left[1 + \frac{200^2}{120^2}\right] = 127.5 \text{ MPa}$$

Tangential stress at common surface at $r = \frac{d_c}{2}$

$$\sigma_{t_c} = \sigma_{t_{oi}} = \frac{p_i d_i^2}{(d_o^2 - d_i^2)}\left[1 + \frac{d_o^2}{d_c^2}\right] = \frac{60 \times 120^2}{(200^2 - 120^2)}\left[1 + \frac{200^2}{160^2}\right] = 86.5 \text{ MPa}$$

Tangential stress at the outer surface of outer member, $r = \frac{d_o}{2}$

$$\sigma_{t_c} = \frac{2 p_i d_i^2}{d_o^2 - d_i^2} = \frac{2 \times 60 \times 120^2}{(200^2 - 120^2)} = 67.5 \text{ MPa} \qquad \text{(Eqn. 12.9 MDHB)}$$

Step-2: Stresses due to contact pressure

Tangential stress at the inner surface of inner member,

$$\sigma_{t_{ii}} = -\frac{2 p_c d_c^2}{d_c^2 - d_i^2} = -\frac{2 \times 8 \times 160^2}{(160^2 - 120^2)} = -36.57 \text{ MPa} \qquad \text{(Eqn. 12.22 MDHB)}$$

Tangential stress at the outer surface of inner member,

$$\sigma_{t_{io}} = p_c\left(\frac{d_i^2 + d_c^2}{d_c^2 - d_i^2}\right) = -8 \times \left(\frac{160^2 + 120^2}{160^2 - 120^2}\right) = -28.57 \text{ MPa} \qquad \text{(Eqn. 12.23 MDHB)}$$

Tangential stress at the inner surface of outer member,

$$\sigma_{t_{oi}} = p_c\left(\frac{d_o^2 + d_c^2}{d_o^2 - d_c^2}\right) = 8 \times \left(\frac{200^2 + 120^2}{200^2 - 120^2}\right) = 36.44 \text{ MPa} \qquad \text{(Eqn. 12.24 MDHB)}$$

Tangential stress at the outer surface of outer member,

$$\sigma_{t_{oo}} = \frac{2 p_c d_c^2}{d_o^2 - d_c^2} = \frac{2 \times 8 \times 160^2}{(200^2 - 120^2)} = 28.44 \text{ MPa} \qquad \text{(Eqn. 12.25 MDHB)}$$

Step-3: Resultant tangential stress distribution

Resultant tangential stress at the inner surface of the inner member,

$$\sigma_1 = \sigma_t + \sigma_{t_{ii}} = 127.5 - 36.57 = 90.93 \text{ MPa} \quad \text{(Tensile)}$$

Resultant tangential stress at the outer surface of inner member,

$$\sigma_2 = \sigma_{t_c} + \sigma_{t_{io}} = 86.5 - 28.57 = 57.93 \text{ MPa} \quad \text{(Tensile)}$$

Resultant tangential stress at the inner surface of outer member,

$$\sigma_3 = \sigma_{t_c} + \sigma_{t_{oi}} = 86.5 + 36.44 = 122.94 \text{ MPa} \quad \text{(Tensile)}$$

Resultant tangential stress at the outer surface of outer member,

$$\sigma_4 = \sigma_{t_o} + \sigma_{t_{oo}} = 67.5 + 28.44 = 95.94 \text{ MPa} \quad \text{(Tensile)}$$

Example 13.7: A cast iron cylinder of internal diameter 200 mm and thickness 50 mm is subjected to a pressure of 5 MPa. Calculate the tangential and radial stresses at the inner, middle and outer surfaces.

Given Data: $d_i = 200$ mm; $t = 50$ mm; $p_i = 5$ MPa

SOLUTION

Step-1: Dimensions

Outer diameter, $d_o = d_i + 2t = 200 + 2 \times 50 = 300$ mm

Mean diameter, $d_m = \dfrac{d_o + d_i}{2} = \dfrac{300 + 200}{2} = 250$ mm

Step-2: Tangential stresses

Tangential stress induced in the cylinder due to internal fluid pressure,

$$\sigma_\theta = \frac{p_i d_i^2}{d_o^2 - d_i^2}\left(1 + \frac{d_o^2}{4r^2}\right) \quad \text{(Eqn. 12.6 MDHB)}$$

Tangential stress at the inner surface of cylinder at $r = d_i/2$

$$\sigma_t = \frac{p_i d_i^2}{d_o^2 - d_i^2}\left(1 + \frac{d_o^2}{d_i^2}\right) = p_i \frac{d_o^2 + d_i^2}{d_o^2 - d_i^2} = \left(\frac{300^2 + 200^2}{300^2 - 200^2}\right) \times 5 = 13 \text{ MPa}$$

Tangential stress at the middle surface of the cylinder at $r = d_m/2$

$$\sigma_{t_m} = \frac{p_i d_i^2}{(d_o^2 - d_i^2)}\left(1 + \frac{d_o^2}{d_m^2}\right) = \frac{5 \times 200^2}{(300^2 - 200^2)}\left(1 + \frac{300^2}{250^2}\right) = 9.76 \text{ MPa} \quad \text{(Tensile)}$$

Tangential stress at the outer surface of the cylinder at $r = d_o/2$

$$\sigma_{t_o} = \frac{2p_i d_i^2}{(d_o^2 - d_i^2)} = \frac{2 \times 5 \times 200^2}{(300^2 - 200^2)} = 8 \text{ MPa} \qquad \text{(Eqn. 12.9 MDHB)}$$

Step-3: Radial stresses

Radial stress at the inner surface of the cylinder,

$$-p_i = -5 \text{ MPa (Compressive)} \qquad \text{(Eqn. 12.8 MDHB)}$$

Radial stress at the middle surface of the cylinder,

$$\sigma_{r_m} = -\frac{p_i d_i^2}{(d_o^2 - d_i^2)}\left(1 + \frac{d_o^2}{d_m^2}\right) \quad \because \left(r = \frac{d_m}{2}\right)$$

$$= -\frac{5 \times 200^2}{(300^2 - 200^2)}\left(1 + \frac{300^2}{250^2}\right) = -1.76 \text{ MPa (Compressive)}$$

Radial stress at the outer surface of the cylinder at $r = d_o/2$

$$\sigma_{r_o} = 0 \qquad \text{(Eqn. 12.10 MDHB)}$$

Example 13.8: Design a shrink-fit joint to join two cylinders of diameter 150 mm $\times$ 200 mm and 200 mm $\times$ 250 mm. Maximum tangential stress in the components due to shrink-fitting is to be limited to 40 MPa. Also determine the axial force necessary to disengage the joint, if the length of the joint is 200 mm and the maximum power that can be transmitted at a rated speed of 1000 rpm. The material of the cylinders has a modulus of elasticity 210 GPa and Poisson's ratio 0.3.

Given Data: $d_i = 150$ mm; $d_c = 200$ mm; $d_o = 250$ mm; $\sigma_{t-oi} = 40$ MPa;

$l = 200$ mm; $n = 1000$ rpm; $E = 210$ GPa $= 210 \times 10^3$ MPa; $\nu = 0.3$

SOLUTION

Step-1: Contact pressure

Tangential stress at the inside diameter of outer cylinder,

$$\sigma_{toi} = p_c\left(\frac{d_o^2 + d_c^2}{d_o^2 - d_c^2}\right) \qquad \text{(Eqn. 12.24 MDHB)}$$

i.e.,

$$40 = p_c\left(\frac{250^2 + 200^2}{250^2 - 200^2}\right)$$

$\therefore$ Solving for $p_c = 8.78$ MPa

Step-2: Change in dimensions

Change in diameter of inner member,

$$\Delta d_i = \frac{p_c d_c}{E}\left[\frac{d_c^2 + d_i^2}{d_c^2 - d_i^2} - \mu_i\right] \qquad \text{(Eqn. 12.27 MDHB)}$$

$$= -\frac{8.78 \times 200}{210 \times 10^3}\left[\frac{200^2 + 150^2}{200^2 - 150^2} - 0.3\right] = -0.027 \text{ mm} \cong -0.03 \text{ mm}$$

Change in diameter of outer member,

$$\Delta d_o = \frac{p_c d_c}{E}\left[\frac{d_o^2 + d_c^2}{d_o^2 - d_c^2} + \mu_o\right] \qquad \text{(Eqn. 12.26 MDHB)}$$

$$= \frac{8.78 \times 200}{210 \times 10^3}\left[\frac{250^2 + 200^2}{250^2 - 200^2} + 0.3\right] = 0.04 \text{ mm}$$

∴ Total interference,

$$\Delta B = \Delta d_o + \Delta d_i = 0.03 + 0.04 = 0.07 \text{ mm} \qquad \text{(Eqn. 12.28 MDHB)}$$

Original inner diameter of outer member,

$$d_h = d_c - \Delta d_o = 200 - 0.04 = 199.96 \text{ mm}$$

Original outer diameter of inner member,

$$d_s = d_c + \Delta d_i = 200 + 0.03 = 200.03 \text{ mm}$$

Step-3: Force necessary to disengage the joint

Axial force necessary to disengage the joint,

$$F_a = \pi d_c L f p_c \qquad \text{(Eqn. 12.30 MDHB)}$$

$$= \pi \times 200 \times 200 \times 0.125 \times 8.78 = 137915.9175 \text{ N}, \ [f = 0.125]$$

Step-4: Torque capacity

Torque of shrink fit, $T = F_a \cdot \dfrac{d_c}{2}$ (Eqn. 12.31 MDHB)

$$= 137915.9175 \times \frac{200}{2} = 13791591.75 \text{ N mm}$$

Step-5: Power transmitted

$$\text{KW} = \frac{TN}{955 \times 10^4} = \frac{13791591.75 \times 1000}{955 \times 10^4} = 1444.15 \text{ kW}$$

Example 13.9: A 440 mm outer diameter, 250 mm inner diameter and 300 mm long steel hub is to be shrunk on to a 250 mm diameter steel shaft. If the torque to be transmitted is 300 kNm and $f = 0.18$, determine the amount of interference required.

Given Data: $d_0 = 440$ mm; $d_c = 250$ mm; $d_i = 0$ ($\therefore$ solid shaft); $l = 300$ mm;

$$T = 300 \text{ kNm} = 300 \times 10^6 \text{ Nmm}; \ f = 0.18$$

SOLUTION

Step-1: Contact pressure

Torque due to shrink fit, $T = \dfrac{\pi d_c^2 L f p_c}{2}$ (Eqn. 12.31 MDHB)

i.e., $$300 \times 10^6 = \frac{\pi \times 250^2 \times 300 \times 0.18 \times p_c}{2}$$

$\therefore$ Contact pressure, $p_c = 56.6$ MPa

Step-2: Total interference

Change in diameter of shaft,

$$\Delta d_i = -\frac{p_c d_c}{E}\left(\frac{d_c^2 + d_i^2}{d_c^2 - d_i^2} - \mu_i\right) = -\frac{p_c d_c}{E}[1 - \mu_i] \qquad \text{(Eqn. 12.27 MDHB)}$$

From Table 2.1 MDHB for steel

Modulus of elasticity, $E = 210 \text{ GPa} = 210 \times 10^3$ MPa

Poisson's ratio, $\mu = 0.3$

$\therefore$ $$\Delta d_i = -\frac{56.6 \times 250}{210 \times 10^3}[1 - 0.3] = 0.047 \text{ mm}$$

Change in inner diameter of hub,

$$\Delta d_o = \frac{p_c d_c}{E}\left[\frac{d_o^2 + d_c^2}{d_o^2 - d_c^2} + \mu_0\right] \qquad \text{(Eqn. 12.26 MDHB)}$$

$$= \frac{56.6 \times 250}{210 \times 10^3}\left[\frac{440^2 + 250^2}{440^2 - 250^2} + 0.3\right] = 0.1518 \text{ mm}$$

$\therefore$ Total interference $\Delta B = \Delta d_i + \Delta d_0 = 0.047 + 0.1518 = 0.1988$ mm

Example 13.10: A steel tube with inner and outer diameters of 50 mm and 75 mm respectively, is jacketed by an outer steel tube with an outer diameter of 100 mm. The compound tube is subjected to an internal pressure of 35 MPa. The shrinkage allowance is such that the maximum tangential stress in each tube has same magnitude. Find,

(i) Shrinkage pressure

(ii) Original dimensions of tubes

Also plot the distribution of tangential stresses.

Given Data: $d_i = 50$ mm; $d_c = 75$ mm; $d_o = 100$ mm; $p_i = 35$ MPa

SOLUTION

Step-1: Lame's equations for tangential and radial stress distribution due to internal fluid pressure p_i,

$$\sigma_t = \frac{p_i d_i^2}{d_o^2 - d_i^2}\left(1 + \frac{d_o^2}{4r^2}\right) \text{ when } p_o = 0 \qquad \text{(Eqn. 12.6 MDHB)}$$

Tangential stress at the inner surface of the inner member due to internal fluid pressure at $r = \frac{d_i}{2}$

$$\sigma_{t_{ai}} = \frac{p_i d_i^2}{d_o^2 - d_i^2}\left(1 + \frac{d_o^2}{d_i^2}\right) = p_i\left(\frac{d_o^2 + d_i^2}{d_o^2 - d_i^2}\right)$$

$$= \left(\frac{100^2 + 50^2}{100^2 - 50^2}\right) \times 35 = 58.33 \text{ MPa}$$

Tangential stress at the inner surface of outer cylinder or common surface due to internal fluid pressure at $r = d_c/2 = 75/2$

$$\sigma_{t_{ac}} = \frac{p_i d_i^2}{d_o^2 - d_i^2}\left(1 + \frac{d_o^2}{d_c^2}\right) = \frac{35 \times 50^2}{100^2 - 50^2}\left(1 + \frac{100^2}{75^2}\right) = 32.41 \text{ MPa} \quad \text{(Tensile)}$$

Tangential stress at the outer surface of outer cylinder due to internal fluid pressure at $r = d_o/2 = 100/2$

$$\sigma_{t_{ao}} = \frac{2 p_i d_i^2}{d_o^2 - d_i^2} = \frac{2 \times 35 \times 50^2}{100^2 - 50^2} = 23.33 \text{ MPa} \quad \text{(Tensile)}$$

The radial stress at $r = d_i/2$,

$$\sigma_{r_{ai}} = -35 \text{ MPa}$$

The radial stress at $r = d_c/2$,

$$\sigma_{r_{ac}} = -p_a \left(\frac{d_i}{d_c}\right)^2 = -35 \times \left(\frac{50}{75}\right)^2 = -15.56 \text{ MPa}$$

The radial stress at $r = d_o/2$

$$\sigma_{r_{ai}} = 0.0$$

Step-2: Tangential stress distribution due to shrinkage fitting

Tangential stress at inside diameter of outer member

$$\sigma_{t_{oi}} = p_c \left(\frac{d_o^2 + d_i^2}{d_o^2 - d_i^2}\right) = p_c \left(\frac{100^2 + 75^2}{100^2 - 75^2}\right) = 3.57 p_c \quad \text{(Eqn. 12.24 MDHB)}$$

Tangential stress at inside diameter of inner member

$$\sigma_{t_{ii}} = \frac{-2p_c d_c^2}{d_c^2 - d_i^2} = \frac{-2p_c \times 75^2}{(75^2 - 50^2)} = 3.6 p_c \text{ (Compressive)} \quad \text{(Eqn. 12.22 MDHB)}$$

Step-3: Shrinkage pressure

As the resultant maximum tangential stress in each member has same magnitude, equating stresses at the inner surfaces of tube and jacket, i.e., $58.33 - 3.6p_c = 32.41 + 3.57p_c$

Solving, $p_c = 3.615$ MPa

Step-4: Original dimensions of tubes

From Table 2.1 MDHB, $E = 206 \text{ GPa} = 210 \times 10^3$ MPa; $\mu = 0.3$

Change in diameter of inner member,

$$\Delta d_i = -\frac{p_c d_c}{E}\left(\frac{d_c^2 - d_i^2}{d_c^2 - d_i^2} - \mu\right) \quad \text{(Eqn. 12.27 MDHB)}$$

$$= -\frac{3.615 \times 75}{210 \times 10^3}\left[\frac{75^2 + 50^2}{75^2 - 50^2} - 0.3\right] = -2.97 \times 10^{-3} \text{ mm}$$

Change in the inner diameter of outer member,

$$\Delta d_o = \frac{p_c d_c}{E}\left(\frac{d_o^2 - d_c^2}{d_o^2 - d_c^2} + \mu\right) \quad \text{(Eqn. 12.26 MDHB)}$$

$$= \frac{3.615 \times 75}{210 \times 10^3}\left[\frac{100^2 + 75^2}{100^2 - 75^2} + 0.3\right] = 4.998 \times 10^{-3} \text{ mm}$$

∴ Original outside diameter of tube,

$$d_t = d_c + \Delta d_i = 75 + 2.97 \times 10^{-3} = 75.00297 \text{ mm}$$

Original inside diameter of jacket,

$$d_j = d_c - \Delta d_o = 75 - 4.998 \times 10^{-3} = 74.995 \text{ mm}$$

Total interference,

$$\Delta B = \Delta d_i + \Delta d_o = 2.97 \times 10^{-3} + 4.998 \times 10^{-3} = 7.968 \times 10^{-3} \text{ mm}$$

Step-5: Resultant tangential stress distribution

Tangential stress due to shrink-fit at outside diameter of jacket

$$\sigma_{t-oo} = 2.57 p_c = 2.57 \times 3.615 = 9.29 \text{ MPa} \quad \text{(Tensile)}$$

Tangential stress due to shrink-fit at inside diameter of jacket,

$$\sigma_{t-oi} = 3.57 p_c = 3.57 \times 3.615 = 12.91 \text{ MPa} \quad \text{(Tensile)}$$

Tangential stress due to shrink-fit at outside diameter of tube,

$$\sigma_{t-oi} = -2.6 p_c = 2.6 \times 3.615 = -9.4 \text{ MPa} \quad \text{(Compressive)}$$

Tangential stress due to shrink-fit at inside diameter of tube,

$$\sigma_{t-ii} = -3.6 p_c = 3.6 \times 3.615 = -13.04 \text{ MPa} \quad \text{(Compressive)}$$

Resultant tangential stresses at outside diameter of jacket,

$$\sigma_{tt-oo} = 23.33 + 9.29 = 32.62 \text{ MPa} \quad \text{(Tensile)}$$

Resultant tangential stress at inside diameter of jacket,

$$\sigma_{tt-oi} = 32.41 + 12.91 = 45.32 \text{ MPa} \quad \text{(Tensile)}$$

Resultant tangential stress at outside diameter of tube,

$$\sigma_{tt-io} = 32.41 - 9.4 = 23.01 \text{ MPa} \quad \text{(Tensile)}$$

Resultant tangential stress at inside diameter of tube,

$$\sigma_{tt-ii} = 58.33 - 13.01 = 45.32 \text{ MPa} \quad \text{(Tensile)}$$

The stress distributions are shown in Fig. 13.9.

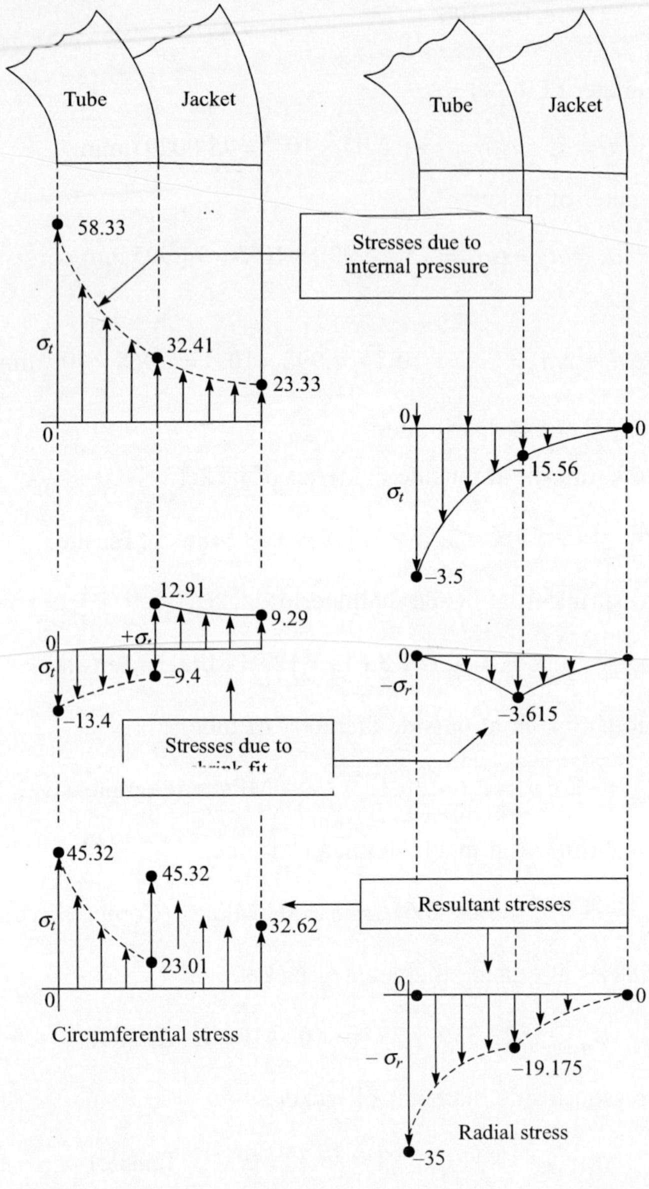

Figure 13.9 Distribution stresses in shrink-fit assembly

13.3 COVER PLATES

Unstayed Flat Heads and Covers

The thickness of flat unstayed circular heads and covers shall be calculated by the formula when the head or cover is attached by bolts causing an edge movement.

$$t = CD\sqrt{\frac{p}{S}} \tag{13.25a}$$

for values of C refer Table 11.13

$$t = D\sqrt{C'p/S} \tag{13.25b}$$

where, $C' = 0.1875$ for fixed edge
$= 0.305$ for supported edge

The thickness of flat unstayed non-circular heads and covers shall be calculated by the formula,

$$t = CZa\sqrt{p/S} \tag{13.26}$$

where Z = the factor for non-circular heads depending on the ratio of short span to long span a/b (Refer Table 13.5).

Spherically Dished Covers

The thickness of spherically dished ends secured to the shell flange connection by means of bolts shall be calculated by the formula (provided that the inside crown radius R of the dished cover does not exceed 1.3 times the shell inside diameter D_i and value of $100t/R$ is not greater than 10),

$$t = \frac{3pD_i}{2SJ} \tag{13.27}$$

where J is joint factor (Table 13.2)

Thickness of spherical head riveted or welded to the cylindrical shell,

$$t = \frac{8.33pR_0}{2Su} \tag{13.28}$$

$$R_0 = \text{Crown radius} < D_i$$

13.4 STAYED AND BRACED PLATES

The thickness of the stayed and braced plates shall be calculated by the formula

$$t = C_1D_1\sqrt{p/S} \tag{13.29}$$

where, D_1 = Diameter of the largest circle in mm which may be inscribed between the supporting points of the plates.

C_1 = Factor depending on the mode of support and is given in Table 2.6.

When the plate is supported in different ways on the circumference, for the inscribed circle of diameter D_1, a mean value is to be adopted for C_1.

Circular Plate Uniformly Loaded

Thickness of the plate with a diameter D supported at the circumference and subjected to a uniformly distributed pressure p over the total area,

$$t = c_1D\sqrt{p/S} \tag{13.30}$$

The maximum deflection,

$$y = \left[\frac{c_2 D^4 p}{Et^3}\right] \tag{13.31}$$

where c_1 and c_2 are coefficients from Table 2.3.

Circular Plate Loaded Centrally

The thickness of a flat cast iron plate supported freely at the circumference with a diameter D and subjected to a load F distributed uniformly over an area $(\pi D_0^2/4)$,

$$t = 1.2\sqrt{(1.670 D_0/D)(F/S)} \tag{13.32}$$

The deflection in mm,

$$y = \frac{0.12 D^2 F}{Et^3} \tag{13.33}$$

Grashof's formula for the thickness of a plate rigidly fixed around the circumference with the above given type of loading,

$$t = 0.65\left[\frac{F}{S}\log_e(D/D_0)\right]^{1/2} \tag{13.34}$$

The deflection in mm,

$$y = \frac{0.055 D^2 F}{Et^3} \tag{13.35}$$

Rectangular Plates

The thickness of a rectangular plate subjected to uniform load (Grashof and Bach),

$$t = abc_3\left[\frac{p}{S(a^2 + b^2)}\right] \tag{13.36}$$

The thickness of a rectangular plate on which a concentrated load F acts at intersection of diagonals,

$$t = c_4\left[\frac{abF}{S(a^2 + b^2)}\right]^{1/2} \tag{13.37}$$

where, a = length of plate, mm

b = breadth of plate, mm

c_3; c_4 = coefficients from Table 13.3

Elliptical Plate

The thickness of uniformly loaded elliptical plate,

$$t = abc_5\left[\frac{P}{S(a^2 + b^2)}\right]^{1/2} \tag{13.38}$$

where, a = minor axis, mm

b = major axis, mm

c_5 = coefficient, (from Table 11.9 MDHB)

13.5 SHAPED CYLINDER HEADS

Thickness of hemispherical and semi-ellipsoidal head, mm,

$$t = \frac{p_i d_i}{4\sigma_t \eta - 0.4 p_i} + CA \tag{13.39}$$

Length of the straight portion of semi-ellipsoid head, mm,

$$l_{st} = 3t \quad \text{or} \quad 20 \text{ mm} \tag{13.40}$$

Thickness of torispherical head, mm,

$$t \frac{0.885 p_i R_h}{\sigma_t \eta - 0.1 p_i} + CA \tag{13.41}$$

R_h inside crown radius $< d_0$

Knuckle s radius, mm

$$r_i = 0.06 R_h \tag{13.42}$$

Thickness of conical head, mm,

$$t = \frac{p_i d_i}{2\cos\alpha(\sigma_t \eta - 0.6 p_i)} + CA \tag{13.43}$$

α half apex angle 30°

d_i internal diameter, mm

p_i internal pressure, N/mm^2

η efficiency of welded joint (Table 11.14 MDHB)

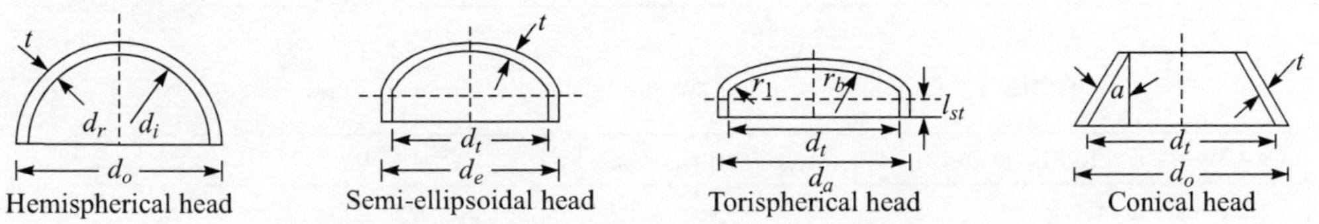

Different forms of cylinder heads

Table 13.3. Coefficients c_1, c_2, c_3, c_4, c_5

Material of cover plate	Method of holding edges	Circular plate		Rectangular plate		Elliptical plate
		c_1	c_2	c_3	c_4	c_5
Cast iron	Supported freely	0.54	0.038	0.75	1.73	1.5
Cast iron	Fixed	0.44	0.010	0.62	1.4; 1.6	1.2
Mild steel	Supported freely	0.42	0.038	0.60	1.38	1.2
Mild steel	Fixed	0.35	0.010	0.49	1.12; 1.28	0.9

Table 13.4. Maximum allowable stresses in stays and stay bolts

	Types of stays	Design stress, MPa for lengths between support	
		Not Exceeding 120 diameter	Exceeding 120 diameter
(a)	Unwelded or flexible stays less than 20 diameter long, screwed through plates with ends riveted over.	53	–
(b)	Hollow steel stays less than 20 dia. long, screwed through plates with ends riveted over.	56	–
(c)	Unwelded stays and unwelded portion of welded stays, except as specified in (a) and (b).	67	60
(d)	Steel through stays exceeding 40 mm, diameter.	73	63
(e)	Welded portions of stays.	42	42

Table 13.5. Value of Z in Eqn. 13.26

Ratio a/b	0.1	0.2	0.3	0.4	0.5	0.6	0.7	0.8	0.9	1.0
Value of Z	1.6	1.58	1.54	1.48	1.43	1.36	1.3	1.24	1.17	1.10

Table 13.6. Values of coefficient C_1 in Eqn. 13.29

Types of stays	Value of C_1	Types of stay	Value of C_1
Flanged head	0.45	Bar stay with washer ($D = 2.5d$)	0.45
Welded brace	0.45		
Welded tube	0.55		
Expanded and bended tabular stay	0.55	Bar stay with the washer and reinforcing plate ($D = 3d$)	0.4

Table 13.7. Values of constant C for the thickness of head

A - Covers riveted or bolted to the flange with gasket	$C = 0.42$
B - Forged heads	$C = 0.45$ to 0.5
C - Flanged flat heads butt welded	$C = 0.35$ to 0.5
D - Head lap welded to shell	$C = 0.45$ to 0.55
E - Plate welded to the inside of the shell	$C = 0.55$ to $0.7\sqrt{(t_c/t_s)}$
F - Plates welded to the end of the shell	$C = 0.7$
G - Plate welded to the shell with fillet weld	$C = 0.7\sqrt{(t_c/t_s)} > 0.55$

*t_c = Calculated thickness

t_s = Actual thickness

Table 13.8. Efficiency, η and maximum thickness, t; mm

Joint	Class	Efficiency	Maximum thickness
Double-butt weld	Class I	I	Any
	Class II	0.85	38 + CA
	Class III	0.56 – 0.70	16 + CA
Single-butt weld with backing strip	Class I	0.9	Any
	Class II	0.85	38 + CA
	Class III	0.55 – 0.65	16 + CA

CA – Corrosion allowance

Example 13.11: A cast steel cylinder of 300 mm internal diameter is to contain liquid at a pressure of 12.5 MPa. It is closed at both ends by unstayed flat cover plates which are attached by bolts. Determine the thickness of the cover plates if the allowable working stress for the cover material is 75 MPa.

Given Data: $D = 300$ mm; $p = 12.5$ MPa; $S_d = 75$ MPa

SOLUTION Minimum thickness of an unstayed flat head or cover plate according to ASME Boiler code,

$$h = D\sqrt{C' p/S} \qquad \text{(Eqn. 11.25a MDHB)}$$

Coefficient, $C' = 0.1875$

$$\therefore \qquad h = 300\sqrt{\frac{0.1875 \times 12.5}{75}} = 53.03$$

$\therefore$ Thickness of cover plate, $h = 53$ mm

Example 13.12: A cast steel cylinder of 350 mm inside diameter is to contain liquid at a pressure of 13.5 MPa. It is closed at both ends by flat cover plates made of alloy steel which are attached by bolts. Determine,

(i) Wall thickness of the cylinder, if the maximum hoop stress in the material is limited to 55 MPa.

(ii) Minimum thickness necessary of the cover plates if the working stress is not to exceed 65 MPa.

Given Data: $p_i = 13.5$ MPa; $d = 350$ mm;
Design stress for cylinder $S = 55$ MPa, for cover plate $S = 65$ MPa

SOLUTION

Step-1: Thickness of cylinder wall

Thickness of cylinder wall by shear theory,

$$h = \frac{d_i}{2}\left[\sqrt{\frac{S}{S - 2xp}} - 1\right] = \frac{350}{2}\left[\sqrt{\frac{55}{55 - 2 \times 13.5}} - 1\right] = 70.268 \text{ mm} \qquad \text{(Eqn. 12.19 MDHB)}$$

Step-2: Thickness of cover plates

Minimum thickness of an unstayed flat head or cover plate according to ASME boiler code,

$$h = D\sqrt{\frac{C'p}{S}} \qquad \text{(Eqn. 11.25a MDHB)}$$

Coefficient, $C' = 0.1875$

$$\therefore \qquad h = 350\sqrt{\frac{0.1875 \times 13.5}{65}} = 69.068 \text{ mm}$$

Take $h = 70$ mm

EXERCISES FOR PRACTICE

13.1 A thin cylinder of 2.5 m diameter is subjected to an internal pressure of 4 MPa. Find the required wall thickness if the hoop stress is limited to 60 MPa and longitudinal stress is limited to 40 MPa.

13.2 A thin cylinder of diameter 1.2 m and thickness 15 mm is to sustain an internal pressure of 1.25 MPa. Find the hoop stress and longitudinal stresses induced in the wall.

13.3 A gas cylinder with an internal diameter of 250 mm is subjected to an operating pressure of 12.5 MPa. It is made of plain carbon steel C 30 and the factor of safety is 2.5. Calculate the cylinder wall thickness assuming it to be a thin cylinder and neglecting the effect of welded joints.

13.4 A cast iron pipe used in a hydraulic circuit is subjected to an internal pressure of 50 MPa. The inner and outer diameters of the pipe are 20 mm and 40 mm respectively. Determine the distribution of principal stresses across the pipe thickness.

13.5 A thick cylinder of diameters 200 mm and 300 mm is subjected to an internal pressure of 15 MPa and external pressure of 5 MPa. Calculate and sketch the stresses distribution across the wall at 10 mm intervals.

13.6 A thick cylinder of internal and external diameters of 150 mm and 250 mm is made of CI having allowable stress of 40 MPa. Find the safe internal pressure that can be carried.

13.7 A thick cylindrical shell having internal diameter 175 mm carries fluid under pressure of 1.5 MPa. Find the required wall thickness taking allowable tangential stress as 95 MPa.

13.8 A 300 mm outside diameter, 200 mm inside diameter cylinder is press-fitted into a 300 mm inside diameter and 400 mm outside diameter hub of 120 mm length. The tangential stress at inside of outer cylinder is not to exceed 60 MPa. Modulus of elasticity is 200 GPa and Poisson's ratio is 0.3.

(i) Find the resulting stresses across the wall and sketch them.

(ii) The original diameters before shrink-fitting.

(iii) The axial force required for assembling of hub if the coefficient of friction is 0.1.

(iv) The power that can be transmitted at 500 rpm by the assembly.

13.9 A hydraulic cylinder with closed ends is subjected to an internal pressure of 15 MPa. The inner and outer diameters of the cylinder are 220 mm and 260 mm respectively. The cylinder material is cast iron FG 300. Determine the factor of safety used in design. If the cylinder pressure is further increased by 50%, what will be the factor of safety?

13.10 A high pressure cylinder consists of an inner cylinder of inner and outer diameters 200 mm and 300 mm respectively. It is jacketed by an outer cylinder of outside diameter 400 mm. The difference between the outer diameter of the inner cylinder and the inner diameter of the jacket before assembly is 0.25 mm (E = 207 GPa). Calculate the shrinkage pressure and the maximum tensile stress induced in any of the cylinders.

13.11 A pressure vessel consists of a cylindrical shell of inside diameter 1750 mm, that is closed by torispherical heads with a crown radius of 1500 mm. The operating pressure inside the vessel is 1.5 MPa. The yield strength of the material used for the shell and head is 260 MPa and the weld joint efficiency may be assumed to be 0.8. The corrosion allowance is 2 mm. Determine the thickness of the cylindrical shell and the torispherical head.

13.12 Determine the thickness of a flat steel plate for a cylinder head of 125 mm diameter subjected to an internal pressure of 10 MPa. It is supported at the circumference and the allowable stress is 150 MPa. Also find the deflection in this plate. What will be the thickness and deflection if the plate is fixed at the circumference?

13.13 A CI plate is used as a cover plate for a cylinder of internal diameter 120 mm. It is supported at circumference and is subjected to a uniformly distributed force of 6 kN over a diameter of 60 mm. Find the thickness of cover and corresponding deflection taking the allowable stress as 18 MPa, E = 96 GPa.

13.14 A cylinder of diameter 240 mm is covered with a CI plate 9 mm thick and is fixed at circumference. It is subjected to a uniform force of 9 kN over a diameter of 200 mm. Find the stress induced in the plate and the corresponding deflection.

13.15 An air receiver, consisting of a 500 mm diameter cylinder closed by hemispherical ends, is made of steel C 30 and the factor of safety is 2.5. The operating pressure is limited to 2.5 MPa. Treating the receiver as a thin cylinder, calculate the thickness of the cylinder wall and the hemispherical ends. Neglect the effect of welded joints. Assume that the cylinder is made of ductile material.

13.16 A pressure vessel, subjected to a design pressure of 0.75 MPa, consists of a cylindrical shell with 2 m inside diameter and 10 mm thickness. An opening of inner diameter 300 mm and wall thickness 10 mm is provided in the shell. The corrosion allowance is 2 mm and the weld joint efficiency is taken as 0.85. The extension of the opening inside the shell is 15 mm. The yield strength of the material used for the shell and the opening is 210 MPa. A reinforcing pad made of a 10 mm thick plate is provided for the opening. Determine the inner and outer diameters of the pad.

SHORT ANSWER QUESTIONS

13.1 What is the difference between thin and thick cylinders?

13.2 What are the kinds of stresses induced in thin cylinders?

13.3 Name some thin cylinder applications.

13.4 What are the kinds of stresses induced in thick cylinders?
13.5 Give the Lame's equations for thick cylinder stresses.
13.6 What are the stresses induced in cylinders subjected to internal fluid pressure only?
13.7 What are the stresses induced in a cylinder subjected to external pressure only?
13.8 What are compound cylinders?
13.9 What are the advantages of compounding?
13.10 What are the other methods of strengthening the cylinders?
13.11 What the different types of cover plates for cylinders?
13.12 Give the equation for the thickness of flat-unstayed cover plate?
13.13 What is the equation for the thickness of the non-circular cover plate?
13.14 Give the equation for spherical head-thickness.
13.15 What is the equation for the thickness of the centrally-loaded circular head cover?
13.16 Give the equation for the thickness of the centrally-loaded rectangular cover plate?
13.17 What is expression for the thickness of a centrally-loaded elliptical cover plate?
13.18 What is the hemispherical cover head for cylindrical pressure vessels?
13.19 What is tori-spherical head? When is it used?
13.20 What is semi-ellipsoidal head? When is it used?

CHAPTER 14

SPRINGS

14.1 INTRODUCTION

A spring is an elastic member, which undergoes deflection under the action of load and returns to its original shape, when the load is removed. The springs are of different shapes, depending upon the applications. The functions and applications of springs are:

1. Store energy and release as required, e.g., clock spring.
2. Absorb shock and impact energy, e.g., suspension springs.
3. Keep the machine in contact, e.g., valve gear mechanism of IC engine, clutches.
4. Control motions, e.g., governors.
5. Control or dampen vibrations, e.g., shock absorbers.
6. To measure the force, e.g., weighing machines.

14.2 MATERIALS FOR SPRINGS

The requirement of spring materials are:

1. High ultimate strength
2. High yield strength
3. Low modulus of elasticity
4. High fatigue strength for the dynamically loaded springs

The materials satisfying these requirements to some extent are:

(i) Medium to high carbon steels C45 to C95;
(ii) Alloy steels 55Si2Mn90; 50Crl; 50CrlV23; 50Si2Min;
(iii) Stainless steel alloys SAE 30302;
(iv) Beryllium copper and phosphor bronze.

Most of the light duty springs are made from cold drawn, round or rectangular wire or thin cold drawn flat strip stock. The heavy duty springs like vehicle suspension springs are typically made from hot rolled or forged formed. The spring materials are typically hardened after forming.

14.3 CLASSIFICATION OF SPRINGS

Depending on the shapes, the springs may be classified as:

(i) Wire springs
(ii) Flat (leaf) springs
(iii) Special shaped springs: disk spring or Belevile springs

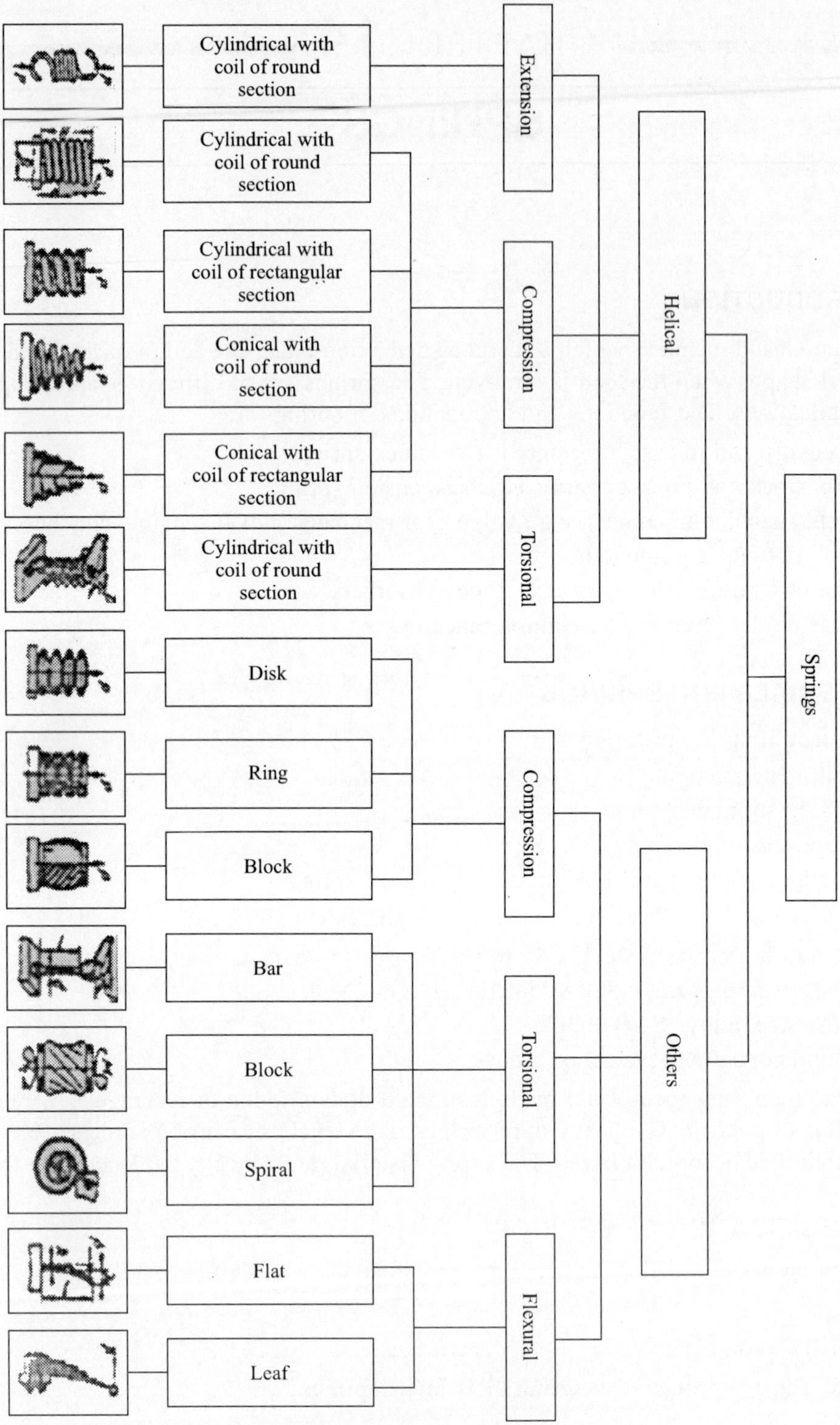

General Classification of Springs

The wire springs may be helically or spirally wound. The shape of the wire may be round, square or rectangular. The helical springs may be cylindrical or conical.

The general classification is shown in the chart given on the previous page.

The various types of springs discussed here are as follows:

(a) Helical coil springs
  (i) Compression (ii) Extension (iii) Torsion

(b) Spiral – coil spring

(c) Leaf springs

(d) Disk spring or Belleville springs

(e) Rubber springs.

14.4 HELICAL COIL SPRINGS

Helical coil springs are made from round or rectangular wire bent in the form of a helix. In compression springs, the coils have large pitch initially and close when full load is applied. The ends are properly finished to receive the load. In tension springs, the coils are close wound initially and extend when the load is applied. The ends are formed in a variety of shapes to facilitate the application of load. In both the types, the load is applied axially.

In torsion springs, the wire is wound in a close helix, the ends are extended tangentially to some length, to support the load.

14.4.1 Stress Induced in Tension or Compression Springs

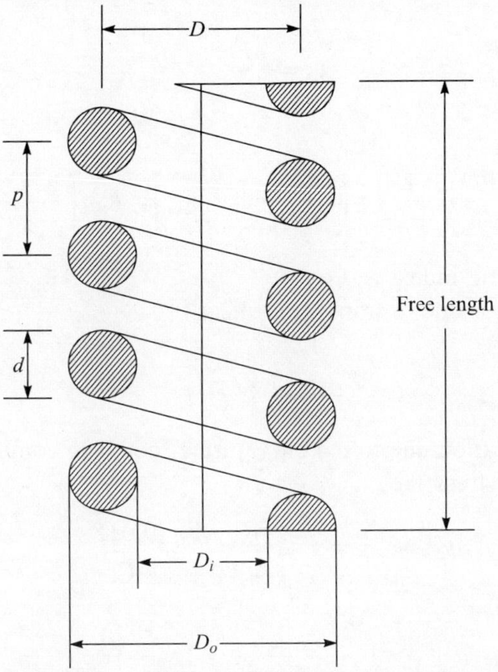

Figure 14.1 Nomenclature of spring

d is the diameter of the wire,

D is the mean diameter of the spring coil

R is the radius of the coils

p is the pitch of the coils. It is the distance between the adjacent coils

F is the load applied along the axis of the coils

$C = d/D$ is called spring index and is in the range of 4 to 12

The spring wire is subjected to torsional and shear loads.

Torsion T = F.R. = FD/2

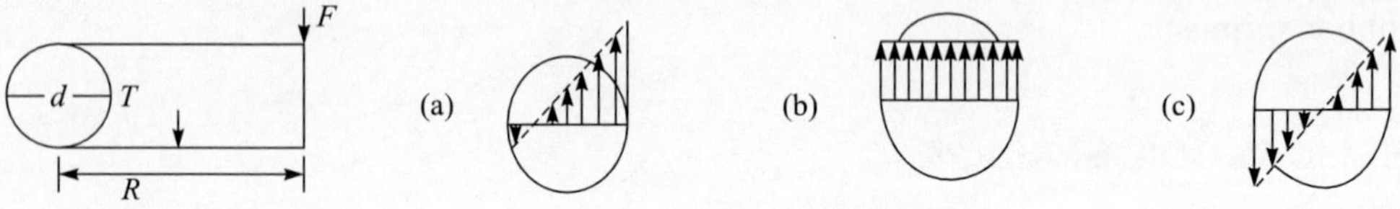

Figure 14.2 Stresses in spring wire

(a) Pure torsional stress
(b) Direct shear stress
(c) Combined torsional, direct and curvature shear stresses

The torsional shear stress,

$$\tau_1 = \frac{16T}{\pi d^3} = \frac{8FD}{\pi d^3} \tag{i}$$

The transverse shear stress,

$$\tau_2 = \frac{4F}{\pi d^3} \tag{ii}$$

The total shear stress,

$$\tau = \frac{8FD}{\pi d^3} + \frac{4F}{\pi d^2} = \left(1 + \frac{1}{2C}\right)\left(\frac{8FD}{\pi d^3}\right) = K_{C1}\left(\frac{8FD}{\pi d^3}\right) \tag{iii}$$

where $C = D/d$ is called spring index.
The stress concentration factor due to transverse shear,

$$K_{C1} = \left(1 + \frac{0.5}{C}\right) \tag{iv}$$

There will be stress concentration due to the curvature effect. The combined effect is accounted for by introducing the Wahl stress factor,

$$K = K_{C1} \cdot K_{C2} = \frac{4C - 1}{4C - 4} + \frac{0.615}{C} \tag{v}$$

The curvature factor K_{C2} can be found as

$$K_{C2} = \frac{K}{K_{C1}} \qquad \text{(vi)}$$

The maximum shear stress, $\tau_{\max} = \dfrac{8FD}{\pi d^3} K$ (14.1)

14.4.2 The Spring Deflection

The spring wire undergoes an angular twist θ given by the torsional equation

$$\frac{T}{J} = \frac{\tau}{r} = \frac{G\theta}{L}$$

The angular deflection is given as $\theta = \dfrac{TL}{JG}$

where, $L = \pi \cdot D \cdot N$ and $N =$ number of coils
The load displacement,

$$y = \theta \cdot R = \theta D/2$$

$$= \left(\frac{TL}{JG}\right)\left(\frac{D}{2}\right) = \frac{(FD/2)(\pi D \cdot N)}{\left(\dfrac{\pi}{32} d^4\right) \cdot G}\left(\frac{D}{2}\right)$$

$$\therefore \qquad = \frac{8FD^3 N}{d^4 G} = \frac{8FC^3 N}{dG} \qquad (14.2)$$

The spring rate or stiffness is defined as

$$K_s = \frac{F}{y} = \frac{d^4 G}{8D^3 N} = \frac{dG}{8C^3 N} \text{ N/mm} \qquad (14.3)$$

The energy stored in the spring (resilience)
$U =$ Work done by the applied load,

$$F \cdot \frac{y}{2} = \frac{4F^2 D^3 N}{d^4 G} = \left(\frac{y^2 d^4 G}{16ND^3}\right) = \frac{V \cdot S_y^2}{4K^2 G} \qquad (14.4)$$

The surge frequency in the spring,

$$f_N = C_1 \sqrt{\left(\frac{K_s}{M_s}\right)} \text{ Hz} = C_1 \left(\frac{2}{\pi N}\right)\left(\frac{d}{D^2}\right)\sqrt{\frac{Gg}{32\rho}} \text{ Hz} \qquad (14.5)$$

where, $C_1 = 7.9$ when one end is at rest

$= 15.8$ for both ends fixed

where, $M_s =$ mass of the spring $= (\pi^2 d^2 DN\rho_m)/4$

ρ_m is mass density in kg/mm^3
K_s is spring rate in N/mm

The above equations are applicable to both tension and compression springs. Now let us consider some features applicable exclusively for tension or compression springs.

14.4.3 Tension Springs

Tension or extension spring is wound solid, (the coils touch one another) initially and end coils are bent at right angle to form a hook or loop for the application of the load. The ends may be shaped to different forms also. A standard extension spring is shown in Fig. 14.3.

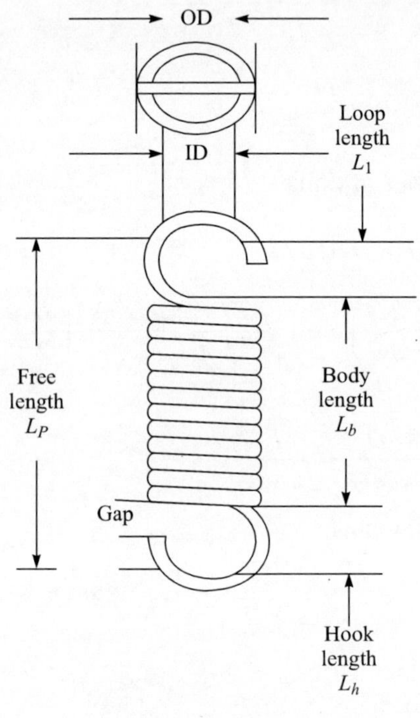

Figure 14.3

Active coils – All the coils in the spring are active coils but one coil is added to get the body length L_b.

$$L_b = d(N+1)$$

The hook length, $L_h \geq D$
The free length, $L_p = L_b + 2L_h$

The deflection of extension spring. The extension spring wire is twisted as it is wound, this introduces an initial pre-load in the coils. This must be overcome before any extension starts. The force extension diagram is as shown in Fig. 14.4.

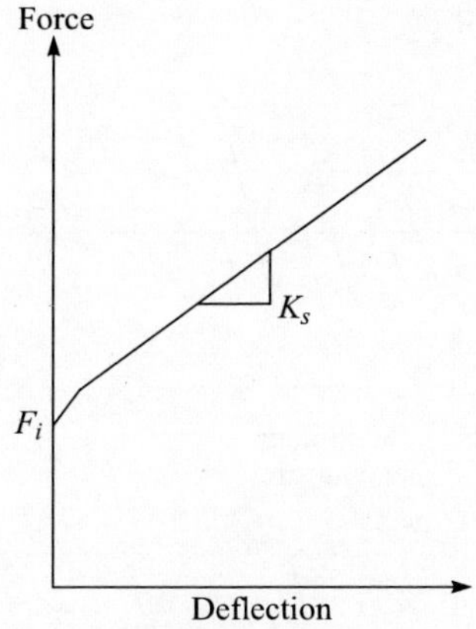

Figure 14.4

Then the coil deflection is given by the equation below

$$y = \frac{8(F - F_i)D^3N}{d^3G} \tag{14.6}$$

Then the spring rate,

$$K_s = \left(\frac{F - F_i}{y}\right) = \frac{d^4G}{8D^3N} = \frac{dG}{8C^3N} \tag{14.7}$$

The spring index $C = (D/d)$
The value of C is in the range of 4 to 12
The initial load, $F_i = (0.25 \text{ to } 0.33)F_{\text{lim}} = (0.275 \text{ to } 0.393)F_{\text{max}}$
In the usual calculation F_i is neglected.

Stresses in the End Hook

The max. direct stress at A in the curved hook,

$$\sigma_A = K_{b1} \cdot \frac{16DF}{\pi d^3} + \frac{4F}{\pi d^2} \tag{14.8}$$

where, $$K_{b1} = \frac{4C_1^2 - C_1 - 1}{4C_1(C_1 - 1)}$$

and $$C_1 = (2R_1/d)$$

The maximum torsional stress at the bend B,

$$\tau_B = K_{b2}\frac{8DF}{\pi d^3} \tag{14.9}$$

where, $$K_{b2} = \frac{4C_2 - 1}{4C_2 - 4}$$

and $$C_2 = (2R_2/d)$$

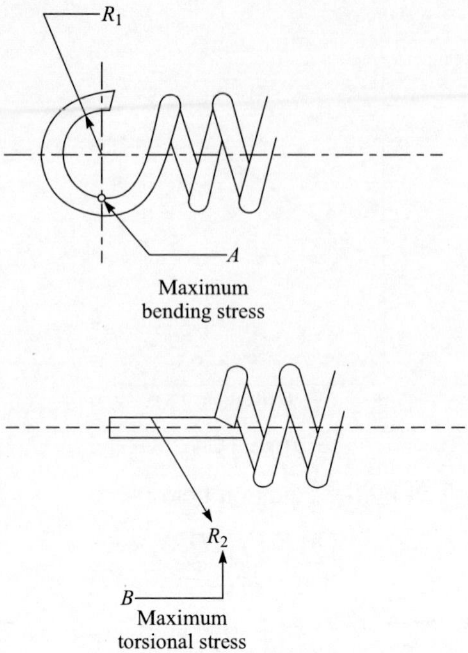

Figure 14.5

14.4.4 Compression Springs

The wire diameter, coil diameter, the number of active coils, coil pitch, free length are the various parameters of a helical compression spring. The various terms used are explained below.

Active Coils: The coils which contribute to spring deflection are active coils. They are less than the actual coils, depending upon the end finish. The ends of the compression springs may be finished in various ways:

(i) Plain ends
(ii) Plain-ground end
(iii) Squared ends
(iv) Squared and ground ends

Coil and finish	Actual no. of coils	N' Solid length	Free length
Plain end	N	$(N+1)d$	$N \cdot p+d$
Ground end	N	N.d	N.p
Squared end	$N+2$	$(N+3)d$	N.p + 3d
Squared and Ground end	$N+2$	$(N+2)d$	N.p + 2d

N = Effective no. of coils p = Pitch

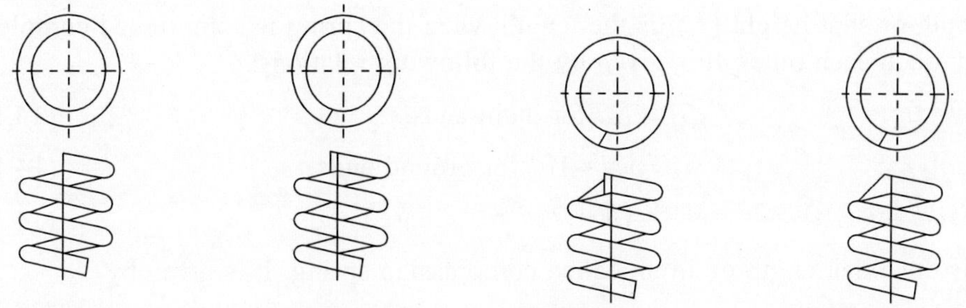

(a) Plain end (b) Ground end (c) Squared end (d) Squared and ground end

Figure 14.6

The free length L_p is the length in the unloaded condition $= (L_d + y_{max} + c)$ (14.10)

c = clash allowance $= 0.15 y_{max}$

Assembled length, L_a is the length after installation to its initial deflection Y_i under the preload F_i,

$$y_i = \frac{8 F_i D^3 N}{d^4 G} \tag{14.11}$$

Working deflection is the deflection y_w due to the working load F_w,

$$y_w = \frac{8 F_w D^3 N}{d^4 G} \tag{14.12}$$

Total deflection $y = y_i + y_w$ (14.13)

The minimum working length, L_{min} is the shortest length to which it is compressed in service. This is also called compressed length.

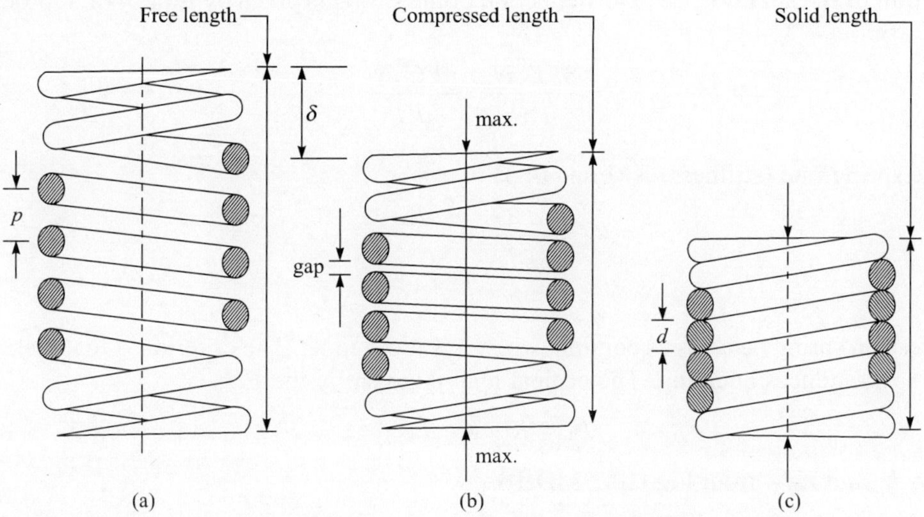

Figure 14.7 Spring length terminology

The solid length or shut height (L_d) is the length when the spring is compressed completely so that the coils touch each other. It is given by the following relations:

$$L_d = Nd \text{ for plain ends.} \tag{14.14a}$$

$$= (N+1)d \text{ for ground ends} \tag{14.14b}$$

$$= (N+2)d \text{ for squared ends} \tag{14.14c}$$

Free length is the axial length of an unloaded compression spring. It is given by

Free length = compressed length + deflection (δ)

= solid length + total axial gap + deflection (δ) (14.14d)

The pitch p is given by

$$p = \frac{\text{free length}}{(N_t - 1)} \tag{14.14e}$$

Stress in the Coils

The stresses induced in the coils are torsional shear and transverse shear and stress concentration due to the curvature effect. They are given by following equation (Eqn. 14.1).

$$\tau_{\max} = K\frac{8F_{\max}D}{\pi d^3} \leq S_{sd}$$

where, $K = \dfrac{4C-1}{4C-4} + \dfrac{0.615}{C}$ and $F_{\max} = F_i + F_w$

The equation is used to determine the wire diameter.

The deflection of the spring: The axial deflection of the spring is given by Eqn. 14.2, reproduced here.

$$y = \frac{8FD^3N}{d^3G} = \frac{8FC^3N}{dG}$$

Then the spring rate (stiffness) is (Eqn. 14.3)

$$K_s = \frac{d^4G}{8D^3N} = \frac{dG}{8C^3N}$$

Buckling

The compression spring behaves as column, when it is too slender. The criterion is that the aspect ratio (Lo/D) > 4 induces buckling. The critical load is given by the Eqn.

$$F_{cr} = K_s \cdot K_L \cdot Lo \tag{14.15}$$

where $Lo = L_p$ and K_L – from Fig. 10.2 MDHB

14.4.5 Design Procedure

1. Select suitable spring index
    - Small value for hard spring
    - Large value for soft spring
2. Select the spring material and design stress.
3. Calculate the wire diameter using Eqn. 14.2 and standardize.
4. Determine the coil dia. $D = C.d$ and check for the specified restrictions, i.e., space requirements.
5. Calculate the spring rate from the deflection requirements.
6. Calculate the number of coils from Eqn. 14.2 or Eqn. 14.3 for the spring rate.
7. Calculate the solid length, free length, etc.
8. Check for buckling of compression spring by Eqn. 14.15.

Design for Fluctuating Loads

We consider the case of fluctuating load, varying between the extreme values of $F_{\max}$ and $F_{\min}$. We define the following terms,

Mean load, $F_m = (F_{\max} + F_{\min})/2$

Amplitude of the load, $F_a = (F_{\max} - F_{\min})/2$

Then the mean shear stress, $\tau_m = \dfrac{8DF_m}{\pi d^3} K_{c1}$

Amplitude of the stress, $\tau_a = \dfrac{8DF_a}{\pi d^3} K$

Where, curvature factor, $K_{c1} = \left(1 + \dfrac{0.5}{C}\right)$ and,

Wahl's factor, $K = \dfrac{4C-1}{4C-4} + \dfrac{0.615}{C}$

It is also recommended to use $K_B = \dfrac{4C+2}{4C-3}$ in place of above two factors.

The stress variation is never reversed. It fluctuates between 0 and S_{sen}, the endurance limit stress in shear is shown in Fig. 14.8.

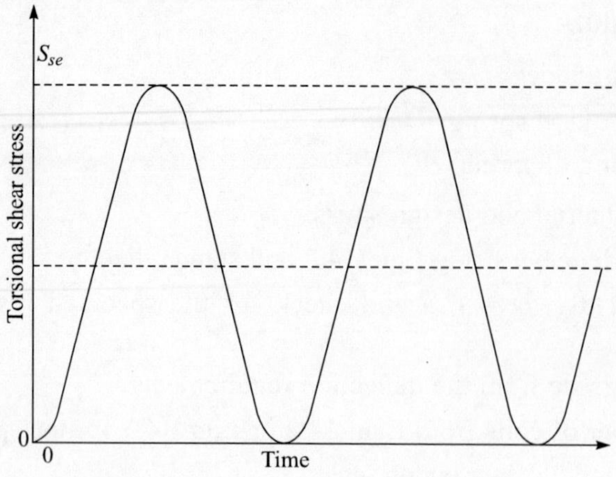

Figure 14.8 Pulsating stress cycle

For the failure analysis, a fatigue failure diagram is considered. This is a modified Soderberg diagram.

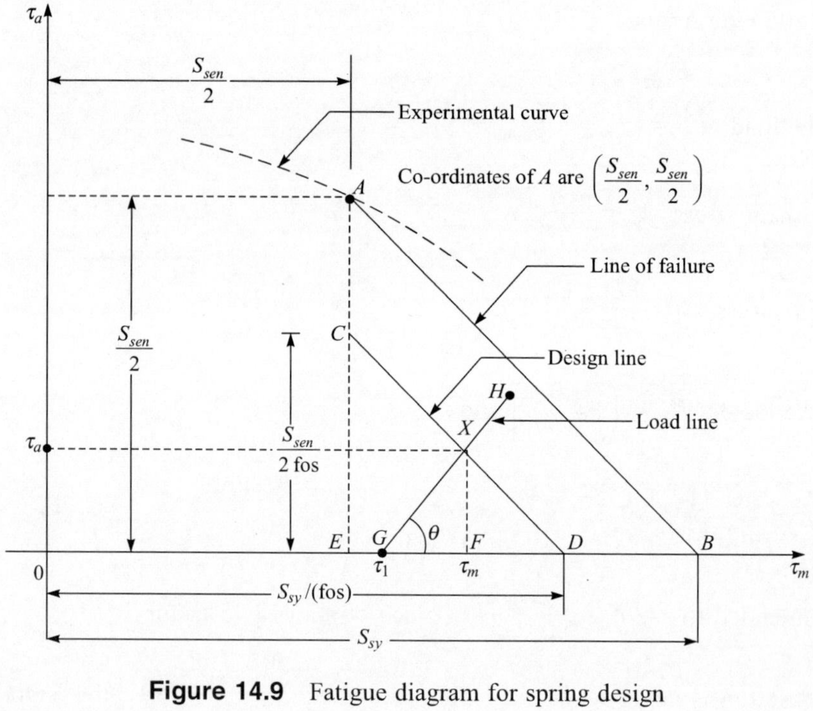

Figure 14.9 Fatigue diagram for spring design

The mean stress τ_m is plotted on abscissa and the stress amplitude τ_a is plotted on the ordinate. The starting point A has the co-ordinates $(S_{sen}/2, S_{sen}/2)$.

This is the failure point, when the spring is subjected to fatigue test. The other point B has the co-ordinates $(S_{sy}, 0)$. This is the failure point of the spring in simple shear test. The line joining A to B is the failure line for other combination variable loads.

For design a factor of safety is used. The point *D* is plotted with its abscissa as S_{sy} /FOS. The line *DC* is drawn parallel to line *AB*.

This is the design line. The equation for this line is derived as follows.

The point *G* is plotted at a distance τ_i representing the initial shear stress in the spring, when the assembly is made. Then a line *GH* is drawn with a slope of tan $\theta = \tau_a/\tau_m$. The line is called the load line. The point of intersection *X* has the co-ordinates (τ_m, τ_a).

Consider the similar triangles, *AEB* and *XFD*. We can write

$$\frac{AE}{EB} = \frac{XF}{FD}$$

$$\therefore \quad \frac{S_{sen}/2}{S_{sy} - S_{sen}/2} = \frac{\tau_a}{S_{sy}/fos - \tau_m}$$

$$\frac{S_{sy}}{\text{FOS}} - \tau_m = \tau_a\left(\frac{2S_{sy}}{S_{sen}} - 1\right)$$

Simplifying we get,

$$\frac{1}{\text{FOS}} = \frac{2\tau_a}{S_{sy}} + \frac{(\tau_m - \tau_a)}{S_{sy}} \tag{14.16}$$

This is modified Soderberg equation, known as Wahl's fatigue design equation.

SOLVED EXAMPLES

Example 14.1: A helical spring is made from steel wire with allowable stress of 700 MPa and modulus of elasticity in shear of 80 GPa. The mean diameter of the spring is 150 mm and a load of 4 kN is applied. Determine the size of the wire and number of coils for spring rate of 10 N/mm.

Given Data: Design stress, S_{sd} = 700 MPa; Shear modulus, G = 80 GPa;

Spring diameter, D = 150 mm; Load, F = 4000 N; Stiffness, K_s = 10 N/mm

SOLUTION

Step-1: Spring dia. for strength

For the shear stress induced

$$\tau_{\max} = \frac{8DF}{\pi d^3}K \le S_{sd} \qquad \text{(Eq. 10.1 MDHB)}$$

where,

$$K = \frac{4C-1}{4C-4} + \frac{0.615}{C}$$

With $\tau_{\max}$ = 700 MPa and neglecting curvature effect,

$$d = \sqrt[3]{\frac{8 \times 4000 \times 150}{\pi \times 700}} = 12.97 \text{ mm} = 13.00 \text{ mm}.$$

Let $d = 14.0$ mm, then $C = \dfrac{150}{14} = 10.7$

$$K = \frac{4 \times 10.7 - 1}{4 \times 10.7 - 4} + \frac{0.615}{10.7} = \frac{42.8 - 1}{42.8 - 4} + \frac{0.615}{10.7} = 1.1347$$

Then the stress induced,

$$\tau = \frac{8 \times 4000 \times 150}{\pi \times 14.0^3} \times 1.1347 = 631.8 \text{ MPa} < S_d = 700 \text{ MPa}$$

The design is safe.

Step-2: Number of coils for stiffness

The spring rate $K_s = 10$ N/mm; the number of coils is evaluated by the equation for stiffness.

$$K_s = \frac{d^4 G}{8 D^3 N}$$

$$\therefore \qquad 10 = \frac{14^4 \times 80 \times 10^3}{8 \times 150^3 \times N}$$

$$\therefore \qquad N = 11.38 \cong 12.0$$

Example 14.2: A spring loaded fly-ball governor with dimensions as shown in Fig. 14.10(a) requires a spring force of 170 N to permit the proper speed with the balls at a 120 mm radius. Assume the limiting position to be that as shown in Fig. 14.10(b) with balls at 180 mm radius. For proper speed regulation, the spring force must be 510 N in that position.

(a) Determine the required spring rate.
(b) For maximum shear stress of 310 MN/m^2 and spring index of 8, determine the required wire and mean coil diameters.
(c) Using a shear modulus G = 81 GN/m^2, determine the number of active turns required.
(d) Allowing one ineffective turn at each end, determine the free length needed so that the spring will be closed in position as shown in Fig. 14.10(b).
(e) What is the spring length in position as shown in Fig. 14.10(a)?

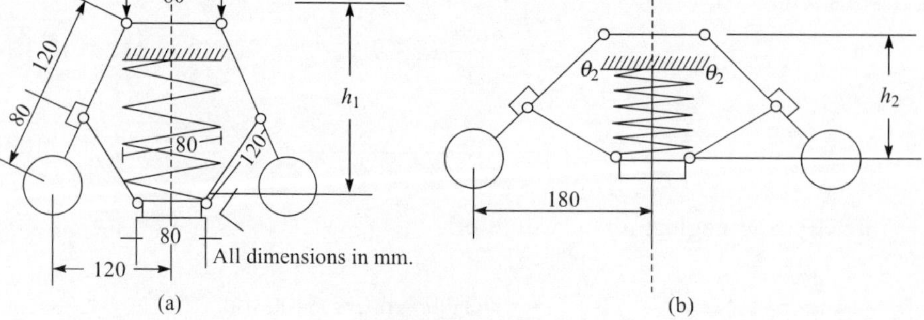

Figure 14.10

Given Data: Min. spring force F_{min} = 170 N; Min. radius r_1 = 120 mm;
Max. spring force F_{max} = 510 N; Max. radius r_2 = 180 mm;
Design stress S_{sd} = 310 MPa; Modulus G = 81 GPa.

SOLUTION

Step-1: Stiffness of the spring

Inclination θ_1 of the arm at position (a)

$$\theta_1 = \sin^{-1}\left[\frac{120-40}{120+80}\right] = \sin^{-1}\left[\frac{80}{200}\right] = 23.6°$$

Then height, $h_1 = 2 \times 120 \times \cos\theta_1 = 220$ mm

The inclination θ_2 of the arm at position (b)

$$\theta_2 = \sin^{-1}\left[\frac{140}{200}\right] = \sin^{-1}(0.7) = 44.427°$$

Then the height, $h_2 = 2 \times 120 \cos 44.427° = 171.4$ mm

The spring compression, $h_1 - h_2 = 220 - 171.4 = 48.6$ mm

$$\text{Then the spring rate, } K_s = \frac{F_2 - F_1}{h_1 - h_2} = \frac{510-170}{48.6} = 7.0 \text{ N/mm}$$

Step-2: Spring dia. for strength

Then the wire dia. for C = 8 and K = 1.18 (Fig. 10.1 MDHB)

$$d = \sqrt{\frac{8F_{max} \times C \times K}{\pi \times S_{sd}}} = \sqrt{\frac{8 \times 510 \times 8 \times 1.18}{\pi \times 310}} = 6.288 \cong 6.3 \text{ mm}$$

Mean coil dia. $D = 6.3 \times 8 = 50.4$ mm

Step-3: No. of coils for stiffness

Number of active coils,

$$N = \frac{d^4 G}{8D^3 K_s} = \frac{d \cdot G}{8C^3 K_s} = \frac{6.3 \times 81 \times 10^3}{8 \times 8^3 \times 7} = 17.8 \cong 18$$

Total no. of coils, $N^1 = 18 + 2 = 20$

Step-4: Length of the spring

Solid length, $L_d = N^1 \times d = 20 \times 6.3 = 126$ mm.

Deflection under max. load, $Y_{max} = 510/7 = 72.857$ mm

Then the free length,

$$L_o = L_d + y + c = 126 + 72.857 + 10 = 208.857 \cong 210 \text{ mm}$$

Step-5: Check for buckling

Buckling load, $F_{cr} = K_s \cdot K_L \cdot L_o$

where $K_s = 7$ and $K_L = 0.6$ (Fig. 10.2 MDHB)

Then $F_{cr} = 7 \times 0.6 \times 210 = 882N > F_{max}$

Hence, design is safe.

Example 14.3: A carbon steel coil spring has a mean diameter of 37.5 mm and is formed of 12.5 mm bar. It has a spring index of 3 and is subjected to a continuously alternating load between a maximum of 7000 N and a minimum of 5000 N. If the torsional yield point of the material is 700 MPa and the endurance limit due to released loading is 300 MPa, determine the factor of safety (FOS) under which the spring is operating.

Given Data: Given $D = 37.5$ mm; $d = 12.5$ mm

$F_{max} = 7000$ N; $F_{min} = 5000$ N;

$S_{sy} = 700$ MPa, $S_{sen} = 300$ MPa

SOLUTION The spring index $= C = D/d = 37.5/12.5 = 3.0$

Step-1: Check for FOS by Soderberg equation

Amplitude of load, $F_a = (7000 - 5000)/2 = 1000$ N

Mean load, $F_m = (7000 + 5000)/2 = 6000$ N

Wahl's factor, $K = \dfrac{4 \times C - 1}{4 \times C - 4} + \dfrac{0.615}{C} = \dfrac{4 \times 3 - 1}{4 \times 3 - 4} + \dfrac{0.615}{3} = 1.58$

Curvature factor, $K_c = 1 + 0.5/C = 1.167$

The stress amplitude,

$$\tau_a = \frac{8 \cdot F_a \times C \times K}{\pi \times d^2} = \frac{8 \times 1000 \times 3 \times 1.58}{\pi \times 12.5^2} = 77.25 \text{ MPa}$$

The mean stress,

$$\tau_m = \frac{8 \cdot F_m \times C \times K_c}{\pi \times d^2} = \frac{8 \times 6000 \times 3 \times 1.167}{\pi \times 12.5^2} = 342.3 \text{ MPa}$$

Modified Soderberg equation

$$\frac{1}{\text{FOS}} = \frac{\tau_m - \tau_a}{S_{sy}} + \frac{2 \cdot \tau_a}{S_{sen}} = \frac{342.34 - 77.25}{700} + \frac{2 \times 77.25}{300} = 0.8937$$

$\therefore$ FOS $= 1.118$

Step-2: Check for FOS by variable load only

Alternatively based on the variable – component only

$$\tau_{max} - \tau_{min} = 2\tau_a = 2 \times 77.25 = 154.5 \text{ MPa}$$

Then $\text{FOS} = \dfrac{S_{en}}{\tau_{max} - \tau_{min}} = \dfrac{300}{154.5} = 1.94$

Step-3: Check for FOS by static theory

Based on yield stress – Maximum stress induced

$$\tau_{max} = \frac{8F_{max}C \cdot K_c}{\pi \times d^2} = \frac{8 \times 7000 \times 3 \times 1.167}{\pi \times 12.5^2} = 399.397 \text{ MPa}$$

FOS based on yield stress

$$= \frac{S_{sy}}{\tau_{max}} = \frac{700}{399.397} = 1.75$$

Example 14.4: A helical spring is set inside another, the outer spring having a free length of 40 mm greater than the inner spring. The dimensions of each spring are as follows:

Dimension	Outer spring	Inner spring
Mean diameter	100 mm	64 mm
Wire diameter	12.5 mm	12.5 mm
Active turns	20	15

Determine the combined spring rate of the two springs after sufficient load has been applied to deflect the outer spring 60 mm and loads shared by them. Use $G = 83\ \text{GN/m}^2$.

Given Data: Note that suffix 1 applies to outer spring, 2 applies to inner spring

$D_1 = 100$ mm $\qquad D_2 = 64$ mm

$d_1 = 12.5$ mm $\qquad d_2 = 12.5$ mm

$N_1 = 20$ $\qquad N_2 = 15$

$N_{i1} = 2$ $\qquad N_{i2} = 2$

$L_{01} - L_{02} = 40$ mm and $G = 83\ \text{GN/m}^2 = 83 \times 10^3$ MPa

SOLUTION

Step-1: Stiffness of each spring

The stiffness of the outer spring,

$$K_{s1} = \frac{d_1^4 G}{8D_1^3 N_1} = \frac{12.5^4 \times 83 \times 10^3}{8 \times 100^3 \times 20} = 12.665 \text{ N/mm}$$

The stiffness of the inner spring,

$$K_{s2} = \frac{d_2^4 G}{8D_2^3 N_2} = \frac{12.5^4 \times 83 \times 10^3}{8 \times 64^3 \times 15} = 64.4 \text{ N/mm}$$

Step-2: Combined stiffness of springs

After the initial compression of the outer spring up to 40 mm, both the springs share the applied load as per their stiffness values, the deflection being the same. They act as springs in parallel, the resultant stiffness,

$$K_s = K_{s1} + K_{s2} = 12.665 + 64.4 = 77.08 \text{ N/mm}$$

Step-3: Loads shared by each spring

For a deflection of 60 mm of the outer spring, the load taken by the outer spring,

$$F_1 = 60 \times 12.665 = 759.9 \text{ N}$$

The corresponding deflection of the inner spring = 60 – 40 = 20 mm
The load taken by inner spring, $F_2 = 20 \times 64.4 = 1288$ N.

Example 14.5: At the bottom of an elevator shaft, a group of 8 identical springs are set in parallel to absorb the shock of the elevator in case of failure. The elevator weight is 28 kN. Assuming that the elevator has a free fall of 1.2 m from rest, determine the maximum stress in each spring, made from 30 mm diameter rod. For each spring, spring index is 6 and the number of active turns is 15. Neglect any effect of counter-weights in the system. Take $G = 83$ GN/m^2.

Given Data: Elevator weight = 28 kN; Fall of the elevator = 1.2 m
Number of springs = 8; Wire diameter d = 30 mm
Spring index C = 6; Coil diameter $D = 30 \times 6 = 180$ mm

SOLUTION

Step-1: Stiffness for energy storage capacity
Let y be deflection of the spring due to the impact
Then impact energy, $E = (1200 + y) \times 28 \times 10^3$ Nmm
The energy stored in each spring, (Eqn. 10.6 MDHB)

$$U_1 = \frac{y^2 dG}{16NC^3} = \frac{30 \times 83 \times 10^3 y^2}{16 \times 15 \times 6^3} = 48.03y^2$$

For energy balance, $E = 8U_1$

$$\therefore \quad (1200 + y)28 \times 10^3 = 8 \times 48.03y^2$$

$$\therefore \quad y^2 - 72.864y - 87440.96 = 0$$

The roots are

$$y = \frac{72.864}{2} \pm \sqrt{\left(\frac{72.864}{2}\right)^2 + 87440.96} = 36.432 \pm 297.9$$

$$= 334.37 \text{ mm, (the –ve value is not feasible)}$$

Step-2: Check for stress induced

The stiffness of each spring, $K_s = \dfrac{dG}{8C^3N} = \dfrac{30 \times 83 \times 10^3}{8 \times 6^3 \times 15} = 96.065$ N/mm

Dynamic load on each spring,

$$F = K_s \times y = 96.065 \times 334.37 = 32.21 \times 10^3\,\text{N}$$

Wahl factor, $K = \dfrac{4 \times 6 - 1}{4 \times 6 - 4} + \dfrac{0.615}{6} = 1.25$

The stress induced in the spring,

$$\tau = K\frac{8FC}{\pi d^2} = \frac{1.25 \times 8 \times 32.121 \times 10^3 \times 6}{\pi \times 30^2} = 681.6\ \text{MPa} < S_d = 700\ \text{MPa}$$

Hence, design is safe.

Example 14.6: A railway wagon weighing 50 kN and moving with a speed of 8 km/hr has to be stopped by four buffer springs in which the maximum compression allowed is 220 mm. Find the number of turns or coils in each spring of mean diameter 150 mm. The diameter of spring wire is 25 mm. Take $G = 84$ GPa. Also find the shear stress.

Given Data: Moving weight, W = 50 kN = 50,000 N

Speed V = 8 km/hr = 8000/3600 = 2.222 m/sec

Number of springs = 4;

y = 220 mm;

D = 150 mm; d = 25 mm

G = 84 GPa = 84,000 MPa

SOLUTION

Step-1: Axial force on each spring

$$\text{Kinetic energy} = \frac{1}{2}MV^2 = \frac{1}{2}\frac{W}{g}V^2 = \frac{1}{2} \times \frac{50000}{9.810} \times (2.222)^2 = 12584.79\ \text{Nm}$$

$$\therefore \text{Energy stored by each spring} = \frac{\text{Total kinetic energy}}{\text{Number of springs}} = \frac{12584.790}{4} = 3146.1975\ \text{Nm}$$

Also, energy absorbed by each spring $= \dfrac{1}{2}Fy = 3146197.5$ Nmm

For deflection, $y = 220$ mm

$$\therefore \text{Axial force on each spring, } F = \frac{3146197.5 \times 2}{220} = 28601.8\ \text{N}$$

Step-2: Number of turns

Deflection, $y = \dfrac{8FD^3N}{d^4G}$ (Eqn. 10.5 MDHB)

$$N = \frac{yd^4G}{8FD^3} = \frac{220 \times 25^4 \times 84000}{8 \times 28601.8 \times 150^3} = 9.348$$

$\therefore$ Choose number of active turns, $N = 10$
Assume squared and ground end
$\therefore$ Number of additional coils, $n = 2$
$\therefore$ Total number of turns, $N^1 = 10 + 2 = 12$

Step-3: Shear stress

Shear stress, $\tau = \dfrac{8FD}{\pi d^3} K \leq S_d$ (Eqn. 10.1 MDHB)

Spring index, $C = \dfrac{D}{d} = \dfrac{150}{25} = 6$

Stress factor, $K = \dfrac{4C-1}{4C-4} + \dfrac{0.615}{C}$ (Eqn. 10.2 MDHB)

$$= \frac{4 \times 6 - 1}{4 \times 6 - 4} + \frac{0.615}{6} = 1.2525$$

$$\therefore \quad \tau = \frac{8 \times 28601.8 \times 150 \times 1.2525}{\pi \times 25^3} = 875.72 \text{ MPa}$$

Example 14.7: A load of 2000 N is dropped axially on a closed coiled helical spring from a height of 250 mm. The spring has 20 effective turns and it is made of 25 mm diameter wire. Find the maximum shear stress induced in the spring and the amount of compression produced. Take $C = 8$ and $G = 84$ GPa.

Given Data: Weight of falling body, $W = 2000$ N; Height of fall, $h = 250$ mm;

$$G = 84 \text{ GPa} = 84 \times 10^3 \text{ MPa}; \; N = 20; \; d = 25 \text{ mm}; \; C = 8$$

SOLUTION

Step-1: Axial force

Potential energy, $E = (h + y)W = (250 + y)2000$

$= 5 \times 10^5 + 2000y$

Also energy stored by the spring $= \dfrac{1}{2} Fy$

Energy stored = P.E. of falling wt.

$$\therefore \quad \frac{1}{2} Fy = 5 \times 10^5 + 2000y$$

i.e., $Fy = 10^6 + 4000y$ (i)

Deflection, $y = \dfrac{8FD^3N}{d^4G} = \dfrac{8FC^3N}{dG} = \dfrac{8 \times F \times 8^3 \times 20}{25 \times 84000} = 0.039\,\text{F}$

$\therefore$ $y = 0.039\,\text{F}$ (ii)

Substitute in equation (i)

$$0.039\,F^2 = 10^6 + 4000 \times 0.039F;$$

$$\therefore \quad F^2 - 4000F - 25.635 \times 10^6 = 0$$

$$\therefore \quad F = \frac{+4000 \pm \sqrt{4000^2 + 4 \times 1 \times 25.635 \times 10^6}}{2 \times 1} = (4000 \mp 10887.61)/2$$

Taking positive value, axial load on the spring, $F = 7443.8$ N

Step-2: Shear stress

Shear stress, $\tau = \dfrac{8FDK}{\pi d^3}$

Stress factor, $K = \dfrac{4C-1}{4C-4} + \dfrac{0.615}{C} = \dfrac{4 \times 8 - 1}{4 \times 8 - 4} + \dfrac{0.615}{8} = 1.184$

$$\therefore \quad \tau = \frac{8 \times 7443.8 \times 200 \times 1.184}{\pi \times 25^3} = 287.274 \text{ MPa}$$

Step-3: Compression or deflection

From (B) $y = 0.039F = 0.039 \times 7443.8 = 290.3$ mm

Example 14.8: The spring used in an automobile engine has to exert 500 N, when the valve is closed and 600 N, when the valve is opened. The displacement of the valve is 5 mm. The engine crankshaft rotates at 8000 rpm. Design the spring if permissible stress in the material of the spring is 300 MPa. The ratio of mean coil diameter to the wire diameter is 6. The specific weight and the modulus of rigidity of the spring material are 7.35×10^{-5} N/mm^3 and 8×10^4 MPa, respectively. The ends of the spring are squared and ground. Inspect the suitability of the spring for this engine. At what speed of the engine does the spring resonate?

Given Data: Minimum load, $F_{min} = 500$ N;

Maximum load, $F_{max} = 600$ N

$$\Delta y' = 5 \text{ mm}; \quad \frac{D}{d} = C = 6$$

Automobile engine, 4-stroke engine

Crankshaft speed = 8000 rpm

Design stress, $S_{sd} = 300$ MPa

Modulus, $G = 8 \times 10^4$ MPa

Specific weight, $\rho = 7.35 \times 10^{-5}$ N/mm^3

Square and ground end, $N^1 = N + 2$

SOLUTION

Step-1: Deflection of spring

$$\text{Maximum deflection, } y_m = \frac{F_2 \Delta y}{F_{max} - F_{min}} = \frac{600 \times 5}{600 - 500} = 30 \text{ mm}$$

Design the spring for maximum load and maximum deflection

Step-2: Diameter of wire

$$\text{Shear stress, } \tau = \frac{8FDK}{\pi d^3} = \frac{8F_{max} C \cdot K}{\pi d^2} = S_{sd} \quad \text{(Eqn. 10.1 MDHB)}$$

$$\text{Wahl's stress factor, } K = \frac{4C - 1}{4C - 4} + \frac{0.615}{C} = \frac{4 \times 6 - 1}{4 \times 6 - 4} + \frac{0.615}{6} = 1.2525$$

$$\text{Hence, } 300 = \frac{8 \times 600 \times 6 \times 1.2525}{\pi d^2} = \frac{36072}{\pi d^2}$$

$$\therefore \text{ Wire diameter } d = \sqrt{\frac{36072}{300\pi}} = 6.186 \text{ mm}$$

The next standard wire diameter from Table 10.5 is $d = 6.3$ mm

Step-3: Diameters of coil

Mean diameter of coil, $D = C \times d = 6 \times 6.3 = 37.8$ mm

Outer diameter of coil, $D_0 = D + d = 37.8 + 6.3 = 44.1$ mm

Inner diameter of coil, $D_i = D - d = 37.8 - 6.3 = 31.5$ mm

Step-4: Number of coils

$$\text{Maximum deflection, } y_{max} = \frac{8F_{max} D^3 N}{d^4 G} = \frac{8F_{max} C^3 N}{dG}$$

i.e., $$30 = \frac{8 \times 600 \times 6^3 \times N}{6.3 \times 8 \times 10^4}$$

i.e., $$N = 14.583$$

$\therefore$ Choose number of active turns, $N = 15$

Total no. of coils, $N^1 = N + 2 = 15 + 2 = 17$

Step-5: Spring stiffness

Required rate of spring or required stiffness of spring,

$$K = \frac{F_{\max}}{Y_{\max}} = \frac{600}{30} = 20 \text{ N/mm}$$

Actual rate of spring,

$$K_s = \frac{d^4 G}{8D^3 N} = \frac{6.3^4 \times 80000}{8 \times 15 \times 37.8^3} = 19.44 \text{ N/mm} \qquad \text{(Eqn. 10.4 MDHB)}$$

Step-6: Solid length of the sprint,

$$L_d = (N^1 - 0.5)\,d = (17.05) \times 6.3 = 103.95 \text{ mm} \qquad \text{(Eqn. 10.18 MDHB)}$$

Step-7: Free length of the coil,

$$L_0 = L_d + y + c \qquad \text{(Eqn. 10.19 MDHB)}$$

$$L_0 = 103.95 + 30 + 0.25 \times 30 = 141.45 \text{ mm}$$

Pitch of the coil, $p_o = D/3 = 37.8/3 = 12.6$ mm (Eqn. 10.20 MDHB)

The helix angle, $\propto = \tan^1(p_0/\pi D) = \tan^1[12.6/(\pi \times 37.6)] = 6.0566°$

Step-8: Length of unwound wire,

$$L_w = \frac{\pi D N^1}{\cos \propto} = \frac{\pi \times 37.8 \times 17}{\cos (6.0566)} = 2030.12 \text{ mm} \qquad \text{(Eqn. 10.21 MDHB)}$$

Step-9: Check for FOS for buckling load

Buckling load, $F_{cr} = K_s K_L L_0$ (Eqn. 10.22 MDHB)

where, $K_s = 20$ N/mm, $L_0 = 141.45$ mm

For $\frac{L_0}{D} = \frac{141.45}{37.8} = 5.74$; $K_L = 0.65$ (Fig. 10.2 MDHB)

$\therefore$

$$F_{cr} = 20 \times 0.6 \times 141.45 = 1838.85 \text{ N}$$

$$\text{FOS} = \frac{F_{cr}}{F_{\max}} = \frac{1838.85}{600} = 3.065$$

This is satisfactory.

Step-10: Check for the critical speed

Natural frequency, $f = 15.8\sqrt{\frac{K_s}{M_s}}$ (Eqn. 10.65 MDHB) where, $K_s = 20$ N/mm ; $M_s = \dfrac{\pi^2 d^2 DN\rho_m}{4}$

where, $\rho_m = \dfrac{7.35 \times 10^{-5}}{9.81} = 0.75 \times 10^{-5}$

$$\therefore \quad M_s = \frac{\pi^2 \times 6.3^2 \times 37.8 \times 15 \times 0.75 \times 10^{-5}}{4} = 41602.77 \times 10^{-5}$$

$$\therefore \quad f = 15.8\sqrt{\frac{20}{41602.77 \times 10^{-5}}} = 15.8\sqrt{\frac{20 \times 10^5}{41602.77}} = 109.55 \text{ (Hert 2)}$$

Critical speed $N_c = 109.55 \times 60 = 6572.98$ rpm. This is greater than cam speed of 8000/2 = 4000 rpm and hence no resonance occurs.

Example 14.9: Design a helical compression spring required for a spring loaded safety valve, mounted on a pressure vessel. The spring is subjected to an initial compression of 50 mm at the time of assembly and will open by 10 mm when the pressure approaches 6 MPa. The diameter of the valve is 25 mm.

Given Data: Initial compression, y_1 = 50 mm,
Maximum (blow-off) pressure, P_2 = 6 MPa
Diameter of the valve = 25 mm
Deflection at blow-off pressure, y^1 = 10 mm

SOLUTION

Step-1: Max. load on the spring

Maximum deflection, $y_2 = y_1 + y' = 50 + 10 = 60$ mm
Max. load, $F_2 = p_2 \times$ area of cross section of the valve

$$= 6 \times \frac{\pi}{4} \times 25^2 = 2945.243 \text{ N}$$

Design the spring for maximum load and maximum deflection.

Step-2: Diameter of spring wire for shear strength

Assume the spring material as chrome-vanadium alloy steel. SAE6150; from Table 10.9 MDHB, $S_{y_s} = 774$ MPa, $G = 84.5$ GPa.

For FOS = 1.5; design stress, $S_{sd} = \dfrac{S_{y_s}}{FOS} = \dfrac{774}{1.5} = 516$ MPa

Max. shear stress, $S_{sd} = \dfrac{8F_{max}DK}{\pi d^3}$

Stress factor, $K = \frac{4C-1}{4C-4} + \frac{0.615}{C} = \frac{4\times6-1}{4\times6-4} + \frac{0.615}{6} = 1.2525$

Spring index, $C = 6$ (assumed value)

$$\therefore \quad D = 6d$$

$$516 = \frac{8\times2945.243\times6d\times1.2525}{\pi d^3} = \frac{177068}{\pi d^2}$$

$$\therefore \quad d = \sqrt{\frac{177068}{\pi\times516}} = 10.45 \text{ mm}$$

Select the standard diameter of wire from Table 10.5 MDHB

∴ Diameter of spring wire, $d = 11.0$ mm

Step-3: Diameters of coil

Mean diameter of coil, $D = 6 \times 11 = 66$ mm

Outer diameter of coil, $D_o = D + d = 66 + 11 = 77$ mm

Inner diameter of coil, $D_i = D - d = 66 - 11 = 55$ mm

Step-4: Number of turns for stiffness

Maximum deflection, $y_2 = \frac{8F_2D^3N}{d^4G} = \frac{8F_2C^3N}{dG}$ (Eqn. 10.5 MDHB)

$$N = \frac{dG}{8FC^3}y_2 = \frac{11\times845\times10^3\times60}{8\times2945.23\times6^3} = 10.9$$

Choose number of active turns, $N = 12$

The total number of coils, $N^1 = 12 + 2 = 14$

Stiffness of the spring, $K_s = \frac{F}{y} = \frac{2945.243}{60} = 49.087$ N/mm

Step-5: Length of the spring

The solid strength $L_d = (N^1 - 0.5)\cdot d$ (Eqn. 10.18 MDHB)

$L_d = (14.0 - 0.5)\times11 = 1485$ mm

Free length of the spring,

$$L_o = L_d + y + c \quad \text{(Eqn. 10.19 MDHB)}$$

$$= 148.5 + 60 + 0.25\times60 = 223.5 \text{ mm}$$

The pitch of the coils, $p_o = D/3 = 66/3 = 22$ mm

The helix angle, $\alpha = \tan^{-1}\left(\frac{p_o}{\pi D}\right) = \tan^{-1}\left(\frac{22}{\pi\times66}\right) = 6.057°$

Length of unwound wire,

$$L_w = \frac{\pi D N^1}{\cos\alpha} = \frac{\pi \times 66 \times 14}{\cos 6.07} = 2919 \text{ mm} \qquad \text{(Eqn. 10.21 MDHB)}$$

Step-6: Check for FOS for buckling load

The buckling load, $F_{cr} = K_s \cdot K_L \cdot L_o$ (Eqn. 10.22 MDHB)

$$K_L \text{ for } \frac{L_o}{D} = \frac{223.5}{66} = 3.39 \text{ from Fig. 10.2, } K_2 = 0.64$$

$$\therefore \qquad F_{or} = 49.087 \times 0.64 \times 223.5 = 7021.45 \text{ N}$$

$$\text{FOS} = \frac{7021.45}{2945.243} = 2.384.$$

This is satisfactory.

Example 14.10: Design a helical compression spring for a service load ranging from 2250 N to 2750 N. The axial deflection of the spring for the load range is 6 mm. Assume a spring index of 5, permissible shear stress of 420 MPa and modulus of rigidity of 84 kN/mm^2.

Given Data: Maximum load, $F_{max} = 2750$ N;

Minimum load, $F_{min} = 2250$ N;

Deflection, $y' = 6$ mm; Spring index, $C = 5$;

$S_{sd} = 420$ MPa; Modulus, $G = 84 \text{ kN/mm}^2 = 84 \times 10^3 \text{ N/mm}^2$

SOLUTION

Step-1: Max. deflection

$$\text{Maximum deflection, } y_{max} = \frac{y' F_{max}}{F_{max} - F_{min}} = \frac{6 \times 2750}{(2750 - 2250)} = 33 \text{ mm}$$

Design the spring for maximum deflection and maximum load.

Step-2: Diameter of wire

$$\text{Shear stress } = \frac{8FD}{\pi d^3} K \leq S_{sd} \qquad \text{(Eqn. 10.1 MDHB)}$$

$$\text{Stress factor, } K = \frac{4C-1}{4C-4} + \frac{0.615}{C} = \frac{4 \times 5 - 1}{4 \times 5 - 4} + \frac{0.65}{5} = 1.3105$$

Substituting the values,

$$420 = \frac{8 \times 2750 \times 5 \times 1.3105}{\pi d^2}$$

∴ Diameter of spring wire, $d = 10.452$ mm

Select standard diameter of wire from Table 10.5, $d = 10.5$ mm

Step-3: Diameters of coil

Mean diameter of coil, $D = 5d = 5 \times 10.5 = 52.5$ mm

Outer diameter of coil, $D_0 = D + d = 52.5 + 5.25 = 57.75$ mm

Inner diameter of coil, $D_i = D - d = 52.5 - 5.25 = 47.25$ mm

Step-4: Number of coils

Maximum, $y_2 = \dfrac{8F_2 D^3 N}{d^4 G}$ (Eqn. 10.5 MDHB)

i.e., $33 = \dfrac{8 \times 2750 \times 52.5^3 \times N}{10.5^4 \times 84 \times 10^3}$

i.e., $N = 10.584$

∴ Number of active coils, $N = 11$

and total no. of coils, $N^1 = N + 2 = 11 + 2 = 13$

Step-5: Stiffness of the spring

Required stiffness, $K_s' = \dfrac{F_2}{y_2} = \dfrac{2750}{33} = 83.33$ N/mm

Actual stiffness,

$$K_s = \frac{d^4 G}{8D^3 N} = \frac{10.5^4 \times 84000}{8 \times 11 \times 52.5^3} = 80.182\text{N/mm}; \cong 83.33 \text{ N/mm}$$

Nearer to the required value and hence design is satisfactory.

Step-6: The length of the spring

Solid length, $L_d = (N^1 - 0.5) \cdot d$ (Eqn. 10.18 MDHB)

$= (13 - 0.5) \times 10.5 = 131.25$ mm

Free length of the spring,

$$L_0 = L_d + y + c \quad \text{(Eqn. 10.19 MDHB)}$$

$$= 131.25 + 33 + 0.25 \times 33 = 172.5 \text{ mm}$$

The pitch of the coils,

$$p = (L_o - 2d)/N \quad \text{(Eqn. 10.20 MDHB)}$$

$$= \frac{(72.5 - 2 \times 10)}{13} = 13.77 \text{ mm}$$

Helix angle, $\propto = \tan^{-1}(p/\pi D) = \tan^{-1}\left(\frac{13.77}{\pi \times 52.5}\right) = 4.77°$

$\therefore \quad \cos \propto = 0.9965$

Length of unwound wire, $L_w = \frac{\pi D N^1}{\cos x}$ (Eqn. 10.21 MDHB)

$$L_w = (\pi \times 52.5 \times 13)/0.9965 = 2151.6 \text{ mm}$$

Step-7: FOS based on critical load

$$F_{cr} = K_s \cdot K_L \cdot L_o \quad \text{(Eqn. 10.22 MDHB)}$$

$$= 80.182 \times 0.67 \times 172.5 \; (\because K_L = 0.67 \text{ from Fig. 10.2}) = 9267.034$$

$$\text{FOS} = F_{cr}/F_{\max} = \frac{9267.03y}{2750} = 3.37 \text{ satisfactory}$$

Example 14.11: Design a helical spring for a spring loaded safety valve for the following data: Operating pressure = 1 MPa; Maximum pressure when the valve blows off = 1.075 MPa. Maximum lift of valve (when pressure is 1.075 MPa) = 6 mm. Diameter of valve seat = 100 mm. Maximum allowable shear stress = 0.4 GPa. Rigidity modulus = 86 GPa. Spring index = 5.5. Assume this to be a Rams bottom safety valve.

Given Data: $P_{\min} = 1$ MPa; $P_{\max} = 1.075$ MPa; $y' = 6$ mm

Diameter of valve seat = 100 mm; $S_{sd} = 0.4$ GPa = 400 MPa;

$$G = 86 \text{ GPa} = 86000 \text{ MPa}; \; C = 5.5$$

SOLUTION

Step-1: Max. force and deflection

As the spring is used in rams bottom safety valve, it is under tension.

Minimum force on the spring,

$$F_{\min} = \text{Area of valve} \times p_{\min} = \frac{\pi}{4} \times 100^2 \times 1 = 7853.98 \text{ N}$$

Maximum force on the spring,

$$F_{\max} = \text{Area of valve} \times p_{\max} = \frac{\pi}{4} \times 100^2 \times 1.075 = 8443.033 \text{ N}$$

Maximum deflection,

$$y_{\max} = \frac{y' F_{\max}}{F_{\max} - F_{\min}} = \frac{6 \times 8443.03}{8443.03 - 7853.98} = 85.999 = 86 \text{ mm}$$

Design the spring for maximum load and maximum deflection.

Step-2: Diameter of wire

Shear stress, $\tau = \dfrac{8F_{\max}D}{\pi d^3}K = \dfrac{8F_{\max}C}{\pi d^2}K$

Spring index, $C = 5.5$

Stress factor, $K = \dfrac{4C-1}{4C-4} + \dfrac{0.615}{c} = \dfrac{4\times5.5-1}{4\times5.5-4} + \dfrac{0.615}{5.5} = 1.2785$

$$\therefore \qquad 400 = \frac{8\times8443.03\times5.5\times1.2785}{\pi\times d^2}$$

$$\therefore \qquad d = 19.44 \text{ mm}$$

Select standard diameter of wire from Table 10.5 MDHB

$$\therefore \qquad d = 20 \text{ mm}$$

Step-3: Diameters of coil

Mean diameter of coil, $D = 5.5\,d = 5.5 \times 20 = 110$ mm

Outer diameter of coil, $D_o = D + d = 110 + 20 = 130$ mm

Inner diameter of coil, $D_i = \text{D} - d = 110 - 20 = 90$ mm

Step-4: Number of coils and stiffness

Maximum deflection, $y_2 = \dfrac{8F_{\max}D^3N}{d^4G}$

$$\therefore \qquad 86 = \frac{8\times8443.03\times110^3\times N}{20^4\times86000}$$

i.e., $\qquad N = 13.16$

$\therefore$ Number of active turns, $N = 14$

Total number, $N^1 = N+2 = 14+2 = 16$

Actual stiffness,

$$K_s = \frac{d^4G}{8D^3N} = \frac{dG}{8C^3N} = \frac{20\times86000}{8\times5.5^3\times14} = 92.3 \text{ N/mm}$$

Step-5: Free length

$\therefore$ Free length of spring recommended,

$$L_o = Nd + 1.1D = 14\times20 + 1.1\times110 = 401 \text{ mm}$$

Step-6: Pitch of the coils,

$$p = \frac{L_0}{N-1} = \frac{401}{14-1} = 30.85 \text{ mm}$$

Step-7: Length of unwound wire,

$$L_w = \frac{\pi D N^1}{\cos\alpha} + \text{hook length}$$

Here, $\propto = \tan^{-1}(p/\pi D) = \tan^{-1}\left(\frac{30.85}{\pi \times 110}\right) = 5.1°$

$$\therefore \quad L_w = \frac{\pi \times 110 \times 16}{\cos(5.1°)} + 2 \times 110 = 5771.18 \text{ mm}$$

14.4.6 Cluster Springs

The cluster springs are formed by placing one spring inside another spring with following conditions:

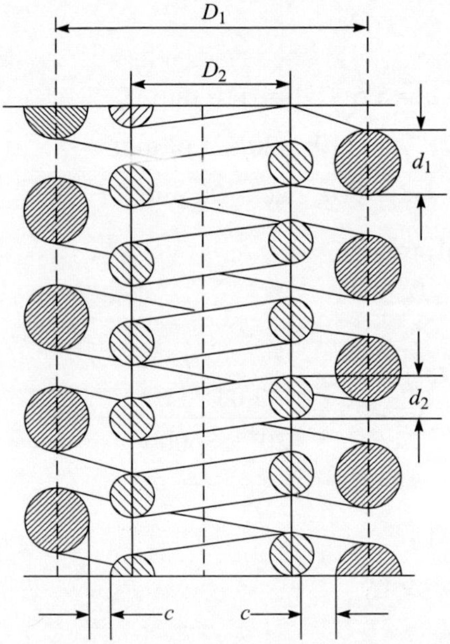

Figure 14.11 Concentric springs

(i) The spring index is the same,

$$D_1/d_1 = D_2/d_2 = C$$

(ii) The material is the same for both springs and the stress induced is the same,

$$F_1/F_2 = (d_1/d_2)^2$$

(iii) Amount of deflection of the spring is the same,

$$d_1 N_1 = d_2 N_2$$

(iv) The solid lengths of the springs is the same,

$$d_1 N_1 = d_2 N_2$$

(v) The wire diameters are related by

$$d_1/d_2 = C/(C-2)$$

Example 14.12: A cluster spring is made up of two springs for a valve gear mechanism. The material of the spring wires is the same. Using the following particulars, estimate the dimensions of the springs:

Maximum load	9750 N
Permissible shear stress	600 MPa
Maximum deflection	45 mm
Spring index	6
Modulus of rigidity	83×10^3 MPa

Adopt standard clearance between two coils and make allowance for the wire curvature.

Given Data: Load applied, $P = 9750$ N; Deflection, $y = 45$ mm;

Spring index = 6; Permissible stress, S_{sd} = 600 MPa;

Modulus of rigidity, $G = 83\times10^3$ MPa

SOLUTION

Step-1: Stiffnesses of springs

The combined stiffness required,

$$K_s = \frac{P}{y} = \frac{9750}{45} = 216.67 \text{ N/mm}$$

$$\therefore \qquad K_{s1} + K_{s2} = 216.67 \qquad (1)$$

The ratio of stiffnesses,

$$K_{s1}/K_{s2} = \frac{d_1/N_1}{d_2/N_2} = \left(\frac{d_1}{d_2}\right)^2$$

But $d_1/d_2 = C/C-2 = 6/4 = 1.5$

$$\therefore \qquad (K_{s1}/K_{s2}) = (1.5)^2 = 2.25 \qquad (2)$$

Solving (1) and (2), $K_{s2} = 216.67/3.25 = 66.67$ N/mm,

$$\therefore \qquad K_{s1} = 2.25 K_{s2} = 150 \text{ N/mm}$$

Step-2: Diameter of the spring wire for strength

The load taken by outer spring,

$$F_1 = y\cdot K_{s1} = 45\times150 = 6750 \text{ N}$$

Stress factor, $K = \dfrac{4C-1}{4C-4} + \dfrac{0.615}{C} = \dfrac{4\times 6-1}{4\times 6-4} + \dfrac{0.615}{6} = 1.25$

Then the dia. of the spring wire

$$d_1 = \sqrt{\frac{8F_1C\cdot K_1}{\pi S_{sd}}} = \sqrt{\frac{8\times 6750\times 6\times 1.25}{\pi\times 600}} = 14.65 \cong 15 \text{ mm.}$$

$\therefore \quad d_2 = d_1/1.5 \cong 10$ mm

Step-3: Number of coils

The number of coils in the outer spring,

$$N_1 = \frac{d_1 G}{8C^3 K_{s1}} = \frac{15\times 83\times 10^3}{8\times 6^3\times 150} = 4.8 = 5 \text{ (say)}$$

Then number of coils in the inner spring,

$$N_2 = 5\times 1.5 = 7.5$$

14.5 TORSIONAL SPRING

It is made from the round or rectangular wire in the same fashion as the tension spring. The ends are extended tangentially and may be bent to form a hook to facilitate the application of load. The loads act perpendicular to the extended arms and cause bending moment on the wire, given, by $M = F\cdot r$

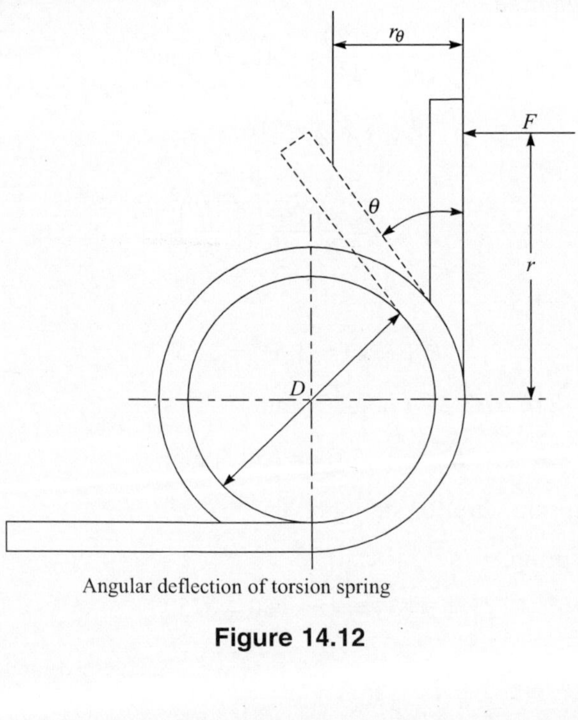

Angular deflection of torsion spring

Figure 14.12

where r is the distance of the load from the coil's axis

The bending stress,

$$\sigma = \frac{32M}{\pi d^3} K_1 \text{ for round wire} \tag{14.16a}$$

$$= \frac{6M}{bt^2} K_2 \text{ for rectangular wire} \tag{14.16b}$$

where, K_1, K_2 are Wahl's correction factors,

$$K_1 = \frac{4C^2 - C - 1}{4C(C-1)} \tag{14.16c}$$

$$K_2 = \frac{3C^2 - C - 0.8}{3C(C-1)} \tag{14.16d}$$

The angular deflection of the end, $\theta = \dfrac{ML}{EI}$

$$= \frac{64 \cdot F \cdot r \cdot D \cdot N}{Ed^4} \text{ in rad for round wire} \tag{14.17a}$$

$$= 3665 \frac{F \cdot D \cdot r \cdot N}{Ed^4} \text{ in deg for round wire} \tag{14.17b}$$

$$= \frac{12\pi F \cdot D \cdot r \cdot N}{Ebt^3} \text{ in rad for rectangular wire} \tag{14.17c}$$

$$= \frac{2160FDrN}{Ebt^3} \text{ in deg for rectangular wire} \tag{14.17d}$$

Linear Deflection

$$Y = \theta \cdot r = \frac{64F \cdot D \cdot r^2 N}{Ed^4} \text{ for round wire} \tag{14.18a}$$

$$= \frac{12\pi FDr^2 N}{Ebt^3} \text{ for rectangular wire} \tag{14.18b}$$

Example 14.13: A torsion spring is wound from a round wire into a coil with mean diameter 50 mm. The torque applied on the spring is 7000 N. mm. Assume the spring index as 10. The permissible stress in the spring is 600 MPa and $E = 0.2 \times 10^6$ MPa. What must be the diameter of the wire and the corresponding deflection in degrees? Take the number of effective coils as 6.5.

Given Data: Torque applied, $M = 7000$ Nmm

Permissible stress $S_d = 600$ MPa

Number of active coils $N = 6.5$

SOLUTION Let spring index, $C = 10$

$$K_1 = \frac{4\times10^2 - 10 - 1}{4\times10\times(10-1)} = 1.08$$

Then stress, $S = \frac{32\cdot M}{\pi d^3} K_1$

$$\therefore \qquad d^3 = \frac{32\times700\times1.08}{\pi\times600} = 127.8$$

$$\therefore \qquad d = 5.029 = 5.0 \text{ mm}$$

Coil dia. $D = 5\times10 = 50$ mm as given

The angular deflection, $\theta = 3665\frac{F\cdot r\cdot D\cdot N}{Ed^4}$ with $(F\cdot r = M)$

$$= \frac{3665\times7000\times50\times6.5}{0.2\times10^6\times5^4} = 66.7°$$

14.6 SPIRAL SPRING

Normally, a spring wire or strip is wound in the form of a spiral with inner end fixed to a spindle at the centre and outer end is fixed to a spindle at the centre and the outer end is fixed to a fixed frame. The spindle is rotated to wind the spring by applying a torque, T.

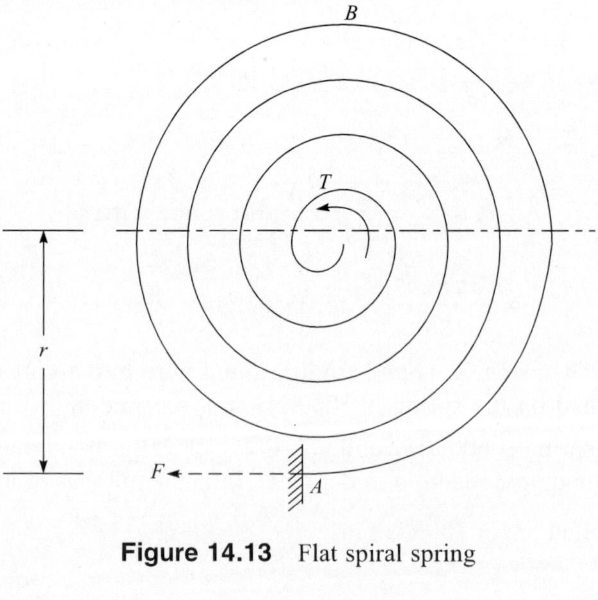

Figure 14.13 Flat spiral spring

The bending stress induced in the strip at A,

$$\sigma_A = \frac{32T}{\pi d^3} K_1 = \frac{32F \cdot r}{\pi d^3} K_1 \text{ for round wire} \tag{14.19a}$$

$$= \frac{6T}{bt^2} K_2 = \frac{6Fr}{bt^2} K_2 \text{ for rectangular strip} \tag{14.19b}$$

The stress at the far end B is double of this value.

where, K_1, K_2 are Wahl's correction factors (Fig. 14.17)

OR refer equations (14.16c) and (14.16d)

The angular twist, $\theta = \dfrac{ML}{EI}$

For round wire, $\theta = \dfrac{64ML}{\pi Ed^4}$ in radians (14.20a)

$$= \frac{1168FL \cdot r}{Ed^4} \text{ in degrees} \tag{14.20b}$$

For rectangular wire, (Eqn. 10.41 and 10.42 MDHB)

$$\theta = \frac{12ML}{Edt^4} = \frac{12F \cdot r \cdot L}{Edt^3} \text{ in rad} \tag{14.20c}$$

$$= \frac{687 \cdot 6F \cdot L \cdot r}{Edt^3} \text{ in deg} \tag{14.20d}$$

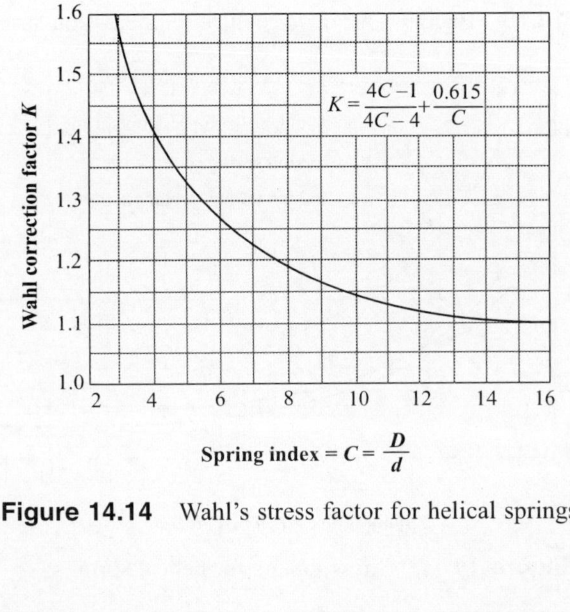

Figure 14.14 Wahl's stress factor for helical springs

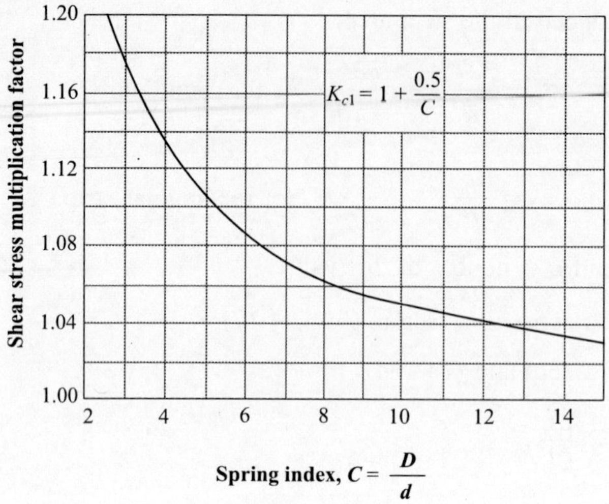

Figure 14.15 Stress concentration factor K_{c1} for shear

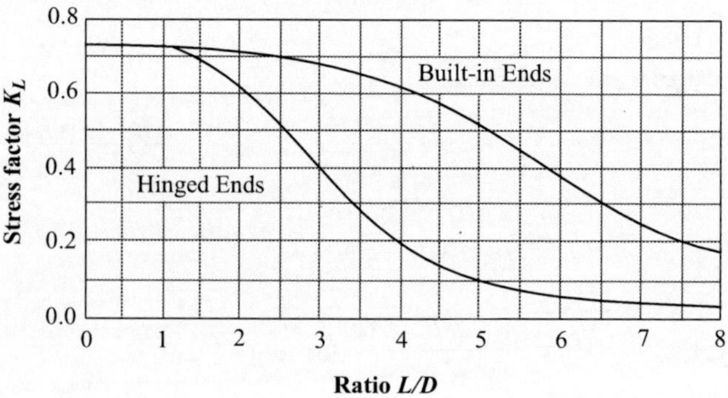

Figure 14.16 Buckling factor for helical (compression spring)

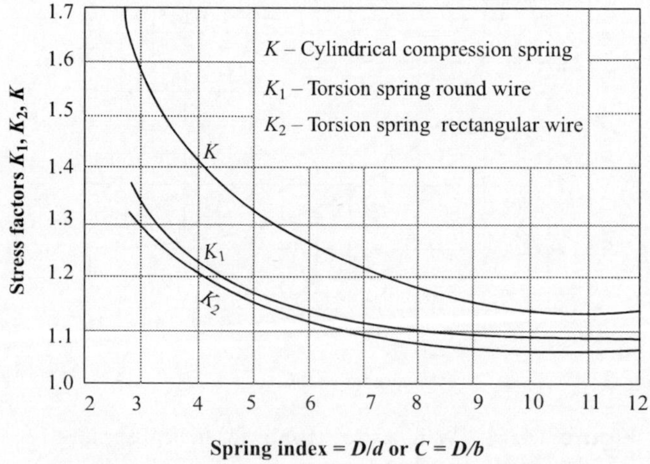

Figure 14.17 Stress factors for helical springs

Example 4.14: A flat spiral spring with rectangular section 0.5 mm × 10 mm is 2000 mm long. One end of it is attached to a small spindle and the other to a fixed point. If the spring is in a state of case of run down, find (a) turning moment on spindle required to give 4 complete turns, (b) strain energy stored, and (c) maximum stress induced. Take $E = 0.2 \times 10^6$ N/mm^2.

SOLUTION Angular twist required, $\theta = 8\pi = 4 \times 360° = 1440°$

Then $1440 = \dfrac{687.6ML}{Ebt^3}$

$$\therefore \qquad M = \frac{1440 \times 0.2 \times 10^6 \times 10 \times 0.5^3}{687.6 \times 2000} = 261.8 \text{ Nmm}$$

The strain energy, $U = M\theta/2 = 261.8 \times 8\pi/2 = 3289.6$ Nmm

The stress at A,

$$\sigma_A = \frac{6.M}{bt^2} = \frac{6 \times 261.8}{10 \times 0.5^2} = 628.3 \text{ MPa}$$

$$\sigma_B = 2 \times \sigma_A = 1256.4 \text{ MPa}$$

14.7 LEAF SPRINGS

The leaf springs find applications in automobiles and railways. The leaf springs are made by placing rectangular flats one above the other and bolting or clamping them together. They are normally made in semielliptical shape. The stress and deflections are analysed on the cantilever beam theory.

Rectangular Plate as Cantilever Beam

Bending stress, $\sigma_b = \dfrac{6M}{Bt^2} = \dfrac{6FL}{Bt^2}$ (14.21a)

If the plate is split in N_f strips and they are placed one above the other and clamped, then also, the stress induced is given by the same equation with $B = N_f \cdot b$

$$\sigma_b = \frac{6 \cdot FL}{N_f \cdot b \cdot t^2} \qquad (14.21b)$$

The deflection,

$$y = \frac{FL^3}{3EI} = \frac{FL^3}{3 \cdot E(Bt^3/12)} = \frac{4FL^3}{EN_f bt^3} \qquad (14.22)$$

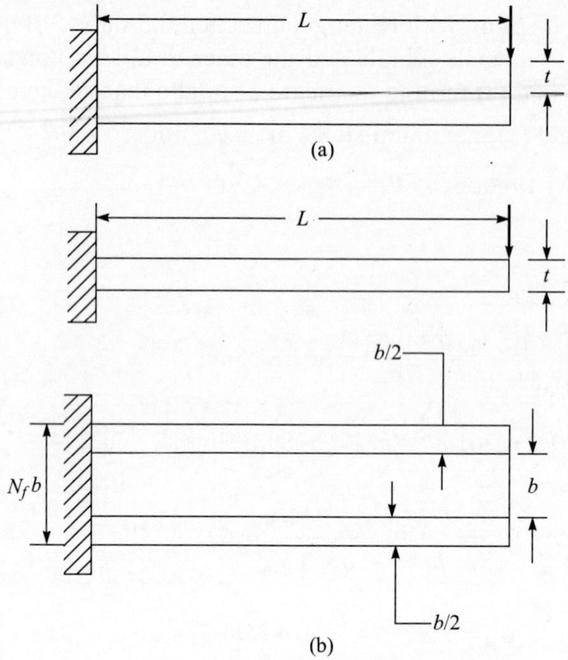

Figure 14.18 Extra full-length leaves as rectangular plate

Triangular Plate as Cantilever Beam

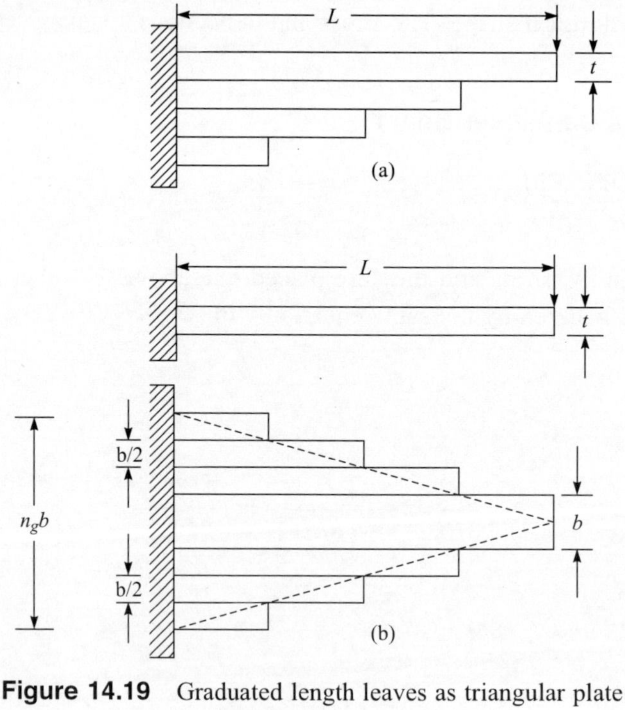

Figure 14.19 Graduated length leaves as triangular plate

The bending stress,

$$\sigma_b = \frac{6 \cdot M}{Bt^2} = \frac{6 \cdot FL}{N_g bt^3} \text{ with } (B = N_g \cdot b) \tag{14.23}$$

The deflection,

$$y = \frac{FL^3}{2 \cdot EI} = \frac{6 \cdot FL^3}{EBt^3} = \frac{6 \cdot FL^3}{EN_g \cdot bt^3} \tag{14.24}$$

The triangular plate is cut in widths of b, and $b/2$ and put together as shown in the last figure.

The triangular plate is more flexible than a rectangular plate but for load application, flat end is more convenient. Hence, two leaves of full length, remaining graduated leaves are clamped together to form a semielliptical leaf spring shown in Fig. 14.20.

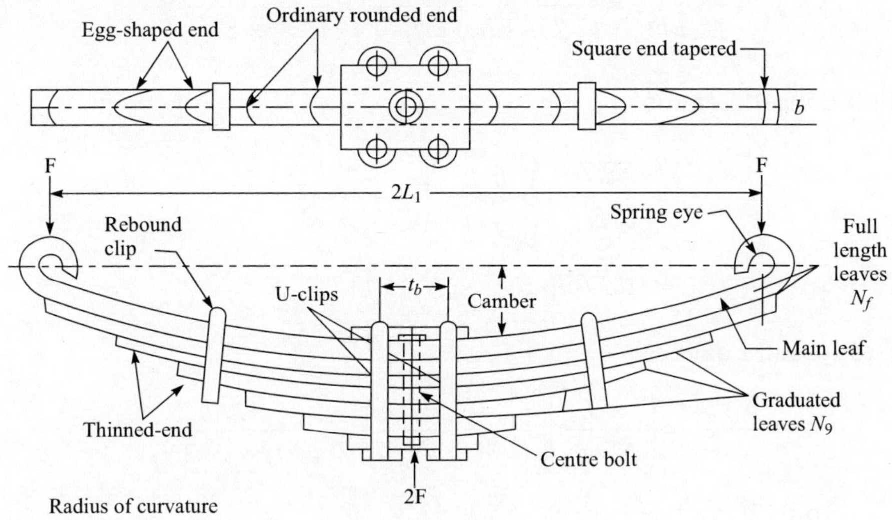

Figure 14.20

Semielliptical Leaf Spring

Radius of curvature $R = (L_1^2/2y)$

The load applied is shared between the full length leaves and graduated leaves such that the deflection is same. The relation is:

$$y = \left(\frac{4F_f \cdot L^3}{N_f \; Ebt^3}\right) = \left(\frac{6F_g \cdot L^3}{N_g \; Ebt^3}\right)$$

$$\therefore \quad \frac{F_f}{N_f} = \frac{3}{2}(F_g/N_g) \tag{14.25}$$

The full load, $F = F_f + F_g = \left(\frac{3}{2}\right)\left(\frac{N_f}{N_g} \cdot F_g\right) + F_g = \frac{(3N_f + 2N_g)}{2N_g} F_g$

$$\therefore \quad F_g = \frac{2N_g F}{(2N_g + 3N_f)} \tag{14.26}$$

$$\text{And } F_f = \left(\frac{3}{2}\right)\left(\frac{N_f}{N_g}\right)\frac{2N_g \cdot F}{(2N_g + 3N_f)} = \frac{3N_f F}{(2N_g + 3N_f)} \tag{14.27}$$

The deflection,

$$y = y_f = \frac{4F_f \cdot L^3}{N_f Ebt^3} = \frac{12N_f FL^3}{(2N_g + 3N_f)\, EN_f\, bt^3} = \frac{12FL^3}{(2N_g + 3N_f)\, Ebt^3} \tag{14.28}$$

Stress in Full Length Leaves

$$\sigma_f = \frac{6F_f L}{N_f bt^2} = \left(\frac{6 \cdot L}{N_f bt^2}\right)\left(\frac{3N_f F}{(2N_g + 3N_f)}\right)$$

$$= (18FL)/[(2N_g + 3N_f)bt^2] \tag{14.29}$$

Stress in Graduated Leaves

$$\sigma_g = \frac{6 \cdot F_g L}{N_g bt^2} = \left(\frac{6 \cdot L}{N_f bt^2}\right)\left(\frac{2N_f F}{(2N_g + 3N_f)}\right) = \frac{12FL}{bt^2(2N_g + 3N_f)} \tag{14.30}$$

Then, $\sigma_g = 2/3\sigma_f$

$$\therefore \quad \sigma_f = 3/2\sigma_g \tag{14.31}$$

Stress equalized springs: The stress in full length leaves is 50% more. For optimum utility, all the leaves should be stressed equally

i.e., $\sigma_f = \sigma_g$

$$\frac{6 \cdot F_f L}{N_f bt^2} = \frac{6 \cdot F_g L}{N_g bt^2}$$

$$\therefore \quad \frac{F_f}{N_f} = \frac{F_g}{N_g} = \frac{F_f + F_g}{N_f + N_g} = \frac{F}{N} \tag{14.32}$$

Then the stress induced in all the leaves,

$$\sigma = \frac{6FL}{Nbt^2} \tag{14.33}$$

The difference in deflections,

$$\delta = \frac{6 \cdot F_g L}{N_g \, Ebt^3} - \frac{4 \cdot F_f L^3}{N_f \, Ebt^3} = \left(\frac{2FL^3}{NEbt^3} \right) \tag{14.34}$$

This difference is provided initially by giving greater curvature to the graduated leaves. The gap is closed by bolting and clamping the leaves during the assembly operation. This puts the full length leaves under compression and graduated leaves in tension initially. When the external load is applied, the initial compression in full length leaves is to be overcome first, then tension starts. The final tensions in the two sets become equal at full load.

The bolt force F_b required to close the gap is found as follows,

δ = the initial gap

= deflection of full length leaves + deflection of graduated leaves

$$\delta = \left(\frac{2FL^3}{NEbt^3} \right) = \left(\frac{4L^3}{N_f \, bt^3} \right) \left(\frac{F_b}{2} \right) + \left(\frac{6L^3}{N_g \, Ebt^3} \cdot \frac{F_b}{2} \right)$$

$$\therefore \qquad F_b = \left[\frac{2N_g \, N_f \cdot F}{N(2N_g + 3N_f)} \right] \tag{14.35}$$

The leaves are assembled to have a radius of curvature, which provides a camber equal to the final deflection.

Then radius, $R = \dfrac{L_1^2}{2y}$ by geometry (14.36)

Example 14.15: A laminated semielliptic leaf spring under a central load of 12 kN is to have an effective length of 1 m and is to deflect not more than 75 mm. The spring has 10 leaves, two of which are full length, and have been pre-stressed so that all leaves have the same stress after full load has been applied. All leaves have the same width and thickness. The maximum stress in the leaves is not to exceed 350 MN/m^2. Calculate the width and thickness of the leaves, the bolt load and the radius of curvature.

Given Data: Central load, $2F$ = 12 kN

Effective length, $2L$ = 1000 mm

Number of full length leaves, $N_f = 2$

Number of graduated leaves, $N_g = 8$

Permissible stress, S = 350 MPa

Maximum deflection, y = 75 mm

SOLUTION

Step-1: The dimensions of the plates

The stress induced (assuming stress equalized springs),

$$S = \frac{6FL}{Nbt^2}$$

$$bt^2 = \frac{6FL}{NS} = \frac{6 \times 6000 \times 500}{10 \times 350} = 5142.8 \quad \text{(i)}$$

The deflection, $y = \dfrac{12FL^3}{bt^3 \cdot E(2N_g + 3N_f)}$

$$bt^3 = \frac{12 \times 600 \times 500^3}{75 \times 0.2 \times 10^6 \times (2 \times 8 + 3 \times 2)} = 27272.72 \quad \text{(ii)}$$

Dividing (ii) by (i)

$\text{(ii)/(i)} = t = \dfrac{27272.72}{51428} = 5.3$

Select standard size, $t = 6.00$ mm

Then, $b = \dfrac{5142.8}{6^2} = 142.8$ mm

Select standard size, $b = 150.0$ mm

The size of the flat selected is $b \times t = 150 \times 6 \text{ mm}^2$

Step-2: Bolt load

The bolt load to close the gap,

$$F_b = \frac{2N_g \cdot N_f \cdot F}{N(2N_g + 3N_f)} = \frac{2 \times 8 \times 2 \times 600}{10 \times 22} = 872.7 \text{ N}$$

Step-3: Spring shape

The radius of curvature, $R = L_1^2/2Y$

Assuming an ineffective length (Steel band) of 100 mm at the center,

$$2L_1 = 1000 + 100 = 1100 \text{ mm}$$

$$L_1 = 55 \text{ mm and } y = 75 \text{ mm}$$

$$R = \frac{(550)^2}{2 \times 75} = 2016.7 \text{ mm}$$

Example 14.16: In a semielliptical laminated spring, the given effective length is 900 mm, the total load is 3600 N, the maximum deflection is 75 mm and the maximum stress due to bending is 360 N/mm^2. Decide the number of leaves and their breadth and thickness. Take $E = 0.2 \times 10^6$ N/mm^2 for the material of the spring.

Given Data: Effective length, $2L = 900$ mm; Central load, $2F = 3600$ N; Deflection, $y = 75$ mm; Permissible stress, $S = 360$ MPa

SOLUTION

Step-1: Plate dimensions-choice I

Choose $N = 10,\ N_f = 2N_g = 8$

From equation for stress,

$$bt^2 = \frac{6FL}{N \cdot S} = \frac{6 \times 1800 \times 450}{10 \times 360} = 1350 \qquad \text{(i)}$$

From deflection equation,

$$bt^3 = \frac{12FL^3}{y \times E(2N_g + 3N_f)}$$

$$= \frac{12 \times 1800 \times 450^3}{75 \times 0.2 \times 10^6 \times (16+6)} = 5964.545 \qquad \text{(ii)}$$

Dividing (ii)/(i) $= t = \dfrac{5964.545}{1350} = 4.4$ mm, choose $t = 4.5$ mm

Then $b = \dfrac{1350}{4.5^2} = 66.67$ mm, choose $b = 70$ mm

Then $(b/t) = 70/4.5 = 15.5$
It is higher.

Step-2: Plate dimensions-choice II

Adopt $N = 12,\ N_f = 2$ and $N_g = 10$

Then

$$bt^2 = (1350 \times 10)/12 = 1125 \qquad \text{(iii)}$$

And $bt^3 = 5964.545 \times (22/26) = 5046.9$ (iv)

Then (iv)/(iii) $= t = \dfrac{5046.9}{1125} = 4.48$, choose $t = 4.5$ mm

Then, $b = \dfrac{1125}{4} \cdot 5^2 = 55.55$ mm, choose $b = 55$ mm

Then $b/t = 55/4.5 = 12.2$

The ratio is acceptable.

Example 14.17: A semielliptical spring with effective length of 1000 mm and central load of 4.5 kN is to be designed to store an impact energy of 165 kNmm. For max. permissible stress of 350 MPa, find the dimensions of the leaves.

Given Data: $2L = 1000$ mm; $2F = 4500$ N;

Energy stored, $U = 165$ kNmm; Max. stress, $S = 350$ MPa;

Modulus, $E = 2.1 \times 10^5$ MPa

SOLUTION From the energy balance

$$U = \frac{PY}{2}, \text{ i.e. } Y = \frac{2U}{P} = \frac{2 \times 165000}{4500} = 73.3 \text{ mm}$$

Choose $N = 12,\ N_f = 2,\ N_g = 10$

Therefore, $2N_g + 3N_f = 26$

Then from stress equation,

$$bt^2 = \frac{6 \times (4500/2) \times (1000/2)}{12 \times 350} = 1607.2 \tag{1}$$

From deflection equation,

$$bt^3 = \frac{12 \times (4500/2) \times 500^3}{73.3 \times 2.1 \times 10^5 \times 26} = 8429.4 \tag{2}$$

Then $t = (2)/(1) = \dfrac{8429.4}{1607.2} = 5.24$ mm, choose $t = 6$ mm

Then $b = 1607.2/6^2 = 44.64$ mm, choose $b = 45$ mm

The ratio $b/t = 45/6 = 7.5$

This is acceptable.

Example 14.18: A locomotive spring has an overall length of 1100 mm and sustains a load of 75 kN at its centre. The spring has 3 full length leaves and 15 graduated leaves with a central band of 100 mm. All the leaves are to be stressed at 0.4 GPa when fully loaded. The ratio of total spring depth to width is 2. Determine: (i) width and thickness of leaves, (ii) initial space that must be provided between full length and graduated leaves before the band is applied, and (iii) load exerted on the band after the spring is assembled.

Given Data: $2F = 75$ kN $= 75000$ N $\therefore F = 37500$ N; $N_f = 3$; $N_g = 15$;

$l_b = 100$ mm; $\sigma = 0.4$ GPa $= 400$ MPa; $2l_1 = 1100$ mm

SOLUTION Assume $E = 206.8 \times 10^3$ MPa

Effective length, $L = \dfrac{2L_1 - l_b}{2} = \dfrac{1100 - 100}{2} = 500$ mm

Total number of leaves, $N = N_f + N_g = 3 + 5 = 18$

Step-1: Plate dimensions

Width and thickness of leaves

Maximum equalized stress, $\sigma = \dfrac{6FL}{Nbt^2}$ (Eqn. 10.47 MDHB)

where, $$F = 37500 \text{ N};\ L = \frac{1100 - 100}{2} = 500 \text{ mm}$$

$$b = \frac{18t}{2} = 9t \text{ and } \sigma = 400 \text{ MPa}$$

i.e., $$400 = \frac{6 \times 37500 \times 500}{18 \times 9t \times t^2}$$

i.e., $$t^3 = \frac{6 \times 37500 \times 500}{18 \times 9 \times 400} = 1736.11$$

$\therefore$ Thickness of each spring leaf, $t = 12.02 \cong 12$ mm

Width of spring leaf, $b = 9t = 9 \times 12 = 108 \text{ mm} \cong 110 \text{ mm}$

Step-2: Initial gap

Initial gap or nipping between full length and graduated leaves,

$$\delta = \frac{2F \cdot L^3}{Nbt^3E} = \frac{2 \times 37500 \times 500^3}{18 \times 206.8 \times 10^3 \times 110 \times 12^3} = 13.25 \text{ mm} \quad \text{(Eqn. 10.45 MDHB)}$$

Step-3: Load on the band

$$\text{Clamping load, } F_b = \frac{2FN_gN_f}{N(2N_g + 3N_f)} = \frac{2 \times 37500 \times 15 \times 3}{18[2 \times 15 + 3 \times 3]} = 4807.7 \text{ N} \quad \text{(Eqn. 10.46 MDHB)}$$

Example 14.19: A truck spring has 12 leaves, two of which are full length leaves. The spring supports are 1.05 m apart and the central band is 85 mm wide. The central load is to be 5400 N with a permissible stress of 0.28 GPa. The ratio of total depth to width of spring is 3 and modulus of elasticity 210 GPa. Determine: (i) thickness and width of steel spring, and (ii) maximum deflection.

Given Data: The depth to width ratio, $Nt/b = 3$

Number of leaves, $N = 12$; Full length leaves, $N_f = 2$

Graduated, $N_g = 10$; Central load, $2F = 5400$ N

Total span, $2L = 1050$ mm; Central band, $l_b = 85$ mm

Design stress, $S_{sd} = 280$ MPa and Modulus, $E = 210$ GPa

SOLUTION

Step-1: Plate dimensions

Load, $F = 5400/2 = 2700$ N

Length, $L = \dfrac{2L_1 - l_b}{2} = \dfrac{1050 - 85}{2} = 482.5$ mm

And width, $b = 12t/3 = 4t$

Maximum stress induced

$$\sigma_f = \frac{18FL}{bt^2(2N_g + 3N_f)} \qquad \text{(Eqn. 10.43 MDHB)}$$

$$= \frac{18 \times 2700 \times 482.5}{4t^3(2 \times 10 + 3 \times 2)} = 280$$

$$\therefore \quad t^3 = \frac{18 \times 2700 \times 482.5}{4 \times 280 \times (20 + 6)} = 805.27$$

$$\therefore \quad t = 9.3 \text{ mm}$$

Choose standard value, $t = 10$ mm

Then width, $b = 4t = 4 \times 10 = 40$ mm

Step-2: Deflection of the spring

$$y = \frac{12FL^3}{bt^3E(2N_g + 3N_f)} \qquad \text{(Eqn. 10.44 MDHB)}$$

$$= \frac{12 \times 2700 \times (482.5)^3}{40 \times 10^3 \times 210 \times 10^3 \times (2 \times 10 + 3 \times 2)} = 16.664 \text{ mm}$$

Example 14.20: A multi-leaf spring with camber is fitted to the chassis of an automobile over a span of 1.2 m to absorb shocks due to a maximum load of 20 kN. The spring material can sustain a maximum stress of 0.4 GPa. All the leaves of the spring were to receive the same stress. The spring is required for at least 2 full-length leaves out of 8 leaves. The leaves are assembled with bolts over a span of 150 mm width at the middle. Design the spring for a maximum deflection of 50 mm.

Given Data: Central load, $2F = 20$ kN

Total span, $2L_1 = 1.2 \text{ m} = 1200$ mm

Maximum stress, $\sigma_{max} = S_d = 0.4 \text{ GPa} = 400$ MPa

Number of leaves, $N = 8$, $N_g = 6$, $N_f = 2$;

Central band, $l_b = 150$ mm

Maximum deflection, $y_{max} = 50$ mm

SOLUTION

Step-1: Plate dimensions
For stress equalized leaf-spring,
The stress induced,

$$\sigma = \frac{6FL}{Nbt^2} \qquad \text{(Eqn. 10.47 MDHB)}$$

where, $F = 20{,}000/2 = 10{,}000$ N; $L = \left(\frac{2l_1 - l_b}{2}\right) = \frac{1200 - 150}{2} = 525$ mm

$N = 8$ and $\sigma = 400$ MPa

$$\therefore \quad 400 = \frac{6 \times 10{,}000 \times 525}{8 \times bt^2}$$

$$\therefore \quad bt^2 = \frac{6 \times 525 \times 10}{8 \times 400} = 9843.75 \qquad (1)$$

Deflection of the spring,

$$y = \frac{12FL^3}{bt^3 E(2N_g + 3N_f)}$$

$$\therefore \quad 50 = \frac{12 \times 10{,}000 \times (525)^3}{bt^3 \times 210 \times 10^3 \times 18}$$

$$bt^3 = \frac{12 \times 10{,}000 \times (525)^3}{50 \times 210 \times 10^3 \times 18} = 91875 \qquad (2)$$

By dividing (2) by (1)

$$\frac{bt^3}{bt^2} = t = \frac{91875}{9843.75} = 9.33 \cong 10 \text{ mm}$$

Then the width, $b = \frac{9843.75}{(10)^2} = 98.4375 \text{ mm} \cong 100 \text{ mm}$

Choose the standard sizes from Table 10.7 MDHB

$t = 10$ mm and $b = 100$ mm

Step-2: Initial gap

Camber, $\delta = \frac{2FL^3}{Nbt^3E} = \frac{2 \times 10000 \times 525^3}{8 \times 100 \times 10 \times 210 \times 10^3} = 16.45$ mm (Eqn. 10.45 MDHB)

Step-3: Load on the central band,

$$F_b = \frac{2N_g N_f F}{N(2N_g + 3N_f)} = \frac{2 \times 10000 \times 6 \times 2}{8(2 \times 6 + 3 \times 2)} = 1666.667 \text{ N}$$

Example 14.21: A semielliptical laminated spring has effective length of 1 m. The spring has to sustain a load of 75 kN. The spring has 3 full length leaves and 16 graduated leaves. If the leaves are pre-stressed such that the stress induced in all the leaves is same and is limited to 400 MPa, when maximum load is applied. The width of the leaves is 9 times the thickness. Assume E = 200 GPa. Determine:

(i) Width and thickness of the leaves.
(ii) Initial space that has to be provided between the full-length leaves before the band is applied.
(iii) Load on the clip to close the initial gap.

Given Data: Effective length, $2L$ = 1000 mm

Central load, $2F = 75 \text{ kN} \therefore F = 37.5 \text{ kN} = 37.5 \times 10^3 \text{ N}$

Full length leaves, $N_f = 3$; Graduated leaves, $N_g = 16$

Maximum equalized stress, $\sigma = 400 \text{ MPa} = 400 \text{ N/mm}^2$

Modulus, $E = 200 \text{ GPa} = 200 \times 10^3 \text{ N/mm}^2$

And $b = 9t$

SOLUTION Number of leaves, $N = N_f + N_g = 3 + 16 = 19$

Step-1: Plate dimensions

Width and thickness of spring leaves from strength equation

Maximum equalized stress, $\sigma = \dfrac{6FL}{Nbt^2}$ (Eqn. 10.47 MDHB)

i.e.,
$$400 = \frac{6 \times 37.5 \times 10^3 \times 500}{19 \times (9) \times t^3}$$

$\therefore$ Thickness of spring leaves, $t = 11.804 \text{ mm} \cong 12 \text{ mm}$

Width of spring leaves, $b = 9t = 108 \text{ mm} \cong 110 \text{ mm}$

Step-2: Initial space between full length and graduated leaves

$$\delta = \frac{2FL^3}{Nbt^3E} = \frac{2 \times 37.5 \times 10^3 \times 500^3}{19 \times 200 \times 10^3 \times 110 \times 12^3} = 12.98 \text{ mm} \quad \text{(Eqn. 10.46 MDHB)}$$

Step-3: Load on the clip to close the initial gap

$$\text{Load on the clip, } F_b = \frac{2N_g N_f F}{N(2N_g + 3N_f)} = \frac{2\times 3\times 16\times 37.5\times 10^3}{19(2\times 16+3\times 3)} = 4621.3094 \text{ N}$$

14.8 BELEVILLE SPRINGS

The Beleville spring is a conical disk. It is like a spring washer. It has characteristic load-deflection curves depending upon the ratio of (h/t). The behaviour is a combination of linear and non-linear relations. The desired load–deflection relation can be obtained by selecting the (h/t) ratio. They can be arranged in series, parallel, parallel-series combinations to get stiffer or softer springs. They provide a wide range of spring constants. They are simple in construction and easy to manufacture. They are compact and occupy small space. They are used in plate clutches, brakes, relief valves and bolted connections.

The load deflection and stress-deflection relations are given below.

Disk Spring

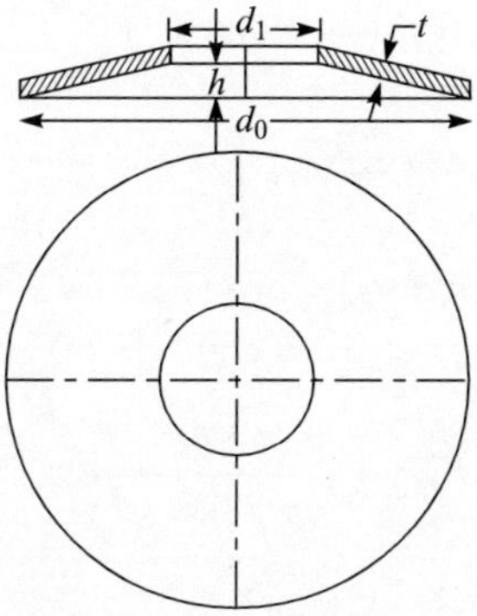

Figure 14.21 Belleville springs

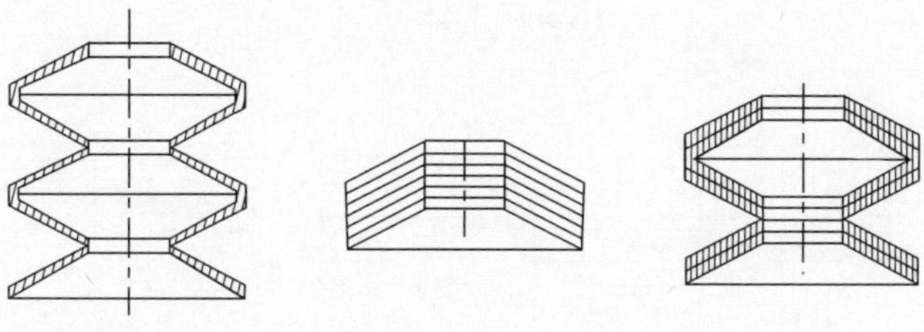

(a) (b) (c)

Figure 14.22 Nesting of Belleville springs

(a) Series combination
(b) Parallel combination
(c) Parallel series combination

Expression for axial load,

$$F=\frac{4E\cdot y}{(1-\mu^2)Md_0^2}\left[(h-y)\left(h-\frac{y}{2}\right)t+t^3\right] \tag{14.37}$$

Maximum stress induced at inner edge,

$$\sigma=\frac{4E\cdot y}{(1-\mu^2)\,Md_0^2}\left[C_1\left(h-\frac{y}{2}\right)+C_2t\right] \tag{14.38}$$

Stress induced at the outer edge,

$$\sigma=\frac{4E\cdot y}{(1-\mu^2)\,Md_0^2}\left[C_1\left(h-\frac{y}{2}\right)-C_2t\right] \tag{14.39}$$

where,

$$M=\frac{6}{\pi\ln(R_d)}\left[\frac{R_d-1}{R_d}\right]^2 \text{ and } R_d=\frac{d_o}{d_i} \tag{14.40a}$$

$$C_1=\frac{6}{\pi\ln(R_d)}\left(\frac{R_d-1}{\ln(R_d)}-1\right) \tag{14.40b}$$

$$C_2=\frac{6}{\pi\ln R_d}\left(\frac{R_d-1}{2}\right) \tag{14.40c}$$

M, C_1, C_2 are constants and can be found from (Fig. 14.24)

Load at flat position $(h=y)$

$$F_{flat}=\frac{4Eht^3}{Md_0^2(1-\mu^2)} \tag{14.41}$$

Thickness required,

$$t=\frac{1}{10}\sqrt[4]{\frac{F_{flat}}{132.4}\left(\frac{d_0^2}{\frac{h}{t}}\right)} \tag{14.42}$$

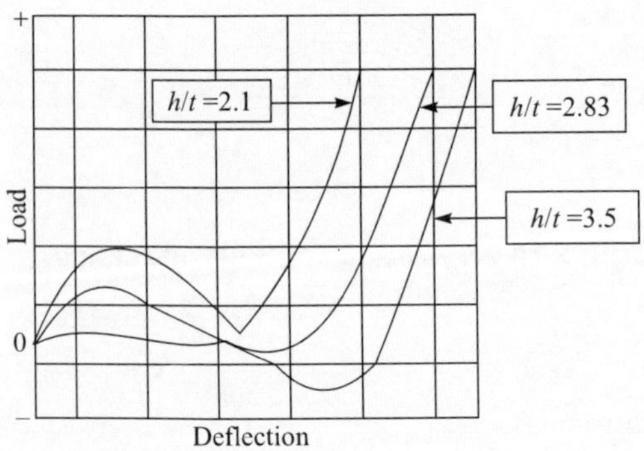

Figure 14.23 Load deflection curve for Belleville springs

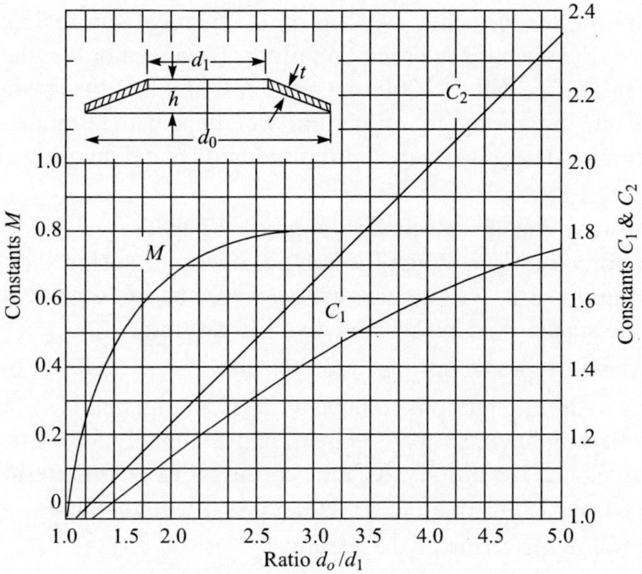

Figure 14.24 Diagram of disk constants

EXERCISES FOR PRACTICE

14.1 (a) Explain the terms: spring rate; spring index, Wahl's stress factor.

(b) The following particulars refer to the valve spring of a petrol engine:

Length of the spring when the valve is open = 41 mm

Length of the spring when the valve is closed = 49 mm

Spring load when the valve is open = 360 N

Spring load when the valve is closed = 220 N

Maximum inside diameter of spring = 25 mm

Evaluate (i) size of wire required for maximum operating stress of 400 MPa and modulus of rigidity 8.3×10^4 MPa, (ii) number of active turns required, (iii) pitch of the spring, and (iv) mean coil diameter.

14.2 (a) What is "Wahl stress factor" ? Explain briefly.

(b) A helical spring is made from a wire of 6 mm diameter and has outside diameter of 75 mm. If the permissible shear stress is 350 MPa and modulus of rigidity of 8.3×10^5 MPa. Find the axial load that the spring can carry and the deflection per active turn: (i) Neglecting the effect of curvature and (ii) Considering the effect of curvature.

14.3 Design a compression coil spring of music wire having squared ends. The spring is to be assembled with a preload of 10 N and exerts force of 50 N, when it is compressed additional 140 mm. Find the wire diameter to the nearest 0.2 mm and the coil diameter to the nearest millimetre using spring index = 12. Do not use the larger wire diameter than is necessary. What is free length and solid length?

14.4 At the bottom of an elevator shaft, a group of 16 identical springs are set in parallel to absorb the shock of the elevator in case of a failure. The elevator weighs 80 kN. Assuming that the elevator has a free fall of 1000 mm from rest and that the springs are compressed by 250 mm. Design the spring for minimum weight and maximum strength using the principles of optimum design. Assume spring index of 6 and an allowable shear stress of 500 Mpa.

14.5 Design a valve spring for an automobile engine. When the valve is closed, the spring should produce a forçe of 45 N. When the valve is open, the spring should produce a force of 55 N. The spring must fit over the valve bush which has an outside diameter of 20 mm and must go inside space of 35 mm. The valve lift is 6 mm.

14.6 (a) State the different types of springs and their uses.

(b) A valve for a hydraulic pump is to have a lift of 10 mm and is to be closed by helical compression spring having a load range of approximately 500 N to 875 N from valve closed to valve open conditions. Assume spring index of 6. Determine wire diameter, outer diameter and number of active coils to be recommended.

14.7 (a) Explain the following terms of the spring:

(i) Free length (ii) Spring rate (iii) Spring index.

(b) A truck spring has 12 leaves, two of which are full length leaves. The spring supports are 1.05 m apart and the central band is 85 mm wide. The central load is to be 5000 N. With a permissible stress of 290 MPa, determine the thicknesses and width of the steel spring leaves. The ratio of the total depth to the width of the spring is 3. Also determine the deflection of the spring.

14.8 (a) Discuss various types of springs used in engineering practice.

(b) The movement of a slow plunger is to be controlled by parallel springs arrangement, consisting of two concentric helical springs having common anchorage. Outer spring is 10 mm longer than the inner. Upward lift of the plunger is 20 mm. The spring force resisting upward movement of the plunger is 200 N up to 10 mm motion. The inner spring now makes contact with the plunger shoulder and further lift of 10 mm changes the spring load to 800 N. Design the spring. Take $S_{sd} = 300 \text{ N/mm}^2$, $G = 8 \times 10^4$ MPa for the spring material.

14.9 A carriage spring 800 mm long is required to carry a proof load of 5000 N at the centre. The spring is made of plates 80 mm wide and 7.5 mm thick. If the maximum permissible stress for the material of the plate is not to exceed 190 MPa, determine:

(i) The number of plates required.

(ii) The deflection of the spring.

(iii) The radius to which the plates must be initially bent.

The modulus of elasticity may be taken as 2.05×10^5 MPa.

14.10 (a) Explain the applications of springs.

(b) The valve spring of a gasoline engine is 40 mm long, when the valve is open and 50 mm long, when the valve is closed. The spring loads are 2300 N with the valve closed and 3650 N with valve open. The inside dia. of the spring cannot be less than 25 mm. Determine:

(i) the wire diameter for a stress of 420 MPa.

(ii) the number of coils required assuming squared and ground ends.

(c) A cantilever spring has an effective length of 525 mm and has two full length leaves and eight graduated leaves each 50 mm wide. The spring is to support a load of 40 kN with a stress of 350 MPa in all leaves.

(i) Determine the thickness of the leaves

(ii) Determine the deflection at full load.

14.11 (a) Design a helical compression spring to sustain an axial load that fluctuates between 1.5 kN and 2 kN with an associated deflection of 15 mm during the fluctuation of load.

(b) An automotive leaf spring is to be designed to consist of 10 graduated leaves and 2 full length leaves. The spring is to support a central load of 5 kN over a span of 1100 mm with the central band width of 100 mm. The ratio of total depth of spring to its width is to be 2.5. Determine the width and thickness of leaves limiting the maximum equalized stress induced in the leaves to 350 MPa. Also determine the initial gap to be provided between full length and graduated leaves before the assembly.

14.12 (a) Design a helical compression spring required for a spring loaded safety valve mounted on a pressure vessel. The spring is subjected to an initial compression of 50 mm at the time of assembly and will open by 10 mm when the pressure approaches 6 MPa. The diameter of the valve is 25 mm.

(b) A semielliptical laminated leaf spring with two full length leaves and ten graduated leaves are to be designed to support a central load of 6 kN over two points 1000 mm apart. The central band width is 100 mm. The ratio of total depth of the spring to its width is 2.5. The design normal stress of the material of the leaves is 400 MPa and the modulus of elasticity is 208 GPa. Determine:

(i) Width and thickness of leaves

(ii) The initial gap between full length and graduated leaves

(iii) The central bolt load.

14.13 (a) Explain stress concentration in helical coil springs and how is it taken care of.

(b) What is surging in springs and how can it be overcome?

(c) The spring used in an automobile engine has to exert 500 N when the valve is closed, and 600 N when the valve is open. The displacement of the valve is 5 mm. The engine crankshaft rotates at 8000 rpm. Design the spring if permissible stress in the material of the spring is 300 MPa. The ratio of mean coil diameter to the wire diameter is 6. The specific weight and the modulus of rigidity of the spring material are 7.35×10^{-5} N/mm^3 and 8×10^4 MPa, respectively. The ends of the spring are square and ground. Inspect the suitability of the spring for this engine. At what speed of the engine does the spring resonate?

14.14 (a) What is surging in helical springs and how can it be eliminated?

(b) A railway wagon weighing 50 kN and moving with a speed of 8 km/hr has to be stopped by four buffer springs in which the maximum compression allowed is

220 mm. Find the number of active turns in each spring of mean diameter 150 mm. The diameter of spring wire is 25 mm. Also determine the maximum shear stress in each spring. Take G = 84 GPa.

(c) A locomotive spring has an overall length of 1100 mm and sustains a load of 75 kN at its centre. The spring has 3 full length leaves and 15 graduated leaves with a central band of 100 mm. All the leaves are to be stressed at 0.4 GPa, when fully loaded. The ratio of total depth of spring to its width is to be 2. Determine: (i) width and thickness of leaves, and (ii) initial gap between full length and graduated leaves.
Take E = 206.8 GPa.

14.15 (a) Derive the equation for energy stored in a helical spring.

(b) Design a helical spring for a spring loaded safety valve (Rams bottom safety valve) for the following conditions:
Diameter of valve seat = 60 mm, Operating pressure = 0.7 N/mm^2, Maximum pressure when the valve blows off freely = 0.75 N/mm^2, Maximum lift of the valve when the pressure rises from 0.7 to 0.75 N/mm^2 = 3.5 mm, Maximum allowable stress = 550 N/mm^2, Modulus of rigidity = 84 kN/mm^2, Spring index = 6.

14.16 (a) Derive an expression for the stress induced in helical coil spring.

(b) Design a helical compression spring for a maximum load of 1000 N and for a deflection of 25 mm. The maximum permissible shear stress for a spring wire is 420 N/mm^2, modulus of rigidity is 0.84×10^5 N/mm^2 and value of spring index is 6.

(c) A one metre long Cantilever spring is composed of 8 graduated leaves and one extra full length leaf. The leaves are 45 mm wide. A load of 2000 N at the end of the spring causes a deflection of 75 mm. Determine the thickness of the leaves and maximum bending stress in the full length leaf assuming that leaves are not pre-stressed.

14.17 (a) A railway wagon weighing 40 kN and moving with a speed of 10 km/hour has to be stopped by four buffer springs in which the maximum compression allowed is 200 mm. Find the number of turns in each spring of mean diameter 150 mm. The diameter of spring wire is 25 mm. Take $G = 82.7 \times 10^3$ MPa.

(b) Design a truck spring that has 12 leaves two of which are full length leaves. The spring supports are 1 metre apart and the central band is 70 mm wide. The central load is to be 6 kN with a permissible stress of 200 MPa. Determine the thickness, width and deflection of spring leaves if the ratio of total depth to width of spring is 3.

14.18 (a) Derive an expression for the maximum strain energy stored in the closed coil spring under axial load in terms of maximum shear stress, modulus of rigidity and volume of the spring.

(b) An automobile helical coil spring is to have a mean diameter of 80 mm and stiffness 200 N/mm. The total axial force is 8000 N and allowable shear stress of spring material is 320 N/mm^2, and $G = 8 \times 10^4$ N/mm^2. Calculate:

(i) Diameter of coil

(ii) Number of effective coils

(iii) Free length of spring

(iv) Maximum energy which can be stored in the spring.

(c) A laminated semielliptical leaf spring under central load of 10 kN is to have an effective length of 1 m and not to deflect more than 75 mm. The spring has 10 leaves, 2 of which are full length and have been pre-stressed so that all leaves have same stress at full load condition. All the leaves have same width and thickness. The maximum allowable stress in the leaves is 350 MPa. Calculate the width and thickness of leaves.

14.19 (a) Derive the expression for the stress induced in helical coil spring.

(b) Design a valve spring for an automobile engine, when the valve is closed, the spring produces a force of 45 N and when it opens, it produces a force of 55 N. They must fit over the valve bush, which has an outside diameter of 20 mm and must go inside a space of 35 mm. The lift of the valve is 6 mm. The spring index is 12. The allowable stress may be taken as 350 MPa and modulus of rigidity $G = 80$ GPa.

14.20 (a) A railway wagon weighing 40 kN and moving with a speed of 10 km/hour has to be stopped by four buffer springs in which the maximum compression allowed is 200 mm. Find the number of turns in each spring of mean diameter 150 mm. The diameter of the wire is 25 mm. Take $G = 82.7 \times 10^3\ \text{MN/m}^2$.

(b) A multileaf spring with camber is fitted to the chassis of an automobile over a span of 1.2 metre to absorb shocks due to a maximum stress of 0.4 GPa. All the leaves of the spring are to receive the same stress. The spring should have at least 2 full length leaves out of 8 leaves. They are assembled with bolts over a span of 150 mm width at the middle. Design the spring for a maximum deflection of 50 mm.

SHORT ANSWER QUESTIONS

14.1 What is a spring?

14.2 What are the functions of springs?

14.3 What are the applications of springs?

14.4 What are the requirements of materials for springs?

14.5 Name some of the materials used for springs.

14.6 How are springs classified?

14.7 What are the types of helical coil springs?

14.8 Define spring index.

14.9 What is the Wahl stress factor?

14.10 Give the expression for the stress in compression-tension springs.

14.11 Define spring rate (spring stiffness).

14.12 What is the expression for resilience of the springs?

14.13 Draw a tension spring and show the terminology.

14.14 Give the equation for the stiffness of extension spring.

14.15 Show the types of end finish for the compression springs.

14.16 Explain with sketch the compression spring length terminology.

14.17 What are active number of coils in compression springs?

14.18 What is the maximum stress induced in compression springs?

14.19 Give the equation for the stiffness of compression springs.

14.20 Draw and show the various terms in the fatigue-diagram for spring design.

14.21 What is the equation for fatigue design of springs?

14.22 What are cluster springs?

14.23 What are the conditions for the cluster springs?

14.24 What is a torsional spring?

14.25 Give the equation for stress induced in torsional springs?

14.26 Where are the torsional springs used?

14.27 What is a spiral spring?

14.28 What are the stresses induced in spiral springs?

14.29 What are the equations for angular twist in spiral springs?

14.30 What are leaf springs?

14.31 Where are the leaf springs used?

14.32 What is the relation between stresses in full length leaves and graduated leaves?

14.33 What is meant by stress-equilised leaves?

14.34 How are the stress equilised leaf springs made?

14.35 What is the load on the clamping bolt?

14.36 What is the camber in leaf springs?

14.37 Explain the utility of centre bolt and rebound clip in leaf spring.

14.38 What is nipping in leaf springs?

14.39 What are the Belleville springs?

14.40 Where are the Belleville springs used?

14.41 What are the advantages of Belleville springs?

CHAPTER 15

SPUR GEARS

15.1 INTRODUCTION

The simplest means of transferring rotary motion from one shaft to another is a pair of friction rollers mounted on the two shafts and pressed to one another. Due to the frictional force between the cylinders, when one cylinder rotates the other also rotates. In the friction-wheel drives, the power transmitted is limited by the maximum frictional force between the surfaces. Another drawback is the possible slip between the surfaces. A positive drive is provided by adding teeth on the two surfaces, one set of teeth meshing with the other set. Then the toothed wheels are called gear set. The smaller one is called pinion and larger one is called gear wheel.

15.2 TYPES OF GEARS

Depending upon the relative orientation of the shaft axes and teeth on the surfaces, the gears are classified as:

(i) Spur gears: The shaft axes are parallel and teeth axes are parallel to the shaft axes.
(ii) Helical gears: The shaft axes are parallel but teeth axes follow the helixes on the pitch surfaces. The pinion has right-hand helix while the gear has left-hand helix or vice versa.
(iii) Bevel gears: The shaft axes are inclined. They may be intersecting or non-intersecting (skew). The pitch surfaces are conical. The teeth may be parallel to the generators or they may follow the spirals on the pitch surface.
(iv) Worm-gears: The shaft axes are non-intersecting. The worm resembles the screw and the gear wheel resembles a helical gear.

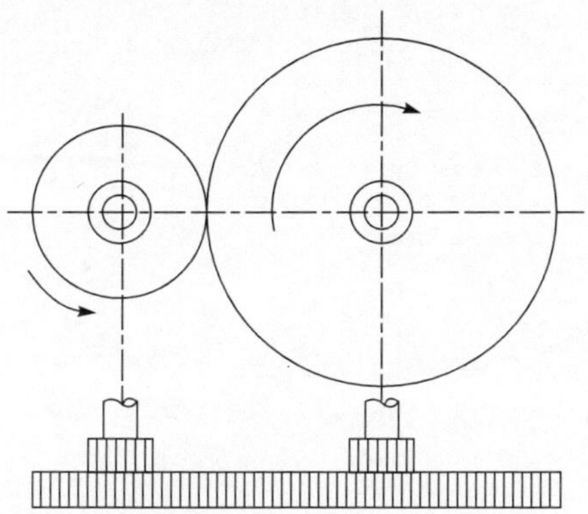

Figure 15.1 Spur gear set

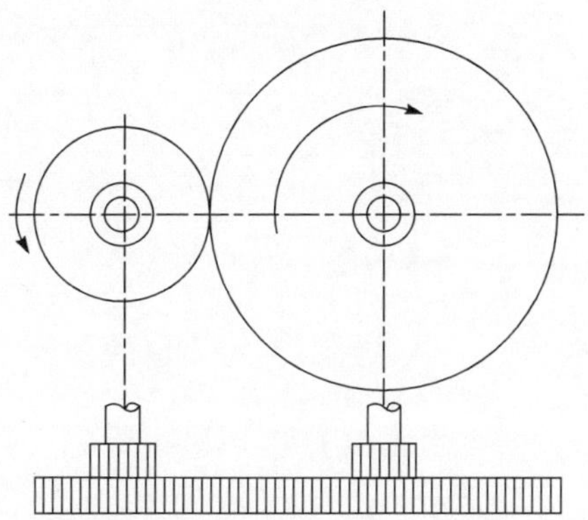

Figure 15.2 Helical gear set

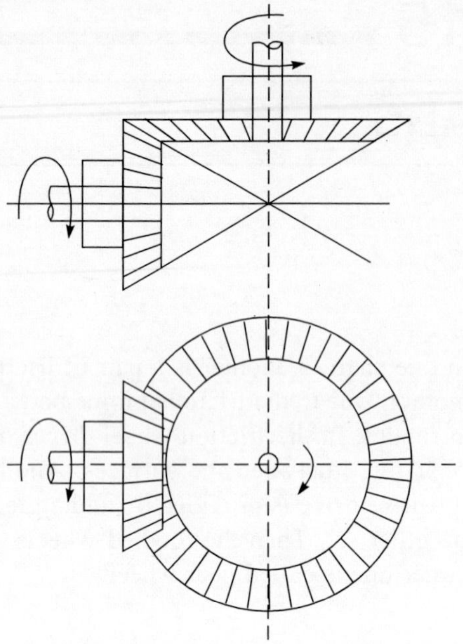

Figure 15.3 Bevel gear set

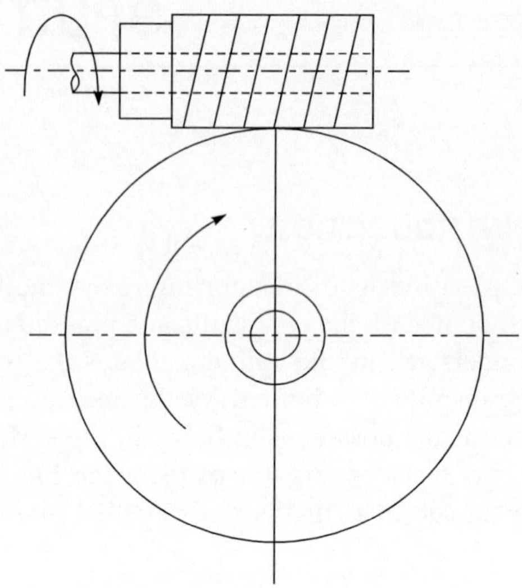

Figure 15.4 Warm gear set

We will study the spur gear design in this chapter.

15.3 THE LAW OF GEARING

Practically, any shape of teeth may be used. But for uniform motion transfer, the fundamental law of gearing must be satisfied. The law states that the angular velocity ratio between the gears of a gear set must remain constant throughout the mesh, mathematically, expressed as

$$\frac{\omega_2}{\omega_1} = \frac{r_1}{r_2} \tag{15.1}$$

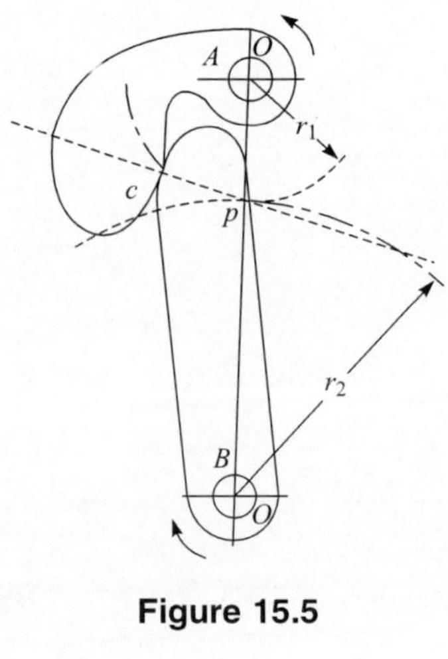

Figure 15.5

where ω_2, and ω_1 are the angular velocities of the output shaft and input shaft respectively and r_1, and r_2 are the radii of the input pitch cylinder and output pitch cylinder respectively. They are called pitch radii. The corresponding diameters are called the **pitch diameters**. The point where the cylinders touch is called **pitch point**. The law of gearing is satisfied when the common normal to the meshing surfaces intersects the line of centres at the pitch point. The cycloidal and involute profiles satisfy this requirement but the involute profile is most widely adopted because the variation in centre distance is not violating this condition. Another important point in favour of this profile is the ease of manufacturing.

15.4 GEAR–TOOTH NOMENCLATURE

Pitch Circle: The circle representing the rolling cylinder or the circle passing through the pitch point is called the pitch circle. The diameter of this circle is called the pitch diameter.

Base Circle: The circle on which the involute profile of the tooth is generated, is called the base circle.

Addendum Circle: The circle passing through the top face of the gear tooth is called addendum circle.

Dedendum Circle: The circle passing through (representing) the bottom surface of the tooth space is called the dedendum circle.

Circular Pitch (P_c): The distance between the central lines of the successive teeth measured along the pitch circle is called the circular pitch and is given by

$$P_c = \pi D/Z \tag{15.2}$$

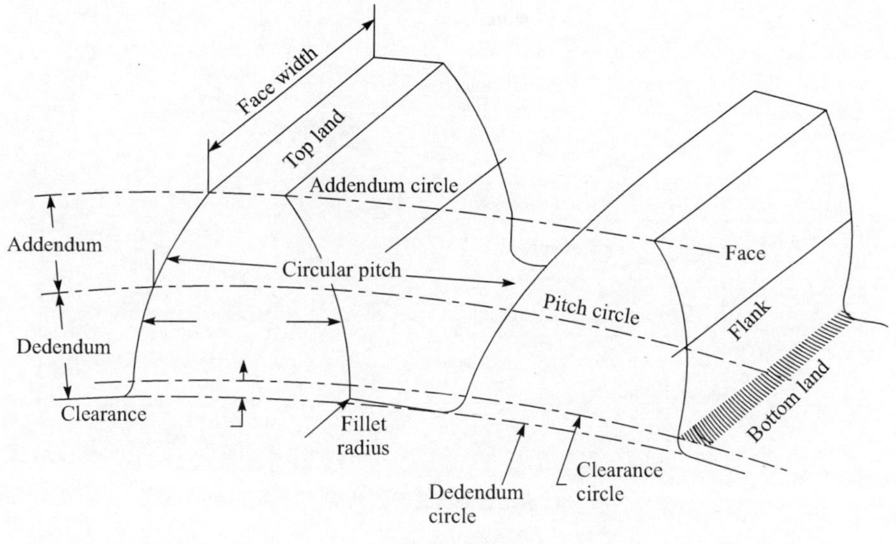

Figure 15.6

where, D = pitch diameter in mm

Z = number of teeth on the gear.

Module: It is the ratio of the pitch diameter to the number of teeth.

$$m = (D/Z) = (P_c/\pi) \tag{15.3}$$

The module is the standard for specifying the gear size in S.I. system of units.

Addendum (a): It is the tooth height above the pitch circle. For a standard tooth, $a = 1.0$ m.

Dedendum (d): It is the radial distance between the dedendum circle and the pitch circle. For a standard tooth, $d = 1.25$ m.

Clearance (c): It is the radial distance between the bottom of the tooth space and the top surface of the engaging tooth. For standard teeth, $c = 0.25$ m.

Face width (b): The tooth length along the axis is called face width.

Pressure Angle (ϕ): When the gears are in engagement, the common normal to the two involute profiles in contact becomes a common tangent to the two base circles of the gears. The contact point moves along this line and driving force is also directed along this line. The angle between this line and common tangent to the pitch circles is called the pressure angle. The standard value is 20°.

Then the relation between the pitch radius and the base radius,

$$r_b = r \cdot \cos\phi \text{ and base pitch } p_b = p_c \cos\phi \tag{15.4}$$

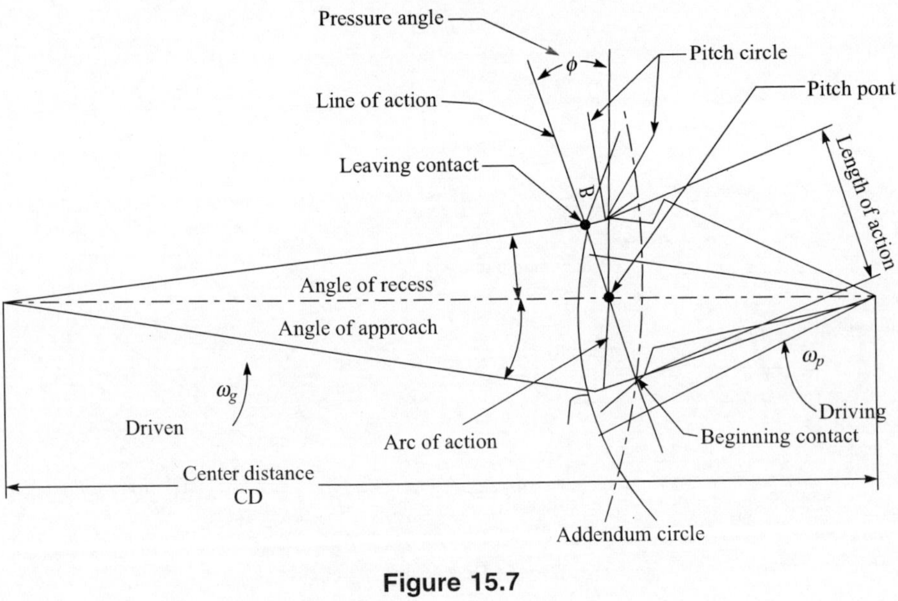

Figure 15.7

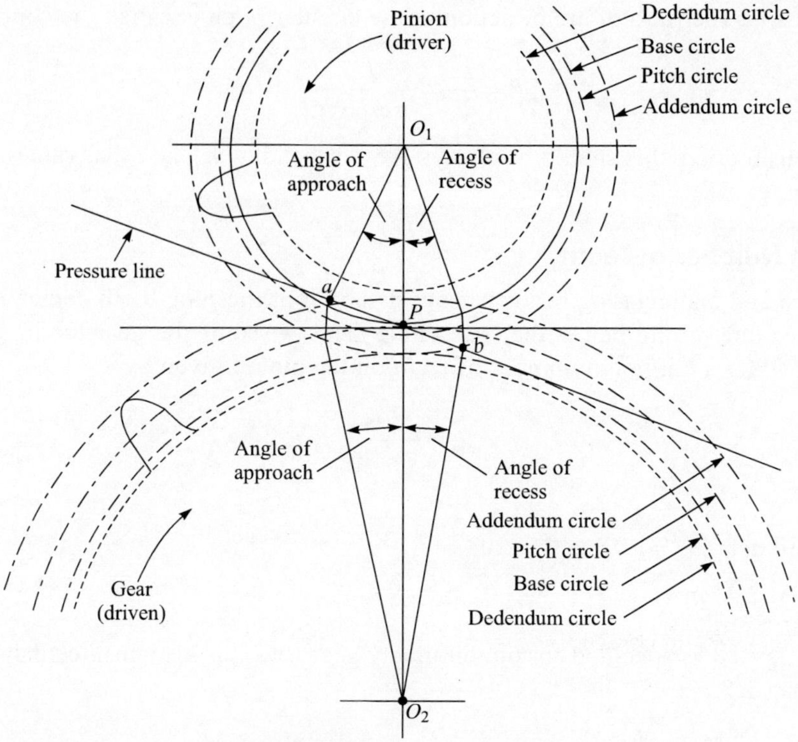

Figure 15.8 Tooth action

Path of Contact (length of action, L_{ab})

The distance along the line of action from the beginning of contact to the end of contact is called path of contact or length of action. The portion from the starting point to the pitch point is called **path of approach** and the portion from the pitch point to the end of contact is called **path of recess**.

$$\text{Path of approach, } L_{ap} = \sqrt{(r_g + a_g)^2 - (r_g \cos \Phi)^2} - r_g \sin \Phi \tag{15.5a}$$

$$\text{Path of recess, } L_{pb} = \sqrt{(r_p + a_p)^2 - (r_g \cos \Phi)^2} - r_p \sin \Phi \tag{15.5b}$$

The Path of Contact: $L_{ab} = L_{ap} + L_{pb}$

$$= \sqrt{(r_p + a_g)^2 - (r_g \cos \Phi)^2} + \sqrt{(r_p + a_p)^2 - (r_g \cos \Phi)^2} \; -(r_g + r_p)\sin \Phi \tag{15.5c}$$

The arc of action: The distance measured along the pitch circle from the beginning of contact to the end of contact is called the arc of action

$$L_r = [L_{ab} / \cos \Phi] \tag{15.6}$$

The contact ratio : The ratio of arc of action to the circular pitch is called the contact ratio.

$$CR = \frac{L_r}{P_c} = \frac{L_{ab}}{\pi m \cos \Phi} \tag{15.7}$$

For continuous motion, this should be always greater than unity. The usual values range from 1.4 to 2.

The Minimum Number of Teeth

The interference and undercutting occur whenever the tip of meshing tooth comes into contact with the non-involute profile below the base circle. This weakens the gear tooth. This can be avoided by providing a minimum number of teeth on the pinion given by

$$Z_{\min} = \frac{2a_p}{\sin^2 \Phi} \tag{15.8}$$

For a standard gear, $a_p = 1.0$ and $\Phi = 20°$

Then $Z_{\min} = 2/\sin^2(20°) = 17.1$

Normally $Z_{\min} = 18$ is specified for pinion in any gear set. This also ensures that the contact ratio is more than 1.5.

15.5 STRESSES IN SPUR GEARS

There are two kinds of stresses that may cause two modes of failures:

(i) Bending stresses at the root of the gear tooth causing bending fatigue fracture.

(ii) The contact stresses, causing surface fatigue (pitting) failure. The abrasive wear (scuffing or scoring) may also occur.

The fatigue fracture can be prevented by keeping the bending stresses within endurance limits. The materials do not exhibit well defined surface endurance limit stress and hence failure by successive or progressive wear cannot be totally prevented.

Bending Stress

The tooth load W acts along the pressure line. The radial component W_r causes uniform compression and the tangential component W_t causes bending stresses.

The bending effect is maximum when the contact is at the tip of the tooth. The bending stress induced is given by

$$\sigma_b = \frac{M}{I} y = \frac{h \cdot W_t}{\frac{b \cdot t^2}{6}} = \frac{W_t}{\frac{b \cdot t^2}{6h}}$$

$$= \frac{W_t}{b \cdot y p_c}, \text{ (substituting } \frac{t^2}{6h} = y p_c\text{)} \tag{15.9a}$$

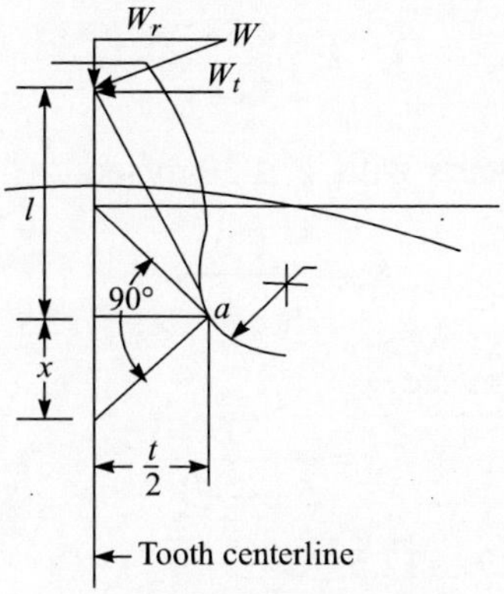

Figure 15.9

$$= \frac{W_t}{b \cdot y \cdot \pi \cdot m}, \text{ (substituting } p_c = \pi \cdot m)$$

$$= \frac{W_t}{b \cdot Ym}, \text{ (substituting } Y = \pi y) \tag{15.9b}$$

where,

W_t = tangential load, N

p_c = circular pitch $= \pi m$

m = module in mm.

b = face width in mm.

$Y = \pi y$ = Tooth form factor

y = Lewis tooth form factor

This equation was modified by Lewis to include the effect of the velocity and is named after him. This is used to estimate the preliminary value of the module as follows:

$$m = \frac{W_t}{Y \cdot bS_d \cdot K_v} = \sqrt{\frac{W_t}{Y \cdot \beta \cdot S_d K_v}} = \left[\frac{2T}{Y\beta \cdot ZS_d \cdot K_v}\right]^{1/3} \tag{15.10}$$

where, S_d = design stress based on the endurance stress in bending as suggested by Lewis (Table 15.1)

K_v = velocity factor which accounts for the dynamic effect of velocity and tooth errors (Fig. 15.12)

Velocity factors (Barth formulae) for ordinary gears with $v \leq 8$ m/sec

$$K_v = \frac{3.0}{3.0 + V} \tag{15.10a}$$

For carefully cut gears with $V \leq 13$ m/sec

$$K_v = \frac{4.58}{4.58 + V} \tag{15.10b}$$

For very accurately cut ground gears with $V \leq 20$ m/sec,

$$K_v = \frac{6.0}{6.0 + V} \tag{15.10c}$$

For precision gears with $V > 20$ m/sec,

$$K_v = \frac{5.6}{5.6 + \sqrt{V}} \tag{15.10d}$$

The face width factor $\beta = b/m = 7$ to 12.5

$$T = \text{Torque in Nmm} = W_t \cdot r = W_t \left(\frac{m \cdot Z}{2} \right)$$

$$Z = \text{Number of teeth}$$

For evaluation of the dynamic load, Sir Buckingham has recommended the following Eqn.

$$W_d = W_t + \frac{21V(bC + W_t)}{21V + \sqrt{bC + W_t}} \tag{15.11}$$

The dynamic beam strength is evaluated by the modified Lewis equation,

$$W_b = S_{dr} \cdot b \cdot Y \cdot m \tag{15.12}$$

The value of $S_{dr} = S_{en}$ - Endurance limit stress in bending.

To prevent the bending failure, $W_b > W_d$

The new AGMA standards recommended the following check against bending failure.

$$\sigma_b = \frac{W_t \cdot K_0 \cdot K_s \cdot K_m}{K_v \cdot m \cdot b \cdot J} \leq S_d. \tag{15.13}$$

where, the permissible stress is calculated by the equation

$$S_d = \frac{S \cdot K_L}{K_T \cdot K_R} \tag{15.14}$$

where, $K_0 =$ Overload correction factor (Table 15.6)

$K_s =$ Size factor = 1.0 (assumed)

$K_m =$ Load distribution factor (Table 15.7)

$K_v =$ Velocity factor (Fig. 15.12)

$J =$ Geometry factor (Fig. 15.13)

$K_L =$ Life factor (Table 15.8)

$K_T =$ Temperature factor = $(273 + t)/273$.

$K_R =$ Reliability factor (Table 15.9)

Contact Stresses: The contact stresses can be evaluated by using the Hertz contact stress theory. The maximum tooth load from the wear consideration as evaluated from the Hertz contact stress equation, applied for pitch point contact, is given by

$$W_w = K \cdot b \cdot D_p \cdot Q \tag{15.15}$$

where, K = load stress factor (Table 15.14)

$$= \frac{S_{fe}^2 \cdot \sin\Phi}{1.4}\left[\frac{1}{E_p} + \frac{1}{E_g}\right] \tag{15.15a}$$

S_{fe} = Surface endurance limit stress $= (2.75\,\text{BHN} - 70)\,\text{MPa}$

$$Q = \text{ load sharing factor } = \frac{2D_g}{D_g + D_p} = \frac{2G}{G \pm 1} \tag{15.15b}$$

(+ for external gear, – for internal gear)

To safeguard against wear failure, $W_w \geq W_d$

Usually, it is better to find the value of K to satisfy this condition and the fix required hardness values from Table 15.14.

The new AGMA recommended the following check equation for contact stress

$$\sigma_H = C_p \sqrt{\frac{W_t C_o \cdot C_s \cdot C_m \cdot C_f}{C_v \cdot D_p b \cdot I}} \leq S_{Hd} \tag{15.16}$$

The permissible contact stress is determined by the relation

$$S_{Hd} = \frac{S_{fe} \cdot C_L \cdot C_H}{C_T \cdot C_R} \tag{15.17}$$

where, C_p = Elastic coefficient (Table 15.10)

$C_o = K_o$ = Overload correction factor (Table 15.6)

$C_s = K_s$ = Size factor = 1

$C_m = K_m$ = Load distribution factor (Table 15.7)

$C_v = K_v$ = Velocity factor (Fig. 15.12)

I = Geometry factor (Fig. 15.14)

$$= \left(\frac{\cos\phi \cdot \sin\phi}{2}\right)\left(\frac{i}{i+1}\right)$$

C_L = Life factor (Table 15.8)

C_H = Hardness ratio factor = 1, if not specified

$C_f =$ Surface condition factor

$C_T = K_T = (273 + t)/273$

$C_R = K_R =$ Reliability factor (Table 15.9)

15.6 GEAR MATERIALS

Steels, cast iron, malleable and nodular irons are the most common choices for gears. Surface hardening is adopted to increase strength and wear resistance. For high corrosion resistance as in marine environment bronzes are often used.

The combination of bronze gear and steel pinion has advantages of material compatibility and conformity and are used in non-marine applications also. The thermoplastics such as nylon and acetal are used for low torque capacity applications.

15.7 DESIGN PROCEDURE

The gear design is an iterative process beginning with a set of initial trial parameters, reasonably estimated on the basis of Lewis equation and then checked for bending strength and wear strength. The following steps may be adopted.

1. Select the materials and design stresses for pinion and gear. The pinion material is to be stronger than the gear material. A steel pinion and cast iron gear are economical to design. The same material for pinion and gear will make the gear the stronger member. For design stress refer Table 5.4.
2. Select the number of teeth (minimum of 18 teeth) for pinion and corresponding number of gears.
3. Determine the tooth form factors Y for pinion and gear from Table 5.3.
4. Calculate product (YS_d) i.e., the strength factors for pinion and gear and compare them to determine the weaker member. Design should be based on the weaker member. This step can be eliminated whenever the same material is used for both pinion and gear. The design is based on the pinion only.
5. **Case (a):** If the centre distance is specified, the pitch radii can be calculated from the gear ratio and the module can be estimated based on the number of teeth assumed in step (2) as $m = (2r_p/Z_p)$ and standardized. The tangential load can be evaluated by the eqn. $W_t = (T_p/r_p)$.

   Pitch line velocity, $V = \dfrac{2\pi r_p N_p}{60,000}$ m/sec.

   Velocity factor, $K_v -$ as per one of the Eqns. 15.10a to 15.10d

   Then face width, $b = \dfrac{W_t}{Y \cdot S_d \cdot K_v \cdot m}$

The value of b/m should be between 7 to 12.5. If not change Z_p suitably and repeat the above steps until the suitable value of *b* is obtained.

Case (b) : Centre distance is not specified

Assume $K_v = 0.3$ (lowest value)

$\beta = 10.0$ (or any value between 7 to 12.5)

Evaluate the module by Lewis Eqn.

$$m = \left[\frac{2T}{Y\beta \cdot Z \cdot S_d \cdot K_v}\right]^{1/3} \tag{15.10}$$

and standardize by Table 5.2.

Then pitch radius, $r_p = (Z_p \cdot m/2)$

Pitch line velocity, $V = \dfrac{2\pi r_p N_p}{60,000}$ m/sec.

Find velocity factor K_v by one of the Eqns. 15.10a to 15.10d and compare it with assumed value. If $K_v <$ assumed value, recalculate the module with the new value of K_v, until current $K_v \geq$ precious K_v.

6. Then find, $r_p = (Z_p \cdot m/2)$; $r_g = (Z_p \cdot m/2)$, $b = \beta \cdot m$
   and load, $W_t = T/r = (T_p/r_p) = (T_g/r_g)$
7. **Calculate the dynamic load by Buckingham equation**

$$W_d = W_t + \frac{21V(bC + W_t)}{21V + \sqrt{bC + W_t}} \tag{15.11}$$

For C refer Table 15.5 (Table 16.5 MDHB)

8. **Calculate dynamic beam strength**

$$W_b = Y \cdot b \cdot m \cdot S_{dr} \tag{15.12}$$

and compare with W_d, if $W_b < W_d$, modify *b* or *m* or both until $W_b > W_d$

9. **Check for wear strength**

$$W_w = K \cdot b \cdot D_p \cdot Q \geq W_d \tag{15.15}$$

If $W_w < W_d$, evaluate load stress Factor $K = \dfrac{W_d}{b \cdot D_p \cdot Q}$

10. Determine the BHN for pinion and gear from Table 15.14 and specify the surface treatment process.

Table 15.1. Spur gear proportions (AGMA)

Quantity	Coarse pitch 1.25 mm		Fine pitch up to 1.25 mm
Pressure angle	20°	25°	20°
Addendum = a	1.00 m	1.00 m	1.00 m
Dedendum = d	1.25 m	1.25 m	1.20 m + 0.05 mm
Working depth (hk)	2.00 m	2.0 m	2.00 m
Whole depth (ht)	2.25 m	2.25 m	2.20 m + 0.05
Circular tooth thickness (t)	$\frac{\pi}{2}$ m	$\frac{\pi}{2}$ m	1.5708 m
Fillet radius of the basic rack r_f	0.30 m	0.30 m	-
Basic clearance c (min)	0.25 m	0.25 m	0.20 m + 0.05
Clearance c (sheaved or ground teeth)	0.35 m	0.35 m	0.35 m + 0.05
Minimum number of pinion teeth	18	12	18
Minimum number of teeth per pair	36	24	-
Minimum width of top land	0.25 m	0.25 m	-

Table 15.2. Recommended module (m) in mm (IS 2535)

Preferred	1	1.25	1.5	2.0	2.5	3.0	4.0	5.0	6.0
	8.0	10.0	12.0	16.0	20.0	25.0	32.0	40.0	50.0
Choice 2	1.125	1.375	1.75	2.25	3.5	4.5	5.5	7.0	9.0
	11.0	14.0	18.0	22.0	28.0	36.0	45		
Choice 3	3.25		3.75						6.5

Table 15.3. Tooth from factor Y

Number of teeth	141/2° composite & involute	20° Involute		No. of teeth	14 1/2° composite & involute	20° Involute	
		Full depth	Stub depth			Full depth	Stub depth
(1)	(2)	(3)	(4)	(1)	(2)	(3)	(4)
12	0.211	0.245	0.312	34	0.327	0.371	0.447
13	0.223	0.261	0.324	38	0.333	0.354	0.457
14	0.236	0.277	0.340	43	0.340	0.397	0.463
15	0.245	0.290	0.350	50	0.346	0.409	0.476
17	0.264	0.303	0.368	60	0.355	0.422	0.485
18	0.270	0.309	0.378	75	0.361	0.435	0.497
20	0.283	0.322	0.394	80	0.362	0.438	0.499
21	0.289	0.328	0.400	90	0.365	0.444	0.502
22	0.292	0.331	0.406	100	0.367	0.447	0.507
24	0.299	0.337	0.416	150	0.374	0.460	0.520
26	0.300	0.346	0.425	300	0.383	0.472	0.535
28	0.314	0.353	0.432	rack	0.390	0.485	0.552
30	0.318	0.359	0.438				

Table 15.4. Allowable static stress for use in Lewis formula

Material	Allowable stress S_d MPa	BHN	Material	Allowable stress S_d MPa	BHN MPa
Ordinary cast iron	56.5	150	Semi-steel	84.5	150
Good grade cast iron	70	250	Cast steel Untreated	140	180
C.I. Gr. 20	47	200	Treated	195	250
25, 35	56.5	220	**Alloy steels**		
35 treated	78.5	300	20 Ni 3-case hardened	350	650
Forge Steels:			45 Cr 2 Nl-heat treated	450	400
SAE 1020 (C-20) case hardened	125	150	45 Cr Va-heat treated	475	450
C30 not treated	140	150	Manganese bronze SAE 43	140	100
C35 not treated	160	200	Gear bronze SAE 62	70	80
C40 not treated	175	140	Phosphor bronze-SAE 65	90	100
C45 not treated	210	200	Aluminium bronze-SAE 68	150	180
C45 hardened	210	250	Raw hide, Fibroli	40	-
C50 hardened	240	240	Bakelite, Micarta	40	-

Table 15.5. Value of C in the dynamic load Eqns. 15.11, 16.14 and 17.11

Material	Tooth form	e-Error in action (mm)					
		0.0125	**0.025**	**0.05**	**0.075**	**0.10**	**0.125**
Cast iron and cast iron	14 1/2°	73	145	290	436	580	730
Cast iron – steel	14 1/2°	100	200	400	600	800	1000
Steel – steel	14 1/2°	145	290	580	875	1160	1450
Cast iron – cast iron	20°	75.5	151	300	455	605	755
Cast iron – steel	Full depth	105	210	415	620	830	1035
Steel – steel	Full depth	150	300	605	910	1210	1510
C.I – C.I	20° stub	78	155	310	470	625	780
C.I – steel	20° stub	107	215	430	645	857	1070
Steel – steel	20° stub	156	312	625	940	1250	1560

Table 15.6. Overload correction factors K_0 and C_0 for all gears

Sources	Driven machinery			
	Uniform	**Light shock**	**Medium shock**	**Heavy shock**
Uniform	1.00	1.25	1.5	1.75 or higher
Light shock	1.10	1.35	1.6	1.85 or higher
Medium shock	1.25	1.50	1.75	2.00 or higher
Heavy shock	1.50	1.75	2.0	2.25 or higher

Table 15.7. Load distribution factor (K_m and C_m) for spur gears

Characteristic of support	Face width (b) in mm			
	Up to 50	150	225	400 and above
Accurate mountings small bearings clearance minimum deflection, precision gears	1.3	1.4	1.5	1.8
Less rigid mountings, less accurate gears, contact across full face	1.6	1.7	1.8	2.2
Accuracy and mounting such that less than full face contact exists	Over 2.2			

Table 15.8. Life factors (K_L C_L) for spur, helical and bevel gears

Total life cycles	Durability life factor C_L	Strength life factor K_L				
		Spur & helical gears				Bevel gears
		160 BHN	250 BHN	450 BHN	RC55	RC55
1000	–	1.6	2.4	3.4	2.7	4.6
10,000	1.5	1.4	1.9	2.4	2.0	3.1
1,00,000	1.3	1.2	1.4	1.7	1.5	2.1
10,00,000	1.1	1.1	1.1	1.2	1.1	1.4
10,000,000	1.0	1.0	1.0	1.0	1.0	1.0

Table 15.9. Reliability factors (K_R and C_R) for all gears

Reliability requirement	C_R	K_R
High 99.9% up	1.25 up	1.50 up
More than 99% to 99.9%	1.00	1.00
67 to 99%	0.80	0.70
50%	0.70	0.70

Table 15.10. Values of elastic coefficient C_p for spur and helical gears with $\mu = 0.30$

Pinion material	Modulus of elasticity E GPa	Gear material					
		Steel	Malleable iron	Nodular iron	Cast iron	Alu bronze	Tin bronze
Steel	200	191	181	179	174	162	158
Malleable iron	170	181	174	172	168	158	154
Modular iron	170	179	172	170	166	156	152
Cast iron	150	174	168	166	163	154	149
Al bronze	120	162	158	156	154	145	141
Tin bronze	110	158	154	152	149	141	137

Table 15.11. Recommended endurance limits for spur helical and bevel gear

Material	(BHN) Minimum hardness	S_{en} Endurance limit MPa		
		Spur gear	Helical gear	Bevel gear
Steels	140	140 – 155	140 – 155	77.5
	180	175 – 195	175 – 195	97.5
S_{en} = 1.76 BHN	300	250 – 280	250 – 315	135
or	450	315 – 350	315 – 450	175
S_{en} = 700 MPa	Rc55	390 – 450	390 – 450	210
Cast irons				
AGMA Grade – 20	140	35	35	19
AGMA Grade – 30	175	65	35	19
AGMA Grade – 40	200	90	90	50

Table 15.12. Service factor C_s

Type of load	Type of service		
	Intermittent or 3 hr per day	8 to 10 hr per day	Continuous 24 hr
Steady	0.80	1.0	1.25
Light shocks	1.00	1.25	1.50
Medium shocks	1.25	1.5	1.80
Heavy shocks	1.50	1.80	2.00

Table 15.13. Surface endurance limits S_{fe} for gear materials

Material	Heat treatment	Min. hardness BHN	Endurance limit S_{fe} MPa
Steels	Case carburized	625	1750
	Case carburized	575	1400
Approximate relation	Flame hardened	500	1330
S_{fe} = 2.8 BHN − 70	Hardened and tampered	440	1330
	-do-	360	1120
	-do-	300	950
	-do-	240	880
	-do-	180	670
Cast irons			
AGMA - 20	As cast	140	210
AGMA - 30	As cast	175	350
AGMA - 40	As cast	200	450

Table 15.14. Loads stress factor - K in Eqns. 15.15, 16.16 and 17.13

Gear		Pinion		Value of K for M.K.S. multiply the values by 10		
Material	**BHN**	**Material**	**BHN**	**Endurance limits S_{fe} of N/mm^2**	**ϕ = 141/2° involute**	**ϕ – 20° involute**
Steel	150	Steel	150	350	0.21	0.288
Steel	150	Steel	200	420	0.302	0.407
Steel	150	Steel	250	490	0.407	0.555
Steel	200	Steel	200	490	0.407	0.555
Steel	200	Steel	250	550	0.535	0.722
Steel	200	Steel	300	632	0.674	0.92
Steel	250	Steel	250	632	0.674	0.92
Steel	250	Steel	300	702	0.835	1.135
Steel	250	Steel	350	770	1.01	1.375
Steel	300	Steel	300	770	1.01	1.375
Steel	300	Steel	350	842	1.20	1.635
Steel	300	Steel	400	877	1.305	1.780
Steel	350	Steel	350	910	1.41	1.93
Steel	350	Steel	400	984	1.65	2.23
Steel	350	Steel	450	1015	1.752	2.400
Steel	450	Steel	450	1195	2.415	3.370
Steel	450	Steel	500	1230	2.560	3.480
Steel	450	Steel	600	1262	2.700	3.70
Steel	450	Steel	500	1335	3.02	4.13
Steel	600	Steel	600	1610	4.42	6.05
C.I.	180	Steel	150	351	0.295	0.42
C.I.	180	Steel	200	492	0.61	0.835
C.I.	180	Steel	250	632	1.01	1.375
Gear	100	Steel	150	351	0.323	0.435
Bronze	100	Steel	200	492	0.6390	0.870
Bronze	100	Steel	250	595	0.945	1.290
C.I.	180	C.I.	180	630	1.350	1.85
Nickel C.I.	-	Steel	150	340	0.304	0.412
	-	Steel	200	490	0.69	0.814
	-	Steel	250	630	0.99	1.344
	-	Steel	300	650	1.059	1.442
Nickel C.I.	-	Nickel C.I.	-	660	1.422	1.942
P-Bronze	-	Nickel C.I.	-	580	1.187	1.62
Metal	-	Non-metallic	(34)	225	1.325	1.81

Table 15.15. Endurance limits for checking the beam strength of the teeth

Material	Core (BHN)	S_{en} MPa
Gray cast iron	160	83.5
Semi-steel (High grade I)	200	124.5
Manganese bronze	100	118.0
Gear bronze	100	165.5
Non-metallic	–	41.0
Cast steel (soft)	110	178.5
Cast steel (medium)	120	206.0
Steels	150	259.0
Steels	200	345.0
Steel normalised	240	414.0
Steels	250	429.0
	280	480.5
	300	515.0
	320	549.0
	350	583.5
	360	618.0
	400	686.5
	450	686.5
	500	686.5
	550	686.5
	600	686.5
Cr, Ni, steel, about 1% Ni and 0.40% C, heat-treated	280	428.5
Cr, Ni, Steel, about 2% Ni and 0.40% C, heat-treated	320	552.0
Steel, Oil tempered	360	621.0
Steel nitralloy	400	906.0

Table 15.16. Tolerance on the adjacent pitch

Grade	e(microns)
1	0.80 + 0.06 ϕ
2	1.25 + 0.10 ϕ
3	2.00 + 0.16 ϕ
4	3.20 + 0.25 ϕ
5	5.00 + 0.40 ϕ
6	8.00 + 0.63 ϕ
7	11.00 + 0.90 ϕ
8	16.00 + 1.25 ϕ
9	22.00 + 1.80 ϕ
10	32.00 + 2.50 ϕ
11	45.00 + 3.55 ϕ
12	63.00 + 5.00 ϕ

These tolerances are calculated by using the following basic equation,

$$\phi = m + 0.25\sqrt{d'}$$

where ϕ = tolerance factor
m = module (mm)
d' = pitch circle diameter (mm)

Table 15.17. Permissible errors in micron

Module mm	Class-1 industrial commercial gears	Class-2 or accurate carefully cutgears	Class-3 or precision gears
4.0	50	25	12.5
5.0	55	27	14.0
6.0	63	32	17.0
8.0	77	38	20.0
10.0	87	45	23.0
12.0	98.0	49	25.0
14.0	105.0	52.0	27.0
16.0	110.0	55.0	27.5
18.0	113.0	57	28.0
20.0	116.0	58	29.0
22.0	118.0	59	30.0
25.0	120.0	60	30.0

Table 15.18. Maximum allowance error in action between gears

Velocity m/s	Error (mm)	Velocity m/s	Error (mm)
1.0	0.0960	8.0	0.0500
2.0	0.0880	10.0	0.0386
2.5	0.0840	12.0	0.0330
3.0	0.0785	15.0	0.0230
4.0	0.0710	20.0	0.0155
5.0	0.0640	25.0	0.0130
6.0	0.0590	26.0	0.0127
		and over	

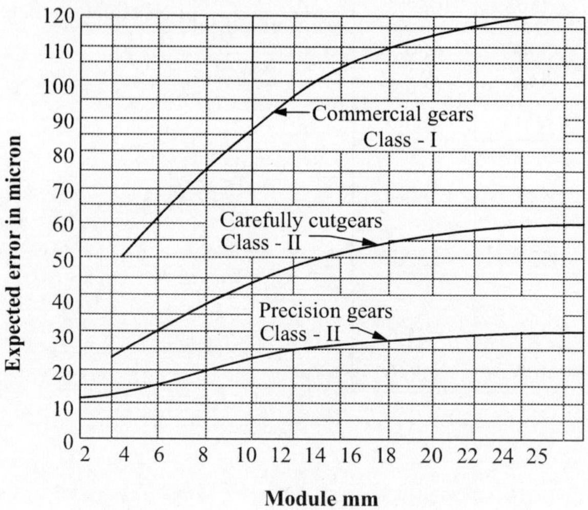

Figure 15.10 Maximum probable error, *e*

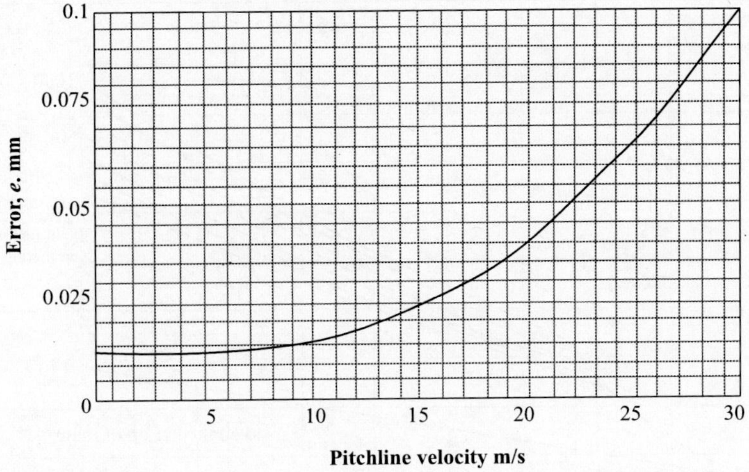

Figure 15.11 Permissible error in action for reasonable noise level

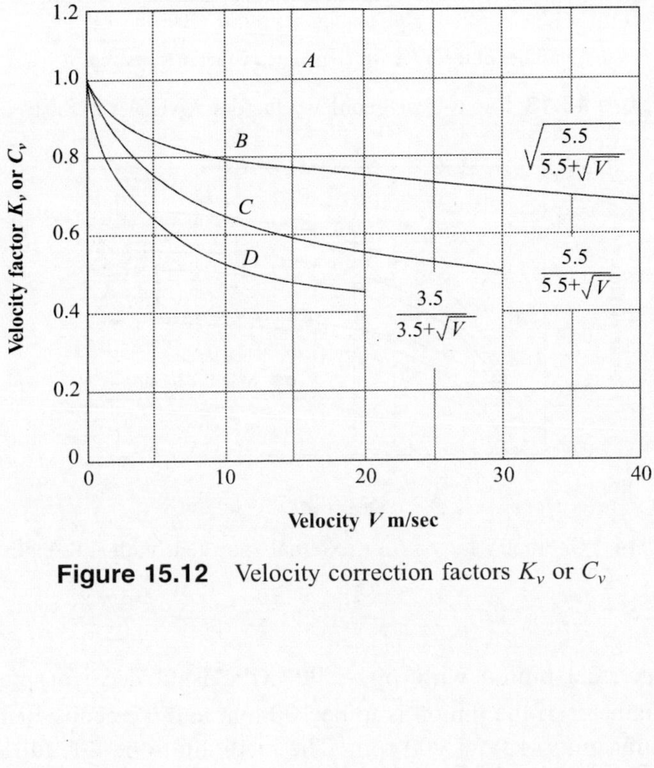

Figure 15.12 Velocity correction factors K_v or C_v

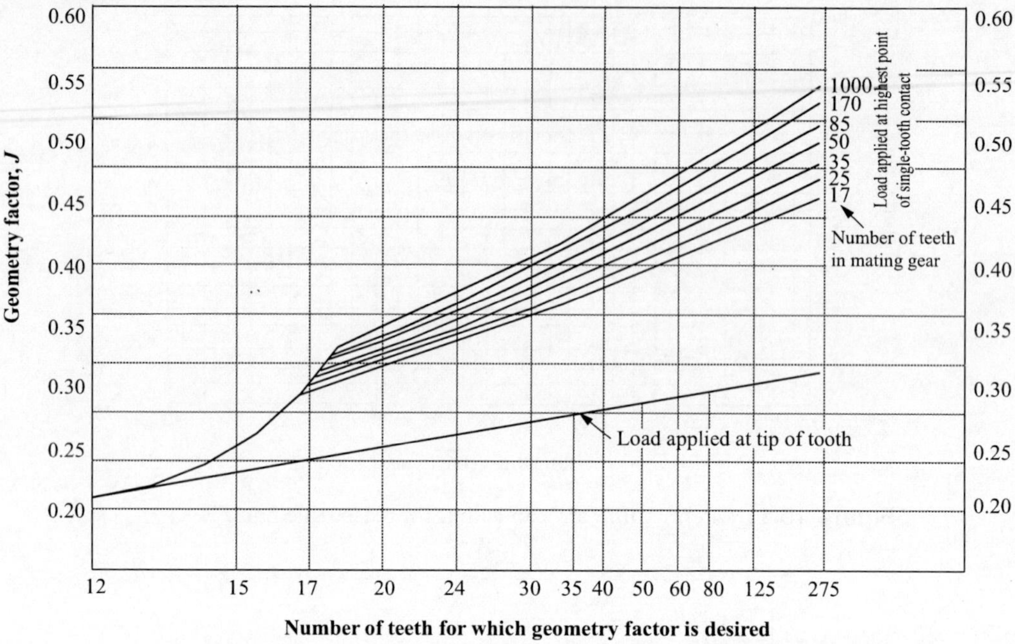

Figure 15.13 Spur-gear geometry factors *J* AGMA 218.01

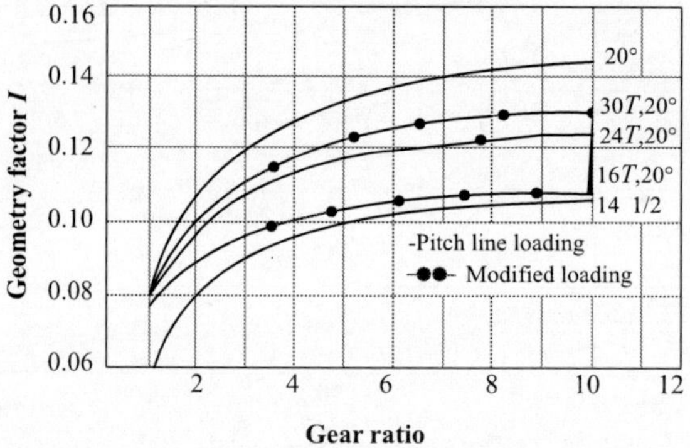

Figure 15.14 Geometry factor *I* for external spur-gear with full depth teeth

SOLVED EXAMPLES

Example 15.1: A spur steel pinion with $S_d = 200$ MPa is to drive a spur steel gear with $S_d = 140$ MPa. The diameter of the pinion is to be 100 mm and the centre distance is to be 200 mm. The pinion is to transmit 5 kW at 900 rpm. The teeth are to be 20° full depth. Determine the necessary module and face width to give greatest number of teeth. Design for strength only using Lewis equation.

Given Data: Power NP = 5 kW @ 900 rpm

Materials, for pinion – steel with $S_d = 200$ MPa

For gear – steel with $S_d = 140$ Mpa

Number of teeth are to be maximum possible

Diameter of pinion $D_p = 100$ mm

Centre – distance $C = 200$ mm

SOLUTION

Step-1: Estimation of module

$\therefore$ Diameter of the gear $= 2 \times C - D_p = 2 \times 200 - 100 = 300$ mm

The gear ratio, $G = 300/100 = 3$

Tentatively let $Z_p = 50$

Then teeth or gear, $Z_g = 150$

Module, $m = \dfrac{D}{Z} = \dfrac{100}{50} = 2.0$ mm

Step-2: Comparison of strength factors

Form factors, $Y_p = 0.409$ and $Y_g = 0.460$ (Table 16.3 MDHB)

Strength factors, $(S_d \cdot Y)_p = 200 \times 0.409 = 81.8$

$$(S_d \cdot Y)_g = 140 \times 0.460 = 64.4$$

$$\therefore \quad (S_d \cdot Y)_p > (S_d \cdot Y)_g$$

The gear is weaker. Hence, design the gear.

Step-3: Face width by Lewis equation

The speed of the gear $= 900/3 = 300$ rpm

$\therefore$ Torque on the gear, $T = \dfrac{955 \times 10^4 \times KW}{N} = \dfrac{955 \times 10^4 \times 5}{300} = 15.92 \times 10^4$ Nmm

Tangential load, $W_t = \dfrac{T}{r} = \dfrac{15.92 \times 10^4}{150} = 1061.1$ N

The pitch line velocity,

$$V = \frac{\pi DN}{60{,}000} = \frac{\pi \times 300 \times 300}{60{,}000} = 4.7 \text{ m/sec}$$

Velocity factor (For $V < 8$ m/sec), $K_v = \dfrac{3.0}{3.0 + V} = \dfrac{3.0}{3.0 + 4.7} = 0.389$

Using Lewis equation (Eqn. 16.10 MDHB)

$$m = \frac{W_t}{Yb \cdot S_d \cdot K_v} = \frac{1061.1}{0.46 \times b \times 140 \times 0.389} = 2$$

$$b = \frac{1061.1}{2 \times 0.46 \times 140 \times 0.389} 21.79$$

Choose $b = 22$ mm, then $(b/m) = 22/2 = 11$

This is within the specified range of 7 to 12.5. Hence,. acceptable. Any trial for $m < 2$ mm, will result in greater value of b/m ratio and hence, not acceptable.

Example 15.2: A pair of spur gears is to be made to the following specifications—20° full depth teeth; S_d for pinion = 80 MPa; S_d for gear = 55 MPa; $Z_p = 20; Z_g = 80$. The pinion is to transmit 5 kW at 200 rpm.

(a) Determine the module and face width for minimum centre distance.
(b) If the dynamic load for the pair is computed to be 3.8 kN, determine whether the design is safe from the standpoint of strength.
(c) Check the design for wear strength if the fatigue constant, $K = 1350$ kN/m^2

Given Data:

$$Z_p = 20, Z_g = 80$$

$$(S_p)_p = 80 \text{ MPa}; (S_d)_g = 55 \text{ MPa}$$

$$\text{Power} = 5 \text{ kW}$$

$$\text{Speed } N_p = 200 \text{ rpm}$$

SOLUTION

Step-1: Comparison of strength factors

The tooth form factors from, (Table 16.3 MDHB)

$$Y_p = 0.322; Y_g = 0.438$$

Strength factors,

$$(YS_p)_p = 0.322 \times 80 = 25.76$$

$$(YS_d)_g = 0.438 \times 55 = 24.09$$

The gear is weaker, hence design the gear.

Step-2: Module by Lewis equation

Gear ratio, $G = 80/20 = 4.0$

Gear speed, $N_g = N_p/G = 200/4 = 50$ rpm

The torque on the gear, $T = \dfrac{955 \times 10^4 \times 5}{50} = 95.5 \times 10^4$ Nmm

For minimum centre distance, the module is to be minimum and hence choose maximum ratio of $b/m = 12.5$.

Substituting the values in Lewis equation, (Eqn. 16.10 MDHB)

$$m = \left[\frac{2.T}{Y \cdot \beta Z S_d K_v}\right]^{1/3}$$

where,

$T = 95.5 \times 10^4$ Nmm; $\beta = 12.5$; $Y = 0.438$; $S_d = 55$ MPa;

$Z = 80$; $K_v = 0.4$ (assumed)

$$m = \left[\frac{2 \times 95.5 \times 10^4}{0.438 \times 12.5 \times 80 \times 55 \times 0.4}\right]^{1/3} = 5.8$$

The nearest standard module, $m = 6.0$ (Table 16.2 MDHB)

The diameter of the gear, $D_g = Z \cdot m = 80 \times 6 = 480$ mm

Pitch line velocity, $V = \dfrac{\pi DN}{60,000} = \dfrac{\pi \times 480 \times 50}{60,000} = 1.26$ m/s

The velocity factor, $K_v = \dfrac{3.0}{3.0 + 1.26} = 0.7 > (K_v = 0.4)$

Design is satisfactory but the centre distance is to be minimum, hence try for smaller value of $m = 5$ mm.

The gear diameter, $D_g = m \cdot Z = 5 \times 80 = 400$ mm

Tooth load, $W_t = T/r = \dfrac{955 \times 10^4}{200} = 4775$ N

Pitch line velocity, $V = \dfrac{\pi DN}{60,000} = \dfrac{\pi \times 400 \times 50}{60,000} = 1.047$ m/sec

The velocity factor, $K_v = \dfrac{3}{3 + 1.047} = 0.74$

Face width, $b = \beta m = 12.5 \times 5 = 62.5$ mm

Then the tooth strength by Lewis equation

$W_t = S_d \cdot Y \cdot bm\, K_v = 55 \times 0.438 \times 62.5 \times 5 \times 0.74 = 5570.8$ N

This is greater than actual required average of $W_t = 4775$ N

Hence, the design with $m = 5$ is acceptable.

Step-3: Check for dynamic beam strength

The dynamic beam strength,

$$W_b = S_{dr} Y \cdot b \cdot m$$

$= 55 \times 0.438 \times 62.5 \times 5 = 7528.125 \text{ N} = 7.528 \text{ kN}$

This is greater than the given value of dynamic load, $W_d = 3.8$ kN
Therefore, the design is safe.

Step-4: Check for wear strength

The wear load is given by

$$W_w = K \cdot b \cdot D_p \cdot Q$$

where, $K = 1350 \text{ kN/m}^2 = 1.350 \text{ MN/m}^2 = 1.350 \text{ MPa}$; $b = 62.5$ mm

$$D_p = 5 \times 20 = 100 \text{ mm}$$

$$Q = \frac{2D_g}{D_p + D_g} = \frac{2 \times 400}{100 + 400} = 1.6$$

Substituting the values

$$W_w = 1.350 \times 62.5 \times 100 \times 1.6 = 13500 \text{ N} = 13.5 \text{ kN} > (W_d = 3.8 \text{ kN})$$

This is greater than the given value of dynamic load. Hence, the design is safe.

Example 15.3: A 15 kW and 1200 rpm motor drives a compressor at 300 rpm through a pair of spur gears having 20° stub teeth. The centre to centre distance between the shafts is 400 mm. The motor pinion is made of forged steel and the gear is made of cast steel. Assuming the drive operates for 8 to 10 hr a day under light shock condition, design the gears.

Given Data: Power NP = 15 kW @ pinion speed = 1200 rpm

Tooth profile—20° stub teeth

Centre distance C = 400 mm

Service—8 to 10 hr per day, light shock

SOLUTION

Step-1: Materials properties

Pinion – Forged steel $C45 - S_d = 210$ MPa (250 HB)

Gear – Cast steel treated, $S_d = 195$ MPa (250 HB)

Step-2: Comparison of strength factors

Teeth for 20° stub; teeth $Z_{\min} = 15$

Gear ratio, $G = 1200/300 = 4$

Choose $Z_p = 16$, then $Z_g = 16 \times 4 = 64$

Tooth form factors, $Y_p = 0.359$, and $Y_g = 0.489$ (Table 16.3 MDHB)

Strength factors $(YS_d)_p = 0.359 \times 210 = 75.39$

$$(YS_d)_g = 0.489 \times 195 = 95.355$$

Pinion is weak, hence design the pinion.

Step-3: Module by geometry

Centre distance, $C = (Z_p + Z_g)m/2$

$$400 = (16 + 64)m/2$$

Preliminary value of $m = 800/80 = 10$ mm

Step-4: Tangential load

Pinion diameter, $D_p = 16 \times 10 = 160$ mm

Gear diameter, $D_g = 64 \times 10 = 640$ mm

Service factor, $K_s = 1.25$ (Table 16.22 MDHB)

The torque, $T = \dfrac{955 \times 10^4 \text{KW}}{N} C_s = \dfrac{955 \times 10^4 \times 15 \times 1.25}{1200} = 149218.75 \text{ Nmm}$

Transmitted load, $W_t = \dfrac{149218.75}{80} = 1865.2 \text{ N}$

Step-5: Face width by Lewis equation

Pitch line velocity, $V = \dfrac{\pi \times 160 \times 1200}{60,000} = 10.05 \text{ m/sec}$

Velocity factor, $K_v = \dfrac{4.58}{4.58 + V} = \dfrac{4.58}{4.58 + 10.05} = 0.313$ (Eqn. 16.12b)

Substituting the values in Lewis equation (Eqn. 16.10 MDHB)

$$m = \frac{W_t}{Y \cdot bS_d \cdot K_v}$$

$$\therefore \quad 10 = \frac{1865.2}{0.359 \times b \times 210 \times 0.313} = \frac{1865.2}{23.6b}$$

$$\therefore \quad b = \frac{1865.2}{23.6 \times 10} = 7.9 \text{ mm}$$

Then $b/m = 7.9/10 = 0.79$

This is very small hence reduce the module.

Step-6: Alternate selection of module

Let m = 2.5 mm, since centre distance does not change, for pitch diameter $D_p = 160$ mm

$$\therefore \quad Z_p = 160/2.5 = 64; \text{ and } Z_g = 64 \times 4 = 256$$

$$\therefore \quad Y_p = 0.489; Y_g = 0.535;$$

$$(Y_p \cdot S_d = 102.69) < (Y_g \cdot S_d = 104.325)$$

The pinion is weak. Design is to be based on pinion.
Then face width,

$$b = \frac{W_t}{mY \cdot S_d \cdot K_v} = \frac{1865.2}{2.5 \times 0.489 \times 210 \times 0.313} = 23.2 \text{ mm}$$

Let $b = 25$ mm

$$b/m = \frac{25}{2.5} = 10 \text{ This is satisfactory.}$$

The preliminary design parameters are:

$D_p = 160$ mm; $m = 2.5$ mm; $b = 25$ mm; $Z_p = 64$ and $Z_g = 256$

Step-7: Dynamic load by Buckingham equation

Tangential load, $W_t = 1865.2$ N

Pitch line velocity, $V = 10.05$ m/sec

For the carefully cut gears, gears error permissible

$e = 25$ mm (From Table 16.28 and 16.29 in MDHB)

The dynamic factor, $C = 312$ (Table 16.5 MDHB)

The dynamic load by Buckingham equation

$$W_d = W_t + \frac{21V(bC + W_t)}{21V + \sqrt{bC + W_t}}$$

$$= 1865.2 + \frac{21 \times 10.05[25 \times 312 + 1865.2]}{11 \times 10.05 + \sqrt{25 \times 312 + 1865.2}}$$

$$= 1865.2 + \frac{21 \times 10.05 \times 7732.2}{309.36} = 8458.9 \text{ N}$$

Step-8: Check for dynamic beam strength

The dynamic beam strength for $S_{dr} = S_{en} = 429$ MPa (Table 16.26 MDHB)

$$W_b = S_{dr} Y \cdot b \cdot m = 429 \times 0.489 \times 25 \times 2.5 = 13111.3N > W_d$$

Hence, the design is satisfactory for bending strength.

Step-9: Check for wear strength

$$\text{Wear strength, } W_w = K \cdot b \cdot D_p \cdot Q$$

For $W_w = W_d$, the strength factor,

$$K = \frac{8458.9}{25 \times 160 \times 1.6} = 1.32$$

Then the surface hardnesses required are

For $K = 1.375$ (Table 16.23 MDHB)

Pinion hardness = 350 BHN; gear hardness = 250 BHN

∴ The pinion is to be surface hardened to 350 BHN.

Example 15.4: Design a spur gear drive required to transmit 55 kW at 800 rpm of pinion. The speed ratio is to be 3.2:1. The teeth are to be 20° full depth involute.

Given Data: Power = 55 kW; Pinion speed = 800 rpm
Gear ratio = 3.2

SOLUTION

Step-1: Material selection

Select pinion material as C45, gear material as cast steel (Table 16.4 MDHB)

Design stress for pinion, $S_d = 210$ MPa (250 BHN)

For gear, $S_d = 195$ MPa (250 BHN)

Step-2: Comparison of strength factors

Number of teeth for pinion, $Z_p = 20$,

Then for gear, $Z_g = 20 \times 3.2 = 64$

Tooth form factors, $Y_p = 0.322$, and $Y_g = 0.489$

Strength factors $(YS_d)_p = 0.322 \times 210 = 67.62$

$$(YS_d)_g = 0.489 \times 195 = 95.355$$

Pinion is weaker hence design the pinion.

Step-3: Design for beam strength

The torque on the pinion, $T_p = \dfrac{955 \times 10^4 \times 55}{800} = 656562.5$ Nmm

Lewis equations (Eqn. 16.10 MDHB)

$$m = \left[\frac{2T}{Y \cdot \beta \cdot Z \cdot S_d \cdot K_v}\right]^{1/3}$$

where, $T = 656562.5$ Nmm; $Y = 0.322$ and $S_d = 210$ MPa

$\beta = 12$ [between 7 to 12.5]; $Z = 20$

$K_v = 0.3$ (assumed tentatively)

Substituting the values

$$m = \left[\frac{2 \times 656562.5}{0.322 \times 12 \times 20 \times 210 \times 0.3}\right]^{1/3} = \left[\frac{2 \times 656562.5}{48686}\right]^{1/3} = 6.34$$

Choose standard value, $m = 6.00$ mm

Diameter of pinion, $D_p = 6 \times 20 = 120$ mm

Dia of gear $D_g = 6 \times 64 = 384$ mm

The pitch line velocity, $V = \dfrac{\pi DN}{60,000} = \dfrac{\pi \times 120 \times 800}{60,000} = 5.02$ m/sec

The velocity factor, $K_v = \dfrac{3.0}{3.0 + V} = \dfrac{3.0}{3.0 + 5.02} = 0.374$

This is greater than the previous value, i.e., assumed value.
Then $m = 6$ acceptable.

Step-4: Alternate choice of module

But let us try an alternative design

$m = 5.0$ mm; $b = 60$ mm; $Z_p = 20$; and $Z_g = 64$

$\therefore D_p = 5.0 \times 20 = 100$ mm, and $D_g = 5 \times 64 = 320$ mm

Tangential load, $W_t = \dfrac{656562.5}{50} N = 13131.25$ N

Pitch line velocity,

$$V = \frac{\pi DN}{60,000} = \frac{\pi \times 100 \times 800}{60,000} = 4.18879 \text{ m/sec} \sim 4.2 \text{ m/sec.}$$

Manufacturing error, $e = 27$ (Table 16.28) for class II

The dynamic factor (Table 16.5 in MDHB), $C = 300$.

Then the dynamic load, (Eqn. 16.14 MDHB)

$$W_d = W_t + \frac{21V(bC + W_t)}{21V + \sqrt{bC + W_t}}$$

$$= 13131.250 + \frac{21 \times 4.2(60 \times 300 + 13131.25)}{21 \times 4.2 + \sqrt{(60 \times 300 + 13131.25)}} = 23506.75 \text{ N}$$

Step-5: Check for dynamic beam strength

$W_b = S_{dr} \cdot Y \cdot b \cdot n,$ where $S_{dr} = S_{en} = 429.0$ (Table 16.25 MDHB)

$= 429 \times 0.322 \times 60 \times 5.0 = 41441.4 \text{ N}$

$W_b > W_d$, hence safe design.

Step-6: Check for wear strength

$W_w = K \cdot b \cdot D_p \cdot Q = 1.25 W_d,$ where $Q = \dfrac{2G}{1+G} = \dfrac{2 \times 3.2}{4.32} = 1.525$

Then the strength factor,

$$\therefore \qquad K = \frac{1.25 \times 23506.75}{60 \times 100 \times 1.525} = 3.214$$

From Table 16.23, surface hardness required,

$$(BHN)_p = 450 \text{ and } (BHN)_g = 450$$

Both pinion and gear are surface hardened to BHN = 450.

Example 15.5: Design a spur gear set for an engine powered portable air compressor gearbox. The gear ratio is 2.5:1. The engine rating is 1.25 kW at 3750 rpm. 20° full depth involute teeth are adopted. The gearbox is to have a life of 10 years at 8 hr per day operation.

Given Data: Power transmitted, $NP = 1.25$ kW;

Speed of the pinion, $N = 3750$ rpm; Gear ratio $G = 2.5$

Tooth form—20° full depth involute;

Service—8 hr per day for 10 years

SOLUTION

Step-1: Material selection

Pinion and gear – C45 Hardened (Table 16.4 MDHB)

Design stress, $S_d = 210$ MPa; Hardness $= 250$ BHN

Step-2: Module by Lewis equation

Select the number of teeth for pinion, $Z_p = 18$

Gear $Z_g = 18 \times 2.5 = 45$

Material being same, design the pinion

Tooth form factor from Table 16.3 MDHB, $Y = 0.309$

Torque on the pinion $T = \dfrac{955 \times 10^4 \times 1.25}{3750} = 3183.3$ Nmm

Assume $\beta = b/m = 10$, and velocity factor $K_v = 0.3$

Then module by Lewis equation,

$$m = \left(\frac{2T}{Y \cdot \beta \cdot ZS_d K_v}\right)^{1/3} = \left[\frac{2 \times 3183.3}{0.309 \times 10 \times 18 \times 210 \times 0.3}\right]^{1/3} = 1.2$$

Choose $m = 2.0$, and $b = 2 \times 10 = 20$ m

Then $D_p = 2.0 \times 18 = 36$ mm, $D_g = 2.0 \times 45 = 90$ mm

Tangential load, $W_t = T/r = \dfrac{3183.3}{(36/2)} = 176.85$ N

Pitch line velocity, $V = \dfrac{\pi \times 36 \times 3750}{60,000} = 7.07$ m/sec

Velocity factor, (Eqn. 16.12a MDHB)

$K_v = \dfrac{3.0}{3.0 + V} = \dfrac{3.0}{3.0 + 7.07} = 0.29 \cong 0.3$, equal to the assumed value.

Hence O.K.

Step-3: Check the design by new AGMA-specifications

Permissible bending stress by AGMA Eqn. (Eqn. 16.18 MDHB)

$$S_d = \frac{SK_L}{K_T \cdot K_R}$$

where, $S = S_{en} = 1.76$ BHN $= 1.76 \times 250 = 440$ MPa (Table 16.20 MDHB)

K_L for life cycles $(10 \times 365 \times 8 \times 60 \times 3750) = 657 \times 10^7$ cycles

$= 1.0$ (Table 16.12 MDHB)

$K_T = 1.0$ for normal temperatures, $K_R = 1.0$ (Table 16.13 MDHB)

Substituting the values,

$$\therefore \quad S_d = \frac{440 \times 1.0}{1.0 \text{x} 1.0} = 440 \text{ MPa}$$

Induced stress by AGMA equation (Eqn. 16.17 MDHB)

$$\sigma_b = \frac{W_t \cdot K_o \cdot K_s \cdot K_m}{K_v \cdot m \cdot b \cdot J}$$

where, $W_t = 176.85$ N; $K_o = 1.5$ (Table 16.10); $K_s = 1.0$ assumed

$K_m = 1.6$ (Table 16.11 MDHB); $K_v = 0.65$ (Fig. 16.5 curve – C MDHB)

$m = 2.0$ mm; $b = 20$ mm; $J = 0.35$ (Fig. 16.6 MDHB)

Substituting the values,

$$\sigma_b = \frac{176.85 \times 1.5 \times 1.6 \times 1.0}{0.65 \times 2.0 \times 20 \times 0.35} = 46.6 \text{ MPa} < (S_d = 440 \text{ MPa})$$

The design is safe.

Step-4: Check for the contact stress by new AGMA specifications

The permissible contact stress,

$$S_{Hd} = \frac{S_{fe} \cdot C_L \cdot C_H}{C_T \cdot C_R} \quad \text{(Eqn. 16.19 MDHB)}$$

where, $S_{fe} = 2.8\ \text{BHN} - 70 = 2.8 \times 250 - 70 = 630$ MPa

$C_L = 1.0$ (Table 16.12 MDHB), $C_H = 1.0$ assumed, $C_T = 1.0$ assumed

$C_R = 1.0$ (Table 16.13 MDHB)

Substituting the values,

$$\therefore \quad S_{Hd} = \frac{630 \times 1.0 \times 1.0}{1.0 \times 1.0} = 630 \text{ MPa}$$

Induced surface stress, (Eqn. 16.20 MDHB)

$$\sigma_H = C_b \sqrt{\frac{W_t \cdot C_o \cdot C_s \cdot C_m \cdot C_f}{C_v \cdot C_p \cdot b \cdot I}}$$

where, $W_t = 176.85$ N; $C_o = K_o = 1.5$ (Table 16.10 MDHB); $C_s = 1.0$ assumed

$C_m = K_m = 1.6$ (Table 16.11 MDHB); $C_f = 1$ assumed

$C_v = K_v = 0.65$ (Fig. 16.5 curve – C); $D_p = 36.0$ mm; $b = 20$ mm;

$$I = \frac{D_g \sin\phi \cos\phi}{2(D_g + D_p)} = \frac{90 \sin 20° \cos 20°}{2 \times (90+36)} = 0.115$$

$$C_p = 191 \qquad \text{(Table 16.14 MDHB)}$$

Substituting the values

$$\sigma_H = 191\sqrt{\frac{176.85 \times 1.5 \times 1 \times 1.6 \times 1.0}{0.65 \times 36 \times 20 \times 0.115}} = 536.4 < (S_{Hd} = 630 \text{ MPa})$$

Design is safe.

Example 15.6: A pair of carefully cut gear spur gear with 20° full depth tooth profile is used to transmit 12 kW at 1200 rpm of pinion. The gear speed is to be 300 rpm. The material for both pinion and gear is medium carbon steel with allowable bending stress of 230 MPa. Take 24 teeth on pinion and modulus of elasticity as 210 GPa. Design the gears.

Given Data: Power transmitted, $NP = 12$ kW

Speed of pinion, $n_1 = 1200$ rpm; Speed of gear $n_2 = 300$ rpm;

Allowable stress, $S_d = 230$ MPa; No. of teeth on pinion = 24;

Modulus of elasticity, $E = 210 \text{ GPa} = 210 \times 10^3$ MPa

SOLUTION

Step-1: Module by Lewis equation

Number of teeth on gear, $Z_2 = \frac{1200}{300} Z_1 = 4 \times 24 = 96$

The torque on the pinion,

$$T = \frac{955 \times 10^4 \times kW}{N} = \frac{955 \times 10^4 \times 12}{1200} = 95500 \text{ Nmm}$$

Since the material for pinion and gear are the same, pinion is weaker hence design the pinion. Applying Lewis equation for bending

$$m = \left[\frac{2 \cdot T}{Y \cdot \beta \cdot Z \cdot S_d \cdot K_v}\right]^{1/3} \qquad \text{(Eqn. 16.10 MDHB)}$$

where, $T = 95500 \times C_s = 95500 \times 1.5$ (from Table 16.21, MDHB $C_s = 1.5$)

$Z = Z_1 = 24$; $Y = 0.337$ (Table 16.3 MDHB), and $\beta = 10$ assumed

$S_d = 230$ MPa, and $K_v = 0.4$ assumed

Then, the module,

$$m = \left[\frac{2\times 95500\times 1.5}{0.337\times 10\times 24\times 230\times 0.4}\right]^{1/3} = 3.37 \text{ mm}$$

Choose $m = 4$ mm

The diameter of pinion, $d_1 = mZ_1 = 4\times 24 = 96$ mm

The diameter of gear, $d_2 = mZ_2 = 4\times 96 = 384$ mm

Pitch line velocity,

$$V = \frac{\pi d_1 n_1}{60,000} = \frac{\pi\times 96\times 1200}{60,000} = 6.032 \text{ m/sec}$$

Then the velocity factor,

$$K_v = \frac{3}{3+6.032} = 0.3321 < K_v = 0.4 \text{ assumed}$$

Then module for $K_v = 0.3321$,

$$m = 3.37\left(\frac{0.4}{0.3321}\right)^{1/3} = 3.58$$

$m = 4$ mm is suitable.

Step-2: Dynamic load by Buckingham equation

The tangential load,

$$W_t = \frac{T}{d/2} = \frac{95500\times 1.5}{(96/2)} = 2984.375 \text{ N}$$

Width of gear, $b = \beta m = 10\times 4 = 40$ mm

Dynamic load by Buckingham equation,

$$W_d = W_t + \frac{21V(b\cdot C + W_t)}{21V + \sqrt{bc + W_t}}$$

where, $W_t = 2984.375$; $b = 40$ mm; $V = 6.032$ m/sec;

$C = 300$ for $e = 25$, class – II gear (Table 16.27 and Table 16.5 MDHB)

Substituting the values,

$$W_d = 2984.375 + \frac{21\times 6.032(40\times 300 + 2984.375)}{21\times 6.032 + \sqrt{40\times 300 + 2984.375}}$$

$$= 2984.375 + \frac{126 \times (14984.375)}{126 + 122.4}$$

$$= 2984.375 + 7600.44 = 10584.82 \text{ N}$$

Step-3: Check for bending strength,

$$W_b = S_{dr} b \cdot Y \cdot m$$

where, S_{dr} = 250 MPa; b = 40 mm; m = 4 mm; Y = 0.337

$$W_b = 230 \times 40 \times 0.337 \times 4 = 12401.6 \text{ N} > (W_d = 10584.82 \text{ N})$$

The design is safe against bending fatigue failure.

Step-4: Check for wear strength,

$$W_w = K b \cdot D_p Q$$

where, $b = 40$ mm; $D_p = 96$ mm;

$$Q = \frac{2D_g}{D_g + D_p} = \frac{2 \cdot D_g / D_p}{D_g / D_p + 1} = \frac{2 \times 4}{4 + 1} = 1.6$$

$K = 0.92$ (Table 16.23, for BHN = 250)

Substituting the values,

$$W_w = 0.92 \times 40 \times 96 \times 1.6 = 5652.48 < W_d$$

The surface hardening is required.
Then the stress factor K required,

$$K = \frac{W_d}{b \cdot D_p \cdot Q} = \frac{10584.82}{40 \times 96 \times 1.6} = 1.723$$

This is achieved by surface hardening the pinion and gear to

$$\text{BHN} = 350 \text{ with } K = 1.93$$

or pinion BHN = 300 gear and BHN = 400 with $K = 1.78$

Example 15.7: Design a pair of spur gears to transmit 20 kW from a shaft rotating at 1000 rpm to a parallel shaft, which is to rotate at 310 rpm. Assume the number of teeth on pinion 31 and 20° full depth form. The material for pinion is $C45$ steel untreated and for gear cast steel 0.25% C untreated.

Given Data: Power transmitted, $NP = 20$ kW;

Speed of pinion, $N_1 = 1000$ rpm; Speed of gear, $N_2 = 310$ rpm

Number of teeth on pinion, $Z_p = 31$; Tooth form = 20° full depth

Material for pinion – C45 untreated

Material for gear – cast steel with 0.25% C.

SOLUTION

Step-1: Module by Lewis equation

The gear ratio $= N_1/N_2 = \dfrac{1000}{310} = 3.226$

Number of teeth on gear, $Z_g = 3.226 \times Z_p = 3.226 \times 31 = 100$

The material properties (Table 16.4 MDHB)

C45 steel, $S_d = 210$ MPa with BHN = 200

Cast steel, $S_d = 140$ MPa with BHN = 180

The tooth form factors (Table 16.3 MDHB)

For pinion, $Y_p = 0.36$; for gear $Y_g = 0.447$

Strength factors,

For pinion $(YS_d)_p = 0.36 \times 210 = 75.6$

For gear $(YS_d)_g = 0.447 \times 180 = 80.46$

$$\therefore \qquad (YS_d)_p < (YS_d)_g$$

The pinion is weaker and hence design pinion.
The torque on the pinion,

$$T = \frac{955 \times 10^4 \text{ kW}}{N} = \frac{955 \times 10^4 \times 20}{1000} = 191000 \text{ Nmm}$$

Lewis equation for module,

$$m = \left[\frac{2.T}{Y \cdot \beta \cdot Z S_d \cdot K_v}\right]^{1/3} \qquad \text{(Eqn. 16.10 MDHB)}$$

where, $T = 191000$ Nmm; $Y = 0.36$; $\beta = 10$ (assumed); $Z = 31$; $S_d = 210$ MPa,

$K_v = 0.4$ assumed

Substituting the values,

$$\therefore \qquad m = \left[\frac{2 \times 191000}{0.36 \times 10 \times 31 \times 210 \times 0.4}\right]^{1/3} = 3.44$$

Choose $m = 4$ (Table 16.2 MDHB)

The pinion diameter, $d_p = 4 \times 31 = 124$ mm

Face width, $b = 10\text{ m} = 10 \times 4 = 40$ mm

Gear diameter, $d_g = 4 \times 100 = 400$ mm

The pitch line velocity,

$$V = \frac{\pi DN}{60,000} = \frac{\pi \times 124 \times 1000}{60,000} = 6.49 \text{ m/sec}$$

Then velocity factor,

$$K_v = \frac{3.0}{3+v} = \frac{3.0}{3.0+6.49} = 0.316 < \text{ assumed}$$

Recalculate the module with new value,

$$m = \left[\frac{2 \times 191000}{0.36 \times 10 \times 31 \times 210 \times 0.316}\right]^{1/3} = 3.72 \text{ mm}$$

∴ The selected module $m = 4$ is O.K.

Then the tangential load,

$$W_t = \frac{T}{d/2} = \frac{2 \times 191000}{124} = 3080.65 \text{ N}$$

Step-2: Dynamic load by Buckingham equation,

$$W_d = W_t + \frac{21V(bc + W_t)}{21V + \sqrt{bc + W_t}} \qquad \text{(Eqn. 16.14 MDHB)}$$

where, $W_t = 3080.65$; $V = 6.49$ m/sec $\cong 6.5$; $b = 40$ mm;

$c = 300$ (Table 16.5 for $e = 25$ μm, (Table 16.27 MDHB)

$$W_d = 3080.65 + \frac{21 \times 6.5(40 \times 300 + 3080.65)}{21 \times 6.5 + \sqrt{40 \times 30 + 3080.65}}$$

$$= 3080.65 + \frac{136 \times (12000 + 3080.65)}{136 + \sqrt{12000 + 3080.65}} = 11005.54 \text{ N}$$

Step-3: Check for the dynamic beam strength

$$W_b = S_{dr} \cdot b \cdot Y \cdot m = 210 \times 40 \times 0.36 \times 4 = 12096 \text{ N} > W_d$$

The design is safe against bending fatigue failure.

Step-4: Check for wear strength,

$$W_w = K \cdot b \cdot D_p Q$$

where, $b = 40$ mm; $D_p = 124$ mm; $Q = \dfrac{2D_g}{D_g + D_p} = \dfrac{2 \times 400}{400 + 124} = 1.53, K = 0.555$ (Table 16.23 MDHB)

Substituting the values,

$$W_w = .555 \times 40 \times 124 \times 1.53 = 4211.8 \text{ N} < W_d$$

This is less than the required. Then the stress factor required,

$$K = \frac{11005.5}{40 \times 124 \times 1.5} = 1.479$$

From Table 16.23 MDHB, for $K = 1.635$

BHN for pinion = 300, BHN for gear = 350

Example 15.8: Design a pair of spur gears to transmit a power of 18 kW from a shaft running at 1000 rpm to a parallel shaft to be run at 250 rpm, with a distance of 160 mm between the shaft centres. Suggest suitable surface hardness for the pair.

Given Data: Power transmitted, $NP = 18$ kW;

Pinion speed, $N_1 = 1000$ rpm; Gear speed, $N_2 = 250$ rpm

The centre distance, $C = 160$ mm

SOLUTION

Step-1: Selection of the materials

(From Table 16.4 MDHB)

For pinion as C45 steel – with $S_d = 210$, BHN = 200

For gear casted with $S_d = 140$, BHN = 180

Step-2: The module by geometry

The gear ratio, $G = \dfrac{1000}{250} = 4 = \dfrac{D_g}{D_p}$

Center distance, $C = \dfrac{D_p + D_g}{2} = 160$

$\therefore \qquad D_p + 4D_p = 320$

$\therefore \qquad D_p = 320/5 = 64$ mm

$D_g = 4 \times 64 = 254$ mm

Assume the number of teeth,

For pinion $Z_1 = 18$ and for gear $Z_2 = 18 \times 4 = 72$

Then the module $m = 64/18 = 3.555$

Choose $m = 3.5$

Then $D_p = 3.5 \times 18 = 63$ mm, $D_g = 3.5 \times 72 = 252$ mm

Then the centre distance, $C = \dfrac{63 + 252}{2} = 157.5 \cong 160$ mm O.K.

Tooth form factors (Table 16.3 MDHB)

For pinion, $Y_p = 0.309$; and for gear $Y_g = 0.435$

Comparison of strength factors

For pinion $(YS_d)_p = 0.309 \times 210 = 64.89$

For gear $(YS_d)_g = 0.435 \times 140 = 60.9$

$$\therefore \quad (YS_d)_g < (YS_d)_p$$

The gear is weaker, hence design the gear.

Step-3: Face width by Lewis equation

The torque transmitted,

$$T = \frac{955 \times 10^4 \text{kW}}{N} = \frac{955 \times 10^4 \times 18}{1000} = 171900 \text{ Nmm}$$

The tangential force on the gear tooth,

$$W_t = \frac{T}{(D_p/2)} = \frac{171900 \times 2}{63} = 5457.14 \text{ N}$$

Pitch line velocity,

$$V = \frac{\pi DN}{60,000} = \frac{\pi \times 252 \times 250}{60,000} = 3.2987 \cong 3.3 \text{ m/sec}$$

The velocity factor, $K_v = \dfrac{3.0}{3.0 + 3.3} = 0.476$

Then from Lewis (Eqn. 16.10 MDHB)

Width of gear, $b = \dfrac{W_t}{m \cdot Y \cdot S_d K_v} = \dfrac{5457.14}{3.5 \times 0.435 \times 140 \times 0.476} = 53.79$ mm

$$\beta = b/m \cong \frac{53.79}{3.5} = 15.37$$

This is more therefore modify the choice of m.
Choose $m = 4.0$ mm and $Z_1 = 16.0$ and $Z_2 = 64$, Stub tooth
The diameters are,

$$D_p = 16 \times 4 = 64 \text{ mm}$$

$$D_g = 64 \times 4 = 256 \text{ mm}$$

The center distance,

$$C = \frac{D_p + D_g}{2} = \frac{64 + 256}{2} = 160 \text{ mm}$$

Tooth form factors (Table 16.3 MDHB)

For pinion $Y_p = 0.359$ and for gear $Y_g = 0.530$

Strength factors

For pinion $(YS_d)_p = 0.359 \times 210 = 75.39$

For gear $(YS_d)_g = 0.530 \times 140 = 74.2$

The gear is weaker, hence design the gear.

The tangential force, $W_t = \dfrac{171900 \times 2}{64} = 5371.875 \text{ N}$

The pitch line velocity,

$$V = \frac{\pi DN}{60,000} = \frac{\pi \times 256 \times 250}{60,000} = 3.36 \text{ m/s}$$

Velocity factor,

$$K_v = \frac{3}{3 + 3.36} = 0.471 \quad \text{(Eqn. 16.12 MDHB)}$$

From Lewis equation 16.10 MDHB

Width of gear, $b = \dfrac{Wt}{m \cdot Y \cdot S_d K_v} = \dfrac{5371.875}{4 \times 0.530 \times 140 \times 0.471} = 38.43 \text{ mm} \cong 40 \text{ mm}$

Choose $b = 40$ mm, then $\beta = b/m = \dfrac{40}{4} = 10 < 12.5$ O.K.

Step-4: Dynamic load by Buckingham equation

The dynamic load,

$$W_d = W_t + \frac{21V(bC + W_t)}{21V + \sqrt{(bC + W_t)}}, \text{ where } C = 312 \quad \text{(Table 16.5 MDHB)}$$

$$= 5371.875 + \frac{(21 \times 3.36)\ (40 \times 312 + 5371.875)}{21 \times 3.36\ + \sqrt{40 \times 312 + 5371.875}}$$

$$= 11541.355 \text{ N}$$

Step-5: Check for the dynamic beam strength,

$$W_b = S_{dr} b \cdot Y \cdot m = 210 \times 40 \times 0.530 \times 4 = 17808N > W_d$$

The design is safe against bending fatigue failure.

Step-6: Check for the wear strength,

$$W_w = K \cdot b \cdot D_p Q \text{ where } Q = \frac{2 \times 4}{4+1} = 1.6$$

∴ Strength factor required,

$$K = \frac{W_d}{b \cdot D_p \cdot Q} = \frac{11541.355}{40 \times 64 \times 1.6} = 2.818$$

From Table 16.23 MDHB, the surface hardness of 450 BHN for both pinion and gear give $K = 3.37$.

Example 15.9: A 12 kW motor running at 1170 rpm drives a fan through a pair of spur gears (forged steel SAE 1030 for pinion & C.I. for gear) with a reduction ratio of 3.9:1. Design the gears and check for dynamic and wear loads.

Given Data: Power, $NP = 12\text{ kW} @ 1170\text{ rpm}$; Reduction ratio, $G = 3.9$

Pinion material is SAE 1030 and gear material is C.I.

SOLUTION

Step-1: Module by Lewis equation

Material properties (Table 16.4 MDHB)

Pinion – SAE 1030; $S_d = 140\text{ MPa (BHN} = 150)$

Gear C.I. – $Gr\,35$; $S_d = 78.5\text{ MPa (BHN} = 300)$

Assume, no. of teeth on pinion, $Z_1 = 20$

Number of teeth on gear, $Z_2 = 20 \times 3.9 = 78$

The tooth form factors (Table 16.3 MDHB)

For pinion, $Y_p = 0.322$, for gear $Y_g = 0.438$

The strength factors

For pinion $(S_d Y)_p = 140 \times 0.322 = 45$

For gear $(S_d Y)_g = 78.5 \times 0.438 = 34.383$

$$(S_d Y)_p > (S_d Y)_g$$

The gear is weaker hence design gear.

The torque on gear,

$$T = \frac{955 \times 10^4 \text{kW}}{N} = \frac{955 \times 10^4 \times 12}{(1170/3.9)} = 382000 \text{ Nmm}$$

Select $\beta = b/m = 10$ and velocity factor $K_v = 0.3$

Module by Lewis equation

$$m = \left[\frac{2 \times T}{Y \times \beta \times Z \times S_d \times K_v}\right]^{1/3} \quad \text{(Eqn. 16.10 MDHB)}$$

$$= \left[\frac{2 \times 382000}{0.438 \times 10 \times 78 \times 78.5 \times 0.3}\right]^{1/3} = 4.5$$

Choose $m = 5.0$ mm

Face width $b = 10 \times m = 10 \times 5 = 50$ mm

Diameter of pinion, $D_p = 5 \times 20 = 100$ mm

Diameter of gear, $D_g = 5 \times 78 = 390$ mm

Pitch line velocity,

$$V = \frac{\pi DN}{60{,}000} = \left(\frac{\pi \times 100 \times 1170}{60{,}000}\right) = 6.126 \text{ m/s}$$

The exact velocity factor,

$$K_v = \frac{3}{3+V} = \frac{3}{3+6.126} = 0.3287$$

This is more than the assumed value hence the choice of $m = 5$ mm is suitable.

Step-2: Dynamic load by Buckingham equation

The tangential load,

$$W_t = \frac{T}{r} = \frac{382000}{(390/2)} = 1959 \text{ N}$$

Specifying precision finish, class III gear,

The permissible error, $e = 14\ \mu\text{m}$ (Table 16.27 MDHB)

The dynamic factor, $C = 120$ (Table 16.5 MDHB)

Then dynamic load,

$$W_d = W_t + \frac{21V(b \cdot C + W_t)}{21V + \sqrt{b \cdot c + W_t}} \quad \text{(Eqn. 16.14 MDHB)}$$

$$= 1959 + \frac{21 \times 6.126 \times (50 \times 120 + 1959)}{21 \times 6.126 + \sqrt{50 \times 120 + 1959}} = 6658.8 \text{ N}$$

Step-3: Dynamic beam strength

The permissible limit stress S_{dr} for C.I. (Table 16.25 MDHB)

$$S_{dr} = 83.5 \text{ MPa}$$

Then $W_b = S_{dr} \cdot b \cdot Y \cdot m = 83.5 \times 50 \times 0.438 \times 5 = 9143.25$ N

$\therefore$ $W_b > W_d$. The design is safe against bending fatigue failure.

Step-4: Wear strength,

$$W_w = K \cdot b \cdot D_p \cdot Q$$

where, $Q = \frac{2 \times D_g}{D_g + D_p} = \frac{2 \times G}{G+1} = \frac{2 \times 3.9}{3.9+1} = 1.59$; $K = 0.42$ (Table 16.23 MDHB)

$\therefore$ $W_w = 0.42 \times 50 \times 100 \times 1.59 = 3339N < W_d$

Hence, the stress factor *K* required,

$$K = \frac{w_d}{b \cdot D_p \cdot Q} = \frac{6658.8}{50 \times 100 \times 1.59} = 0.8375$$

The required hardness of the pinion = 250 BHN and of gear = 180 BHN which gives *K* = 1.375.

Example 15.10: Design a pair of super gears to transmit 20 kW of power at a pinion speed of 1000 rpm. The required velocity ratio is 3.5:1. 20° involute stub tooth profile is to be used. The static design-stress for the pinion is 100 MPa and for the gear is 70 MPa, and the pinion has 16 teeth. Determine the module, face width, and pitch circle diameters based on a service factor of 1.25.

Given Data: Power $NP = 20$ kW @ speed $N_1 = 1000$ rpm

Speed ratio, $G = 3.5$; No. of teeth $Z_1 = 16$

Allowable stress for pinion, $S_d = 100$ MPa; for gear $S_d = 70$ MPa

Tooth form 20° involute stub; Service factor $C_s = 1.25$

SOLUTION

Step-1: Comparison of strength factors

Number of teeth on pinion, $Z_1 = 16$

Number of teeth on gear, $Z_2 = 16 \times 3.5 = 56$

The tooth form factor (Table 16.3 MDHB)

For pinion $Y_p = 0.359$; for gear $Y_g = 0.481$

Strength factors

For pinion $(S_d Y)_p = 100 \times 359 = 35.9$

For gear $(S_d Y)_g = 70 \times 481 = 33.67$

The gear is weaker hence design the gear.

Step-2: Module by Lewis equation

The speed of the gear, $N_2 = N_1/G = 1000/3.5 = 285.7$ rpm

$$T_d = \frac{955 \times 10^4 \text{ kW}}{N_2} C_s = \frac{955 \times 10^4 \times 20}{285.7} \times 1.25 = 668500 \text{ Nmm}$$

Assume $\beta = b/m = 10$; and the velocity factor $K_v = 0.3$

Then by Lewis equation 16.10 MDHB

$$m = \left[\frac{2 \times T}{Y \cdot \beta \cdot Z \cdot S_d \cdot K_v}\right]^{1/3} = \left[\frac{2 \times 668500}{0.481 \times 10 \times 56 \times 70 \times 0.3}\right]^{1/3} = 6.183$$

Choose $m = 6.0$ mm

Step-3: Dimension of gears

Diameter of pinion, $D_p = 16 \times 6 = 96$ mm

Diameter of gear, $D_p = 56 \times 6 = 336$ mm

The face width, $b = \beta \cdot m = 10 \times 6 = 60$ mm

Step-4: Dynamic load by Buckingham equation

$$V = \frac{\pi DN}{60,000} = \frac{\pi \times 96 \times 1000}{60,000} = 5.02 \text{ m/s}$$

The tangential load,

$$w_t = \frac{T}{r} = \frac{665500}{(336/2)} = 3979.17 \text{ N}$$

Specifying class-III accuracy, the permissible error

$e = 17$ μm (Table 16.27 MDHB); the dynamic factor $C = 150$ (Table 16.5 MDHB)

Then dynamic load by Buckingham equation

$$w_d = w_t + \frac{21V(bc + w_t)}{21V + \sqrt{(b \cdot c + w_t)}}$$

$$= 3979.17 + \frac{21 \times 5.02 \times (60 \times 150 + 3979.17)}{21 \times 5.02 + \sqrt{60 \times 150 + 3979.17}} = 10217.1 \text{ N}$$

Step-5: Dynamic beam strength,

$$w_b = S_{dr} \cdot b \cdot Y \cdot m \qquad \text{(Eqn. 16.15 MDHB)}$$

For C-I gear $S_{dr} = 83.5$ MPa (Table 16.25 MDHB)

$$\therefore \quad w_b = 83.5 \times 60 \times 0.481 \times 6 = 14458.86 \text{ N} > W_d$$

This is more than the dynamic load, hence design is safe against bending failure.

Step-6: Wear strength,

$$W_w = K \cdot b \cdot D_p \cdot Q \qquad \text{(Eqn. 16.16 MDHB)}$$

where,

$$Q = \frac{2.G}{G+1} = \frac{2 \times 3.5}{3.5 + 1} = 1.56$$

The stress factor K for $W_w = W_d$

$$K = \frac{W_d}{b \cdot D_p \cdot Q} = \frac{10217.1}{60 \times 96 \times 1.56} = 1.137$$

From Table 16.23 MDHB the surface hardness required for pinion is BHN = 250 and for gear BHN = 180 with K = 1.375.

Example 15.11: Two spur gears are to be used for a rock crusher drive and are to be of minimum size. The gears are to be designed for the following requirements. The power to be transmitted is 20 kW. Speed of the pinion is 1200 rpm. Velocity ratio is 3.5:1. Tooth profile is 20° involute stub. Determine the module, face width for strength requirement only.

Given Data: Power, $NP = 20$ kW; Pinion speed, $N_1 = 1200$ rpm
Velocity ratio, $G = 3.5$; Tooth form – 20° involute stub

SOLUTION

Step-1: Materials data (Table 16.4 MDHB)

Material for pinion cast steel width, S_d = 195 MPa

Hardness = 250 BHN;

For gear–C.I. Gr 35 treated with S_d = 78.5 MPa

Hardness = 300 BHN

Step-2: Strength factors

The number of teeth for pinion, $Z_1 = 16$

Number of teeth for gear, $Z_2 = 16 \times 3.5 = 56$

The tooth form factors (Table 16.3 MDHB)

For pinion, $Y_p = 0.359$; For gear $Y_g = 0.481$

Strength factors

For pinion $(S_d Y)_p = 195 \times 0.359 = 70$

For gear $(S_d Y)_g = 78.5 \times 0.481 = 37.76$

The gear is weak, hence design the gear.

Step-3: Module by Lewis equation

The gear speed, $N_2 = N_1/G = \dfrac{1200}{3.5} = 342.85$ rpm

The torque on the gear,

$$T = \frac{955 \times 10^4 \, \text{kW}}{N} = \frac{955 \times 10^4 \times 20}{342.85} = 557083.3 \text{ Nmm}$$

The ratio, $\beta = \dfrac{b}{m} = 10$ assumed

The velocity factor, $K_v = 0.3$ assumed

$$\text{Then module, } m = \left[\frac{2T}{Y \cdot \beta \cdot Z \cdot S_d \cdot K_v}\right]^{1/3}$$

$$= \left[\frac{2 \times 557083.3}{0.481 \times 10 \times 56 \times 78.5 \times 0.3}\right]^{1/3} = 5.6$$

Choose $m = 6.0$ mm (Table 16.2 MDHB)

Then face width, $b = \beta \cdot m = 10 \times 6 = 60$ mm

Diameter of the pinion, $D_p = 6 \times 16 = 96$ mm

Diameter of the gear, $D_g = 6 \times 56 = 312$ mm

Pitch line velocity,

$$V = \frac{\pi DN}{60{,}000} = \frac{\pi \times 96 \times 1200}{60{,}000} = 6.032 \text{ m/s}$$

The exact velocity factor,

$$K_v = \frac{3}{3 + V} = \frac{3}{3 + 6.032} = 0.332$$

This is more the assumed value of $K_v = 0.3$ and the selection of module is correct.

Example 15.12: It is required to transmit 15 kW power from a shaft running at 1200 rpm to a parallel shaft with speed reduction of 3.0. The centre distance between the shafts is to be 300 mm. The material used for pinion is steel with $S_d = 200$ MPa and for gear is C.I. with $S_d = 140$ MPa. The service factor is 1.25 and tooth profile is 20° involute full depth. Design the spur gears and check the design for dynamic and wear loads.

Given Data: Power, $NP = 15\text{ kW}$ @ speed, $N_1 = 1200$ rpm;

Speed ratio, $G = 3.0$; Centre distance, $C = 300$ mm;

Material for pinion steel with $S_{d_1} = 200$ MPa

Material for gear C.I. with $S_d = 140$ MPa

Tooth form – 20° involute full depth

SOLUTION

Step-1: Diameters of gears by geometry

Center distance, $C = (D_p + D_g)/2 = 300$ mm

i.e., $$D_p + D_g = 600 \text{ mm} \tag{i}$$

Speed ratio, $G = 3$, i.e., $\dfrac{D_g}{D_p} = 3$

$$\therefore \quad D_g = 3D_p \tag{ii}$$

Substituting (ii) in (i)

$$D_p + 3D_p = 600$$

$\therefore$ Pinion diameter, $D_p = 600/4 = 150$ mm

Gear diameter, $D_g = 150 \times 3 = 450$ mm

Step-2: Module by geometry

Number of teeth on pinion, $Z_1 = 30$

Number of teeth on gear, $Z_2 = Z_1 \cdot G = 30 \times 3 = 90$

The module, $m = \dfrac{D_p}{Z_1} = \dfrac{150}{30} = 5$ mm

Choose, $m = 5.0$ mm (Table 16.2 MDHB)

Step-3: Comparison of strength factors

Tooth form factors

For pinion, $Y_p = 0.359$; for gear, $Y_g = 0.444$

Strength factor for pinion $(S_d Y)_p = (200 \times 0.359) = 71.8$

For gear $(S_d Y)_g = (140 \times 0.444) = 62.16$

∴ The gear is weaker, hence design the gear.

Step-4: Face width by Lewis equation

The torque on the gear,

$$T = \frac{955 \times 10^4 \text{kW}}{N} = \frac{955 \times 10^4 \times 15}{(1200/3)} = 358125 \text{ Nmm}$$

The tangential load on tooth,

$$w_t = (T/r) = \frac{358125}{(450/2)} = 1591.67 \text{ N}$$

The pitch line velocity,

$$V = \frac{\pi DN}{60{,}000} = \frac{\pi \times 150 \times 1200}{60{,}000} = 9.43 \text{ m/s}$$

The velocity factor, $K_v = \dfrac{4.58}{4.58+V} = \dfrac{4.58}{4.58+9.43} = 0.33$ (Eqn. 16.12a MDHB)

Face width by Lewis equation, (Eqn. 16.10 MDHB)

$$b = \frac{w_w}{Y \cdot mS_d \cdot K_v} = \frac{1591.67}{0.444 \times 5 \times 140 \times 0.33} = 15.52 \text{ mm}$$

Then, $b/m = \dfrac{15.52}{5} = 3.104$

This is too small, hence try alternate solution.

Step-5: Alternate choice

Try alternate value of module, $m = 3.0$ mm

Then, $Z_1 = \dfrac{150}{3} = 50$ and $Z_2 = \dfrac{450}{3} = 150$

Tooth form factor for gear, $Y_g = 0.460$

Then face width,

$$b = \frac{w_t}{Y \cdot m \cdot S_d \cdot K_v} = \frac{1591.67}{0.46 \times 3 \times 140 \times 0.33} = 24.97 \text{ mm} \cong 25 \text{ mm}$$

Then $\beta = b/m = \dfrac{25}{3} = 8.32$

This is within the specified range of 7 to 12.5 and hence, acceptable.

Step-6: Dynamic load by Buckingham equation

The permissible error for class – III accuracy (Fig. 16.3 MDHB) $e = 11\,\mu\text{m}$ the dynamic factor from (Table 16.5 MDHB) $C = 90$ (by extrapolation)

Dynamic load,

$$w_d = 1591.67 + \frac{21 \times 9.43 \times (25 \times 90 + 1591.67)}{21 \times 9.43 + \sqrt{25 \times 90 + 1591.67}} = 4517.57 \text{ N}$$

Step-7: Dynamic beam strength,

$$w_b = S_{dr} \cdot b \cdot Y \cdot m = 140 \times 25 \times 0.444 \times 3 = 4662N > W_d$$

Hence, design is safe against bending failure.

Step-8: Wear strength,

$$W_w = K \cdot b \cdot D_p \cdot Q$$

where, $Q = \dfrac{2.G}{G+1} = \dfrac{2 \times 3}{3+1} = 1.5$; $b = 25$ mm; $D_p = 150$ mm

The strength factor required,

$$K = \frac{w_d}{b \cdot D_p Q} = \frac{4517.57}{25 \times 150 \times 1.5} = 0.803$$

The surface hardness required (Table 16.23 MDHB)

For pinion $= 200$ BHN; for gear $= 180$ BHN

Example 15.13: A cast steel 24 teeth spur pinion operating at 1150 rpm transmits 3 kW of power to cast-steel, 56 teeth spur gear. The gears have the following specifications:

Module, $m = 3$ mm, face width, $b = 35$ mm,

The allowing stress, $S_d = 100$ MPa

Tooth form—$14\frac{1}{2}$ full depth profile

The factor of dynamic loading $C = 350$ N/mm.

Wear load factor, $K = 0.28$ MPa. Determine the induced stress in the weaker member. Also determine the dynamic load and wear load. Comment on the results.

Given Data: Number of teeth on pinion = 24; No. of teeth on gear = 56

Power, $NP = 3$ kW @ speed 1150 rpm

Module, $m = 3$; Face width $b = 35$ mm

Allowable stress, $S_d = 100$ MPa; Tooth profile: $14\frac{1}{2}$ full depth

Dynamic load factor, $C = 350$ N/mm;

Wear load factor, $K = 0.28$ MPa

SOLUTION

Step-1: Induced stress by Lewis equation

The torque on pinion,

$$T = \frac{955 \times 10^4 \text{ kW}}{N} = \frac{955 \times 10^4 \times 3.0}{1150} = 24913 \text{ Nmm}$$

Diameter of pinion, $D_p = 24 \times 3 = 72$ mm

Diameter of gear, $D_g = 56 \times 3 = 168$ mm

The tangential load,

$$w_t = \frac{T}{r} = \frac{24913}{(72/2)} = 692.03 \text{ N}$$

Tooth form factor, $Y = 0.299$

The pitch line velocity, $V = \dfrac{\pi DN}{60,000} = \dfrac{\pi \times 72 \times 1150}{60,000} = 4.34$ m/s

Velocity factor, $K_v = \dfrac{3}{3+v} = \dfrac{3}{3+4.34} = 0.41$

The stress induced by Lewis equation,

$$\sigma_b = \frac{w_t}{b \cdot Y \cdot m \cdot K_v} = \frac{692.03}{35 \times 0.299 \times 3 \times 0.41} = 53.76 \text{ MPa} < (S_d)$$

Step-2: Dynamic load by Buckingham equation

$$w_d = w_t + \frac{21v(b \cdot c + w_t)}{21v + \sqrt{bc + w_t}}$$

$$= 692.03 + \frac{21 \times 4.34 \times (35 \times 350 + 692.03)}{21 \times 4.34 + \sqrt{35 \times 350 + 692.03}} = 6448.59 \text{ N}$$

Step-3: Dynamic beam strength,

$$w_b = Y \cdot b \cdot m \cdot S_{dr} = 0.299 \times 35 \times 3 \times 100 = 3139 \text{ N}$$

This is less than the dynamic load. Hence, the material with higher strength value is to be selected or the material is to be heat treated to increase the hardness and strength.

Step-4: The wear strength,

$$w_w = K \cdot b \cdot D_p \cdot Q$$

where $Q = \dfrac{2 \times 56}{56 + 24} = 1.4$

Then the wear strength factor,

$$K = \frac{W_d}{b \cdot D_p Q} = \frac{6448.59}{35 \times 72 \times 1.4} = 1.829$$

The surface hardnesses required are:

For pinion BHN = 350; for gear BHN = 350.

Example 15.14: A compressor running at 300 rpm is driven by 15 kW at 1200 rpm motor through a pair of 20° full depth system spur gears. The centre distance between gears is 380 mm. Motor pinion is made of 0.3% carbon forged steel (heat treated) and gear is made of 0.2% carbon heat treated cast steel. Assuming medium shock conditions, determine module, face width, number of teeth on pinion and gear and hardness.

Given Data: Power, $NP = 15$ kW at speed $N_1 = 1200$ rpm;

Driven speed, $N_2 = 300$ rpm;

Tooth profile—20° full depth; Centre distance, $C = 380$ mm;

Material for pinion -0.3% carbon forged steel;

For gear $-$ 0.2% carbon-cast steel;

Service: Moderate shock

SOLUTION

Step-1: Material properties (Table 16.4 MDHB)

For pinion $-$ $C45$ treated $S_d = 210$ MPa, BHN = 250

For gear $-$ cast steel treated $S_d = 195$ MP; BHN = 250

Step-2: Diameters of the gears

The speed ratio, $\frac{N_1}{N_2} = \frac{1200}{300} = 4$

$\therefore$ The ratio of diameters, $D_2/D_1 = 4$ (1)

The centre distance, $C = \frac{D_1 + D_2}{2} = 380$

$\therefore$ $D_1 + D_2 = 760$ (2)

Substitution for D_2 from (1)

$$D_1 + 4D_1 = 760$$

$\therefore$ $D_1 = 760/5 = 152$ mm

$$D_2 = 152 \times 4 = 608 \text{ mm}$$

Step-3: Estimation of module by Lewis equation

The torque on pinion,

$$T = \frac{955 \times 10^4 \times kW}{N} = \frac{955 \times 10^4 \times 15}{1200} = 119375 \text{ Nmm}$$

Service factor, $C = 1.5$ (Table 16.21 MDHB)

Design torque, $T_d = T \times C_s \ = 119375 \times 1.5 = 179062.5$ Nmm

The tangential force on pinion,

$$w_t = \frac{T_d}{r} = 179062.5/76 = 2356.1 \text{ N}$$

Pitch line velocity, $V = \dfrac{\pi DN}{60,000} = \dfrac{\pi \times 152 \times 1200}{60,000} = 9.55$ m/s

The velocity factor, $K_v = \dfrac{4.58}{4.58 + V} = \dfrac{4.58}{4.58 + 9.55} = 0.324$

Tooth form factor, $Y = 0.30$ (assumed)

Select $b/m = \beta = 7$

The module by Lewis equation,

$$m = \sqrt{\frac{w_t}{Y \cdot \beta \cdot S_d \cdot K_v}} = \sqrt{\frac{2356.1}{0.30 \times 7 \times 210 \times 0.324}} = 4.06$$

Choose $m = 4$ mm

Number of teeth on pinion, $Z_1 = 152/4 = 38$

Number of teeth on gear, $Z_2 = 38 \times 4 = 152$

The face width, $b = \beta, m = 7 \times 4 = 28$ mm

The tooth form factors (Table 16.3 MDHB)

For pinion, $Y_p = 0.384$; for gear, $Y_g = 0.46$

The bending stress induced in pinion,

$$\sigma_b = \frac{w_t}{Y_p \cdot b \cdot m K_v} = \frac{2356.1}{0.384 \times 28 \times 4 \times 0.324} = 169.08 \text{ MPa} < S_d$$

The bending stress in gear tooth,

$$\sigma_{bg} = \frac{2356.1}{0.46 \times 28 \times 4 \times 0.324} = 141.15 \text{ MPa} < (S_d = 195 \text{ MPa})$$

Step-4: Dynamic load by Buckingham equation

The permissible error for class-II gear, e = 25 μm (Table 16.27 MDHB)

The dynamic factor, $C = 150$ (Table 16.5 MDHB)

Then dynamic load,

$$w_d = w_t + \frac{21V(b \cdot c + w_t)}{21V + \sqrt{b \cdot c + w_t}}$$

$$= 2356.1 + \frac{21 \times 9.55 \times (28 \times 150 + 2356.1)}{21 \times 9.55 + \sqrt{28 \times 150 + 2356.1}} = 7026.56 \text{ N}$$

Step-5: Dynamic beam strength,

$$W_b = S_{dr} b \cdot m \cdot Y = 210 \times 28 \times 4 \times 0.384 = 9031.7\,N > W_d$$

This is greater than the dynamic load and hence design is safe against bending failure.

Step-6: Wear strength,

$$w_w = K \cdot b \cdot D_p Q$$

where,

$$Q = \frac{2 \times 4}{4 + 1} = 1.6$$

The strength factor for $W_w = W_d$

$$K = \frac{7026.56}{28 \times 152 \times 1.6} = 1.032$$

From Table 16.23 MDHB, surface hardness

For pinion = 300 BHN; for gear = 250 BHN.

EXERCISES FOR PRACTICE

15.1 (a) By the inspection of the Lewis equation, state what simple product determines whether the pinion tooth or the gear tooth is weaker. Explain.

(b) It is required to determine the proportions of a spur gear to transmit 20 kW power from a shaft running at 600 rpm to a parallel shaft with a speed reduction of 3 : 1. The centre-to-centre distance of the shaft be near about 300 mm. Assuming that teeth are 20° involute stub and the pinion is made of forged steel untreated C-30, select the material for the gear to obtain a compact design and determine the module, number of teeth, face width and pitch diameter of gears.

15.2 (a) Derive Lewis equation for the design of a pair of spur gears.

(b) Design a pair of spur gears to transmit a power of 30 kW from a shaft rotating at 1,500 rpm to another shaft to be run at 450 rpm. Give detailed specifications including the hardness to which gear surfaces must be treated.

15.3 A cast-iron spur gear is driven by a cast steel untreated pinion, transmitting a power of 4 kW at a speed of 1500 rpm. The gear rotates at a speed of 500 rpm. The drive operates for 8 to 10 hours per day with medium shock. Design stress for pinion is 190 MPa and for gear is 100 MPa. The pinion is of 200 BHN and gear of 180 BHN. The teeth are 20° full depth involute. Design the gears and check for dynamic load and wear load.

15.4 (a) Derive the Lewis equation for bending stress in gear tooth.

(b) Two shafts running at 100 rpm and 135 rpm and transmitting 60 kW, are connected by cut spur-gears of high grade cast-iron. Determine the necessary module, number of teeth, face width and exact centre distance. The service is 18 hours per day.

15.5 A compressor running at 400 rpm, is driven by a 20 kW, 1200 rpm motor through a pair of 14.5° full depth involute spur gears. The centre distance is to be 40 cm. The motor pinion is to be of forged steel with an allowable stress of 210 MPa and 250 BHN and the gear is to be of cast steel with an allowable stress of 195 MPa and 250 BHN. Design the gear drive for safe and continuous operation. Also check the gears for dynamic and wear loads.

15.6 Design a spur gear drive, from consideration of dynamic load and wear load, to transmit 90 kW at 600 rpm to a shaft running at 150 rpm. Assume 10 mm module and 20° full depth involute system. The pinion is to have 20 teeth and is made of heat treated cast steel, having value of $E = 2 \times 10^5$ MPa and safe static stress of 200 MPa. Assume suitable material for gear.

15.7 A pair of spur gear with 20° full depth teeth transmits 14.5 kW at 1500 rpm of the pinion. The speed reduction is 4:1. Select the suitable materials for the gears and design the gear drive. The minimum number of teeth on the pinion should not be less than 20.

15.8 (a) Define: (i) Module (ii) Addendum (iii) Dedendum (iv) Working depth (v) Pressure angle.

(b) A power of 20 kW is to be transmitted at a pinion speed of 900 rpm and with a gear ratio of 2.3 to 1. The pinion is made of cast steel with permissible design stress of $S_d = 140$ MPa in bending. The gear is made of cast iron with a design stress of $S_d = 55$ MPa. The centre distance is to be approximately 245 mm. Design the gear set for bending and check for wear strength.

15.9 Design a pair of spur gears to transmit a power of 20 kW from a shaft running at 1500 rpm to another shaft to be run at 350 rpm maintaining a distance of (80 ± 5) mm between the shaft centres.

15.10 In the layout of the drive for a packing machine, a pair of 20° full depth spur gears is to transmit 4 kW at a transmission ratio of 3:1. The pinion operates at 1500 rpm. Design the system of pair of gears. Assume carefully cut teeth.

15.11 (a) Explain the terms : Static bending load, dynamic load and wear load in case of spur gear design. State the importance of each.

(b) A driving shaft *P* rotates at 300 rpm and is placed in alignment with a driven shaft *Q* rotating at 25 rpm. A spur gear on *P* drives a gear *B* on a back shaft *S*. Another gear *C* on shaft *S* drives a gear *D* on shaft *Q*. The centre distance between shafts *P* and *S* is to be not less than 225 mm and not more than 250 mm. Gears *C* & *D* are of module

6 mm and *A* and *B* are of the module 5 mm. Find the number of teeth of the gears, if no gear is to have more than 75 teeth and less than 12 teeth. The drive transmits 1.5 kW. Choosing your own materials for the gears, design the gears and sketch the arrangement.

15.12 Design a pair of spur gears to transmit a power of 25 kW from a motor running at 1440 rpm to a pump running at 360 rpm. The maximum centre distance is to be 100 mm. Choose the suitable materials for the gears. Check the design for dynamic beam strength and wear loads.

15.13 A 45 kW motor running at 1000 rpm is required to drive a rotary mill at 120 rpm through a pair of spur gears. The distance between the axes of the motor and the mill shaft is to be about 1.0 m. Design the spur gear to go with the mill shaft. Assume a shaft diameter of 100 mm for the mill. Assume a suitable material for gear and pinion.

15.14 It is desired to transmit 15 kW from a motor shaft rotating at 1440 rpm to a low speed shaft with a speed reduction of 3:1. The teeth are 14.5° involute with 25 teeth on the pinion. The starting torque is 5% higher than the running torque. Both the pinion and gear are made of forged steel heat treated for which maximum allowable static stress may be taken as 210 MPa. The shaft on which the gear is to be mounted, and the key are of mild-steel for which the safe stress may be taken as 45 MPa. Design the suitable spur gear drive.

15.15 Design a spur gear drive, from consideration of dynamic load and wear load to transmit 100 kW at 600 rpm to a shaft running at 150 rpm. Assume 10 mm module and 20° full depth involute system. The pinion is to have 20 teeth and is made of heat-treated cast-steel having value of *E* as 2×10^5 MPa and safe static stress of 200 MPa. Assume suitable material for gear.

15.16 Design a pair of spur gears to transmit a power of 18 kW from a shaft running at 1000 rpm to a parallel shaft to be run at 250 rpm, maintaining a distance of 160 mm between the shaft centres. Suggest suitable surface hardness for the gear pair.

15.17 A shaft rotating at a speed of 1000 rpm is to transmit a power of 40 kW to a parallel shaft to be rotated at 350 rpm. The distance between the shaft centres is 160 mm. Design a pair of spur gears to connect these two shafts.

15.18 (a) Give reasons for the selection of involute tooth profile for gears more commonly. While generating less number of teeth on gear wheels, the problem of interference prevails in involute tooth gearing. Does such a problem really exist in cycloidal tooth gearing?

(b) A pair of spur gears is required to transmit 5 kW of power. The pinion rotates at a speed of 1620 rpm and the gear is required to run 420 rpm. Determine the least number of teeth on the gear and pinion such that velocity ratio does not deviate at all. The tooth form is 20° full depth involute. Therefore, the number of teeth selected on the pinion should not be less than the theoretical minimum. The permissible stress in the material of the gear and the pinion are 55 MPa and 65 MPa respectively. Design the gears for beam strength only and determine all the proportions of the gearing.

15.19 (a) Derive an expression for beam strength of a spur gear tooth with standard notations.

(b) A pair of spur gears has to transmit 20 kW from a shaft rotating at 1000 rpm to a parallel shaft which is to rotate at 310 rpm. Number of teeth on pinion is 31 with 20°

full depth involute tooth form. The material for pinion is steel SAE 1040 untreated with allowable static stress of 206.81 MPa and the material for gear is cast steel 0.20% C untreated with allowable static stress 137.34 MPa. Determine the module and face width of the gear pair. Also find the dynamic tooth load on the gears. Take service factor as 1.5.

15.20 (a) Derive an expression for beam strength of a spur gear tooth.

(b) Design a pair of spur gears to transmit 15 kW from a shaft rotating at 1000 rpm to a parallel shaft which is to rotate at 310 rpm. Assume number of teeth on pinion 31 and 20° full depth tooth form. The material for pinion is C-40 steel untreated and for gear, cast steel 0.20% C untreated.

15.21 A cast steel pinion rotating at 900 rpm is to drive a cast iron gear at 144 rpm. The static design stresses for pinion and gear materials are 103 MPa and 55 MPa respectively. The teeth are to have standard 20° stub involute profiles and the maximum power to be transmitted is 25 kW. Design the spur gears completely and check for the dynamic and wear loads. The gear surfaces are hardened to BHN 250. Use 16 teeth on the pinion.

15.22 A machine running at 360 rpm is driven by 12 kW, 1440 rpm motor through a 14½ involute gear. The centre distance between the drive is 250 mm. The pinion is made up of heat treated cast steel with allowable stress of 191.2 MPa and 450 BHN. The gear is made of untreated cast steel with allowable static stress of 137.3 MPa and BHN of 300. Assuming gears are working 8 hours/day and subjected to light shocks, determine the module, face width and number of teeth on each gear. Also check the design for wear.

15.23 It is required to transmit 25 kW of power from shaft running at 1000 rpm to a parallel shaft with speed reduction 2.5 to 1. The centre-to-centre distance of shafts is to be about 300 mm. The material used for pinion is steel ($\sigma_d = 200\ \text{N/mm}^2$, BHN = 250), and for gear is cast iron ($\sigma_d = 180\ \text{N/mm}^2$, BHN = 200). Considering class-II gears with tooth profile 20° full depth involute, design the spur gear and check for dynamic load and wear load.

SHORT ANSWER QUESTIONS

15.1 What is a gear set?

15.2 What are the types of gears?

15.3 What is the law of gearing?

15.4 What is the tooth profile?

15.5 Which are the profiles satisfying the law of gearing?

15.6 What are the advantages of the involute profile?

15.7 What are the merits of cycloidal profile?

15.8 Which one is the preferred profile and why?

15.9 What is pitch circle?

15.10 What is base circle?

15.11 What is addendum circle?

15.12 What is dedendum circle?

15.13 What is circular pitch?
15.14 What is module?
15.15 What is diametral pitch?
15.16 What is addendum?
15.17 What is dedendum?
15.18 What is clearance?
15.19 What is face width?
15.20 What is pressure angle?
15.21 What is the path of approach?
15.22 What is the path of recess?
15.23 What is the path of contact?
15.24 What is the arc of action?
15.25 What is contact ratio?
15.26 What is interference?
15.27 How do you avoid it?
15.28 What is the minimum number of teeth on pinion to avoid interference?
15.29 What are the stresses induced in spur-gear tooth?
15.30 Give the equation for bending stress.
15.31 What is the Lewis equation for the module based on bending strength?
15.32 What is velocity factor?
15.33 Give the Barth formulae for velocity factors.
15.34 What is dynamic load?
15.35 Give the Buckingham dynamic load equation.
15.36 What is wear-strength equation based on Hertz contact stress?
15.37 Give the new AGMA equation for bending stress in spur gear tooth.
15.38 Give the new AGMA equation for contact stress in spur gear tooth.
15.39 Explain the various correction factors in the above equations.
15.40 List materials suitable for gears.
15.41 What is backlash? List the factors that influence backlash.
15.42 List the reasons for gear failure.
15.43 What is the effect of increasing the pressure angle in gears?

CHAPTER 16

HELICAL GEARS

In helical gears, the teeth are formed along a helix on the pitch cylinder instead of parallel to the axis as in spur gears. If the pinion has right-hand helical teeth, the gear has left-hand helical teeth, the helix angle being the same.

16.1 TERMINOLOGY OF HELICAL GEAR

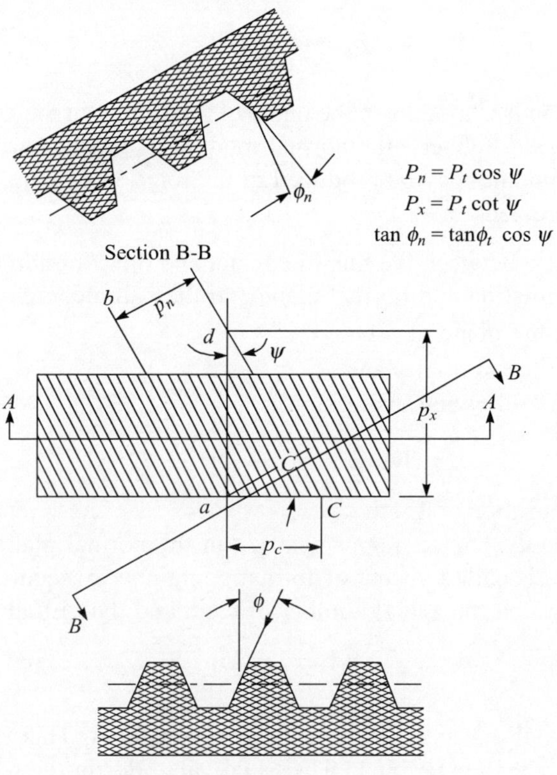

Figure 16.1 Terminology of helical gears

Helix angle (Ψ) is the angle which the tooth central line makes with the axis of gear

Module is the plane of rotation, $m = (D/Z)$

Normal module is the module in the normal plane, which is perpendicular to the helix. It is the module of the equivalent spur gear, given by

$$m_n = m \cos \Psi \tag{16.1}$$

Circular pitch (p_c) is the pitch measured in the plane of rotation or transverse plane.

$$p_c = \frac{\pi D}{Z} = \pi m \tag{16.2}$$

Normal pitch is the circular pitch measured in the plane normal to the helix or it is the circular pitch of the equivalent spur gear.

$$p_n = p_c \cos \Psi \tag{16.3}$$

Axial pitch is the distance between the central lines of the successive teeth measured in the axial direction.

$$p_a = p_c / \tan \Psi \tag{16.4}$$

The gears may have standard module in the plane of rotation or in the normal plane. Whenever the teeth are formed by gear hobbers, the normal module is the standard value. If form cutters are used, then standard module may be produced in the rotational plane. In design practice, the normal module is standardized.

Pressure angle is the angle between the line of contact and the common tangent. The values are different in the plane of rotation and normal plane and they are denoted as follows:

ϕ_t = Pressure angle in the plane of rotation

ϕ_n = Pressure angle in the normal plane

They are related by the following expression,

$$\tan \phi_n = \tan \phi_t \cos \Psi \tag{16.5}$$

Virtual or Formative Number of Teeth

The helical gear set is equivalent to a spur gear set in the normal plane. The number of teeth on these equivalent gears is called virtual or formative number of teeth. Geometrically, we can establish the relation between the actual number of teeth and the virtual number of teeth as

$$Z_v = Z / \cos^3 \Psi \tag{16.6}$$

Face width (b) is the axial distance between the two end faces. This is the width of the gear blank. It is always greater than axial pitch. The usual value lies in the range,

$$b = 1.15\, p_a \text{ to } 2p_a \tag{16.7}$$

16.2 TOOTH LOADS

The actual load between the engaging teeth is normal to the teeth profile in the normal plane. It will have tangential, axial and radial component as shown in Fig. 16.2. They are given below.

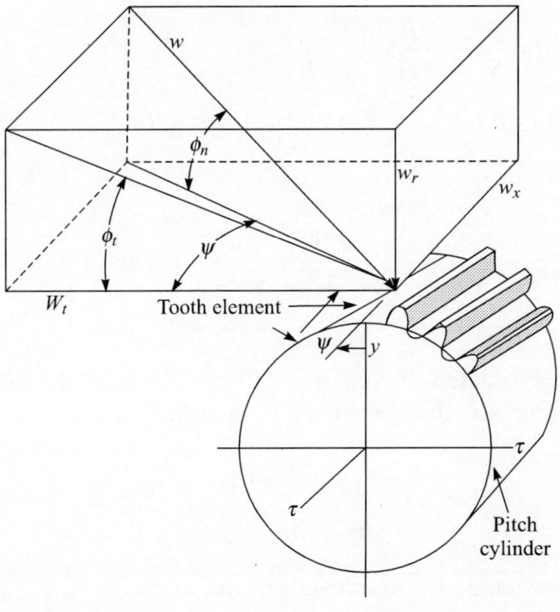

Figure 16.2 Tooth forces acting on a right-hand helical gear

Axial thrust, $W_a = W_t \tan \Psi$ (16.8)

Radial load, $W_r = W_t \tan \Phi_t$ (16.9)

Normal load, $W = \dfrac{W_t}{\cos \Psi \cos \Phi_t}$ (16.10)

These are expressed in terms of W_t, because the value of W_t can be evaluated from the power transmitted by the gear set using equation,

$$W_t = T/r = \frac{955 \times 10^4 \text{ kW}}{N \cdot r} \qquad (16.11)$$

16.3 DESIGN PROCEDURE

In the normal plane, helical gear set is equivalent to spur gear set. The Lewis equation can be adopted as

$$W_{tn} = S_d \left(\frac{b}{\cos \Psi} \right) \cdot Y m_n$$

Since $W_t = W_{tn} \cos \Psi$, then $W_t = S_d \cdot b \cdot Y \cdot m_n$

Correcting for dynamic effect, $W_t = S_d \cdot b \cdot Y \cdot m_n \cdot K_v$

Normal module, $$m_n = \frac{W_t}{S_d \cdot b \cdot K_v \cdot Y} \tag{16.12}$$

or Tangential module, $$m = \frac{W_t}{S_d \cdot b \cdot Y \cdot K_v \cdot \cos \Psi} \tag{16.12a}$$

To account for the nature of lubrication, a lubrication factor C_w is introduced.

Then, $$m_n = \frac{W_t C_w}{S_d \cdot bY \cdot K_v} \tag{16.13}$$

or $$m = \frac{W_t C_w}{S_d \cdot b \cdot Y \cdot K_v \cdot \cos \Psi} \tag{16.13a}$$

These equations are used to determine the preliminary values of the module and face width.

The Buckingham equation for dynamic load is modified to account for inclination of the helical teeth as follows:

$$W_d = W_t + \frac{21V(b \cdot C \cdot \cos^2 \Psi + W_t) \cos \Psi}{[21V + \sqrt{b \cdot C \cos^2 \Psi + W_t})]} \tag{16.14}$$

Check equation for the dynamic beam strength of the tooth,

$$W_b = S_{dr} \cdot Y \cdot b \cdot m \cdot \cos \Psi = S_{dr} \cdot b \cdot Y \cdot m_n \tag{16.15}$$

This value should be more than the dynamic load, i.e., $W_b > W_d$.

This ensures that the teeth do not break due to the fluctuating bending stresses at the root of the teeth.

Wear Strength: The equation has a modification due to the inclination of the teeth as follows:

$$W_w = \frac{K \cdot b \cdot D_p \cdot Q}{\cos^2 \Psi} \tag{16.16}$$

To safeguard against pitting and scoring failure, this value should be more than the dynamic load, i.e., $W_w \geq W_d$.

Alternatively the strength factor is calculated to satisfy this condition,

$$K = \frac{W_d \cos^2 \Psi}{b \cdot D_p \cdot Q} \tag{16.16a}$$

The surface hardnesses required to obtain this strength value is determined from Table 15.14 MDHB and the treatment process is specified.

As per new AGMA standard the check for induced bending stress and the contact stress, is done by the following equations instead of Buckingham equation.

Bending Stress: The permissible bending stress is calculated by the equation

$$S_d = \frac{S \cdot K_L}{K_T \cdot R_R} \tag{16.17}$$

where, $S =$ Endurance limit stress (Table 15.11)

$K_L =$ Life factor (Table 15.8)

$K_T =$ Temperature factor = 273/(273 + t)

$K_R =$ Reliability factor (Table 15.9)

The induced stress is evaluated by the modified Lewis equation

$$\sigma_b = \frac{W_t \cdot K_o \cdot K_s \cdot K_m}{K_v \cdot m \cdot b \cdot J} \leq S_d \tag{16.18}$$

where, $K_o =$ Overload factor (Table 15.6)

$K_s =$ Size factor = 1.0

$K_m =$ Load distribution factor (Table 16.4)

$K_v =$ Velocity factor (Fig. 15.12) (Fig. 16.5 MDHB)

Velocity Factor equations are,

For low helical angle gear with $V < 5$ m/sec,

$$K_v = \frac{4.5}{4.5 + V} \tag{16.18a}$$

For all helical and Herringbone gears, for $V < 10$ to 20 m/sec,

$$K_V = \frac{6}{6 + V} \tag{16.18b}$$

For precision gears with $V < 20$ m/sec,

$$K_v = \frac{5.5}{5.5 + \sqrt{V}} \tag{16.18c}$$

For non-metallic gears,

$$K_v = \frac{0.75}{1 + V} + 0.25 \tag{16.18d}$$

$J =$ Geometry factor (Fig. 16.3 and Fig. 16.4a & 16.4b)

Contact Stress: The permissible contact stress is calculated by the equation,

$$S_{Hd} = \frac{S_{fe} \cdot C_L \cdot C_H}{C_T \cdot C_R} \tag{16.19}$$

where, S_{fe} = Surface endurance limit (Table 15.13)

$= 2.8\ \text{BHN} - 70$ for steels

$C_L = K_L$ = Life factor (Table 15.8)

$C_T = K_T$ = Temperature factor

$C_R = K_R$ = Reliability factor (Table 15.9)

The induced contact stress is calculated by Hertz contact equation modified with various modifying factors as given below:

$$\sigma_H = C_p \sqrt{\frac{W_t \cdot C_o \cdot C_s \cdot C_m \cdot C_f}{C_v \cdot D \cdot b \cdot I}} \leq S_{Hd} \tag{16.20}$$

where, C_p = Elastic coefficient (Table 15.10)

C_o = Overload factor (Table 15.6)

C_s = Size factor = 1.0 (assumed)

$C_m = K_m$ = Load distribution factor (Table 16.4)

C_f = Surface correction factor = 1.0 (assumed)

C_v = Velocity factor (Fig. 15.12)

I = Geomtry factor = 0.225 $[i/(1 + i)]$

i = Gear ratio $= Z_g / Z_p$

Table 16.1. Helical & herringbone gears standard proportions

Pressure angle in the plane of rotation	Maximum Minimum	25° 15°
Helix angle	Maximum Minimum	45° 20°
Addendum = a	Maximum Minimum	m 0.7 m
Clearance = c	Maximum Minimum	0.3 m 0.157 m
Dedendum = b		$a + c$

Table 16.2. Standard tooth proportions for helical gears

Quantity*	Formula	Quantity*	Formula
Addendum	$[1.0\ m_n]$	**External gears**	
Dedendum	$[1.25\ m_n]$	Standard centre distance	$\frac{D+d}{2}$

Contd.

Pinion pitch diameter	$\left[\frac{Z_p m_n}{\cos \psi}\right]$	Gear outside diameter	$D + 2a$
Gear pitch diameter	$\left[\frac{Z_G m_n}{\cos \psi}\right]$	Pinion outside diameter	$d + 2a$
Normal arc tooth thickness†	$\left[\pi m_n - \frac{B_n}{2}\right]$	Gear root diameter	$D - 2b$
Pinion base diameter	$d \cos \phi_t$	Pinion root diameter Internal gears	$d - 2b$
Gear base diameter	$D \cos \phi_t$	Centre distance	$\frac{D-d}{2}$
Base helix angle	$\tan^{-1}(\tan \psi \cos \phi_t)$	Inside diameter Root diameter	$D - 2a$ $D + 2b$

†B_n is the normal backlash.

Table 16.3. Tooth proportion for crossed helical gears

Driven helix angle $\psi_1°$	**Minimum tooth number Z**	**Driven helix angle $\psi_2°$**	**Normal pressure angle ϕ_n**
45	20	45	14° – 30′
60	9	30	17° – 30′
75	4	15	19° – 30′
86	1	4	20° – 0

Table 16.4. Load distribution factors K_m and C_m for helical gears

	Face width mm			
Characteristics	**50 mm**	**150 mm**	**225 mm**	**400 and up**
Accurate mounting small bearing clearance, minimum clearance, minimum deflection, precision gears	1.2	1.3	1.4	1.7
Less rigid mounting, less accurate gears contact across full face exists	1.5	1.6	0.17	2.0
Less accuracy and mounting such that less than full face contact exists			Over 2.0	–

Table 16.5. Wear and lubrication factor K_w

Nature of lubrication	K_w
(i) For enclosed gears, continuously lubricated with proper oil	1.15
(ii) Scant lubrication with regular inspection	1.25
(iii) Indifferent lubrication	1.35

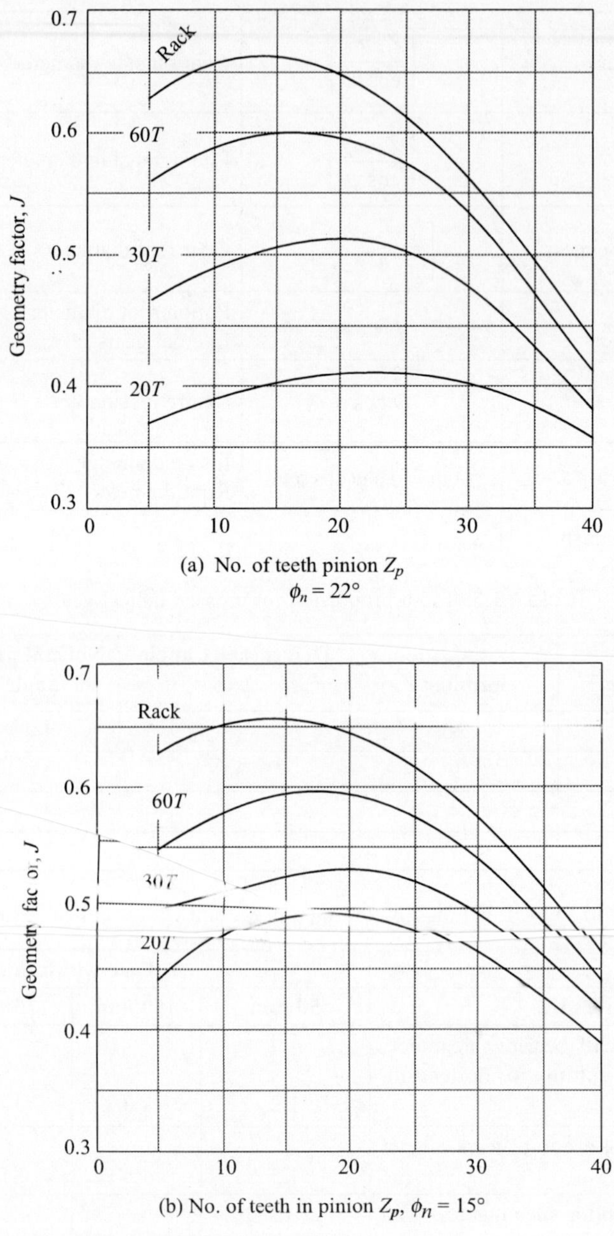

Figure 16.3 Geometry factor J for helical and herringbone gears (top) 22 deg. normal pressure angle (bottom) 15 deg. normal pressure angle

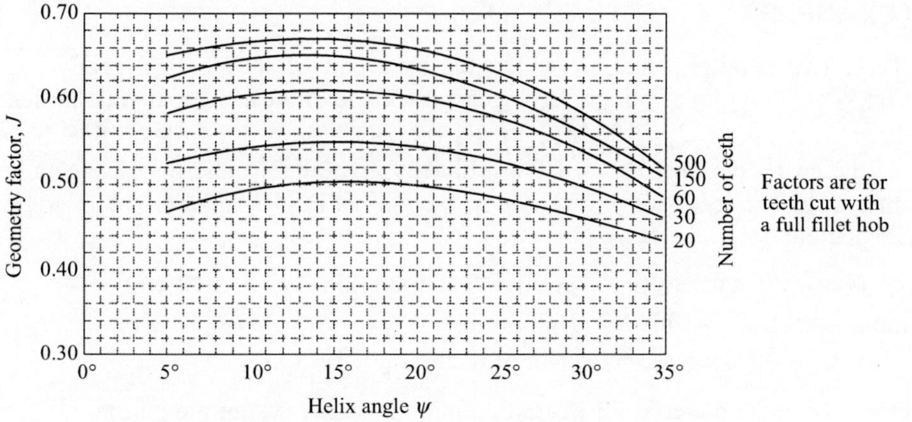

Figure 16.4(a) Helical-gear geometry factors J AGMA 218.01

The modifying factor can be applied to the J factor when other than 75 teeth are used in the mating element

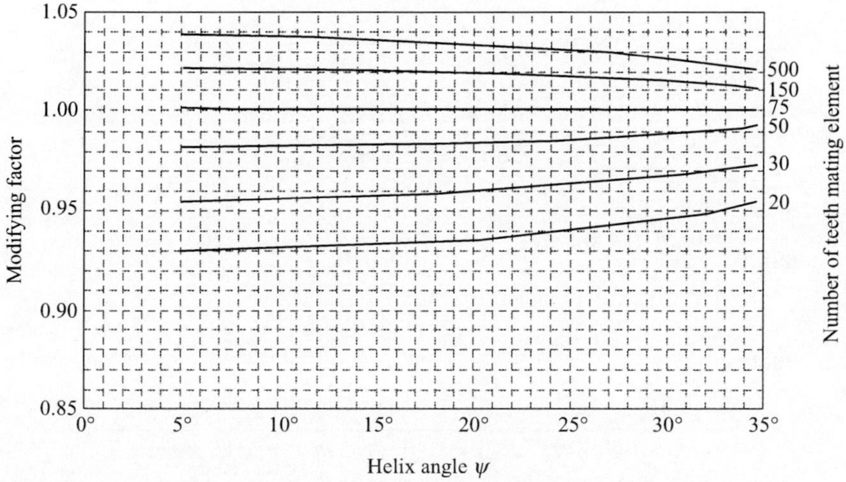

Figure 16.4(b) J'-factor multipliers for use with Fig. 16.4(a) to find J. AGMA 218.01

Equation for geometry factor I,

$$I = \begin{cases} \dfrac{\cos\phi_1 \sin\phi_1}{2m_N} \dfrac{m_G}{m_G+1} & \text{external gears} \\ \dfrac{\cos\phi_2 \sin\phi_2}{2m_N} \dfrac{m_G}{m_G-1} & \text{internal gears.} \end{cases} \tag{16.21}$$

where $m_N = 1$ for spur gears

$m_N = p_n/(0.957)$ for helical gears

m_G = gear ratio

SOLVED EXAMPLES

Example 16.1: Two parallel shafts are connected by a pair of steel helical gears. The pinion transmits 10 kW at 4000 rpm of pinion. Both gears are made of the same material, hardened steel with an allowable stress of $S_d = 100$ MPa. If the velocity ratio is 4.5:1, determine the smallest diameters of the gears that may be used for sufficient strength. No less than 30 teeth are to be used on either gear. The teeth are of 20° stub in the diametric plane and the helix angle is 45°. Use Lewis equation.

Given Data: Power -10 kW at Pinion speed, $N_p = 4000$ rpm
Speed ratio = 4.5:1, $S_d = 100$ MPa; Helix angle, $\Psi = 45°$;
Tooth form, $\Phi_t = 20°$ involute stub; No. of teeth, $Z_p = 30$

SOLUTION Since the material for gear and pinion is same, design the pinion.
Virtual number of teeth for pinion,

$$Z_{pv} = 30/\cos^3\Psi = \frac{30}{\cos^3 45} = 84.85$$

Tooth form factor, $Y = 0.5015$

$$\text{Torque, } T = \frac{955 \times 10^4 \times 10}{4000} = 23875 \text{ Nmm}$$

$$\text{Tangential load, } W_t = \frac{T}{r} = \frac{23875}{\frac{(m \cdot Z)}{2}} = \frac{23875 \times 2}{30m} = \frac{1592}{m}\text{N}$$

$$\text{Let } b = 2p_x = \frac{2 \cdot \pi m}{\tan \Psi} = 6.28 \text{ m} \cong 6.5 \text{ m}$$

Assume, $K_v = 0.4$ and $C_w = 1.25$

Using Lewis equation (Eqn. 16.31 MDHB)

$$m = \frac{W_t C_w}{S_d \cdot K_v \cdot b \cdot Y \cos \Psi} = \frac{(1592/m) \times 1.25}{100 \times 0.4 \times 6.5m \times 0.5015 \times \cos 45}$$

$$\therefore \quad m^3 = 21.57, \text{ i.e., } m = 2.75 \cong 3.0$$

$$D_p = 3.0 \times 30 = 90 \text{ mm}$$

To reduce the diameter, let $b = 15$ m

$$m = \frac{(1592/m) \times 1.25}{100 \times 0.4 \times 15m \times 0.5015 \times \cos 45}$$

$$m^3 = \frac{1592 \times 1.25}{100 \times 0.4 \times 15 \times 0.5015 \times 0.707}$$

$$\therefore \quad m = \sqrt[3]{9.354} = 2.10$$

Let $m = 2.0$ mm

Then $b = 2.0 \times 15 = 30$ mm ; $D_p = 2.0 \times 30 = 60$ mm

Pitch line velocity, $V = \dfrac{\pi DN}{60{,}000} = \dfrac{\pi \times 60 \times 4000}{60{,}000} = 12.6$ m/sec.

Then velocity factor, $K_v = 15/(15 + V) = 15/(15 + 12.6) = 0.54$
It is more than the assumed value of 0.4
Hence, the module selected is OK.

Example 16.2: Two precision cut forged steel helical gears have the following specifications: $S_d = 70$ MN/m^2, $S_{fe} = 630$ MN/m^2; $m = 3$ mm; $b = 30$ mm. The pinion makes 600 rpm. The transmission ratio is 25/9, helix angle, $\Psi = 23°$ and pressure angle, $\phi_t = 20°$ full depth. Determine the transmitted load W_t, wear load W_w, and dynamic load W_d, and end thrust load W_a.

Given Data: Module m = 3 mm; face width b = 30 mm; speed ratio = 25/9;
Helix angle, $\Psi = 23°$; Pressure angle, $\phi_t = 20°$;
Design stresses, $S_d = 70$ MN/m^2, $S_{fe} = 630$ MN/m^2

SOLUTION

Step-1: Tangential load by Lewis equation

Let $Z_p = 18$; $Z_g = 18 \times 25/9 = 50$

Then virtual number of teeth, $Z_{pv} = \dfrac{18}{\cos^3 \Psi} = 23.078$

The tooth form factor, $Y_p = 0.334$ (Table 16.3 MDHB)

Diameter, $D_p = 18 \times 3 = 54$ mm

Pitch line velocity, $V = \dfrac{\pi DN}{60{,}000} = \dfrac{\pi \times 54 \times 600}{60{,}000} = 1.69$ m/sec

$$\text{Velocity factor } K_v = \frac{4.5}{4.5 + V} = \frac{4.5}{4.5 + 1.69} = 0.7269$$

Then using Lewis equation,

$$m = \frac{W_t C_w}{S_d \cdot K_v \cdot b \cdot Y \cdot \cos \Psi} \quad \text{with } C_w = 1.25$$

$$W_t = (3 \times 70 \times 0.7269 \times 30 \times 0.334 \times \cos 23)/1.25 = 1126.4 \text{ N}$$

Step-2: Dynamic load,

$$W_d = W_t + \frac{21V(C \cdot b \cdot \cos^2 \Psi + W_t) \cos \Psi}{21V + \sqrt{Cb \cos^2 \Psi + W_t}}$$

where, $W_t = 1126.4\text{ N};\ V = 1.69\text{ m/s};\ b = 30\text{ mm}.$

and $C = 150$ for $e = 12.5\ \mu\text{m}.$ (Table 16.5 MDHB)

$$\therefore \quad W_d = 1126.4 + \frac{21 \times 1.69(30 \times 150\cos^2 23 + 1126.4)\cos 23}{21 \times 1.69 + \sqrt{30 \times 150\cos^2 23 + 1126.4}} = 2652\text{ N}$$

Step-3: Dynamic wear strength, (Eqn. 16.39 MDHB)

$$W_w = \frac{K \cdot b \cdot D \cdot Q}{\cos^2 \Psi}$$

where, $$K = \frac{S_{fe}^2 \sin\phi}{1.4}\left[\frac{1}{E_p} + \frac{1}{E_g}\right] = \frac{630^2 \times \sin(20)}{1.4} \times \frac{2}{210000} = 0.923$$

$$Q = \frac{2 \times 50}{50 + 18} = 1.47$$

Substituting the values,

$$\therefore \quad W_w = \frac{0.923 \times 30 \times 54 \times 1.47}{\cos^2 23} = 2585.6\text{ N} \le W_d$$

Since the difference is small, the design may be accepted.

The axial thrust, $W_a = W_t \tan\Psi = 1126.4\tan 23 = 478.12\text{ N}.$

Example 16.3: The 20° full depth, involute gear set has a velocity ratio of 8. The pinion has 25 teeth and rotates at 4000 rpm. The module is 5 mm and the active face width is 100 mm. The permissible stress for the material of the gears is 70 MPa. The helix angle is 23°. Estimate the power that can be transmitted by the gears.

Given Data: Tooth form 20° full module depth; Helix angle $= 23°$;

Module m = 5 mm; Face width b = 30 mm;

Velocity ratio G = 8; Number of teeth on pinion = 25;

Speed of the pinion N = 4000 rpm; Design stress = 70 MPa

SOLUTION Let number of teeth on the pinion, $Z_p = 25$

The diameter of pinion $= 25 \times 5 = 125$ mm

Since the material is the same for both pinion and gear, the calculations are based on the pinion

Equivalent no. of teeth, $Z_{pv} = 25/\cos^3 23 = 32.05$

The tooth form factor, $Y = 0.365$

Then Lewis equation for transmitted load,

$w_t = (K_v \cdot S_d \cdot Y \cdot b \cdot m \cdot \cos\Psi)/C_w$ (Eqn. 16.31 MDHB)

where, $S_d = 70$ MPa; $b = 100$ mm; $m = 5$; $\Psi = 23°$; $C_w = 1.25$

The pitch line velocity, $V = \dfrac{\pi DN}{60,000} = \dfrac{\pi \times 125 \times 4000}{60,000} = 26.18$ m/sec

The velocity factor, (Eqn. 16.35 MDHB)

$$K_v = \frac{5.5}{5.5 + \sqrt{V}} = \frac{5.5}{5.5 + \sqrt{26.18}} = 0.52$$

Substituting the values,

$$W_t = (0.52 \times 70 \times 0.365 \times 100 \times 5 \times \cos 23)/1.25 = 4892 \text{ N}$$

Then the torque, $T = 4892 \times \dfrac{125}{2} = 305745.7$ Nmm

The power transmitted,

$$\text{Power} = \frac{TN}{955 \times 10^4} = \frac{305745.7 \times 4000}{955 \times 10^4} = 128 \text{ kW}$$

Example 16.4: Design a pair of helical gears with the following data, power transmitted = 80 kW; speed of the pinion = 720 rpm.
The speed ratio is 4:1

SOLUTION

Step-1: Selection of material properties
From Table 16.4 MDHB,
For pinion steel – C45 with $S_d = 210$ MPa, and BHN = 250
For gear – C.I. Gr 35 with $S_d = 78.5$ MPa, and BHN = 300

Step-2: Comparison of strength factors
Tooth form -20° involute full depth in normal plane
Helix angle = 30°
Number of teeth on pinion, $Z_p = 15.0$
Number of teeth on gear, $Z_g = 60.00$
Formative no. of teeth for pinion, $Z_{pv} = 15/\cos^3\Psi = 23.0$
for gear, $Z_{gv} = 60/\cos^3\Psi = 92.37$
Tooth form factors, (Table 16.3 MDHB)

$$Y_p = 0.334, \text{ and } Y_g = 0.445$$

Strength factors, $(YS_d)_p = 0.334 \times 210 = 70.14$ MPa

$$(YS_d)_g = 0.445 \times 78.5 = 34.93 \text{ MPa}$$

The gear is weak, hence design the gear.

Step-3: Module by Lewis equation

The torque on the gear,

$$T = \frac{955\times10^4\,\text{kW}}{N} = \frac{955\times10^4\times80}{180} = 424.4\times10^4\ \text{Nmm}$$

The tangential load,

$$w_t = T/r = \frac{424.4\times10^4}{\left(\dfrac{Z_g m}{2}\right)} = \frac{141481.5}{m}\,\text{N}$$

Assume $K_v = 0.4;\ C_w = 1.25,\ b = 20$ m

Substituting in Lewis equation,

$$m = \frac{(141981.5/m)\times1.25}{0.4\times34.93\times20m\times\cos30}$$

$\therefore$ $$m^3 = 730.8,\ \therefore m = 9.007$$

$\therefore$ $$m_n = m\cos\Psi = 7.8$$

Choose $m_n = 8.0$ (Table 16.2 MDHB)

Then $m = 8.0/\cos30 = 9.24$

The dimensions are

$$D_g = mZ_g = 9.24\times60 = 554.2\ \text{mm}$$

$$b = 20\times9.24 = 185\ \text{mm}$$

$$D_p = 15\times9.24 = 138.6\ \text{mm}$$

Pitch line velocity, $V = \dfrac{\pi DN}{60{,}000} = \dfrac{\pi\times584.2\times180}{60{,}000} = 5.5$ m/sec

The actual, $K_v = \dfrac{6}{6+V} = \dfrac{6}{6+5.25} = 0.52$

It is more than the assumed value of 0.4 and module selected is OK.

Step-4: The dynamic load by Buckingham equation

Tangential load, $W_t = 141481.5/\text{m} = 15311.8$ N
Dynamic constant, C for class-II gear
The error, $e = 45\ \mu\text{m} < 64\ \mu\text{m}$ max. allowed (from Tables 16.27 and 16.28 MDHB)
Then C (from Table 16.5 MDHB) $= 375$
The dynamic load, (Eqn. 16.37 MDHB)

$$W_d = W_t + \frac{21V(bc\cos^2\Psi + W_t)\cos\Psi}{21V + \sqrt{b\cdot C\cdot\cos^2\Psi + W_t}}$$

$$= 15311.8 + \frac{21 \times 5.5 (185 \times 375 \cos^2 30 + 15311.8) \cos 30}{21 \times 5.5 + \sqrt{185 \times 375 \cos^2 30 + 15311.8}}$$

$$= 15311.8 + 25957.27 = 41269.07 \text{ N}$$

Step-5: Check for beam strength

Dynamic beam strength

$$W_b = S_{dr} \cdot Y \cdot b \cdot m \cdot \cos \Psi$$

$$= S_{dr} \cdot Y \cdot b \cdot m_n$$

$$= 102.4 \times 0.445 \times 185 \times 8$$

$$= 67440.64 \text{ N} > W_d$$

where,

$$S_{dr} = S_{en} = 0.4 S_u$$

$$= 0.4 \times 256 \text{ (Table 2.4 MDHB)} = 102.4 \text{ Mpa}$$

$$Y = 0.445, b = 185 \text{ mm}, m_n = 8 \text{ mm}$$

Hence, design is safe against bending failure.

Step-6: Check for wear strength

Wear strength

$$W_w = \frac{K \cdot b \cdot D_p Q}{\cos^2 \Psi}$$

$$= \frac{1.375 \times 185 \times 138.6 \times 1.6}{\cos^2 30}$$

$$= 75213.6 > W_d$$

where $K = 1.375$ (Table 16.23 MDHB)

$$Q = \frac{2 \times 92.37}{92.37 + 23} = 1.6$$

Hence, design is safe against wear failure.

Example 16.5: Design a pair of helical gears with the following data:
Power transmitted = 80 kW; Speed of the pinion = 720 rpm.
Centre distance = 400 mm; Gear ratio = 4:1

SOLUTION

Step-1: Module by geometry

An additional restriction on centre distance is to be satisfied. There is change in the preliminary estimate of the module, face width and number of teeth.

Since $\dfrac{D_p + D_g}{2} = \dfrac{D_p + 4D_p}{2} = C = 400$

$$\therefore \quad D_p = \frac{2C}{5} = \frac{2 \times 400}{5} = 160 \text{ mm}$$

$$\therefore \quad D_g = 4 \times D_p = 4 \times 160 = 640 \text{ mm}$$

Let us select $Z_p = 23$, and helix angle $\Psi = 30°$
Then module, $m = 160/23 = 6.956$
The normal module, $m_n = m \cos \Psi = 6.02$
Then choose $m_n = 6.0$
To suit this combination, a slight change in helix angle is required

$$\cos \Psi = \frac{6.00}{6.956} = 8.86505, \text{ therefore } \Psi = 30.11°$$

Step-2: Face width by Lewis equation

Selecting steel for both the pinion and gear, (Table 16.4 MDHB)

$$S_d = 195 \text{ MPa, and BHN} = 250$$

The pinion is to be designed because it is weaker of the two.

The formative number of teeth, $Z_{pv} = \dfrac{23}{\cos^3 30.11} = 35.53$

The tooth form factor, $Y = 0.375$ (Table 16.3 MDHB)

The pitch line velocity, $V = \dfrac{\pi DN}{60{,}000} = \dfrac{\pi \times 160 \times 720}{60{,}000} = 6.03 \text{ m/sec}$

Then velocity factor, (Eqn. 16.33 MDHB)

$$K_v = 6/(6 + 6.03) = 0.50$$

The torque on pinion,

$$T = \frac{955 \times 10^4 \text{kW}}{N} = \frac{955 \times 10^4 \times 80}{720} = 106.1 \times 10^4 \text{ Nmm}$$

The tangential load,

$$W_t = (T/r) = (106.1 \times 10^4/80) = 13263.8 \text{ N}$$

Then by Lewis equation, (Eqn. 16.31 MDHB)

$$W_t = S_d \cdot Y \cdot b \cdot K_v \cdot m \cdot \cos \Psi = S_d \cdot y \cdot b \cdot K_v \cdot m_n$$

$$13263.8 = 195 \times 0.375 \times b \times 0.50 \times 6$$

$$b = 60.46 \cong 60 \text{ mm} > (1.15 p_a = 43.53 \text{ mm})$$

$$b/m_n = 60.0/6 = 10. \text{ This is OK.}$$

Step-3: Dynamic load by Buckingham equation

The dynamic constant for $e = 17$ mm (Table 16.28 MDHB)

$C = 210$ (Table 16.5 MDHB)

Substituting the values, dynamic load,

$$W_d = 13263.8 + \frac{21 \times 6.03 (60 \times 210 \cos^2 30 + 13263.8) \cos 30}{21 \times 6.03 + \sqrt{(60 \times 210 \cdot \cos^2 30 + 13263.8)}} = 22245 \text{ N}$$

Step-4: Check for beam strength

Dynamic beam strength,

$$W_b = S_{dr} \cdot b \cdot Y \cdot m_n$$

$$= 440 \times 60 \times 0.375 \times 6$$

$$= 59400 \text{ N} > W_d$$

Design is safe against bending.

where,

$$S_{dr} = S_{en} = 1.76 \text{ BHN}$$

$$= 1.76 \times 250 = 440 \text{ MPa}$$

$$b = 60 \text{ mm}, Y = 0.375, \; m_n = 6$$

Step-5: Check for wear strength

Wear strength,

$$W_w = \frac{K \cdot b \cdot D_p \cdot Q}{\cos^2 \Psi}$$

$$= (0.92 \times 60 \times 160)/\cos^2 30$$

$$= 1888.5 \text{ N} < W_d$$

Design is not safe against wear.

where,

$K = 0.92$ (Table 16.23 MDHB)

$b = 60$ mm, $D_p = 160$ mm

$Q = (2 \times 4)/(4+1) = 1.6$

Hence, surface hardening is necessary to get $(W_w = W_d)$

$$K = \frac{22245 \times \cos^2 30.1}{60 \times 160 \times 1.6} = 1.083$$

From Table 16.23 MDHB

$$(\text{BHN})_p = 300, \text{ and } (\text{BHN})_g = 250 \text{ with } K = 1.13$$

Pinion is to be surface hardened.

Example 16.6: A single stage helical gear reducer is to receive power from a 25 kW at 1440 rpm induction motor. The tooth form is 20° full depth involute in the normal plane. The helix angle is 23°. Gear ratio is to be 3:1. Design the gear set.

Given Data: Power = 25 kW at speed = 1440 rpm,

Helix angle, Ψ = 23°; Tooth form, ϕ_n = 20° involute full depth

SOLUTION

Step-1: Module by Lewis equation

Torque transmitted by pinion,

$$T = \frac{955 \times 10^4 \text{ kW}}{N} = \frac{955 \times 10^4 \times 25}{1440} = 16.58 \times 10^4 \text{ Nmm}$$

Material – Steel C 45 – for both pinion and gear

$$S_d = 210 \text{ MPa; BHN} = 250$$

Number of teeth, $Z_p = 16$, and $Z_g = (16 \times 3) = 48$

Virtual number of teeth, $Z_{pv} = \dfrac{16}{\cos^3 23} = 20.5$

Form factor, $Y = 0.322$

Tangential load,

$$W_t = T/r = \frac{16.58 \times 10^4}{\left(\dfrac{16m}{2}\right)} = \frac{20725}{m}\text{ N}$$

Let $b = 20$ m, $C_w = 1.25$ and $K_v = 0.4$

Substituting in Lewis Eqn. 16.31 MDHB,

$$m = \frac{(20725/m)1.25}{210 \times 0.322 \times 20m \times 0.4 \times \cos 23}$$

$$\therefore \qquad m = \sqrt[3]{52.04} = 3.728$$

Choose standard module in normal plane, $m_n = 4.0$ mm,

$$\text{Then, } m = 4.0/\cos 23 = 4.345$$

Diameter of pinion, $D_p = 16 \times 4.345 = 69.5$ mm

$$\text{Dia. of gear } D_g = 48 \times 4.345 \text{ mm} = 208.5 \text{ mm}$$

Face width, $b = 20 \times 4.345 = 86.9 \cong 90$ mm

The pitch line velocity, $V = \dfrac{\pi DN}{60{,}000} = \dfrac{\pi \times 69.5 \times 1440}{60{,}000} = 5.25$ m/sec

Velocity factor, $K_v = 6/(6 + 5.24) = 0.534$ (Eqn. 16.33 MDHB)

It is more than the assumed values of $K_v = 0.4$. Therefore, module, $m_n = 4.0$ is OK.

Step-2: Dynamic load by Buckingham equation

The tangential load, $W_t = 20725/4.345 = 4769.85$ N

For class-II gear finish (Table 16.28 & Table 16.29 MDHB)

The gear error, $e = 25$ μm (Table 16.29 MDHB)

The dynamic factor, $C = 300$ (Table 6.5 MDHB)

Then the dynamic load, (Eqn. 16.37 MDHB)

$$W_d = W_t + \frac{21V\,(b \cdot C \cdot \cos^2 \Psi + W_t)\cos \Psi}{21V + \sqrt{b \cdot C \cdot \cos^2 \Psi + W_t}}$$

$$= 4769.85 + \frac{21 \times 5.24 (90 \times 300 \times \cos^2 23 + 4769.85) \cos 23°}{21 \times 5.24 + \sqrt{90 \times 300 \times \cos^2 23 + 4769.85}} = 14904.7 \text{ N}$$

Step-3: Check for beam strength

Dynamic beam strength,

$$W_b = S_{dr} \cdot b \cdot Y \cdot m_n$$
$$= 440 \times 90 \times 0.322 \times 4$$
$$= 51004.8 > W_d$$

where,

$S_{dr} = S_{en} = 440$ MPa

$b = 90$ mm, $Y = 0.322, m_n = 4.0$

This is greater than W_d, and design is safe against bending fatigue failure.

Step-4: Check for wear strength

Dynamic wear strength,

$$W_w = \frac{K \cdot b \cdot D_p \cdot Q}{\cos^2 \Psi}$$
$$= \frac{0.92 \times 90 \times 69.5 \times 1.5}{\cos^2 23}$$
$$= 10187 < W_d$$

where,

$K = 0.92$ (Table 16.23 MDHB),

$b = 90, D_p = 69.5$

$Q = (2 \times 3)/(3 + 1) = 1.5$

Design is unsafe against wear failure and surface hardening is required. Then stress factor required,

$$K = \frac{14905 \times \cos^2 23}{90 \times 69.5 \times 1.5} = 1.35$$

The hardeness values required (from Table 16.23 MDHB),

$$(\text{BHN})_p = 600, \text{ and } (\text{BHN})_g = 450$$

Alternate Solution, as per the new AGMA standards

Step-5: Check for bending strength

The permissible bending stress,

$$S_d = \frac{S \cdot K_L}{K_T K_R} = 440 \text{ MPa}$$

where,

$S = S_{en} = 440$ MPa

$K_L = 1.0, K_T = 1.0, K_R = 1.0$

Induced bending stress,

$$\sigma_b = \frac{W_t \cdot K_o \cdot K_s \cdot K_m}{K_v \cdot b \cdot m \cdot J}$$

$$= \frac{4769.85 \times 1.2 \times 1 \times 1.6}{0.75 \times 90 \times 4.345 \times 0.44}$$

$$= 70.96 \text{ MPa} < (S_d = 440)$$

The teeth are strong in bending.

where,

$W_t = 4769.85$ N

$K_o = 1.2$ (Table 16.11), $K_s = 1.0$

$K_m = 1.6$ (Table 16.15 MDHB)

$K_v = 0.75$ (Fig. 16.5 MDHB)

$J = 0.44$ (From Fig 16.9a MDHB)

Step-6: Check for contact strength

The permissible contact stress,

$$S_{Hd} = \frac{S_{fe} \cdot C_L \cdot C_H}{C_T \cdot C_R}$$

$$= \frac{630 \times 1 \times 1}{1 \times 1} = 630 \text{ MPa}$$

where,

$S_{fe} = 2.8\text{ BHN} - 70$

$= 2.8 \times 250 - 70 = 630$ MPa

$C_L = 1.0, C_L = 1.0$

$C_T = 1.0, C_R = 1.0$

Induced contact stress,

$$\sigma_H = C_p \sqrt{\frac{W_t \cdot C_o \cdot C_s \cdot C_m \cdot C_f}{C_v \cdot D \cdot b \cdot I}}$$

$$= 191 \times \sqrt{\frac{4769.85 \times 1.2 \times 1 \times 1.6 \times 1}{0.75 \times 69.5 \times 90 \times 0.16875}}$$

$$= 649.6 > (S_{Hd} = 630.0)$$

where,

$C_p = 191$ (Table 16.14)

$C_o = 1.2$ (Table 16.10), $C_s = 1.0$

$C_m = 1.6$ (Table 16.15), $C_f = 1$

$C_v = K_v = 0.75$

$I = 0.225\left(\frac{3}{3+1}\right) = 0.16875$ (Eqn. 16.44 MDHB)

Hence, pinion is to be hardened to $\text{BHN} = \frac{649.6 + 70}{2.8} = 257$

The comparison of the two methods shows that design based on Buckingham dynamic load equation as per old AGMA specification is more conservative and results in over design or extra strong design.

Example 16.7: A pair of helical gears with helix angle of 30° is used to transmit 20 kW at 10,000 rpm of pinion. The velocity ratio is 4:1. Both the gears are made of hardened steel with 200 BHN. The tooth form is 20° full-depth in the normal plane. The number of teeth on the pinion is to be 16 and the face width is to be 15 times the module. Design the gears.

Given Data: Power transmitted, NP = 20 kW; Pinion speed = 10000 rpm;

Speed ratio = 4; Tooth form -20° full – depth in normal plane;

Number of teeth on pinion = 16; Face width, b = 15 m

Material hardened steel with BHN = 200

SOLUTION

Step-1: Module by Lewis equation

Torque transmitted by pinion,

$$\therefore \quad T = \frac{955\times10^4\,\text{KW}}{N} = \frac{955\times10^4\times20}{10000} = 19100\text{ Nmm}$$

Material – hardened steel – for both pinion and gear

$\text{BHN} = 200;\ S_{en} = 1.76\text{ BHN} = 1.76\times200 = 352\text{ MPa}$

The design stress for FOS = 2, $S_d = 352/2 = 176\text{ MPa}$

Number of teeth, $Z_p = 16$ and $Z_g = (16\times4) = 64$

Virtual number of teeth, $Z_{pv} = \dfrac{16}{\cos^3 30} = 24.6$

Form factor $Y = 0.337$ (Table 16.3 MDHB)

Tangential load, $W_t = T/r = 19100/(8\text{ m}) = 2387.5/\text{m}$

For Lewis equation (Eqn. 16.31) assume $Cw = 1.25$ and $K_v = 0.4$,

$$m = \frac{(2387.5/m)1.25}{176\times0.337\times15m\times0.4\times\cos30}$$

$$\therefore \quad m = \sqrt[3]{9.68} = 2.12\text{ mm}$$

Choose standard module $m_n = 2.0$ mm,

Then $m = \dfrac{2.0}{\cos 30} = 2.31$

$$D_p = 16\times2.31 = 36.95\text{ mm};\ D_g = 64\times2.31\text{ mm} = 147.84\text{ mm}$$

$$b = 15\times2.31 = 34.65 \cong 35\text{ mm}$$

Pitch line velocity,

$$V = \pi DN/60000 = \pi\times36.95\times10000/60000 = 19.35\text{ m/s}$$

The velocity factor, (Eqn. 16.34 MDHB)

$K_v = 15/(15+V) = 15/(15+19.35) = 0.43 >$ assumed value & hence the preliminary design is acceptable.

Tangential load, $W_t = 2387.5/2.31 = 1033.55\text{ N}$

Step-2: Check for bending strength (new AGMA standards)

The permissible bending stress,

$$S_d = \frac{S \cdot K_L}{K_T K_R} = 352 \text{ MPa}$$

where,

$S = S_{en} = 352$ MPa

$K_L = 1.0$ (Table 16.12 MDHB), $K_T = 1.0$

$K_R = 1.0$ (Table 16.13 MDHB)

Induced bending stress

$$\sigma_b = \frac{W_t \cdot K_o \cdot K_s \cdot K_m}{K_v \cdot b \cdot m \cdot J}$$

$$= \frac{1033.55 \times 1.5 \times 1 \times 1.2}{0.75 \times 35 \times 2.31 \times 0.44}$$

$$= 71.31 \text{ MPa} < (S_d = 352 \text{ MPa})$$

where, $W_t = 1033.55$ N

$K_o = 1.5$ (Table 16.10 MDHB), $K_s = 1.0$

$K_m = 1.2$ (Table 16.15 MDHB), $K_v = .75$ (Fig. 16.5 MDHB)

$J = 0.44$ (Fig. 16.9 MDHB)

Step-3: Check for contact strength (new AGMA standards)

The permissible contact stress,

$$S_{Hd} = \frac{S_{fe} \cdot C_L \cdot C_H}{C_T \cdot C_R}$$

$$= \frac{490 \times 1 \times 1}{1 \times 1} = 490 \text{ MPa}$$

where,

$S_{fe} = 2.8$ BHN $- 70$

$= 2.8 \times 200 - 70 = 490$ MPa

$C_L = 1.0$, $C_L = 1.0$

$C_T = 1.0$, $C_R = 1.0$

Induced contact stress,

$$\sigma_b = C_p \frac{W_t \cdot C_o \cdot C_s \cdot C_m \cdot C_f}{C_v \cdot D \cdot b \cdot I}$$

$$= 191 \sqrt{\frac{1033.55 \times 1.5 \times 1 \times 1.2 \times 1}{0.75 \times 36.95 \times 35 \times 0.18}}$$

$$= 623 > (S_{Hd} = 490 \text{ MPa})$$

where,

$C_p = 191$ (Table 16.14), $C_o = 1.5$

$C_s = 1.0$; $C_m = K_m = 1.2$

$C_f = 1$; $C_v = K_v = 0.75$

$$I = 0.225 \times \frac{4}{4+1} = 0.18$$

The design is not safe against contact surface failure.

Then take $m_n = 2.5$ mm and $m = 2.5/\cos 30 = 2.89$ mm

$b = 15 \times 2.89 = 45$ mm (rounded); $D_p = 16 \times 2.89 = 46.24$ mm

$W_t = 2387.5/2.89 = 826.125$ N, then contact stress is

$$= 191 \times \sqrt{\frac{826.125 \times 1.5 \times 1 \times 1 \times 1.2 \times 1}{0.75 \times 46.24 \times 45 \times 0.18}} = 439.45 \text{ MPa} < 490 \text{ MPa}$$

This is acceptable.

Example 16.8: Design a pair of helical gears to transmit a power of 20 kW from a shaft running at 1500 rpm to a parallel shaft to be run at 450 rpm. Suggest suitable surface hardness for the gear pair.

Given Data: Power transmitted, $NP = 20$ kW; Pinion speed, $N_1 = 1500$ rpm
Gear speed, $N_2 = 450$ rpm

SOLUTION

Step-1: Module by Lewis equation

Assume the tooth form as 20° full depth in normal plane
Helix angle, $\Psi = 25°$ (Table 16.6a MDHB)
Material C45 – hardened $S_d = 210$ MPa and BHN = 250 (Table 16.4 MDHB)
Number of teeth on the pinion, $Z_p = 15$

$$\text{The gear ratio, } G = \frac{1500}{450} = 3.33$$

No. of teeth on gear, $Z_g = GZ_p = 50$
Virtual number of teeth

$$\text{On pinion, } Z_{vp} = \frac{15}{\cos^3 25} = 20.5$$

$$\text{On gear, } Z_{vg} = \frac{50}{\cos^3 25} = 67.165$$

Tooth form factors

For pinion, $Y_p = 0.322$ and for gear, $Y_g = 0.359$

Torque transmitted,

$$T = \frac{955 \times 10^5 \text{kW}}{N} = \frac{955 \times 10^5 \times 20}{1500} = 127333.3 \text{ Nmm}$$

The tangential load,

$$W_t = \frac{T}{(D_p/2)} = \frac{2T}{m \cdot Z_p} = \frac{2 \times 127333.3}{15m} = (16977.8/m)\,\text{N}$$

$$\text{Assume } b = 2p_x = 2p_c/\tan\Psi = \frac{2\pi m}{\tan\Psi} = \frac{2 \times \pi \times m}{\tan 25°} = 13.5 \text{ m}$$

Assume $K_v = 0.4$ and $C_w = 1.25$

Substituting the values in Lewis equation,

$$m = \frac{W_t C_w}{S_d \cdot K_v \cdot b \cdot Y \cdot \cos\Psi}$$

$$= \frac{(16977.8/\text{m}) \times 1.25}{210 \times 0.4 \times 13.5m \times 0.322 \times \cos 25°}$$

$$\therefore \qquad m^3 = \frac{16977.8 \times 1.25}{210 \times 0.4 \times 13.5 \times 0.322 \times 0.91} = 64.13$$

$$\therefore \qquad m = \sqrt[3]{64.13} = 4.0027 \text{ mm}$$

The normal module, $m_n = m \cos \Psi = 4.0027 \times \cos 25 = 3.64$

Choose $m_n = 4.0$ mm

Then $m = m_n / \cos \Psi = 4.4$

Width, $b = 13.5 \text{ m} = 13.5 \times 4.4 = 59.4 \cong 60$ mm

Diameter of pinion, $D_p = m \cdot Z_p = 4.4 \times 15 = 66$ mm

Diameter of gear, $D_g = mZ_g = 4.4 \times 50 = 220$ mm

Pitch line velocity,

$$V = \frac{\pi DN}{60,000} = \frac{\pi \times 66 \times 1500}{60,000} = 1.518 \text{ m/sec}$$

Velocity factor, $K_v = \dfrac{6}{6+V}$ (Eqn. 16.33 MDHB)

$$= \frac{6}{6 + 5.18} = 0.536 > 0.4 \text{ assumed value}$$

Thus selection of module, $m_n = 4.0$ mm is O.K.

Step-2: Dynamic load by Buckingham equation

The tangential load,

$$w_t = 16977.8/\text{m} = \frac{16977.8}{4.4} = 3858.6 \text{ N}$$

The dynamic load by Buckingham equation,

$$W_d = W_t + \frac{21V(b \cdot C \cdot \cos^2\Psi + W_t)\cos \Psi}{21V + \sqrt{b \cdot C \cdot \cos^2\Psi + W_t}}$$

where, $w_t = 3858.6$ N, $b = 60$ mm; $V = 5.18\dfrac{\text{m}}{\text{s}}$; $\Psi = 25°$

$C = 150$ for $e = 0.125$, class III gear (Tables 16.5 & 16.27 MDHB)

Substituting the values,

$$W_d = 3858.6 + \frac{21 \times 5.18[60 \times 150 \times \cos^2 25 + 3858.6]\cos 25}{21 \times 5.18 + \sqrt{60 \times 150 \times \cos^2 25 + 3858.6}}$$

$$= 3858.6 + \frac{108.78[1311.5]}{108.78 + \sqrt{1311.5}} = 9578.1 \text{ N}$$

Step-3: Check for the dynamic beam strength,

$$W_b = S_{dr} b \cdot Y \cdot m \cos\Psi, \text{ where } S_{dr} = 210 \text{ MPa}$$
$$= 210 \times 60 \times 0.322 \times 4.4 \times \cos 25° = 16245.03 \text{ N} > w_d$$

The teeth are safe in bending.

Step-4: Check for wear strength

The wear load,

$$W_w = \frac{K \cdot b \cdot D_p Q}{\cos^2\Psi}, \text{ where } Q = \frac{2 \times G}{G+1} = \frac{2 \times 4}{4+1} \doteq 1.6$$

The stress factor for $W_w = W_d$

$$K = \frac{w_d \cos^2\Psi}{b \cdot D_p \cdot Q} = \frac{9578.1 \times \cos^2 25}{60 \times 66 \times 1.6} = 1.242$$

Then hardness required,

$$(\text{BHM})_p = 350; \ (\text{BHN})_g = 250 \qquad \text{(Table 16.23 MDHB)}$$

Example 16.9: A pair of steel helical gears is to transmit 15 kW at 5000 rpm of pinion. Both the gears are made of the same material, hardened steel with allowable bending stress of 120 MPa. The gears have to operate at a centre distance of 200 mm. The speed reduction ratio is 4:1. The teeth are 20° full depth involutes on the normal plane. Helix angle is 45°. The gears are manufactured to class–III accuracy (precision class). The face width can be taken as 16 times the normal module. If the wear strength has to be more than the dynamic load, determine the following:

(a) Normal module
(b) Transverse module
(c) Pressure angle in transverse – plane
(d) Number of teeth on pinion and gear
(e) Face width
(f) Required surface endurance limit

Given Data: Power, NP = 15 kW; Speed of pinion, $N_1 = 5000$ rpm
Gear ratio $4:1$; Speed of gear, $N_2 = 5000/4 = 1250$ rpm
Centre, $C = 200$ mm; Helix angle, $\Psi = 45°$; Face width $b = 16 m_n$
Tooth form – 20° involute full depth on normal plane
Quality – precision gear – class–III; Allowable stress, $S_d = 120$ MPa

SOLUTION

Step-1: Module by Lewis equation

Diameters of the gears,

$$C = \left(\frac{D_p + D_g}{2}\right) = 200$$

$$\therefore \qquad D_p + D_g = D_p + 4D_p = 400$$

$$\therefore \qquad D_p = \frac{400}{5} = 80 \text{ mm}$$

$$D_g = 80 \times 4 = 320 \text{ mm}$$

Assume $Z_p = 14$ and $Z_g = 4 \times 14 = 56$
Virtual number of teeth $Z_{vp} = 14/\cos^3 45 = 39.59$
Tooth form factor $Y_p = 0.367$
The Lewis equation,

$$m = \frac{w_t c_w}{S_d \cdot K \cdot b \cdot Y \cdot \cos \Psi} \qquad \text{(Eqn. 16.31 MDHB)}$$

$$\therefore \qquad m_n = m \cos \Psi = \frac{w_t c_w}{S_d \cdot K \cdot b \cdot Y}$$

where,

$$w_t = \frac{2T}{D_p} = \frac{2 \times 955 \times 10^4 \times 15}{80 \times 5000} = 716.25 \text{ N}$$

$c_w = 1.25, S_d = 120$ MPa, $K_v = 0.4$ assumed, $b = 16m_n, Y = 0.367$

$$\therefore \qquad m_n = \frac{716.25 \times 1.25}{120 \times 0.4 \times 16 m_n \times 0.367}$$

$$m_n^2 = \frac{716.25 \times 1.25}{120 \times 0.4 \times 16 \times 0.367} = 3.126$$

$$\therefore \qquad m_n = 1.78$$

Choose $m_n = 2.0$ mm
Transverse module, $m = m_n/\cos \Psi = 2/\cos 45° = 2.83$
Number of teeth on pinion, $Z_p = 80/2.83 = 28.3 \cong 28$
Revised, $Z_p = 28$
Then correct diameter of pinion, $D_p = Z_p \cdot m = 28 \times 2.83 = 79.24$ mm
The diameter of gear, $D_g = 4 \times D_p = 316.96$ mm

The centre distance, $C = \left(\dfrac{D_p + D_g}{2} \right) = \dfrac{79.24 + 316.96}{2} = 198.1$ mm

This is nearer to $C = 200$ mm and is acceptable
Then the correct tangential load,

$$w_t = \frac{2T}{D_p} = \frac{2 \times 955 \times 10^4 \times 15}{79.24 \times 5000} = 723.12 \text{ N}$$

Pitch line velocity,

$$V = \frac{\pi DN}{60,000} = \frac{\pi \times 79.24 \times 5000}{60,000} = 20.745 \text{ m/s}$$

Velocity factor, $K_v = \dfrac{5.5}{5.5+\sqrt{v}} = \dfrac{5.5}{5.5+\sqrt{20.745}} = 0.547 > 0.4$ assumed value

Hence, $m_n = 2$ mm is O.K.

Gear width, $b = 16m_n = 16 \times 2 = 32$ mm

Step-2: The dynamic load by Buckingham equation

with $C = 150$ for $e = 0.0125$ (Table 16.5 MDHB)

$$w_d = w_t + \frac{21V\,(b \cdot C \cdot \cos^2 \Psi + w_t)\cos \Psi}{21V + \sqrt{b \cdot C \cdot \cos^2 \Psi + w_t}}$$

$$= 723.12 + \frac{21 \times 20.745(32 \times 150 \times \cos^2 45 + 723.12)\cos 45}{21 \times 20.745 + \sqrt{(32 \times 150 \times \cos^2 45 + 723.12)}}$$

$$= 723.12 + \frac{435.645 \times (3123.12)}{435.645 + 55.885} = 723.12 + 1957.3 = 2680.42\,\text{N}$$

Step-3: Check for dynamic beam strength

Dynamic beam strength,

$$w_b = S_{dr} \cdot b \cdot Y \cdot m \cos \Psi$$

where, Y for $Z_{pv} = \dfrac{28}{\cos^3 45} = 79.2$; $Y = 0.438$

$\therefore$ $w_b = 120 \times 32 \times 0.438 \times 2 \times \cos\ 45 = 3363.84\ \text{N} > w_d = 2680.42\ \text{N}$

The tooth is safe against bending failure.

Step-4: Check for wear strength

The wear strength,

$$w_w = \frac{K \cdot b \cdot D_p \cdot Q}{\cos^2 \Psi}$$

Then the stress factor for $w_w = w_d$

$$K = \frac{w_d \cdot \cos^2 \Psi}{b \cdot D_p \cdot Q}$$

where, $w_d = 2680.12\,\text{N}$; $b = 32$ mm; $D_p = 79.24$; $Q = \dfrac{2 \times 4}{4+1} = 1.6$

Substituting the values,

$$K = \frac{2680.12 \times \cos^2 45}{32 \times 79.24 \times 1.6} = 0.3303$$

But $K = \dfrac{S_{fe}^2 \cdot \sin \Phi}{1.4}\left[\dfrac{1}{E_g} + \dfrac{1}{E_p}\right] = \dfrac{2S_{fe}^2 \cdot \sin \Phi}{1.4 \cdot E}$ (Eqn. 16.16a MDHB)

$$\therefore \quad S_{fe}^2 = \frac{1.4}{2}\frac{KE}{\sin \phi} = \frac{0.7 \times 0.3303 \times 210 \times 10^3}{\sin 20} = 141.96 \times 10^3$$

$\therefore$ The surface endurance limit stress, $S_{fe} = 376.8$ MPa

Example 16.10: The following data refer to a helical gear drive:

(a) Power transmitted 34 kW at 2800 rpm of pinion
(b) Speed reduction ratio 4.5
(c) Helix angle, $\Psi = 25°$
(d) Material for both pinion and gear is medium carbon steel with allowable bending stress as 230 MPa and BHN = 275
(e) The pinion diameter is limited to 125 mm

Determine module face, width, and check the design for wear strength against dynamic loading. Also determine the axial thrust on the shaft.

Given Data: Power, $NP = 34$ kW; Speed of pinion $= 2800$ rpm;
Speed ratio, $G = 4.5$; Helix angle, $\Psi = 25°$;
Design stress, $S_d = 230$ MPa and BHN = 275;
Diameter of pinion, $D_p \leq 125$ mm

SOLUTION

Step-1: Module by geometry

Assume $Z_p = 56$

Then $Z_g = 56 \times 4.5 = 252$

Then $m = 125/56 = 2.23$

Choose normal module, $m_n = 2.00$ mm

Modified transverse module,

$$m = m_n / \cos \Psi = 2/\cos(25) = 2.21$$

The modified pinion diameter,

$$D_p = m \cdot Z_p = 2.21 \times 56 = 123.76 \text{ mm} < 125 \text{ mm O.K.}$$

Then, $$D_g = 123.76 \times 4.5 = 556.92 \text{ mm}$$

Step-2: The face width by Lewis equation

Virtual number of teeth,

$$Z_{vp} = Z_p/\cos^3\Psi = \frac{56}{\cos^3 25} = 75.22$$

Tooth form factor, $Y = 0.435$
The torque transmitted by pinion,

$$T = \frac{955\times10^4 \times \text{kW}}{N} = \frac{955\times10^4\times34}{2800} = 115964.3\ \text{N}\cdot\text{mm}$$

The tangential load on the tooth,

$$w_t = \frac{T}{(D_p/2)} = \frac{2T}{D_p} = \frac{2\times115964.3}{123.76} = 1874.02\ \text{N}$$

The pitch line velocity,

$$V = \frac{\pi DN}{60{,}000} = \frac{\pi\times123.76\times2800}{60{,}000} = 18.144\ \text{m/sec}$$

The velocity factor,

$$K_v = \frac{15}{15+v} = \frac{15}{15+18.144} = 0.45 \qquad \text{(Eqn. 16.34 MDHB)}$$

From the Lewis equation (Eqn. 16.31 MDHB)

Width of the pinion, $b = \dfrac{w_t\cdot c_w}{S_d\cdot K_v\cdot Y\cdot m_n}$

$$= \frac{1874.02\times1.25}{230\times0.45\times0.435\times2} = 26.015\ \text{mm}$$

Choose b = 26 mm
$b/m_n = 26.0/2 = 13.0$; the ratio is acceptable.

Step-3: The dynamic load by Buckingham equation

$$w_d = w_t + \frac{21V\,(b\cdot c\cos^2\Psi + w_t)\cos\Psi}{21V + \sqrt{b\cdot c\cos^2\Psi + w_t}} \qquad \text{(Eqn. 16.37 MDHB)}$$

(where, c = 150 for e = 0.025 Table 16.5 MDHB)

$$= 1874.02 + \frac{(21\times18.14)(26\times150\times\cos^2 25+1874.02)\cos 25}{21\times18.14+\sqrt{(26\times150\times\cos^2 25+1874.02)}}$$

$$= 1874.02 + \frac{380.94\times(5077.46)\times\cos 25}{380.94+\sqrt{5077.46}} = 5750.63\,\text{N}$$

Step-4: Check for dynamic beam strength

Endurance limits stress, $S_{dr} = 480$ MPa (Table 16.25 MDHB)

$$w_b = S_{dr} \cdot b \cdot Y \cdot m \cos \Psi \qquad \text{(Eqn. 16.38 MDHB)}$$

$$= 480 \times 26 \times 0.435 \times 2 = 10857.6 \text{ N} > w_d$$

The tooth is safe against bending failure.

Step-5: Check for wear strength,

$$w_w = \frac{K \cdot b \cdot D_p \cdot Q}{\cos^2 \Psi} \geq w_d \qquad \text{(Eqn. 16.39 MDHB)}$$

The strength factor for $w_w = w_d$

$$K = \frac{w_d \cdot \cos^2 \Psi}{b \cdot D_p \cdot Q}$$

where, $$b = 26,\ D_p = 123.76,\ Q = \frac{2 \times 4.5}{4.5 + 1} = 1.64$$

$$= \frac{5750.63 \times \cos^2 25}{26 \times 123.76 \times 1.64} = 0.8951$$

The hardness required from Table 16.23 MDHB

$$(\text{BHN})_p = (\text{BHN})_g = 250 \text{ with } K = 0.92 \text{ is suitable}$$

Therefore hardening is not required.

Step-6: Tooth loads

The axial thrust on the shaft,

$$w_a = w_t \tan \Psi = 1874.02 \tan (25) = 873.87 \text{ N} \qquad \text{(Eqn. 16.28 MDHB)}$$

The radial load,

$$w_r = w_t \tan \Phi_t = w_t \frac{\tan \Phi_n}{\cos \Psi} = \frac{1874.02 \times \tan 20}{\cos 25} = 752.6 \text{ N}$$

Normal load,

$$w = w_t / (\cos \Psi \cdot \cos \Phi_n) \qquad \text{(Eqn. 16.30 MDHB)}$$

$$= 1874.02 / (\cos 25 \cdot \cos 20) = 2200.46 \text{ N}$$

Example 16.11: A 55 kW motor running at 450 rpm is geared to a pump by means of double helical gearing. The forged steel pinion on motor shaft has PCD of around 200 mm and it drives a good grade C.I gear on the pump shaft at 120 rpm. The allowable stress for pinion and gear materials should be taken as 224 MPa and 56 MPa respectively. Assuming $14\frac{1}{2}°$ angle tooth form with helix angel $\Psi = 20°$ and number of teeth on pinion $Z_1 = 24$, design the gears completely.

Given Data: Power transmitted, $NP = 55$ kW; Pinion speed, $N_1 = 450$ rpm

Gear speed, $N_2 = 120$ rpm; PCD of pinion, $D_p \cong 200$ mm

Allowable stresses

For pinion $(S_d)_p = 224$ MPa; for gear $(S_d)_g = 56$ MPa

Tooth form, $\Phi = 14.5°$; Helix angle $\Psi = 20°$

No. of teeth on pinion, $Z_1 = 24$.

SOLUTION

Step-1: Estimation of module

Speed ratio, $G = \dfrac{N_1}{N_2} = \dfrac{450}{120} = 3.75$

$\therefore$ Number of teeth on gear, $Z_2 = G \cdot Z_1 = 3.75 \times 24 = 90$

The gear diameter, $D_g = G \times D_p = 3.75 \times 200 = 750$ mm

The transverse module, $m = D_p / Z_1 = \dfrac{200}{24} = 8.33$

The normal module, $m_n = m \cdot \cos\Psi$ (Eqn. 16.21 MDHB)

$$\therefore \quad m_n = 8.33 \cos 20° = 7.83$$

Choose $m_n = 8$ (Table 16.2 MDHB)

The transverse module, $m = m_n / \cos\Psi = 8/\cos 20° = 8.5$

The diameter of the pinion,

$$D_p = 2 \cdot Z_1 = 8.5 \times 24 = 204 \text{ mm}$$

The gear diameter, $D_g = G \times D_p = 204 \times 3.75 = 765$ mm

Pitch line velocity, $V = \dfrac{\pi DN}{60{,}000} = \dfrac{\pi \times 204 \times 450}{60{,}000} = 4.81$ m/sec

Step-2: Tangential load

The torque transmitted by pinion,

$$T = \frac{955 \times 10^4 \text{ kW}}{N} = \frac{955 \times 10^4 \times 55}{450} = 1167222.2 \text{ Nmm}$$

The tangential load on the tooth,

$$w_t = \frac{T}{(D_p/2)} = \frac{2 \times 1167222.2}{204} = 11443.4 \text{ N}$$

Step-3: Width of the pinion gear by Lewis equation

Virtual number of teeth for pinion, $Z_{1v} = 24/\cos^3 20 = 28.92$

Virtual number of teeth for gear, $Z_{2v} = 90/\cos^3 20 = 108.46$

Tooth form factors

For pinion $Y_p = 0.318$

For gear $Y_g = 0.368$

The strength factors

For pinion $(YS_d)_p = 0.318 \times 224 = 71.23$

For gear $(YS_d)_g = 0.368 \times 56 = 20.61$

The gear is weaker, hence design the gear.

Using the Lewis Eqn. (16.31 MDHB), the width of the gear,

$$b = \frac{w_t c_w}{S_d \cdot K_v \cdot Y \cdot m \cdot \cos\Psi}$$

where, $w_t = 11443.4$ N, $S_d = 56$ MPa, $Y = 0.368$, $c_w = 1.25$ assumed

$$\text{Velocity factor, } K_v = \frac{4.5}{4.5 + v} = \frac{4.5}{4.5 + 4.81} = 0.484$$

Substituting the values,

$$\therefore \quad b = \frac{11443.4 \times 1.25}{56 \times 0.484 \times 0.368 \times 8} = 179.26 \text{ mm}$$

Choose $b = 180$ mm

$$b > \frac{2.3\text{ m}}{\tan\Psi} = 53.7 \text{ mm, but } b < \frac{30\text{ m}}{\tan\Psi} = 700 \text{ mm}$$

The choice is acceptable.

Step-4: Checking for dynamic load by Buckingham equation,

$$w_d = w_t + \frac{21V\,(b \cdot c \cdot \cos^2\Psi + w_t)\cos\Psi}{21V + \sqrt{b \cdot c \cdot \cos^2\Psi + w_t}}$$

Dynamic factor C = 150 from Table 16.5 for e = 0.020 from Table 16.27 MDHB.

$$w_d = 11443.4 + \frac{21 \times 4.81(180 \times 150 \times \cos^2 20 + 11443.4)\cos 20}{21 \times 4.81 + \sqrt{(180 \times 150 \times \cos^2 20 + 11443.4)}}$$

$$= 11443.4 + \frac{101.01(35285)}{101.01 + \sqrt{35285}} = 23038.2 \text{ N}$$

Step-5: Check for dynamic beam strength,

$$w_b = S_{dr} \cdot b \cdot Y \cdot m_n$$

$$= 56 \times 180 \times 0.368 \times 8 = 29675.52 \text{ N} > w_d$$

The tooth is safe in bending.

Step-6: Check for wear strength,

$$w_w = \frac{K \cdot b \cdot D_p \cdot Q}{\cos^2 \Psi} \geq w_d$$

The strength factor for $w_w = w_d$

$$K = \frac{w_d \cos^2 \Psi}{b \cdot D_p \cdot Q} \quad \text{where } Q = \frac{2 \times 3.75}{3.75 + 1} = 1.58$$

$$= \frac{23038.2 \times \cos^2 20}{180 \times 204 \times 1.58} = 0.351$$

Then surface hardness (Table 16.23 MDHB)

For gear BHN $= 180$, and pinion BHN $= 150$, with $K = 0.42$ is a suitable choice.

Example 16.12: Design a single-reduction parallel helical gear speed reducer having speed ratio of 5:1. It is to be capable of transmitting the full load rating of 18.75 kW at 1200 rpm of pinion. Use nickel chrome steel for pinion and C45 steel for gear. Select class-III precision gears. The helix angle of the gear wheels is 23°. The normal pressure angle of the tooth is 20° and the profile is full depth involute. Check the design for dynamic and wear loads.

Given Data: Power $NP = 18.75$ kW at $N_1 = 1200$ rpm;
Gear ratio, $G = 5:1$;
Material for pinion–Nickel chrome steel; for gear – C45 steel
Helix angle $\Psi = 23°$; Normal pressure angle $\Phi_n = 20°$

SOLUTION

Step-1: Material properties
(From Table 16.4 MDHB)
Pinion Nickel-chrome steel – 45Cr 2Ni – $S_d = 450$ MPa, BHN = 400
Gear – C45 steel $S_d = 210$ MPa, BHN = 250

Step-2: Comparison of strength factors
Assume the no. of teeth on pinion, $Z_1 = 20$
The number of teeth on gear, $Z_2 = G \cdot Z_1 = 5 \times 20 = 100$
The virtual number of teeth
For pinion, $Z_{1v} = 20/\cos^3(23) = 25.642$

For gear, $Z_{2v} = Z_2/\cos^3 \Psi = \dfrac{100}{\cos^3(23)} = 128.21$

Tooth form factors
For pinion, $Y_1 = 0.346$; for gear $Y_2 = 0.455$
Strength factors
For pinion $(Y_1 S_d) = 0.346 \times 450 = 155.7$ MPa

For gear $(Y_2 S_d) = 0.455 \times 210 = 95.55$ MPa
The gear is weaker, hence design the gear.

Step-3: Module by Lewis equation

Gear speed, $N_2 = 1200/5 = 240$ rpm
The torque on gear,

$$T = \frac{955 \times 10^4 \times \text{kW}}{N} = \frac{955 \times 10^4 \times 18.75}{(240)} = 746093.75 \text{ Nmm}$$

The tangential load on the tooth,

$$W_t = \frac{T}{r} = \frac{746093.75}{\left(\frac{100\text{ m}}{2}\right)} = \frac{14921.875}{m} \text{ N}$$

The face width, $b = \dfrac{2\pi m}{\tan \Psi} = 14.8$ m

Assume the velocity factor, $K_v = 0.3$, and lubrication factor, $c_w = 1.25$
Substituting the values in Lewis equation,

$$m = \frac{w_t c_w}{S_d \cdot K_v \cdot b \cdot Y \cdot \cos \Psi} = \frac{(14921.875/\text{m}) \times 1.25}{210 \times 0.3 \times 14.8 \text{ m} \times 0.455 \times \cos(23)}$$

$$\therefore \quad m = \sqrt[3]{47.76} = 3.63$$

Normal module, $m_n = m \cdot \cos \Psi = 3.63 \times \cos 23 = 3.34$
Choose $m_n = 4$

Then $m = \dfrac{4}{\cos 23} = 4.35$

The diameter of pinion, $D_p = mZ_1 = 4.35 \times 20 = 87$ mm

The diameter of gear, $D_g = mZ_2 = 4.35 \times 100 = 435$ mm

The pitch line velocity, $V = \dfrac{\pi DN}{60{,}000} = \dfrac{\pi \times 87 \times 1200}{60{,}000} = 5.47$ m/s

The velocity factor (Eqn. 16.33), $K_v = \dfrac{6}{6+v} = \dfrac{6}{6+5.47} = 0.52 > 0.3$

This is greater than the assumed value and hence the selection $m = 4.0$ is suitable.

Then face width, $b = 14.8\text{ m} = 14.8 \times 4.35 = 64.39 \cong 65$ mm

Step-4: Dynamic load by Buckingham equation

The tangential load,

$$w_t = \frac{14921.872}{m} = \frac{14921.872}{4.35} = 3430.3 \text{ N}$$

For class-III precision gear, Permissible error (Table 16.27 MDHB) $e = 12.5\,\mu\text{m}$
The dynamic factor (Table 16.5 MDHB), $C = 150$
Then substitute the value in the Buckingham equation, (Eqn. 16.37 MDHB)

$$w_d = w_t + \frac{21V[b\cdot C\cos^2\Psi + w_t]\cos\Psi}{21V + \sqrt{b\cdot C\cdot\cos^2\Psi + w_t}}$$

$$\therefore \quad = 3430.3 + \frac{21\times 5.47\times[65\times 150\times\cos^2(23) + 34303]\cos 23}{21\times 5.47 + \sqrt{65\times 150\times\cos^2(23) + 3430.3}} = 8974.15\,\text{N}$$

Step-5: Check for dynamic beam strength,

$$w_b = S_{dr}\cdot b\cdot Y\cdot m\cdot\cos\Psi$$

$$= 210\times 65\times 0.455\times 4.35\times\cos 23 = 24869.06\,\text{N} > w_d$$

This is more than the dynamic load, hence design is safe.

Step-6: Check for wear strength

The ratio factor, $Q = \dfrac{2\times 5}{5+1} = 1.67$

The strength factor, $K = \dfrac{w_d}{b\cdot D_p\cdot Q} = \dfrac{8974.15}{65\times 87\times 1.67} = 0.95,$

From Table 16.23, $(\text{BHN})_p = 300$ and $(\text{BHN})_g = 250$ with $K = 1.135$

Example 16.13: Design a helical gear pair to transmit a power of 15 kW from a shaft running at 1000 rpm, to another shaft to be run at 360 rpm. Assume involute profile with pressure angle of 20°. The material for the pinion is forged steel SAE 1030, whose $\sigma_d = 172.375$ MPa and the material for the gear is cast steel 0.20% C untreated with $\sigma_d = 137.34$ MPa. The gears operate under a condition of medium shock for a period of 10 hr per day. Check for dynamic load, if the load factor, $C = 580$ N/mm, also for wear load.

Given Data: Power, $MP = 15$ kW at speed, $N_1 = 1000$ rpm,
Speed of gear, $N_2 = 360$ rpm,
Tooth profile - 20° full depth involute,
Material for pinion SAE 1030 with $S_d = 172.375$ MPa
Material for gear 0.20% C- cast steel with $S_d = 137.34$ MPa
Service – medium shock, 10 hr per day
Dynamic factor, $C = 580$ N/mm.

SOLUTION

Step-1: Design torque,

$$T = \frac{955\times 10^4\times kW}{N}C_s$$

where kW = 15; $N = 1000$; and $C_s = 1.5$ (Table 16.21 MDHB)

Substituting the values,

$$\therefore \quad T = \frac{955 \times 10^4 \times 15 \times 1.5}{1000} = 214875 \text{ Nmm}$$

Step-2: Compression of strength factors

Gear ratio, $G = \dfrac{1000}{360} = 2.78$

Assume number of teeth on pinion, $Z_1 = 18$

Number of teeth on gear, $Z_2 = Z_1 G = 18 \times 2.78 = 50$

Assume helix angle, $\Psi = 25°$

Virtual number of teeth

For pinion, $Z_{v_1} = Z_1/\cos^3 \Psi = \dfrac{18}{\cos^3 (25)} = 24.18$

For gear, $Z_{v_2} = Z_2/\cos^3 \Psi = \dfrac{50}{\cos^3 (25)} = 67.22$

The tooth form factors (Table 16.3 MDHB)

For pinion, $Y_1 = 0.337$; for gear, $Y_2 = 0.429$

The strength factors

For pinion $(Y_1 S_d)_p = (0.337 \times 172.375) = 58.09$

For gear $(Y_2 S_d)_g = (0.429 \times 137.34) = 58.92$

∴ The pinion is weaker, hence design the pinion.

Step-3: Module by Lewis equation

The face width, $b = 2p_x = \dfrac{2\pi m}{\tan \Psi} = \dfrac{2\pi m}{\tan 25} = 13.5 \text{ m}$

The tangential load,

$$W_t = \frac{T}{r} = \frac{214875}{\left(\dfrac{18 \text{ m}}{2}\right)} = \frac{23875}{m}$$

Assume velocity factor $K_v = 0.3$ and lubrication factor, $C_w = 1.25$

Substitute the values in Lewis equation, (Eqn. 16.31 MDHB)

$$m = \frac{w_t \cdot c_w}{S_d \cdot K_v \cdot b \cdot Y \cdot \cos \Psi} = \frac{(23875/m) \times 1.25}{172.375 \times 0.3 \times 13.5m \times 0.337 \times \cos 25}$$

$$\therefore \quad m^3 = \frac{23875 \times 1.25}{172.375 \times 0.3 \times 13.5 \times 337 \times \cos 25} = 139.36$$

$\therefore \quad m = 5.19$

Choose $m_n = 5.0$

Then $m = m_n/\cos 25 = 5/\cos 25 = 5.52$

The diameters are:

For pinion $D_1 = Z_1 \cdot m = 18 \times 5.52 = 99.36$ mm

For gear $D_2 = Z_2 \cdot m = 50 \times 5.52 = 276$ mm

The pitch line velocity,

$$V = \frac{\pi DN}{60,000} = \frac{\pi \times 99.36 \times 1000}{60,000} = 5.2 \text{ m/s}$$

The velocity factor, $K_v = \dfrac{4.5}{4.5+V} = \dfrac{4.5}{4.5+5.2} = 0.464 >$ assumed value of 0.3

Hence, the module $m_n = 5.0$ is suitable.

Then face width $b = 13.5 \text{ m} = 13.5 \times 5.52 = 74.52 \cong 75$ mm

Step-4: Dynamic load by Buckingham equation

The tangential load,

$$W_t = \frac{23875}{m} = \frac{23875}{5.52} = 4325.18 \text{ N}$$

Dynamic load by Buckingham equation, (Eqn. 16.37 MDHB)

$$W_d = W_t + \frac{21V(b \cdot C \cdot \cos^2\Psi + W_t)\cos\Psi}{21V + \sqrt{b \cdot C \cdot \cos^2\Psi + W_t}}$$

$$= 4325.18 + \frac{21 \times 5.2(75 \times 580 \times \cos^2 25 + 4325)\cos 25}{21 \times 5.2 + \sqrt{75 \times 580 \times \cos^2 25 + 4325.18}} = 17140.47 \text{ N}$$

Step-5: Check for dynamic beam strength, (Eqn. 16.38 MDHB)

$$W_b = S_{dr} \cdot b \cdot Y \cdot m \cdot \cos\Psi$$

$$= 172.375 \times 75 \times 0.337 \times 5.52 \times \cos 25 = 21796.17 \text{ N} > W_d$$

Design is safe against bending fatigue failure.

Step-6: Check for wear strength, (Eqn. 16.39 MDHB)

$$W_w = \frac{K \cdot b \cdot D_p \cdot Q}{\cos^2\Psi}$$

where, $Q = \dfrac{2 \cdot G}{G+1} = \dfrac{2 \times 2.78}{2.78+1} = 1.47$, $b = 75$ mm, and $D_p = 99.36$

For $w_w = w_d = 17140.47\ \text{N}$, the surface strength factor,

$$K = \frac{w_d \cos^2\Psi}{b \cdot D_p \cdot Q} = \frac{17140.47 \times \cos^2 25}{75 \times 99.36 \times 1.47} = 1.284$$

From Table 16.23 MDHB, the surface hardnesses required are:

For pinion $\text{BHN} = 350$, for gear $\text{BHN} = 250$, with $K = 1.375$.

Example 16.14: A pair of mating gears have 20° pressure angle in the normal plane. The normal module is 5 mm and the module in the diameter plane is 5.7735 mm. The pitch diameter of the smaller gear is 115.47 mm. If the transmission ratio is 4.1, calculate (i) helix angle, (ii) normal pitch, (iii) transverse pitch, (iv) number of teeth on each gear, (v) addendum, (vi) whole depth, (vii) clearance, (viii) tooth thickness, (xi) working depth (x) outside diameter, (xi) centre distance, (xii) root circle diameter and (xiii) base circle diameters.

Given Data: Normal module, $m_n = 5.00\ \text{mm}$;

Transverse module, $m = 5.7735\ \text{mm}$

Diameter of the pinion, $D_p = 115.47\ \text{mm}$

Gear ratio, $G = 4$

SOLUTION From Eqn. 16.21 MDHB, the helix angle,

$$\cos\Psi = \frac{m_n}{m} = \frac{5}{5.7735} = 0.866$$

$$\therefore \qquad \Psi = \cos^{-1}(0.866) = 29.999 \cong 30°$$

(i) Normal pitch, $p_n = \pi m_n = \pi \times 5 = 15.708\ \text{mm}$

(ii) Transverse pitch, $p_t = \pi m = \pi \times 5.7735 = 18.138\ \text{mm}$

(iii) Number of teeth, $Z_1 = D_p/m = (115.47/5.7735) = 20$

Number of teeth for gear, $Z_2 = GZ_1 = 4 \times 20 = 80$

(From Table 16.6b MDHB)

(iv) Addendum, $a = 1.0m_n = 5\ \text{mm}$

(v) Addendum, $b = 1.25m_n = 1.25 \times 5 = 6.25\ \text{mm}$

(vi) Whole depth $= a + b = 5 + 6.25 = 11.25\ \text{mm}$

(vii) Clearance, $c = b - a = 6.25 - 5 = 1.25\ \text{mm}$

(viii) Tooth thickness, $t = \dfrac{\pi m_n}{2} = \dfrac{\pi \times 5}{2} = 7.85\ \text{mm}$

(ix) Working depth, $h = 2a = 10\ \text{mm}$

(x) Outside diameter for pinion,

$$D_{p_o} = D_p + 2a = 115.47 + 10 = 125.47\ \text{mm}$$

For gear $D_{g_o} = D_g + 2a = 115.47 \times 4 + 10 = 471.88$ mm

(xi) Centre distance,

$$C = \frac{D_p + D_g}{2} = \frac{115.47 + 461.88}{2} = 288.675 \text{ mm}$$

(xii) Root circle diameter

Pinion diameter $D_{p_i} = D_p - 2b = 115.47 - 2 \times 6.25 = 102.97$ mm

Gear diameter $D_{g_i} = D_g - 2b = 461.88 - 2 \times 6.25 = 449.38$ mm

(xiii) Base circle diameters

The pressure angle in transverse plane

$$\tan \Phi_t = \frac{\tan \Phi_n}{\cos \Psi} \text{ (Eqn. 16.24)} = \frac{\tan 20}{\cos 30} = 0.42$$

$$\Phi_t = \tan^{-1}(0.42) = 22.79°$$

Pinion base circle diameter,

$$D_{p_b} = D_p \cos \Phi_t = 115.47 \times \cos(22.79) = 106.45 \text{ mm}$$

Gear base circle diameter

$$D_{g_b} = D_g \cos \Phi_t = 461.88 \times \cos(22.79) = 425.821 \text{ mm}$$

EXERCISES FOR PRACTICE

16.1 The pair of helical gears for transmitting power is to be designed for the following data:

(i) Velocity Ratio = 8

(ii) R.P.M. of pinion = 4000

(iii) Power transmitted = 80 kW

(iv) Helix angle = 23°

Use full depth involute gear with 20° pressure angle in normal plane.

16.2 Design the helical gears for a single reduction speed reducer to have a velocity ratio of 4.0. The gearbox is to transmit 25 kW at 3000 rpm of the pinion. The helical angle is to be 20°. The teeth are 20° involute in the normal plane. The face width is to be approximately two times the pitch diameter. The material for the gears is to be high carbon steel heat- treated to have yield strength, $S_y = 400$ MPa and BHN = 450.

16.3 A pair of double helical gears is to be designed for a turbine with a transmission ratio of 10:1. The teeth are to be 20° full depth. The pinion is to rotate at 5000 rpm and to have 25 teeth. Gear must be of forged steel, 0.3% carbon, untreated. Power transmitted is 125 kW. Design the gears and specify:

(i) Material of the pinion, so that both gear and pinion have almost equal strength.

(ii) Number of teeth, diameters and normal module.

(iii) Helix angle, face width and centre distance.

16.4 A helical pinion with 20° full depth teeth, helix angle of 23°, pitch diameter of 100 mm and face width of 40 mm, transmits 20 kW at 1100 rpm. If the pressure angle is measured in the plane of rotation, determine the tangential force, the radial force and the thrust force.

16.5 Two parallel shafts are connected by a pair of steel helical gears. The pinion transmits 12 kW at 3000 rpm. Velocity ratio is 4:1. Static design stress for material of gears is 130 MPa. If the teeth are to be 20° stub in the diametral plane, design the gears with a helix angle of 30°. Check your design for wear load.

16.6 A pair of helical gears transmit 15 kW at 4000 rpm of the pinion. Both the gears are made of same material, hardened steel with an allowable static stress of 100 MPa. If the velocity ratio is 4.5 to 1, determine the smallest diameter gears that may be used having sufficient strength. Not less than 30 teeth are to be used on either gear, the teeth are of 20° involute in diametral plane and the helix angle is 45°.

16.7 Two helical gears, having a 20° normal pressure angle and a helix angle of 23°, connect two parallel shafts. The pinion has 20 teeth, the gear has 100 teeth, and the module is 5 mm. The face width of the gear is 80 mm. Both gears are made of steel, C30 heat treated to 300 BHN. Speed of pinion is 2500 rpm. Assume class-III gears for dynamic and wear load.

16.8 A pair of helical gears is to be designed for a speed ratio of 10:1. Design the gears and specify:

(i) Material of pinion

(ii) Number of teeth, diameters and normal module.

(iii) Helix angle, face width and centre distance.

16.9 A helical cast steel gear drive is required to transmit 30 kW at 2800 rpm of the pinion, having 24 teeth. The helix angle is 30° and the teeth are cut with 20° full depth involute form in the normal place. The design stress is to be 80 MPa. Design and specify the details of the gear drive.

16.10 A pair of helical gears transmits 15 kW power. The driver and driven gears run at 1800 rpm and 600 rpm respectively. The pinion has 24 teeth and the helix angle is 30°. Both the gears are made of cast steel with an allowable static stress of 60 MPa. Design the gears assuming all other necessary data.

16.11 (a) Derive an expression for virtual number of teeth in terms helix angle and actual number of teeth of a helical gear.

(b) A pair of helical gears are to transmit 15 kW of power. The teeth are 20° stub in diametral plane and have a helix angle of 45°. The pinion has 80 mm pitch diameter and operates at 900 rpm. The gear has 320 mm pitch diameter. If the gears are made of cast steel with allowable static stress of 100 MPa. Design the gears. Pinion and gear are heat-treated to 300 and 200 BHN respectively.

16.12 Design a pair of helical gears to transmit a power of 20 kW from a shaft running at 1500 rpm to a parallel shaft to be run at 450 rpm. Suggest suitable surface hardness for the gear pair.

16.13 Design a pair of helical gears to transmit a power of 30 kW from a shaft rotating at 1500 rpm to a parallel shaft to be rotated at 450 rpm.

16.14 A pair of herringbone gears is used to transmit 50 kW power. The pinion rotates at 2800 rpm. The number of teeth on the pinion and gear are 21 and 109 respectively. The tooth form is 20° full depth involute and the helix angle is 25°. The material for the gear is cast steel with a hardness of 150 BHN and for the pinion is steel. The wear and lubrication factor may be taken as 1.15. The normal module employed for the gears is 4 mm and the face width of the gears is 20 times the normal module. Determine the required hardness for the pinion for continuous operation of the drive. Also recommend the class of gears.

16.15 A pair of carefully cut helical gears for a turbine has a transmission ratio of 10:1. The teeth are 20° stub involute in the normal plane. Pinion has 25 teeth and rotates at 5000 rpm. Material for pinion and gear is 0.4% carbon steel untreated with allowable static stress of 70 MPa. Helix angle = 30°. Power transmitted = 90 kW. Service factor = 1.25. Wear and lubrication factor = 1.25. Determine the module in normal plane and face width of the gears. Suggest suitable surface hardness for the gear pair.

16.16 (a) Explain formative number of teeth in helical gears.

(b) A pair of helical gears are to transmit 16 kW. The teeth are 20° stub in diametral plane and have a helix angle of 45°. The pinion runs at 1000 rpm and has 80 mm pitch diameter. The gear has 320 mm pitch diameter. If the gears are made of cast steel having allowable static strength of 100 MPa. Determine the module and face width from static strength considerations and check the gears for wear, given $\sigma_{es} = 618$ MPa.

16.17 A pair of helical gears with a 23° helix angle is to transmit 25 kW at 10,000 rpm of the pinion. The velocity ratio is 4:1. Both gears are to be made of hardened steel with an allowable stress of 100 MPa for each gear. The gears are 20° stub and the pinion is to have 24 teeth. Design the gears and determine the required BHN.

16.18 Design a steel helical gear pair from the following data:

Power transmitted = 30 kW

Speed of pinion = 1500 rpm

Velocity ratio = 1:4

Number of teeth on pinion = 24

Helix angle, $\beta = 30°$

Static stress for steel $\sigma_{d1} = \sigma_{d2} = 50.7$ MPa. Brinell hardness number for gear material = 350. Check the design from wear point of view also.

16.19 Design a single reduction parallel helical gear speed reducer having speed ratio 5:1. It is to be capable of transmitting the full load rating of 18.75 kW at 1200 rpm of pinion. Use nickel chrome steel for pinion and C45 steel for gear. Select class III precision gears. The helix angle for the gear wheels is 23°. The normal pressure angle of the tooth is 20° and the profile is full depth involute. Check the design for dynamic and wear load.

16.20 Design a pair of helical gears to transmit 30 kW at 2000 rpm of the pinion. The speed reduction ratio is 4:1. The helix angle and normal pressure angle are to be 25° and 20° respectively. The materials for pinion and gear are steels with permissible stresses as 55 MPa and 40 MPa, hardness as 340 BHN and 300 BHN respectively. The minimum number of teeth on pinion are 30.

SHORT ANSWER QUESTIONS

16.1 What is a helical gear set?
16.2 What are the directions of helices on the mating helical gears?
16.3 What is profile of gear tooth of a helical gear?
16.4 How are teeth produced on the gear blank?
16.5 What is helix angle?
16.6 What is circular (transverse) pitch?
16.7 What is normal pitch?
16.8 How are they related?
16.9 What is circular module?
16.10 What is normal module?
16.11 How are they related?
16.12 What is axial pitch?
16.13 How is it related to circular pitch?
16.14 What is pressure angle in the plane rotation?
16.15 What is the pressure angle in the normal plane?
16.16 How are they related?
16.17 What is the formative number of teeth?
16.18 What is the AGMA recommendation for face width?
16.19 What is the recommended contact ratio?
16.20 What are the loads on the helical gear tooth?
16.21 How are they related?
16.22 What is the modified Lewis equation for the module of helical gear?
16.23 Explain the dynamic load equation for the helical gear set.
16.24 Why are they preferred over spur gears for heavy high speed applications?
16.25 What is the dynamic beam strength equation?
16.26 What is the wear-strength equation?
16.27 Give the new AGMA equations for checking bending strength of gear teeth.
16.28 What are the new AGMA equations for checking the contact strength of gear teeth?
16.29 What are herringbone gears?
16.30 What are double helical gears?
16.31 What are the advantages of these gears over single helical gears?
16.32 What are the disadvantages of these gears?
16.33 What are crossed helical gears?
16.34 Where are they used?
16.35 What are the disadvantages of the crossed helical gears?

CHAPTER 17

BEVEL GEARS

17.1 INTRODUCTION

When the shaft axes intersect, the power can be transmitted from one to the other by putting two contacting friction cones on the shafts. When one cone rotates, the other cone rotates because of frictional force between them. The drive can be made positive by providing teeth on the surface of the truncated cones. Then we have bevel gears. Depending upon the orientation of the teeth, there are mainly two types of bevel gears:

(a) **Straight bevel gears:** The tooth axis coincides with a generator.

(b) **Spiral bevel gears:** The tooth axis runs along a helix on the conical surface, i.e., along a spiral.

The straight bevel gears are common. Here we consider the design of only straight bevel gears.

17.2 GEAR GEOMETRY: TERMINOLOGY

Shaft angle θ – The angle between the axes of the shafts.

Pitch angle (α, β) – The half cone angle of the equivalent friction cone.

Pitch diameter (D_p, D_g) – The diameter of the cone at the larger end of teeth.

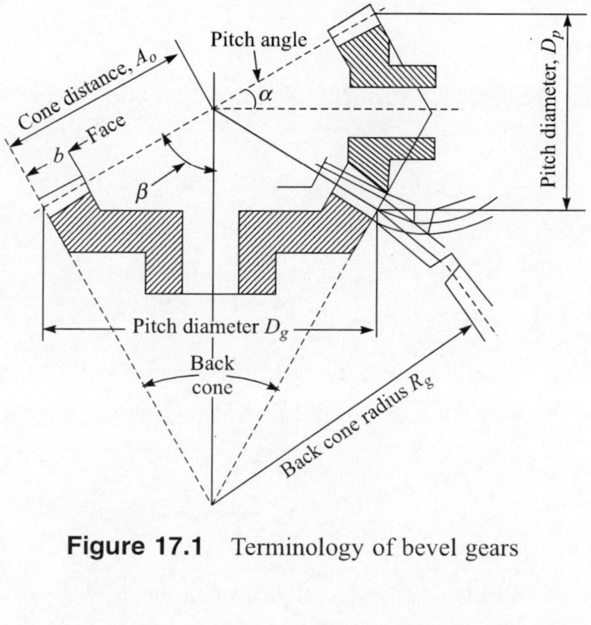

Figure 17.1 Terminology of bevel gears

Back- Cone – It is the cone formed by the outer end faces of the teeth. The slant height of the back-cone or the length of the back-cone generator is called the back cone radius and is given by:

For pinion, $R_{bp} = R_p/\cos\alpha$ (17.1a)

For gear, $R_{bg} = R_g/\cos\beta$ (17.1b)

Module – The tooth height reduces towards the apex. It is proportional to the pitch cone radius at any position. For specification purposes, the module is defined on the basis of the larger end of tooth i.e., outer pitch diameter.

$$m = \frac{D}{Z} = \frac{D_p}{Z_p} = \frac{D_g}{Z_g} \tag{17.2}$$

Cone-distance (A_o) – It is length of the generator of the pitch cone and is given by the relations,

$$A_o = R_p/\sin\alpha = R_g/\sin\beta \tag{17.3a}$$

$$= \sqrt{R_p^2 + R_g^2} \text{ for shaft angle } \theta = 90° \tag{17.3b}$$

When viewed along the contact line, the engagement resembles the spur gear set with centres at the apexes of the back cones. The imaginary gears are equivalent gears and the number of teeth on these gears are called virtual or formative number of teeth expressed by the relations:

For pinion, $Z_{pv} = \dfrac{2R_{bp}}{m} = \dfrac{2R_p}{m\cos\alpha} = Z_p/\cos\alpha$ (17.4a)

For gear, $Z_{gv} = \dfrac{2R_{bg}}{m} = \dfrac{2R_g}{m\cos\beta} = Z_g/\cos\beta$ (17.4b)

Face width (b) – It is length of a tooth measured along the pitch generator.

17.3 TOOTH LOADS

The force acting between the contact surfaces is normal to the contact surfaces. It has the following components.

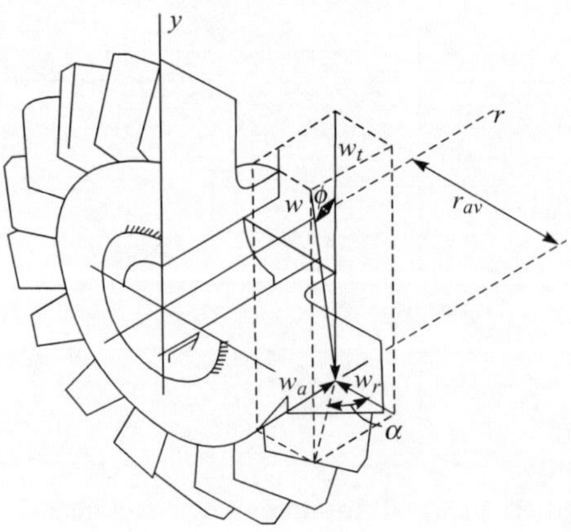

Figure 17.2 Bevel-gear tooth forces

where,

(i) $W_t =$ tangential component. It acts along the tangent to the pitch cones. (T/r_{av}) (17.5)

(ii) $W_r =$ radial component. It is the force acting along the radial line perpendicular to the cone axis

$$W_{rp} = W_t \tan\phi \cos\alpha \text{ for pinion} \tag{17.6a}$$

$$W_{rg} = W_t \tan\phi \cos\beta \text{ for gear} \tag{17.6b}$$

(iii) $W_a =$ axial thrust. It is the force acting parallel to the cone axis

$$W_{ap} = W_t \tan\phi \sin\alpha \text{ for pinion} \tag{17.7a}$$

$$W_{ag} = W_t \tan\phi \sin\beta \text{ for gear} \tag{17.7b}$$

For gear the tangential force remains the same but the radial and axial components get interchanged.

For $\theta = 90°$, $W_{rp} = W_{ag}$ and $W_{ap} = W_{rg}$

17.4 DESIGN PROCEDURE

Modification to Lewis Equation

The tooth height varies from inside edge to the outside edge and the load is distributed along its length. Evaluation method of the bending strength has to consider this effect.

We consider an elemental width dx of the tooth at a distance 'x' from the apex.

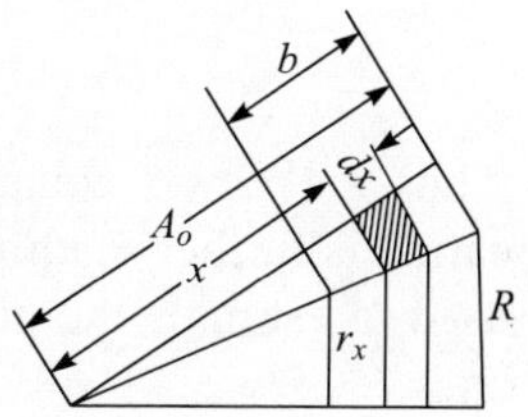

Figure 17.3

Let m_x be module at the position $= \dfrac{m}{R} \cdot r_x$

$r_x =$ Pitch radius at the position

The beam strength of this tooth element,

$$dW_t = S_d \cdot dx \cdot Y \cdot m_x \tag{i}$$

The torque transmitted by this elemental portion,

$$dT = dW_t \cdot r_x = S_d \, dx \cdot Y \cdot m_x \cdot r_x = S_d Y \cdot \left(\frac{m}{R}\right)\left(\frac{R}{A_o} x\right)^2 \cdot dx$$

$$= S_d Y \cdot \left(\frac{m}{R}\right)\left(\frac{R}{A_o}\right)^2 x^2 \cdot dx \tag{ii}$$

The total torque transmitted by the tooth is the sum of all such element portions from inside edge to outside edge.

$$T = \int dT = \int_{A_o - b}^{A_o} \left(S_d \cdot Y \cdot \frac{mR}{A_0^2}\right) x^2 \cdot dx = \left(\frac{S_d \cdot Y \cdot mR}{A_0^2}\right) \int_{A_o - b}^{A_o} x^2 \cdot dx$$

$$= \left(\frac{S_d \cdot Y \cdot mR}{A_0^2}\right)\left[\frac{x^3}{3}\right]_{A_o - b}^{A_o} = \left(\frac{S_d \cdot Y \cdot mR}{3A_0^2}\right)[A_0^3 - (A_o - b)^3]$$

$$= \left(\frac{S_d \cdot Y \cdot mR}{3A_0^2}\right)[A_0^3 - A_0^3 + 3A_0^2 \cdot b - 3A_o b^2 + b^3]$$

$$= S_d \cdot Y \cdot m \cdot Rb \left[1 - \frac{b}{A_o} + \frac{1}{3}\left(\frac{b}{A_o}\right)^2\right] \quad \text{(iii)}$$

Since $b/A_o \leq 1/3$, the last term is very small compared to the second and hence, it is neglected.

Then the torque, $T = (S_d \cdot Y \cdot m \cdot b \cdot R)\left(1 - \frac{b}{A_o}\right)$ (iv) (17.8)

The tangential force referred to the larger end,

$$W_t = \frac{T}{R} = S_d \cdot Y \cdot m \cdot b \cdot \left(1 - \frac{b}{A_o}\right) \quad \text{(v)} \qquad (17.9)$$

Thus the Lewis equation is modified to account for the varying height of the tooth. This equation is further modified by introducing the velocity factor to account for dynamic effect and is used to estimate the preliminary value of the module and other dimensions, in the following form,

$$W_t = S_d \cdot C_v \cdot Y \cdot b \cdot m \cdot \left(1 - \frac{b}{A_o}\right) \qquad (17.10)$$

where,

Y = tooth form factor based on the virtual no. of teeth (Table 15.3)

C_v = Velocity factor

$= \frac{3}{3+V}$ for machine cut gears

$= \frac{6}{6+v}$ for precision generated gears

And $A_o \cong 18$ m to 30 m and $b = A_o/3 = 6$ m to 10 m

The dimensions so estimated are checked for dynamic effect by using the Buckingham dynamic load equation,

$$W_d = W_t + \frac{21V(b \cdot C + W_t)}{21V + \sqrt{b \cdot C + W_t}} \qquad (17.11)$$

It is same as for spur gears.

The dynamic load can also be calculated by the Eqn.,

$$W_d = W_t + \frac{e \cdot N_p \cdot Z_p \cdot b \cdot R_p \cdot R_g}{K_1 \cdot \sqrt{R_p^2 + R_g^2}} \qquad (17.11a)$$

where, $R_p = \dfrac{D_p}{2} - \dfrac{b \sin \propto}{2}$ and $R_g = \dfrac{D_g}{2} - \dfrac{b \sin \beta}{2}$

$$K_1 = 2530 \text{ for steel-steel}$$
$$= 3785 \text{ for CI-CI}$$
$$= 3260 \text{ for steel-CI}$$

Check for dynamic beam strength,

$$W_d = S_{dr} \cdot b \cdot Y \cdot m \cdot \left(1 - \frac{b}{A_o}\right) \geq W_d \tag{17.12}$$

where, $S_{dr} \cong S_{en} \cong 1.76$ BHN for steels

Check for wear strength, $W_w = \dfrac{K \cdot b_e \cdot D_p \cdot Q}{\cos \alpha} > W_d$ (17.13)

where,

$$Q = \frac{2Z_{vg}}{Z_{vp} + Z_{vg}} \tag{17.13a}$$

$$b_e = 0.75b \text{ for out board gears}$$
$$= 1.00b \text{ for in board gears}$$

According to the new AGMA standard, check is done for induced bending stress and contact stress, evaluated as follows:

Induced bending stress is given by

$$\sigma_b = \frac{W_t \cdot K_o \cdot K_s \cdot K_m}{K_v \cdot m \cdot b \cdot J} \leq S_d \tag{17.14}$$

The design stress in bending S_d is given by

$$S_d = \frac{S \cdot K_L}{K_T \cdot K_R} \tag{17.15}$$

Induced contact stress is given by

$$\sigma_H = C_p \sqrt{\frac{W_t \cdot C_o C_s \cdot C_m \cdot C_f}{C_v \cdot D \cdot b \cdot I}} \leq S_{Hd} \tag{17.16}$$

where the design stress for contact is given by

$$S_{Hd} = \frac{S_{fe} \cdot C_L}{C_T \cdot C_R} \tag{17.17}$$

The coefficients have the same meaning as defined for spur gears.

Table 17.1. Tooth proportions for 20° straight bevel - gear

Working depth					
$h_k = 2.0$ m Clearance: $c = 0.188$ m + 0.05 Addendum of gear: $a_G = 0.54\text{ m} + \dfrac{0.460\text{ m}}{(i_{90})^2}$ Gear ratio: $i = Z_G \mid Z_p$	Root angle $\alpha_r = \alpha - \tan^{-1}\left(\dfrac{d}{A_o}\right)$ Face angle $\alpha_F = \alpha + \tan^{-1}\left(\dfrac{a}{A_o}\right)$ d = dedendum a = addendum		Equivalent 90° gear ratio $i_{90} = i$ for $\theta = 90°$ $= \sqrt{i\dfrac{\cos\alpha}{\cos\beta}}$ when $\theta \neq 90°$ $A_o = \sqrt{R_G^2 + R_P^2}$ Smaller of the following $b = \dfrac{A_o}{3}$ or $b = 10$ m		
Min. number of teeth (Z)	Pinion	16	15	14	13
	Gear	16	17	20	30

Table 17.2. Spiral gear proportions

Pressure angle, $\phi = 20°$	Working depth, $hk = 0.7$ m
Spiral angle, $\psi = 35°$	Clearance, $c = 0.188$ m

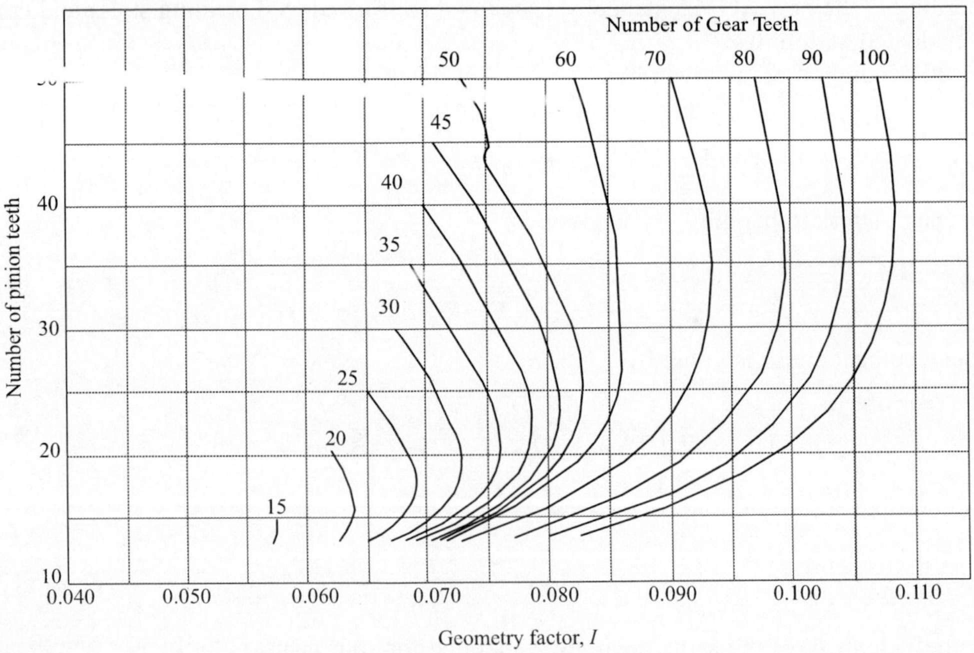

Figure 17.4 Geometry factor I for straight bevel gears with $\phi = 20°$ and $0.120/p_d$ tool-edge radius

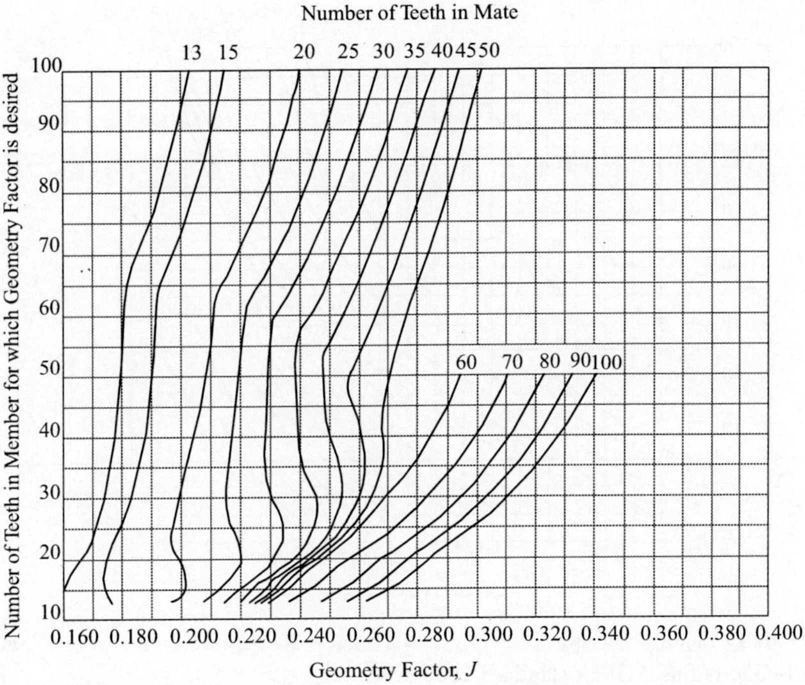

Figure 17.5 Geometry factor J for straight bevel gears with $\phi = 20°$ and $0.120/p_d$ tool-edge radius

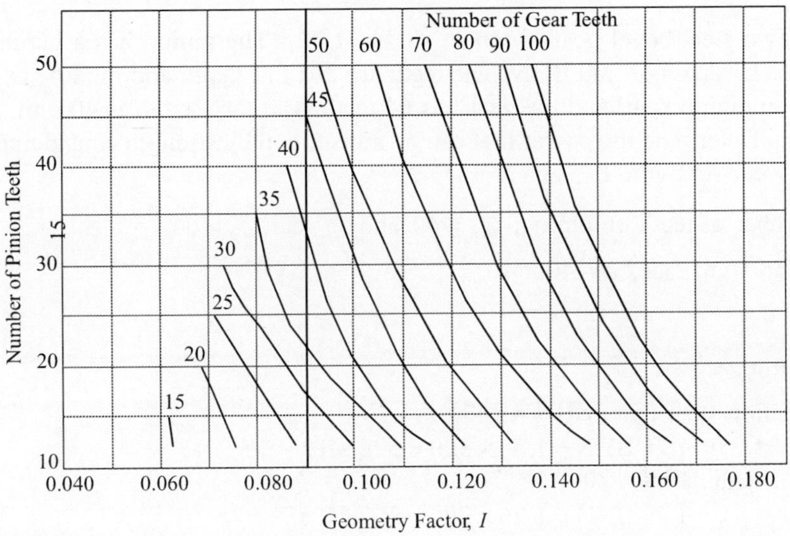

Figure 17.6 Geometry factor I for spiral bevel gears with $\phi = 20°$, spiral angle, $\psi = 35°$, and $0.240/p_d$ tool-edge radius AGMA 200-A86

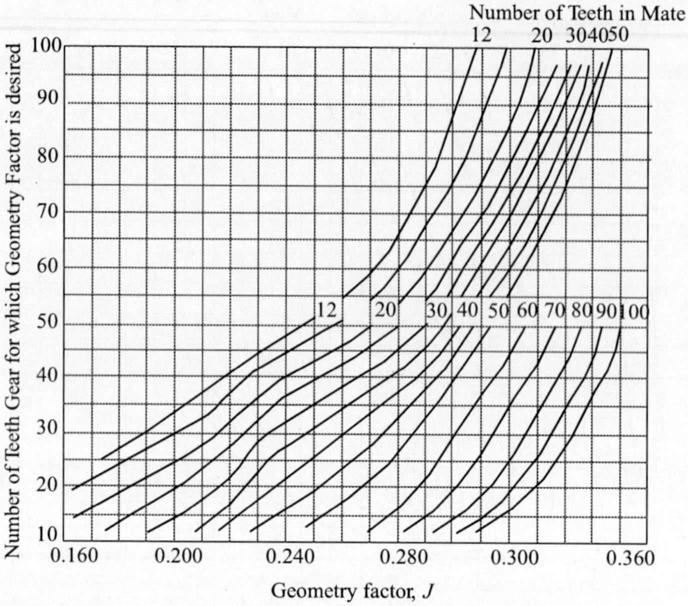

Figure 17.7 Geometry factor J for spiral bevel gears with $\phi = 20°$, spiral angle, $\psi = 35°$, and $0.240/p_d$ tool-edge radius AGMA standard 2003-A86

SOLVED EXAMPLES

Example 17.1: Two steel bevel gears connect shafts at 90°. The pinion has a hardness of 300 BHN and gear has hardness of 200 BHN. The teeth are 20° full depth and module is 3 mm. The number of teeth on pinion is 40 and the gear has 64 teeth. The face width is 40 mm. The pinion rotates at 900 rpm. Determine the power that can be transmitted by strength consideration. Check for the dynamic wear strength.

Given Data: Number of teeth on pinion, $Z_p = 40$, and on gear, $Z_g = 64$

Module, $m = 3.0$ mm, Face, $b = 40$ mm

SOLUTION

Step-1: Design stresses

Material for pinion steel with $\text{BHN} = 300$,
Endurance stress, $S = 1.76\ \text{BHN} = 1.76 \times 300 = 528\ \text{MPa}$
for gear – steel with $\text{BHN} = 200$
Endurance stress, $S = 1.76\ \text{BHN} = 1.76 \times 200 = 352\ \text{MPa}$
Let factor of safety $= 2.0$

Design stress for pinion, $S_d = \dfrac{528}{2} = 264\ \text{MPa}$

For gear, $S_d = \dfrac{352}{2} = 176\ \text{MPa}$

Step-2: Virtual number of teeth

Gear ratio $= 64/40 = 1.6$

Pitch cone angle of gear, $\beta = \tan^{-1}(1.6) = 58°$

Pitch angle of pinion, $\alpha = 90 - 58° = 32°$

Virtual number of teeth, $Z_{vp} = \dfrac{Z_p}{\cos\alpha} = \dfrac{40}{\cos 32} = 47.16$

$$Z_{Vg} = \frac{Z_g}{\cos\beta} = \frac{64}{\cos 58} = 120.8$$

Tooth form factor, $Y_p = 0.405$, and $Y_g = 0.455$

Step-3: Comparison of strength factors

For pinion $(Y_p \cdot S_d)_p = 0.405 \times 264 = 106.32$

For gear $(Y_g \cdot S_d)_g = 0.455 \times 176 = 80.08$

The gear is weaker, hence base the calculations on gear.

Step-4: Power capacity by Lewis equation

Gear diameter, $D_g = 64 \times 3 = 192$ mm

Gear speed, $N_g = 900/1.6 = 562.5$ rpm

Velocity factor, $V = \dfrac{\pi DN}{60{,}000} = \dfrac{\pi \times 192 \times 562.5}{60{,}000} = 5.65$ m/sec

Velocity factor, $K_v = C_v = \dfrac{6}{6+V} = \dfrac{6}{(6+5.65)} = 0.514$

The pinion diameter, $D_p = 40 \times 3 = 120$ mm

The cone distance, $A_o = \sqrt{R_g^2 + R_p^2} = \sqrt{\left(\dfrac{192}{2}\right)^2 + \left(\dfrac{120}{2}\right)^2} = 113.2$ mm

Then transmitted load by Lewis equation,

$$W_t = S_d \cdot C_v \cdot Y \cdot b \cdot m \cdot \left(1 - \frac{b}{A_o}\right) \quad \text{(Eqn. 16.60 MDHB)}$$

$$= (176 \times .514 \times 0.455 \times 40 \times 3) \times \left(1 - \frac{40}{113.2}\right) = 3194\,\text{N}$$

The torque on the gear, $T = W_t \times R_G = 3194 \times \dfrac{192}{2} = 306624$ Nmm

Power transmitted, $NP = \dfrac{TN}{955 \times 10^4} = \dfrac{306624 \times 562.5}{955 \times 10^4} = 18.06$ kW

Step-5: Dynamic load by Buckingham equation

The gear error, $e = 12.5\ \mu m$ (Table 16.28 MDHB) (precision gear)

The dynamic factor, $C = 150$ (Table 16.5 MDHB)

By Buckingham equation, (Eqn. 16.58 MDHB)

$$W_d = W_t + \frac{21V(b \cdot C + W_t)}{21V + \sqrt{b \cdot C + W_t}}$$

$$= 3194 + \frac{21 \times 5.65 \times (40 \times 150 + 3194)}{21 \times 5.65 + \sqrt{40 \times 150 + 3194}} = 8278.8\ \text{N}$$

Step-6: Check for dynamic beam strength

Endurance limit stress, $S_{dr} = 352$ MPa

Beam strength, $W_b = S_{dr} \cdot Y \cdot b \cdot m \cdot \left(1 - \frac{b}{A_o}\right)$

$$= (352 \times 0.455 \times 40 \times 3) \times \left(1 - \frac{40}{113.2}\right) = 12428\ \text{N} > W_d$$

Design is safe against bending.

Step-7: Check for wear strength

The wear strength,

$$W_w = Kb_e \cdot D_p \cdot Q / \cos\alpha$$

$$= \frac{0.92 \times 40 \times 120 \times 1.44}{\cos 32}$$

$$= 7498.44\ \text{N} < W_d$$

where,

$K = 0.92$ (Table 16.23 MDHB)

$b_e = b = 40$ mm

$D_p = 120$ mm

$$Q = \frac{2 \times 120.8}{120.8 + 47.16} = 1.44$$

Then the surface hardening is required.

The load stress factor required, $K = \frac{8278.8 \times \cos 32}{40 \times 120 \times 1.44} = 1.016$

Then BHN for gear = 250; BHN for pinion = 300 with $K = 1.135$ is suitable.

Example 17.2: A pair of straight bevel gears is required to transmit 10 kW at 500 rpm of motor shaft to another shaft at 250 rpm. The pinion has 24 teeth, the pitch angle is 20°. If the shaft axes are at right angles to each other, find the module, face width, and pitch diameters. The static design stress for the material is 60 MPa. The Lewis form factor may be taken as $y = 0.154 - \frac{912}{Z}$; Velocity factor, $K_v = \frac{4.5}{4.5 + V}$, the face width is to be one-fourth the cone distance.

Given Data: Power, $NP = 10$ kW at speed, $N = 500$ rpm;
Driven speed = 250 rpm; No. of teeth on pinion = 24;
Pitch angle of pinion, $\alpha = 20°$; Shaft angle, $\theta = 90°$;
Design stress, $S_d = 60$ MPa; Face width, $b = \dfrac{A_0}{4}$

SOLUTION

Step-1: Module by Lewis equation

The torque on the pinion,

$$T = \frac{955 \times 10^4 \text{ kW}}{N} = \frac{955 \times 10^4 \times 10}{500} = 191000 \text{ Nmm} = 191 \text{ Nm}$$

Pitch diameter of the pinion, $D_p = Zm = 24$ m

The tangential load on the pinion,

$$W_t = \frac{T}{r} = \frac{191 \times 10^3}{\frac{24\text{ m}}{2}} = \frac{15916.7}{m}\text{N}$$

Virtual number of teeth, $Z_{pv} = \dfrac{24}{\cos 20°} = 25.54$

Tooth form factor, $Y = \pi\left(0.154 - \dfrac{0.912}{Z_v}\right) = 0.372$

Pitch line velocity,

$$V = \frac{\pi DN}{60{,}000} = \frac{\pi \times (24 \cdot m) \times 500}{60{,}000} = (0.628 \text{ m}) \text{ m/sec}$$

Velocity factor, $C_v = \left(\dfrac{4.5}{4.5 + 0.628 \text{ m}}\right)$

Cone distance, $A_o = R_p / \sin\alpha = \left(\dfrac{24 \text{ m}}{2}\right) / \sin 20° = 35.086$ m

Face width, $b = A_o/4 = 8.77$ m

Substituting these in Lewis equation

$$W_t = S_d \cdot C_v \cdot Y \cdot b \cdot m \cdot \left(1 - \frac{b}{A_o}\right)$$

$$\frac{15916.7}{m} = 60 \times \left(\frac{4.5}{4.5 + 0.628 \text{ m}}\right) \times 0.372 \times 8.77 \text{ m} \times m\left(1 - \frac{1}{4}\right)$$

$$= \frac{660 \text{ m}^2}{4.5 + 0.628 \text{ m}}$$

$\therefore$ $$660\ m^3 = 15916.7\ (4.5 + 0.628\ m)$$

$\therefore$ $$m^3 = 24.1\ (4.5 + 0.628\ m) = 108.4 + 15.13\ m$$

$\therefore$ $$m^3 - 15.13\ m = 108.4$$

Try $m = 4.0$, LHS $= 3.48 <$ RHS

Try $m = 5.0$, LHS $= 49.35 <$ RHS

Try $m = 6.0$, LHS $= 125.2 >$ RHS

The solution is between $m = 5$ and $m = 6$

Hence, take $m = 6$ mm

Step-2: Dimensions of the gears

The cone distance $A_o = 35.086 \times 6 = 210.5$ mm
Face width, $b = 8.77 \times 6 = 52.63\ \text{mm} \cong 55.0$ mm (chosen)
Pitch diameter, $D_p = 24 \times m = 144$ mm

Example 17.3: A pair of bevel gears is required to transmit 30 kW at 500 rpm. The output shaft runs at 200 rpm and is at right angles to the input shaft. The gears are supported in housing close to them. The gear is of cast iron and pinion is of cast steel. The teeth are of 20° stub involute form. The static strength of cast iron is 70 MPa and of cast steel is 142 MPa. Determine the proportions of gears based on strength only.

Given Data: Power $= 30$ kW; Pinion speed, $N_p = 500$ rpm;

Gear speed, $N_G = 200$ rpm; Shaft angle = 90°;

Material for pinion cast steel with, $S_d = 142$ MPa;

Material for gear–cast iron with, $S_d = 70$ MPa;

Tooth form - 20° stub involute

SOLUTION

Step-1: Pitch cone angles

$$\text{Gear ratio} = \frac{500}{200} = 2.5$$

The pitch cone angle for pinion, $\alpha = \tan^{-1}(1/2.5) = 21.8°$

for gear, $\beta = \tan^{-1}(2.5) = 68.2°$

Step-2: Comparison of strength factors

Let $Z_p = 20$ and $Z_g = 50$

Formative number of teeth

$$\text{For pinion, } Z_{vp} = \frac{Z_p}{\cos\alpha} = \frac{20}{\cos 21.8} = 21.55$$

For gear, $Z_{vg} = \dfrac{50}{\cos 68.2} = 134.64$

Form factors for pinion, $Y_p = 0.406$; for gear, $Y_g = 0.516$

Strength factors

For pinion $(YS_d)_p = (0.406 \times 142) = 57.6$

For gear $(YS_d)_g = (0.516 \times 70) = 36.12$

The gear is weaker, hence base the calculations on gear.

Step-3: Module by Lewis equation

Torque on gear, $T = \dfrac{955 \times 10^4 \times 30}{200} = 143.25 \times 10^4$ Nmm

The tangential load, $W_t = \dfrac{143.25 \times 10^4}{25\ \text{m}} = \dfrac{5.73 \times 10^4}{m}$ N

The velocity factor, $K_v = 0.4$ (assumed)

The pinion diameter, $D_p = 20$ m

The cone distance, $A_o = \sqrt{R_p^2 + R_g^2} = \sqrt{(10\ \text{m})^2 + (25\ \text{m})^2} = 26.9258$ m

Face width, $b = \dfrac{A_o}{3} = \dfrac{26.9258\ \text{m}}{3} = 8.975$ m

Then using Lewis equation for strength,

$$W_t = S_d \cdot C_v \cdot Y \cdot b \cdot m \cdot \left(1 - \frac{b}{A_o}\right)$$

$$\frac{5.73 \times 10^4}{m} = 70 \times 0.4 \times 0.516 \times 8.975\ \text{m} \times m\left(1 - \frac{1}{3}\right) = 86.88\ \text{m}^2$$

$$\therefore \qquad m = \sqrt[3]{\frac{5.73 \times 10^4}{86.88}} = 8.7$$

Choose $\qquad m = 9.0$

Then face width, $b = 8.975 \times m = 8.975 \times 9 = 80.775\ \text{mm} \cong 81$ mm

The diameter of gear, $D_g = 50 \times 9 = 450$ mm

The diameter of pinion, $D_p = 20 \times 9 = 180$ mm

Pitch line velocity

$$V = \frac{\pi DN}{60,000} = \frac{\pi \times (180) \times 500}{60,000} = 4.71 \text{ m/sec}$$

$$\text{Velocity factor } C_v = \frac{6}{6+4.71} = 0.56$$

This value is more than the assumed value of 0.4 and design is safe.

Example 17.4: A right angle speed reducer uses hardened steel alloy precision cut bevel gears. The transmission ratio is 5 to 1 and pinion rotates at 900 rpm and transmits 40 kW. If the teeth are of 20° full depth form and the pinion has a diameter of 115 mm, what must be the module and face width of the gears. For both gears, the same material with $S_d = 200$ MPa is used. What hardness is required for satisfactory wear strength?

Given Data: Power transmitted $= 40$ kW at speed $= 900$ rpm,

Gear ratio $5:1$, Shaft angle, $\theta = 90°$, Pinion diameter, $D_p = 115$ mm

Design stress, $S_d = 200$ Mpa

SOLUTION

Step-1: Dimensions by geometry

Diameter of the gear, $D_g = 115 \times 5 = 575$ mm
Pitch cone angle of the pinion, $\alpha = \tan^{-1}(1/5) = 11.31°$
Pitch cone angle of the gear, $\beta = \tan^{-1}(5) = 78.69°$

$$\text{The cone distance, } A_o = R_p/\sin\alpha = \frac{115/2}{\sin 11.31} = 293.2 \text{ mm}$$

$$\text{Choose face width, } b = \frac{A_o}{3} = 97.73 \text{ mm} \cong 100 \text{ mm}$$

Step-2: Module by Lewis equation

Torque on the pinion,

$$T = \frac{955 \times 10^4 \times 40}{900} = 42.44 \times 10^4 \text{ Nmm}$$

$$\text{Tangential load, } W_t = \frac{42.44 \times 10^4}{57.5} \text{ N} = 7381.64 \text{ N}$$

Using Lewis equation,

$$W_t = S_d \cdot Y \cdot b \cdot m \cdot C_v \left(1 - \frac{b}{A_o}\right)$$

where, $W_t = 7381.64$ N, $S_d = 200$ MPa, $b = 100$ mm, $b/A_o \cong 1/3$

Pitch line velocity, $V = \dfrac{\pi DN}{60,000} = \dfrac{\pi \times 115 \times 900}{60,000} = 5.42$ m/sec

Velocity factor, $C_v = \dfrac{6}{6 + 5.42} = 0.525$

Substituting the values,

$$7381.64 = 200 \times Y \times 100 \times m\,(1 - 1/3) \times 0.525 = 7035(Y \cdot m)$$

$$\therefore \quad (Y \cdot m) = 1.05$$

Let $m = 2.5$, $z = \dfrac{115}{2.5} = 46$,

Virtual number of teeth, $Z_v = \dfrac{46}{\cos 11.31} = 46.9$, then $Y = 0.403$

Then $Y \cdot m = 2.5 \times 0.403 = 1.0075$

It is nearly equal to the required value of 1.05.

The design parameters acceptable are:

$$m = 2.5, \text{ and } b = 100 \text{ mm}$$

Then by Lewis equation,

$$W_t = S_d \cdot C_v \cdot b \cdot m\left(1 - \frac{b}{A_o}\right) = 200 \times 0.525 \times 0.403 \times 100 \times 2.5 \times 2/3$$

$$= 7052.5 \text{ N} \cong W_t = 7381.64 \text{ N}$$

The values are nearly equal and can be accepted.

Step-3: Dynamic load by Buckingham equation,

$$W_d = W_t + \frac{21V\,(b \cdot C + W_t)}{21V + bC + W_t} \text{ for } C = 150$$

$$= 7381.64 + \frac{21 \times 5.42 + (100 \times 150 + 7381.64)}{21 \times 5.42 + \sqrt{100 \times 150 + 7381.64}} = 17052.24 \text{ N}$$

Step-4: Hardness for wear strength

In wear load equation

$$W_w = (Kb_e \cdot D_p \cdot Q)/\cos\alpha$$

where, $W_w = W_d = 17052.24$, $b_e = b = 100$ mm, $D_p = 115$ mm

$$Z_{Vg} = \frac{Z_V}{\cos\beta} = \frac{230}{\cos 78.69} = 1172.7, \; Z_{VP} = 46.911$$

$$Q = \frac{2 \times Z_{Vg}}{Z_{Vg} + Z_{VP}} = \frac{2 \times 1172.7}{1172.7 + 469} = 1.922$$

$$\therefore \qquad K = \frac{17052.24 \times 0.98}{100 \times 115 \times 1.92} = 0.757$$

Hardness required is BHN = 250 for both with $K = 0.92$.

Example 17.5: Two shafts at right angles are to be connected with a pair of bevel gears having 20° full depth teeth. The velocity ratio is to be 4.5:1. The pinion is to be made of steel ($S_d = 100$ MPa) and the gear is to be made of semi-steel ($S_d = 85$ MPa). The pinion is to transmit 4 kW at 900 rpm. Determine the minimum diameters, module and face width based on strength.

Given Data: Power = 4 kW; Pinion speed, $N_p = 900$ rpm;
Speed ratio = 4.5; Gear speed, $N_g = (900/4.5) = 200$ rpm;
Design stress for pinion = 100 MPa, and for gear = 85 MPa

SOLUTION

Step-1: Geometric relations

The pitch angle of pinion, $\alpha = \tan^{-1}(1/4.5) = 12.53°$

The pitch angle of gear, $\beta = \tan^{-1}(4.5) = 77.47°$

Let number of teeth on pinion, $Z_p = 16$

Number of teeth on gear, $Z_g = 16 \times 4.5 = 72$

Virtual number of teeth for pinion, $Z_{pv} = 16/\cos(12.53) = 16.4$

Virtual number of teeth for gear, $Z_{gv} = 72/\cos(77.47) = 331.9$

Step-2: Comparison of strength factors

Tooth form factors: for pinion, $Y_p = 0.303$; for gear, $Y_g = 0.475$
Strength factors are:
For pinion $(YS_d)_p = 0.303 \times 100 = 30.3$
For gear $(YS_d)_g = (0.475 \times 85) = 40.375$
The pinion is weaker, hence design the pinion.

Step-3: Module for strength by Lewis equation

Cone distance, $A_o = \sqrt{R_p^2 + R_g^2} = \sqrt{\left(\frac{16\text{ m}}{2}\right)^2 + \left(\frac{72\text{ m}}{2}\right)^2} = 36.89\text{ m}$

Face width, $b = A_o/3 = 12.29$ m
Assume velocity factor, $C_v = 0.4$

Torque on the pinion, $T = \frac{955 \times 10^4 \text{ kW}}{N} = \frac{955 \times 10^4 \times 4}{900} = 42444.4$ Nmm

Tangential load, $W_t = \frac{T}{r} = \frac{42444.4}{8\text{ m}}$

Substituting the values in Lewis equation,

$$W_t = S_d \cdot C_v \cdot Y \cdot b \cdot m(1 - b/A_o)$$

$$\therefore \quad \frac{42444.4}{8\,\text{m}} = 100 \times 0.4 \times 0.303 \times 12.29\ \text{m} \times m \times 2/3$$

$$\therefore \quad m = \sqrt[3]{53.4} = 3.76$$

Let $m = 4.0$; and $D_p = 4 \times 16 = 64$ mm

Pitch line velocity, $V = \dfrac{\pi DN}{60{,}000} = \dfrac{\pi \times 64 \times 900}{60{,}000} = 3.016$ m/sec

The velocity factor for precision gears

$$C_v = \frac{6}{6+V} = \frac{6}{6+3.016} = 0.66$$

It is more than the assumed value of 0.4. It is OK. But we may get smaller module with this larger value of C_v. Let us try $m = 3.0$

The pinion diameter, $D_p = 3 \times 16 = 48$ mm

Pitch line velocity, $V = \dfrac{\pi DN}{60{,}000} = \dfrac{\pi \times 48 \times 900}{60{,}000} = 2.26$ m/sec

Velocity factor, $C_v = 6/(6 + 2.26) = 0.726$

With this new value in Lewis equation, we get

$$m^3 = \frac{42444.4 \times 3}{8 \times 100 \times 0.726 \times 0.303 \times 12.29 \times 2} = 29.44$$

$$\therefore \quad m = \sqrt[3]{29.44} = 3.09\ \text{mm}$$

$\therefore$ Module $m = 3.0$ is OK

$$b = 12.29 \times 3 = 36.9 \cong 35.0\ \text{mm (chosen)}$$

Step-4: Dynamic load by Buckingham equation

Error, $e = 12.5\ \mu\text{m} = 0.0125$ mm; Dynamic factor, $C = 150$ (Table 16.5 MDHB)

$$W_t = \frac{42444.4}{8 \times 3} = 1768.5\ \text{N}$$

Substituting the values in Buckingham equation (Eqn. 16.58 MDHB)

$$W_d = W_t + \frac{21V(bC + W_t)}{21V + \sqrt{b \cdot C + W_t}}$$

$$= 1768.5 + \frac{21 \times 2.26(35 \times 150 + 1768.5)}{21 \times 2.26 + \sqrt{35 \times 150 + 1768.5}} = 4306.6\ \text{N}$$

Step-5: The strength factor for wear strength for $W_w = W_d$

$$\text{And } Q = 2\times\frac{331.9}{331.9+16.4} = 1.91$$

$$K = \frac{W_d \times \cos\alpha}{b\cdot D_p Q} = \frac{4306.6\times\cos 12.53}{35\times 48\times 1.91} = 1.31$$

Then the surface hardening to the following values is specified

$(\text{BHN})_p = 300$, $(\text{BHN})_g = 300$ with $K = 1.375$ (from Table 16.23 MDHB)

Example 17.6: The shaft of a ventilating fan which absorbs 30 kW at 750 rpm is driven by a motor running at 1000 rpm. The shaft axes are at right angles. Design the bevel gears for the drive.

Given Data: Power = 30 kW; Speed of the fan = 750 rpm;
Speed of the motor = 1000 rpm

SOLUTION

Step-1: Geometric relations

The speed ratio, G = 1000/750 = 4/3 = 1.33

Let number of teeth on pinion, $Z_p = 18$,

On gear, $Z_g = 18\times 1.33 = 24$

The pitch cone angle

For pinion, $\alpha = \tan^{-1}(3/4) = 36.87°$

For pinion, $\beta = \tan^{-1}(4/3) = 53.13°$

Step-2: Module estimation by Lewis equation

Material–Cast steel treated for both (Table 16.4 MDHB)

$$S_d = 195 \text{ MPa (BHN 250)}$$

Since the material is the same, design the pinion.

$$\text{Torque on pinion, } T = \frac{955\times 10^4 \times 30}{1000} = 286500 \text{ Nmm} = 286.5 \text{ Nm}$$

$$\text{The tangential load, } W_t = T/r = \frac{286500}{\frac{18\text{ m}}{2}} = \left(\frac{31833.33}{m}\right)\text{N}$$

$$\text{Virtual number of teeth for pinion, } Z_{v_p} = \frac{18}{\cos 36.87} = 22.5$$

Tooth form factor, $Y = 0.331$ (for 20° full depth)
The diameter of pinion, $D_p = 18$ m
The diameter of gear, $D_g = 24$ m

The cone distance, $A_o = \sqrt{R_p^2 + R_g^2} = \sqrt{(9\text{ m})^2 + (12\text{ m})^2} = 15\text{ m}$

Face width, $b = \dfrac{A_o}{3} = 15\text{ m}/3 = 5\text{ m}$

The velocity factor, $C_v = 0.4$ (assumed).

Then substituting the values in Lewis equation,

$$\frac{31833.3}{m} = 195 \times 0.311 \times 5\text{ m} \times m\,(1 - 1/3) \times 0.4$$

$$\therefore \qquad m^3 = \frac{31833.3 \times 3}{195 \times 0.331 \times 5 \times 2 \times 0.4} = 369.7$$

$$\therefore \qquad m = 7.18 \cong 7\text{ mm}$$

Pinion diameter $D_p = 7 \times 18 = 126$ mm

Pitch line velocity, $V = \dfrac{\pi DN}{60,000} = \dfrac{\pi \times 126 \times 1000}{60,000} = 6.59$ m/sec

Actual velocity factor, $C_v = \dfrac{6}{6+v} = \dfrac{6}{6+6.59} = 0.47$

This is greater than the assumed value and hence module selected is OK. The preliminary design parameters,

$$m = 7;\ b = 35\text{ mm};\ D_p = 126\text{ mm};\ D_g = 168\text{ mm}.$$

The tangential load, $W_t = 31833.3/\text{m} = 4547.6$ N

Step-3: Check for bending stress (new AGMA Eqn.)

Design stress in bending, $S_d = \dfrac{S \cdot K_L}{K_T \cdot K_R}$

where, $S = 1.76\text{ BHN} = 1.76 \times 250 = 440$ MPa,

$K_L = 1;\ K_T = 1;\ K_R = 1.5;$

$$\therefore \qquad S_d = \frac{440 \times 1.0}{1.0 \times 1.5} = 293.3\text{ MPa}$$

Bending stress induced

$$\sigma_b = \frac{W_t K_o K_s \cdot K_m}{K_v m \cdot b \cdot J} = \frac{4547.6 \times 1.5 \times 0.75 \times 1.1}{0.7 \times 35 \times 7 \times 0.24}$$

$$= 136.7\text{ MPa} < S_d$$

where

$W_t = 4547.6$ N

$K_o = 1.5$ (Table 16.10 MDHB)

$K_s = 0.75$ (Table 16.17 MDHB)

$K_m = 1.1$ (Table 16.16 MDHB)

$K_v = 0.7$ (Fig. 16.5c MDHB)

$J = 0.24$ (Fig. 16.13 MDHB)

It is less than the design stress and gear is safe against bending fatigue failure.

Step-4: Check for contact stress

Permissible contact stress,

$$S_{Hd} = \frac{S_{fe} \cdot C_L}{C_T \cdot C_R} = \frac{630 \times 1.0}{1.0 \times 1.25}$$

$$= 504 \text{ MPa}$$

where,

$S_{fe} = 2.8 \text{ BHN} - 70$ (Table 16.22 MDHB)

$= 2.8 \times 250 - 70 = 630$ MPa

$C_L = 1.0$, $C_R = 1.25$ (Table 16.13 MDHB)

Induced contact stress

$$\sigma_H = C_p \sqrt{\frac{W_t C_o \cdot C_s \cdot C_m \cdot C_f}{C_v \cdot b \cdot D \cdot I}}$$

where,

$W_t = 4547.6$ N, $C_o = K_o = 1.5$ (Table 16.10)

$C_s = K_s = 0.75$, (Table 16.17); $C_m = K_m = 1.1$ (Table 16.16 MDHB);

$C_v = K_v = 0.7$ (Fig. 16.5 MDHB); $I = 0.068$ (Fig. 16.12 MDHB);

$C_p = 237$ (Table 16.18 MDHB), $C_f = 1.0$ (assumed)

Substituting the values,

$$= 237\sqrt{\frac{4547.6 \times 1.5 \times 0.75 \times 1.1 \times 1.0}{0.7 \times 35 \times 126 \times 0.068}} = 1227 \text{ MPa} > S_{Hd}$$

∴ Wear failure occurs

To have sufficient wear strength, pinion has to be surface hardened to BHN $= \dfrac{1227 + 70}{2.8} \cong 465$

Example 17.7: Design a pair of bevel gears to transmit 12 kW at 300 rpm of gear, and 1470 rpm of pinion. The angle between the shafts is 90°. The pinion has 20 teeth and the material of gears is cast steel with allowable stress in bending, $S_d = 183.33$ MPa; BHN = 320. Take the service factor as 1.25. Cheek the gears for dynamic load and wear load. Suggest the suitable hardness for the gear pair.

Given Data: Power transmitted, $NP = 12$ kW;

Speed of pinion, $N_1 = 1470$ rpm; Speed of gear, $N_2 = 300$ rpm;

Number of teeth on pinion, $Z_p = 20$; Shaft angle, $\theta = 90°$

The material is cast steel with allowable stress in bending $S_d = 183.33$ MPa and hardness BHN = 320

Service factor $= 1.25$

SOLUTION

Step-1: Geometric relations

The gear ratio, $N_1/N_2 = \dfrac{1470}{300} = 4.9$

Number of teeth on gear, $Z_g = 4.9Z_p = 4.9 \times 20 = 98$

The pitch angle of the gear

$\beta = \tan^{-1}(Z_g/Z_p) = \tan^{-1}(4.9) = 78.47°$ (Eqn. 16.46a MDHB)

The pitch angle of pinion, $\propto = \theta - \beta = 90 - 78.47 = 11.53°$

The formative number of teeth for pinion,

$$Z_{vp} = \frac{Z_p}{\cos\alpha} = \frac{20}{\cos 11.53°} = 20.412$$

For gear, $Z_{Vg} = \dfrac{Z_g}{\cos\beta} = 98/\cos(78.57) = 490.29$

The radius of the pinion, $R_p = \dfrac{Z_p \cdot m}{2} = 10\text{ m}$

The radius of the gear, $R_g = \dfrac{Z_g \cdot m}{2} = \dfrac{98\text{ m}}{2} = 49\text{ m}$

The centre distance,

$A_o = \sqrt{R_g^2 + R_P^2} = (\sqrt{49^2 + 10^2}) \cdot m = 50\text{ m}$ (Eqn. 16.50 MDHB)

The gear width, $b = 10\text{ m for } A_o > 30\text{ m}$ (Eqn. 16.57 MDHB)

Step-2: Module by Lewis equation

The torque transmitted,

$$T = \frac{955 \times 10^4\text{ kW}}{N} C_s = \frac{955 \times 10^4 \times 12 \times 1.25}{1470} = 97450\text{ Nmm}$$

The tangential force on the tooth,

$$W_t = T/R_p = \frac{97450}{10\text{ m}} = (9745/\text{m})$$

The Lewis equation,

$$W_t = S_d \cdot C_v \cdot Y \cdot b \cdot m\left[1 - \frac{b}{A_o}\right]$$ (Eqn. 16.55 MDHB)

where velocity factor $C_v = 0.4$ (assumed) $Y = 0.322$ (Table 16.3 MDHB)

Substituting the values,

$$\frac{9745}{m} = 183.33 \times 0.4 \times 0.322 \times 10\ \text{m} \times m\left[1 - \frac{10\ \text{m}}{50\ \text{m}}\right]$$

$$\therefore \qquad m^3 = \frac{9745 \times 5}{183.33 \times 0.4 \times 0.322 \times 10 \times 4} = 51.59$$

$$m = \sqrt[3]{51.59} = 3.72$$

Choose $m = 4$ (Table 16.3 MDHB)

Then pitch diameter of pinion, $D_p = 20\ \text{m} = 20 \times 4 = 80$ mm
The pitch line velocity,

$$V = \frac{\pi DN}{60{,}000} = \frac{\pi \times 80 \times 1470}{60{,}000} = 6.16\ \text{m/sec}$$

Velocity factor,

$$C_v = \frac{6}{6+V} = \frac{6}{6+6.16} = 0.49 > 0.4\ \text{(assumed)} \qquad \text{(Eqn. 16.56b MDHB)}$$

Hence $m = 4$ mm is satisfactory

Face width, $b = 10\ \text{m} = 10 \times 4 = 40$ mm

Long distance, $A_o = 50\ \text{m} = 50 \times 4 = 200$ mm

The tangential load, $W_t = 9745/m = 9745/4 = 2436.25$ N

Step-3: Dynamic load

The Buckingham equation for dynamic load,

$$W_d = W_t + \frac{21V\left(b \cdot C + W_t\right)}{21V + \sqrt{(b \cdot c + w_t)}} \qquad \text{(Eqn. 16.58 MDHB)}$$

where $C = 150$ for $e = 0.0125$ (Table 16.5 MDHB)

$\because$ $e = 12.5\ \mu\text{m}$ (from Table 16.27 MDHB)

Substituting the values,

$$W_d = 2436.25 + \frac{21 \times 6.16 \times (40 \times 150 + 2436.25)}{21 \times 6.16 + \sqrt{(40 \times 150 + 2436.25)}}$$

$$= 2436.25 + \frac{129.36 \times 8436.25}{129.36 + \sqrt{(8436.25)}} = 7369.65\ \text{N}$$

Step-4: Check for dynamic beam strength,

$$W_b = S_{dr} \cdot Y \cdot b \cdot m\left[1 - \frac{10\ \text{m}}{50\ \text{m}}\right] \qquad \text{(Eqn. 16.60 MDHB)}$$

$$=183.33\times 0.322\times 40\times 4\times\left(1-\frac{40}{200}\right)=7556.13\text{ N}>W_d$$

The teeth are safe against bending failure.

Step-5: Check for wear strength,

$$W_w=\frac{K\cdot b_e\cdot D_p\cdot Q}{\cos\propto}\geq W_d \qquad \text{(Eqn. 16.61 MDHB)}$$

For $W_w=W_d$, the strength factor, $K=\dfrac{W_d\cos\propto}{b_e\cdot D_p\cdot Q}$

where, $b_e=0.75b=30$ mm,

$$Q=\frac{Z_{VG}}{Z_{vp}+Z_{vG}}=\frac{2\times 490.29}{20.4+490.29}=1.92$$

Substituting the values,

$$K=\frac{7369.65\cos(11.57)}{30\times 80\times 1.92}=1.57$$

∴ The hardnesses for gear pair,

$$(\text{BHN})_p=350;\ (\text{BHN})_g=300\text{ with }K=1.635$$

are suggested (from Table 16.23 MDHB)

Example 17.8: A pair of bevel gear wheels with 20° pressure angle consists of 20 teeth pinion meshing with 30 teeth gear. The module is 4 mm while face width is 20 mm, the surface hardness of both pinion and gear is 400 BHN. The pinion rotates at 500 rpm and receives power from an electrical motor. The starting torque of the motor is 150% of the rated torque. Determine the safe power that can be transmitted considering the dynamic load, wear strength and endurance strength. The allowable bending stress may be taken as 240 MPa.

Given Data: Number of teeth on pinion, $Z_p=20$; Number of teeth on gear, $Z_g=30$;

The speed of pinion, $N_1=500$ rpm;

The module, $m=4$ mm; Face width, $b=20$ mm;

Surface hardness $(\text{BHN})_p=(\text{BHN})_g=400$;

Starting torque $=1.5\times$ Average torque;

Allowable bending stress, $S_d=240$ MPa;

Determine the power transmitted.

SOLUTION

Step-1: Geometric relations

The gear ratio, $G=\dfrac{Z_g}{Z_P}=\dfrac{30}{20}=1.5$

Speed of gear, $N_2 = \dfrac{500}{1.5} = 333.33$ rpm

The pitch angle of gear, $\tan\beta = Z_g/Z_P$ (Eqn. 16.46a MDHB)

$$\therefore \quad \beta = \tan^{-1}(Z_g/Z_P) = \tan^{-1}(1.5) = 56.31°$$

The pitch angle of pinion, $\propto = \theta - \beta = 90 - 56.31 = 33.69°$

Diameter of pinion, $D_p = Z_p m = 20 \times 4 = 80$

Diameter of gear, $D_g = Z_g m = 30 \times 4 = 120$ mm

Cone distance, $A_o = \sqrt{R_g^2 + R_P^2} = \sqrt{60^2 + 40^2} = 72.11$ mm

For pinion, $Z_{vp} = Z_p/\cos\propto = 20/\cos(33.69) = 24.04$

For gear, $Z_{Vg} = \dfrac{Z_g}{\cos\beta} = \dfrac{30}{\cos 56.31} = 54.08$

The pitch line velocity,

$$V = \frac{\pi DN}{60,000} = \frac{\pi \times 80 \times 500}{60,000} = 2.1 \text{ m/sec}$$

Step-2: Dynamic beam strength,

$$W_d = S_d \cdot Y \cdot b \cdot m \cdot \left[1 - \frac{b}{A_o}\right] \quad \text{(Eqn. 16.60 MDHB)}$$

where, $Y = 0.337$ (Table 16.3), $S_d = 240$ MPa;

$m = 4$ mm, $b = 20$ mm, $A_o = 72.11$ mm

Substituting the values,

$$\therefore \quad W_b = 240 \times 0.337 \times 20 \times 4\left[1 - \frac{20}{72.11}\right] = 4675.81 \text{ N}$$

Step-3: Dyamic wear strength,

$$W_w = \frac{K \cdot b_e \cdot D_p \cdot Q}{\cos\propto} \quad \text{(Eqn. 16.61 MDHB)}$$

where, $K = 2.3$ from Table 16.23 MDHB, taken between 2.23 and 2.40

$b_e = 0.75 \times b = 0.75 \times 20 = 15$, $D_p = 80$ mm

$$Q = \frac{2 \times Z_{Vg}}{Z_{VP} + Z_{Vg}} = \frac{2 \times 54.08}{24.04 + 54.08} = 1.38$$

$$\cos\propto = \cos(33.69) = 0.832$$

Substituting the values,

$$W_w = \frac{2.3 \times 15 \times 80 \times 1.38}{0.832} = 4577.6\,\text{N}$$

Step-4: Tangential load by Buckingham dynamic load equation

The smaller of the two is $W_d = 4577.6\,\text{N}$
Assuming precision gear, the permissible error, $e = 12.5\,\mu\text{m}$ (from Table 16.27 MDHB)
The dynamic factor $C = 150$ (from Table 16.5 MDHB)
Substituting the values in Buckingham equation,

$$W_d = W_t + \frac{21V(b \cdot c + W_t)}{21V + \sqrt{b \cdot c + W_t}}$$

$$4577.6 = W_t + \frac{21 \times 2.1(20 \times 150 + W_t)}{21 \times 2.1 + \sqrt{20 \times 150 + W_t}} = W_t + \frac{44.1 \times (3000 + W_t)}{44.1 + \sqrt{3000 + W_t}}$$

By trial and error method, $W_t \cong 2550\,\text{N}$

The average value $(W_t)_{av} = W_t/1.5 = \dfrac{2550}{1.5}$

∴ The average torque,

$$T = (W_t)_{av} \times R_p = \frac{2550}{1.5} \times 40 = 68000\,\text{Nmm}$$

The power transmitted, $NP = \dfrac{TN}{955 \times 10^4} = \dfrac{68000 \times 500}{955 \times 10^4} = 3.56\,\text{kW}$

Wear Power Rating

$$NP = 0.8 C_A \cdot C_B \cdot b \times 10^{-3}\,\text{kW} \qquad \text{(Eqn. 16.68a MDHB)}$$

where, $C_A = 0.5$ (Table 16.29 MDHB),

$$C_B = \frac{D_p^{1.5} N_p}{1012}\left(\frac{5.6}{5.6 + \sqrt{v}}\right) = \frac{80^{1.5} \times 500 \times 0.794}{1012} = 280.7$$

Substituting the values,

$$NP = 0.8 \times 0.5 \times 280.7 \times 20 \times 10^{-3} = 2.245\,\text{kW}$$

The power capacity is the smaller of the two; **NP = 2.245 kW**

Example 17.9: Two steel bevel-gears connect shafts at 90°, the pinion has a surface hardness of 300 BHN and the gear has a surface hardness of 200 BHN, the tooth profile is to be 14.5° full depth involute and module is 4 mm. The number of teeth on pinion is 30 and the gear has

48 teeth. The face width is 40 mm. Determine the wear strength of the pair. What would be the power that can be transmitted based on the above wear strength, if the pinion rotates at 1440 rpm?

Given Data: Shaft angle, $\theta = 90°$;

No. of teeth on pinion, $Z_p = 30$; No. of teeth on gear, $Z_g = 48$;

Tooth profile, $\phi = 14.5^{\circ}$ involute full depth;

Module, $m = 4$ mm; Face width, $b = 40$ mm;

The surface hardness,

For pinion $(\text{BHN})_p = 300$, for gear $(\text{BHN})_g = 200$

Pinion speed, $N_p = 1440$ rpm

SOLUTION

Step-1: Geometric relations

The gear ratio, $G = \dfrac{Z_g}{Z_p} = \dfrac{48}{30} = 1.6$

The pitch angle for gear, $\tan\beta = \dfrac{Z_g}{Z_p} = 1.6$ (Eqn. 16.48a MDHB)

$\therefore$ $$\beta = \tan^{-1}(1.6) = 58°$$

The pitch angle for pinion,

$$\propto = \theta - \beta = 90° - 58° = 32° \quad \text{(Eqn. 16.49 MDHB)}$$

The pitch diameters

For pinion, $D_p = m \cdot Z_p = 4 \times 30 = 120$ mm

For gear, $D_p = m \cdot Z_g = 4 \times 48 = 192$ mm

The formative number of teeth

For pinion, $Z_{vp} = Z_p / \cos\propto = 30/\cos 32° = 35.38$

For gear, $Z_{vg} = Z_g / \cos\beta = 48/\cos 58° = 90.58$

The ratio factor,

$$Q = \frac{2Z_{VG}}{Z_{VG} + Z_{VP}} = \frac{2 \times 90.58}{90.58 + 35.38} = 1.44 \quad \text{(Eqn. 16.62 MDHB)}$$

The strength factor $K = 0.92$ (Table 16.23 MDHB)

The effective face width $b_e = 0.75b$ for out board $= 0.75 \times 40 = 30$

Wear strength,

$$W_w = \frac{K \cdot b_e D_p \cdot Q}{\cos \propto} = \frac{0.92 \times 30 \times 120 \times 1.44}{\cos 32} = 5623.83\,\text{N} \qquad \text{(Eqn. 16.61 MDHB)}$$

The dynamic load,

$$W_d = W_t + \frac{e \cdot N_p \cdot Z_p \cdot b \cdot R_p \cdot R_g}{K_1 \cdot \sqrt{R_p^2 + R_g^2}} \qquad \text{(Eqn. 16.59 MDHB)}$$

where,

$$e = 50\ \mu\text{m for commercial gear} \qquad \text{(Table 16.27 MDHB)}$$

$$N_p = 1440\ \text{rpm},\ Z_p = 30.0,\ b = 40\ \text{mm}$$

$$R_p = \frac{D_p}{2} - \frac{b \sin \propto}{\propto} = 60 - \frac{40 \times \sin 32}{2} = 49.4\ \text{mm}$$

$$R_g = \frac{D_g}{2} - \frac{b \sin \beta}{\propto} = 96 - \frac{40 \times \sin 58}{2} = 79.04\ \text{mm}$$

$$K_1 = 2530\ \text{for steel-steel}$$

$$W_d = W_w = 5623.83\ \text{N}$$

Then substituting the values in

$$W_t = W_d - \frac{e \cdot N_p \cdot Z_p b R_p \cdot R_g}{K_1 \sqrt{R_p^2 + R_g^2}}$$

$$= 5623.83 - \frac{50 \times 1440 \times 30 \times 40 \times 49.4 \times 79.04}{2530 \sqrt{79.04^2 + 49.4^2}} = 4193.23\,\text{N}$$

The torque, $T = W_t \dfrac{D_p}{2} = 4193.23 \times \dfrac{120}{2} = 251593.8\ \text{Nmm}$

The power transmitted,

$$NP = \frac{TN}{955 \times 10^4} = \frac{251593.8 \times 1440}{955 \times 10^4} = 37.94\ \text{kW}$$

Example 17.10: A pair of mitter gears have pitch diameter of 280 mm and face width of 36 mm and run at 250 rpm. The teeth are $14\frac{1}{2}^\circ$ involute and accurately cut and transmit 6 kW. Neglecting friction angle find the following:

(i) Outside diameter of gears
(ii) Resultant tooth load tangent to pitch cone
(iii) Radial load on the pinion
(iv) Thrust load on the pinion.

Given Data: Pitch diameters $D_p = D_g = 280$ mm; Face width = 36 mm;
Power transmitted, NP = 6 kW at 250 rpm;
Tooth form 14.5° full depth, accurately cut

SOLUTION

Step-1: Geometric relations

Pitch cone angle, $\propto = \beta = 45°$

Pitch radii, $R_p = R_g = 140$ mm

Cone distance, $A_o = \sqrt{R_p^2 + R_g^2} = \sqrt{140^2 + 140^2} = 198$ mm

The torque transmitted,

$$T = \frac{955 \times 10^4 \text{ kW}}{250 \text{ N}} = \frac{955 \times 10^4 \times 6}{250} = 229200 \text{ Nmm}$$

The radius at midpoint, $R_{av} = \frac{D_p}{2} - \frac{b \sin \propto}{2}$ (Eqn. 16.52a MDHB)

$$= 140 - \frac{36 \times \sin 45}{2} = 127.27 \text{ mm}$$

Then, tangential load at the midpoint,

$$W_t = \frac{T}{R_{av}} = \frac{229200}{127.27} = 1800.87 \text{ N.}$$ (Eqn. 16.52 MDHB)

Radial force $W_r = W_t \tan \phi \cos \propto$, (Eqn. 16.53 MDHB)

$$= 1800.87 \times \tan(14.5) \times \cos 45° = 323.33 \text{ N}$$

Axial load (thrust load), $W_a = W_t \tan \phi \sin \propto$ (Eqn. 16.54 MDHB)

$$= 1800.87 \times \tan(14.5) \times \sin(45°) = 323.33 \text{ N}$$

Assume the number of teeth, $Z_p = Z_G = 70$

Then module, $m = D_p / Z_p = \frac{280}{70} = 4.0$ mm

The formative no. of teeth, $Z_{VP} = Z_{VG} = \frac{Z_p}{\cos \propto} = \frac{70}{\cos 45} = 99.$ (Eqn. 16.51 MDHB)

The tooth form factor (Table 16.2 MDHB) $Y = 0.367$

The pitch line velocity,

$$V = \frac{\pi DN}{60,000} = \frac{\pi \times 280 \times 250}{60,000} = 3.67 \text{ m/sec}$$

The velocity factor,

$$C_v = \frac{3}{3+V} = \frac{3}{3+3.67} = 0.45 \qquad \text{(Eqn. 16.56a MDHB)}$$

Material property – (Table 16.4 MDHB)

Material C45 hardened, $S_d = 210$ MPa, BHN = 250

Using Lewis equation the permitted tangential load,

$$W_t^1 = S_d \cdot C_v \cdot Y \cdot b \cdot m \cdot \left[1 - \frac{b}{A_o}\right]$$

$$= 210 \times 0.45 \times 0.367 \times 36 \times 4 \times \left[1 - \frac{36}{198}\right] = 4086.11 \text{ N}$$

This is much more than the actual tangential load of $W_t = 1800.87$ N and hence $m = 4$ mm chosen is O.K.

The addendum of the gear, (Table 16.7 MDHB)

$$a_G = 0.54 \text{ m} + \frac{0.46 \text{ m}}{i^2} = m = 4 \text{ mm for } i = 1$$

Face angle,

$$\propto_F = \propto + \tan^{-1}\left(\frac{a}{A_o}\right) = 45 + \tan^{-1}\left(\frac{4}{198}\right) = 46.16°$$

Outside diameter,

$$D_o = 2A_o \sin \propto_F = 2 \times 198 \times \sin(46.16) = 285.63 \text{ mm}$$

Example 17.11: A pair of straight-tooth bevel gears at right angles is to transmit 5 kW at 1500 rpm of the pinion at a speed ratio of 3. Diameter of the pinion is 75 mm. The tooth form is $14\frac{1}{2}°$ involute. Pinion is made of steel ($S_d = 160$ MPa) and gear of CI ($S_d = 80$ MPa). Design the gear pair and check the design for dynamic load and wear.

Given Data: Power, $NP = 5$ kW at 1500 rpm; Speed ratio, $G = 3$

Diameter of pinion, $D_p = 75$ mm; Tooth form, 14.5° involute

Material for pinion-steel $S_d = 160$ MPa; for gear, CI $S_d = 80$ MPa

SOLUTION

Step-1: Module by geometric relations

Minimum number of teeth on pinion,

$$Z_p = 48/(\sqrt{1+G^2}) = 8/(\sqrt{1+3^2}) = 15.17 \qquad \text{(Eqn. 16.46 MDHB)}$$

Choose $Z_p = 15$

The number of teeth on the gear, $Z_g = G \cdot Z_p = 3 \times 15 = 45$

The pitch angle of gear, $\tan\beta = \dfrac{Z_g}{Z_p} = 3$ (Eqn. 16.46a MDHB)

$$\therefore \quad \beta = \tan^{-1}(3) = 71.57°$$

The pitch angle of pinion, $\propto = 90 - \beta = 90 - 71.57 = 18.43°$

The module, $m = \dfrac{D_p}{Z_p} = \dfrac{75}{15} = 5.0$ mm

The diameter of gear, $D_g = G \cdot D_p = 75 \times 3 = 225$ mm

The cone distance,

$$A_o = \sqrt{R_p^2 + R_g^2} = \sqrt{\left(\frac{75}{2}\right)^2 + \left(\frac{225}{2}\right)^2} = 118.59 \text{ mm} \quad \text{(Eqn. 16.50 MDHB)}$$

Step-2: Face width by Lewis equation

The torque transmitted,

$$T = \frac{955 \times 10^4 \text{ kW}}{N} = \frac{955 \times 10^4 \times 5}{1500} = 31833.33 \text{ Nmm}$$

The tangential load, $W_t = \dfrac{T}{r} = \dfrac{31833.33}{(75/2)} = 848.9$ N

The formative number of teeth of pinion,

$$Z_{pv} = \frac{Z_p}{\cos\propto} = \frac{15}{\cos(18.43)} = 15.811$$

The formative no. of teeth of gear,

$$Z_{vg} = \frac{Z_g}{\cos\beta} = \frac{45}{\cos(71.57)} = 142.34$$

Tooth form factors (Table 16.3 MDHB)

For pinion, $Y_p = 0.255$ and for gear, $Y_g = 0.374$

Strength factors

For pinion $(S_d \cdot Y)_p = (160 \times 0.255) = 40.8$

For gear $(S_d \cdot Y)_g = (80 \times 0.374) = 29.92$

The gear is weaker, hence design the gear.

The face width by Lewis equation,

$$W_t = S_d \cdot Y \cdot C_v \cdot b \cdot m \cdot \left(1 - \frac{b}{A_o}\right)$$

The pitch line velocity,

$$V = \frac{\pi DN}{60,000} = \frac{\pi \times 75 \times 1500}{60,000} = 5.89 \text{ m/s}$$

The velocity factor, (Eqn. 16.56b MDHB)

$$C_v = \frac{6}{6+V} = \frac{6}{6+5.89} = 0.505$$

Substituting the values,

$$848.9 = 80 \times 0.374 \times 0.505 \times b \times 5 \times \left(1 - \frac{b}{118.59}\right)$$

$$= 75.55b\left(1 - \frac{b}{118.59}\right)$$

$\therefore$ $b^2 - 118.59b + 1332.5 = 0$

$\therefore$ $b = 12.57$ mm

By proportions, $b = A_o/3 = 118.59/3 = 35.53$

Choose $b = 40$ mm

Step-3: Dynamic load by Buckingham equation,

$$W_d = W_t + \frac{21V(bc + w_t)}{21V + \sqrt{bc + w_t}}$$ (Eqn. 16.58 MDHB)

The dynamic factor for class-III gear

With permissible error e = 14 μm (Table 16.27 MDHB)

and $C = 112$ (Table 16.5 MDHB)

Substituting the values,

$$W_d = 848.9 + \frac{21 \times 5.89 \times (40 \times 112 + 848.9)}{21 \times 5.89 + \sqrt{40 \times 112 + 848.9}} = 4200 \text{ N}$$

Step-4: Check for dynamic beam strength,

$$W_b = S_{dr} \cdot Y \cdot b \cdot m\left[1 - \frac{b}{A_0}\right]$$

$$= 83.5 \times 0.374 \times 40 \times 5 \times \left[1 - \frac{40}{118.9}\right] = 4144.6 \text{ N} \approx W_d$$

This is nearly equal to the dynamic load, hence design is acceptable.

Step-5: Check for wear strength

$$W_w = \frac{K \cdot b_e \cdot D_p Q}{\cos \propto}$$

where,
$$Q = \frac{2 \times 142.34}{15.811 + 142.34} = 1.8$$

∴
$$b_c = 1.00 \times 40 = 40,\ \cos \propto = \cos 18.43 = 0.949$$

∴ The strength factor for,
$$W_w = W_d$$

$$K = \frac{4200 \times 0.949}{40 \times 75 \times 1.8} = 0.738$$

The hardness values for pinion = 200 BHN, for gear = 180 BHN, with $K = 0.814$.

Example 17.12: Design a pair of bevel gears to connect two shafts at 60°. The power transmitted is 25 kW at 900 rpm of pinion. The reduction ratio is 5:1. The teeth are 20° full depth involute and pinion has 24 teeth. Check the design for dynamic and wear considerations.

Given Data: Power- 25 kW at $N_1 = 900$ rpm;
Reduction ratio, $G = 5:1$; Shaft angle $\theta = 60°$;
Number of teeth on pinion, $Z_1 = 24$;
Tooth profile, 20° full depth involute

SOLUTION

Step-1: Selection of materials (Table 16.4 MDHB)

Select material for pinion

As C45 treated width, $S_d = 210$ MPa and BHN = 250

For gear cast steel treated with $S_d = 195$ MPa and BHN = 250

Step-2: Comparison of strength factors

Number of teeth on gear, $Z_g = 24 \times 5 = 120$
Cone angle of gear,

$$\tan \beta = \frac{\sin \theta}{\left[\frac{Z_p}{Z_g} + \cos \theta\right]} = \frac{\sin 60}{\left[\frac{1}{5} + \cos 60\right]} = 1.237 \qquad \text{(Eqn. 16.47 MDHB)}$$

$$\beta = \tan^{-1}(1.257) = 51.05°$$

Cone angle for pinion, $\propto = \theta - \beta = 60 - 51.05 = 8.95°$

Formative number of teeth, for pinion, $Z_{pv} = \frac{Z_p}{\cos \propto} = \frac{24}{\cos 8.95} = 24.30$

For gear, $Z_{GV} = \dfrac{Z_g}{\cos\beta} = \dfrac{120}{\cos 51.05} = 190.89$

The tooth form factors, $Y_p = 0.337$, and $Y_g = 0.464$ (Table 16.2 MDHB)

The strength factors

For pinion $(Y \cdot S_d)_p = 0.337 \times 210 = 70.77$

For gear $(Y \cdot S_d)_g = 0.464 \times 195 = 90.48$

The pinion is weaker, hence design the pinion.

Step-3: Module by Lewis equation

The diameter of pinion, $D_p = Z_p \times m = 24\,\text{m}$

The diameter of gear, $D_g = 120 \times m = 120\,\text{m}$

The torque on pinion,

$$T = \frac{955 \times 10^4 \times \text{kW}}{N} = \frac{955 \times 10^4 \times 25}{900} = 265277.8 \text{ Nmm}$$

The tangential load on pinion tooth,

$$W_t = \frac{T}{r} = \frac{26577.8}{12\text{ m}} = \frac{22106.5\text{ N}}{m}$$

The cone distance, $$A_o = \frac{R_p}{\sin \propto} = \frac{12\text{ m}}{\sin(8.95)} = 77.14\text{ m}$$

The face width $b = A_o/3 = 25.71\text{ m}$

Assume $C_v = 0.3$

Substituting the values of Lewis equation,

$$w_t = s_d \cdot c_v \cdot y \cdot b \cdot m\left[1 - \frac{b}{A_o}\right]$$

$$\frac{22106.5}{m} = 210 \times 0.3 \times 0.337 \times 25.71 \times m\left[1 - \frac{1}{3}\right]$$

$$m = \sqrt[3]{\frac{22106.5}{210 \times 0.3 \times 0.337 \times 25.71 \times 0.67}} = 3.93$$

Choose $m = 4$ mm (Table 16.2 MDHB)

The diameter of pinion, $D_p = 24 \times 4 = 96$ mm

The pitch line velocity, $V = \dfrac{\pi DN}{60{,}000} = \dfrac{\pi \times 96 \times 900}{60{,}000} = 4.5$

Then velocity factor $C_v = \frac{6}{6+v} = \frac{6}{6+4.5} = 0.57 > 0.3$ (assumed value)

This is more than assumed value of 0.3.
Hence, $m = 4$ mm is satisfactory
The face width, $b = 25.71 \times 4 = 102.84$
Choose $b = 100$ mm

Step-4: Dynamic load by Buckingham equation

Tangential load, $W_t = \frac{22106.5}{4} = 5526.625$ N

For class-III precision gear

The permissible error, $e = 12.5$ μm (Table 16.27 MDHB)

The dynamic factor (Table 16.5 MDHB), $C = 150$

Then by Buckingham equation,

$$W_d = W_t + \frac{21V(b \cdot C + W_t)}{21V + \sqrt{bC + W_t}}$$

$$= 5526.625 + \frac{21 \times 4.5 \times [100 \times 150 + 5526.625]}{21 \times 4.5 + \sqrt{100 \times 150 + 5526.625}} = 13684.75 \text{ N}$$

Step-5: Check for dynamic beam strength,

$$w_b = S_{dr} \cdot b \cdot Y \cdot m \cdot \left[1 - \frac{b}{A_o}\right]$$

$$= 210 \times 100 \times 0.337 \times 4 \times \left[1 - \frac{1}{3}\right] = 18966.36 \text{ N} > w_d$$

The design is safe against bending failure.

Step-6: Check for wear strength

The load sharing factor, $Q = \frac{2 \times Z_{vg}}{Z_{vg} + Z_{vp}}$

$$= (2 \times 190.69)/(24.3 + 19069) = 1.77$$

The strength factor, $K = W_d \cos \alpha / (b \cdot D_p \cdot Q)$

$$= \frac{13684.75 \times \cos(8.95)}{100 \times 96 \times 1.77} = 0.7855$$

Then hardnesses required are $(\text{BHN})_p = (\text{BHN})_g = 250$ with $K = 0.92$

Example 17.13: A pair of straight bevel gears are used to transmit 15 kW at 1500 rpm input speed. The number of teeth on pinion is 20 and the speed ratio is 5:1. Design the gears for strength only. Assume 14.5° full depth tooth form.

Given Data: Power, $NP = 15$ kW at $N_1 = 1500$ rpm, Speed ratio, $G = 5$

Number of teeth on pinion, $Z_p = 20$

SOLUTION Assume the shaft angle, $\theta = 90°$

Step-1: Material selection (Table 16.4 MDHB)

Material for pinion $C40$ with $S_d = 175$ MPa; BHN = 140

Material for gear—Cast steel untreated, $S_d = 140$ MPa; BHN = 180

Step-2: Pitch cone angles

For gear, $\tan\beta = \dfrac{Z_g}{Z_p} = G = 5$

$$\therefore \qquad \beta = \tan^{-1}(5) = 78.69°$$

For pinion, $\propto = \theta - \beta = 90° - 78.69 = 11.31°$

Step-3: Comparison of strength factors

Number of teeth on pinion, $Z_p = 20$

Number of teeth on gear, $Z_g = 5 \times 20 = 100$

The formative number of teeth,

For pinion, $Z_{pv} = \dfrac{Z_p}{\cos\propto} = \dfrac{20}{\cos 11.31} = 21.02$

For gear, $Z_{gv} = \dfrac{Z_g}{\cos\beta} = \dfrac{100}{\cos(78.69)} = 509.90$

Tooth form factors,

For pinion, $Y_p = 0.289$; for gear, $Y_g = 0.39$

Strength factors,

For pinion $(S_d \cdot Y)_p = 175 \times 0.289 = 50.57$

For gear $(S_d \cdot Y)_g = (140 \times 0.39) = 54.6$

The pinion is weaker and the design is to be based on pinion.

Step-4: Module by Lewis equation

The torque on pinion,

$$T = \frac{955 \times 10^4 \times \text{kW}}{N} = \frac{955 \times 10^4 \times 15}{1500} = 95500 \text{ Nmm}$$

The pinion diameter, $D_p = Z_p m = 20\,\text{m}$

The tangential load on the tooth,

$$W_t = \frac{T}{r} = \frac{95500}{10\,\text{m}} = \frac{9550}{m}\,\text{N}$$

The cone distance, $A_o = \frac{R_p}{\sin\propto} = \frac{10\,\text{m}}{\sin(11.31)} = 50.99\,\text{m}$

Face width, $b = \frac{A_o}{3} = 17\,\text{m}$

Assume, $C_v = 0.3$

Substituting the values in Lewis equation

$$\frac{9550}{m} = 175 \times 0.289 \times 0.3 \times 17\,\text{m} \times m\left[1 - \frac{1}{3}\right]$$

$$m = \sqrt[3]{\frac{9550}{175 \times 0.289 \times 0.3 \times 17 \times 0.67}} = 3.809$$

Choose $m = 4$ mm

Diameter of pinion, $D_p = 20 \times 4 = 80$ mm

The pitch line velocity, $V = \frac{\pi DN}{60{,}000} = \frac{\pi \times 80 \times 1500}{60{,}000} = 6.28$ m/s

Velocity factor, $C_v = \frac{6}{6+v} = \frac{6}{6+6.28} = 0.488$

This is greater than assumed value of 0.3 and $m = 4$ mm is OK.

The face width, $b = 17\,\text{m} = 17 \times 4 = 68$ mm

The design parameters are

Module, $m = 4.0$ mm, Pinion dia., $D_p = 80$ mm, Gear dia., $D_g = 400$ mm

Face width, $b = 68\text{ mm} \cong 70$ mm

The cone angle for pinion, $\propto = 11.31$

The cone angle for gear, $\beta = 78.69°$

Example 17.14: A pump is driven by a 30 kW motor through a pair of right angled bevel gears. The speed of the motor is 1200 rpm. The pinion on the motor shaft has a pitch diameter of 150 mm and carries 30 teeth and the gear on the pump shaft has 40 teeth. The pinion is made of C45 steel–untreated, whereas the gear is made of 0.2% carbon steel, untreated. The teeth are generated to have 20° full depth involute. Check whether the gear pair is safe from the stand point of bending strength.

Given Data: Power transmitted = 30 kW at speed of the motor = 1200 rpm;

Pitch diameter of pinion, $D_p = 150$ mm; Shaft angle, $\theta = 90°$;

Number of teeth on pinion, $Z_p = 30$; Number of teeth on gear, $Z_g = 40$;

Material for pinion – C45 untreated; for gear C20 untreated

SOLUTION

Step-1: Material properties (Table 16.4 MDHB)

Pinion – $C45 - S_d = 210$ MPa, BHN = 200

Gear $C20 - S_d = 125$ MPa, BHN = 150

Step-2: Pitch cone angles

Pitch cone angle of gear, $\tan\beta = \dfrac{Z_g}{Z_P} = \dfrac{40}{30} = 1.33$

$$\therefore \quad \beta = \tan^{-1}(1.33) = 53.13°$$

Pitch cone angle of pinion, $\propto = 90° - 53.13 = 36.87°$

Step-3: Comparison of strength factors

The formative number of teeth

For pinion, $Z_{pv} = \dfrac{Z_p}{\cos\propto} = \dfrac{30}{\cos(36.87)} = 37.5$

For gear, $Z_{gv} = \dfrac{Z_g}{\cos\beta} = \dfrac{40}{\cos(53.13)} = 66.67$

The tooth form factors (Table 16.3 MDHB)

For pinion, $Y_p = 0.384$ and for gear, $Y_g = 0.428$

Strength factors are,

For pinion $(Y \cdot S_d)_p = 0.384 \times 210 = 80.64$

For gear $(Y \cdot S_d)_g = 0.428 \times 125 = 53.5$

The gear is weaker, hence the design is to be based on gear.

Step-4: Check for bending stress by Lewis equation

The torque of the motor,

$$T = \frac{955 \times 10^4 \text{ kW}}{N} = \frac{955 \times 10^4 \times 30}{1200} = 238750 \text{ Nmm}$$

The tangential load on the tooth, $w_t = \left(\dfrac{T}{r}\right) = \dfrac{238750}{(150/2)} = 3183.33$ N

The cone distance, $A_o = \dfrac{R_p}{\sin \propto} = \dfrac{75}{\sin(36.87)} = 125 \text{ mm}$

The face width, $b = A_o/3 = 125/3 \cong 42 \text{ mm}$

The module, $m = \dfrac{D_p}{Z_p} = \dfrac{150}{30} = 5 \text{ mm}$

The pitch line velocity, $V = \dfrac{\pi DN}{60{,}000} = \dfrac{\pi \times 150 \times 1200}{60{,}000} = 9.42 \text{ m/s}$

The velocity factor, $C_v = \dfrac{6}{6+v} = \dfrac{6}{6+9.42} = 0.39$

The stress induced in the gear tooth,

$$\sigma_b = \frac{w_t}{C_v \cdot Y \cdot b \cdot m\left[1 - \dfrac{b}{A_o}\right]} = \frac{3183.33}{0.39 \times 0.428 \times 42 \times 5 \times 0.67}$$

$$= 135.54 \text{ MPa} > (S_d = 125 \text{ MPa})$$

The stress induced is more than the allowable stress and hence the material for the gear has to be changed to C30 with $S_d = 140$ MPa.

Example 17.15: A pair of mitter gears have pitch diameter of 280 mm and face width of 36 mm and runs at 250 rpm. The teeth are 14.5° involute and accurately act and transmit 6 kW. Neglecting friction angle, find the following: (i) outside diameters of gears, (ii) resultant tooth load tangent to the pitch cone, (iii) radial load on pinion, and (iv) thrust on the pinion. Assume low carbon cast steel 0.2%C heat treated as material for both gears.

Given Data: Pitch diameter $= 280$ mm; Face width, $b = 36$ mm;

Power transmitted $= 6$ kW; $N = 250$ rpm;

Material C20; Tooth form – 14.5° involute.

SOLUTION

Step-1: Check for bending stress

Material property (Table 16.4 MDHB)

C20 treated $S_d = 125$ MPa

Assume number of teeth, $Z_1 = Z_2 = 70$

The module by proportions (geometry),

$$m = \frac{D}{Z_1} = \frac{280}{70} = 4 \text{ mm}$$

The torque transmitted,

$$T = \frac{955 \times 10^4 \text{ kW}}{N} = \frac{955 \times 10^4 \times 6}{250} = 229200 \text{ Nmm}$$

The tangential load, W_t = T/R = 229200/140 = 1637.14 N

Cone distance, $A_o = \sqrt{140^2 + 140^2} = 198 \text{ mm}$

Pitch line velocity,

$$V = \frac{\pi DN}{60,000} = \frac{\pi \times 280 \times 250}{60,000} = 3.67$$

Velocity factor, $C_v = \frac{3}{3+v} = \frac{3}{3+3.67} = 0.45$

Formative number of teeth, $Z_{pv} = \frac{Z_p}{\cos \propto} = \frac{70}{\cos 45} = 98.99$

Tooth form factor $Y = 0.367$
By Lewis equation, bending stress induced

$$\sigma_b = \frac{W_t}{C_v \cdot Y \cdot b \cdot m \left[1 - \frac{b}{A}\right]}$$

$$= \frac{1637.14}{0.45 \times 0.367 \times 36 \times 4 \times \left[1 - \frac{36}{198}\right]} = 84.14 \text{ MPa} < S_d = 125 \text{ MPa}$$

The design is safe. The module $m = 4$ mm and other parameters are satisfactory.

Step-2: Outside diameter

The addendum, $a = 1.0 \times m = 1.0 \times 4 = 4$ mm

The outside diameter, $D_o = D + 2a \cos \propto$

$$= 280 + 2 \times 4 \times \cos 45 = 285.66$$

Step-3: Tooth loads

The tangential load, $W_t = 1637.14$ N

Radial load, $W_r = W_t \tan \phi \cos \propto$ (Eqn. 16.53 MDHB)

$$= 1637.14 \times \tan 14.5 \times \cos 45 = 299.38 \text{ N}$$

Axial load, $W_a = W_r = 299.38$ N

Example 17.16: A pair of bevel gear-wheels with 20° pressure angle consists of 20 teeth pinion meshing with 30 teeth gear. The module is 4 mm and width is 20 mm. The surface hardness of both pinion and gear is 400 BHN. The pinion rotates at 500 rpm and receives power from an electric motor. The starting torque is 150 per cent of the rated torque. Determine the safe power that can be transmitted considering the dynamic load, wear strength and endurance strength. The allowable bending stress may be taken as 240 MPa.

Given Data: Tooth form – 20° involute;

Number of teeth on pinion, $Z_p = 20$; Number of teeth on gear, $Z_g = 30$;

Module, $m = 4$ and face width, $b = 20$ mm;

Speed of pinion, $N_1 = 500$ rpm; Service factor, $C_s = 1.5$;

Allowable bending stress, $S_d = 240$ MPa

SOLUTION

Step-1: Cone distance A

The pitch cone angle of gear, (Eqn. 16.46a MDHB)

$$\beta = \tan^{-1}\left(\frac{Z_G}{Z_p}\right) = \tan^{-1}\left(\frac{30}{20}\right) = 56.31^\circ$$

The pitch cone angle of pinion, $\propto = 90 - \beta = 90 - 56.31 = 33.69^\circ$

The pitch diameter of pinion, $D_p = mZ_p = 4 \times 20 = 80$ mm

The pitch diameter of gear, $D_g = mZ_g = 4 \times 30 = 120$ mm

The cone distance,

$$A = \frac{R_p}{\sin \propto} = \frac{R_g}{\sin \beta} = 40/\cos 33.69 = 72.11 \text{ mm} \qquad \text{(Eqn. 16.50a MDHB)}$$

Step-2: The allowable tangential load by Lewis equation

The formative number of teeth for pinion,

$$Z_{pv} = \frac{Z_p}{\cos \propto} = \frac{20}{\cos 33.69} = 24.04$$

For gear,

$$Z_{gv} = \frac{30}{\cos 56.31} = 54.083$$

The tooth form factor for pinion and gear,

$$Y_p = 0.337; \text{ and } Y_g = 0.415$$

The pitch line velocity,

$$V = \frac{\pi DN}{60{,}000} = \frac{\pi \times 80 \times 500}{60{,}000} = 2.09 \text{ m/s}$$

The velocity factor, $C_v = \dfrac{3}{3+v} = \dfrac{3}{3+2.09} = 0.59$

The permissible tangential load,

$$w_t = S_d \cdot C_v \cdot Y \cdot b \cdot m\left[1-\frac{b}{A}\right] \qquad \text{(Eqn. 16.55 MDHB)}$$

$$= 240 \times 0.59 \times 0.337 \times 20 \times 4 \times \left[1-\frac{20}{72.11}\right] = 2758.73 \text{ N}$$

Step-3: Dynamic load by Buckingham equation,

$$w_d = w_t + \frac{21V(b \cdot c + w_t)}{21V + \sqrt{b \cdot c + w_t}} \qquad \text{(Eqn. 16.58 MDHB)}$$

where $C = 150$ from Table 16.5 MDHB,

$$\therefore \quad W_d = 2758.73 + \frac{21 \times 2.09 \times (20 \times 150 + 2758.73)}{21 \times 2.09 + \sqrt{20 \times 150 + 2758.73}}$$

$$= 4868.92 \text{ N}$$

Step-4: Check for dynamic beam strength

$$W_b = S_{dr} \cdot Y \cdot b \cdot m\left[1-\frac{b}{A}\right] \qquad \text{(Eqn. 16.60 MDHB)}$$

where $S_{dr} = 686.5$ MPa (Table 16.25 MDHB)

Substituting the values,

$$W_b = 686.5 \times 0.337 \times 20 \times 4 \times \left[1-\frac{20}{72.11}\right] = 13374.76 \text{ N} > W_d$$

Step-5: Check for wear strength,

$$w_w = \frac{K \cdot b_e \cdot D_p \cdot Q}{\cos \propto} \qquad \text{(Eqn. 16.61 MDHB)}$$

where $K = 2.23$ (Table 16.23 BHN = 400); $b_e = 20$ mm and $D_p = 80$ mm,

$$Q = \frac{z \times z_{vG}}{z_{VG} + Z_{VP}} = \frac{20 \times 54.083}{54.083 + 24.04} = 1.385$$

$$\cos \propto = \cos 33.69° = 0.832$$

Substituting the values,

$$w_w = \frac{2.23 \times 20 \times 80 \times 1.385}{0.832} = 5939.52 \text{ N} > w_d$$

Step-6: Permissible power transmitted,

$$KW = \frac{TN}{955 \times 10^4 \times c_s}$$

where, $T = W_t R_p = 2758.73 \times 40 = 110349.2$ Nmm

$N = 500$ rpm; $C_s = 1.5$

$$\therefore \quad \text{kW} = \frac{110349.2 \times 500}{955 \times 10^4 \times 1.5} = 3.8576$$

The safe power, $NP = 3.85$ kW

EXERCISES FOR PRACTICE

17.1 (a) Explain the forces acting on a bevel gear.

(b) A pair of 20° full height involute bevel gears transmit power between two shafts at right angles with a speed ratio 4:1. The pinion transmits 40 kW at 750 rpm. Both the gears are made of steel having an allowable static stress of 100 MPa. Design and specify the details of bevel gears.

17.2 Design a pair of bevel gears to transmit a power of 25 kW from a shaft rotating at a speed of 1000 rpm to a perpendicular shaft to be run at a speed of 350 rpm. Give detailed specifications including the BHN to which gear surfaces must be treated.

17.3 Design a pair of bevel gears to transmit a power of 25 kW from a shaft running at 750 rpm to a perpendicular shaft to be run at 350 rpm.

17.4 Design a pair of bevel gears for two shafts whose axes are at right angles. Speed of pinion shaft is 300 rpm and that of gear shaft is 120 rpm. The power transmitted is 80 kW at gear shaft.

17.5 A pair of straight bevel gears is required to transmit 15 kW at 1000 rpm of the motor shaft to another shaft running at 250 rpm. The pressure angle is 20°. If the shaft axes are at right angles to each other, find the module, face width, addendum, outside diameter and slant height. The gears are capable of withstanding a static stress of 60 MPa. The tooth form factor may be taken as $y = 0.514 - 0.912/NE$, where NE is the equivalent number of teeth. Assume velocity factor as $C_v = 4.5/(4.5 + v)$, where v is the pitch line speed in m/sec. The face width may be taken as 1/4 of the slant height of the pitch cone.

17.6 A right angle speed reducer uses hardened alloy steel bevel gears. Transmission ratio is 5 to 1 and the pinion rotates at 900 rpm while transmitting 70 kW. Teeth are 200 full depth form and the pinion has a dimeter of 100 mm. Determine the module, face width, number of teeth, and pitch cone angles of the gears.

17.7 Two shafts at right angles to each other are connected by a pair of straight teeth bevel gears. The velocity ratio is 3:1. Tooth profile is 20° full depth involute. Pinion is made of steel C 30, heat treated and gear is made of cast steel 0.20% C, heat treated. The pinion transmits 32 kW at 1200 rpm. If the pinion has 30 teeth, determine the mean module, face width and the pitch diameters of gear wheels. Specify the cone angles and slant height.

17.8 A pair of bevel gears is used to transmit 10 kW power with the following details:

Speed of the pinion = 1200 rpm

Speed ratio = 3:1

Minimum number of teeth on the pinion = 20 C-15 steel is to be used for both wheels. Design the gears and draw the arrangement.

17.9 A pair of bevel gears is required to transmit 40 kW at 500 rpm. The output shaft is running at 200 rpm. If the gear and pinion are to be cast steel 0.20% C, untreated and C 40 untreated carbon steel respectively, determine the proportion of gear wheels, with 20° involute stub teeth.

17.10 A vertical drilling machine requires a set of bevel gears to reduce the speed from 2500 rpm to 1250 rpm at the spindle. Design the bevel gear drive with 20° full depth form. The power to be transmitted is 15 kW. Design the bevel gear drive assuming both the gears to be made of high grade cast iron.

17.11 Design a right angle speed reducer to transmit 50 kW power. The velocity ratio is 5:1 and pinion rotates at 900 rpm. The operation is considered to be continuous. Select the materials to obtain a compact design and cost should be kept low. Use 20° full depth involute teeth.

17.12 Design the full depth 14½° teeth of a pair of straight bevel gears to transmit 20 hp. at 1250 rpm of the pinion. The velocity ratio should be about 3.5 and the pinion should have 18 teeth. Determine the module, face width, pitch diameters and pitch cone angles for the gears. Cast steel having an allowable stress of 1970 kgf /cm^2 may be used for both gears.

17.13 A pair of bevel gears is to transmit 45 (kW) with a pinion speed of 240 rpm and gear speed of 120 rpm. Pinion has 20 teeth of 20° involute full depth and a module of 15 mm. Both wheels are made of cast steel. What face width do you recommend for the gear wheels? What should be the hardness of pinion and gear for dynamic and wear resistance?

17.14 Design a pair of straight cast iron bevel gears to transmit 75 kW power from a shaft running at 200 rpm to another running at 65 rpm. The 18 teeth pinion are cut with a standard 20° stub. Allowable static strength of cast iron may be taken as 55 N/mm^2. The service is continuous with light shocks. Determine the module, face width and the pitch dimater of the gears.

17.15 Design a pair of right-angled bevel gears to transmit a power of 15 kW from a shaft running at a speed of 750 rpm to a perpendicular shaft to be run at 250 rpm. Suggest suitable surface hardness for the gear pair.

17.16 Design a pair of bevel gears to transmit a power of 25 kW from a shaft rotating at 1200 rpm to a perpendicular shaft to be rotated at 400 rpm.

17.17 Determine the cone pitch angles. Pitch diameters for the following bevel gear pairs:

(i) For shaft angle 77° (Acute angle bevel gearing)

(ii) For shaft angle 147° (Obtuse angle bevel gearing)

The module is 5 mm and the number of teeth on the pinion and the gear are 14 and 42, respectively. Draw the sketches of gearing.

17.18 (a) A pair of straight bevel gears transmits 15 kW at 1250 rpm of 120 mm diameter pinion. The speed reduction is 3.5. Use 14.5° involute tooth system. The angle between the shaft axes is 90°. The pinion is made of case hardened alloy steel with allowable static stress 343.34 MPa and gear is cast steel of 0.20% C heat treated with allowable static stress 191.295 MPa. Determine module, face width and number of teeth on pinion and gear. Suggest suitable surface hardness for the gear pair. Take the service factor as 1.5 and assume the teeth are generated.

(b) Explain formative number of teeth in bevel gears.

17.19 A pair of 20° full depth involves teeth bevel gears connect two shafts at right angles, having velocity ratio 3:1. The gear is made of cast steel, 0.20% untreated and the pinion material is of steel, C30 heat-treated. The pinion has 20 number of teeth and transmits 40 kW at 750 rpm. Determine: (i) Module (ii) Face width and (iii) Pitch diameters. Assume width of gear face as one-third of the length of pitch cone.

17.20 A pair of straight tooth bevel gears at right angles is to transmit 5 kW at 1200 rpm of the pinion. The diameter of the pinion is 80 mm and the velocity ratio is 3.5. The tooth form is 14½°. Both the pinion and gear are made of cast iron with allowable stress of 55 MN/m^2. Determine the module, face width from the stand point of strength and also check the design from the standpoint of dynamic load and wear load.

SHORT ANSWER QUESTIONS

17.1 What are bevel gears?
17.2 Where are they used?
17.3 What are straight bevel gears?
17.4 What are spiral bevel gears?
17.5 Which of these gears are widely used?
17.6 What is shaft angle in straight bevel gears?
17.7 What is pitch angle?
17.8 What is pitch diameter?
17.9 Define module for bevel gears.
17.10 What is cone distance?
17.11 What is face width?
17.12 What is back cone?
17.13 What is back cone radius?
17.14 What are equivalent gears?
17.15 What are the virtual or formative number of teeth?
17.16 What is the equation for the formative number of teeth for pinion?
17.17 What is the equation for formative number of teeth for gear?
17.18 What are the loads on the bevel gear tooth?
17.19 How do you determine their directions?
17.20 What are the relations between the loads on pinion tooth and gear tooth?
17.21 Give the modified Lewis equation for bevel gears?
17.22 What is the dynamic load on bevel gear tooth?

17.23 What is dynamic beam strength equation?
17.24 Give the wear strength equation.
17.25 What is the bending stress as per the new AGMA standards?
17.26 What is the new AGMA standards equation for the contact stress in bevel gears?
17.27 Explain the modifying factors in both above equations.
17.28 What are meter gears? What are their characteristics?
17.29 What are crown gears? List their characteristics.

CHAPTER 18

WORM GEARS

18.1 INTRODUCTION

Worm gears consist of a gear and a worm. The gear is somewhat similar to helical gear. The shape of the teeth is slightly different to suit the shape of the thread on worm. The worm is like a screw with trapezoidal shaped thread. The threads may be one or two or three. Accordingly it is called single threaded, double threaded or triple threaded worm.

The gear ratio G is given as $G = Z_G/Z_W$ (18.1)

where, Z_G = No. of teeth on the gear

Z_W = No. of threads on the worm.

The axes of the gear and worm are non-intersecting (skew) and usually perpendicular.

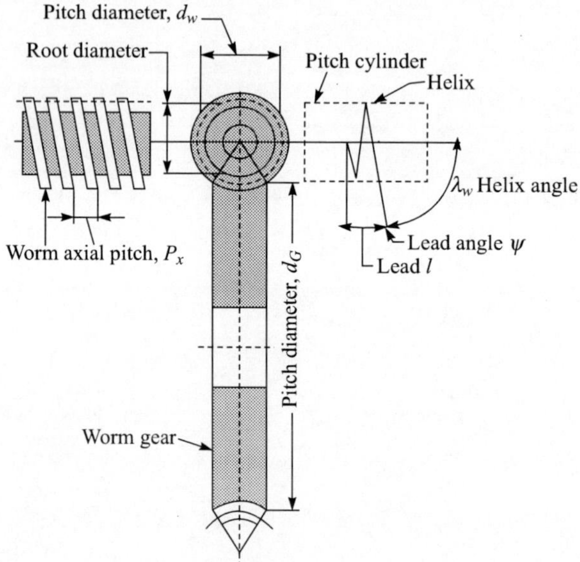

Figure 18.1

The gear diameter D_G is given by the relation $D_G = mZ_G$ (18.2)

where, m = tangential (circular) module of the gear

Such a relation does not exist between diameter of the worm and thread pitch.

The axial pitch P_x of the worm thread must be equal to circular pitch of the gear for proper engagement.

$$P_x = P_c = \pi m \tag{18.3}$$

The lead of the threads, $L = Z_W \cdot P_x = \pi m \cdot Z_W$ (18.4)

The recommended number of threads are as follows:

Velocity ratio	No. of threads
20 and above	Single start
12 – 20	Double start
8 – 12	Triple start
6 – 12	Quadruple start
4 – 10	Sextuple start

The lead angle of threads Ψ is given by the relation:

$$\tan \Psi = \frac{L}{\pi D_W} = \frac{\pi \cdot m \cdot Z_W}{\pi \cdot D_W} = \frac{m \cdot Z_W}{D_W} = \frac{D_G}{Z_G} \cdot \frac{Z_W}{D_W} = \left(\frac{D_G}{D_W}\right)\left(\frac{Z_W}{Z_G}\right) \tag{18.5}$$

$$\therefore \quad \Psi = \tan^{-1}\left(\frac{L}{\pi D_W}\right) = \tan^{-1}\left(\frac{D_G}{D_W}\frac{Z_W}{Z_G}\right) \simeq \tan^{-1}\left(\sqrt[3]{\frac{Z_W}{Z_G}}\right) \tag{18.5a}$$

For compact design, the values of Ψ = (5° to 45°)

The distance between the two axes,

$$C = \frac{D_G + D_W}{2} = \frac{1}{2}\left[\frac{Z_G}{Z_W} \cdot D_W \tan \Psi + D_W\right] = \frac{L}{2\pi}\left[G + \frac{1}{\tan \Psi}\right] \tag{18.6}$$

The AGMA standard recommendation,

Minimum worm diameter, $D_W = 0.68C^{0.875}$ mm (18.7)

The face width, $b = 0.5C^{0.875} = 0.73D_W$ (18.8)

It cannot be greater than the worm diameter.

$$\text{Recommended, } P_c = \left(\frac{D_W}{3}\right) = \pi m \tag{18.9}$$

The recommended pressure angles for wormset are 14.5°, 17.5°, 20°, 22.5°, 25°, 27.5° and 30°. The minimum number of teeth on the worm gear are related to pressure angle as follows:

Table 18.1

ϕ	14.5°	17.5°	20°	22.5°	25°	27.5°	30°
Z_{min}	40	27	21	17	14	12	10
Ψ	0 – 16°	16 – 25°		25 – 35°		35 – 45°	
y	0.100	0.125				0.175	

18.2 MATERIALS

Worm: It is highly stressed and requires a strong material. Low and medium carbon steels hardened to BHN 650 – 660 are usually used.

Worm Gear: It is comparatively less loaded and subjected to wear. A soft material suitable for high sliding strength is required. Usually, bronzes as cast or forged condition is used. The phosphor or tin bronzes are used for high power applications. The manganese bronze is used for small power, slow speed applications. Cast iron, soft steel and plastics are used for light loads and low speed applications.

18.3 FORCES IN WORM GEARS

The load acting on the gear tooth and worm thread has three components: tangential, axial and radial. There is also frictional force, which can be resolved in these directions.

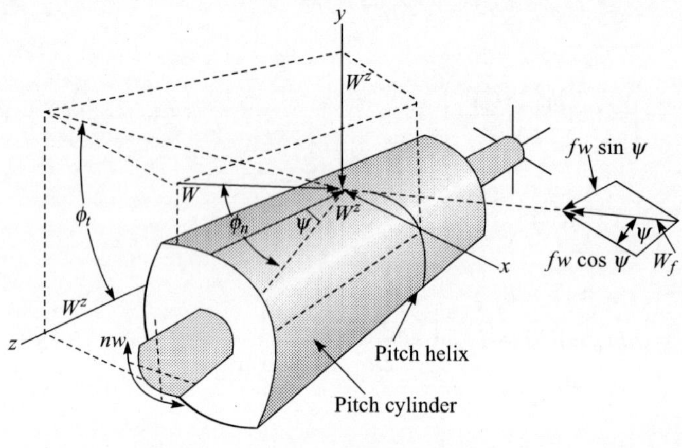

$$\left.\begin{aligned} W_W^t &= -W_G^a = W^x \\ W_W^r &= -W_G^r = W^y \\ W_W^a &= -W_G^t = W^z \end{aligned}\right\}$$

$$\left.\begin{aligned} W^x &= W = (\cos\phi_n \sin\psi + f\cos\psi) \\ W^y &= W\sin\phi_n \\ W^z &= W(\cos\phi_n \cos\psi - f\sin\psi) \end{aligned}\right\}$$

Figure 18.2 Forces on the gears

The axial load on the worm, W_W^a = The tangential load on the gear W_G^T

$$= \frac{2T_G}{D_G} = W[\cos\phi_n \cdot \cos\Psi] \text{ without friction} \tag{18.10}$$

$$= W[\cos\phi_n \cdot \cos\Psi - f \cdot \sin\Psi] \text{ with friction} \tag{18.10a}$$

where, W is the normal force on the thread or tooth

The axial force on the gear, W_G^a = The tangential force on the worm

$$W_W^T = \frac{2T_W}{D_W} = W[\cos\phi_n \cdot \sin\Psi] \text{ - without friction} \tag{18.11}$$

$$= W[\cos\phi_n \cdot \sin\Psi + f \cdot \cos\Psi] \text{ with friction} \tag{18.11a}$$

$$= W_G^T \cdot \frac{[\cos\phi_n \cdot \sin\Psi + f \cdot \cos\Psi]}{[\cos\phi_n \cdot \cos\Psi - f\sin\Psi]} \tag{18.11b}$$

The radial load on the gear = Radial load on the worm,

$$= W \sin \phi_n = \frac{W_G^T \tan \phi_n}{\cos \Psi} = W_G^T \tan \phi_t \tag{18.12}$$

$$\therefore \qquad \tan \phi_n = \tan \phi_t \cdot \cos \Psi \tag{18.13}$$

The efficiency of the worm gearing,

$$\eta = \frac{T_G N_G}{T_W \cdot N_W} = \left[\frac{\cos \phi_n - f \tan \Psi}{\cos \phi_n + f \cot \Psi}\right] \tag{18.14}$$

For maximum efficiency,

$$\tan \Psi = \sqrt{1 + f^2} - f \tag{18.15}$$

If the efficiency is less than 50%, then the device will be self locking. It can transmit motion from worm to wheel. The other way (reverse) is not possible. This fact can be included as a safety feature in some applications like cranes.

18.4 DESIGN PROCEDURE

Step-1: Module by AGMA standards – The module of the gear is estimated from AGMA recommendation Eqns. 18.7 and 18.9.

If the centre distance C is specified, or assumed,

Then worm diameter, $D_w = 0.68C^{0.875}$

Then module, $m = (D_w/3\pi)$

The module so evaluated is standardized using Table 15.2 and worm diameter and face width of worm gear are calculated by following equations:

The worm diameter, $D_w = 3\pi \cdot m$

The face width, $b = 0.73 D_w = 0.5C^{0.875}$

The Ψ, ϕ, y combination are selected from Table 18.1

Z_W = Number of threads on the worm selected for the application.

The number of teeth, Z_G is determined by the relation, $Z_G = G \cdot Z_W$

The diameter of the gear, $D_G = m \cdot Z_G$

The other dimensions are calculated using equations in Tables 8.3a and 8.3b.

Step-2: Efficiency of the drive

The coefficient of friction f is assumed suitably or calculated from equations,

$$f = \frac{0.042}{V_s^{0.28}} \quad \text{for } (0.2 < V_s < 3.8)$$

$$= 0.025 + \frac{V_s}{300} \quad \text{for } V_s > 3.8$$

The sliding velocity, $V_s = \dfrac{V_w}{\cos \Psi} = \dfrac{\pi N_w D_w}{60000 \cos \Psi}$

Efficiency of the worm gearset, $\eta = \dfrac{\cos \phi_n - f \tan \Psi}{\cos \phi_n + f \cot \Psi}$

Step-3: Bending strength

(i) Check for stress by the Lewis beam strength equation,

$$S = \frac{W_G^T K_V}{\pi \cdot m \cdot b \cdot y \cdot \cos \Psi} \leq S_d \tag{18.16}$$

where, design stress S_d is selected from Table 18.4

and $$W_G^T = \frac{T_G}{R_G} = \frac{2T_G}{m \cdot Z_G} = \frac{2T_W \cdot N_W \cdot \eta}{N_G \cdot m Z_G} = \frac{2T_W \eta}{m Z_W}$$

The torque on gear T_G is evaluated from the output power KW_o,

$$= \frac{955 \times 10^4 \, KW_o}{N_G} = \frac{955 \times 10^4 \times \eta \times KW_i}{N_G}$$

Torque on worm, $T_W = \dfrac{955 \times 10^4 \, KW_i}{N_w}$

Velocity factor, $K_v = (3 + V_g)/3$

The velocity, $V_g = \dfrac{\pi D_G}{60000} N_G$ m/s

Step-4: Wear strength

(ii) The design is checked for surface strength by the equation

$$S_c = \frac{W_G^T \cdot K_v \cdot C_s}{A \cos \Psi} \leq S_{cd} \tag{18.17}$$

where, $$A = \frac{h_W D_W B}{114.6}, \quad K_v = (3 + V_g)/3$$

S_{cd} is taken from Table 8.6.

(iii) Check by limiting wear load equation,

$$W_w = D_G \cdot b \cdot K \geq W_d \tag{18.18}$$

where $W_d = K_v \cdot W_G^T$ and the value of K is taken from Table 18.7.

Step-5: Check for heat balance

An additional check for heat dissipating capacity is necessary for worm gears set.

The heat generated due to friction,

$$Q_g = W_f \cdot V_s = 1000(1-\eta)KW_i \text{ J/sec} \tag{18.19}$$

The heat dissipation capacity,

$$Q_d = 0.035(A_g + A_w)(t_g - t_r) > Q_g \tag{18.20}$$

where, A_g = Projected area of the gear = $(\pi D_G^2/4)$

A_w = Projected area of the worm = $FL \cdot D_w$ (sometimes $A_w = \pi D_G^2$ is taken)

FL = Face length of the worm = $\left(4.5 + \frac{Z_G}{50}\right)P_x$

The heat dissipating capacity should be greater than the heat generated.

AGMA-Power rating equations

For wear strength

$$\text{Power in kW} = N_g \cdot K_1 \cdot C_v \cdot Q \tag{18.21}$$

N_g = speed in rpm of the gear

$Q = G/(G + 2.5)$

C_v = velocity factor

$$= \frac{2.3}{(2.3 + V_w + 3V_w/G)}$$

V_w = pitchline velocity of the worm in m/s

K_1 = constant depending upon the centre distance (Table 18.2)

For heat dissipation

$$\text{Power in kW} = \frac{0.029\,C^{1.7}}{(G+5)} \tag{18.22}$$

C = Centre distance in mm.

Table 18.2. Pressure constant K_1 in kW/rpm. in Eqn. (18.21)

Center distance C (mm)	25	50	75	100	125	150	175	200	225
Constant K_1	0.0092	0.0184	0.0294	0.0661	0.125	0.213	0.330	0.485	0.727

Centre distance C (mm)	250	375	500	750	1000	1200	1500	1750	2000
Constant K_1	881	2.95	5.87	21.3	48.5	88.1	147	235	235

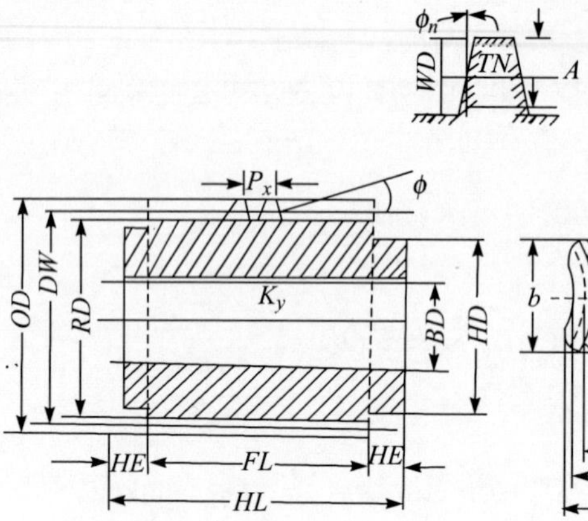

Figure 18.3a Worm shaft

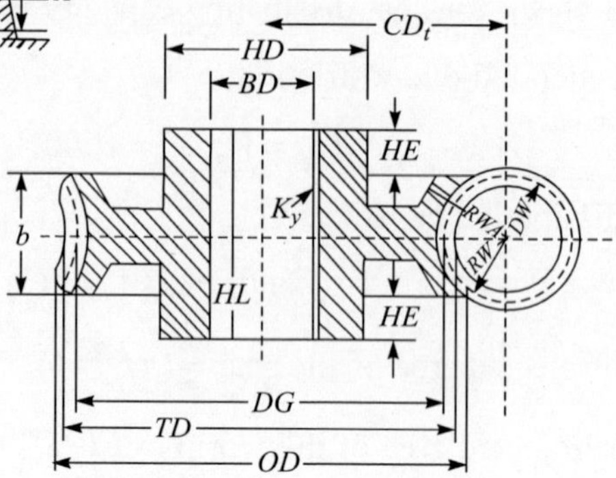

Figure 18.3b Worm shaft

Table 18.3. Worms mounted shafts: (Fig. 18.3a)

Dimension	Single and double threads	Triple and quadruple threads
ϕ_n – normal pressure angle	14.5°	20°
Dw – pitch diameter	$2.4\ p_x + 28$ mm	$2.4\ p_x + 28$ mm
A – addendum	$0.318\ p_x$	$0.286\ p_x$
WD – whole depth of tooth	$0.686\ p_x$	$0.623\ p_x$
FL – face length	$\left(4.5 + \frac{Z_G}{50}\right) p_x$	$\left(4.5 + \frac{Z_G}{50}\right) p_x$
HD – hub diameter	$1.66\ p_x + 25$ mm	$1.726\ p_x + 25$ mm
HE – hub extension	$(p_x + 16)$ OR p_x	$(p_x + 16)$ OR p_x
TRc – top round	$0.05\ P_x$	$0.05\ p_x$

Table 18.3a. Proportions of worm gears (Fig. 18.3b)

Dimension	Single and double threads	Triple threads
ϕ_n – Normal pressure angle	14.5°	20°
b – Face width	$2.38\ p_x + 6.35$	$2.15\ p_x + 5.08$
TD – Throat diameter	$D_G + 0.636\ p_x$	$D_G + 0.572\ p_x$
OD – Outside diameter	$TD + 0.475\ p_x$	$TD + 0.3183\ p_x$
HD – Hub diameter	$1.875 \times BD$	$1.875 \times BD$
HE – Hub extension	$0.25 \times BD$	$0.25 \times BD$
RWF – Radius of wheel face	$0.881\ p_x + 14$ mm	$0.914\ p_x + 14$ mm
RWR – Radius of wheel rim	$2.2\ p_x + 14$ mm	$2.1\ p_x + 14$ mm
ER – Edge round	$0.25\ p_x$	$0.25\ p_x$

p_x = Axial pitch (mm)

Table 18.4. Beam stress for worm gears

Materials	Safe stress S_d MPa	Ultimate stress S_u MPa	Yield stress S_y MPa
Cast iron good grade	70	210	–
Semi-steel	184	250	–
Cast steel	140	420	190
Manganese bronze SAE 43	140	490	175
Bronze SAE = lead gun metal	56	210	84
Phosphorus bronze SAE 65	105	246	140

Table 18.5. Service factors for worm gears

Type of service	C_s
Intermittent with light shock	1.0-1.5
Continuous with medium shock line-hushers, etc.	1.5-2.0
Continuous with heavy shock reciprocating pumps Paper and rubber mills, etc.	2.0-2.6
Continuous with frequent and very heavy shock Main-line drivers, steel mills, etc.	2.5-3.0

Table 18.6. Permissible surface pressure S_c MPa for 14.5° pressure angle

Materials		Number of teeth in gear							
Worm	Gear	10	20	30	40	50	60	70	80 and above
C 20 steel untreated	Cast iron	0.5	1.6	3.0	5.0	6.5	7.5	9	9.5
C 40 steel untreated	Bronze SAE63 Sand cast	0.8	2.5	4.5	7.5	9.5	11.5	13.5	14
C 40 steel heat Treated ground	Bronze SAE63 Sand cast	1.2	3.5	6.5	11.0	14	17	20	21
0.10C alloy steel Carburized, hardened	Bronze SAE65 Sand cast	1.6	47.5	9.0	15	19	23	27	28

Contd.

Ground Bronze	SAE 65	2.2	6.5	12.0	21	26	31.5	37	38.5
	Chill cast								
	Nickel bronze								
	Sand cast	2.6	8.0	15.0	25	31.5	38.5	45.0	47.0
	Nickel Bronze								
	Chill cast	3.2	9.5	17.5	30	38	46	53.5	56.5

(*Note:* 20° pressure angle, values are multiplied by 1.05)

Table 18.7. Load stress factors for worm gear

Materials		**Load stress factors – *K* (MPa)**		
Worm	**Gear**	$\psi = 0 - 10°$	$\psi = 10 - 25°$	$\psi \geq 25°$
Steel 250 BHN	Phosphor bronze	0.423	0.528	0.634
Hardened steel	Phosphor bronze	0.564	0.704	0.845
Hardened steel	Chilled phosphor bronze	0.845	1.057	1.268
Hardened steel	Antimony bronze	0.845	1.057	1.268
Cast iron	Phosphor bronze	1.057	1.317	1.585

SOLVED EXAMPLES

Example 18.1: A double threaded worm has a lead angle of 20°. For pitch of 30 mm, what is the diameter of the worm?

SOLUTION We have, $\tan \Psi = \dfrac{L}{\pi D_w}$,

$$\therefore \qquad D_w = \frac{L}{\pi \tan \Psi} = \frac{P_c \cdot Z_w}{\pi \tan \Psi} = \frac{30 \times 2}{\pi \tan 20} = 52.5 \text{ mm}$$

Example 18.2: A worm gear reducer unit is to have a 250 mm centre distance. What should be the worm diameter and axial module of the worm according to AGMA recommendations?

SOLUTION As per AGMA recommendations, the worm diameter D_w is given by equation,

$$D_w = 0.68C^{0.875} = 0.68(250)^{0.875} = 85.25 \text{ mm}$$

The recommended module, $m = \dfrac{D_w}{3\pi} = \dfrac{85.25}{3\pi} = 9.04$

Choose $m = 9.0$ mm

Example 18.3: A worm gear reducer unit has a centre distance of 220 mm and a transmission ratio of 20. What is its approximate power input rating in order to prevent over heating?

SOLUTION The power rating equation based on the heat dissipation,

$$\text{Power in kW} = \frac{0.029C^{1.7}}{(G+5)} = \frac{0.029\times(220)^{1.7}}{(20+5)} = 11.132\text{ kW} \qquad \text{(Eqn. 17.24 MDHB)}$$

Example 18.4: A hardened steel worm and phosphor bronze gear reducer unit has transmission ratio of 20 to 1. The speed of the worm is 1200 rpm and its diameter is 68 mm. What is the approximate permissible power from the wear consideration?

SOLUTION The AGMA equation for wear

$$\text{kW} = N_g \cdot K_1 \cdot C_v \cdot Q \qquad \text{(Eqn. 17.23 MDHB)}$$

where,

$$N_g = \frac{N_w}{G} = \frac{1200}{20} = 60\text{ rpm}$$

$$Q = G/(G+2.5) = 20/(20+2.5) = 20/(20+2.5) = 0.889$$

$$V_w = \frac{\pi DN}{60000} = \frac{\pi\times 68\times 1200}{60000} = 4.27\text{ m/s}$$

$$C_v = \frac{2.3}{(2.3+V_w+3V_w/G)} = \frac{2.3}{2.3+4.27+3\left(\frac{4.27}{20}\right)} = 0.319$$

From Table 17.2, for $C = (D_w/0.68)^{1/0.875} = 193.2,\ K_1 = 0.442$

$\therefore$ Power NP $= 60\times 0.442\times 0.319\times 0.889 = 7.5$ kW

Example 18.5: A hardened steel worm transmits power to phosphor bronze gear with a velocity ratio of 20 to 1. The centre distance is 400 mm. Determine the axial module and lead angle for proportions as close as possible to AGMA recommendations.

SOLUTION The AGMA recommendation for worm diameter

$$D_w = 0.68\,C^{0.875} = 0.68(400)^{0.875} = 128.6\text{ mm}$$

Then $3P_c = D_w = 128.6$, i.e., $3\pi m = 128.6$

$\therefore \quad m = (128.6/3\pi) = 13.65$

Next standard value, $m = 14$

For double start thread, lead, $L = \pi m Z_W = 2\pi m$

The angle of helix, $\Psi = \tan^{-1}\left(\frac{L}{\pi D_w}\right) = \tan^{-1}(2\pi m/\pi D_w)$

$$= \tan^{-1}(2\pi m/3\pi^2 m) = \tan^{-1}(2/3\pi) = 11.98 = 12°$$

Examplc 18.6: A hardened steel worm rotating at 1250 rpm, transmits power to phosphor bronze gear with a transmission ratio of 15 to 1. The centre distance is 225 mm. Determine the remaining design parameters and give estimated power-input ratings from the standpoint of endurance strength, wear strength and heat dissipation. The teeth are of 14.5° involute full depth form.

Given Data: Worm speed = 1250 rpm; gear ratio = 15,

Centre distance, C = 225 mm; tooth form-14.5° full depth;

Material for worm: hardened steel; for gear: P-bronze

SOLUTION

Step-1: Module and dimensions by AGMA standards

By AGMA recommendations, worm diameter for C = 225 mm, is given by

$$D_w = 0.68C^{0.875} = 0.68(225)^{0.875} = 77.7 \simeq 80 \text{ mm (say)}$$

Again by AGMA recommendations $3P_c = 3\pi m = D_w = 80$

$$\therefore \qquad m = 80/3\pi = 8.48 \text{ mm}$$

Let $m = 8.0$ mm

The worm requires to have triple start to satisfy the minimum teeth condition,

$$Z_G = 40 \text{ for } \phi = 14.5° \text{ i.e., } Z_w = 3$$

So that $Z_G = G \cdot Z_w = 15 \times 3 = 45$

Then gear diameter $= 45 \times 8 = 360$ mm

Then worm diameter to satisfy the centre distance condition C = 225,

$$D_w = 2C - D_G = 2 \times 225 - 360 = 450 - 360 = 90 \text{ mm}$$

This is more than minimum specified by AGMA equations. $D_w = 80$ mm

The face width, $b = 0.73D_w = 0.73 \times 90 = 65.7 = 65.0$ mm (say)

The speed of the gear, $N_g = 1200/15 = 80$ rpm

The lead of the worm, $L = \pi m \cdot Z_w = \pi \times 8 \times 3 = 75.4$ mm

Then the helix angle, $\Psi = \tan^{-1}\left(\dfrac{L}{\pi D_w}\right) = \tan^{-1}\left(\dfrac{75.4}{\pi \times 90}\right) = 15°$

Centre distance, $C = \dfrac{90 + 360}{2} = \dfrac{450}{2} = 225.0$ mm

Step-2: For beam strength,

$$W_G^T = \frac{\pi S_d \cdot m \cdot b \cdot Y \cdot \cos\Psi}{K_v}$$

$$= \frac{\pi \times 105 \times 8 \times 65 \times \cos 15}{1.48}$$

$$= 11195.0 \text{ N}$$

Where pitch line velocity,

$$V_g = \frac{\pi DN}{60000} = \frac{\pi \times 360 \times 80}{60000} = 1.5 \text{ m/sec}$$

$$K_v = \frac{3.1 + V_g}{3.1} = 1.48$$

$S_d = 105$ MPa (Table 17.4 MDHB)

$y = 0.100$ (Table 17.1 MDHB)

Step-3: Power rating

(i) Power rating by beam strength consideration,

$$= \frac{W_G^T \times R_G \times N_g}{955 \times 10^4} = \frac{11195 \times 180 \times 80}{955 \times 10^4} = 16.9 \text{ kW}$$

(ii) Power rating by wear strength

$$= N_g \cdot K_1 \cdot C_v \cdot Q$$

$$= 80 \times 0.727 \times 0.559 \times 0.857$$

$$= 27.86 \text{ kW}$$

where, $N_g = 80$ rpm.

$K_1 = 0.727$

$$C_v = \frac{2.3}{2.3 + 1.5 + \frac{3 \times 1.5}{15}}$$

$$= 0.559$$

$Q = 15/(15 + 2.5) = 0.857$

(iii) Power rating by heat dissipation,

$$\text{Power} = \frac{0.092 C^{1.7}}{G + 5} = \frac{0.092 (225)^{1.7}}{15 + 5} = 14.45 \text{ kW}$$

The power rating of gear set is the minimum of all three, i.e., $NP = 14.45$ kW

Example 18.7: Design a speed reducer unit of worm and worm-wheel for an input of 1 kW with transmission ratio of 25. The speed of the worm is 1600 rpm. The worm is made of hardened steel and wheel of phosphor bronze for which the material combination factor is 0.7 N/mm^2. The static stress for the wheel material is 56 MPa. The worm is made of double start and the centre distance between the axes of the worm and wheel is 120 mm. The tooth form is 14.5° involute. Check the design for strength, wear and heat dissipation.

Given Data: Power $= 1.0$ kW @ speed $N_w = 1600$ rpm.

Gear ratio $= 25$, speed of the gear $= \frac{1600}{25} = 64$ rpm.

SOLUTION

Step-1: Module and dimensions by AGMA standards

The centre distance $C = 120$ mm

Then $D_w = 0.68C^{0.875} = 0.68(120)^{0.875} = 44.85$ mm

$$\therefore \quad 3P_c = 3\pi m = D_w = 44.85$$

$$\therefore \quad m = \frac{44.85}{3\pi} = 4.76 \text{ mm}$$

Select $m = 5.0$ mm, $D_w = 3\pi m = 47.1$ mm, choose $D_w = 50$ mm

The lead, $L = \pi m \cdot Z_w = \pi \times 5 \times 2 = 31.42$ mm

The helix angle, $\Psi = \tan^{-1}\left(\frac{L}{\pi D_w}\right) = \tan^{-1}\left(\frac{31.42}{\pi \times 50}\right) = 11.3°$

The correct centre distance,

$$C = \frac{L}{2\pi}\left(\frac{1}{\tan\Psi} + G\right) = \frac{31.42}{2\pi}\left(\frac{1}{\tan 11.3} + 25\right) = 150.0 \text{ m}$$

The number of teeth on gear, $Z_G = 2 \times 25 = 50$

The gear diameter, $D_G = 5 \times 50 = 250$ mm

Centre distance, $C = \frac{D_w + D_G}{2} = \frac{50 + 250}{2} = 150.00$

This is more than the specified value and acceptable.

$\therefore$ The module selected is geometrically suitable.

The face width, $b = 0.73D_w = 0.73 \times 50 = 36.5 \text{ mm} \cong 35$ mm (Say)

Step-2: Check for efficiency

Pitch line velocity of the worm,

$$V_w = \frac{\pi D_w N}{60000} = \frac{\pi \times 50 \times 1600}{60000} = 4.19 \text{ m/sec.}$$

Sliding velocity, $V_s = V_w/\cos\Psi = 4.27$ m/sec

Coefficient of friction $= 0.025 + \frac{V_s}{300} = 0.025 + \frac{4.27}{300} = 0.039 \simeq 0.04$

The efficiency, $\eta = \frac{\cos\Phi_n - f\tan\Psi}{\cos\Phi_n + f\cot\Psi} = \frac{\cos 14.5 - 0.04\tan 11.3}{\cos 14.5 + 0.04\cot 11.3}$

$$= \frac{0.97 - 0.008}{0.97 + 0.200} = 0.82 = 82\%$$

Step-3: Check for beam strength

Input power $= 1.0$ kW @ 1600 rpm.

The torque on the worm, $T_w = \dfrac{955 \times 10^4 \times 1.0}{1600} = 5968.75$ Nmm

Torque on the gear,

$$T_G = \frac{T_w N_w}{N_G} \eta = T_w \cdot G \cdot \eta = 5968.75 \times 25 \times 0.82 = 122672.0 \text{ Nmm}$$

The tangential load on gear,

$$W_G^T = T_G / R_G = \frac{122672.0}{125.0} = 981.4 \text{ N}$$

$$V_G = \frac{\pi D_g N}{60000} = \frac{\pi \times 250 \times 64}{60000} = 0.84 \text{ m/s}$$

Velocity factor, $K_v = \dfrac{3.1 + V_G}{3.1} = \dfrac{3.1 + 0.84}{3.1} = 1.27$

Form factor, $y = 0.10$

The bending stress induced,

$$S = \frac{W_t \cdot K_v}{\pi \cdot m \cdot b \cdot y \cos \Psi} = \frac{981.4 \times 1.27}{\pi \times 5 \times 35 \times 0.1 \times \cos 11.3^\circ}$$

$$= 23.12 \text{ MPa} < (S_d = 56 \text{ MPa})$$

Hence, the design is safe.

Step-4: Check for wear srength

The dynamic load, $W_d = W_G^T \times K_v = 1246.38$ N

The limiting wear load, $W_w = D_G \cdot b \cdot K$ (with $K = 0.7$ N/mm^2)

$$= 250 \times 35 \times 0.7 = 6125 \text{ N} > (W_d = 1246.38 \text{ N})$$

The design is safe against wear.

Step-5: Check for heat dissipation

Heat generated,

$$Q_g = 1000(1 - \eta) KW = 1000(1 - 0.82) \times 1.0 = 180 \text{ J/sec.}$$

Heat dissipating capacity,

$$Q_d = 3.35(A_g + A_w)(t_g - t_r)$$

$$= 0.035 \times (49087 + 4319.7) + 17$$

$$= 5490.5 \text{ J/sec} > Q_g$$

where, $A_g = \frac{\pi}{4} D_G^2 = 49087.4 \text{ mm}^2$

$$A_w = FLD_w = \left(4.5 + \frac{50}{50}\right)\pi \times 5 \times 50 = 4319.7 \text{ mm}^2$$

$$(t_g - t_r) = (40 - 23) = 17°$$

This is more than the heat generated. Hence, the design is satisfactory from heating consideration.

Example 18.8: Design worm gear speed reducer to transmit 2 kW at a speed of 1440 rpm. The desired velocity ratio is 24:1. Assume the worm is made of hardened steel and gear of phosphor bronze.

Given Data: Power NP = 2 kW @1440 rpm; velocity ratio, $G = 24$

SOLUTION

Step-1: Module and dimensions by AGMA recommendation

Tooth form 14.5° involute, Double start, $Z_w = 2$

Number of teeth on the gear, $Z_G = 2 \times 24 = 48$

Assume the centre distance, $C = 400$ mm

The worm diameter, $D_w = 0.68C^{0.875} = 128.6$ mm

$$3P_c = 3\pi m = D_w = 128.6$$

$$\therefore \quad m = \frac{128.6}{3\pi} = 13.65 \cong 14.0$$

$$D_w = 3\pi \times 14 = 132.0 \text{ mm}$$

Lead $L = \pi m Z_w = \pi \times 14 \times 2 = 87.96$ mm

$$\Psi = \tan^{-1}\left(\frac{L}{\pi D_w}\right) = \tan^{-1}\left(\frac{87.96}{\pi \times 132}\right) = 11.97 \simeq 12$$

Gear diameter, $D_G = 14 \times 48 = 672.0$ mm

The modified centre distance $= \frac{D_w + D_G}{2} = \frac{132.0 + 672.0}{2} = 402$ mm

It is nearly equal to the assumed value of 400 mm

Hence, the estimated of the module is correct.

Step-2: Check for efficiency

The pitch line velocity, $V_w = \frac{\pi D_w \cdot N_w}{60000} = \frac{\pi \times 132 \times 1440}{60000} = 9.95$ m/sec

Sliding velocity, $V_s = \dfrac{V_w}{\cos\Psi} = \dfrac{9.95}{\cos 12} = 10.17$ m/sec

Coefficient of friction, $f = 0.025 + \dfrac{10.17}{300} = 0.058$

Then efficiency,

$$\eta = \frac{\cos\Phi_n - f\tan\Psi}{\cos\Phi_n + f\tan\Psi} = \frac{\cos 14.5 - 0.04\tan 12.5}{\cos 14.5 + 0.04\cot 12.4}$$

$$= \frac{0.968 - 0.1286}{0.968 + 0.2016} = 0.77 < (\text{desired value of } 0.8)$$

Hence, modify the design.

Step-3: Modified module and dimensions

Let us assume $C = 250$ mm

Then $D_w = 0.68C^{0.875} = 85.25$ mm

Module $m = D_w/3\pi = 9.04$

Let $m = 9.0$

The modified worm diameter, $D_w = 3\pi m = 3\times\pi\times 9 = 85.00$

Then lead, $L = \pi m Z_w = \pi\times 9\times 2 = 56.55$ mm

Helix angle, $\Psi = \tan^{-1}\left[\dfrac{L}{\pi D_w}\right] = \tan^{-1}\left[\dfrac{56.55}{\pi\times 85}\right] = 12°$

The gear diameter, $D_G = 9\times 48 = 432$ mm

The correct center distance, $C = \dfrac{432+85}{2} = 258.5$ mm

Step-4: Check for efficiency

Pitch line velocity, $V_w = \dfrac{\pi D_w N}{60000} = \dfrac{\pi\times 85\times 1440}{60000} = 6.4$ m/sec

Sliding velocity, $V_s = \dfrac{V_w}{\cos\Psi} = \dfrac{6.4}{\cos 12} = 6.55$ m/sec

Coefficient of friction, $f = 0.025 + \dfrac{V_s}{300} = 0.025 + \dfrac{6.55}{300} = 0.047$

The efficiency, $\eta = \dfrac{0.968 - 0.047\times 0.212}{0.968 + 0.047\times 4.7} = \dfrac{0.958}{11189} = 0.806$

This is equal to desired value and hence the design is acceptable.

Step-5: Check for strength

The torque on the worm,

$$T_w = \frac{955 \times 10^4 \times 2}{1440} = 1.33 \times 10^4 \text{ Nmm}$$

The tangential load, $W_w^T = T_w / R_w = 2 \times 1.33 \times 10^4 / 85 = 312.1 \text{ N}$

Normal load (Eq. 17.8), $W = W_W^T / (\cos \Phi_n \sin \Psi + f \cos \Psi)$

$$= 312.1/(\cos 14.5 \times \sin 12 + 0.047 \times \cos 12) = 1261.8 \text{ N}$$

The tangential load on gear, $W_G^T = W \cdot (\cos \Phi_n \cos \Psi - f \sin \Psi)$

$$= 1261.8 \times (\cos 14.5 \times \cos 12 - 0.047 \times \sin 12) = 1182.58 \text{ N}$$

The pitch velocity of the gear,

$$V_g = \frac{\pi D N}{60000} = \frac{\pi \times 432 \times 60}{60000} = 1.357 \text{ m/sec}$$

$$\text{Velocity factor, } K_v = \frac{3.1 + V_g}{3.1} = \frac{3.1 + V_g}{3.1} = 1.44$$

Check for bending strength

Induced stress,

$$S = \frac{W_G^T K_v}{\pi m \cdot b \cdot Y \cos \Psi}$$

$$= \frac{1182.58 \times 1.44}{\pi \times 9 \times 62 \times 0.1 \times 0.978}$$

$$= 9.93 \text{ MPa}$$

where, $b = 0.73 D_w = 62.0$ mm

$Y = 0.10$

$K = 1.44$

$W_G^T = 1182.58$ N

$m = 9.0$

but $S_d = 105$ MPa (Table 17.4 MDHB) The induced stress is less and can be acceptable

Step-6: Check for wear strength

Dynamic load, $W_d = W_G^T \times K_v = 1182.58 \times 1.44 = 1702 \text{ N}$

Wear load, $W_w = D_G \cdot b \cdot K$

$$= 432 \times 62 \times 0.704$$

$$= 18855.94 \text{ N} > W_d$$

where,

$D_G = 432$ mm, $b = 62.0$ mm

$K = 0.704$ (Table 17.7 MDHB)

This is more than the dynamic load W_d and hence the design is safe against wear failure.

Step-7: Check for heat balance

Heat generated, (Eqn. 17.21 MDHB)

$$Q_g = 1000(1-\eta)\,\text{kW} = 1000(1-0.80)\times 2 = 400\ \text{J/sec} = 0.4\ \text{kW}$$

Heating dissipating capacity, (Eqn. 17.22 MDHB)

$$Q_d = 0.035(A_g + A_w)(t_g - t_r)$$

where, $A_g = \dfrac{\pi}{4}D_G^2 = \dfrac{\pi}{4}(432)^2 = 146574\ \text{mm}^2$

$$A_w = FL\cdot D_w = (4.5 + 84/50)\times\pi\times 9\times 85 = 13122.1\ \text{mm}^2$$

Assume $t_g = 40°, Ct_r = 22°\text{C}, t_g - t_r = 18°\text{C}$

$\therefore \quad Q_d = 0.035\times(146574 + 13122.1)\times 18$

$= 17498\ \text{J/sec} > Q_g\ \therefore$ Safe

This is more than the heat generated. The operating temperature will be much less than specified 40°C. Hence, the design is acceptable.

Example 18.9: A worm gear reducer unit is to be designed for a centre distance of 200 mm. The worm is made of hardened steel and gear is made of phosphorus bronze. The transmission ratio is 10. The speed of the worm is 1750 rpm. The gear has 20° involute stub-teeth. Determine the input capacity for the phosphorus bronze gear strength of 55 MPa.

Given Data: Worm gear – reducer unit

Centre distance, C = 200 mm, Gear ratio, G = 10

Speed of worm, $N_w = 1750$ rpm, Tooth form –20° involute stub

Material for gear - P bronze, and for worm – hardened steel

The design stress for gear, $S_d = 55$ MPa

SOLUTION

Step-1: Design parameters by AGMA standards

Lead angle, $\Psi = \tan^{-1}(\sqrt[3]{Z_w/Z_G})$ (Eqn. 17.5 MDHB)

$$= \tan^{-1}(\sqrt[3]{1/G}) = \tan^{-1}(\sqrt[3]{0.1}) = 24.89 \cong 25°$$

The worm diameter,

$$D_w = 0.68C^{0.875} = 0.68\times(200)^{0.875} = 70.13 \simeq 70\ \text{mm} \quad \text{(Eqn. 17.2 MDHB)}$$

Gear width, $b = \dfrac{C^{875}}{2} = 51.57\ \text{mm} \simeq 52.0\ \text{mm}$ (Eqn. 17.3 MDHB)

The gear diameter, $D_G = 2C - D_w = 2\times 200 - 70 = 330$ mm

The module from the relation, $D_w = 3P_c = 3\,\pi m$ (Eqn. 17.2 MDHB)

$$\therefore \qquad m = \frac{D_w}{3 \times \pi} = \frac{70}{3 \times \pi} = 7.44$$

Choose standard value, $m = 8$ mm

The number of teeth on the gear, $Z_G = D_G / m = \dfrac{330}{8} = 41.25$

Choose $Z_G = 42.0$

The modified gear diameter, $D_G = m \cdot Z_G = 8 \times 42 = 336$ mm

The exact centre distance, $C = \dfrac{D_w + D_G}{2} = \dfrac{70 + 336}{2} = 203$ mm (acceptable)

From the relation – Eqn. 17.4 and 17.5

Lead, $L = \pi m \cdot Z_w = \pi D_w \tan \Psi$

The number of threads, $Z_w = \dfrac{D_w}{m} \tan \Psi = \dfrac{70}{8} \tan (25°) = 4.08$

Choose $Z_w = 4.0$

Then correct lead, $L = \pi m Z_w = \pi \times 8 \times 4 = 100.53$ mm

Step-2: Power output by beam strength

The pitch line velocity of gear,

$$V_g = \frac{\pi DN}{60000} = \frac{\pi \times 336 \times 175}{60000} = 3.079 \text{ m/s}$$

The velocity factor, $K_v = \dfrac{3.1 + V_g}{3.1} = \dfrac{3.1 + 3.097}{3.1} = 1.993$

The form factor from Table 17.1 MDHB, $y = 0.125$

Then by beam strength (Eqn. 17.16 MDHB)

$$S = \frac{W_G^T \cdot K_v}{\pi m \cdot b \cdot y \cdot \cos \Psi}$$

The tangential load on the gear,

$$W_G^T = \frac{S \pi \cdot m \cdot b \cdot y \cdot \cos \Psi}{K_v} = \frac{55 \times \pi \times 8 \times 52 \times 0.125 \times \cos (25)}{1.993} = 4085.87 \text{ N}$$

Torque on the gear,

$$T_G = W_G^T \times R_G = 4085.87 \times \frac{336}{2} = 686425.8 \text{ Nmm}$$

The power output,

$$NP = \frac{T \cdot N}{955 \times 10^4} = \frac{686425.8 \times 175}{955 \times 10^4} = 12.58 \text{ kW}$$

Step-3: The efficiency of the drive,

$$\eta = \frac{\cos Q_n - f \tan \psi}{\cos Q_n + f \cos \psi} \qquad \text{(Eqn. 17.13 MDHB)}$$

Where f friction factor, is evaluated as follows
The sliding velocity,

$$V_s = \frac{V_w}{\cos \psi} = \frac{\pi D_w N_w}{60000 \cos \psi} = \frac{\pi \times 70 \times 1750}{60000 \times \cos(25)} = 7.077 \text{ m/s}$$

The coefficient of friction,

$$f = 0.025 + \frac{V_s}{300} = 0.025 + \frac{7.077}{300} = 0.049 \qquad \text{(Eqn. 17.14a MDHB)}$$

and normal pressure angle, $\phi_n = 20°$

$$\therefore \text{ Efficiency, } \eta = \frac{\cos 20° - 0.049 \tan 25°}{(\cos 20° + 0.049 \cot 25°)} = \frac{0.94 - 0.049 \times 0.47}{0.94 + 0.049 \times 2.145} = 0.88 \text{ (acceptable)}$$

Step-4: The power input by strength,

$$NP = \frac{\text{output}}{\eta} = \frac{12.58}{0.88} = 14.34 \text{ kW}$$

Step-5: The input power by AGMA rating equation,

$$NP = N_g \cdot K_1 \cdot C_v \cdot Q \qquad \text{(Eqn. 17.23 MDHB)}$$

where,

$$Q = G/(G + 2 - 5) = 10/(10 + 2.5) = 0.8$$

$$C_v = \frac{2.3}{2.3 + V_w + \frac{3V_w}{G}} = \frac{2.3}{2.3 + 6.4 + \frac{3 \times 6.4}{10}} = 0.216$$

$$K_1 = 0.485 \qquad \text{(Table 17.2 MDHB)}$$

$$\therefore \quad NP = 175 \times 0.485 \times 0.216 \times 0.8 = 14.67 \text{ kW}$$

Step-6: The correct power input capacity

The smaller of the two is the input power capacity of the reducer set = 14.34 kW.

Example 18.10: The following data refer to a worm gear drive (i) centre distance, $C = 200$ mm, (ii) pitch circle diameter of the worm = 80 mm, (iii) number of starts = 4, (iv) axial module = 8 mm, (v) transmission ratio = 20, (vi) The worm gear is made of phosphor bronze with an allowable stress = 55 MPa, (vii) the worm is made of hardened steel, and (viii) the tooth form is 20° full depth involute.

Determine (i) the number of teeth on the worm gear, (ii) the lead angle, and (iii) face width of the gear to transmit 15 kW at 1750 rpm of the worm, based on the beam strength of the gear.

Given Data: Worm diameter, $D_w = 80$ mm; Number of starts, $Z_w = 4$;

Centre distance, $C = 200$ mm; Axial module, $m = 8$ mm;
The gear ratio, $G = 20$
Material – worm-hardened steel
Gear – P bronze, Allowable stress, $S_d = 55$ MPa
Power transmitted, $NP = 15$ kW at speed of worm $N_1 = 1750$ rpm
Find number of teeth on gear Z_G, face width b and lead angle ψ

SOLUTION

Step-1: The lead angle by geometry

Gear ratio, $G = \dfrac{N_1}{N_2} = \dfrac{Z_G}{Z_w} = \dfrac{Z_G}{4} = 20$

$\therefore \quad Z_G = 4 \times 20 = 80$

Gear diameter, $D_G = m \cdot Z_G = 8 \times 80 = 640$ mm

The lead, $L = \pi m Z_w = \pi \times 8 \times 4 = 100.53$ mm

The lead angle, $\psi = \tan^{-1}\left(\dfrac{L}{\pi D_w}\right) = \tan^{-1}\left(\dfrac{100.53}{\pi \times 80}\right) = 21.8°$

Step-2: Face width by strength

The speed of the gear, $N_2 = N_1/G = \dfrac{1750}{20} = 87.5$ rpm

The torque on the gear,

$$T = \frac{955 \times 10^4 \text{ kW}}{N} = \frac{955 \times 10^4 \times 15}{87.5} = 167.71 \times 10^4 \text{ Nmm}$$

The tangential load on the gear,

$$W_G^T = T/R_G = \frac{163.71 \times 10^4}{(640/2)} = 5116.07 \text{ N}$$

The pitch line velocity of the gear,

$$V_g = \frac{\pi D_G N_G}{60000} = \frac{\pi \times 640 \times 87.5}{60000} = 2.93 \text{ m/s}$$

The velocity factor, $K_v = \dfrac{3.1 + V_1}{3.1} = \dfrac{3.1 + 2.93_1}{3.1} = 1.95$ (Eqn. 17.17 MDHB)

Tooth form factor, $y = 0.125$ (Table 17.1 MDHB)

The face width of the gear for beam strength, (Eqn. 17.16 MDHB)

$$b = \frac{W_G^T \cdot K_v}{S \cdot \pi m \cdot y \cdot \cos\psi} = \frac{5116.07 \times 1.95}{55 \times \pi \times 8 \times 0.125 \times \cos(21.8)} = 62.185 \text{ mm}$$

Choose $b = 62.0$ mm

Example 18.11: A speed reducer unit is to be designed for an input power of 0.75 kW with a transmission ratio of 27. The speed of the hardened steel worm is 1750 rpm. The worm wheel is made of phosphor bronze. The tooth form is 14.5° involute. The allowable stress for the wheel may be taken as 80 MPa.

Given Data: Input power, $NP = 0.75$ kW at worm speed, $N_1 = 1750$ rpm;
Speed ratio, $G = 27$; Tooth form – 14.5° involute;
Material- worm-hardened steel; Gear- P. bronze;
Design stress, $S_d = 80$ MPa

SOLUTION

Step-1: Geometric relations for dimensions as per AGMA

The lead angle,

$$\psi = \tan^{-1}(\sqrt[3]{Z_w/Z_G}) = \tan^{-1}(\sqrt[3]{1/G}) = \tan^{-1}(\sqrt[3]{1/27}) = 18.435^6$$

Worm thread – Double start $Z_w = 2.0$,
and the number of teeth on gear $Z_G = G \cdot Z_W = 27 \times 2 = 54$
Diameter of the gear, $D_G = 54$ m
Speed of gear, $N_G = 1750/27 = 64.815$ rpm
Diameter of worm from relations 17.4 and 17.5 MDHB

$$\tan\psi = \frac{L}{\pi D_w} = \frac{\pi m \cdot Z_w}{\pi D_w} = \frac{m \cdot Z_w}{D_w}$$

$$\therefore \quad D_w = m \cdot Z_w / \tan\psi = \frac{2 \text{ m}}{\tan(18.435)} = 5.999 \text{ m}$$

The gear face width,

$$b = 0.73 D_w = 0.73 \times 5.999 \text{ m} = 4.38 \text{ m} \quad \text{(Eqn. 17.3 MDHB)}$$

Step-2: Module from strength equation

The torque on the gear,

$$T_G = \frac{955 \times 10^4 \times 0.75}{64.815} = \frac{955 \times 10^4 \times 0.75}{64.815} = 110506.83 \text{ Nmm}$$

The tangential load on the gear, $W_G^T = \dfrac{T_G}{R_G} = \dfrac{110506.83}{\left(\dfrac{54 \text{ m}}{2}\right)} = \left(\dfrac{4092.86}{m}\right) N$

Assume velocity factor $K_v = 1.2$, tooth form factor $y = 0.100$ (Table 17.1 MDHB). Then by beam strength Eqn. 17.16 MDHB,

$$S = \frac{W_G^T \cdot K_v}{\pi m \cdot b \cdot y \cdot \cos\psi}$$

$$\therefore \quad 80 = \frac{\left(\frac{4092.86}{m}\right) \times 1.2}{\pi m \times 4.38\ \text{m} \times 0.1 \times \cos(18.435)}$$

$$\therefore \quad m^3 = \frac{4092.86 \times 1.2}{(80 \times \pi \times 4.38 \times 0.1 \times \cos 18.435)} = 45.39$$

$$\therefore \quad m = \sqrt[3]{45.39} = 3.56$$

Choose $m_n = 4.0$

Then $m = m_n \cos\Psi = 4 \times \cos(18.435) = 3.795$

Then gear diameter, $D_G = 3.795 \times 54 = 205$ mm

Worm diameter, $D_w = 5.999\ \text{m} = 5.999 \times 3.795 = 22.74 \cong 23$ mm

The centre distance, $C = \dfrac{D_G + D_w}{2} = \dfrac{205 + 23}{2} = 114.0$ mm

The face width $b = 4.38\ \text{m} = 4.38 \times 3.795 = 16.62 \simeq 17$ mm

The tangential load

$$W_G^T = \frac{4092.86}{m} = \frac{4092.86}{3.795} = 1078.49\ \text{N}$$

The tangential velocity of gear,

$$V_g = \frac{\pi D_G N_G}{60000} = \frac{\pi \times 205 \times 64.815}{60000} = 0.696\ \text{m/sec}$$

Velocity factor, $K_v = \dfrac{3.1 + V_g}{3.1} = \dfrac{3.1 + 0.696}{3.2} = 1.2244$

The bending stress induced in gear,

$$S = \frac{W_G^T \cdot K_v}{\pi m b \cdot \cos y} = \frac{1078.49 \times 1.2244}{\pi \times 3.795 \times 17 \times 0.1 \times \cos C18.435}$$

$$= 68.64\ \text{MPa} < S_d = 80\ \text{MPa}$$

Hence, design is safe.

Step-3: Check for wear load

The load stress factor, $K = 0.704$ from Table 17.7 MDHB

The wear load,

$$W_w = D_G \cdot b \cdot K = 205 \times 17 \times 0.704 = 2453.44 \text{ N} > W_G^T \qquad \text{(Eqn. 17.20 MDHB)}$$

The design is safe against wear failure.

Step-4: Check by AGMA rating equations

(i) For wear strength

Power in $\text{kW} = N_g K_1 \cdot C_v \cdot Q$ (Eqn. 17.23 MDHB)

where $N_g = 64.815 \text{ rpm}$, $K_1 = 0.125$ (Table 17.2 MDHB)

The worm speed, $V_w = \dfrac{\pi D_w N_w}{60000} = \dfrac{\pi \times 23 \times 1750}{60000} = 2.11 \text{ m/s}$

Velocity factor, $C_v = \dfrac{2.3}{2.3 + 2.11 + \dfrac{3 \times 2.11}{27}} = 0.495$

$$Q = G/(G+2.5) = \frac{27}{27+2.5} = 0.915$$

$NP = 64.815 \times 0.125 \times 0.495 \times 0.915 = 3.67 \text{ kW} > \text{input power}$

Design is safe for wear strength.

(ii) For heat dissipation

Power, $NP = \dfrac{0.029 C^{1.7}}{(G+5)}$ (Eqn. 17.24 MDHB)

$$= \frac{0.029(114)^{1.7}}{(27+5)} = 2.844 \text{ kW} > 0.75 \text{ kW}$$

Design is safe for heat dissipation.

Example 18.12: A hardened steel worm rating at 1250 rpm, transmits power to a phosphor bronze gear with transmission ratio of 15:1. The centre distance is 225 mm. Design the gear drive and estimate the power rating of the gear drive from the standpoint of strength and heat dissipation. The tooth form is 14.5° full depth involute.

Given Data: Worm speed, $N_w = 1250$ rpm, speed ratio, $G = 15$

Centre distance, $C = 225$ mm

Material- worm – hardened steel, Gear – P- bronze

Power rating - ?

SOLUTION

Step-1: Module by AGMA equations

The worm diameter, $D_w = 0.68 C^{0.875}$ (Eqn. 17.2 MDHB)

$$= 0.68 \times (225)^{0.875} = 77.74 \text{ mm} = 78.0 \text{ mm}$$

We have by Eqn. 17.2, $D_w = 3P_c = 3 \times \pi \times m$

$$\therefore \quad m = \frac{D_w}{3\pi} = \frac{78}{3\pi} = 8.276$$

Choose $m = 8$

The diameter of the gear, $D_G = 2C - D_w = 2 \times 225 - 78 = 372.0$ mm

Number of teeth, $Z_G = D_G / m = 372/8 = 46.25 \cong 45.0$

The modified gear dia., $D_G = m \cdot Z_G = 8 \times 45 = 360.0$ mm
The modified centre distance,

$$C = \frac{D_w + D_G}{2} = \frac{78 + 360}{2} = 219 \text{ mm}$$

The gear face width $b = 0.73D_w = 0.73 \times 78 = 56.94 = 57$ mm

The lead, $L = \pi m Z_w = \pi m \dfrac{Z_G}{G} = \pi \times 8 \times \dfrac{45}{15} = 75.4$ mm

Then the lead angle, $\Psi = \tan^{-1}\left(\dfrac{L}{\pi D_w}\right) = \tan^{-1}\left(\dfrac{75.4}{\pi \times 78}\right) = 17.1°$

Step-2: The power by strength equation

The safe stress, $S_d = 105$ MPa (Table 17.4 MDHB)
The pitch line velocity of gear,

$$V_g = \frac{\pi D_G N_g}{60000} = \frac{\pi \times 360 \times (1250/15)}{60000} = 1.57 \text{ m/s}$$

The velocity factor, $K_v = \dfrac{3.1 + V_g}{3.1} = \dfrac{3.1 + 1.57}{3.1} = 1.51$ (Eqn. 17.17 MDHB)

The tooth form factor $y = 0.100$ (from Table 17.1 MDHB)
Then from beam strength Eqn. 17.16 MDHB
The tangential load,

$$W_G^T = \frac{S_d \cdot \pi \cdot m \cdot b \cdot y \cdot \cos \Psi}{K_v} = \frac{105 \times \pi \times 8 \times 57 \times 0.1 \times \cos(17.1)}{1.51} = 9521.18 \text{ N}$$

The torque, $T = W_G^T \cdot R_G = 9521.18 \times \dfrac{360}{2} = 1713812.97$ Nmm

The power, $NP = \dfrac{TN}{955 \times 10^4} = \dfrac{1713812.97 \times (1250/15)}{955 \times 10^4} = 14.9547$ kW

Step-3: Power by AGMA rating equations

(i) For wear strength,

$$NP = N_g \cdot K_1 \cdot C_v Q \qquad \text{(Eqn. 17.23 MDHB)}$$

where, $N_g = \dfrac{1250}{15} = 83.33; K_1 = 0.727$ (Table 17.2 MDHB)

$$Q = G/(G+2.5) = 15/(15+2.5) = 0.857$$

The pitch line velocity of worm,

$$V_w = \frac{\pi D_w N_w}{60000} = \frac{\pi \times 78 \times 1250}{60000} = 5.105 \text{ m/s}$$

Velocity factor,

$$C_v = (2.3)/\left[2.3 + V_w + \frac{3V_w}{G}\right] = (2.3)/\left[2.3 + 5.105 + \frac{3 \times 5.105}{15}\right] = 0.273$$

Substituting the values,

$$NP = 83.33 \times 0.727 \times 0.273 \times 0.873 = 14.44 \text{ kW}$$

(ii) For heat dissipation,

$$NP = \frac{0.029C^{1.7}}{(G+5)} = \frac{0.029 \times (219)^{1.7}}{(15+5)} = 13.81 \text{ kW} \qquad \text{(Eqn. 17.24 MDHB)}$$

Step-4: The feasible power capacity

The least power capacity is the feasible power capacity.

The power capacity, $NP = 13.81$ kW

Example 18.13: Design a worm gear drive for a speed of 500 rpm of the worm transmitting 20 kW. The velocity ratio is 25:1. The material for gear is phosphorus bronze and worm is hardened steel. Determine the efficiency of the drive also.

Given Data: Power, $NP = 20$ kW, Speed of worm $N_w = 500$ rpm

The velocity ratio, $G = 25:1$,

Material for gear $\rightarrow$ P bronze, for worm hardened steel

SOLUTION

Step-1: Geometric relations

Number of threads, $Z_w = 2$

Number of teeth on gear, $Z_G = Z_w \times G = 2 \times 25 = 50$

The lead angle, $\Psi = \tan^{-1}(\sqrt[3]{Z_w/Z_G})$ (Eqn. 17.5 MDHB)

$$= \tan^{-1}(\sqrt[3]{1/25}) = 18.88°$$

Choose, $\Psi = 19°$

The lead, $L = \pi m \cdot Z_w = 2\pi m = 6.28\text{ m}$

The worm diameter, $D_w = \dfrac{L}{\pi \tan \Psi} = \dfrac{2\pi m}{\pi \tan 19°} = 5.81\text{ m}$ (Eqn. 17.5 MDHB)

Face width, $b = 0.73 D_w = 0.73 \times 5.81\text{ m} = 4.24\text{ m}$ (Eqn. 17.3 MDHB)

The gear diameter, $D_G = m \cdot Z_G = 50\text{ m}$

Step-2: Module by beam strength equation

Speed of gear $N_G = 500/25 = 20$ rpm
The torque on the gear,

$$T = \frac{955 \times 10^4 \times 20}{N} = \frac{955 \times 10^4 \times 20}{20} = 955 \times 10^4 \text{ Nmm}$$

The tangential force on gear,

$$W_G^T = \frac{T}{R_G} = \frac{955 \times 10^4}{25\text{ m}} = \frac{38.2 \times 10^4}{m}\text{ N}$$

Velocity factor, $K_v = 1.5$ (assumed)

Tooth form factor, $y = 0.1$ (Table 17.1 MDHB),

The strength, $S = 105$ (Table 17.4 MDHB)

Then, substituting in beam strength Eqn. 17.16 MDHB

$$105 = \frac{(38.2 \times 10^4/\text{m}) \times 1.5}{\pi \times m \times 4.24\text{ m} \times 0.1 \times \cos 19°}$$

$$\therefore \quad m^3 = \frac{38.2 \times 10^4 \times 1.5}{\pi \times 105 \times 4.24 \times 0.1 \times \cos 19°} = 4332.9$$

$$\therefore \quad m = \sqrt[3]{4332.9} = 16.3$$

Choose $m = 16$

Step-3: Check for the stress induced

The tangential load, $W_G^T = \dfrac{38.2 \times 10^4}{16} = 23875\text{ N}$

The gear diameter, $D_G = 50 \times 16 = 800$ mm

The pitch line velocity, $V_g = \dfrac{\pi \times 800 \times 20}{60000} = 0.84$ m/s

The velocity factor, $K_v = \dfrac{3.1 + V_g}{3.1} = \dfrac{3.1 + 0.84}{3.1} = 1.27$

Face width, $b = 4.24\text{ m} = 4.24\text{ m} = 4.24 \times 16 = 67.84$

Choose $b = 68$ mm

Worm diameter, $D_w = 5.81 \times m = 5.81 \times 16 = 92.96 = 93.0$ mm

Lead $L = \pi m Z_w = \pi \times 16 \times 2 = 100.53$ mm

Then the bending stress induced,

$$S = \frac{W_G^T \cdot K_v}{\pi m b y \cdot as\, y} \qquad \text{(Eqn. 17.16 MDHB)}$$

$$= \frac{23875 \times 1.27}{\pi \times 16 \times 68 \times 0.1 \times \cos(19°)} = 93.82\text{ N/mm}^2 < S_d = 105\text{ MPa}$$

Hence, the design is safe.

Step-4: AGMA power rating for strength

$$NP = N_g K_1 C_v Q \qquad \text{(Eqn. 17.25 MDHB)}$$

where, $N_g = 20$ rpm

$$Q = G/(G + 2.5) = 25/(25 + 2.5) = 0.91$$

The worm velocity, $V_w = \dfrac{\pi D_w \cdot N}{60000} = \dfrac{\pi \times 93 \times 500}{60000} = 2.435$ m/sec

Velocity factor, $C_v = \dfrac{2.3}{2.3 + V_w + \dfrac{3V_w}{G}} = \dfrac{2.3}{2.3 + 2.435 + \dfrac{3 \times 2.435}{25}} = 0.4575$

The centre distance, $C = \dfrac{D_w + D_G}{2} = \dfrac{93 + 800}{2} = 446.5$ mm

$\therefore \quad K_1 = 4.64$ (Table 17.2 MDHB)

$\therefore \quad NP = 20 \times 4.64 \times 0.4575 \times 0.91 = 38.63$ kW

Step-5: AGMA power rating for heat dissipation

$$NP = \frac{0.029 C^{1.7}}{G + 5} \qquad \text{(Eqn. 17.24 MDHB)}$$

$$= \frac{0.029 \times (446.5)^{1.7}}{(25 + 5)} = 30.9\text{ kW} > \text{ (Given power = 20 kW)}$$

This means that the heat dissipation capacity is more than the heat generated. Hence design is safe.

Example 18.14: Design a worm gear drive to transmit 12 kW at 1200 rpm. Speed reduction ratio is 30:1 The worm is made of hardened steel with $S_d = 210$ MPa and gear of phosphor bronze with $S_d = 90$ MPa. The teeth are 14.5° involute. Check the heating capacity of the gears.

Given Data: Power $NP = 12$ kW at $n_1 = 1200$ rpm; Reduction ratio $= 30$;
Material for worm – steel – $S_d = 210$ MPa
Material for gear – P – bronze – $S_d = 90$ MPa

SOLUTION

Step-1: Number of teeth on gear

Number of threads on worm, $Z_1 = 1$ Single start

Number of teeth on gear, $Z_2 = G \cdot Z_1 = 30 \times 1 = 30$

Step-2: The helix angle,

$$\Psi = \tan^{-1}(\sqrt[3]{Z_w/Z_G}) = \tan^{-1}(\sqrt[3]{1/30}) = 17.84°$$

Select $\Psi = 16°$ for $\phi = 14.5°$ (Table 17.1 MDHB)

Step-3: The tangential load on the gear tooth

Torque on gear, $T = \dfrac{955 \times 10^4 \times kW}{N} C_s$

where, $kW = 12$ and $N = 1200/30 = 40$

The service factor, $C_s = 1.5$ (Table 17.5)

Substituting the values, $T = \dfrac{955 \times 10^4 \times 12 \times 1.5}{40} = 4297500$ Nmm

The gear diameter, $D_G = m \cdot Z_G = 30$ m

$\therefore$ The tangential load, $w_t = \dfrac{T}{r} = \dfrac{4297500}{15\text{ m}} = \dfrac{286500\text{ N}}{m}$

Step-4: Module by Lewis equation

$$S = \frac{W_G^T \cdot K_v}{\pi m \cdot b \cdot y \cdot \cos\Psi} \quad \text{(Eqn. 17.16)}$$

i.e.

$$W_t = \frac{\pi \cdot m \cdot s \cdot b \cdot y \cdot \cos\Psi}{K_v}$$

where, $S = S_d = 90$ MPa, $b = 10$ m, $y = 0.10$ (Table 17.1), and $\Psi = 16°$ and $K_v = 1.2$ (assumed)

Substituting the value,

$$\frac{286500}{m} = \frac{\pi \times m \times 90 \times 10\ \text{m} \times 0.10 \times \cos 16°}{1.2}$$

$$m = \sqrt[3]{\frac{286500 \times 1.2}{90 \times \pi \times 10 \times 0.1 \times \cos 16°}} = \sqrt[3]{1264.96} = 10.81$$

Choose $m = 12$ (Table 16.2 MDHB)

Diameter of gear, $D_G = m \cdot Z_2 = 12 \times 30 = 360$ mm

The pitch line velocity, $V_g = \dfrac{\pi DN}{60000} = \dfrac{\pi \times 360 \times 40}{60000} = 0.75$ m/s

The velocity factor, $K_v = \dfrac{3.1 + V_g}{3.1} = \dfrac{3.1 + 0.75}{3.1} = 1.2419$

Then module with the actual value of velocity factor,

$$m = \sqrt[3]{1264.96 \times \frac{1.2419}{1.2}} = 10.93$$

Therefore, $m = 12$ mm is OK.

Step-5: Dimensions

Gear wheel - diameter $D_G = 12 \times 30 = 360$ mm

Face $b = 10\ \text{m} = 10 \times 12 = 120$ mm

or from Table 17.3 and 17.3a MDHB

The face width, $b = 1.28P_n + 6.35 = 96$ mm

The worm diameter, $D_w = 2.4P_n + 28 = 2.4 \times \pi \times 12 + 28 = 118.48 = 120.00$

The face width, $b = 96$ mm is acceptable because it should be less than D_w

Whole depth of tooth, $h_w = 0.686P_n = 0.686 \times \pi \times 12 = 25.86$ mm

The lead, $L = \pi m \cdot Z_w = \pi \times 12 \times 1 = 37.7$ mm

The modified helix angle, $\Psi = \tan^{-1}\left(\dfrac{L}{\pi D_w}\right) = \tan^{-1}\left(\dfrac{37.7}{\pi \times 120}\right) = 5.7°$

Step-6: Efficiency of the drive

Sliding velocity of the drive, (Eqn. 17.15 MDHB)

$$V_s = \frac{\pi N_w D_w}{60000 \cos \Psi} = \frac{\pi \times 1200 \times 120}{60000 \times \cos(5.7)} = 7.5 \text{ m/s}$$

The coefficient of friction, (Eqn. 17.14a MDHB)

$$f = 0.025 + V_s/300 = 0.025 + 75/300 = 0.05$$

The efficiency of the drive, (Eqn. 17.13 MDHB)

$$\eta = \frac{\cos\phi_n - f\tan\Psi}{\cos\phi_n + f\cot\Psi} = \frac{\cos 14.5 - 0.05\tan 5.7}{\cos 14.5 + 0.05\cot 5.7} = 0.6845 = 68.45\%$$

Step-7: Heat balance

Heat generated, $Q_g = 1000(1-\eta)\,\text{kW}$ (Eqn. 17.21 MDHB)

$$= 1000 \times (1 - 0.6845) \times 12 = 3786.0 \text{ J/s}$$

Heat dissipated, $Q_d = 0.035(A_g + A_w)(t_g - t_r)$ (Eqn. 17.22 MDHB)

where, $A_w = FL \cdot D_w$ and $A_g = \frac{\pi}{4} D_G^2$

$$A_w = \left(4.5 + \frac{Z_G}{50}\right) p_x \cdot D_w = \left(4.5 + \frac{30}{50}\right) \times \pi \times 12 \times 120 = 23071.87 \text{ mm}^2$$

$$A_g = \frac{\pi}{4} \times 360^2 = 101787.6 \text{ mm}^2$$

$$t_g - t_f = 60° - 30° = 30°\text{C}$$

$\therefore$ $$Q_d = 0.035 \times (23071.87 + 101787.6) \times 30 = 131102.5 \text{ J/s} > Q_g$$

The heat dissipating capacity is more than the heat generated. Hence, the design is safe.

Example 18.15: A hardened steel worm at 1250 rpm transmits power to phosphor bronze gear with a transmission ratio of 20. The centre distance is 200 mm. Determine the input power capacity. Assume $\propto = 14.5°$ FDI.

Given Data: Speed of worm, $N_w = 1250$ rpm

Transmission ratio, $G = 20$

Centre distance, $C = 200$ mm

Material for gear- $P-$ bronze for worm – steel

Tooth form, $\Phi = \propto = 14.5°$ FDI

SOLUTION

Step-1: Module and dimensions by AGMA specifications

From Table 17.1 MDHB, for $\Phi = 14.5°$; tooth form factor $y = 0.100$;
Lead angle, $\Psi = 0 - 16°$ and $Z_G = 40$

Number of threads on worm, $Z_w = Z_G/G = \frac{40}{20} = 2$

The diameter of worm by AGMA, (Eqn. 17.2 MDHB)

$$D_w = 0.68C^{0.875} = 0.68(200)^{0.875} = 70.13 \text{ mm}$$

Module by Eqn. 17.2 MDHB, $D_w = 3p_c = 3\times\pi\times m$

$$\therefore \quad m = \frac{D_w}{3\pi} = \frac{70.13}{3\times\pi} = 7.44$$

Choose $m = 8$ mm (Table 16.2 MDHB)

Then modified worm diameter

$$D_w = 3\pi m = 3\times\pi\times 8 = 75.4 \text{ mm}$$

Lead, $L = \pi m\cdot Z_w = \pi\times 8\times 2 = 50.27$

The helix angle,

$$\Psi = \tan^{-1}\left(\frac{L}{\pi D_w}\right) = \tan^{-1}\left(\frac{50.27}{\pi\times 75.4}\right) = 11.98 \simeq 12°$$

The gear diameter, $D_G = m\cdot Z_G = 8\times 40 = 320$ mm

The centre distance, $C = \dfrac{D_w + D_G}{2} = \dfrac{75.4+320}{2} = 197.7$ mm

Very near to given value, hence O.K.

The face width of gear, $b = 0.73D_w = 0.73\times 75.4 = 55.042$

Choose, $b = 55.0$ mm

Step-2: The power output by beam strength consideration

The gear speed, $N_g = 1250/20 = 62.5$ rpm

The pitch line velocity, $V_g = \dfrac{\pi D_G\cdot N_g}{60000} = \dfrac{\pi\times 320\times 62.5}{60000} = 1.047$ m/s

The velocity factor, $K_v = \dfrac{3.1+V_g}{3.1} = \dfrac{3.1+1.047}{3.1} = 1.34$

The allowable stress (Table 17.4 MDHB), $S_d = 105$ MPa

The permissible tangential load from Eqn. 17.16 MDHB

$$W_G^T = \frac{S\cdot\pi\cdot m\cdot b\cdot y\cdot\cos\Psi}{K_v} = \frac{105\times\pi\times 8\times 55\times 0.100\times\cos 12°}{1.34} = 10594.77 \text{ N}$$

The permissible torque,

$$T_G = W_G^T\times R_G = 10594.77\times 160 = 1695163.2 \text{ Nmm}$$

The power output,

$$NP_o = \frac{TN}{955 \times 10^4} = \frac{1695163.2 \times 62.5}{955 \times 10^4} = 11.094 \text{ kW}$$

Step-3: Efficiency of the drive

The worm pitch line velocity,

$$V_w = \frac{\pi D_w N_w}{60000} = \frac{\pi \times 75.4 \times 1250}{60000} = 4.935 \text{ m/s velocity}$$

The sliding velocity, (Eqn. 17.15 MDHB)

$$V_s = \frac{V_w}{\cos \Psi} = \frac{4.935}{\cos 12^\circ} = 5.045 \text{ m/s}$$

The coefficient of friction, (Eqn. 17.14a MDHB)

$$f = 0.025 + \frac{V_{st}}{300} = 0.025 + \frac{5.045}{300} = 0.0418$$

The efficiency of the drive, (Eqn. 17.13 MDHB)

$$\eta = \frac{\cos \varphi_n - f \tan \Psi}{\cos \Phi_n + f \cot \Psi} = \frac{\cos 14.5 - 0.0418 \tan 12^\circ}{\cos 14.5 + 0.0418 \cot 12^\circ} = 0.8235 = 82.35\%$$

Step-4: The permissible input power,

$$NP_i = \frac{NP_o}{\eta} = \frac{11.094}{0.8235} = 13.47 \text{ kW}$$

Step-5: Power by AGMA

Equation for wear strength,

$$NP_i = N_g, k_1, C_v, Q \quad \text{(Eqn. 17.23 MDHB)}$$

where, $N_g = 62.5 \text{ rpm};\; Q = G/(G+2.5) = \frac{20}{20+2.5} = 0.89$

$$C_v = \frac{2.3}{2.3 + V_w + \left(\frac{3V_w}{G}\right)} = \frac{2.3}{2.3 + 4.935 + \left(\frac{3 \times 4.935}{20}\right)} = 0.2884$$

$K_1 = 0.485$ (Table 17.2 MDHB)

∴ Substituting the values,

$$NP_i = 62.5 \times 0.485 \times 0.2884 \times 0.89 = 7.78 \text{ kW}$$

Step-6: AGMA equation for heat dissipation,

$$NP_i = \frac{0.029 \cdot C^{1.7}}{(G+5)} = \frac{0.029 \times (200)^{1.7}}{(20+5)} = 9.47 \text{ kW}$$

Step-7: Permissible input power

The least of all the three values is the input power capacity of the gear set

∴ Input power capacity, $NP = 7.78$ kW

EXERCISES FOR PRACTICE

18.1 An elevator cage is lifted at the rate of 90 m/min. The elevator drum is 60 cm in diameter and the load lifted is 18 kN. A worm gear is keyed to the drum shaft. The speed of driving motor is 1000 rpm. Determine (i) the worm and worm gear proportions, and (ii) the required power of the motor.

18.2 Design a worm reducer to transmit 7.5 kW at 100 rpm. The input speed from the motor is 1800 rpm. Calculate the efficiency of the drive.

18.3 A worm gear drive is to be designed for transmitting 2 kW power at a speed reduction ratio of 27:1. Worm rotates at 1750 rpm. Tooth profile is to be 14.5° involute. Select standard materials for the worm and heel. Centre distance between the shaft must be around 100 mm. Check the design for wear.

18.4 Design a worm and worm gear drive for a speed reduction ratio of 1:5, worm rotates at 1500 rpm transmitting 25 kW. Select suitable materials for worm and worm wheel. Check heating effect.

18.5 Design a worm reducer to transmit 10 kW at 60 rpm. The input speed from the motor is 1440 rpm. Calculate the efficiency of the drive.

18.6 A worm gear drive is to be designed for transmitting 5 kW power at a speed reduction ratio of 24:1. Worm rotates at 1800 rpm. Tooth profile is to be 14.5° involute. Select standard materials for the worm and worm-heel. Centre distance between the shafts must be around 150 mm. Check the design for wear and heat dissipation.

18.7 An elevator cage is lifted at the rate of 90 m/min. The elevator drum dia. is 0.6 m and the load to be lifted is 18 kN. A worm gear is keyed to the drum shaft. Assuming speed of the driving motor is 900 r.p.m., determine:

(i) The worm and gear proportions

(ii) The efficiency of the drive and

(iii) The required kW of the motor if the efficiency of the hoist itself is 90%.

18.8 Design a worm gear speed reducer unit to transmit 10 kW at a speed of 1440 rpm of the worm. The desired velocity ratio is 24:1. Worm is to be made of hardened steel and the gear is to be phosphor bronze. An efficiency of at least 80 per cent is desired. Check the designed drive for efficiency and wear load.

18.9 A hardened steel worm rotating at 1250 rpm transmits power to a phosphor bronze gear with a transmission ratio of 15 : 1. The centre distance is close to 225 mm. Determine the remaining design and give estimated power input ratings from the standpoint of strength, endurance and heat dissipation, the teeth are of 14° full depth involute form.

18.10 A hardened steel worm of speed 1250 rpm has to transmit 20 kW to a phosphor bronze gear with a transmission ratio of 15 : 1. The centre distance is to be about 250 mm. Design the worm gear drive. What is its efficiency?

18.11 Design a worm drive for a speed reducer to transmit 25 kW at worm speed of 600 rpm. The desired velocity ratio is 25:1. The efficiency of the drive is to be at least 85%. Assume that the worm is made of hardened steel and select the material for the gear.

18.12 A worm and worm gear speed reducer is to be designed to transmit 2 kW power at 1450 rpm and with a velocity ratio 9:1. The worm is to be made of hardened steel (BHN 250) having a static strength of 120 N/mm^2 and the worm gear is made of phosphor bronze (BHN 100) having a static strength of 55 N/mm^2.

(a) Derive an expression for velocity ratio in worm and worm gear.

18.13 Design a worm gear drive for a speed of 500 rpm of the worm transmitting 20 kW. The velocity ratio is 25:1. The material of the gear is phosphorus bronze and the worm is hardened steel. Determine the efficiency of the drive also.

18.14 A hardened steel worm rotating at 1250 rpm transmits power to a phosphor bronze gear with transmission ratio of 15:1. The centre distance of drive is 225 mm. Design the gear drive, and estimate the power rating of the drive from stand point of strength and heat dissipation. The teeth are 14½° full depth involute form.

SHORT ANSWER QUESTIONS

18.1 What are worm gears?

18.2 Where are they used?

18.3 What is tooth form on gear?

18.4 What is the form of the thread on the worm?

18.5 What is gear ratio?

18.6 What is relation between the gear ratio and number of starts on the worm?

18.7 What is the axial pitch of worm thread?

18.8 What is the relation between the axial pitch of thread and the circular pitch of gear?

18.9 What is the lead of the thread?

18.10 What is the relation between the lead of thread and module of the gear?

18.11 What is recommended value of lead angle?

18.12 What is equation for the centre distance?

18.13 What is the AGMA recommendation for worm diameter?

18.14 What is the AGMA recommendation for face width of gear?

18.15 What is the recommendation for the circular pitch?

18.16 What is the material for the worm?

18.17 What is the material for the gear?

18.18 Why the two materials should not be the same?

18.19 What are the loads on the worm?

18.20 What are the loads on the gear wheel?

18.21 What is the frictional force between the threads?

18.22 What is the recommended coefficient of friction?

18.23 What is the bending stress in gear tooth?
18.24 What is the wear strength equation?
18.25 What is the heat generated in worm gear set?
18.26 Give the heat dissipating capacity of the worm gear set.
18.27 What is the AGMA power rating equation for wear strength?
18.28 What is the AGMA power rating equation for heat dissipation?

CHAPTER 19

CLUTCHES AND BRAKES

19.1 INTRODUCTION

Theoretically, the clutches and brakes are essentially the same devices. Each provides a frictional, mechanical, hydraulic, or magnetic connection between two elements of a machine. If both connected elements (shafts) rotate, it is called a clutch. A clutch provides an interruptible connection between two rotating shafts as in automobiles. It is used to connect the transmission shaft to the engine crank shaft intermittently. The disconnection is done at the will of the driver or automatically as per the speed of the engine. The clutch is used to transmit power.

If one element rotates and the other is fixed, it is called a brake. A brake provides an interruptible connection between the rotating element and a non-rotating (fixed) ground plane. An example is the wheel of an automobile and its chasis.

19.2 CLASSIFICATION

It is based on the following factors.

19.2.1 Based on the Mode of Actuation

(i) Mechanical (ii) Pneumatic or hydraulic
(iii) Electrical (iv) Automatic

(i) In mechanical actuation, the contact between the elements is controlled by pedal and lever mechanism. This is a widely used device.

(ii) In pneumatic or hydraulic systems, the lever mechanism is replaced by pneumatic or hydraulic cylinders and control valves with connecting pipes or tubes. The operation is fast and smooth.

(iii) Electrical or magnetic system operates on electromagnetic principle. The operation is fast and smooth. The control is easy.

(iv) The automatic system operates on the speed difference principle. It is simple and is used in small machines like two wheelers, e.g., the centrifugal clutch.

19.2.2 Based on the Operating Principle

(i) **Positive contact clutch:** The character of engagement is the mechanical interference between the jaws or teeth on the contacting surfaces. These devices are not useful as brakes, because they cannot dissipate large quantities of heat. At the most, they can be used as holding devices. They are used as clutch at low speed applications (about 60 rpm for jaw clutches and 300 rpm for toothed clutches). Their advantage is positive engagement and large torque capacity. They are sometimes used in combination with friction clutch as in synchromesh clutch. The friction clutch brings the two elements to nearly the same speed before the jaws or teeth engage.

(ii) **Friction clutches and brakes:** In these devices, two or more surfaces are pressed together with a normal force to create frictional coupling between the surfaces. The friction surfaces may be flat and perpendicular to the axis of rotation. Examples are disk clutches or brakes. They may be cylindrical with radial normal force. The examples are drum brakes or clutches. This arrangement is more common in brakes. The contacting surfaces may be conical with axial actuating force. Examples are conical clutches and brakes.

In these devices one of the surfaces is metal mostly cast iron or steel, and the other is a high friction material, usually referred as lining, fixed on metal surface. In multi-disk clutches, many pairs of surfaces are provided to achieve higher torque capacities. The friction clutches may operate dry or wet. The wet operation is provided by oil bath. The oil bath reduces the coefficient of friction but increases heat dissipating capacity. The oil bath protects the friction surfaces from dust, dirt and water. For automatic transmission wet operation is preferred, while for manual operations, dry operation is adopted.

(iii) **Over running clutches:** They operate automatically based on the relative velocity of the two elements. They act on the circumference (cylindrical surface) and allow relative motion in one direction only. If rotation tends to reverse, the clutch mechanism grabs the shaft and locks up. The back stop clutches operating on this principle, are used in hoists to prevent the fall in the event of power failure. They are used in indexing mechanisms, where input shaft oscillates back and forth, while the output shaft turns intermittently, in only one direction. Another application is the free-wheel on rear hub of a bicycle. This allows the wheel motion when the drive sprocket is not operated for rest.

(iv) **Magnetic clutches and brakes:** There is no mechanical contact between the surfaces. The coupling is attained by the magnetic field setup between the surfaces. There are three types of magnetic clutches and brakes:

(a) Magnetic particle clutches and brakes
(b) Magnetic hysteresis clutches and brakes
(c) Eddy current clutches.

(a) **Magnetic particle type:** In these, the gap between the surfaces is filled with fine ferrous powder. When the magnetic coil is energized, the magnetic field orients the ferrous powder particles in the form of chains and couple the surfaces with no slip. The torque is controlled by the coil current. The device slips when the torque exceeds the limit set by the coil current.

(b) **Magnetic hysteresis type:** There is an air gap between the surfaces. The rotor called drag cup is dragged along the magnetic field setup by the field coil or permanent magnet. They exert a constant torque. These are used in winding machines. These devices are extremely smooth, quiet and durable.

(c) **Eddy current type:** In this type, the coil sets up eddy currents in the engaging elements and cause the magnetic coupling between the surfaces. There is always a slip between the rotor and the pole to generate the eddy currents supplying the coupling force. They are used in winding machine to maintain a constant tension in the wire or filament.

(v) **Fluid (coupling) clutches:** The coupling between the input and output shafts is a fluid, oil or water. The input shaft carries an impeller with blades and output shaft carries a turbine with similar blades. The two are covered by a casing containing the fluid. The impeller churns the fluid, which impinges on the turbine blades and rotates it. The output shaft is gradually accelerated to almost the full speed of the impeller. But some amount of slip remains. It absorbs the shock. With output shaft fixed, it absorbs the energy of the input shaft and converts it into heat. It acts as dynamometer or brake.

Now let us consider a few of the commonly used clutches and brakes.

19.3 CLUTCHES

19.3.1 Disk Clutch

The main parts of this clutch are a disk and pressure plate carried by the driving flange and a disk with friction lining attached to the driven flange. The driven shaft is splined and the driven hub can be moved axially on them. The spring mounted on a sleeve on the driven shaft exerts force on the pressure plate through levers and keeps the friction surfaces pressed together. The sleeve can be pulled out by a proper actuating mechanism and the friction surfaces can be disengaged.

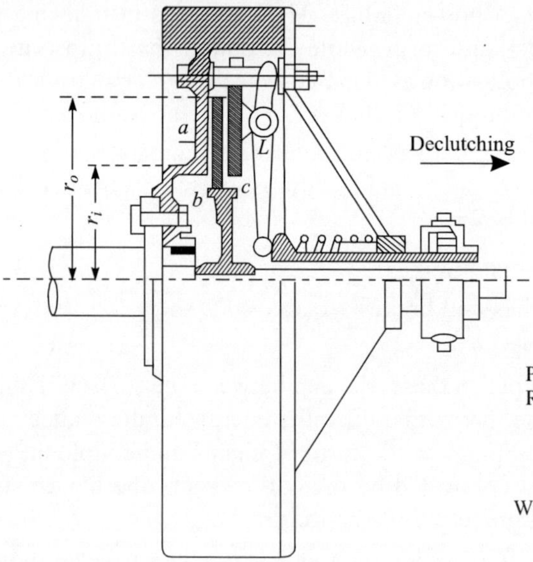

a = Driving plate
b = Driven plate
c = Pressure plate
L = Levers

Proportions of the disk–
Relation between radii

$$\frac{r_1}{r_0} = 0.6 \text{ to } 0.8$$

Width of the friction face
$b = 0.5\, r_m$ to $0.2\, r_m$

Figure 19.1

The torque transmitted, due to the friction between the contacting surfaces is evaluated as follows.

Consider a ring of width dr of the friction surface at radius r. Let p be the normal pressure between the surfaces.

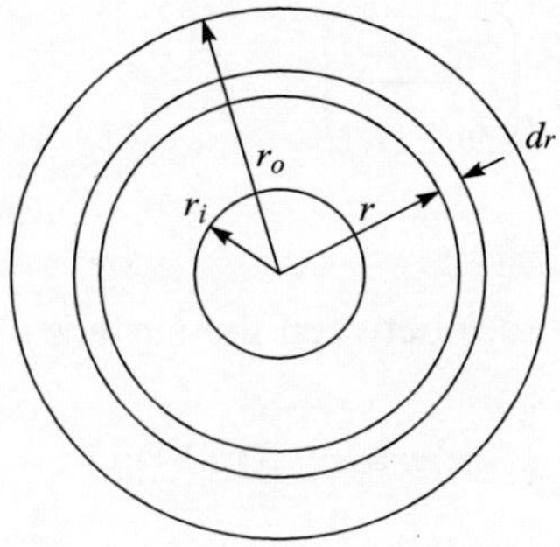

Figure 19.2

The normal force on the elemental ring is given by

$$df_n = 2\pi\ r \cdot dr \cdot p.$$

The torque due to the friction,

$$dT = 2\pi fp \cdot r^2 \cdot dr, \qquad \text{where } f \text{ is the coefficient of friction.}$$

The distribution of pressure across the surface may be uniform or variable. Accordingly, two assumptions are made:

(A) The pressure is uniform: p = constant = C

(B) The wear is uniform and variation of pressure follows the law

$$p \cdot r = \text{constant} = C$$

(A) **Uniform Pressure:** p = Const. = C

The total normal force between the surfaces,

$$F_n = \int_{r_i}^{r_0} 2\pi p \cdot r \cdot dr = 2\pi p \left[\frac{r^2}{2}\right]_{r_i}^{r_0} = 2\pi p \frac{(r_0^2 - r_i^2)}{2}$$

Then axial force to be exerted by the spring (or springs),

$$F_a = F_n = \pi (r_0^2 - r_i^2) p \tag{19.1}$$

The total torque transmitted by each pair of surfaces,

$$T = \int dT = \int_{r_i}^{r_0} 2\pi \cdot f \cdot p \cdot r^2 \cdot dr = 2\pi \cdot p \cdot f \left[\frac{r^3}{3}\right]_{r_i}^{r_0} = 2\pi \cdot p \cdot f \left(\frac{r_0^3 - r_i^3}{3}\right)$$

Substituting for p,

$$T = \left(\frac{2}{3}\right)\left[\frac{r_0^3 - r_i^3}{r_0^2 - r_i^2}\right] f \cdot F_a = f \cdot F_a \cdot r_m \tag{19.2}$$

where, $$r_m = \left(\frac{2}{3}\right)\left(\frac{r_0^3 - r_i^3}{r_0^2 - r_i^2}\right)$$

(B) **Uniform wear,** $p \cdot r$ = constant = C

Then the normal force (axial force) between the surfaces,

$$F_a = \int_{r_i}^{r_0} (2\pi rp)\, dr = 2\pi C \cdot (r_0 - r_i) \tag{19.3}$$

The torque transmitted by each pair of surfaces,

$$T = \int_{r_i}^{r_0} (2\pi rp) f \cdot r \cdot dr = 2\pi Cf\left(\frac{r_0^2 - r_i^2}{2}\right) = \left(\frac{F_a}{r_0 - r_i}\right) f \cdot \left(\frac{r_0^2 - r_i^2}{2}\right)$$

$$= f \cdot F_a\left(\frac{r_0 + r_i}{2}\right) = f \cdot F_a \cdot r_m \tag{19.4}$$

where, $$r_m = \left(\frac{r_0 + r_i}{2}\right)$$

For many pairs of surfaces in contact, the torque transmitted,

$$T = n \cdot f \cdot F_a \cdot r_m \tag{19.5}$$

where, $n = 2$ for single plate clutch with both sides effective

$= (n_1 + n_2 - 1) = 2n_2$, for a multidisc clutch

n_1 = number of driver plates = n_2 + 1

n_2 = number of driven plates

Then the power transmitted, $NP = \dfrac{N \cdot T}{955 \times 10^4 \times K_L}$

$$NP = \frac{n \cdot f \cdot F_a \cdot r_m \cdot N}{955 \times K_L} 10^{-4} = \frac{n \cdot f \cdot F_a \cdot r_m \cdot N}{K_L} \times 1.05 \times 10^{-7} \tag{19.6}$$

where, r_m = Mean friction radius in mm

K_L = Load factor (Table 19.1)

Table 19.1. Load factors for various machines (couplings) K_L

Driver	Driven machinery	Factor K_L
Steam turbine	Electric generator, steady load, turbine blower	1.00
	Electric generator, uneven load, centrifugal pump	1.25
	Induced draft fan, line shaft, gear drive	1.5
	Rolling mill, gear drive	2.00
Electric motor	Turbine blower, metal working machinery	1.25
	Centrifugal pump, wood working machinery	1.50
	Line shaft, ship propeller, double acting pump	1.75
	Triplex single acting pump, elevator, crane	1.75
	Compressor	2.50
Steam engine	Values of electric drive multiplied by 1.2 to 1.5	
Gas and oil engines	Values of electric drive multiplied by 1.3 to 1.6, the factor depending on the coefficient of steadiness of the flywheel	

Table 19.2. Service factor for clutches K_s

Type of service	K_s	Type of service	K_s
Driving machine		**Pumps**	
Electric motor steady load	1.00	Centrifugal	1.0
Fluctuating load	1.5	Single acting	2.0
Gas engine single cylinder	1.5	Double acting	1.5
Multiple cylinders	1.0		
Diesel engine		Line shaft	1.5
High speed	1.5	Wood working machinery	1.75
Large low speed	2.0		
Driven machine		Hoists elevators	2.0
Generator steady load	1.0	Cranes, shovels	2.0
Fluctuating load	1.5	Hammer mills	2.0
Blower	1.0	Ball mills, crushers	2.0
Compressor	2 to 2.5		
		Brick machinery	3.0
		Rock crushers	3.0

Table 19.3. Design data for clutch facings

Facing material		Coefficient of friction f	Unit pressure (p) MPa
Cast iron on cast iron	Dry Oily	0.210 0.07	0.275 – 0.412 –
Cast iron on steel	Dry Oily	0.30 0.10	0.275 – 0.412 –
Leather on cast iron	Dry Oily	0.5 0.15	0.0687 – 0.10434 –
Asbestos fabric on metal	Dry Oily	0.35 – 0.46 0.25	0.626 – 1.206 0.206 – 0.412
Moulded asbestos	–	0.25 – 0.35	0.206 – 0.412
Wood on cast iron	–	0.30	0.171 – 0.343

Table 19.4. Friction materials for clutches

Materials		Friction Coefficient		Max. temperature °C	Max. pressure MPa	Relative cost	Comment
Wearing	Opposing	Wet	Dry				
Cast bronze	Cast iron or steel	0.05	–	150	0.56-0.85	Low	Subject to seizure.
Cast iron	Cast iron	0.05	0.15-0.2	315	1.02-1.2	Very low	Good at low speeds.
Cast iron	Steel	0.06	–	260	0.85-1.4	Very low	Fair at low speeds
Hard steel	Hard steel	0.05	–	260	0.7	Moderate	subject to falling.
Hard steel	Hard steel chromium plated	0.03	–	260	1.4	High	Durable combination.
Hard-drawn phosphor bronze	Hard steel, chromium plated	0.03	–	260	1.0	High	Good wearing qualities.
Powder metal:	Cast iron or steel	0.05-0.1	0.1-0.4	540	1.0	High	Good wearing qualities.
Powder metal:	Hard steel chromium Plated	0.05-0.11	0.1-0.3	540	2.1	Very high	High energy absorption.
Wood	Cast iron or steel	0.16	0.2-0.35	150	0.4 to 0.6	Lowest	Unsuitable at high speed.
Leather	Cast iron or steel	0.12-0.15	0.3-0.5	90	0.07 to 0.3	Very low	Subject to glazing.
Cork	Cast iron or steel	0.15-0.25	0.3-0.5	90	0.06 to 0.1	Very low	Cork insert type. preferred
Felt	Cast iron or steel	0.18	0.22	90	0.035 to 0.07	Low	Resinent engagement

Contd.

Vulcanized fibre or paper	Cast iron or steel	–	0.3-0.5	90	0.070 to 0.28	Very low	Low speed, light duty
Woven asbestos	Cast iron or steel	0.1-0.2	0.13-0.6	260	0.070 to 1.4	Low	This rating is for short infrequent engagements.
Woven asbestos	Hard steel chromium plated	0.1	–	–	1.4	Very high	Used in Napier sabre engine.
Moulded asbestos	Cast iron or steel	0.8-0.12	0.2-0.5	260	0.35 to 1.0	Very low	Wide field of applications
Impregnated asbestos	Cast iron or steel	0.12	0.32	260-400	1.0	Moderate	For demanding applications
Carbon Graphite	Steel	0.5-0.1	0.25	370-540	2.1	High	For critical requirements
Moulded Phanotic plastic, Macerated cloth base	Cast iron or steel	0.1-0.15	0.25	150	0.7	Low	For light special service

SOLVED EXAMPLES

Example 19.1: A disk clutch consists of two steel disks in contact with one asbestos fabric faced disk having an outside diameter of 250 mm and an inside diameter of 200 mm. Determine the power that can be transmitted at 1000 rpm, if the coefficient of friction is 0.35 and the disks are pressed together by an axial force of 8000 N.

Given Data: $f = 0.35;\ r_0 = 250/2 = 125;\ r_i = 200/2 = 100$ mm;
Speed $N = 1000$ rpm; $F_a = 8000$ N; Friction pairs $n = 2$

SOLUTION

Step-1: Power transmitted for uniform pressure distribution

When uniform pressure distribution holds, frictional mean radius,

$$r_m = \frac{2}{3}\left[\frac{r_0^3 - r_i^3}{r_0^2 - r_i^2}\right] = \frac{2}{3}\left[\frac{125^3 - 100^3}{125^2 - 100^2}\right] = \frac{2}{3}\left[\frac{953125}{5625}\right] = 112.96 \text{ mm}$$

$$\text{Power, } NP = \frac{nf \cdot F_a \cdot r_m \cdot N}{955 \times 10^4 K_L} \text{ kW} = \frac{2 \times 0.35 \times 8000 \times 112.96 \times 1000}{955 \times 10^4 \times 1.0} = 66.24 \text{ kW}$$

Step-2: Power transmitted for uniform wear condition

When uniform wear holds, the frictional mean radius,

$$r_m = \frac{r_0 + r_i}{2} = \frac{125 + 100}{2} = 112.5 \text{ mm}$$

$$\text{Power, } NP = \frac{2\times 0.35\times 8000\times 112.5\times 1000}{955\times 10^4\times 1.0}\ \text{kW} = 65.97 \simeq 66.0\ \text{kW}$$

The two values are almost equal.

Example 19.2: A multiple plate clutch is used to transmit 15 kW at 1000 rpm. The inner and outer diameters of the contact surfaces are 90 mm and 150 mm respectively. The coefficient of friction is 0.1 and the pressure intensity is 0.35 MPa. Find the total number of steel and bronze plates.

Given Data: Power, NP = 15 kW, at speed = 1000 rpm,

Outer radius $r_0 = 150/2 = 75.0$ mm, Inner radius $r_i = 90/2 = 45.0$ mm,
Pressure p = 0.35 MPa; Coefficient of friction = 0.1

SOLUTION

Step-1: Assuming uniform pressure distribution, frictional mean radius,

$$r_m = \left(\frac{2}{3}\right)\left[\frac{r_0^3 - r_i^3}{r_0^2 - r_i^2}\right] = \left(\frac{2}{3}\right)\left[\frac{75^3 - 45^3}{75^2 - 45^2}\right] = \left(\frac{2}{3}\right)\left[\frac{330750}{3600}\right] = 61.25\ \text{mm}$$

Step-2: The axial force,

$$F_a = \pi p(r_0^2 - r_i^2) = \pi \times 0.35 \times (75^2 - 45^2) = 3958.4\ \text{N}$$

Step-3: The number of contact pairs from power transmitted,

$$NP = \frac{n\cdot f\cdot F_a\cdot r_m\cdot N}{955\times 10^4\times K_L} \qquad \text{(Eqn. 6.26 MDHB)}$$

$$\text{Contact pairs, } n = \frac{15\times 955\times 10^4\times 1.0}{0.1\times 3958.4\times 61.25\times 1000} = 5.9 \cong 6.0\ (\text{with, } K_L = 1.0)$$

Total plates are $n_1 + n_2 = n + 1 = 7$
Steel plates $n_1 = 4$ and Bronze plates $n_2 = 3$

Example 19.3: A single plate clutch is to transmit 100 kW at 1200 rpm. The ratio of face width to the mean radius is 1/5. The coefficient of friction is 0.25 and mean pressure $p_m = 0.2$ MPa. Both sides are effective. Find the dimensions of the friction surfaces. Also design the spring.

Given Data:	
Power = 100 kW	$b/r_m = 1/5$
Speed = 1200 rpm	$p_m = 0.2$ MPa
$f = 0.25$	$n = 2.0$

SOLUTION

Step-1: The mean radius by geometrical relation

$$b/r_m = \frac{r_0 - r_i}{\frac{r_0 + r_i}{2}} = 1/5$$

i.e.,
$$10(r_0 - r_i) = r_0 + r_i$$

i.e.,
$$r_0 = \frac{11}{9} r_i = 1.22 r_i$$

$$r_m = \frac{r_0 + r_i}{2} = \frac{\left(\frac{11}{9} r_i + r_i\right)}{2} = \frac{10}{9} r_i$$

Step-2: The axial force

$$F_a = \pi p_m (r_0^2 - r_i^2) \qquad \text{(Eqn. 6.21 MDHB)}$$

$$= \pi \times 0.2 \times (1.22^2 r_i^2 - r_i^2) = \pi \times 0.2 \times (0.4938) r_i^2 = 0.31026 r_i^2$$

Step-3: Dimensions of the friction plate

The inner radius by power equation with $K_L = 1.2$

$$\text{kW} = \frac{2 \times f \times F_a \times r_m \times N}{955 \times 10^4 \times K_L}$$

$$= \frac{2 \times 0.25 \times 0.31026 r_i^2 \times (10/9) r_i \times 1200}{955 \times 10^4 \times 1.2}$$

∴
$$r_i^3 = \frac{100 \times 955 \times 10^4 \times 1.2 \times 9}{2 \times 0.25 \times 0.31026 \times 10 \times 1200} = 5540.44 \times 10^3$$

∴
$$r_i = 176.4 \text{ mm} \cong 175.0 \text{ mm (adopted value)}$$

and $r_0 = \frac{11}{9} r_i = 213$ mm, adopt $r_0 = 215$ mm

Step-4: Force on the springs

The axial force, $F_a = \pi \times 0.2 \times (215^2 - 175^2) = 9801.77$ N

If there be six springs exerting this force, then force per spring,

$$F_1 = 9801.77/6 = 1633.6 \text{ N}$$

Assuming 5% increase for further compression of 2 mm, during disengagement, the maximum spring force,

$$F_{max} = 1633.6 \times 1.05 = 1715.3 \text{ N}$$

Step-5: The spring stiffness required,

$$K_s = \frac{F_{max} - F_{min}}{2} = \frac{1715.3 - 1633.6}{2} = \frac{81.7}{2} \cong 41 \text{ N/mm}$$

Step-6: Spring design for strength

Choose material for the spring as 50 CrV23 with yield stress

$$S_y = 1800 \text{ MPa, then } S_{sy} = 900 \text{ MPa}$$

Let $FS = 2.0$, then $S_{sd} = 900/2 = 450$ MPa

Maximum force, $F_{max} = 1715.3$ N

With the assumed value of spring index $C = 6.0$,

$$K = \frac{4C-1}{4C-4} + \frac{0.615}{C} = 1.2525 \qquad \text{(Eqn. 10.2 MDHB)}$$

The diameter of the spring for the strength, (Eqn. 10.1 MDHB)

$$S_{sd} = K\frac{8FD}{\pi d^3} = K\frac{8FC}{\pi d^2}$$

$$\therefore \quad 450 = \frac{8 \times 1715.3 \times 6 \times 1.2525}{\pi d^2}$$

$$\therefore \quad d^2 = \frac{8 \times 1715.3 \times 6 \times 1.2525}{\pi \times 450} = 72.95$$

$$\therefore \quad d = 8.54 \text{ mm}$$

Choose $d = 8.5$ mm and $D = 8.5 \times 6 = 51.0$ mm

Step-7: The number of coils from stiffness equation,

$$K_s = \frac{d^4 G}{8D^3 N} = \frac{d \cdot G}{8C^3 N} \qquad \text{(Eqn. 10.4 MDHB)}$$

$$\therefore \quad N = \frac{dG}{8C^3 K_s} = \frac{8.5 \times 81 \times 10^3}{8 \times 6^3 \times 41.0} = 9.7, \text{ choose } N = 10$$

For squared and ground ends, total number of coils = 10 + 2 = 12

Example 19.4: A power of 75 kW at 3000 rpm is to be transmitted by a multiplate friction clutch. The plates have friction surfaces of steel and phosphor bronze alternatively and run in oil. Design the clutch for 20% overload and sketch the arrangement of the plates.

Given Data: Power transmitted, *NP* = 75 kW at 3000 rpm with overload = 20%.

SOLUTION

Step-1: Design torque,

$$T_d = \frac{955 \times 10^4 \times kW \cdot Cs}{N} = \frac{955 \times 10^4 \times 75 \times 1.2}{3000} = 286500 \text{ Nmm}$$

Step-2: Selection of clutch data

Referring Table 6.9 MDHB for steel-bronze combination,
Friction coefficient, $f = 0.03$
Maximum pressure, $p = 1.0$ MPa

For proportions, $\frac{r_i}{r_0} = 0.6$, and $b = 0.5r_m$

Step-3: The axial force for uniform wear

$$F_a = 2\pi C(r_0 - r_i) \text{ and } C = p \cdot r \qquad \text{(Eqn. 6.24 MDHB)}$$

$$\therefore \quad F_a = 2\pi(p \cdot r)(r_0 - r_i) = 2\pi p_{\max} \cdot r_i (r_0 - r_i)$$

$$= 2\pi \times 1.0 \times 0.6 \cdot r_0 \times (r_0 - 0.6r_0) = 2\pi \times 0.24 r_0^2 = 1.51 r_0^2$$

Step-4: Mean radius,

$$r_m = \left(\frac{r_0 + r_i}{2}\right) = \frac{r_0 + 0.6r_0}{2} = 0.8r_0$$

Step-5: The dimensions of the friction plate for torque capacity

The torque transmitted per pair of contact surfaces,

$$T = f \cdot F_a \cdot r_m = 0.03 \times 1.51 r_0^2 \times 0.8 r_0 = 0.036 r_0^3$$

Assuming 6 pairs of contact surface, total torque transmitted,

$$T_d = n \times T$$

$$\therefore \quad 286500 = 6 \times 0.036 r_0^3$$

$$\therefore \quad r_0 = \sqrt[3]{\frac{286500}{6 \times 0.036}} = 109.87$$

Choose $r_0 = 110$ mm

Then $r_i = 0.6 \times r_0 = 0.6 \times 110 = 66$ mm
The face width $b = r_0 - r_i = 110 - 66 = 44$ mm

Step-6: Check for the average pressure

Axial force, $F_a = 1.57 \times r_0^2 = 1.57 \times (110)^2 = 13927$ N

The average pressure, $p = \dfrac{13927}{\pi(110^2 - 66^2)} = 0.5725$ MPa (acceptable)

This is less than the maximum specified and hence acceptable.

Example 19.5: Design a single plate clutch consisting of two pairs of contacting surfaces for a torque capacity of 200 Nm. Due to space limitations, the outside diameter of the clutch is to be 250 mm.

Given Data: Single plate clutch; $M_t = 200\,\text{Nm} = 2 \times 10^5$ Nmm;
$D_2 = 250$ mm; $n = 2$ (Since two pairs of contacting surfaces)

SOLUTION

Step-1: Selection of clutch data

Assume leather as the friction material. From Table 6.9 MDHB for leather

$$f = 0.3 \text{ to } 0.5 \text{ and } p = 0.07 \text{ to } 0.3 \text{ MPa}$$

$$\text{Select } f = 0.4 \text{ and } p = 0.15 \text{ MPa}$$

Step-2: Dimensions for uniform wear

The mean radius, $r_m = \dfrac{r_0 + r_i}{2} = \dfrac{r_0 + r_i}{2} = \left(\dfrac{125 + r_i}{2}\right)$

Axial force, $F_a = 2\pi C (r_0 - r_i)$ (Eqn. 6.24 MDHB)

$$= 2\pi p_i r_i (r_0 - r_i) \; (\therefore pr = C)$$

$$= 2\pi \times 0.15 \times r_i \times (125 - r_i) = 0.94 \times r_i \times (125 - r_i)$$

Torque transmitted per pair, $T = f \cdot F_a \cdot r_m$ (Eqn. 6.20 MDHB)

$$= 0.4 \times 0.94 \times r_i \times (125 - r_i) \times \left(\frac{125 + r_i}{2}\right)$$

$$= 0.188 \times r_i \times (125^2 - r_i^2) = 200 \times 10^3/2 = 10^5$$

$$\therefore \qquad r_i (125^2 - r_i^2) = \frac{10^5}{0.188} = 531914.9$$

By trial, we get three roots,

$$r_i = 37.5 \text{ mm}, r_i = 102 \text{ mm, and } r_i = -139.5 \text{ mm}$$

The last one is infeasible and $r_i = 37.5$ mm is small, hence choose $r_i = 102 \text{ mm} \cong 100$ mm

with $\dfrac{r_i}{r_0} = \dfrac{100}{125} = 0.8$ and $r_m = \dfrac{125 + 100}{2} = 112.5$ mm

$b/r_m = 25/112.5 = 0.22$ (Acceptable)

Step-3: Dimensions for uniform pressure

Mean radius, $r_m = \left(\dfrac{2}{3}\right)\left(\dfrac{r_0^3 - r_i^3}{r_0^2 - r_i^2}\right)$

Axial force, $F_a = \pi \cdot p \cdot (r_0^2 - r_i^2)$

The torque/pair, $T = fF_a \cdot r_m = \frac{2}{3} f\pi p \cdot (r_0^3 - r_i^3)$

$$= 2 \times 0.4 \times \pi \times 0.15 \times (125^3 - r_i^3)/3 = 100 \times 10^3 \text{ Nmm}$$

$$(125^3 - r_i^3) = \frac{100 \times 10^3 \times 3}{2 \times 0.4 \times \pi \times 0.15} = 7.95775 \times 10^5 = 795775$$

$$\therefore \quad r_i^3 = 125^3 - 795775 = 1157350$$

The solution, $r_i = 105$ mm

Width $b = 125 - 105 = 20$ mm and $r_m = (125 + 105)/2 = 115$ mm

$b/r_m = 20/115 = 0.174.$

This is nearer to the minimum specified and hence acceptable.

Example 19.6: A multiple disc clutch has five plates having four pairs of active friction surfaces. If the intensity of pressure is not to exceed 0.127 N/mm², find the power transmitted at 500 rpm. The outer and inner radii of friction surfaces are 125 mm and 75 mm respectively. Assume uniform wear and take coefficient of friction as 0.3.

Given Data: $n = 4; p = 0.127 \text{ N/mm}^2$, Speed $N = 500$ rpm; $f = 0.3$;
$r_0 = 125$ mm, and $r_i = 75$ mm

SOLUTION For uniform wear condition,

The power transmitted,

$$NP = \frac{n \cdot f \cdot F_a \cdot r_m \cdot N}{955 K_s} \times 10^{-4} \qquad \text{(Eqn. 6.26 MDHB)}$$

$$= \frac{n \cdot f \cdot 2\pi p_i r_i (r_o - r_i)(r_o + r_i) N \times 10^{-4}}{955 \times K_s \times 2}$$

$$= \frac{n \cdot f \cdot \pi p_i r_i (r_o^2 - r_i^2) N \times 10^{-4}}{955 \times K_s}$$

$$= \frac{4 \times 0.3 \times \pi \times 0.127 \times 75 \times (125^2 - 75^2) \times 500 \times 10^{-4}}{955 \times 1} = 18.8 \text{ kW}$$

Example 19.7: Design a single plate clutch having both sides effective from the following data: Power transmitted = 30 kW; speed of shaft = 1500 rpm, allowable limiting pressure, $p = 0.147$ MPa; maximum diameter of the clutch, $D_0 = 300$ mm, service factor, $C_s = 1.5$; number of springs $n = 9$; compression of the spring during engagement, $y = 2.5$ mm.

SOLUTION

Step-1: The design torque,

$$T_d = \frac{955 \times 10^4 \times kW \times Cs}{N} = \frac{955 \times 10^4 \times 30 \times 1.5}{1500} = 286500 \text{ Nmm}$$

Step-2: Axial force,

$$F_a = \pi \cdot (r_0^2 - r_i^2) \cdot p = \pi \cdot (150^2 - r_i^2) \times 0.147 = 0.462 \times (150^2 - r_i^2)$$

Step-3: Torque capacity per pair,

$$\text{The mean radius, } r_m = \frac{2}{3}\frac{(r_0^3 - r_i^3)}{r_0^2 - r_i^2} = \left(\frac{2}{3}\right) \times \left(\frac{150^3 - r_i^3}{150^2 - r_i^2}\right)$$

$$T = f \cdot F_a \cdot r_m = 0.3 \times 0.462 \times (150^2 - r_i^2) \times \left(\frac{2}{3}\right) \times \left(\frac{150^3 - r_i^3}{150^2 - r_i^2}\right)$$

$$= 0.0924 \times (150^3 - r_i^3)$$

Step-4: The inside radius for the torque capacity

$$T_d = 2 \times T \text{ i.e., } 286500 = 2 \times 0.0924 \times (150^3 - r_i^3)$$

$$\therefore \qquad r_i = \sqrt[3]{\left(150^3 - \frac{286500}{2 \times 0.0924}\right)} = 122.2 \text{ mm} \cong 122.0 \text{ mm}$$

The width of the friction face, $b = r_0 - r_i = 150 - 122.0 = 28$ mm

Step-5: Axial force on each spring

$$\text{Axial force on springs, } F_a = \pi (r_0^2 - r_i^2) \cdot p$$

$$= \pi \times (150^2 - 122^2) \times 0.147 = 3517.18 \text{ N}$$

$$\text{Force on each spring } F = \frac{3517.18}{9} = 390.8 \text{ N}$$

Assuming 10% increase in the axial load during disengagement

$$F_{max} = 1.1 \times 390.8 = 429.9 \text{ N}$$

Step-6: The spring diameter

Spring material – C45 with Sy = 1050 MPa (Table 10.3 MDHB)

$$\text{The design stress in shear for } FS = 2, S_{sd} = \frac{\left(\frac{1050}{2}\right)}{2} = 262.5 \text{ MPa}$$

For spring index $C = 10$ assumed.

$$\text{Wahl factor } K = \frac{4\times 10-1}{4\times 10-4} - \frac{0.615}{10} = 1.45$$

The spring diameter for the strength, (Eqn. 10.1 MDHB)

$$\tau = \frac{8FD}{\pi d^3}K = \frac{8FC}{\pi d^2}K = S_{sd}$$

Substituting the values,

$$\frac{1.45\times 8\times 429.9\times 10}{\pi\cdot d^2} = 262.5$$

$$\therefore \quad d = \sqrt{\frac{1.45\times 8\times 429.9\times 10}{\pi\times 262.5}} = 7.776$$

Choose $d = 8.0$ mm and $D = 10\times 8 = 80$ mm.

Step-7: The number of coils for stiffness

The spring stiffness,

$$K_s = \frac{\Delta F}{Y} = \frac{429.9-390.8}{2.5} = 15.96 \text{ N/mm} = 16.0 \text{ N/mm}$$

Number of coils by the stiffness equation $K_s = \dfrac{d^4\cdot G}{8D^3\cdot N} = \dfrac{d\cdot G}{8C^3\cdot N}$

$$\therefore \text{ Number of active coils, } N = \frac{d\cdot G}{8C^3\cdot K_s} = \frac{8\times 80\times 10^3}{8\times 10^3\times 16.0} = 5.00$$

Total number of coils $= 5+2 = 7$

Example 19.8: A multiplate clutch consists of five steel plates and four bronze plates. The inner and outer diameters of the friction disks are 75 mm and 150 mm respectively. The coefficient of friction is 0.1 and the intensity of pressure is limited to 0.3 MPa. Assuming uniform wear theory, calculate:

(i) The required operating force
(ii) Power transmitting capacity at 750 rpm.

Given Data: Dimensions of the friction disks

Outside diameter, $D_0 = 150$ mm; Inner diameter, $D_i = 75$ mm

Coefficient of friction, $f = 0.1$, Pressure $p_{max} = 0.3$ MPa

Uniform wear condition exists; No. of disks $n_1 = 5$; $n_2 = 4$

SOLUTION

Step-1: Operating axial force

$$F_a = 2\pi C(r_0 - r_i) \qquad \text{(Eqn. 6.24 MDHB)}$$

and $C = p \cdot r = p_{\max} \cdot r_i$ (Eqn. 6.23 MDHB)

$\therefore \; F_a = (2\pi \cdot p_{\max} \cdot r_i)(r_0 - r_i)$

$= 2\pi \times 0.3 \times 37.5 \times (75 - 37.5) = 2650.72 \text{ N}$

Step-2: Mean friction radius,

$$r_m = \frac{75 + 37.5}{2} = 56.25 \text{ mm}$$

Step-3: Power transmitted

Number of active faces $n = 2n_2 = 2 \times 4 = 8$

$$NP = \frac{n \cdot f \cdot f_a \cdot r_m \cdot N}{955K_s} \times 10^{-4} \quad \text{(Eqn. 6.26 MDHB)}$$

$$= \frac{8 \times 0.1 \times 2650.72 \times 56.25 \times 750}{955 \times 1.0} \times 10^{-4} = 9.3677 \text{ kW}$$

Example 19.9: In a multiplate disk clutch, the radial width of the friction material is to be 0.2 of the maximum radius. The coefficient of friction is 0.25. The clutch is to transmit 60 kW at 3000 rpm. Its maximum diameter is 250 mm and the axial force is limited to 600 N. Determine: (i) number of driving and driven plates, and (ii) mean unit pressure on each contact surface. Assume uniform wear condition.

Given Data: Width of friction surface, $b = 0.2 \times r_0$

Power transmitted, $NP = 60$ kW @ 3000 rpm; Maximum diameter, $D_0 = 250$ mm

Coefficient of friction, $f = 0.25$, Maximum axial force $F_{\max} = 600$ N

SOLUTION

Step-1: Dimensions of friction face

Outside radius, $r_o = D_o/2 = \frac{250}{2} = 125$ mm

The face width, $b = 0.2 \times r_0 = 0.2 \times 125 = 25$ mm

Inner diameter, $r_i = r_o - b = 125 - 25 = 100$ mm

Mean friction radius, $r_m = \frac{r_o + r_i}{2} = \frac{125 + 100}{2} = 112.5$ mm

Step-2: Number of friction surface pairs for power capacity

From Eqn. 6.26 MDHB, number of friction pairs,

$$n = \frac{955 \text{ kW} \cdot K_s}{f \cdot F_a \cdot r_m \cdot N} \times 10^4 = \frac{955 \times 60 \times 1 \times 10^4}{0.25 \times 600 \times 112.5 \times 3000} = 11.32$$

Choose $n = 12$ (next even number)

The number of driven plates, $n_2 = \dfrac{n}{2} = 6$

The number of driving plates, $n_1 = \dfrac{n}{2} + 1 = 6 + 1 = 7.$

Step-3: The average unit pressure

From expression for axial force, (Eqn. 6.24 MDHB)

$$C = \frac{F_a}{2\pi(r_o - r_i)} = \frac{600}{2\pi \times (125 - 100)} = 3.82$$

Then unit pressure, $p = c/r$

Unit pressure at outer radius, $p_o = C/r_o = \dfrac{3.82}{125} = 0.03056$ MPa

Unit pressure at inner radius, $p_i = C/r_i = \dfrac{3.82}{100} = 0.0382$ MPa

The average pressure, $p_m = \dfrac{p_i + p_o}{2} = \dfrac{0.0382 + 0.03056}{2} = 0.0344$ MPa

Example 19.10: A multiplate clutch has steel on bronze, and is to transmit 8 kW at 1440 rpm. The inner diameter of contact is 80 mm and outer diameter of contact is 140 mm. The clutch operates in oil with expected coefficient of friction of 0.1. The average allowable pressure is 0.35 MPa. Assume uniform wear theory and determine the following: (i) number of steel and bronze plates, (ii) axial force required, and (iii) actual maximum pressure.

Given Data: Power transmitted, NP = 8 kW @ 1440 rpm

Inner diameter, $d_i = 80$ mm, Outer diameter, $d_o = 140$ mm

Coefficient of friction, $f = 0.1$, Averge pressure, $P_a = 0.35$ MPa

SOLUTION

Step-1: Number of plates

From Eqn. 6.26 MDHB, number of plates

$$n = \frac{955 \times 10^4 \text{ kW} \cdot K_s}{f F_a \cdot r_m \cdot N}$$

where, kW = 8, $K_s = 1.5$ (assumed), $f = 0.1$ and $N = 1440$ rpm

$$F_a = \pi(r_o^2 - r_i^2)p = \pi(70^2 - 40^2) \times 0.35 = 3628.54 \text{ N}$$

$$r_m = \frac{r_o + r_i}{2} = \frac{70 + 40}{2} = 55 \text{ mm}$$

Substituting the value, $n = \dfrac{955 \times 8 \times 1.5 \times 10^4}{0.1 \times 3628.54 \times 55 \times 1440} = 3.987$

Choose $n = 4$

The number of bronze plates (both sides effective), $n_2 = \dfrac{n}{2} = 2$

The number of steel plates, $n_1 = \dfrac{n}{2} + 1 = 3$

Step-2: Axial force required

The average pressure, $p = \dfrac{p_i + p_0}{2} = \dfrac{C/r_i + C/r_0}{2}$

$$0.35 = \frac{C}{2}\left[\frac{1}{r_i} + \frac{1}{r_o}\right] = \frac{C}{2}\left[\frac{1}{40} + \frac{1}{70}\right]$$

$$\therefore \quad C = \frac{0.35 \times 2 \times 40 \times 70}{(70 + 40)} = 17.82$$

$$\therefore \quad F_a = 2\pi \cdot C \cdot (r_o - r_i) = 2\pi \times 17.82 \times (70 - 40) = 3359 \text{ N}$$

Step-3: The actual maximum pressure,

$$P_{\max} = C/r_i = \frac{17.82}{40} = 0.4455 \text{ MPa}$$

Example 19.11: A single plate clutch with maximum diameter of 600 mm, has maximum lining pressure of 0.35 MPa. The power to be transmitted at 400 rpm is 135 kW and $f = 0.3$. Find inside diameter and spring force to engage the clutch. If six springs with spring index of 6 are used, find the spring diameter.

Given Data: Power, $N = 135$ kW at 400 rpm, coefficient of friction, $f = 0.3$
Outer diameter $D_d = 600$ mm, Lining pressure $p = 0.35$ MPa
Number of springs used, $N = 6$, Spring index, $C = 6$

SOLUTION

Step-1: Inner diameter of friction face

The power transmitted, $\text{kW} = \dfrac{n \cdot f \cdot F_a \cdot r_m \cdot N}{955 K_s} \times 10^{-4}$ (Eqn. 16.26 MDHB)

where, for uniform pressure, $F_a = \pi (r_0^2 - r_i^2) p$ (Eqn. 16.21 MDHB)

$$= \pi \times (300^2 - r_i^2) \times 0.35 = 1.099 \times (300^2 - r_i^2)$$

$r_m = \left(\dfrac{2}{3}\right)\left(\dfrac{300^3 - r_i^3}{300^2 - r_i^2}\right)$; $f = 0.3$ and $N = 400$ rpm, $K_s = 1.2$ assumed, kW = 135

Substituting the values,

$$135 = \left(\frac{2 \times 0.3 \times 1.099}{955 \times 1.2}\right)(300^2 - r_i^2)\left(\frac{2}{3}\right)\left(\frac{300^3 - r_i^3}{300^2 - r_i^2}\right) \times 400 \times 10^{-4}$$

$$= 0.153438 \times (300^3 - r_i^3) \times 10^{-4}$$

$$r_i = \sqrt[3]{300^3 - \frac{135 \times 10^4}{0.153438}} = 263 \text{ mm}$$

∴ The inside diameter $d_i = 2 \times 263 = 526$ mm

Step-2: The axial force

$$F_a = \pi \cdot p \cdot (r_0^2 - r_i^2) = 1.099 \times (300^2 - 263^2) = 22893.27 \text{ N}$$

Step-3: Diameter of spring wire

Let the increase in spring force during disengagement be 20%

The maximum force on each spring,

$$F_{max} = \frac{F_a \times 1.2}{6} = \frac{22893.27 \times 1.2}{6} = 4578.65 \text{ N}$$

Select the material of spring as chrome vanadium steel with $s_{sd} = 450$ MPa

Assuming a spring index $C = 6$,

$$\text{Wahl factor } K = \frac{4.C - 1}{4.C - 4} + \frac{0.615}{C} = \frac{4 \times 6 - 1}{4 \times 6 - 4} + \frac{0.165}{6} = 1.25$$

Then spring diameter (from Eqn. 10.1 MDHB),

$$d = \sqrt{\frac{8F \cdot C \cdot K}{\pi \cdot S_{sd}}} = \sqrt{\frac{8 \times 4578.365 \times 6 \times 1.25}{\pi \times 450}} = 14 \text{ mm}$$

19.3.2 Cone Clutches

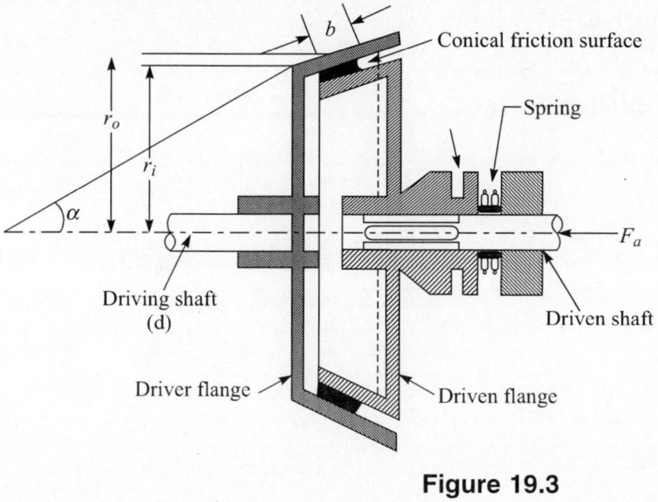

Proportions:
The mean radius $r_m = 2.5d$ to $5d$
The width of the friction face
$b = 0.25\ r_m$ to $0.4\ r_m$

Figure 19.3

Proportions

The mean radius, $r_m = 2.5d$ to $5d$

The width of the friction face, $b = 0.25\,r_m$ to $0.4r_m$.

The driving and driven flanges are shaped in the form of frustrum of a cone. The driven flange surface is lined with friction material and fits inside the driving flange cone. They are held pressed by the spring on the driven shaft. The contact can be disengaged by pulling back the driven flange hub against the spring force by the actuating mechanism.

The torque transmitted is evaluated as follows (Fig. 19.4).

Consider conical ring of the contact surface of width 'dx' (between radii r and $r + dr$).

p = unit normal pressure between the surfaces

$\propto$ = half cone angle, f = coefficient of friction

The normal force on the element, $dF_n = (2\pi \cdot r \cdot p \cdot dx)$

The tangential frictional force,

$$dF_t = f dF_n = 2\pi p \cdot f \cdot r \cdot dx = \left(\frac{2\pi p \cdot f \cdot r \cdot dr}{\sin \alpha}\right)$$

The corresponding frictional torque $dT = dF_t \cdot r = \left(\frac{2\pi p \cdot f \cdot r \cdot dr}{\sin \alpha}\right) \cdot r$

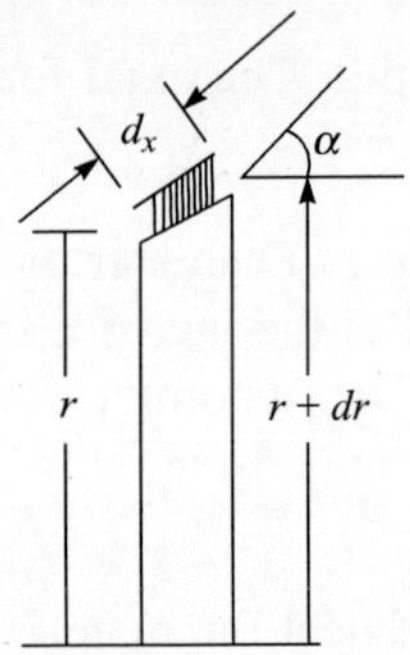

Figure 19.4

The total frictional torque, $$T = \int_{r_i}^{r_0} dT = \int_{r_i}^{r_0} \frac{2\pi p \cdot f \cdot r^2 \cdot dr}{\sin \alpha} \tag{19.7}$$

The axial force,

$$F_a = \int dF_n \sin \propto = \int 2\pi r \cdot p \cdot dx \sin \propto = \int_{r_i}^{r_0} 2\pi r \cdot p \cdot dr \tag{19.8}$$

The integrals depend upon the pressure distribution. Two situations exist:

(a) Uniform pressure, p = constant = C

(b) Uniform wear, p.r. = constant = C

Accordingly, the expressions are different.

(a) **Uniform pressure** –p = const. = C

The normal force,

$$F_n = \int \left(\frac{2\pi p \cdot r \cdot dr}{\sin \alpha}\right) = \left(\frac{2\pi p}{\sin \alpha}\right) \int_{r_i}^{r_0} r dr = \left(\frac{\pi p}{\sin \alpha}\right)(r_0^2 - r_i^2) = 2\pi p r_m b \tag{19.9}$$

The axial force,

$$F_a = 2\pi p \cdot \int_{r_i}^{r_0} r dr = \pi p (r_0^2 - r_i^2)$$

$$= 2\pi p \cdot r_m \cdot b \sin\alpha = F_n \cdot \sin\alpha \tag{19.10}$$

The friction torque,

$$T = \int_{r_i}^{r_0} \left(\frac{2\pi pf}{\sin\alpha}\right) r^2 \cdot dr = \left(\frac{2\pi pf}{\sin\alpha}\right)\left(\frac{r_0^3 - r_i^3}{3}\right)$$

Substituting for p from (19.10),

$$T = \left(\frac{fF_a}{\sin\alpha}\right)\left(\frac{2}{3}\right)\left(\frac{r_o^3 - r_i^3}{r_o^2 - r_i^2}\right) = \frac{f \cdot F_a r_m}{\sin\alpha} \tag{19.11}$$

where, $r_m = \left(\frac{2}{3}\right)\left(\frac{r_o^3 - r_i^3}{r_o^2 - r_i^2}\right)$ is frictional mean radius.

(b) **Uniform wear** p.r. = C

The normal force,

$$F_n = \int \left(\frac{2\pi p \cdot r \cdot dr}{\sin\alpha}\right) = \left(\frac{2\pi C}{\sin\alpha}\right)\int_{r_i}^{r_0} dr = \left(\frac{\pi p \cdot r_m}{\sin\alpha}\right)(r_o - r_i) = 2\pi p \cdot r_m \cdot b$$

The axial force,

$$F_a = \int_{r_i}^{r_0} 2\pi C \cdot dr = 2\pi C(r_o - r_i)$$

$$= 2\pi p r_m \cdot b \cdot \sin\alpha = F_n \sin\alpha \tag{19.12}$$

The force required to engage the clutch,

$$F_{a1} = F_n(\sin\propto + f\cos\propto) \tag{19.13}$$

The torque transmitted,

$$T = \int_{r_i}^{r_0} \frac{2\pi C \cdot f}{\sin\propto} \cdot r dr = \frac{2\pi C \cdot f}{\sin\propto}\left(\frac{r_0^2 - r_i^2}{2}\right)$$

$$T = \frac{2\pi C \cdot f}{\sin\propto}\left[\frac{(r_o + r_i)(r_o - r_i)}{2}\right] = \frac{fF_a}{\sin\propto}\left[\frac{(r_o + r_i)}{2}\right] = \frac{fF_a \cdot r_m}{\sin\propto} \tag{19.14}$$

where, $r_m = \frac{(r_o + r_i)}{2}$ – is frictional mean radius for uniform wear

The expressions are similar to the disk clutch with f being replaced by ($f/\sin\propto$). We can introduce virtual coefficient of friction $f_1 = f/\sin\propto$ and use all the disk clutch equations. But it is not possible to have multiple pairs of frictional surfaces in cone clutch.

The power transmitted is given by,

$$NP = \frac{f \cdot pr_m^2 \cdot b \cdot N}{15K_L} \times 10^{-5} \qquad (19.15)$$

Example 19.12: A soft cone clutch has cone pitch angle of $\alpha = 10°$, a mean diameter of 300 mm and a face width of 100 mm. Using a coefficient of friction of 0.2 and an average pressure of 70 kN/m^2 find the force required to engage the clutch. If uniform wear exists, for a speed of 500 rpm, determine the power transmitted.

Given Data: Cone face angle, $\alpha = 10°$, Mean diameter, D_m = 300 mm,

Face width, b = 100 mm, Coefficient of friction, f = 0.2,

The average pressure, p = 70 kN/m^2 = 0.07 MPa; Speed = 500 rpm

SOLUTION

Step-1: Forces on the cone

The normal force

$$F_n = 2\pi r_m p \cdot b = 2\pi \times 150 \times 0.07 \times 100 \text{ N} = 6597.34 \text{ N} \qquad \text{(Eqn. 6.28 MDHB)}$$

The force required to engage the clutch,

$$F_{a1} = F_n(\sin\alpha + f\cos\alpha) \qquad \text{(Eqn. 6.29 MDHB)}$$

$$= 6597.34 \times (\sin 10° + 0.2 \times \cos 10°)$$

$$= 6597.34 \times (0.1736 + 0.1969) = 2444.7 \text{ N}$$

The force required to maintain the engagement

$$F_a = F_n \sin\alpha = 6597.34 \times \sin 10° = 1145.6 \text{ N} \qquad \text{(Eqn. 6.27 MDHB)}$$

Step-2: The power transmitted by the clutch

$$NP = \frac{f \cdot pr_m^2 \cdot b \cdot N}{15K_L} \times 10^{-5} \qquad \text{(Eqn. 6.32 MDHB)}$$

$$= \frac{0.2 \times .07 \times 150^2 \times 100 \times 500 \times 10^{-5}}{15 \times 1.0} = 10.5 \text{ kW}$$

Example 19.13: An engine developing 30 kW at 1250 rpm is fitted with a cone clutch built into the fly-wheel. The cone has a face angle of 12.5° and a mean diameter of 350 mm. The coefficient of friction is 0.2 and the normal pressure on the friction surface is not to exceed 80 kN/m^2. Determine the required face width and the force required to engage the clutch, assuming uniform wear.

Given Data: Power = 30 kW at Speed = 1250 rpm
Cone pitch angle, 2 $\propto$ = 12.5°, Mean dia., d_m = 350 mm
Coefficient of friction = 0.2; The normal pressure, p = 80 kN/m^2 = 0.08 MPa

SOLUTION

Step-1: The face width for power capacity

Assume uniform wear condition

Mean radius, $r_m = \dfrac{r_0 + r_i}{2} = d_m/2 = 350/2 = 175$ mm

Service factor, $K_L = 1.0$ assumed

From the equation for power transmitted, (Eqn. 6.32 MDHB)

$$\text{kW} = \frac{f \cdot p \cdot r_m^2 b \cdot N}{15 \times K_L} \times 10^{-5}$$

$$b = \frac{15 \times K_L \times \text{kW} \times 10^5}{f \cdot p \cdot r_m^2 \cdot N} = \frac{15 \times 1.0 \times 30 \times 10^5}{0.2 \times 0.08 \times 175^2 \times 1250} = 73.47 \text{ mm}$$

Choose $b = 75$ mm

Step-2: The forces on the cone

The normal force $F_n = 2\pi r_m \cdot b \cdot p = 2\pi \times 175 \times 75 \times 0.08 = 6597.34$ N

The axial force to engage the clutch, $F_a = F_n(\sin \propto + f \cos \propto)$

$$= 6597.34(\sin 6.25° + 0.2 \times \cos 6.25°) = 2029.86 \text{ N}$$

Example 19.14: A cone clutch transmits 4.5 kW at 225 rpm. The small diameter of the cone is 230 mm and the face width is 50 mm. The cone face makes an angle of 15° with the horizontal. The coefficient of friction is 0.25. The uniform wear exists. Determine the max. contact pressure and force to engage the clutch.

Given Data: Power = 4.5 kW, @ Speed = 225 rpm; Face width b = 50 mm,
The cone pitch angle $\propto = 15°$, Coefficient of friction f = 0.25

SOLUTION

Step-1: The dimensions of the clutch

From the geometry, large diameter

$$D_0 = D_i + 2b\sin \propto = 230 + 2 \times 50 \times \sin 15° = 255.88 \text{ mm}$$

Let $D_0 = 256$ mm, $r_0 = 256/2 = 128$ mm, $r_i = 230/2 = 115$ mm

The frictional mean radius,

$$r_m = \frac{r_o + r_i}{2} = \frac{128+115}{2} = 121.5 \text{ mm}$$

Step-2: Average pressure by power equation

From the power equation, (Eqn. 6.32 MDHB)

$$\text{kW} = \frac{f \cdot p \cdot r_m^2 b \cdot N \times 10^{-5}}{15 \times K_L}$$

Average or mean pressure,

$$p = \frac{KW \times 15 \times K_L}{f \cdot r_m^2 b \cdot N} \times 10^5 = \frac{4.5 \times 15 \times 1.0 \times 10^5}{0.25 \times 121.5^2 \times 50 \times 225} = 0.1626 \text{ MPa}$$

Step-3: Maximum pressure by wear consideration

For uniform wear,

$$C = p_m r_m = 0.1626 \times 121.5 = 19.75$$

$$\therefore \quad p_{\max} = \frac{C}{r_{\min}} = \frac{19.75}{115} = 0.1717 \text{ MPa}$$

Step-4: Forces on the cone

The normal force,

$$F_n = 2\pi r_m \cdot b \cdot p_m = 2\pi \times 121.5 \times 50 \times 0.1626 = 6206.5 \text{ N}$$

Then the force required to engage the clutch,

$$F_{ai} = F_n (\sin \propto + f \cos \propto)$$

$$= 6026.5 \times (\sin 15° + 0.25 \times \cos 15°) = 3105.114 \text{ N}$$

Example 19.15: A leather faced cone-clutch transmits 15 kW at 1200 rpm. The total cone angle is to be 20° and the average pressure $p = 70$ kN/m^2. The coefficient of friction $f = 0.25$. Design the clutch.

Given Data: Power = 15 kW at Speed = 1200 rpm

Cone angle $2\propto = 20°$, i.e., $\propto = 10°$,

Coefficient $f = 0.25$; Contact pressure = 0.07 MPa

SOLUTION

Step-1: The dimensions of the clutch for power capacity

Assume (i) $b = 0.3\, r_m$ and (ii) Uniform wear exists, $r_m = \dfrac{r_o + r_i}{2}$

Then in power equation $\text{kW} = \dfrac{f \cdot p \cdot r_m^2 b \cdot N}{15 \times K_L} \times 10^{-5}$ (Eqn. 6.32 MDHB) kW = 15, $N = 1200, f = 0.25,\ p = 0.07$ Mpa, $b = 0.3r_m,\ K_L = 1.0$

Substituting the values,

$$\therefore \quad 15 = \frac{0.25 \times 0.07 \times r_m^2 \times 0.3r_m \times 1200 \times 10^{-5}}{15 \times 1.0}$$

$$\therefore \quad r_m^3 = \frac{15 \times 15 \times 1.0 \times 10^5}{0.25 \times 0.07 \times 0.3 \times 1200} = 35.715 \times 10^5$$

$$\therefore \quad r_m = 152.85 \text{ mm and } b = 0.3 \times r_m = 45.85 \text{ mm}$$

$$\therefore \quad r_o - r_i = b \cdot \sin \propto = 7.963$$

$$\therefore \quad r_o + r_i = 2 \times 152.85 = 305.7$$

Solving, we get

$$\therefore \quad r_o = 156.83 \text{ mm and } r_i = 148.87 \text{ mm}$$

Choose $r_i = 150$ mm and $r_o = 160$ mm

Then
$$b = \frac{r_o - r_i}{\sin \propto} = \frac{(160 - 150)}{\sin 10^\circ} = 57.6 \text{ mm}$$

$$r_m = \frac{r_o + r_i}{2} = \frac{160 + 150}{2} = 155 \text{ mm}$$

The power capacity of the clutch will be more than required or the contact pressure will be less than maximum of 0.07 MPa at the rated power.

Step-2: Forces on the cone

The normal force, $F_n = 2\pi r_m \cdot p \cdot b = 2\pi \times 155 \times 0.07 \times 57.6 = 3926.75$ N

The axial force, $F_a = F_n \sin \propto = 3926.75 \times \sin 10^\circ = 681.87$ N

The force required to engage the clutch,

$$F_{ai} = F_n(\sin \propto + f \cos \propto)$$

$$= 3926.75 \times (\sin 10^\circ + 0.25 \times \cos 10^\circ) = 1648.65 \text{ N}$$

$\therefore$ Minimum spring force, $F_{\min} = 681.87$ N and maximum spring force, $F_{\max} = 1648.65$ N

Step-3: Spring design

Assuming axial displacement during disengagement y = 10.0 mm

$$\text{The stiffness} = \frac{1648.65 - 681.8}{10} = 96.6 \text{ N/mm}$$

Assume spring index, $C = 5.0$, then $K = 1.3$ from Fig. 10.1 MDHB

The spring material; Cr-V Steel, S_y = 1800 MPa (Table 10.3 MDHB)

$$\therefore \qquad S_{sy} = 0.57 \times S_y = 1026 \text{ MPa}$$

For $FS = 1.5$, design stress, $S_{sd} = \dfrac{1026}{1.5} = 684.0$ MPa

From the strength equation, $S_{sd} = \dfrac{8 \cdot F \cdot D \cdot K}{\pi d^3} = \dfrac{8FCK}{\pi d^2}$

$$d = \sqrt{\frac{8F \cdot C \cdot K}{\pi S_{sd}}} = \sqrt{\frac{8 \times 1648.6 \times 5 \times 1.3}{\pi \times 684.0}} = 6.316 \text{ mm}$$

Choose $d = 6.5$ mm (Table 10.5 MDHB)

$$D = 6.5 \times 5 = 32.5 \text{ mm}$$

Inner diameter, $D_i = 32.5 - 6.5 = 26.0$ mm,
Outer diameter, $D_0 = 32.5 + 6.5 = 39$ mm

Then from stiffness equation, $K_s = \dfrac{d^4 G}{8D^3 N} = \dfrac{dG}{8C^3 N}$ (Eqn. 10.4 MDHB)

$$N = \frac{dG}{8C^3 K_s} = \frac{6.5 \times 81 \times 10^3}{8 \times 5^3 \times 96.6} = 5.45 \cong 6$$

Total number of coils, $N_1 = 6 + 2 = 8$

Step-4: Design of the shaft

The torque on the shaft, $T = \dfrac{955 \times 10^4 \text{ kW}}{\text{N}}$

$$= \frac{955 \times 10^4 \times 15}{1200} = 119375 \text{ Nmm}$$

With design stress, $S_{sd} = 350$ MPa for the shaft

The shaft diameter, $d = \left[\dfrac{16T}{\pi S_{sd}}\right]^{\frac{1}{3}} = \left[\dfrac{16 \times 119375}{\pi \times 350}\right]^{1/3} = 12$ mm

The value is very small. To account for weakening effect of the spline or keys, the shaft diameter has to be increased. Considering the other dimensions of the clutch, the diameter $d = 25$ mm is a reasonable choice. The spring with inner diameter $D_i = 26.00$ mm can be placed on this shaft.

Example 19.16: Design a cone clutch to transmit a power of 40 kW at rated speed of 750 rpm. Also determine (i) axial force necessary to transmit the torque, and (ii) axial force necessary to engage the clutch.

Given Data: Power, $NP = 40$ kW @ Speed = 750 rpm

SOLUTION Assume uniform wear

Step-1: Selection of friction material

Select leather as the friction material

From Table 6.9 MDHB for leather on C.I. or steel

Select $f = 0.4$; $p = 0.2$ N/mm^2 and $\alpha = 12.5°$

Step-2: The mean radius by power,

$$\text{kW} = \frac{f \cdot p \cdot r_m^2 \cdot b \cdot N}{15K_s} \times 10^{-5} \qquad \text{(Eqn. 6.32 MDHB)}$$

where, $f = 0.4$; $p = 0.2$ MPa, $b = 0.3r_m$ (assumed)

$K_s = 1.2$ (assumed), kW = 40, $N = 750$ rpm

Substituting the values,

$$40 = \frac{0.4 \times 0.2 \times r_m^2 \times 0.3r_m \times 750}{15 \times 1.2} \times 10^{-5}$$

$$\therefore \quad r_m^3 = \frac{40 \times 15 \times 1.2 \times 10^5}{0.4 \times 0.2 \times 0.3 \times 750} = 40 \times 10^5$$

$$\therefore \quad r_m = 158.74 \text{ mm}$$

Step-3: Dimensions of the clutch

Maximum width, $b = 0.3r_m = 0.3 \times 158.74 = 47.62$ mm

Mean diameter, $D_m = 2r_m = 2 \times 158.74 = 317.48$ mm

The outer diameter, $D_0 = 2\left[r_m + \frac{b}{2}\sin \propto\right] = D_m + b\sin \propto$

$= 317.48 + 47.62 \sin 12.5 = 327.78$

Choose $D_0 = 330.0$ mm

The inner diameter, $D_i = D_m - b\sin \propto = 317.48 - 47.62\sin 12.5 = 307.17$

Choose $D_i = 310$ mm

$\therefore$ Then $D_m = \frac{330 + 310}{2} = 320$ mm and $r_m = 160.0$ mm

Face width, $b = \frac{D_0 - D_i}{2\sin\alpha} = \frac{330 - 310}{2 \times \sin 12.5} = 46.2$ mm

Step-4: The axial force

The normal force, $F_n = 2 \cdot \pi \cdot r_m \cdot p \cdot b$ (Eqn. 6.28 MDHB)

$$= 2 \times \pi \times 160 \times 0.2 \times 46.2 = 9289.06 \text{ N}$$

The axial force, $F_a = F_n \sin \propto$ (Eqn. 6.27 MDHB)

$$= 9289.06 \times \sin 12.5 = 2010 \text{ N}$$

Step-5: Axial force to engage the clutch

$$F_a = F_n (\sin \propto + f \cos \propto)$$

$$= 9289.06 \times (\sin 12.5° + 0.4 \times \cos 12.5°) = 5638.07 \text{ N}$$

Example 19.17: An engine developing 50 kW at 1000 rpm is fitted with a cone clutch. The cone has a face angle of 12.5° and width of the face is one-fourth of mean diameter of the friction lining. If the normal intensity of the pressure between the contact surfaces is not to exceed 0.1 N/mm^2, assuming uniform wear criterion and taking $f = 0.2$, calculate the dimensions of the clutch.

Given Data: Power *NP* = 50 kW @ 1000 rpm

Cone face angle, $\alpha = 12.5°$; Width of friction face, $b = \dfrac{D_m}{4}$

The normal pressure, $p = 0.1$ MPa; Coefficient of friction, $f = 0.2$

SOLUTION

Step-1: The mean friction radius

The power equation for uniform wear, (Eqn. 6.32 MDHB)

$$\text{kW} = \frac{f \cdot p \cdot r_m^2 \cdot b \cdot N}{15 \cdot K_s} \times 10^{-5}$$

where, kW = 50; $N = 1000$; $f = 0.2$; $p = 0.1$ MPa,

$$b = \frac{D_m}{4} = \left(\frac{r_m}{2}\right), \text{ and } K_s = 1.2 \text{ (assumed)}$$

Substituting the values,

$$50 = \frac{0.2 \times 0.1 \times r_m^2 \times (r_m/2) \times 1000 \times 10^{-5}}{15 \times 1.2}$$

$$\therefore \quad r_m^3 = \frac{50 \times 15 \times 1.2 \times 2}{0.2 \times 0.1} \times 10^2 = 9 \times 10^6$$

$$\therefore \quad r_m = 208.0 \text{ mm}$$

Step-2: The dimension of friction lining

$\therefore$ Face width, $b = r_m/2 = \dfrac{208}{2} = 104.0$ mm

$$\text{The outer radius, } r_o = r_m + \frac{b}{2}\sin\propto \; = 208 + \frac{104}{2}\sin 125 = 219.255 \text{ mm}$$

$$\text{Inner radius, } r_i = r_m - \frac{b}{s}\sin\propto = 208 - \frac{104}{2}\sin 125 = 196.745 \text{ mm}$$

$$\text{Outer diameter, } D_0 = 438.51 \text{ mm and Inner dia. } D_i = 393.49 \text{ mm}$$

Example 19.18: A cone clutch with face angle of 12.5°, is to transmit 7.5 kW at 900 rpm. The width of the face is half the mean radius and the normal pressure between the contact faces is not to exceed 0.09 N/mm². Assuming uniform wear and the coefficient of friction between the contact faces as 0.2; determine the main dimensions of the clutch and the axial force required to engage the clutch.

Given Data: Power, NP = 7.5 kW @900 rpm

The cone face angle, $\alpha = 12.5°$, Face width, $b = r_m/2$

Normal pressure, $p = 0.09$ MPa, Coefficient of friction, $f = 0.2$

SOLUTION

Step-1: Mean radius from power equation,

$$\text{kW} = \frac{f \cdot p \cdot r_m^2 b \cdot N}{15K_s} \times 10^{-5} \qquad \text{(Eqn. 6.32 MDHB)}$$

where, $f = 0.2'$, $p = 0.09$ MPa, $b = 0.5r_m$, $N = 900$ rpm; $K_s = 1.2$ (assumed)
Substituting the values,

$$7.5 = \frac{0.2 \times 0.09 \times r_m^2 \times 0.5r_m \times 900 \times 10^{-5}}{15 \times 1.2}$$

$$\therefore \quad r_m^3 = \frac{7.5 \times 15 \times 1.2 \times 10^5}{0.2 \times 0.09 \times 0.5 \times 900} = 16.67 \times 10^5$$

$$\therefore \quad r_m = 118.56 \text{ mm}$$

Step-2: Main dimensions

The face width $b = \dfrac{r_m}{2} = \dfrac{118.56}{2} = 59.28$

The outer diameter,

$$D_o = 2\left[r_m + \frac{b}{2}\sin\propto\right] = 2\left[118.56 + \frac{59.28}{2}\sin 12.5\right] = 249.95 \cong 250 \text{ mm}$$

The inner diameter,

$$D_i = 2\left[r_m - \frac{b}{2}\cdot\sin\propto\right] = 2\left[118.56 - \frac{59.28}{2}\cdot\sin 12.5\right]$$

$$= 224.3 \text{ mm} \cong 225.0 \text{ mm}$$

The exact face width, $b = \left(\frac{D_0 - D_i}{2}\right)/\sin\propto$

$$= \left(\frac{250-225}{2}\right)/(\sin 12.5) = 57.75 \text{ mm}$$

Step-3: Axial force

$$F_a = 2\pi C(r_0 - r_i) \qquad \text{(Eqn. 6.24 MDHB)}$$

$$= 2\pi\cdot p\cdot r_i\cdot(r_0 - r_i) = 2\pi\times 0.09\times\left(\frac{225}{2}\right)\left(\frac{250-225}{2}\right) = 795.22 \text{ N}$$

Example 19.19: A cone-clutch is to be designed to transmit 40 Nm of torque. The semi-cone angle is 12.5°. The mean radius of the clutch is twice the face width; the coefficient of friction is $f = 0.18$; the normal pressure intensity between the contacting surfaces should not exceed 0.1 N/mm^2. Considering uniform wear, calculate:

(i) The inner and outer diameter of friction layer.
(ii) Face width of friction layer.
(iii) Force required to engage the clutch.

Given Data: Torque, $T = 40$ Nm = 40000 N mm; Face angle, $\propto = 12.5°$

Mean radius, $r_m = 2\cdot b$; Coefficient of friction, $f = 0.18$
Pressure, $p = 0.1 \text{ N/mm}^2$

SOLUTION

Step-1: Mean radius by power equation

The torque, $T = \frac{955\times 10^4\times KW}{N}K_s$

$$= \left(\frac{955\times 10^4}{\text{N}}\right)\left(\frac{f\cdot p\cdot r_m^2\cdot b\cdot N}{15}\right)\times 10^{-5} \qquad \text{(Eqn. 6.32 MDHB)}$$

$$= \frac{955 f\cdot p\cdot r_m^2\cdot b}{150} = \frac{955\times 0.18\times 0.1\times r_m^3}{300} = 0.0573 r_m^3$$

but $T = 40$ Nm = 40000 Nmm

$$\therefore \qquad 0.0573 r_m^3 = 40000$$

$$\therefore \qquad r_m^3 = \frac{40000}{0.0573} = 698.08 \times 10^3$$

$$\therefore \qquad r_m = 88.71 \text{ mm}$$

Step-2: Main dimensions of the clutch

The face width, $b = \dfrac{r_m}{2} = \dfrac{88.71}{2} = 44.355 \cong 45.0$ mm

The mean diameter, $D_m = 2 \times 88.71 = 177.42$ mm

The outer diameter,

$$D_0 = D_m + b \sin \propto = 177.42 + 45 \times \sin(12.5°) = 187.16 \cong 190 \text{ mm}$$

The inner diameter,

$$D_i = 177.42 - 45 \times \sin 12.5° = 167.68 \cong 170 \text{ mm}$$

The exact face width, $b = \left(\dfrac{D_0 - D_i}{2}\right) \times \dfrac{1}{\sin \propto} = \dfrac{(190 - 170)}{2 \times \sin 12.5°} = 46.20$ mm

Step-3: The axial force

The average pressure, $p = \dfrac{p_i}{2}\left(\dfrac{r_i}{r_o} + 1\right) = \left(\dfrac{0.1}{2}\right)\left(\dfrac{170}{190} + 1\right) = 0.0947$ MPa

The normal force, $F_n = 2\pi \cdot r_m \cdot p \cdot b$

$$= 2 \times \pi \times 90 \times 0.0947 \times 46.2 = 2475 \text{ N}$$

The axial force to engage the clutch

$$F_a = F_n (\sin \alpha + f \cos \alpha) \qquad \text{(Eqn. 6.29 MDHB)}$$

$$= 2475 (\sin 12.5° + 0.18 \times \cos 12.5°) = 970.65 \text{ N}$$

Example 19.20: Determine the dimensions of a simple cone-clutch to transmit 20 kW at 1000 rpm. The minimum diameter is to be 300 mm and the cone angle 20°. Assume f = 0.2 and permissible pressure = 0.1 N/mm^2. Also determine the axial force required to engage the clutch.

Given Data: Power NP = 20 kW at 1000 rpm, Inner diameter, $D_i = 300$ mm

The cone angle, $2\alpha = 20°$, Coefficient of friction, $f = 0.2$

Pressure, $p = 0.1$ N/mm^2

SOLUTION Assume uniform pressure

Step-1: Torque transmitted,

$$T = \frac{955 \times 10^4 \times \text{kW}}{N} = \frac{955 \times 10^4 \times 20}{1000} = 191000 \text{ Nmm}$$

Step-2: Dimensions of clutch for torque capacity,

$$T = \frac{f \cdot F_a \cdot r_m}{\sin \propto}$$

where, $f = 0.2; \propto = 10°$

$$F_a = \pi(r_o^2 - r_i^2) \cdot p = \pi \times (r_o^2 - r_i^2) \times 0.1 = 0.3142 \times (r_o^2 - r_i^2)$$

$$r_m = \left(\frac{2}{3}\right)\left(\frac{r_o^3 - r_i^3}{r_o^2 - r_i^2}\right), \text{ and } T = 191000 \text{ Nmm}$$

Substituting the values,

$$191000 = \left(\frac{0.2 \times 0.3142}{\sin 10°}\right)(r_o^2 - r_i^2) \times \left(\frac{2}{3}\right)\left(\frac{r_o^3 - r_i^3}{r_o^2 - r_i^2}\right)$$

$$= 0.241254(r_o^3 - r_i^3)$$

$$\therefore \quad r_o^3 - r_i^3 = \frac{191000}{0.241254} = 791696.42$$

$$\therefore \quad r_o^3 = 791696.42 + 150^3 = 4166696.42$$

$$\therefore \quad r_o = 160.91$$

The outer diameter, $D_o = 2r_o = 2 \times 160.91 \cong 322$ mm

The face width, $b = \dfrac{D_o - D_i}{2\sin \propto} = \dfrac{322 - 300}{2 \times \sin 10°} = 63.35$ mm

Mean diameter, $D_m = \dfrac{D_o + D_i}{2} = \dfrac{322 + 300}{2} = 311$ mm

Mean radius, $r_m = \dfrac{311}{2} = 155.5$ mm

Step-3: Axial force to maintain the contact

$$F_a = F_n \cdot \sin \propto = 2\pi r_m \cdot p \cdot b \sin \alpha$$

$$F_n = 2 \times \pi \times 155.5 \times 0.1 \times 63.35 = 6189.52 \text{ N}$$

$$\therefore \quad F_a = 6189.52 \times \sin 10 = 1074.8 \text{ N}$$

Step-4: Axial force to engage the clutch,

$$F_a = F_n(\sin \propto + f \cos \propto) = 6189.52 \times (\sin 10° + 0.2 \times \cos 10°) = 2293.9 \text{ N}$$

Example 19.21: A friction cone-clutch has to transmit a torque of 200 Nm at 1440 rpm. The larger diameter of the cone is 350 mm. The cone pitch angle is 6.25°. The face width is 65 mm. The coefficient of friction is 0.2. Determine (i) the axial force to transmit the torque, and (ii) the average normal pressure on the contact surface when maximum torque is transmitted.

Given Data: The torque transmitted, $T = 200$ Nm at Speed, $N = 1440$ rpm
Outer diameter of the cone, $D_o = 350$ mm; Face width, $b = 65$ mm
Cone pitch angle, $\propto = 6.25°$; Coefficient of friction, $f = 0.2$

SOLUTION

Step-1: Axial force to transmit the torque

The torque equation, $T = \dfrac{f \cdot F_a \cdot r_m}{\sin \propto}$ (Eqn. 6.31 MDHB)

where, $r_m = r_0 - \dfrac{b}{2} \sin \propto = \dfrac{350}{2} - \dfrac{65}{2} \sin 6.25° = 171.46$ mm

$f = 0.2$, $T = 200$ Nm $= 200000$ Nmm

$$F_a = \frac{T \sin \propto}{f \cdot r_m} = \frac{2 \times 10^5 \times \sin 6.25°}{0.2 \times 171.46} = 634.93 \text{ N}$$

Step-2: Average normal pressure

The normal force, $F_n = \dfrac{F_a}{\sin \propto} = \dfrac{634.93}{\sin 6.25°} = 5832.2$ N

But the equation for normal force is $F_n = 2 \cdot \pi \cdot r_m \cdot p \cdot b$

∴ The normal pressure, $p = \dfrac{F_n}{2\pi r_m \cdot b}$

$$= \frac{5832.2}{2 \times \pi \times 171.46 \times 65} = 0.0833 \text{ MPa}$$

Example 19.22: A cone-clutch has a semi-cone angle of 12°. It is to transmit 10 kW power at 750 rpm. The width of the face is one-fourth of mean diameter of friction lining. If the normal intensity of pressure between the contacting surfaces is not to exceed 0.085 N/mm^2 and the coefficient of friction is 0.2, assuming the uniform wear conditions, calculate the dimensions of the clutch.

Given Data: Power, $NP = 10$ kW at 750 rpm; Face width, $b = \dfrac{D_m}{4}$

The maximum normal pressure, $p = 0.085$ N/mm^2

Coefficient of friction, $f = 0.2$; Uniform wear conditions exist

SOLUTION

Step-1: The mean radius by power equation,

$$\text{kW} = \frac{f \cdot p \cdot r_m^2 \cdot b \cdot N \times 10^{-5}}{15 \cdot K_s} \qquad \text{(Eqn. 6.32 MDHB)}$$

where, $\text{kW} = 10$; $N = 750$; $f = 0.2$, $p = 0.085$

$$b = D_m/4 = r_m/2 \text{ and } K_s = 1.2 \text{ (assumed)}$$

Substituting the values,

$$10 = \frac{0.2 \times 0.085 \times r_m^2 \times \left(\frac{r_m}{2}\right) \times 750 \times 10^{-5}}{15 \times 1.2}$$

$$\therefore \quad r_m^3 = \frac{10 \times 15 \times 1.2 \times 10^5 \times 2}{0.2 \times 0.085 \times 750} = 2.8235 \times 10^6$$

$$\therefore \quad r_m = 141.33 \text{ mm}$$

Step-2: Dimensions of the clutch

The face width, $b = \frac{r_m}{2} = 70.67 \text{ mm} \cong 71 \text{ mm}$

Mean diameter, $D_m = 2r_m = 282.66$ mm

The outside diameter, $D_o = D_m + b\sin \propto = 282.66 + 71\sin 12° = 297.4$

Choose $D_o = 300$ mm

The inner diameter,

$$D_i = D_m - b\sin \propto = 282.66 - 71 \times \sin 12° = 267.89 \text{ mm}$$

choose $D_i = 270$ mm

The exact face width,

$$b = \left(\frac{D_o - D_i}{2 \sin \propto}\right) = \frac{300 - 270}{2 \times \sin 12°} = 72.15 \text{ mm}$$

19.3.3 Centrifugal Clutch

In this type, the driving shaft carries a spider which houses four heavy shoes (or 'blocks') in radial guides. They are held in place by flat springs. The outer surface of the shoe is lined with friction material. This assembly is placed inside a rim-casing mounted on the driving shaft. When the engine is not rotating, the springs hold the shoes in position and there is no contact between the shoe face and the rim. When the engine starts rotating, the shoes are subjected to centrifugal force and are displaced outward. At a certain speed they touch the under surface of the casing-rim and are pressed against it by further increase in speed. Due to the friction between the surfaces, the casing also rotates along with spider. This is basically a friction clutch with cylindrical contact surfaces. Only actuation is due to the centrifugal action. Design of the clutch involves mainly, the determination of the mass and size of the shoes and sizes of the spider and the casing. The following analysis helps in determining them.

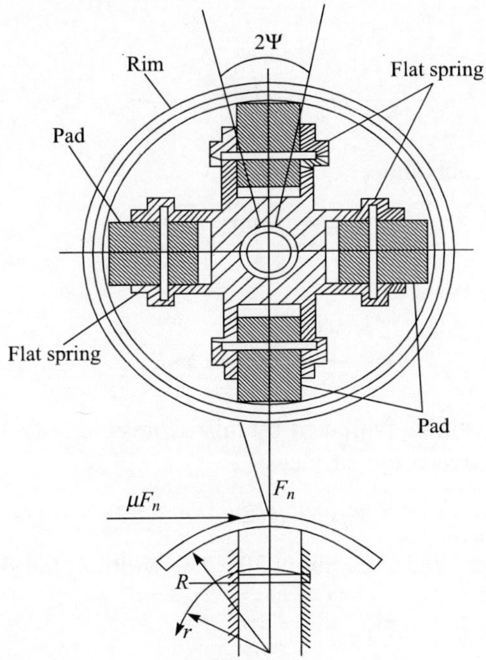

Figure 19.5

Let m = mass of each shoe.

r = radius of the C.G. of the shoe mass.

R = radius of the inner surface of the rim.

b = axial width of the shoe.

2Ψ = angle subtended by the shoe face at the centre.

N = Normal speed of the engine shaft in rpm.

ω = Normal angular speed of the engine shaft = $\left(\dfrac{2\pi N}{60}\right)$ rad/sec

ω_i = The angular speed at which the engagement starts

= ¾ ω (in most of the cases)

f = coefficient of friction

The centrifugal force on each shoe at the normal speed, $F_c = m\omega^2 \cdot r$

The C.F. at moment of engagement, $F_i = m\omega_1^2 r = m(3\omega/4)^2 r = \dfrac{9}{16} m\omega^2 r$

The radial force on the rim due to each shoe,

$$F_n = (F_c - F_i) = m\omega^2 r - \frac{9}{16} m\omega^2 r = \frac{7}{16} m\omega^2 r \tag{19.16}$$

Friction force per shoe,

$$F_t = fF_n = f\left(\frac{7}{16} m\omega^2 r\right) \tag{19.17}$$

The torque transmitted by each shoe,

$$T_1 = \frac{7}{16} fm\omega^2 r \cdot R \tag{19.18}$$

Total torque transmitted by n-shoes,

$$T = \frac{7}{16} n \cdot f \cdot m \cdot \omega^2 r \cdot R \tag{19.19}$$

The mass, m of each shoe can be evaluated by this expression.
Let p = normal pressure between the surfaces.

Then
$$F_n = b \cdot R(2\Psi) \cdot p \tag{19.20}$$

The angle Ψ is normally specified to be about 30°. The width of the shoe, b can be determined from the above expression.

The circumferential length, $L = 2R\Psi$.

From these the spider and casing dimensions are fixed to accommodate the shoe in place.

Example 19.23: A centrifugal clutch is to transmit 25 kW at 750 rpm. The engagement begins at 75% of the normal speed. The inner diameter of the drum is 300 mm and the radial distance of the C.G. of the shoe mass is 125 mm from the shaft centre. The coefficient of friction is to be 0.25. Determine the mass and size of the shoe and the normal pressure is limited to 0.1 N/ mm^2.

Given Data: Power, NP = 25 kW at 750 rpm; Drum diameter = 300 mm;
CG of the mass = 125 mm;
Coefficient of friction, f = 0.25; Normal pressure, p = 0.1 MPa

SOLUTION

Step-1: Mass of the shoe

The torque transmitted,

$$T = \frac{955 \times 10^4 \text{kW}}{N} = \frac{955 \times 10^4 \times 25}{750} = 318.3 \times 10^3 \text{Nmm} = 318.3 \text{ Nm}.$$

Torque per shoe, $T = \dfrac{318.3}{4} = 79.575$ Nm

The expression for torque, $T_1 = \dfrac{7}{16} f \cdot m \cdot \omega^2 \cdot r \cdot R$

where, $T_1 = 79.575$ Nm, $f = 0.25$, $r = 125$ mm $= 0.125$ m,

$$R = 150 \text{ mm} = 0.15 \text{ m}, \omega = \frac{2\pi N}{60} = \frac{2\pi \times 750}{60} = 78.54 \text{ rad/sec}$$

$$\therefore \quad 79.575 = \frac{7}{16} \times 0.25 \times m \times (78.54)^2 \times 0.125 \times 0.150 = 12.65 \text{ m}$$

$$\therefore \quad m = \frac{79.575}{12.65} = 6.29 \text{ kg.}$$

Mass of each shoe = 6.29 kg.

Step-2: The shoe dimensions

The tangential force, $F_t = T_1/R = \dfrac{79575}{150} = 530.5$ N

The normal force, $F_n = \dfrac{530.5}{0.25} = 2122$ N

Contact area of the shoe, $A = \dfrac{F_n}{p} = \dfrac{2122}{0.1} = 21220 \text{ mm}^2$

But $A = 2b \cdot R\Psi$, where, $R = 150$ mm, $\Psi = 30° = 0.5236$ rad

Shoe width, $b = \dfrac{21220}{2 \times 150 \times 0.5236} = 135.1$ mm

Length, $L = 2R\Psi = 2 \times 150 \times 0.5236 = 157.0$ mm

19.4 BRAKES

The brakes may be classified as

(i) Block brakes (ii) Shoe brakes

(iii) Band brakes (iv) Band and block brakes

19.4.1 Block Brakes

A typical block brake is shown in Fig. 19.6. A block with curved lined surface is pressed against the brake-drum by lever arrangement. The frictional force between the surfaces resists the rotation of the drum and slows it down. The braking torque can be evaluated as follows:

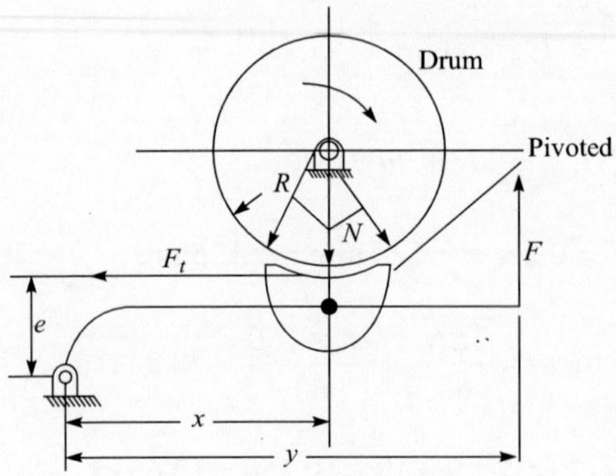

Figure 19.6

Consider an elemental area dA of the contact surface at angle ϕ, from the central axis of the block,

$d\phi$ = small angle subtended by the element
b = width of the block
p = normal pressure between the surfaces
r = radius of the drum.

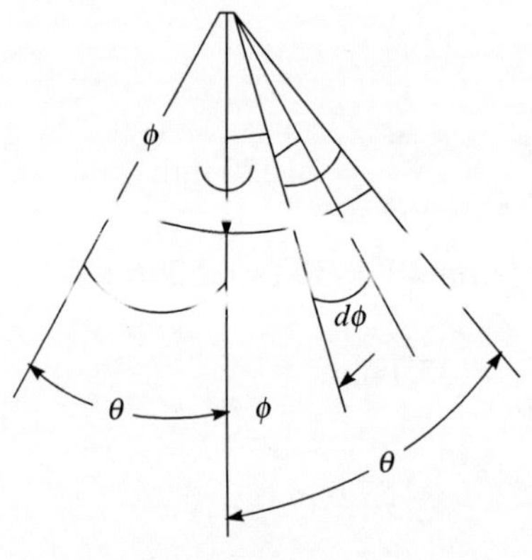

Figure 19.6a

The normal force on the element, $dF_n = p \cdot dA = p \cdot b \cdot r \cdot d\phi$.
The component along the axis, $dN = dF_n \cos\phi = p \cdot b \cdot r \cdot \cos\phi \cdot d\phi$.
The total force,

$$N = \int dN = \int_{-\theta}^{\theta} p \cdot b \cdot r \cdot \cos\phi \, d\phi = (br) \int_{-\theta}^{\theta} p \cdot \cos\phi \, d\phi \tag{19.21}$$

The frictional force on the element, $dF_t = f \cdot dF_n = f \cdot p \cdot b \cdot r \cdot d\phi$

The torque due to the element, $dT = r \cdot dF_t = f \cdot p \cdot b \cdot r^2 \cdot d\phi$

∴ The total frictional torque,

$$T = \int_{-\theta}^{\theta} f \cdot p \cdot b \cdot r^2 \cdot d\phi \tag{19.22}$$

The integrals depend upon the variation of pressure. The block is pressed down and the wear of the lining is constant along the axis. Then the pressure variation is given by $p = C \cos \phi$ where C is a constant.

The normal force,

$$N = \int_{-\phi}^{\phi} brC\cos^2\phi \cdot d\phi = (C \cdot b \cdot r)\int_{-\phi}^{\phi} \cos^2\phi \cdot d\phi = (Cbr)\left[\frac{2\theta + \sin 2\theta}{2}\right] \tag{19.23}$$

The braking torque,

$$T = \int_{-\theta}^{\theta} f \cdot b \cdot C \cdot r^2 \cdot \cos\phi\, d\phi = 2f \cdot b \cdot C \cdot r^2 \sin\theta \tag{19.24}$$

Eliminating C,

$$T = f \cdot N \cdot r\left[\frac{4\sin\theta}{2\theta + \sin 2\theta}\right] \tag{19.25}$$

The tangential force,

$$F_t = \frac{T}{r} = f \cdot \left[\frac{4\sin\theta}{2\theta + \sin 2\theta}\right] N = f_1 N \tag{19.26}$$

where, $f_1 = f\left[\dfrac{4\sin\theta}{2\theta + \sin 2\theta}\right]$ is called the virtual coefficient of friction.

The relation between the operating force F and T is derived as follows:

Taking moment about the hinge, O_1.

$$F \cdot y - x \cdot N + F_t \cdot e = 0$$

∴
$$F \cdot y - x \cdot N + f_1 \cdot N \cdot e = 0$$

Solving
$$N = \left[\frac{F \cdot y}{x - f_1 e}\right] \tag{19.27}$$

The torque,

$$T = f_1 N \cdot r = \frac{f_1 F \cdot y \cdot r}{x - f_1 e} \tag{19.28}$$

If $x = f_1 e$, the brake is called self energizing brake.

Double-block brake

In this arrangement two identical blocks are provided on the diametrically opposite positions. A common spring connected between the two lever ends, provides the braking force.

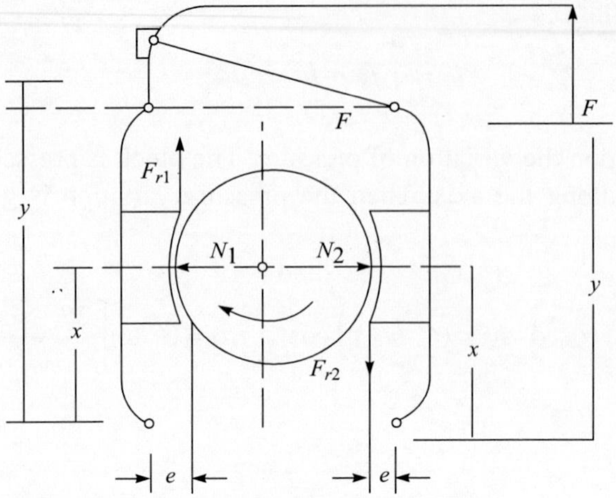

Figure 19.7

Considering the equilibrium of the two levers separately,

$$F \cdot y - F_{t1} e - N_1 x = 0 \text{ and } N_1 = F_{t1}/f_1 \text{ for the left block}$$
$$F \cdot y + F_{t2} e - N_2 x = 0 \text{ and } N_2 = F_{t2}/f_1 \text{ for the right block}$$

Solving the equations, we get, $F_{t1} = \dfrac{f_1 F \cdot y}{x + e f_1}$ and $F_{t2} = \dfrac{f_1 F \cdot y}{x - e f_1}$ (19.29)

Total torque, $T = (F_{t1} + F_{t2}) \cdot r$ (19.30)

Table 19.5. Block-brake formulae for pivoted block-brake 20 > 60°, replace f by $f_o = (4f \sin\theta)/(2\theta + \sin 2\theta)$

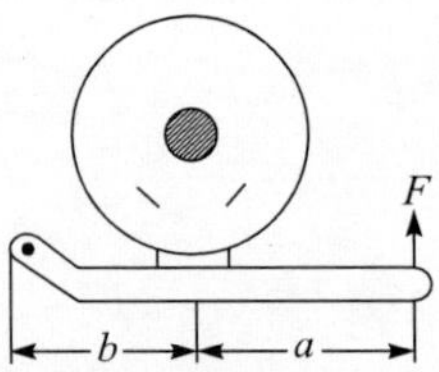	$F = \dfrac{Ft}{f}\left[\dfrac{b}{a+b}\right]$ for rotation in either directions	F = Operating force at the end of the lever N Ft = Tangential force at the rim of the brake drum N
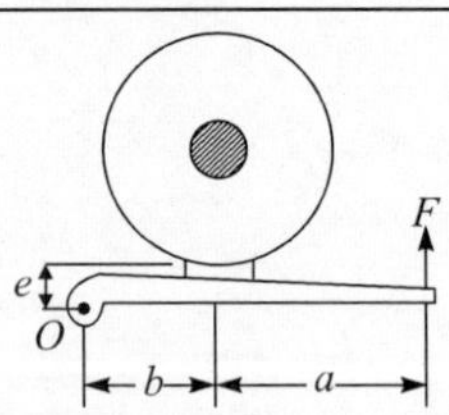	Clockwise rotation $F = \left(\dfrac{Ft\,b}{a+b}\right)\left(\dfrac{1}{f} - \dfrac{e}{b}\right)$	Anticlockwise rotation $F = \left(\dfrac{Ft\,b}{a+b}\right)\left(\dfrac{1}{f} + \dfrac{e}{b}\right)$
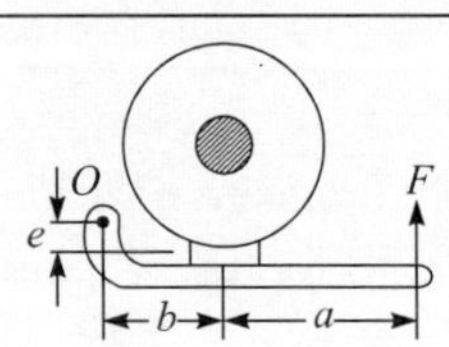	Clockwise rotation $F = \left(\dfrac{Ft\,b}{a+b}\right)\left(\dfrac{1}{f} + \dfrac{e}{b}\right)$	Anticlockwise rotation $F = \left(\dfrac{Ft\,b}{a+b}\right)\left(\dfrac{1}{f} - \dfrac{e}{b}\right)$

Example 19.24: Determine the braking torque of a single block-brake shown in Fig. 19.8 for the coefficient of friction = 0.3.

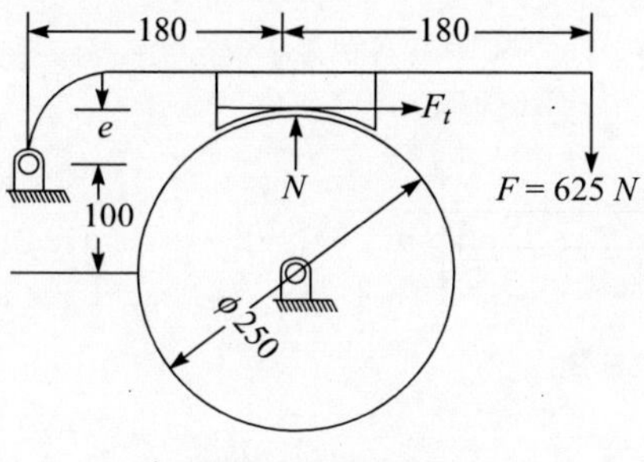

Figure 19.8

$F = 625$ N,

$r = 125$ mm,

$e = 125 - 100 = 25$ mm,

$X = 180$ mm,

$Y = 360$ mm

$f_1 = f$, for very small θ

SOLUTION

Taking moments about the hinge point,

$$F \cdot y - Nx + eF_t = F \cdot y - Nx + Nf_1e = 0$$

$$\therefore \quad N = \frac{F \cdot y}{x - f_1e} \text{ and } F_t = f_1N = \frac{f_1F \cdot y}{x - f_1e}$$

The brake torque, $T = \dfrac{f_1F \cdot Y \cdot r}{x - f_1e} = \dfrac{0.3 \times 625 \times 360 \times 125}{180 - 0.3 \times 25}$

$= 48913$ Nmm $= 48.913$ Nm

Example 19.25: A double block-brake is actuated as below. The drum rotates at 80 rpm. When the applied force, $F = 2350$ N, and the coefficient of friction = 0.3, determine the braking torque and the amount of heat per unit time.

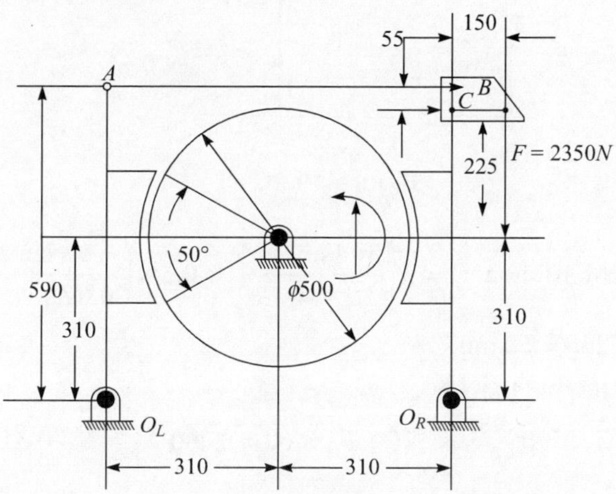

Figure 19.9

SOLUTION Free body diagram of the operating lever

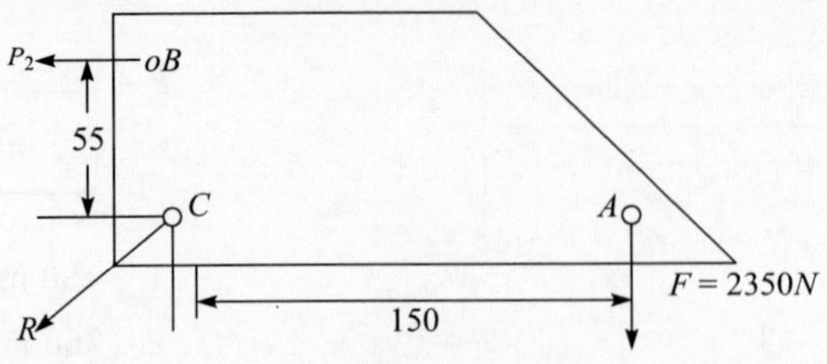

Figure 19.9a

Taking moment about C,

$$P_2 \times 55 = 2350 \times 150$$

$$\therefore \quad P_2 = \frac{2350 \times 150}{55} = 6409.1 \text{ N}$$

The force on the pin at C

$$R = \sqrt{F^2 + P_2^2} = \sqrt{2350^2 + 6409.1^2} = 6826.3 \text{ N}$$

The actuating force on the left lever $P_1 = P_2 = 6409.1$ N

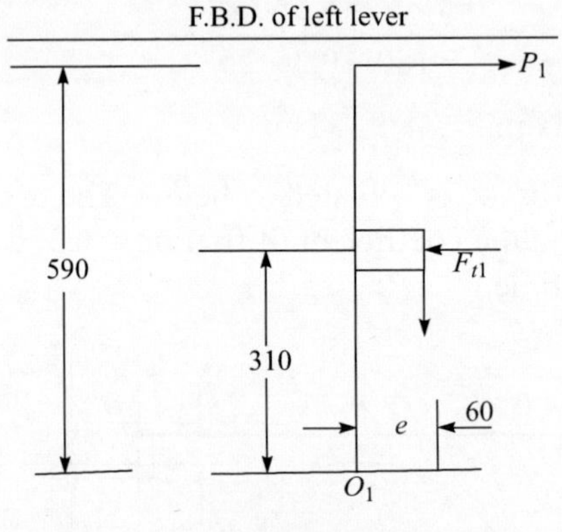

Figure 19.9b

$$\text{Virtual coefficient of friction } f_1 = f\left(\frac{4\sin\theta}{2\theta + \sin 2\theta}\right) = 0.3 \times \left(\frac{4 \times \sin 25^\circ}{0.87 + \sin 50^\circ}\right) = 0.311$$

Eccentricity $e = 310 - 250 = 60$ mm

Taking moment about the hinge O_1,

$$6409.1 \times 590 - N_1 \times 310 + F_{t1} \times 60 = 0 \text{ and } N_1 = F_{t1}/0.311$$

$$\therefore \quad F_{t1} = \frac{6409.1 \times 590}{\left(\dfrac{310}{0.311} - 60\right)} = 4022.7 \text{ N}$$

Considering free body diagram of right lever,

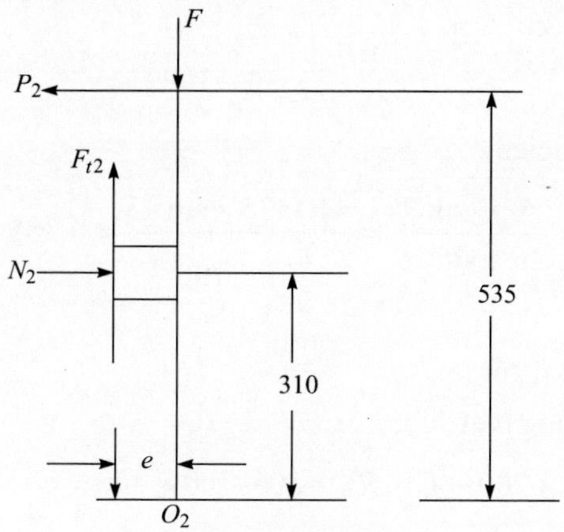

Figure 19.9c

Moments about O_2,

$$F_{t2} \times 60 + N_2 \times 310 - 6409.1 \times 535 = 0$$

$$\therefore \qquad F_{t2} = \frac{6409.1 \times 535}{\left(60 + \dfrac{310}{0.311}\right)} = 3234.8 \text{ N}$$

The total braking torque,

$$T = (F_{t1} + F_{t2}) \times r = (4022.7 + 3234.8) \times 225 \text{ Nmm} = 1632937.5 \text{ Nmm}$$

Heat generated /sec.,

$$H_g = T\omega = 1632.94 \times \frac{2\pi \times 80}{60} = 13680.1 \text{ Nm/sec} = 13.68 \text{ kW} = 3.27 \text{ KCl/sec}$$

Example 19.26: The diameter of the drum of a single block brake shown in Fig. 19.10 is 200 mm and the angle of contact is 90°. If the operating force of 700 N is applied at the end of a lever and the coefficient of friction between the drum and lining is 0.35, determine the torque that may be transmitted by the block brake.

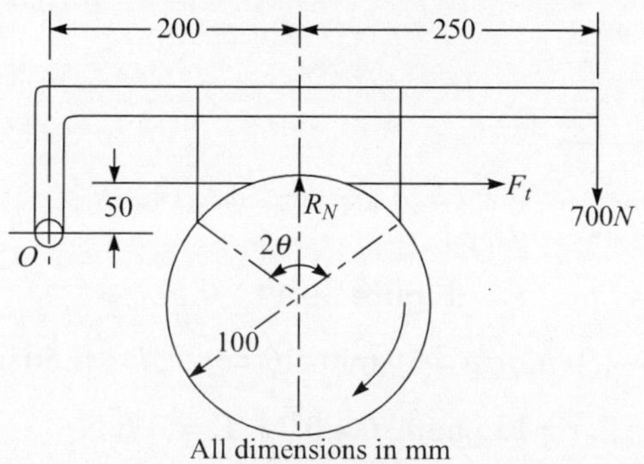

Figure 19.10

Given Data: Drum diameter $D = 200$ mm; Block angle $2\theta = 90°$
Coefficient of friction $f = 0.35$

SOLUTION

Step-1: Effective friction coefficient,

$$f_1 = \frac{4 \cdot f \cdot \sin\theta}{2\theta + \sin 2\theta} = \frac{4 \times 0.35 \times \sin 45}{\frac{\pi}{2} + \sin 90°} = 0.385$$

Step-2: Determination of normal force R_N.
By taking moments about the pivot,

$$R_N \times 200 - f_0 \times R_N \times 50 - 700 \times 450 = 00$$

$$\therefore \quad (200 - 0.385 \times 50)R_N = 700 \times 450$$

$$\therefore \quad R_N = 1742.75 \text{ N}$$

Friction force, $F_t = f_1 R_N = 0.385 \times 1742.75 = 671.0$ N

Step-3: Braking torque, $T = r \cdot F_t = 100 \times 671 = 67100$ Nmm

Example 19.27: A single block brake is shown in Fig. 19.11. The drum diameter is 250 mm. The contact angle is 90°. If an operating force of 700 N is applied at the end of the lever and the coefficient of friction is 0.35, determine the torque that may be sustained by the brake.

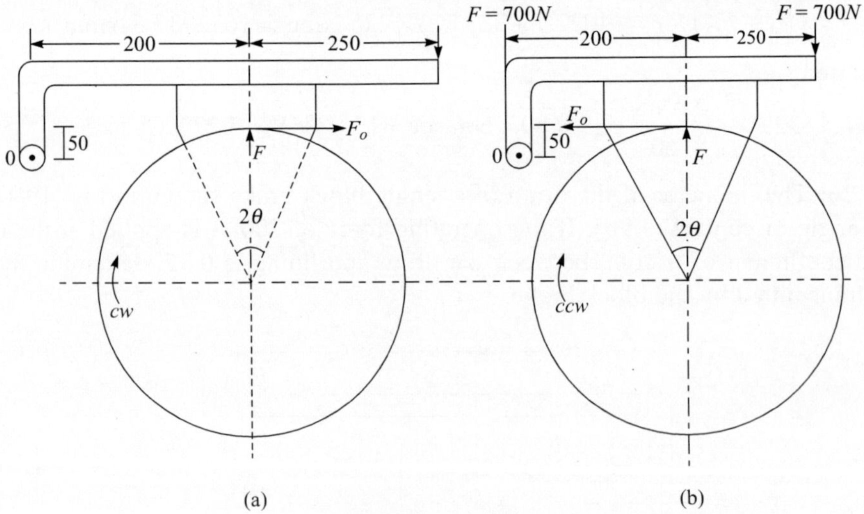

Figure 19.11

Given Data: $b = 200$ mm; $a = 250$ mm; $e = 50$ mm; $2\theta = 90°$; $D = 250$ mm

$$\therefore \quad r = 125 \text{ mm}; f = 0.35; F = 700 \text{ N}$$

SOLUTION

Step-1: Effective friction coefficient

Since $2\theta > 60°$, equivalent coefficient of friction,

$$f_1 = \frac{4f \sin\theta}{2\theta + \sin 2\theta} = \frac{4 \times 0.35 \times \sin 45°}{\dfrac{\pi}{2} + \sin 90°} = 0.385$$

For clockwise rotation since F_t is away from the fulcrum, use formula 6.89 for ccw from the table for block brakes (in MDHB).

Applied effort or actuating force, $F = \dfrac{F_t \cdot a}{a+b}\left(\dfrac{1}{f} - \dfrac{e}{b}\right)$ (Eqn. 6.89 MDHB)

i.e.,

$$700 = \frac{F_t \times 200}{200 + 250}\left[\frac{1}{0.385} - \frac{50}{200}\right]$$

∴ Tangential force, $F_t = 670.954$ N

∴ Torque, $M_t = F_t \cdot r = 670.954 \times 125 = 83869.3$ Nmm

For ccw rotation F_t is towards the fulcrum (Fig. 9.11b), use the formula 6.88 for cw from the table for block brakes (in MDHB).

Applied effort, $F = \dfrac{F_t a}{a+b}\left(\dfrac{1}{f} + \dfrac{e}{a}\right)$ (Eqn. 6.88 MDHB)

i.e.,

$$700 = \frac{F_t \times 200}{200 + 250}\left(\frac{1}{0.385} + \frac{50}{200}\right)$$

∴ Tangential force, $F_t = 553.136$ N

∴ Torque, $M_t = F_t \cdot r = 553.136 \times 125 = 69142$ Nmm

Hence, torque capacity of the given brake is the larger of the two.

∴ Torque capacity, $M_t = 83869.3$ Nmm

Example 19.28: A single block-brake with a torque capacity of 250 Nm is shown in Fig. 19.12. The brake drum rotates at 100 rpm and the coefficient of friction is 0.35. Calculate:

(i) The actuating force and the hinge pin reaction.

(ii) Rate of heat generated during the braking action.

(iii) The dimensions of the block, if the intensity of pressure between the block and brake drum is 1 MPa. The length of the block is twice its width.

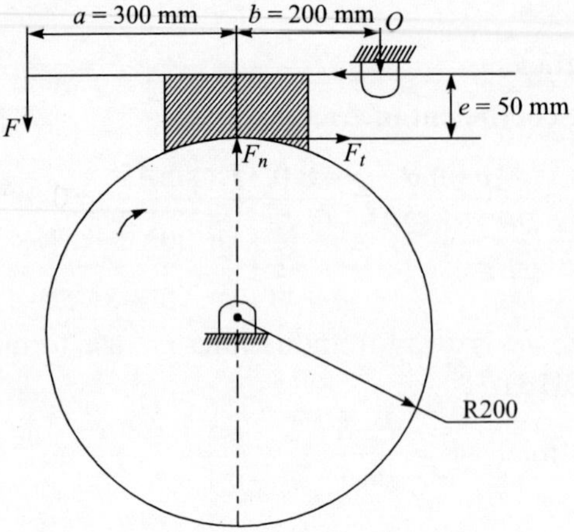

Figure 19.12

Given Data: $M_t = 250\text{ Nm} = 250 \times 10^3\text{ Nmm}$; Speed $N = 100$ rpm

a = 200 mm; b = 300 mm; c = 50 mm; $f = 0.3$

$p = 1\text{ MPa} = 1\text{ N/mm}^2$; $L = 2b$; r = 200 mm.

SOLUTION For the given direction of rotation (i.e., *cw* direction) the various forces acting on the shoe and the pin hinge are shown in Fig. 19.12.

Step-1: Actuating force for the torque capacity

Torque capacity, $M_t = F_t r$

i.e., $$250 \times 10^3 = F_t \times 200$$

$\therefore$ Tangential force, $F_t = \dfrac{250 \times 10^3}{200} = 1250\text{ N}$

Normal force on the shoe, $F_n = \dfrac{F_t}{f} = \dfrac{1250}{0.35} = 3571.43\text{ N}$

Taking moments about the hinge O

$$F \times (a+b) + F_t \times e = F_n \times b;$$

i.e., $$F(a+b) + F_t \cdot e = \frac{F_t}{f} \cdot b$$

i.e., $$F = \frac{\dfrac{F_t}{f} \cdot b - F_t \cdot e}{(a+b)} = \frac{F_t \cdot b}{(a+b)}\left(\frac{1}{f} - \frac{e}{b}\right)$$

$$= \frac{1250 \times 200}{(200+300)}\left(\frac{1}{0.35} - \frac{50}{200}\right) = 1303.6\text{ N}$$

$\therefore$ Actuating force, $F = 1303.6$ N

Step-2: Hinge reaction

For the equilibrium of lever in the vertical direction, sum of the vertical forces must be zero

$$F - F_n + F_y = 0$$

i.e., $$1303.6 - 3571.43 + F_y = 0;$$

$\therefore$ $$F_y = 2267.83 \text{ N}$$

For equilibrium in the horizontal direction, $F_t - F_x = 0$

i.e., $$1250 - F_x = 0;$$

$\therefore$ $$F_x = 1250 \text{ N}$$

$\therefore$ Resultant hinge pin reaction, $F_R = \sqrt{F_x^2 + F_y^2}$

$$= \sqrt{1250^2 + 2267.83^2} = 2589.5 \text{ N}$$

Step-3: Rate of heat generated

Heat generated, $H_g = f \cdot F_n \cdot V$

$$= 0.35 \times 3571.43 \times \left(\frac{\pi \times 400 \times 100}{60000}\right) = 2618 \text{ j/sec.}$$

Step-4: Dimensions of the block

Normal force, $F_n = p \times$ Projected area of the block $= p \cdot b \cdot L$

i.e., $$3571.43 = 1 \times 2b \times b \; (\therefore L = 2b')$$

$\therefore$ Width of shoe $b = 42.26$ mm, and length of shoe $L = 84.52$ mm.

Example 19.29: Determine the torque that may be resisted by the single block brake shown in Fig. 19.13 for a coefficient of friction 0.3.

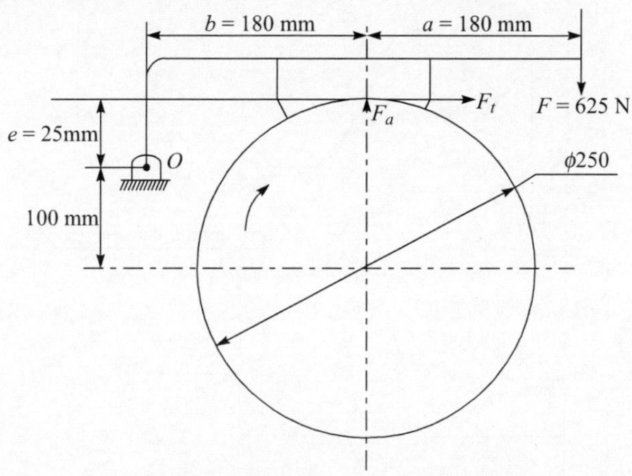

Figure 19.13

Given Data: $a = 180$ mm; $b = 180$ mm; $e = 125 - 100 = 25$ mm;

$$D = 250 \text{ mm}; r = \frac{250}{2} = 125 \text{ mm}; f = 0.3; F = 625 \text{ N}$$

SOLUTION For the given direction of rotation (*cw*), the various forces acting on the shoe are as shown in Fig. 9.13. As F_t is above the centre of drum and fulcrum and its direction is away from the fulcrum, use *ccw* formula 6.89 from MDHB.

$$\text{Actuating force, } F = \frac{F_t \cdot b}{(a+b)}\left(\frac{1}{f} - \frac{e}{b}\right) \qquad \text{(Eqn. 6.89 MDHB)}$$

i.e.,

$$625 = \frac{F_t \cdot 180}{(180+180)}\left(\frac{1}{0.3} - \frac{25}{180}\right)$$

$\therefore$ Tangential force, $F_t = 391.30435$ N

$\therefore$ Braking torque, $M_t = F_t \cdot r = 391.30435 \times 125 = 48913$ Nmm

Example 19.30: The block-brake shown in Fig. 19.14 below has two shoes each subtending an angle 80°. The coefficient of friction of the brake material is 0.3. The brake drum diameter is 800 mm while the rope drum measures 1200 mm in diameter. It is required to stop the load of 20 kN lowering at a velocity of 5.0 m/s in a distance of 20 metres. Determine the effort *E* required at the end of the lever, the width of the brake shoes given the permissible pressure of 0.7 MPa and the amount of heat generated.

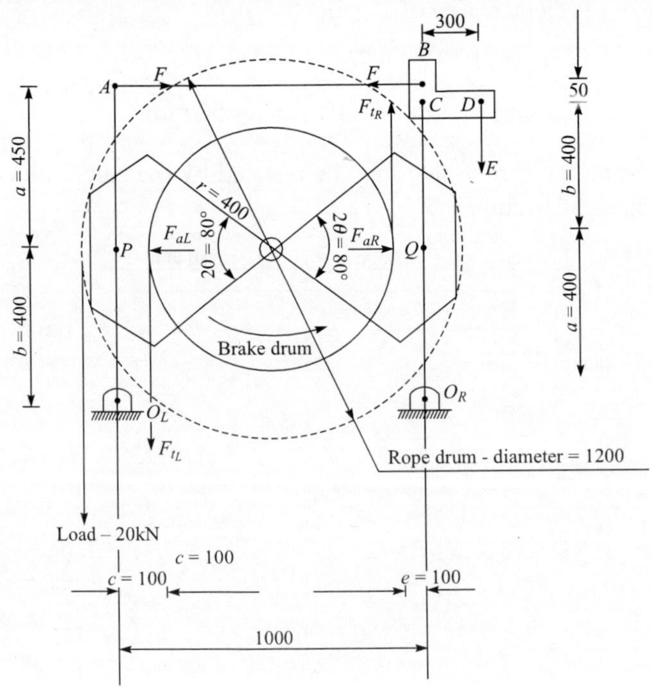

Figure 19.14

Given Data: $2\theta = 80°$, $\therefore \theta = 40°$; $f = 0.3$; $D = 800$ mm, $\therefore r = 400$ mm;
Diameter of rope drum = 1200 mm; Load on the rope drum = 20 kN
Velocity, $V = 5$ m/sec; Distance moved by the load = 20 m
Permissible pressure, $p = 0.7$ MPa

SOLUTION

Step-1: Torque transmitted

Torque on the rope drum shaft = Load on the rope drum × Radius of rope drum

$$= 20{,}000 \times \frac{1200}{2} = 12 \times 10^6 \text{ Nmm}$$

Since the rope drum and the brake drum are on the same shaft, braking torque,

$$M_t = 12 \times 10^6 \text{ Nmm}$$

Step-2: Equilibrium of the left shoe

The various forces acting on the shoes for the given direction of rotation are shown in Fig. 19.14. Since $2\theta > 60°$, equivalent coefficient of friction,

$$f_1 = \frac{4 f \sin\theta}{2\theta + \sin 2\theta} = \frac{4 \times 0.3 \times \sin 40°}{80 \times \frac{\pi}{180} + \sin 80°} = 0.324$$

F_t lies between the centre of drum and the fulcrum and acting towards the fulcrum, use eqn. 6.86 MDHB from Table for block brake.

$$\therefore \text{ Actuating force, } F = \frac{F_{tL} \cdot b}{(a+b)}\left[\frac{1}{f_1} - \frac{e}{b}\right]$$

i.e.,

$$F = \frac{F_{tL} \times 400}{(400+450)}\left[\frac{1}{0.324} - \frac{100}{400}\right] = 1.3348\, F_{tL}$$

$$\therefore \qquad F_{tL} = 0.7492 F$$

Step-3: Equilibrium of the right shoe

F_{t_R} lies between the centre of drum and the fulcrum, and is acting away from the fulcrum, use eqn. 6.87 MDHB from Table for block-brakes.

$$\therefore \text{ Actuating force, } F = \frac{F_{\theta_R} \cdot b}{(a+b)}\left[\frac{1}{f_1} + \frac{e}{b}\right] \qquad \text{(Eqn. 6.87 MDHB)}$$

i.e.,

$$F = \frac{F_{t_R} \times 400}{(400+400)}\left[\frac{1}{0.324} + \frac{100}{400}\right] = 1.66821\, F_{t_R}$$

$\therefore \quad F_{t_R} = 0.59944F$

Step-4: Effort, E required

Braking torque, $M_t = (F_{tL} + F_{tR})r$

i.e., $\quad 12 \times 10^6 = (0.7492\,\text{F} + 0.59944\,\text{F}) \times 400$

$\therefore$ Actuating force, $F = 12 \times \dfrac{10^6}{1.34864 \times 400} = 22244.6$ N

Taking moments about, C

$$F \times 50 = E \times 300$$

i.e.,
$$22244.6 \times 50 = E \times 300$$

$\therefore$ Effort, E required at the end of the lever = 3707.4 N

Step-5: Width of brake shoe

Tangential force on the left shoe,

$$F_{tL} = 0.7492\,F = 0.7492 \times 22244.6 = 16665.65 \text{ N}$$

Tangential force on the right shoe,

$$F_{tR} = 0.59944\,F = 0.59944 \times 22244.6 = 13334.3 \text{ N}$$

Normal force on the left shoe, $F_{nL} = \dfrac{F_{tL}}{f} = \dfrac{16665.65}{0.324} = 51437.2$ N

Normal force on the right shoe, $F_{tR} = \dfrac{F_{\theta R}}{f_i} = \dfrac{13334.3}{0.324} = 41155.25$ N

$\therefore$ Maximum normal force, $F_{\max} = 51437.2$ N

Projected length of shoe, $L = r\theta = 600 \times \dfrac{80 \times \pi}{180} = 558.5$ mm

$\therefore$ Width of shoe, $b' = \dfrac{F_{\max}}{L} = \dfrac{51437.2}{558.5} = 131.57 \text{ mm} = 132 \text{ mm}$

Step-6: Heat generated

Total heat generated per unit time = Heat generated on the left shoe + Heat generated on the right shoe $= (F_{\theta L} + F_{n_R})V$

$$= (16665.65 + 13334.3) \times 5 \times \frac{800}{1200} = 99999.83 \text{ Watts} = 100 \text{ kW or kJ/sec}$$

Time taken by the drum to come to rest, $t = \dfrac{20}{v} = \dfrac{20}{5} = 4$ sec.

$\therefore$ Total amount of heat generated $= 100 \times 4 = 400$ kJ.

Example 19.31: A double shoe brake is shown in Fig. 19.15. The braking torque required is 1200 Nm. The brake drum is 200 mm in diameter and the coefficient of friction between the shoe and drum is 0.3. Find the operating force P and the size of the shoe, if the permissible pressure does not exceed 1.5 MPa. Assume the length of the shoe as twice the width. Analysis may be made considering it as short shoe.

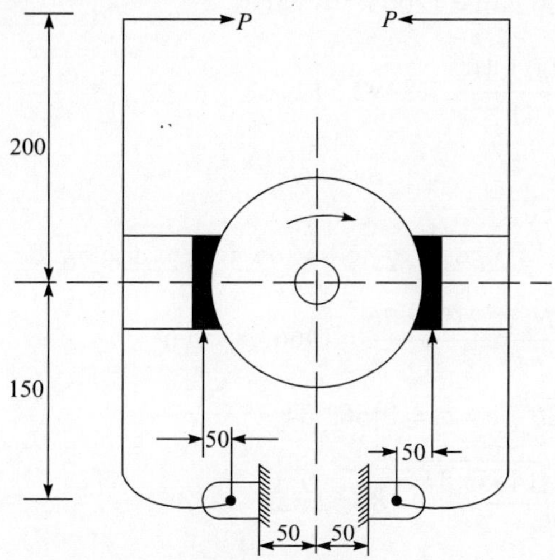

Figure 19.15

Given Data: Torque = 1200 Nm; Drum dia. = 200 mm; Coefficient of friction, $f = 0.3$
Maximum contact pressure, p = 1.5 MPa

SOLUTION

Step-1: The equilibrium of left lever

Taking moment about O,

$$150N_1 - 50 \times F_{t_1} - 350P = 0$$

$$150N_1 - 50 \times 0.3N_1 = 350P$$

$$\therefore \qquad N_1 = \frac{350}{135} p = 2.59P \tag{1}$$

Step-2: The equilibrium of right lever

Taking moments about O_2,

$$150N_2 + 50F_{t2} - 350P = 0$$

$$150N_2 + 50 \times 0.3N_2 = 350P$$

$$N_2 = \frac{350}{165} p = 2.12P \tag{2}$$

Step-3: The operating force

The torque due to left shoe, $T_1 = F_{t1}r = 0.3N_1 \times 100 = 30N_1$

The torque due to the right shoe, $T_2 = r \cdot F_{t_2} = 100 \times 0.3 N_2 = 30 N_2$.

The total torque, $T = T_1 + T_2 = 30(N_1 + N_2)$

$$= 30(2.56 + 2.12)P = 141.3 \tag{3}$$

The given torque, $T = 1200 \text{ Nm} = 1200 \times 10^3 \text{ Nmm}$.

$$\therefore \quad P = \frac{1200 \times 10^3}{141.3} = 8492.57 \text{ N}$$

Step-4: Size of the shoes

The maximum normal force,

$$N_1 = 2.59P = 2.59 \times 8492.57 = 21995.76 \text{ N}$$

The area of the shoe, $A = \dfrac{N_1}{p} = \dfrac{21995.76}{1.5} = 14663.84 \text{ mm}^2$

$$\therefore \quad L \times b = 2b \times b = A = 14663.84$$

$$\therefore \quad b = \sqrt{\frac{14663.84}{2}} = 85.62 \text{ mm}$$

Choose b = 86 mm, and length $L = 2b$ = 172 mm.

19.4.2 Shoe Brakes

These are of two types:

(a) Internal expanding shoe brake

(b) External contracting shoe brake

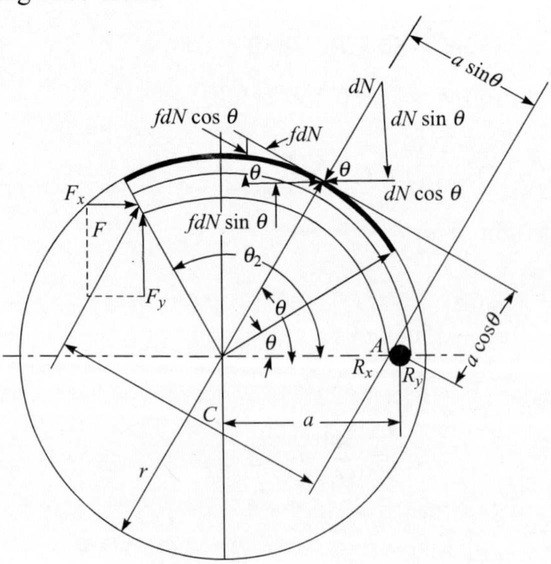

Figure 19.16 Internal expanding shoe

The general analysis is as follows.
Consider an element subtending an angle $d\theta$
Normal force on the element,

$$dN = p \cdot b \cdot r \cdot d\theta = \frac{p_m}{\sin \theta_m} b \cdot r \cdot \sin \theta \cdot d \cdot \theta$$

where, p_m = Maximum pressure, θ_m = position of maximum pressure

The frictional force on the element,

$$dF_t = fdN = \frac{f \cdot p_m \cdot br}{\sin \theta_m} \sin \theta \cdot d\theta$$

The frictional torque on the element,

$$dT = dF_t \cdot r = f \cdot dN \cdot r = \frac{f \cdot p_m \cdot b \cdot r^2}{\sin \theta_m} \sin \theta \cdot d\theta$$

The total frictional torque,

$$T = \int_{\theta_1}^{\theta_2} dT = \int_{\theta_1}^{\theta_2} \frac{f \cdot p_m \cdot br^2}{\sin \theta_m} \sin \theta \cdot d\theta = \left(\frac{f \cdot p_m \cdot br^2}{\sin \theta_m} \right) \int_{\theta_1}^{\theta_2} \sin \theta \cdot d\theta$$

$$T = \left(\frac{f \cdot p_m \cdot br^2}{\sin \theta_m} \right) \int [\cos \theta_1 - \cos \theta_2] = (f \cdot R_N r)(\cos \theta_1 - \cos \theta_2) \tag{19.31}$$

where, $R_N = \dfrac{p_m \cdot b \cdot r}{\sin \theta_m}$ – unit normal force

Moment of normal force about hinge point,

$$M_N = \int dM_N = \int dF_n \cdot a \cdot \sin \theta = \frac{\int p_m \cdot b \cdot r \cdot a}{\sin \theta_m} \sin^2 \theta \cdot d\theta$$

$$= R_N a \int_{\theta_1}^{\theta_2} \sin^2 \theta \cdot d\theta \tag{19.32}$$

Moment of frictional forces about hinge point,

$$M_f = \int dM_f = \int f dF_n (r - a \cos \theta) = \int \frac{f \cdot p_m \cdot b \cdot r}{\sin \theta_m} (r - a \cos \theta) \sin \theta d\theta$$

$$= f R_N \int_{\theta_1}^{\theta_2} (r - a \cos \theta) \sin \theta d\theta \tag{19.33}$$

The actuating force, $F = \dfrac{M_N \pm M_f}{C}$ (19.34)

For internal shoe, –ve for clockwise rotation & +ve for anticlockwise rotation
For external shoe, +ve for clockwise rotation & –ve for anticlockwise rotation
Normally the shoes are used in pairs, and total torque is,

$$T = (f \cdot R_{N1} r + f \cdot R_{N2} r)(\cos\theta_1 - \cos\theta_2)$$

$$= \frac{f \cdot b \cdot r^2}{\sin\theta_m}(p_{m1} + p_{m2})(\cos\theta_1 - \cos\theta_2) \qquad (19.35)$$

p_{m1} = Maximum pressure for right shoe

$$p_{m2} = \text{Maximum pressure for left shoe} = \frac{C \cdot F \cdot p_m}{M_N + M_f} \qquad (19.36)$$

θ_m = Location of max. pressure $= 90°$ if $\theta_2 > 90°$

$= \theta_2$ if $\theta_2 < 90°$

Example 19.32: For internal expanding shoe brake shown in Fig. 19.17 below, $\theta_1 = 15°, \theta_2 = 105°$. The value of maximum shoe pressure is 0.7 MPa and the coefficient of friction $f = 0.25$. The width of the lining $b = 50$ mm. The inner radius of the drum is 125 mm. Find the torque capacity.

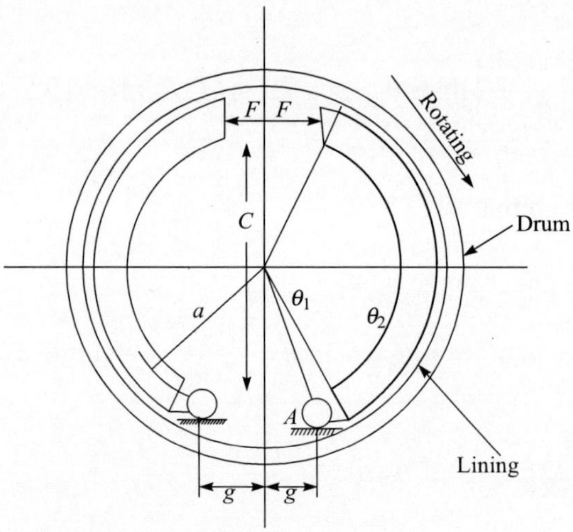

Figure 19.17 Internal expanding shoe brake

Given Data:

$f = 0.25$	$C = 180$ mm
$b = 50$ mm	$g = 40$ mm
$r = 125$ mm	

$$a = \sqrt{90^2 + 40^2} = 98.5 \text{ mm}$$

$p_{m1} = 0.7$ MPa for right shoe

$$p_{m2} = \frac{CFp_{m1}}{M_{N1} + M_{f1}} \text{ for the left shoe}$$

$$\theta_m = 90°(\theta_2 > 90°)$$

SOLUTION

Step-1: Moments on the right shoe

For the right shoe

Unit normal force, $R_N = \frac{p_m b \cdot r}{\sin \theta_m} = \frac{0.7 \times 50 \times 125}{\sin 90°} = 4375 \text{ N}$

The moment of normal forces,

$$M_N = R_N a \int_{\theta_1}^{\theta_2} \sin^2 \theta \, d\theta \qquad \text{(Eqn. 6.70 MDHB)}$$

$$= aR_N \int_{\theta_1}^{\theta_2} \frac{1 - \cos 2\theta}{2} d\theta = aR_N \left[\frac{\theta}{2} - \frac{\sin 2\theta}{4} \right]_{\theta_1}^{\theta_2}$$

$$= aR_N \left[\frac{\theta_2 - \theta_1}{2} - \frac{\sin 2\theta_2 - \sin 2\theta_1}{4} \right] = aR_N \left[\frac{\pi}{4} - \frac{\sin 210° - \sin 30°}{4} \right]$$

$$= aR_N \left[\frac{\pi}{4} + \frac{2 \sin 30°}{4} \right] = aR_N \left[\frac{\pi}{4} + \frac{1}{4} \right]$$

$$= 98.5 \times 4375 \times 1.035 = 446191.9 \text{ Nmm}$$

The moments of the frictional forces,

$$M_f = f \cdot R_N \int \sin \theta (r - a \cos \theta) \, d\theta \qquad \text{(Eqn. 6.71 MDHB)}$$

$$= f \cdot R_N \left[r(\cos \theta_1 - \cos \theta_2) + \frac{a(\cos 2\theta_2 - \cos 2\theta_1)}{2} \right]$$

$$= 0.25 \times 4375(+153.1 - 85.3) = 74156 \text{ Nmm}$$

Step-2: Actuating force,

$$F = \frac{M_N - M_f}{C} = \frac{446191.9 - 74156}{180} = 2066.86 \text{ N}$$

Step-3: The maximum pressure on left shoe,

$$p_{m2} = \frac{CFp_{m1}}{M_N + M_f} = \frac{M_N - M_f}{M_N + M_f} p_{m1} = \frac{372035.9}{520347} \times 0.7 = 0.5 \text{ MP}$$

Step-4: Total torque,

$$T = f \cdot b \cdot r^2 (\cos\theta_1 - \cos\theta_2)(p_{m1} + p_{m2})$$

$$T = (0.25 \times 50 \times 125^2)(7 + 0.50)(\cos 15° - \cos 105°) = 287049.58 \text{ Nmm}$$

Example 19.33: Figure 19.18 shows internally expanding shoe brake, subjected to loads P and P' as shown. The width of the brake lining is 30 mm and the intensity of pressure at any point is given by $p = 0.5\sin\theta$ MPa, where θ is measured from either pivot as shown. Determine the braking torque and the magnitude of forces P and P'. Assume the coefficient of friction as 0.3.

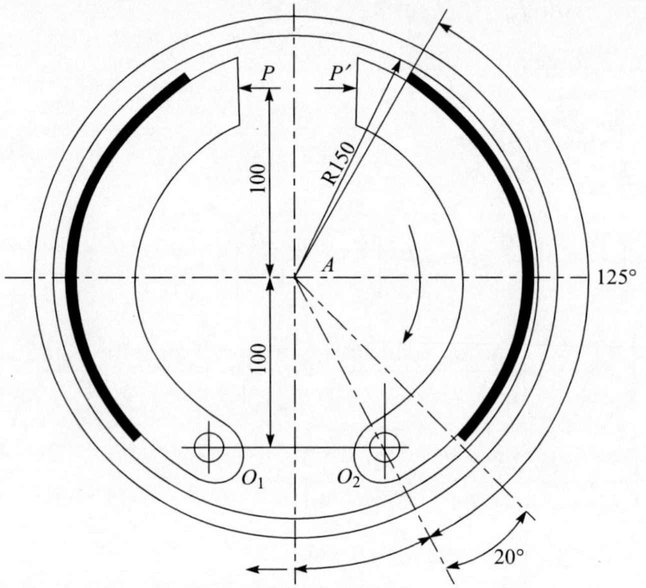

Figure 19.18

Given Data: The radius of contact surface, $r = 150$ mm

The distance of actuating force from pivot,

$$C = 100 + 100 = 200 \text{ mm}$$

Width of brake-lining, $b = 30$ mm

Contact pressure, $p = 0.5\sin\theta$ MPa

Coefficient of friction, $f = 0.3$

Location of pivot point, $\theta_0 = 25°$

Starting position of lining, $\theta_1 = 20°$

Subtended angle of lining, $\theta_2 = 125°$

SOLUTION

Step-1: Braking torque applied

Braking due to each shoe

$$T = (fR_N \cdot r)(\cos\theta_1 - \cos\theta_2) \qquad \text{(Eqn. 6.72 MDHB)}$$

where, $f = 0.3;\ r = 150\ \text{mm},\ \theta_1 = 20°, \text{ and } \theta_2 = 125°$

$$R_N = \frac{p_m b \cdot r}{\sin\theta_m} = 0.5 \times 30 \times 150 = 2250\ \text{N, because } \theta_m = 90°$$

Substituting the values,

$$T_1 = 0.3 \times 2250 \times 150 \times (\cos 20° - \cos 125°) = 153218.5\ \text{Nmm}$$

The total braking torque by two shoes,

$$T = 2T_1 = 2 \times 153218.5 = 306437\ \text{Nmm} = 306.437\ \text{Nm.}$$

Step-2: The moment of normal forces about the hinge point,

$$M_N = R_N a \int_{\theta_1}^{\theta_2} \sin^2\theta\, d\theta \qquad \text{(Eqn. 6.70 MDHB)}$$

$$= R_N a \left[\frac{\theta}{2} - \frac{\sin 2\theta}{4}\right]_{\theta_1}^{\theta_2}$$

$$= 2250 \times \frac{100}{\cos 25°} \times \left[\frac{105}{2} \times \frac{\pi}{180} - \frac{\sin 250° - \sin 40°}{4}\right]$$

$$= 248260.0 \times [0.92 + 0.3956] = 326610.9\ \text{Nmm}$$

Step-3: The moment of frictional forces about the hinge point,

$$M_f = fR_N \int_{\theta_1}^{\theta_2} \sin\theta \cdot (r - a\cos\theta) \cdot d\theta$$

$$= fR_N \left[\int_{\theta_1}^{\theta_2} r \cdot \sin\theta\, d\theta - \int_{\theta_1}^{\theta_2} a \sin\theta \cos\theta\, d\theta\right]$$

$$= fR_N \left[r(\cos\theta_1 - \cos\theta_2) - a\left(\frac{\sin^2\theta_2 - \sin^2\theta_1}{2}\right)\right]$$

$$= 0.3 \times 2250 \left[150 \times (\cos 20° - \cos 125°) - 110.34 \times \left(\frac{\sin^2 125° - \sin^2 20°}{2}\right)\right]$$

$$= 132583.5\ \text{Nmm}$$

Step-4: The actuating force on the right shoe,

$$P = \frac{M_N - M_f}{C} \qquad \text{(Eqn. 6.73 MDHB)}$$

$$= \frac{326610.9 - 132583.5}{200} = 970.12 \text{ N}$$

Step-5: The actuating force on the left show,

$$P^1 = \frac{M_N + M_f}{C} \qquad \text{(Eqn. 6.74 MDHB)}$$

$$P^1 = \frac{326610.9 + 132583.5}{200} = 2295.97 \text{ N}$$

19.4.3 Band Brakes

They consist of a steel band wrapped on the drum with the two ends attached to a pivoted lever. Depending upon the way the band ends are fixed to the lever, these are of two types:

(a) Simple band brakes
(b) Differential band brakes

Simple band brakes: The band is fixed to the pivot at one end and the other end is fixed to the lever away from the pin joint on one side left or right.

Simple band brakes

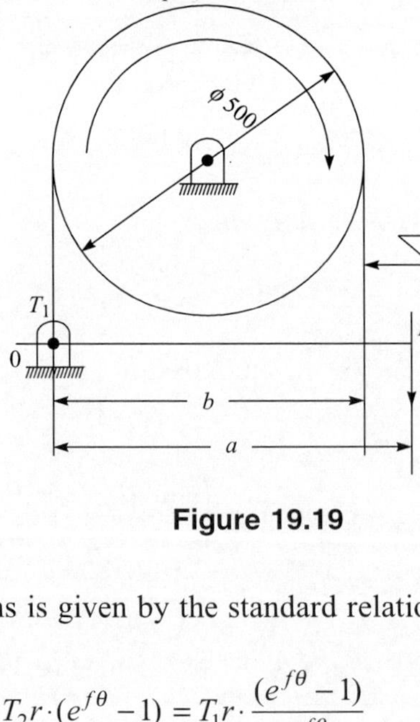

Figure 19.19

The ratio of band tensions is given by the standard relation, $\left(\frac{T_1}{T_2}\right) = e^{f\theta}$

$$\text{Torque, } T = (T_1 - T_2)\cdot r = T_2 r\cdot(e^{f\theta} - 1) = T_1 r\cdot\frac{(e^{f\theta} - 1)}{e^{f\theta}} \qquad (19.37)$$

For equilibrium of the lever for clockwise rotation, taking moments about 'O', $b \cdot T_2 - a \cdot F = 0$

$$F = \frac{b}{a}T_2 = \left(\frac{b \cdot T}{a \cdot r}\right)\left(\frac{1}{e^{f\theta} - 1}\right) \tag{19.38}$$

For anticlockwise rotation, $b \cdot T_2 - a \cdot F = 0$

$$F = \frac{b \cdot T_1}{a} = \left(\frac{b \cdot T}{a \cdot r}\right)\left(\frac{e^{f\theta}}{e^{f\theta} - 1}\right) \tag{19.39}$$

These relations get exchanged if the band is fixed to the other side of the pivot.

Differential Band Brakes

Here the two ends are separately fixed to the lever away from the pivot point either on both sides or on one side of it. Accordingly, there are two types. The basic relations are given below.

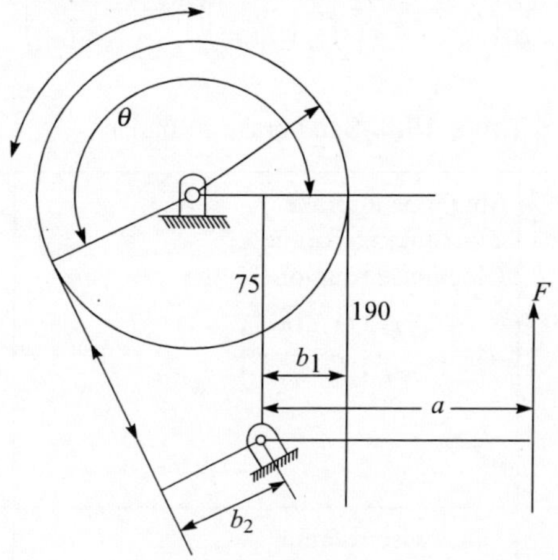

Figure 19.20

Ratio of tensions, $T_1/T_2 = e^{f\theta}$
Torque applied,

$$T = (T_1 - T_2) \cdot r = T_2 \cdot r(e^{f\theta} - 1) = T_1 \cdot r\frac{e^{f\theta} - 1}{e^{f\theta}} \tag{19.40}$$

When the ends are fixed one on each side of the pivot, for equillibrium for anticlockwise rotation,

$$Fa + T_1 \cdot b_1 - T_2 \cdot b_2 = 0$$

$$F = \frac{T_2}{a}\left(b_2 - b_1\frac{T_1}{T_2}\right) = \frac{T_2}{a}(b_2 - b_1 e^{f\theta}) = \frac{T}{a \cdot r}\left(\frac{b_2 - b_1 e^{f\theta}}{e^{f\theta - 1}}\right) \tag{19.41}$$

For clockwise rotation, the expression,

$$F \cdot a + T_2 \cdot b_1 - T_1 b_2 = 0$$

$$F = \frac{T_2}{a}\left(b_2 \frac{T_1}{T_2} - b_1\right) = \left(\frac{T}{a \cdot r}\right)\left(\frac{b_2 e^{f\theta} - b_1}{e^{f\theta} - 1}\right) \tag{19.42}$$

When both the ends are fixed on one side only,

For anticlockwise rotation, $F \cdot a = b_2 T_2 + b_1 T_1$

$$\therefore \quad F = \frac{T_2}{a}\left(b_2 + b_1 \frac{T_1}{T_2}\right) = \frac{T}{a \cdot r}\left(\frac{b_2 + b_1 e^{f\theta}}{e^{f\theta} - 1}\right) \tag{19.43}$$

For clockwise rotation, $F \cdot a = b_1 T_2 + b_2 T_1$

$$\therefore \quad F = \frac{T_2}{a}\left(b_1 + b_2 \frac{T_1}{T_2}\right) = \left(\frac{T}{a \cdot r}\right)\left(\frac{b_2 e^{f\theta} + b_1}{e^{f\theta} - 1}\right) \tag{19.44}$$

Table 19.6. Band brake formulae

Band brake formulae 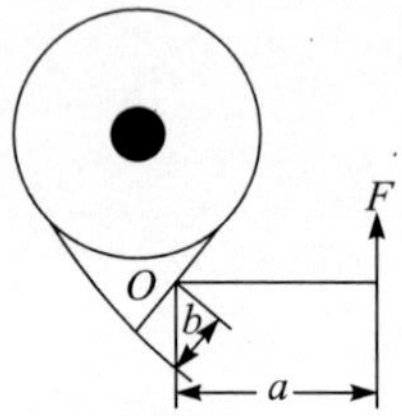	Single band brake T = Braking Torque Clockwise rotation $F = \left(\frac{Tb}{a \cdot r}\right)\left(\frac{e^{f\theta}}{e^{f\theta} - 1}\right)$	Anticlockwise rotation $F = \left(\frac{Tb}{a\,r}\right)\left(\frac{1}{e^{f\theta} - 1}\right)$
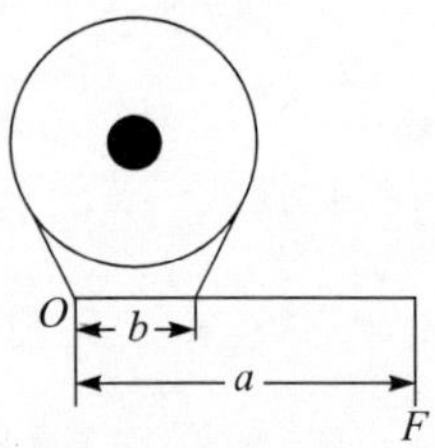	Clockwise rotation $F = \left(\frac{Tb}{a \cdot r}\right)\left(\frac{1}{e^{f\theta} - 1}\right)$	Anticlockwise rotation $F = \left(\frac{Tb}{a \cdot r}\right)\left(\frac{e^{f\theta}}{e^{f\theta} - 1}\right)$
Differential band brake 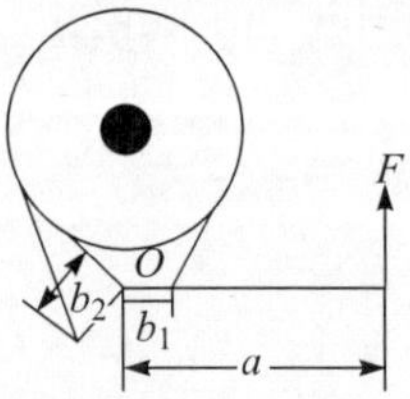	Clockwise rotation $F = \left(\frac{T}{a \cdot r}\right)\left[\frac{b_2 e^{f\theta} - b_1}{e^{f\theta} - 1}\right]$	Anticlockwise rotation $F = \left(\frac{T}{a \cdot r}\right)\left[\frac{b_2 - b_1 e^{f\theta}}{e^{f\theta} - 1}\right]$

Contd.

Differential band brake	Clockwise rotation	Anticlockwise rotation
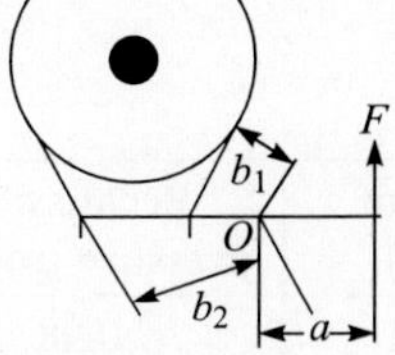	$F = \left(\frac{T}{a \cdot r}\right)\left[\frac{b_2\, e^{f\theta} + b_1}{e^{f\theta} - 1}\right]$	$F = \left(\frac{T}{ar}\right)\left[\frac{b_2 + b_1\, e^{f\theta}}{e^{f\theta} - 1}\right]$

19.4.4 Band and Block Brake

Here the steel band carries a number of blocks fixed on its underside. They come in contact with the drum.

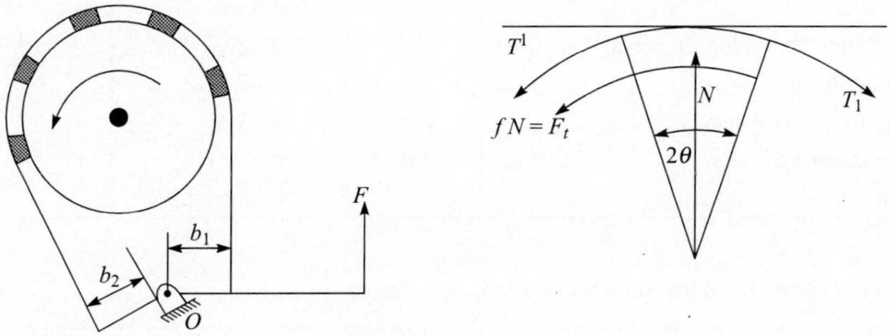

Figure 19.21

Consider the first block. For equilibrium

$(T_1 + T_1^1) \sin \theta = N$ in radial direction

$(T_1 - T_1^1) \cos \theta = F = fN$, in tangential direction

Combining the two, we get

$$(T_1 - T_1^1) \cos \theta = f(T_1 + T_1^1) \sin \theta$$

$$\therefore \qquad f \tan \theta = \frac{T_1 - T_1^1}{T_1 + T_1^1}$$

$$\therefore \qquad \frac{T_1}{T_1^1} = \frac{1 + f \tan \theta}{1 - f \tan \theta} \qquad (19.45)$$

For n-blocks from the first to the last

$$\frac{T_1}{T_2} = \frac{T_1}{T_1} \times \frac{T_1^1}{T_2^1} \times \frac{T_2^1}{T_3^1} \times \frac{T_{n-2}^1}{T_{n-1}^1} \times \frac{T_{n-1}}{T_2} = \left[\frac{1 + f \tan \theta}{1 - f \tan \theta}\right]^n \qquad (19.46)$$

In all the band equations $e^{f\theta}$ is to be replaced by the expression $\left[\dfrac{1+f\tan\theta}{1-f\tan\theta}\right]^n$

Table 19.7. Design data for brakes

Material	Max. temp °C	Coefficient of friction (f)	Permissible pressure (p) MPa
Metal on metal	315	0.2 to 0.25	0.98
Wood on metal	65	0.2 to 0.25	0.343
Leather on metal or wood	65	0.3 to 0.4	0.098
Cork on iron	–	0.35	0.069
Wood on wood	–	0.25 to 0.35	0.171
Asbestos in rubber compound			
Compressed on metal	200	0.3 to 0.4	0.49
Asbestos in resin binder moulded on metal	315	0.3 to 0.4	0.49
Asbestos embedded in oil	–	0.10	4.12
Asbestos, flexible woven on metal in oil	260	0.35 to 0.45	0.343
Sintered metal on cast iron	200	0.2 to 0.4	2.75
Sintered metal on oil	–	0.05 to 0.08	

Table 19.8. Values recommended by Hutle for brakes

Service P-MPa; V-m/s	PV
(1) Intermittent operation with long rest periods and poor heat-radiation as with wood blocks	1.95
(2) Continuous service with short rest period with poor radiation	1.00
(3) Continous service with good radiation as with oil bath	2.90

Table 19.9. Radiating factor C_r for brakes

Temperature difference (t_r)(°K)	Radiating factor C_r
55	30×10^{-6}
110	38×10^{-6}
165	41×10^{-6}
220	45×10^{-6}

Table 19.10. Working pressure for block brakes

Rubbing velocity m/sec.	Pressure (MPa)		
	Wood block	Asbestos fabric	Asbestos block
1	0.56	0.7	1.1
2	0.45	0.56	1.05
3	0.35	0.42	0.9
4	0.24	0.28	0.7
5	0.175	0.21	0.49
10	0.175	0.21	0.28

Table 19.11. Comparison of hoist brakes

Brake characteristic	Block brakes		Band brakes		Axial brakes	
	Double block	**V-grooved disk**	**Simple**	**Both directions of rotation**	**Cone**	**Multi-disk**
Force ratio, F/Ft	$\frac{b}{fa}$	$\frac{b\sin\alpha}{fa}$	$\frac{b}{a(e^{f\theta}-1)}$	$\frac{b(e^{f\theta}+1)}{a(e^{f\theta}-1)}$	$\frac{b\sin\alpha}{fa}$	$\frac{b\sin\alpha}{nfa}$
Average numerical value	0.667	0.282	0.0323	0.165	0.161	0.097
Relative value	20.6	87	1	5.1	5.0	3.0
Travel at lever end	$\frac{ha}{b}$	$\frac{ha}{b\sin\alpha}$	$\frac{ha\theta}{2\pi b}$	$\frac{ha\theta}{4\pi b}$	$\frac{ha}{b\sin\alpha}$	$\frac{ih'a}{b}$
Average travel, mm	0.785	18.5	73.6	36.8	32.5	5.5
Maximum capacity, kW	1500	18	225	75	38	90

Example 19.34: The diameter of a simple band brake is 500 mm. The coefficient of friction is 0.25. For the dimensions shown in the figure below, determine the maximum power rating @ 300 rpm. If V-belt is used in place of flat belt, what is the increase in the power rating, for groove angle 2 β = 45°?

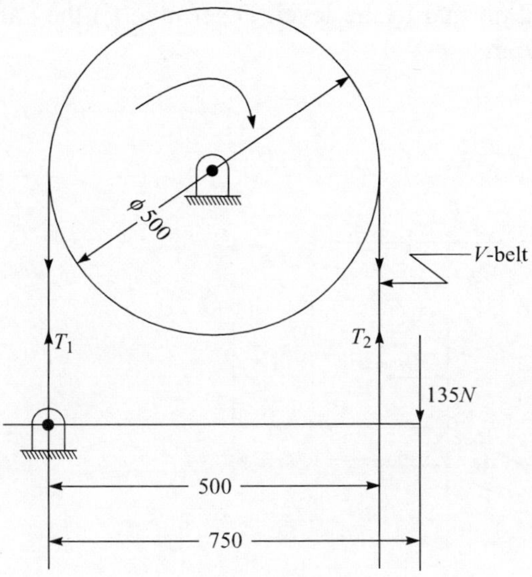

a = 750 mm
b = 500 mm
$\theta = \pi$ = 180°
f = 0.25
r = 250 mm

Figure 19.22

SOLUTION

Step-1: Applied force,

$$F = \frac{T_2 \cdot b}{a} = \left(\frac{T \cdot b}{a \cdot r}\right)\left(\frac{1}{e^{f\theta}-1}\right) \qquad \text{(Eqn. 6.91 MDHB)}$$

$$\text{Torque, } T = \left(\frac{F \cdot a \cdot r}{b}\right)(e^{f\theta} - 1) = \left(\frac{135 \times 750 \times 250}{500}\right)(e^{0.25 \times \pi} - 1) = 60409.8 \text{ Nmm}$$

Step-2: The power rating,

$$NP = \frac{TN}{955 \times 10^4} = \frac{60409.8 \times 300}{955 \times 10^4} = 1.8976 \text{ kW}$$

Step-3: For V-belt with groove angle, $2\beta = 45°$

$$\text{Coefficient of friction, } f_1 = f/\sin\beta = \frac{0.25}{\sin 22.5°} = 0.65$$

$$\text{Torque, } T = \left(\frac{135 \times 750 \times 250}{500}\right)(e^{0.65 \times \pi} - 1) = 339505.3 \text{ Nmm}$$

$$\text{Power, } NP = \frac{TN}{955 \times 10^4} = \frac{339505.3 \times 300}{955 \times 10^4} = 10.66 \text{ kW}$$

There is a large increase in the power rating.

Example 19.35: A band brake has dimensions as shown in the figure. The ratio of the band tensions is 1.75. The belts are fastened normal to the lever. Determine (i) the value of b_2, and (ii) the power rating for speed of 150 rpm.

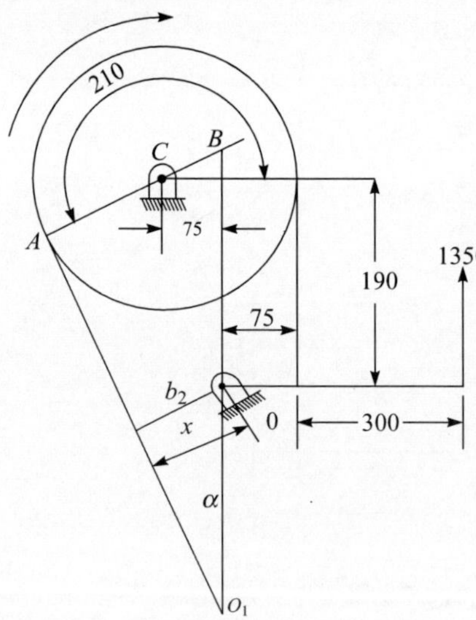

Figure 19.23

SOLUTION

Step-1: Dimensions

From the Fig., the drum radius $R = AC = 75 + 75 = 150$ mm

The distance, $b_2 = \dfrac{AB}{O_1B} \times O_1O = \dfrac{AB}{O_1B}(O_1B - OB) = (AB - OB\sin\propto)$

where, $AB = 150 + \dfrac{75}{\cos\propto} = 150 + \dfrac{75}{\cos 30^\circ} = 236.6$ mm

and $OB = 190 + 75\tan\propto = 190 + 75\tan 30^\circ = 233.3$ mm

Then distance $b_2 = 236.6 - 233.3\sin 30^\circ = 119.95 \cong 120$ mm

Tension ratio, $\dfrac{T_1}{T_2} = e^{f\theta} = 1.75$

Step-2: The torque applied,

$$T = F \cdot a \cdot r\left[\frac{e^{f\theta} - 1}{b_2 e^{f\theta} - b_1}\right] \qquad \text{(Eqn. 6.94 MDHB)}$$

$$= (1350 \times 375 \times 150)\left[\frac{1.75 - 1}{120 \times 1.75 - 75}\right] = 421875 \text{ Nmm}$$

Step-3: Power rating,

$$NP = \frac{TN}{955 \times 10^4} = \frac{421875 \times 150}{955 \times 10^4} = 6.63 \text{ kW}$$

Example 19.36: The band brake, with drum diameter of 200 mm (shown in Fig. 19.24 below) is applied to a shaft carrying a flywheel of 360 kg with radius of gyration of 0.45 metre, running at 300 rpm. Find (i) the torque applied for an applied force of $F = 90$ N and coefficient of friction $f = 0.2$, and (ii) the number of rotations of the wheel before it is brought to rest.

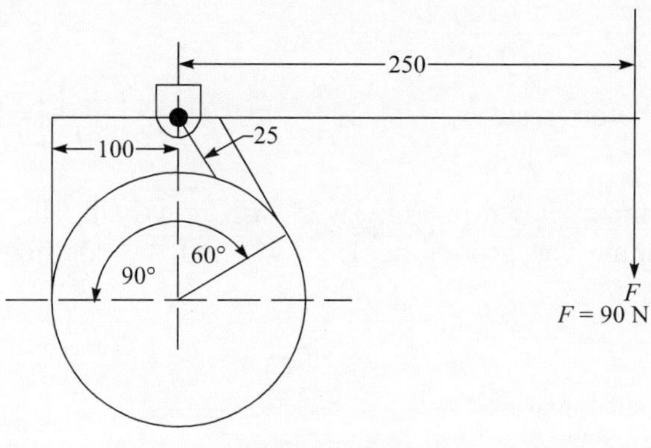

Figure 19.24

Given Data: Applied force, $F = 90$ N, Coefficient of friction $f = 0.2$
Brake drum diameter, $D = 200$ mm

SOLUTION

Step-1: Belt tensions,

$$f\theta = 0.2 \times \frac{210\pi}{180} = 0.733$$

The belt tension ratio, $T_1/T_2 = e^{f\theta} = e^{.733} = 2.08$

$\therefore$ $T_1 = 2.08T_2$

For equilibrium of the lever, taking moment about the hinge point,

$$F \times a + T_1 \times b_1 - T_2 \times b_2 = 0$$

$\therefore$ $$90 \times 250 + 25 \times T_1 - 100 \times T_2 = 0$$

$\therefore$ $$90 \times 250 + 52T_2 - 100T_2 = 0$$

$\therefore$ $$T_2 = 469 \text{ N and}$$

$\therefore$ $$T_1 = 469 \times 2.08 = 975.7 \text{ N}$$

Step-2: Torque transmitted,

$$T = (T_1 - T_2)r = (975.7 - 469) \times 100 = 50670 \text{ Nmm} = 50.67 \text{ Nm}$$

Step-3: Retardation

Moment of inertia of the wheel, $I = MR^2 = 360 \times 0.45^2 = 72.9 \text{ kgm}^2$

Then retardation, $\propto = \frac{T}{I} = \frac{50.67}{72.9} = 0.695 \text{ rad/sec}^2$

Step-4: Number of revolutions

Time taken to stop, $t = \frac{\omega}{\alpha} = \frac{2\pi \times 300}{60 \times 0.695} = 45.2 \text{ sec}$

Number of revolutions before rest, $N_1 = t \cdot N/2 = (48.2/60) \times \left(\frac{360}{2}\right) = 144.6$

Example 19.37: A band brake shown in Fig. 19.25, uses a V-belt. The pitch diameter of the v-grooved pulley is 400 mm. The groove angle is 45° and the coefficient of friction is 0.3. Determine the power rating.

SOLUTION

Step-1: Determination of belt tensions

Effective friction coefficient, $f_i = f/\sin\beta = 0.3/\sin 22.5° = 0.784.$ (Eqn. 18.31 MDHB)

The ratio of belt tensions, $\frac{T_1}{T_2} = e^{f_i\theta} = e^{0.784 \times \pi} = 11.74$ (1)

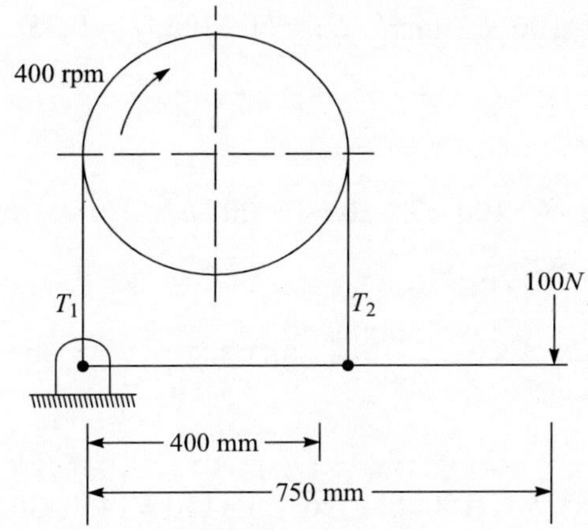

Figure 19.25

Taking moment about the pivot, $T_2 \times 400 - 100 \times 750 = 0$

$$T_2 = \frac{75000}{400} = 187.5 \text{ N} \tag{2}$$

By substituting (2) in (1),

$$T_1 = 11.74 \times T_2 = 11.74 \times 187.5 = 2201.25 \text{ N}$$

Step-2: Braking torque,

$$T = (T_1 - T_2) \cdot r = (2201.25 - 187.5) \times 200 = 402750 \text{ Nmm}$$

Step-3: Power rating,

$$NP = \frac{TN \times 10^{-4}}{955} = \frac{402750 \times 400 \times 10^{-4}}{955} = 16.87 \text{ kW}$$

Example 19.38: A band brake is shown in Fig. 19.26; the width and thickness of the band are 100 mm and 3 mm respectively. The permissible stress in the band material is limited to 50 N/mm^2. Coefficient of friction between the band and drum is 0.25.

Calculate (i) tensions in the band, (ii) actuating force, (iii) torque capacity of the brake and (iv) check for self-locking.

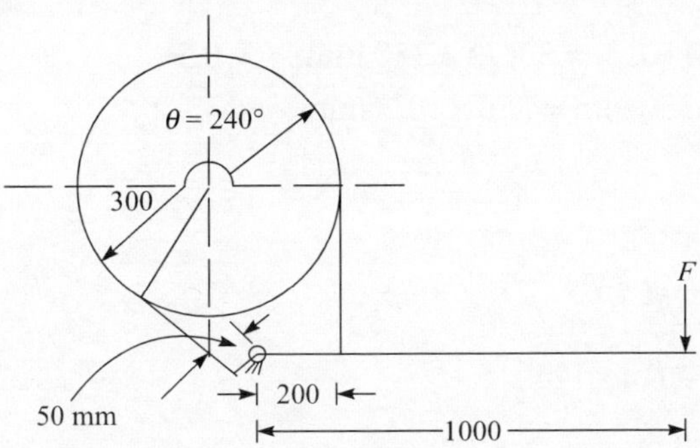

Figure 19.26

Given Data: Band size $w \times t = 100 \times 3$ mm^2, $S_d = 50$ MPa, $f = 0.25$

SOLUTION

Step-1: Band tensions

Tension on tight side, $T_1 = A \cdot \sigma = 100 \times 3 \times 50 = 15000$ N.

Tension ratio, $T_1/T_2 = e^{f\theta} = e^{0.25 \times \frac{240}{180} \times \pi} = 2.85$

Tension on slack side, $T_2 = T_1/2.85 = \dfrac{15000}{2.85} = 5263.2$ N

Step-2: Braking torque,

$$M_t = (T_1 - T_2) \cdot R = (15000 - 5263.2) \times 150 = 1460430 \text{ Nmm}$$

Step-3: Actuating force

For clockwise rotation,

$$F = \left(\frac{T}{a \cdot r}\right)\left(\frac{b_2 e^{f\theta} - b_1}{e^{f\theta} - 1}\right)$$

$$= \left(\frac{1460430}{1000 \times 150}\right)\left[\frac{50 \times 2.85 - 200}{2.85 - 1}\right] = -302.612 \text{ N}$$

For anticlockwise rotation,

$$F = \frac{T}{ar}\left[\frac{b_2 - b_1 e^{f\theta}}{e^{f\theta} - 1}\right] = \left(\frac{1460460}{1000 \times 150}\right)\left[\frac{50 - 200 \times 2.85}{2.85 - 1}\right] = -2736.66$$

Example 19.39: A differential band brake as shown in Fig. 19.27, has an angle of contact of 225°. The band has a compressed woven lining and bears against a cast iron drum of 350 mm diameter. The brake is to sustain a torque of 350 Nm and the coefficient of friction between the band and the drum is 0.3. Find:

(i) The necessary force P for the clockwise and anticlockwise rotation of the drum.

(ii) The value of OA for the brake to be self-locking, when the drum rotates clockwise.

Given Data: $\theta = 225°$; $b_2 = 150$; $b_1 = 35$; $D = 350$ mm; $f = 0.3$

The braking torque $M_t = 350$ Nm $= 350 \times 10^3$ Nmm

SOLUTION

Step-1: Band tensions

Braking torque, $M_t = (T_1 - T_2)R$

i.e., $350 \times 100^3 = (T_1 - T_2) \times 175$

$\therefore$ $T_1 - T_2 = 2000$ N (i)

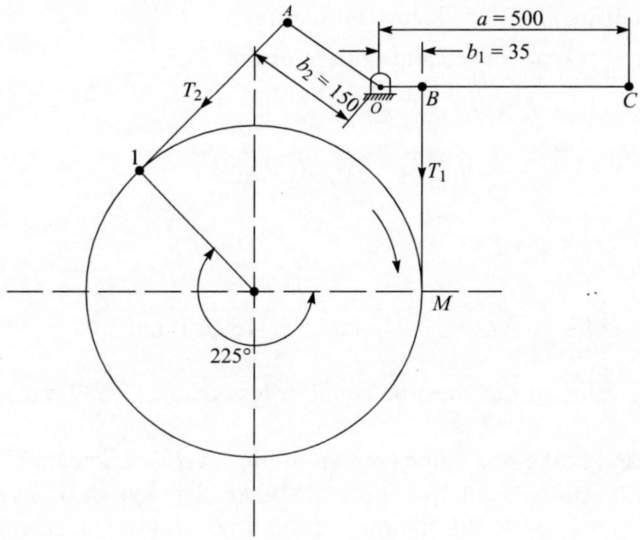

Figure 19.27

The ratio of tensions, $\frac{T_1}{T_2} = e^{\mu\theta} = e^{0.3\times 225\times\frac{\pi}{180}} = 3.2482$

i.e., $T_1 = 3.2482T_2$ (ii)

Substituting in equation (i), $3.2482T_2 - T_2 = 2000$

$\therefore$ Tension on slack side, $T_2 = \frac{2000}{2.2482} = 889.6$ N

Tension on tight side, $T_1 = 3.2482 \times T_2 = 3.2482 \times 889.6 = 2889.6$ N

Step-2: Operating force for clockwise rotation

Taking moment about 'O' for cw rotation of the drum,

$$P \times 500 + T_1 \times 35 = T_2 \times 150$$

$\therefore$ Actuating force $P = \frac{(889.6\times 150 - 2889.6\times 35)}{500} = 64.61$ N

Step-3: Operating force for anticlockwise rotation

Taking moment about 'O' for ccw rotation of the drum,

$$P \times 500 + T_2 \times 35 = T_1 \times 150$$

$\therefore$ Actuating force, $P = \frac{(2889.6\times 150 - 889.6\times 35)}{500} = 804.61$ N

Step-4: Condition for self-locking for clockwise rotation

Taking moments about '*O*' for cw rotation of the drum,

$$P \times OC + T_1 \times OB = T_2 \times OA$$

i.e.,
$$P = \frac{T_2 \times OA - T_1 \times OB}{OC}$$

For self-locking $T_2 \times OA \leq T_1 \times OB$,

i.e.,
$$OA \leq \frac{T_1}{T_2} \times OB \leq 3.2482 \times 35 = 113.687 \text{ mm}$$

∴ For self-locking, the value of OA must be equal or less than 113.687 mm.

Example 19.40: The band brake of a crane (shown in Fig. 19.28) is actuated by a lever, the free end of which is pulled upwards in order to apply the brake. The length of lever = 440 mm. The tight end of the band is attached to the fulcrum of the lever and the slack end to a pin 50 mm from fulcrum, i.e., 390 mm from the free end of the lever, where pull is applied. The diameter of the brake drum = 1000 mm and that of barrel = 650 mm. Find the necessary pull at the end of the brake lever in order to hold a load of 20 kN. Take $f = 0.35$ and angle of lap = 300°.

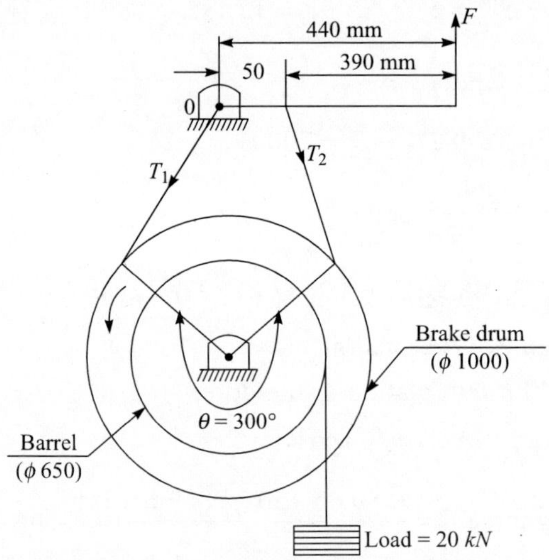

Figure 19.28

Given Data: $D = 1000$ mm ∴ $R = 500$ mm; $f = 0.35$; $\theta = 300°$

Diameter of barrel = 650 mm; Load on the barrel = 20 kN = 20×10^3 N

SOLUTION

Step-1: Belt tensions

Torque on the barrel shaft $= 20 \times 10^3 \times \dfrac{650}{2} = 65 \times 10^5$ Nmm

Since the barrel and the brake drum are on the same shaft, braking torque $M_t = 65 \times 10^5$ Nmm.

Also $M_t = (T_1 - T_2)R$

$$\therefore \quad T_1 - T_2 = (65 \times 10^5)/500 = 13000 \text{ N} \qquad \text{(i)}$$

$$\frac{T_1}{T_2} = e^{u\theta} = e^{0.35 \times 300 \times \pi/180} = 6.25 \qquad \text{(ii)}$$

$$\therefore \quad T_1 = 6.25 T_2$$

Substituting in equation (i) $6.25T_2 - T_2 = 13000$

$$\therefore \quad T_2 = 2476.2 \text{ N and } T_1 = 6.25 \times 2476.2 = 15476.2 \text{ N}$$

Step-2: Operating force F

Taking moments about O, $F \times 440 = T_2 \times 50$

i.e., $F = 2476.2 \times 50/440 = 281.4$

$\therefore$ Force required to hold the load $F = 281.4$ N

Example 19.41: A differential band brake is shown in Fig. 19.29. The width and the thickness of the steel band are 100 mm and 3 mm respectively and the maximum tensile stress in the band is 50 MPa. The coefficient of friction between the friction lining and the brake drum is 0.25. Calculate:

(i) Tensions in the band
(ii) Actuating force
(iii) Torque capacity of the brake.

Check whether the brake is self-locking.

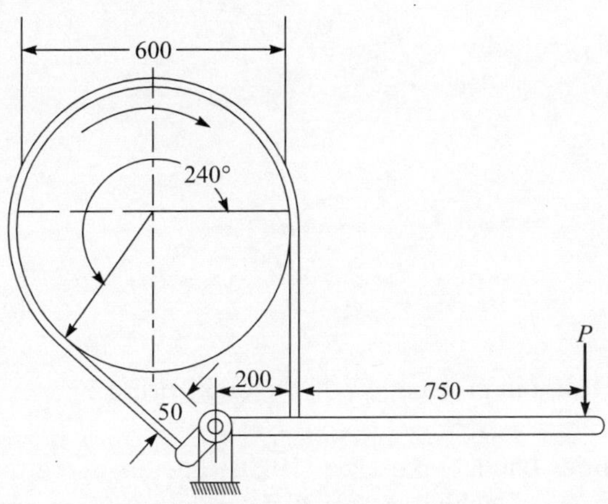

Figure 19.29

Given Data: $R = 300$ mm; $\theta = 240°$; $t = 3$ mm; $w = 100$ mm; $f = 0.25$

$\sigma_d = 50$ N/mm^2; $b_2 = 50$ mm; $b_1 = 200$ mm; $a = 750 + 200 = 950$ mm

SOLUTION

Step-1: Tensions in the band

Maximum tension in the band, $T_1 = wtS_d$

$$= 10 \times 3 \times 50 = 15000 \text{ N}$$

i.e., Tension on the tight side, $T_1 = 15000$ N

$$\text{Ratio of tensions, } \frac{T_1}{T_2} = e^{\mu\theta} = e^{0.25 \times 240 \times \frac{\pi}{180}} = 2.85$$

$$\therefore \text{ Tension on the slack side, } T_2 = \frac{T_1}{2.85} = \frac{15000}{2.85} = 5263.2 \text{ N}$$

Step-2: Actuating force

Taking moments about 0, $P \times a + T_1 \times b_2 = T_2 \times b_1$

$$\text{i.e., } P = \frac{T_2 b_1 - T_1 b_2}{a} = \frac{5263.2 \times 200 - 15000 \times 50}{950}$$

$\therefore$ Actuating force, $P = 318.57$ N

Step-3: Torque capacity of the brake,

$$M_t = (T_1 - T_2)R = (15000 - 5263.2) \times 300 = 2921040 \text{ Nmm}$$

Step-4: Check for self-locking

For the given direction, i.e., clockwise $P = \dfrac{T_2 b_1 - T_1 b_2}{a}$

For self-locking $T_2 b_1 \leq T_1 b_2$

$$\text{i.e., } \frac{b_1}{b_2} \leq \frac{T_1}{T_2}$$

But $\dfrac{200}{50} = 4 > 2.85$

Hence, there is no self-locking in *cw* rotation of the brake drum.

Example 19.42: A differential band brake (Fig. 19.30) has an operating lever 225 mm long. The ends of the brake band are attached so that their operating arms are 38 mm and 127 mm long. Brake drum diameter = 600 mm. Arc of contact is 300° and $f = 0.22$. The band section is 3.2 mm × 100 mm.

(i) Find the least force required at the end of operating lever when the band is subjected to a stress of 55 N/ mm^2.
(ii) What is the torque applied to the brake drum shaft?
(iii) Is this brake self-locking? Prove your answer.

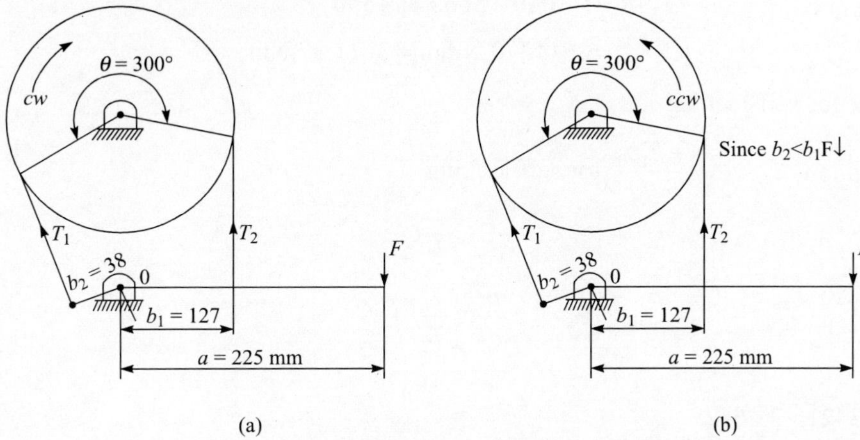

Figure 19.30

Given Data: $D = 600$ mm; $\therefore R = 300$ mm; $f = 0.22$; $\theta = 300°$;
$h = 3.2$ mm; $w = 100$ mm; $\sigma_d = 55$ N/mm^2

SOLUTION

Step-1: Band tensions

Maximum tension in the band, $T_1 = \sigma_d wh = 55 \times 100 \times 3.2 = 17600$ N

The ratio of band tensions, $\frac{T_1}{T_2} = e^{\mu\theta} = e^{0.22 \times 300 \times \frac{\pi}{180}} = 3.16425$

$$\therefore \quad T_2 = \frac{T_1}{3.16425} = \frac{17600}{3.16425} = 5562.1 \text{ N}$$

Step-2: Least actuating force

Taking moments about O for *cw* rotation [Fig. 19.30(a)]

$$F \times a + T_1 \times b_2 = T_2 \times b_1$$

i.e.,
$$F \times 225 + 17600 \times 38 = 5562.1 \times 127$$

$\therefore$ Actuating force, $F = 167.1$ N

Taking moments about O for *ccw* rotation [Fig. 9.30(b)]

$$F \times a + T_2 \times b_2 = T_1 \times b_1$$

i.e.,
$$F \times 225 + 5562.1 \times 38 = 17600 \times 127$$

i.e., $$F = 8994.8 \text{ N}$$

$\therefore$ Least actuating force, $F = 167.1$ N

Step-3: Braking torque

Braking torque, $M_t = (T_1 - T_2)R = [17600 - 5562.1] \times 300$

$$= 3611370 \text{ Nmm} = 3611.37 \text{ Nm}$$

Step-4: Check for self-locking

For *cw* rotation, $F = \dfrac{T_2 b_1 - T_1 b_2}{a}$ for self-locking

$$T_2 b_1 \leq T_1 b_2$$

i.e., $$\frac{b_1}{b_2} \leq \frac{T_1}{T_2}$$

but, $\dfrac{127}{38} = 3.3421 > 3.16425$

Hence, no self-locking in cw rotation.

For *ccw* rotation, $F = \dfrac{T_1 b_1 - T_2 b_2}{a}$

For self-locking,

$$T_1 b_1 \leq T_2 b_2$$

i.e., $$\frac{T_1}{T_2} \leq \frac{b_2}{b_1}$$

but, $3.16425 > \dfrac{38}{127} = 0.2992$

Hence, no self-locking in *ccw* rotation also.

Example 19.43: The torque absorbed in the band brake shown in Fig. 19.31 is 400×10^3 Nmm. Design the band and lever, taking $f = 0.27$ and diameter of drum as 400 mm. The allowable stress in band may be taken as 70 MPa.

Given Data: $M = 400 \times 10^3$ Nmm; $f = 0.27$; $D = 400$ mm; $R = 200$ mm;

$\sigma_d = 70$ N/mm² $\theta = 270°$ (refer figure)

SOLUTION

Step-1: The belt tensions

Braking torque, $M_t = (T_1 - T_2)R$

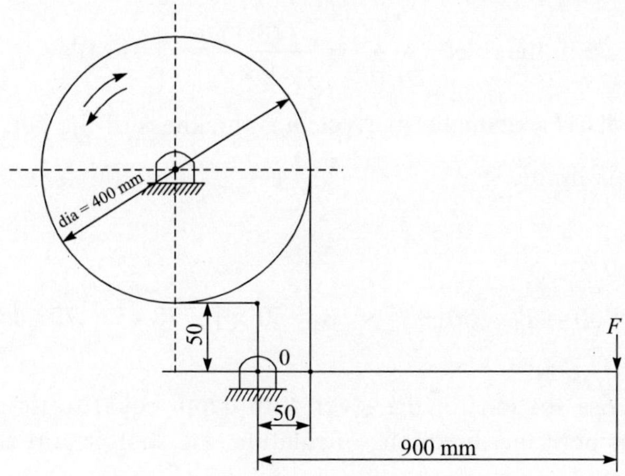

Figure 19.31

i.e., $400 \times 10^3 = (T_1 - T_2) \times 200$

$$T_1 - T_2 = 2000 \text{ N} \quad \text{(i)}$$

Also, $\dfrac{T_1}{T_2} = e^{\mu\theta} = e^{0.27 \times 270 \times \frac{\pi}{180}} = 3.5692$

i.e., $T_1 = 3.5692 T_2$ (ii)

Substituting in equation (i),

$$3.5692 T_2 - T_2 = 2000$$

$\therefore$ $$T_2 = \frac{2000}{2.5692} = 778.45 \text{ N}$$

and $$T_1 = 3.5692 \times 778.45 \text{ N} = 2778.45$$

Step-2: Design of the band

Thickness of band, $h = 0.005D = 0.005 \times 400 = 2$ mm

Width of band, $w = \dfrac{T_1}{h\sigma_d} = \dfrac{2778.45}{2 \times 70} = 19.846 \text{ mm} \approx 20 \text{ mm}$

Assume four rivets of 8 mm size is used and these are arranged in two rows. Hence, the band will not be weakened by more than two rivet holes.

Considering the rivet holes, width of band, $w = 20 + 2x$ and diameter of rivet hole $= 20 + 2 \times 9 = 38$ mm

$\therefore$ Adopt width of band, $w = 40$ mm

Check for the stresses

Maximum shear stress in the rivet $= \dfrac{T_1}{n \cdot A} = \dfrac{2778.45}{4 \times \frac{\pi}{4} \times 8^2} = 13.82$ MPa

Maximum crushing stress in the rivet $= \dfrac{T_1}{n \cdot dh} = \dfrac{2778.45}{4 \times 8 \times 2} = 43.41$ MPa

where, n = number of rivets; d = diameter of rivet; h = thickness of the belt

These stresses are within limits.

Step-3: Design of lever

Taking moment about 0 for *cw*

$$F \times 900 = T_1 \times 50 + T_2 \times 50 = 50 \times [2778.45 + 778.45]$$

∴ Actuating force, F = 197.6 N

Assume rectangular cross section for the lever with depth equal to thrice the width. Since the brake is to operate for both the directions of rotation, the shorter arm is to be designed for maximum bending moment.

∴ Maximum BM $M_b = T_1 \times 50 = 2778.45 \times 50 = 138922.5$ Nmm

Select the material of the lever as C40 steel with factor of safety = 3

From Table 2.6 (MDHB) for C40 steel, $S_y = 360$ MPa

$$\therefore \qquad S_d = \frac{S_y}{FOS} = \frac{360}{3} = 120 \text{ MPa}$$

We have the bending equation, $\dfrac{M_b}{I} = \dfrac{S_d}{c}$

where, $I = \dfrac{b_l h_l^3}{12} = \dfrac{b_l (3b_l)^3}{12} = \dfrac{27}{12} b_l^4$ and $c = \dfrac{h_1}{2} = \dfrac{3b_1}{2} = 1.5b_1$

$$\therefore \qquad \frac{I}{c} = \left(\frac{27}{12} b_l^4\right)/(1.5b_1) = 1.5\, b_l^3 = 138922.5/120 = 1157.69$$

$$b_1 = \sqrt[3]{1157.69/1.5} = 9.2$$

Adopt width of lever $b_1 = 9.5$ mm

∴ Thickness of lever, $h_1 = 3b_l = 3 \times 9.5 = 28.5$ mm

EXERCISES FOR PRACTICE

19.1 (a) Enumerate the various types of clutches.

(b) Design a single plate friction clutch having both sides effective from the following data:

Power transmitted	- 30 kW
Speed of shaft	- 1500 rpm
Allowable lining pressure	- 0.15 MPa
Maximum diameter of the clutch	- 30 cm

Service factor - 1.5

Number of springs - 9

Compression of spring during disengagement - 2.5 mm

19.2 (a) What is the significance of the service factor or engagement factor in the design of a friction clutch? Discuss briefly the factors on which the factor is dependent and illustrate the selection of service factor on a relative basis for the following cases: Centrifugal pumps and fans, machine tools, reciprocating pumps, winding and coiling machine.

(b) Design a cone clutch to transmit 150 kW at 1000 rpm. It has a mean radius of 25 cm and the maximum pressure on the lining is limited to 0.6 MPa. The coefficient of friction is taken as 0.2. Assume other data.

19.3 (a) State the application of the following clutches:

(i) Single-plate clutch (ii) Multi-plate clutch

(iii) Cone clutch (iv) Centrifugal clutch.

(b) A drum having a diameter of 152.4 cm is used to support a wire that is attached to a mass of 3570 kg. The drum is rotating at 60 rpm, has a radius of gyration of 61 cm, and weighs 4460 kg. A brake is to bring the drum to rest in 457 cm. Draw a neat sketch of the arrangement. If drum rotates at 90 rpm, calculate average braking torque. If the brake rim is made of steel, has a 26 cm diameter, 1.25 cm thickness, and has 10 cm face width, what will be the temperature rise consuming all energy is absorbed by rim?

19.4 (a) A single plate clutch is to have a max. capacity of 50 kW at 1800 rpm. The clutch facing has a coefficient of friction of 0.40 and a permissible pressure of 0.2 MPa. Determine the diameters of the clutch facing and the axial force required to keep the clutch engaged.

(b) The rope drum of an elevator hoist is 600 mm diameter and the speed of the elevator is 3 m/s. This drum is fitted with a brake drum 900 mm in diameter, having two C.I. brake shoes, each subtending an angle of 90° on the brake drum. The elevator weighs 18 kN when loaded. The brake is to have sufficient capacity to stop the elevator in 3-6 minutes. The coeff. of friction for C.I on C.I. may be taken as 0.2. The allowable pressure on the brake shoes is 0.35 MPa. Determine (i) the radial force at each brake, (ii) width of the shoe, and (iii) the heat generated in stopping the elevator.

19.5 (a) A conical clutch of included angle 60° has inner and outer radii of 25 mm and 50 mm and the axial load is 500 N. If the variation of normal pressure with radius is given by $a + b/r$ and the maximum pressure is 1.5 times the minimum, calculate the torque transmitted for a value of coefficient of friction 0.35.

(b) In the winding drum band brake, the braking force P is applied to one end of a lever ABC pivoted at F, between A and B, where AF = 25 mm, BF = 100 mm and CF = 625 mm. If the drum diameter is 1 m, the arc of contact 225° and the coefficient of friction 0.3, calculate the value of P required to develop a braking torque of 5000 Nm when the drum is rotating clockwise.

19.6 (a) Explain with example: (i) Self-energizing brake, and (ii) self-locking brake.

(b) The ends of a differential band brake are attached on either sides of the pivot so that the operating arms are 60 mm and 150 mm long. The shorter operating arm is between pivot and point of application of effort. The distance between pivot and effort is 350 mm. The diameter of brake drum is 600 mm. The ankle of contact is 270° and coefficient of friction is 0.25. If the effort is 1500 N acting upwards, what is the torque that can be absorbed by the brake drum when it is rotating clockwise? Design the lever if the allowable stress in lever is limited to 60 MPa. Make a neat sketch of brake.

19.7 (a) Compare the clutches and couplings.

(b) Design the main dimension of a cone clutch to transmit 40 kN at 750 rpm. Assume the suitable material properties.

19.8 (a) Design a centrifugal clutch from the following data:

Power transmitted	= 15 kW
Speed	= 1,000 rpm
Number of shoes	= 4
Engagement speed	= 75% of runnings speed
Coefficient of friction	= 0.25
Inner radius of pulley	= 125 mm
Radius at which the centre of gravity of shoes revolves	= 100 mm
Arc of contact of each shoe	= 60°

(b) A differential band brake is mounted on a drum of diameter 1000 mm. The rope of rope drum carries 15 kN. Determine:

(i) The force F for clockwise rotation.

(ii) What is the width of the band required, if the stress in the band is not to exceed 55 MPa?

The thickness of the band may be taken as 2.5 mm.

19.9 (a) What do you mean by self-energizing brake? When does a brake becomes self-energizing?

(b) List the important factors upon which the capacity of brake depends.

(c) A centrifugal friction clutch has a driving member consisting of a spider carrying four shoes which are kept from contact with the clutch case by means of springs until increase of centrifugal force overcomes the resistance of the springs and the power is transmitted by the friction between the shoe and the case.

Determine the necessary mass and size of each shoe if 25 kW is to be transmitted at 800 rpm with engagement beginning at 75% of the running speed. The inside diameter of the drum is 310 mm and the radial distance of the centre of gravity of each shoe from the shaft axis is 127 mm. Assume rest of the data from data book.

19.10 (a) Explain the terms self-energising and self-locking in brakes and clutches with respect to the various types of brakes and clutches. Draw the sketches if necessary.

(b) Design a multi-plate motor cycle type clutch from following data:

Power transmitted	= 7 kW
Speed	= 3000 rpm
Maximum outside diameter of plate	= 125 mm
Number of springs	= one at centre
Spring compression for disengagement	= 2 mm

For spring compression for variation of 15%.

Allowable one-way shear stress for spring steel = 450 MPa

Choose your own friction material, sketch the clutch assembly and show arrangement of plates and spring.

19.11 (a) What is the function of a brake? How does it differ from that of a clutch?

(b) Design a single-plate automobile clutch to go with a six-cylinder engine that has a maximum torque of 240 Nm at 1900 rpm. The flywheel diameter is large enough so that a 24 cm outside diameter disc clutch can be used. After the clutch dimensions have been decided upon, we have to analyse the characteristics of the drive system, assuming the clutch is engaged when the car is travelling at 50 km per hour after a shift from second to third gear. Determine the number of revolutions of clutch slip during engagement and the energy of friction work to be dissipated by the clutch for each engagement.

Engine speed at beginning of engagement	= 1900 rpm
Engine torque constant during engagement	= 78.4 Nm
Automobile weight when loaded	= 13725 N
Automobile wheel diameter	= 70 cm
Combined moment of inertia	= 1.08 Nm sec^2
Gear reduction at differential	= 7:1
Torque at rear wheels available for accelerating	= 137.2 Nm
Coefficient of friction of clutch material	= 0.3
Design pressure at clutch material	= 0.25 MN/m^2

19.12 (a) What are the common basic principles adopted in design of brakes and clutches? What is the effect of pressure distribution of friction surface on design torque?

(b) An automobile vehicle weighing 15 kN is moving on a level road at a speed of 100 km/hr. When the brakes are applied, it is subjected to certain negative acceleration of 10 m/sec^2. There are brakes on all four wheels. The tyre diameter is 75 mm. The kinetic energy of the rotating part is 10% of the K.E. of the moving vehicle. The mass of each brake drum assembly is 12 kg and the specific heat capacity is 500 J/kg°C.

Calculate:

(i) Braking time,

(ii) The braking distance,

(iii) The total energy absorbed by each brake,

(iv) The torque capacity of each brake

(v) The temperature rise of brake drum assembly.

19.13 A four-wheeled automobile car has a total mass of 1000 kg. The moment of inertia of each wheel about a transverse axis through its centre of gravity is 0.5 kg-m^2. The rolling radius of the wheel is 0.35 m. The rotating and reciprocating parts of the engine and the transmission system are equivalent to a moment of inertia of 2.5 kg-cm^2 rotating at five times the road-wheel speed. The car is travelling at a speed of 100 km/hr on a plane road. When the brakes are applied, the car decelerates at 0.5 g. There are brakes on all four wheels. Calculate:

(i) The energy absorbed by each brake

(ii) The torque capacity of each brake.

19.14 (a) What are the advantages and disadvantages of cone clutch? What do you mean by self-locking in case of cone clutch?

(b) Design and draw dimensional sketch of a cone clutch from the following data:

Power to be transmitted	= 40 kW
Speed	= 500 rpm
Mean diameter of the clutch	= 300 mm
Shock factor on driver side	= 1.2
Shock factor on driven side	= 1.8
Semi-cone angle	= 12.5°

Assume other data.

19.15 (a) Differentiate various types of friction clutches and their limitations. Is there any self-locking phenomena in any clutch?

(b) Design a multiple motor cycle type clutch from the following data:

Power transmitted	= 6.0 kW
Speed	= 2000 rpm
Ratio of outer to inner radius of plate	= 2.00
Clutch displacement	= 2 mm

For spring force variation 15%

Assume other data from data book. Sketch the clutch assembly and show arrangement of plates and spring.

19.16 (a) Design a cone clutch to transmit a power of 40 kW at a rated speed of 750 rpm. Also determine:

(i) The axial force capacity

(ii) The axial force necessary to transmit the torque

(iii) The axial force necessary to engage the cone clutch.

A simple band brake shown in Fig. 19.32 is to be designed to absorb a power of 30 kW at a rated speed of 750 rpm. Determine:

(i) The effort required to stop clockwise rotation of the brake drum.

(ii) The effort required to stop counterclockwise rotation of the brake drum.

(iii) The dimensions of the rectangular cross section of the brake lever assuming its depth to be twice the width.

(iv) The dimension of the cross section of the band assuming its width to be ten times the thickness.

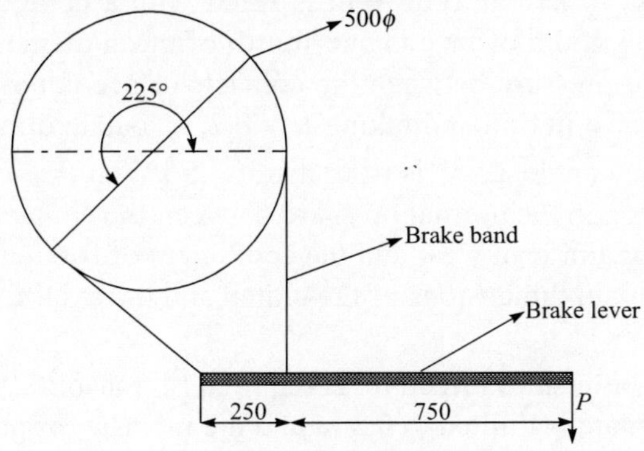

Figure 19.32

19.17 (a) Design a cone clutch to transmit a power of 40 kW at a rated speed of 750 rpm. Also determine:

(i) Axial force necessary to transmit torque

(ii) Axial force necessary to engage the cone clutch.

(b) A simple band brake shown in Fig. 19.33 transmitting a power of 45 kW at rated determine is to be designed to stop the rotation of a shaft speed of 500 rpm. Selecting suitable materials, determine:

(i) Dimensions of the rectangular cross section of the band

(ii) Dimensions of the rectangular cross section of the brake lever

(iii) Diameter of the fulcrum pin.

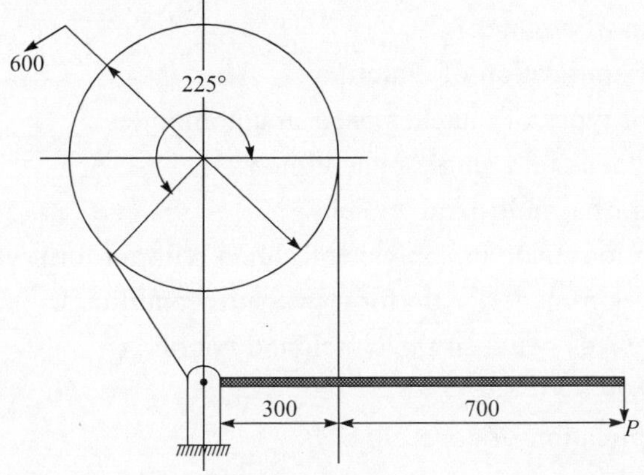

Figure 19.33

19.18 A multiple-plate clutch with steel on bronze is to transmit 8 kW at 1440 rpm. The inner diameter of the contact is 80 mm and the outer diameter of contact is 140 mm. The clutch operates in oil with expected coefficient of friction of 0.1 and allowable pressure of 0.35 MPa. Assume uniform wear theory and determine the number of steel and bronze plates.

19.19 An engine developing 50 kW at 1000 rpm is fitted with a cone clutch. The cone has a face angle of 12.5° and width of face is one-fourth of mean diameter of friction lining. If the normal intensity of pressure between the contact surface is not to exceed 0.1 N/mm^2, assuming uniform wear criterion and taking $\mu = 0.2$, calculate dimensions of the clutch.

19.20 A cone clutch with face angle 12.5° is to transmit 7.5 kW at 900 rpm. The width of face is half the mean radius and the normal pressure between the contact faces is not to exceed 0.09 MN/m^2. Assuming uniform wear and the coefficient of friction between contact faces as 0.2, determine the main dimensions of the clutch and the axial force required to engage the clutch.

19.21 A single-plate clutch, both sides effective, is required to transmit 25 kW at 1600 rpm. The outer diameter of the plate is limited to 0.3 m and the intensity of pressure is not to exceed 0.07 MPa. Assuming uniform wear and the coefficient of friction 0.3, determine the inner diameter of the plate and the axial force necessary to engage the clutch.

19.22 (a) With the help of a neat sketch derive an equation for torque transmitting capacity of single plate clutch, considering uniform wear.

(b) A cone clutch is to be designed to transmit 40 Nm of torque. The semi-cone angle is 12.5°. The mean radius of clutch is twice the face width, coefficient of friction $\mu = 0.18$, normal intensity of pressure between contacting surfaces should not exceed 0.1 N/mm^2. Considering uniform wear calculate:

(i) The inner and outer diameter of friction layer

(ii) Face width of friction layer

(iii) Force required to engage the clutch.

SHORT ANSWER QUESTIONS

19.1 What is a clutch?

19.2 What is the function of a clutch?

19.3 What is the field of application of clutch?

19.4 Specify the different types of clutches used in automobiles.

19.5 What are the advantages of a single plate clutch?

19.6 What are the merits of a multi-plate clutch?

19.7 What is the assumption made in analysis of clutch with uniform wear considerations?

19.8 What is the assumption made for uniform pressure condition?

19.9 What are the differences between the clutch and coupling?

19.10 What is cone clutch?

19.11 What is field of application of cone clutch?

19.12 What are the advantages of plate clutch over cone clutch?

19.13 Explain the principle of operation of centrifugal clutch.
19.14 When is the use of single plate clutch recommended?
19.15 Where are the multiple plate clutches used?
19.16 Distinguish between a clutch and a brake.
19.17 What is a shoe break?
19.18 Name the types of shoe break configurations that can be used.
19.19 What are band brakes?
19.20 What are the types of band brakes?
19.21 Obtain a relation for torque transmitted by a simple band break.
19.22 What PV values are recommended for break with continuous service with good radiation?
19.23 What are the friction materials used in clutches and brakes?
19.24 Why are splines used in clutches?
19.25 Why are clutches immersed in oil?

CHAPTER 20

FLEXIBLE MECHANICAL ELEMENTS (DRIVES)

20.1 INTRODUCTION

The drives are of two types: rigid and flexible.

In the rigid drives, the driving and driven elements are in direct contact. The gear drives are rigid drives. In flexible drives, the driving and driven elements are joined by a flexible link. The driving element gives motion to the flexible link and the link, in turn, moves the driven element. The belt drives, rope drives, chain drives are flexible drives.

Advantages of flexible drives

(i) They transmit power over long distances.
(ii) The center distance between the shafts is flexible.
(iii) They are simple in design and construction.
(iv) They are cheap.
(v) The maintenance cost is low.

Disadvantages of flexible drives

(i) The drive speed may not be uniform.
(ii) The drive speed is affected by slip.
(iii) The speed ratios are low.
(iv) The occupy a large space.

20.2 BELT DRIVES

The belt drive consists of a driving pulley on driving shaft, a driven pulley on driven shaft and a belt wrapped tightly on them. As the driving pulley rotates, it pulls the belt along it, the belt, in turn, pulls the driven pulley, due to the friction between the belt and pulley. Thus, the driven shaft starts rotating. Belt drives are used to transmit power between shafts placed at distances. There are three types of belt drives, based on the shape of the belt, namely:

(a) Flat belt
(b) V-belt
(c) Toothed belt

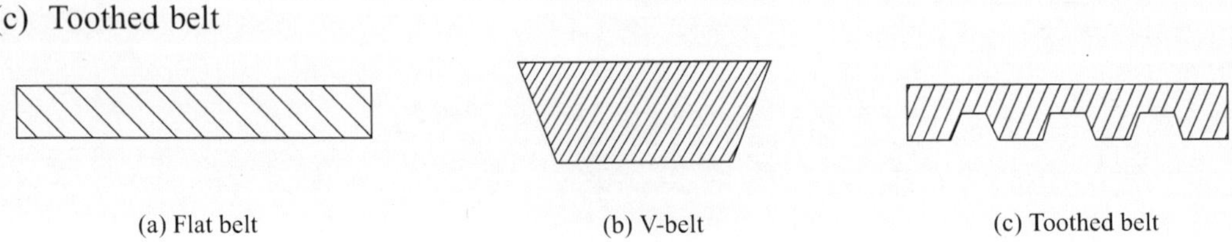

(a) Flat belt (b) V-belt (c) Toothed belt

Figure 20.1

20.2.1 Flat Belts Drives

These belts are rectangular in cross section and are wrapped on the flat pulleys mounted on the shafts being connected. The ends are joined by riveting or stitching or using metallic links. Because of the joint, the efficiency is less than 100%. Materials for these belts are leather, canvas, rubber, fabric with rubber. They are available in standard widths and thicknesses

Leather Belts: The leather belt has an ultimate stress of approximate value of 30 MPa but the design stress used is limited to about 2.8 MPa. They are suitable for speeds upto 45 mps. They are available in widths from 20 to 300 mm.

Rubber Belts: They are made from a number of layers of fabric impregnated with rubber. The ultimate strength of these belts ranges from 37.0 to 44.0 MPa. The strength reduces with the number of plies. They operate at speeds up to 30 m/s. They are available in widths from 20 to 1200 mm.

Canvas Belts: They are suitable for lower power. They are less durable and not suitable for higher speeds. They are not suitable for shock loads also.

Woollen Belts: They are flexible and strong. They are suitable for medium and heavy loads, can operate in moist, dusty and hot atmosphere. The tensile strength is 30 MPa.

Advantages of flat belt drives over V-belt drives

(i) They are simple in design and construction.
(ii) They have low initial and maintaince cost.
(iii) They can be used in dusty and abrasive atmosphere.
(iv) They do not need any casing.
(v) The speed ratios can be varied by step pulley arrangement.
(vi) They can be used as clutch by loose and tight pulley arrangement.
(vii) The drive efficiency is more than V-belt drive.

Disadvantages of flat belt drives over V-belt drives

(i) The ratios are low.
(ii) The power transmitting capacity is less.
(iii) They occupy more space.
(iv) They generate more noise.

The flat belts are used in textile machinery, clay machinery, baking machinery, saw mills, brick industry, crushers, line shafts, as belt conveyors, bucket elevators, etc.

Belt Drive Geometry

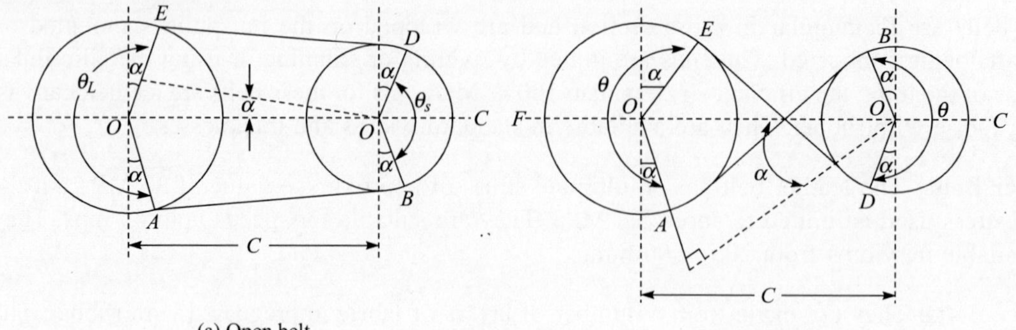

Figure 20.2

Open Belts

Angle of contact for smaller pulley, $\theta_s = \pi - 2\alpha$

$$\theta_s = \pi - \left\{2\sin^{-1}\left(\frac{D-d}{2C}\right)\right\} \tag{20.1}$$

Angle of contact for bigger pulley, $\theta_l = \pi + 2\alpha$

$$\theta_l = \pi + \left\{2\sin^{-1}\left(\frac{D-d}{2C}\right)\right\} \tag{20.2}$$

Length of Open Belt

L = length of straight portion + length wrapped on the pulleys

$$L = \sqrt{4C^2 - (D-d)^2} + 0.5\,(D\theta_L + d\theta_s) \tag{20.3}$$

$$= 2C\left\{\sqrt{1-\frac{(D-d)^2}{4C^2}}\right\} + 0.5\left[\pi D + \left\{2\sin^{-1}\left(\frac{D-d}{2C}\right)\right\}D + \pi d - \left\{2\sin^{-1}\left(\frac{D-d}{2C}\right)\right\}d\right]$$

$$= 2C\left\{\sqrt{1-\frac{(D-d)^2}{4C^2}}\right\} + 0.5\left[\pi\,(D+d) + \left\{2\sin^{-1}\left(\frac{D-d}{2C}\right)\right\}(D-d)\right]$$

$$\cong 2C\left\{1-\frac{(D-d)^2}{8C^2}\right\} + 0.5\left[\pi\,(D+d) + \left\{2\left(\frac{D-d}{2C}\right)\right\}(D-d)\right]$$

$$= 2C - \frac{(D-d)^2}{4C} + 05\,\pi\,(D+d) + \frac{(D-d)^2}{2C}$$

$$\therefore\quad L = 2C + 1.57\,(D+d) + \frac{(D-d)^2}{4C} \tag{20.3a}$$

Crossed belts

Angle of contact for both pulleys,

$$\theta = \theta_s = \theta_l = \pi + \left\{2\sin^{-1}\left(\frac{D+d}{2C}\right)\right\} \tag{20.4}$$

Length of crossed belt

L = Length of straight portion + length wrapped on the pulleys

$$L = \sqrt{4C^2 - (D+d)^2} + \frac{\theta}{2}(D+d) \tag{20.5}$$

$$= 2C\left\{\sqrt{1 - \frac{(D+d)^2}{4C^2}}\right\} + \frac{1}{2}(D+d)\left[\pi + \left\{2\sin^{-1}\left(\frac{D+d}{2C}\right)\right\}\right]$$

$$\cong 2C\left\{1 - \frac{(D-d)^2}{8C^2}\right\} + \frac{\pi}{2}(D+d) + \frac{1}{2}(D+d)\left(\frac{D+d}{C}\right)$$

$$= 2C - \frac{(D+d)^2}{4C} + \frac{\pi}{2}(D+d) + \frac{(D+d)^2}{2C}$$

$$\therefore \quad L = 2C + 1.57\,(D+d) + \frac{(D+d)^2}{4C} \tag{20.5a}$$

Belt Tensions

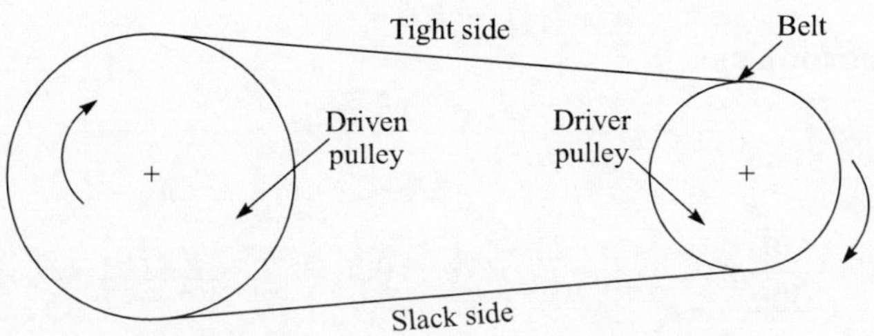

Figure 20.3 Tight and slack sides of belt

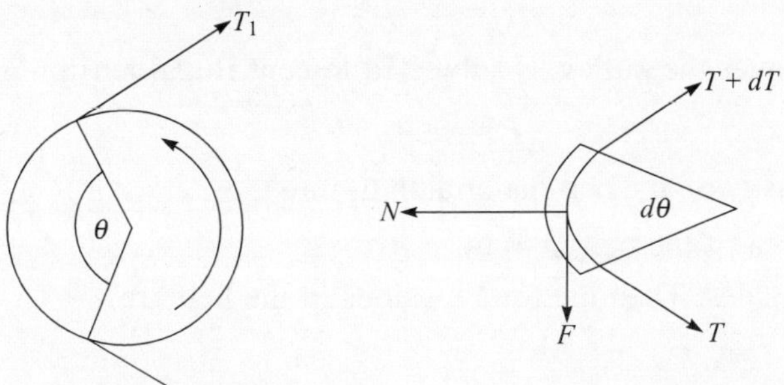

Figure 20.4 Variation of belt tension on the pulley

The belt is held tightly on the pulley and as it is pulled, the friction between the belt and pulley causes the pulley to rotate. There is gradual change in the belt tension. For an elemental length subtending an angle $d\theta$, the incremental tension is derived as follows.

Considering the components in the tangential direction,

$$(T + dT)\cos(d\theta/2) - T\cos(d\theta/2) - F = 0$$

$$(T + dT)\cos(d\theta/2) - T\cos(d\theta/2) - f \cdot N = 0$$

$$\therefore \quad dT\cos(d\theta/2) - f \cdot N = 0$$

$$\therefore \quad dT - f \cdot N = 0 \qquad \text{(i) (because } \cos(d\theta/2) \cong 1)$$

Considering the components in the radial direction,

$$N - (T + dT)\sin(d\theta/2) - T\sin(d\theta/2) = 0$$

$$\therefore \quad N - T \cdot d\theta - dT \cdot \frac{d\theta}{2} = 0$$

$$\therefore \quad N - T \cdot d\theta - = 0 \qquad \text{(ii)}$$

Substituting (ii) in (i),

$$\therefore \quad dT - f \cdot T \cdot d\theta = 0$$

$$\therefore \quad \frac{dT}{T} = f d\theta \qquad \text{(iii)}$$

Integrating for the angle of lap,

$$T_1/T_2 = e^{f\theta} \qquad (20.6)$$

Then effective driving torque,

$$T = (T_1 - T_2)R$$

The power transmitted in kW,

$$NP = \frac{T\omega}{1000}\,\text{kW} = (T_1 - T_2)R \cdot \omega/1000 = (T_1 - T_2)\,V/1000 \qquad (20.7)$$

where, ω is speed of the pulley in rad/sec;

V is belt speed in m/s

The belt, as it passes over the pulley, is subjected to centrifugal tension given by

$$P_c = m \cdot A \cdot V^2 \qquad (20.8)$$

where, m is the mass density of the belt material in g/mm^3;

A is cross-sectional area of the belt in $\text{mm}^2 = b \cdot t$

This is equal on both sides. Then the total tensions in the belt are:

on tight side $P_1 = T_1 + P_c$

on slack side $P_2 = T_2 + P_c$

Then the belt tension ratio

$$\frac{P_1 - P_c}{P_2 - P_c} = e^{f\theta} \tag{20.9}$$

And the power transmitted is

$$NP = \left[\frac{P_1 - P_2}{1000}\right] V \cdot \text{kW} \tag{20.10}$$

In order to transmit motion, there has to be initial tension P_o in the belt. The expression for the initial tension can be derived by making the following assumptions:

(i) The length of the belt is constant.

(ii) The belt material is linearly elastic.

When the driving pulley starts rotating, there is increase in the tension to P_1 from P_o. The belt extension is $dl = k \cdot (P_1 - P_o)$.

The tension on the loose side decreases from P_o to P_2.

The belt contraction is $dc = k \cdot (P_o - P_2)$

Since the belt length remains costant, $dl = dc$

$$\therefore \quad k \cdot (P_1 - P_o) = k \cdot (P_0 - P_2)$$

$$\therefore \quad P_o = (P_1 + P_2)/2 \tag{20.11a}$$

An empirical equation suggested by C.G. Barth,

$$2\sqrt{P_o} = \sqrt{P_1} + \sqrt{P_2} \tag{20.11b}$$

Condition for maximum power

For full tension in the belt, $P_1 = P_{max}$

The power transmited,

$$\begin{aligned} NP &= (T_1 - T_2)\,V/1000\ \text{kW} = (T_1 - T_2)\,V \text{ watts} \\ &= T_1(1 - T_2/T_1)\cdot V = T_1(1 - e^{-f\theta})\cdot V = (P_{max} - P_c)\cdot V \cdot (1 - e^{-f\theta}) \\ &= (V \cdot P_{max} - m \cdot A \cdot V^3)\cdot C;\ \text{where } C = (1 - e^{-f\theta}) \text{ is a constant} \end{aligned}$$

For maximum power,

$$\frac{d(NP)}{dV} = (P_{max} - 3\,\text{m} \cdot A \cdot V^2)\cdot C = 0$$

$$\therefore \quad P_{max} = 3\,m \cdot A \cdot V^2 = 3 \cdot P_c$$

$$\therefore \quad P_c = \frac{P_{max}}{3} \tag{20.12}$$

$$\therefore \text{ Optimum belt velocity, } V = \sqrt{\frac{P_{max}}{3\,mA}} \tag{20.12a}$$

The stress induced in the belt due to the tension,

$$\sigma_1 = [P_1/A] \tag{20.13}$$

There is additional bending stress in the belt due to bending of the belt on the pulley and is given by

$$\sigma_b = (Et/d) \tag{20.14}$$

∴ The total stress in the belt should be less than the design stress i.e.,

$$\sigma_{max} = \sigma_1 + \sigma_b < \sigma_d \tag{20.15}$$

Then for the belt cross section, we can write

$$A = bt = P_1/[\sigma_d - \sigma_b] \tag{20.16}$$

This equation is standardized in the following form,

$$A = bt = P/[k_{uo} \cdot C_1 \cdot C_2 \cdot C_3 \cdot C_4] \tag{20.16a}$$

where, $P = P_1 - P_2$ = effective pull = 1000 kW/V

k_{uo} = strength factor from Table 20.13 (Table 18.17 MDHB)

$C_1 \cdot C_2 \cdot C_3 \cdot C_4$ are modifying coefficients.

C_1 = coefficient for angle of contact = $1 - 0.17(\pi - \theta)$

C_2 = coefficient of belt velocity = $1.04 - 0.0004\,V^2$

C_3 = coefficient of operative condition = 1 to 0.9 for constant load

= 0.8 to 0.7 for considerable variations

= 0.7 to 0.6 for extremely uneven load

C_4 = belt inclination coefficient (Table 20.14) (Table 18.18 MDHB)

For a prescribed thickness, the belt width is given by

$$b = P/[\sigma_d - \sigma_b]t \tag{20.17}$$

The belt width b can also be calculated from the power – rating data, by the standardised equation: (by Eqn. 18.19 MDHB)

$$b = KW/[KW_r \cdot C_m \cdot C_p \cdot C_f] \tag{20.17a}$$

The value of power rating $(\text{kW})_r$, is taken from Table 20.10 (Table 18.14 MDHB). The value of the coefficients C_m, C_p and C_f are chosen from Tables 20.7, 20.8, & 20.9 respectively

Pulley Design: The pulley diameter is either specified or determined by speed ratio consideration. The minimum pulley diameters are specified in Tables 18.3 & 18.8 MDHB. The pulley rim width is to be wider than the belt width and evaluated by the C.G. Barth equations,

$B = 1.167b + 10$ for heavy duty (20.18a)

$B = 1.09375b + 5$ for light duty (20.18b)

The crawning is to be provided as per the following equations:

$h - B/200$ for well aligned shaft (20.19a)

$h = B/12$ for poorly aligned shaft (20.19b)

$h = 0.004568\, B^{2/3}$ for wide belt (20.19c)

The recommended rim thickness at the edge,

$$t = 0.25\sqrt{D} + 1.5 \text{ for light duty pulley} \quad (20.20a)$$

$$t = 0.35\sqrt{D} + 2.5 \text{ for heavy duty pulley} \quad (20.20b)$$

The diameter of the hub,

$$D_1 = 1.5d + 25 \quad (20.21)$$

The arms cross section is elliptical or rectangular.
The section modulus for bending strength is given by,

$= 2T_q/nS_d$ where T_q is the design torque and

S_d is the design stress

$= ba^2/6$ for rectangular section

$= \pi ba^2/32$ for elliptical section

The details of the arm and hub are calculated by using equations with Fig. 20.5.

Cast Iron Pulley Proportions

(a) Number of arms

Diameter	n
1. $D < 200$ mm	webs
2. $200 < D \le 450$	4 arms
3. $D > 450$ mm	6 arms

(b) Section of arm - Elliptical arm

$$a = 2.94\sqrt[3]{\frac{BD}{4n}} \quad \text{For single belt}$$

$$a = 2.94\sqrt[3]{\frac{BD}{2n}} \quad \text{For double belt}$$

Taper on arm 4 mm in 100 mm

Hub length, $L = 3B/4$

Hub diameter

($D_1 = d_2 + 0.84\sqrt[3]{BD} + 12$) for single belt.

($= d_2 + 1.06\sqrt[3]{BD} + 12$) for double belt.

Or $D_1 = 1.7d$ to $2d$

$r_1 = r_2 = a/2$

Rim thickness,

$t = D/200 + 3$ for single belt

$= D/200 + 6$ for double belt

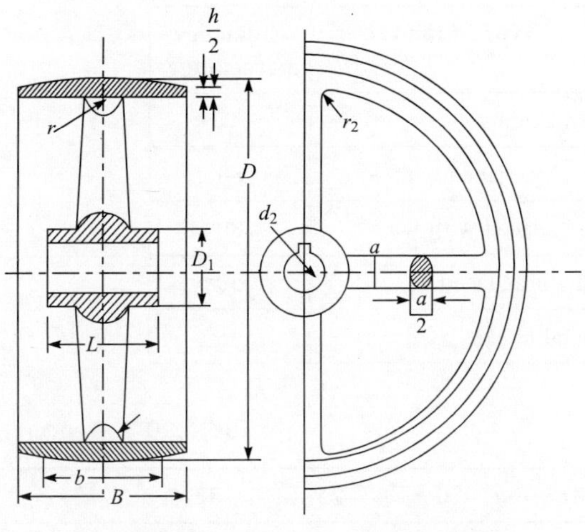

Figure 20.5 Cast iron pulley proportions

Table 20.1. Coefficient of friction for belts (flat)

Belt (material)	Pulley material						
	Cast iron & steel			Wood	Compressed paper	Leather faced	Rubber faced
	Dry	Wet	Greasy				
Leather, oak tanned	0.25	0.20	0.15	0.30	0.33	0.38	0.40
Leather, chrome tanned	0.35	0.32	0.22	0.40	0.45	0.48	0.50
Canvas, stitched	0.20	0.15	0.12	0.23	0.25	0.27	0.30
Cotton, woven	0.22	0.15	0.12	0.25	0.28	0.27	0.30
Camel hair, woven	0.35	0.25	0.20	0.40	0.45	0.45	0.45
Rubber	0.30	0.18	-	0.32	0.35	0.40	0.42
Balata	0.32	0.20	-	0.35	0.38	0.40	0.42

Table 20.2. Average efficiency of belt joint

Type of joint	Efficiency percentage
Cemented	100
Machine laced with wire	88
Hand laced with wire	80
Raw hide laced	50
Metal hooks	35

Table 20.2a. Density of belt material

Material of belt	Density in mg/mm^3
Leather	0.9
Canvas	1.2
Rubber	1.14
Woven belt, single	1.17
double	1.25

Table 20.3. Standard widths of belts (mm)

16	20	25	32	40	50	63	71	80	90
112	100	125	140	160	180	200		224	250
315	289	355	400	450	500				

Table 20.4. Minimum pulley diameter (mm)

	Single ply		Double ply			Triple ply	
Belt speed m/sec	**4.5 mm**	**5 mm**	**7 mm**	**8 mm**	**9 mm**	**12 mm**	**13.5 mm**
Up to 12.5	56	57	100	125	200	400	500
12.5–20.0	75	87.5	110	150	225	450	550
20.0–30.0	87.5	100	125	175	250	500	600

Table 20.5. Effective belt tensions $P = (P_1 - P_2)N$ per mm of belt width for arc of contact of 180°

Number of plies	Belt material				
	Rubber	**Balata**	**Canvas**	**Woven with cotton**	**Woven with camel hair**
3	5.5	8.75	5.5	Single 10.00	Single 11.00
4	7.25	11.80	7.25	Double 14.5	Double 15.5
5	9.0	14.50	9.00	Triple 20.0	Triple 22.5
6	11.0	17.00	11.00		
7	12.70	20.00	12.70		
8	14.50	23.00	14.50		
10	18.00	29.00	18.00		

Table 20.6. Maximum belt speeds

Type of pulley	Belt speed, m/sec
Cast iron pulley	
(1) Solid with flat or crown face	25
(2) Split pulley	16.7
Mild steel pulley solid or split	25

Table 20.7. Correction factors for type of motor and starting C_m

Type of drive	C_m
Any except electric motor	1.0
Squirrel cage, compensator starting	0.67
Squirrel cage, line starting	0.5
Slip ring, high starting torque	0.4

Table 20.8. Correction factor for the size of the small pulley C_p

Pulley size, mm	C_p
Upto 100	0.5
105 – 200	0.6
225 – 300	0.7
325 – 400	0.8
425 – 750	0.9
above 750	1.0

Table 20.9. Correction factors for service condition C_f

Operating conditions	C_f
Oily net or dusty atmosphere	0.74
Vertical drives	0.83
Jerky loads	0.83
Shock and reversing loads	0.71

Table 20.10. $(kW)_r$–power rating per mm width of oak tanned leather belt

Belt speed m/sec	Single ply		Double ply			Triple ply	
	4.5 mm medium	5 mm heavy	7 mm light	8 mm medium	9 mm heavy	12 mm medium	13.5 mm heavy
3.0	0.032	0.035	0.044	0.053	0.065	0.074	0.082
4.0	0.041	0.050	0.059	0.071	0.085	0.097	0.106
5.0	0.053	0.062	0.079	0.091	0.106	0.118	0.132
6.0	0.062	0.074	0.091	0.109	0.127	0.144	0.156
7.0	0.074	0.085	0.103	0.127	0.144	0.168	0.185
8.0	0.082	0.097	0.118	0.144	0.165	0.191	0.209
9.0	0.094	0.109	0.132	0.159	0.183	0.215	0.235
10.0	0.103	0.121	0.144	0.177	0.203	0.238	0.262
12.0	0.124	0.144	0.174	0.209	0.241	0.280	0.305
14.0	0.144	0.165	0.200	0.241	0.280	0.324	0.355
16.0	0.159	0.185	0.224	0.271	0.313	0.360	0.368
18.0	0.174	0.203	0.244	0.295	0.344	0.386	0.434
20.0	0.188	0.218	0.264	0.320	0.371	0.427	0.471
22.0	0.203	0.233	0.283	0.346	0.394	0.453	0.497
24.0	0.212	0.244	0.298	0.360	0.412	0.475	0.523
26.0	0.221	0.253	0.309	0.375	0.431	0.493	0.545
28.0	0.227	0.261	0.317	0.383	0.442	0.508	0.559
30.0	0.230	0.262	0.320	0.390	0.445	0.519	0.567

Table 20.11. Properties of flat belts

Belt material	Ultimate strength, N/mm² S_u	Endurance limit, N/mm² S_{en}	Factor of safety n_f	Maximum velocity in m/sec v	Density, mg/mm³ p	d/t	Width, mm b	Thickness, mm t	Modulus of elasticity, N/mm² E
Leather oak tanned	20-30	4	8-10	20-40	0.999	25-35	20-300	3-7.5	100-350
Rubber	37-44	8	6-10	20-30	1.27-1.53	30-40	20-500	2.5-14	80-12
Cotton woven	35-40	8	6-10	15-25	0.765-1.53	25-40	30-25	4.5-8.5	30-60
Woollen woven	30	6	8-10	20-30	0.917-1.27	25-30	50-500	2-10	-

Generalised values for flat belts are:

v = 20 m/sec, S_u = 25 N/mm² and E = 80 N/mm² n_f = 10, d/t = 30, and S_{en} = 6 N/mm².

Table 20.12. Rated kW per mm width of rubberised canvas belt

Belt speed, m/sec **No. of plies**	**3**	**4**	**5**	**6**	**8**	**10**	**12**
5	0.035	0.050	0.060	0.080	0.103	0.130	0.162
10	0.070	0.094	0.121	0.153	0.203	0.253	0.303
15	0.103	0.138	0.174	0.221	0.294	0.365	0.442
20	0.127	0.208	0.221	0.277	0.365	0.458	0.548
25	0.144	0.194	0.247	0.318	0.415	0.448	0.624

Table 20.13. Effective tension K_{uo} N/mm^2

Type of belt	**Maximum allowable ratio [*d*/*t*]**	***d*/*t* ratio**				
		25	30	40	50	100
Leather	25	1.7	1.9	2.1	2.4	2.6
Rubber	30	-	2	2.1	2.2	2.2
Cotton	25	1.5	1.6	1.7	1.8	1.9
Woollen	20	1.2	1.3	1.5	1.6	1.7

Table 20.14. Values of factor C_4

Drive	**Inclination θ°**		
	0°-60°	**60°-80°**	**80°-90°**
Open belt drive with periodic adjustment of tension	1.0	0.9	0.3
Half twist belt drive	0.9	0.8	0.7
Quarter twist belt drive	0.8	0.7	0.6

Table 20.15. Rubber transmission – flat belt

Flat belt rating at 80° arc of contact and 10 m/s belt speed

Type of belt	**Mass density, W_m**	**Rating per mm per ply, $KW_r(P)_r$**	**Use**
Dunlop Full speed 878 fabric belting	0.878 mg/mm^3	0.023 KW	Light duty and medium duty and for high belt speeds
Dunlop "Fort" 949 fabric belting hard fabric	0.949 mg/mm^3	0.0289 KW	For heavy duty and for medium belt speeds

Note: Rating at belt speed V, $(P)_r = KW_r\left(\frac{V}{10}\right)$

$$\text{Millimetre - plies of belt} = \frac{\text{Corrected load}}{\text{Load rating per mm per ply at belt speed}} = \frac{P \cdot K \cdot K_c}{(P)_r}$$

Table 20.16. Load correction factor, K_l

	Load classification	Load factor K_l
Normal load	Where peak load is known	1.0
Steady load	Screens, centrifugal pumps and fans, evaporators, agitators, belt conveyor, light machine tools: laundry, printing machinery and textile machinery	1.2
Intermittent load	Heavy duty fans and blowers, brickwork machinery reciprocating pumps, compressors, saw mill and paper mill machinery, heavy machine tools, strokers, elevators and line shafts	1.3
Shock loads	Vaccum pumps tube and ball mills, rolling mills, crushing machinery disintegrators, stamp presses, grinder hammer, mills, and calendar and automatic machinery	1.5

Table 20.17. Arc of contact factor K_c

Arc of contact $\theta = 180° - 60° \times (D + d)/C$								
Arc of contact θ	90°	120°	130°	140°	150°	160°	170°	180°
K_c	1.68	1.33	1.36	1.90	1.13	1.08	1.04	1.0
Arc of contact θ	190	200	210	220	230	240	250	
K_c	0.97	0.94	0.91	0.88	0.86	0.84	0.82	

Table 20.18. Standard belt (fabric) widths (mm)

3 ply (3.9 mm thick)	25	32*	40	44*	50	63	76	90*	100*
4 ply (5.1 mm thick)	25*	32*	40	44	50	63	76	90	100
	112	125	140*	152	200*				
5 ply (6.4 mm thick)	76	90*	100	112	125	180*	200*	224	250*
6 ply (7.7 mm thick)	100*	112	125	150	180	250			
8 ply (10.4 mm thick)	200**	250**	305**	355	400**				

*Available Hi-Speed only **Available in FORT only

Table 20.19. Recommended pulley diameter (mm) (IS - 1691-1968)

40, 45, 50, 56, 63, 71, 80, 90, 100, 112, 125, 140, 160, 180,
200, 224, 250, 280, 315, 355, 400, 450, 500, 560, 630, 710, 800, 900,
1000, 1120, 1250, 1400, 1800, 2000

SOLVED EXAMPLES

Example 20.1: A pulley is driven by a flat belt running at 600 m/min. The coefficient of friction between the pulley and belt is 0.3. The centre distance between the shafts is 1.425 m and the diameters of the pulleys are 1000 mm and 500 mm. The maximum tension in the belt is 700 N. Determine the power transmitted.

Given Data: Belt speed, V = 600 m/min, coefficient of friction, f = 0.3, centre distance, C = 1.425 m, D = 1000 mm, d = 500 mm, maximum tension, $P_1 = 700$ N

SOLUTION

Step-1: Belt tensions

Assuming open belt, the angle of lap on the smaller pulley, (Eqn. 18.12 MDHB)

$$\theta_s = \pi - 2\sin^{-1}\left(\frac{D-d}{2C}\right)$$

$$= \pi - 2\sin^{-1}\left(\frac{1000-500}{2\times 1425}\right) = 2.79 \text{ rad}$$

The ratio of belt tensions, (Eqn. 18.1 MDHB)

$$\frac{P_1 - P_c}{P_2 - P_c} = e^{f\theta} = e^{0.3\times 2.79} = 2.31$$

$\therefore$
$$P_2 - P_c = \frac{P_1 - P_c}{2.31} = 0.433\times(P_1 - P_c)$$

Neglecting centrifugal tension,

$$P_2 = 0.433\,P_1$$

but P_1 = max. tension = 700 N.

$\therefore$ $P_2 = 0.433 \times 700 = 303$ N.

Step-2: Power transmitted

$\therefore$ The belt speed V = 600/60 = 10 m/sec

$\therefore$ The power transmitted (Eqn. 18.6)

$$NP = \left(\frac{P_1 - P_2}{1000}\right)V = \frac{(700-303)\times 10}{1000} = 3.97\,\text{kW}$$

Example 20.2: Calculate the centrifugal tension in the belt running over two pulleys at 30 m/s. The belt is 200 mm wide and 8 mm thick and has a mass of 1000 kg/m^3. If the coefficient of friction is 0.28 and the angle of lap on the larger pulley is 190°, find the maximum power that can be transmitted at the given speed, given that the maximum permissible stress is 2.2 MPa. What is the percentage increase in the power, when the belt runs at the optimum speed?

Given Data: Belt speed = 30 m/s, Belt cross section $b \times t = 200 \times 8$ mm

Belt density, m = 1000 kg/m^3, Friction coefficient, f = 0.28,

Angle of lap, $\theta_l = 190°$, Maximum stress, S_d = 2.2 MPa

SOLUTION

Step-1: Belt tensions

The centrifugal tension, (Eqn. 18.4 MDHB)

$$P_c = \text{mbt V}^2$$

where, belt area $A = bt = 200 \times 8\ \text{mm}^2$

$m = 1000\ \text{kg/m}^3 = 1000 \times 10^3\ \text{gm}/1000^3\ \text{mm}^3 = 10^{-3}\ \text{gm/mm}^3$

$\therefore P_c = 10^{-3} \times 200 \times 8 \times (30)^2 = 1440\ \text{N}$

The maximum pull in the belt

$P_1 = A \cdot S_d = 200 \times 8 \times 2.2 = 3520\ \text{N}$

$\therefore P_1 - P_c = 3520 - 1440 = 2080\ \text{N}$

The angle of lap on the smaller pulley

$\theta_s = 2\pi - \theta_l = 360° - 190° = 170° = 2.967\ \text{rad}$

Then the belt tension ratio, $\dfrac{P_1 - P_c}{P_2 - P_c} = e^{f\theta}$ (Eqn. 18.1 MDHB)

$$\therefore \quad \frac{P_1 - P_c}{P_2 - P_c} = e^{0.28 \times 2.967} = 2.295$$

$$P_2 - P_c = \frac{P_1 - P_c}{2.295} = 906.3$$

$$\therefore \quad P_2 = 906.3 + 1440 = 2346.3\ \text{N}$$

Step–2: Power transmitted,

$$NP = \frac{(P_1 - P_2)V}{1000} = \frac{(3520 - 2346.3) \times 30}{1000} = 35.21\ \text{kW} \quad \text{(Eqn. 18.6 MDHB)}$$

Step–3: Power capacity for optimum speed,

$$P_c = P_1/3 = 3520/3 = 1173.3\ \text{N}$$

$$P_2 - P_c = (3520 - 1173.3)/2.295 = 1022.5\ \text{N}$$

$$\therefore \quad P_2 = 1022.5 + 1173.3 = 2195.8\ \text{N}$$

$\therefore$ Optimum belt speed,

$$V = \sqrt{(P_c/\text{mbt})} = \sqrt{\frac{1173.3}{10^{-3} \times 200 \times 8}} = 27.1\ \text{m/sec.}$$

$\therefore$ Power transmitted,

$$NP = \frac{(3520 - 2195.8) \times 27.1}{1000} = 35.88\ \text{kW}$$

There is an increase of **0.67 kW** for optimum belt speed.

Example 20.3: In an open belt drive, the smaller pulley is 300 mm diameter and rotates at 300 rpm. The bigger pulley has a diameter of 500 mm and is situated at a distance of 576 mm centre to centre from the smaller pulley. The coefficient of friction is 0.25. When 4 kW is being transmitted, which of the following alternatives is more effective to increase the power to be transmitted?

(a) Increasing initial tension by 10%
(b) Increasing coefficient of friction by 10%
(c) Increasing angle of lap to $190°$ by using idler pulley.

Given Data: Diameter of smaller pulley, $d = 300$ mm
Diameter of larger pulley, $D = 500$ mm
Centre distance, $C = 576$ mm
Speed of the smaller pulley = 300 rpm
Coëfficient of friction, $f = 0.25$
Power transmitted, $NP = 4$ kW

SOLUTION

Step-1: Belt tensions

Angle of lap on the smaller pulley, (Eqn. 18.12 MDHB)

$$\theta_s = \pi - 2\sin^{-1}\left(\frac{D-d}{2C}\right) = \pi - 2\sin^{-1}\left(\frac{500-300}{2\times 576}\right) = 2.79 \text{ rad}$$

$\therefore$ Tension ratio, $\dfrac{P_1}{P_2} = e^{f\theta} = e^{0.25\times 2.79} = 2.01$

Belt speed, $V = \dfrac{\pi dN}{60{,}000} = \dfrac{\pi \times 300 \times 300}{60{,}000} = 4.7$ m/s

Then power transmitted,

$$NP = \frac{(P_1 - P_2)}{1000} V \text{ kW}$$

$$P_1 - P_2 = 1000\ NP/V = (1000\times 4)/4.7 = 851.1 \text{ N}$$

$\therefore$ $2.01\,P_1 - P_2 = 851.1$ N

$\therefore$ $P_2 = 842.64$ N

$\therefore$ $P_1 = 842.64 \times 2.01 = 1693.7$ N

Initial tension, $P_0 = \dfrac{P_1 + P_2}{2} = 1268.17$ N

Step-2: Power capacity by increasing initial tension by 10%

$$P_0^1 = 1.1\,P_0 = 1.1 \times 1268.17 = 1394.98 \text{ N}$$

$\therefore$ $P_1 + P_2 = 2.P_0^1 = 2789.96$ N

$\therefore$ $2.01P_2 + P_2 = 2789.96$

$\therefore$ $P_2 = 926.9$ N

$\therefore$ $P_1 = 926.9 \times 2.01 = 1863.1$ N

New power transmitted, $NP = \dfrac{(1863.1 - 926.9)\times 4.7}{1000} = 4.4$ kW

Step-3: Power capacity by increasing coefficient of friction by 10%

$$\frac{P_1}{P_2} = e^{f\theta} = e^{1.1 \times .25 \times 2.79} = 2.154 \text{ (neglecting centrifugal action)}$$

Initial tension, $P_o = 1268.17$ N

Then $P_1 + P_2 = 2 \times P_0 = 2 \times 1268.17 = 2536.34$ N

$$P_2 = 2536.34/3.154 = 804.17 \text{ N},$$

$$P_1 = 804.17 \times 2.154 = 1732.17 \text{ N}$$

$$NP = \frac{(P_1 - P_2)V}{1000} = \frac{(1732.17 - 804.17) \times 4.7}{1000} = 4.36 \text{ kW}$$

Step-3: Power capacity by increasing, $\theta_s = 190° = 3.316$ rad

$$P_1/P_2 = e^{0.25 \times 3.316} = 2.291$$

but $P_1 + P_2 = 2 \times 1268.17 = 2536.34$ N

$\therefore$ $P_2 = 2536.34/3.291 = 770.68$ N

$\therefore$ $P_1 = 2.291 \times P_2 = 1765.65$ N

The power, $NP = \dfrac{(P_1 - P_2)V}{1000} = \dfrac{(1765.65 - 770.68) \times 4.7}{1000} = 4.676$ kW

The last option gives highest increase in power and hence idler is to be used.

Example 20.4: Design a flat belt drive to transmit 100 kW at a belt speed of 25 m/s between two pulleys of diameter 250 mm and 400 mm with centre distance of 1000 mm. The allowable belt stress is 8.5 MPa and the coefficient of friction is 0.3. The belt material density is 1000 kg/ m^3 and the thickness to width ratio is 0.1.

Given Data: Power transmitted, $NP = 100$ kW, Speed, $V = 25$ m/s,

Diameter of small pulley, $d = 250$ mm,

Diameter of bigger pulley, $D = 400$ mm,

Centre distance, $C = 1000$ mm

Allowable stress, $S_d = 8.5$ MPa

Coefficient of friction, $f = 0.3$

Belt material density, $m = 1000$ Kg/m^3

Thickness/width ratio = 0.1

SOLUTION

Step-1: Belt tensions

The effective pull, $P = (P_1 - P_2)$

But from the power equation (Eqn. 18.6 MDHB), effective pull is:

$$P_1 - P_2 = 1000\ \text{kW}/V = 1000 \times 100/25 = 4000\ \text{N} \tag{1}$$

The angle of lap on the smaller pulley, (Eqn. 18.12 MDHB)

$$\theta_s = \pi - 2\sin^{-1}\left(\frac{D-d}{2C}\right)$$

$$= \pi - 2\sin^{-1}\left(\frac{400-250}{1000\times 2}\right) = 171.4° = 2.99\ \text{rad}$$

The ratio of belt tensions,

$$\frac{P_1 - P_c}{P_2 - P_c} = e^{f\theta} = e^{0.3\times 2.99} = 2.45 \tag{2}$$

$$\therefore \quad \frac{P_1 - P_c}{P_1 - P_2} = \frac{2.45}{1.45} \quad \text{(by ratio proportions)}$$

$$\therefore \quad P_1 - P_c = \frac{2.45}{1.45}\times(P_1 - P_2) = \frac{2.45}{1.45}\times 4000 = 6758.62\ \text{N}$$

$$\sigma_1 = S_d = 8.5\ \text{MPa}$$

Centrifugal stress,

$$\sigma_c = mV^2 = 10^{-3}\times(25)^2 = 0.625\ \text{MPa}$$

Step-2: The cross section of the belt,

$$A = \frac{P_1 - P_c}{\sigma_1 - \sigma_c} = \frac{6758.62}{8.5 - 0.625} = 858.24\ \text{mm}^2$$

Width, $t/b = 0.1$

$\therefore \quad A = bt = 0.1\,b^2 = 858.24$

Width, $b = 92.64$ mm and $t = 9.264$ mm

Choose $-b = 95.0$ mm and $t = 9.5$ mm

Step-3: Pulley dimensions

The pulley width (Eqn. 18.21a)

$$B = 1.167\text{b} + 10 = 1.167\times 95 + 10 = 121\ \text{mm}$$

Rim thickness, $t = 0.35\sqrt{D} + 2.5 = 0.35\sqrt{250} + 2.5 = 8.00$ mm

Example 20.5: An exhaust fan fitted with 900 mm diameter pulley is driven by a flat belt from 30 kW, 950 rpm squirrel cage motor. The diameter of the pulley on the motor shaft is 250 mm. The centre distance is 2.25 m. Design the drive.

Given Data: Power, NP = 30 kW, Speed N_1 = 950 rpm,

Diameter of small pulley, d = 250 mm,

Diameter of bigger pulley, D = 900 mm,

Center distance, C = 2.25 m = 2250 mm

SOLUTION

Step-1: The belt speed,

$$V = \frac{\pi DN}{60,000} = \frac{\pi \times 250 \times 950}{60,000} = 12.4 \text{ m/s}$$

Step-2: The angle of lap on the smaller pulley,

$$\theta_s = \pi - 2\sin^{-1}\left(\frac{D-d}{2C}\right)$$

$$= \pi - 2\sin^{-1}\left(\frac{900-250}{2\times 2250}\right) = 2.85 \text{ rad}$$

Step-3: The length of the belt required,

$$L = 2C + 1.57\,(D+d) + \frac{(D-d)^2}{4C} \qquad \text{(Eqn. 18.14a MDHB)}$$

$$= 2\times 2250 + 1.57\,(900+250) + \frac{(900-250)^2}{4\times 2250}$$

$$= 4500 + 1805.5 + 46.9 = 6352.4 \text{ mm}$$

Step-4: Belt cross section

Select double ply, 9 mm thick leather belt.

The width of the belt by the rating Eqn. $KW = (kW)_r \cdot b \cdot C_m \cdot C_p \cdot C_r$

where $kW_r = 0.241$ kW/mm, (Table 20.10, double ply, 9 mm)

$C_m = 0.67$, Table 20.7 (Table 18.11 MDHB)

$C_p = 0.7$ Table 20.8 (Table 18.12 MDHB)

$C_f = 0.74$ Table 20.9 (Table 18.13 MDHB)

and kW = 30 (given)

$$b = 30/(0.241\times 67\times 0.7\times 0.74) = 358.6 \text{ mm}$$

Nearest available size Table 20.3 (Table 18.7 MDHB) is,

$$b = 355.0 \text{ mm}$$

Step-5: Pulley dimensions

Then pulley width, $B = 1.167b + 10 = 424.3 \text{ mm} \cong 425 \text{ mm}$

The rim thickness, $t = 0.35\sqrt{250} + 2.5 = 8$ mm

The number of arms (Fig. 18.5) $n = 4$

Major axis of the eliptical section of the arm,

$$a = 2.94\sqrt[3]{\frac{BD}{4n}} == 2.94\sqrt[3]{\frac{425 \times 250}{4 \times 4}} = 55.0 \text{ mm}$$

Minor axis, $b = a/2 = 55/2 = 27.5$ mm

Example 20.6: Design a belt drive to transmit 20 kW @ 960 rpm from a motor to a centrifugal pump running at 300 rpm. The centre distance is 1.2 m. The pulley diameter should not be less than 150 mm, the coefficient of friction is 0.35 and the mass of the belt is 9 kg/m-width/m-length. The drive is to be designed for 20% over load.

Given Data: Power, $NP = 20$ kW, Speed, $N_1 = 950$ rpm, Driven speed, $N_2 = 300$ rpm,
Centre distance, $C = 1.2$ m $= 1200$ mm,
Diameter of pulley, $d > 150$ mm,
Coefficient of friction, $f = 0.35$
Mass of the belt, $mm_1 = 9$ kg/m width/m length

SOLUTION

Step-1: Pulley diameters

Select the belt material as oak-tanned leather belt, double ply 8 mm thick.

Then choose $d/t \geq 25$

$d = 8 \times 25 = 200 \text{ mm} > d_{mim} = 150 \text{ mm}$, and hence acceptable

The diameter of the larger pulley,

$$D = d \times \frac{960}{300} = (200 \times 960)/300 = 640 \text{ mm}$$

Step-2: The angle of lap on the smaller pulley,

$$\theta_s = \pi - 2\sin^{-1}\frac{(D-d)}{2C}$$

$$= \pi - 2\sin^{-1}\left(\frac{640-200}{2 \times 1200}\right) = 158.9° = 2.773 \text{ rad}$$

Step-3: Belt cross section

$$\text{The belt velocity, } V = \frac{\pi dN}{60,000} = \frac{\pi \times 200 \times 960}{60,000} = 10.05 \text{ m/s}$$

$$\text{The effective pull, } P = P_1 - P_2 = \frac{1000 \text{ kW}}{V} = \frac{1000 \times 20}{10.05} = 1989.4 \text{ N}$$

The cross-sectional area of the belt, (Eqn. 18.20 MDHB)

$$A = \left[\frac{P}{K_{uo} C_1 \cdot C_2 \cdot C_3 \cdot C_4}\right]$$

where $P = 1989.4$ N

$$K_{uo} = \frac{1.7\text{N}}{\text{mm}^2} \qquad \text{Table 20.13 (Table 18.17 MDHB)}$$

$$C_1 = 0.17\,(\pi - \theta_s) = 0.93734$$

$$C_2 = 1.04 - 0.0004\,(10.05)^2 = 0.9996$$

$$C_3 = 0.8 \text{ (uneven load)}$$

$C_4 = 1.0$ for horizontal drive Table 20.14 (Table 18.18 MDHB)

$\therefore \quad A = [(1989.4)/1.7 \times 0.93734 \times 0.9996 \times 8.1.0)] = 1561.0 \text{ mm}^2$

The width of the belt, $b = A/t = 1561.0/8 = 195.13 \text{ mm} \cong 200 \text{ mm}$

The size of the belt, $A = (200 \times 8) \text{ mm}^2$

Step-4: Pulley dimensions

Pulley – rim width,

$B = 1.167\,b + 10 = 1.167 \times 200 + 10 = 243.4 \cong 245$ mm

The rim thickness for smaller pulley,

$t = 0.35\sqrt{D} + 2.5 = 0.35 \times \sqrt{200} + 2.5 = 7.5$ mm

The rim thickness of the bigger pulley,

$t = 0.35 \times \sqrt{640} + 2.5 = 11.35 \cong 12.0$ mm

The number of arms from Fig. 20.5 (Fig. 18.5 MDHB)

$n = 4.0$ for smaller pulley

$n = 6.0$ for bigger pulley

The section of the arm

For smaller pulley,

$$a = 2.94\sqrt[3]{\frac{BD}{4n}} = 2.94\sqrt[3]{\frac{245 \times 200}{4 \times 4}} = 42.7 \text{ mm} \cong 45.0 \text{ mm}$$

Minor axis, $b = a/2 = 22.5$ mm

For bigger pulley,

$$a = 2.94\sqrt[3]{\frac{245 \times 640}{4 \times 6}} = 55.0 \text{ mm}$$

Minor axis, $b = a/2 = 27.5$ mm

The torque on driving shaft (small pulley),

$$M_1 = \frac{955 \times 10^4 \times 20}{960} = 198958 \text{ Nmm.}$$

The torque on the driven shaft,

$$M_2 = 198958 \times \frac{960}{300} - 636666.7 \text{ Nmm.}$$

The shaft diameter for smaller pulley,

$$d_1 = \sqrt[3]{\frac{16M_1}{\pi S_{sd}}} = \sqrt[3]{\frac{16 \times 198958}{\pi \times 50}} = 27.26$$

$$d_1 = 30.0 \text{ mm}$$

For the bigger pulley,

$$d_2 = \sqrt[3]{\frac{16 \times 63666.7}{\pi \times 50}} = 40.17$$

Choose $d_2 = 40.00$ mm

Hub diameter for smaller pulley, (Eqn. 18.27 MDHB)

$$D_{b1} = 1.5d + 25 = 1.5 \times 30 + 25 = 70 \text{ mm}$$

Hub diameter for bigger pulley,

$$D_{b2} = 1.5 \times 40 + 25 = 85 \text{ mm}$$

Step-5: Check for strength of the arms

For smaller pulley arms,

$$Z_1 = \pi b h^2/32 = \pi\left(\frac{22.5 \times 45^2}{32}\right) = 4473.087 \text{ mm}^2$$

The bending stress in arms,

$$\sigma = \frac{2Tq}{nZ} = \frac{2 \times 198958}{4 \times 4473.087} = 22.24 \text{ MPa} < (S_d = 25.0 \text{ MPa for C.I})$$

For bigger pulley arms,

$$Z_2 = \pi \times 27.5 \times 55^2/(32) = 8166.9 \text{ mm}^3$$

The bending stress in arms,

$$\sigma = \frac{2Tq}{nZ} = \frac{2 \times 636666.7}{6 \times 8166.9} = 25.98 \text{ MPa} \simeq (S_d = 25.0 \text{ MPa})$$

Marginally equal. Hence, the design may be taken to be safe.

Example 20.7: A nylon core flat belt 200 mm wide, weighing 20 N/m, connecting 300 mm diameter pulley to a 900 mm diameter driven pulley at a shaft spacing of 6 m, transmits 55.2 kW at a belt speed of 25 m/sec. (i) Calculate the belt length and the angles of wrap and (ii) compute the belt tensions based on a coefficient of friction of 0.38.

Given Data: $b = 200$ mm; $d = 300$ mm; $D = 900$ mm and $C = 6$ m $= 6000$ mm;
$NP = 55.2$ kW; $V = 25$ m/sec; $f = 0.38$;
Weight per metre length = 20 N/m

SOLUTION

Step-1: Angles of lap for open belt

Angle of contact for larger pulley,

$$\theta_L = \pi + \left\{2\sin^{-1}\left(\frac{D-d}{2C}\right)\right\} \qquad \text{(Eqn. 18.12 MDHB)}$$

$$= \pi + \left[2\sin^{-1}\left(\frac{900-300}{2\times 6000}\right)\right] = 3.2416 \text{ radian } = 185.732^\circ$$

Angle of contact for smaller pulley,

$$\theta_s = \pi - \left\{2\sin^{-1}\left(\frac{D-d}{2C}\right)\right\} \qquad \text{(Eqn. 18.12 MDHB)}$$

$$= \pi + \left[2\sin^{-1}\left(\frac{900-300}{2\times 6000}\right)\right] = 3.04155 \text{ radian } = 174.268^\circ$$

Length of open belt,

$$L = \sqrt{4C^2 - (D-d)^2} + 0.5\,(D\theta_L + d\theta_s) \qquad \text{(Eqn. 18.14 MDHB)}$$

$$= \sqrt{4\times 6000^2 - (900-600)^2} + \frac{1}{2}(900\times 3.2416 + 300\times 3.04155)$$

$$= 13895.893 \text{ mm} \approx 13900 \text{ mm}$$

Step-2: Belt tensions

The power transmitted by the belt,

$$kW = \frac{(\sigma_1 - \sigma_c)}{1000} A\left(\frac{e^{f\theta}-1}{e^{f\theta}}\right)V = \frac{P_1 - P_c}{1000}\left(\frac{e^{f\theta}-1}{e^{f\theta}}\right)V \quad \text{(Eqn. 18.6a MDHB)}$$

where, $V = 25$ mps,

$$P_c = m\cdot b\cdot t\,V^2 = \frac{w}{g}V^2 = \frac{20\times 25^2}{9.81} = 1274.21 \text{ N},$$

$$e^{f\theta} = e^{0.38\times 3.04155} = 3.1765, \text{ and kW} = 55.2,$$

$$\therefore \quad 55.2 = \left(\frac{P_1 - 1274.21}{1000}\right)\left(\frac{3.1765-1}{3.1765}\right)\times 5 = \left(\frac{P_1 - 11274.21}{1000}\right)\times 17.13$$

$$P_1 = \frac{55.2\times 1000}{17.13} + 1274.21 = 4469.64 \text{ N and}$$

$$P_2 = (P_1 - P_c)/e^{f\theta} + P_c$$

$$= (4496.64 - 1274.21)/3.1765 + 1274.21 = 2288.67 \text{ N}$$

Example 20.8: A leather belt 125 mm wide and 6 mm thick, transmits power from a pulley of 750 mm diameter, which runs at 500 rpm. The angle of lap is 150° and the coefficient of friction between the belt and the pulley is 0.3. If the belt material density is 1000 kg/m^3 and the stress in the belt is not to exceed 2.75 MPa, find the power transmitted by the belt. Also find the initial tension in the belt.

Given Data: $b = 125$ mm; $t = 6$ mm; $d = 750$ mm; $n = 500$ rpm; $\theta = 150°$;

$f = 0.3$; $m = 1000\ kg/m^3 = 10^{-6}\ kg/mm^3$; $\sigma_1 = \sigma_d = 2.75$ MPa

SOLUTION

Step-1: Power transmitted by the belt drive

$$\text{Velocity of belt} = \frac{\pi(d+t)n}{60,000} = \frac{\pi(750+6)\times 500}{60,000} = 19.8 \text{ m/sec}$$

Tension on right side, $P_1 = \sigma_1 bt = 2.75 \times 125 \times 6 = 2062.5$ N

Centrifugal tension, $P_c = m A V^2$ (Eqn. 18.4 MDHB)

$$P_c = 1000 \times 10^{-6} \times 125 \times 6 \times 19.8^2 = 293.8 \text{ N}$$

Ratio of belt tension,

$$\frac{P_1 - P_c}{P_2 - P_c} = e^{f\theta} = e^{0.3} \times \frac{150}{180}\pi = 2.1938 \qquad \text{(Eqn. 18.1 MDHB)}$$

i.e.,
$$\frac{2062.5 - 293.8}{P_2 - 293.8} = 2.1938$$

$\therefore$ Tension on slack side, $P_2 = 1100.2$ N

$\therefore$ Power transmitted by the belt, $NP = \left(\frac{P_1 - P_2}{1000}\right) V$ kW

$$= \frac{(2062.5 - 1100.2)}{1000} \times 19.8 = 19.146 \text{ kW}$$

Step-2: Initial tension in the belts,

$$2\sqrt{P_0} = \sqrt{P_1} + \sqrt{P_2} = \sqrt{2065} + \sqrt{1100.2} \qquad \text{(Eqn. 18.17 MDHB)}$$

$\therefore$ Initial tension, $P_0 = 1543.862$ N

Example 20.9: A compressor is driven by 900 rpm motor by means of 250 × 10 mm flat belt. The motor pulley is 0.3 m in diameter and the compressor pulley is 1.5 m diameter. The distance between the centres of the pulleys is 2 m. A jockey pulley is used to make the angle of wrap on the smaller pulley 220° and the larger pulley 270°. The coefficient of friction between the belt and the smaller pulley is 0.3 and between the belt and the larger pulley is 0.22. The maximum allowable belt stress is 2 MN/m^2 and the specific weight of the belt material is 9.515 kN/m^3. Determine the power that can be transmitted by the belt drive.

Given Data: $n_1 = 900$ rpm; $b = 250$ mm; $t = 10$ mm; $d = 0.3$ m;

$$D = 1.5 \text{ m}; C = 2 \text{ m}; \theta_s = 220°; \theta_L = 270°; f_s = 3; f_L = 0.22;$$

$$\sigma_1 = \sigma_d = 2 \text{ MN/m}^2 = 2 \text{ N/mm}^2; w = 9.515 \text{ kN/m}^3 = 9515 \text{ N/m}^3$$

SOLUTION

Step-1: Belt tensions

Belt pull on tight side,

$$\therefore \quad P_1 = \sigma bt = 2 \times 250 \times 10 = 5000 \text{ N}$$

Centrifugal tension,

$$P_c = m \cdot btV^2 \qquad \text{(Eqn. 18.4 MDHB)}$$

$$\text{Belt speed, } V = \frac{\pi(d+t)n_1}{60000} = \frac{\pi \times (300+10) \times 900}{60,000} = 14.61 \text{ m/sec}$$

where, $m = 9.575 \text{ kN/m}^3 = 9515 \text{ N/m}^3$

$$= \frac{9515}{9.81} \text{ kg/m}^3 = 969.93 \text{ kg/m}^3 = 969.93 \times 10^{-6} \text{ gm/mm}^3$$

Then substituting the values,

$$\therefore \quad P_c = 969.93 \times 10^{-6} \times 250 \times 10 \times 14.61^2 = 517.58 \text{ N}$$

Step-2: Power transmitted

$$\text{The belt tension ratio} = e^{f\theta} = e^{.22} \times \frac{270}{180} \times \pi = e^{1.036} = 2.82$$

Power transmitted,

$$\text{kW} = \frac{(\sigma_1 - \sigma_c)A}{1000}\left(\frac{e^{f\theta} - 1}{e^{f\theta}}\right)V, = \left(\frac{P_1 - P_C}{1000}\right)\left(\frac{e^{f\theta} - 1}{e^{f\theta}}\right)V$$

$$= \left(\frac{5000 - 517.58}{1000}\right)\left(\frac{2.82 - 1}{2.82}\right) \times 14.61 = 42.26 \text{ kW}$$

Example 20.10: The layout of the leather belt drive transmitting 15 kW power is shown in Fig. 20.10. The centre distance between the pulleys is twice the diameter of the big pulley. The belt should operate at a velocity of 20 m/sec and the stresses in the belt should not exceed 2.25 MPa. The density of the leather belt is 0.95 gm/cc and the coefficient of friction is 0.35. The thickness of the belt is 5 mm. Calculate:

(i) Diameter of the pulleys

(ii) The length and width of belt

(iii) Belt tensions.

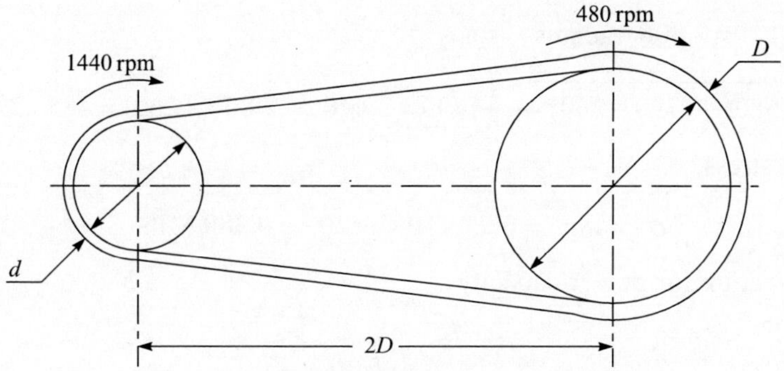

Figure 20.6

Given Data: $NP = 15$ kW; $C = 2$ D; $v = 20\dfrac{\text{m}}{\text{sec}}$; $\sigma_d = \sigma_1 = 2.25\dfrac{\text{N}}{\text{mm}^2}$;

$f = 0.35$; $\rho = 0.95$ gm/cc $= 0.95 \times 10^{-3}$ g/mm^3

SOLUTION

Step-1: Diameter of the pulleys

Velocity of the belt drive, $v = \dfrac{\pi(d+t)n_1}{60{,}000}$ m/sec

i.e., $$= 20 = \frac{\pi(d+5)\times 1440}{60{,}000}$$

$\therefore$ Diameter of smaller pulley, $d = 260.258 \text{ mm} \cong 260 \text{ mm}$

For the same belt speed, $n_1(d+t) = n_2(D+t)$

$$D = \frac{n_1}{n_2}(d+t) - t = \frac{1440}{480}(260+5) - 5 = 790 \text{ mm}$$

$\therefore$ Diameter of larger pulley, $D = 790$ mm

Step-2: Length of belt

Centre distance, $C = 2D = 2 \times 790 = 1580$ mm

Length of open belt,

$$L = \sqrt{4C^2 - (D-d)^2} + \frac{1}{2}(D\theta_L + \theta_S d) \qquad \text{(Eqn. 18.14 MDHB)}$$

$$\theta_L = \pi + \left\{2\sin^{-1}\left(\frac{D-d}{2C}\right)\right\} = \pi + \left\{2\sin^{-1}\left(\frac{780-260}{2\times 1560}\right)\right\}$$

$$= 3.4765 \text{ radians}$$

$$\theta_S = \pi - \left\{2\sin^{-1}\left(\frac{D-d}{2C}\right)\right\} = \pi + \left\{2\sin^{-1}\left(\frac{780-260}{2\times 1560}\right)\right\}$$

$$= 2.806 \text{ radians}$$

$$\therefore \quad L = \sqrt{4 \times 1580^2 - (790 - 260)^2} + \frac{1}{2}[790 \times 3.4765 + 2.806 \times 260] = 4853.79 \text{ mm}$$

Step-3: Centrifugal stress,

$$\sigma_c = mv^2 = 0.95 \times 10^{-3} \times 20^2 = 0.380 \text{ MPa}$$

Step-4: Belt cross section for power capacity

Power equation is:

$$\text{kW} = \frac{(\sigma_1 - \sigma_c)A}{1000}\left[\frac{e^{f\theta} - 1}{e^{f\theta}}\right]V \qquad \text{(Eqn. 18.6a MDHB)}$$

where, $\sigma_1 = 2.25$ MPa, $\sigma_c = 0.38$ MPa

$$e^{f\theta} = e^{0.35 \times 2.804} = 2.668\,; V = 20 \text{ m/sec}\,; \text{kW} = 15 \text{ kW}$$

$$\therefore \quad 15 = \frac{(2.25 - 0.38)}{1000} \times A \times \left(\frac{2.668 - 1}{2.668}\right) \times 20$$

$$A = 641.52 \text{ mm}^2$$

$$b = A/t = 641.52/5 = 128.30$$

Choose $b = 130$ mm

Step-5: Belt tensions

Centrifugal stress, $\sigma_c = 0.38$ MPa

$$\text{Now,} \quad \frac{\sigma_1 - \sigma_c}{\sigma_2 - \sigma_c} = e^{f\theta}$$

$$\text{i.e.,} \quad \frac{2.25 - 0.38}{\sigma_2 - 0.38} = 2.668$$

$$\therefore \quad \sigma_2 = 1.08 \text{ MPa}$$

Tension on tight side,

$$P_1 = \sigma_1 A = \sigma_1 \times b \times t = 2.25 \times 130 \times 5 = 1462.5 \text{ N}$$

Tension on slack side,

$$P_2 = \sigma_2 A = \sigma_2 \times b \times t = 1.08 \times 130 \times 5 = 702.65 \text{ N}$$

20.2.2 V-Belts

V-belts are specially made endless belts with reinforcements impregnated with rubber. They are standardized and are available in five standard shapes designated by letters *A*, *B*, *C*, *D* and *E*. They vary in the width and thickness proportions. They are specified by a letter followed by a number indicating the inside length of the belt.

The pitch length is calculated by, (derived for flat belt)

$$L_p = 2C + 1.57\,(D+d) + \frac{(D-d)^2}{4C} \tag{20.22}$$

The relation between the pitch length and inside length is given the Table 20.35 (Table 18.46 MDHB). The proportions and other data for the different sections are given in Table 20.20 (Table 18.29 MDHB) and the standard dimensions of pulleys are given in Table 20.33 (Table 18.44 MDHB). The Tables 20.22 and 20.20 (Tables 18.33 and 18.29 MDHB) and Fig. 20.7 are useful in selecting the suitable section for a particular application. The design of the *V*-belt drive based on strength, requires the calculation of the maximum tension P_1 in the belt by using the following relations, drived in case of flat belt.

The tension ratio, $$\frac{P_1 - P_c}{P_2 - P_c} = e^{f_1\theta} \tag{20.23}$$

where, $f_1 = f/\sin\beta$ is the effective coefficient of friction and β is the semi-groove angle.

This equation is derived as follows:

The frictional force on the belt element,

$$F = 2\,N\cdot f \tag{i}$$

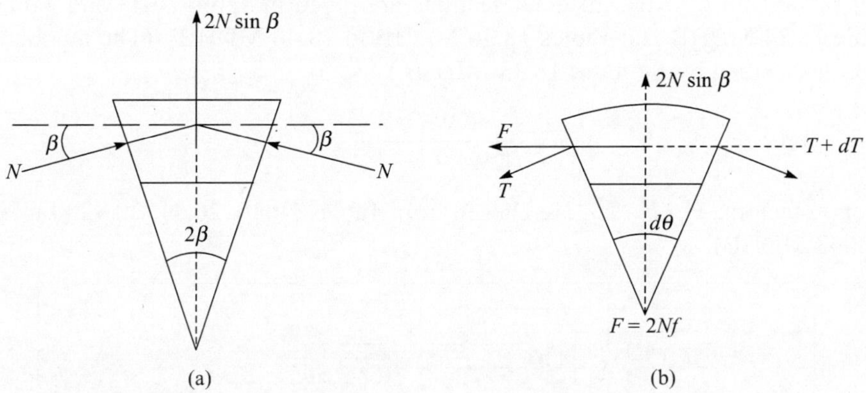

Figure 20.7(a) Forces on belt **Figure 20.7(b)** Variation of tension

For equilibrium in tangential direction,

$$(T + dT)\cos(d\theta/2) - T\cos(d\theta/2) - F = 0$$

∴ Simplifying we get, $$dT - 2Nf = 0 \tag{ii}$$
(because $\cos(d\theta/2) \approx 1$)

For force balance *N* radial direction,

$$(T + dT)\sin(d\theta/2) + T\sin(d\theta/2) - 2N\sin\beta = 0$$

Simplifying we get

$$T\cdot d\theta - 2\mathrm{N}\sin\beta = 0 \quad (\sin(d\theta/2) \approx d\theta/2)$$

$$\therefore \qquad 2N = \frac{T\,d\theta}{\sin\beta} \tag{iii}$$

Substituting (iii) in (ii), $dT - f\dfrac{T\,d\theta}{\sin\beta} = 0$

$$\frac{dT}{T} = \frac{f}{\sin\beta}\,d\theta = f_1 d\theta, \text{ where } f_1 = \frac{f}{\sin\beta}$$

Integrating over θ, $\dfrac{T_1}{T_2} = e^{f_1 d\theta}$ (iv)

$$\therefore \qquad \frac{P_1 - P_c}{P_2 - P_c} = e^{f_1 d\theta} \tag{v}$$

And the power is $NP = \left[\dfrac{P_1 - P_2}{1000}\right] V$ kW (20.24)

The number of belts required is by the relation,

$$n = P_1 / P_d, \text{ where } P_d = A \times S_d$$

The design procedure based on the power rating is as follows.

The power rating $(KW)_r$ of the different sections are given in Table 20.23 and Tables 20.25 to 20.29 (Table 18.34 MDHB and Tables 18.36 MDHB to 18.40 MDHB). (The number of belts to be used are determined by the Eqn. 18.33 MDHB.)

$$n = \frac{\text{KW}\,F_a}{(\text{KW})_r \cdot F_c \cdot F_d} \tag{20.25}$$

The correction factors, $F_a \cdot F_c \cdot F_d$, are chosen from Tables 20.30, 20.31, 20.32 (Tables 18.41, 18.42, and 18.43 MDHB).

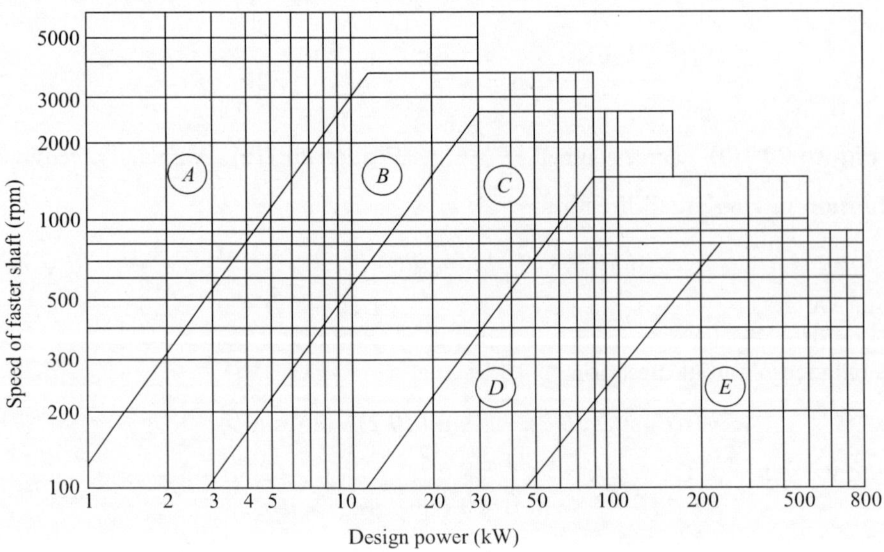

Figure 20.8 Selection of cross section of *V*-belt

Table 20.20. Data on standard *V*-belt section* IS: 2494-1964

Cross section symbol	Usual load of drive		Recommended minimum pulley pitch diameter, mm	Normal top width W, mm	Nominal thickness T, mm	Mass per metre, kg	Max no. of strands
	kW	HP					
A	0.75-4	1-5	75	13	8	0.106	6
B	2-15	3-20	125	17	11	0.189	9
C	7.5-75	10-100	200	22	14	0.343	12
D	22-150	30-200	355	32	19	0.596	14
E	30-190	40-250	500	38	23	0.885	20

*Indian standard specification for *V*-belt industrial purpose

Table 20.21. Suggested service factors K_s for *V*-belt drives (Eqn. 18.35 MDHB)

Driven machinery	Source of power	
	Normal torque characteristic	High of non-uniform torque
Uniform	1.0 to 1.2	1.1 to 1.3
Light shock	1.1 to 1.3	1.2 to 1.4
Medium shock	1.2 to 1.4	1.4 to 1.6
Heavy shock	1.3 to 1.5	1.5 to 1.8

Table 20.22. Section of *V*-belts

Normal kW	Motor speed in rpm			
	1800	1200	900	720 and below
1	*A*	*A*	*A*	*A*
1.51 – 3 .75	*A* or *B*	*A* or *B*	*A* or *B*	*A* or *B*
7.5	*A* or *B*	*A* or *B*	*B*	*B* or *C*
8 – 10	*B* or *C*	*B* or *C*	*B* or *C*	*C*
15.0	*B*	*C*	*C*	*C*
15 – 22 .5	*C*	*C*	*C*	*C*
30.0	*C*	*C* or *D*	*C* or *D*	*D*
37.5 – 45 .0	*C* or *D*	*C* or *D*	*C* or *D*	*D* or *E*

Contd.

55	*C* or *D*	*C* or *D*	*C* or *D*	*D* or *E*
75	*C*	*D*	*D*	*D* or *E*
90	-	*D*	*D*	*D* or *E*
110 – 150	-	*D*	*D*	*E*
185	-	*E*	*E*	*E*
225 and above	-	-	*E*	*E*

Table 20.23. Power transmitting capacity of *V*-belts in kW

Cross section	Formula for kW	Max. d_e mm
A	$\text{kW} = \left[0.45V^{-0.09} - \frac{19.62}{d_e} - 0.765 \times 10^{-4}V^2\right]V$	125
B	$\text{kW} = \left[0.79V^{-0.09} - \frac{50.8}{d_e} - 1.32 \times 10^{-4}V^2\right]V$	175
C	$\text{kW} = \left[1.47V^{-0.09} - \frac{142.7}{d_e} - 2.34 \times 10^{-4}V^2\right]V$	300
D	$\text{kW} = \left[3.22V^{-0.09} - \frac{506.7}{d_e} - 4.78 \times 10^{-4}V^2\right]V$	425
E	$\text{kW} = \left[4.58V^{-0.09} - \frac{952}{d_e} - 7.05 \times 10^{-4}V^2\right]V$	700

kW = Maximum power in kW for 180° contact angle.

V = Belt speed in m/sec

d_e = Equivalent pitch diameter (mm)

$= d_p \times F_b$

d_p = Pitch diameter of smaller pulley mm

F_b = Diameter factor for smaller pulley. (Fig. 18.5)

Table 20.24. Diameter factor F_b

Speed ratio range	F_b	Speed ratio range	F_b
1.0 – 1.019	1.00	1.179 – 1.222	1.07
1.02 – 1.032	1.01	1.223 – 1.274	1.08
1.033 – 1.055	1.02	1.275 – 1.340	1.09

Contd.

1.056 – 1.081	1.03	1.341 – 1.429	1.10
1.082 – 1.109	1.04	1.43 – 1.562	1.11
1.11 – 1.142	1.05	1.563 – 1.814	1.12
1.143 – 1.178	1.06	1.815 – 2.948	1.13
		2.949 up	1.14

Table 20.25. Rating for *V*-belts in kilowatt cross section A

Belt speed (m/s) (1)	**Equivalent pitch diameter d_e in mm**					
	80 (2)	**90 (3)**	**100 (4)**	**110 (5)**	**120 (6)**	**125 & over (7)**
0.5	0.13	0.13	0.14	0.14	0.15	0.15
1	0.22	0.24	0.25	0.27	0.28	0.29
2	0.37	0.40	0.43	0.46	0.49	0.51
3	0.51	0.58	0.64	0.68	0.72	0.74
4	0.58	0.74	0.81	0.88	0.93	0.96
5	0.74	0.85	0.95	1.04	1.13	1.18
6	0.81	0.94	1.05	1.15	1.25	1.32
7	0.88	1.08	1.25	1.39	1.50	1.54
8	0.96	1.17	1.35	1.50	1.62	1.69
9	1.03	1.32	1.54	1.69	1.77	1.84
10	1.10	1.40	1.62	1.77	1.91	1.99
11	1.18	1.47	1.69	1.91	2.06	2.13
12	1.25	1.54	1.84	2.06	2.21	2.28
13	1.32	1.62	1.91	2.13	2.35	2.43
14	1.32	1.69	1.99	2.28	2.50	2.50
15	1.32	1.77	2.06	2.35	2.57	2.65
16	1.40	1.84	2.13	2.43	2.65	2.79
17	1.40	1.84	2.21	2.50	2.79	2.87
18	1.40	1.84	2.28	2.57	2.87	2.94
19	1.40	1.84	2.28	2.65	2.94	3.02

Contd.

20	1.32	1.91	2.35	2.72	3.02	3.09
21	1.32	1.91	2.35	2.72	3.02	3.16
22	1.25	1.91	2.35	2.79	3.09	3.24
23	1.25	1.84	2.35	2.79	3.16	3.31
24	1.18	1.84	2.35	2.79	3.16	3.31
25	1.10	1.77	2.28	2.79	3.16	3.31
26	1.03	1.69	2.28	2.72	3.16	3.31
27	0.88	1.62	2.20	2.72	3.09	3.31
28	0.81	1.54	2.13	2.65	3.09	3.24
29	0.66	1.47	2.06	2.57	3.02	3.24
30	0.51	1.32	1.99	2.50	2.92	3.16

Note as given below the next table

Table 20.26. Ratings for *V*-belts in kilowatts cross section *B*

Belt speed (m/s) (1)	**Equivalent pitch diameter d_e in mm**					
	130 (2)	**140 (3)**	**150 (4)**	**160 (5)**	**170 (6)**	**180 & over (7)**
0.5	0.22	0.22	0.22	0.22	0.29	0.29
1	0.37	0.44	0.44	0.44	0.51	0.51
2	0.66	0.74	0.81	0.81	0.88	0.88
3	0.96	1.03	1.10	1.17	1.25	1.39
4	1.18	1.32	1.40	1.47	1.50	1.62
5	1.47	1.54	1.69	1.84	1.91	1.99
6	1.62	1.84	1.99	2.06	2.21	2.28
7	1.91	2.13	2.21	2.35	2.50	2.65
8	2.06	2.28	2.43	2.65	2.79	2.94
9	2.21	2.43	2.65	2.87	3.02	3.16
10	2.35	2.65	2.87	3.09	3.31	3.46
11	2.50	2.79	3.09	3.31	3.53	3.75
12	2.65	3.02	3.31	3.53	3.75	3.97
13	2.79	3.16	3.46	3.75	3.97	4.19
14	2.87	3.16	3.60	3.90	4.19	4.56

Contd

15	2.94	3.38	3.75	4.05	4.34	4.63
16	3.02	3.46	3.90	4.19	4.49	4.78
17	3.09	3.60	3.97	4.34	4.71	4.92
18	3.16	3.60	4.04	4.49	4.78	5.07
19	3.16	3.68	4.19	4.56	4.92	5.22
20	3.16	3.75	4.19	4.63	5.00	5.44
21	3.16	3.75	4.27	4.71	5.07	5.44
22	3.16	3.75	4.27	4.78	5.15	5.52
23	3.09	3.75	4.27	4.78	5.22	5.59
24	3.02	3.68	4.27	4.78	5.22	5.66
25	2.94	3.60	4.19	4.78	5.22	5.66
26	2.79	3.53	4.19	4.71	5.22	5.66
27	2.65	3.46	4.12	4.63	5.15	5.66
28	2.50	3.31	3.97	4.56	5.15	5.59
29	2.43	3.16	3.82	4.49	5.00	5.52
30	2.13	2.94	3.68	4.34	4.92	5.44

Note: 1. Belt speed is calculated using the pitch diameter and revolution per second of one of the pulleys on the drive.

2. If the belt speeds are over 25 metres per second, the pulleys may require special construction or special material. They may also require dynamic balancing. Such applications should be referred to the manufacturer.

Table 20.27. Rating for *V*-belts in kilowatt cross section *C*

Belt speed (m/s)	**Equivalent pitch diameter d_e in mm**						
	180	**200**	**220**	**240**	**260**	**280**	**300 & over**
0.5	0.37	0.44	0.44	0.51	0.51	0.51	0.51
1	0.64	0.73	0.81	0.88	0.88	0.96	0.96
2	1.15	1.32	1.47	1.50	1.69	1.77	1.84
3	1.62	1.84	2.06	2.21	2.35	2.50	2.57
4	2.06	2.35	2.57	2.79	3.02	3.16	3.31
5	2.35	2.79	3.09	3.38	3.60	3.75	3.97
6	2.72	3.16	3.60	3.90	4.19	4.41	4.63
7	3.02	3.60	4.05	4.41	4.71	5.00	5.22
8	3.31	3.97	4.49	4.85	5.22	5.59	5.81
9	3.60	4.27	4.85	5.37	5.81	6.10	6.40
10	3.82	4.56	5.22	5.81	6.25	6.62	6.99

Contd.

11	4.04	4.92	5.59	6.18	6.69	7.13	7.58
12	4.19	5.15	5.96	6.62	7.13	7.72	8.00
13	4.34	5.44	6.25	6.91	7.58	8.16	8.46
14	4.49	5.59	6.55	7.28	7.94	8.53	8.97
15	4.63	5.81	6.77	7.58	8.31	8.90	9.41
16	4.71	5.96	7.06	7.87	8.61	9.27	9.78
17	4.78	6.10	7.21	8.09	8.90	9.56	10.15
18	4.78	6.25	7.35	8.38	9.19	9.93	10.51
19	4.78	6.33	7.58	8.53	9.41	10.15	10.81
20	4.78	6.33	7.65	8.68	9.63	10.44	11.11
21	4.71	6.33	7.72	8.83	9.86	10.66	11.33
22	4.56	6.33	7.80	8.90	10.00	10.81	11.62
23	4.34	6.25	7.80	8.97	10.08	11.03	11.77
24	4.19	6.18	7.72	9.05	10.15	11.11	11.91
25	4.05	6.03	7.71	9.05	10.15	11.18	11.99
26	3.82	5.88	7.58	8.97	10.15	11.18	12.06
27	3.60	5.66	7.43	8.90	10.15	11.18	12.13
28	3.24	5.44	7.35	8.75	10.08	11.18	12.13
29	2.79	5.15	7.06	8.60	9.93	11.03	12.06
30	2.43	4.78	6.77	8.38	9.71	10.80	11.99

Note: 1. Belt speed is calculated using the pitch diameter and revolution per second of one of the pulleys on the drive.

2. If the belt speeds are over 25 metres per second, the pulleys may require special construction or special material. They may also require dynamic balancing. Such applications should be referred to the manufacturer.

Table 20.28. Rating for *V*-belts in kilowatt cross section *D*

Belt speed (m/s)	**Equivalent pitch diameter d_e in mm**						
	300	**320**	**340**	**360**	**380**	**400**	**420**	**430 & over**
0.5	0.81	0.88	0.96	0.96	1.03	1.03	1.10	1.1
1	1.47	1.54	1.69	1.77	1.84	1.91	1.91	1.99
2	2.50	2.79	2.94	3.09	3.24	3.38	3.53	3.53
3	3.46	3.75	4.04	4.34	4.56	4.71	4.92	5.00
4	4.34	4.78	5.15	5.44	5.74	6.03	6.25	6.00
5	5.07	5.66	6.10	6.55	6.91	7.21	7.58	7.65

Contd.

6	5.81	6.55	7.06	7.51	7.94	8.38	8.75	8.90
7	6.47	7.28	7.94	8.46	8.97	9.41	9.86	10.08
8	7.13	7.94	8.68	9.34	10.00	10.52	10.96	11.25
9	7.65	8.68	9.49	10.22	10.96	11.47	12.06	12.28
10	8.23	9.34	10.22	11.03	11.84	12.43	13.02	13.31
11	8.68	9.86	10.88	11.84	12.58	13.39	14.05	14.34
12	9.12	10.37	11.47	12.50	13.39	14.19	14.93	15.30
13	9.49	10.88	12.06	13.16	14.05	15.00	15.74	16.11
14	9.79	11.25	12.58	13.75	14.56	15.74	16.55	16.99
15	10.00	11.62	13.09	14.34	15.44	16.40	17.28	17.72
16	10.30	11.91	13.46	14.78	15.96	16.99	18.02	18.46
17	10.44	12.21	13.83	15.22	16.55	17.58	18.61	19.05
18	10.51	12.43	14.12	15.59	16.99	18.09	19.20	19.71
19	10.51	12.50	14.27	15.89	17.36	18.53	19.86	20.23
20	10.51	12.58	14.49	16.11	17.65	18.98	20.15	20.74
21	10.37	13.31	14.56	16.40	17.87	19.27	20.52	21.11
22	10.22	12.58	14.56	16.47	18.02	19.49	20.81	21.48
23	9.93	12.43	14.56	16.40	18.17	19.71	21.11	21.70
24	9.63	12.21	14.34	16.40	18.24	19.78	21.26	21.99
25	9.19	11.91	14.27	16.25	18.17	19.78	21.33	22.06
26	8.68	11.47	13.97	16.11	17.87	19.78	21.33	21.99
27	8.16	11.03	13.53	15.81	17.80	19.56	21.26	21.99
28	7.51	10.51	13.09	15.44	17.43	19.34	21.11	21.84
29	6.77	9.86	12.58	15.00	17.14	19.12	20.89	21.62
30	5.96	9.12	11.91	14.42	16.70	18.68	20.52	21.33

Note: 1. Belt speed is calculated using the pitch diameter and revolution per second of one of the pulleys on the drive.

2. If the belt speeds are over 25 metres per second, the pulleys may require special construction or special material. They may also require dynamic balancing. Such applications should be referred to the manufacturer.

Table 20.29. Rating for *V*-belts in kilowatt cross section *E*

Belt speed (m/s)	Equivalent pitch diameter d_e in mm					
	450	500	550	600	650	700 & over
0.5	1.40	1.47	1.54	1.62	1.69	1.77
1	2.43	2.65	2.87	2.94	3.09	3.24
2	4.34	4.78	5.15	5.44	5.66	5.88
3	6.03	6.69	7.21	7.65	8.02	8.38
4	7.58	8.46	9.19	9.78	10.22	10.74
5	9.05	10.15	11.03	11.77	12.36	13.02
6	10.44	11.69	12.80	13.68	14.49	15.07
7	11.77	13.32	14.49	15.51	16.40	17.14
8	13.02	14.78	16.11	17.28	18.31	19.05
9	14.12	16.11	17.72	18.98	20.15	21.11
10	15.22	17.43	19.12	20.59	21.85	22.95
11	16.25	18.61	20.59	22.20	23.61	24.71
12	17.21	19.78	21.92	23.68	25.15	26.40
13	18.02	20.81	23.17	25.08	26.69	28.02
14	18.84	21.85	24.35	26.40	28.10	29.57
15	19.56	22.80	25.45	27.65	29.49	31.04
16	20.15	23.61	26.40	28.83	30.82	32.44
17	20.74	24.42	27.36	29.86	31.99	33.76
18	21.18	25.08	28.24	30.89	33.10	35.01
19	21.55	25.60	28.97	31.77	34.13	36.19
20	21.77	26.11	29.71	32.58	35.08	37.22
21	21.99	26.48	30.23	33.17	35.97	38.17
22	22.06	26.77	30.74	33.90	36.70	38.98
23	21.99	27.07	31.04	34.42	37.29	39.72
24	21.92	27.07	31.33	34.79	37.88	40.38
25	21.71	26.99	31.48	35.08	38.25	40.89
26	21.25	26.92	31.48	35.30	38.54	41.26
27	20.74	26.62	31.41	35.38	38.69	41.56
28	20.15	25.89	31.18	35.30	38.83	41.78
29	19.49	25.74	30.89	35.08	38.76	41.78
30	18.61	25.08	30.45	34.79	38.54	41.70

Note: 1. Belt speed is calculated using the pitch diameter and revolution per second of one of the pulleys on the drive.

2. If the belt speeds are over 25 metres per second, the pulleys may require special construction or special material. They may also require dynamic balancing. Such applications should be referred to the manufacture.

Table 20.30. Correction factors for industrial service, F_a is 2494-1964

Severity of service	Types of driver machines service	Types of driving units					
		AC Motors; Normal Torque, Squirrel Cage, Synchronous and Split Phase. DC Motors: Shunt Wound, Multiple Cylinder Internal Combustion Engines Over 600 rpm			AC Motors; High Torque, High Slip Repulsion- Induction, Single Phase Service Wound and Slip Ring DC Motors: Series Wound and Compound wound. Single Cylinder Internal Combustion Engines < 600 rpm		
		Upto 10 hr	Over 10 hr to 16 hr	Over 16 hr & continuous service	Over 10 hr	Over 10 hr to 16 hr	Over 16 hr & continuous service
Light-duty	Agitators for liquid blowers and exhausters, centrifugal pumps, compressor fans up to 7.5 kW (10 hp) and light-duty conveyors.	1.0	1.1	1.2	1.1	1.2	1.3
Medium-duty	Belt conveyors for sand, grain etc.; rough mixes, fans over 7.5 kW (10 hp); generators; line shafts; laundry machinery; machine tools; punches, presses and shears, printing machinery; positive displacement rotary pumps; and revolving and vibrating screens.	1.1	1.2	1.3	1.2	1.3	1.4
Heavy-duty	Brick machinery bucket elevators, exciters, piston compressors, conveyors (dragpanscrew), hammer mills, paper mill beaters, piston pumps, positive displacement blowers, pulverizers, saw mill and wood, working machinery and textile machinery.	1.2	1.3	1.4	1.4	1.5	1.6
Extra heavy	Crushers (gyratory-jaw, roll), mills (ballrod tube), boists, and rubber (calenders extruders mills).	1.3	1.4	1.5	1.5	1.6	1.8

Note: 1. The table gives only a few examples of particularly representative machines.

2. If idler pulley is used, the following values must be added to the service factor:

Idler pulley on the Slack side: Inside: 0; Outside: 0.1

Idler pulley on the tight side: Inside: 0.1; Outside: 0.2

Table 20.31. Correction factors for belt length, F_c (IS 2494-1964)

Belt cross section symbols									
Normal inside length, mm	**A**	**B**	**C**	**Normal inside length, mm**	**A**	**B**	**C**	**D**	**E**
610	0.80	-	-	2057	1.04	0.98	0.89	-	-
660	0.81	-	-	2159	1.05	0.99	0.90	-	-
711	0.82	-	-	2286	1.06	1.00	0.91	-	-
787	0.84	-	-	2438	1.08	-	0.92	-	-
813	0.85	-	-	2464	-	1.02	-	-	-
889	0.87	0.81	-	2540	-	1.03	-	-	-
914	0.87	-	-	2667	1.10	1.04	0.94	-	-
965	0.88	0.83	-	2845	1.11	1.05	0.95	-	-
991	0.88	-	-	3048	1.13	1.07	0.97	0.86	-
1016	0.89	0.84	-	3150	-	-	0.97	-	-
				3251	1.14	1.08	0.98	0.87	
1067	0.90	0.85	-	3404	-	-	0.99	-	-
1092	0.90	-	-	3658	-	1.11	1.00	0.90	-
1168	0.92	0.87	-	4013	-	1.13	1.02	0.92	-
1219	0.93	0.88	-	4115	-	1.14	1.03	0.92	-
1295	0.94	0.89	0.80						
				4394	-	1.15	1.04	0.93	
1372	-	0.90	-	4572	-	1.16	1.05	0.94	-
1397	0.96	0.90	-	4953	-	1.18	1.07	0.96	-
1422	0.96	0.90	0.81	5334	-	1.19	1.08	0.96	0.94
1473	0.97	-	- -	6045	-	-	1.11	1.00	0.96
1524	0.98	0.92	0.82						
				6807	-	-	1.14	1.03	0.99
1600	0.99	-	-	7569	-	-	1.16	1.05	1.01
1626	0.99	-	-	8331	-	-	1.19	1.07	1.03
1651	1.00	0.94	-	9093	-	-	1.21	1.09	1.05
1727	1.00	0.95	0.85	9855	-	-	1.23	1.11	1.07
1778	1.01	0.95	-						
				10617	-	-	1.24	1.12	1.09
1905	1.02	0.97	0.87	12141	-	-	-	1.16	1.12
1981	1.03	0.98	-	13665	-	-	-	1.18	1.14
2032	1.04	-	-	15189	-	-	-	1.20	1.17
				16713	-	-	-	1.23	1.19

Table 20.32. Correction factors for arc of contact, F_d

Arc of contact on smaller pulley (in degrees)	Correction factor (proportion of 180° rating)		Arc of contact on smaller pulley (in degrees)	Correction factor (proportion of 180° rating)	
	V-V	*V*-Flat		*V-V*	*V*-Flat
180	1.00	0.75	133	0.87	0.85
177	0.99	0.76	130	0.86	0.86
174	0.99	0.76	127	0.85	0.85
171	0.98	0.77	123	0.83	0.83
169	0.97	0.78	120	0.82	0.82
166	0.97	0.79	117	0.81	0.81
163	0.96	0.79	113	0.80	0.80
160	0.95	0.80	110	0.78	0.78
157	0.94	0.81	106	0.77	0.77
154	0.93	0.81	103	0.75	0.75
151	[illegible]	0.82	99	0.73	0.73
148	0.92	0.83	95	0.72	0.72
145	0.91	0.83	91	0.70	0.70
142	0.90	0.84	87	0.68	0.68
			83	0.65	0.65

Note: 1. Arcs of contact below 120° should not be used unless full drive details are first submitted to the *V*-drive manufacturer concerned for confirmation.

2. It should be noted that the advantage of using *V*-flat drives diminishes increasingly for arcs of contact greater than 133° and the use of such drives is usually not found to be practical for arcs of contact greater than 151°.

Table 20.33. Recommended standard pulley pitch diameter (mm) for *V*-belts

40, 44, 50, 56, 71, 80, 90, 100, 112, 125, 140, 160, 180, 200, 224, 250,
280, 315, 400, 450

Table 20.34. Recommended centre distance

Speed ratio, $i = D/d$	1	2	3	4	5	6 to 9
Recommended C/D ratio	1.5	1.2	1.0	0.95	0.9	0.85

Table 20.35. Nominal pitch length of the *V*-belts

Nominal inside length, mm (1)	Standard section of *V*-belts					Pitch length variation	
	Cross section A, mm (2)	Cross section B, mm (3)	Cross section C, mm (4)	Cross section D, mm (5)	Cross section E, mm (6)	Pitch length limits mm (7)	Maximum variation in length with a matched set mm (8)
610	645	-	-	-	-		
660	696	-	-	-	-	+11.4 −6.4	
711	747	-	-	-	-		
787	828	-	-	-	-		
830	848	-	-	-	-		
889	925	932	-	-	-		2.5
914	950	-	-	-	-	+12.5 −5.7	
				-	-		
965	1001	1008	-	-	-		
991	1026	-	-	-	-		
1016	1051	1059	-	-	-		
1067	1102	1110	-	-	-		
1092	1128	-	-	-	-	+14.0 −8.9	
1168	1204	1212	-	-	-		
1219	1255	1262	-	-	-		
1295	1331	1339	1351	-	-		
1372	-	1415	-	-	-	+16.0 −9.0	
1397	1433	1440	-	-	-		
1422	1458	1466	-	-	-		5.0
1473	1509	-	-	-	-		
1524	1560	1567	1580	-	-		
1600	1636	-	-	-	-		
1626	1661	-	-	-	-	+17.8 −12.5	
1651	1687	1694	-	-	-		
1727	1763	1770	1783	-	-		
1778	1814	1824	-	-	-		
1905	1914	1948	1961				
1981	2017	2024	-	-	-		7.5
2032	2068	-	-	-	-	+17.8 −12.8	
2057	2093	2101	2113	-	-		
2159	2195	2202	2215	-	-		
2286	2322	2329	2342	-	-	+30 −60	
2438	2474	-	2494		-		

Contd.

(1)	(2)	(3)	(4)	(5)	(6)	(7)	(8)
2464	-	2507	-	-	-		
2540	-	2583	-	-	-		
2667	2703	2710	2723	-	-	+34	
2845	2880	2888	2901	-	-	−18	10
3048	3084	3091	3104	3127	-		
3150	-	-	3205	-	-		
3251	3287	3294	3307	3330	-	+38.0	
3404	-	-	3459	-	-	−21	
3658	3693	3701	3713	3736	-		
4013	-	4056	4069	4092	-		
4115	-	4158	4171	4194	-	+43	
4394	-	4437	4450	4473	-	−28	
4572	-	4615	4628	4651	-		12.5
4953	-	4996	5009	5032	-		
5334	-	5377	5390	5413	5426	+49	
6045	-	-	6101	6124	6137	−28	
6807	-	-	6863	6886	6899	+56	
7569	-	-	7625	7648	7661	−32	
8331	-	-	8387	8410	8423	+65	15.0
9093	-	-	9143	9172	9185	−37	
9855	-	-	-	9334	9947		17.5
10617	-	-	-	10696	10709	+76	
12141	-	-	-	12220	12233	−43	
13665	-	-	-	13744	13757	+89	
15189	-	-	-	15268	15281	−50	
16713	-	-	-	16792	16805	+105	
						−59	

Table 20.36. Dimensions of standard *V*-groove & pulley

Groove cross section	Pitch width L_p, mm	Minimum distance down to pitch line b	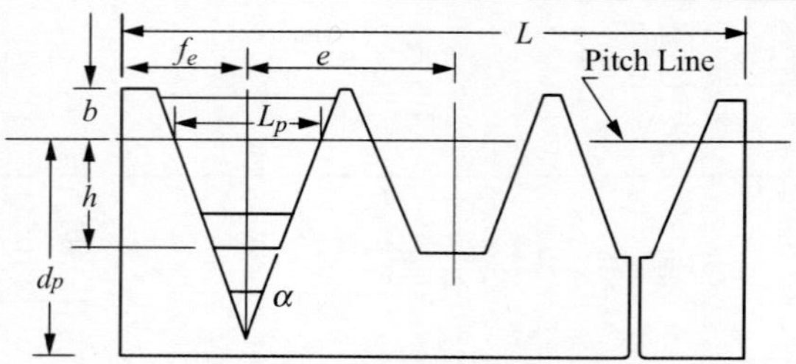 **Pulley pitch diameter *dp***	Angle α tolerance $\pm 1/2°$ in degree,	Minimum depth below pitch line (h), mm	Centre to centre distance of grooves (e), mm	Edge of pulley to first groove centre, mm (f)
A	11	3.3	67 mm to 75 mm Recommended min 75 mm and under 125 mm 125 mm and over	32 34 38	8.7	15 ± 0.3	10^{+2}_{-1}
B	14	4.2	117 mm to 125 mm Recommended min 125 and under 200 mm 200 mm and over	32 34 38	10.8	19 ± 0.4	12.5^{+2}_{-1}
C	19	5.7	175 mm to 200 mm Recommended min 200 mm and under 300 mm 300 mm and over	34 36 38	14.3	25.5 ± 0.5	17^{+3}_{-1}
D	27	8.1	300 mm to 355 mm Recommended min 355 and under 500 mm 500 mm and over	34 36 38	19.9	37 ± 0.6	24^{+3}_{-1}
E	32	9.6	500 mm to 630 mm 630 mm and over	36 38	23.4	44.5 ± 0.7	29^{+4}_{-1}

Note: (1) The recommended minimum is for standard use, but the largest possible pulleys should be used for long belt life, provided that the belt speed does not exceed 25 metres/second.

(2) The values given above have been calculated subject to certain approximations, from the following formulae:

$b = 0.13L_p$ $e = 1.35\ L_p$ $f = 0.9\ L_p$ $h = 7\ L_p + 1$ (expressed in millimetres)

(3) Maximum $L = (X - 1)e + 2f$ where X = number of grooves

Example 20.11: Design a *V*-belt drive for connecting a 20 kW, 1440 rpm motor to a compressor running at 480 rpm for service of 15 hr/day. The centre distance is approximately 1.2 m.

Given Data: Power transmitted, $NP = 20$ kW @ Speed = 1440 rpm,
Compressor speed = 480 rpm; Centre distance, $C = 1.2$ m = 1200 mm

SOLUTION

Step-1: Belt selection

Referring to Table 18.33 MDHB, the type C-section is suitable for the application. From Table 18.29 MDHB, the dia. of the smaller pulley $d = 200$ mm. The belt section is 22×14 mm and the mass per meter length $m_1 = 0.343$ kg/m.

Step-2: Diameter of the larger pulley,

$$D = d\left(\frac{n_1}{n_2}\right) = \frac{200 \times 1440}{480} = 600 \text{ mm}$$

Step-3: The angle of lap on the small pulley,

$$\theta_s = \pi - 2\sin^{-1}\left(\frac{600 - 200}{2 \times 1200}\right) = 2.807 \text{ rad} = 160.8°$$

Step-4: The pitch length, (Eqn. 18.34 MDHB)

$$L = 2C + 1.57\,(D + d) + \frac{(D-d)^2}{4C} \tag{20.22}$$

$$L = 2 \times 1200 + 1.57(600 + 200) + \frac{(600 - 200)^2}{4 \times 1200}$$

$$= 2400 + 1256 + 33.33 \ = 3689.3 \text{ mm}$$

From Table 18.46 MDHB, the nearest available pitch length, $L = 3713$ mm with inside length $L_i = 3658$ mm. So the belt specification is C3658.

Step-5: The number of belts

The belt speed, $V = \dfrac{\pi dN}{60,000} = \dfrac{\pi \times 200 \times 1440}{60,000} = 15.08$ m/sec.

The power rating from Table 20.27 (Table 18.38 MDHB);

(for $d_e = 200 \times 1.14 = 228$)

$(KW)_r = 7.17$ (by interpolation)

∴ The number of belts, (Eqn. 18.33 MDHB)

$$n = (KW\,F_a)/[(KW)_r F_c \cdot F_d]$$

where, $KW = 20$; and $(KW)_r = 7.17$

$$F_a = 1.1 \text{ Table 20.30} \qquad \text{(Table 18.41 MDHB)}$$

$$F_c = 1.0 \text{ Table 20.31} \qquad \text{(Table 18.42 MDHB)}$$

$$F_d = 0.95 \text{ Table 20.32} \qquad \text{(Table 18.43 MDHB)}$$

$$\therefore \qquad n = (20 \times 1.1) / (7.17 \times 1.0 \times 0.95) = 3.229$$

Then select $n = 4.0$ for safe design.

Alternatively we can choose $d = 300$ mm and $D = 900$ mm

Step-3: Angle of lap on smaller pulley,

$$\theta_s = \pi - 2\sin^{-1}\left(\frac{900-300}{2 \times 1200}\right) = 151.1°$$

Step-4: Belt length,

$$L = 2 \times 1200 + 1.57\,(900 + 300) + \frac{(900-300)^2}{4 \times 1200}$$

$$= 2400 + 1884 + 75 = 4359.0 \text{ mm}$$

From Table 18.46 MDHB, the next available length is $L = 4450$ with $L_i = 4394$. Then the belt specification in C 4394.

Step-5: The number of belts

The belt speed, $V = \pi \times 300 \times 1440/60{,}000 = 22.62$ m/sec.

and $d_e = d \times F_b = 300 \times 1.14$ (Table 20.24 for F_b) $= 342.0$ mm

The power rating $(KW)_r = 11.71$, and $F_a = 1.04, F_c = 1.04, F_d = 0.93$

$\therefore$ The number of belts, $n = (20 \times 1.1) / (11.71 \times 1.04 \times .93) = 1.94$

Choose $n = 2.0$ belts

Example 20.12: A *V*-belt drive is used to transmit power from 100 kW motor running at 475 rpm. The belt has a mass of 0.6 kg/m. The maximum permissible tension is 900 N. The groove angle is 38° and the angle of contact is 160°. The coefficient of friction is 0.2. Find the number of belts and the pulley diameter.

Given Data: Power transmitted, *NP* =100 kW @ Speed = 475 rpm,

Mass of the belt/met, $m_1 = 0.6$ kg/m

Maximum permissible tension, $P_1 = 900$ N

The groove angle, $2\beta = 38°$

The angle of contact, $\theta = 160° = 2.8$ rad

The coefficient of friction, $f = 0.2$

SOLUTION Equations for flat belt hold good for *V*-belt also with *f* replaced by f_1 given by $f_1 = f \operatorname{cosec} \beta$

Step-1: Pulley diameter for optimum power

For optimum power transmission, $P_c = \dfrac{P_1}{3}$. (Eqn. 18.4 MDHB)

Since $P_1 = 900$ N;

$$P_c = \text{mAV}^2 = \frac{P_1}{3} = 300 \text{ N}$$

where, $mA = 0.6 \text{ kg/m} = 0.6 \text{ g/mm}$

$$\therefore \quad 0.6 \times V^2 = 300$$

$$\therefore \quad V = \sqrt{\frac{300}{0.6}} = 22.36 \text{ m/sec.}$$

$$\therefore \quad \pi\, dN/60{,}000 = V = 22.36$$

$$\therefore \quad d = \frac{22.36 \times 60{,}000}{\pi \times 475} = 899.0 \text{ mm}$$

Choose the pulley diameter as $d = 900.0$ mm

Step-2: Number of belts

The ratio of belt tensions,

$$\frac{P_1 - P_c}{P_2 - P_c} = e^{f_1\theta} = e^{f\theta\,\text{cosec}\beta}$$

$$e^{(0.2 \times 2.8 \times \text{cosec}19)} = e^{1.7} = 5.56$$

$$P_2 - P_c = \frac{(P_1 - P_c)}{5.56} = \frac{900 - 300}{5.56} = 107.9\text{N}$$

$$\therefore \quad P_2 = 107.9 + 300 = 407.9\text{N}$$

The power per belt $= \dfrac{(P_1 - P_2)V}{1000}$ (Eqn. 18.6 MDHB)

$$= (900 - 407.9) \times 22.36/1000 = 11.0 \text{ kW}$$

Number of belts required,

$$n = \text{total power/power per belt} = 100/11 = 9.08 \cong 9$$

∴ Nine belts are sufficient.

Example 20.13: A *V*-belt drive is to transmit 20 kW power from a 250 mm pitch diameter sheave to a 900 mm diameter flat pulley. The centre distance between the shafts is 1000 mm. The groove angle is $40°$ and the coefficient of friction for the belt and sheave is 0.2 and that for flat pulley is 0.25. The cross section of the belt is 40 mm wide at the top and 20 mm wide at the bottom and 25 mm deep. The density of the belt material is 1000 kg/m^3. The allowable stress is 3.0 MPa. Find the number of belts required.

Given Data: Power transmitted, $NP = 20$ kW;

Diameter of small pulley, $D = 250$ mm,

Diameter of bigger pulley, $D = 900$ mm,

Center distance, $C = 1000$ mm.

Belt material density = 1000 kg/ m^3 ; Allowable stress, $S_d = 3.0$ MPa

Coefficient of friction, smaller pulley, $f = 0.2$, bigger pulley, $f = 0.25$

Belt cross section, $A = \left(\dfrac{40+20}{2}\right)\times 25$ mm

SOLUTION

Step-1: Belt tensions

The cross-sectional area of the belt $A = \left(\dfrac{40+20}{2}\right)\times 25 = 750\ \text{mm}^2$

Maximum tension in the belt,

$$P_1 = A\sigma_d = 750\times 3 = 2250\ \text{N}$$

At the optimum speed, $P_c = P_1/3$

$\therefore$ $P_c = mAV^2 = P_1/3 = 2250/3 = 750$ N

Step-2: Optimum belt speed,

$$V = \sqrt{(P_c)/mA}$$

$$= \sqrt{750/[(10^{-3})\times 750]} = 31.62\ \text{m/sec}$$

Step-3: The virtual coefficient of friction for V-belt,

$$f_1 = f\ \text{cosec}\ \alpha = 0.2\times \text{cosec}\ 20° = 0.585$$

Step-4: The angle of lap for the sheave,

$$\theta_s = \pi - 2\sin^{-1}\left(\frac{D-d}{2C}\right)$$

$$= \pi - 2\sin^{-1}\left(\frac{900-250}{2\times 1000}\right) = 2.48\ \text{rad}$$

Step-5: The belt tension ratio,

$$e^{f1\theta} = e^{0.585\times 2.48} = 4.26$$

Step-6: Power capacity for sheave

The maximum power capacity per belt on the sheave,

$$NP_1 = \left[\frac{(P_1 - P_c)}{1000}\right]\left[\frac{e^{f_1\theta}-1}{e^{f_2\theta}}\right]V \qquad \text{(Eqn. 18.32 MDHB)}$$

$$NP_1 = \left[\frac{(2250-750)\times 31.62}{1000}\right]\left[\frac{4.26-1}{4.26}\right] = 36.32\ \text{kW}$$

Step-7: For the flat pulley, the angle of lap,

$$\theta_s = \pi + 2\sin^{-1}\left(\frac{D-d}{2C}\right) = \pi + 2\sin^{-1}\left(\frac{900-250}{2000}\right) = 3.8 \text{ rad}$$

Step-8: Belt tension ratio for flat pulley,

$$\therefore \qquad e^{f\theta} = e^{0.25\times 3.8} = 2.58$$

Step-9: Max. power capacity per belt on flat pulley,

$$NP_2 = \left[\frac{P_1 - P_c}{1000}\right]\left[\frac{e^{f\theta}-1}{e^{f\theta}}\right]V$$

$$= \left[\frac{(2250-750)\times 31.62}{1000}\right]\left[\frac{2.58-1}{2.58}\right] = 29.19 \text{ kW}$$

Step-10: The power capacity of the drive

The minimum of the two values is the power that can be transmitted by the drive

$\therefore$ Power transmitting capacity $= 29.19$ kW

Hence, one belt is sufficient for transmitting 20 kW.

Example 20.14: Design *V*-belt for transmitting 15 kW at 1440 rpm from an electric motor to centrifugal pump running at 360 rpm. The centre distance is to be 1000 mm approximately. The service is for 24 hours per day.

Given Data: Power, $NP = 15$ kW at speed $N_1 = 1440$ rpm,

Pump speed $N_2 = 360$ rpm; Centre distance $C = 1000$ mm;

Service 24 hours per day

SOLUTION

Step-1: Selection of the belt

From Table 20.22 (Table 18.33 MDHB),

B- or *C*-type is suitable; select *B*-type

From Table 20.20 (Table 18.29 MDHB), for *B*-type

Minimum pulley diameter = 125 mm

Select $d = 150$ mm

Step-2: Bigger pulley diameter

Then for bigger pulley for speed ratio = 1440/360 = 4

$$\therefore \qquad D = 150 \times 4 = 600 \text{ mm}$$

Step-3: The angle of lap on the smaller pulley,

$$\theta_s = \pi - 2\sin^{-1}\left(\frac{D-d}{2C}\right) = \pi - 2\sin^{-1}\left(\frac{600-150}{2\times 1000}\right) = 2.69 \text{ rad}$$

Step-4: The length of the belt, (Eqn. 18.34 MDHB)

$$L = 2C + 1.57(D+d) + \frac{(D-d)^2}{4C}$$

$$= 2\times1000 + 1.57(600+150) + \frac{(600-150)^2}{4\times1000}$$

$$= 2000 + 1177.5 + 50.625 = 3228.125 \text{ mm}$$

From Table 18.46 MDHB, nearest available size is L = 3294 mm with inside length L_i = 3251 mm.

∴ The belt specification is B 3251.

Step-5: Number of belts

The belt speed,

$$V = \pi dN/60{,}000 = (\pi\times150\times1440)/60{,}000 = 11.31 \text{ m/sec}$$

Equivalent diameter $d_e = d\times F_b = 150\times1.14 = 171$ mm

For the number of belts required.

The power rating $(KW)_r = 3.598$ (Table 18.37 MDHB)

The service factor, F_a = 1.2 (Table 18.41 MDHB)

Belt length correction factor, $F_a = 1.08$ (Table 18.42 MDHB)

Area of contact correction factor, $F_d = 0.93$ (Table 18.43 MDHB)

∴ The number belts required, (Eqn. 18.33 MDHB)

$$n = (KW\cdot F_a)/(KW_r F_c\cdot F_a)$$

$$= (15\times1.2)/(3.598\times1.08\times0.93) = 4.98$$

∴ Choose $n = 5$ belts.

Step-6: The sheave dimensions

The width of the sheave, (Table 18.47 MDHB)

$$L = (n-1)\,e + 2f = (5-1)\,19 + 2\times12.5 = 101 \text{ mm}$$

The rim thickness $t > b + h$

i.e.,
$$t > (4.2 + 10.8 = 15)$$

Choose $t = 20$ mm

For torque, $T_q = \frac{(955\times10^4\times15)}{1440} = 99500$ Nmm

And for design stress S_{sd} = 60 MPa, the shaft diameter is,

$$d_{s1} = \sqrt[3]{\frac{16\times99500}{\pi\times60.0}} = 20.36$$

Choose $d_{s1} = 20$ mm

$\therefore$ Hub diameter, $d_{n1} = 2 \times d = 40$ m

For the larger pulley, torque $= (4 \times 99500)$ Nmm

$\therefore$ The shaft diameter $= \sqrt[3]{(16 \times 4 \times 99500)/(\pi \times 60)} = 32.33$ mm

Choose $d_{s2} = 35.0$ mm

and hub diameter, $d_{h2} = 2 \times 35 = 70$ mm

Example 20.15: A compressor requiring 90 kW is to run at 250 rpm. The drive is by *V*-belt from an electric motor running at 750 rpm. The diameter of the pulley on the compressor shaft is 1 m, while the centre distance between the pulleys is limited to 1.75 m. The belt speed should not exceed 1600 m/min. Determine the number of *V*-belts required to transmit the power, if each belt has a cross-sectional area of 375 mm^2. The groove angle of the pulley is $35°$ and the coefficient of friction between the belt and the pulley is 0.25.

Given Data: Power, $NP = 90$ kW; Motor speed, $n_1 = 750$ rpm;

Compressor speed, $n_2 = 250$ rpm, $A = 375\ \text{mm}^2$;

Larger diameter, $D = 1\ \text{m} = 1000\ \text{mm}$, $C = 1.75\ \text{m} = 1750\ \text{mm}$;

Belt speed $= 1600\ \text{m/min} = \dfrac{1600}{60} = 26.67\ \text{m/sec}$

$m = 1000\ \text{kg/m}^3 = 10^{-3}\ \text{gm/mm}^3$; Design stress $\sigma_d = \sigma_1 = 2.5$ MPa

SOLUTION

Step-1: Diameter of smaller pulley for equal belt speed, $n_1 d = n_2 D$

$$\therefore \qquad 750 \times d = 250 \times 1000$$

$$\therefore \qquad d = 250 \times 1000/750 = 333.33\ \text{mm}$$

i.e., Diameter of smaller pulley, $d = 333.33$ mm

Step-2: Centrifugal stress and centrifugal tension

Centrifugal stress,

$$\sigma_c = mv^2 = 10^{-3} \times 26.667^2 = 0.711\ \text{MPa} \qquad \text{(Eqn. 18.5 MDHB)}$$

Centrifugal tension, $P_c = \sigma_c \times A = 0.711 \times 375 = 266.67$ N

Step-3: Tension on the tight side,

$$P_1 = \sigma_1 A = \sigma_d A = 2.5 \times 375 = 937.5\ \text{N}$$

Step-4: Ratio of belt tensions

Angle of contact on smaller pulley,

$$\theta_s = \pi - \left\{2 \sin^{-1}\left(\frac{D-d}{2C}\right)\right\} \qquad \text{(Eqn. 18.12 MDHB)}$$

$$= \pi - \left\{ 2 \sin^{-1} \left(\frac{1000 - 333.33}{2 \times 1750} \right) \right\} = 2.7583 \text{ radian}$$

The ratio of belt tensions $= e^{\frac{f\theta}{\sin\alpha}} = e^{\frac{f\theta}{\sin\alpha}} = e^{\frac{0.25 \times 2.7583}{\sin 17.5}} = 9.9065$

Step-5: Power transmitted/belt,

$$kW = \left[\frac{(P_1 - P_c)}{1000} \right] \left[\frac{e^{f_1\theta} - 1}{e^{f_1\theta}} \right] \qquad \text{(Eqn. 18.32 MDHB)}$$

$$= \left[\frac{(937.5 - 266.67) \times 26.67}{1000} \right] \left[\frac{9.9065 - 1}{9.9065} \right] = 16.055 \text{ kW}$$

Step-6: Number of V-belts,

$$n = \frac{\text{Total power transmitted}}{\text{Power transmitted / belt}} = \frac{90}{16.055} = 5.596$$

∴ Number of *V*-belts, $n = 6$.

Example 20.16. Two shafts, 1 metre apart are connected by a *V*-belt to transmit 90 kW at 1200 rpm of a driver pulley of 300 mm effective diameter. The driven pulley rotates at 400 rpm. The angle of groove is 40° and the coefficient of friction between the belt and the pulley rim is 0.25. The area of the belt section is 400 mm^2 and the permissible stress is 2.1 MPa. Density of belt material is 1100 kg/m^3. Calculate the number of belts required and the length of the belt.

Given Data: Power, $NP = 90$ kW: $n_1 = 1200$ rpm; Pulley dia., $d = 300$ mm
Centre distance, $C = 1$ m $= 1000$ mm; Speed of driven pulley, $n_2 = 400$ rpm

$$2\alpha = 40°; \alpha = 20°; f = 0.25; A = 400 \text{ mm}^2; \sigma_d = 2.1 \text{ MPa}$$

$$m = 1100 \text{ kg/m}^3$$

SOLUTION

Step-1: Diameter of larger pulley

For no slip, $$n_1 d = n_2 D$$

∴ $$1200 \times 300 = 400 \times D$$

∴ Diameter of larger pulley, $$D = \frac{1200 \times 300}{400} = 900 \text{ mm}$$

Step-2: Belt speed,

$$V = \frac{\pi d_1 n_1}{60{,}000} = \frac{\pi \times 300 \times 1200}{60{,}000} = 18.85 \text{ m/sec}$$

Step-3: Ratio of belt tensions

Angle of contact on smaller pulley, $\theta_s = \pi - \left\{2\sin^{-1}\left(\frac{D-d}{2C}\right)\right\}$

$$= \pi - \left\{2\sin^{-1}\left(\frac{900-300}{2x1000}\right)\right\} = 2.5322 \text{ radian}$$

Ratio of belt tension $= e^{\frac{f\theta_s}{\sin\alpha}} = e^{\frac{0.25\times 2.5322}{\sin 20}} = 6.3656$

Step-4: The centrifugal stress,

$$\sigma_c = mV^2$$

where, $m = 1100 \text{ kg/m}^3 = 1100\times 10^{-6} \text{g/mm}^3$

$V = 18.55 \text{m/sec}$

$\sigma_c = 1100\times 10^{-6}\times(18.55)^2 = 0.38 \text{ MPa}$

Step-5: Power transmitted/belt,

$$kW = \left[\frac{A(\sigma_1 - \sigma_c)}{1000}\right]\left[\frac{e^{f_1\theta}-1}{e^{f_2\theta}}\right]V \qquad \text{(Eqn. 18.32a MDHB)}$$

$$= \left[\frac{400\times(2.1-0.391)}{1000}\right]\left[\frac{6.3656-1}{6.3656}\right]\times 18.85 = 10.86$$

Step-6: Number of belts required, $n = \frac{\text{Total power}}{\text{Power per belt}} = \frac{90}{10.86} = 8.286$

Choose the number of belts $n = 9$.

Step-7: Length of belt

$$L = 2C + \frac{\pi}{2}(D+d) + \frac{(D-d)^2}{4C} \qquad \text{(Eqn. 18.34 MDHB)}$$

$$= 2\times 1000 + 1.57\times(900+300) + \left(\frac{900-300}{4\times 1000}\right)^2 = 3975 \text{ mm}$$

Example 20.17: Select a *V*-belt drive to connect a 15 kW at 2880 rpm motor to a centrifugal pump, running at approximately 2400 rpm, for a service of 18 hr/day. The centre distance should be approximately 400 mm. Assume the pitch diameter of driving pulley as 125 mm.

Given Data: $NP = 15$ kW; $n_1 = 2880$ rpm; $n_2 = 2400$ rpm; $C = 400$ mm;
$d = 125$ mm

SOLUTION

Step-1: Diameter of larger pulley,

$$D = (n_1 d)/n_2 = \frac{2880 \times 125}{2400} = 150 \text{ mm}$$

∴ Diameter of larger pulley, $D = 150$ mm

Step-2: Select the cross section of belt
Equivalent pitch diameter of smaller pulley, $d_e = d_p F_b$
where, $d_p = d = 125$ mm

Speed ratio, $\frac{n_1}{n_2} = \frac{2880}{2400} = 1.2$

From Table 18.35 MDHB for $\frac{n_1}{n_2} = 1.2$,

Smaller diameter factor, $F_e = 1.07$
∴ Equivalent pitch diameter of smaller pulley,

$$d_e = 125 \times 1.07 = 133.75 \text{ mm}$$

From Table 18.33 MDHB, the cross section of the belt selected corresponding to the equivalent pitch diameter $d_e = 133.75$ mm is 'B' cross section belt.

Step-3: Belt velocity,

$$v = \frac{\pi d_1 n_1}{60000} = \frac{\pi \times 125 \times 2880}{60000} = 18.82 \text{ m/sec.}$$

Step-4: Power capacity (Table 18.34 MDHB)

For 'B' Cross section belt,

$$(kW)_r = \left[0.79V^{-0.09} - \frac{50.8}{d_e} - 1.32 \times 10^{-4} V^2\right] V$$

$$= 18.85\left(\frac{0.79}{18.85^{0.09}} - \frac{50.8}{133.75} - \frac{1.32 \times 18.85^2}{10^4}\right) = 3.39 \text{ kW}$$

Step-5: Number of belts

Number of V-belts, $$n = \frac{KWF_a}{(kW)_r F_c F_d}$$ (Eqn. 18.33 MDHB)

From Table 18.41 MDHB for the driven machine-centrifugal pump and a service of 18 hours per day, load application factor or correction factor for service $F_a = 1.2$

Pitch length of belt,

$$L = 2C + 1.57(D + d) + \frac{(D - d)^2}{4C}$$ (Eqn. 18.34 MDHB)

$$= 2 \times 400 + 1.57 \times (150 + 125) + \frac{(150 - 125)^2}{4 \times 400} = 1232.36 \text{ mm}$$

From Table 18.46 MDHB, the nearest standard value of nominal pitch length for the selected 'B' cross section belt, L = 1212 mm and Nominal inside length = 1168 mm.

From Table 18.42 MDHB, for nominal inside length = 1168 mm and 'B' cross section belt, Correction factor for length, $F_c = 0.87$

Angle of contact, $\theta = \pi - 2\sin^{-1}\left(\frac{D-d}{2c}\right)$ (Eqn. 18.12 MDHB)

$$= \pi - \sin\left(\frac{150-125}{2\times 400}\right) = 176.42 \text{ deg}$$

From Table 18.43 MDHB, for $\theta = 176.42°$ and *V-V* belt (assume) correction factor for angle of contact $F_d = 0.99$

Number of belts,

$$\therefore \quad n = \frac{15\times 1.2}{3.32\times 0.87\times 0.99} = 6.295$$

Choose number of *V*-belts, $n = 7$

(i) Correct centre distance, (Eqn. 18.34a MDHB)

$$C = 0.25\left\{\left[L_p - 1.57(D+d)\right] + \sqrt{\left[L_p - 1.57(D+d)\right]^2 - 2(D-d)^2}\right\}$$

$$= 0.25\left\{\left[1212 - 1.57(150+125)\right] + \sqrt{\left[1212 - 1.57(150+125)\right]^2 - 2\times(25)^2}\right\}$$

$$= 389.925 \text{ mm}$$

Step-6: Specification of *V*-belt

The *V*-belt selected is B-1168

From Table 18.29 for 'B' cross section belt,

Nominal thickness, T = 11 mm

Nominal top width, W = 17 mm

Example 20.18: Following data is given for a *V*-belt drive connecting a 20 kW motor to a compresser.

	Motor pulley	Compressor pulley
Pitch diameter (in mm)	300	900
Speed (in rpm)	1440	480
Coefficient of friction	0.2	0.2

The centre distance between pulleys is 1 m. Cross section of belt is trapezoidal with parallel sides being 12 mm and 22 mm respectively and depth 14 mm. The density of the belt material is 0.97 gm/cc and the allowable tension per belt is 850 N. Determine the number of belts required for this application.

Given Data: $d = 300$ mm; $D = 900$ mm; $n_1 = 1440$ rpm; $n_2 = 480$ rpm;

$$f_s = f_l = 0.2;\ C = 1\ \text{m} = 1000\ \text{mm};\ NP = 20\ \text{kW};$$
$$m = 0.97\ \text{gm/cc} = 0.97 \times 10^{-3}\ \text{gm/mm}^3$$

SOLUTION

Step-1: Velocity of the belt,

$$v = \frac{\pi d n_1}{60{,}000} = \frac{\pi \times 300 \times 1440}{60{,}000} = 22.62\ \text{m/sec}$$

Step-2: Ratio of tensions

Angle of contact on smaller pulley,

$$\theta = \pi - \left\{2 \sin^{-1}\left(\frac{D-d}{2C}\right)\right\} \qquad \text{(Eqn. 18.12 MDHB)}$$

$$= \pi - \left\{2 \sin^{-1}\left(\frac{900-300}{2 \times 100}\right)\right\} = 2.532\ \text{radian}$$

The semi groove angle of belt is given by

$$\tan\alpha = \frac{22-12}{2 \times 14} = \frac{5}{14};$$

∴ Half groove angle, $\alpha = 19.654°$

∴ Tension ratio $= e^{\frac{0.2 \times 2.532}{\sin 19.654}} = 4.507$

Step-3: Centrifugal tension

Cross-sectional area of the belt,

$$A = \frac{1}{2}(22+12) \times 14 = 238\ \text{mm}^2$$

Centrifugal tension, $P_1 = mAV^2$ (Eqn. 18.4 MDHB)

$$= 97.10^{-3} \times 238 \times 22.62^2 = 118.122\ \text{N}$$

Tension on tight side, $P_1 = 850$ N

Step-4: Power transmitted per belt,

$$= \left[\frac{(P_1 - P_c)V}{1000}\right]\left[\frac{e^{f_1\theta} - 1}{e^{f_2\theta}}\right] = \left[\frac{(850 - 118.122) \times 22.62}{1000}\right]\left[\frac{4.507 - 1}{4.507}\right] = 12.882\ \text{kW}$$

Step-5: Number of *V*-belts,

$$n = \frac{\text{Total power transmitted}}{\text{Power transmitted per belt}} = \frac{20}{12.85} = 1.5528$$

∴ Choose number of *V*-belts $n = 2$

Example 20.19: Select a standard *V*-belt to transmit 30 kW from an AC induction motor rotating at 1500 rpm to a centrifugal pump rotating at 750 rpm. The drive operates continuously for 8 hr/day. Calculate the number of belts.

Given Data: $NP = 30$ kW; $n_1 = 1500$ rpm; $n_2 = 750$ rpm

SOLUTION

Step-1: Selection of belt cross section

As the given power of motor is 30 kW at 1500 rpm from Table 18.33 MDHB, select 'C' cross section belt.

Step-2: Diameter of two pulleys

From Table 18.29 MDHB, for 'C' cross section belt, min. pitch diameter of smaller pulley $d = 200$ mm

For diameter of large pulley, $n_1 d = n_2 D$

$$1500 \times 200 = 750 \times D$$

$\therefore$ Pitch diameter of larger pulley, $D = 400$ mm

Step-3: From Table 20.30 (Table 18.41 MDHB), for the driver AC induction motor, driven centrifugal pump and operation of 8 hr/day continuously, factor of application or Correction factor for service $F_a = 1.3$

Step-4: Power rating

$$\text{Belt speed, } V = \frac{\pi dn}{60000} = \frac{\pi \times 200 \times 1500}{60,000} = 15.708 \text{ m/sec}$$

Power rating, from Table 18.38, $(kW)_r = 7.1$ kW

Step-5: Length correction factor

Recommend, $C/D = 1.2$

The centre distance, $C = 1.2 \times 400 = 480$ mm

$$\text{Belt length, } L = 2C + 1.57(D + d) + \frac{(D - d)^2}{4C} \qquad \text{(Eqn. 18.14a MDHB)}$$

$$= 2 \times 480 + 1.57 \times (400 + 200) + \frac{(400 - 200)^2}{4 \times 480} = 1922.83 \text{ mm}$$

Nearest standard length (Table 18.46 MDHB)

$L = 1961$ mm and nominal inside length, $L_i = 1905$ mm

Then length correction factor, $F_c = 0.87$ (Table 18.42 MDHB)

Step-6: Angle correction factor

$$\text{The angle of lap, } \theta = \pi - 2\sin^{-1}\left(\frac{400 - 200}{2 \times 480}\right) = 155.95°$$

Correction factor for contact angle (Table 20.32), $F_d = 0.94$

Step-7: Number of belts,

$$n = \frac{kW \cdot F_a}{(kW)_r F_c F_d} = \frac{30 \times 1.3}{7.1 \times 0.87 \times 0.94} = 6.7$$

Choose $n = 7$

20.2.3 Toothed Belt Drives

The allowable specific peripheral force on the belt (N/mm),

$$F_0 = F_0 C_s C_{sr} C_{ip} C_w \tag{20.26}$$

The required belt width,

$$b = \frac{F}{F_0 - qv^2} \tag{20.27}$$

where, $F = \frac{1000\,\text{kW}}{v}$

$v =$ velocity of the belt m/s

$q =$ mass of 1 m of the belt of the width equal to 1 mm kg/mm

The force acting on the shafts from the drives, $Q = (1 \text{ to } 1.2)F$ (20.28)

The belt module is selected in accordance with the power to be transmitted.

Table 20.37. Recommended belt module

P, kW. . .	up to 0.4	0.4 to 3	3 to 5.5	5.5 to 10	10 to 22	22 and over
Module m, mm. . .	2 and 3	3 and 4	4 and 5	4, 5 and 7	5 and 7	7 and 10

Table 20.38. Unit strength f_o N/mm width

Module m, mm. . .	2	3	4*	4**	5	7	10
f_o, N/mm. . .	5	10	15	25	35	45	60

*Belt with cable core diameter 0.4 mm.

**Ditto, but diameter 0.75 mm.

Table 20.39. The service factor C_s

Peak loads in % of normal loads. . .	up to 150	150 to 250	250 to 400
C_s in single-shift operation. . .	0.6 to 0.75	0.55 to 0.7	0.5 to 0.65

Note: The over lower values are for machines with severe service conditions; the higher ones are for light service. For two-shift operation, C_s is reduced by 0.05; for three-shift operation, by 0.1.

Table 20.40. The speed ratio factor C_{sr} for over drives

$u = \frac{n_1 \ldots}{n_2}$	1 to 0.8	0.8 to 0.6	0.6 to 0.4	0.4 to 0.3	0.3 and less
$C_{sr} \ldots$	1.0	0.95	0.9	0.85	0.8

Factor C_{ip} is introduced when the drives have idler pulley or a guide pulley: with one pulley $C_{ip} = 0.9$; with two $C_{ip} = 0.8$.

Table 20.41. The belt width factor C_w

Width b	10	16	25	40	63	100
C_w	0.8	0.9	1.0	1.05	1.1	1.2

Table 20.42. Unit mass of the belt kg/m/mm width in Eqn. 20.27

Module m, mm. . .	2	3	4	5	7	10
q, kg/mm. . .	0.003	0.004	0.006	0.007	0.008	0.011

Table 20.43. Recommended value of the belt width mm

3,	4,	5,	8	10,	12.5,	16,	20,	25,	32,
40,	50,	63,	80,	100,	125,	160,	and 200 mm.		

Table 20.44. Recommended lengths of the belt, $L = \pi m z_b$, where z_b is the number of teeth mm

40,	42,	45	48	50	53,	56,	60,	63,	67,
71,	75	80	85	90	95	100	105	112	115
125	130	140	150	160	170	180,	190,	200,	210
220,	235 and 250.								

Table 20.45

Recommended maximum belt velocity	
Module m	**Velocity v m/sec**
2	25
3	30
4	35
5-10	40

Table 20.46

Recommended min. number of teeth on small pulley	
Module m	**Number of teeth**
2-5	16-20
7-10	20-26

Example 20.20: Design a toothed belt-drive for transmitting power from 10 kW at 1440 rpm motor to a punch press running at 480 rpm.

SOLUTION Belt module (Table 20.37)

Choose $m = 5$ mm

Number of teeth on the smaller pulley, $Z_p = 40$

The diameter of pulley, $d = mZ_p = 5 \times 40 = 200$ mm

The belt speed, $V = \dfrac{\pi dN}{60,000} = \dfrac{\pi \times 200 \times 1440}{60,000} = 15.1\ \text{m/sec}$

This is less than the max. specified in Table 20.45 MDHB and the selection of diameter is O.K. Then the transmitted force,

$$F = \frac{1000\ \text{kW}}{V} = 1000 \times 10 / 15.1 = 662.25\ \text{N}$$

Allowable specific force (Eqn. 20.26 MDHB),

$$F_o = f_o \cdot C_s \cdot C_{sr} \cdot C_{ip} \cdot C_w$$

where, $f_o = 35.0$ N/mm from Table 20.38 MDHB,

$$C_s = 0.7, \text{ for } 150\ \%\ \text{load, Table 20.39 MDHB}$$

Speed ratio factor (Table 20.40) for $u = \dfrac{1440}{480} = 3$
$C_{sr} = 1.0$ and $C_{cp} = 1.0$ for
Width factor, $C_w = 1.0$ (assumed)

$$F_o = 35 \times 0.7 \times 1.0 \times 1.0 \times 1.0 = 24.5\ \text{N/mm}$$

Unit mass, $q = 0.007$ (Table 20.42)
∴ The belt width, b (Eqn. 20.27)

$$b = \frac{F}{F_o - qV^2} = (662.25)/(24.5 - 0.007 \times 15^2) = 28.887\ \text{mm}$$

Choose the next recommended value (Table 20.43), $b = 32$ mm.

20.3 CHAIN DRIVES

A chain drive consists of driving and driven sprockets and a power chain which runs over the sprockets and meshes with their teeth. A chain consists of links connected by pin joints.

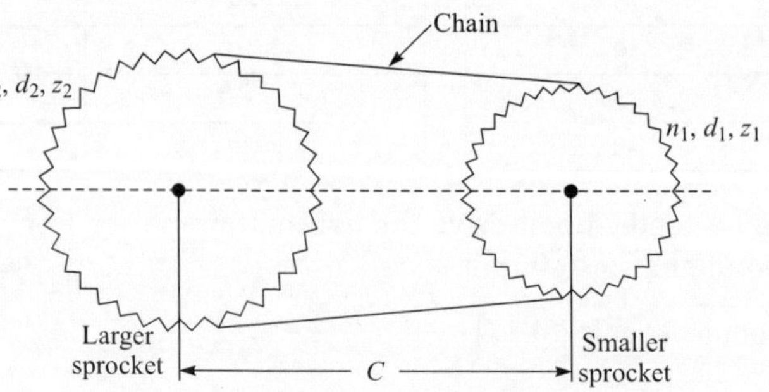

Figure 20.9 Chain drive

Advantages

(1) No slippage occurs
(2) Smaller overall size
(3) High efficiency up to 98%
(4) No initial tension, hence there are small forces on the shaft.
(5) The chain drive can be installed easily; assembly tolerances are flexible.
(6) Easy to redesign and replacement of chain
(7) The operating loads are spread on many teeth.
(8) Several sprockets can be driven by a single chain.
(9) Chains can operate over wide range of centre distances.
(10) The atmospheric conditions do not affect the performance of chain drives.
(11) They are cheaper than gear drives.

Disadvantages

(1) The chain velocity is not uniform due to polygonal effect. They are unsuitable for precise motion applications.
(2) They are subjected to excessive wear at the joints.
(3) The pitch increases with wear. Hence, adjustment for slackness is required.
(4) Tensioning devices are required.
(5) Housing is required to protect against dust and grit.
(6) Chain drives generate noise.
(7) Chain drives are not suitable for non-parallel shafts.
(8) They require proper lubrication.
(9) Precise alignment of the shafts is required.

The chains are used in motor cycles, bicycles, automobiles, conveyors, agricultural machinery, oil-well drilling rigs, machine tools, material handling equipments, etc.

Chain drives are recommended for low velocity ratios less than 10:1 and low velocities up to 25 m/s. They are capable to transmit power up to 100 kW.

Classification: Depending upon the use, they are classified as:

(1) Load chains (link chains)
(2) Hauling chains (block chains)
(3) Power transmission chains (roller & bush chains)

Load Chains

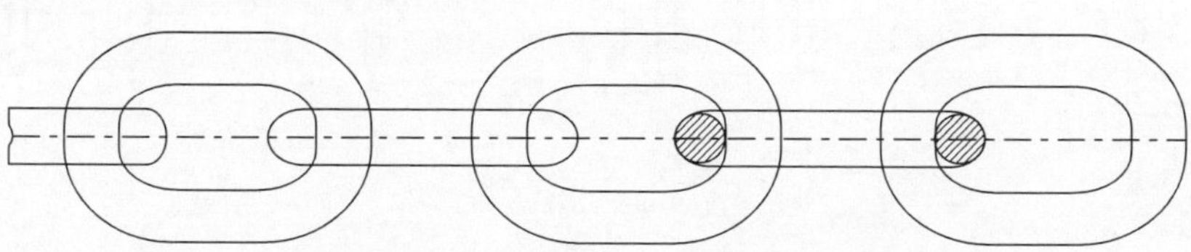

Figure 20.10 Link chain

They are formed by the chain links jointed to one another.

The load chains operate at low speeds and high loads. Mostly they are used in cranes.

Advantages

1. They are flexible.
2. They can operate with small pulleys.
3. They produce less noise.
4. They are manufactured easily.

Hauling chains

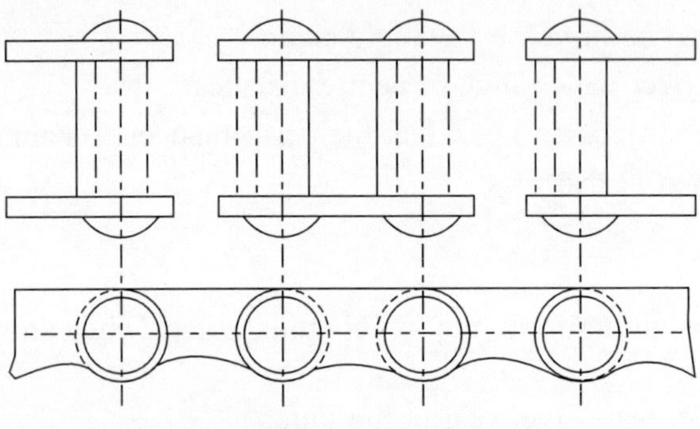

Figure 20.11 Block chain

Hauling chains are constructed by side plates and pins. They have long pitches, and are long and mesh with sprockets.

Hauling chains are suitable for medium speeds of 2 to 4 m/s and medium loads. Mostly they are used for conveyors. They are noisy. They are subjected to rapid wear.

Power Transmission Chains

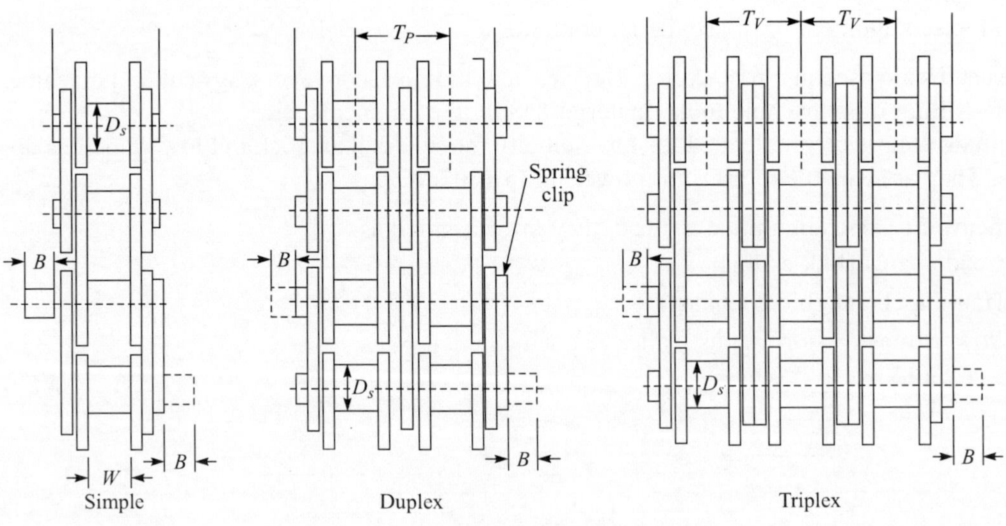

(a) Roller chains

Contd.

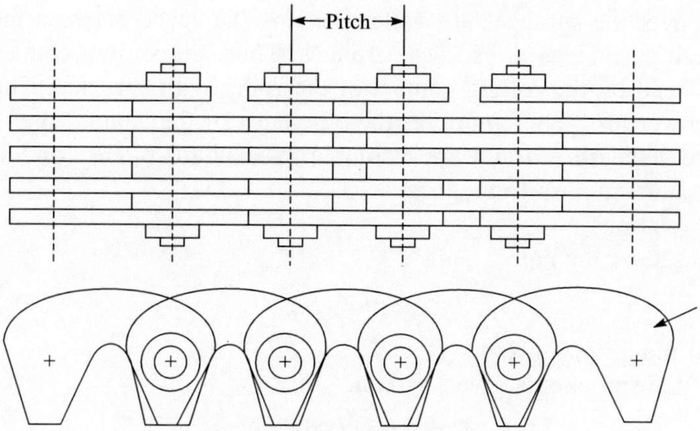

(b) Silent chain or inverted tooth chains

Figure 20.12 Power transmission chains

The power transmission chains operate over a wide range of velocities. They are designed with smaller pitches to reduce the dynamic effect. They are used in automobiles and other power transmission devices.

There are three types of power chains namely, (1) roller chains, (2) bush chains and (3) plate or salient chains.

Roller Chain Construction

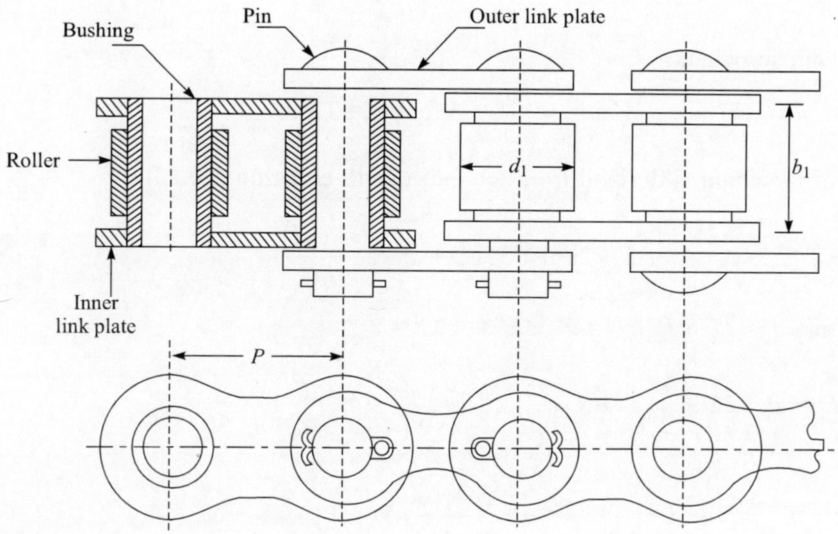

Figure 20.12a Construction of roller chain

The chain is made of alternate inner and outer link plates jointed by pin and bush. The bush with the roller on it is press-filled in the link plates. The pin is passed through the bush fitted in the outer link plates. The outer link plates are free to swivel with respect to inner link plates as the chain is moved over the sprocket, the rollers occupy the space between the sprocket teeth. And roll on the tooth surface as the sprocket rotates. Thus, the sliding contact between chain and sprocket is replaced by the rolling contact of the roller with sprocket. The rolling friction reduces wear and power loss. This improves the efficiency of the chain drive.

These chains are standardized and are available in standard pitches and widths. They are selected from the manufacturer's catalogues.

The geometrical relations:

The pitch is choosen by the eqn.,

$$P \le 25(900/N)^{2/3} \tag{20.29}$$

where N is speed of the driver in rpm.

The pitch dia. of the sprocket is given by eqn.

$$D = P/\sin(180/z) \tag{20.30}$$

The average chain speed is given by eqn.

$$V = (PZN)/60{,}000 \tag{20.31}$$

Minimum number of teeth considering the sprocket bore diameter d_s

For roller chain sprocket,

$$Z\,\text{min} = \frac{4d_s}{p} + 5 \text{ for } p < 25\text{ mm} \tag{20.32}$$

$$Z\,\text{min} = \frac{4d_s}{p} + 4 \text{ for } 30\text{ mm} \le p \le 65\text{ mm} \tag{20.33}$$

For silent chain sprocket,

$$Z\,\text{min} = \frac{4d_s}{p} + 6 \text{ for } p \le 50\text{ mm} \tag{20.34}$$

The length of the chain is derived from the belt length equation (20.3a)

$$L = 2C + 1.57(D+d) + \frac{(D-d)^2}{4C}$$

By substituting $L = L_p \times P, \pi D = Z_2 \times P$ and $\pi d = Z_1 \times P$

$$\therefore \quad L_p \times P = 2C + \frac{1}{2}(Z_2 \times P + Z_1 \times P) + \left(\frac{Z_2 P}{\pi} - \frac{Z_1 P}{\pi}\right)^2 \times \frac{1}{4C}$$

$$\therefore \quad L_p = 2\left(\frac{C}{P}\right) + \left(\frac{Z_1 + Z_2}{2}\right) + \frac{(Z_2 - Z_1)^2}{4\pi^2}\left(\frac{P}{C}\right)$$

The chain length in pitch lengths is given by

$$L_p = 2C_p + \frac{Z_1 + Z_2}{2} + \frac{K(Z_1 - Z_2)^2}{C_p} \tag{20.35}$$

where, $C_p = \dfrac{C}{P}$ and $K = \dfrac{1}{4\pi^2}$

L_p is the number of links rounded to the nearest even number.

The exact centre distance is calculated by arranging the above equation as follows:

$$C = \frac{P}{4}\left[L_p - \left(\frac{Z_1 + Z_2}{2}\right) + \sqrt{\left(L_p - \frac{Z_1 + Z_2}{2}\right)^2 - 8\left(\frac{Z_1 - Z_2}{2\pi}\right)^2}\right] \quad (20.36)$$

Power transmitted and power rating

Power transmitted, $$NP = \frac{F_w \cdot V}{1000\,K_s}\ \text{kW} \quad (20.37)$$

where, K_s = service factor = $k_1, k_2, k_3, k_4, k_5, k_6$

The values of the various factors (Table 20.51) (Table 20.5 MDHB)

F_w = Working load or permissible load. It is evaluated by one of the following equations:

(a) Working load by empirical equation:

$$F_w = 100\frac{A}{V+3} \quad (20.38)$$

(b) Working load based on bearing stress:

$$F_w = S_b \cdot A \quad (20.39)$$

S_b from Table 20.54

(c) Working load based on breaking load Q:

$$F_w = Q/n \quad (20.40)$$

n = Factor of safety from Table 20.53 (Table 20.7 MDHB)

Check for actual factor of safety:

$$n = \frac{Q}{P} \quad (20.41)$$

where, $$P = P_t + P_c + P_s$$

$$= \frac{1000\ \text{KW}}{V} + \frac{wV^2}{g} + K_g wC \quad (20.42)$$

w = Weight per metre of chain N

C = Centre distance m.

K_g = Coefficient of sag. Table 20.56

Q = Breaking load in N.

Load on shafts due to chain drive N,

$$Q_0 = K_L \frac{1000\ \text{kW}}{V} \quad (20.43)$$

where, K_L = Load factor – Table 20.55

Empirical equation for rated KW for roller chain/strand,

$$kW = 0.75p^2\left[\frac{V}{74}-\left(\frac{V}{67}\right)^{1.41}\times\left\{1+50\sin^2\left(\frac{90}{Z_1}\right)\right\}\right] \quad (20.44)$$

The manufacture of chains has been standardized; they are available off the self of the manufacturers. They can be selected by the use of the manufacturer's catalogues containing the data in the form of tables and charts. For a given application the *kW*, rating required can be evaluated by the relation given below:

$$kW_r \text{ rating} = \frac{(kW \text{ to be transmitted}) \times K_s}{K_1 K_2} \quad (20.45)$$

where, K_s – service factor

K_1 – Tooth correction factor – Table 20.64 (Table 20.18 MDHB)

K_2 – Multiple strand factor – Table 20.65 (Table 20.19 MDHB)

The chain number is selected from the power rating Tables 20.59, 20.62 or 20.63.

If the chain number is selected from Table 20.60 or 20.61 and rating per strand from Table 20.59, 20.62 or 20.63, the number of strands are selected from Table 20.65, for the required valve of K_2 as given by the equation:

$$K_2 = \frac{kW}{kW_r \times K_1}$$

Failure Modes for the chain

(a) Due to fatigue of side plates, roller and bush.

(b) Due to impact of roller on the sprocket.

(c) Wear of pin and other parts due to insufficient lubrication.

Polygonal action of chain

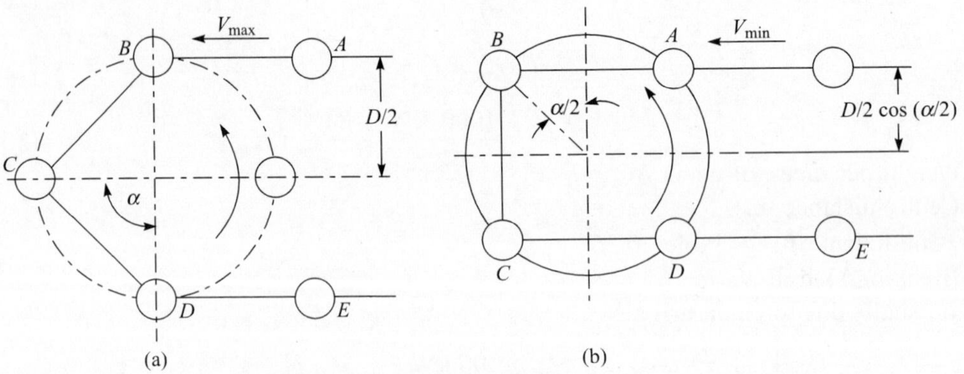

Figure 20.13

As the sprocket rotates, the chain link takes two extreme positions as in Fig. 20.10(a) and 20.10(b). When the link is at the position shown in Fig. (a), the chain has maximum velocity given by $V_{max} = \dfrac{\omega D}{2} = \dfrac{\pi Dn}{60000}$ m/s

When the link is at the position shown in Fig. (b), the link is at a distance of $\dfrac{D}{2}\cos\left(\dfrac{\alpha}{2}\right)$ from the centre. And the chain has minimum velocity given by, $V_{min} = \dfrac{\pi Dn \dfrac{D}{2}\cos\left(\dfrac{\alpha}{2}\right)}{60000}$ m/s

where, $\propto$ is the pitch angle $= \left(\dfrac{2p}{D}\right) = \dfrac{360^\circ}{Z}$.

Thus, the linear velocity of the chain is not uniform but undergoes a cycle variation from V_{max} to V_{min} for every cycle of tooth engagement. The variation in the velocity,

$$V_{max} - V_{min} = \frac{\pi Dn}{60000}\left[1-\cos\left(\frac{\alpha}{2}\right)\right]$$

$$\therefore \quad V_{max} - V_{min} \propto \left[1-\cos\left(\frac{180}{Z}\right)\right] \tag{20.46}$$

This variation can be decreased by increasing Z the number of teeth on sprocket. It is observed that for $Z = 11$, the variation is 4%, for $Z = 17$, it is 1.6% and for $Z = 24$, it is 1%. It is advisable to have minimum number of teeth i.e., 17 on the driving sprocket.

Selection procedure based on breaking strength Q

1. The type of chain to be used is selected with help of Table 20.51.
2. The pitch is chosen as recommended by the Eqn. 20.29.
3. The number of teeth on smaller sprocket is chosen from Table 20.57 or by Eqns. (20.32, 20.33 or 20.34). The dia. of the sprocket is evaluated by the Eqn. 20.30.
4. Chain speed is evaluated by the Eqn. 20.31.
5. Chain number and the breaking load Q, are chosen from Table 20.51 or 20.58 or 20.60 or 20.61 depending on the type of the chain chosen.
6. The factor of safety is chosen from Table 20.52.
7. The permissible tangential F_w is evaluated by Eqns. 20.38, 20.39 or 20.40.
8. The power transmitted per strand is evaluated by Eqn. 20.37. The service factor being chosen from Table 20.50.
9. The number of strands required,

$$n = \frac{\text{Actual power to be transmitted } KW_a}{\text{Power per strand } NP}$$

Alternatively (after Step 6)

(7) Find the factor of safety (FOS) by Eqn. 20.41 assuming the single strand is used.

(8) Compare the value with the values specified in Table 20.52 MDHB. If FOS is more than the specified, the single strand is O.K.

(9) If it is less than the specified, use more than one strand so as to get a satisfactory factor of safety.

Procedure based on the power rating (applicable to roller chains)

1. Calculate recommended pitch from Eqn. 20.29.
2. Select the number of teeth on the small sprocket from Table 20.57 or calculate from Eqns. 20.33 or 20.34 or 20.35.
3. Calculate the sprocket diameter from Eqn. 20.30.
4. Select the chain number based on pitch from Tables 20.58 or 20.60 or 20.61.
5. Then find the power rating per strand, kW_r from Tables 20.59 or 20.62 or 20.63.
6. Select a suitable value of K_1 from Table 20.64.
7. For a suitable value of K_2 in equation.

$$KW = K_1 \cdot K_2 \cdot KW_r \text{ i.e., } K_2 = \left[\frac{KW}{K_1 KW_r}\right]$$

The number of strands are selected from Table 20.65

Table 20.47. Strength of hoisting chains as percentage of bar strength

Type of chain link		Strength
Standard close link		138%
Coil chain	–	120%
B - B crane chain	–	145%
Stud chain	–	165%

Table 20.48. Values of K in equation (Eqn. 20.35)

$\frac{Z_1 - Z_2}{C_p}$	0.1	0.1	2.0	3.0	4.0	5.0	6.0
K	0.02533	0.02538	0.02555	0.02584	0.02631	0.02704	0.02828

Table 20.49. Tentative selection factors

Chain speed m/sec.	Power kW	Types of chain
1.75	15	Detachable
2.25	30	Pintle type
5.00	190	Offset side bar
12.50	1140	Roller
20.00	1900	Silent

Table 20.50. Minimum centre distance *C* min

Transmission ratio (i)	Min. centre distance C min
3	a' + 30 to 50 mm
3 - 4	$1.2a'$
4 - 5	$1.3a'$
5 - 6	$1.4a'$
6 - 7	$1.5a'$

Table 20.51. Service factor K_s

Service factor K_s	$K_s = k_1, k_2, k_3, k_4, k_5, k_6$
Load factor, k_1 = 1.0 for constant load = 1.25 for variable load with mild shock = 1.5 for variable load with heavy shock	Centre distance of sprocket factor = k_3 $k_3 = 1.25$ for $C_p < 25$ or $\frac{L_p}{Z_1+Z_2} > 1$
Distance regulation factor, k_2 = 1.0 for adjustable supports = 1.1 for drive using idler sprockets = 1.25 for fixed centre distance	= 1.0 for C_p = 30 to 50 or $\frac{L_p}{Z_1+Z_2} > 1.5$ = 0.8 for C_p = 60 to 80 or $\frac{L_p}{Z_1+Z_2} > 200$
Sprockets position factor, k_4 = 1.0 for inclination of line joining the centres of sprockets to horizontal up to 60° = 1.25 for inclination over 60°	$C_p = \frac{C}{P} = \frac{\text{Centre distance}}{\text{Pitch}}$
Lubrication factor, k_5 = 0.8 for continuous lubrication = 1.0 for drop lubrication = 1.5 for periodic lubrication	Rating factor, k_6 k_6 = 1.0 for 8 hours/day = 1.25 for 16 hours/day = 1.5 for continuous running

Table 20.52. Chain application

Type of chain	Pitch, mm	Pitch velocity, m/s	Ultimate strength, kN	Velocity ratio	Application
Roller	12.5 – 65	25	7.50 – 400	8	Motorcycles, general-purpose field machinery
Silent	12.5 – 32	35	1.9 – 700	12	Main drive in various machines
Bush	8 – 100	25	5 – 500	8	General-purpose, motorcycles
Bush roller with bent plates	40 – 100	5	8.0 – 500	4	Heavy duty
Hooked link	30 – 38	1.75	6 – 9	2.5	Agriculture machinery
Stud link	42	–	2.5	–	Drives under impact load

Of the six types of driving chains, the two most commonly used chains are: (a) roller or bush roller chains and (b) silent or inverted tooth chain.

Table 20.53. Factor of safety for chain drives

Types of Chain	R.P.M. of pinion or small sprocket for $K_s = 1$, $Z_1 = 15\text{-}30$								
	50	**200**	**400**	**600**	**800**	**1000**	**1200**	**1600**	**2000**
Bush Roller Chain:									
P = 12 - 15 mm.	7	7.8	8.55	9.35	10.2	11.0	11.7	13.2	14.8
= 20 - 25 mm.	7	8.2	9.35	10.3	11.7	12.9	14.0	16.3	-
= 30 - 35 mm.	7	8.55	10.2	13.2	14.8	16.3	19.5	-	
Silent Chain:									
P = 12.7 - 15.87 mm.	20	22.2	24.4	28.7	29.0	31.0	33.4	37.8	42.0
= 19.05 - 25.5 mm.	20	23.4	26.7	30.0	33.4	36.8	40.0	46.5	53.5

Table 20.54. Allowable bearing stress S_b MPa

(for K_s = 1, and Z_1 = 15 to 30)

Pitch p, mm	Speed of small sprocket										
	50	**200**	**400**	**600**	**800**	**1000**	**1200**	**1600**	**2000**	**2400**	**2800**
9.525 12.7 15.875	35	31.5	28.7	26.2	24.2	22.4	21	18.5	16.5	15	13.7

Table 20.55. Load factor K_L in (Eqn. 20.43)

Load factor	Position of chain drive			
	Inclination 40°		**Inclination above 40°**	
	constant	**shock**	**constant**	**shock**
K_L	1.15	1.3	1.05	1.15

Table 20.56. Coefficient of sag - K_g (Eqn. 20.42)

Sag factor	Position of chain drive			
	Horizontal	**Up to 40°**	**Above 40°**	**Vertical**
K_s	6	4	2	1

Table 20.57. Preferred transmission ratio - *i*

1,	1.12,	1.25,	1.4,	1.6,	1.8,	2,	2.25,	3.15,	4,	4.5,	5,	5.6,	6.3,	7.1

Table 20.57a. Recommended minimum teeth in small sprocket

Transmission ratio	1 - 2	2 - 3	3 - 4	4 - 5	5 - 7
No. of teeth ZI	3 - 27	27 - 25	25 - 23	23 - 21	21 - 17

Table 20.58. Characteristics of roller and bush chains

Designation as per IS 2403	**Pitch *P* mm**	**Roller dia. D_r, max. (mm)**	**Width between inner plates *W*, min. (mm)**	**Breaking load, kN**	**Weight per metre Wt, N/m**
Roller chains					
5.0*S*	6.0	4.0	2.8	3	1.2
6.1*S*	8.0	5.0	3.1	5	1.8
	9.52	6.35	5.9	10	4.1
8.0*S*	12.7	7.75	6.45	15	4.9
	12.7	8.51	7.85	20	7.1
10.1*S*	15.875	10.16	9.85	25	9.1
12.0*S*	19.05	12.07	11.7	31	11.7
	19.05	11.9	12.7	33	14.7
16.0*S*	25.4	15.88	17.1	45	26.3
6.1*D*	9.525	6.35	5.9	18	7.4
8.3*D*	12.7	8.51	7.85	35	13.6
10.1*D*	15.875	10.16	9.85	45.8	18.2
12.0*D*	19.05	12.07	11.7	60	23.6
16.0*D*	25.4	15.88	17.1	87	52.2
6.1*T*	9.525	6.35	5.9	27	10.9
8.3*T*	12.7	8.51	9.85	51	20.2
10.1*T*	15.875	10.16	9.85	71	27.3
12.0*T*	19.05	12.07	11.70	89	35.4
16.0*T*	25.4	15.88	17.10	130	78.5
Bush chains	(Rolon NO)				
GB 63	9.525	5.08	4.77	11	4
GB 65	9.525	5.00	7.55	13	5.3
GB 66	9.525	6.00	9.55	12	6.2
GB 66H	15.875	6.00	8.00	15.5	7.6
DB 63	9.525	5.08	4.77	21	7.1
Letters *D* & *T* denote double and triplex chains respectively. Other chains are simple chains.					

Table 20.59. kW power rating of chains (per strand)

Pinion speed (rpm)	Chain number								
	Roller (IS. Number)						Bush (Rolon Number)		
	5	6	8	10	12	16	*GP* 63	*GP* 65	*GP* 66
100	-	0.265	0.625	1.03	1.62	5.152	0.293	0.338	0.530
200	0.118	0.478	1.178	1.84	2.797	9.568	0.442	0.626	1.030
300	0.184	0.662	1.693	2.650	3.901	12.512	0.648	0.957	1.398
400	0.228	0.883	2.134	3.362	5.005	14.72	0.957	1.178	1.84
600	0.331	1.251	2.870	4.416	6.624	19.136	1.251	1.693	2.576
800	0.405	1.620	3.827	5.888	8.832	23.55	1.472	2.061	3.238
1000	0.478	1.914	4.048	6.256	9.568	25.76	1.776	2.502	3.827
1200	0.530	2.208	4.563	6.992	11.04	26.50	1.987	2.944	4.416
1600	0.662	2.576	5.152	8.096	11.78	27.23	2.355	3.312	3.152
2000	0.736	2.944	5.888	9.568	11.77	22.08	2.723	4.048	5.888
24.00	0.883	3.312	6.109	8.832	10.304	14.72	2.944	4.416	6.624

Table 20.60. Dimensions of single-strand American standard roller chain (ISO type A)

Chain size	Pitch, mm	Width, mm	Breaking load, kN	Mass, kg/m	Roller diameter, mm	Strand spacing, mm
35 (06*C*-1)	9.52	4.78	10.2	0.33	5.08	10.13
40 (08*A*-1)	12.70	7.92	17.2	0.61	7.92	14.38
50 (10*A*-1)	15.88	9.53	28.3	0.98	10.16	18.11
60 (12*A*-1)	19.05	12.70	38.5	1.59	11.91	22.78
80 (16*A*-1)	25.40	15.88	65.8	2.56	15.88	29.29
100 (20*A*-1)	31.75	19.05	108.9	3.78	19.05	35.76
120 (24*A*-1)	38.10	25.40	154.2	5.82	22.23	45.44
140 (28*A*-1)	44.45	25.40	208.7	7.62	25.40	48.87
160 (32*A*-1)	50.80	31.75	263.1	9.88	28.58	58.55
200 (40*A*-1)	63.50	38.10	430.9	15.91	39.67	71.55

*First number is ANSI size designation; parentheses enclosed is the ISO size designation
Detachable; mass is slightly less if riveted.

Table 20.61. Dimensions of single-strand British standard roller chain (ISO type *B*)

Chain size	Pitch, mm	Width, mm	Breaking load, kN	Mass, kg/m	Roller diameter, mm	Strand spacing, mm
06 *B* - 1	9.52	5.72	10.7	0.39	6.35	10.24
08 *B* - 1	12.70	7.75	18.2	0.68	8.51	13.92
10 *B* - 1	15.88	9.65	22.7	0.85	10.16	16.59
12 *B* - 1	19.05	11.68	29.5	1.16	12.07	19.46
16 *B* - 1	25.40	17.02	65.0	2.71	15.88	31.88
20 *B* - 1	31.75	19.56	98.1	3.70	19.05	36.45
24 *B* - 1	38.10	25.40	108.9	6.70	25.40	48.36
28 *B* - 1	44.45	30.99	131.5	8.25	27.94	59.56
32 *B* - 1	50.80	30.99	172.4	9.22	29.21	68.55
40 *B* - 1	63.50	38.10	272.2	15.48	39.37	72.29

Table 20.62. Power ratings for medium-pitch single-strand roller chain based on a 17-teeth driving wheel

Pinion speed, rev./ min.	Power kW									
	ISO Type A					ISO Type B				
	35	40	50	60	80	06B	08B	10B	12B	16B
50	0.12	0.28	0.53	0.94	2.06	0.14	0.34	0.64	1.07	2.59
100	0.22	0.53	0.98	1.74	4.03	0.25	0.64	1.18	2.01	4.83
200	0.41	0.98	1.83	3.40	7.34	0.47	1.18	2.19	3.75	8.94
300	0.58	1.34	2.68	4.56	11.63	0.61	1.70	3.15	5.43	13.06
500	0.98	2.24	4.34	7.69	11.99	1.09	2.72	5.01	8.53	20.57
700	1.29	2.95	5.91	10.73	23.26	1.48	3.66	6.71	11.63	27.73
1000	1.76	3.94	8.05	14.32	28.63	2.03	5.09	8.97	15.65	34.89
1400	2.42	5.28	11.18	14.32	18.49	2.73	6.81	11.67	18.15	38.47
1800	3.07	6.98	8.05	10.44		3.44	8.10	13.03	19.85	
2000	3.40	6.26	7.16	8.50		3.80	8.67	13.49	20.57	
3000	4.74					4.65	10.02			
4000	3.04					5.19				

Table 20.63. Power ratings for large-pitch single-strand roller chain based on a 17-teeth driving wheel

Pinion speed, rev./ min.	Power kW								
	ISO Type A					ISO Type B			
	100	120	140	160	200	20B	24B	28B	32B
25	2.24	3.75	5.81	9.08	11.88	2.68	4.83	6.80	10.20
50	4.29	7.15	11.18	15.65	22.37	5.10	8.95	12.52	19.24
100	7.96	13.42	20.57	30.42	42.05	9.66	16.99	23.20	35.79
200	14.31	25.05	38.47	56.36	79.63	17.94	31.31	43.84	67.10
300	21.47	35.79	54.57	80.52	116.31	26.31	45.03	63.22	98.42
400	26.84	46.53	71.57	111.84	134.20	34.27	58.15	82.31	120.78
500	34.00	56.36	89.47	134.20		42.05	71.57	96.63	129.70
600	39.37	67.10	107.30			47.41	78.28	103.78	
700	44.73	76.05	80.52			52.78	84.99	109.15	
800	49.21	57.86				54.87	89.02		
900	40.26	48.31				56.96	93.05		
1000	34.00					59.05			

Table 20.64. Tooth correction factors K_1 based upon the number of teeth Z on the driving sprocket

Z_1	K_1	Z_1	K_1	Z_1	K_1	Z_1	K_1
11	0.53	17	1.00	23	1.35	35	1.95
12	0.62	18	1.05	24	1.41	40	2.15
13	0.70	19	1.11	25	1.46	45	2.37
14	0.78	20	1.18	27	1.57	50	2.51
15	0.85	21	1.26	29	1.68	55	2.66
16	0.92	22	1.29	31	1.77	60	2.80

Table 20.65. Multiple strand factor, K_2

Number of strands	K_2
1	1.0
2	1.7
3	2.5
4	3.3

The corrected power is obtained by applying these two factors to the rated power from Tables 20.16 and 20.17 as follows: $KW = K_1\ K_2\ KW_r$, where KW is the fully corrected rating.

Example 20.21: Design a roller chain to transmit 20 kW from a motor to a reciprocating pump. The pump operates continuously for 24 hours a day. The speed of the motor is 960 rpm and that of the pump is 300 rpm.

Given Data: Power, NP = 20 kW at Speed N_1 = 960 rpm,
Driven speed, N_2 = 300 rpm, Service: 24 hours per day

SOLUTION

Step-1: Selection of chain

The recommended pitch (Eqn. 20.5 MDHB),

$$p \le 25\left[\frac{900}{N}\right]^{\frac{2}{3}} = 25\left[\frac{900}{960}\right]^{\frac{2}{3}} = 23.94 \text{ mm}$$

From Table 20.12 MDHB, the chain is 16.0 S, with pitch = 25.4 mm.

Breaking load, $Q = 45$ kN; $wt/m = 26.3$ N/m; Width, $w = 17.1$ mm.

Step-2: Diameters of sprockets

The transmission, $i = 960/300 = 3.2$

Number of teeth on the small sprocket (Table 20.57 MDHB) $Z_1 = 25$

$\therefore$ On bigger sprocket, $Z_2 = 25 \times 3.2 = 80$

The diameter of the smaller sprocket,

$D_1 = P/\sin(180/Z_1) = 25.4/\sin(180/25) = 202.66$ mm

Step-3: Evaluation of number of strands

The chain speed,

$V = \pi D_1 N_1/60{,}000 = (\pi \times 202.66 \times 960)/60{,}000 = 10.20$ m/s

The FOS from Table 20.7 MDHB, $n = 12.9$

$\therefore$ Permissible load, $F_w = Q_1/n = 45000/12.9 = 3488.4$ N

The service factor, $K_s = K_1 \cdot K_2 \cdot K_3 \cdot K_4 \cdot K_5 \cdot K_6$

where (from Table 20.50 MDHB) $K_1 = 1.25$

$K_2 = 1.0, K_3 = 1.0$ for $C_p = 50$

$K_4 = 1.0; K_5 = 1.5$ and $K_6 = 1.5$

$K_s = 1.25 \times 1.0 \times 1.0 \times 1.5 \times 1.5 = 2.8125$

Power transmitted per strand,

$$NP = \frac{F_w V}{1000 K_s} = \frac{3488.4 \times 10.2}{1000 \times 2.8125} = 12.65 \text{ kW}$$

Power to be transmitted = 20 kW

$\therefore$ Number of strands $20/12.65 = 1.58$

Choose $n = 2.0$ strands

A duplex chain 16.0*D* is suitable.

Step-4: Exact centre distance

The centre distance for $C_p = 50$

$$\therefore \quad C = 20 \times 25.4 = 1270 \text{ mm} = 1.27 \text{ m}$$

The pitch length of the chain,

$$L_p = 2C_p + \frac{Z_1 + Z_2}{2} + \frac{K(Z_1 - Z_2)^2}{C_p} \quad \text{(Eqn. 20.10 MDHB)}$$

$$L_p = 2 \times 50 + \left(\frac{25 + 80}{2}\right) + 0.025 \times \frac{(80 - 25)^2}{50} = 154$$

The exact centre distance, (Eqn. 20.11 MDHB)

$$C = \frac{P}{4}\left[L_p - \left(\frac{Z_1 + Z_2}{2}\right) + \sqrt{\left(L_p - \frac{Z_1 + Z_2}{2}\right)^2 - 8\left(\frac{Z_2 - Z_1}{2\pi}\right)^2}\right]$$

$$C = \frac{25.4}{4}\left[154 - \frac{25 + 80}{2} + \sqrt{\left(154 - \frac{25 + 80}{2}\right)^2 - 8\left(\frac{80 - 25}{2\pi}\right)^2}\right]$$

$$= 1269.62 \text{ mm}$$

Example 20.22: Design a chain drive to transmit 15 kW at 1440 rpm of motor to transmission shafting running at 360 rpm. The operation involves moderate shock.

Given Data: Power, NP =15 kW, at Speed N_1 = 1440 rpm,
Driven speed, N_2 = 360 rpm

SOLUTION

Step-1: Design power

Load factor, $K_1 = 1.25$ (Table 20.50)

$\therefore$ The design power $= 1.25 \times 15 = 18.85$ kW

Step-2: Selection of pitch

The recommended pitch, $p < 25\left(\frac{900}{1440}\right)^{2/3}$

$$p < 25(900/1440)^{2/3} = 18.24 \text{ mm}$$

The available pitch, $p = 19.05$ mm, for 12.0*S* chain (Table 20.58 MDHB)

Step-3: Diameters of sprokets

The speed ratio, $i = 1440/360 = 4$

The number of teeth on small sprocket, (Table 20.57 MDHB)

$$Z_1 = 25, \text{ then } Z_2 = 100$$

Diameter of small sprocket,

$$D_1 = p/\sin(180/Z) = 19.05/\sin(180/25) = 152 \text{ mm}$$

Diameter of the larger sprocket,

$$D_2 = 19.05/\sin(180/100) = 606.5 \text{ mm}$$

Step-4: Centre distance

Minimum centre distance, $C_{\min} = 1, 2a^1$ (Table 20.49 MDHB)

$$a^1 = \frac{d_{a1} + d_{a2}}{2} = \frac{1}{2}\left[\frac{p}{\tan(180/Z_1)} + \frac{p}{\tan(180/Z_2)} + 1.2p\right]$$

$$= \frac{19.05}{2}[7.9 + 31.82 + 1.2] = 389.77$$

$$C_{\min} = 1.2 \times 389.77 = 467.7 \text{ mm}$$

$$C_{\max} = 80p = 80 \times 19.05 = 1524 \text{ mm}$$

Choose $C = 50p = 50 \times 19.05 = 952.5$ mm.

Step-5: Selection of chain

From Table 20.12 MDHB; choose 12.0*S* chain

With breaking load, $Q = 31$ kN, and wt/m = 11.7 N/m

Step-6: Factor of safety

The chain speed,

$$V = PZN/60{,}000 = 19.05 \times 25 \times 1440/60{,}000 = 11.43 \text{ m/sec}$$

The pull in the chain,

$$P_t = 1000 \text{ kW}/V = 1000 \times 18.85/11.43 = 1649.17 \text{ N}$$

$$P_c = \frac{wV^2}{2} = 11.7 \times 11.43^2/9.81 = 155.815 \text{ N}$$

$$P_s = K_g wC = 6 \times 11.7 \times 0.9525 = 66.87 \text{ N}$$

The total pull in the chain,

$$P = P_t + P_c + P_s = 1649.17 + 155.815 + 66.87 = 1871.855 \text{ N}$$

∴ The factor of safety,

$$n = Q/P = 31000/1535 = 20.2$$

Recommended value from Table 20.52, is FOS = 13

∴ The actual value is much more and hence acceptable.

Step-7: Exact centre distance

The chain pitch length,

$$L_P = 2C_P + \left(\frac{Z_1 + Z_2}{2}\right) + \frac{K(Z_1 - Z_2)^2}{C_p}$$

$$= 2 \times 50 + \left(\frac{25 + 100}{2}\right) + \frac{0.025\,(75)^2}{50} = 165$$

The exact centre distance,

$$C = \frac{P}{4}\left[L_p - \left(\frac{Z_1 + Z_2}{2}\right) + \sqrt{\left(L_p - \frac{Z_1 + Z_2}{2}\right)^2 - 8\left(\frac{Z_2 - Z_1}{2\pi}\right)^2}\right]$$

$$C = \frac{19.05}{4}\left[165 - \frac{125}{2} + \sqrt{(165 - 62.5)^2 - 8\left(\frac{75}{2\pi}\right)^2}\right] = 949.1 \text{ mm}$$

Example 20.23: Design a roller chain drive for transmitting 20 kW at 1440 rpm to a pump running at 480 rpm. The service is continuous with medium shock.

Given Data: Power, NP = 20 kW at Speed N_1 = 1440 rpm, and N_2 = 480 rpm

SOLUTION

Step-1: Design power

The service factor or load factor, $K_1 = 1.25$

∴ Design power, $kW_d = 1.25 \times 20 = 25$ kW

Step-2: Selection of chain

The recommended pitch, (Eqn. 20.5 MDHB)

$$P < 25\left(\frac{900}{1440}\right)^{2/3} = 18.25 \text{ mm}$$

Choose the available pitch, $P = 19.05$ mm

Then choose chain $12B-1$ from Table 20.15 MDHB

with P = 19.05 mm, breaking load, $Q = 29.5$ kN and mass, $m_1 = 1.16 \frac{\text{kg}}{\text{m}}$, width = 11.68 mm.

Step-3: Diameters of sprockets

The speed ratio $= 1440/480 = 3$

The number of teeth on small sprocket, $Z_1 = 25$

∴ Number of teeth on bigger sprocket, $Z_2 = 75$

∴ Diameter of small sprocket,

$$D_1 = 19.05/\sin(180/25) = 152 \text{ mm}$$

Diameter of the bigger sprocket,

$$D_2 = 19.05/\sin(180/75) = 454.92 \text{ mm}$$

Step-4: Verification by power rating equation

The power rating of $12B$ chain at 1440 rpm (Table 20.16 MDHB)

$$KW_r = 18.25 \text{ kW}$$

Then the correction factor, $K_1 = 1.46$ (Table 20.18 MDHB)

The multiple strand factor for single strand, $K_2 = 1.0$

The rated power for the chain, $KW_r = 18.25$ kW

Then the correction factor, $K_1 = 1.46$ (Table 20.18 MDHB)

The multiple strand factor for single strand, $K_2 = 1.0$

The rated power for the chain, $KW = 1.46 \times 1.0 \times 18.25 = 26.645$

This is more than the required, $KW = 25$. Hence, the design is safe.

Step-5: Check for the factor of safety

Centre distance, $C = 50p = 50 \times 19.05 = 952.5$ mm

The speed of the chain,

$$V = PZN/60{,}000 = (19.05 \times 25 \times 1440)/60{,}000 = 11.43 \text{ m/s}$$

The driving pull,

$$P_t = \frac{1000 \text{ kW}}{V} = 1000 \times 25/11.43 = 2187.23 \text{ N}$$

Centrifugal pull, $P_c = \dfrac{wV^2}{g} = 1{,}16 \times 11.43^2 = 151.5$ N

The sag pull, $P_s = K_g\, wC = 6 \times 1.16 \times 9.81 \times 0.9525 = 65$ N

The total pull (Eqn. 20.23 MDHB),

$$P = P_t + P_c + P_s = (2187.23 + 151.5 + 65.0) = 2403.8 \text{ N}$$

Then factor of safety, $n = Q_p = \dfrac{29500}{2403.8} = 12.25$

This is acceptable.

Example 20.24: A roller chain is to transmit 66.24 kW from a 17-tooth sprocket to a 34-tooth sprocket at a pinion speed of 300 rpm. The loads are with moderate shock. The equipment is to run for 18 hours/day. Specify the length and size of the chain required for a center distance of about 25 pitches.

Given Data: $N = 66.24$ kW; $z_1 = 17$; $z_2 = 34$; $n_1 = 300$ rpm;

Moderate shock and runs 18 hours/day; $C_p = 25$

SOLUTION

Step-1: Selection of chain

Pitch of chain

$$p \le 25\left(\frac{900}{N}\right)^{2/3} = 52.2 \text{ mm} \quad \text{(Eqn. 20.5 MDHB)}$$

From Table 20.21 MDHB, the nearest standard value of pitch

$$p = 50.8 \text{ mm}$$

Select chain number: 216A
Breaking load, $F_u = 56.7 \text{ kN} = 56700 \text{ N}$
Measuring load, $w = 510 \text{ N}$

Step-2: Speed of larger sprocket,

$$\frac{n_1}{n_2} = \frac{z_1}{z_2} = \frac{300}{n_2} = \frac{34}{17} \quad \therefore \; n_2 = 150 \text{ rpm}$$

Step-3: Pitch diameter of the sprocket,

$$D = \frac{p}{\sin\left(\frac{180}{z}\right)} \qquad \text{(Eqn. 20.6 MDHB)}$$

Pitch diameter of smaller sprocket,

$$D_1 = \frac{p}{\sin\left(\frac{180}{z_1}\right)} = \frac{50.8}{\sin\left(\frac{180}{17}\right)} = 276.46 \text{ mm}$$

Pitch diameter of larger sprocket,

$$D_2 = \frac{p}{\sin\left(\frac{180}{z_2}\right)} = \frac{50.8}{\sin\left(\frac{180}{34}\right)} = 550.57 \text{ mm}$$

Step-4: Velocity of chain,

$$V = \frac{pNz_1}{60,000} = \frac{50.8 \times 17 \times 300}{60,000} = 4.318 \text{ m/sec} \qquad \text{(Eqn. 20.8 MDHB)}$$

Step-5: Pull in the chain,

$$\text{Power } NP = \frac{F_w V}{1000 K_s} \text{ kW} \qquad \text{(Eqn. 20.18 MDHB)}$$

From Table 20.5 $K_s = K_1 \cdot K_2 \cdot K_3 \cdot K_4 \cdot K_5 \cdot K_6$

$$K_s = 1.5 \times 1.25 \times 1.25 \times 1.0 \times 1.5 \times 1.5 = 5.273$$

$$\therefore \quad 66.24 = \frac{F_w \times 4.318}{1000 \times 5.273}$$

$$\therefore \quad F_w = 80890 \text{ N}$$

Step-6: Allowable pull,

$$F_a = \frac{F_u}{n_o}$$

From Table 20.7 MDHB for $n_1 = 300$ rpm and $p = 50.8$ mm

Working factor of safely, $n_o = 10$

$$\therefore \qquad F_a = \frac{56700}{10} = 5670 \text{ N}$$

Step-7: Number of strands,

$$n = \frac{F_w}{F_a} = \frac{80890}{5670} = 14.3$$

$\therefore$ Number of strands, $n = 15$

Step-8: Check for actual factor of safety,

$$n_a = \frac{Q}{P} = \frac{Q}{P_t + P_c + P_s} \qquad \text{(Eqn. 20.22 MDHB)}$$

where, $P_t = \dfrac{1000 \times \text{kW}}{V} = \dfrac{1000 \times 66.24}{4.318} = 15340.435$ N

$$P_c = \frac{wV^2}{g} = \frac{510.0 \times 4.318^2}{9.81} = 969.32$$

$$P_s = K_g wC$$

From Table 20.10 MDHB for horizontal drive, $K_g = 6$

$$\therefore \qquad P_s = 6 \times 10.0 \times \left(\frac{25 \times 50.8}{1000}\right) = 3886.2 \text{ N}$$

$\therefore$ Actual FOS, $n_a = \left(\dfrac{56700 \times 15}{15340.435 + 969.32 + 3886.2}\right) = 41.37$

Since $n_a > n_o$, design is safe.

Step-9: Length of the chain,

$$L_p = 2C_p + \frac{z_1 + z_2}{2} + \frac{K(z_2 - z_1)^2}{C_p} \qquad \text{(Eqn. 20.10 MDHB)}$$

where, $\quad C_p = 25,\ z_1 = 17;\ z_2 = 34,\ K = \dfrac{0.25}{\pi^2} = 0.0253$

$$\therefore \qquad L_p = 2 \times 25 + \frac{17 + 34}{2} + \frac{0.0253(34 - 17)^2}{25} = 75.5$$

Choose, $L_p = 76$

Step-10: The exact center distance,

$$C = \frac{p}{4}\left[L_p - \frac{z_1 + z_2}{2} + \sqrt{\left(L_p - \frac{z_1 + z_2}{2}\right)^2 - 8\left(\frac{z_2 - z_1}{2\pi}\right)^2}\right]$$

$$= \frac{50.8}{4}\left[76 - \frac{51}{2} + \sqrt{\left(76 - \frac{51}{2}\right)^2 - 8\left(\frac{17}{2\pi}\right)^2}\right] = 1275.29 \text{ mm}$$

Example 20.25: A roller chain 208B is used to drive the cam shaft of an I.C. engine, both shafts rotate at 360 rpm. The centre distance between the shafts is approximately 550 mm. Calculate:

(i) Number of teeth on each sprocket
(ii) Number of chain links
(iii) Correct centre distance
(iv) The rated power transmitted.

Given Data: Chain number – 208B

Speed, $n_1 = n_2 = 360$ rpm, Approx., centre distance $C = 550$ mm

SOLUTION

Step-1: Chain data

For chain 208B from Table 20.21 MDHB, $p = 25.4$ mm
Measuring load, $F_m = 130$ N
Breaking load, $F_u = 18.2$ kN

Step-2: Diameter sprockets

The speeds, $n_1 = n_2 = 360$ rpm

The speed ratio $= \dfrac{n_1}{n_2} = 1$

The number of teeth, $Z_1 = Z_2 = 30$ (Table 20.11a MDHB)

The diameter sprockets, $d_1 = d_2 = \dfrac{p}{\sin\left(\dfrac{180}{z}\right)} = \dfrac{25.4}{\sin\left(\dfrac{180}{30}\right)} = 243$ mm

Step-3: Chain length in pitches (links),

$$L_p = C_p + \left(\frac{Z_1 + Z_2}{2}\right) + \frac{K\left(Z_2 - Z_1\right)^2}{C_p} \quad \text{(Eqn. 20.10 MDHB)}$$

where, $C_p = \dfrac{C}{p} = \dfrac{550}{25.4} = 21.65$, $K = \dfrac{0.25}{\pi^2} = 0.02533$

$$L_p = 2\times 21.65 + \left(\frac{30+30}{2}\right) + \frac{0.02533(30-30)^2}{21.65} = 73.3$$

Choose $L_p = 74$

Step-4: The correct centre distance,

$$C = \frac{p}{4}\left[L_p - \frac{Z_1+Z_2}{2} + \sqrt{\left(L_p - \frac{Z_1+Z_2}{2}\right)^2 - 8\left(\frac{Z_2-Z_1}{2\pi}\right)^2}\right]$$

$$= \frac{25.4}{4}\left[74 - \frac{30+30}{2} + \sqrt{(74-30)^2 - 8(0)2}\right] = 558.8 \text{ mm}$$

Step-5: The rated kW power,

$$kW = 0.75p^2\left[\frac{V}{74} - \left(\frac{V}{67}\right)^{1.41}\left\{1 + 50\sin^2\left(\frac{90}{Z_1}\right)\right\}\right] \quad \text{(Eqn. 20.25 MDHB)}$$

where, $V \frac{PZN}{60,000} = \frac{25.4\times 30\times 360}{60,000} = 4.572 \text{ m/sec}$

$$kW = 0.75\times(25.4)^2\left[\frac{4.572}{74} - \left(\frac{4.572}{67}\right)^{1.41}\left\{1 + 50\sin^2\left(\frac{90}{30}\right)\right\}\right] = 17.4 \text{ kW}$$

20.4 ROPE DRIVES

There are two types of ropes:

(1) Fibre ropes and (2) wire ropes

20.4.1 Fibre Ropes

They are made from fibrous material like hemp, manila and cotton. The hemp ropes are subjected to sliding and wear in sheaves. To prevent this wear, they are lubricated by tar, tallow or graphite. The cotton ropes are smooth and need no lubrication. The ropes are circular in cross section. For manila ropes the strength is given by $F_u = 500d^2$ kN and for cotton rope, the tensile strength is given by $F_u = 350\,d^2$ kN, d being the diameter in mm.

The sheave for the ropes is grooved with groove angle of $45°$. The rope rests on the sides of the groove and there is gap at the bottom. Then the tensions in the rope are related by a relation, similar to V-belt namely,

$$\frac{T_1}{T_2} = e^{f\theta\,\text{cosec}\beta}, \text{ where } 2\beta \text{ is the groove angle.}$$

Hence, all the equations derived for V-belt drives, apply here also.

20.4.2 Wire Ropes

They are used to transmit large amount of power over long distances (up to 150 metres). They are used in elevators, mine-hoists, conveyors, cranes, and hauling devices. The wire ropes run over grooved pulleys or drums. They rest on the bottom side of the groove and not wedged between the groove sides.

Wire rope constructions: Depending upon the way the wire are wound and arranged in the rope, the ropes have different constructions as designated in Table 20.68. This table gives various characteristics of the wire ropes and is very much useful in designing wire rope drives.

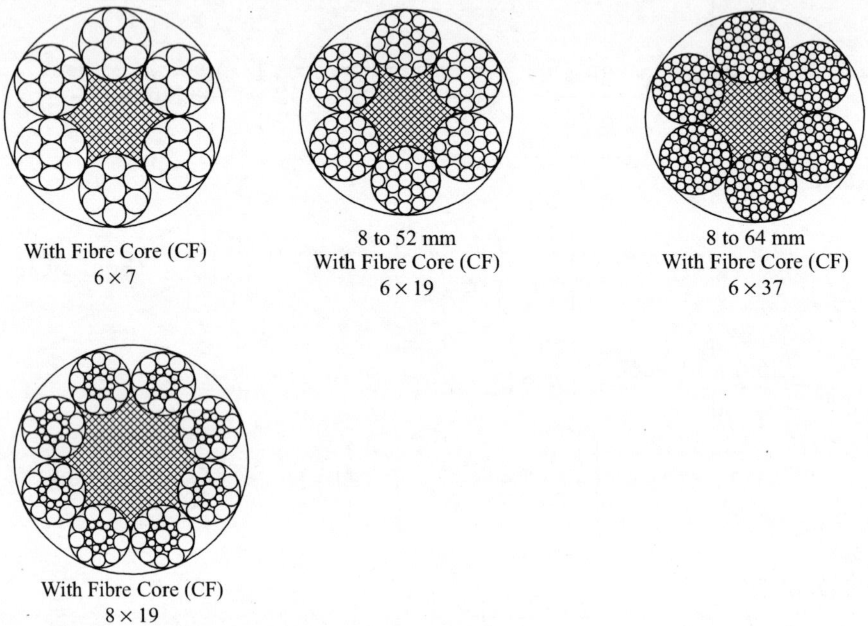

Figure 20.14 Steel wire ropes

The wire ropes are contracted by twisting the number of steel wires up to strands and then twisting the strands around fiber or steel core. The popular contracting are shown in the figure above.

In 6 × 7 contraction, there are 7 steel wires in each strand and there are 6 strands twisted around the core.

In 6 × 19 contraction, there are 19 wires in each strand and there are 6 strands in the ropes, wound around steel or fiber core.

In 6 × 37 contraction, there are 37 wires in each strand and there are 6 strands in the ropes.

The core can be of three types: (1) fiber core, (2) wire strand core and (3) independent wire rope core (IWRC).

In fibre cores are most flexible and elastic but get easily crushed. The second types are strong and are used for suspension, the third type is most durable. It has the following advocates: (1) It increases the strength, (2) It has greater resistance to crushing and (3) It is resistant to excessive heat.

There are subclassifications depending upon the arrangement of wires in the strands.

Depending upon the direction of twisting wires in the strands and strands to make the wires, the ropes are classified as regular lay ropes or lang's lay ropes.

In the regular lay ropes, the direction of twisting of strands is opposite to the direction of twisting of wires in the strand.

In the Lang's lay ropes, the direction of twisting of strands and the wires in the strands are the same.

The regular lay ropes are more popular due to the following advantages:

1. They have smaller tendency to rotate under load.
2. There is reduced possibility of kinking.
3. They are strong against crushing and distortion.
4. They are structurally stable.
5. They can be installed easily.

The end of the ropes tends to fray readily and cannot be connected to the plant and equipment. To prevent the fraying and facilitate the connection to the equipment, the most commonly used method is to turn the end back to form a loop tumble to maintain the proper shape of the loop. The loose end is fixed back on the rope by U-bolts. The thimble in addition prevents the direct contact of the load with the wires.

Rope Sheaves and Rope Drums

The rope sheaves and drums are used to drive the rope. The hoisting tackle consists of number of sheaves. The rope passes over the sheaves. The end of the rope is attached to a drum connected to the driving motor. The rope is wound around the drum as the load is lifted, and unwound as the loads are lowered. As the rope passes over, the sheaves and drum are subjected to bending stress. The smaller the diameter of the sheave, greater is the bending stress. To reduce bending stress, the diameter of sheaves and drums should be larger. But the larger diameter results in greater centrifugal force and greater cost. Hence, an optimum size is to be used. Depending upon the type of rope, the sheave diameter rope diameter ratios are recommended in Table 20.75 MDHB.

The rope sheaves are cast in cast-iron or fabricated from steel plate. The life of the rope is more for cast iron sheaves and hence is preferred. Normally, FG200 cast-iron is used. Small sheaves are cast as one piece as solid disk. The larger sheaves are provided with ribs. The rim of the sheaves should have proper shape to reduce the likely damage to the wire ropes. The proportions are shown in figure ahead.

Rope drums may be plain cylindrical drum or cylindrical drum with proper helical grooves on its surface. Usually, the grooved drums are preferred, because the wear on rope is less and rope life is more. They are made of cast-iron. In some cases, they may be fabricated from steel plate. The drums with one coiling rope have only one right-hand helical groove. When two ropes are used, there are two helical grooves one right-hand and the other left-hand with a space at the center. The grooves should have a proper radius to accommodate the rope without any damage to the wires. The pitch of the helical groove,

$$S = d_r + (2 \text{ to } 3\text{ mm})$$

The drum wall thickness, $w = 0.02D + 6 \text{ to } 10\text{ mm}$

Proportions of rope drum and sheaves

(A) Proportions of rope sheaves

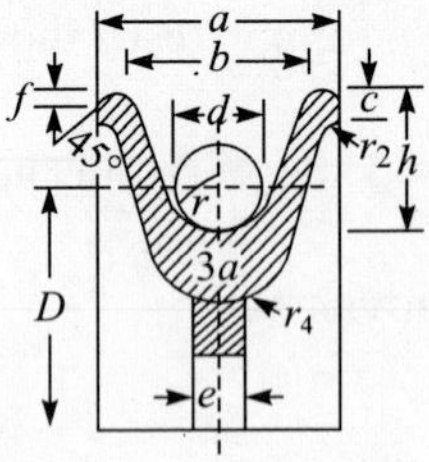

Sheaves diameter

$D \geqslant 10d \qquad c = 0.4d$
$a = 2.7d \qquad n = 1.6d$
$b = 21d \qquad e = 0.75d$

(B) Proportions of rope drum diameter of groove

Rope dia. *d*, mm	Min. groove dia. *di*, mm
6 to 8	$d + 0.8$
9 to 19	$d + 1.6$
21 to 29	$d + 2.4$
over 29	$d + 3.2$

Min. clearance between the rope

Dia. of rope, mm		Clearance
From	To	
–	12	1.5
12	29	2.5
29	over	3.0

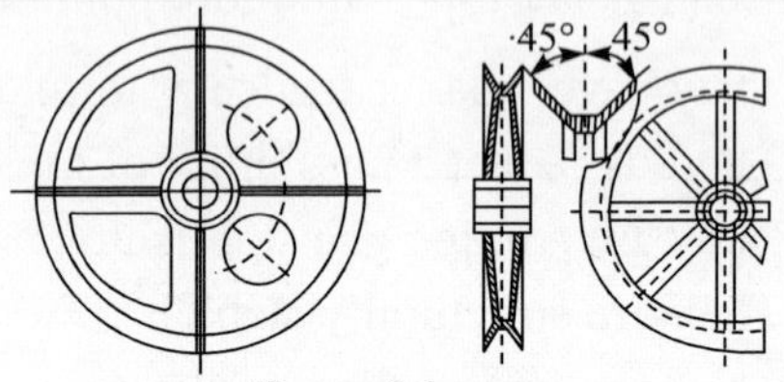

Types of sheaves

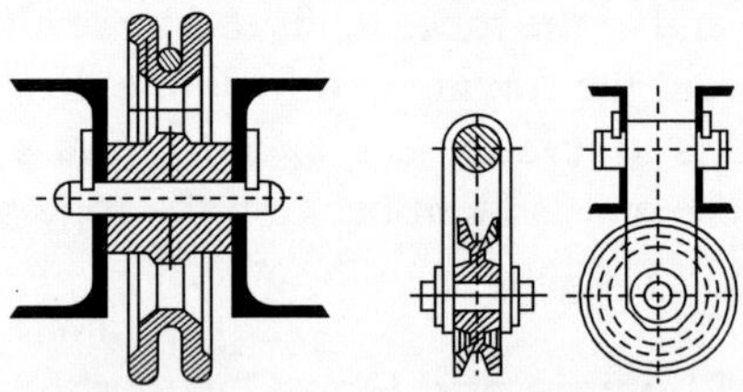

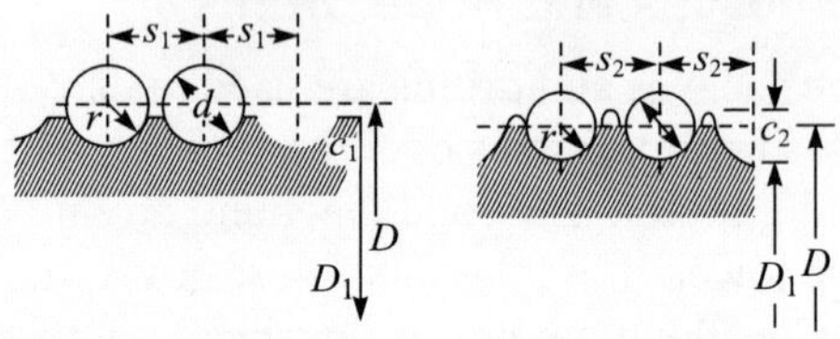

(a) Standard groove (b) Deep groove

Figure 20.15 Proportions of rope sheaves

Hoisting Tackle

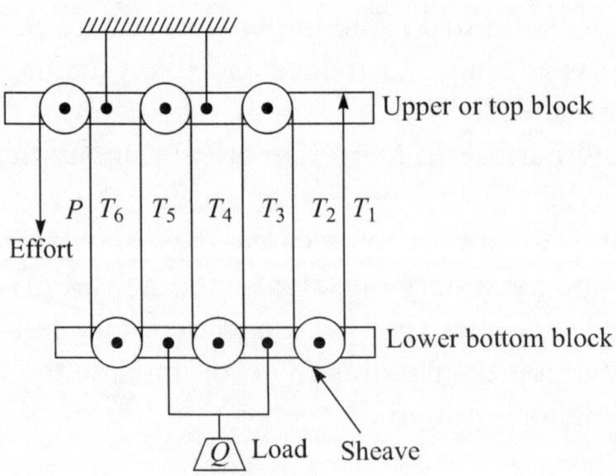

Figure 20.16

The relation between the pulls on both sides,

$$P = CQ \tag{20.47}$$

The value of C is given in Table 20.66. Normally, in hoisting machinery, the wire ropes pass over a number of pulleys in top and bottom pulley blocks, and the effort to be applied is given by the following relations:

The effort required to raise the load Q, by hoist with n pulleys is given by the relation,

$$P = \frac{C^n (C-1)}{(C^n - 1)} Q \tag{20.48}$$

Effort required for lowering the load Q,

$$P = \frac{(C-1)}{C(C^n - 1)} Q \tag{20.49}$$

The efficiency of the hoist to raise the load,

$$\eta = \frac{(C^n - 1)}{nC^n (C-1)} \tag{20.50}$$

Differential block

Effort required to raise the load without friction,

$$P_0 = \frac{Q}{2}(1 - n_1) \tag{20.51}$$

Effort required to raise the load with friction,

$$P = \frac{Q}{(1+C)}(C^2 - n_1) \tag{20.52}$$

Raising efficiency,

$$\eta = \left[\frac{(1-n_1)}{2}\right]\left[\frac{(1+C)}{(C^2 - n_1)}\right] \tag{20.53}$$

Effort required to lower the load with friction,

$$P = \frac{Q}{C}\left[\frac{1 - n_1 C^2}{1+C}\right] \tag{20.54}$$

Lowering efficiency,

$$\eta = \frac{2}{C}\left[\frac{(1 - n_1 C^2)}{(1+C)(1-n_1)}\right] \tag{20.55}$$

Loads on the Wire Ropes

The wire is subjected to service load,

$$Q = F_s = W + W_c + W_r \tag{20.56}$$

where,

W = Load lifted, N

W_c = Cage weight, N

W_r = Weight of the rope, N

In addition to service load Q, there is bending load as calculated below.

As the wire rope passes over the pulley (sheave), the individual wires are subjected to bending stress. The stress induced is given by the equation,

$$\frac{M}{I} = \frac{\sigma_b}{y} = \frac{E}{r}$$

where, $$y = \frac{d_w}{2} \text{ and } r = \frac{D}{2}$$

d_w = wire diameter and D = drum diameter.

$$\therefore \quad \frac{2\sigma_b}{d_w} = \frac{2E}{D}$$

$$\therefore \quad \sigma_b = E\frac{d_w}{D} \tag{20.57}$$

The equivalent **bending load** is given by eqn.

$$F_b = \sigma_b A = \frac{AEd_w}{D}, \tag{20.58}$$

where, $E = 0.85 \times 10^5$ MPa, $A = 0.38\, d_1^2$

There is **acceleration load** during acceleration period given by

$$F_a = \frac{F_s}{g}\, a, \text{ where } a = \text{max. acceleration} = (V/t_a)\,\text{m/s}^2$$

There is **impact load** during starting given by eqn.

$$F_{st} = F_s\left[1 + \sqrt{1 + \frac{2a \cdot h \cdot E \cdot A}{L \cdot F_s \cdot g}}\right] \tag{20.59}$$

Total Load on the Rope

$$F = F_s + F_b \text{ during normal working} \tag{20.60a}$$

$$= F_s + F_b + F_a \text{ during acceleration} \tag{20.60b}$$

$$= F_{st} + F_b \text{ during starting} \tag{20.60c}$$

F_{max}, the maximum of the three values given by these equations, is considered for finding the diameter of the rope.

Table 20.66. Values of C in equations 20.47 to 20.55

Manila rope C = 1.14

Wire rope

Rope diameter (mm)	*C*	Rope diameter (mm)	*C*
10	1.09	18	1.055
12	1.076	22	1.045
14	1.070	24	1.0415
16	1.062	25	1.04

Table 20.67. Ultimate strength of Manila Rope (*Fu*)

d = Rope dia. in mm			*Fu* = Ultimate strength (kN)		
d, mm	*Fu*, kN	*d*, mm	*Fu*, kN	*d*, mm	*Fu*, kN
10	6	22	29	38	76
12	9	24	34	41	90
14	12	25	36	44	102
16	16	29	48	48	122
18	20	32	56	51	136
20	24	35	66	54	150

Table 20.68. Wire rope data (all sizes in mm)

Rope *designation	Mass, kg/100 m	Min. dia., sheave	Size range, *d*	Strength *Fu*, kN	Elast. mod. *E*, GPa
6 × 7 *FC*	$0.35\ d^2$	42*d*	6.5 - 38	$0.608\ d^2$	80.6
6 × 7 *IWRC* μ	$0.39\ d^2$	42*d*	6.5 - 38	$0.656\ d^2$	68.6
6 × 19 *FC*	$0.39\ d^2$	34*d*	6.5 - 70	$0.643\ d^2$	74.4
6 × 19 *IWRC*	$0.43\ d^2$	34*d*	6.5 - 70	$0.688\ d^2$	93.0
6 × 19 *IWRC* †	$0.43\ d^2$	34*d*	6.5 - 70	$0.795\ d^2$	93.0
6 × 37 *FC*	$0.39\ d^2$	23*d*	6.5 - 83	$0.643\ d^2$	68.2
6 × 37 *IWRC*	$0.43\ d^2$	23*d*	6.5 - 103	$0.688\ d^2$	86.8
6 × 37 *IWRC* †	$0.43\ d^2$	23*d*	6.5 - 103	$0.798\ d^2$	86.8
8 × 19 *FC*	$0.37\ d^2$	27*d*	6.5 - 38	$0.520\ d^2$	55.8
8 × 19 *IWRC*	$0.42\ d^2$	27*d*	13 - 29	$0.579\ d^2$	-
8 × 19 *IWRC* †	$0.42\ d^2$	27*d*	13 - 29	$0.668\ d^2$	-

* Material is improved plow steel except as noted.

† Material is extra-improved plow steel.

Table 20.69. Minimum factors of safety for wire rope

		Passenger elevators, speed in m/s:	
Track cables	3.2	0.25	7.60
Guys	3.5	1.50	9.20
Mine shafts, depth in m		4.00	11.25
Up to 150 m	8.0	6.00	11.80
300 - 600 m	7.0	7.50	11.90
600 - 900 m	6.0	Freight elevators, speed in m/s:	
Over 900	5.0	0.25	6.65
Hoisting	5.0	1.50	8.20
Haulage	6.0	4.00	10.00
Cranes and derricks	6.0	6.00	10.50
Electric hoists	7.0	7.50	10.55
Hand elevators	5.0	Powered dumbwaiters, speed in m/s:	
Private elevators	7.5	0.25	4.8
Hand dumbwaiter	4.5	1.50	6.6
Grain elevator	7.5	2.50	8.0

* Use of these factors does not preclude a fatigue failure.

Table 20.70. Common wire rope application and data rope dia. d(mm) = 1.5 $d_w \sqrt{i}$; d_w = wire dia. mm; i = no. of wires in the rope

Type of service	Rope type	Ultimate strength, *Fu* N.	Mass, kg/m.	Wire dia. *dw* (mm)	Area of rope A(mm^2)	Recommended sheave dia. (mm)	
						Average	Minimum
Standard hoisting rope, e.g., mine hoists, quarries, ore docks, cargo hoists, car pullers, cranes, derricks, dredges, tramways, well drilling, elevators	6 × 19	$500d^2$	$0.0038d^2$	$0.063d$	$0.38d^2$	$45d$	$30d$
Special flexible hoisting rope, steel mill ladles, cranes. High-speed elevators, service where sheave diameters are limited	6 × 37	$475d^2$	$0.0037d^2$	$0.045d$	$0.38d^2$	$27d$	$18d$
Extra flexible hoisting rope	8 × 19	$435d^2$	$0.0036d^2$	$0.05d$	$0.38d^2$	$31d$	$21d$
Haulage rope, e.g., mine haulage, factory yard haulage, inclined plane, transways power transmission, guywires	6 × 7	$475d^2$	$0.0035d^2$	$0.106d$	$0.38d^2$	$72d$	$42d$

Table 20.71. Efficiency of wire rope fasteners

Methods of fastening	Efficiency %
Wire rope socket with zinc	100
Thimble with four or five wire tucks	90
Special offset thimble with clips	90
Regular thimble with clips	85
Three bolt wire clamps	75

Table 20.72. Preferred millimetres size of wire rope

6.5,	8,	9.5,		11.5,	13,	14.5,	16,	19,	22,	26,	29,	32,	35,	38
42,	45,	48,	51,	54,	57,	61,	64,	67,	70,	74,	77,	80,	83,	86,
90,	96,	103												

Table 20.73. Mass and breaking force for 6 × 7 (6-1) construction ropes (IS-2266-2002 Clauses 1, 4 and 5)

Nominal Diameter	Approximate Mass		Minimum Breaking Force Corresponding to Rope Grade of					
			1570		1770		1960	
	Fibre core (CF)	Steel core[1] (CWR)	Fibre core (CF)	Steel core[1] (CWR)	Fibre core (CF)	Steel core[1] (CWR)	Fibre core (CF)	Steel core[1] (CWR)
(1)	(2)	(3)	(4)	(5)	(6)	(7)	(g)	(9)
mm	kg/100 m		kN	kN	kN	kN	kN	kN
8	22.9	25.2	33	36	38	41	42	45
9	28.9	31.8	42	46	48	51	53	57
10	35.7	39.3	52	56	59	64	65	70
11	43.2	47.6	63	68	71	77	79	85
12	51.5	56.6	75	81	85	91	94	101

Note: To calculate the aggregate breaking force, multiply the figures given in col 4, 6 and 8 by 1.111 and in col 5, 7 and 9 by 1.193.

[1]Wire strand core (CWS) may be used for rope diameter 12 mm and below.

Table 20.74. Mass and breaking force for 6 × 19 M (12/6-1) construction ropes (IS-2266-2002 Clauses 1, 4 and 5)

Nominal Diameter	Approximate Mass		Minimum Breaking Force Corresponding to Rope Grade of					
			1570		1770		1960	
	Fibre core (CF)	Steel core[1] (CWR)	Fibre core (CF)	Steel core[1] (CWR)	Fibre core (CF)	Steel core[1] (CWR)	Fibre core (CF)	Steel core[1] (CWR)
(1)	(2)	(3)	(4)	(5)	(6)	(7)	(g)	(9)
mm	kg/100 m		kN	kN	kN	kN	kN	kN
8	22.1	24.4	31	33	35	38	39	42
9	28,0	30.8	39	42	44	48	49	53
10	34.6	38.1	48	52	54	59	60	65
11	41.9	46.1	58	63	66	71	73	79
12	49.8	54.8	69	75	78	85	87	94
13	58.5	64.3	82	88	92	99	102	110
14	67.8	74.6	95	102	107	115	118	128
16	88.6	97.4	124	133	139	150	154	167
18	112	123	156	169	176	190	195	211
19	125	137	174	188	196	212	217	235
20	138	152	193	208	218	235	241	260
22	167	184	234	252	263	284	292	315
24	199	219	278	300	313	338	347	375
26	234	257	326	352	368	397	407	440
28	271	298	378	409	426	461	472	510
32	354	390	494	534	537	602	617	666
36	448	493	625	675	705	761	781	843
38	500	550	697	732	785	848	870	939
40	554	609	772	834	870	940	964	1041
44	670	—	934	—	1053	—	I 166	—
48	797	—	1 112	—	1 253	—	1 388	—
52	936	—	1 305	—	1 471	—	1 629	—

Note: To calculate the aggregate breaking force, multiply the figures given in col 4, 6 and 8 by 1.163 and in col 5, 7 and 9 by 1.25.

[1]Wire strand core (CWS) may be used for rope diameter 12 mm and below.

Table 20.75. Mass and breaking force for 6 × 37 M (18/6-1) construction ropes (IS-2266-2002 Clauses 1, 4 and 5)

Nominal Diameter	Approximate Mass		Minimum Breaking Force Corresponding to Rope Grade of					
			1570		1770		1960	
	Fibre core (CF)	Steel core[1] (CWR)	Fibre core (CF)	Steel core[1] (CWR)	Fibre core (CF)	Steel core[1] (CWR)	Fibre core (CF)	Steel core[1] (CWR)
(1)	(2)	(3)	(4)	(5)	(6)	(7)	(g)	(9)
mm	kg/100 m		kN	kN	kN	kN	kN	kN
8	22.1	24.4	30	32	33	36	37	40
9	28.0	30.8	37	40	42	46	47	51
10	34.6	38.1	46	50	52	56	58	62
11	41.9	46.1	56	60	63	68	70	76
12	49.8	54.8	67	72	75	81	83	90
13	58.5	64.3	78	84	88	95	98	105
14	67.8	74.6	91	98	102	110	113	122
16	88.6	97.4	118	128	134	144	148	160
18	112	123	150	162	169	183	187	202
19	125	137	167	180	188	203	209	225
20	138	152	185	200	209	225	231	250
22	167	184	224	242	253	273	280	302
24	199	219	267	288	301	325	333	359
26	234	257	313	338	353	381	391	422
28	271	298	363	392	409	442	453	489
32	354	390	474	512	534	577	592	639
36	448	493	600	648	676	730	749	809
38	500	550	668	722	753	814	834	901
40	554	609	741	800	835	902	924	999
44	670	737	896	968	1010	1091	1119	1208
48	797	877	1066	1152	1202	1298	1331	1438
52	936	1029	1252	1352	1411	1524	1562	1687
56	1085	-	1451	-	1636	-	1812	-
60	1246	-	1666	-	1878	-	2080	-
64	1417	-	1896	-	2137	-	2367	-

Note: To calculate the aggregate breaking force, multiply the figures given in col 4, 6 and 8 by 1.212 and in col 5, 7 and 9 by 1.302.

[1]Wire strand core (CWS) may be used for rope diameter 12 mm and below.

Table 20.76. Mass and breaking force for 6 × 17 S (8-8-1) and 6 × 19 S (9-9-1) construction ropes (IS-2266-2002 Clauses 1, 4 and 5)

Nominal Diameter	Approximate Mass		Minimum Breaking Force Corresponding to Rope Grade of					
			1570		1770		1960	
	Fibre core (CF)	Steel core[1] (CWR)	Fibre core (CF)	Steel core[1] (CWR)	Fibre core (CF)	Steel core[1] (CWR)	Fibre core (CF)	Steel core[1] (CWR)
(1)	(2)	(3)	(4)	(5)	(6)	(7)	(g)	(9)
mm	kg/100 m		kN	kN	kN	kN	kN	kN
8	23.8	26.2	33	36	37	40	42	45
9	30.2	31.2	42	45	47	51	53	57
10	37.3	41.0	52	56	59	63	65	70
11	45.1	49.6	63	68	71	77	78	85
12	53.7	59.0	75	81	84	91	93	101
13	63.0	69.3	88	95	99	107	110	118
14	73.0	80.3	102	110	115	124	127	137
16	95.4	105	133	144	150	162	166	179
18	121	133	168	182	190	205	210	227
19	135	148	188	203	211	228	234	253
20	149	164	208	224	234	233	260	280
22	180	198	252	272	284	31]6	314	339
24	215	236	299	323	337	364	374	403
26	252	277	351	379	396	428	439	474
28	292	321	407	440	459	496	509	549
32	382	420	532	675	600	648	664	717
36	483	531	673	727	759	820	841	908
38	538	592	750	810	846	913	937	1012
40	596	656	831	898	937	1012	1038	1121
44	721	794	1006	1086	1134	1225	1256	1356
48	858	944	1197	1293	1350	1458	1495	1611
52	1008	1108	1405	1517	1584	1711	1754	1894

Note: To calculate the aggregate breaking force, multiply the figures given in col 4, 6 and 8 by 1.163 and in col 5, 7 and 9 by 1.25.

[1] Wire strand core (CWS) may be used for rope diameter 12 mm and below.

Table 20.77. Mass and breaking force for 8 × 19 S (9-9-1) construction ropes (IS-2266-2002 Clauses 1, 4 and 5)

Nominal Diameter	Approximate Mass		Minimum Breaking Force Corresponding to Rope Grade of					
			1570		1770		1960	
	Fibre core (CF)	Steel core[1] (CWR)	Fibre core (CF)	Steel core[1] (CWR)	Fibre core (CF)	Steel core[1] (CWR)	Fibre core (CF)	Steel core[1] (CWR)
(1)	(2)	(3)	(4)	(5)	(6)	(7)	(g)	(9)
mm	kg/100 m		kN	kN	kN	kN	kN	kN
8	22.3	27.2	29	34	33	38	36	42
9	28.2	34.4	36	43	41	49	46	54
10	34.9	42.5	45	53	51	60	56	66
11	42.2	51.4	55	64	61	73	68	80
12	50.2	61.2	65	77	73	36	81	96
13	58.9	71.8	76	90	86	101	95	112
14	683	83.3	88	104	100	117	110	130
16	89.2	109	115	136	130	153	144	170
18	113	138	146	172	165	194	182	215
19	126	153	163	192	183	216	203	240
20	139	170	180	213	203	240	225	265
22	169	206	218	257	246	290	272	321
24	201	245	260	306	293	345	324	382
26	236	287	305	359	343	405	380	449
28	273	333	353	417	198	470	441	520
32	357	435	461	544	520	614	576	680
36	452	551	584	689	658	777	729	860
38	503	614	651	768	73-4	865	812	958
40	558	680	721	851	813	959	900	1062
44	675	823	872	1029	983	1160	1039	1285
48	803	979	1038	1225	1170	1381	12.96	1529
52	942	1149	1218	1437	1374	1621	1521	1795

Note: To calculate the aggregate breaking force, multiply the figures given in col 4, 6 and 8 by 1.19 and in col 5, 7 and 9 by 1.332.

[1]Wire strand core (CWS) may be used for rope diameter 12 mm and below.

Example 20.26: A rope drive is required to transmit 150 kW from a pulley of 1 m diameter running at 450 rpm. The safe pull in each rope is 2250 N and the mass of the rope is 1 kg/m length. The angle of lap and groove angle are 150° and 45° respectively. Find the number of ropes required for the drive if the coefficient of friction between the pulley and rope is 0.3.

Given Data: Power, $NP = 150$ kW at $N_1 = 450$ rpm, pulley diameter = 1.0 m,

Maximum safe load on each rope, $F_d = 2250$ N

Mass per metre, $m_1 = 1.0$ kg/m

Angle of lap, $\theta_s = 150°$, Groove angle = 45°

SOLUTION

Step-1: Tensions in the rope

The ratio of rope tensions is,

$$\frac{P_1 - P_c}{P_2 - P_c} = e^{f\theta \operatorname{cosec} \alpha} = e^{0.3\times(180\times150/\pi)\times \operatorname{cosec} 22.5} = 7.786$$

The centrifugal tension, $P_c = mbtV^2$

where, $m \cdot b \cdot t = (1\times10^3/1000)$ g/mm $= 1.0$ g/mm

And rope speed

$$V = (\pi DN)/60,000 = (\pi\times1000\times450)/60,000 = 23.56 \text{ mps}$$

$$\therefore \quad P_c = 1\times(23.56)^2 = 555.16 \text{ N}$$

Maximum tension allowed, $P_1 = 2250$ N

Substituting the values,

$$\frac{2250 - 555.16}{P_2 - 555.16} = 7.786$$

$$P_2 = 1694.8/7.786 + 555.16 = 772.83 \text{ N}$$

Step-2: The power capacity of each rope

The power transmitted by each rope,

$$= \frac{(P_1 - P_2)V}{1000} = \frac{(2250 - 772.83)}{10,000} \times 23.56 = 34.8 \text{ kW}$$

Step-3: The number of ropes required

The power transmitted = 150 kW

$$n = 150/34.8 = 4.3$$

Choose n = 5 ropes

Example 20.27: A block and tackle consists of one 3-sheaves block at the top and 4-sheaves block at the bottom. Assume friction loss at each sheave to be 3%. The pulleys are made of C. I. and other parts of mild steel. The permissible load on the rope may be taken to be $52.5d^2N$, where d is the diameter of rope in mm. The hook for suspension has round section. The load to be lifted is 5.0 kN. The allowable tensile stress in mild steel is 80 MPa. Design the rope, pulley and cross arm.

Given Data: Number of sheaves on top, $n_1 = 3$, at bottom, $n_2 = 4$,

Friction loss = 3%, $F_d = 52.5\,d^2$

Load to be lifted = 5.0 kN

Allowable tensile strees, $S_d = 80$ MPa

SOLUTION

Step-1: Pull in the rope

The value of $C = 1.03$, (because of 3% loss at each sheave)

The effort P required (Eqn. 19.3 MDHB)

$$P = \frac{C^n (C-1)}{C^n - 1} Q$$

where, $C = 1.03$ and $n = 7$

$$Q = 5 \times 10^3 \text{ N}$$

$$\therefore \quad P = \frac{1.03^7 (1.03 - 1)}{(1.03^7 - 1)} \times 5 \times 10^3 = \frac{1.2298 \times 0.03 \times 5 \times 10^3}{0.2298} = 802.5 \text{ N}$$

Step-2: Loads on the rope

Service load, $F_s = 802.5$ N

The bending load, $F_b = \dfrac{AEd_w}{D}$ (Eqn. 19.12 MDHB)

where, $A = 0.38\,d^2$, $E = 0.85 \times 10^5$ MPa

Wire diameter, $d_w = 0.054\,d$ and sheave diameter, $D = 45\,d$ (Table 19.5 MDHB)

$$\therefore \quad F_b = \frac{0.38\,d^2 \times 0.85 \times 10^5 \times 0.045\,d}{45d} = 32.3d^2$$

$\therefore$ Total load, $F = F_s + F_b = 802.5 + 32.3d^2$

Step-3: Diameter of the rope

The strength of the rope $= \dfrac{F_u}{n} = F_d = 52.5d^2 =$ Total load F

$$\therefore \quad 52.5\,d^2 = 802.5 + 32.3\,d^2$$

$$\therefore \quad d = \sqrt{802.5/20.2} = 6.3 \text{ mm}$$

Available dia. $d = 6.5$ mm

Step-4: The pulley dimensions

The pulley diameter $= 45d = 45 \times 6.5 = 292.5$ mm

Width of the pulley $a = 2.7\,d = 17.55$ mm (From Plate 19.1 MDHB)

Step-5: Dimension of the cross bar

There are four such pulleys in the bottom block. Therefore width of the cross bar,

$$L_{\text{mm}} = 4 \times 17.55 = 70.2 \text{ mm}$$

Allowing for clearance and end bearings, choose $L = 100$ mm

$$\text{The B.M. on the cross bar} = \frac{5 \times 10^3 \times 100}{4} = 125000 \text{ Nmm}$$

The section modules of the cross bar,

$$Z = \frac{M}{S_d} = 1{,}25{,}000/80 = 1562 \text{ mm}^3$$

The diameter of the cross bar,

$$d = \sqrt[3]{\frac{32Z}{\pi}} = \sqrt[3]{\frac{32 \times 1562}{\pi}} = 25 \text{ mm}$$

The shearing stress in the bar

$$\tau = (F/2)/\left(\frac{\pi}{4}d^2\right) = \frac{\dfrac{5 \times 10^3}{2}}{\left[\dfrac{\pi}{4}(25^2)\right]} = 5.093 \text{ MPa}$$

It is safe against shear failure.

Step-6: Dimensions of the hook

The cross section of the hook,

$$A = \frac{Q}{S_d} = 5000/80 = 62.5 \text{ mm}^2$$

$$\therefore \text{ The diameter of the hook shank, } d_1 = \sqrt{\frac{4 \times 62.5}{\pi}} = 9.12 \text{ mm}$$

Choose $d_1 = 10.0$ mm

Example 20.28: A mine hoist employs 6×37 wire-rope and 6 falls of the rope. The max. lifting capacity is 25 kN. The maximum velocity of 2 m/sec., is attained by the hoist trolley within 1.5 seconds. The dead weight accounts for 10% of the service load. Design the wire rope for the above application using the following data in terms of rope diameter *d*.

Braking load, $F_u = 470d^2$

Wire diameter, $d_w = 0.045d$
Effective area of the rope, $A = 0.4d^2$
Minimum sheave diameter, $D = 25d$
Choose a suitable factor of safety.

SOLUTION

Step-1: The loads on the rope

From Table 19.4 MDHB, the factor of safety for hoist is 5.0

Then the permissible load for the rope,

$$F_a = 470\,d^2/5 = 94.0\,d^2$$

Choose a friction factor, $C = 1.06$ (approx). (Table 19.1 MDHB)

The service load, $F_s = 25 \times 1.1 = 27.5$ kN

Acceleration, $a = v/t = (2.0/1.5)$ mm/ s^2

The acceleration load,

$$F_a = \frac{F_s}{g} a = \frac{25 \times 1.1}{9.81} \times \frac{2.0}{1.5} = 3.74 \text{ kN}$$

∴ Total load being overcome, $Q = 27.5 + 3.74 = 31.24$ kN

Maximum effort to be applied to the rope at the driving end for approximate value of $C = 1.06$

$$P = \frac{C^n (C-1)}{(C^n - 1)} Q = \frac{1.06^6 (1.06-1)}{(1.06^6 - 1)} \times 31.24 = 6.353 \text{ kN} = 6353 \text{ N}$$

The bending load (Eqn. 19.12 MDHB)

$$F_b = (A.E.d_w)/D = (0.4d^2 \times 0.85 \times 10^5 \times 0.045\,d)/25\,d \; = 61.2d^2$$

Step-2: Diameter of the rope

For equilibrium, the permissible load = the load to be resisted

$$\therefore \quad 94.0\,d^2 = 6353 + 61.2\,d^2$$

$$\therefore \quad d^2 = (6353/32.8) = 193.69$$

$$d = \sqrt{193.69} = 13.92 \text{ mm}$$

Choose next available size, $d = 14.0$ mm

∴ The sheave diameter $= 25 \times 14 = 350.0$ mm

There are three pulleys in each of top and bottom blocks.

Example 20.29: Select a wire rope for a vertical mine hoist to lift 1800 tonnes of ore in each 8 hour shift from a depth of 780 m. Assume a two-compartment shaft with hoisting skips in balance, use a maximum velocity of 720 rpm with acceleration and retardations of 15 sec. each and rest period of 10 sec. for discharging and loading of skips.

A hoisting skip weighs approximately 0.6 of the load capacity.

Given Data: Quantity of ore lifted = 1800 Tonnes/shift

Shift period = 8 hours, Depth = 780 m, Maximum velocity = 720 rpm

Acceleration and retardation periods = 15 sec. each

Rest period = 10 sec., Skip weight = 0.6 of load capacity

Two-compartment shaft with hoisting skips in balance

SOLUTION

Step-1: Number of trips per shift

Maximum skip velocity = 720 rpm = $720/60$ m/s $= 12$ m/s.

The acceleration & retardation period = 15 sec. each

∴ Acceleration/retardation, $a = 12/15 = 0.8$ m/sec^2

The distance traversed during acceleration and retardation

$$S_1 + S_3 = 2\left(\frac{12}{2}\times 15\right) = 180 \text{ m}$$

∴ Distance travelled with uniform velocity,

$$S_2 = (780 - 180)m = 600 \text{ m}$$

Time taken for this distance, $t_2 = 600/12 = 50$ sec

The total time for one cycle of operation

= acc. + uniform speed + ret. + rest at top and bottom

$= (15 + 50.0 + 15 + 2\times 10.00)$ sec = 100 sec

∴ The number of cycles per shift, $n = (8\times 60\times 60/100) = 288$

Step-2: Mass lifted per trip

The ore lifted in each cycle $= \left(\frac{1800}{288}\right)$ Tonnes = 6.25 Tonnes = 6250 kg

∴ The ore per trip per skip $= 6250/2 = 3125$ kg

∴ The skip mass $= 0.6\times 3125$ kg $= 1875$ kg

∴ Total mass lifted $= 3125 + 1875 = 5000$ kg

Step-3: Selection of rope data

Choose 6×19 rope (Table 19.5 MDHB)

From Table 19.3 for 6×19 WRC, breaking load $F_u = 0.688d^2$ kN

Sheave diameter, $D = 34d$, Mass, $m = 0.43d^2$ kg/100 m $= 0.0043d^2$ kg/m and Elastic modulus, $E = 93$ GPa, Wire diameter $d_w = 0.045d$ (approx.)

The rope area $A = 0.38d^2$ (approx.)

Step-4: Permissible load

The factor of safety (Table 19.4 MDHB) = 6

The permissible load in the rope,

$$F = F_u/\text{FOS} = \frac{0.688d^2}{6.0} = 0.1147d^2\ \text{kN} = 114.7d^2\ \text{N}$$

Step-5: Loads on the rope

The service load,

$$F_s = (5{,}000 + 0.0043d^2 \times 780)\,\text{kg} = (5{,}000 + 3.354d^2)\,\text{kg}$$
$$= (5{,}000 + 3.354d^2) \times 9.81\,\text{N} = (49050 + 32.9d^2)\text{N}$$

The acceleration load,

$$F_a = (F_s/g) \times a = (5{,}000 + 3.354d^2) \times 0.8\ \text{N} = (4{,}000 + 2.682d^2)N$$

The bending load,

$$F_b = AE\,d_w/D = 0.38d^2 \times 93 \times 10^3 \times 0.045d/34d = 46.77d^2N$$

∴ Total load, $F = F_s + F_a + F_b$

$$F = (49050 + 32.9d^2) + (4{,}000 + 2{,}682d^2) + 46.77d^2$$
$$= (53050 + 82.355d^2)\ \text{N}$$

Step-6: Rope diameter

Equating total load to strength,

∴ $$114.7d^2 = 53050 + 82.355d^2$$

∴ $$d = \sqrt{53050/32.34} = \sqrt{1640.16} = 40.5\ \text{mm}$$

Choose the available diameter, $d = 42$ mm

Example 20.30: Select a suitable rope to lift a load of 10 kN of derbis from a well 60 m deep. The rope should have a factor of safety equal to 5. The weight of the bucket is 5 kN. The load is lifted with a maximum speed of 150 rpm, which is attained in 1.5 sec. There is an initial slackness of 250 mm.

SOLUTION

Step-1: Selection of rope data

From Table 19.3 MDHB, the data available for 6 × 7 FC are

$F_u = 0.608d^2$ kN, $M = 0.35d^2$ kg/100 m $= 0.0035d^2$ kg/m

Sheave diameter, $D = 42d$ and $E_r = 80.6$ GPa,

Assume wire diameter, $d_w = 0.45d$

Step-2: Rope diameter

The live load, $Q = 10 + 5 = 15$ kN

Then we can write, $F_u/\text{FOS} = Q$ or $F_u = Q \cdot FS$, with FOS = 15

$$\therefore \quad 0.608d^2 = 15 \times 15$$

$$d = \sqrt{\frac{15 \times 15}{0.608}} = 19.24 \text{ mm}$$

Select available diameter, d = 22 mm Table 3.79 (Table 19.11 MDHB)

Step-3: Rope data for, d = 22 mm

The ultimate strength, $F_u = 0.608 \times 22^2 \text{ kN} = 294.3 \text{ kN}$

The mass, $M = 0.0035d^2 = 0.0035 \times 22^2 = 1.69 \text{ kg/m}$

Sheave diameter $= 42 \times 22 = 924 \text{ mm}$

The cross-sectional area, $A = 0.38d^2 = 0.38 \times 22^2 = 184 \text{ mm}^2$

The mass of the rope $= 60 \times 1.69 = 101.4 \text{ kg}$

Step-4: Loads on the rope

$\therefore$ The service load, $F_s = 15 \text{ kN} + \dfrac{101.4 \times 9.81}{1000} = 15.995 \text{ kN}$

Acceleration, $a = (150/60)/1.5 = 1.67 \text{ m/s}^2$

$\therefore$ Inertia load (Eqn. 19.14), $F_a = (15.995/9.81) \times 1.67 = 2.72 \text{ kN}$

The staring load, (Eqn. 19.15 MDHB)

$$F_{st} = F_s\left[1 + \sqrt{1 + \frac{2a \cdot h \cdot EA}{LF_s g}}\right]$$

$$= 15.995\left[1 + \sqrt{1 + \frac{2 \times 1.67 \times 0.250 \times 80.6 \times 10^3 \times 184}{60 \times 9.81 \times 15995}}\right] = 40.28 \text{ kN}$$

Bending load, (Eqn. 19.12 MDHB)

$$F_b = \frac{AEd_w}{D} = \frac{184 \times 80.6 \times 10^3 \times 0.045d}{42d} = 15.889 \text{ kN}$$

Total load on the rope (Eqn. 19.16 MDHB)

Normal working load, $F = F_s + F_b = 15.995 + 15.889 = 31.885 \text{ kN}$

During acceleration, $F = F_s + F_b = 5.995 + 15.889 + 2.72 = 34.6 \text{ kN}$

During starting, $F = F_{st} + F_b = 40.25 + 15.889 = 56.139 \text{ kN}$

Maximum, F = 56.139 kN

Step-5: The actual factor of safety,

$$F_u/F = 294.3/56.139 = 5.24$$

This is more than the specified value of 5 from Table 19.4 MDHB and hence, acceptable.

Example 20.31: Select a wire rope to lift a load of 10 kN through height of 600 m from a mine. The weight of bucket is 2.5 kN. The load should attain a maximum speed of 50 m/min in 2 seconds.

Given Data: Load lifted = 10 kN, Height = 600 m, Hoist speed = 50 m/min,

Period of acceleration = 2 sec

SOLUTION

Step-1: Selection of rope data

From Table 19.5 MDHB, select type of rope as 6 × 19

$$F_u = 500.0d^2 N \text{ where } d \text{ is in mm}$$

Mass per metre length, $m = 0.0038\,d^2$ kg/m

Wire diameter, $d_w = 0.063d$, mm

Area of c/s, $A = 0.38\,d^2$, mm^2

Sheave diameter, $D = 45d$, mm

FOS = n = 5 (assumed)

Total load lifted,

$$W = \text{Load to be lifted} + \text{weight of skip} = 10 + 2.5 = 12.5 \text{ kN}$$

Total mass of rope, M_R = Mass per meter length × length of rope

$$= 0.0038\,d^2 \times 600 = 2.28\,d^2 \text{ kg}$$

Step-2: Acceleration,

$$v = u + at = at \text{ Since } u = 0$$

$$\therefore \quad a = \frac{v}{t} = \frac{50}{60 \times 2} = 0.417 \text{ m/sec}^2$$

Step-3: Load due to acceleration,

$$\therefore \quad F_a = \left(\frac{12500}{9.81} + 2.28d^2\right) \times 0.417 = (531.35 + 0.951d^2)N$$

Step-4: Bending load, $F_b = \dfrac{AEd_w}{D}$ (Eqn. 19.12 MDHB)

where, $E = 0.85 \times 10^5 \text{ MPa}, \; A = 0.38\,d^2, \; d_w = 0.063\,d, \; D = 45d$

$$\therefore \quad F_b = [0.38\,d^2 \times 0.85 \times 10^5 \times 0.063.d]/(45\,d) = 45.22\,d^2\text{N}$$

Step-5: Service load, F_s = Live load + wt of cage + wt of rope

$$= 10000 + 2500 + 9.81 \times 2.24\,d^2 = (12500 + 22.37d^2)\text{ N}$$

Step-6: Starting load,

$$F_{st} = F_s\left[1+\sqrt{1+\frac{2ahEA}{L\cdot F_s g}}\right] \simeq 2F_s \text{ (impact neglected)}$$

$$= 25000 + 44.74d^2$$

Step-7: The total load during starting,

$$F = F_{st} + F_b$$

$$= 25000 + 44.74d^2 + 45.22d^2 = 25000 + 89.96d^2$$

Step-8: Total load during acceleration,

$$F = F_s + F_a + F_b$$

$$= 12500 + 22.37d^2 + 531.35 + 0.951d^2 + 45.22d^2$$

$$= 13031.35 + 68.5431d^2$$

Step-9: Maximum load

The max. load occurs during starting, $F_{max} = 25000 + 89.96d^2$

Step-10: Diameter of the rope

Equating the maximum load to allowable load $\frac{500d^2}{5} = 25000 + 89.96d^2$

$$d = \sqrt{\left(\frac{25000}{10.04}\right)} = 49.9 \text{ mm}$$

Choose the next value, $d = 51$ mm (From Table 19.7b)

Example 20.32: Select a 6× 19 steel rope to lift 15 kW of debris from a tunnel 200 m deep. The bucket weighs 8 kN. The velocity of the rope is 100 m/min to be attained in 20 seconds. What will be the maximum load on the rope when there is a slack of 10 m in the rope?

Given Data: 6× 19 steel rope; Load to be lifted $= 15 \text{ kN} = 15\times10^3 \text{ N}$

$H = 200$ m; Weight of bucket $8 \text{ kN} = 8000$ N;

$$V = 10 \text{ m/min} = \frac{100}{60} = 1.667 \text{ m/sec}; t = 20 \text{ seconds}$$

Slack $h = 10$ m

SOLUTION

Step-1: Selection of rope data

From Table 19.5 MDHB, for 6× 19 rope, $F_u = 500.0d^2$ N where d is in mm

Mass per metre length $= 0.0038d^2$ kg/m

Wire diameter, $d_w = 0.063d$, mm

Area of cross section, $A = 0.38d^2$
Average sheave diameter, $D = 45d$, mm
From Table 19.4 MDHB for 200 m depth, select FOS = 7

Step-2: Service load

Live load = Load to be lifted + Weight of the bucket

$$= 15000 + 8000 = 23000 \text{ N}$$

Weight of rope per metre $W_{R/m}$ = mass per meter 9.81

$$= .0038d^2 \times 9.81 = 0.03728d^2$$

$$W_r = \text{Weight per meter length} \times \text{length of rope}$$

$$= 0.03728d^2 \times (200+10) = 7.828d^2 \text{ } N$$

Service load, F_s = Live load + weight of rope

$$= W + W_r = 23000 + 7.828d^2$$

Step-3: Inertial load due to acceleration

Velocity, $v = u + at$

i.e.,
$$1.667 = 0 + a \times 20$$

$$\therefore \text{ Acceleration, } a = \frac{1.667}{20} = 0.0833 \text{ m/sec}^2$$

$$F_a = \frac{F_s}{g} = \left(\frac{23000 + 7.828d^2}{9.81}\right)(0.0833) = (195.3 + 0.0665d^2)N$$

Step-4: Bending load,

$$F_b = \frac{AEd_w}{D}, \text{ where } A = 0.38d^2, E = 0.85 \times 10^5 \qquad \text{(Eqn. 19.12 MDHB)}$$

$$= (0.38d^2 \times 0.85 \times 10^5 \times 0.063d)/45d = 45.22d^2 N$$

Step-5: Starting load

Neglecting slack:

$$F_{st} = 2 \cdot F_s = 2 \times [23000 + 7.828d^2] = 46000 + 15.656d^2$$

Step-6: Maximum load

Load during uniform velocity, $F = F_s + F_b$

$$= (23000 + 7.828d^2) + 45.22d^2 = (23000 + 53.048d^2)\,N$$

Load during acceleration,

$$F = F_s + F_a + F_b = 23000 + 7.828d^2 + 195.3 + 0.0665d^2 + 45.22d^2$$

Load during starting, $F = F_{st} + F_b$

$$F = 46000 + 15.656d^2 + 45.22d^2 = 46000 + 60.876d^2$$

The maximum value occurs at the start,

$$F_{max} = 46000 + 60.876d^2$$

i.e.,
$$\frac{500\,d^2}{7} \geq 46000 + 60.876d^2$$

$\therefore$
$$d \geq 66.0, \text{ choose } d = 70 \text{ mm}$$

Step-7: Considering slack, max. load is evaluated as follows:

Taking diameter of rope, $d = 70$ mm, $A = 0.38d^2 = 1862$ mm

Service load, $F_s = 23000 + 7.828 \times 70^2 = 61357.2$ N

(i) Considering slack, starting load,

$$F_{st} = F_s\left[1 + \sqrt{1 + \frac{2ahEA}{LF_s g}}\right] \quad \text{(Eqn. 19.15 MDHB)}$$

$$= 61357.2 \times \left[1 + \sqrt{1 + \frac{2 \times 0.0833 \times 10 \times 0.85 \times 10^5 \times 1862}{200 \times 61357.2 \times 9.81}}\right] = 61357.2 \times 2.763 = 169529.93 \text{ N}$$

Load during starting, $F = F_{st} + F_b$

$$= 169529.93 + 45.22 \times 70^2 = 391107.94 \text{ N}$$

Step-8: Checking for FOS

Breaking strength of the rope,

$$F_u = 500d^2 = 500 \times 70^2 = 2450000 \text{ N}$$

Maximum load with 10 m slack, $F_{max} = 391107.94$ N,

$$\text{FOS } n = \frac{F_u}{F_{max}} = 2450000/391107.94 = 6.264$$

This is nearer to '7' and is satisfactory

Hence, diameter of rope $d = 70$ mm is acceptable.

Example 20.33: A 25 mm 6×37 steel wire rope is used in a mine of 80 m depth. The velocity of the cage is 2 m/sec and the time required to accelerate the cage to the desired velocity is 10 sec. The diameter of the drum is 1.25 m. Determine the safe load that the hoist can handle by assuming a factor of safety as 8. Neglect the impact load on the rope.

Given Data: $d = 25$ mm; Type of steel wire rope 6×37; $H = 80$ m;

$V = 2$ m/sec; $t = 10$ sec.; $D = 1.25$ m $= 1250$ mm; FOS $n = 8$

SOLUTION

Step-1: Selection of rope data

From Table 19.5 for 6×37 rope,

$$F_u = 475d^2 = 475 \times 25^2 = 296875 \text{ N}$$

Mass per metre length of rope = $0.0037\,d^2$ kg/m = $0.0037 \times (25)^2 = 2.3125$ kg/m

Weight per metre, $w = m \times g = 2.3125 \times 9.81 \text{ N} = 22.686 \text{ N}$

Wire diameter, $d_w = 0.045d = 0.045 \times 25 = 1.125$ mm

Area of cross section, $A = 0.38\,d^2 = 0.38 \times 25^2 = 237.5 \text{ mm}^2$

Average sheave diameter, D = 1250 mm (given)

Step-2: Service load

Total load to be lifted = W

Total weight of rope W_R = Weight per meter length × Length of rope = $22.686 \times 80 = 1815$ N

Service load, $F_s = W + W_R = W + 1815$

Step-3: Inertia load due to acceleration,

$$F_a = \frac{(W + W_R)}{g} a = \left(\frac{W + 1815}{9.81}\right) \times 0.2 = 0.0204\,W + 37$$

Step-4: Bending load,

$$F_b = \frac{AEd_w}{D} = \frac{23.75 \times 0.85 \times 10^5 \times 1.125}{1250} = 18169 \text{ N} \quad \text{(Eqn. 19.12 MDHB)}$$

Step-5: Maximum load

Neglecting impact,

$$\begin{aligned} F_{\max} &= W + W_R + F_a + F_b \\ &= W + 1815 + 0.0204W + 37 + 18169 \\ &= 1.0204W + 20021 \end{aligned}$$

Now, $F_a \geq F_{\max}$,

$$\frac{F_u}{n} \geq F_{\max}$$

i.e.,

$$\frac{29687.5}{8} \geq 1.0204\,W + 20021$$

i.e.,

$$W = \frac{37109.4 - 20021}{1.0204} = 16746 \text{ N}$$

Example 20.34: Select a wire rope for a vertical mine hoist to lift a load of 55 kN from a depth of 300 metres. A rope speed of 500 m/min is to be attained in 10 seconds.

Given Data: Load lifted = 55 kN, Depth = 300 m,
Speed of 500 m/min, acceleration time = 10 sec.

SOLUTION

Step-1: Selection of rope data

From Table 19.5 MDHB, select the most commonly used type of rope, i.e., 6 × 19.

From Table 19.5 for 6×19 rope,

Ultimate strength, $F_u = 500d^2$ where d is in millimeters and F_u is N,

Weight per meter length $= 9.81 \times 0.0037\, d^2$ N/m $= 0.03630\, d^2$ N/m

Wire diameter, $d_w = 0.063d$ where d in mm

Area of cross section, $A = 0.38\, d^2$

Average sheave diameter, $D = 45\, d$ where d in mm

From Table 19.4 MDHB, for 300 m depth,

Select

Factor of safety, $n = 7$

Step-2: Service load

Live load, $W = 55$ kN $= 55000$ N

Total weight of rope

$W_R =$ Weight per meter length × Length of rope

$= 0.03630 \times d^2 \times 300 = 10.89\, d^2$

Service load, $F_s = W + W_R = 55000 + 10.89d^2$

Step-3: Inertia load

Acceleration, $a = \dfrac{V}{t} = \dfrac{500}{60 \times 10} = 0.833$ m/sec^2

Inertia load due to acceleration, $F_a = Ma = F_a = \dfrac{F_s}{g} a$

$$= \frac{(55000 + 10.89d^2)}{9.81} \times 0.833 = 4670.234 + 0.9247d^2$$

Step-4: Bending load,

$$F_b = EA \cdot \frac{d_w}{D} = \frac{85 \times 10^5 \times 0.38d^2 \times 0.063d}{45d} = 45.22d^2 \quad \text{(Eqn. 19.12 MDHB)}$$

Step-5: Starting load

Neglecting shock,

$$F_a = 2(W + W_R) = 2[55000 + 10.89\, d^2] = 11 \times 10^4 + 21.78\, d^2$$

Step-6: Maximum load

Assume impact load is neglected

$\therefore$ Maximum load on the rope, $F_{max} = F_s + F_a + F_b$

$$= 55000 + 10.89 d^2 + 4670.234 + 0.9247 d^2 + 45.22 d^2$$

$$= 59670.234 + 57.035 d^2$$

Step-7: Diameter of rope

The allowable load $\geq$ Maximum load,

i.e., $$\frac{F_u}{n} \geq F_{max}$$

Substituting, $$\frac{500 \cdot d^2}{7} \geq 59670.234 + 57.035 d^2$$

Solving, $d \geq 64.4$

Choose diameter of rope $= 65$ mm

EXERCISES FOR PRACTICE

20.1 (a) What are the various types of belts and on which factors do their selection depend?

(b) An open belt drive connects two pulleys *A* of 1250 mm diameter and *B* of 625 mm diameter on parallel shafts 4.5 metres apart. The belt weighs 0.893 kg per metre and the tensile strength is 2.5 MPa. The value of friction coefficient is 0.3. The driving pulley *A* runs at 220 rpm. Calculate torque on each of the shafts and the power transmitted.

20.2 (a) What are the common belt materials?

(b) A belt is required to transmit 20 kW of power from a pulley of 900 mm diameter, running at 250 rpm to another pulley running at 450 rpm. The centre distance is 2500 mm. The coefficient of friction for the pulley and belt is 0.25. The permissible stress for the belt material is 1.75 MPa and the mass density is 1000 kg/ m^3. The belt thickness is 10 mm. Find the width of the belt and design the pulley.

20.3 (a) What are the various types of flat belt drives and what are their applications?

(b) A leather belt, 125 mm wide and 6 mm thick, transmits power from a 750 mm pulley running at 500 rpm. The angle of lap is 150° and friction coefficient is 0.3. The mass density of belt material is 1.0 mg per cubic metre and the tensile strength is 2.75 MPa. Find the max. power that can be transmitted.

20.4 (a) Derive the equation for the effective friction factor in *V*-belt drives.

(b) A compressor requiring 50 kW at 250 rpm is driven by an electric motor running at 760 rpm through a *V*-belt drive. The centre distance between the pulleys is 1.75 metres. The max. dia. of the pulley on the compressor shaft is to be 1.2 metres. The belt speed should not exceed 30 m/s. Determine the number of belts if each belt has a cross-sectional area of 375 mm^2 and mass density of 0.001 gm per cubic millimetre. The permissible tensile strength is 2.5 MPa. The groove angle is 35° and coefficient of friction is 0.25.

20.5 (a) Explain the merits and demerits of *V*-belt drives.

(b) A 30 kW, 1440 rpm motor is to drive a compressor by means of *V*-belts. The diameters of the pulleys are 220 mm and 750 mm. The centre between the pulleys is 400 mm. Design (i) motor shaft, (ii) number of belts and (iii) the pulley.

20.6 (a) Explain the merits and demerits of chain drives.

(b) Design roller chain drive to transmit 150 kW from an engine running at 500 rpm to a generator running at 960 rpm. Determine the pitch and number of strands of the chain, and the sprocket size.

20.7 (a) How are the chains classified? Explain each type briefly.

(b) A 15 kW, 1440 rpm motor drives a line shaft at 300 rpm.

The shaft centre is to be about 500 mm. The motor shaft diameter is 40 mm. The starting torque is 2.5 times the running torque. The load is applied with moderate shock. Select the roller chain for the drive and determine (i) the sprocket size, (ii) the chain length, and (iii) the exact centre distance.

20.8 (a) Explain the chordal action in chain drives.

(b) Design a chain drive to run a blower at 600 rpm. The power is available from a 10 kW motor at 1440 rpm. The centre distance is to be 1200 mm approximately.

20.9 (a) Explain the wire rope constructions with neat sketches.

(b) A 6 × 19 wire rope of steel is used in a lift carrying maximum of 10 persons of average mass of 60 kg. An overload of 75% allows for cabin and inertia loads. The rope has four falls. Design the rope and the sheave.

20.10 (a) What are the different types of stresses in a wire rope used in hoisting?

(b) Select a wire rope for a vertical mine hoist to lift 1800 tonnes of ore in each 8-hour shift from a depth of 900 m. Assume a two-compartment shaft with hoisting skips in balance The max. velocity is 15 m/s with acceleration and deceleration of 15 sec each and a rest period of 10 sec. for discharging and loading of the skips. The hoisting skip weighs 0.6 of its load capacity. The factor of safety is to about 6.

20.11 (a) Select a *V*-belt drive to transmit a power of 6 kW from a shaft rotating at 1500 rpm to a parallel shaft to be run at 375 rpm. The distance between the shaft centres is 500 mm. The pitch diameter of the smaller grooved pulley can be taken to be 150 mm. The factor of application is to be taken as 1.2.

(b) Select a wire rope to lift a load of 10 kN by 200 m. The desired velocity of 23.20 m/min is to be achieved while travelling through a distance of 10 m.

20.12 (a) Select a *V*-belt drive to transmit a load of 6 kW from a shaft rotating at 1000 rpm to a parallel shaft to be rotated at 350 rpm. The space limits the center distance between shafts to 500 mm. The pitch diameter of the smaller pulley could be assumed to be 150 mm.

(b) Select a suitable wire rope of a standard strand to raise a load of 10 kN through 400 m. The load has to achieve a desired linear speed of 20 m/mm while traversing through a distance of 15 m from the start.

20.13 (a) Show that in flat belt drives, the ratio of belt tensions is given by $= \frac{T_1}{T_2} = e^{\mu\theta}$, where T_1 and T_2 are belt tensions, μ is the coefficient of friction and θ is the angle of wrap.

(b) Select the type and number of *V*-belts required to drive a crusher, which works 8 hours a day. The power transmitted is 65 kW. The motor shaft runs at 900 rpm and carries a pulley of 250 mm diameter; the crusher shaft rotates at 300 rpm and the centre distance is 700 mm. Determine the pitch length of the belt required.

20.14 (a) Select a *V*-belt drive to transmit 10 kW of power from a pulley of 200 mm diameter mounted on an electric motor running at 720 rpm to another pulley mounted on compressor running at 200 rpm. The approximate centre distance between the two pulleys is 600 mm. The correction factor for service is 1.3. Find the number of belts and the correct centre distance.

(b) Select a suitable wire rope of a standard strand to lift a load of 10 kN through a height of 600 m from a mine. The weight of bucket is 2.5 kN. The load should attain a maximum speed of 50 m/min in 2 seconds.

20.15 (a) Derive the equation $\frac{T_1}{T_2} = e^{\mu\theta}$, where T_1 = tension in the belt on the tight side, T_2 = tension in the belt on the slack side, μ = the coefficient of friction between the belt and pulley and θ = Angle of contact in radians.

(b) A compressor, requiring 85 kW, is to run at 250 rpm. The drive is by *V*-belts from an electric motor running at 800 rpm. The diameter of the pulley on the compressor shaft must not be greater than 1.0 m while centre distance between the pulleys is limited to 1.8 m. The belt speed should not exceed 1500 m/min. Determine the number of *V*-belts required to transmit the power if each belt has a cross-sectional area of 360 mm^2, density 1000 kg/m^3 and an allowable tensile stress of 2.5 N/mm^2. The groove angle of pulley is 35°. The coefficient of friction between the belt and the pulley is 0.25.

20.16 (a) Derive an expression for the ratio of tension in *V*-belt drive.

(b) Two shafts one metre apart are connected by a *V*-belt drive to transmit 90 kW at 1200 rpm of a driver pulley of 300 mm effective diameter. The driver pulley rotates at 400 rpm. The angle of groove is 40° and coefficient of friction between the belt and the pulley rim is 0.25. Area of the belt section is 400 mm^2 and the permissible stress is 2.1 MPa, and density of belt material = 1100 kg/ mm^2. Determine the number of belts required and the length of the belt.

20.17 (a) A 25 mm diameter 6×37 steel rope is used in a mine 80 m deep. The velocity of the cage is 2 m/sec and the time required to accelerate the cage to the desired velocity is 10 seconds. The diameter of the drum is 1.25 m. Determine the safe load that the hoist can handle by assuming a factor of safety as 8.

20.18 (a) A flat belt drive is required to transmit 20 kW of power at 1440 rpm. The driver and driven shafts are approximately 3.0 m apart. The dimensions of pulley mounted on driver and driven shafts are 300 mm and 450 mm respectively. The weight density of selected belt material is 9.7×10^4 N/m^3, allowable design stress, $\sigma_d = 2$ MPa and coefficient of friction = 0.3.

Determine:

(i) Width of the belt if thickness = 6 mm

(ii) Length of the belt

(iii) Initial tension in the belt

(iv) Centrifugal tension in the belt.

(b) Select type of *V*-belt and number of belts required for 10 kW, 750 rpm induction motor to drive an exhaust fan in a steel plant at 250 rpm. The minimum centre distance between shafts is 1.2 m. Pitch diameter of motor pulley is 200 mm.

20.19 Design a chain drive to actuate a compressor from a 10 kW motor running at 960 rpm. The compressor speed is to be 320 rpm. The centre distance should not be less than 500 mm. The compressor operates in two shifts of 8 hours each. The chain tension can be adjusted by shifting the motor on the slides. Give the complete details of the drive.

20.20 Select a wire rope for a vertical mine hoist to lift 25000 kN of ore in each 8 hours shift from a depth of 1200 m. Assume a two-compartment shaft with hoisting skips in balance. Use maximum velocity of 900 m/min with acceleration and retardation periods of 15 sec each and rest period of 15 sec for discharging and loading the skips. The hoisting skip weighs approximately 0.6 of its load capacity.

SHORT ANSWER QUESTIONS

20.1 What are flexible drives?

20.2 List the advantages and disadvantages of flexible drives.

20.3 What are different types of belt drives used in transmitting power?

20.4 What is the effect of centrifugal tension on a power transmitted by a belt?

20.5 What are the advantages of flat belt over V-belt drive?

20.6 What is the length of a open belt drive with pulleys separated by a center distance ‘C’?

20.7 What is the length of crossed v-belt drive with pulleys separated by a center distance ‘C’?

20.8 What is the optimal belt velocity for maximum power transmission?

20.9 What is the range of coefficient of friction for cast iron or steel pulley?

20.10 What is the range of correction factors used considering service condition?

20.11 What load factors are used in steady and shock loads?

20.12 What is crowing of belt pulley?

20.13 How are the V-belts specified?

20.14 On what basis is the number of V-belts to be used decided?

20.15 What are advantages of a chain drive over belt drive?

20.16 List five disadvantages of chain drives.

20.17 Distinguish between single, duplex and triplex roller chains.

20.18 Name the modes in which chain drives can fail.

20.19 Give a comparison between a chain and rope drive?

20.20 What is effort required to lower the load in a rope drive?

20.21 What is range of factor of safety recommended for passenger elevators?

CHAPTER 21

BEARINGS AND LUBRICATION

21.1 INTRODUCTION

Whenever two parts have relative motion, they constitute a bearing, regardless of their shape or configuration. There are two types of bearings: (i) plain (sliding) bearings and (ii) antifriction bearings.

Plain (Sliding) Bearings

A plain bearing is the one formed by two elements, one sliding on the other. Examples are a sleeve around a shaft or a flat surface under a slider. The transmission shafts, links and pins are of the first category and machine tool table guides are of the second category. The moving parts like shaft is made of steel or cast-iron to have the required strength and hardness. The other part, called bearing, is made of softer materials like bronze, babbitt or a non-metallic polymer. This may be in the form of a complete bush or split bush mounted in the housing for support. This is also called journal bearing.

Another type of plain bearing is thrust bearing, wherein the bearing surface is floating flat or conical ring. The shaft has integral collars or conical end.

Antifriction bearing has hardened steel balls or rollers held between hardened steel-race ways. These bearings are designed and manufactured to support radial or axial or a combination of both the loads. They are selected from the manufacturer's catalogue to suit a particular application.

We shall discuss the sliding contact or plain bearing in this chapter and the rolling contact bearings in the next chapter.

The sliding bearings consist of two parts, one of hard material like steel and the other one as soft and wear resistant material like bronzes and brasses. The two are separated by a lubricant.

21.2 MATERIALS FOR BEARINGS

The characteristics of the bearing material are:

1. Relatively soft
2. Reasonably strong
3. Machinability
4. Lubricity
5. Corrosion resistant
6. Porosity
7. High fatigue strength

8. High thermal conductivity
9. Anti-weld and anti-scoring property

Some of the materials satisfying these requirements are as follows.

Bronze: These are copper alloys like leaded bronze, tin bronze, aluminium bronze and beryllium copper. The bronzes are well suited to work against ferrous materials. They have strength, good machinability and corrosion resistance.

Babbitts: The family of alloys based on lead and tin in combination with other elements are known as babbitts. They are good for hydrodynamic or hydrostatic lubrication. They are used as bushes or electroplated thin film on steel surface.

Gray Cast-Iron: The free graphite in C.I. imparts lubricity. They are used in the bush form and used for low speeds.

Cintered materials: These are made from metal powder and are microscopically porous. The porosity helps them to absorb the lubricant and release during the operation.

Non-metallic material: These materials are self-lubricating and require no additional lubricant. The graphite, thermoplastics like nylon, acetal or teflon are some of the non metals used as bearing materials. These can operate well only at low speeds and loads. Table 21.1 shows the usable combination of shaft and bearings.

21.3 LUBRICANTS

They are materials with low shear strength between the intermolecular layers. The layers in contact with the two parts get adhered to them. The intermediate layers get sheared. It becomes the weakest material in the interface and reduces the coefficient of friction.

The functions of lubricants are:

(i) To reduce the friction and wear.

(ii) To carry away the heat generated due to friction.

(iii) To protect the journal and vary from corrosions.

The lubricants may be gaseous, liquid or solid. The liquid and solid lubricants are more common. Some of the important types are listed in Tables 21.2 and 21.3. The petroleum based oils are the most widely used liquid lubricants. The graphite and phosphate coatings are the commonly used solid lubricants.

Table 21.1. Recommended bearing materials for sliding against steel or cast iron

Bearing Material	Hardness, kg/m^2	Minimum Shaft Hardness, kg/mm^2	Hardness Ratio
Lead-based babbitt	15-20	150	8
Tin-based babbitt	20-30	150	6
Alkali-hardened lead	22-26	200-250	9

Contd.

Copper-lead	20-23	300	14
Silver (overplated)	25-50	300	8
Cadmium base	30-40	200-250	6
Aluminium alloy	45-50	300	6
Lead bronze	40-80	300	5
Tin bronze	60-80	300-400	5

VISCOSITY: Viscosity is the most important property of the lubricant. It is the measure of a fluid's resistance to internal shear. The coefficient of viscosity is given by the expression,

$$\mu' = \frac{\tau}{\left(\frac{du}{dy}\right)}$$

μ' is called absolute viscosity measured in Ns/m^2 or kg/ms.

$$\tau = \text{Shear stress in the layer N/m}^2,$$
$$u = \text{Velocity of sliding surface in m/s}$$

Expressed in Nmm-S

μ is expressed in N-s/mm^2 or MPa-s

$(\mu')\,\text{kg/ms} = (\mu')\text{N-s/m}^2 = 10^{-6}\,(\mu)\,\text{N-s/mm}^2$

$Z(\text{CP}) =$ Viscosity in centipoises

$$\frac{Z(\text{POISE})}{100} = \frac{1}{100}\frac{\text{dyn-se}}{\text{cm}^2} = \frac{1}{100}\left(\frac{N}{10^2}\right)\left(\frac{\text{S}}{10^2\,\text{mm}^2}\right)$$

$$= \frac{1}{10^9}\frac{N-S}{\text{mm}^2} = 10^{-9}\,\text{MPa}-\text{s}$$

$$\therefore \qquad \mu = Z/10^9$$

The kinematic viscosity $Z_K = \left(\frac{z}{\rho}\right)$ in stokes

Table 21.2. Types of liquid lubricants

Type	Properties	Typical Uses
Petroleum oils (mineral oils)	Fair lubrication ability, but additives produce great improvement. Poor lubrication action at high temperatures.	Very wide and general
Polyglycols	Quite good lubricants; do not form sludge on oxidizing.	Brake fluid

Contd.

Silicones	Poor lubrication ability, especially against steel. Good thermal stability.	Rubber seals, mechanical dampers
Chlorofluorocarbon	Good lubricants; good thermal stability.	Oxygen compressors, chemical processing equipment.
Polyphenyl ethers	Very wide liquid range; excellent thermal stability; fair lubricating ability.	High temperature sliding systems
Phosphate esters	Good lubricants—EP action	Hydraulic fluid lubricant
Dibasic esters	Good lubricating properties; can stand higher temperatures than mineral oils.	Jet engines.

Table 21.3. Types of solid film lubricants

Type	Properties	Typical uses
Graphite and/or MoS_2 + binder	Best general-purpose lubricants. Low friction (0.12-0.06), reasonably long life (10^4-10^6 cycles)	Locks and other intermittent mechanisms.
Teflon + binder	Life not quite as long as previous type, but resistance to some liquids better.	As above
Rubbed graphite or MoS_2 film	Friction very low (0.10-0.04), but life quite short (10^2) cycles	Deep drawing and other metal working operations.
Soft metal (lead, indium, cadmium)	Friction higher (0.30-0.15) and life not as long as resin bonded types	Running-in protection (temporary)
Phosphate, anodized films, other chemical coatings.	Friction is high (~ 0.20). Galling preventatives leave "spongy" surface layer.	Undercoating for resin-bonded film.

21.4 THEORY OF LUBRICATION

Lubrication can be classified as thin film lubrication and thick film lubrication.

Thin film lubrication occurs when the layer of the lubricant is very thin. The physical contact between the surfaces is barely prevented. The inter molecular forces are in action. This is a case of boundary lubrication.

Thick film lubrication occurs when the surfaces are separated by a thick layer of lubricant. The thick film lubrication are classified as: (1) hydrodynamic lubrication and (2) hydrostatic lubrication.

In hydrodynamic lubrication the thick film lubrication is created by rotation of the shaft, the lubricant is kept in the confined wedge-shaped space between the moving part (shaft) and stationary bearing surface. The parabolic pressure created in the film of oil is responsible to keep the moving surface floating on the film of oil.

In hydrostatic lubrication, the thick film of oil between the surfaces is created by the supply of pressured fluid by an external pump.

The lubrication in a journal bearing, as the shaft speeds of from rest to full speed, passes through three types of lubrications, namely:

(i) Boundary lubrication
(ii) Mixed film lubrication
(iii) Full film lubrication.

This is illustrated in the figures 21.1a, 21.1b, and 21.1c.

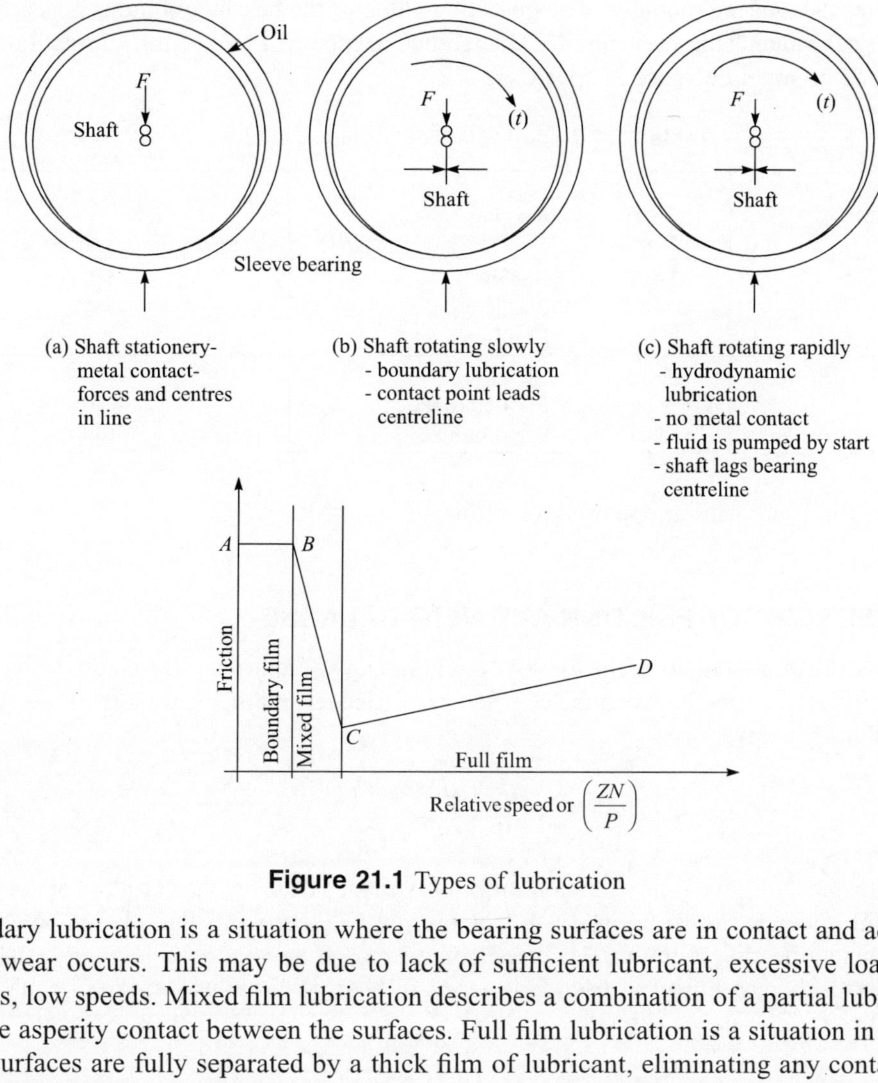

Figure 21.1 Types of lubrication

Boundary lubrication is a situation where the bearing surfaces are in contact and adhesive or abrasive wear occurs. This may be due to lack of sufficient lubricant, excessive load, surface roughness, low speeds. Mixed film lubrication describes a combination of a partial lubricant film plus some asperity contact between the surfaces. Full film lubrication is a situation in which the bearing surfaces are fully separated by a thick film of lubricant, eliminating any contact.

Figure 21.1 shows the relation between the friction and sliding speed in bearing. At slow speeds, boundary lubrication occurs. It is characterized by the high friction (*AB*).

As the sliding speed is increased beyond point *B*, a hydrodynamic fluid-film begins to form, reducing the asperity contact. This is the mixed film lubrication (*BC*). The friction is rapidly reduced upto *C*. At higher speeds beyond *C*, a full film is formed between the surfaces and pressure is developed in the film (*CD*). A lateral load on the floating member can be supported. In a rotating journal bearing, all three regimes are experienced during start-up and shut-down. As the shaft begins to turn, it is in boundary lubrication. As the speed is increased, it passes through the mixed lubrication to full film lubrication. At the operating speeds, the full film lubrication exists. The *X*-axis of the Fig. 21.1, can be converted to dimensionless factor $\left(\frac{ZN}{p}\right)$ or $\left(\frac{\mu N}{p}\right)$ which is called bearing modulus. The bearing modulus corresponding to minimum friction (*C*) is called the critical bearing modulus. The operating value of the bearing modulus is nearly three times the critical value. This prevents the chance occurrence of mixed film lubrication due to increase in load or small decrease in speed.

Table 21.4. Critical values of bearing modulus

Shaft	Bearing	$C = \left(\frac{ZN}{p}\right)$
Hardened ground	Babbit	2540
	Plastic bronze	4260
	Rigid bronze	5700
Machined soft steel	Babbit	3550
	Plastic bronze	4980
	Rigid bronze	7100

Note: For the fluid lubrication, operating modulus $\left(\frac{ZN}{p}\right)_0 \cong 3C$.

21.5 COEFFICIENT OF FRICTION AND HEAT BALANCE

For light loads, the journal floats in the film of oil without any eccentricity. Based on the Newton's law of viscosity, the relation between the coefficient of friction and the bearing modulus for light load due to Petroff is given by,

$$f = \left(\frac{\pi^2}{30}\right)\left(\frac{\mu N}{p}\right)\left(\frac{r}{c}\right) \tag{21.1}$$

Considering the hydrodynamic behaviour of heavily loaded bearing, the variation of coefficient of friction with speed is shown in Fig. 21.1. The portion *CD* represents the relation between the friction coefficient bearing modulus. Mackee Brothers, based on experimental work established an expression for the coefficient of friction in terms of bearing modulus. Mackee's equation for friction coefficient for heavy load,

$$f = 0.326\left(\frac{\mu N}{p}\right)\left(\frac{r}{c}\right) + \Delta f \tag{21.2}$$

$$f = K_a\left(\frac{ZN}{p}\right)\left(\frac{r}{c}\right) \times 10^{-10} + \Delta f \tag{21.2a}$$

The value of Δf is a function of L/d as shown in Fig. 21.2. In the usual range of L/d = 0.75 to 2.8, the average value of $\Delta f = 0.002$ is used.

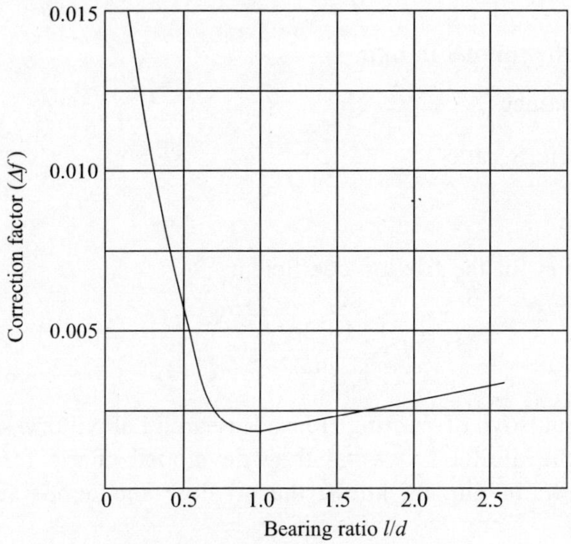

Figure 21.2 Correction Factor

When the bearings are partially lubricated, the coefficient of friction, given by Louis Illmer,

$$f = \frac{C_1 C_2}{97.3} \sqrt[4]{\left(\frac{p}{V}\right)} \tag{21.3}$$

C_1 = constant from Table 21.6

C_2 = constant from Table 21.7

Raynold Theory of Lubrication

The theoretical analysis of fluid film lubrications in bearings was done by Sir Osborne Raynold. He developed the following differential equation to describe the behaviour of the fluid film in bearing.

$$\frac{\partial}{\partial x}\left[h^3 \frac{\partial p}{\partial x}\right] + \frac{\partial}{\partial Z}\left(h^3 \frac{\partial p}{\partial Z}\right) = 6\mu \cup \left(\frac{\partial h}{\partial x}\right) \tag{21.4}$$

The solution for this equation,

$$S = \frac{(2+\epsilon^2)(1+\epsilon^2)^{\frac{1}{2}}}{12\pi^2 \epsilon} \tag{21.5}$$

where, Sommerfeld number $\qquad S = \left(\frac{r}{c}\right)^2 \left(\frac{\mu N_s}{p}\right)$

where, μ = Viscosity of the lubricant in Ns/mm^2 (MPa-s)

$N_s =$ Speed in rps

$p =$ Bearing pressure in MPa $= \dfrac{W}{Ld}$

$r =$ Radius of the journal in mm

$c =$ Radial clearance

$\in = \dfrac{e}{c} =$ Eccentricity ratio

$e =$ Eccentricity

The Sommerfeld solution for the friction coefficient,

$$f = \left[\frac{1+2\in^2}{3\in}\right]\left(\frac{c}{r}\right) \tag{21.6}$$

A. A. Raimondi and John Boyd of Wasting House Research Laboratory, solved the differential equation and based on the digital solutions, they developed charts for various parameters like fiction variable, minimum film thickness, the oil flow and temperature rise in terms of Sommerfeld number.

Using the chart (Fig. 21.5) the coefficient of friction can be evaluated by the relations,

$$f = \left(\frac{c}{r}\right)(FV) \tag{21.7}$$

For large values of Sommerfeld number, the coefficient of friction from the charts is nearly equal to the values obtained from Petroff equation and Mackee's equation.

The heat generated due to friction is given by

$$H_g = fWV_s = fW\frac{(\pi DN)}{60{,}000}\ \text{J/Sec.} = \frac{N\cdot f\cdot W\cdot r}{9550}\cdot \text{J/sec} \tag{21.8}$$

The heat dissipated by the bearing due to radiation is evaluated by the following empirical relations,

(a) Heat dissipated from the bearing based on area of housing:

$$H_d = K_1A(t_B - t_A) \geq H_g \tag{21.9}$$

$$K_1 = 11.4\times10^{-6}\ \text{W mm}^2\text{/K for still air}$$

$$= 15.3\times10^{-6}\ \text{W mm}^2\text{/K for average design}$$

$$= 33.5\times10^{-6}\ \text{W mm}^2\text{/K for air moving at 2.5 m/s}$$

$A =$ Area of surface of bearing housing or pedestal = 20 dL mm^2

(b) Based on projected area of bearing: Heat dissipated from the bearing,

$$H_d = K_2Ld(t_B - t_A) \geq H_g \tag{21.10}$$

$$K_1 = 280\times10^{-6}\ \text{W mm}^2\text{/K for well-ventilated}$$

$$= 170\times10^{-6}\ \text{W mm}^2\text{/K for unventilated}$$

(c) Another formula to determine heat dissipated from the bearing,

$$H_d = \frac{Ld(T+18)^2}{K_3} \geq H_g \tag{21.11}$$

where, $K_3 = 0.273 \times 10^{+6}$ °C mm²/watt for bearings of heavy construction and well-ventilated, $= 0.484 \times 10^{+6}$ °C mm²/watt for bearings of light construction in still air, i.e. unventilated and

$$T = (t_B - t_A)°\text{C}$$

Check for – $H_d \geq H_g$

For safe operation ($H_d > H_g$), otherwise artificial cooling by circulation of the lubricant is to be adopted. The chart in Fig. 21.8 gives the rise in temperature on the assumption that entire heat generated is taken away by the lubricant.

Bearing design involves the proper selection of μ, L, c to satisfy the above conditions. Another condition to be satisfied is the minimum film thickness to avoid the seizure. The minimum film thickness h_0 is found from the chart (Fig. 21.7).

$$h_0 = c\,(MFT) = C(1 - \in) \tag{21.12}$$

It should be greater than minimum from the surface roughness consideration, i.e.,

$$h_0 > 3R_0 \tag{21.13}$$

where R_0 is the combined roughness value of the two surfaces (in relative motion) of the bearing, namely shaft and bush.

21.6 DESIGN PROCEDURE

Step-1: For a given type of bearing and machine, select the following recommended design dated from Table 21.10.

(i) Maximum average pressure $[p]_r$

(ii) Viscosity $[Z]_r$

(iii) Minimum bearing modulus $\left[\frac{ZN}{p}\right]_r$

(iv) Clearance ratio $\left[\frac{c}{r}\right]_r$

(v) Length to diameter ratio $\left[\frac{l}{d}\right]_r$

Normally, the speed N and diameter d are pre-determined (or specified) by other design considerations.

Step-2: From the selected L/d ratio, determine length L, and check for bearing pressure,

$$p = \frac{W}{L.d} \leq \text{Recommended } [P]_r$$

If the diameter is not specified, determine diameter $d = \sqrt{\dfrac{[p]_r}{\left[\dfrac{L}{d}\right]_r}}$ and length, $L = d \cdot \left[\dfrac{L}{d}\right]_r$

Step-3: Determine the operating value of bearing modulus

$$\left(\frac{ZN}{p}\right)_o = 3\left[\frac{ZN}{p}\right]_r$$

Then evaluate viscosity

$$Z = \frac{p}{N}\left(\frac{ZN}{p}\right)_0$$

For operating temp T_0 (Assume $T_0 = 60°C$ if not given), determine the type of oil from chart Fig. 21.3 or Fig. 21.4 and the actual value of Z (or μ) and $p\left(=\dfrac{W}{Ld}\right)$

Step-4: Evaluate the actual operating bearing modulus $\left(\dfrac{ZN}{p}\right)$ and Sommerfeld number

$$S = \left(\frac{\mu \cdot N_s}{p}\right)\left(\frac{r}{c}\right)$$

Step-5: Determine the friction coefficient f from the Mackee's equation (21.2),

$$f = 0.326\left(\frac{\mu \cdot N}{p}\right)\left(\frac{r}{c}\right) + \Delta f$$

or from the Eqn. 10.2a

$$f = K_a\left(\frac{ZN}{p}\right)\left(\frac{r}{c}\right) \times 10^{-10} + \Delta f$$

or determine the coefficient of friction from the chart Fig. 21.5.

$$f = \frac{c}{r}(FV), \text{ take (FV) from Fig. 21.5}$$

Step-6: Determine the power lost in friction (heat-generated)

$$H_g = \frac{NT}{955\times10^4} = \frac{N\cdot f\cdot W\cdot r}{955\times10^4}\, kW = \frac{N\cdot f\cdot W\cdot r}{9550}\cdot\frac{J}{s}$$

Step-7: Evaluate the heat dissipation capacity of the bearing by Eqn. (21.9) or Eqn. (21.10) or Eqn. (21.11).

Check for $H_d \geq H_g$

If not, modify $\left(\frac{c}{r}\right)$ ratio or specify artificial cooling.

Step-8: Calculate the quantity of oil flow by the equation,

$$Q = (FLV)\,(r\cdot C\cdot N_s\cdot L)$$

Take (FLV) from Fig. 21.6

Step-9: Determine temperature rise ΔT

$$\Delta T = \frac{(TRV)\cdot p}{\rho\cdot C_p}, \text{ take } (TRV) \text{ from Fig. 21.8}$$

where, $C_p = 1675\times10^3$ Nmm/kg°C

$\rho =$ density of the oil $= 9\times10^{-7}$ kg/mm^3

It can be evaluated by the alternate equation given below,

$$\Delta T = \left(\frac{4\pi p}{\rho\cdot C_p}\right)\left(\frac{FV}{FLV}\right) = 8.3p\left(\frac{FV}{FLV}\right), \text{ for } \rho = 0.86 \text{ and } C_p = 1.76 \text{ kJ/kg°C}$$

Determine the oil temperature by,

$$T_o = T_A + (\Delta T)$$

Determine the bearing temperature by,

$$T_B = T_A + \left(\frac{\Delta T}{2}\right)$$

Compare this with the given value. If the value is more, artificial cooling is necessary.

Step-10: Determine the minimum film thickness h_0

$$h_0 = (MFT)\cdot c, \text{ take } (MFT) \text{ from Fig. 21.7}$$

Step-11: Check for the combined surface roughness value

$$R_o \leq (h_o/3)$$

If not, specify the surface finish required or change the clearance ratio.

Table 21.5. Viscosity - Temperature chart (Fig. 21.3)

Curve number Fig. 14.1	Type of oil	Specific gravity at 15°C
1.	Light oil for light service and high speeds	0.875
2.	Turbine oil (oil rings) for light service and high speeds	0.88
3.	Turbine oil (oil rings) for light service and high speeds	0.89
4.	Extra-light motor oil for ring-oiled bearings, transmission shafting small generators, motors, and high-speed engines	0.935
5.	S.A.E. 20 light transmission oil for gears	0.925
6.	S.A.E. 40 medium transmission oil for large generators, motors steam turbines, high-speed gears, heavy motor oil	0.93
7.	Airplane 100 G light cylinder oil	0.89
8.	S.A.E. 110 light steam cylinder oil; heavy duty gears	0.93
9.	Medium cylinder oil – slow speeds, worm gear	0.91
10.	SAE 160 heavy cylinder at heavy duty low speed gearing	0.935
11.	Heavy steam cylinder oil	0.93

Table 21.6. Value of the type constant C_1 (Eqn. 21.3)

Lubrication	Workmanship	Attendance	Location	Constant C_1
Oil bath flooded	High grade	First class	Clean and protected	1
Oil free drop (const. feed)	Good	Fairly good	Favourable	2
Oil cup or grease (intermittent feed)	Fair	Poor	Exposed to dirt, or other unfavourable condition	4

Table 21.7. Value of the type constant C2 (Eqn. 21.3)

	Type of bearing	C_2
1.	Rotating journals such as rigid bearings and crankpins	1.
2.	Oscillating journals, such as rigid wrist pins and pintle block	1.
3.	Rotating bearings locking ample rigidity such as eccentricity and the like	2.
4.	Rotating flat surface lubricated from the centre to the circumference, such as annular step or pivot bearings	2.
5.	Sliding flat surface wiping over the guide ends such as reciprocating cross head shoes, long guides	2.
6.	Sliding or wiping surface lubricated from the periphery or outer wiping edge such as marine thrust bearings and worm gears	3-4
7.	Long power screw nuts and similar wiping parts over which it is difficult to effect a uniform distribution of lubricant or load.	4-6

Table 21.8. Values of h_o/c for max. load capacity or minimum power loss

	Bearings arc $\beta°$			
	360°	**180°**	**150°**	**120°**
For max. load capacity h_o/c	0.331	0.3275	0.3025	0.280
For minimum power loss h_o/c	0.317	0.315	0.295	0.255

Table 21.9. Materials and applications of journal bearing

Material	Characteristics	Limitations	Applications
Wood (maple Ligium vital)	Self-lubricating, low cost, long life	Light loads at high speeds, under 65°C	Conveyors
Cast iron	Low friction, low cost	Not over 3.5 MPa and 40 m/min	Camshafts, light transmission
Steel	Low cost	Light loads 45 m/min	Guides
Bronze bushing	Low cost, simple construction	Loads up to 21 MPa 270 m/min	All equipment
Heavy babbit liner on steel or C.I.	Long life, low friction	Steady loads under 7 MPa	Motor, turbine shafting
Light liner on steel or bronze backing	Heavy duty, general purpose, good for dynamic loads	About 35 MPa and 1200 m/min	Gas and diesel engines, compressors
Rubber	Low friction, resists abrasion, shock absorbent, long life	About 0.55 MPa needs water lubrication	Marine propellers, pumps, turbines
Carbon graphite	No lubrication needed, light duty applications	Under 45°C 4 MPa at low speeds	Electric motors, meters, conveyors
Moulded plastic laminate	Low friction, stronger than babbit when water lubricated	About 120 m/min and 17.5 MPa; must be well cooled	Pumps, propellers
Moulded plastic	Low friction, clean	Low loads if used at high speeds	Dairy, textile and food machinery
Sintered bearings	Self-lubricating	Low loads, high rpm	All inaccessible places, typewriter, electric fans instruments, piston rings

Table 21.10. Design data for bearings

Machinery	Bearing	Maximum p MPa	Suitable Z CP	Minimum ZN/p	$\frac{c}{r}$	$\frac{L}{d}$
1	2	3	4	5	6	7
1. Automobile and aircraft engines	Main Crankpin Wrist pin	5.6 – 11.9 10.5 – 24.6 16.1 – 35.1	7 8 8	2135 1422 1140	- - -	0.8 – 1.8 0.7 – 1.4 1.5 – 2.2
2. Gas and oil engines four-stroke	Main Crankpin Wrist pin	4.9 – 8.4 9.8 – 12.6 12.6 – 15.5	20 40 65	2845 1422 711	0.001 < 0.001 < 0.001	0.6 – 2.0 0.6 – 1.5 1.5 – 2.0
3. Gas and oil engines two-stroke	Main Crankpin Wrist pin	3.5 – 5.6 7 – 10.5 8.4 – 12.6	20 40 45	3560 1708 1422	0.001 < 0.001 < 0.001	0.6 – 2.0 0.6 – 1.5 1.5 – 2.0
4. Marine steam engines	Main Crankpin Wrist pin	3.5 4.2 10.5	30 40 50	2845 2135 1422	< 0.001 < 0.001 < 0.001	0.7 – 1.5 0.7 – 1.2 1.2 – 1.7
5. Stationary slow speed steam engine	Main Crankpin Wrist pin	2.8 4.2 12.6	60 80 60	2845 854 712	< 0.001 < 0.001 < 0.001	1.0 – 2.0 0.9 – 1.3 1.2 – 1.5
6. Stationary high speed steam engine	Main Crankpin Wrist pin	1.8 4.2 12.6	15 30 25	3560 854 712	< 0.001 < 0.001 < 0.001	1.5 – 3.0 0.9 – 1.5 1.3 – 1.7
7. Reciprocating pump and compressors	Main Crankpin Wrist pin	1.8 4.2 7	30 - 80	4270 2845 1422	< 0.001 < 0.001 < 0.001	1.0 – 2.2 0.9 – 1.7 1.5 – 2.0
8. Steam locomotives	Driving axel Crankpin Wrist pin	3.8 14 28	100 40 30	4270 712 712	< 0.001 < 0.001 < 0.001	1.6 – 1.8 0.7 – 1.1 0.8 – 1.3
9. Railway cars	Axle	3.1	100	711	< 0.001	1.8 – 2.0
10. Steam turbines	Main	0.7 – 1.9	2 - 16	14220	< 0.001	1.0 – 2.0
11. Gyroscope	Rotor	6	30	7825	0.0013	–
12. Transmission shafting	Light fixed Self-aligning Heavy	0.2 1 1	- 25-60	14220 4270 4270	0.005 0.001 0.001	2.0 – 3.0 2.5 – 3.0 2.5 – 4.0
13. Cotton mill	Spindle	0.007	2	1422000	0.001	2.0 – 3.0
14. Machine tool	Main	2.1	40	5690	0.001	1.0 – 4.0
15. Punching and shearing machines	Main Crankpin	28 26	100 100	- -	0.001 0.001	1.0 – 2.0 1.0 – 2.0
16. Rolling mills	Main	21	50	1425	0.0015	1.1 – 1.5
17. Generators motor centrifugal pumps	Rotor	0.7 – 1.4	25	28450	0.0013	1.0 – 2.0

Note: $\left(\frac{\mu N}{p}\right) = \left(\frac{ZN}{p}\right) \times 10^{-9}$; μ in Ns/mm^2 and $\left(\frac{\mu' N}{p}\right) = \left(\frac{ZN}{p}\right) \times 10^{-3}$, μ' in Ns/m^2 (kg/ms)

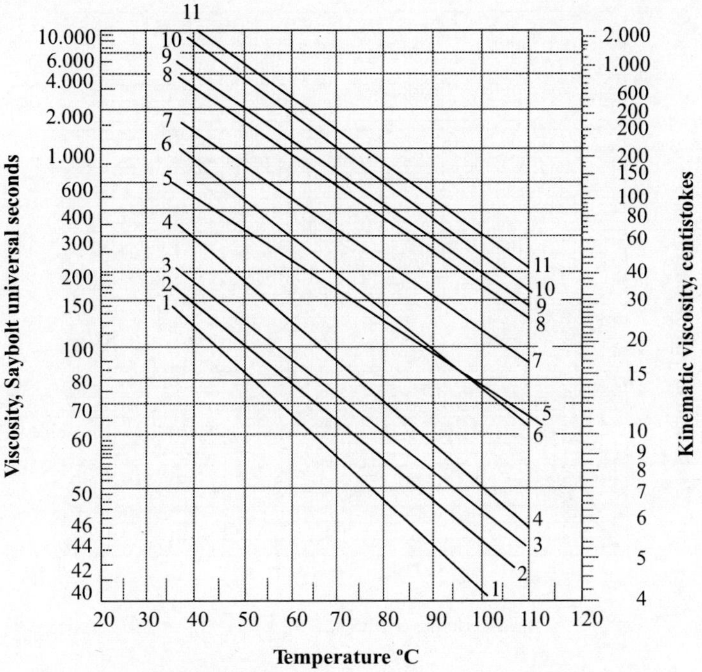

Figure 21.3 Viscosity temperature chart

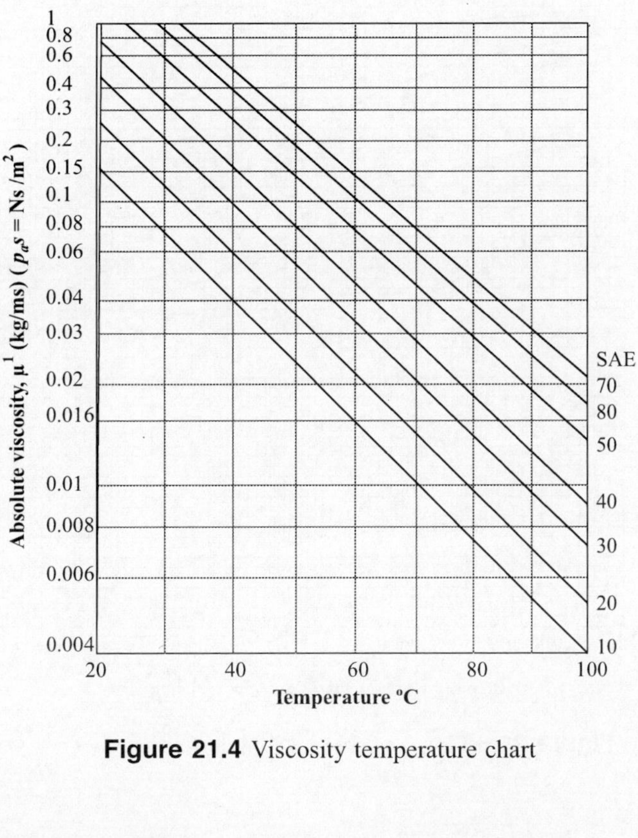

Figure 21.4 Viscosity temperature chart

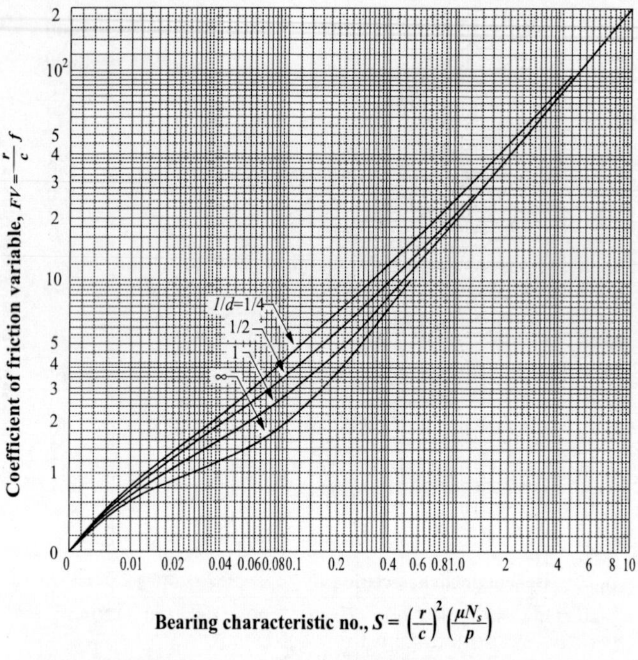

Figure 21.5 Friction variable vs Sommerfeld number chart

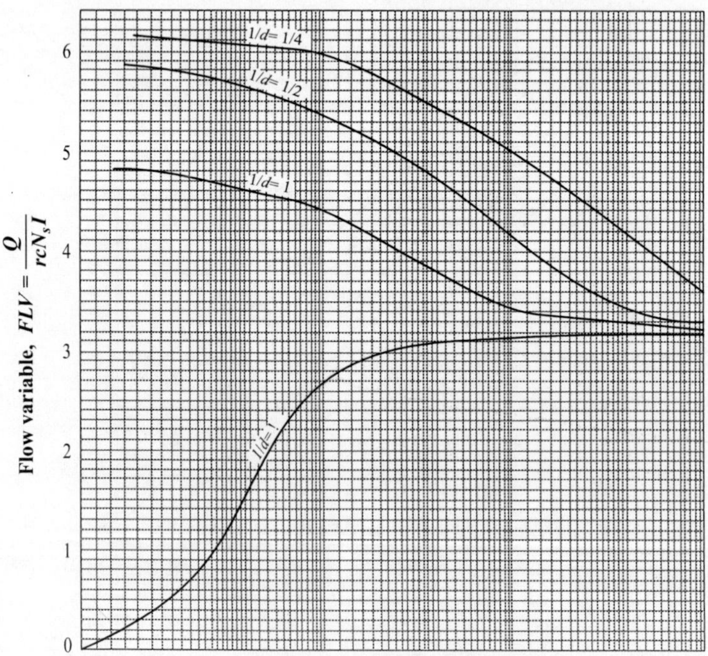

Figure 21.6 Flow variable vs Sommerfeld number

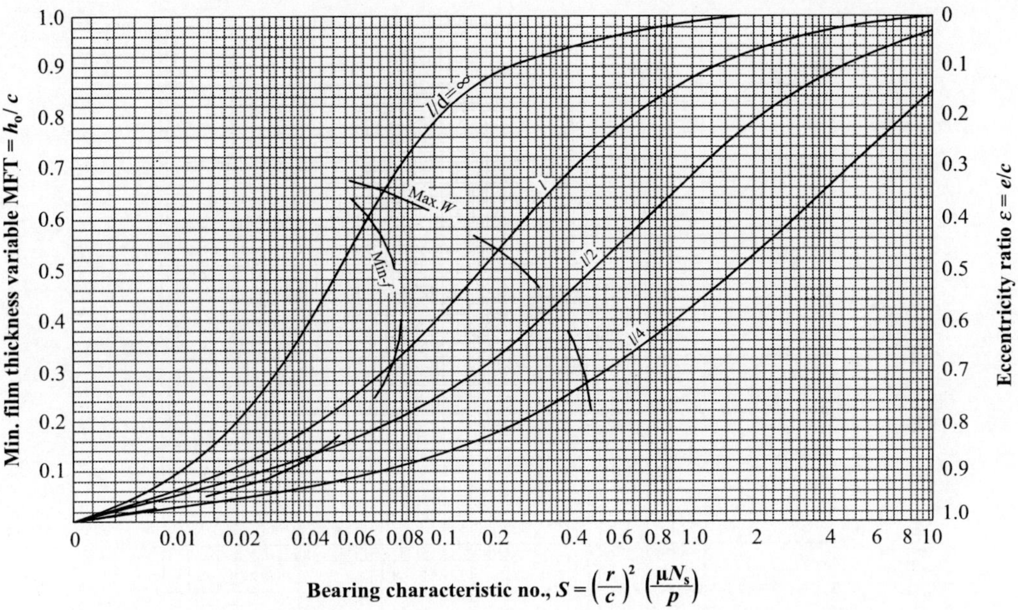

Figure 21.7 Minimum film thickness vs Sommerfeld number

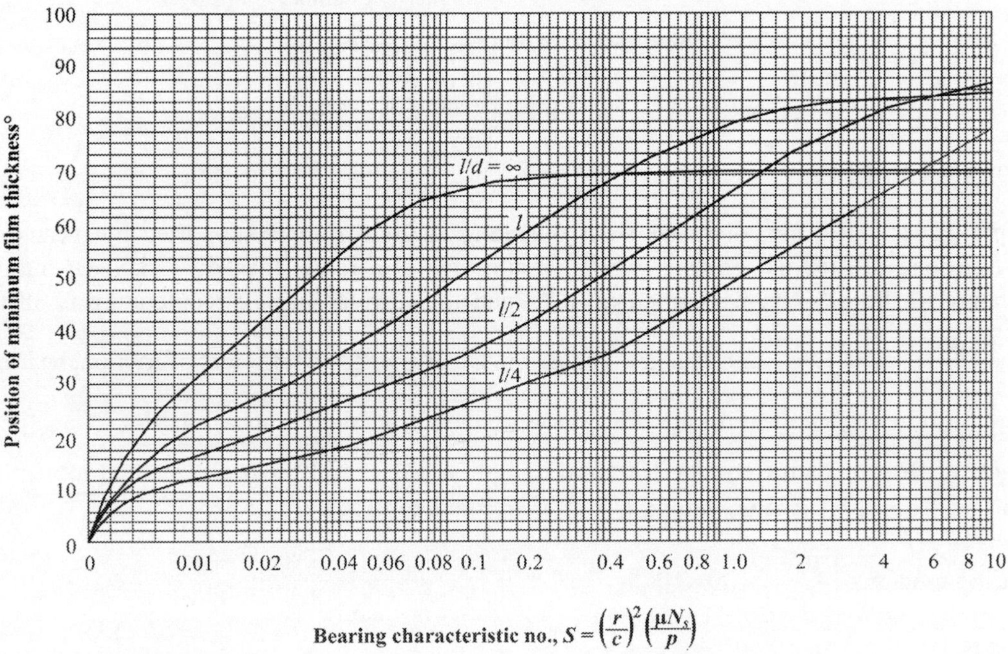

Figure 21.7a Position of minimum film thickness vs Sommerfeld number

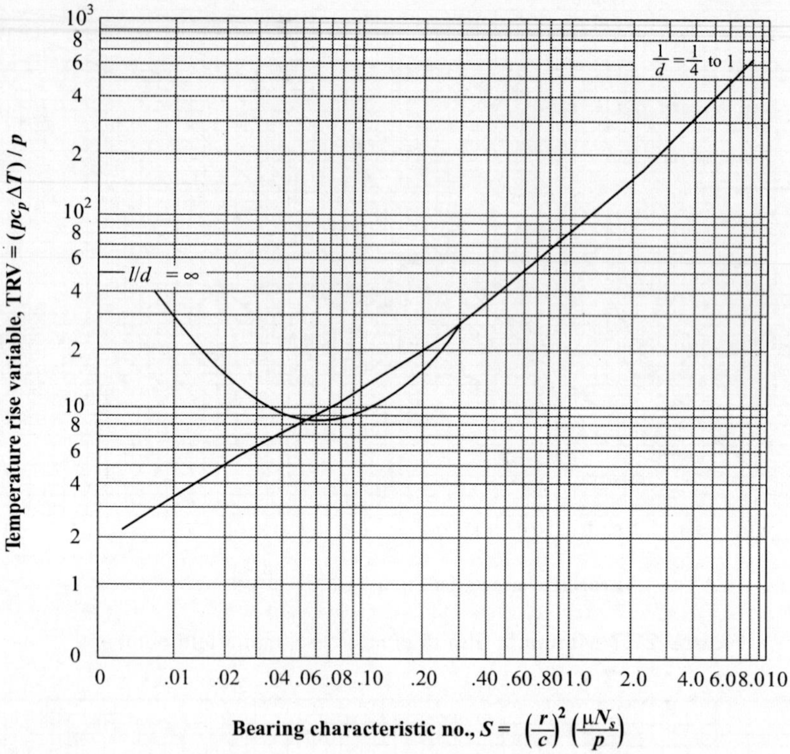

Figure 21.8 Chart for temperature-rise variable

SOLVED EXAMPLES

Example 21.1: A 100 mm diameter shaft operating at 2000 rpm is supported by means of a 150 mm long full journal bearing, which is subjected to a radial load of 43 kN. Operating temperature of the oil is to be limited to 80°C and the surrounding air temperature is 38°C. Assume $\left(\frac{\mu N}{p}\right) = 30 \times 10^{-6}$ and Mackee's equation. Determine the coefficient of friction, bearing pressure, heat generated, heat dissipated, oil to use and artificial cooling required.

Given Data: Bearing diameter, D = 100 mm; Length, L = 150 mm; Speed, N = 2000 rpm
Load, W = 43 kN; Oil temperature, $t_0 = 80°C$; Room temperature, $t_A = 38°C$;

Bearing modulus $\left(\frac{\mu N}{p}\right) = 30 \times 10^{-6}$.

SOLUTION

Step-1: Coefficient of friction

Mackee's equation, $f = 0.326\left(\frac{\mu N}{p}\right)\left(\frac{r}{c}\right) + \Delta f$

Assume $\frac{r}{c} = 1000; \frac{\mu N}{p} = 30 \times 10^{-6}; \Delta f = 0.002$

Substituting the values,

$$\therefore \quad f = 0.326 \times 30 \times 10^{-6} \times 1000 + 0.002 = 0.00978 + 0.002 = 0.01178$$

Step-2: Heat generated

The peripheral velocity of the shaft,

$$V_s = \frac{\pi DN}{60{,}000} = \frac{\pi \times 100 \times 2000}{60{,}000} = 10.47 \text{ m/sec}$$

Heat generated,

$$H_g = fWV_s = 0.01178 \times 43 \times 10.47 = 5.3 \text{ k watts} = 5300 \text{ watts}$$

Step-3: Heat dissipating capacity

Temperature of oil $t_0 = 80°\text{C}$, and surronding temp, $t_A = 38°\text{C}$

$$T = t_B - t_A - \frac{(t_0 - t_A)}{2} = \frac{(80 - 38)}{2} = 21°\text{C}$$

The heat dissipating capacity,

$$H_d = \frac{LD(T+18)^2}{K_3} = \frac{150 \times 100(21+18)^2}{0.484 \times 10^6} = 47.13 \text{ watts}$$

Step-4: Artificial cooling required

Since $H_d < H_g$, artificial cooling is necessary

$$H = 5300 - 47.13 = 5252.87 \text{ watts}$$

Step-5: Viscosity of the oil

The average pressure, $p = \frac{W}{LD} = \frac{43 \times 10^3}{100 \times 150} = 2.87 \text{ MPa} = 2.87 \times 10^6 \text{ Pa}$

Substituting the value in $\frac{\mu N}{p} = 30 \times 10^{-6}$

$$\mu = \frac{30 \times 10^{-6} \times p}{N} = \frac{30 \times 10^{-6} \times 2.87 \times 10^{+6}}{2000}$$

$$= 0.043 \text{ Ns/m}^2 \text{ @ } 80°\text{C} = 0.043 \text{ kg/ms}$$

From the graph in Fig. 14.2 MDHB, the oil is SAE 60.

Example 21.2: A 60 mm journal bearing, 100 mm long supports a load of 2200 N at 600 rpm. If the room temperature is 24°C, what viscosity of oil should be used to limit the bearing surface temperature to 60°C?

Given Data: Journal diameter, D = 60 mm; Length, L = 100 mm;

Load, W = 2200 N; Speed, N = 600 rpm;

Bearing temperature, t_B = 60°C; Room temperature, t_A = 24°C

SOLUTION

Step-1: Heat generated

Heat generated, $H_g = fWV_s$ watts

where, $W = 2200$ N, $V_s = \dfrac{\pi DN}{60,000} = \dfrac{\pi \times 60 \times 600}{60,000} = 1.885$ m/sec

$$\therefore \quad H_g = f \times 2200 \times 1.885 = (4146.9.\text{f}) \text{ watts}$$

Step-2: Heat dissipating capacity

Heat dissipated $H_d = K_1 A\,(t_B - t_A)$

where, $K_1 = 11.4 \times 10^{-6}$ w/mm/^{0}K,

$$A = 20 \times 60 \times 100 = 12 \times 10^4 \text{ mm}^2$$

$$t_B - t_A = 60 - 24 = 36°\text{C}$$

$$H_d = 11.4 \times 10^{-6} \times 12 \times 10^4 \times 36 = 49.24 \text{ watts}$$

Step-3: Coefficient of friction by heat balance

For heat balance, Heat generated = Heat dissipated

$$4146.9f = 49.24$$

$$f = 0.0119$$

Step-4: Viscosity of the oil

The Mackee's equation for friction,

$$f = 0.326\left(\frac{\mu N}{p}\right)\left(\frac{r}{c}\right) + \Delta f$$

where, N = 600 rpm, r/c = 1000 (assumed)

$$p = \frac{W}{LD} = \frac{2200}{60 \times 100} = 0.367 \text{ MPa}, \text{ and } \Delta f = 0.002 \text{ (assumed)}$$

$$\therefore \quad 0.0119 = 0.326\left(\frac{\mu \times 600}{0.367 \times 10^6}\right) \times 1000 + 0.002$$

$$\therefore \qquad \mu = \frac{(0.0119 - 0.002) \times 0.367 \times 10^6}{0.326 \times 600 \times 1000} = 0.0186 \text{ kg/ms}$$

At temperature, $t_0 = 2 \times 36 + 24 = 96°\text{C}$

∴ SAE 50 oil can be used. (Fig. 14.2 MDHB)

Example 21.3: A 2.8 kN load is supported by a journal bearing 75 mm long on a 65 mm shaft. The bearing has a clearance of 0.05 mm and is lubricated with an oil of viscosity 0.0207 kg/ms at its operating temperature. Determine the maximum speed of rotation if the bearing is capable of dissipating 80 watts by means of heat transfer. Use Mackee's equation.

SOLUTION Speed of the shaft for heat balance

Heat generated = Heat dissipated

$$\therefore \qquad f \cdot W \cdot V_S = 80$$

Using Mackee's equation for *f*, we get

$$\left[0.326\left(\frac{\mu N}{p}\right)\left(\frac{r}{c}\right) + \Delta f\right] \times W \times V_s = 80$$

where, $\mu = 0.02070 \text{ kg/m.s} = 0.0207 \times 10^{-6} \text{ N/mm}^2$

$$p = \frac{W}{LD} = \frac{2.8 \times 10^3}{75 \times 65} = 0.574 \text{ MPa}, \quad \frac{r}{c} = \frac{D}{C} = \frac{65}{0.05} = 1300$$

$$V_s = \frac{\pi DN}{60{,}000} = \frac{\pi \times 65 \times N}{60{,}000} = (3.4 \times 10^{-3} N) \text{ m/s}$$

Substituting these values, we get

$$\left[0.326\left(\frac{.0207 \times 10^{-6} N}{0.574}\right) \times 1300 + .002\right] \times 2800 \times (3.4 \times 10^{-3} N) = 80$$

On simplification, $N^2 + 131N = 0.55 \times 10^6$

$$\text{Try } N = 700, \text{ LHS} = 0.58 \times 10^6 > \text{RHS}$$

$$\text{Try } N = 675, \text{ LHS} = 544050 < \text{RHS}$$

$$\text{Try } N = 680, \text{ LHS} = 551480 \simeq 0.55 \times 10^6 = \text{RHS}$$

$\therefore N = 680$ rpm

Example 21.4: A full journal bearing is 90 mm in diameter and 150 mm long and has radial load of 2 MN/m^2 of projected area. The shaft speed is 500 rpm and H9 - e9 fit is to be used. The surrounding air temperature is 20°C and the oil used has a viscosity of 0.0099 kg/ms at the operating temperature. Determine the probable temperature of the bearing surface, assuming all the heat generated is dissipated by radiation heat transfer according to

$$H_d = \frac{(T_B - T_A + 18)^2}{0.484} \times LD$$

Given Data: Bearing diameter, $D = 90$ mm $= 0.09$ m,

Length, $L = 150$ mm $= 0.15$ m

Average pressure $p = 2$ MN/m^2 $= 2$ MPa

Speed $N = 500$ rpm, $\mu = 0.0099$ kg/ms;

Fit H9 - e9

SOLUTION

Step-1: Clearance ratio from fit specified

$H9$ is $- - -\ 90^{0.087}$ (Table 13.11 MDHB)

and $e9$ is $- - \rightarrow 90^{-0.072}_{-0.159}$ (Table 13.12 MDHB)

Diametral clearance, $C = 0.072$ to 0.246 mm

For min. clearance D/C $= \dfrac{90}{0.072} = 1250$

Step-2: Coefficient of friction

Using Mackee's equation, $f = 0.326\left(\dfrac{\mu N}{p}\right)\left(\dfrac{r}{c}\right) + \Delta f$

$$f = 0.326 \times \frac{0.0099 \times 500}{(2.0 \times 10^6)} \times 1250 + 0.002 = 0.00301$$

Step-3: Heat generated

Heat generated, $H_g = f \cdot W \cdot V_s$

where, $f = 0.00301, W = 2 \times 90 \times 150 = 27000$ N

$$V_s = \frac{\pi DN}{60{,}000} = \frac{\pi \times 90 \times 500}{60{,}000} = 2.356 \text{ m/s}$$

$\therefore$ $$H_g = 0.00301 \times 27000 \times 2.356 = 191.5 \text{ watts}$$

Step-4: Bearing temperature for heat balance

By the heat balance, $H_d = H_g$

$$H_d = \frac{(T_B - T_A + 18)^2}{0.484} \times 0.150 \times 0.09 = 191.5$$

$\therefore$ $$T_B - T_A + 18 = \sqrt{\frac{191.5 \times 0.484}{0.150 \times 0.09}} = 82.85$$

$\therefore$ $$T_B = 82.85 - 18 + 20 = 84.85°\text{C}$$

Example 21.5: A 100 mm dia. by 100 mm long full bearing supports a load of 1300 N. The shaft rotates at 1000 rpm. The diametral clearance is 0.1 mm and the oil has viscosity of 0.0003 kg/ms. Assume specific heat as $C_p = 4200$ Nm/kg°C and density $\rho = 1000$ kg/m^3. Using Raimondi and Boyd curves, determine the coefficient of friction, the min. film thickness, the flow of the lubricant into the bearing, end leakage of the lubricant, temperature rise of oil and power loss due to friction.

Given Data: $D = 100$ mm; $L = 100$ mm; $N = 1000$ rpm.;

$W = 1300$ N; $C = 0.1$ mm; $\mu = 0.0003$ kg/ms;

$C_P = 4200$ Nm/kg°C; $\rho = 1000$ kg/m^3.

SOLUTION

Step-1: Sommerfeld number

$$S = \left(\frac{\mu N_S}{P}\right)\left(\frac{D}{C}\right)^2$$

where,
$$N_s = \frac{1000}{60} = 16.67 \text{ rps}$$

$$p = \frac{1300}{0.1 \times 0.1} = 13 \times 10^4 \text{ N/m}^2$$

$$\left(\frac{D}{C}\right) = \frac{100}{0.1} = 1000$$

∴
$$S = \frac{0.0003 \times 16.67}{13 \times 10^4} \times 10^6 = 0.0385$$

Step-2: Coefficient of friction

From charts,

Friction variable, $FV = f\left(\frac{r}{c}\right) = 1.6$ (Fig. 14.3 MDHB)

∴ Coefficient of friction, $f = \frac{FV}{\left(\frac{r}{c}\right)} = \frac{1.6}{1000} = 0.0016$

Step-3: Minimum film thickness variable

From (Fig. 14.5 MDHB), $MFT = \frac{h_0}{c} = 0.18$

∴
$$h_0 = 0.18 \times 0.05 = 0.009 \text{ mm} = 9\ \mu\text{m}$$

Step-4: Quantity of lubricant

Flow variable (Fig. 14.4 MDHB) $\dfrac{Q}{rcN_s L} = 4.6$

$$\therefore \qquad Q = 4.6 r \cdot c \cdot N_s \cdot L$$

$$= 4.6 \times 0.05 \times (0.05/1000) \times 16.67 \times 0.1 = 19.17 \times 10^{-6} \text{ m}^3/\text{s}$$

Side flow ratio (Fig. 14.6 MDHB) $\dfrac{Q_0}{Q} = 0.86$

$$Q_0 = 0.86 \times Q = 0.86 \times 19.17 \times 10^{-6} = 16.5 \times 10^{-6} \text{m}^3/\text{s}$$

Step-5: Temperature rise

Temp. variable (Fig. 16.8 MDHB) $TRV = \dfrac{\rho C_p \Delta T}{P} = 7.5$

$$\Delta T = \frac{7.5 \times p}{\rho \cdot C_p} = \frac{7.5 \times 13 \times 10^4}{1000 \times 4200} = 0.23°\text{C}$$

Step-6: Power loss

Heat generated, $H_g = f \cdot W \cdot V_s$

$$= 0.0016 \times 1300 \times \frac{\pi \times 100 \times 1000}{60{,}000} = 10.89 \text{ watts}$$

Example 21.6: A full journal bearing 200 mm dia. and 200 mm long supports a radial load of 45 kN. The journal rotates at 1200 rpm and D/C ratio is 1000. Viscosity of oil at the operating temperature of 80°C is 0.018 kg/ms. Ambient temperature is 20°C. Using Raimondi & Boyd Charts, determine the film thickness, heat generated, heat dissipated by radiation (use Lache equation)

$$H_d = \frac{(\Delta T + 18)^2}{0.273} LD,$$

and the heat removed by the end leakage of oil.

Given Data: The load supported, W = 45 kN; Speed, N = 1200 rpm;
Journal dimensions, D = 200 mm; L = 200 mm; D/C = 1000;
Viscosity, μ = 0.018 kg/ms; Temperature, t = 80°C

SOLUTION

Step-1: Sommerfeld number

Average pressure, $p = \dfrac{45 \times 10^3}{0.2 \times 0.2} = 1125 \times 10^3 \text{Pa}$

Sommerfeld number, $S = \dfrac{\mu N_s}{p}\left(\dfrac{D}{C}\right)^2 = \dfrac{0.018 \times 20}{1125 \times 10^3} \times (1000)^2 = 0.32$

Step-2: From charts, minimum film thickness

Minimum film thickness variable (Fig. 14.5 MDHB) $\frac{h_0}{c} = 0.62,$

$$\therefore \quad h_0 = 0.62 \times 0.1 = 0.062 \text{ mm}$$

Step-3: Friction coefficient from chart

Friction variable (Fig. 14.3 MDHB) $f(r/c) = 7.5$

$$\therefore \quad f = 7.5/\left(\frac{r}{c}\right) = 7.5 \times 0.001 = 0.0075$$

Step-4: Heat generated

$$\text{Heat generated, } H_g = fWV_s = 0.0075 \times 45 \times 10^3 \times \frac{\pi \times 200 \times 1200}{60{,}000} = 4.24 \times 10^3 \text{ watts} = 4.24 \text{ kW}$$

Step-5: Quantity of oil flow

Oil flow variable (Fig. 14.4 MDHB) $\left(\frac{Q}{rcN_S L}\right) = 3.8$

$$\therefore \quad Q = 3.8 \times \frac{100}{1000} \times \frac{0.1}{1000} \times 20 \times 0.2 = 152 \times 10^{-6} \text{ m}^3/\text{sec}$$

Side leakage variable $\left(\frac{Q_0}{Q}\right) = 0.48$ (Fig. 14.6 MDHB)

$$\therefore \quad Q_0 = 0.48 \times Q = 0.48 \times 152 \times 10^{-6} = 72.9 \times 10^{-6} \text{ m}^3/\text{sec}$$

Step-6: Heat carried by the leaking oil

Temperature variable, $\rho C_p \frac{\Delta T}{p} = 26$ (Fig. 14.8 MDHB)

Heat taken away by the leaking oil for $\rho \cdot C_p\, \Delta T = 26 \times p$

$$H_d = Q_0 \cdot \rho \cdot C_p \cdot \Delta T = 26 \times p \times Q_o$$

$$= 26 \times 1125 \times 10^3 \times 72.9 \times 10^{-6} = 2132 \text{ watts} = 2.132 \text{ kW}$$

Step-7: Heat dissipated by radiation,

$$H_d = \frac{(\Delta T + 18)^2}{0.273} \times LD \quad [\text{with, } \Delta T = (80 - 20)/2 = 30]$$

$$= \frac{(30 + 18)^2}{0.273} \times 0.2 \times 0.2 = 337.6 \text{ watts}$$

Example 21.7: The following data refer to a 360° hydrodynamic bearing

Radio load, $W = 3.2$ kN, Journal speed, $N = 1500$ rpm

Journal diameter, $d =$ bearing length, $L = 50$ mm

The viscosity of lubricant = 25 cp

Assuming that the total heat generated is carried away by the oil flow in bearing, calculate: (i) Coefficient of friction, (ii) Power lost in friction, (iii) Minimum film thickness, (iv) Oil flow requirement, and (v) Temperature rise T.

Given Data: Radial load, $W = 3.2$ kN, Speed, $N = 1500$ rpm

Bearing diameter, $d =$ bearing length, $L = 50$ mm

Radial clearance, $c = 0.05$ mm, Viscosity, $Z = 25$ cp

SOLUTION

Step-1: Coefficient of friction by Mackee's equation (Eqn. 14.18 MDHB)

$$f = Ka\left(\frac{ZN}{p}\right)\left(\frac{r}{c}\right)\times 10^{-10} + \Delta f$$

where, $Ka = 3.326$; $Z = 25$ cp; $N = 1500$ rpm

$$p = \frac{W}{L.d} = \frac{3200}{50\times 50} = 1.28 \text{ MPa}, \frac{r}{c} = \frac{25}{.05} = 500 \text{ and } \Delta f = 0.002$$

Substituting the values,

$$f = 3.326\left(\frac{25\times 1500}{1.28}\right)(500)\times 10^{-10} + 0.002$$

$$= 0.004872 + 0.002 = 0.00687$$

Step-2: Power lost in friction (Eqn. 14.47 MDHB)

$$H_g = (f \cdot W \cdot V_s) \text{ watts}$$

$$\text{where, } V_s = \frac{\pi DN}{60,000} = \frac{\pi\times 50\times 1500}{60,000} = 3.93 \text{ m/s}$$

$W = 3200$ N and $f = 0.00687$

Substituting the values,

$$H_g = 0.00687\times 3200\times 3.93 = 86.33 \text{ watts}$$

Step-3: Minimum film thickness

$$\text{Sommerfeld number, } S = \left(\frac{r}{c}\right)^2\left(\frac{\mu N_s}{p}\right)$$

where, $\frac{r}{c} = 500, \mu = 25 \times 10^{-9} \cdot \frac{\text{Ns}}{\text{mm}^2}$,

$$N_s = \frac{1500}{60} = 25 \text{ rps and } p = 1.28 \text{ MPa}$$

Substituting the value, $S = (500)^2 \left(\frac{25 \times 10^{-9} \times 25}{1.28} \right) = 0.12207$

From chart Fig. 14.5 MDHB, MFT = 0.4

Then minimum film thickness, (Eqn. 14.32 MDHB)

$$h_0 = (\text{MFT}) \cdot C = 0.4 \times 0.05 = 0.02 \text{ mm}$$

Step-4: The oil flow Q

The flow variable from Fig. 14.4 MDHB, $FLV = 4.4$

The flow rate, $Q = (FLV) \cdot r \cdot c N_s L$

$$= 4.4 \times 25 \times 0.05 \times 25 \times 50 = 6875 \text{ mm}^3/\text{sec}$$

Step-5: The temperature rise (ΔT)

The temperature rise variable (Fig. 14.8) $TRV = 15$

Then the temperature rise, $\Delta T = \frac{(TRV) \cdot p}{\rho \cdot C_p}$ (Eqn. 14.38 MDHB)

$$= \frac{15 \times 1.28}{9 \times 10^{-7} \times 1675 \times 10^3} = 12.74°\text{C}$$

Example 21.8: A full journal bearing 50 mm in diameter and 50 mm long, operates at 1000 rpm and carries a load of 5 kN. The radial clearance is 0.025 mm. The bearing is lubricated with SAE 30 oil and the operating temperature of the oil is 80°C. Assume the attitude angle 60°, and determine: (i) Bearing pressure, (ii) Sommerfeld number, (iii) Attitude, (iv) Minimum film thickness, (v) Heat generated, (vi) Heat dissipated, if the ambient temperature is 20°C and (vii) Amount of artificial cooling, if necessary. Use Mackee's and Peterson's equations.

Given Data: Bearing diameter, $d = 50$ mm; Bearing length, $L = 50$ mm

The load on the bearing, $W = 5$ kN; Radial clearance, $c = 0.025$ mm

Lubricating oil, SAE 30; Temperature of oil, $t_0 = 80°\text{C}$

Angle of attitude, $\beta = 60°$

SOLUTION

Viscosity of the oil SAE 30 (Fig. 14.2) $\mu' = 0.015$ kg/ms (Ns/m^2)

Step-1: Bearing pressure,

$$p = \frac{W}{Ld} = \frac{5000}{50 \times 50} = 2 \text{ MPa}$$

Step-2: Sommerfeld number,

$$S = \left(\frac{r}{c}\right)^2 \left(\frac{\mu Ns}{p}\right)$$

where, $\mu = \mu' \times 10^{-6} = 0.015 \times 10^{-6}\, \text{Ns/mm}^2$,

$$N_s = \frac{1000}{60} = 16.67 \text{ rps}, \; p = 2 \text{ MPa}, \; \frac{r}{c} = \left(\frac{25}{.025}\right) = 1000$$

$$\therefore \text{ Sommerfeld number, } S = (1000)^2 \left(\frac{0.015 \times 10^{-6} \times 16.67}{2}\right) = 0.125$$

Step-3: Minimum film thickness,

$$h_0 = (\text{MFT}) \cdot C \qquad \text{(Eqn. 14.32 MDHB)}$$

MFT coefficient from Fig. 14.5 is 0.42, and $C = 0.025$

$$\therefore \qquad h_0 = 0.42 \times 0.025 = 0.0105 \text{ mm}$$

Step-4: The coefficient of friction, f

Mackee's equation for friction coefficient of friction

$$f = (1950/10^8)\left(\frac{\mu' N_s}{p}\right)\left(\frac{r}{c}\right) + \Delta f$$

$$f = \frac{1950}{10^8}\left(\frac{.015 \times 16.67}{2}\right)\left(\frac{25}{.025}\right) + 0.002 = 0.00444$$

Step-5: Heat generated,

$$H_g = fW \cdot V_s \qquad \text{(Eqn. 14.47 MDHB)}$$

where, $V_s = \dfrac{\pi DN}{60{,}000} = \dfrac{\pi \times 50 \times 1000}{60{,}000} = 2.62 \text{ m/s}$

$$f = 0.0044 \quad \text{and} \quad W = 5000 \text{ N}$$

Substituting the values,

$$\therefore \qquad H_g = 0.00444 \times 5000 \times 2.62 = 58.119 \text{ watts}$$

Step-6: Heat dissipating capacity,

$$H_d = K_1 \cdot A \cdot (t_B - t_A) \qquad \text{(Eqn. 14.51 MDHB)}$$

where $K_1 = 33.5 \times 10^{-6}$ W/mm^2/K (moving air)

$$A = 20 \cdot d \cdot L = 20 \times 50 \times 50 = 50000 \text{ mm}^2$$

$$(t_B - t_A) = \frac{1}{2}(t_0 - t_A) = \frac{1}{2}(80 - 20) = 30°$$

$\therefore \quad H_d = 33.5 \times 10^{-6} \times 5 \times 10^4 \times (30) = 50.25$ watts

The artificial cooling required,

$$H = H_g - H_d = 58.119 - 50.25 = 7.869 \text{ watts}$$

Example 21.9: A 75 mm long full journal bearing of diameter 75 mm, supports a radial load of 12 kN at speed of 1800 rpm. Assume the ratio of diameter to diameteral clearance as 1000. The viscosity of the oil is 0.01 Ns/m^2 at the operating temperature.

Determine: (i) Sommerfeld number, (ii) Coefficient of friction and (iii) Amount of heat generated.

Given Data: Diameter, $d = 75$ mm; Length, $L = 75$ mm;

The diameteral ratio $\left(\frac{d}{C}\right) = 1000$; Viscosity $\mu' = 0.01$ Ns/m^2

Load on bearing, $W = 12$ kN, Speed, $N = 1800$ rpm

SOLUTION

Step-1: Sommerfeld number

$$S = \left(\frac{r}{c}\right)^2 \left(\frac{\mu N_s}{p}\right)$$

where $\frac{r}{c} = \frac{d}{c} = 1000$

$$\mu = \mu' \times 10^{-6} \text{ Ns/mm}^2 = 0.01 \times 10^{-6} \text{ Ns/mm}^2$$

$$N_s = \frac{N}{60} = \frac{1800}{60} = 30 \text{ rps}, \; p = \frac{W}{ld} = \frac{12000}{75 \times 75} = 2.13 \text{ MPa}$$

Substituting the values,

$$S = (1000)^2 \times \left(\frac{0.01 \times 10^{-6} \times 30}{2.13}\right) = 0.1406$$

Step-2: Coefficient of friction,

$$f = (Fv)\left(\frac{c}{r}\right) \qquad \text{(Eqn. 14.27 MDHB)}$$

where $\left(\frac{c}{r}\right) = \frac{1}{1000}$, $(Fv) = 3.0$ (Fig. 14.3 MDHB)

$$\therefore \qquad f = \frac{3.0}{1000} = 0.003$$

Step-3: The amount of heat generated

$$H_g = f \cdot W \cdot V_s \qquad \text{(Eqn. 14.47 MDHB)}$$

where $f = 0.003$ and $W = 12000$ N

$$V_s = \frac{\pi DN}{60,000} = \frac{\pi \times 75 \times 1800}{60,000} = 7.07 \text{ m/s}$$

$$\therefore \qquad H_g = 0.003 \times 12000 \times 7.07 = 254.52 \text{ watts}$$

Example 21.10: Design a journal bearing to withstand a load of 5886 N, at the speed of 1000 rpm. The journal is made of hardened steel and bearing is made of babbit; operating temperature is 70°C and ambient temperature is 30°C. Check the design for the thermal equilibrium and also determine the power loss at the bearing. The lubricant used is of grade *SAE* 40 and $l/d = 1.5$.

Given Data: Load on the bearing, $W = 5886$ N, Speed, $N = 1000$ rpm

Material for journal – Hardened steel, for bush – babbit

Operating temperature, $t_0 = 70°\text{C}$, Ambient temperature, $t_a = 30°\text{C}$, $\frac{l}{d}$ ratio $= 1.5$

SOLUTION

Step-1: Bearing data

$$\text{Bearing modulus} \left(\frac{ZN}{p}\right)_c = 2540 \qquad \text{(Table 14.9 MDHB)}$$

$$\text{The operating values of bearing modulus} \left(\frac{ZN}{p}\right)_0 = 3 \times 2540 = 7620$$

$$\text{The oil pressure, } p = \frac{W}{l \cdot d} = \frac{5886}{1.5d^2} = \left(\frac{3924}{d^2}\right)$$

From Fig. 14.2, the viscosity of oil $\mu' = 0.028 \text{ NS/m}^2$, $Z = 28Cp$
The bearing speed $N = 1000$ rpm
Substituting these values in operating modulus,

$$\frac{28 \times 1000}{(3924/d^2)} = 7620$$

$$\therefore \qquad d^2 = \frac{7620 \times 3924}{28 \times 1000} = 1067.89$$

$$\therefore \qquad d = 32.67 \text{ mm}$$

Choose $d = 35$ mm

$$\therefore \quad l = 1.5 \times d = 1.5 \times 35 = 52.5$$

Choose $l = 55$ mm; assume $\frac{r}{c} = 1000$

Step-2: Sommerfeld number

$$S = \left(\frac{r}{c}\right)^2 \left(\frac{\mu N_s}{p}\right)$$

where, $\mu = 0.028 \times 10^{-6}$ Ns/mm^2, $N_s = 1000/60 = 16.67$ rps

$$p = \frac{5886}{35 \times 55.0} = 3.06 \text{ MPa}$$

$$\therefore \quad S = (1000)^2 \times \left(\frac{0.028 \times 10^{-6} \times 16.67}{3.06}\right) = 0.1525$$

Step-3: Coefficient of friction

The friction variable (Fig. 14.3 MDHB) $FV = 3$

Then friction coefficient, $f = (FV)\frac{C}{r} = 3/1000 = 0.003$

Step-4: Minimum film thickness

$$h_0 = (\text{MFT}) \cdot c \qquad \text{(Eqn. 14.32 MDHB)}$$

where, MFT = 0.42, (Fig. 14.5 MDHB)

$$c = \frac{r}{1000} = \frac{35}{2000} = 0.0175 \text{ mm}$$

$$\therefore \quad h_0 = 0.42 \times 0.017 = 0.00735 \text{ mm} = 7.35 \text{ µm}$$

The combined surface roughness, $R_0 = \frac{7.5}{3} = 2.45$ µm

Step-5: Power lost in friction or heat generated

$$H_g = f \cdot W \cdot V_s \qquad \text{(Eqn. 14.47 MDHB)}$$

where, $f = 0.003$ and $W = 5886$ N

$$V_s = \frac{\pi dN}{60000} = \frac{\pi \times 35 \times 1000}{60000} = 1.8325 \text{ m/s}$$

$$\therefore \quad H_g = 0.003 \times 5886 \times 1.8325 = 32.36 \text{ watts}$$

Step-6: The heat dissipating capacity

$$H_d = \frac{L \cdot d\,(T+18)^2}{K_3} \qquad \text{(Eqn. 14.53 MDHB)}$$

where, $$T = (t_B - t_A) = \frac{1}{2}(t_0 - t_A) = \frac{1}{2}(70 - 30) = 20$$

$$K_3 = 0.273 \times 10^6 \text{ mm}^2/\text{watt}$$

$$\therefore \qquad H_d = \frac{55 \times 35 \times (20+18)^2}{0.273 \times 10^6} = 10.182 \text{ watts}$$

$\therefore$ $H_g > H_d$, hence artificial cooling is required.

Example 21.11: The following data refer to a full journal bearing:

Radial load, $W = 25$ kN; $\frac{L}{d}$ ratio 1 : 1; Class of fit – H7e7

Unit bearing pressure, $p = 2.5$ MPa; Viscosity of the oil, $Z = 20$ CP

Calculate (i) Dimensions of the bearing

(ii) Minimum film thickness

(iii) Required oil flow

SOLUTION

Step-1: Dimensions of the bearing

The unit pressure, $p = \frac{W}{l \cdot d} = \frac{W}{d^2}$

$\therefore$ Diameter of bearing, $d = \sqrt{\frac{W}{p}} = \sqrt{\frac{25000}{2.5}} = 100$ mm

Length of the bearing, $l = d = 100$ mm

For fit class – H7e7

The exact diameters –

Hole, upper limit, $U = 100.035$ mm; lower limit, $L = 100$ mm

Shaft, upper limit, $u = (100 - 0.072)$ mm, lower limit, $l = (100 - 0.107)$ mm

Maximum clearance, $C_g = U - l = 100.035 - (100 - 0.107) = 0.142$ mm

Minimum clearance, $C_s = L - u = 100 - (100 - 0.072) = 0.072$ mm

Diametral clearance ratio, $\frac{d}{C} = \frac{100}{0.072} = 1388.89$

Step-2: Sommerfeld number

$$S = \left(\frac{r}{c}\right)^2 \left(\frac{\mu N_s}{p}\right) \qquad \text{(Eqn. 14.23 MDHB)}$$

where $\left(\frac{r}{c}\right) = \left(\frac{d}{C}\right) = 1388.89,\ \mu = 20 \times 10^{-9}\ \text{Ns/mm}^2,\ p = 2.5\ \text{MPa}$

$$N_S = \frac{N}{60} = \frac{1200}{60} = 20\ \text{rps}\ (N = 1200\ \text{rpm assumed})$$

$\therefore$ Sommerfeld number, $S = (1388.89)^2 \times \left(\frac{20 \times 10^{-9} \times 20}{2.5}\right) = 0.309$

Step-3: Minimum film thickness

$$h_0 = (MFT) \cdot c \qquad \text{(Eqn. 14.32 MDHB)}$$

where, MFT (from Fig. 14.5) $= 0.65,\ c = .072/2 = 0.036$ mm

$\therefore \qquad h_0 = 0.65 \times 0.036 = 0.0234$ mm

Step-4: Oil flow required,

$$Q = (FLV) \cdot (r \cdot c \cdot N_s \cdot L) \qquad \text{(Eqn. 14.36 MDHB)}$$

where, (FLV) (from Fig. 14.4) $= 4$

$r = 50$ mm; $c = 0.036,\ N_s = 20$ rps; and $L = 100$

Substituting the values,

$$Q = 4 \times 50 \times 0.036 \times 20 \times 100 = 14400\ \text{mm}^3/\text{s}$$

Example 10.12: A 75 mm long journal bearing of diameter 75 mm, supports a load of 15 kN. The ratio of $\frac{d}{C} = 1000$ and the viscosity of the oil is 25×10^{-3} Pas. The heat generated in bearing is 442 watts. Determine the maximum speed of the journal using Mackee's equation.

Given Data: Load, $W = 15$ kN,

Bearing diameter, d = length of bearing $L = 75$ mm

The clearance ratio, $\frac{d}{C} = \left(\frac{r}{c}\right) = 1000,$

Viscosity of oil, $\mu' = 25 \times 10^{-3}$ P

Heat generated, $H_g = 442$ watts

SOLUTION

Step-1: Coefficient of friction

The unit pressure pressure, $p = \frac{W}{Ld} = \frac{15000}{75 \times 75} = 2.67$ MPa

Then by Mackee's equation, the friction coefficient,

$$f = \frac{32.6}{10^8} \times \left(\frac{25 \times 10^{-3} \times N}{2.67} \right) \times 1000 + 0.002 = \frac{305.24}{10^8} \cdot N + 0.002 \quad (1)$$

Step-2: The heat generated,

$$H_g = f \cdot W \cdot V_s \quad \text{(Eqn. 14.47 MDHB)}$$

where, $W = 15000$ N; $H_g = 442$ watts

$$V_s = \frac{\pi dN}{60,000} = \frac{\pi \times 75 \times N}{60,000} = 3.93 \times 10^{-3} \times N$$

$$\therefore \quad 442 = f \times 15000 \times 3.93 \times 10^{-3} \text{ N} = (58.9 f) \times N$$

$$\therefore \quad f \times N = \frac{442}{58.9} = 7.504 \quad (2)$$

Step-3: Speed of the bearing

Substituting for f; from (1) in (2)

$$\left(\frac{305.243}{10^8} \cdot N + 0.002 \right) \times N = 7.504$$

$$\frac{305.243}{10^8} \cdot N^2 + 0.002 \cdot N = 7.504$$

Substituting $N = 1000N'$

$$\frac{305.24}{100} \cdot N'^2 + 2 \cdot N' = 7.504$$

$$\therefore \quad N'^2 + 0.655N' = 2.46 \quad (3)$$

Solving $N' = 1.9298$ and $N = 1929.8 \simeq 1930$ rpm

The maximum speed of the journal $N = 1930$ rpm

Example 21.13: The main bearing of a steam engine runs at 1500 rpm and has a diameter of 40 mm. The load on the bearing is estimated to be 3 kN. Assume clearance ratio of 0.001, length to diameter ratio is 1.5 and is well ventilated. The operating temp. of the oil film is 60°C and the oil used is turbine oil SAE 10. Determine whether (i) Fluid film lubrication is expected, (ii) Is artificial cooling necessary? and (iii) Amount of oil flow.

Given Data: The load on the bearing, $W = 3$ kN; Speed $N = 1500$ rpm

Diameter of bearing, $d = 40$ mm; $\frac{L}{d}$ ratio $= 1.5$,

The clearance ratio, $c/r = 0.001$.

Oil used–turbine oil SAE 10, Operating temperature, $t_0 = 60°C$

SOLUTION

Step-1: Bearing modulus

Average pressure, $p = \dfrac{W}{Ld} = \dfrac{3000}{1.5 \times 40 \times 40} = 1.25$ MPa

The viscosity of the oil (Fig. 14.2), $\mu' = 0.016$ Ns/mm^2, $Z = 16$ cp

Bearing modulus $\left(\dfrac{ZN}{p}\right)_0 = \dfrac{16 \times 1500}{1.25} = 19200$

The modulus recommended (from Table 14.7 MDHB) is $\left(\dfrac{ZN}{p}\right)_{min} = 3560$

$$\therefore \qquad \left(\frac{ZN}{p}\right)_0 \bigg/ \left(\frac{ZN}{p}\right)_{min} = 5.4$$

Hence, the fluid film lubrication is possible.

Step-2: The coefficient of friction

Mackee's equation, $f = \dfrac{32.6}{10^8}\left(\dfrac{\mu' N}{p}\right)\left(\dfrac{r}{c}\right) + \Delta f$ (Eqn. 14.17a MDHB)

$$= 0.00626 + 0.002 = 0.00826$$

Step-3: Heat balance

The heat generated, $H_g = f \cdot W \cdot V_s$ (Eqn. 14.47 MDHB)

where, $f = 0.00826$, and $W = 3000$ N

$$V_s = \frac{\pi dN}{60,000} = \frac{\pi \times 40 \times 1500}{60,000} = 3.14 \text{ m/s}$$

$$\therefore \qquad H_g = 0.00826 \times 3000 \times 3.14 = 77.81 \text{ watts}$$

The heat dissipating capacity,

$$H_d = \frac{Ld\,(T+18)^2}{K_3} \qquad \text{(Eqn. 14.53 MDHB)}$$

where, $K_3 = 0.273 \times 10^6$;

$$T = (t_b - t_a) = \frac{1}{2}(t_0 - t_a) = \frac{(60-20)}{2} = 20°C;$$

$$\therefore \quad H_d = \frac{60 \times 40 \times (20+18)^2}{0.273 \times 10^6} = 20.255 \text{ watts}$$

$\therefore \quad H_g > H_d$ Hence, artificial cooling is required.

Step-4: Sommerfeld number

$$S = \left(\frac{r}{c}\right)^2 \left(\frac{\mu N_s}{p}\right)$$

where, $\mu = \mu' \times 10^{-6} = 0.016 \times 10^{-6} \text{ Ns/mm}^2, N_s = \frac{1500}{60} = 25 \text{ rps}$

$$p = 1.25 \text{ MPa}; \ \frac{r}{c} = 1000$$

$$\therefore \quad S = (1000)^2 \times \left(\frac{0.016 \times 25}{1.25}\right) \times 10^{-6} = 0.32$$

Step-5: Quantity of out flow

$$Q = (FLV)\,(rcN_sL) \qquad \text{(Eqn. 14.26 MDHB)}$$

where, (FLV) (from 14.4) = 3.2

$$\therefore \quad Q = 3.2 \times 20 \times \frac{20}{1000} \times 25 \times 60 = 1920 \text{ mm}^3/\text{s}.$$

Example 21.14: The following data are given for a $360°$ hydrodynamic bearing:

Bearing diameter, $d_h = 50.02$ mm; Journal diameter, $d_l = 49.93$ mm

Bearing length = 50 mm; Journal speed, $N = 1440$ rpm

Radial load, $W = 8$ kN; Viscosity of lubricant, $Z = 12$ Cp

The bearing is machined on a lathe from bronze casting, while the steel journal is hardened and ground. The surface roughness values for turning and grinding are 0.8 and 0.4 microns respectively. For the thick film lubrication, the minimum film thickness should be five times the sum of the roughness values for the journal and bearing. Calculate: (i) Permissible minimum film thickness, (ii) Actual film thickness under operating conditions, (iii) Power loss in friction and (iv) Flow requirement.

Given Data: Bearing diameter, $d_h = 50.02$ mm; Journal diameter, $d_l = 49.93$ mm
Bearing length = 50 mm; Journal speed, $N = 1440$ rpm
Radial load, $W = 8$ kN; Viscosity of lubricant, $Z = 12$ Cp

Roughness for shaft, $R_{01} = 0.4$ micron, for bearing $R_{02} = 0.8$ micron

SOLUTION

Step-1: Permissible minimum film thickness

The surface roughness for journal = 0.4 μm

The surface roughness for bush = 0.8 μm

The combined roughness, $R_0 = 0.4 + 0.8 = 1.2\ \mu\text{m}$

To avoid peak to peak contact of the surfaces, the film thickness,

$$h_0 = 5 \times 1.2 = 6.0\ \mu\text{m}$$

Step-2: Actual minimum film thickness

Actual clearance, $C = 50.02 - 49.93 = 0.09\text{ mm} = 90\ \mu\text{m}$

Radial clearance, $c = \frac{1}{2}C = \frac{1}{2} \times 90 = 45\ \mu\text{m} = .045\text{ mm}$

Clearance ratio $\left(\frac{r}{c}\right) = \frac{25}{0.045} = 555.56$

The average fluid pressure, $p = \frac{W}{Ld} = \frac{8 \times 10^3}{50 \times 50} = 3.2\text{ MPa}$

Speed, $N_s = \frac{N}{60} = \frac{1440}{60} = 24\text{ rps},$

Viscosity $\mu = Z \times 10^{-9} = 12 \times 10^{-9}\text{ Ns/mm}^2$

Sommerfeld number

$$S = \left(\frac{r}{c}\right)^2 \left(\frac{\mu N_s}{p}\right) \qquad \text{(Eqn. 14.23 MDHB)}$$

$$= (555.56)^2 \times \left(\frac{12 \times 10^{-9} \times 24}{3.2}\right) = 0.028$$

The minimum film thickness for MFT (from Fig. 14.5) = 0.15

$$(h_0)_a = (\text{MFT}) \cdot c \qquad \text{(Eqn. 14.32 MDHB)}$$

$$= 0.15 \times 0.045 = 6.75 \times 10^{-3}\text{ mm} = 6.75\ \mu\text{m} > h_0 = 6\ \mu\text{m Specified}$$

Hence, thick film lubrication is possible.

Step-3: The power loss in friction

The coefficient of friction variable (from Fig. 14.3), $FV = 1.6$

Then $f = (fv) \cdot \frac{r}{c} = 1.6 \times \frac{1}{555.56} = 0.00288$

The journal speed, $V_s = \frac{\pi dN}{60{,}000} = \frac{\pi \times 50 \times 1440}{60{,}000} = 3.77\text{ m/s}$

The power loss in friction = Heat generated

$$H_g = fWV_s \qquad \text{(Eqn. 14.47 MDHB)}$$

$$= .00288 \times 8000 \times 3.77 = 86.86 \text{ watts}$$

Step-4: Quantity of oil flow

The flow variable (from Fig. 14.4 MDHB) is $FLV = 1.4$.

The oil flow $Q = (FLV)\,(V \cdot c \cdot N_s L)$ (Eqn. 14.36 MDHB)

$$= 1.4 \times 25 \times 0.045 \times 24 \times 50 = 1890 \text{ mm}^3/\text{s}$$

Example 21.15: Design the main bearing of a steam turbine that runs at 1800 rpm and 70°C. The load on the bearing is estimated to be 2500 N.

Given Data: Load on the bearing, $W = 2500$ N

Speed of the journal, $N = 1800$ rpm; Operating temperature, $t_0 = 70°C$

SOLUTION

Step-1: Basic dimensions of bearing

From Table 14.7 MDHB, the data for steam turbine bearing are

Average Pressure, $p = 0.7$ to 1.9 MPa; Viscosity, $Z = 2$ to 16 CP

Modulus, $\frac{ZN}{p} = 14220$, Clearance ratio, $\frac{c}{r} = 0.001$

Select viscosity, $Z = 16$ CP

The corresponding oil from Fig. 14.2 MDHB, is SAE 20

The operating value of modulus is

$$\left(\frac{ZN}{p}\right)_0 = 3 \times 14220 = 42660$$

Then operating pressure, $p = \dfrac{16 \times 1800}{42660} = 0.675$ MPa

Assuming $\dfrac{L}{d} = 1.5$, $p = \dfrac{W}{Ld} = \dfrac{2500}{1.5d^2} = 0.675$ MPa

$$\therefore \quad d = \sqrt{\frac{2500}{1.5 \times 0.675}} = 49.68 \text{ mm}$$

Choose $d = 50$ mm. Then $L = 1.5d = 1.5 \times 50 = 75$ mm

The actual operating pressure, $p = \dfrac{2500}{75 \times 50} = 0.667$ MPa

This is acceptable.

Select $\frac{c}{r} = 0.001$, then $c = 0.001 \times 25 = 0.025$ mm

Shaft fit H7f7 gives this clearance.

Step-2: The coefficient of friction by Mackee's equation,

$$f = \frac{32.6}{10^8}\left(\frac{\mu' N}{p}\right)\left(\frac{r}{c}\right) + \Delta f$$

$$= \left(\frac{32.6}{10^8}\right)\left(\frac{.016 \times 1800}{0.667}\right) \times 1000 + 0.002$$

$$= 0.014 + 0.002 = 0.016$$

Step-3: Heat generated,

$$H_g = f \cdot W \cdot V_s \qquad \text{(Eqn. 14.47 MDHB)}$$

$$= 0.016 \times 2500 \times \left(\frac{\pi \times 50 \times 1800}{60{,}000}\right) = 188.5 \text{ watts}$$

Step-4: The heat dissipating capacity,

$$H_d = K_1 A (t_B - t_A) \qquad \text{(Eqn. 14.51 MDHB)}$$

where, $K_1 = 33.5 \times 10^{-6}$ watts/mm^2/K

$$A = 20dL = 20 \times 50 \times 75 = 75000 \text{ mm}^3$$

$$t_B - t_A = \frac{1}{2}(t_0 - t_A) = \frac{1}{2}(70 - 20) = 25°\text{C}$$

$$\therefore \qquad H_d = 33.5 \times 10^{-6} \times 75 \times 10^3 \times 25 = 62.812 \text{ watts}$$

$\therefore$ $H_d < H_g$. Hence, artificial cooling is required.

Step-5: Minimum film thickness

Sommerfeld number, $S = \left(\frac{r}{c}\right)^2 \left(\frac{\mu N_s}{p}\right)$

where, $N_s = \frac{1800}{60} = 30$ and $\mu = 16 \times 10^{-9}$ and $p = 0.667$

$$\therefore \qquad S = (1000)^2 \times \left(\frac{16 \times 10^{-9} \times 30}{0.667}\right) = 0.7164$$

The MFT factor from Fig. 14.5 = 0.85

$\therefore$ $h_0 = \text{MFT}\, c = 0.85 \times 0.025 = 0.02125 \text{ mm} = 21.25\ \mu\text{m}$

The combined surface roughness value, $R = 21.25/3 = 7.08\ \mu\text{m}$

This can be achieved by grinding of shaft, – $R_s = 6.3\ \mu\text{m}$

and honing of bush, $R_h = 0.4\ \mu\text{m}$ (Table 13.6 MDHB)

Step-6: Quantity of oil flow

Flow variable (Fig. 14.4) $FLV = 3.6$

$$\therefore \quad Q = (FLV)(r \cdot c \cdot N_s L) \qquad \text{(Eqn. 14.36 MDHB)}$$

$$= 3.6 \times 25 \times 0.025 \times 30 \times 75 = 5062\ \text{mm}^3/\text{s}.$$

21.7 HYDROSTATIC LUBRICATIONS

Hydrostatic step bearing

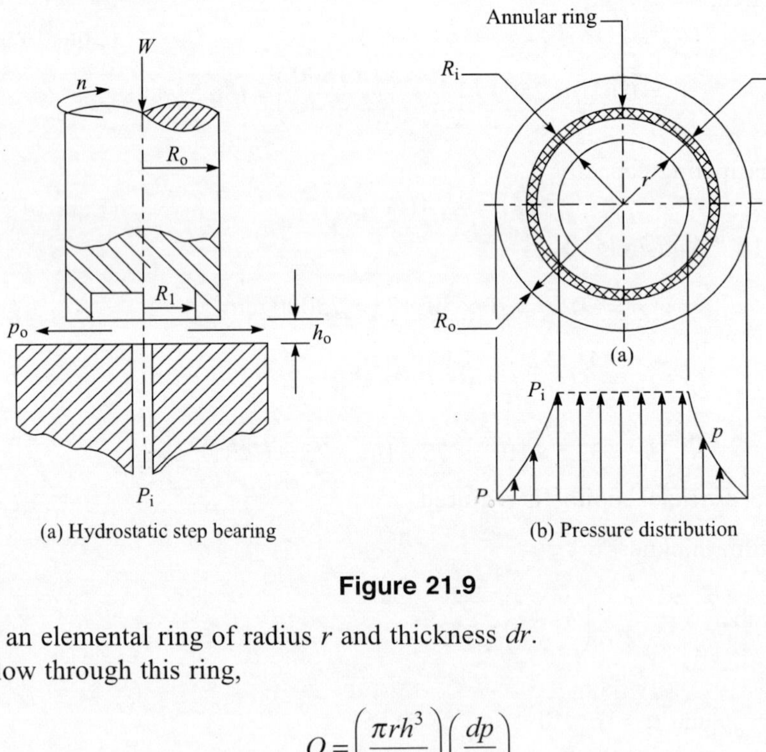

(a) Hydrostatic step bearing (b) Pressure distribution

Figure 21.9

Consider an elemental ring of radius r and thickness dr.
The oil flow through this ring,

$$Q = \left(\frac{\pi r h^3}{6\mu}\right)\left(\frac{dp}{dr}\right)$$

Rearranging the terms,

$$dp = -\frac{6\mu Q}{\pi h^3}\frac{dr}{r}$$

Integrating, we get

$$p = -\left(\frac{6\mu Q}{\pi h^3}\right)\log_e r + C$$

C is the constant of integration and is evaluated by the boundary condition that

$$r = R_o, \; p = 0$$

∴ $$C = +\frac{6\mu Q}{\pi h^3}\log_e R_0$$

Substituting the value, the pressure distribution is

$$p = \frac{6\mu Q}{\pi h^3}\log\left(\frac{R_o}{r}\right) \tag{21.14}$$

Since $p = P_i$ at $r = R_i$, we get

$$P_i = \frac{6\mu Q}{\pi h^3}\log\left(\frac{R_0}{R_i}\right) \tag{21.15}$$

Then oil flow,

$$Q = \frac{\pi p_i h^3}{6\mu \log\left(\frac{R_0}{R_i}\right)} \tag{21.16}$$

Load Capacity

The load supported is the sum of the load supported by the constant pressure in central portion plus that supported by the variable pressure portion.

$$W = (\pi R_i^2 P_i) + \int_{R_i}^{R_o} 2\pi r \cdot p \cdot dr$$

$$= \pi R_i^2 P_i + \int_{R_i}^{R_o} 2\pi\left(\frac{6\mu Q}{\pi h^3}\right)\log\left(\frac{R_0}{r}\right) rdr$$

$$= \pi R_i^2 P_i + \left[\left(\frac{12\mu Q}{h^3}\right)\right]\left[\frac{r^2}{2}\log\left(\frac{R_0}{r}\right) + \frac{r^2}{4}\right]_{R_i}^{R_0}$$

$$= \pi R_i^2 P_i + \frac{12\mu Q}{h^3}\left[\frac{R_0^2 - R_i^2}{4} - \frac{R_i^2}{2}\log\frac{R_0}{R_i}\right]$$

$$= \pi R_i^2 P_i + \frac{2\pi P_i}{\log\left(\frac{R_0}{R_i}\right)}\left[\frac{R_0^2 - R_i^2}{4} - \frac{R_i^2}{2}\log\frac{R_0}{R_i}\right]$$

$$W = \frac{\pi P_i (R_0^2 - R_i^2)}{2\log\frac{R_0}{R_i}} = \left(\frac{\pi P_i}{2}\right)\left(\frac{R_0^2 - R_i^2}{log\frac{R_0}{R_i}}\right) \tag{21.17}$$

Power Loss in the Bearing

Consider the ring of the oil film, the viscous resistance by the ring by Newton's law of viscosity,

$$dF = \mu A\left(\frac{U}{h}\right) = \mu(2\pi r\,dr)\left(\frac{2\pi r\cdot n}{60}\right)\frac{1}{h} = \left(\frac{4\pi^2 n\mu}{60h}\right)r^2 dr$$

The torque due to the viscous drag,

$$dT_f = \left(\frac{4\pi^2 n\mu}{60h}\right)r^3 d_r$$

The total torque by integration,

$$T_f = \left(\frac{4\pi^2}{60}\right)\left(\frac{\mu n}{h}\right)\left(\frac{r^4}{4}\right)_{R_i}^{R_0}$$

$$= \left(\frac{\pi^2}{60}\right)\left(\frac{\mu\cdot n}{h}\right)(R_0^4 - R_i^4) \tag{21.18}$$

Power loss is given by,

$$kW_f = T_f\frac{\omega}{10^6} = \left(\frac{2\pi n}{60}\right)\left(\frac{\pi^2}{60}\right)\left(\frac{\mu n}{h}\right)\frac{R_0^2 - R_i^2}{10^6}$$

$$= \frac{2\pi^3 n^2\mu}{3600\times 10^6}\times\left(\frac{R_0^4 - R_i^4}{h}\right)$$

$$kW_f = \frac{n^2\mu}{58.05\times 10^6}\times\left(\frac{R_0^4 - R_i^4}{h}\right) \tag{21.19}$$

The power required to pump the oil,

$$kW_p = Q(P_i - P_o)\times 10^{-6} \tag{21.20}$$

The total power loss,

$$kW_T = kW_p + kW_f \tag{21.21}$$

Conical Thrust Bearing

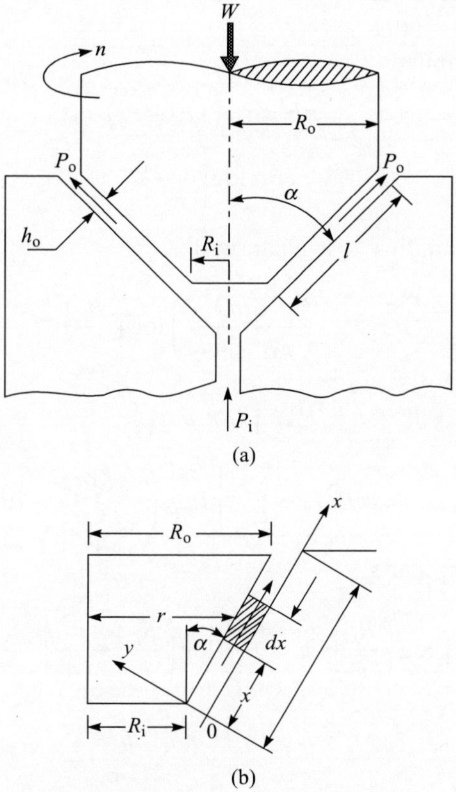

Figure 21.10 Conical thrust bearing

Consider an elemental ring of thickness *dx* at a distance *x* from the bottom end as shown in figure.

The radius is, $r = R_1 + x \sin \propto$

The oil flow through the ring,

$$Q = \frac{\Delta P(2\pi r)h^3}{12\mu dx}$$

$$\Delta P = \left(\frac{12\mu Q}{2\pi h^3}\right)\frac{dx}{r}$$

Integrating,

$$\int_{P_s}^{P_a} dp = -\int_{R_i}^{R_o} \left(\frac{6\mu Q}{\pi h^3}\right)\frac{1}{\sin \propto}\left(\frac{d_r}{r}\right)$$

$$(P_s - P_a) = +\left(\frac{6\mu Q}{\pi h^3 \sin \propto}\right)\log\left(\frac{R_0}{R_i}\right)$$

P_s = the oil supply pressure at inlet
P_a = the oil pressure at the outlet
The quantity of the oil supplied,

$$\therefore \qquad Q = \frac{\pi h^3 \sin \propto (P_s - P_a)}{6\mu \log\left(\frac{R_o}{R_i}\right)} \tag{21.22}$$

For the oil pressure at any radius 'r' we can write,

$$P = P_s - \left(\frac{6\mu Q}{\pi h^3 \sin \propto}\right) \log\left(\frac{R_i}{r}\right) \tag{21.23}$$

Load capacity,
$$W = \pi R_i^2 P_s + \int 2\pi r\, pdr$$

$$= \pi R_i^2 P_s + \int 2\pi \left[P_s - \left(\frac{6\mu Q}{\pi h^3 \sin \propto}\right) \log\left(\frac{R_i}{r}\right)\right] rdr$$

Where, using integration by parts,

$$\int \log\left(\frac{R_i}{r}\right) rdr = \left[\frac{r^2}{2}\log\left(\frac{R_i}{r}\right) + \frac{r^2}{4}\right]_{R_i}^{R_0}$$

$$= \frac{R_0^2 - R_i^2}{4} - \frac{R_o^2}{2}\log\left(\frac{R_o}{R_i}\right)$$

$$\frac{12\mu Q}{h^3 \sin \propto} = \frac{(P_s - P_a)}{\log\left(\frac{R_o}{R_i}\right)} = \frac{2\pi (P_s - P_a)}{\log\left(\frac{R_o}{R_i}\right)}$$

Substituting the values,

$$\therefore \qquad W = \pi(P_s - P_a)R_i^2 + \left[\frac{2\pi(P_s - P_a)}{\log\left(\frac{R_o}{R_i}\right)}\right]\left[\frac{R_0^2 - R_i^2}{4} - \frac{R_1^2}{2}\log\left(\frac{R_o}{R_i}\right)\right]$$

$$= \frac{\pi (P_s - P_a)(R_0^2 - R_i^2)}{2 \log\left(\frac{R_o}{R_i}\right)} \tag{21.24}$$

Power loss due to friction in the bearing

Consider the ring of the oil film, the viscous resistance by the ring, by Newton's law of viscosity is,

$$dF = \mu A\left(\frac{U}{h}\right) = \mu(2\pi r dx)\left(\frac{2\pi r \cdot n}{60}\right)\frac{1}{h} = \left(\frac{4\pi^2 n\mu}{60h\sin\alpha}\right)r^2 dr$$

The torque due to the viscous drag,

$$dT_f = \left(\frac{4\pi^2 n\mu}{60h\sin\alpha}\right)r^3 d_r$$

The total torque by integration,

$$T_f = \left(\frac{4\pi^2}{60}\right)\left(\frac{\mu n}{h\sin\alpha}\right)\left(\frac{r^4}{4}\right)_{R_i}^{R_0}$$

$$= \left(\frac{\pi^2}{60}\right)\left(\frac{\mu \cdot n}{h\sin\alpha}\right)(R_0^4 - R_i^4)$$

Power loss is given by

$$kW_f = T_f\frac{\omega}{10^6} = \frac{2\pi^3 n^2\mu}{3600\times 10^6}\times\left(\frac{R_o^4 - R_i^4}{h\sin\alpha}\right)$$

$$kW_f = \frac{n^2\mu}{58.05\times 10^6}\times\left(\frac{R_o^4 - R_i^4}{h\sin\alpha}\right) \qquad (21.25)$$

The power required to pump the oil is the same as step bearing.

Example 21.16: The following are the data for a hydrostatic thrust bearing:

Shaft diameter = 400 mm; Recess diameter = 250 mm

Film thickness, h = 0.15 mm; Supply pressure = 5 MPa

Viscosity of lubricant = 30 cP; Specific gravity = 0.86

Specific heat = 1.75 kJ/kg°C; Shaft speed = 720 rpm

Calculate (i) load carrying capacity of the bearing, (ii) flow requirement, (iii) pumping power loss, (iv) frictional power loss, and (v) temperature rise.

Given Data:

Shaft diameter = 400 mm, Recess diameter = 250 mm,

Film thickness, h = 0.15 mm; Supply pressure = 5 MPa,

Viscosity of lubricant = 30 cP, Specific gravity = 0.86,

Specific heat = 1.75 kJ/kg°C, Shaft speed = 720 rpm

SOLUTION

The load capacity,

$$W = \left(\frac{\pi P_i}{2}\right)\left(\frac{R_o^2 - R_i^2}{\log\frac{R_o}{R_i}}\right) = \left(\frac{\pi\times 5}{2}\right)\left(\frac{400_o^2 - 250_o^2}{\log\frac{400}{250}}\right)$$

$$= 407318 \text{ N} = 407.318 \text{ kN}$$

Flow required,

$$Q = \frac{\pi p_i h^3}{6\mu \log\left(\frac{R_0}{R_i}\right)} = \frac{\pi \times 5 \times 0.15^3}{6 \times 30 \times 10^{-9} \log\left(\frac{400}{250}\right)}$$

$$= 6.26643 \times 10^5 \text{mm}^3/\text{s} = 37.6 \text{ litres/min}$$

Pumping power,

$$kW_p = Q(P_i - P_o) \times 10^{-6} = 6.26643 \times 10^5 \times 5 \times 10^{-6} = 3.13 \text{ kW}$$

Frictional power loss,

$$kW_f = \frac{n^2 \mu}{58.05 \times 10^6} \times \left(\frac{R_o^4 - R_i^4}{h}\right) = \frac{720^2 \times 30 \times 10^{-9}}{58.05 \times 10^6} \times \left(\frac{400_o^4 - 250_i^4}{0.15}\right) = 2.42 \text{ kW}$$

The total power loss,

$$kW_t = kW_p + kW_f = 3.133 + 2.42 = 5.553 \text{ kW}$$

The generated, $H = 5.553 \text{ kW} = 5.553 \text{ kJ/s}$

The mass of the oil flow,

$$M = \rho Q = 0.86 \times 6.26643 \times 10^5 \times 10^{-6} \text{ kg/s} = 0.5389 \text{ kg/s}$$

The temperature rise,

$$\Delta t = \frac{\text{heat generated}}{MC_p} = \frac{5.553}{0.5389 \times 1.75} = 5.88° \text{C}$$

Example 21.17: The following data refer to a hydrostatic conical thurst bearing:

The shaft speed = 250 rpm; Shaft diameter = 400 mm

Bottom end diameter of the shaft = 200 mm; Cone angle, $2\alpha = 150°$

The film thickness = 0.20 mm; Supply pressure = 5 MPa

Viscosity of the oil = 30 cP; Mass density, $\rho = 0.86 \times 10^{-6}$ kg/ mm^3

Calculate (i) load capacity, (ii) quantity of oil flow, (iii) pumping power loss, (iv) frictional loss, and (v) temperature rise.

SOLUTION

The load capacity, $W = \dfrac{\pi(P_s - P_a)(R_o^2 - R_i^2)}{2 \log\left(\frac{R_o}{R_i}\right)} = \dfrac{\pi(5-0)(400_o^2 - 200_i^2)}{2 \log\left(\frac{400}{200}\right)} = 1359.71 \text{ kN}$

The quantity of oil flow, $Q = \dfrac{\pi h^3 \sin \propto (P_s - P_a)}{6\mu \log\left(\frac{R_o}{R_i}\right)}$

$$= \frac{\pi \times 0.2^3 \sin(75)(5-0)}{6 \times 30 \times 10^{-9} \log\left(\frac{400}{200}\right)} = 9.72872 \times 10^5 \text{ mm}^3/\text{s}$$

The pumping power loss, $kW_p = Q\,(P_i - P_o) \times 10^{-6}$

$$kW_p = 9.72872 \times 10^5 \times (5-0) \times 10^{-6} = 4.864 \text{ kW}$$

Energy lost in friction, $kW_f = \dfrac{n^2 \mu}{58.05 \times 10^6} \times \left(\dfrac{R_o^4 - R_i^4}{h \sin \alpha}\right)$

$$kW_f = \frac{250^2 \times 30 \times 10^{-9}}{58.05 \times 10^6} \times \left(\frac{400_o^4 - 200_i^4}{0.2 \times \sin(75)}\right) = 4.013 \text{ kW}$$

The temperature rise,

$$\Delta T = \frac{H}{m \cdot C_p}$$

where, H = heat generated = Total power loss = 4.864 + 4.01 = 8.874 kW = 8.874 KJ/s

m = mass of the oil = $Q\rho = 9.72872 \times 10^5 \times 0.86 \times 10^{-6} = 0.8367$ kg/s

C_p = specific heat of the oil = 1.75 kJ/kg

Substituting the values,

$$\Delta T = \frac{8.874}{0.8367 \times 1.75} = 6.061°\text{C}$$

EXERCISES FOR PRACTICE

21.1. (a) Explain the terms, bearing modulus and Sommerfeld number.

(b) A full journal bearing 200 mm diameter by 200 mm length supports a radial load of 45 kN. The journal rotates at 1200 rpm and the diametral clearance ratio is 1000. The viscosity of the oil at its operating temperature of 80°C is 18 MPa.S. (.018 kg/m.s). Ambient temperature is 20°C. Using Raimondi and Boyd curves, determine the minimum film thickness, the heat generated in the bearing, and the natural heat loss from the bearing assuming,

$$Hd = \frac{(\Delta T + 18)^2}{K} \times (\text{proj. area})$$

$K = 0.273°\text{C}\ M^2/W$, and the amount of oil leaking out of both ends.

21.2 (a) Describe how the friction between the journal and bearing varies from the instant at which rotation of journal begins until high speed is attained. Also sketch the curve showing how the coefficient of friction varies with bearing characteristic number.

(b) Design a hydrodynamic journal bearing for the following particulars:

Diameter of journal	- 70 mm
Speed of shaft	- 250 r.p.m.
Load on shaft	- 7,500 N
Viscosity of oil	- Saybolt of 270 seconds at 35°C
and Specific gravity	- 0.905 at 35°C.

Determine: (i) Length of bearing, (ii) Running clearance, (iii) Minimum film thickness, (iv) Loss of power due to friction, (v) Heat dissipation capacity of bearing, and (vi) Whether artificial cooling is required.

21.3 (a) Explain hydrodynamic lubrication in a journal bearing.

(b) A journal bearing 60 mm diameter, rotates at 175 rpm. The total load on the bearing due to belt pull and pulley weight is 7500 N. The load is steady. The lubricating oil has a viscosity of 270 Saybolt cc. at 40°C. The oil has a density of p 0.905 gm/cc at 15°C. Determine the bearing length, clearance, the power loss and the temperature of the oil film.

21.4 A journal bearing is to be designed for the main bearing of a four-stroke oil engine to sustain a load of 1500 kN. The diameter of the shaft is 40 mm and the shaft is to rotate at 1000 r.p.m. Determine the (i) dimensions of the bearing assuming suitable materials for the shaft and the bearing bush, (ii) viscosity of an oil to be used as lubricant and hence suggest a suitable oil, (iii) coefficient of friction, (iv) heat generated, (v) heat dissipating capacity of the bearing, and (vi) amount of heat to be removed by artificial cooling, if necessary.

21.5 The operating conditions of a journal bearing for a centrifugal pump are as follows:

Load on each journal	= 11500 N
Speed of the journal	= 1,440 rpm
Diameter of journal	= 7.5 mm
Operating temperature of bearing	= 70°C
Ambient temperature	= 40°C

Design the bearing and specify the suitable grade of lubricating oil to be used.

21.6 A journal bearing is to be designed, for main bearing of a four stroke oil engine, to sustain a radial load of kN and support a shaft of diameter of 40 mm. The shaft operates at a speed of 1000 rpm. The ambient temperature is 35°.

(i) Selecting suitable material for bearing bush, determine the length of bearing bush.
(ii) Choose an appropriate lubricant for the bearing.
(iii) Determine the coefficient of friction of bearing surfaces.
(iv) Determine the heat generated.
(v) Determine the heat dissipated.
(vi) Check whether artificial cooling is required. If required, determine the amount of heat to be removed by artificial cooling.

21.7 (a) Explain with sketches principles of hydrodynamic lubrication.

(b) Design the journal bearing for a centrifugal pump running at 1440 rpm. Diameter of the journal is 10 cm and the load on each bearing is 20 kN.

The factor ZN/P may be suitably assumed for centrifugal pump bearing.

21.8 (a) Explain the theory of hydrodynamic lubrication. Clarify the terms: (i) bearing modulus, and (ii) bearing characteristic number.

(b) A journal bearing with a diameter of 200 mm and length of 200 mm has to carry a load of 30 kN at the speed of 200 rpm. It is advised to run the bearing with slack running. If possible, the bearing is to operate at 32°C ambient temperature without external cooling with a maximum oil temperature of 90°C. If external cooling is required, it is to be as little as possible to minimize the required oil flow rate and heat exchanger size.

(i) What type of oil do you recommend?

(ii) Will the bearing operate without external cooling?

(iii) If the bearing operates without external cooling, determine operating oil temperature.

(iv) If the bearing operates with the external cooling, determine the amount of oil in kg/mm required to carry the excess heat generated over heat dissipated, when the oil temperature rises from 85°C to 90°C, when passing through the bearing.

21.9 (a) Define and explain the following:

(i) Bearing characteristic number

(ii) Sommerfeld number

(iii) Reliability factor

(iv) Critical characteristic number, in case of hydrodynamically lubricated bearings.

(b) A partial 120° centrally loaded bearing has the following specifications:

Diameter of the journal	= 100 mm
Radius of the bearing	= 50.0725 mm
L/D ratio	= 1.25
Speed of the journal	= 1000 rpm
Expected oil temperature	= 70°C
Minimum film thickness	= 0.0125 to 0.015 mm
Lubricating oil	SAE 60.

Calculate and comment on: (i) Load on bearing, (ii) Attitude with which bearing should operate, (iii) Sommerfeld number and (iv) Power loss in the bearings.

21.10 (a) State the importance of Sommerfeld number in design of journal bearings. Elaborate Sommerfeld number versus coefficient of friction variable, minimum film thickness variable, flow variable, flow ratio, temperature rise variable.

(b) A full formal bearing 200 mm diameter having L/D ratio as 1, supports a radial load of 45 kN. The journal rotates at 1200 rev./min and the D/C ratio is 1000. The viscosity of oil at the its operating 80°C temperature is 0.018 kg/ms. Ambient temperature is

20°C. Using Raimondi and Boyd curves, determine the minimum film thickness, the heat generated in the bearing, the amount of oil leaking out of both ends and heat removal from the bearing due to end leakage.

21.11 (a) Explain with sketches theory of hydrodynamic lubrication.

(b) A journal bearing is required to be designed for a rotary compressor for operation at a speed of 1500 rpm. The bearing is to sustain a load of 4500 N and the diameter of the main shaft is 50 mm. Determine:

(i) Length and inner diameter of the bearing bush.

(ii) Viscosity of the oil to be used as a lubricant and hence suggest a lubricating oil.

(iii) The coefficient of friction.

(iv) Heat generated.

(v) Heat dissipating capacity.

(vi) Amount of heat to be removed by artificial cooling.

(vii) Sommerfeld number.

21.12 (a) Derive Petroft's equation for coefficient of friction of a lightly loaded journal bearing.

(b) It is required to design a main bearing of a four-stroke oil engine to sustain a load of 6 kN over a shaft of diameter 50 mm. The operating speed of the shaft is 1000 rpm and the operating temperature is 50°C. Determine:

(i) Dimensions of the bearing.

(ii) The viscosity of the oil to be used for the bearing and hence suggest the appropriate oil.

(iii) Coefficient of friction.

(iv) Heat generated.

(v) Heat dissipated.

(vi) Heat to be removed by the artificial cooling if necessary.

(vii) Sommerfeld number.

21.13 (a) Derive the Petroff's equation for frictional power loss of a lightly loaded journal rotating at high speed concentric to the bearing.

(b) Determine the power loss in a bearing, the diameter of journal is 60 mm and length 80 mm. The diametral clearance is 0.12 mm. The bearing supports a load of 5000 N and the journal rotates at a speed of 2500 rpm. The kinematic viscosity and the specific gravity of the oil used at the operating temperature of the bearing are 50 centi-stokes and 0.9, respectively.

21.14 (a) Explain the bearing modulus.

(b) Derive Petroft's equation for a lightly loaded bearing.

(c) A 75 mm long full journal bearing of diameter 75 mm supports a radial load of 12 kN at the shaft speed of 1800 rpm. Assume ratio of diameter to the diametral clearance as 1000. The viscosity of oil is 0.01 Pas at the operating temperature. Determine:

(i) Sommerfeld number.

(ii) Coefficient of friction based on Mackee's equation.

(iii) Amount of heat generated.

21.15 (a) Explain the significance of the bearing characteristic number in the design of sliding contact bearings.

(b) A full journal bearing of 60 mm diameter and 100 mm length has a bearing pressure of 1.4 MPa. The speed of the journal is 800 rpm and the ratio of journal diameter to the diametral clearance is 1000. The bearing is lubricated with oil whose absolute viscosity at the operating temperature of 75°C may be taken as 0.011 kg/ms. The room temperature is 30°C. Find: (i) Amount of artificial cooling required, and (ii) Mass of the lubricating oil required, if the difference between the outlet and inlet temperatures of the oil is 10°C. Take specific heat of the oil as 1850 J/kg/°C.

21.16 Design a full journal bearing subjected to 6000 N at 1000 rpm of the journal. The journal is of hardened steel and the bearing is of babbit metal. The bearing is operating with SAE 40 oil at 70°C and the ambient temperature is 30°C. Also determine the amount of artificial cooling required.

21.17 (a) What is bearing modulus? Explain the significance of bearing modulus in the design of bearing.

(b) Design a journal bearing for a centrifugal pump from the following data:

Load on the journal = 10 kN

Speed of the journal = 900 rpm

Ambient temperature = 15°C.

21.18 (a) Explain with a neat sketch, the importance of bearing characteristic number in design of journal bearing.

(b) Design a journal bearing for a centrifugal pump running at 1200 rpm. Diameter of journal is 100 mm and load on bearing is 15 kN. Take = 1.5, bearing temperature 50°C and ambient temperature 30°C. Find whether artificial cooling is required.

21.19 Design a bearing and journal to support a load of 10 kN at 1000 rpm using a hardened steel journal and a bronze-backed babbit bearing. An abundance of oil, having a viscocity of 250 sec-Saybolt at 40°C and specific gravity of 0.9 at 15°C is provided. Assume an oil temperature of 85°C and clearance ratio of 0.0015.

21.20 Design a suitable journal bearing a soft steel machined shaft of 80 mm diameter, rotating at 1500 rpm. The load at the rigid bronze bearing is 15 kN. The operating temperature is 70°C and the ambient temperature is 25°C. The lubricant used has a viscosity of 27 cp at 70°C and specific heat of 0.45.

(i) Suggest the bearing dimensions.

(ii) Estimate the Sommerfeld number and coefficient of friction.

(iii) Compare this value with the value by Mackee's equation.

(iv) Determine the minimum film thickness.

(v) Estimate the heat generated and check the heat balance.

(vi) In case artificial cooling is required, determine the quantity of oil flow rate for temperature rise of 10°C.

21.21 The following data are given for a hydrostatic step bearing:

Thrust load = 500 kN; Shaft speed = 750 rpm

Shaft diameter = 500 mm; Recess diameter = 300 mm

Oil film thickness = 0.15 mm; Viscocity of oil =160 SUS

Specific gravity of the oil = 0.86; Specific heat = 2 kJ/Kg/C

Calculate: (i) Supply pressure, (ii) Fluid flow in litres/min, (iii) Power loss in pumping, (iv) Frictional power loss, and (v) Temperature rise.

21.22 Following data refers to hydrostatic step bearing of a vertical turbo generator:

Thrust load = 500 kN; shaft diameter = 500 mm; recess diameter = 250 mm; shaft speed = 75 rpm; viscocity of the lubricant = 30 cP. Calculate the optimum film thickness for minimum power loss.

SHORT ANSWER QUESTIONS

21.1 What constitutes a bearing?

21.2 What is a journal bearing? Explain.

21.3 What is a lubricant?

21.4 Name two principal characteristics of a lubricant.

21.5 What is viscosity? Indicate the coefficient of absolute viscosity by mathematical relation.

21.6 Give the classification of lubrication.

21.7 Give the mathematical representation of Petroff equation for lightly loaded journal bearing.

21.8 What is bearing characteristic number?

21.9 Give the outline of a journal bearing design.

21.10 What is the effect of temperature on kinematic viscosity? What is it's consequence on load carrying capacity?

21.11 How does the surface roughness of a bearing effect its working life?

21.12 List any five materials used in manufacturing journal bearing.

21.13 Name any five machineries where journal bearings are used.

21.14 Name the factors that account for heat generation in a journal bearing.

21.15 What is bearing pressure in a journal bearing?

21.16 What is the load carrying capacity of hydrostatic bearing in terms of the pressure and its geometry?

21.17 Differentiate between hydrostatic and hydrodynamic bearing from operational point of view.

21.18 Give the relation between oil flow, pressure and geometry of a hydrostatic bearing.

21.19 Give a comparison of load carrying capacity of hydrostatic step and conical bearing.

21.20 Give a comparison of power required to pump oil into hydrostatic step and conical bearing.

CHAPTER 22

BALL AND ROLLER BEARINGS

22.1 INTRODUCTION

The ball and roller bearings are also called antifriction bearings, because of the small coefficient of friction. The basic elements of these bearings are:

1. Outer and inner races with ball or roller raceways.
2. The rolling elements (ball or rollers) placed in the raceways between the two races.
3. Cage or the retainer which holds the rolling elements in their proper relative positions and guides them.

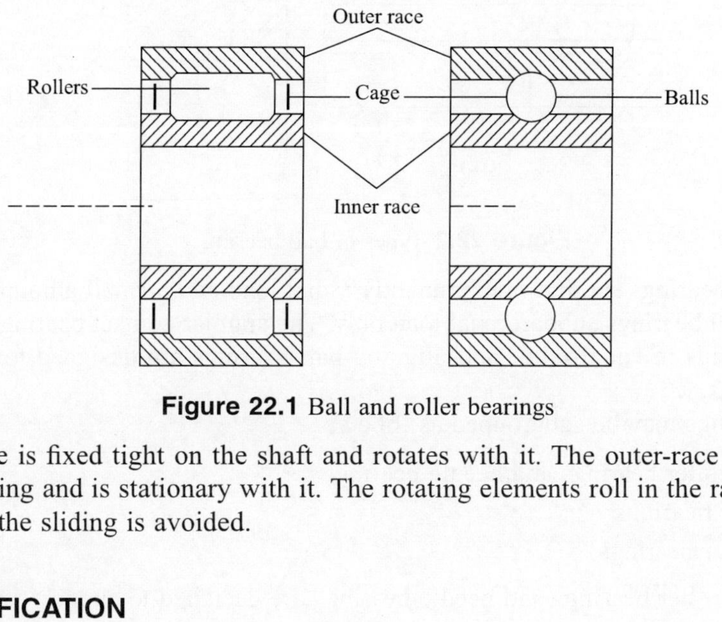

Figure 22.1 Ball and roller bearings

The inner-race is fixed tight on the shaft and rotates with it. The outer-race is fixed tight in the bearing housing and is stationary with it. The rotating elements roll in the races as the shaft rotates and thus the sliding is avoided.

22.2 CLASSIFICATION

The ball bearings may be subgrouped as follows:

(a) Radial ball bearings
(b) Thrust ball bearings
(c) Angular contact ball bearings
(d) Self-aligning ball bearings

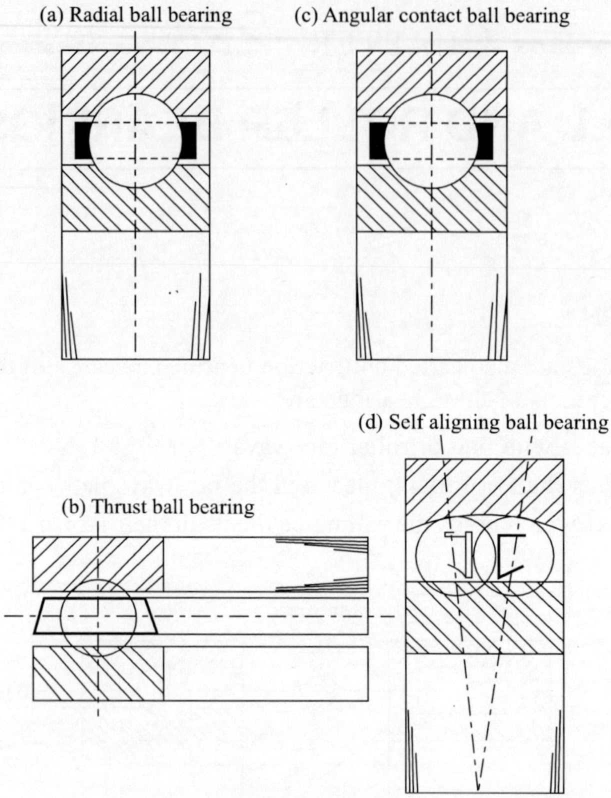

Figure 22.2 Types of ball bearing

The radial ball bearings support predominantly radial load and a small amount of axial load also. The thrust ball bearings support axial load only. The angular contact bearings support both radial and axial loads in equal ratio. Self-aligning ball bearings are designed to tolerate small shaft misalignments.

The roller bearings may be subgrouped as follows:

1. Cylindrical roller bearings and needle bearings
2. Thrust roller bearings
3. Tapered roller bearings

The cylindrical roller bearings and needle bearings are designed to support radial load only. The cylindrical roller bearings may support a small amount of axial load.

The tapered roller bearings are designed to take considerable amount of axial load also in addition to radial load. The thrust roller bearings are designed to support axial load only. The roller bearings, in general, are used for heavy loads.

The antifriction bearings are manufactured in a variety of types and sizes. They are standardized and can be selected from the manufacturer's catalogue.

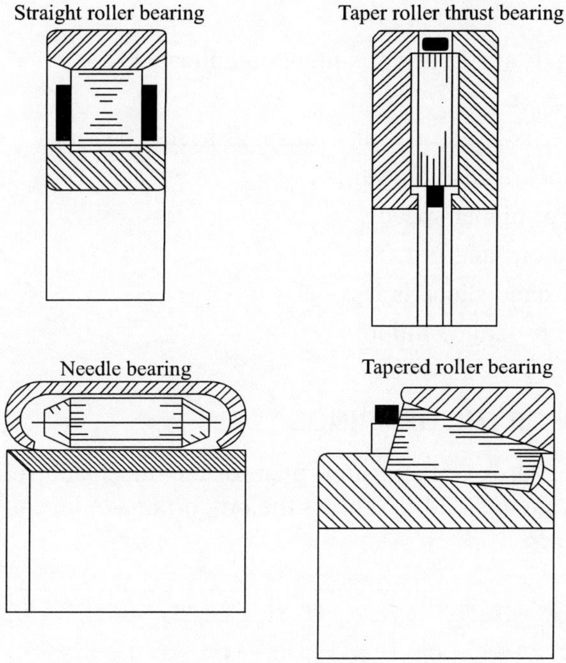

Figure 22.3 Types of roller bearings

22.3 ADVANTAGES AND DISADVANTAGES OF ANTIFRICTION BEARINGS

Advantages

1. Reduced friction and heat generation.
2. The coefficient of friction does not depend upon the running speed of the shaft.
3. The starting torque is considerably less than in sleeve bearings.
4. Consumption of lubricant is less than in sleeve bearings.
5. The maintenance is much simpler.
6. They have higher load carrying capacity.
7. They can carry heavy momentary overloads without failure.
8. They are particularly suited for low speeds.
9. They maintain accurate shaft alignment over long periods of time.
10. Replacement, in case of failure, is easy.
11. They occupy less axial space compared to diameter.
12. They can support both axial and radial loads.
13. They can be pre-loaded to make the shaft stiffer.
14. The lubrication is simple with grease package or with simple oil system.
15. The increased noise is the early warning of impending failure.

Disadvantages

1. The design of the shaft and housing is more complicated.
2. The initial cost is higher.
3. The housing diameter is larger except with small needle bearings.
4. The resistance to shock loads is lower.
5. There is more noise at higher speeds.
6. They are sensitive to dirt and grit.
7. The capacity to withstand shock is less.
8. The life is finite due to fatigue failure.

22.4 IDENTIFICATION OF THE BEARINGS

The ISO (International Standards Organization) plan for the dimension series of the bearings is illustrated in Fig. 22.4 (a). Figure 22.4 (b) shows the proportionate dimensions of SKF bearings with the same inner diameter.

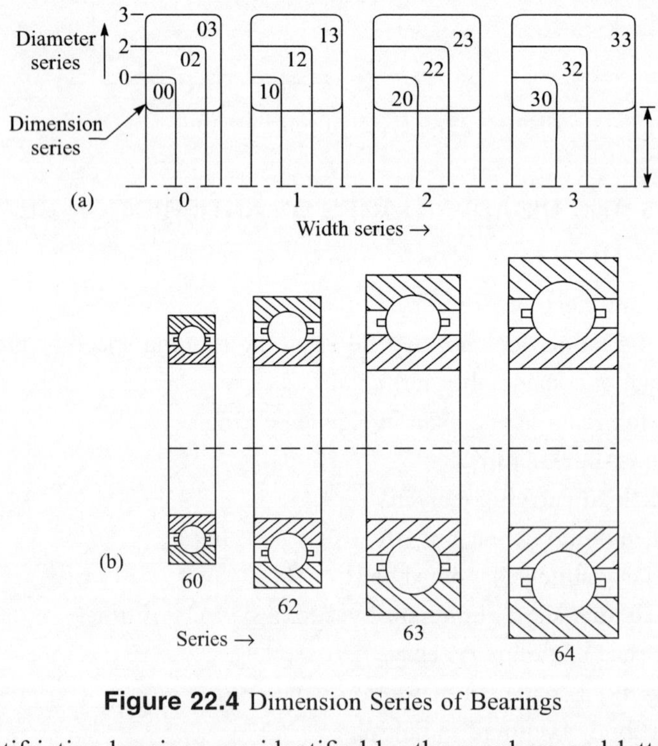

Figure 22.4 Dimension Series of Bearings

The standard antifriction bearings are identified by the numbers and letters that indicate the type of bearing and its dimensions. The first two digits counting from the right specify the bore diameter of the bearing. For bearings with a bore diameter from 20 mm to 495 mm, the digits indicate the bore diameter divided by 5. The third digit from the right indicates load series.

The extra light series by	1
Light series by	2
Medium series by	3
Heavy series by	4
Light wide series by	5
Medium wide series by	6
Heavy wide series by	7

The fourth digit from the right specifies the type of the bearings.

Single row radial ball bearing	0
Double row radial self-aligning ball bearing	1
Radial cylindrical roller bearings with short rollers	2
Double row self-aligning spherical roller bearing	3
Radial cylindrical roller bearing with long rollers or needles	4
Flexible roller bearings	5
Angular contact deep groove ball bearings	6
Tapered roller bearings	7
Thrust ball bearings	8
Thrust roller bearings	9

The fifth or fifth and sixth digits from the right are included for certain kinds of bearing only to indicate certain design features of the bearing.

The single row radial ball bearings with bore diameter of 50 mm are identified as follows:

Light series	210
Medium series	310
Heavy series	410

The roller bearings with short rollers are indicated as follows (diameter = 50 mm):

Light series	2210
Medium series	2310
Heavy series	2410

The groove ball bearings are indicated as (bore diameter = 50 mm):

Light series	6210
Medium series	6310
Heavy series	6410

Tapered roller bearings are indicated as (bore diameter = 50 mm):

Light series	7210
Medium series	7310
Light wide series	7510
Medium wide series	7610

22.5 THE STATIC CAPACITY OF BEARINGS

The static load is the load acting on the bearing when the shaft is stationary. This load causes deformation in the ball and races. The deformation is limited to 0.0001 of the ball diameter.

The basic static load rating is defined as that radial load which causes a total permanent deformation of ball and race at the most heavily stressed contact; of 0.0001 of the ball diameter.

The initial work done by Stribeck serves as basis for static load capacity of the bearings. He developed an equation for the static load capacity based on the Hertz's contact equation.

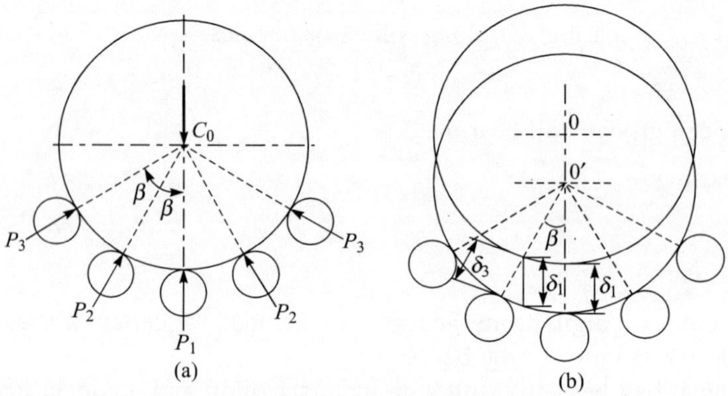

Figure 22.5 Deflection of balls

Applying the Hertz's equation for the ball and race contact, we can write a relationship between the deflection and load at each ball contact as given below:

$$\delta \propto (P)^{2/3} = c\,(P)^{2/3}$$

Applying the relation for each ball at the bottom of the shaft,

For the first ball, $\delta_1 = c\,(P_1)^{2/3}$

For the second ball, $\delta_2 = c\,(P_2)^{2/3}$

From the geometry, $\delta_2 = \delta_1 \cos\beta$

$$\therefore \qquad \frac{\delta_2}{\delta_1} = \left(\frac{P_1}{P_2}\right)^{2/3} = \cos\beta$$

$$\therefore \qquad P_2 = P_1\,(\cos\beta)^{3/2}$$

Similarly, $P_3 = P_1(\cos 2\beta)^{3/2}$

The total load capacity is the sum of all such loads,

$$\begin{aligned} C_o &= P_1 + 2P_2 + 2P_3 + \ldots \\ &= P_1 + 2P_1(\cos\beta)^{3/2} + 2P_1(\cos 2\beta)^{3/2} + \ldots \\ &= P_1(1 + 2(\cos\beta)^{3/2} + 2(\cos 2\beta)^{3/2} + \ldots) \\ &= P_1 \cdot M \end{aligned}$$

where, $M = (1 + 2(\cos\beta)^{3/2} + 2(\cos 2\beta)^{3/2} + \ldots)$

The angle between the balls $\beta = 360/Z$

The number of balls in one row = Z

Substituting the value of β in terms of z, we can show that,

$M = Z/5$

Then $C_o = Z \cdot P_1/5$

Experimentally it is shown that

$$P_1 = K \cdot D^2$$

where, D is the diameter of the balls and K is a constant, depending upon the curvature at the contact point and the modulii of the ball and race materials.

Then Stribeck equation for static capacity for ball bearings is:

$$C_o = \frac{KZD^2}{5} \tag{22.1}$$

where, K is a constant depending upon the ball diameter and the ball and race materials.

Z = Number of balls in the bearings.

D = The diameter of the balls.

For roller bearing, the static capacity is:

$$C_o = \frac{KZDL}{5} \tag{22.2}$$

where, K = a constant; Z = no. of rollers

D = diameter of the roller; L = length of the roller.

The BIS standards equation for the basic load capacity is:

$$C_o = f_o i \cdot Z \cdot D^2 \cdot \cos\alpha \tag{22.3}$$

where, f_o = constant; i = number of rows of ball

Z = number of balls in a row; D = ball diameter

α = angle of contact

If the actual loading is a combination of radial and axial loads, the static equivalent load is defined as that static radial load, which, if applied, would cause the same total permanent deformation at the most heavily stressed ball and race contact; as that which occurs under the actual loading condition. It is evaluated as follows:

$$P_o = X_o F_r + Y_o \cdot F_a \tag{22.4}$$

or

$$P_o = F_r \tag{22.4a}$$

whichever is greater

where, X_0 = radial factor, Y_o = thrust factor

(The values are chosen from Table 22.6 MDHB)

F_r = the radial load; F_a = axial load.

22.6 DYNAMIC CAPACITY OF BEARINGS

It is based on the fatigue life of the bearings. The life may be expressed in hours at specified speed or in terms of number of revolutions that the bearing can run before the occurrence of the first fatigue crack in balls or races. Since the performance of the bearings is not uniform, the dynamic capacity is determined by fatigue test conducted on a group of bearings. The bearings are rated on either of the two following bases depending upon the manufacturers:

(i) The average life of a group of bearings, i.e., 50% survival.

(ii) The life which 90% bearings will reach or exceed.

AFBMA and I.S.O specify the 90% survival-life, defined as:

"The rating life of a group of identical bearings is defined as the number of revolutions that 90% of the bearings complete or exceed before the first evidence of fatigue crack."

Normally, the average life is five times the 90% survival life.

The basic dynamic load rating is that constant stationary radial load which a group of apparently identical ball bearings with stationary outer ring, can endure for a rating life of one million (cycles) revolutions of the inner ring. Theoretically, it is calculated by the equations:

$$C = f_c (i \cos \alpha)^{0.7} Z^{2/3} D^{1.8} \text{ for } D \leq 25.4 \text{ mm} \tag{22.5a}$$

$$= 3.647 f_c (i \cos \alpha)^{0.7} Z^{\frac{2}{3}} (0.23 D^{1.4}) \text{ for } D > 25.4 \text{ mm} \tag{22.5b}$$

The values for different bearings are tabulated in the manufacturer's catalogues.

Bearing Load Capacity

If two groups of identical bearings are tested under loads P_1 and P_2 and they record lives of L_1 and L_2 revolutions respectively, they are related by the expression,

$$\frac{L_1}{L_2} = (P_2/P_1)^k$$

where, k is a constant = 3 for ball bearings

= 10/3 for roller bearings

For basic load rating, the rating life is one million cycles. We can substitute,

$L_2 =$ one million cycles

P_2 = Basic load capacity = C

P_1 = Actual load on the bearing = P

$L_1 = L_n =$ life in millions of cycles

Then the life in million cycles of a group of bearings under load P, can be given by,

$$L_n = (C/P)^k \tag{22.6}$$

where, P = equivalent radial = $XVF_r + Y \cdot F_a$ (22.7)

and $X =$ radial factor; $Y =$ thrust factor; $V =$ rotation factor

These values are given in Table 22.5 MDHB.

By rearranging the terms, we can write for the basic load required as

$$C = (L_n)^{\frac{1}{k}} \cdot P = f_N \cdot P \tag{22.8}$$

where, $$f_N = \sqrt[k]{\frac{N \times T}{10^6}}$$

Since in the load rate testing, one million cycles are achieved by running the test for 500 hours at 33.3 rpm, we can write,

$$f_N = \sqrt[k]{\frac{N \times T}{10^6}} = = \sqrt[k]{\frac{N \times 60 \times L_h}{500 \times 60 \times 33.3}} \tag{22.9a}$$

where, L_h = expected life of bearing in hours.

We can rewrite the equation as

$$f_N = \sqrt[k]{\frac{L_h/500}{33.3/N}} = \frac{f_h}{f_n} \tag{22.9b}$$

where, Life factor, $f_h = \sqrt[k]{(L_h/500)}$ (22.10)

Speed factor, $f_n = \sqrt[k]{(33.3/\text{N})}$ (22.11)

The values of, f_N, f_h, f_n can be chosen from Figs. 22.1 to 22.4.

22.7 PROBABILITY OF SURVIVAL-RELATION

The probability of failure of bearings is governed by the Wiebull distribution. The relationship between the bearing life and reliability is given by

$$R = e^{-(L/a)^b} \quad \text{or} \quad \frac{1}{R} = e^{(L/a)^b}$$

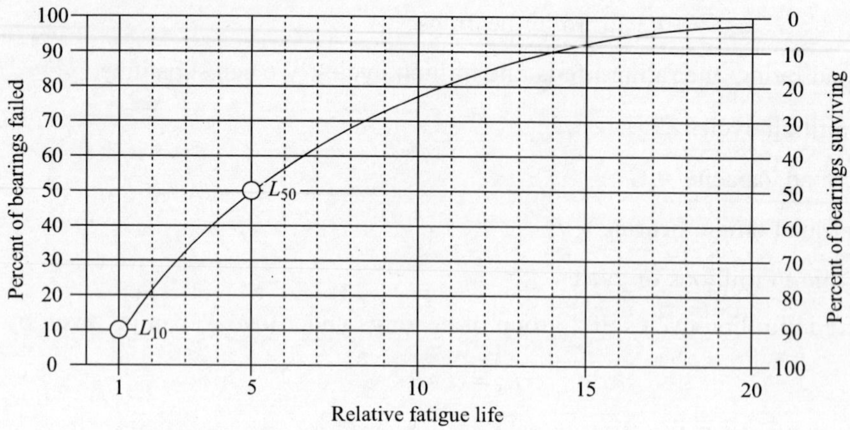

Figure 22.6 Life distribution in antifriction bearings

Taking log of both sides,

$$\text{Ln}\left(\frac{1}{R}\right) = (L/a)^b$$

Solving for life $-L = a \cdot \left[\text{Ln}\left(\frac{1}{R}\right)\right]^{\frac{1}{b}}$

Applying this equation for 50%

$$L_{50} = a \cdot \left[\text{Ln}\left(\frac{1}{0.50}\right)\right]^{\frac{1}{b}}$$

and 90% survival

$$L_{90} = a \cdot \left[\text{Ln}\left(\frac{1}{0.90}\right)\right]^{\frac{1}{b}}$$

Taking the ratio of the two,

$$\frac{L_{50}}{L_{90}} = \frac{\ln\left(\frac{1}{0.50}\right)^{\frac{1}{b}}}{\ln\left(\frac{1}{0.90}\right)} = 5$$

Solving for b,

$$b = 1.17$$

and solving for a,

$$a = \frac{L_{90}}{\left[\text{Ln}\left(\frac{1}{0.90}\right)\right]^{\frac{1}{b}}} = 6.85 \text{ for } b = 1.17$$

Then life expectancy and reliability of survival relation is

$$(L/L_{90}) = 6.85\,[\log_e(1/R)]^{0.855} \tag{22.12}$$

where, L_{90} – life for 90% survival, and L – life for reliability R.

22.8 DESIGN FOR CYCLIC LOADS AND SPEEDS

In many applications, the load and speed may not remain constant throughout the life. There may be cyclic variations of both loads and speeds.

Let the loading system contain n-loads at n-different revolutions, like

P_1 for N_1 revolutions

P_2 for N_2 revolutions

P_n for N_n revolutions

The life L_1 for load P_1 is given by $L_1 = \left(\dfrac{C}{P_1}\right)^k \times 10^6$ rev.

The life L_2 for load P_2 is given by $L_2 = \left(\dfrac{C}{P_2}\right)^k \times 10^6$ rev.

The life L_n for load P_n is given by $L_n = \left(\dfrac{C}{P_n}\right)^k \times 10^6$ rev.

The damages caused by the loads are

$$D_1 = \frac{N_1}{L_1} = \frac{N_1}{10^6} \times \left(\frac{P_1}{C}\right)^k$$

$$D_2 = \frac{N_2}{L_2} = \frac{N_2}{10^6} \times \left(\frac{P_2}{C}\right)^k$$

$$D_n = \frac{N_n}{L_n} = \frac{N_n}{10^6} \times \left(\frac{P_n}{C}\right)^k$$

The total damage per cycle,

$$D_T = \frac{N_1}{10^6} \times \left(\frac{P_1}{C}\right)^k + \frac{N_2}{10^6} \times \left(\frac{P_2}{C}\right)^k + \ldots + \frac{N_n}{10^6} \times \left(\frac{P_n}{C}\right)^k$$

Let P_e be the equivalent load for the complete cycle of total N_T revolutions. Then we can write for total damage as,

$$D_T = \frac{N_T}{10^6} \times \left(\frac{P_e}{C}\right)^k$$

Equating the two,

$$\frac{N_T}{10^6} \times \left(\frac{P_e}{C}\right)^k = \frac{N_1}{10^6} \times \left(\frac{P_1}{C}\right)^k + \frac{N_2}{10^6} \times \left(\frac{P_2}{C}\right)^k + \ldots + \frac{N_n}{10^6} \times \left(\frac{P_n}{C}\right)^k$$

Solving for the equivalent load,

$$P_e = \sqrt[k]{\frac{N_1 P_1^k \pm N_2 P_2^k \ldots \mp N_n P_n^k}{N_T}}$$

Then for variable loads, the equivalent static mean value is by

$$P_e = \left[\sum \frac{P_i^k N_i}{N_T}\right]^{\frac{1}{k}} \text{ for varying speed} \quad (22.13)$$

$$= \left[\sum \frac{P_i^k t_i}{T}\right]^{\frac{1}{k}} \text{ for time varying} \quad (22.14)$$

where, $k = 3$ for ball bearings

$= 10/3$ for roller bearings

N_T = total number of revolutions

T = total time of the load cycle

If the loads consist of radial and axial loads, then P_i are the equivalent radial loads, calculated as,

$$P_i = \text{X.V. } F_{ri} + Y \cdot F_{ai} \quad (22.15)$$

Table 22.1. Ball and roller bearings

Designation	Application	Type and figure
Deep groove ball bearing	Can carry considerable thrust load apart from radial load – high speed	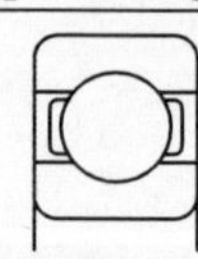
Self-aligning ball bearing	Minor angular displacements of shafts will not affect	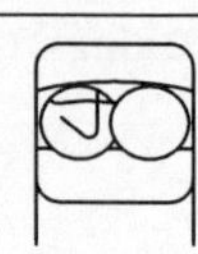

Contd.

*Single row angular contact ball bearing	For heavy axial loads	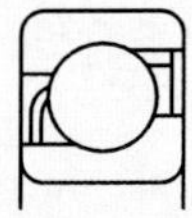
Double row angular contact ball bearing	Can carry radial loads with heavy thrust in both directions	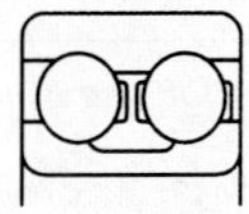
Spherical roller bearing	High carrying capacity, self-åligning for heavy radial loads with considerable axial load in both directions	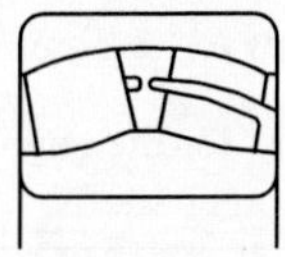
Cylindrical roller bearing	For heavy radial loads at high speeds, permit slight axial displacement	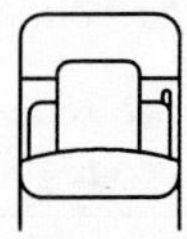
*Taper roller bearing	For combined radial and axial loads	
Single thrust ball bearing	For axial load in one direction only	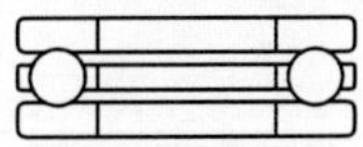
Double thrust ball bearing	For axial loads acting on both directions	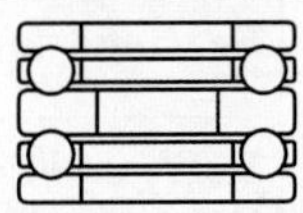
Spherical roller thrust bearing	For heavy axial loads, high speed, self-aligning	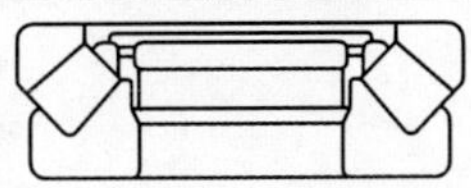

*****Note:** Always adjusted towards another bearing capable of carrying axial load in the opposite direction.

Table 22.2. Bearing life

Class of machine	Life in working hours, Lh
Instruments and apparatus that are used only seldom: Demonstration apparatus: mechanism for operating sliding doors. Machines used for short period or intermittently and whose breakdown would not have serious consequences.	500
Hand tools: lifting tackle in workshops, hand operated machines—generally, agricultural machines, cranes in erecting shops, domestic machines	4000 - 8000

Contd.

Machines working intermittently and whose breakdown would have serious consequences: Auxiliary machinery in power stations: conveyor plant for flow production lifts: cranes for piece goods: machine tools used infrequently	8000 - 12000
Machines for use 8 hours per day and not always fully utilised: Stationary electric motors: general-purpose gear units.	12000 - 20,000
Machines for use - 8 hours per day and fully utilised Machines for the engineering industry generally cranes for bulk goods; ventilating fans; counter shafts.	20000 - 30000
Machines for continuous use 24 hours per day: Separators; compressors, pumps, mine hoists, stationary electric machines, machine in continuous operation on board naval vessels.	40000 - 60000
Machines required to work, with a high degree of reliability 24 hours a day. Pulp and paper making machinery, public power plants, mine pumps, water works; machines in continuous operation on board merchant ships.	100000 - 200000

Table 22.3. Reduction in carrying capacity with temperature

Bearing temperature in	• C 125 • F 225	150 300	175 345	200 400	225 435	250 480
Decrease in carrying capacity	5%	10%	15%	25%	35%	40%

Table 22.4. Coefficient of friction for antifriction bearings

S.No.	**Type of bearing**	***f***
1	Self-aligning bearing (radial load)	0.0010
2	Cylindrical roller bearing (radial load)	0.0011
3	Thrust ball bearing (thrust)	0.0013
4	Single row deep groove ball bearing (radial)	0.0015
5	Tapered and spherical roller bearing (radial)	0.0018
6	Needle bearings (radial)	0.0045

Note: Values apply to run in bearings, without seal and with good lubrication.

Table 22.5. Factors X, V and Y for ball bearings

Bearing type			**In relation to the load, the inner ring is**		**Single row bearings**[(2)]		**Double row bearings**[(3)]				***e***
			Rotating	**Stationary**	$\frac{Fa}{VFr} > e$		$\frac{Fa}{VFr} < e$		$\frac{Fa}{VFr} > e$		
			V	***V***	***X***	***Y***	***X***	***Y***	***X***	***Y***	
Radial contact groove ball bearings[(4)]	$\frac{Fa^{(1)}}{Co}$	$\frac{Fa}{iZD^2}$									
		Units kPa									
	0.014	172				2.30				2.30	0.19
	0.028	345				1.90				1.99	0.22
	0.056	689				1.71				1.71	0.26
	0.084	1030	1	1.2	0.56	1.55	1	0	0.56	1.55	0.28
	0.11	1380				1.45				1.45	0.30
	0.17	2070				1.31				1.31	0.34
	0.28	3450				1.15				1.15	0.38
	0.42	5170				1.04				1.04	0.42
	0.56	6890				1.00				1.00	0.44
	$\frac{iFa^{(1)}}{Co}$	$\frac{Fa}{ZD^2}$									
		Unit kPa			For this type use the X, Y and e values applicable to single row radial contact bearings						
Angular contact groove ball bearings with contact angle[(4)] 5°	0.014	172						2.78		3.74	0.23
	0.028	345						2.40		3.23	0.26
	0.056	689						2.07		2.78	0.30
	0.085	1030						1.87		2.52	0.34
	0.11	1380					1	1.75	0.78	2.36	0.36
	0.17	2070	1	1.2				1.58		2.13	0.40
	0.28	3450						1.39		1.87	0.45
	0.42	5170						1.26		1.69	0.50
	0.56	6890						1.21		1.63	0.52

Contd.

Bearing type			In relation to the load the inner ring is		Single row bearings[(2)]		Double row bearings[(3)]				e
			Rotating	Stationary	$\frac{Fa}{VFr} > e$		$\frac{Fa}{VFr} < e$		$\frac{Fa}{VFr} > e$		
			V	V	X	Y	X	Y	X	Y	
10°	0.014	172	1	1.2	0.46	1.88	1	2.18	0.75	3.06	0.29
	0.029	345				1.71		1.98		2.78	0.32
	0.057	689				1.52		1.76		2.47	0.36
	0.086	1030				1.41		1.63		2.29	0.38
	0.11	1380				1.34		1.55		2.18	0.40
	0.17	2070				1.23		1.42		2.00	0.44
	0.29	3450				1.10		1.27		1.79	0.49
	0.43	5170				1.01		1.17		1.64	0.54
	0.57	6890				1.00		1.16		1.63	0.54
15°	0.015	172	1	1.2	0.44	1.47	1	1.65	0.72	2.39	0.38
	0.029	345				1.40		1.57		2.28	0.40
	0.058	689				1.30		1.46		2.11	0.43
	0.087	1030				1.23		1.38		2.00	0.46
	0.12	1380				1.19		1.34		1.93	0.47
	0.17	2070				1.12		1.26		1.82	0.50
	0.29	3450				1.02		1.14		1.66	0.55
	0.44	5170				1.00		1.12		1.63	0.56
	0.58	6890				1.00		1.12		1.63	0.56
20°			1	1.2	0.43	1.00	1	1.09	0.70	1.63	0.57
25°			1	1.2	0.41	0.87	1	0.92	0.67	1.41	0.68
30°			1	1.2	0.39	0.76	1	0.78	0.63	1.24	0.80
40°			1	1.2	0.37	0.66	1	0.66	0.60	1.07	0.95
			1	1.2	0.35	0.57	1	0.55	0.57	0.93	1.14
Self-aligning ball bearing			1	1	0.40	0.4 cot α	1	0.42 cot α	0.65	0.65 cot α	1.5 tan α

Note: For Table 22.5 (MDHB).

(1) C_o is the static basic load rating.

(2) For single row bearings, when $\frac{F_a}{VF_r} \le e$, use $X = 1$ and $Y = 0$.

Two similar single row angular contact ball bearings mounted "face-to-face or "back-to-back" are considered one double row angular contact bearing. For two or more similar single row ball bearings mounted "in tandem", use the values of *X*, *Y* and *e* which apply to one single row ball bearing. When α is smaller than 20°, F_r and F_a are not total loads but the loads per single row bearing. C_o and *i* also refer to one single row bearing.

(3) Double row bearings are presumed to be symmetrical.

(4) Permissible maximum value of F_a/C_o depends on the bearing design.

(5) Values of *X*, *Y* and *e* for a load or contact angle other than shown in Table 22.5 (MDHB) are obtained by linear interpolation.

Table 22.6. Values for factors X_o and Y_o IS: (Part I) - 1966

Bearing type		**Single row bearings**[§]		**Double row bearings**[†]	
		X_o	Y_o	X_o	Y_o
Radial contact groove ball bearing[§‡]		0.6	0.5	0.6	0.5
Angular contact groove ball bearing[‖]	$\propto = 20°$	0.5	0.42	1	0.84
	$\propto = 25°$	0.5	0.38	1	0.76
	$\propto = 30°$	0.5	0.33	1	0.66
	$\propto = 35°$	0.5	0.29	1	0.58
	$\propto = 40°$	0.5	0.26	1	0.52
Self-aligning ball bearing		0.5	$0.22 \cot \propto$	1	$0.44 \cot \propto$
Self-aligning and tapered roller bearings	$\propto \neq 0°$	0.5	$0.22 \cot \propto$	1	$0.44 \cot \propto$

Note:

§ P_o is always $\geq F_r$.

† Double row bearings are presumed to be symmetrical.

± Permissible maximum value F_a/C_o depends on the bearing design (groove depth and internal clearance).

‖ For two similar single row angular contact ball bearings, mounted 'face-to-face' or 'back-to-back' use the values of X_o and Y_o which apply to a double row angular contact ball bearings. For two or more similar single row angular contact ball bearings, mounted 'in tandem', use the value of X_o and Y_o which apply to single row angular contact ball bearing.

Table 22.7. Factors *X* and *Y* for roller bearings

Type of bearing	**Single row bearing**				**Double row bearing**				*e*
	$F_a/VF_r < e$		$F_a/VF_r > e$		$F_a/VF_r \leq e$		$F_a/VF_r > e$		
	X	*Y*	*X*	*Y*	*X*	*Y*	*X*	*Y*	
Taper roller bearing	1	0	0.4	$0.4 \cot \alpha$	1	$0.45 \cot \alpha$	0.67	$0.67 \cot \alpha$	$1.5 \tan \alpha$
Steep angle roller bearing			$\tan \alpha$	1	$1.5 \tan \alpha$	0.67	$\tan \alpha$	1	$1.5 \tan \alpha$
Special roller bearing	1	2.1 to 2.9	0.67	3.1 to 4.9	1	2.1 to 2.9	0.67	3.1 to 4.9	

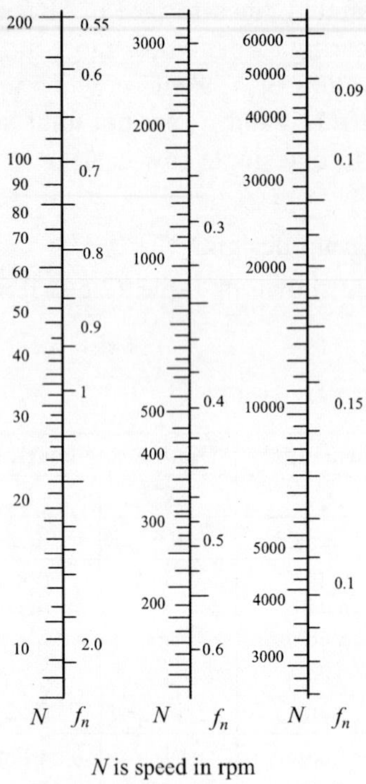

Figure 22.7 Speed factor

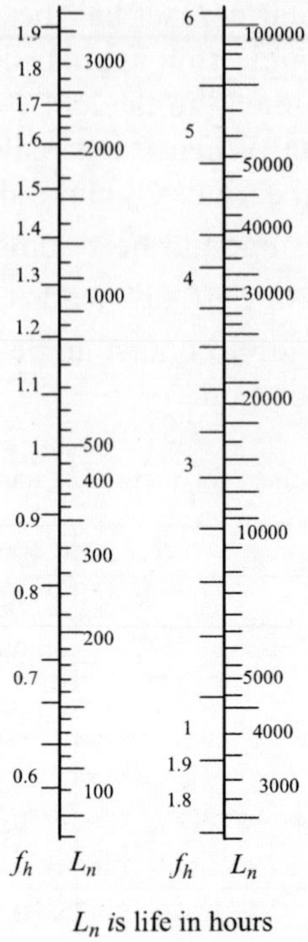

Figure 22.8 Life factor

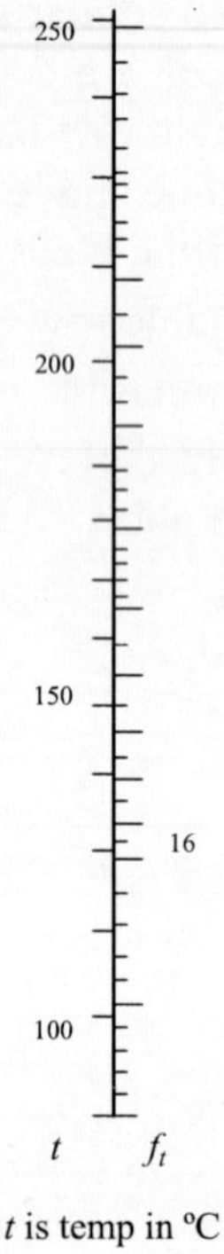

Figure 22.9 Temperature factor

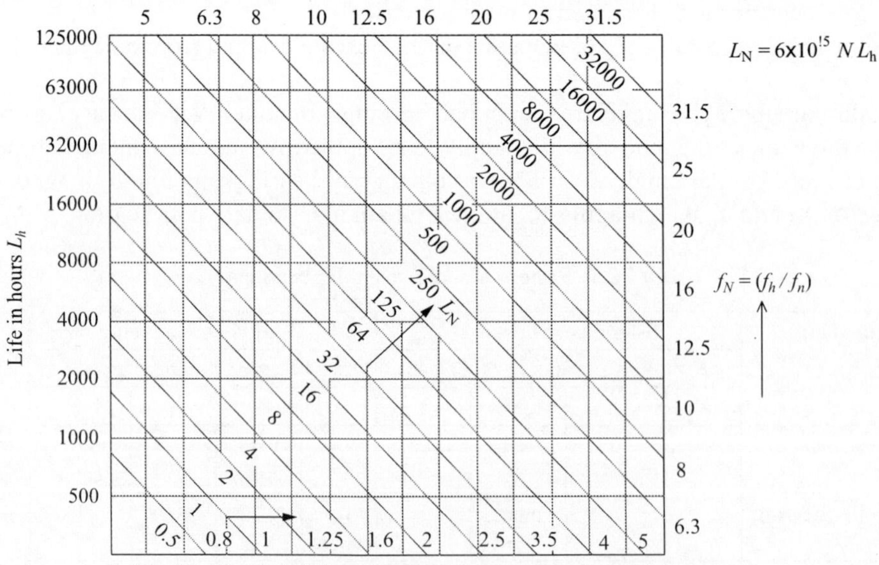

Figure 22.10 Life factor, f_N

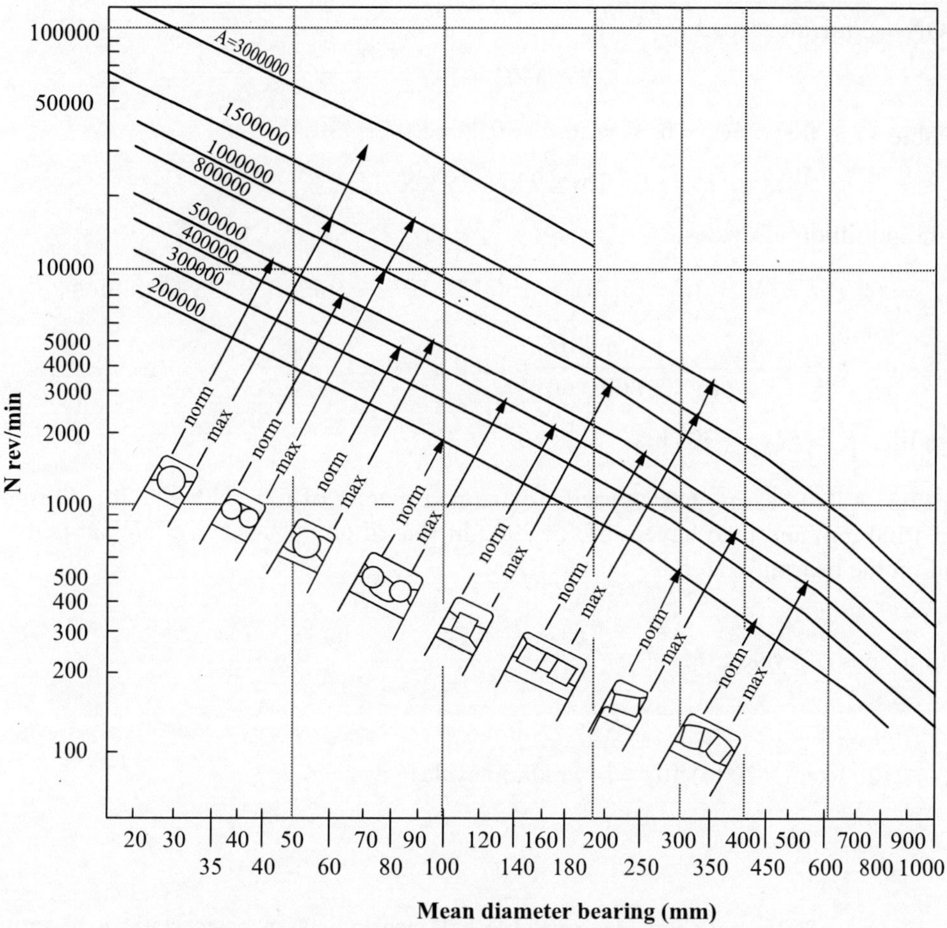

Figure 22.11 Limiting speeds for antifriction bearing

SOLVED EXAMPLES

Example 22.1: A deep groove radial ball bearing is subjected to a radial load of $F_r = 9$ kN and axial load of $F_a = 9$ kN. Using AFBMA equation, determine the equivalent static radial load P_o.

SOLUTION

AFBMA equation for equivalent load

$P_o = X_o F_r + Y_o F_a$ or $P_o = F_r$ whichever is larger

From Table 15.6 MDHB, $X_o = 0.6$; $Y_o = 0.5$

$P_o = 0.6 \times 9 + 0.5 \times 9 = 9.9$ kN

Example 22.2: A single row deep groove ball bearing has a specific dynamic capacity of 46.3 kN. The actual radial load is $F_r = 9$ kN. The speed of rotation is 1800 rpm. What is the life (a) in cycles of operation, (b) in hours, and (c) what is average life?

SOLUTION Equivalent load

$$P = XVF_r + Y \cdot F_a$$

From Table 11.5, for $F_a/C_o = 0, X = 1.0, Y = 0, V = 1.0$

$$\therefore \qquad P = 1.0 \times 1.0 \times 9 \text{ kN} = 9 \text{ kN}$$

Then life in million of cycles,

$$L_n = (C/P)^3 = (46.3/9)^3 = (5.14)^3 = 136.15 \times 10^6 \text{ revolution} = 136.15 \text{ millions}$$

Life in hours, $L_h = \dfrac{L_n}{(N \times 60)} = \dfrac{136.1 \times 10^6}{1800 \times 60} = 1260.6$ hrs

Average life, $L_a = 5L_h = 6303$ hrs

Example 22.3: A ball bearing has a specific dynamic capacity of $C = 24.5$ kN. It is to rotate at a speed of 1200 rpm and is to have a life of 2000 hr (based on 90% survival). What is the safe radial load on the bearing?

SOLUTION

Life in millions of cycles, $L_n = (C/P)^3$

$$P = C/\sqrt[3]{L_n}$$

where, $L_n = (1200 \times 60 \times 2000)/10^6 = 144$ million cycles

$$\therefore \qquad P = \frac{24.5}{\sqrt[3]{144}} = 4.682 \text{ kN}$$

Example 22.4: A 6203 single row deep groove ball bearing has a basic static load rating of $C_o = 4500$ N and a basic dynamic capacity of $C = 7350$ N. What is life expectancy for 90% survival for a radial load of $F_r = 1350$ N and a thrust load of $F_a = 1260$ N? The outer ring is stationary.

SOLUTION From Table 11.5 MDHB

For $\dfrac{F_a}{C_o} = \dfrac{1260}{4500} = 0.28; e = 0.38$

For $F_a/VF_r = \dfrac{1260}{1 \times 1350} = 0.933 > e$

$$\therefore \qquad X = 0.56, Y = 1.15$$

Equivalent load,

$$P = VXF_r + YF_a = 1 \times 0.56 \times 1350 + 1.15 \times 1260 = 2205 \text{ N}$$

The life $L_n = (C/P)^3 = \left(\frac{7350}{2205}\right)^3 = 37.0$ million rev. $= 37.0 \times 10^6$ rev.

Example 22.5: A ball bearing is to be selected for an application with radial load $F_r = 2000$ N for 90% of the time and 8000 N for the remaining 10% of the time. The shaft is to rotate at 150 rpm. Determine the minimum value of dynamic capacity for 5000 hrs of operation with 90% survival probability.

SOLUTION The mean cubic load

$$F_m = \left[\frac{\Sigma F_i^3 t_i}{T}\right]^{1/3} = [\Sigma(2000^3 \times 0.9 + 8000^3 \times 0.1)]^{1/3} = 3.88 \text{ kN}$$

Equivalent load $P = XVF_r + YF_a$

For $F_a = 0, X = 1, Y = 0, V = 1.0$

$$\therefore \qquad P = 1 \times 1 \times 3.88 + 0 \times 0 = 3.88 \text{ kN}$$

The capacity required $C = (\sqrt[3]{L_n}) \cdot P$

$$= \left[\sqrt[3]{\frac{150 \times 60 \times 5000}{10^6}}\right] \times 3.88 = 13.8 \text{ kN}$$

Example 22.6: A ball bearing subjected to a radial load of $F_r = 5.0$ kN is expected to have a life of 8000 hrs at 1450 rpm with reliability of 99%. Calculate the dynamic capacity of the bearing @ 90% survival reliability.

SOLUTION The life relations are

$$\frac{L}{L_{90}} = 6.85[\log_e(1/R)]^{0.855} = 6.85\left[\log_e \frac{1}{0.99}\right]^{0.855} = 0.134$$

$$\therefore \qquad L_{90} = L/0.134 = \frac{8000 \times 60 \times 1450}{0.134 \times 10^6} \text{ millions } = 5194.0 \text{ millions}$$

Equivalent load $P = XVF_r + YF_a$
Since $F_a = 0, P = F_r = 5$ kN
The dynamic capacity

$$C = (\sqrt[3]{L_n}) \cdot P = (\sqrt[3]{5194}) \times 5 = 86.34 \text{ kN}$$

Example 22.7: Select a ball bearing for 50 mm shaft running at 1000 rpm and subjected to radial load of 8000 N and axial load of 6000 N. The required life in hours is 2500 hrs.

SOLUTION

Radial load, $F_r = 8000\text{ N} = 8\text{ kN}$

Axial load, $F_a = 6000\text{ N} = 6\text{ kN}$

Equivalent load, $P = XVF_r + YF_a$

Assume, $X = 1.0, Y = 1.5, V = 1$

Equivalent load, $P = 1.0 \times 1.0 \times 8 + 1.5 \times 6.0 = 17\text{ kN}$

Speed factor, $f_n = 1/\sqrt[3]{0.03 \times 1000} = 0.325$

Life factor for life in hr (Eqn. 15.16 MDHB)

$$f_h = \sqrt[3]{L_h/500} = \sqrt[3]{2500/500} = 1.7$$

Then the dynamic capacity required (Eqn. 15.14),

$$C = \frac{f_h}{f_n} P = \frac{1.7}{0.325} \times 17 = 88.9\text{ KN}$$

From Table 15.10 MDHB, the double row angular contact ball bearing no. 3310, has dynamic capacity, $C = 81.5\text{ kN}$ and static capacity, $C_o = 73.5\text{ kN}$.
We select this bearing and check for exact values of X and Y

From Table 15.5 MDHB, for $\frac{F_a}{C_o} = \frac{6}{73.5} = 0.0815, e = 0.34,$

and $F_a/VFr = 0.75 > 0.34,\ X = 0.78, Y = 2.52$

The equivalent load,

$$P = 0.78 \times 1 \times 8 + 2.52 \times 6 = 21.24\text{ kN}$$

The dynamic capacity required,

$$C = \left(\frac{f_h}{f_n}\right) P = \frac{1.7}{0.325} \times 21.24 = 111.1\text{ kN} > 81.5\text{ kN}$$

The bearing capacity is less than the required and hence not suitable. Then from Table 15.12 MDHB for roller bearing, Taper roller bearing no. 32310 with $C = 151.00\text{ kN}$ and $C_o = 145\text{ kN}$ may be selected and checked for exact values of X and Y from Table 15.7 MDHB.

For $F_a/VF_r = 0.75, \alpha = 30°, e = 0.86,\ X = 1.0, Y = 0$

$$P = 1 \times 1.0 \times 8 + 0 \times 6 = 8\text{ kN}$$

Then $$C = \frac{1.7}{0.325} \times 8 = 41.85\text{ kN}$$

This is much less than the actual capacity of $C = 151.0$ kN and hence over design. We can select the cylindrical roller bearing no. NU2210 with $C = 46.5$ kN which is slightly more than the required.

Example 22.8: Design a suitable ball bearing for a radial load of 2000 N and a thrust load of 700 N at an operating speed of 3000 rpm. The load is steady without shock. The life of the bearing should be 15000 hr. The nominal diameter of the shaft is 50 mm. Sketch neatly the bearing assembly in the housing.

SOLUTION

Speed factor, $f_n = \dfrac{1}{\sqrt[3]{0.03\,\text{N}}} = \dfrac{1}{\sqrt[3]{0.03 \times 3000}} = 0.225$

Life factor, $f_h = \sqrt[3]{L_h/500} = \sqrt[3]{15000/500} = 3.07$

Equivalent radial load, $P = XVF_r + YFa$

Assume, $X = 1, Y = 1.5, V = 1$

$$\therefore \qquad P = 1 \times 1 \times 2000 + 1.5 \times 700 = 3050\text{ N} = 3.05\text{ kN}$$

The dynamic capacity required,

$$C = \frac{f_h}{f_n} \cdot P = \frac{3.07}{0.2265} \times 3.05\text{ kN} = 41.34\text{ kN}$$

Select bearing no. 6310 (Table 15.9 MDHB)

with $C = 48.00$ kN and $C_o = 35.5$ kN

Check for X and Y (Table 15.5 MDHB) and exact value of C.

$$Fa/Co = \frac{700}{35500} = 0.02;\ e = 0.25,\ X = 0.56, Y = 1.63$$

$\therefore$ Equivalent load,

$$P = 1 \times 0.56 \times 2000 + 1.63 \times 700 = 2261\text{ N} = 2.261\text{ kN}$$

The dynamic capacity required,

$$C = \frac{3.07}{0.2265} \times 2.261 = 30.64\text{ kN} < 48.0\text{ kN}$$

Hence, O.K.

Example 22.9: A rolling contact bearing is subjected to the work cycle as follows: (a) radial load of 6 kN at 150 rpm for 25% of the time, (b) radial load of 7.5 kN at 600 rpm for 20% of the time, and (c) radial load of 2 kN at 300 rpm for 55% of the time. The ring rotates and the loads are steady. Select a bearing for an average life of 2500 hr.

SOLUTION

(a) Load $F_1 = 6$ kN for the duration $t_1 = 0.25 \times 2500$ hrs. The no. of revolutions at this load

$$n_1 = 150 \times 60 \times 0.25 \times 2500 = 5.625 \times 10^6 \text{ rev.}$$

(b) Load $F_2 = 7.5$ kN for the duration $t_2 = 0.2 \times 2500$ hr.

The number of revolutions at this load, $n_2 = 600 \times 60 \times 0.2 \times 2500 = 18 \times 10^6$ rev.

(c) Load $F_3 = 2$ kN for the duration $t_3 = 0.55 \times 2500$ hr.

The number of revolutions at this load $n_3 = 300 \times 60 \times 0.55 \times 2500 = 24.75 \times 10^6$ rev.

Then the total number of revolutions $N_t = (5.625 + 18 + 24.75) \times 10^6$ rev.

$$= 48.375 \times 10^6 \text{ rev.}$$

The equivalent cubic load is given by

$$F_m = [(F_1^3 n_1 + F_2^3 n_2 + F_3^3 n_3)/N_t]^{1/3}$$

$$[(6_1^3 5.625 + 7.5_2^3 18 + 2_3^3 24.75) \times 10^6 / 48.375 \times 10^6]^{1/3} = 5.688 \text{ kN}$$

The equivalent radial load,

$$P = vxF_r + yF_a = 1.0 \times 1.0 \times 5.688 + 0.0 \times 0.0 = 5.688 \text{ kN}$$

The bearing life in millions of rev, $L_N = N_t / 10^6 = 48.375$
The bearing capacity required,

$$C = f_N \cdot P = (L_N)^{1/3} \cdot P = (48.375)^{1/3} \times 5.688 = 20.698 \text{ kN}$$

Then from Tables 15.9 and 15.10 MDHB, the suitable bearing numbers are 6208; 6306; 6404; 7207B; 7306B; 3305; 2211K; 2306K.

Example 22.10: A ball bearing is subjected to the work cycle as follows: (a) radial load of 3 kN at 720 rpm for 30% of the time, (b) radial load of 7.0 kN at 1440 rpm for 40% of the time and (c) radial load of 5 kN at 900 rpm for 30% of the time. The bearing has a dynamic load capacity of 16.6 kN. Calculate (i) average speed, (ii) equivalent load, and (iii) bearing life in hours.

SOLUTION Let the bearing life = T hr.

(a) Load $F_1 = 3$ kN for the duration $t_1 = 0.3T$ hr.

The number of rev. at this load $n_1 = 720 \times 60 \times 0.3 \times T = (12960 \times T)$ rev.

(b) Load $F_2 = 7.0$ kN for the duration $t_2 = 0.4 \times T$ hr.

The number of revolutions at this load

$$n_2 = 1440 \times 60 \times 0.4 \times T = (34560 \times T) \text{ rev.}$$

(c) Load $F_3 = 5$ kN for the duration $t_3 = 0.30 \times T$ hr.

Then number of revolutions at this load,

$$n_3 = 900 \times 60 \times 0.30 \times T = (16200 \times T) \text{ rev.}$$

Then the total no. of revolutions

$$N_t = (12960 + 34560 + 16200)T \text{ revs.} = (63720 \times T) \text{ rev.}$$

The average speed in rpm, $N = (63720 \times T)/(60 \times T) = 1062$ rpm

The equivalent cubic load is given by

$$F_m = [(F_1^3 n_1 + F_2^3 n_2 + F_3^3 n_3)/N_t]^{1/3}$$

$$= [(3^3 \times 12960 + 7.0^3 \times 34560 + 5^3 \times 16200)/63720]^{1/3} = 6.056 \text{ kN}$$

The equivalent radial load,

$$P = V \cdot X \cdot F_r + Y \cdot F_a = 1.0 \times 1.0 \times 6.056 + 0.0 \times 0.0 = 6.056 \text{ kN}$$

The bearing life in millions of rev, $L_N = N_t/10^6 = 0.063720 \times T$

The bearing capacity, $C = f_N \cdot P = (L_N)^{1/3} P$

Hence, life factor, $f_N = C/P = 16.6/6.056 = 2.741$

But the life, $L_N = (f_N)^3 = (2.741)^3 = 20.59$ millions of rev.

i.e., $0.063720 \times T = 20.59$

The bearing life in hours, $T = 20.59/0.063720 = 323$ hr.

EXERCISES FOR PRACTICE

22.1 (a) Explain the advantages of the antifriction bearing.

(b) Explain: (i) static capacity, (ii) dynamic capacity, (iii) life factor, (iv) speed factor, and (v) application factor.

(c) A helical gear used in a paper making machine is mounted on a shaft supported on two bearings 300 mm apart. The load on the left bearing is radial and axial both. Radial load is 25 kN and the axial load is 8 kN. The right hand bearing carries only radial load of 10 kN. The shaft rotates at 150 rpm and transmits 50 kW of power. The shaft material is commercial steel. Select suitable load bearings for the shaft and sketch the mounting arrangement.

22.2 A bearing for an axial flow compressor is to carry an axial thrust of 1500 N and a radial load of 2500 N. The bearing will be in use for 40 hours/week for five years. The speed of the shaft is 1000 rpm. Select a suitable deep grove ball bearing.

22.3 A spur gear drive transmits power of 15 kW at 900 rpm. The diameter of the gear is 200 mm. Select a suitable deep groove ball bearing. The gear drive operates for 10 hours a day, 6 days a week and 46 weeks per year. The bearing has to be replaced once in 2 years. The gear shaft is made of steel with allowable shear stress of 45 MPa.

22.4 A ball bearing is to be used for a drill press operating at 3000 rpm. The bearing is subjected to a thrust load of 1500 N and a radial load of 3500 N. The press is operated for 8 hours per day and 6 days a week. The expected life is two years. Select a suitable bearing.

22.5 (a) With the help of neat sketches, explain the basic difference between deep-grooved and self-aligning ball bearings.

(b) The bearings for a worm shaft of worm gearing are to be selected. The worm is to rotate in both the directions and transmits 15 kW at 1500 rpm. The forces at pitch point of the worm are 3600 N radial, 2600 N tangential and 6600 N axial. The pitch diameter of the worm is 70 mm and the shaft at bearings is 40 mm in diameter. The worm is at the middle of the bearings. Select the bearings for a required life of 2000 hours.

22.6 Select suitable standard ball bearing to support a radial load of 7000 N and an axial thrust load of 5000 N to be mounted on a shaft of diameter 50 mm, operating at a speed of 1000 rpm. The desired life of the bearing is 8000 hours.

22.7 (a) Explain the meaning of "dynamic load capacity" and "static load capacity" as applied to rolling contact bearings.

(b) A ball bearing having a dynamic capacity of 48 kN is successfully used in a particular machine. An improved model of the machine indicates that the bearing load will increase by 40%. If the useful life of the bearing is to be increased by 50%, recommend the minimum value of dynamic load capacity for the bearing for the new machine.

(c) Ball bearings are to be selected for the application in which shaft rotates at 2100 rpm and the radial load variation during each revolution can be approximated by 9 kN for 90°, 27 kN for 260° and 68 kN for the remainder of the revolution. What should be the basic dynamic load rating for failure in 500 hours?

22.8 Specify a suitable ball bearing for a radial load of 2.5 kN and thrust load of 0.8 kN at 1500 rpm. The load is steady. The bearing is to operate for five years at 8 hr per day, 6 days a week. Nominal shaft diameter is to be about 50 mm.

22.9 In an industrial application, the radial load is 2000 N during 90% of the time and 8000 N during the remaining period. The axial thrust remains constant at 5000 N during the entire life. The shaft runs at 150 rpm and a life of 5000 working hr is required. Select a suitable deep groove ball bearing for the above conditions.

22.10 Select a suitable ball bearing for the spindle of a wood working machine revolving at 1200 rpm, one bearing is subjected to a radial load of 3 kN and a thrust load of 2.5 kN. The other bearing carries only a radial load of 3.25 kN. The machine is to be used 8 hr a day and 6 days a week and a service life of 10 yr is desired. The diameter of the spindle is 50 mm.

22.11 A Timken bearing supporting one end of a worm shaft of a speed reducer turning at 1800 rpm is to carry a radial load of 800 N and a thrust load of 1950 N. The desired life of the bearing is 12,000 hours. Determine the rated capacity of the bearing to be selected. Application factor may be taken as unity.

22.12 (a) What are the factors which are to be considered while selecting ball and roller bearings?

(b) The shaft of an electric motor, which rotates at 1440 rpm is to be supported by a pair of Timken roller bearings. Each bearing is to sustain a pure radial load of 5 kN and is to have a desired life of 12,000 hours. Design a suitable bearing.

22.13 (a) Define the following:

(i) Rating life.

(ii) Basic-load rating.

(iii) Equivalent load.

Explain how they are determined.

(b) A deep groove ball bearing is required to carry a radial load of 4.5 kN and a trust load of 2.26 kN at 1000 rpm. The life should be at least 10,000 working hours. Select approximate bearing number.

22.14 (a) What is Stribeck's equation? Derive it from fundamentals, in case of rolling contact bearing.

(b) A taper roller bearing has a dynamic load capacity of 26 kN. The desired life for 90% of the bearings is 8000 hr and the speed is 300 rpm. Calculate the equivalent radial load that the bearing can carry.

22.15 Select a deep groove ball bearing to support a radial load of 5kN and a trust load of 2 kN at an operating speed of 1500 rpm. The average life of the bearing is expected to be 5 years at the rate of 10 hr per day and 350 days per year. The shaft diameter is to be 50 mm. Estimate the actual life of the bearing.

SHORT ANSWER QUESTIONS

22.1 What are the functions of a ball bearing?

22.2 Why is a roller bearing called antifriction bearing?

22.3 Name the basic elements of a ball bearing?

22.4 Give the classification of ball bearings.

22.5 Give the classification of roller bearings.

22.6 Which of the loads can be resisted by a tapered roller bearing?

22.7 List three advantages of antifriction bearings.

22.8 Explain the ISO code to identify a ball bearing.

22.9 What is the static load carrying capacity of a bearing?

22.10 Write down the Stribeck's equation of ball bearing. Indicate what each term represents?

22.11 Write down the Stribeck's equation roller bearing.

22.12 How is the static load capacity of a ball bearing evaluated?

22.13 On what basis are the bearings rated for their dynamic load carrying capacity?

22.14 Name the statistical distribution that estimates probability of a roller bearing.

22.15 In a variable speed of operation, write the equation used to estimate static mean value of load.

22.16 Under what loading conditions are self aligning ball bearings recommended?

22.17 Under what loading conditions are double thrust ball bearings recommended?

22.18 Under what loading conditions are cylindrical roller bearings recommended?

22.19 How does the bearing load carrying capacity vary with increasing operating temperature?

22.20 Which configuration of ball bearing has lowest coefficient of friction?

CHAPTER 23

DESIGN OF I.C. ENGINE – PARTS

23.1 PISTON

It is a cylindrical member, reciprocating inside a engine cylinder. Its functions are (i) to seal the high pressure combustion gases in the combustion space, (ii) to support the side thrust from the connecting rod, and (iii) support the piston pin, which connects the piston to connecting rod for converting the reciprocating motion to rotary motion.

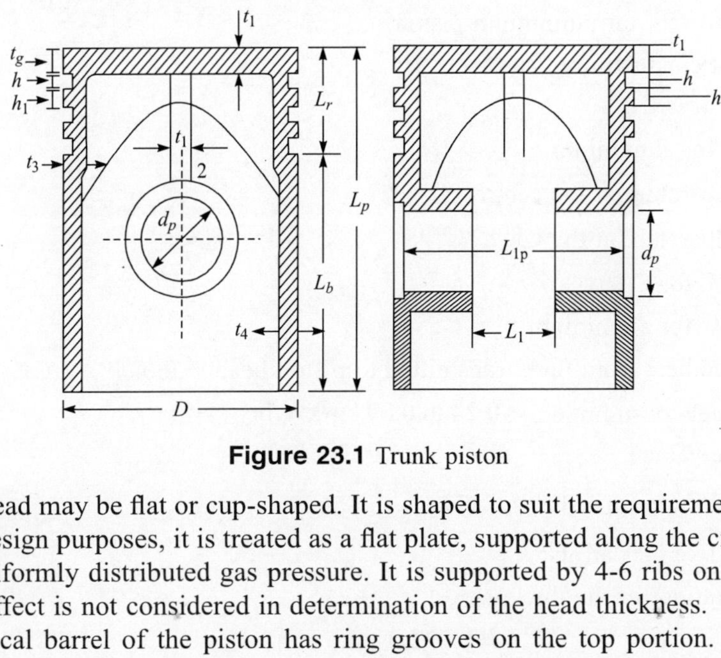

Figure 23.1 Trunk piston

The piston head may be flat or cup-shaped. It is shaped to suit the requirement of combustion chamber. For design purposes, it is treated as a flat plate, supported along the circumference and subjected to uniformly distributed gas pressure. It is supported by 4-6 ribs on underside. Their strengthening effect is not considered in determination of the head thickness.

The cylindrical barrel of the piston has ring grooves on the top portion. The piston rings are placed into the grooves. The rings prevent the leakage of the high pressure gases, in the combustion chamber. The bottom portion of the barrel is called skirt and acts as bearing surface to support the side thrust from the connecting rod. In this portion, bosses are provided to support the piston pin. The gas force is transmited to connecting rod through the piston pin. The piston pin is treated as a beam supported in the end bosses.

Design Procedure

The piston–head thickness: The thickness of the head is determined by treating it as a circular flat plate subjected to uniformly distributed lateral pressure and fixed at the edge. Applying

Grashoff's formula, $t_1 = \sqrt{\dfrac{3 \cdot p_g D^2}{16 S_t}} = 0.43D\sqrt{\dfrac{p_g}{S_t}}$ (23.1)

where, D is the diameter of the piston mm

p_g is maximum gas pressure in MPa

S_t is design stress for piston material in MPa

The piston head has to conduct the heat received from the hot gases to the cylinder wall. From the heat flow consideration, the required thickness of the head is given by

$$t_1 = \frac{D^2 q}{16K(T_c - T_e)} = \frac{Q}{4\pi K(T_c - T_e)} \tag{23.2}$$

where, $q =$ Heat flow from gases $\text{J/mm}^2/\text{hr}/^\circ\text{C}$

$= 115$ to 460 for C.I. pistons

$= 230$ to 920 for aluminium piston

$K =$ Heat conductivity J/mm/hr/°C

$= 270$ for C.I.

$= 650$ for aluminium

$T_c =$ Temperature at the centre, °C

$T_e =$ Temperature at the edge, °C

$T_c - T_e = 222$°C for C.I.

$= 111$°C for aluminium

$Q =$ Portion of the heat from fuel, transmitted to piston head $= 0.05\, W_f \cdot HCV \cdot kW$ (23.2a)

$W_f =$ Specific fuel consumption $= 0.24$ to 0.29 kg/kW/hr

$HCV =$ Calorific value

$= 45 \times 10^6$ J/kg for diesel

$= 47 \times 10^6$ J/kg for petrol

The empirical relation for the thickness,

$$t_1 = 0.032D + 1.5 \text{ mm} \tag{23.3}$$

The larger of the three values is chosen for the thickness of the head plate.

Piston Rings

The radial thickness of the piston ring,

$$t_r = D\sqrt{3p_r/S} \tag{23.4}$$

where, $S =$ allowable stress $= 84$ to 112 MPa for C.I.

$p_r =$ Radial pressure (Table 23.7)

The axial depth of piston ring, $h = 0.7t_r$ to $t_r \leq 0.1D/i$
where, $i =$ number of piston rings.

The lands between the ring grooves, $h_l = 0.75h$ to h

The total depth over the ring grooves, $h_t = i(h + h_l)$

$$= \frac{D}{7} + 0.6 \text{ mm for gas and oil engines}$$

$$= \frac{D}{5.5} \text{ for petrol engines.}$$

The distance from the top to first groove, $t_g = t_1$ to $1.2t_1$

The length of the piston over the piston rings,

$$L_f = t_g + h_t = t_g + i(h + h_l) \tag{23.5}$$

Bearing length of the piston,

$$L_b = \frac{F_N}{D \cdot p_b} = \frac{0.1F_{max}}{D \cdot p_b} \tag{23.6}$$

where, $p_b = 0.2$ to 0.25 MPa; and $F_{max} = \frac{\pi}{4} D^2 p_g$

The empirical relation for L_b is

$L_b = D$ for high speed engines

$= 1.25D$ to $1.45D$ for slow speed engines

The larger of the two values be chosen for L_b

The total length of the piston, $L_p = L_f + L_b$ (23.7)

The maximum thickness of the piston barrel,

$$t_3 = 0.03D + b + 4.5 \text{ mm} \tag{23.8}$$

where, $b =$ depth of ring groove $= t_r + 0.4$ mm

The thickness of the shirt, $t_4 = 0.25t_3$ to $0.35t_3$ (23.9)

Piston Pin

The diameter of the pin is determined from the bearing pressure consideration. The pin may be fixed in the piston hub and oscillates in the connecting rod. Then the bearing length, $L_1 = 0.45D$ to $0.5D$. The pin may be fixed in connecting rod and oscillate in the piston hub, then the bearing length, $L_1 = 0.62D$.

Maximum load on the bearing surface, $F_{max} = \frac{\pi}{4} D^2 p_g$

The bearing pressure, $p_b = \frac{F_{max}}{L_1 d} = \frac{\pi D^2 \cdot p_g}{4L_1 \cdot d}$

∴ The diameter of the pin, $$d = \left[\frac{\pi D^2 p_g}{4 \cdot L_1 \cdot p_b}\right] \quad (23.10)$$

Treating the pin as beam supported in the piston bosses, the bending stress induced,

$$\sigma_b = \frac{F_{\max} \cdot D}{8Z} \leq S_d \quad (23.11)$$

Table 23.1. Allowable tensile stress, S_t for cylinder and piston

Material	Stress MPa	Material	Stress MPa
Ordinary C.I.	25-30	Low carbon steel	84
Close grained C.I.	39	Case hardned steel	84
Nickle C.I.	56	Heat treated alloy steel	140
AI. alloy	39		

Table 23.2. Reboring factor, k

Cylinder diameter mm	75	100	150	200	250	300	350	400	450	500
Reboring factor k	1.5	2.25	3.57	6	7.5	9	10.5	12	12	12

Table 23.3. Recommended piston speed

Class of engines	V_m speed of piston in m/sec
Ordinary direct-acting pump engines	0.45 - 0.65
Ordinary horizontal engines	1 - 2
Horizontal compound and triple-expansion engines	2 - 40
Ordinary marine engines	2 - 3.25
Locomotive engines (mail)	4 - 5
Internal combustion engines	3.25 - 18
Low speed engines (100-400 rpm)	4.0 - 5.5
Medium speed engines (400-1000 rpm)	5.5 - 8.5
High-speed engines (1000-2000 rpm)	8.5 - 12

Table 23.4. Stroke-bore ratio Ls/D

Class of engines	Ratio L_s/D
Ordinary horizontal slow speed engines	1.5 - 2.0
Vertical quick-running engine	1.25 - 1.6
Locomotive engines	0.9 - 1.9
Internal combustion engines	0.9 - 1.9
Low speed engines	1.25 - 1.9
Medium speed engines	1.00 - 1.5
High-speed engines	1.1 - 1.3
Air-cooled aircraft engines	1.0

Table 23.5. Mass of reciprocating parts

Types of engines	Mass of piston per unit area, W_r/A kg /cm^2	
	Spark ignition	Compression ignition
Single acting engine		
Aircraft	0.0154 to 0.028	0.0210 - 0.0315
Automobile	0.0210 to 0.035	-
Truck and tractor	0.0245 to 0.0525	0.0245 - 0.0525
Medium speed stationary engine	0.1050 to 0.2800	0.1050 - 0.0315
Low speed short stroke (L < 1.5D)	0.28 to 0.49	0.35 - 0.56
Low speed long stroke (L > 1.5D)	0.42 to 0.70	0.56 - 0.84
Double acting engines single cylinder per crank	0.84 - 1.4	0.98 - 1.54
Tandem	1.4 - 1.82	-

Table 23.6. Thickness of piston head

Type of engine	Piston material	Four-stroke	Two-stroke
Compression ignition engines	Cast iron	0.11D - 0.13D	0.16D - 0.18D
-do-	Aluminium	0.13D - 0.16D	0.17D - 0.20D
Spark ignition engines	Cast iron	0.12D - 0.14D	0.20D - 0.23D

Table 23.7. Magnitude of radial pressure of the piston rings on the cylinder barrel (according to Unwin)

Type of engine	Steam engine			Gas or oil engines	Petrol engines
	H.P.	I.P.	L.P		
P_r MPa	0.028	0.21	0.014	0.0245	0.028-0.035

Table 23.8. IC Engine design data

Air-fuel ratio		Compression ratios of various engines	
Diesel engines	1.8 - 2	Diesel engines	
Petrol engines	1.5 - 1.8	High speed	13 - 14
Experimental value of charge efficiencies		Medium speed	14 - 15
Four-stroke engines	0.83 - 0.86	Low speed	13 - 14
Two-stroke engines	0.85 - 1.10	Automotive carburettor engines	5.6 - 6.5
Excess air coefficient $= \dfrac{\text{Actual air}}{\text{Theoretical air}}$		**Indicated mean effective pressure at full load, p_m MPa** Diesel engines	

Contd.

Low-speed engines	1.7-2	Four-stroke airless injection, unsupercharged	0.65 - 0.75
High-speed diesel engines	1.3 - 1.7	Four-stroke airless injection, supercharged	0.85 - 1.7
Carburettor engines	0.85 - 1.05	Two-stroke airless injection with unflow scavenging	0.66 - 0.85
Mechanical efficiency at full load		Two-stroke airless injection scavenging pressure from 0.2 to 0.6 N/mm^2	0.9 - 1.8
Four-stroke naturally-aspirated	0.78 - 0.83		
Turbocharged	0.8 - 0.88	Two-stroke air injection	0.6 - 0.65
Two-stroke unsupercharged	0.73 - 0.86	Carburettor engines	0.7 - 1.1
Supercharged	0.83 - 0.9	**Permissible bearing pressure in piston's skirt MPa**	
Carburettor four-stroke engines	0.8 - 0.85	Four-stroke low speed engines	0.84 - 0.28
		Four-stroke medium speed engines	0.3 - 0.35
		Four-stroke high-speed engines	0.57
		Two-stroke medium speed engines	0.32 - 0.35

SOLVED EXAMPLES

Example 23.1: Design a cast iron piston for a single acting diesel engine with the following data: Cylinder bore = 200 mm; length of stroke = 250 mm; speed = 600 rpm; brake mean effective pressure = 0.60 MPa; max. gas pressure p_{max} = 4 MPa. The fuel consumption, w_f = 2.5 kg /kW/ hr; l/d ratio for small end of connecting road = 1.5.

SOLUTION

Step-1: Brake power of the engine,

$$BP = \frac{PLAN}{60,000} = \frac{0.6 \times 250 \times \frac{\pi}{4} \times 200^2 \times 600}{60,000} = 47123.88 \text{ W} = 47.124 \text{ kW}$$

Step-2: Heat transferred to head,

$$Q = 0.05 \cdot W_f \cdot HCV \cdot kW = 0.05 \times 0.25 \times 45 \times 10^6 \times 47.124 = 26.51 \times 10^6 \text{ J/hr}$$

Step-3: The thickness of the head plate

(i) The thickness by heat flow consideration,

$$t_1 = \frac{Q}{4\pi \cdot K \cdot (T_c - T_e)} \qquad \text{(Eqn. 21.32 MDHB)}$$

where, $Q = 26.51 \times 10^6$ J/hr ; K = 270 J/hr/° C, and $T_c - T_e = 222$°C

$$\therefore \qquad t_1 = \frac{26.51 \times 10^6}{4 \times \pi \times 270 \times 222} = 35.2 \text{ mm}$$

(ii) The thickness of head for strength,

$$t_1 = 0.43D\sqrt{p_g / S} \qquad \text{(Eqn. 21.30 MDHB)}$$

where, $D = 200$ mm; $p_g = 4$ MPa and $S = 39$ MPa (Table 21.1 MDHB)

$$t_1 = 0.43 \times 200\sqrt{4/39} = 27.54 \text{ mm}$$

(iii) The thickness by empirical equation,

$$t_1 = 0.032D + 1.5 \text{ mm} = 0.032 \times 200 + 1.5 = 7.9 \text{ mm}$$

Choose $t_1 = 35.2$ mm as the largest of the three.

Step-4: Piston ring proportions,

$$t_r = D\sqrt{3p_r / S} \qquad \text{(Eqn. 21.35 MDHB)}$$

where, $D = 200$ mm; $p_r = 0.0245$ MPa (Table 21.7 MDHB); $S = 90$ MPa

$$\therefore \qquad t_r = 200\sqrt{\frac{3 \times 0.0245}{90}} = 5.72 \text{ mm}$$

Choose $t_r = 6$ mm

The axial depth of the ring, $h = 0.7t_r = 0.7 \times 6 \cong 4.0$ mm

Step-5: Thickness of the barrel

Maximum thickness of piston barrel,

$$t_3 = 0.03D + b + 4.5 \qquad \text{(Eqn. 21.33 MDHB)}$$

$$= 0.03D + t_r + 0.4 + 4.5 = 0.03 \times 200 + 6 + 4.9 = 16.9 \text{ mm}$$

The thickness at skirt (open end),

$$t_4 = 0.3t_3 = 0.3 \times 16.9 \cong 5 \text{ mm} \qquad \text{(Eqn. 21.34 MDHB)}$$

Step-6: Length of the piston

The landing between the rings,

$$h_l = 0.75 \cdot h = 0.75 \times 4 = 3.0 \text{ mm} \qquad \text{(Eqn. 21.37 MDHB)}$$

The distance from the top $t_g = t_1 = 35.2$ mm

The length of the piston over the piston rings,

$$L_r = t_g + i(h + h_l) \qquad \text{(Eqn. 21.40 MDHB)}$$

$$= 35.2 + 5\,(4 + 3) = 70.2 \text{ mm } (i = 5 \text{ assumed})$$

The bearing length, $L_b = \dfrac{0.1F_{max}}{Dp_b}$ (Eqn. 21.41 MDHB)

$$= \frac{0.1 \times \frac{\pi}{4} D^2 \times p_g}{D p_b} = \frac{0.1 \times \frac{\pi}{4} \times 200^2 \times 4}{200 \times 0.25} = 251.33 \text{ mm}$$

The total length, $L = L_r + L_b = 70.2 + 251.33 = 321.53$ mm

By proportions, $L = D = 200$ mm

Choose $L = 322$ mm

Step-7: Piston pin design

The piston pin diameter for bearing pressure,

$$d = \frac{\pi D^2 p_g}{4 L_1 \cdot p_b} \qquad \text{(Eqn. 21.44 MDHB)}$$

where, $D = 200$ mm; $L_1 = 0.45D = 90$ mm

$p_g = 4$ MPa and $p_b = 12.5$ MPa (Table 21.10)

Then $d = \dfrac{\pi \times 200^2 \times 4}{4 \times 90 \times 12.5} = 111.7$ mm

Choose $d = 112$ mm

Step-8: Check for bending stress

The bending stress is given as $\sigma_b = \dfrac{F_p \cdot D}{8 \cdot Z}$

where, $F_p = \dfrac{\pi}{4} D^2 \cdot p_g = \dfrac{\pi}{4} \times 200^2 \times 4 = 125663.7$ N

$$Z = \frac{\pi}{32} d^3 = \frac{\pi \times 112^3}{32} = 137928.5 \text{ mm}^3$$

$$\therefore \qquad \sigma_b = \frac{125663.7 \times 200}{8 \times 137928.5} = 22.77 \text{ MPa}$$

This value is smaller than permissible value and hence the design is safe.

Example 23.2: Design a trunk piston for an I.C. engine. The piston is made of cast iron with an allowable stress of 38.5 MPa. The bore diameter of the cylinder is 200 mm and the max. explosion pressure is 4.0 MPa; the permissible bending stress of the material of the gudgeon is 100 MPa. The bearing pressure in the gudgeon pin bearing of the connecting rod is to be taken as 20 MPa.

Given Data: Cylinder diameter, $D = 200$ mm; Maximum gas pressure, $p_g = 4.0$ MPa;

Allowable stress, $S = 38.5$ MPa; Permissible bending stress for piston pin, $S_{d1} = 100$ MPa; Permissible bearing pressure, $p_b = 20$ MPa

SOLUTION

Step-1: Piston head thickness

(i) Thickness by strength consideration,

$$t_1 = 0.43D\sqrt{P_g/S_t} \qquad \text{(Eqn. 21.30 MDHB)}$$

$$= 0.43 \times 200 \times \sqrt{4/38.5} = 27.72 \text{ mm}$$

(ii) Thickness by heat flow consideration,

$$t_1 = \frac{D^2 q}{16.K\,(T_c - T_e)} \qquad \text{(Eqn. 21.32 MDHB)}$$

$$= \frac{200^2 \times 250}{16 \times 270 \times 222} = 10.43 \text{ mm}$$

(iii) Thickness by empirical relation,

$$t_1 = 0.032D + 1.5 = 0.032 \times 200 + 1.5 = 7.9 \text{ mm} \qquad \text{(Eqn. 21.31 MDHB)}$$

Choose $t_1 = 28.0$ mm as largest of the three.

Step-2: Piston ring proportions

The radial thickness of the piston ring,

$$t_r = D\sqrt{3p_r/s} \qquad \text{(Eqn. 21.35 MDHB)}$$

where, $D = 200$ mm; $p_r = 0.028$ MPa; $S = 85$ MPa

$$\therefore \qquad t_r = 200 \times \sqrt{\frac{3 \times 0.028}{85}} = 6.287 \text{ mm}$$

Choose $t_r = 6.5$ mm

The axial depth of the ring,

$$h = t_r = 6.5 \text{ mm} \qquad \text{(Eqn. 21.36 MDHB)}$$

Step-3: The thickness of the barrel

Maximum thickness in the ring portion,

$$t_3 = 0.03D + b + 4.5 \qquad \text{(Eqn. 21.33 MDHB)}$$

$$= 0.03D + t_r + 0.4 + 4.5 = 0.03 \times 200 + 6.5 + 4.9 \cong 17.5 \text{ mm}$$

Thickness at the skirt,

$$t_4 = 0.3t_3 = 0.3 \times 17.5 = 5.22 \text{ mm} \qquad \text{(Eqn. 21.34 MDHB)}$$

Step-4: Length of the barrel

Distance from the top to the first ring,

$$t_g = t_1 = 28 \text{ mm} \qquad \text{(Eqn. 21.39 MDHB)}$$

Depth over piston rings,

$$h_t = \frac{D}{7} + 0.6 = \frac{200}{7} + 0.6 = 29.17 \qquad \text{(Eqn. 21.38 MDHB)}$$

The length of the piston over the piston rings,

$$L_r = t_g + h_t = 28 + 29.17 = 57.17 \qquad \text{(Eqn. 21.40 MDHB)}$$

Choose $L_r = 58$ mm

The bearing length,

$$L_p = \frac{0.1F_{\max}}{D \cdot P_b} \qquad \text{(Eqn. 21.41 MDHB)}$$

where, $F_{\max} = \frac{\pi}{4} D^2 \cdot p_g = \frac{\pi}{4} \times 200^2 \times 4 = 125663.7 \text{ N}$

$D = 200$ mm and $p_b = 0.25$ MPa

$$\therefore \quad L_p = \frac{0.1 \times 125663.7}{200 \times 0.25} = 251.33 \text{ mm}$$

Total length, $L = L_r + L_p = 58 + 251.33 = 309.33$ mm

By empirical relation, $L = D = 200$ mm

$\therefore$ Choose $L = 310$ mm

Step-5: Piston pin diameter

Bearing length of pin,

$$L_1 = 0.45D = 0.45 \times 200 = 90 \text{ mm} \qquad \text{(Eqn. 21.45 MDHB)}$$

The diameter of the pin by bearing pressure consideration,

$$d_p = \frac{\pi D^2 p_g}{4 \cdot L_1 \cdot p_b} = \frac{F_{\max}}{L_1 \cdot p_b} = \frac{125663.7}{90 \times 20} = 69.8 \text{ mm} \qquad \text{(Eqn. 21.44 MDHB)}$$

By empirical relation (Plate 23.1 MDHB),

$$d_p = 0.3D = 0.3 \times 200 = 60 \text{ mm}$$

Choose $d_p = 70$ mm

Inner diameter, $d_i = 0.5d_p = 0.5 \times 70 = 35$ mm

The section modulus,

$$Z = \frac{\pi}{32}\left(\frac{d_p^3 - d_i^4}{d_p}\right) = \frac{\pi}{32} \times \left(\frac{70^4 - 35^4}{70}\right) = 31569.3 \text{ mm}^3$$

The bending stress,

$$\sigma_b = \frac{F_p \cdot D}{8Z} = \frac{125663.7 \times 200}{8 \times 31569.3} = 99.5 \text{ MPa} \qquad \text{(Eqn. 21.46 MDHB)}$$

This is less than the permissible value of $S_d = 100$ MPa and design is safe.

Example 23.3: Design a cast iron piston for a single acting four-stroke diesel engine from the following data: Cylinder bore = 100 mm; Length of stroke = 125 mm; Speed = 2000 rpm. Maximum gas pressure, $p_g = 5$ MPa; Brake mean effective pressure, $P_m = 0.5$ MPa;

Fuel consumption, $w_f = 0.25$ kg/kw/hr

SOLUTION

Step-1: Power output,

$$NP = \frac{P_m LAN}{60,000} = \frac{0.5 \times 0.125 \times \frac{\pi}{4}(100)^2 \times (2000/2)}{60,000} = 8.12 \text{ kW}$$

Step-2: Piston head thickness

(i) From heat transfer consideration,

$$t_1 = \frac{Q}{4\pi K (T_c - T_e)} \qquad \text{(Eqn. 21.32 MDHB)}$$

where, $Q = 0.05 w_f \cdot HCV \cdot kW$

$$= 0.05 \times 0.25 \times 45 \times 10^6 \times 8.12 = 4.6 \times 10^6 \text{ J/hr}$$

$K = 270$ and $T_c - T_e = 222°\text{C}$

Substituting the values,

$$t_1 = \frac{4.6 \times 10^6}{4 \times \pi \times 270 \times 222} = 6.11 \text{ mm}$$

(ii) From strength consideration with s_t = 39 MPa

$$t_1 = 0.43 D \sqrt{p_g / S_t} \qquad \text{(Eqn. 21.30 MDHB)}$$

$$= 0.43 \times 100 \times \sqrt{5/39} = 15.39 \text{ mm}$$

(iii) Empirical relation,

$$t_1 = 0.032D + 1.5 = 0.032 \times 100 + 1.5 = 4.7 \text{ mm} \qquad \text{(Eqn. 21.31 MDHB)}$$

Choose the largest value $t_1 = 15.39 \cong 15.5$ mm

Step-3: Piston ring proportions

The radial thickness of the ring,

$$t_1 = D\sqrt{3p_r/S} \quad \text{(Eqn. 21.35 MDHB)}$$

where, $S = 85$ MPa and $p_r = 0.0245$ MPa (Table 21.7 MDHB)

$$\therefore \quad t_r = 100\sqrt{\frac{3 \times 0.0245}{85}} = 2.94 \text{ mm}$$

Choose $t_r = 3$ mm

The axial depth of the ring,

$$h = 0.8t_r = 0.8 \times 3 = 2.4 \cong 2.5 \text{ mm}$$

Step-4: Length of piston over piston rings

The landing between the rings,

$$h_l = (0.75h \text{ to } h) = h = 2.5 \text{ mm}$$

The distance from the top to the first ring,

$$t_g = (t_1 \text{ to } 1.2t_1) = t_1 = 15.5 \text{ mm} \quad \text{(Eqn. 21.39 MDHB)}$$

Assume number of rings = 5

The total length over the rings,

$$L_r = t_g + i(h + h_l) \quad \text{(Eqn. 21.40 MDHB)}$$

$$= 15.5 + 5 \times (2.5 + 2.5) = 40.5 \text{ mm}$$

Step-5: Length for bearing pressure,

$$L_b = \frac{0.1 \cdot F_{\text{max}}}{D \cdot P_b} \quad \text{(Eqn. 21.41 MDHB)}$$

where, $F_{\text{max}} = \frac{\pi}{4} D^2 \cdot p_g = \frac{\pi}{4} \times 100^2 \times 5 = 3.93. \times 10^4 \cdot N$

$p_b = 0.57$ MPa (Table 23.8 MDHB)

$$\therefore \quad L_b = \frac{0.1 \times 3.93 \times 10^4}{100 \times 0.57} = 68.94 \text{ mm}$$

Step-6: Total length of the piston,

$$L_p = L_r + L_b = 40.5 + 68.94 = 109.44 \cong 110 \text{ mm}$$

Empirical relation, $L_p = D = 100$ (Eqn. 21.43 MDHB)

Choose $L_p = 110$ mm

Step-7: The diameter of the pin

(i) For bearing pressure,

$$d = \frac{\pi D^2 p_g}{4 \cdot L_1 p_b} \qquad \text{(Eqn. 21.44 MDHB)}$$

where, $p_g = 5$ MPa, and $L_1 = 0.45D = 45$ mm;

And $p_b = 12.5$ (Table 21.10 MDHB)

$$\therefore \qquad d = \frac{\pi \times 100^2 \times 5}{4 \times 45 \times 12.5} = 69.8$$

Choose $d = 70$ mm

Step-8: Check for bending strength of the pin,

$$\sigma_b = \frac{F_p D}{8Z} \qquad \text{(Eqn. 21.46 MDHB)}$$

where, $F_p = F_{\max} = 3.93 \times 10^4$ N;

$$Z = \frac{\pi}{32} \cdot d^3 = \frac{\pi}{32} \times 70^3 = 33673.95 \text{ mm}^3$$

$$\therefore \qquad \sigma_b = \frac{3.93 \times 10^4 \times 100}{8 \times 33673.95} = 14.59 \text{ MPa}$$

This is less than the allowable stress, $S_d = 100$ MPa for piston pin. Hence, design is safe.

Example 23.4: Design an aluminium alloy piston with a flat head for an I.C. engine, having 100 mm bore dia. The max. explosion pressure many be taken as 4.5 MPa. Draw the dimensioned sketch of the piston.

Given Data: Bore diameter, $D = 100$ mm;

Gas pressure, $p_g = 4.5$ MPa

SOLUTION

Step-1: The thickness of the head

(i) Thickness for strength,

$$t_1 = 0.43D\sqrt{p_g/s_t} \qquad \text{(Eqn. 21.30 MDHB)}$$

where, $p_g = 4.5$ MPa and $s_t = 39$ MPa (Table 21.1 MDHB)

$$\therefore \qquad t_1 = 0.43 \times 100 \times \sqrt{\frac{4.5}{39}} = 14.61 \text{ mm}$$

(ii) Thickness for heat flow,

$$t_1 = \frac{D^2 \cdot q}{16.K\,(T_c - T_e)} \qquad \text{(Eqn. 21.32 MDHB)}$$

where, $q = 500\ (230 \text{ to } 920)$; $K = 650$ and $T_c - T_e = 111°\text{C}$

$$\therefore \qquad t_1 = \frac{100^2 \times 500}{16 \times 650 \times 111} = 4.33 \text{ mm}$$

(iii) Thickness by empirical relation,

$$t_1 = 0.032D + 1.5 \qquad \text{(Eqn. 21.31 MDHB)}$$

$$= 0.032 \times 100 + 1.5 = 4.7 \text{ mm}$$

Choose the largest value of 14.61 mm, and let $t_1 = 15$ mm.

Step-2: Piston ring proportions

The radial thickness,

$$t_1 = D\sqrt{3p_r/S} \qquad \text{(Eqn. 21.35 MDHB)}$$

where, $p_r = 0.0245$ MPa (Table 23.7); and $S = 85$ MPa

$$\therefore \qquad t_r = 100 \times \sqrt{\frac{3 \times 0.0245}{85}} = 2.94 \text{ mm}$$

Choose t_r = 3 mm.

The axial depth,

$$h = (0.7t_r \text{ to } t_r) = t_r = 3 \text{ mm} \qquad \text{(Eqn. 21.36 MDHB)}$$

Step-3: Piston barrel thickness

The maximum thickness,

$$t_3 = 0.03D + b + 4.5 \qquad \text{(Eqn. 21.33 MDHB)}$$

$$= 0.03D + t_r + 0.4 + 4.5 = 0.03 \times 100 + 3 + 4.9 = 10.9 \cong 11 \text{ mm}$$

Thickness of the skirt,

$$t_4 = 0.3t_3 = 0.3 \times 11 = 3.3 \text{ mm} \cong 3.5 \text{ mm} \qquad \text{(Eqn. 21.34 MDHB)}$$

Step-4: Length of piston over the piston rings

The landing between the rings,

$$h_l = 0.7h = 0.7 \times 3 = 2.1 \text{ mm} \qquad \text{(Eqn. 21.37 MDHB)}$$

The distance from top to the first ring,

$$t_g = (1.2t_1) = 1.2 \times 15 = 18 \text{ mm}$$

Let the number of rings, $i = 5$

Then length over the rings,

$$L_r = t_g + i(h + h_l) - 18 + 5(3 + 2.1) - 43.5 \text{ mm} \qquad \text{(Eqn. 21.40 MDHB)}$$

Step-5: Length for bearing pressure,

$$L_b = \frac{0.1 \cdot F_{max}}{D \cdot p_b} \qquad \text{(Eqn. 21.41 MDHB)}$$

where, $F_{max} = \frac{\pi}{4} D^2 p_g = \frac{\pi}{4} \times 100^2 \times 4.5 = 3.53 \times 10^4 \text{ N}$

$$p_b = 0.35 \text{ MPa} \qquad \text{(Table 21.8 MDHB)}$$

$$\therefore \quad L_b = \frac{0.1 \times 3.53 \times 10^4}{100 \times 0.35} = 100.86 \text{ mm}$$

Step-6: Total length of the piston,

$$L = L_r + L_b = 43.5 + 100.86 = 144.35$$

Choose $L = 145$ mm

Step-7: The piston pin design

The diameter of the pin for bearing pressure,

$$d = \frac{\pi D^2 p_g}{4 L_1 p_b} = \frac{F_{max}}{L_1 p_b}$$

where, $F_{max} = 3.53 \times 10^4 \text{ N}; L_1 = 0.45D = 0.45 \times 100 = 45 \text{ mm}$

$p_b = 12.5$ MPa (Table 21.10 MDHB)

$$\therefore \quad d = \frac{3.53 \times 10^4}{45 \times 12.5} = 62.76 \text{ mm}$$

Choose $d = 65$ mm

Check for bending stress,

$$\sigma_b = \frac{F_p \cdot D}{8Z} \qquad \text{(Eqn. 21.46 MDHB)}$$

where, $F_p = F_{max} = 3.53 \times 10^4 \text{ N}$

$$Z = \frac{\pi}{32} \times (d^3) = \frac{\pi}{32} \times 65^3 = 26961.25 \text{ mm}^3$$

$$\therefore \quad \sigma_b = \frac{3.53 \times 10^4 \times 10}{8 \times 26961.25} = 16.36 \text{ MPa}$$

This is within the safe limits and hence the design is safe.

Example 23.5: Design a trunk type piston for a single cylinder, four-stroke cycle engine developing 5 kW at 600 rpm. The diameter of the piston is 120 mm and the maximum explosion pressure is 4.0 MPa. The heat supplied to the engine is 19000 kJ/kWhr; about 6% of this heat is conducted through the crown. The heat conduction factor is 60 kW/hr/m/°C. The temperature difference between the center and edge of the crown is 250°C. Assume necessary data.

Given Data: Power $= 5$ kWat $N = 600$ rpm; Diameter of piston $=$ 120 mm;

Maximum gas pressure, $p_g = 4.0$ MPa; Heat supplied, $H = 19000$ kJ/kW/h$_r$; The portion to crown $= 6\%$; Conductivity, $K = 60$ kW/h$_r$/m/°C ; Temperature gradient, $\Delta T = T_c - T_e = 250$°C

SOLUTION

Material – C.I. with $S_t = 39$ MPa

Step-1: Thickness of the crown

(i) Thickness for strength,

$$t_1 = 0.43D\sqrt{\frac{p_g}{S_t}} \qquad \text{(Eqn. 21.30 MDHB)}$$

$$= 0.43 \times 120 \times \sqrt{\frac{4}{39}} = 16.52 \text{ mm}$$

(ii) Thickness for heat flow,

$$t_1 = \frac{Q}{4\pi K \cdot (T_c - T_e)} \qquad \text{(Eqn. 21.32 MDHB)}$$

where, $Q = 0.06 \times 19000 \times 5 = 5700$ kJ/h$_r$; $K = 60$ kW/h$_r$/m/°C

$$t_1 = \frac{5700}{4 \times \pi \times 60 \times 250} = 0.030 \text{ m} = 30 \text{ mm}$$

(iii) Thickness by empirical relation,

$$t_1 = 0.032D + 1.5 = 0.032 \times 120 + 1.5 = 5.34 \text{ mm} \qquad \text{(Eqn. 21.31 MDHB)}$$

Choose the highest, $t_1 = 30.0$ mm

Step-2: Piston ring proportions

The radial thickness of the piston ring,

$$t_r = D\sqrt{\frac{3p_r}{S}} \qquad \text{(Eqn. 21.35 MDHB)}$$

where, $p_r = 0.0245$ MPa (Table 21.7 MDHB); $S = 85$ MPa

$$\therefore \qquad t_r = 120\sqrt{\frac{3 \times 0.0245}{85}} = 3.53 \text{ mm}$$

Choose $t_r = 4$ mm

The axial depth of the piston ring,

$$h = (0.7t_r \text{ to } t_r) \;= 0.7 \times 4 = 2.8 \text{ mm} \qquad \text{(Eqn. 21.36 MDHB)}$$

Choose $h = 3$ mm

Step-3: Piston barrel thickness

(i) The thickness in the ring portion,

$$t_3 = 0.03D + b + 4.5 \qquad \text{(Eqn. 21.33 MDHB)}$$

$$= 0.03D + t_r + 4.9 = 0.03 \times 120 + 4 + 4.9 = 12.5 \text{ mm}$$

(ii) The thickness towards skirt,

$$t_4 = (0.25t_3 \text{ to } 0.35t_3)\,) = 0.3 \times 12.5 = 3.75 \text{ mm} \qquad \text{(Eqn. 21.34 MDHB)}$$

Choose $t_4 = 4$ mm

Step-4: The length of the barrel

(i) The length over piston rings,

$$L_r = t_g + h_l = t_1 + \frac{D}{7} + 0.6 = 30 + \frac{120}{7} + 0.6 = 47.75 \text{ mm}$$

(ii) The bearing length,

$$L_b = \frac{0.1F_{\max}}{D \cdot p_b}$$

where, $F_{\max} = \dfrac{\pi}{4} D^2 \cdot p_g = \dfrac{\pi}{4} \times (120)^2 \times 4.5 = 50893.8 \text{ N}$

$$p_b = 0.35 \text{ MPa} \qquad \text{(Table 21.8 MDHB)}$$

$$L_b = \frac{0.1 \times 50893.8}{120 \times 0.35} = 121.2 \text{ mm}$$

(iii) Total length of barrel,

$$L = L_r + L_b = 47.75 + 121.2 = 168.9 \text{ mm}$$

Choose $L = 170$ mm

Step-5: The piston pin diameter

(i) The piston pin diameter for bearing pressure,

$$d = \frac{\pi D^2 p_g}{4L_1 p_b} = \frac{F_{\max}}{L_1 \cdot p_b}$$

where, $F_{\max} = 50893.8 \text{ N}; \; L_1 = 0.45D = 0.45 \times 120 = 54 \text{ mm}$

$$p_b = 15 \text{ MPa} \qquad \text{(Table 21.10 MDHB)}$$

$$\therefore \quad d = \frac{5093.8}{54 \times 15} = 62.83 \text{ mm}$$

Choose $d = 65$ mm

(ii) Check for the bending stress,

$$\sigma_b = \frac{F_p \cdot D}{8Z} = \frac{F_{\max} D}{8Z} = \frac{50893.8 \times 120}{8 \times \left(\frac{\pi}{32} \times 6.5^3\right)} = 28.32 \text{ MPa (Eqn. 21.46 MDHB)}$$

This is smaller than the permissible value, $S_d = 100 \text{ MPa}$ for the piston pin. Hence the design is safe.

Example 23.6: Design a cast iron piston for a four-stroke internal combustion engine for the following data:
Cylinder bore = 150 mm; Stroke = 200 mm; Indicated mean effective pressure = 6 bars; Fuel consumption = 4 kg/hr; Max. gas pressure = 5 MPa; Higher calorific value of fuel = 4100 kJ/kg; Speed of the engine = 600 rpm; Mechanical efficiency = 75%; Allowable stress for piston = 30 N/mm^2; Allowable tensile stress for piston ring = 90 MPa; Allowable bending stress for piston pin = 80 MPa.

SOLUTION

Step-1: Piston head thickness

(i) Thickness for strength,

$$t_1 = 0.43D\sqrt{(p_g/S)} \quad \text{(Eqn. 21.30 MDHB)}$$

$$= 0.43 \times 150 - \sqrt{5/30} = 26.33 \text{ mm}$$

(ii) By empirical relation,

$$t_1 = 0.032D + 1.5 \quad \text{(Eqn. 21.31 MDHB)}$$

$$= 0.032 \times 150 + 1.5 = 6.3 \text{ mm}$$

(iii) By heat conduction consideration,

$$t_1 = \frac{Q}{4\pi K(T_c - T_c)} \quad \text{(Eqn. 21.32 MDHB)}$$

where, $Q = 0.05 W_f HCV\ KW = 0.05(W_t \cdot kW)HCV$

$$= 0.05 \times 4 \times 4100 = 820 \text{ kJ/h}_r$$

$$K = 270 \text{ and } T_c - T_e = 222°\text{C}$$

$$t_1 = \frac{820 \times 10^3}{270 \times 4 \times \pi \times (222)} = 1.1$$

Choose the largest value of $t_1 = 26.33 \text{ mm} \cong 30 \text{ mm}$

Step-2: Piston ring proportions

The radial thickness, $t_1 = D\sqrt{3p_r/S}$

where, $p_r = 0.0245$ MPa (Table 21.7 MDHB); and $S = 90$ MPa

$$\therefore \qquad t_r = 150 \times \sqrt{\frac{3 \times 0.0245}{90}} = 4.3 \text{ mm}$$

Choose $t_r = 4.5$ mm

The axial depth of the ring,

$$h = (0.7t_r \text{ to } t_r) = 0.8 \times 4.5 = 3.6 \text{ mm} \qquad \text{(Eqn. 21.36 MDHB)}$$

Choose $h = 3.5$ mm

Step-3: Piston barrel thickness

(i) Thickness in the ring portion,

$$t_3 = 0.03D + b + 4.5 \qquad \text{(Eqn. 21.33 MDHB)}$$

$$= 0.03D + t_r + 0.4 + 4.5 = 0.03 \times 150 + 4.5 + 4.9 \cong 14 \text{ mm}$$

(ii) Thickness towards open end (skirt),

$$t_4 = (0.25t_3 \text{ to } 0.35t_3) \qquad \text{(Eqn. 21.34 MDHB)}$$

$$= 0.3t_3 = 0.3 \times 14 = 4.2 \text{ mm}$$

Step-4: The piston length

(i) Length over the piston rings

The distance from the top to the first groove,

$$t_g = (t_1 \text{ or } 1.2t_1) = t_1 = 30 \text{ mm} \qquad \text{(Eqn. 21.39 MDHB)}$$

The landing between the rings,

$$h_l = (0.75h \text{ to } h) = h = 3.5 \text{ mm} \qquad \text{(Eqn. 21.37 MDHB)}$$

Assume the no. of rings, $i = 4$

The length over the piston rings,

$$L_r = t_g + i(h + h_l) \qquad \text{(Eqn. 21.40 MDHB)}$$

$$= 30 + 4 \times (3.5 + 3.5) = 58 \text{ mm}$$

(ii) The length for bearing pressure,

$$L_b = \frac{0.1F_{\max}}{D \cdot p_b} \qquad \text{(Eqn. 21.41 MDHB)}$$

where, $F_{\max} = \frac{\pi}{4} D^2 \times p_g = \frac{\pi \times 150^2 \times 5}{4} = 88357.3$ N; $p_b = 0.25$ MPa

$$\therefore \qquad L_b = \frac{0.1 \times 88357.3}{150 \times 0.25} = 235.62 \text{ mm}$$

(iii) The total length of the barrel,

$$L = L_r + L_b = 58 + 235.62 = 293.62$$

Choose $L = 300$ mm

Step-5: Design of piston pin

(i) The dia. of the pin by bearing pressure consideration,

$$d = \frac{\pi D^2 \cdot p_g}{4 \cdot L_1 \cdot p_b} = \frac{F_{max}}{L_1 \cdot p_b} \qquad \text{(Eqn. 21.44 MDHB)}$$

$$= \frac{88357.3}{0.5 \times 150 \times 12.5} = 94.25 \text{ mm}$$

Choose $d = 95$ mm and $L_1 = 0.5 \times D = 0.5 \times 150 = 75$ mm

(ii) Check for the bending stress,

$$\sigma_b = \frac{F_p \cdot D}{8.Z} \qquad \text{(Eqn. 21.46 MDHB)}$$

where, $F_p = F_{max} = 88357.3\,\text{N}$; $Z = \frac{\pi}{32} d^3 = \frac{\pi \times 95^3}{32} = 84172.6 \text{ mm}^3$

$$\therefore \qquad \sigma_b = \frac{88357.3 \times 150}{8 \times 84172.6} = 19.68 \text{ MPa}$$

This is less than the permissible. The design is safe.

23.2 CONNECTING ROD

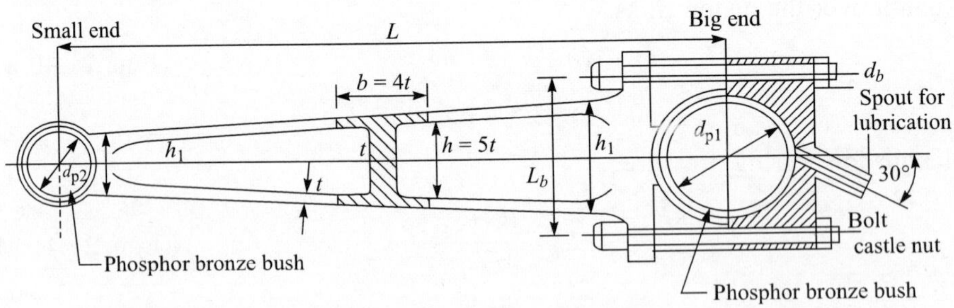

Figure 23.2 Connecting rod

Stresses in the connecting rod: The connecting rod is a link connecting piston to crankshaft. It converts the linear (reciprocating) motion of piston to the rotary motion of crankshaft. The small end covers the piston pin and the big end envelopes the crankpin. The small end reciprocates and the big end rotates. The body is subjected to combination of linear and rotary motions. It is subjected to accelerations and retardations. These motions give rise to the inertia forces in axial and lateral directions. These add to the axial force due to gas pressure. At the I.D.C. the load is axial both due to the gas pressure and inertia. The net axial load is given by

$$Q_1 = F_{\max} - F_i \tag{23.12}$$

where, $F_{\max}$ = Maximum gas load $= \dfrac{\pi}{4} D^2 \cdot p_g$ and inertia load $F_i = W_r \cdot a$

W_r = The reciprocating mass

$$a = \text{Acceleration} = \left(\frac{\omega^2 R}{1000}\right)\left[1 - \frac{1}{n}\right]$$

The connecting rod is long and hence there is column action. The stress induced is given by the Rankine equation,

$$\sigma_c = \frac{Q_1}{A}\left[1 + a\left(\frac{L}{K}\right)^2\right] \tag{23.13}$$

At inclined position of the crank, the connecting rod is subjected to both axial and lateral inertia forces. The lateral inertia forces are maximum, when the crank and connecting rod are at right angles. At this position the crankpin end is subjected to maximum lateral acceleration of $a = \omega^2 \cdot R$. The piston-pin end has no lateral acceleration. The inertia force in the lateral direction varies linearly from zero at the small end to a maximum value at the big end. The variation is shown in Fig. 23.3.

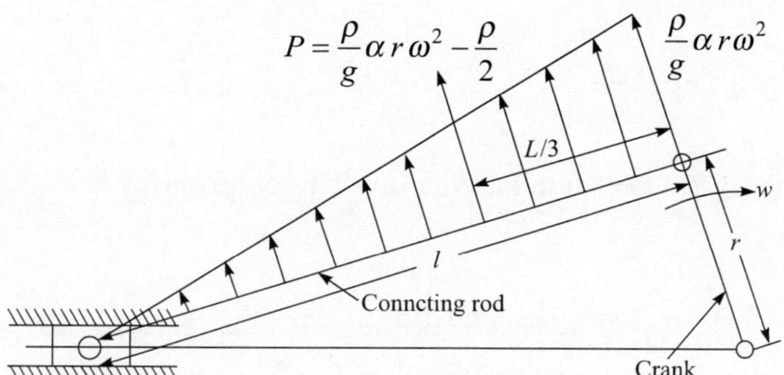

Figure 23.3 Inertia loading on a connecting rod in its plane of oscillation

The inertia force per unit length at the big end, $F_2 = \dfrac{\rho}{g} A\omega^2 R$

where, ρ = Weight density of the material
A = Cross-sectional area
R = Radius of the crank

The inertia force at a distance x from the small end,

$$F_x = \frac{F_2}{L}x = \left(\frac{\rho A\omega^2 R}{gL}\right)x$$

The total lateral force on the connecting rod,

$$F_i = \int_0^L F_x \cdot dx = \left(\frac{\rho A\omega^2 R}{gL}\right)\int_0^L x dx = \left[\frac{\rho A\omega^2 RL}{2g}\right] \tag{23.14}$$

The bending moment at x,

$$M_x = R_1 x - \left(F_x \frac{x}{2}\right)\frac{x}{3} = \frac{F_i}{3}x - \frac{F_i}{L^2}\frac{x^3}{3} = \left(\frac{F_i}{3}\right)\left(x - \frac{x^3}{L^2}\right)$$

For maximum bending moment,

$$\frac{dM_x}{d_x} = 0 = \frac{F_i}{3}\left(1 - \frac{3x^2}{L^2}\right)$$

$$\therefore \qquad x = \frac{L}{\sqrt{3}}$$

$\therefore$ Maximum bending moment, $M_{\max} = \dfrac{F_i}{3}\left(\dfrac{L}{\sqrt{3}} - \dfrac{L}{3\sqrt{3}}\right) = \dfrac{2\cdot\sqrt{3}}{27}F_i \cdot L$ (23.15)

The bending stress, $\sigma_b = \left(\dfrac{M_{\max}}{Z}\right)$ (23.16)

At this position, the gas force introduces an axial force given by the relation (Eqn. 21.7 MDHB)

$$Q_2 = F_{p2}\left[1 - \frac{\sin^2\theta}{n^2}\right]^{-1/2} = F_{p_2}/\cos\beta \tag{23.17}$$

where, $\theta = \tan^{-1}\left(\dfrac{L}{R}\right)$ and $\beta = \tan^{-1}\left(\dfrac{R}{L}\right)$; $n = \dfrac{L}{R}$;

F_{p_2} = Gas force at this position

Then compressive stress due to this force,

$$\sigma_{c2} = \frac{Q_2}{A}\left[1 + a\left(\frac{L}{K}\right)^2\right] \tag{23.18}$$

The net stresses are:
In the inside layer (compressive),

$$\sigma_i = (\sigma_b + \sigma_{c2}) \tag{23.19}$$

In the outer layer (tensile),

$$\sigma_o = (\sigma_b - \sigma_{c2}) \tag{23.20}$$

Cap bolts: In two-stroke engines, the connecting rod is always in compression. In four- stroke engines, during suction stroke, the piston is to be pulled out and hence there is a tensile force equal to the inertia force.

$$F_t = F_i = \frac{W_r \cdot \omega^2 R}{1000}\left[1 - \frac{1}{n}\right] \tag{23.21}$$

This force is exerted by the crankpin on the big end cap and transmitted to the connecting rod through the bolts. The bolts are subjected to the tensile stresses. Normally, there are two bolts sharing this load. The bolt diameter is given by

$$d_b = \sqrt{\frac{2F}{\pi S_d}} \tag{23.22}$$

Materials for the connecting rod: The connecting rod is subjected to fatigue loads, and the materials used must be resistant to fatigue failure. The medium carbon steels and alloy steel forgings are used. The design stresses adopted are as follows:

Carbon steel, $S_d = 80$ to 120 MPa

Alloy steel, $S_d = 120$ to 185 MPa

The bolts of the connecting rod are made of high grade medium carbon alloy steels, heat treated. The permissible stresses are as follows:

Carbon steel, $S_d = 50$ to 60 MPa

Alloy steels, $S_d = 90$ to 100 MPa

The small end bushes are made of phosphor bronze. The big end bearing bushes are split and made of steel, lined with babbit or lead bronze.

Cross sections: There are three sections normally used for connecting rods:

(i) Circular section
(ii) Rectangular section
(iii) I–section

The connecting rod is subjected to column action. The end conditions are different in the two planes. In the plane of rotation, they are free and hinged. In the lateral planes, the rod ends have restricted motion due to the support provided by the end bosses. They are treated as fixed ends. Then the basic condition to be satisfied is: $I_{xx} = 4 \cdot I_{yy}$

The circular section does not satisfy this condition. But it is simple and easy to manufacture. This is adopted for slow speed steam engines.

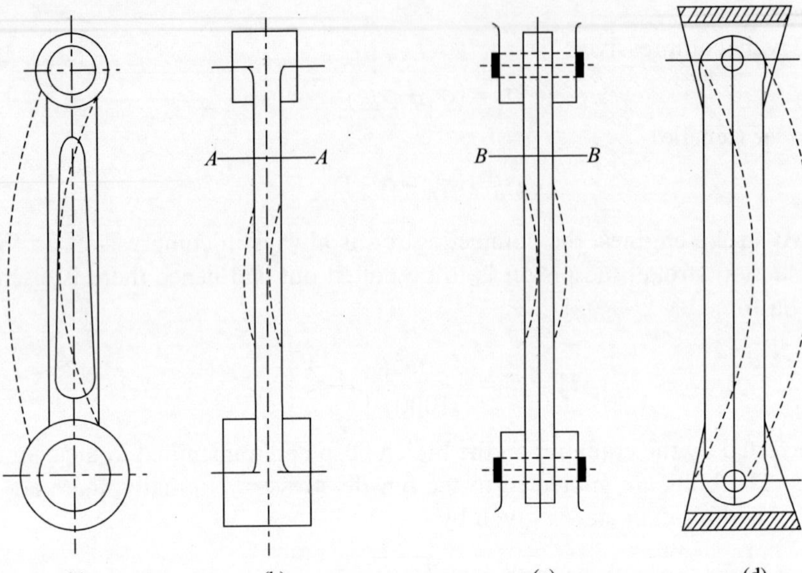

Figure 23.4 Buckling of links in two perpendicular planes

The rectangular section with $\frac{b}{h} = \frac{1}{2}$ satisfies this condition.

$$I_{xx} = \frac{bh^3}{12} \text{ and } I_{yy} = \frac{h \cdot b^3}{12}$$

$$\frac{I_{xx}}{I_{yy}} = \frac{(b \cdot h^3/12)}{(b \cdot b^3/12)} = \left(\frac{h}{b}\right)^2$$

$$\therefore \quad \text{for } \frac{h}{b} = 2, \frac{I_{xx}}{I_{yy}} = 4$$

This section has larger mass and produces larger inertia force. The rectangular section is suitable for slow speed engines, but not suitable for high speeds.

The I-section with proper proportions is suitable for high speed I-C engines. The proportions prescribed are shown in Fig. 23.5.

$$I_{xx} = \frac{1}{12}[4 \times 5.35^3 - 3 \times 3.35^3]t^4 = \frac{518}{12}t^4$$

$$I_{yy} = \frac{1}{12}[2 \times 4^3 + 3.35 \times 1^3]t^4 = \frac{131.35}{12}t^4$$

$$\therefore \quad \frac{I_{xx}}{I_{yy}} = \left(\frac{518}{131.3}\right) = 3.95$$

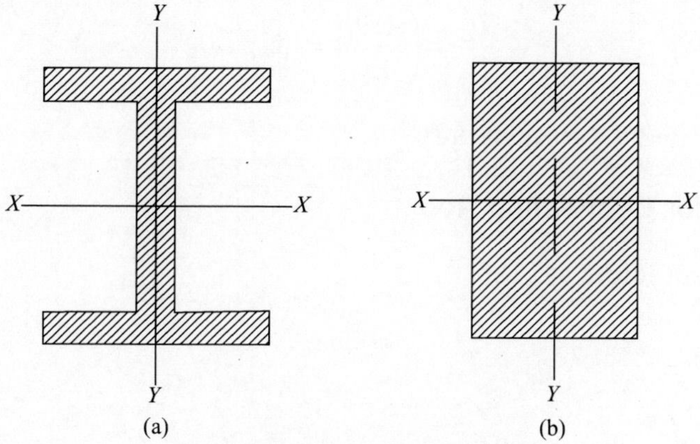

X-X = Neutral axis, free ended column
Y-Y = Neutral axis, fixcd ended column

Figure 23.5

This section is suitable theoretically, but in practice for the fixed ended columns, the end fixity coefficient is taken to be between 3 and 3.5. The second alternative section shown in Fig 23.6 is usually chosen.

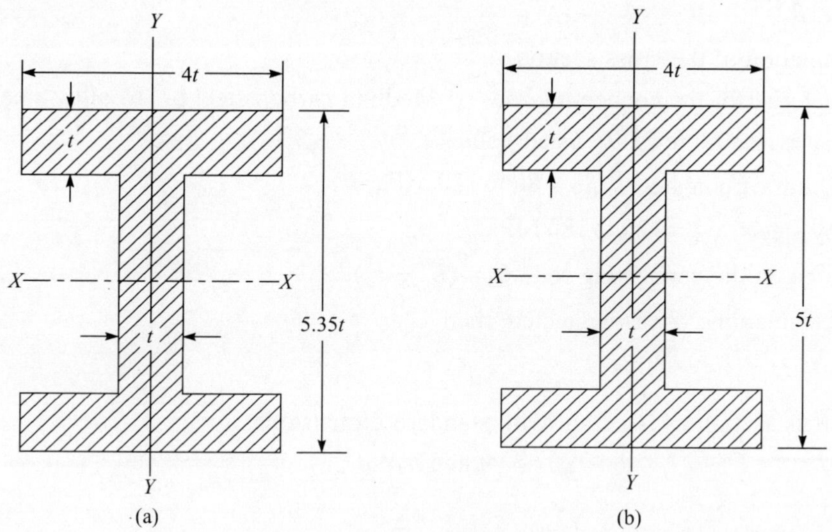

Figure 23.6

$$I_{xx} = \frac{1}{12}[4 \times 5^2 - 3 \times 3^3]t^4 \;= \frac{411}{12}t^4$$

$$I_{yy} = \frac{1}{12}[2 \times 1 \times 4^3 + 3. \times 1^3]t^4 \;= \frac{131}{12}t^4$$

$$\therefore \qquad \frac{I_{xx}}{I_{yy}} = \left(\frac{419}{13}\right) = 3.2$$

This value is considered to be satisfactory.

Design Procedure for Connecting Rod

Step-1: Determination of forces at I.D.C.

(i) The gas force, $F_g = \frac{\pi}{4} \cdot D^2 \cdot p_g$ (1)

(ii) The intertia force, $F_i = \frac{W_r \cdot \omega^2 R}{1000}\left(1 + \frac{1}{n}\right)$ (2)

where, W_r = Mass of reciprocating parts in kg

ω = Angular speed of the crankshaft in rad/sec

R = Radius speed of the crank in mm

$n = \frac{L}{R}$ ratio

Step-2: Determination of the cross section

Select the material for the connecting rod: (i) Medium carbon steel or (ii) alloy steel.

The design stresses (Table 21.9), usual values

for medium carbon steels, $S_d = 80 \text{ to } 120$ MPa;

for alloy steels, $S_d = 120 \text{ to } 180$ MPa

The axial force on the connecting rod, $Q_1 = (F_g - F_i)$

For design calculations, F_i is neglected, then

$$Q_1 = F_g$$

Select the cross section as I-section with standard dimensions,

$$b = 4t \text{ and } h = 5t$$

Then, $A = 11t^2; k^2 = \frac{419}{132}t^2; I_{xx} = \left(\frac{419t^4}{12}\right)$

Then applying the Rankine equation,

$$\sigma_c = \frac{Q_1}{A}\left[1 + a\left(\frac{L}{K}\right)^2\right] \leq S_d \qquad (3)$$

$$\therefore \quad \frac{A}{1+a\left(\frac{L}{K}\right)^2} = \left(\frac{Q_1}{S_d}\right)$$

$$\therefore \quad \frac{11t^2}{1+a\left(\frac{L^2}{3.17t^2}\right)} = \frac{Q_1}{S_d}$$

$$\therefore \quad \frac{34.92 \cdot t^4}{3.17t^2 + a \cdot L^2} = \frac{Q_1}{S_d} \tag{4}$$

This equation is solved for t. Then the dimensions of the cross section are determined. Alternatively, the cross-sectional area A is determined by the simple compression equation,

$$A_C = \left(\frac{Q_1}{S_d}\right)$$

To account for the column action, choose the actual area A as $2.\ A_C$ and determine t

$$t = \sqrt{\frac{A}{11}} = \sqrt{\frac{2}{11}\left(\frac{Q_1}{S_d}\right)} \tag{5}$$

Then determine

$$K^2 = \frac{419}{132}t^2 \quad \text{and} \quad I_{xx} = \frac{419}{12}t^4 \quad \text{and} \quad Z = \frac{419}{30}t^3$$

and check for the stress induced

$$\sigma_c = \frac{Q_1}{A}\left[1+a\left(\frac{L}{K}\right)^2\right] \le S_d \tag{6}$$

If this condition is not satisfied, increase t suitably.

Step-3: Check for the stress in the position of maximum bending

The crank angle at this position,

$$\theta = \tan^{-1}\left(\frac{L}{R}\right) \text{ and } \beta = \tan^{-1}\left(\frac{R}{L}\right)$$

The axial force, $Q_1 = F_g/\cos\beta$

The axial stress, $\sigma_c = \dfrac{Q_1}{A}\left[1+a\left(\dfrac{L}{K}\right)^2\right]$ (7)

The inertia force on the rod,

$$F_i = \frac{A\rho\omega^2 RL}{2000} \tag{8}$$

Maximum bending moment $M_{max} = \frac{2\sqrt{3}F_i}{27}$ (9)

The bending stress,

$$\sigma_b = \frac{M_{max}}{Z} \tag{10}$$

The total stress,

$$\sigma = \sigma_c + \sigma_b \le S_d \tag{11}$$

If this condition is not satisfied increase, i.e., t.

Step-4: Diameters of end bearings

The diameter of the bearing is evaluated by the relation,

$$d_p = \sqrt{\frac{Q_1}{p_b \cdot k}} \tag{12}$$

where, $p_b =$ bearing pressure from Table 23.10 MDHB

$K = 0.6$ to 1.5 for big end bearing

$= 1.5$ to 2.0 for small end bearing

The diameters of piston pin (small end) and crankpin (big end) are determined by this relation.

Step-5: The design of the cap bolts

The load on the cap bolts is the inertia force of the reciprocating mass given by the relation

$$F = \frac{W_r \cdot \omega^2 \cdot R}{1000}\left[1 - \frac{1}{n}\right] \tag{13}$$

There are two bolts resisting this force and the diameter,

$$d = \sqrt{\frac{2.F}{\pi S_d}} \tag{14}$$

For determining S_d the data in Table 23.9 MDHB may be used by assuming a factor of safety of 1.5 to 2.

The approximate values are

$$S_d = 50 \text{ to } 60 \text{ MPa for carbon steels}$$
$$= 90 \text{ to } 100 \text{ MPa for alloy steels}$$

Step-6: The cap design

B.M. on the cap $M = F \cdot L_b / 4$

The big end cap thickness is determined for the bending strength by the equation,

$$\sigma_b = \frac{6M}{Bt^2} \le S_d \tag{15}$$

where, B is equal to the length of the crankpin.

Table 23.9. Strength of some varieties of steel

Material	Ultimate strength MPa	Yield point MPa	Elastic limit MPa
Carbon steel (C - 0.05% to 0.15%)	420	273	217
Carbon steel (C - 0.15% to 0.25%)	469	301	245
Carbon steel (C - 0.25% to 0.35%)	490	322	294
Carbon steel (C - 0.35% to 0.45%)	595	385	350
Nickel steel (Ni - 3.25% to 3.75%)			
(a) water quenched at 434°C	539	350	315
(b) water quenched at 700°C	980	770	700

Table 23.10. Bearing (design data)

Types of engine	Kind of bearing	Highest allowable pressure P_b MPa	Suitable Z centipoise	Minimum ZN/p	Clearance C/r	Length to diameter ratio L/d
Automotive and aircraft engines	Main	5.5 - 12	7 to 9	2150	0.001	0.6 - 1.8
	Crankpin	10.5 - 24.5	"	1425	0.001	0.6 - 1.4
	Piston pin	16 - 35	"	1150	0.001	1.5 - 2.0
Gas and oil engines, four-stroke	Main	5 - 8.5	15 to 65	2850	0.001	0.6 - 2.0
	Crankpin	10 - 14	"	1425	< 0.001	0.6 - 1.5
	Piston pin	12.5 - 17	"	850	< 0.001	1.5 - 2.0
Gas and oil engine two-stroke	Main	3.5 - 5.5	20 to 65	3550	< 0.001	0.6 - 2.0
	Crankpin	7 - 10.5	"	1700	< 0.001	0.6 - 1.5
	Piston pin	8.5 - 12.5	"	1150	< 0.001	1.5 - 2.0

Table 23.11. Empirical design of crankshaft

Engines type	L/D	d_{cp}/D	L_{cp}/D	d_s/D	L_s/D**
Carburettor engines in line	1.20-1.28	0.60-0.70	0.45-0.65	0.60-0.80	0.45 – 0.60 0.74 – 0.84
Vee-type with connecting rods attached to one crankpin	1.25-1.35	0.56-0.66	0.8-1.0	0.63-0.75	0.50 – 0.70 0.70 – 0.88
Diesel engines in line	1.25-1.30	0.64-0.75	0.7-1.0	0.70-0.90	0.45 – 0.60 0.75 – 0.85
Vee-type with connecting rods attached to one crankpin	1.47-1.55	0.65-0.72	0.8-1.0	0.70-0.75	0.50 – 0.65 0.65 – 0.86

**The upper figures are for intermediate bearings
The lower figures are for the main end bearings

Table 23.12. Coefficient *a* in the American Bureau of shipping formula

Type	No. of cylinder		Ratio of stroke to distance over crank webs = *L*/*C*							
	Four-stroke	**Two-stroke**	**0.7**	**0.8**	**0.9**	**1.0**	**1.1**	**1.12**	**1.13**	**1.14**
Explosion engines	1, 2, 4	1, 2	1.17	1.17	1.17	1.17	1.17	1.17	1.17	1.17
	3, 5, 6	3	1.17	1.17	1.17	1.17	1.19	1.20	1.22	1.24
	8	4	1.17	1.19	1.21	1.23	1.25	1.28	1.30	1.32
	10, 11, 12,	5, 6	1.18	1.20	1.23	1.25	1.28	1.31	1.33	1.35
Air injection diesel engines	1, 2, 4	1, 2	1.17	1.19	1.22	1.25	1.28	1.31	1.34	1.36
	3, 5, 6	3	1.19	1.22	1.25	1.28	1.32	1.35	1.38	1.41
	8	4	1.20	1.24	1.27	1.30	1.33	1.37	1.40	1.43
	12	5, 6	1.22	1.25	1.29	1.32	1.36	1.39	1.42	1.45
	16	8	1.25	1.29	1.33	1.36	1.40	1.44	1.47	1.50

Table 23.13. Constant for torsion of rectangular bars

b*/*h	**1.0**	**1.20**	**1.50**	**1.75**	**2.0**	**2.50**	**3.00**
α_1	0.208	0.219	0.231	0.239	0.246	0.258	0.267
α_2	0.208	0.235	0.269	0.291	0.309	0.336	0.355
β	0.1406	0.1666	0.196	0.214	0.229	0.249	0.263
	b*/*h	**4.00**	**5.00**	**6.00**	**8.00**	**10.00**	∞
	α_1	0.282	0.291	0.299	0.307	0.312	0.333
	α_2	0.378	0.392	0.402	0.414	0.421	-
	β	0.281	0.291	0.299	0.307	0.312	0.333

Example 23.7: Design a connecting rod for a diesel engine with the following data:

Maximum gas pressure, $p_g = 3.7$ MPa

Cylinder diameter, $D = 300$ mm; Stroke $L_s = 450$ mm

Connecting rod length, $L = 1000$ mm;

The speed of the engine, $N = 250$ rpm

SOLUTION

Step-1: The force on the connecting rod at I.D.C

$$\text{Gas force, } F_g = \frac{\pi}{4} D^2 p_g = \frac{\pi}{4} \times (300)^2 \times 3.7 = 261537.6 \text{ N}$$

Neglecting the inertia force, the max. force on the connecting rod $Q_1 = F_g = 261537.6\,\text{N}$.

Step-2: The cross section of the rod

The material for the rod be carbon steel with permissible stress, $S_d = 80$ MPa. The rod section be round of diameter d.

Then cross-sectional area, $A = \left(\frac{\pi}{4}d^2\right)$; radius of gyration $k = \frac{d}{4}$; end constant $a = \frac{1}{7500}$

(Table 1.1 MDHB)

Then substituting the values in Rankine equation

$$\sigma_c = \frac{Q_1}{A}\left[1 + a\left(\frac{l}{k}\right)^2\right] \le S_d$$

$$\frac{261537.6}{\left(\frac{\pi}{4}d^2\right)}\left[1 + \frac{1}{7500}\left(\frac{1000\times 4}{d}\right)^2\right] = 80$$

$$\therefore \quad d^4 - \frac{261537.6\times 4}{\pi\times 80}\left[d^2 + \frac{16\times 10^6}{7500}\right] = 0$$

$$\therefore \quad d^4 - 4162.5d^2 - 8.88\times 10^6 = 0$$

Solving $d = 75.6\text{ mm} \cong 76\text{ mm}$

Step-3: Check for bending stress

The lateral inertia force, $F_i = \frac{A\rho\omega^2 R.L}{2000}$ (Eqn. 21.48 MDHB)

where, $A = \frac{\pi}{4}d^2 = \frac{\pi}{4}\times 76^2 = 4536.46\text{ mm}^2$; $\rho = 8\times 16^{-6}\text{ kg/mm}^3$

$$\omega = \left(\frac{2\pi N}{60}\right) = \frac{2\pi\times 250}{60} = 26.18\text{ rad/sec}$$

$$R = \frac{L_s}{2} = \frac{450}{2} = 225\text{ mm};\ L = 1000\text{ mm}$$

$\therefore$ Substituting the values,

$$F_i = \frac{4536.46\times 8\times 10^{-6}\times (26.18)^2\times 225\times 10^3}{2000} = 2798.32\text{ N}$$

$$B.M = \frac{2\sqrt{3}F_i L}{27}$$ (Eqn. 21.49 MDHB)

$$= \frac{2 \times \sqrt{3} \times 2798.32 \times 1000}{27} = 359 \times 10^3 \text{ Nmm}$$

$$\text{Bending stress, } \sigma_b = \frac{M}{2} = \frac{359.0 \times 10^3}{\frac{\pi}{32} \times (76)^3} = 8.33 \text{ MPa}$$

This is less than the design stress and safe.

Step-4: The diameter of small end bearing,

$$d_{p1} = \sqrt{\frac{Q}{p_b \cdot k}} \qquad \text{(Eqn. 21.52 MDHB)}$$

where, $Q = 261537.6\,\text{N}$; $p_b = 15.0\,\text{MPa}$ and $k = 1.5$ (Table 21.10 MDHB)

$$\therefore \qquad d_{p1} = \sqrt{\frac{261537.6}{15.0 \times 1.5}} = 107.8$$

Choose $d_{p1} = 110.0$ mm

Step-5: The diameter of the big end bearing

$$\text{The crankpin diameter, } d_{p2} = \sqrt{\frac{Q_1}{P_b \cdot k}} \qquad \text{(Eqn. 21.52 MDHB)}$$

where, $Q_1 = 261537.6\,\text{N}$; $p_b = 12\,\text{MPa}$; $k = 1.0$

∴ The crankpin diameter,

$$d_{p2} = \sqrt{\frac{261537.6}{12 \times 1.0}} = 147.63 \text{ mm}$$

Choose $d_{p2} = 150$ mm

Example 23.8: Design a connecting rod for a petrol engine running at 750 rpm. The cylinder diameter is 70 mm and stroke is 90 mm. The connecting rod length is 180 mm. The maximum gas pressure is 2.7 MPa.

Given Data: Speed of the engine $N = 750$ rpm;

Cylinder diameter, $D = 70$ mm; Stroke $L_s = 90$ mm;

The length of the connecting rod length, $L = 180$ mm;

Maximum gas pressure, $P_g = 2.7$ MPa; $n = \frac{L}{R}$ ratio $= \frac{180}{45} = 4$

SOLUTION

Step-1: Maximum force on the rod at I.D.C.

The max. gas pressure force,

$$F_g = \frac{\pi}{4} \cdot D^2 \cdot p_g = \frac{\pi}{4} \times (70)^2 \times 2.7 = 10390.84 \text{ N}$$

Neglecting the inertia force, the maximum load on the connecting rod at I.D.C. $Q_1 = F_g = 10390.84 \text{ N}$

Step-2: Cross section of the rod

Assume that the section is rectangular with $h/b = 2$

Area of the section, $A = bh^2 = 2b^2 \text{ mm}^2$

The moment of inertia, $I = \frac{bh^3}{12} = \frac{b(2b)^3}{12} = \left(\frac{2}{3}\right) b^4$

The radius of gyration, $k^2 = I/A = \left(\frac{2}{3} b^4\right) / 2b^2 = \frac{1}{3} b^2$

Material for the rod C 45 steel with $S_d = 80$ MPa

Using Rankine equation, $\sigma_c = \frac{Q}{A}\left[1 + a\left(\frac{l}{k}\right)^2\right] \leq S_d$

$$\left(\frac{10390.84}{2b^2}\right)\left[1 + \frac{1}{7500}\left(\frac{180^2}{b\frac{2}{3}}\right)\right] = 80$$

$\therefore \qquad b^4 - 65b^2 - 841.6 = 0$

$\therefore \qquad b = 8.7 \text{ mm} \cong 9.0 \text{ mm and } h = 2 \times 9 = 18 \text{ mm}$

Step-3: Check for bending stress

The lateral inertia force,

$$F_i = \frac{A \cdot \rho \cdot \omega^2 R \cdot L}{2000} \qquad \text{(Eqn. 21.48 MDHB)}$$

where, $A = b \cdot h = 9 \times 18 = 162 \text{ mm}^2$; $\rho = 8 \times 10^{-6} \text{ kg/mm}^3$

$$\omega = \frac{2\pi N}{60} = \frac{2 \times 75}{60} = 78.54 \frac{\text{rad}}{\text{sec}}; R = 90/2 = 45 \text{ mm}; L = 180 \text{ mm}$$

$$\therefore \qquad F_i = \frac{162 \times 8 \times 10^{-6} \times (78.54)^2 \times 180 \times 45}{2000} = 32.38 \text{ N}$$

Maximum bending moment, $M_{\max} = \frac{2V_3 F_i L}{27} = \frac{2 \times \sqrt{3} \times 32.38 \times 180}{27} = 747.7 \text{ Nmm}$

Bending stress, $\sigma_b = \dfrac{M}{Z} = \dfrac{747.7}{\left(\dfrac{9 \times 18^2}{6}\right)} = 1.54 \text{ MPa}$

This is too small and hence design is safe.

Step-4: The small end bearing

The diameter of the pin by the bearing pressure consideration,

$$d_{p_1} = \sqrt{\frac{Q}{p_b \cdot k}} \qquad \text{(Eqn. 21.52 MDHB)}$$

where, $Q = 10390.84 \text{ N}; p_d = 12.5 \text{ MPa}; k = 1.5$ (Table 21.10 MDHB)

$$\therefore \quad d_{p1} = \sqrt{\frac{10390.84}{12.5 \times 1.5}} = 23.54 \cong 25 \text{ mm}$$

Step-5: The big end bearing

Diameter for bearing pressure, $d_{p2} = \sqrt{\dfrac{Q}{p_b \cdot k}}$

where, $p_b = 10 \text{ MPa}$ and $k = 0.75$

$$d_{p2} = \sqrt{\frac{10390.84}{10 \times 0.75}} = 37.22 \cong 40 \text{ mm}$$

Step-6: Design of cap bolts

The load on the cap bolts is the inertia force, $F = \dfrac{W_r \cdot \omega^2 R}{1000}\left(1 - \dfrac{1}{n}\right)$

where, $W_r =$ mass of reciprocating parts

$$\therefore \quad F = \frac{19 \times (78.54)^2 \times 45}{1000}\left(1 - \frac{1}{4}\right) = 3955.57 \text{ N}$$

For steel bolt, $S_d = 50$ MPa, the bolt diameter,

$$d = \sqrt{\frac{2 \times 3955.57}{\pi \times 50}} = 7.1 \text{ mm}$$

Choose $d = 10$ mm

Step-7: The big end cap design

From Plate 23.2 MDHB

Width of the cap $B = l_p = 0.75 \times d_p = 0.75 \times 40 = 30$ mm

Distance between the bolts,

$$L_b = 1.5 d_p = 1.5 \times 40 = 60 \text{ mm}$$

Maximum bending moment on the cap,

$$M = \frac{F L_b}{4} = \frac{3955.57 \times 60}{4} = 59333.55 \text{ Nmm}$$

The bending stress in the cap,

$$\sigma_b = \frac{M}{Z} = \frac{6M}{Bt^2} \le S_d$$

$$\therefore \quad \frac{6 \times 59333.55}{30 \times t^2} = 80$$

$$\therefore \quad t = 12.18 \cong 12.5 \text{ mm}$$

Example 23.9: Design the connecting rod for a diesel engine running at 1000 rpm. The cylinder diameter is 200 mm and stroke length, L_s = 200 mm. The connecting rod length $L = 400$ mm and the maximum gas pressure, $p_g = 8.5$ MPa.

Data: Speed of the engine, N = 1000 rpm;
Cylinder diameter, D = 200 mm; Stroke length, L_s = 200 mm;
Connecting rod length, L = 400 mm; Maximum gas pressure = 8.5 MPa

SOLUTION Assumptions

(i) The gas pressure at the position of max. B.M.,

$$P_{g_1} = 0.40 \times p_g = 0.4 \times 8.5 = 3.4 \text{ MPa}$$

(ii) Material for the connecting rod
Alloy steel with, $S_d = 120$ Mpa

(iii) Bearing pressures (Table 21.10)
for small end (piston pin), $P_{b_1} = 15.0$ MPa
for big end (crankpin), $P_{b_2} = 12.0$ Mpa
The mass of reciprocating parts (Table 21.5 MDHB),

$$W_r = 0.0315 \times A = 0.0315 \times \frac{\pi}{4} \times 200^2 = 9.896 \text{ kg}$$

Step-1: The force on the connecting rod at I.D.C.
The piston force at I.D.C.,

$$F_p = F_g \pm F_i \qquad \text{(Eqn. 21.6 MDHB)}$$

Maximum gas force,

$$F_g = \frac{\pi}{4} D^2 \times p_g = \frac{\pi}{4} \times (200)^2 \times 8.5 = 267035.38 \text{ N}$$

The inertia force,

$$F_i = \frac{W_r \cdot \omega^2 \cdot R}{1000}\left[1+\frac{1}{n}\right] \qquad \text{(Eqn. 21.5 MDHB)}$$

where, $W_r = 9.896\text{ kg}\,;\ \omega = \dfrac{2\pi \times 1000}{60} = 104.72\text{ rad/sec}$

$$R = \frac{L_s}{2} = \frac{200}{2} = 100\text{ mm};\ n = 400/100 = 4$$

$$\therefore \quad F_i = \frac{9.896\times(104.72)^2\times 100}{1000}\left[1+\frac{1}{4}\right] = 13565.29\text{ N}$$

$$F_p = 267035.80 - 13565.2 = 253470.16\text{ N}$$

Step-2: The cross section of connecting rod

Select the I-section with proportions shown in Fig. 23.6.

Cross-sectional area, $A = 11t^2$; Moment of inertia $= \dfrac{419}{12}t^4$

The radius of gyration, $K^2 = \dfrac{419}{132}t^2$; Section modulus $Z = \dfrac{419}{30}t^3$

The connecting rod is treated on as a column. Using the Rankine equation, the compressive induced,

$$\sigma_c = \frac{Q_1}{A}\left[1 + a\left(\frac{L}{K}\right)^2\right] \le S_d$$

where, $Q_1 = F_b = 253470.16\text{ N};\ A = 11t^2$ and $K^2 = \dfrac{419}{132}t^2$

$$L = 400\text{ mm};\ S_d = 120\text{ MPa};\ a = \frac{1}{7500}$$

Substituting the values,

$$\left(\frac{253470.16}{11t^2}\right)\left[1+\frac{1}{7500}\left(\frac{400^2}{\frac{419}{132}t^2}\right)\right] = 120$$

$$\therefore \quad t^4 - 192t^2 - 1290.54 = 0$$

Solving, $t = 14.1$ mm

Choose $t = 15$ mm; and $b = 4\times 8 = 60$ mm; $h = 5 \times t = 75$ mm

$$A = 11t^2 = 11\times 11^2 = 1331\text{ mm}^2;\ k^2 = \frac{419}{132}(11)^2 = 661.$$

$$Z = \frac{419}{30}(11)^3 = 18589.63 \text{ mm}^3$$

Step-3: Check for the bending stress

The lateral inertia force,

$$F_i = \frac{A\rho\omega^2 R \cdot L}{2000} \qquad \text{(Eqn. 21.48 MDHB)}$$

$$= \frac{1331 \times 8 \times 10^{-6} \times (104.72)^2 \times 100 \times 400}{2000} = 2335.38 \text{ N}$$

Maximum bending moment, $M_{\max} = \dfrac{2 \times \sqrt{3} \times F_i \cdot L}{27}$
(Eqn. 21.49 MDHB)

$$= \frac{2 \times \sqrt{3} \times 2335.38 \times 400}{27} = 119851.69 \text{ Nmm}$$

The bending stress,

$$\sigma_b = \frac{M}{Z} = \frac{119851.69}{18589.63} = 6.45 \text{ MPa}$$

The angle obliquity, $\beta = \tan^{-1}(R/L) = \tan^{-1}(100/400) = 14°$

The gas force on C.R. at this position,

$$Q_2 = \frac{F_{g2}}{\cos\beta} = \frac{\pi}{4} D^2 P_{g_2} / \cos\beta = \frac{\pi}{4} \times \frac{200^2 \times 3.4}{\cos 14°} = 110084.12 \text{ N}$$

Then the direct compressive stress,

$$\sigma_c = \frac{Q_2}{A} = \left[1 + \frac{1}{a}\left(\frac{l}{k}\right)^2\right] = \left(\frac{110084.12}{1331}\right)\left[1 + \frac{1}{7500} \times \frac{400^2}{661.0}\right] = 85.38 \text{ MPa}$$

The total stress, $\sigma = \sigma_b + \sigma_c = 6.45 + 85.38 = 91.83$ MPa

This is less than the design stress and design is safe.

Step-4: The small end bearing

The bearing diameter for bearing pressure,

$$p_b = 15 \text{ MPa} \quad \text{and} \quad k = 1.5$$

$$d_{p_1} = \sqrt{\frac{Q}{p_b \cdot k}} = \sqrt{\frac{253470.12}{15 \times 1.5}} = 106.14 \text{ mm}$$

Choose $d_{p1} = 110.0$ mm

Length of the bearing, $l_{p1} = 1.5 \times 110 = 165.0$ mm

Step-5: The big end bearing

The bearing diameter for bearing pressure,

$$d_{p_2} = 12.0 \text{ MPa} \quad \text{and} \quad k = 1$$

$$d_{p_2} = \sqrt{\frac{253470.12}{12 \times 1}} = 145.34 \text{ mm}$$

Choose $d_{p2} = 145.0$ mm

Length of the bearing, $l_{p2} = 145.0$ mm

Step-6: Design of cap bolts

Bolt load, $F = F_i = 13565.29$ N

Number of bolts, $N = 2$

The bolt diameter for $S_d = 90$ MPa

$$d = \sqrt{\frac{2 \times 13565.29}{\pi \times 90}} = 9.79$$

Choose $d = 10$ mm

Step-7: Cap design

The distance between the bolts $L_b = 1.4 \times d_p = 1.4 \times 145 = 203$ mm

$$B.M = \frac{F \cdot L_b}{4} = \frac{13565.29 \times 203}{4} = 688438.47 \text{ Nmm}$$

Bearing width, $B = L_p = 145.0$ mm and for $S_d = 120$ Mpa

The cap thickness, (Eqn. 23.56 MDHB)

$$t = \sqrt{\frac{6M}{B \cdot S_d}} = \sqrt{\frac{688438.47 \times 6}{145 \times 120}} = 15.41$$

Choose $t = 16.0$ mm

Example 23.10: Design a connecting rod for an I.C. engine running at $N = 1500$ rpm. The cylinder diameter is 125 mm and stroke is 150 mm; max. gas pressure is 6.5 MPa. Select the other data suitably.

Given Data: Maximum gas pressure, $p_g = 6.5$ MPa ; Speed, $N = 1500$ rpm

Cylinder diameter, $D = 125$ mm ; Stroke, $L_s = 150$ mm

SOLUTION Assumptions

$$\frac{L}{R} \text{ ratio} = n = 4$$

Reciprocating mass, $w_r = 0.105 \times A = 0.105 \times \frac{\pi}{4} \times (12.5)^2 = 12.9$ kg

The connecting material Al alloy with design stress, $S_d = 125$ MPa

The specific mass, $\rho = 2.7 \times 10^{-6}$ kg/mm^3

Step-1: The force on connecting rod at I.D.C.

The gas force, $F_g = \frac{\pi}{A} \cdot D^2 \cdot p_g = \frac{\pi}{4}(125)^2 \times 6.5 = 79767$ N

The inertia load due to the reciprocating mass is not considered in the design of connecting rod. Hence, the load on the connecting rod,

$$Q_1 = F_g = 79767 \text{ N}$$

Step-2: The cross section of the connecting rod

Select the I-section with proportions shown in Fig. 23.6

Then for the cross section,

$$A = 11t^2;\ I = \frac{419}{12}t^4;$$

$$Z = \frac{419}{30}t^3;\ k^2 = \frac{419}{132}t^2$$

The length of the connecting rod, $L = 4 \cdot r = 4 \times \left(\frac{150}{2}\right) = 300$ mm

Then using Rankine equation, the compressive stress,

$$\sigma_c = \frac{Q_1}{A}\left[1 + a\left(\frac{L}{K}\right)^2\right] \le S_d \qquad \text{(Eqn. 23.47 MDHB)}$$

Substituting the values,

$$\frac{79767}{11t^2}\left[1 + \frac{1}{700} \times \frac{300^2}{\left(\frac{419}{132}t^2\right)}\right] = 125$$

$$\therefore \qquad t^4 - 58.00t^2 - 219.31 = 0$$

Solving for $t = 7.85$ mm

Choose $t = 8$ mm

Then $A = 11 \times t^2 = 11 \times 8^2 = 704$ mm^2; $k^2 = \frac{419}{132} \times (8)^2 = 203.15$

$$Z = \frac{419}{30}t^3 = \frac{419}{30} \times (8)^2 = 7150.93 \text{ mm}^3$$

Step-3: Check for bending stress

The lateral inertia force,

$$F_i = \frac{A\rho\omega^2 RL}{2000}$$

where, $A = 704 \text{ mm}^2$; $\rho = 2.7 \times 10^{-6}$ kg/mm^3

$$\omega = \frac{2\pi \times N}{60} = \frac{2\pi \times 1500}{60} = 157.08 \text{ rad/s}$$

$R = 75$ mm; $L = 300$ mm

Substituting the values,

$$F_i = \frac{704 \times 2.7 \times 10^{-6} (157.08)^2 \times 75 \times 300}{2000} = 527.63 \text{ N}$$

The maximum bending moment $= \dfrac{2\sqrt{3} F_i L}{27} = \dfrac{2\sqrt{3} \times 527.63 \times 300}{27} = 6082.56$ Nmm

The bending stress,

$$\sigma_b = \frac{M}{Z} = \frac{6082.56}{7150.93} = 0.85 \text{ MPa}$$

The gas pressure at this position,

$$P_{g_2} = 0.5 P_{g_1} = 0.5 \times 6.5 = 3.25 \text{ MPa}$$

∴ The gas force on the piston,

$$F_{g_2} = \frac{\pi}{4} D^2 p_{g_2} = \frac{\pi}{4} \times (125)^2 \times 3.25 = 39883.5 \text{ N}$$

The obliquity angle, $\beta = \tan^{-1}\left(\dfrac{R}{L}\right) = \tan^{-1}\left(\dfrac{75}{300}\right) = 14°$

The force on the connecting rod,

$$Q_2 = \frac{F_{g2}}{\cos\beta} = \frac{39883.5}{\cos 14°} = 41104.5 \text{ N}$$

The direct compressive stress,

$$\sigma_c = \frac{Q_2}{A}\left[1 + a\left(\frac{L}{K}\right)^2\right] = \frac{41104.5}{704}\left[1 + \frac{1 \times 300^2}{7500 \times 203.15}\right] = 61.84 \text{ Mpa}$$

The total stress, $\sigma = \sigma_c + \sigma_b$

$$= 61.84 + 0.85 = 62.69 \text{ MPa} \leq S_d$$

This is less than the design stress and hence the design is safe.

Step-4: Small end bearing

The permissible bearing pressure,

$$p_{b_1} = 17.0 \text{ MPa} \quad \text{and} \quad k = 1.5 \qquad \text{(Table 23.10 MDHB)}$$

The diameter of the piston pin,

$$P_{b_1} = \sqrt{\frac{Q_1}{P_{b_1} \cdot k}} = \sqrt{\frac{79767}{17 \times 1.5}} = 55.93$$

Choose $d_{p_1} = 55$ mm and $l_{b_1} = 1.5 \times 55 = 82.5$ mm

Step-5: Big end bearing

The permissible bearing pressure,

$$p_{b_1} = 14.0 \text{ MPa; } k = 1 \qquad \text{(Table 23.10 MDHB)}$$

Then the crankpin diameter,

$$d_{p_2} = \sqrt{\frac{79767}{14 \times 1.0}} = 75.5$$

Choose $d_{p2} = 75.5$ mm and $l_{p2} = 75.5$ mm

Step-6: Design of cap bolts

Select the number of bolts $n = 2$

Inertia force, $F = \dfrac{w_r \cdot \omega^2 \cdot R}{1000}\left[1 - \dfrac{1}{n}\right]$

where, $w_r = 12.9$ kg; $\omega = 157.08$ rad/s; $R = 75$ mm; $n = 4$

$$\therefore \qquad F = \frac{w_r \cdot \omega^2 R}{1000}\left[1 - \frac{1}{n}\right]\omega = \frac{12.9 \times 17.08^2 \times 75}{1000}\left[1 - \frac{1}{4}\right] = 17904.16 \text{ N}$$

The bolt diameter, for design stress $S_d = 90$ MPa,

$$\therefore \qquad d = \sqrt{\frac{2 \cdot F}{\pi S_d}} = \sqrt{\frac{2 \times 17904.16}{\pi \times 90}} = 11.25$$

Choose $d = 12$ mm (Table 9.7 MDHB)

Step-7: Cap design

The cap width $B = l_p = 75.5$ mm

The distance between the bolts, $L_b = 1.5 l_p = 1.5 \times 75.5 = 113.25$

Choose $L_b = 115$ mm

Bending moment on the cap,

$$M = \frac{F \cdot L_b}{4} = \frac{17904.16 \times 115}{4} = 514744.6 \text{ Nmm}$$

Then the cap thickness (From Eqn. 21.56 MDHB),

$$t = \sqrt{\frac{6M}{B \cdot S_d}} = \sqrt{\frac{6 \times 514744.6}{115 \times 120}} = 14.9 \cong 15 \text{ mm}$$

23.3 CRANKSHAFT

The crankshaft is a part of the slider crank mechanism, which converts the reciprocating motion of the piston into rotary motion of the shaft. The crankshaft consists of shaft parts supported in the main bearings, the crankpin(s) enclosed in the big end of the connecting rod, and crank arm(s) or webs, connecting the crankpin to the shaft parts. Depending upon the position of crankpin, the crankshafts may be classified in two types.

(i) Side crank or overhung crankshaft

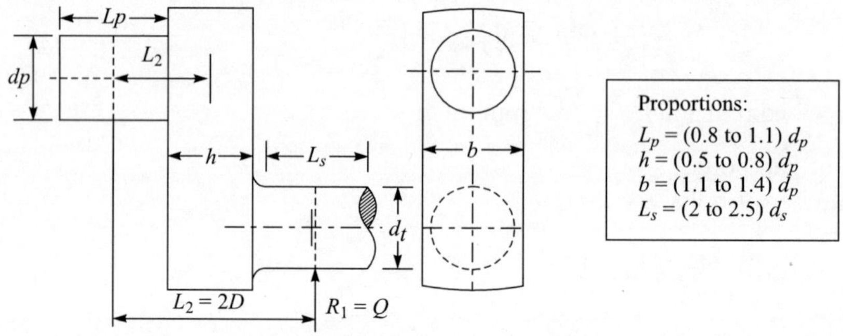

Figure 23.7 Side crankshaft

(ii) Centre crankshaft

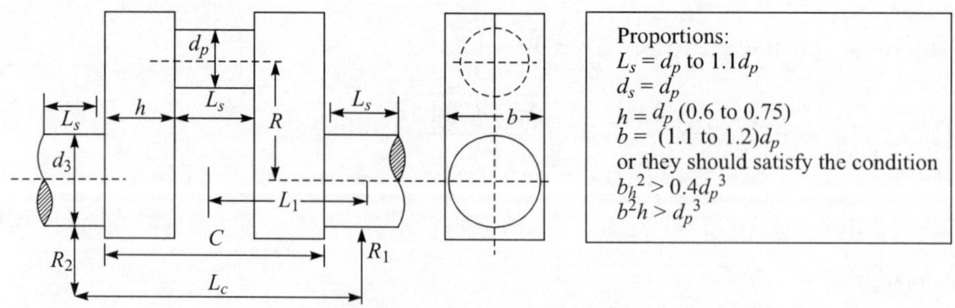

Figure 23.8 Central crankshaft

The centre crankshaft is further subdivided into single crankshafts and multi-crankshafts (Fig. 23.9 and Fig. 23.10). The multi-cylinder engines use the multi-crankshafts. The main design feature of the multi-crankshaft is the number of main bearings. This depends upon the number of cylinders and their configuration. It is necessary to use the minimum number of bearings to reduce the size of the engine.

Figure 23.9 Multithrough crankshaft for a high speed diesel engine

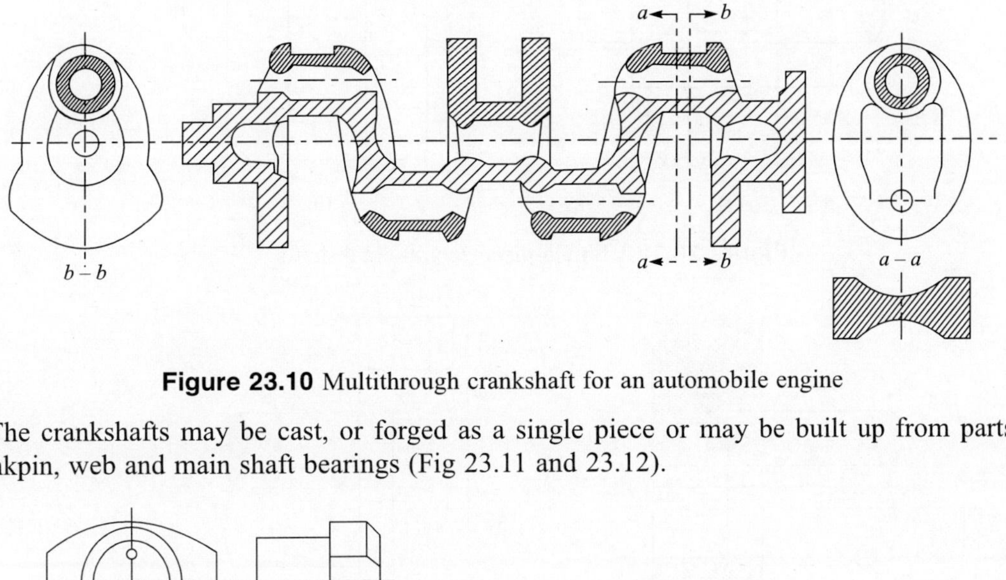

Figure 23.10 Multithrough crankshaft for an automobile engine

The crankshafts may be cast, or forged as a single piece or may be built up from parts of crankpin, web and main shaft bearings (Fig 23.11 and 23.12).

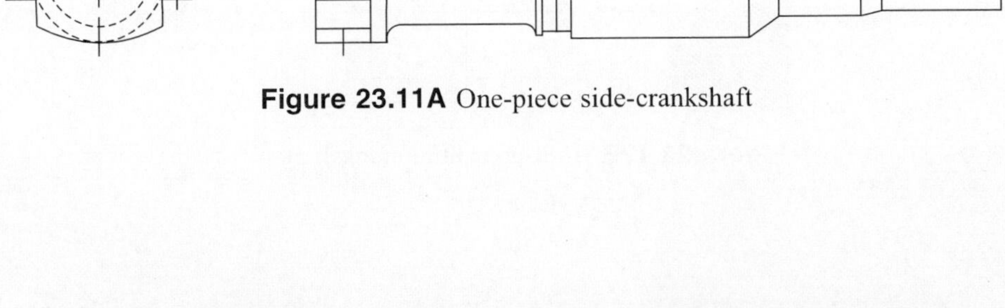

Figure 23.11A One-piece side-crankshaft

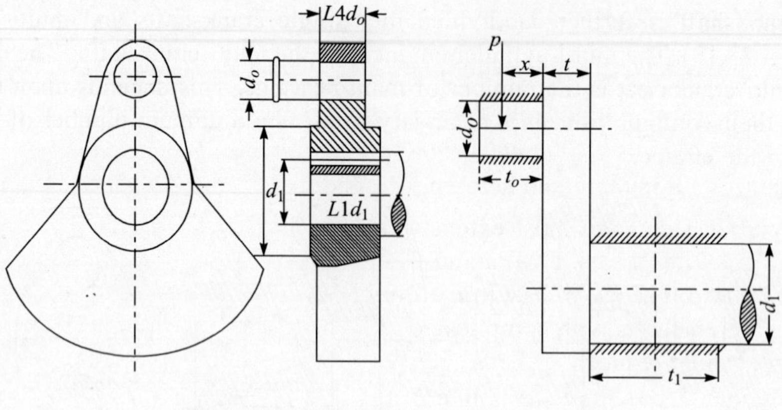

Figure 23.11B Built-up side crankshaft

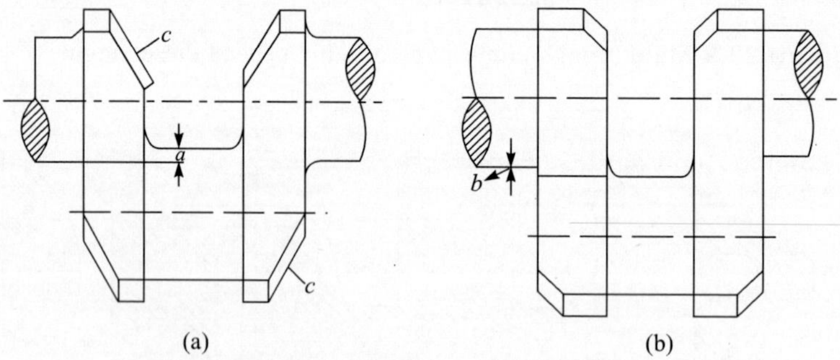

Figure 23.12A Single-piece central crankshafts

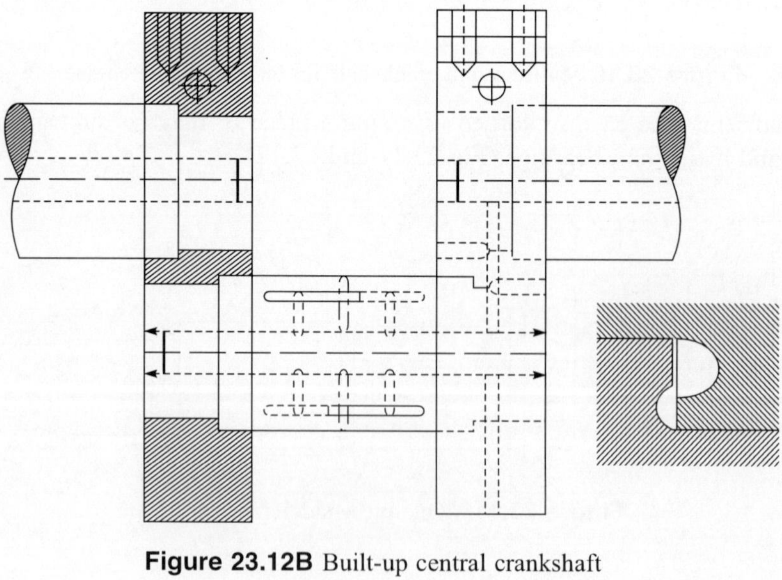

Figure 23.12B Built-up central crankshaft

Materials for the Crankshaft

The material used should have the following properties:

1. High strength to resist the high stresses due to the bending and twisting moments.
2. High rigidity to resist distortion and vibration.
3. Light weight for automobile and aero engines.

The medium carbon steels are most extensively used in industrial engines. C35 steel with ultimate strength of 500 MPa to 525 MPa and C45–steels with ultimate strength of 627 to 780 MPa are commonly used. The use of cast iron alloys for large and high speed engines, is steadily increasing. The cast iron shafts with complicated shapes can be easily casted with economy and the machining cost can be reduced.

In automobile engines, the alloy steels like manganese steel with $S_u = 785 \text{ to } 940 \text{ MPa}$ and crome-nickel steel (SAE 3140) with $S_u = 1070 \text{ MPa}$ and $S_y = 910 \text{ MPa}$, are used. For aero engines, crome-nickel steels with $S_u = 910 \text{ to } 1100 \text{ MPa}$ are used.

Permissible Stresses

The stresses induced in the crankshaft are due to the fluctuating bending and torsional moments. Hence, the fatigue failure is a possibility. The design is to be based on the endurance limit stress. The factor of safety in the range of 3 to 4 should be used. The allowable stresses are given below.

<u>For carbon steels</u>

S_d in bending $= 55 \text{ to } 75 \text{ MPa}$

S_{sd} in shear $= 30 \text{ to } 40 \text{ MPa}$

<u>For alloy cast irons</u>

S_d in bending $= S_{sd}$ in shear $= 35 \text{ to } 45 \text{ MPa}$

<u>For alloy steel</u> (like chrome-nickel steel)

S_d in bending $= 130 \text{ to } 175 \text{ MPa}$

S_{sd} in shear $= 70 \text{ to } 97 \text{ MPa}$

To reduce the effect of stress concentration, the fillet radius should not be less than $0.2d$, where d is the diameter of shaft.

Bearing Pressures

The bearing pressures are very important for the design of the crankshaft. The permissible values depend upon the maximum cylinder pressure, the journal velocity, the change of direction of the applied load and method of lubrication. They are different for different bearings of the crankshafts. The representative values of the bearing data, are given in Table 23.10 MDHB. The permissible pressures are selected from this table.

Forces on the Crankshaft

The gas load and inertia forces due to reciprocating parts are transmitted to crankshaft through the connecting rod. The angle between the connecting rod and crank changes as the crank rotates and tangential and radial components of the force acting through connecting rod continuously change. Accordingly, the bending and twisting moments on the crankshaft and the corresponding stresses also change. We have to locate the position of crank, when the effect of these forces is critical. Normally, two positions are considered to be critical and considered for the design calculations.

Position 1: Crank at I.D.C. when $\theta = 0$

The gas force, $F_g = \dfrac{\pi}{4} D^2 \,.\, p_g$ (23.23)

At this position, the crank and connecting rod are collinear and the force is fully radial and maximum. The bearing pressures are maximum for both crankpin and main shaft bearings.

The crankpin and shaft are subjected to bending stresses. The web (crank-arm) is subjected to bending stresses and torsional shear stresses.

The preliminary dimensions of the crankshaft are determined based on the forces in the position.

Position 2: Maximum torque position

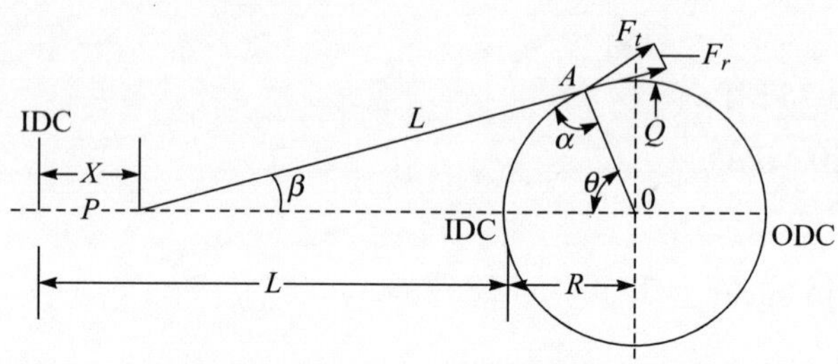

Figure 23.13 Maximum torque position

The force through the connecting rod,

$$Q = F_p/\cos\beta = F_p\left[1 - \frac{\sin^2\theta}{n^2}\right]^{\frac{-1}{2}} \tag{23.24}$$

The radial force on the crankpin,

$$F_r = F_p \cos(\theta + \beta)/\cos\beta \tag{23.25}$$

The tangential force on the crankpin,

$$F_t = F_p \sin(\theta + \beta)/\cos\beta \tag{23.26}$$

The torque on the crankshaft,

$$T = F_t \cdot R = R \cdot F_p \sin(\theta + \beta)/\cos\beta \tag{23.27}$$

The value of this torque varies due to the variation of angles θ and β and the gas load F_p. The maximum torque position can be determined graphically by taking the pressures from net-effort diagram. Usually, the crank angle for maximum torque position is between $30°$ to $40°$ for diesel engines and $25°$ to $35°$ for petrol engines. The crankshaft is subjected to maximum torsional shear stress and bending stress. The check for the combined stress is to be done.

Design Procedure

At the I. D. C. Position

Step-1: Determine the load on the crankpin and shaft from the maximum gas pressure, neglecting the intertia force due to the reciprocating mass,

$$Q = F_p = F_g = \frac{\pi}{4} D^2 p_g$$

Step-2: Select the bearing data from the Table 21.10 and select the material and design stress from Table 21.9 MDHB

Step-3: Determine diameter of the crankpin, d_p, for bearing pressure using equation (Eqn. 21.61 MDHB)

$$d_p = \sqrt{\frac{Q}{k \cdot p_b}}$$

For both, side crankshaft and central crankshaft
Check for the strength by Eqn. 21.60 for side crankshaft,

$$d_p = \sqrt{\frac{5Qk}{S_d}}$$

or determine the optimum diameter by,
Equation 21.62 for overhung crankshaft,

$$d_p = \sqrt{\frac{16}{\pi}\left(\frac{Q^2}{S_d P_b}\right)}$$

Equation 21.67 for central crankshaft,

$$d_p = \sqrt{\frac{5Q \cdot k}{2S_d}}$$

Step-4: Determine the diameter of the main bearing by bearing pressure consideration by,

$$d_s = \sqrt{\frac{R_1}{k \cdot P_b}}$$

for both overhung and central crankshafts where R_1 is the reaction at the bearing

Step-5: Determine the web dimensions by proportions

Below Fig. 23.7 for over hung crank

$$L_p = (0.8 \text{ to } 1.1)d_p;\ h = (0.5 \text{ to } 0.8)d_p$$

$$b = (1.1 \text{ to } 1.4)d_p;\ L_s = (2 \text{ to } 2.5)d_s$$

Below Fig 23.8 for central crankshaft

$$L_S = d_p \text{ to } 1.1d_p\ ;\ d_s = d_p$$

$$h = (0.6 \text{ to } 0.75)d_p\ ;\ b = (1.1 \text{ to } 1.2)d_p$$

Or they should satisfy the conditions

$$bh^2 > 0.4d_p^3 \text{ and } b^2h > d_p^3$$

Step-6: Determine the distances between the bearings

Maximum Torque Position

Step-7: Determine the position of maximum torque from the net-effort diagram (i.e., crank angle vs gas pressure diagram) or assume

$$\theta = 25° \text{ to } 35° \text{ for petrol engine}$$

$$= 30° \text{ to } 40° \text{ for decibel engine}$$

Then the angle obliquity is given by

$$\cos\beta = \left[1 - \frac{\sin^2\theta}{n^2}\right]^{\frac{1}{2}}$$

and the angle between connecting rod and crank,

$$\boxed{\propto = \theta + \beta}$$

Step-8: Determine the force through the connecting rod,

$$Q = F_p/\cos\beta = F_p\left[1 - \frac{\sin^2\theta}{n^2}\right]^{\frac{-1}{2}}$$

where, $$F_p = \frac{\pi}{4}D^2 \cdot P_{g2}$$

P_{g_2} is the gas pressure at this position [from Eqn. $PV^\gamma = C$]

Radial component,

$$P_r = F_r = Q\cos\propto = F_p\cos(\theta + \beta)/\cos\beta$$

The tangential component,

$$P_t = F_t = Q \sin \propto = F_p \cos(\theta + \beta)/\sin\beta$$

Step-9: Then find the bearing reactions by treating the crankshaft as simply supported beam and assuming that the loads are shared by the neghbouring two bearings.

Step-10: Check for the strength of the shaft by equation 21.64 MDHB,

$$d_s = \left[\frac{\sqrt{(Q\cdot L_1)^2 + (P_t R)^2}}{0.2S_{sd}}\right]^{1/3}$$

and Eqn. 21.65 MDHB for overhung crank,

$$d_s = \left[\frac{(Q\cdot L_1) + \sqrt{(Q\cdot L_1)^2 + (P_t R)^2}}{0.2S_d}\right]^{1/3}$$

and Eqn. 21.69 MDHB for central crank,

$$d_s = \left[\frac{\sqrt{\left(\frac{Q\cdot L_1}{2}\right)^2 + (P_t R)^2}}{0.2S_{sd}}\right]^{1/3}$$

Step-11: Check for the stresses in the web by the following equations.

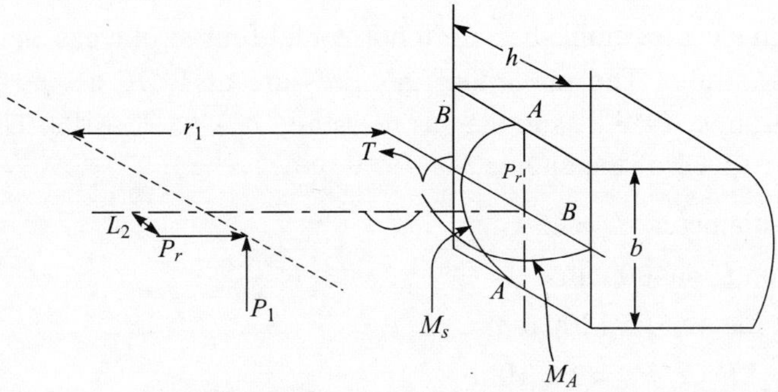

Figure 23.14 The loads on the web

Stresses at B

Shear stress, $\tau_1 = \dfrac{T}{\alpha_1 b\cdot h^2} + \left(\dfrac{3P_1}{2bh}\right)$

Total direct stress, $\sigma_B = \dfrac{P_r}{bh} + \left(\dfrac{6M_A}{bh^2}\right)$

Resultant stress

Maximum shear stress, $\tau_{\max} = \sqrt{\left(\frac{\sigma_B}{2}\right)^2 + \tau_1^2} < S_{sd}$

Maximum direct stress, $\sigma_{\max} = \frac{\sigma_B}{2} + \sqrt{\left(\frac{\sigma_B}{2}\right)^2 + \tau_1^2} < S_{sd}$

Stresses at A

Shear stress, $\tau_2 = \frac{T}{\alpha_2 b \cdot h^2}$

Total direct stress, $\sigma_A = \frac{P_r}{bh} + \left(\frac{6M_A}{hb^2}\right)$

Check the resultant stress

Maximum shear stress, $\tau_{\max} = \sqrt{\left(\frac{\sigma_A}{2}\right)^2 + \tau_2^2} < S_{sd}$

Maximum direct stress, $\sigma_{\max} = \frac{\sigma_A}{2} + \sqrt{\left(\frac{\sigma_A}{2}\right)^2 + \tau_2^2} < S_{sd}$

Example 23.11: Design a centre crankshaft for a horizontal four-stroke gas engine with 250 mm diameter and 400 mm stroke. The maximum gas pressure at 1 DC is $p_g = 2.27$ MPa. The maximum torque position is $\theta = 35°$ and the gas pressure, $p_{g_2} = 1.15$ MPa. The connecting rod length is $l = 1200$ mm; speed of the engine, $N = 250$ rpm.

Given Data: Cylinder diameter, $D = 250$ mm

Stroke of the engine, $L_s = 400$ mm

The connecting rod length, $l = 1200$ mm

The gas pressure at I.D.C., $P_{g_1} = 2.4$ MPa

The position of maximum torque, $\theta = 35°$

The gas pressure at this position, $P_{g_2} = 1.15$ MPa

Speed of the engine, $N = 250$ rpm

SOLUTION

Step-1: Maximum gas force,

$$F_p = F_g = \frac{\pi}{4} D^2 \cdot P_{g_2} = \frac{\pi}{4} \times (250)^2 \times 2.4 = 117809.73 \text{ N} \cong 117810 \text{ N}$$

Step-2: Select the bearing data

Bearing pressure p_b, from Table 21.10 MDHB

for crankpin $p_b = 14$ MPa; for main shaft $p_b = 8.5$ MPa

The $\frac{l}{d}$ ratio, for crankpin $k = 0.6$; for main shaft $k = 0.6$

Select the material as carbon steel C 40 with $S_y = 350$ MPa (Table 21.9 MDHB)

Assuming the factor of safety as 5,

The design stress, $S_d = \frac{350}{5} = 70$ MPa

Design stress in shear, $S_{sd} = 0.57 S_d = 40$ MPa

Step-3: Diameter for pin

The diameter of the crankpin for bearing pressure,

$$d_p = \sqrt{\frac{Q}{p_b \cdot k}} = \sqrt{\frac{117810}{14 \times 0.6}} = 118.5 \qquad \text{(Eqn. 21.66 MDHB)}$$

Choose $d_p = 120$ mm and $l_p = 0.6 \times 120 = 72$ mm

Check for the strength of the pin from Eqn. 23.67 MDHB,

$$S = \frac{5 \cdot Q \cdot k}{2d_p^2} = \frac{5 \times 117810 \times 0.6}{2 \times 120^2} = 12.27 \text{ MPa} < S_d = 70 \text{ MPa. Hence, safe.}$$

Choose the shaft diameter, $d_s = d_p = 120$ mm and $l_s = 0.6 \times 120 = 72$ mm

Step-4: Web dimensions

The web thickness, $h = 0.75 d_p = 0.75 \times 120 = 90$ mm

The web width, $b = 1.1 d_p = 1.1 \times 120 = 132$ mm

Choose b = 132 mm

Step-5: The distance between the shaft bearings,

$$L_c = l_p + 2 \cdot h + l_s + \text{clearance} = 72 + 2 \times 90 + 72 + 10$$

$$= 334 \cong 335 \text{ mm}$$

Step-6: The position of maximum torque

The crank angle, $\theta = 35°$ and gas pressure $p_{g_2} = 1.15$ MPa

The gas force, $F_{g_2} = \frac{\pi}{4} \cdot D^2 \cdot p_{g_2} = \frac{\pi}{4} \times (250)^2 \times 1.15 = 56450.5$ N

The piston force, $F_p = F_{g_2} = 56450.5$ N

Step-7: The force through the connecting rod,

$$Q = F_p\left[1 - \frac{\sin^2\theta}{n^2}\right]^{-\frac{1}{2}} = F_p/\cos\beta \qquad \text{(Eqn. 21.7 MDHB)}$$

where, $F_p = 56450.50\ \text{N}$; $\theta = 35°$; $n = \frac{l}{r} = \frac{1200}{200} = 6$

$$\cos\beta = \left[1 - \frac{\sin^2\theta}{n^2}\right]^{\frac{1}{2}} = \left[1 - \frac{\sin^2 35}{6^2}\right]^{\frac{1}{2}} = 0.99754$$

$\therefore \quad \beta = 5.48°$

$\therefore \quad Q = 56450.5/0.9954 = 56711.37\ \text{N}$

$\therefore \quad \propto = \theta + \beta = 35 + 5.48 = 40.48° = 40.5$

$\therefore$ Radial force,

$$P_r = Q\cos\propto = 56711.37 \times \cos(40.5) = 43123.67\ \text{N}$$

The tangential force,

$$P_t = Q\sin\propto = 56711.37 \times \sin(40.5) = 36831.09\ \text{N}$$

Bearing reactions,

$$R_1 = \frac{Q}{2} = \frac{56711.37}{2} = 28355.685\ \text{N}$$

The bearing pressure in the pin,

$$p_b = \frac{Q}{d_p l_p} = \frac{56711.37}{120 \times 72} = 6.56\ \text{MPa} < 14.0\ \text{MPa}$$

This is less than permissible value of 14 MPa and hence safe.
Bearing pressure in the main bearing,

$$p_b = \frac{R_1}{d_s l_s} = \frac{28355.685}{120 \times 72} = 3.28\ \text{MPa}$$

This value is less than permissible value of $p_b = 8.5$ MPa, and hence the design is safe.

Step-8: Check for the strength of shaft,

$$d_s = \left[\frac{\sqrt{\left(\frac{QL_1}{2}\right)^2 + (P_t R)^2}}{0.2 S_d}\right]^{\frac{1}{3}}$$

where, $Q = 56711.37$ N and $L_1 = \frac{1}{2}L_c = \frac{335}{2} = 167.5$ mm

$$P_t = 36831.09 \text{ N and } R = 200 \text{ mm}; \ d_s = 120.0 \text{ mm}$$

The stress induced is

$$S_s = \left[\frac{\sqrt{(56711.37 \times 167.5)^2 + (36831.09 \times 200)^2}}{0.2 \times (120)^3}\right] = 34.78 \text{ MPa}$$

This is less than the permissible value of $S_{sd} = 40.0$ MPa. The design is safe.

Step-9: Check for the stresses in web

Refer Fig. 23.14 for loads on the web

Torsional moment,

$$T = P_t l_2 = P_t\left(\frac{l_p}{2} + \frac{h}{2}\right) = 36831.09 \times \left(\frac{72}{2} + \frac{90}{2}\right) = 2983318.29 \text{ Nmm}$$

Moment about *A-A*,

$$M_A = P_r \cdot L_2 = \frac{43123.67}{2} \times 81 = 174608.635 \text{ Nmm}$$

Moment about *B-B*,

$$M_B = P_t \cdot r = 36831.09 \times 200 = 7366218 \text{ Nmm}$$

<u>Stresses at *B*</u>

Direct stresses, $\sigma_1 = \dfrac{43123.67}{132 \times 90} = 3.63$ MPa

Bending stress, $\sigma_b = \dfrac{6M_A}{bh^2} = \dfrac{6 \times 1746508.635}{132 \times 90^2} = 9.8$ MPa

Total direct stress, (Eqn. 21.76 MDHB)

$$\sigma = 3.63 + 9.8 = 13.43 \text{ MPa}$$

Shear stress, $\tau = \dfrac{T}{\propto_1 bh^2} + \dfrac{3P_t}{2bh}$ (Eqn. 21.75 MDHB)

$$= \frac{2983318.29}{0.23 \times 132 \times 90^2} + \frac{3 \times 36831.09}{2 \times 132 \times 90} = 12.131 + 4.65 = 16.78 \text{ MPa}$$

Maximum shear stress, (Eqn. 21.78 MDHB)

$$\tau_{\max} = \sqrt{\left(\frac{\sigma}{2}\right)^2 + \tau^2} = \sqrt{\left(\frac{13.43}{2}\right)^2 + 16.78^2}$$

$$= 18.07 \text{ MPa} < S_{sd} = 40 \text{ MPa safe}$$

Maximum direct stress, (Eqn. 21.79 MDHB)

$$\sigma_{\max} = \frac{\sigma}{2} + \sqrt{\left(\frac{\sigma}{2}\right)^2 + \tau^2} = \frac{13.43}{2} + \sqrt{\left(\frac{13.43}{2}\right)^2 + 16.78^2}$$

$$= 24.79 \text{ MPa} < S_d = 75.0 \text{ MPa safe}$$

<u>Stresses at *A*</u>

Direct stress, (Eqn. 21.83 MDHB)

$$\sigma = \frac{P_r}{bh} + \frac{6M_B}{hb^2} = \frac{(43123.67/2)}{132 \times 90} + \frac{6 \times 7366218}{90 \times 132^2} = 30 \text{ MPa}$$

Shear stress, (Eqn. 21.81 MDHB)

$$\tau_2 = \frac{T}{\propto_2 bh^2} = \frac{2983318.29}{0.269 \times 132 \times 90^2} = 10.37 \text{ MPa}$$

Maximum shear stress, (Eqn. 21.84 MDHB)

$$\tau_{\max} = \sqrt{\left(\frac{\sigma}{2}\right)^2 + \tau^2} = \sqrt{\left(\frac{30}{2}\right)^2 + 10.37^2} = 18.24 \text{ MPa}$$

Maximum direct stress, (Eqn. 21.85 MDHB)

$$\sigma_{\max} = \frac{\sigma}{2} + \sqrt{\left(\frac{\sigma}{2}\right)^2 + \tau^2} = \frac{30}{2} + 18.24 = 33.23 \text{ MPa}$$

Both the values are less than the allowable values and hence the design is safe.

Example 23.12: Design an overhung crank of an engine having the cylinder diameter of 250 and stroke 400 mm. The connecting rod length is 5 times the crank length. The maximum gas pressure at I.D.C. is 1.7 MPa and runs at 200 rpm. The max. torque position is at $\theta = 30°$ and the gas pressure at this position is 0.8 MPa.

Given Data: The cylinder diameter, $D = 250$ mm; The stroke, $L = 400$ mm

The ratio, $n = \dfrac{l}{r} = 5$; Gas pressure, p_{g_1} at IDC $= 1.7$ MPa

For maximum torque position, $\theta = 30°$

The gas pressure at this position $= 0.8$ MPa

SOLUTION

Step-1: Selection of bearing data

Bearing data from Table 21.10 MDHB

Crankpin, $p_b = 10$ MPa and $\dfrac{l}{d} = 1.2$

Main shaft, $p_b = 5.0$ MPa and $\dfrac{l}{d} = 2.0$

Material: Carbon steel – C 45 with $S_d = 85$ MPa and

$$S_{sd} = 0.57S_d = 0.57 \times 85 = 48.5 \text{ MPa}$$

The ideal $\dfrac{l}{d}$ for the pin,

$$\frac{l}{d} = \sqrt{\frac{\pi}{16}\left(\frac{S_d}{p_b}\right)} = \sqrt{\frac{\pi \times 85}{16 \times 10}} = 1.3$$

Choose $\dfrac{l}{d} = 1.3$

Step-2: Maximum gas load at I.D.C.

$$F_g = \frac{\pi}{4} d^2 p_{g_1} = \frac{\pi}{4} \times (250)^2 \times 1.7 = 83448.55 \text{ N}$$

Design the crankshaft based on this load.

Step-3: Crankpin dimensions for load at I.D.C.

The pin diameter for equal strength,

$$d_p = \sqrt[4]{\frac{16}{\pi}\left(\frac{Q^2}{S_d \cdot p_b}\right)} = \sqrt[4]{\frac{16}{\pi}\left(\frac{83448.55^2}{85 \times 10}\right)} = 80.37 \text{ mm} \qquad \text{(Eqn. 21.62 MDHB)}$$

Choose $d_p = 80.0$ mm and $l_p = 1.3 \times 80 = 104 \cong 105$ mm

Step-4: Dimensions of web

By proportions, $h = 0.8d_p = 0.8 \times 80 = 64 \cong 70$ mm and $b = 1.4d_p = 1.4 \times 80 \cong 112$ mm

Choose $b = 120$ mm

The maximum stress at the web at B $\sigma_B = \dfrac{P_r}{b \cdot h} + \dfrac{6M_A}{bh^2}$

where, $P_r = Q = 83448.55$ N, $b = 120$ mm and $h = 70$ mm

$$M_A = P_r \cdot L_2 = 83448.55 \times \left(\frac{105}{2} + \frac{70}{2}\right) = 7301748.125 \text{ Nmm}$$

$$\therefore \qquad \sigma_B = \frac{83448.55}{120 \times 70} + \frac{6 \times 7301748.125}{(120 \times 70^2)} = 84.44 \text{ MPa} < S_d = 85 \text{ MPa}$$

Step-5: Main shaft bearing dimensions

The shaft diameter for bearing pressure,

$$d_s = \sqrt{\frac{R_1}{kp_b}} = \sqrt{\frac{83448.55}{1.5 \times 5.0}} = 105.5 \text{ mm} \cong 105 \text{ mm} \qquad \text{(Eqn. 21.63 MDHB)}$$

And $l_s = 1.5 \times 105 = 157.5 \cong 160.0$ mm

The shaft diameter for strength,

$$d_s = \left[\frac{\sqrt{(QL_1)^2 + (P_t R)^2}}{0.2 \times S_{sd}}\right]^{\frac{1}{3}}$$

where, $Q = 83448.55$ N; $P_t = 0$, and $S_{sd} = 48.5$ MPa

$$L_1 = \frac{105}{2} + 70 + \frac{160}{2} + 10 = 212.5 \text{ mm}$$

$$\therefore \qquad d_s = \left[\frac{83448.55 \times 212.5}{0.2 \times 48.5}\right]^{\frac{1}{3}} = 122.27 \text{ mm}$$

∴ Choose d_s = 125 mm and L_s = 160 mm

Then modify web width as

$$b = 1.1 d_s = 11 \times 125 = 137.5 \text{ mm} \cong 140 \text{ mm}$$

Step-6: The forces at maximum torque position

The gas force, $F_p = \frac{\pi}{4} D^2 \cdot p_{g2} = \frac{\pi}{4} (250)^2 \times 0.8 = 39269.91$ N

The force through connecting rod,

$$Q = \frac{F_p}{\cos \beta} = \frac{F_p}{\left[1 - \frac{\sin^2 \theta}{n^2}\right]^{\frac{1}{2}}} \qquad \text{(Eqn. 21.7 MDHB)}$$

where, β = the angle of obliquity

$$\cos \beta = \left[1 - \frac{\sin^2 \theta}{n^2}\right]^{\frac{1}{2}} = \left[1 - \frac{\sin^2 \theta}{5^2}\right]^{\frac{1}{2}} = 0.995$$

$$\therefore \qquad \beta = 5.74°$$

$$\therefore \qquad Q = \frac{39269.91}{\cos 5.74^\circ} = \frac{39269.91}{0.995} = 39467.25 \text{ N}$$

The angle between the connecting rod and crank,

$$\propto = \theta + \beta = 30 + 5.74 = 35.74^\circ$$

The radial force,

$$P_r = Q \cos \propto = 39467.25 \times \cos 35.74 = 32034.62 \text{ N}$$

The tangential force,

$$P_t = Q \sin \propto = 39467.25 \times \sin 35.74 = 23053.14 \text{ N}$$

Step-7: Check for the strength of the shaft,

$$d_s = \left[\frac{\sqrt{(QL_1)^2 + (P_t R)^2}}{0.2 S_{sd}} \right]^{\frac{1}{3}}$$

$$= \left[\frac{\sqrt{(39467.25 \times 212.5)^2 + (23053.14 \times 200)^2}}{0.2 \times 48.5} \right]^{\frac{1}{3}} = 99.55 \text{ mm}$$

This is less than selected value of 120 mm. Hence, design is safe.

Step-8: Check the stresses in web

Revised web dimensions are $b = 140$ mm and $h = 70$ mm

Refer Fig. 23.13 for loads on the web

The torsional moment, (Eqn. 21.71 MDHB)

$$T = P_t \cdot L_3 = 23053.14 \times \left(\frac{105}{2} + \frac{70}{2} \right) = 2017149.75 \text{ Nmm}$$

Moment about *A-A*, (Eqn. 21.72 MDHB)

$$M_A = P_r L_2 = 32034.62 \times 87.5 = 2803029.25 \text{ Nmm}$$

Moment about *B-B*, (Eqn. 21.73 MDHB)

$$M_B = P_t r_1 = 23053.14 \times 200 = 4610628 \text{ Nmm}$$

The stresses at B

The direct stress, (Eqn. 21.77 MDHB)

$$\sigma_B = \frac{P_r}{bh} + \frac{6M_A}{bh^2} = \frac{32034.62}{140 \times 70} + \frac{6 \times 2803029.25}{140 \times 70^2} = 27.79 \text{ MPa}$$

The shear stress, (Eqn. 21.75 MDHB)

$$\tau_B = \frac{T}{\alpha_1 bh^2} + \left(\frac{3}{2}\frac{P_t}{bh}\right) = \frac{2017149.75}{0.246 \times 140 \times 70^2} + \frac{3 \times 23053.14}{2 \times 140 \times 70} = 15.48 \text{ MPa}$$

Maximum shear stress, (Eqn. 21.78 MDHB)

$$\tau_{\max} = \sqrt{\left(\frac{\sigma}{2}\right)^2 + \tau^2} = \sqrt{\left(\frac{27.79}{2}\right)^2 + (15.48)^2} = 20.8 \text{ MPa} < S_{sd} \therefore \text{ Safe}$$

Maximum direct stress, (Eqn. 21.79 MDHB)

$$\sigma_{\max} = \frac{\sigma}{2} + \sqrt{\left(\frac{\sigma}{2}\right)^2 + \tau^2} = \frac{27.79}{2} + \sqrt{\left(\frac{27.79}{2}\right)^2 + 15.48^2}$$

$$= 34.7 \text{ MPa} < S_d = 85 \text{ MPa}$$

<u>The stresses at A</u>

Direct stress, (Eqn. 21.83 MDHB)

$$\sigma = \frac{P_r}{b \cdot h} + \frac{6M_B}{hb^2} = \frac{32034.62}{140 \times 70} + \frac{6 \times 4610628}{70 \times 140^2} = 23.43 \text{ MPa}$$

Shear stress, (Eqn. 21.81 MDHB)

$$\tau = \frac{T}{\alpha_2 bh^2} = \frac{2017149.75}{0.309 \times 140 \times 70^2} = 9.52 \text{ MPa}$$

∴ Maximum shear stress, (Eqn. 21.84 MDHB)

$$\tau_{\max} = \sqrt{\left(\frac{\sigma}{2}\right)^2 + \tau^2} = \sqrt{\left(\frac{23.43}{2}\right)^2 + 9.52^2} = 15.1 \text{ MPa} < S_{sd} \therefore \text{ Safe}$$

Maximum direct stress, (Eqn. 21.85 MDHB)

$$\sigma_{\max} = \frac{\sigma}{2} + \sqrt{\left(\frac{\sigma}{2}\right)^2 + \tau^2} = \left(\frac{23.43}{2}\right)^2 + \sqrt{\left(\frac{23.43}{2}\right)^2 + 9.52^2}$$

$$= 26.93 \text{ MPa} < S_d \therefore \text{ Safe}$$

Example 23.13: Design the three-bearing centre crankshaft for a horizontal two-stroke gas engine with cylinder diameter of 200 mm and stroke length of 500 mm. The connecting rod length is 1250 mm. The gas pressure at I.D.C. is 2.5 MPa. The flywheel weight is 1600 N. The belt pull $(F_1 + F_2) = 3200$ N. The distance between the bearing centres, as shown in Fig. 23.14, are $a = a' = 250$ mm and $b = b' = 300$ mm. Check for the stresses at maximum torque position, when the crank angle $\theta = 30°$ and the gas pressure $p_{g_2} = 1.2$ MPa.

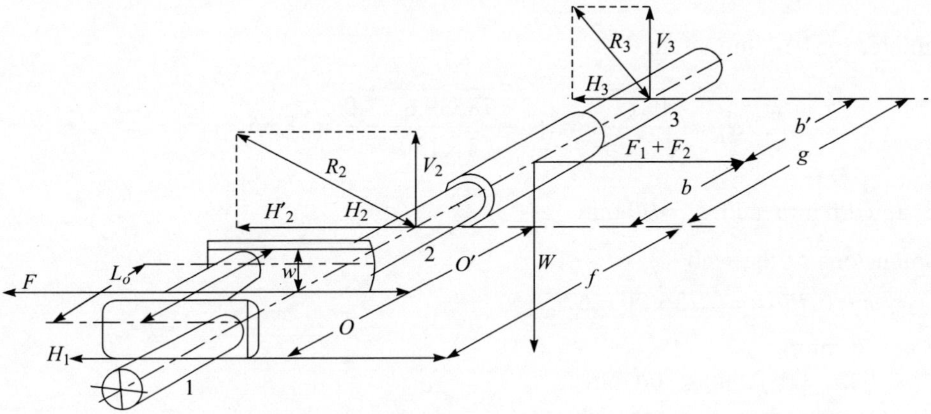

Figure 23.15 Forces on crankshaft at I.D.C.

Given Data: Cylindia diameter, $D = 200$ mm; Stroke length, $L = 500$ mm;

The gas pressure at I.D.C, $p_{g_1} = 2.5$ MPa;

Fly wheel weight, $W = 16000$ N;

The belt pull, $P = F_1 + F_2 = 3200$ N;

The maximum torque position $= 30°$;

The gas pressure at this position, $p_{g_1} = 1.2$ MPa

SOLUTION

Step-1: Selection of basic data

Bearing data from Table 21.10 MDHB,

For the pin, bearing pressure, $p_b = 10.0$ MPa and $k = \frac{l}{d} = 1.0$

For the main shaft, $p_b = 5.5$ MPa and, $k = \frac{l}{d} = 1.2$

The material for the crankshaft – Nickel steel with $S_d = 130$ MPa and $S_{s_d} = 75$ MPa

Step-2: Gas force at I.D.C.,

$$Q = F_p = \frac{\pi}{4} D^2 \times p_{g_1} = \frac{\pi}{4} (200)^2 \times 2.5 = 78539.8 \text{ N}$$

Step-3: Crankpin dimensions

The crankpin diameter for bearing pressure,

$$d_p = \sqrt{\frac{Q}{p_b \cdot k}} = \sqrt{\frac{78539.8}{10 \times 1}} = 88.62 \text{ mm}$$

The diameter for bending strength,

$$d_p = \sqrt{\frac{5 \cdot Q \cdot k}{2S_d}} = \sqrt{\frac{5 \times 78539.8 \times 1.0}{2 \times 130}} = 38.86$$

Choose $d_p = 90$ mm and $l_p = 90$ mm

Step-4: Dimensions of the web

Thickness, $h = 0.75d_p = 0.75 \times 90 = 67.5$

Choose h = 70 mm

Width, $b = 1.1d_p = 1.1 \times 90 \cong 100$ mm

Check for the stresses at B in the web,

$$\sigma_B = \frac{P_r}{bh} + \frac{6M_A}{bh^2}$$

where, $P_r = \frac{Q}{2} = \frac{78539.8}{2} = 39269.9$ N (because the load is by two webs)

and $M_A = P_r \cdot l_2 = 39269.9 \times \left(\frac{90}{2} + \frac{70}{2}\right) = 3141592$ Nmm

Substituting the value,

$$\sigma_B = \frac{39269.9}{100 \times 70} + \frac{6 \times 3141592}{100 \times 70^2} = 44.08 \text{ MPa} < S_d \quad \therefore \text{ Safe}$$

Step-5: Reactions at main bearings

To make the calculations simpler, it is customary to assume that the influence of the bending forces does not extend beyond the two bearings between which the force is situated.

Reactions at the two main bearings due to the gas-force,

$$R_{H_1} = R_{H_2} = \frac{Q}{2} = \frac{78539.8}{2} = 39269.9 \text{ N}$$

The reactions at bearings 2 and 3 due to the fly wheel weight,

$$R_{v_2} = R_{v_3} = \frac{16000}{2} = 8000 \text{ N}$$

The reactions due to the belt-pull,

$$R^1_{H_2} = R^1_{H_3} = \frac{3200}{2} = 1600 \text{ N}$$

The resultant reaction at bearing 2,

$$R_2 = \sqrt{(39269.9 + 1600)^2 + 8000^2} = 41645.512 \text{ N}$$

The reaction at the bearing 3,

$$R_3 = \sqrt{R_{H_3}^2 + R_{V_3}^2} = \sqrt{1600^2 + 8000^2} = 8158.43 \text{ N}$$

Step-6: Dimensions of the shaft bearings

The diameter of main bearing '2' for bearing pressure,

$$d_s = \sqrt{\frac{R_2}{p_2 k}} = \sqrt{\frac{41645.512}{5.5 \times 1.2}} = 79.44 \text{ mm}$$

But by proportions

$d_s = d_p = 90$ mm; and $l_s = 1.2 \times 90 = 108.0$ mm

The shaft diameter at bearing 3

The diameter for bearing pressure

$$d_{s_3} = \sqrt{\frac{8158.43}{5.5 \times 1.2}} = 35.156$$

Choose $d_{s_3} = 40$ mm and $l_{s_3} = 40 \times 1.2 = 48 \text{ mm} \cong 50$ mm

Step-7: The shaft diameter under the fly wheel

Bending moment on the shaft under fly wheel,

$$M = R_3 \times b = 8158.43 \times 300 = 2447529 \text{ Nmm}$$

The diameter of the shaft,

$$d = \sqrt[3]{\frac{32M}{\pi s_d}} = \sqrt[3]{\frac{32 \times 2447529}{\pi \times 130}} = 57.66 \text{ mm}$$

Choose $d = 60$ mm

Step-8: Forces on the crankshaft at maximum torque position

The forces on the crankshaft at maximum torque position

The gas force, $F_p = \frac{\pi}{4} D^2 p_{g_2} = \frac{\pi}{4} \times (200)^2 \times 1.2 = 37699.11$ N

The angle of obliquity is given by the relation,

$$\cos\beta = \left[1 - \frac{\sin^2\theta}{n^2}\right]^{\frac{1}{2}} = \left[1 - \frac{\sin^2 30}{S^2}\right]^{\frac{1}{2}} = 0.995$$

$$\beta = \cos^{-1}(0.995) = 5.74°$$

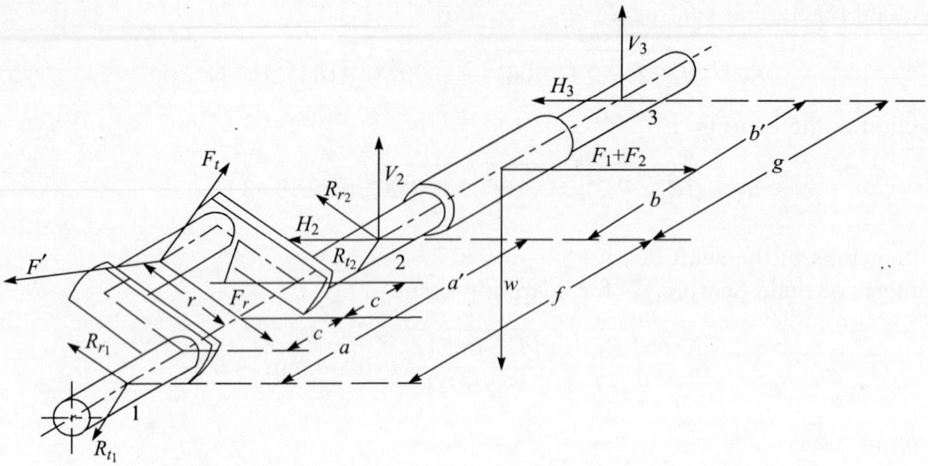

Figure 23.16 Forces on crankshaft at maximum torque position

The angle between the connecting rod and crank $\propto$ is given by

$$\propto = \theta + \beta = 30 + 5.74 = 35.74°$$

The force through the connecting rod,

$$Q = \frac{F_p}{\cos\beta} = \frac{37699.11}{0.995} = 37888.55 \text{ N}$$

The radial component,

$$P_R = Q\cos\propto = 37888.55 \times \cos(35.74) = 30753.23 \text{ N}$$

The tangential component,

$$P_T = Q\sin\propto = 37888.55 \times \sin(35.74) = 22131.01 \text{ N}$$

The torque on shaft,

$$T = P_t \times R = 22131.01 \times \left(\frac{500}{2}\right) = 5532752.5 \text{ Nmm}$$

Step-9: Reactions at the bearings at this position

The reactions due to the flywheel weight and belt pull do not change. The reactions due to the gas force at bearing '1' and '2' change.

The horizontal components due to gas load are

$$R_{H_1} = R_{H_2} = \frac{Q\cos\beta}{2} = \frac{F_p}{2} = \frac{37699.11}{2} = 18849.56 \text{ N}$$

Net horizontal reaction, R_{H_2} = 18849.56 + 1600 = 20449.56 N

The vertical components are

$$R_{v_1} = R_{v_2} = \frac{Q\sin\beta}{2} = \frac{37888.55 \times \sin 5.74°}{2} = 1894.7 \text{ N}$$

These are acting vertically upward. The net vertical component at bearing 2,

$$R_{v_2} = 8000 - 1894.7 = 6105.2 \text{ N}$$

The resultant reaction on bearing 2,

$$R_2 = \sqrt{(20449.56)^2 + (6105.2)^2} = 21341.41 \text{ N}$$

Step-10: Check for the stresses in the shaft

This value of reaction R_2 is less than the reaction at I.D.C.

$R_2 = 41645.512$ N and hence the bearing pressure is less than the permissible.

The shear stress induced in the shaft,

$$\tau = \left[\frac{\sqrt{\left(\frac{QL_1}{2}\right)^2 + (P_t R)^2}}{0.2d_s^3}\right] = \left[\frac{\sqrt{\left(\frac{37888.55 \times 350}{2}\right)^2 + (5532752.5)^2}}{0.2 \times 90^3}\right]$$

$$= 59.23 \text{ MPa} < S_{sd} = 75 \text{ MPa} \therefore \text{ Safe}$$

The stress in the shaft under the flywheel,

$$\tau = \frac{\sqrt{M^2 + T^2}}{0.2d^3} \frac{\sqrt{2447529^2 + 5532752.5^2}}{0.2 \times 60^3}$$

$$= 140.04 \text{ MPa} > S_{sd} = 75 \text{ MPa}$$

The stress is more than the permissible and the diameter is to be increased. The desired diameter is given by

$$d_3 = \left[\frac{\sqrt{2447529^2 + 5532752.5^2}}{0.2 \times 75}\right]^{\frac{1}{3}} = 73.88 \text{ mm}$$

Choose $d_3 = 75$ mm

Step-11: Check for the stresses in the web

Refer Fig. 23.14 for loads

The torsion on the web,

$$T = P_t \cdot L_2 = 22131.01 \times 80 = 1770480.8 \text{ Nmm}$$

The shear stress at B,

$$\tau_B = \frac{T}{\propto_1 bh^2} + \frac{3P_t}{2bh} = \frac{1770480.8}{0.236 \times 100 \times 70^2} + \frac{3 \times 22131.01}{2 \times 100 \times 70} = 20.38 \text{ MPa}$$

The bending stress at B,

$$\sigma_B = \frac{P_r}{bh} + \frac{6M_A}{bh^2} = \frac{P_r}{bh} + \frac{6 \cdot P_r l_2}{bh^2}$$

$$= \frac{30753.23}{100 \times 70} + \frac{6 \times 30753.23 \times 80}{100 \times 70^2} = 34.52 \text{ MPa}$$

The maximum shear stress, (Eqn. 21.78 MDHB)

$$\tau_{\max} = \sqrt{\left(\frac{\sigma}{2}\right)^2 + \tau^2} = \sqrt{\left(\frac{34.52}{2}\right)^2 + (20.38)^2} = 26.71 \text{ MPa} < S_{sd} \therefore \text{ Safe}$$

Maximum direct stress, (Eqn. 21.79 MDHB)

$$\sigma_{\max} = \left(\frac{\sigma}{2}\right) + \sqrt{\left(\frac{\sigma}{2}\right)^2 + \tau^2} = \left(\frac{34.52}{2}\right) + 26.71$$

$$= 43.97 \text{ MPa} < S_d = 130 \text{ MPa} \therefore \text{ Safe}$$

<u>The stress at A</u>

Torsional shear stress, $\tau_2 = \dfrac{T}{\propto_2 bh^2} = \dfrac{1770480.8}{0.269 \times 100 \times 70^2} = 13.43 \text{ MPa}$

The direct stress, $\sigma_2 = \dfrac{P_r}{bh} + \dfrac{6M_B}{bh^2} = \dfrac{P_r}{bh} + \dfrac{6P_t R}{hb^2}$

$$= \frac{30753.23}{100 \times 70} + \frac{6 \times 22131.01 \times 250}{70 \times 100^2} = 51.82 \text{ MPa}$$

Maximum shear stress, (Eqn. 21.84 MDHB)

$$\tau_{\max} = \sqrt{\left(\frac{\sigma}{2}\right)^2 + \tau^2} = \sqrt{\left(\frac{51.82}{2}\right)^2 + (13.43)^2} = 29.18 \text{ MPa} < S_{sd} \therefore \text{ Safe}$$

Maximum direct stress, (Eqn. 21.85 MDHB)

$$\sigma_{\max} = \left(\frac{\sigma}{2}\right) + \sqrt{\left(\frac{\sigma}{2}\right)^2 + \tau^2} = \left(\frac{51.82}{2}\right) + 29.18$$

$$= 55.1 \text{ MPa} < S_d \therefore \text{ Safe}$$

Example 23.14: Design the crankshaft for a horizontal gas engine having the cylinder diameter of 450 mm and stroke of 500 mm. The connecting rod length is 1200 mm, the other dimensions are $a = 375$ mm; $b = 1000$ mm; $c = 470$ mm. The weight of the flywheel is 50 kN. The gas pressure at I.D.C. is $p_{g_1} = 2.5$ MPa. The maximum torque position is at $\theta = 40°$ and the gas pressure at this position is $p_{g_2} = 0.9$ MPa. No belt pull.

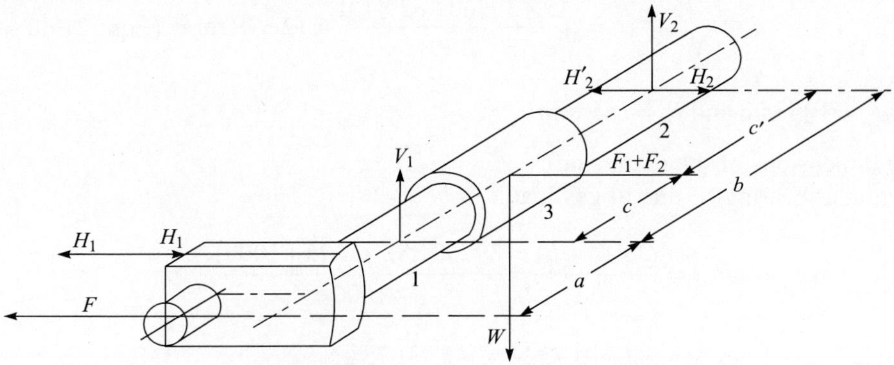

Figure 23.17 Forces on crankshaft at I.D.C

Given Data: The cylinder dia., $D = 450$ mm; stroke length, $L = 500$ mm; connecting rod length, $l = 1200$ mm; distances between bearings, $a = 375$ mm, $b = 1000$ mm, and $C = 470$ mm; the weight of the fly wheel, $W = 50$ kN; gas pressure at I.D.C., $P_{g_1} = 2.5$ MPa; maximum torque position, $\theta = 40°$; gas pressure at this position, $P_{g_2} = 0.9$ MPa.

SOLUTION

Step-1: Selection of basic data

The bearing data from Table 23.10 MDHB

For crankpin, $p_b = 12$ MPa and $k = \dfrac{l}{d} = 1.0$;

For main bearing, $p_b = 7.5$ MPa and $\dfrac{l}{d} = 1.5$

Material for the crankshaft- Alloy steel

with $S_d = 120$ MPa and $S_{sd} = 70$ MPa

Step-2: The gas load at I.D.C.,

$$F_p = \frac{\pi}{4} D^2 \cdot p_g = \frac{\pi}{4} \times (450)^2 \times 2.5 = 397607.82 \text{ N}$$

The force through connecting rod,

$$Q = F_p = 397607.82 \text{ N}$$

Step-3: The crankpin dimensions

The crankpin diameter for bearing pressure,

$$d_p = \sqrt{\frac{Q}{p_b k}} = \sqrt{\frac{397607.82}{12\times 1.0}} = 182\ \text{mm} \qquad \text{(Eqn. 21.61 MDHB)}$$

The pin diameter for strength,

$$d_p = \sqrt{\frac{5\cdot Q\cdot k}{S_d}} = \sqrt{\frac{5\times 397607.82\times 1.0}{120}} = 128.79\ \text{mm} \quad \text{(Eqn. 21.60 MDHB)}$$

Choose $d_p = 185$ mm and $l_p = 185$ mm

Step-4: The dimensions of the main shaft

The reaction at bearing '1' due to gas load,

$$R_{H_1} = \frac{Q\times(a+b)}{b} = \frac{397607.82\times(375+1000)}{1000}$$

$$= 546710.75\ \text{N} = 546.71075\ \text{kN}$$

The reaction at bearing 2,

$$R_{H_2} = \frac{Q\times a}{b} = \frac{397607.82\times 375}{1000} = 149102.93\ \text{N} = 149.10293\ \text{kN}$$

The vertical reactions due to flywheel weight,

$$R_{V_1} = \frac{W\cdot C'}{b} = \frac{50\times(1000-475)}{1000} = 26.25\ \text{kN} = 26250\ \text{N}$$

And $R_{V_2} = 50 - 26.25 = 23.75\ \text{kN} = 23750\ \text{N}$

The total reactions,

$$R_1 = \sqrt{R_{H_1}^2 + R_{V_1}^2} = \sqrt{546710.75^2 + 26250^2} = 547340.58\ \text{N}$$

$$R_2 = \sqrt{R_{H_2}^2 + R_{V_2}^2} = \sqrt{149102.93^2 + 23750^2} = 150982.60\ \text{N}$$

The diameter of shaft at bearing '1' for bearing pressure,

$$d_s = \sqrt{\frac{k_1}{k\cdot p_b}} = \sqrt{\frac{547340.58}{1.5\times 7.5}} = 220.57\ \text{mm} \qquad \text{(Eqn. 21.63 MDHB)}$$

The diameter for strength,

$$d_s = \left[\frac{\sqrt{(Q\cdot L_1)^2 + (P_t R)^2}}{0.2S_{sd}}\right]^{\frac{1}{3}}$$

where, $Q = 397607.82$, and $L_1 = a = 375$

$P_t = 0$ and $R = 250$ mm ; $S_{sd} = 70$ MPa,

$$\therefore \qquad d_s = \left[\frac{397607.82 \times 375}{0.2 \times 70}\right]^{\frac{1}{3}} = 220.015 \text{ mm}$$

Choose $d_s = 220$ mm and $L_s = 1.5 d_s = 1.5 \times 220 = 330$ mm

Shaft diameter at bearing '2',

$$d_{p2} = \sqrt{\frac{R_2}{kp_b}} = \sqrt{\frac{150982.60}{1.5 \times 7.5}} = 115.8 \text{ mm}$$

Choose $d_{s_2} = 120$ mm

The bending moment on the shaft under the flywheel,

$$M = R_3 \times C' = 150982.6 \times 525 = 79265865.4 \text{ mm}$$

The shaft diameter for bending strength,

$$d = \left[\frac{32M}{\pi S_d}\right]^{\frac{1}{3}} = \left[\frac{32 \times 79265865}{\pi \times 120}\right]^{\frac{1}{3}} = 188.78 \text{ mm}$$

To account for torque to be transmitted choose $d = 200$ mm

Step-5: Dimensions of the web

Thickness, $h = 0.8 d_p = 0.8 \times 185 = 148 \cong 150$ mm

Width, $b = 1.4 \cdot d_p = 1.4 \times 185 = 259 \cong 260$ mm

The bending stress at B,

$$\sigma_B = \frac{P_r}{bh} + \frac{6MA}{bh^2} = \frac{P_r}{bh} + \frac{6 \times P_r \times L_2}{bh^2}$$

where, $$L_2 = \frac{l_p}{2} + \frac{h}{2} = \frac{185}{2} + \frac{150}{2} = 167.5 \text{ mm}$$

$$\sigma_B = \frac{397607.82}{260 \times 150} + \frac{6 \times 397607.82 \times 167.5}{260 \times 150^2}$$

$= 78.5$ MPa $< S_d = 120$ MPa $\qquad \therefore$ Safe

Step-6: Forces on the crankshaft at maximum torque position

The forces on crankshaft at maximum torque position

The crank angle, $\theta = 40°$

$$l/R \text{ ratio}, \; n = \frac{1200}{250} = 4.8$$

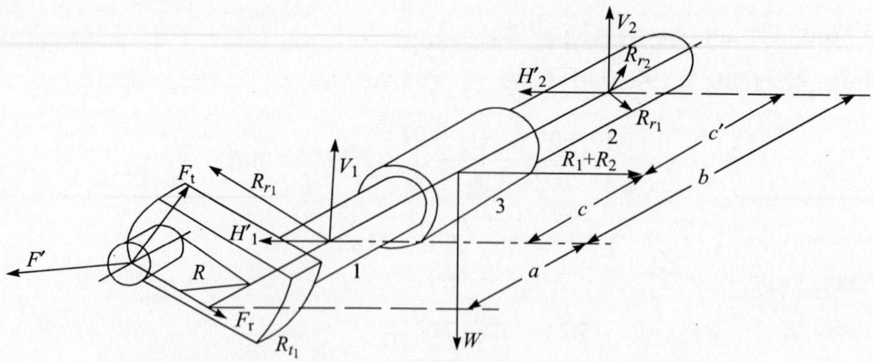

Figure 23.18 Forces on crankshaft at max. torque position

The angle of obliquity β, is given by

$$\cos\beta = \left[1 - \frac{\sin^2\theta}{n^2}\right]^{\frac{1}{2}} = \left[1 - \frac{\sin^2 40}{4.8^2}\right]^{\frac{1}{2}} = 0.991$$

$\therefore$ $$\beta = \cos^{-1}(0.991) = 7.7°$$

The angle between connecting rod and crank

$$\propto = \theta + \beta = 40 + 7.7 = 47.7°$$

The gas load,

$$F_p = \frac{\pi}{4} \cdot D^2 \cdot P_{g2} = \frac{\pi}{4} \times (450)^2 \times 0.9 = 143138.82 \text{ N}$$

The force through connecting rod,

$$Q = F_p / \cos\beta = \frac{143138.82}{0.991} = 144438.76 \text{ N}$$

The radial component,

$$P_r = Q \cos \propto = 144438.76 \times \cos 47.7 = 97209.1 \text{ N}$$

The tangential component,

$$P_t = Q \sin \propto = 144438.76 \times \sin 47.7 = 106831.4 \text{ N}$$

The torque on the shaft,

$$T = P_t R = 106831.4 \times 250 = 26707849.55 \text{ Nmm}$$

Step-7: Check for stresses in the shaft

Check for strength of the shaft,

$$d_s = \left[\frac{\sqrt{(Q \cdot L_1)^2 + (P_t R)^2}}{0.2 S_{sd}}\right]^{\frac{1}{3}}$$

Shear stress induced, $\tau = \dfrac{\sqrt{(Q \cdot L_1)^2 + (P_t R)^2}}{0.2.d_s^3}$

$$= \frac{\sqrt{(144438.76 \times 375)^2 + (26707849.55)^2}}{0.2 \times 220^3}$$

$$= 28.36 \text{ MPa} < S_{sd} = 70 \text{ MPa}$$

The maximum shear stress in the shaft under flywheel,

$$\tau = \frac{\sqrt{M^2 + T^2}}{0.2 \times d^3}$$

where, M is due to the new reactions
Reactions due to gas load,

$$R_{H_2} = 375 \times 143138.82/1000 = 53677.1 \text{ N}$$

$$R_{v_2}^1 = \frac{Q_v \times a}{5} = \frac{144438.76 \times \sin 7.7 \times 375}{1000} = 7257.3 \text{ N}$$

Net vertical reactions,

$$R_{v_2} = 23750 - 7257.3 = 16492.7 \text{ N}$$

The total reaction, $R_2 = \sqrt{53677.1^2 + 16492.7^2} = 56153.72$ N

Bending moment at flywheel,

$$M = R_2 \cdot c^1 = 56153.72 \times 525 = 29480702.76 \text{ Nmm}$$

The shear stress,

$$\tau = \frac{\sqrt{M^2 + T^2}}{0.2d^3} = \frac{\sqrt{29480702.76^2 + 26707849.53^2}}{0.2 \times 200^3}$$

$$= 24.66 \text{ MPa} < S_{sd} \therefore \text{ Safe}$$

Step-8: Check for stresses in web

The torsion on the web,

$$T = P_t \cdot l_2 = 106831.4 \times 167.5 = 17894259.5 \text{ Nmm}$$

Moment about *A-A*,

$$M_A = P_r \cdot L_2 = 97209.1 \times 167.5 = 16282524.25 \text{ Nmm}$$

Moment about *B-B*, (r_1 = R-d/2 = 250-220/2 = 140 mm)

$$M_B = P_r \cdot r_1 = 106831.4 \times 140 = 14956396 \text{ Nmm}$$

Stresses at B,

Shear stress,

$$\tau = \frac{T}{\propto_1 bh^2} + \frac{3P_t}{2bh} = \frac{17894259.5}{0.239 \times 260 \times 150^2} + \frac{3 \times 106831.4}{2 \times 260 \times 150} = 16.91 \text{ MPa}$$

Direct stress,

$$\sigma = \frac{P_r}{bh} + \frac{6.M_A}{bh^2} = \frac{97209.1}{260 \times 150} + \frac{6 \times 16282524.25}{260 \times 150^2} = 19.19 \text{ MPa}$$

Resultant maximum shear stress, (Eqn. 21.78 MDHB)

$$\tau_{max} = \sqrt{\left(\frac{\sigma}{2}\right)^2 + \tau^2} = \sqrt{\left(\frac{19.19}{2}\right)^2 + (16.91)^2} = 19.44 \text{ MPa} < S_{sd} \therefore \text{ Safe}$$

Resultant maximum direct stress, (Eqn. 21.79 MDHB)

$$\sigma_{max} = \left(\frac{\sigma}{2}\right) + \sqrt{\left(\frac{\sigma}{2}\right)^2 + \tau^2} = \frac{19.19}{2} + 19.44 = 29.04 \text{ MPa} < S_d \therefore \text{ Safe}$$

Stresses at A

Shear stress, $\tau = \frac{T}{\propto_2 bh^2} = \frac{17894259.5}{0.291 \times 260 \times 150^2} = 10.51 \text{ MPa}$

Direct stress,

$$\sigma = \frac{P_r}{bh} + \frac{6M_B}{hb^2} = \frac{97209.1}{260 \times 150} + \frac{6 \times 14956396}{150 \times 260^2} = 11.34 \text{ MPa}$$

Resultant maximum shear stress, (Eqn. 21.84 MDHB)

$$\tau_{max} = \sqrt{\left(\frac{\sigma}{2}\right)^2 + \tau^2} = \sqrt{\left(\frac{11.34}{2}\right)^2 + (10.51)^2}$$

$$= 11.94 \text{ MPa} < S_{sd} = 70 \text{ MPa} \quad \therefore \text{ Safe}$$

Maximum direct stress, (Eqn. 21.85 MDHB)

$$\sigma_{max} = \left(\frac{\sigma}{2}\right) + \sqrt{\left(\frac{\sigma}{2}\right)^2 + \tau^2} = \left(\frac{11.34}{2}\right) + 11.94$$

$$= 17.61 \text{ MPa} < S_d = 120 \text{ MPa} \therefore \text{ Safe}$$

EXERCISES FOR PRACTICE

23.1 A trunk type cast iron piston for an internal combustion engine has a diameter of 100 mm and a length of 150 mm. The maximum pressure is 3.5 MPa. Maximum permissible stress for cast iron for the design of head thickness is 30 MPa, and for piston pin material is 45 MPa. Bearing pressure for piston pin should not exceed 20 MPa. Determine (a) thickness of the head treating it as a flat plate fixed at the periphery and uniformly loaded, and (b) dimensions of the piston pin.

Sketch the cross section of the piston through the piston bosses.

23.2 Design a trunk type piston for a single cylinder, four-stroke cycle engine developing 5 kW at 600 rpm. Draw to scale a fully dimensioned working drawing of the piston with the piston rings, scraper rings and the piston pin in position. Show clearly the arrangement for fitting the gudgeon pin.

Diameter of piston is 120 mm and the maximum explosion pressure is 4 MPa. Heat supplied to the engine is 19000 kJ/kWh. About 6% of the heat is conducted through the piston crown. The heat conduction factor for cast iron may be taken as 60 kW hour metre°C. The temperature difference between the centre of the crown and the edge of the crown may be taken as $250°C$.

Assume any data necessary for the design.

23.3 Design an aluminium alloy piston with a flat head for an I.C. engine having 100 mm bore. Draw a neat dimensioned sketch of the piston to bring out the details clearly. The maximum gas pressure may be taken as 4.5 MPa.

23.4 The critical section of a connecting rod for a diesel engine is at a $\frac{2}{5}$th distance from the gudgeon pin. It is an I-section having the following proportions:

Width of the flange 40 mm; thickness of the flange 4 mm; depth of the section 50 mm and thickness of the web 10 mm.

The diameter and stroke of the piston are 120 mm and 220 mm respectively and the length of the connecting rod is 450 mm. The maximum gas pressure in the cylinder is 4.7 MPa. The speed of the engine is 1200 r.p.m. Check whether the factor of safety is sufficient or not.

23.5 The high speed diesel engine has the following particulars: Bore of the cylinder = 90 mm, stroke = 140 mm, speed = 1500 rpm, compression ratio = 16, maximum pressure = 4.5 MPa upto 7% of the stroke, mass of the reciprocating parts is 3 kg and length of the connecting rod is 350 mm. Design a suitable connecting for the given duty.

23.6. Design an overhung crank for a 500 × 600 mm four-stroke cycle gas engine having the following specifications:

Speed of the engine is 300 rpm.

Length of the connecting rod is five times the crank length.

Maximum torque on the shaft occurs when the crank has turned through an angle of 36° from the inner dead centre, the gas pressure at this position being 0.9 MPa.

Maximum explosion pressure = 2.5 MPa

23.7 Design and draw the connecting rod of a petrol engine having the following specifications: Piston diameter = 120 mm, weight of reciprocating parts = 2.5 N, length of rod centre to centre = 380 mm, stroke = 150 mm, normal speed = 2000 rpm and over speed is 2800 rpm, compression ratio = 5.5 and probable maximum explosion pressure = 3 MPa.

23.8 Design the big end of connecting rod for a vertical diesel engine for the following data: Load along the rod 45 kN, and the length of the connecting rod measures 1000 mm between crankpin and gudgeon pin axes.

Choose the suitable value for the $\frac{l}{d}$ ratio for the crankpin. Also select the suitable materials for the connecting rod and cap bolts.

23.9 Check the suitability of the I–section 15 mm × 12 mm × 3 mm for designing the connecting rod in the case of single cylinder motor cycle engine of 40 mm bore by 55 mm stroke wherein the maximum pressure produced is 3.6 MPa and in which the connecting rod is four times the crank. Draw a neat sketch of the connecting rod.

23.10 State the various forces which affect the design of the connecting rod, and explain their influence in choosing the suitable materials for various parts and adopting suitable stress values.

Design a connecting rod of I-section for a diesel engine running at 1500 rpm. Bore of the engine is 120 mm and stroke is 150 mm. Length of the rod is 350 mm. Maximum gas pressure is 5 MPa, which is constant upto 5% of the stroke. Weight of reciprocating parts is 0.015 N/sq mm of piston area. Law of expansion is $PV^{1.25}$ = constant

Check the design for the transverse inertia stresses.

23.11 The crankpin of a forged side crank is acted upon by a force of 25 kN at right angles to the crank. The length of the crank is 150 mm. The distance of the plane of rotation of the centre of the crankpin from the centre of the adjacent bearing is 0.2 metre. Safe bearing pressure for pin is 7 MPa; safe bearing pressure for the crankshaft journal is 3.5 MPa; permissible tensile and shear stress intensities for crank material are 70 MPa. Design the crank and give two views of the same. Mark important dimensions on your sketch.

23.12 A two-stroke diesel engine develops 20 kW at 450 r.p.m. and is directly connected to a generator through a flexible coupling. The stroke of the engine is 150 mm and connecting rod is 700 mm long. The maximum torque occurs at 60° from I.D.C. position, when the thrust of the connecting rod is 1.8 kN. A flywheel weighing 18 kN forms one half of the coupling and is situated centrally between main bearings of the engine and the generator. The main bearings of the crankshaft are situated at 15 mm and 160 mm from the centre line of the cylinder. Design the crankshaft if the permissible shear stress is 30 MPa. Balance weights may be omitted. Assume suitable bearing pressures. Draw a dimensioned sketch and indicate lubricating arrangements.

23.13 Design an overhung crank for a 500 mm × 600 mm four-stroke cycle gas engine having the following specifications:

Speed of the engine 300 r.p.m.

Length of the connecting rod is five times the crank length.

Maximum torque on the shaft occurs when the crank has turned through an angle of 36° from the inner dead centre, the gas pressure at this position being 0.9 MPa.

Maximum explosion pressure 2.5 MPa.

Choose your own material and suitable values for stresses.

23.14 Design a forged overhung crankshaft of an I.C. engine having the bore diameter of 250 mm and the stroke of 400 mm. The connecting rod length is five times the crank length. The maximum explosion pressure is 3 MPa and the engine runs at 500 rpm. The maximum torque occurs when the crank is at 30° from I.D.C. position. The gas pressure at this position is 1.25 MPa. Select the suitable materials and design stresses.

23.15 The following data are for a four-stroke vertical diesel engine: Cylinder bore diameter = 75 mm; stroke length = 100 mm; the connecting rod length = 250 mm; the maximum gas pressure = 5 MPa (it is constant for 6% of the stroke); the law of expansion is $PV^{1.3}$ = constant; mass of the reciprocating parts is 0.04 kg/ cm^2 of the piston area.

Design the connecting rod for the following design stresses:

Nickel-steel body, S_d = 150 MPa;

Cap material, S_d = 80 MPa;

The bolt material, S_d = 75 MPa.

SHORT ANSWER QUESTIONS

23.1 Name the different stokes in a two-stroke internal combustion engine.

23.2 Name the different strokes in a four-stroke internal combustion engine.

23.3 What is the recommended range of air-fuel ratio in petrol engines?

23.4 What is the recommended range of air-fuel ratio in diesel engines?

23.5 What are the functions of a piston?

23.6 What are the considerations in a piston design?

23.7 What is the function of piston skirt?

23.8 What is the function of a piston ring?

23.9 What is the consideration to design a piston pin?

23.10 What is the range of piston speed of an internal combustion engine?

23.11 What is the range of efficiency of super charged internal combustion engine?

23.12 What is brake horsepower of an internal combustion engine?

23.13 What is the function of a connecting rod in an internal combustion engine?

23.14 List the materials used in construction of a connecting rod of an internal combustion engine in an automobile.

23.15 What type of motion do the big end and small end of a connecting rod undergo?

23.16 Name the column theory generally adopted in design of a connecting rod.

23.17 List the cross sections generally used in design of connecting rod of an internal combustion engine.

23.18 What forces are considered in determining the cross section of a connecting rod in an internal combustion engine?

23.19 What are the functions of a crankshaft?

23.20 Name the different types of crankshafts used in internal combustion engine of an automobile.

23.21 What manufacturing processes are used in manufacturing a crankshaft?

23.22 What properties the material used in crankshaft should possess?

23.23 What type of stresses are induced in a crankshaft?

23.24 What is angular position of the crank at inner dead centre?

23.25 What is the angular position of the crank for the maximum torque output?

CHAPTER 24

FLYWHEEL

24.1 INTRODUCTION

A flywheel is an energy storage device. It absorbs mechanical (kinetic) energy, when the supply of energy is more than the requirement and releases the stored energy, whenever the energy supply is less than required. The flywheel smoothens the speed variation of the machine. Whenever there is a mismatch between the supply-torque and the load-torque, there is speed variation. The flywheel mounted on the shaft of the power source machine ensures that the out speed is almost constant.

In case of internal combustion engines, there is power output during expansion stroke only, while during the remaining strokes, power is to be supplied to drive the piston. The flywheel mounted on the engine shaft, absorbs the excess energy during the expansion stroke and releases the required energy during the remaining strokes. It ensures that the output torque and speed are fairly constant.

In case of punching press, the power is required during the punching operation only. The driving torque of the driving motor is almost uniform. The flywheel mounted on the driven shaft absorbs the excess energy during the non-punching period of the work cycle and releases the required energy during the punching operation. The reciprocating pumps, rock-crushers are other examples of flywheel use.

24.2 FUNCTIONS OF FLYWHEEL

1. It stores the energy, when the supply is more than the demand and releases it when the demand is more than the supply during the work cycle.
2. It reduces the speed variation during a work cycle, but it does not affect the mean speed.
3. It reduces the average power capacity of driving machine.

24.3 THE FLYWHEEL MATERIAL

The commonly used material is cast iron. It is cheap and can be casted into any desired shape. The cast iron flywheels have excellent vibration damping capacity. The disadvantage is the low tensile strength. The flywheel failure is sudden and total. Hence for high-speed applications, cast steel is preferred. Nowadays the composite like graphite-fibre-reinforced polymer (GFRP) is becoming more popular in modern engines.

24.4 TYPES OF FLYWHEELS

There are two types of flywheels:

1. Disk flywheel
2. Rim flywheel.

The disk flywheels are solid circular disks with hub for mounting on the shaft. They are used on small engines. The rim flywheels have a cylindrical rim attached to the hub by radial arms (spokes) or thin solid circular disk. It may be cast as a single piece or it may be split into two halves. The each half is casted separately to avoid the built-in shrinkage stresses and then assembled into one by bolting them together.

24.5 TORQUE-TIME (ROTATION) DIAGRAM

The typical torque-time diagram for a four-stroke engine is shown in figure below. During the suction and compression strokes, the piston is to be driven and the torque is to be supplied and there is a negative torque output. During the expansion (explosion) stroke, the piston is being driven by the high pressure gases and there is positive torque output. During the exhaust stroke the burnt gases are to be driven out by the piston and there is negative torque output. The net power output is the sum total of the positive power during the working stroke and negative power during the remaining strokes. Mathematically, we can find the total output per cycle by the expression,

$$E = \int_0^{4\pi} T d\theta$$

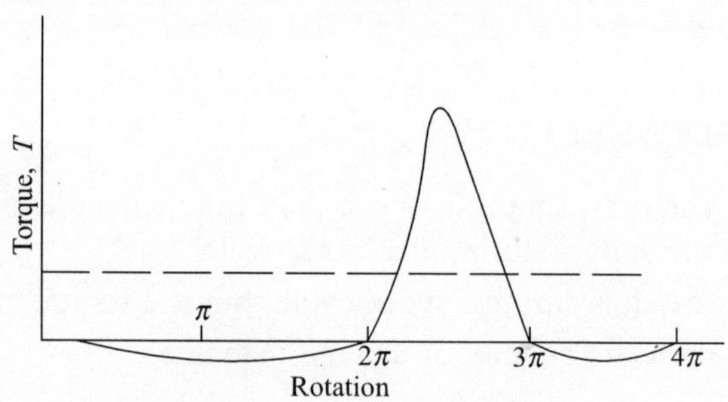

Figure 24.1 Torque-rotation diagram

The mean torque output,

$$T_m = E/4\pi = \frac{1}{4\pi} \int_0^{4\pi} T d\theta$$

Then the excess torque during the expansion stroke,

$$\Delta T = (T - T_m)$$

This value is positive from A to B as shown in figure. The excess energy is evaluated by the expression,

$$\Delta\text{E} = \int_A^B (T - T_m)\,d\theta \tag{24.1}$$

This excess energy is stored in the flywheel by increasing its speed to a maximum value $\omega_{\max}$ at B from a minimum value $\omega_{\min}$ at A.

Then we can write the energy balance equation as

$$\Delta\text{E} = \frac{I\omega_{\max}^2}{2} - \frac{I\omega_{\min}^2}{2} = I\left(\frac{\omega_{\max} + \omega_{\min}}{2}\right) \times \omega \times \left(\frac{\omega_{\max} - \omega_{\min}}{\omega}\right) = I \cdot \omega^2 \cdot C_s \tag{24.2}$$

where, $\omega = \left(\dfrac{\omega_{\max} + \omega_{\min}}{2}\right)$ = mean angular speed rad/s

C_s is called coefficient of speed fluctuation.

$$= \left(\frac{\omega_{\max} - \omega_{\min}}{\omega}\right)$$

The reverse of this is defined as the coefficient of steadiness,

$$m = \frac{1}{C_s} = \frac{\omega}{(\omega_{\max} - \omega_{\min})} = \frac{n}{(n_{\max} - n_{\min})}$$

The coefficient of energy fluctuation is defined as the ratio of the excess energy to the energy produced during a cycle as

$$C_E = \left(\frac{\Delta E}{E}\right)$$

The typical values of the coefficient of speed fluctuation and the coefficient of energy fluctuation are given in Table 24.1 (Table 21.15 MDHB) and Table 24.2 (Table 21.16 MDHB) respectively.

24.6 MOMENT OF INERTIA AND DIMENSIONS OF THE FLYWHEEL

The required moment of inertia of the flywheel is

$$I = \frac{\Delta E}{C_S\omega^2} \tag{24.3}$$

For a disk flywheel, the mass of the flywheel,

$$M = \frac{I}{K^2} = \frac{2I}{R^2} \tag{24.4}$$

where, R is the radius of the disk and
K is the radius of gyration of the disk.
The thickness of the disk,

$$t = \frac{M}{2\pi R\rho} \tag{24.5}$$

For a rim flywheel the mass of the rim is given by

$$M = \frac{I}{K^2} = \frac{I}{R^2} \tag{24.6}$$

(assuming that 100% of the inertia is provided by the rim).

It is customary to assume that 10% of the inertia is provided by the hub and arms. Then the mass of the rim is given by

$$M1 = \frac{0.9I}{K^2} = \frac{0.9I}{R^2} \tag{24.7}$$

The dimensions of the rim can be evaluated by the expression,

$$M1 = 2\pi R \cdot B \cdot t \cdot \rho \tag{24.8}$$

where, ρ is the mass density of the material,

$$= 7100 \text{ to } 7200 \text{ kg/m}^3 \text{ for C.I.}$$

$$= 7800 \text{ kg/m}^3 \text{ for steel}$$

The thickness to width ratio is assumed to be $t/B = 0.5$ to 1.25

When the flywheel is used as belt pulley, the width,

$$B = b + 25 \text{ to } 50 \text{ mm}$$

where, b is the width of the belt.

The rim radius R should be the largest possible under the limiting conditions of speed and space availability. For cast iron the limiting speed is 30 m/s, then the rim radius,

$$R = (30/\omega) \text{ m} = (30000/\omega) \text{ mm}$$

Table 24.1. Recommended rim velocities-V

For stationary engines
V = 25 m/sec. for C.I. wheels of engines under 75 kW
= 30 m/sec. for larger engines
= 35 m/sec. with special rim design without blow holes in the rim
= 50 m/sec. for cast steel wheels
For automobile engines
V = 50 m/sec. C.I. flywheel
= 75 m/sec. semi-steel wheels
= 100 m/sec. steel wheels

Table 24.2. Coefficient of speed fluctuation, C_s

Driven Machinery	C_s
Punch press	1/3 to 1/15
Pumps	1/20 to 1/30
Machine shop	1/35 to 1/40
Looms and paper mills	1/40
Spinning mills	1/60
D.C. Generator (lighting)	1/150
A.C. Generator	1/300
D.C. Generator, direct drives	1/100 to 1/120
D.C. Generator, belt driven	1/70 to 1/200
A.C. Generator, belt driven	1/125 to 1/150
Automobile normal speed	1/10
Automobile (Idling)	1/5

Table 24.3. Fluctuation of energy for steam engine C_E

Cutoff percentage	Single cylinder engines	Two cylinder with cranks 90° apart	Three cylinders with crank 120° apart
10	0.35	0.088	0.040
20	0.33	0.082	0.037
40	0.31	0.078	0.034
60	0.29	0.072	0.032
80	0.28	0.070	0.031
100	0.27	0.068	0.030

Table 24.4. Fluctuation of energy for I.C. engines, C_E

Type of engines	No. of cylinders	Angle between cranks	Fluctuation on energy	
			4-stroke cycle	2-stroke cycle
Single	1	-	2.35 – 2.40	0.95 – 1.00
Acting	2	180°	1.50 – 1.60	0.20 – 0.25
Vertical	3	120°	0.60 – 0.75	0.15 – 0.18
Vertical	4	180° and 90°	0.15 – 0.20	0.075 – 0.10
Vertical	6	180° and 60°	0.10 – 0.12	0.016 – 0.12

24.7 STRESSES INDUCED IN THE FLYWHEEL

Disk Flywheel

The disk flywheel may be treated as a solid circular rotating disk and stresses induced are,

$$\text{Radial stress, } \sigma_r = K_1 \left[1-\left(\frac{r}{R}\right)^2\right]$$

Tangential stress, $\sigma_t = K_1\left[1-\left(\frac{3\mu+1}{\mu+3}\right)\left(\frac{r}{R}\right)^2\right]$

where, $K_1 = 1000\,\rho_m\,V^2\left(\frac{\mu+3}{8}\right)$

ρ_m = mass density kg/mm^3

V = peripheral velocity

= $(\omega R/1000)$ m/s

μ = Poisson's ratio

The maximum stress at $r = 0$,

$$\sigma_{r\max} = \sigma_{t\max} = K_1 = 1000\,\rho_m\,V^2\left(\frac{\mu+3}{8}\right) \qquad (24.9)$$

This should be less than the design stress S_d for flywheel material.

If it is treated as rotating disk with central hole, the maximum stress occurs at the inner radius and is given by,

$$\sigma_{t\max} = 2\,K_1\left[1-\left(\frac{\mu-1}{\mu+3}\right)\left(\frac{R_1}{R_2}\right)^2\right] \qquad (24.10)$$

where, R_1 = inner radius

R_2 = outer radius

If the maximum stress is more than the design stress, the outer radius of the disk should be reduced appropriately and thickness be increased.

Rim Flywheel

The rim is subjected to uniformly distributed centrifugal force P_c, acting radially outwards. The outward displacement of the rim is restricted by the arms. The rim between the arms is subjected to bending moment. Thus, the rim is subjected to combined stress due to centrifugal action and bending action.

(A) An approximate approach given by Prof. Lanza

(a) The centrifugal stress when unrestrained by the arms:

Consider an element of the rim, subtending an angle $\partial\theta$ at the centre.

The centrifugal force per unit length of the rim,

$$P_c = \frac{mV^2}{R}$$

where, m = mass per unit length = $A\rho$

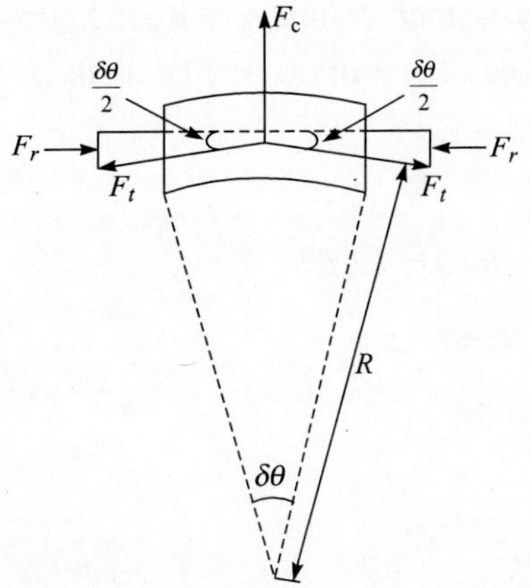

Figure 24.2 Forces acting on the element of the rim

ρ = Mass density, kg/m^2

A = Cross-sectional area of the rim, m^2

R = Mean radius of the rim, m

V = Peripherial velocity of the rim, m/s

The length of the element, $\partial L = R\partial\theta$

The centrifugal force on the element,

$$F_c = P_c \partial L = \frac{mV^2}{R} R\,\partial\theta = m\,V^2\,\partial\theta \quad \text{(i)}$$

The resisting force on the side faces of the element,

$$F_t = A \cdot \sigma_c$$

The radial component,

$$F_r = 2\,F_t\,\partial\theta/2 = F_t\partial\theta = A\,\sigma_c\partial\theta \quad \text{(ii)}$$

For equilibrium, equate the two, $F_r = F_t$

$$A\,\sigma_c\partial\theta = m\,V^2\,\partial\theta$$

Then the stress in the rim,

$$\sigma_c = \frac{m}{A}V^2 = \rho V^2 \text{ Pa} \quad \text{(24.11a)}$$

When the mass density is expressed in kg/mm^3, the centrifugal stress in MPa is given by,

$$\sigma_c = 1000\,\rho_m\,V^2 \text{ MPa} \quad \text{(24.11b)}$$

(b) The bending stress, when restrained completely by the arms.

The portion of the rim between the arms is treated as fixed beam with uniform load.

The rate of loading, $w = F_c = \dfrac{mV^2}{R}$

The length of the beam, $L = R \cdot \theta = 2\pi R/n$

where, m = mass per unit length

n = the number of arms

The maximum bending,

$$M = \frac{wL^2}{12} = \left(\frac{mV^2}{12R}\right)\left(\frac{2\pi R}{n}\right)^2 = \left(\frac{mR\pi^2V^2}{3n^2}\right)$$

The bending stress,

$$\sigma_b = \frac{6M}{bt^2} = \left(\frac{6}{bt^2}\right)\left[\frac{mR\pi^2V^2}{3n^2}\right] = \frac{2\rho R\pi^2V^2}{t \cdot n^2}\text{ Pa} \tag{24.12a}$$

When the mass density is expressed in kg/mm^2, the bending stress in MPa is given by

$$\sigma_b = \frac{2000\rho_m R\pi^2V^2}{t \cdot n^2}\text{ MPa} \tag{24.12b}$$

where, ρ_m = Mass density in kg/mm^3

R = Radius in mm

t = Rim thickness in mm

(c) The resultant stress, assuming a partial restraint,

$$\sigma_t = 0.75\,\sigma_c + 0.25\,\sigma_b \tag{24.13}$$

(B) The analysis given by S. Timoshenko

The stress is a combination of direct stress due to centrifugal force plus the stress due to the constraint by the arms and is given by

$$\sigma_t = \frac{P}{B \cdot t} + \frac{6M}{B \cdot t^2}$$

where, P at $\varphi = 1000\text{ m }V^2 - \dfrac{P_1 \cos\varphi}{2\sin\alpha}$

$$M \text{ at } \varphi = \left(\frac{1000P_1R}{2}\right)\left[\frac{\cos\varphi}{\sin\alpha} - \frac{1}{\alpha}\right]$$

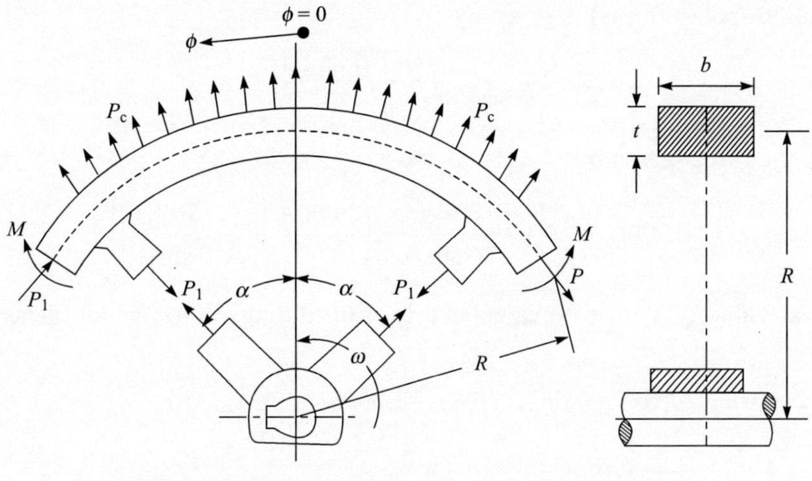

Figure 24.3 Forces and moments in rimmed flywheel

The force through the arm,

$$P_1 = \left(\frac{2}{3}\right)\left[\frac{1000\ \text{mV}^2}{C}\right]$$

The constant C is given by,

$$C = (12\times10^6)\left(\frac{R^2}{t^2}\right)f_2(\alpha) + f_1(\alpha) + \frac{A}{A_1}$$

$$f_1(\alpha) = \left(\frac{0.5}{\sin\alpha^2}\right)\left[\frac{\sin 2\alpha}{4} + \frac{\alpha}{2}\right]$$

$$f_2(\alpha) = \left(\frac{0.5}{\sin\alpha^2}\right)\left[\frac{\sin 2\alpha}{4} + \frac{\alpha}{2}\right] - \frac{0.5}{\alpha}$$

where, m = mass of the rim per unit length kg/mm = $\rho_m \cdot A$

R = the mean radius of the rim, m

t = thickness of the rim, mm

B = width of the rim, mm

A = the cross-sectional area of the rim, mm^2

A_1 = the cross-sectional area of the spoke, mm^2

2α = angle between two consecutive arms, rad

n = the number of arms

V = rim velocity, m/s

The stress in the spokes (arms) is given by

$$-\sigma_1 = P_1/A_1 = \left(\frac{2}{3}\right)\left[\frac{1000\,\text{mV}^2}{CA_1}\right] \tag{24.14}$$

The stress in the rim is given by

$$\sigma_t = (1000\rho_m V^2)\left\{1-\frac{\cos\alpha}{3C\sin\alpha} \pm \left(\frac{2000R}{C\cdot t}\right)\left[\frac{1}{\alpha}-\frac{\cos\varphi}{\sin\alpha}\right]\right\} \tag{24.15}$$

The numerical value of C can be calculated by substituting for α, which depends on the number of arms.

For 4-arms, $2\alpha = \frac{\pi}{2}$; $f_1(\alpha)$ =0.643, $f_2(\alpha)$ =0.0068

$$C = (72960)\left(\frac{R^2}{t^2}\right) + 0.643 + \frac{A}{A_1}$$

For 6-arms, $2\alpha = \frac{\pi}{3}$; $-f_1(\alpha) = 0.957$; $f_1(\alpha) = 0.00169$

$$C = (20280)\left(\frac{R^2}{t^2}\right) + 0.957 + \frac{A}{A_1}$$

SOLVED EXAMPLES

Example 24.1: A single cylinder single acting four-stroke gas engine develops a power of 15 kW at 300 rpm. The work done by the gases during the power stroke is 1.4 times the average work done during the whole cycle. If the total fluctuation of speed is not to exceed +2% of the mean speed and the turning moment diagram during the power stroke is assumed to be triangular in shape, find the cross section of the rim of the flywheel, assuming the width of the rim to be twice the thickness and radius of the wheel not to exceed 500 mm. Also design the shaft, hub, and arms of the flywheel assuming the shear stress for the shaft at 40 MPa and tensile stress for cast-iron as 15 MPa.

Given Data:

Power of the engine = 15 kW, at speed = 300 rpm,

The work done during expansion stroke $W1$ = 1.4 during the whole cycle

The turning moment diagram is triangular during the working stroke.

The total fluctuation of speed $\leq \pm$ 2%

The radius of the wheel, $R \leq 500$ mm

The rim width to thickness ratio = 2

SOLUTION

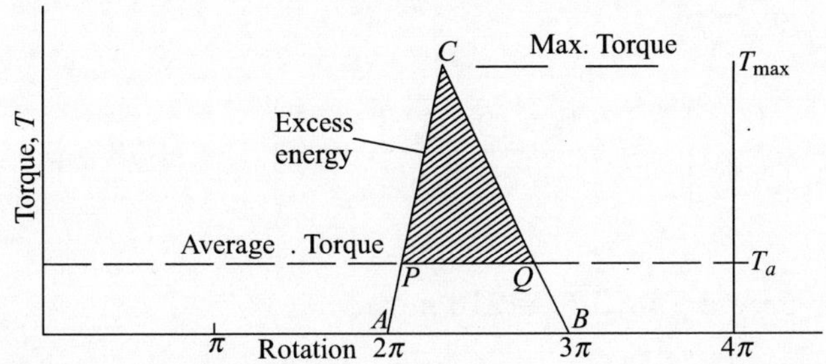

Figure 24.4 Torque rotation diagram for four-stroke engine

The average torque, $T = 955 \times 10^6$ kW/N = 9550000 × 15/300 = 477500 Nmm

The work done per cycle, $W = T \times 4\pi$ = 477500 × 4 × 3.14 = 6000000 Nmm = 6000 Nm

Work done during the power stroke, $W_1 = 1.4 \times W$ = 1.4 × 6000 = 84000 Nm

Assuming triangular torque-moment diagram, the maximum torque,

$$T_{max} = 4W_1/\pi = 4 \times 84000/\pi = 10695.21 \text{ Nm}$$

The excess energy = area of triangle PQC = $W_1 \times (T_{max} - T)/T_{max} = W_1(1 - T/T_{max})$

$$\Delta E = 10695.21 \times (1 - 477.5/10695.21) = 8025 \text{ Nm}$$

The angular speed of the engine, $\omega = 2\pi N/60 = 2\pi \times 300/60$ = 31.42 rad/s
The moment of inertia of the flywheel,

$$I = \Delta E/(C_s \times \omega^2) = 8025/(0.04 \times 31.42 \times 31.42) = 203.28 \text{ kg m}^2$$

The mass of the rim (assuming 90% is provided by the rim),

$$M_1 = 0.9 \times I/R^2 = 0.9 \times 203.28/0.5^2 = 731.79 \text{ kg}$$

The cross section of the rim,

$$A = M1/(2\pi R\rho) = 731.79/(2 \times 31.4 \times 0.5 \times 7200) = 0.0323532 \text{ m}^2 = 32352.32 \text{ mm}^2$$

But width to thickness ratio $B/t = 2$

$$A = B \times t = 2t \times t = 2t^2$$

Thickness, $t = \sqrt{(A/2)} = \sqrt{(32352.32/2)}$ = 127.18 mm ≅ 130 mm

Width of the rim = $2t$ = 2 × 130 = 260 mm

The arm cross section $A1$

The rim velocity, $V = \omega R$ = 31.42 × 0.5 = 15.71 m/s

The stress in the arm, $\sigma_1 = 1000 \times \rho_m \times v^2 = 1000 \times 72000 \times 10^{-9} \times 15.71^2$ = 1.777 MPa.

Assuming design stress for arm = 15 MPa, the cross-sectional area of the arm,

$A1$ = 1.777/15 = 0.1185 mm^2

This is too small and assume an elliptical section of 10 × 20 mm

The shaft diameter is given by

$$d = \left[\frac{16T}{\pi S_d}\right]^{1/3} = \left[\frac{16 \times 10695.21 \times 10^3}{\pi \times 40}\right]^{1/3} = 117.8 \text{ mm}$$

Choose $d = 120$ mm

Hub diameter $D = 1.75d = 1.75 \times 120 = 210$ mm

Example 24.2: The turning moment diagram of a multicylinder engine is drawn with a scale of 1 mm = 2° on abscissa and 1 mm = 1250 Nmm on the ordinate. The intercepted areas between the torque developed by the engine and the mean resisting torque of the machine taken in order from one end are –30, +400, –270, +330, –310, +230, –380, +270, and –240 mm². The engine is running at a mean speed of 240 rpm. The coefficient of speed fluctuation is limited to 0.02. A rimmed flywheel made of grey cast-iron FG200 with $P = 7100$ kg/m³ is provided. The rim contributes 90% of the inertia. The rim has rectangular cross section with width to thickness ratio of 1.5. Determine the dimensions of the flywheel.

Given Data:

Torque-moment diagram with scale x-axis –1 mm = 2° y-axis 1 mm = 1250 Nmm

Areas in mm² from right, –30; +400; –270; +330; –310; +230; –380; +270; –240.

The engine speed = 240 rpm; Coefficient of speed fluctuation $C_s = 0.02$

Material—C.I. FG200 with $\rho = 7100$ kg/m³

Width to thickness ratio = 1.5

SOLUTION

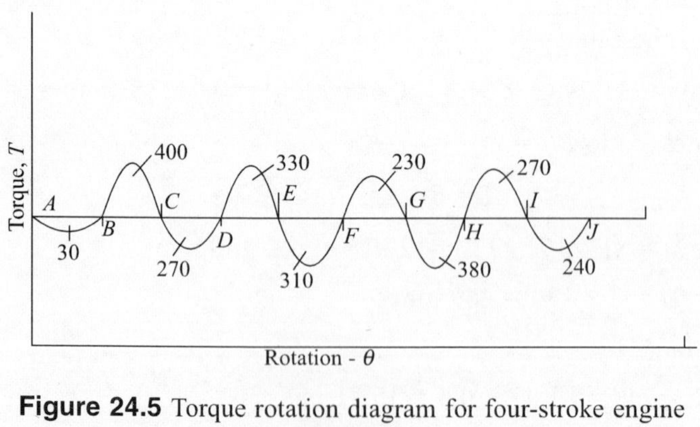

Figure 24.5 Torque rotation diagram for four-stroke engine

The energy levels at various change-over points
Let E_A be the energy level at A,
The energy level at B, $E_B = E_A - 30$;

The energy level at C, $E_C = E_A - 30 + 400 = E_A + 370$

The energy level at D $E_D = E_A + 370 - 270 = E_A + 100$

The energy level at E $E_E = E_A + 100 + 330 = E_A + 430$

The energy level at F $E_F = E_A + 430 - 310 = E_A + 120$

The energy level at G $E_G = E_A + 120 + 230 = E_A + 350$

The energy level at H $E_H = E_A + 350 - 380 = E_A - 30$

The energy level at I $E_I = E_A - 30 + 270 = E_A + 240$

The energy level at J $E_J = E_A + 240 - 240 = E_A$

The maximum and minimum values are:
$E_{max} = E_A + 430$ at E and $E_{min} = E_A - 30$ at H
Energy fluctuation, $\Delta E = E_{max} - E_{min} = E_A + 430 - E_A + 30 = 460\ mm^2 = 460 \times 1250 \times 2 \times \pi/180 = 20071.3$ Nm
The mean speed = 240 rpm,
The angular speed, $\omega = 2\pi \times 240/60 = 25.133$ rad/s
The moment of inertia of the rim,

$$I = \frac{K \times \Delta E}{C_s \omega^2} = \frac{0.9 \times 20071.3}{0.02 \times 25.133^2} = 1429.9\ \text{kgm}^2$$

The radius of the rim,

$$R = \frac{30}{\omega} = \frac{30}{25.133} = 1.194\ \text{m} = 1194\ \text{mm}$$

The mass of the rim,

$$M = \frac{1429.9}{1.194^2} = 1003.6\ \text{kg}$$

The cross section of the rim,

$$A = \frac{M}{2\pi R\rho} = \frac{1003.6}{2 \times \pi \times 1.194 \times 7100} = 0.018841\ \text{m}^2 = 18841\ \text{mm}^2$$

For $B/t = 1.5$, the thickness is given by, $t = \sqrt{A/1.5} = \sqrt{18841/1.5} = 112.1\ \text{mm} \cong 115\ \text{mm}$

$$B = 1.5 \times 115 = 168 \cong 170\ \text{mm}$$

Example 24.3: Design the flywheel for a punching press with the following specifications.

The punch-press has a capacity of 24 holes per minute of 20 mm diameter in 18 mm thick mild steel plates having a shear strength of 350 MPa. The punching operation lasts for 1/10th of a revolution of the press crankshaft through a reduction gearing of 1:9. Assume that,

(i) The mechanical efficiency of the press is 80%.

(ii) During the punching operation, the speed of the flywheel is reduced by a maximum of 10%.

(iii) The space restrictions limit the diameter of the flywheel to 900 mm.

Given Data:

Punch capacity = 24 holes per minute

Diameter, d = 20 mm; Plate thickness = 18 mm; Plate strength = 350 MPa

The punching duration = 1/10th of a revolution

Coefficient of speed fluctuation = 0.1

Gear ratio of the drive = 1:9

Mechanical efficiency of the drive = 80%

The flywheel diameter D = 900 mm

SOLUTION

The punch-shaft speed = Number of punching operations per minute = 24 rpm

The speed of the driving machine, $N = 24 \times 9 = 216$ rpm

The angular speed, $\omega = \dfrac{2\pi N}{60} = \dfrac{2\pi \times 216}{60} = 22.62$ rad/s

The energy required to punch one hole, $W = (\pi d t S_u)t = \pi \times 20 \times 18 \times 350 \times 18$

$= 7125132.138$ Nmm $= 7125.132$ Nm

The energy required per minute, $W1 = 24.\ W = 24 \times 7125.132 = 171003.1713$ Nm

The power capacity of the driving machine, $P = \dfrac{171003.1}{0.8 \times 60} = 3562.56$ Watts $= 3.563$ kW

The average torque on punch shaft, $T_a = W/(2\pi) = 7125.132/(2 \times \pi) = 134$ Nm

The maximum torque, $T_{max} = \dfrac{W}{(2\pi/10)} = \dfrac{10 \times 7125.132}{2 \times \pi} = 11340$ Nm

The excess energy required during punching operation,

$$\Delta E = (T_{max} - T_a)\left(\frac{2\pi}{10}\right) = (11340 - 1134)\left(\frac{2\pi}{10}\right) = 6412.62$$

The excess energy at flywheel on driving shaft,

$$\Delta E_1 = \frac{\Delta E}{0.8} = \frac{6412.62}{0.8} = 8015.8 \text{ Nm}$$

The moment of inertia of the flywheel,

$$I = \frac{\Delta E_1}{C_s\omega^2} = \frac{8015.8}{0.1\times 22.62^2} = 156.7 \text{ kgm}^2$$

The mass of the flywheel rim,

$$M = 0.9\ I/R^2 = 0.9 \times 156.7/0.45^2 = 696.4 \text{ kg}$$

The cross-sectional area of the rim,

$$A = M/(2\pi R\rho) = 696.4/(2\times\pi\times 0.45\times 7200) = 0.0154 \text{ m}^2 = 15400 \text{ mm}^2$$

Since $B/t = 2$, the thickness of the rim, $t = \sqrt{(15400/2)} = 87.75$

Choose thickness, $t = 90$ mm, then width $B = 2 \times 90 = 180$ mm

The velocity, $V = \omega \times R = 22.62 \times 0.45 = 10.18$ m/s

The centrifugal stress in the arm,

$$\sigma_1 = 1000\ \rho_m V^2 = 1000\times 7200\times 10^{-9}\times 10.18^2 = 0.746 \text{ MPa}$$

This is very small stress, choose a 10 mm × 20 mm elliptical section.
The maximum torque on the driving shaft,

$$T_{m1} = \frac{T_{max}}{9} = 11340/9 = 1260 \text{ Nm}$$

The design stress for the shaft $S_d = 55$ MPa is chosen.

The shaft diameter, $d = \left[\frac{16T}{\pi S_d}\right]^{1/3} = \left[\frac{16\times 1260000}{\pi\times 55}\right]^{1/3} = 48.8 \text{ mm}$

Choose $d = 50$ mm
The key dimensions for rectangular key, $b = d/4 = 50/4 = 12.5$ mm; $t = d/6 = 50/6 = 8.3$ mm
Choose the standard dimensions from table as 14 mm × 9 mm
Check for shear stress,

$$\tau_1 = \frac{2T}{bdL} = \frac{2\times 1260000}{14\times 50\times 90} = 40.0 \text{ MPa} \quad \therefore \text{ Safe}$$

Check for crushing stress,

$$\sigma_c = \frac{4T}{hdL} = \frac{4\times 1260000}{9\times 50\times 90} = 124.5 \text{ MPa} \quad \therefore \text{ Safe}$$

Example 24.4: An electric motor is employed to drive a rolling mill. The power is being transmitted by cotton ropes. The motor is rated at 190 kW with an overload capacity of 40% The mill is liable to require excess power upto 300 kW for a maximum duration of 10 seconds.

The mill shaft rope pulley must supply the excess power with a drop of speed not exceeding 10% of the mean speed.The normal speeds of the motor and mill shaft are 400 rpm and 90 rpm respectively. The ropes of 40 mm nominal diameter with a friction coefficient between rope and pulley of 0.2. Angle of pulley groove is 45°.

Design the flywheel, which is to be in two halves, showing details of the joints at the rim and boss. Assume pulley diameter = 30 × rope diameter.

Given Data:

The driving motor power = 190 kW with overload of 40%

The speed of motor, $N_1 = 400$ rpm

The speed of the mill shaft, $N_2 = 90$ rpm

The excess power during rolling = 300 kW for a duration of 10 seconds

The rope diameter, $d = 40$ mm

The coefficient of friction, $f = 0.2$

The pulley groove angle, $2\beta = 45°$.

The smaller pulley diameter, $D_1 = 30 \times d = 30 \times 40 = 1200$ mm = 1.2 m

SOLUTION

The speed ratio = 400/90 = 4.44

The diameter of the larger pulley, $D_2 = 1200 \times 4.44 = 5333.33$ mm = 5.333 m

Assume a center distance, $C = 10{,}000$ mm = 10 m

The angle of lap on smaller pulley,

$$\theta = \pi - 2\sin\left(\frac{D_2 - D_1}{2C}\right) = \pi - 2\sin\left(\frac{5.333 - 1.2}{2 \times 10}\right) = 3.134 \text{ rad}$$

Virtual coefficient of friction = $f/\sin\beta = 0.2/\sin 22.5° = 0.5226$

The rope tension ratio,

$$\frac{T_1}{T_2} = e^{f_1\theta} = e^{0.5226 \times 3.134} = 5.1444$$

The maximum tension in the rope $T_1 = 1750$ N

The minimum tension, $T_2 = 1750/5.1444 = 340.3$ N

The speed of the smaller pulley = 400 rpm

The angular speed, $\omega = 2 \times \pi \times 400/60 = 41.89$ rad/s

The belt speed, $V = \omega \times R = 41.89 \times 1.2/2 = 25.133$ m/s

The power transmitted per rope, kW1 = $\dfrac{(T_1 - T_2) \times V}{1000}$ = (1750 – 340.3) × 25.133 /1000

= 35.44 kW

Number of ropes required = 300/35.44 = 8.5, Choose $n = 9$
The excess energy during rolling,

ΔE = (maximum power required – power supplied by the motor) × duration

$$= (300 - 190 \times 1.4) \times 10^3 \times 10 = 34 \times 10^4 \text{ Nm}$$

The speed of mill shaft = 90 rpm
The angular speed, $\omega = 2\pi N/60 = 2 \times \pi \times 90/60 = 9.425$ rad/s
The moment inertia required,

$$I = \frac{\Delta E_1}{C_s \omega^2} = \text{I} = \frac{34 \times 10^4}{0.1 \times 9.425^2} = 38275.1 \text{ kg m}^2$$

The mass of the rim (assuming 90% is provided by the rim)

$$M = \frac{0.9 \cdot I}{R^2} = \frac{0.9 \times 38275}{2.667^2} = 4842.97 \text{ kg}$$

The cross section of the rim,

$A = M/(2\pi R\rho) = 4842.97/(2 \times \pi \times 2.667 \times 7200) = 0.04014 \text{ m}^2 = 40140 \text{ mm}^2$

Width assume thickness = 2, the thickness $t = \sqrt{A/2} = \sqrt{(40140/2)} = 142$ mm

The width of the pulley $B = 2 \times t = 2 \times 142 = 184$ mm

The rim speed, $V = \omega \times R = 9.425 \times 2.667 = 25.136$ m/s

The centrifugal stress in the rim,

$$\sigma_c = \rho V^2 = 7200 \times 25.136^2 = 4549265.1 \text{ Pa} = 4.55 \text{ MPa}$$

The centrifugal force, $F = A\,\sigma_c = 40140 \times 4.55 = 182607.5$ N

Using 4 bolts with design stress = 45 MPa, the cross-sectional area,

$$\text{As} = 182607.5/(45 \times 4) = 1014.5 \text{ mm}^2$$

The bolts M 39 × 3 mm width As = 1030 mm^2 are suitable (Table 9.7 MDHB)

Example 24.5: The following data refers to a rim flywheel made of gray cast-iron:
Mean radius of the rim, $R = 1.0$ m, Thickness of the rim, $t = 100$ mm, Width of the rim, $b = 200$ mm

Number of spokes, $n = 4$; Cross-sectional area of spoke $A1 = 6500$ mm^2

Speed of rotation, $N = 720$ rpm; Mass density of the material = 7200 kg/m^2

Calculate (i) Maximum tensile stress in the rim; (ii) Axial stress in each spoke.

Given Data:

Flywheel rim, radius = 1.0 m; Thickness, $t = 100$ mm; Width, $b = 200$ mm

Number of spokes = 4; C S area = 6500 mm^2; Material density, $\rho = 7200$ kg/m^2

Speed of rotation, $N = 720$ rpm

SOLUTION

The rim velocity, $V = \dfrac{\pi DN}{60} = \dfrac{\pi \times 2 \times 720}{60} = 75.4$ m/s

(A) Prof. Lanza's Approach

The tensile stress in rim, if unrestrained,

$$\sigma_c = \rho V^2 = 7200 \times 75.4^2 = 40933152 \text{ Pa} = 40.93 \text{ MPa}$$

The bending stress, when restrained by spokes,

$$\sigma_b = \frac{2000\rho_m R\pi^2 V^2}{t \cdot n^2} = \frac{2000 \times 7200 \times 10^{-9} \times \pi^2 \times 1000 \times 75.4^2}{100 \times 4^2} = 505 \text{ MPa}$$

The resultant stress as per Prof. Lanza' approach,

$$\sigma_t = 0.75\ \sigma_c + 0.25\ \sigma_b = 0.75 \times 40.93 + 0.25 \times 505 = 157 \text{ MPa}$$

(B) Timoshenko's Analysis

The stress in the rim is given by, (for $\alpha = \varphi = \pi/4$)

$$\sigma_t = (1000\rho_m V^2)\left\{1 - \frac{\cos\alpha}{3C\sin\alpha} \pm \left(\frac{2000R}{C \cdot t}\right)\left[\frac{1}{\alpha} - \frac{\cos\varphi}{\sin\alpha}\right]\right\}$$

$$= (1000 \times 7200 \times 10^{-9} \times 75.4^2)\left\{1 - \frac{\cos(\pi/4)}{3 \times 11.06 \times \sin(\pi/4)} \pm \left(\frac{2000 \times 1}{11.06 \times 100}\right)\left[\frac{1}{(\pi/4)} - \frac{\cos(\pi/4)}{\sin(\pi/4)}\right]\right\}$$

$$= 40.933 \times (1.0 \pm 0.464) = 59.9 \text{ MPa}$$

The numerical value of C can be calculated by substituting for α, which depends upon the number of arms.

For 4-arms, $2\alpha = \dfrac{\pi}{2}$; $f_1(\alpha) = 0.643$, $f_2(\alpha) = 0.0068$

$$C = (72960)\left(\frac{R^2}{t^2}\right) + 0.643 + \frac{A}{A_1}$$

$$C = (72960)\left(\frac{1^2}{100^2}\right) + 0.643 + \frac{100 \times 200}{6500} = 7.297 + 0.643 + 3.077 = 11.016$$

Mass per unit length, $m = B \times t \times \rho_m = 200 \times 100 \times 7200 \times 10^{-9}$ kg/mm

The stress in the spokes (arms) is given by

$$-\sigma_1 = \left(\frac{2}{3}\right)\left[\frac{1000\,mV^2}{CA_1}\right] = \left(\frac{2}{3}\right)\left[\frac{1000 \times 100 \times 200 \times 7200 \times 10^{-9} \times 75.4^2}{11.06 \times 6500}\right] = 7.59 \text{ MPa}$$

EXERCISES FOR PRACTICE

24.1 A machine shaft running at a mean speed of 200 rpm, requires a torque, which varies uniformly from 1000 Nm to 3500 Nm during the first half revolution, remains constant for the following revolution, decreases uniformly to 1000 Nm during the next revolution and remains constant for the next two revolutions, the cycle being repeated. It is directly coupled to a motor, which exerts a constant torque and has a rotor of mass 40 kg with radius of gyration of 230 mm. It is desired to limit the fluctuation of speed by addition of a flywheel to +2% of the mean speed. Assuming the flywheel rim to contribute 90% of the inertia, design a suitable rim hub and arms. The rim speed should not exceed 1250 m/min. Suggest suitable size of the shaft.

24.2 A machine is driven by motor width constant torque output. The resisting torque of the machine increases uniformly from 500 Nm at the beginning of a revolution to 2000 Nm at the end of the revolution. The mean angular velocity of the machine is 40 rad/s. The coefficient of speed fluctuation is 0.2. A solid circular steel disk is used as a flywheel. The mass density of the steel is 7800 kg/m^3 and the Poisson's ratio is 0.3. Calculate the outer radius of the disk and the maximum stress induced in the disk.

24.3 The flywheel of a rolling mill engine weighs 300 kN, and has a mean radius of 2.5 m and an average speed of 60 rpm. It is secured to a shaft 30 cm in diameter. During the rolling operation, which lasts for 2 seconds, the speed drops from 60 rpm to 45 rpm. The flywheel has 10 arms of elliptical cross section, the major axis being twice the minor axis. Suggest a suitable cross section for the arm near the hub for a permissible stress of 12 MPa.

24.4 A flywheel is to be designed for a four-stroke engine developing 75 kW at 300 rpm. The total fluctuation of speed is to be limited to 5% of the mean speed. The work done during the expansion stroke is 1.5 times the average work done in the full cycle. Assuming the stress in the rim is not to exceed 5.0 MPa, allowable shear stress for the shaft to be 40 MPa. The tensile stress for the cast iron is not to exceed 20 MPa and the mass density of cast iron as 7200 kg/m^3, determine cross sections of rim and arms, and the diameters of shaft and hub. The width to thickness ratio for rim is 2. There are six elliptical arms with major axis twice the minor axis.

24.5 A flywheel is required for a punching machine capable of punching 20 mm diameter holes through a steel plate of 15 mm thickness. The punching rate is 25 holes per minute. The hole is punched during 1/4th of a revolution of the crankshaft. The ultimate shear strength of the plate is 350 MPa. Calculate the dimensions of the flywheel and the size of the motor, running at a speed of 1450 rpm.

24.6 The torque developed by a three-crank engine is given by the expression: $Te = 1900 + 7000 \sin (3\theta)$ Nm. The resisting torque is $Tr = 1900 + 3000 \sin(\alpha)$ Nm, where θ is the crank angle. The engine is running at a mean speed of 300 rpm and the coefficient of speed fluctuation is limited to 0.03.. Assume that 90% of the moment of inertia is provided by the rim. The material of the flywheel is gray cast-iron FG 200 with a mass density of 7100 kg/ m^3. Determine the dimensions of the flywheel.

24.7 Design a flywheel for a punch press, which must be brought to rest by one punching operation, if the power has been shut off. The maximum work required consists in punching 30 mm hole in 20 mm thick mild steel plate with ultimate shear strength of 350 MPa. The punch capacity is 24 holes per minute and the speed ratio of the driving shaft to the eccentric shaft operating the punch is 9:1. In order to clear the floor, the wheel diameter is limited to 1.05 m. The wheel is keyed to the shaft and mechanical efficiency of the press and driving gear is 72%. Determine (a) the weight of the flywheel rim if the density of cast iron is 7000 kg/cub m, (b) the maximum stress in the rim and (c) the coefficient of steadiness of the driving shaft if the duration of the working stroke is one-half that of the idle stroke.

24.8 A machine shaft running at a mean speed of 200 rpm, requires torque which increases uniformly from 1000 Nm to 3500 Nm during the first half of the revolution, remains constant for the following revolution, decreases uniformly to 1000 Nm during the next revolution and then remains constant again for next two revolutions, the cycle is repeated. It is directly coupled to a motor which exerts a constant torque and has a rotor of mass 40 kg and a radius of gyration 230 mm. It is desired to limit the fluctuation of speed to $\pm 2\%$ of the mean speed. Assuming the flywheel rim to contribute 90% of the inertia, design suitable rim, arms and shaft. The rim speed is not to exceed 1250 m/min.

24.9 It is to design a rimmed flywheel for an engine consisting of three single acting cylinders with their cranks set equally at 120° to each other. The torque crank angle diagram for each cylinder consists of a triangle with following values:

Crank angle (deg)	0	60	180	180 to 360
Torque (Nm)	0	300	0	0

The engine runs at a mean speed of 240 rpm and the coefficient is limited to 0.03. The resisting torque of the machine is constant throughout the cycle. From the consideration of space, the radius of the flywheel rim should not exceed 0.25 m. The material for the flywheel is grey cast-iron FG220 with a mass density of 7150 kg/m^3.

24.10 The following data are given for a rimmed flywheel made of gray cast-iron FG200:

The mean radius of the rim = 1.5 m

Thickness of rim = 200 mm

Width of rim = 300 mm

Number of spokes = 6

Cross-sectional area of spokes = 12000 mm^2

Speed of rotation = 720 rpm

The mass density of the material FG200 = 7100 kg/m^3

Calculate the maximum stress in the rim and spoke.

SHORT ANSWER QUESTIONS

24.1 What is a flywheel?

24.2 List the functions of a flywheel.

24.3 Which of the materials are used in construction of a flywheel?

24.4 Name different types of flywheels based on their construction.

24.5 Why is torque negative during suction stroke in torque vs time diagram?

24.6 Why is torque positive during expansion stroke in torque vs time diagram?

24.7 Why is torque negative during exhaust stroke in torque vs time diagram?

24.8 Define the coefficient of speed fluctuation.

24.9 What is the coefficient of energy fluctuation?

24.10 What are different types of stresses induced in a disk flywheel?

24.11 What is a rimmed flywheel?

24.12 What are different types of stresses induced in a rimmed flywheel?

CHAPTER 25

PROBABILISTIC APPROACH TO DESIGN

In the determinist design the approximate average values of material properties and the approximate average values of the stresses are used. The factor of safety is introduced to safeguard against failure. This provides a margin of safety to account for the uncertainties and unknowns. A large factor of safety ensures a robust design but not a secure design. The failure is likely to occur. Moreover the design is uneconomical. This kind of design is not suitable for aerospace and nuclear fields.

In the probabilistic design the statistically variable nature of the material properties and the stresses evaluated are considered. The safeguard against the possible failure is provided by constraining the probability of failure or by the choice of suitable reliability factor. This approach results in safe and economical design.

25.1 STATISTICAL FUNDAMENTALS-FREQUENCY DISTRIBUTIONS

The statistics deals with the collection of data about a characteristic of interest, processing the data and analysis of the data. The characteristic may be yield strength of a material, the diameter of a bolt manufactured in a shift. The number of observations made about the characteristic constitute population. Sometimes it is not possible to observe all elements, like all bars produced or all bolts manufactured in shift, if the number is very large. In such cases the data is collected for a few representative products called sample. The sample is a part of the population. The data so collected is called raw data.

The raw data is arranged in a systematic manner. The whole range from minimum to maximum is split into smaller intervals and number of occurrences in each interval are recorded and plotted in the form of graph with the class interval as base along x-axis and a rectangle on the interval representing the number of occurrences in the interval. The set of rectangles is called histogram. The lines joining the midpoints of the top end of the rectangles form a polygon called frequency polygon. When the interval size is reduced, the polygon becomes a curve called frequency distribution curve. The shape of the curve may be symmetrical about a central mean or it may be skewed to the right or to the left.

The symmetric curve with central mean is called normal distribution or Gaussian distribution. Most of the material properties, loads and stresses follow the normal distribution. The shape of the distribution curve is determined by the central tendency, represented by the mean or median or mode and dispersion represented by mean deviation or standard deviation.

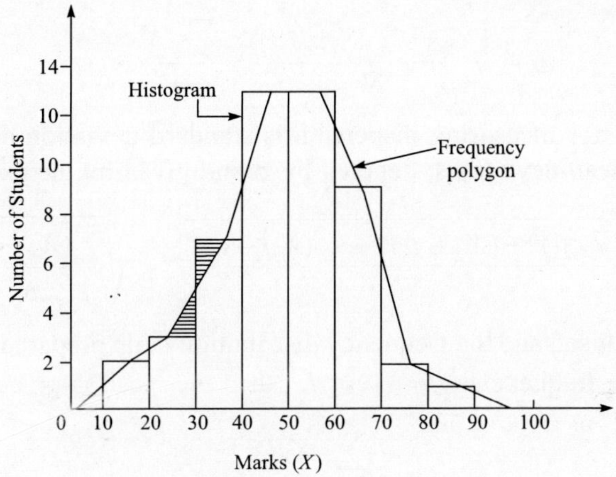

Figure 25.1 Histogram

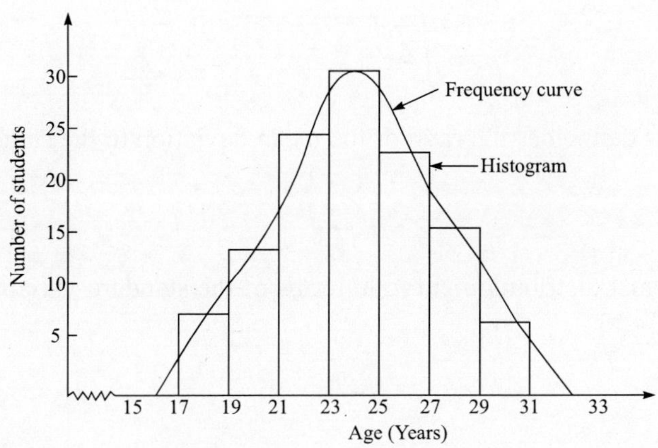

Figure 25.2 Frequency curve

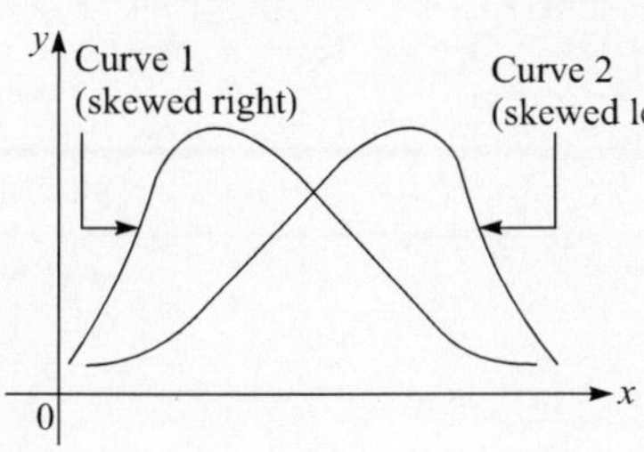

Figure 25.3 Skewed distributions

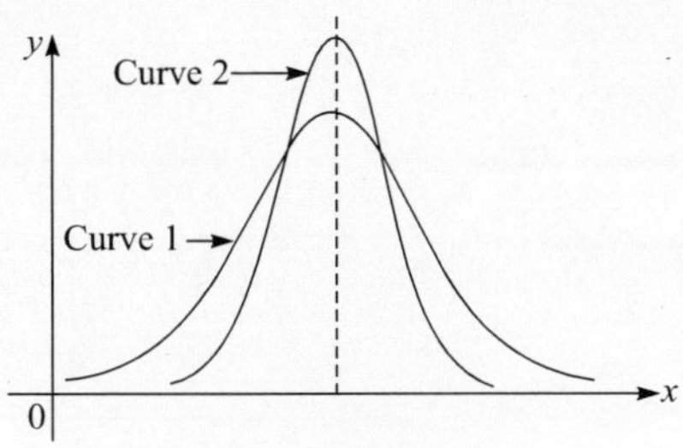

Figure 25.4 Central distributions

The most popular unit for central tendency is the arithmetic mean denoted by μ and defined as the average of N number of observations X_i as follows:

$$\mu=\frac{[X_1+X_2+\dots X_N]}{N}=\frac{\sum_1^N X_i}{N}$$

The most popular unit for measuring dispersion is standard deviation defined as the root of mean of squares of the mean deviations, denoted by σ and given by the expression:

$$\sigma=\sqrt{\frac{[(X_1-\mu)^2+(X_2-\mu)^2+\dots(X_N-\mu)^2]}{N}}=\sqrt{\frac{\sum(X_i-\mu)^2}{N}}$$

If the population is analysed and the frequency distribution table prepared, let the observations $X_1, X_2, \dots X_n$ occur with frequencies $f_1, f_2, \dots f_n$, then the mean is given by the equation,

$$\mu=\frac{\sum_1^N X_i f_i}{N}$$

The standard deviation is given by the equation,

$$\sigma=\sqrt{\frac{\sum f_i(X_i)^2}{N}-\mu}$$

A standard variable is defined as the ratio of the mean deviation to the standard deviation and is given by equation:

$$Z=\frac{X-\mu}{\sigma}$$

The equation for normal distribution curve in terms of the standard variable is given as

$$f(z)=\frac{1}{\sqrt{2\pi}}e^{-Z^2/2}$$

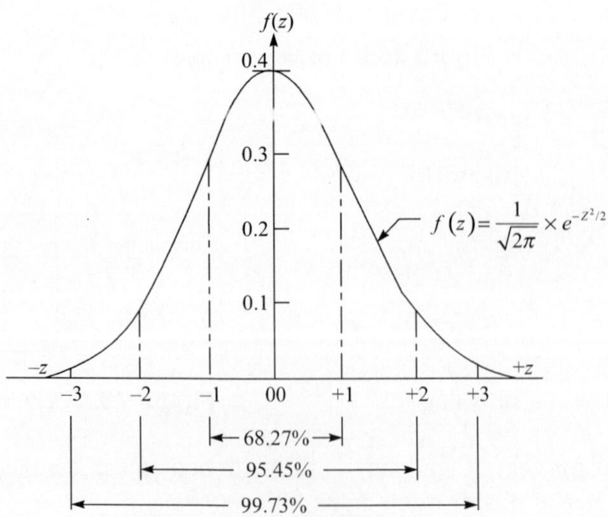

Figure 25.5 Normal distribution

25.2 THE COMBINATIONS OF THE POPULATIONS

In statistics, the populations can be combined. They can be added or subtracted to get a third population. For example, the population representing the data of the inside diameters of the bearings can be subtracted from the population representing the data of the diameters of the shafts and a third population representing the data of the possible clearances can be obtained. Then the characteristics like mean and standard deviation, are given by the following expressions,

Mean, $\mu_c = \mu_b - \mu_s$

Standard deviation, $\sigma_c = \sqrt{(\sigma_b)^2 + (\sigma_s)^2}$

When two populations *X, Y* are added to get a third population *Z*, the mean and standard deviation of the third population are as follows:

Mean $\mu_Z = \mu_X + \mu_Y$

Standard deviation $\sigma_Z^c = \sqrt{(\sigma_X)^2 + (\sigma_Y)^2}$

When they are multiplied, the characteristics of the resulting population are

Mean $\mu_Z = \mu_X \cdot \mu_Y$

Standard deviation, $\sigma_Z = \sqrt{(\mu_Y)^2\,(\sigma_X)^2 + (\sigma_X)^2 \cdot (\sigma_Y)^2 + (\mu_X)^2\,(\sigma_Y)^2}$

When they are divided, the characteristics of the resulting population are

Mean $\mu_Z = \mu_X / \mu_Y$

$$\text{Standard deviation, } \sigma_Z = \frac{1}{\mu_Y}\left[\frac{(\mu_Y)^2 \cdot (\sigma_X)^2 + (\mu_X)^2 \cdot (\sigma_Y)^2}{(\mu_Y)^2 + (\sigma_Y)^2}\right]^{1/2}$$

When a population is squared, the characteristics of the resulting population are

Mean $\mu_Z = (\mu_X)^2 + (\sigma_X)^2$

$$\text{Standard deviation, } \sigma_Z = \frac{1}{2}\left(\frac{\sigma_X}{\mu_X}\right)[4(\mu_X)^2 + (\sigma_X)^2]$$

When a population is multiplied thrice, the characteristics of resulting population are

Mean $\mu_Z = (\mu_X)^3 + 3 \cdot \mu_X (\sigma_X)^2$

Standard deviation, $\sigma_Z = 3 \cdot (\mu_X)^2 \cdot \sigma_X + 3 \cdot (\sigma_X)^3$

When the population is inversed, the characteristics of the resulting population are

$$\text{Mean } \mu_Z = \frac{!}{\mu_X} \cdot \left[1 + \left(\frac{\sigma_X}{\mu_X}\right)^2\right]$$

$$\text{Standard deviation, } \sigma_Z = \frac{\sigma_X}{(\mu_X)^2}\left[1+\frac{(\sigma_X)^2}{(\mu_X)^2}\right]$$

25.3 PROBABILISTIC APPROACH TO DESIGN

Traditional deterministic design, the mean values of strength (yield or ultimate), and stresses induced are used for design of mechanical components. The safety factor is used to account for uncertainties in the various factors influencing product safety. This approach may lead to robust but uneconomical, unreliable design. In the modern design, specially in space, aerospace and nuclear fields the reliability is an important factor. In the probability approach, the stochastic nature of various factors like material properties, the manufacturing methods, the loads applied, are considered. The probability of failure is controlled by introduction of reliability factor. Probabilistic approach is a technique to design a product with specified reliability.

According to International Standards Organization, reliability is defined as the ability of a product to perform a required function under stated conditions for a stated period of time. The definition specifies four factors, namely:

1. The product is able to perform its intended function (strong enough to sustain the loads)
2. The operating conditions are specified (resist the environmental effects)
3. The service life of the product is specified (resist the time defendant failures)
4. The reliability of the product is a probability function (stochastic nature of reliability due the stochastic nature of the above factors).

25.4 RELIABILITY IN STRENGTH DESIGN

In the probabilistic approach, the strength properties (yield or ultimate) are assumed to follow normal distribution shown by the curve (1) in Fig. 25.6, as represented by S_y. The stresses induced also conform to the normal distribution shown by curve (2) and are represented by σ_y.

The overlapping area represents the region of unreliability. Some failures are likely to occur in such situations. There is a probability that the stress induced being greater than strength and hence, the component fails.

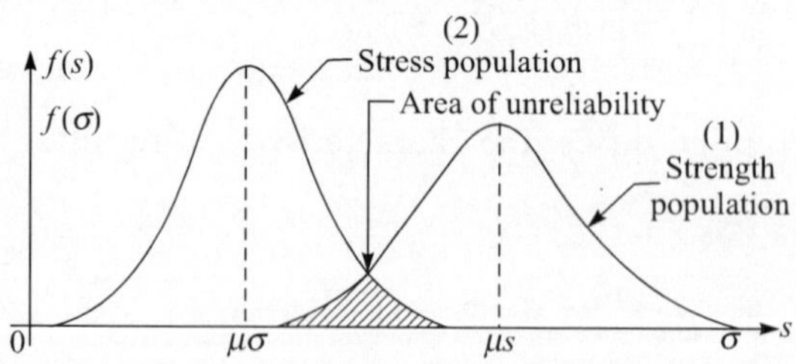

Figure 25.6 Normal distribution of strength and stress populations

To study the situation a third population called margin of safety is defined by deducting the stress population from the strength population as given below,

$$\text{Margin of safety, } m = S_y - \sigma_y$$

With the mean $\mu_m = \mu_s - \mu_\sigma$

And standard deviation, $\sigma_m = \sqrt{(\sigma_\sigma)^2 + (\sigma_s)^2}$

We define a standard variable Z as follows,

$$Z = \frac{m - \mu_m}{\sigma_m}$$

The normal distribution of the margin of safety as a function of standard variable is shown in Fig. 25.7.

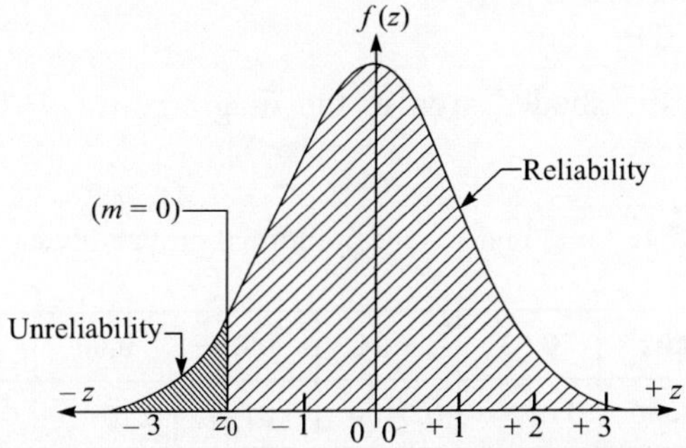

Figure 25.7 Normal curve for population of margin of safety

The condition for failure is $m = 0$, and the value of Z for failure,

$$Z_0 = \frac{-\mu_m}{\sigma_m} = \frac{\mu_s - \mu_\sigma}{\sqrt{(\sigma_\sigma)^2 + (\sigma_s)^2}}$$

The area under the curve from -00 to Z_0 represents the probability of failure P:

$$= P(\sigma > S_y) = \int_{-\infty}^{z_0} f(Z)dZ$$

And the reliability, $R = 1 - P = \int_{Z_0}^{\infty} f(Z)dZ$

This is the area under the curve from Z_0 to ∞.

The region of unreliability can be reduced by decreasing Z_0. That is to increase mean μ_m, and decrease standard deviation σ_m. The first part can be achieved by improving material properties and/or decreasing the stresses by increasing the sizes of component. The second part can be achieved by reducing the standard deviations of both strength population and stress population. This can be achieved by controlling incoming material and the manufacturing methods.

Normal probability curve is given by:

$$f(x) = \frac{1}{\sigma\sqrt{2x}} \exp\left\{-\frac{1}{2}\left(\frac{x-\mu}{\sigma}\right)^2\right\}; -\infty < x < \infty$$

and standard normal probability curve is given by:

$$f(z) = \frac{1}{\sqrt{2\pi}} \exp\left(-\frac{1}{2}z^2\right), -\infty < z < \infty$$

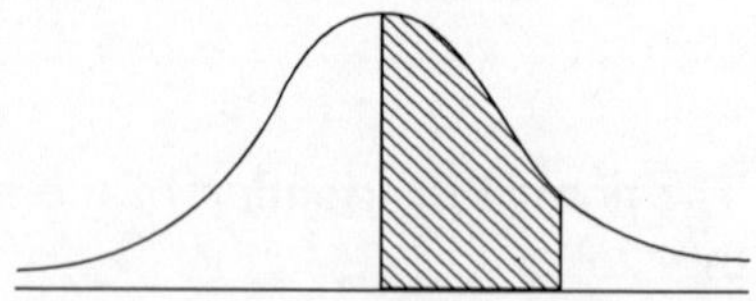

where, $$Z = \frac{X - E(X)}{\sigma_x} = \frac{X-\mu}{\sigma} \sim N(0, 1)$$

The following table gives the shaded area in the diagram, *viz.* $P(0 < Z < z)$ for different values of z.

Table 25.1. Areas under standard normal probability curve

z \ z	0.00	0.01	0.02	0.03	0.04	0.05	0.06	0.07	0.08	0.09
0.0	0.0000	0.0040	0.0080	0.0120	0.0160	0.0199	0.239	0.0279	0.0319	0.0359
0.1	0.0398	0.0438	0.0478	0.0517	0.0557	0.0596	0.0636	0.0675	0.0714	0.0753
0.2	0.0793	0.0832	0.0871	0.0910	0.0948	0.0987	0.1026	0.1064	0.1103	0.1141
0.3	0.1179	0.1217	0.1255	0.1293	0.1331	0.1368	0.1406	0.1443	0.1480	0.1517
0.4	0.1554	0.1591	0.1628	0.1664	0.1700	0.1736	0.1772	0.1808	0.1844	0.1879
0.5	0.1915	0.1950	0.1985	0.2019	0.2054	0.2088	0.2123	0.2157	0.2190	0.2224
0.6	0.2257	0.2291	0.2324	0.2357	0.2389	0.2422	0.2454	0.2486	0.2517	0.2549
0.7	0.2580	0.2611	0.2642	0.2673	0.2704	0.2734	0.2764	0.2794	0.2823	0.2852
0.8	0.2881	0.2910	0.2939	0.2967	0.2995	0.3023	0.3051	0.3078	0.3106	0.3133
0.9	0.3159	0.3186	0.3212	0.3238	0.3264	0.3289	0.3315	0.3340	0.3365	0.3389
1.0	0.3413	0.3438	0.3461	0.3485	0.3508	0.3531	0.3554	0.3577	0.3599	0.3621
1.1	0.3643	0.3665	0.3686	0.3708	0.3729	0.3749	0.3770	0.3790	0.3810	0.3830
1.2	0.3849	0.3869	0.3888	0.3907	0.3925	0.3944	0.3962	0.3980	0.3997	0.4015
1.3	0.4032	0.4049	0.4066	0.4082	0.4099	0.4115	0.4131	0.4147	0.4162	0.4177
1.4	0.4192	0.4207	0.4222	0.4236	0.4251	0.4265	0.4279	0.4292	0.4306	0.4319
1.5	0.4332	0.4345	0.4357	0.4370	0.4382	0.4394	0.4406	0.4418	0.4429	0.4441

Contd.

z \ z	0.00	0.01	0.02	0.03	0.04	0.05	0.06	0.07	0.08	0.09
1.6	0.4452	0.4463	0.4474	0.4484	0.4495	0.4505	0.4515	0.4525	0.4535	0.4545
1.7	0.4554	0.4564	0.4573	0.4582	0.4591	0.4599	0.4608	0.4616	0.4625	0.4633
1.8	0.4641	0.4649	0.4656	0.4664	0.4671	0.4678	0.4686	0.4693	0.4625	0.4633
1.9	0.4713	0.4719	0.4726	0.4732	0.4738	0.4744	0.4750	0.4756	0.4761	0.4767
2.0	0.4772	0.4778	0.4783	0.4788	0.4793	0.4798	0.4803	0.4808	0.4812	0.4817
2.1	0.4821	0.4826	0.4830	0.4834	0.4838	0.4842	0.4846	0.4850	0.4854	0.4857
2.2	0.4861	0.4864	0.4868	0.4871	0.4875	0.4878	0.4881	0.4884	0.4887	0.4890
2.3	0.4893	0.4896	0.4898	0.4901	0.4904	0.4906	0.4909	0.4911	0.4913	0.4916
2.4	0.4918	0.4920	0.4922	0.4925	0.4927	0.4929	0.4931	0.4932	0.4934	0.4936
2.5	0.4938	0.4940	0.4941	0.4943	0.4945	0.4946	0.4948	0.4949	0.4951	0.4952
2.6	0.4953	0.4955	0.4956	0.4957	0.4959	0.4960	0.4961	0.4962	0.4963	0.4964
2.7	0.4965	0.4966	0.4967	0.4968	0.4969	0.4970	0.4971	0.4972	0.4973	0.4974
2.8	0.4974	0.4975	0.4976	0.4977	0.4977	0.4978	0.4979	0.4979	0.4980	0.4981
2.9	0.4981	0.4982	0.4982	0.4983	0.4984	0.4984	0.4985	0.4985	0.4986	0.4986
3.0	0.4987	0.4987	0.4987	0.4988	0.4988	0.4989	0.4989	0.4989	0.4990	0.4990
3.1	0.4990	0.4991	0.4991	0.4991	0.4992	0.4992	0.4992	0.4992	0.4993	0.4993
3.2	0.4993	0.4993	0.4994	0.4994	0.4994	0.4994	0.4994	0.4992	0.4993	0.4995
3.3	0.4995	0.4995	0.4995	0.4996	0.4996	0.4996	0.4996	0.4995	0.4995	0.4997
3.4	0.4997	0.4997	0.4997	0.4997	0.4997	0.4997	0.4997	0.4997	0.4997	0.4998
3.5	0.4998	0.4998	0.4998	0.4998	0.4998	0.4998	0.4998	0.4998	0.4998	0.4998
3.6	0.4998	0.4998	0.4999	0.4999	0.4999	0.4999	0.4999	0.4999	0.4999	0.4999
3.7	0.4999	0.4999	0.4999	0.4999	0.4999	0.4999	0.4999	0.4999	0.4999	0.4999
3.8	0.4999	0.4999	0.4999	0.4999	0.4999	0.4999	0.4999	0.4999	0.4999	0.4999
3.9	0.5000	0.5000	0.5000	0.5000	0.5000	0.5000	0.5000	0.5000	0.5000	0.5000

SOLVED EXAMPLES

Example 25.1: A mechanical component is subject to an axial stress. The stress is found to be normally distributed with a mean value of 200 MPa and standard deviation of 25 MPa. The material of the component has yield strength, normally distributed with a mean value of 400 MPa and standard deviation of 15 MPa. Find the reliability in design.

SOLUTION *Given Data:*

The mean of the stress, $\mu_\sigma = 200$ MPa

The standard deviation, $\sigma_\sigma = 25$ MPa

The material yield strength, S

The mean of yield strength, $\mu_s = 400$ MPa

Standard deviation, $\sigma_s = 15$ MPa

The margin of safety population, $m = S - \sigma$ is also normally distributed

The mean, $\mu_m = \mu_S - \mu_\sigma = 400 - 200 = 200$ MPa

The standard deviation, $\sigma_m = \sqrt{(\sigma_S)^2 + (\sigma_\sigma)^2}$

The standard deviation, $\sigma_m = \sqrt{(15)^2 + (25)^2} = 29.155$ MPa

The standard variable for $m = 0$,

$$Z = \frac{m - \mu_m}{\sigma_m} = \frac{0 - 200}{29.155} = -3.43$$

The area under the normal curve from $Z = 0$ to $Z = 3.43$ (Table 25.1)

$$A = 0.4997$$

The area under the curve for $Z = -\infty$ to $Z - 3.43$ is,

$$A1 = 0.5 - 0.4997 = 0.0003$$

Then the probability of failure, $P = 0.0003 = 0.03\%$
The reliability of design, $R = 1 - P = 0.97\%$

25.5 RELIABILITY IN TIME DEPENDENT FAILURE

Let N number of items put under test be:
$N_F(t)$ be the number of items that have failed at time t
$N_S(t)$ be the number of items that have failed at time t
We state as,

$$N_S(t) + N_F(t) = N$$

On dividing by N,

$$\frac{N_S(t)}{N} + \frac{N_F(t)}{N} = 1$$

We define reliability and probability of failure at time t as follows,

$$\text{Reliability } R(t) = \frac{N_S(t)}{N}$$

Probability of failure, $F(t) = \dfrac{N_F(t)}{N}$

Then we get,

$$R(t) + F(t) = 1$$

The frequency distribution of time to failure is given by

$$f(t) = \frac{dF(t)}{dt} = \frac{d[1 - R(t)]}{dt} = -\frac{dR(t)}{dt}$$

The instantaneous failure rate is the number of units failing per unit time in the time interval t_1 and $t_1 + dt_1$ and is given by expression,

$$h(t) = P(t_1 \le t \le t_1 + dt_1)$$

$$= \frac{f(t)}{1 - F(t)} = \frac{f(t)}{R(t)} = -\frac{1}{R(t)} \frac{dR(t)}{dt}$$

Integrating, we get

$$R(t) = \text{Exp}\left[-\int_0^t h(t) \cdot dt\right]$$

For a given rate of failure, the reliability can be evaluated by this expression. The typical failure rate curve is shown in Fig. 25.8.

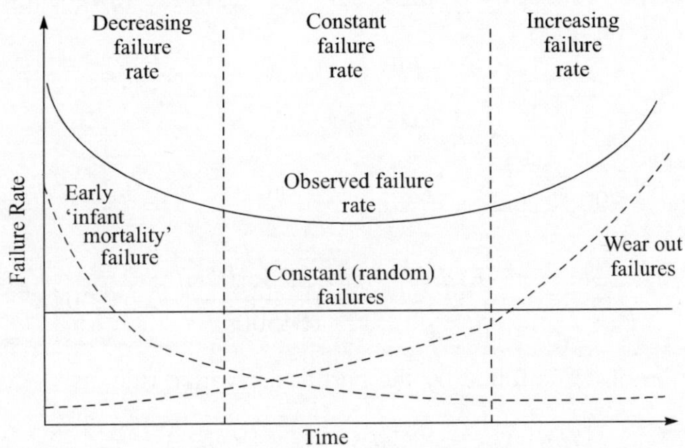

Figure 25.8 Time dependent failure rate

Example 25.2: A ball bearing has normally distributed time to failure data. The mean value is 16000 hr and the standard deviation is 1000 hr. A group of 100 bearings are fitted at a time. How many may be expected to survive for 17500 hr?

SOLUTION *Given Data:*

Time to failure data – normally distributed

The mean $\mu = 16000$ hrs
The standard deviation, $\sigma = 1000$ hrs
Then the standard variable,

$$Z = \frac{t-\mu}{\sigma} = \frac{17500-16000}{1000} = 1.5$$

Then from Table 25.1, the area under the normal curve upto $Z = 1.5$,

$$A = 0.5 + 0.4332 = 0.9332$$

The probability of failure, $P = 0.9332 = 93.32\%$

This means 94 bearings fail during the service of 17500 hr. The number of bearings still in service is only 6.

Example 25.3: A connecting rod of length *l*, and diameter *d*, is subjected to an axial compressive load *F*. The Euler buckling load of the connecting rod P_c, assuming hinged ends, is given by

$$P_c = \frac{\pi^2 \cdot E \cdot I}{l^2}$$

where *E* is Young's modulus, and *I* is the inertia of the cross section.

The parameters are given as $l = N(500, 12.5)$ mm, $d = N(d_m, 0.1\, d_m)$,

$F = N(9000, 900)$ kN, $E = N(2.1 \times 10^5, 2.1 \times 10^4)$ MPa. Design the connecting rod to achieve a reliability of $R = 0.99$ against buckling.

SOLUTION The specified reliability of 0.99, corresponds to the lower limit of the standard variable $Z_1 = -2.33$ (refer Table 25.1)

Then we can write, $$Z_1 = -2.33 = \frac{P_c - F}{\sqrt{\sigma_p^2 + \sigma_F^2}} \tag{1}$$

where, $F = 9000$ N, $\sigma_F = 900$ N, and

$$P_c = \frac{\pi^2 \cdot E \cdot I}{l^2} = \frac{\pi^3 \cdot E \cdot d^4}{64 l^2} = \frac{\pi^3 \times 2.1 \times 10^5 \cdot d^4}{64500^2} = 0.407\, d^4 \tag{2}$$

The standard deviation of P_c is found by the partial derivative rule, as

$$\sigma_{P_c}^2 = \left[\left(\frac{\partial P_c}{\partial E}\right)^2 \cdot \sigma_E^2 + \left(\frac{\partial P_c}{\partial d}\right)^2 \cdot {\sigma_d}^2 + \left(\frac{\partial P_c}{\partial l}\right)^2 \cdot {\sigma_l}^2\right] \tag{3}$$

where, $\left(\frac{\partial P_c}{\partial E}\right) = \frac{P_c}{E}$, $\left(\frac{\partial P_c}{\partial d}\right) = 4\,\frac{P_c}{d}$. $\left(\frac{\partial P_c}{\partial l}\right) = -2\frac{P_c}{l}$

And by defining the coefficient of variation as,
γ = (standard deviation/mean value)

And substituting as follows

$$\gamma_{P_c} = \frac{\sigma_{P_c}}{P_c},\ \gamma_E = \frac{\sigma_E}{E},\ \gamma_d = \frac{\sigma_d}{d},\ \gamma_l = \frac{\sigma_l}{l}$$

Equation (3) becomes

$$\gamma_{P_c}^2 = \gamma_E^2 + 16\cdot\gamma_d^2 + 4\cdot\gamma_l^2 = (0.1)^2 + 16\cdot(0.1)^2 + 4\cdot(0.1)^2 = 21\cdot(0.1)^2$$

$$\therefore\ \gamma_{P_c} = 0.4582$$

Then $\sigma_{P_c} = \gamma_{P_c}\cdot P_c = 0.4582 \times 0.407.\,d^4$ (4)

Substituting the values in equation (1), we get

$$-2.33 = \frac{P_c - F}{\sqrt{\sigma_p^2 + \sigma_F^2}} = \frac{0.407d^4 - 9000}{\sqrt{(0.1865d^4)^2 + (900)^2}}$$

Putting $y = d^4$ and solving we get

$$Y^2 + 326328.50\cdot Y - 3408246928 = 00 \tag{5}$$

$$\therefore\quad Y = \frac{326328.50}{2} \mp \sqrt{\left(\frac{326328.50}{2}\right)^2 + (3408246928)^2}$$

$$= -163164.25 \mp 173233.79 = 10129.54$$

Then $d = \sqrt[4]{Y} = \sqrt[4]{10129.54} = 10.03$ mm

The diameter of the rod is 10 mm.

EXERCISES FOR PRACTICE

25.1 A machine component is subjected to a normally distributed axial load of $F = 250$ kN, with a standard deviation of 25 kN. The cross-sectional area is 1000 mm^2. The yield strength of the material of the component is also normally distributed with a mean value of 300 MPa and standard deviation of 30 MPa. Find the probability of failure.

25.2 A cantilever beam is a bending stress with a mean value of 300 MPa and standard deviation of 25 MPa. The material of the beam has a yield strength with a mean of 360 MPa and standard deviation of 30 MPa. Find the reliability and the mean factor of safety.

25.3 A rotating shaft with 40 mm diameter is made of steel with mean yield strength of 580 MPa and standard deviation of 30 MPa. The shaft is subjected to a torque with mean value of 250 Nm and standard deviation of 25 Nm. Find the mean factor of safety and the reliability.

25.4 A rectangular beam with 25 mm × 50 mm cross section is subjected to a bending moment with a mean of 1250 Nm and standard deviation of 150 Nm. The material of the shaft has a mean yield strength of 600 MPa and standard deviation of 40 MPa. Determine the reliability of the design.

25.5 A ball bearing has normally distributed time of failure with a mean of 20000 hr and standard deviation of 1500 hr. There are 200 bearings. How many of them will serve for 15000 hr?

SHORT ANSWER QUESTIONS

25.1 What are the considerations on the material and stress in a probabilistic design?

25.2 What is raw data?

25.3 What is frequency distribution?

25.4 What is a histogram?

25.5 What is the frequency distribution curve?

25.6 Defines the terms, mean and mode.

25.7 What does standard deviation indicate?

25.8 What is variance?

25.9 What is a normal distribution curve?

25.10 Write down the probability distribution equation of normal distribution and indicate what each term represents.

25.11 Given the mean inside diameter of the bearing and mean of outer diameter of a shaft, what is the mean clearance of the assembly?

25.12 What is reliability of a product?

25.13 What is margin of safety?

25.14 Define the standard variable factor Z.

25.15 How can the region of unreliability be reduced?

INDEX